CW01198220

RAILS

A GUIDE TO THE RAILS,
CRAKES, GALLINULES AND COOTS
OF THE WORLD

RAILS

A GUIDE TO THE RAILS, CRAKES, GALLINULES AND COOTS OF THE WORLD

Barry Taylor

Illustrated by Ber van Perlo

PICA PRESS
SUSSEX

© 1998 Barry Taylor and Ber van Perlo

Pica Press
(an imprint of Helm Information Ltd.)
The Banks, Mountfield,
Nr. Robertsbridge,
East Sussex TN32 5JY

ISBN 1-873403-59-3

A catalogue record for this book is available from the British Library.

All rights reserved. No part of this publication may be reproduced, stored in a retrieval system, or transmitted, in any form or by any means, electronic, mechanical, photocopying, recording, or otherwise, without prior permission of the publisher.

Published in the Netherlands and Belgium as *Dutch Birding Vogelgids 11* by
  Ger Meesters Boekprodukties
  Vrijheidsweg 86,
  2033 CE Haarlem,
  The Netherlands

ISBN 90-74345-20-4
(Netherlands and Belgium only)

Distributed in Southern Africa by
  Russel Friedman Books CC
  PO Box 73,
  Halfway House,
  1685,
  South Africa.

Edited by Nigel Collar
Series Editor: Nigel Redman
Production and Design: Julie Reynolds

WWF®

This project was supported by the World Wide Fund for Nature - South Africa

Production and design by Fluke Art, Bexhill on Sea, East Sussex
Colour separation by The Scanning Gallery, Tonbridge, Kent
Printed by Midas Printing, Hong Kong

Contents

	Page	Plate
Introduction	9	
Acknowledgements	10	
Layout and Scope of the Book	13	
Topography	24	
Phylogeny and Classification	26	
Morphology and Flightlessness	33	
Habitat	36	
Food and Foraging	39	
Voice	42	
Behaviour	44	
Breeding	49	
Movements	52	
Conservation and Extinction	56	
Colour Plates	63	
Systematic Section	151	

		Page	Plate
White-spotted Flufftail	*Sarothrura pulchra*	151	1
Buff-spotted Flufftail	*Sarothrura elegans*	154	1
Red-chested Flufftail	*Sarothrura rufa*	159	2
Chestnut-headed Flufftail	*Sarothrura lugens*	163	2
Streaky-breasted Flufftail	*Sarothrura boehmi*	165	2
Striped Flufftail	*Sarothrura affinis*	168	1
Madagascar Flufftail	*Sarothrura insularis*	171	3
White-winged Flufftail	*Sarothrura ayresi*	172	3
Slender-billed Flufftail	*Sarothrura watersi*	175	3
Nkulengu Rail	*Himantornis haematopus*	177	4
Grey-throated Rail	*Canirallus oculeus*	179	4
Madagascar Wood-rail	*Canirallus kioloides*	181	4
Swinhoe's Rail	*Coturnicops exquisitus*	183	5
Yellow Rail	*Coturnicops noveboracensis*	184	5
Speckled Rail	*Coturnicops notatus*	187	5
Ocellated Crake	*Micropygia schomburgkii*	190	5
Chestnut Forest-rail	*Rallina rubra*	192	6
White-striped Forest-rail	*Rallina leucospila*	194	6
Forbes's Forest-rail	*Rallina forbesi*	195	6
Mayr's Forest-rail	*Rallina mayri*	196	6
Red-necked Crake	*Rallina tricolor*	197	7
Andaman Crake	*Rallina canningi*	200	7
Red-legged Crake	*Rallina fasciata*	201	7
Slaty-legged Crake	*Rallina eurizonoides*	203	7
Chestnut-headed Crake	*Anurolimnas castaneiceps*	206	8
Russet-crowned Crake	*Anurolimnas viridis*	207	8
Black-banded Crake	*Anurolimnas fasciatus*	209	8
Rufous-sided Crake	*Laterallus melanophaius*	210	9
Rusty-flanked Crake	*Laterallus levraudi*	213	9
Ruddy Crake	*Laterallus ruber*	214	9
White-throated Crake	*Laterallus albigularis*	216	9
Grey-breasted Crake	*Laterallus exilis*	217	10
Black Rail	*Laterallus jamaicensis*	220	10
Galapagos Rail	*Laterallus spilonotus*	224	10

Red-and-white Crake	*Laterallus leucopyrrhus*	226	11
Rufous-faced Crake	*Laterallus xenopterus*	228	11
Woodford's Rail	*Nesoclopeus woodfordi*	230	11
Bar-winged Rail	*Nesoclopeus poecilopterus*	231	11
Weka	*Gallirallus australis*	233	12
New Caledonian Rail	*Gallirallus lafresnayanus*	237	12
Lord Howe Rail	*Gallirallus sylvestris*	238	12
Gilbert Rail	*Gallirallus conditicius*	242	12
Okinawa Rail	*Gallirallus okinawae*	243	15
Barred Rail	*Gallirallus torquatus*	245	13
New Britain Rail	*Gallirallus insignis*	247	13
Buff-banded Rail	*Gallirallus philippensis*	248	14
Roviana Rail	*Gallirallus rovianae*	256	14
Guam Rail	*Gallirallus owstoni*	258	13
Wake Rail	*Gallirallus wakensis*	260	15
Tahiti Rail	*Gallirallus pacificus*	261	16
Dieffenbach's Rail	*Gallirallus dieffenbachii*	262	16
Chatham Rail	*Gallirallus modestus*	263	16
Sharpe's Rail	*Gallirallus sharpei*	264	16
Slaty-breasted Rail	*Gallirallus striatus*	265	15
Clapper Rail	*Rallus longirostris*	269	17
King Rail	*Rallus elegans*	278	17
Plain-flanked Rail	*Rallus wetmorei*	283	18
Virginia Rail	*Rallus limicola*	284	18
Bogotá Rail	*Rallus semiplumbeus*	289	18
Austral Rail	*Rallus antarcticus*	291	18
Water Rail	*Rallus aquaticus*	293	19
African Rail	*Rallus caerulescens*	299	19
Madagascar Rail	*Rallus madagascariensis*	302	19
Brown-banded Rail	*Lewinia mirificus*	304	20
Lewin's Rail	*Lewinia pectoralis*	305	20
Auckland Rail	*Lewinia muelleri*	310	20
White-throated Rail	*Dryolimnas cuvieri*	312	21
African Crake	*Crex egregia*	316	27
Corncrake	*Crex crex*	320	27
Rouget's Rail	*Rougetius rougetii*	327	21
Snoring Rail	*Aramidopsis plateni*	329	21
Inaccessible Rail	*Atlantisia rogersi*	331	21
Little Wood-rail	*Aramides mangle*	334	22
Rufous-necked Wood-rail	*Aramides axillaris*	335	22
Grey-necked Wood-rail	*Aramides cajanea*	337	22
Brown Wood-rail	*Aramides wolfi*	341	23
Giant Wood-rail	*Aramides ypecaha*	342	23
Slaty-breasted Wood-rail	*Aramides saracura*	344	23
Red-winged Wood-rail	*Aramides calopterus*	345	23
Uniform Crake	*Amaurolimnas concolor*	346	24
Bald-faced Rail	*Gymnocrex rosenbergii*	349	24
Bare-eyed Rail	*Gymnocrex plumbeiventris*	350	24
Talaud Rail	*Gymnocrex talaudensis*	352	24
Brown Crake	*Amaurornis akool*	353	25
Isabelline Bush-hen	*Amaurornis isabellinus*	355	25
Plain Bush-hen	*Amaurornis olivaceus*	356	25
Rufous-tailed Bush-hen	*Amaurornis moluccanus*	357	25
Talaud Bush-hen	*Amaurornis magnirostris*	361	25
White-breasted Waterhen	*Amaurornis phoenicurus*	363	26
Black Crake	*Amaurornis flavirostris*	366	26
Sakalava Rail	*Amaurornis olivieri*	371	26

Black-tailed Crake	*Amaurornis bicolor*	372	**26**
Little Crake	*Porzana parva*	374	**28**
Baillon's Crake	*Porzana pusilla*	379	**28**
Laysan Crake	*Porzana palmeri*	387	**32**
Spotted Crake	*Porzana porzana*	389	**28**
Australian Crake	*Porzana fluminea*	396	**29**
Sora	*Porzana carolina*	399	**29**
Dot-winged Crake	*Porzana spiloptera*	406	**29**
Ash-throated Crake	*Porzana albicollis*	407	**29**
Hawaiian Crake	*Porzana sandwichensis*	409	**32**
Ruddy-breasted Crake	*Porzana fusca*	411	**30**
Band-bellied Crake	*Porzana paykullii*	414	**30**
Spotless Crake	*Porzana tabuensis*	416	**30**
Kosrae Crake	*Porzana monasa*	422	**32**
Henderson Crake	*Porzana atra*	422	**31**
Yellow-breasted Crake	*Porzana flaviventer*	424	**31**
White-browed Crake	*Porzana cinerea*	427	**31**
Striped Crake	*Aenigmatolimnas marginalis*	431	**27**
Zapata Rail	*Cyanolimnas cerverai*	435	**33**
Colombian Crake	*Neocrex colombianus*	436	**33**
Paint-billed Crake	*Neocrex erythrops*	438	**33**
Spotted Rail	*Pardirallus maculatus*	440	**34**
Blackish Rail	*Pardirallus nigricans*	443	**34**
Plumbeous Rail	*Pardirallus sanguinolentus*	445	**34**
Chestnut Rail	*Eulabeornis castaneoventris*	448	**35**
Invisible Rail	*Habroptila wallacii*	451	**35**
New Guinea Flightless Rail	*Megacrex inepta*	452	**34**
Watercock	*Gallicrex cinerea*	454	**38**
Purple Swamphen	*Porphyrio porphyrio*	458	**36**
White Gallinule	*Porphyrio albus*	470	**37**
Takahe	*Porphyrio mantelli*	471	**37**
Allen's Gallinule	*Porphyrio alleni*	476	**38**
American Purple Gallinule	*Porphyrio martinica*	480	**38**
Azure Gallinule	*Porphyrio flavirostris*	485	**37**
Samoan Moorhen	*Gallinula pacifica*	488	**37**
San Cristobal Moorhen	*Gallinula silvestris*	489	**40**
Tristan Moorhen	*Gallinula nesiotis*	490	**39**
Common Moorhen	*Gallinula chloropus*	492	**39**
Dusky Moorhen	*Gallinula tenebrosa*	503	**39**
Lesser Moorhen	*Gallinula angulata*	509	**39**
Spot-flanked Gallinule	*Gallinula melanops*	512	**40**
Black-tailed Native-hen	*Gallinula ventralis*	515	**40**
Tasmanian Native-hen	*Gallinula mortierii*	518	**40**
Red-knobbed Coot	*Fulica cristata*	522	**41**
Common Coot	*Fulica atra*	527	**41**
Hawaiian Coot	*Fulica alai*	535	**41**
American Coot	*Fulica americana*	536	**42**
Caribbean Coot	*Fulica caribaea*	543	**42**
White-winged Coot	*Fulica leucoptera*	544	**42**
Andean Coot	*Fulica ardesiaca*	546	**42**
Red-gartered Coot	*Fulica armillata*	548	**43**
Red-fronted Coot	*Fulica rufifrons*	551	**43**
Giant Coot	*Fulica gigantea*	553	**43**
Horned Coot	*Fulica cornuta*	555	**43**

BIBLIOGRAPHY 558

INDEX OF SCIENTIFIC AND ENGLISH NAMES 593

*To my children Richard and Peter;
to all those who love marshes, forests, and enigmatic birds;
and to Sonny.*

Barry Taylor

INTRODUCTION

The family Rallidae comprises a large and rather homogeneous assemblage of small to large, terrestrial, marsh and aquatic birds, which are variously referred to as rails, crakes, wood-rails, gallinules, moorhens and coots, although the term 'rail' is also applied generally to the whole family. It is by far the largest family in the Gruiformes, comprising 150 species, 133 extant and 15 recently extinct (since c. 1600) and two almost certainly extinct. Rails are of worldwide distribution, being absent only from polar regions, completely waterless deserts, and mountainous regions above the permanent snow line. They occupy a very diverse range of habitats, including forests, wetlands and grasslands, and even scrub-covered remote oceanic islands and coral cays. Apart from the coots, and to a lesser extent some moorhens, which live in relatively open aquatic habitats, most species live on the ground in dense vegetation and are very difficult to observe.

Part of the fascination of rails is that so many species are rarely seen and are very poorly known; however, despite their apparently secretive behaviour, many rails are remarkably confiding and allow the patient observer fascinating glimpses of their often complex behaviour, social organisation and family life. Rails also capture the imagination because of the apparent paradox that such poorly flying birds can undertake intercontinental migrations and are widespread and successful colonists of remote oceanic islands. They are also remarkable because many species on such islands have evolved rapidly into flightless forms in the absence of predators, a fact which has led to the rapid extinction of numerous island taxa following the arrival of man and his commensals.

Despite the worldwide occurrence of rails and the intrinsic interest of this group of enigmatic and challenging birds, the only existing major treatise dealing exclusively with the family is Ripley's (1977) *Rails of the World*. Commenting on the production of that work, as quoted in *Discovery* 12(3): 44 (1977), Ripley explained that "if ever I could write a monograph on rails, not only had one never been written before, but the likelihood was that one would never be written again, and almost anything I managed to say about the rails was likely not to be very strongly corrected in the future, so that I had clear sailing by taking on a family that is as obscure, and little known, and enigmatic as the rails". No such claims are made for the present volume. In the last twenty years much of great significance has been published on rails and there is a need for a comprehensive review of our current knowledge. What is most urgently needed in the near future is a critical re-appraisal of the classification of the family, which should be carried out when DNA studies have been made of a sufficient number of species.

This book describes all 150 living and recently extinct rail species, including two which have only just been discovered, and all recently extinct races of surviving species. All living species, and 12 of those recently extinct, or almost certainly extinct, are illustrated in the 43 colour plates. The text is compiled from over 1,800 references and also includes new and unpublished observations from all over the world. This is the first work to give comprehensive information on field identification (including voice) covering all species and races for which details are known, and it describes for the first time the immature and juvenile plumages of many species. It also provides a detailed summary of our current knowledge of all aspects of rail biology and behaviour, and emphasises gaps in that knowledge: for example, immature and juvenile plumages are still poorly known and the nest and eggs of some species remain undescribed.

Rail habitats are disappearing rapidly and many rails are becoming endangered before enough is known about them to enable effective conservation measures to be planned. This book deals in detail with distribution, status, habitat requirements and current threats, points out where critical knowledge is lacking, and summarises conservation priorities for threatened species. It is designed to be a comprehensive guide and handbook to the rails of the world, and hopefully it will also stimulate further study in areas where knowledge is seriously lacking, as well as providing a detailed source of information on all aspects of the biology and ecology of rails.

ACKNOWLEDGEMENTS

Although extensive and detailed studies have been made of a few rail species, such as the Common Moorhen, the Purple Swamphen, the Clapper Rail, the Corncrake and some of the coots, many members of the family are poorly known and the literature on them is relatively sparse. Even basic information on the natural history of some species is either lacking or, if available, difficult to locate, being published in obscure or unexpected places. For some species, especially those which have not been recorded for many years or have been seen very rarely, little if anything of significance has been published since the original descriptions. For these reasons the collection of information for this book has relied heavily on the help of other people in tracing and obtaining obscure references, and on the willingness of rail watchers to supply unpublished observations of poorly known species.

I must express my sincere thanks to the staff of many museums who, as well as providing loans of specimens and details of birds in their collections, also made available a great deal of reference material and helped in tracing other sources of information. A tremendous amount of assistance in these respects was provided by Robert Prys-Jones, Michael Walters and Effie Warr at the Natural History Museum, Tring, while Mark Adams and Don Smith also gave assistance during my visits to the museum. Most of the illustrations in the book are based on specimens from this museum and from the National Museum of Natural History, Leiden, where René Dekker provided great assistance during visits, as well as supplying reference material and plumage descriptions. I am also very grateful to Paul Sweet and Ana Luz Porzecanski (American Museum of Natural History), Michael Largen (Liverpool Museum), David Willard (Field Museum of Natural History, Chicago), Brian Schmidt and Rebecca Siegel (Smithsonian Institution, Washington), Don Baepler (Marjorie Barrick Museum, University of Nevada), Robin Restall and Philip Precey (Coleccion Ornitologica Phelps, Caracas, Venezuela), David Oren (Museu Goeldi, Belem, Para, Brazil), David Allan (Durban Natural History Museum), and John Gerwin (North Carolina State Museum of Natural Sciences, Raleigh). Nigel Collar very kindly provided information on Asian rail specimens held in museums around the world.

Library staff at many institutions have also provided much reference material, and I must express my particular thanks to the staff at the University of Natal in the Life Sciences Library and the Inter-Library Loans section of the University Library, especially to Annette Anderson, Leonie Prozesky, Mrs R. A. Metcalfe, N. A. Gani and O. Deoparsad. At the library of the Durban Museum, Belinda Eisenhower provided very valuable assistance, as did Sharynne Hearn and John Timms at the Natal Museum Library, Pietermaritzburg.

I have been very fortunate in obtaining a great deal of help and information (especially reference material) from the staff of BirdLife International (Cambridge), throughout the project as well as during my visit to Cambridge in October 1997, and I must particularly thank Gary Allport, John Fanshawe, Lincoln Fishpool, Katharine Gotto, James Lowen, Hugo Rainey and David Wege.

For help in tracing and obtaining references I am also very grateful to Juan Mazar Barnett, Bob Behrstock, Bill Belton, Adolfo Beltzer, Lori Bronas, Tom Brooks, Rob Clay, Courtney Conway, Jorge Cravino, René Dekker, Rafael Dias, Bill Eddleman, Lois Jammes, Nagahisa Kuroda, Frank Lambert, Sandra Loor-Vela, Manuel Marin, Rick Nuttall, Fabio Olmos, Nigel Redman, Craig Robson, Chris Sharpe, Andy Simpson and John Wall. For help with translations of papers and books from German, Dutch and Spanish, I am very grateful to Juan Mazar Barnett, Bill Belton, Rafael Dias, Gordon Maclean and Fabio Olmos.

The basic groundwork for the text of this book was done during my preparation of the rail family and species accounts for Lynx Edicion's *Handbook of the Birds of the World*, Volume 3. I owe a great deal to Josep del Hoyo and Andy Elliott, editors of that excellent work, for their help and encouragement, for their critical review of my texts, and for allowing me to use information from the family account as the basis for some of the introductory sections in the present volume. I am also grateful to Sheila Hardie for much help during that project, and to Teresa Pardoe for providing many references.

I am very grateful to all those who provided unpublished observations and other information, namely John Ash, John Atkins, Dylan Aspinwall, Juan Mazar Barnett, John Bates, A. J. Beakbane, Bill Belton, Peter Boesman, Robin Brace, Richard Brooke, Tom Brooks, Ian Burrows, Roselle Chapman, Rob Clay, Deon Coetzee, Nigel Collar, Charles Collins, Courtney Conway, Jorge Cravino, Rafael Dias, Guy Dutson, David Eades, Bill Eddleman, Christian Erard, Ilay Ferrier, Lincoln Fishpool, Jon Fjeldså, Rosendo Fraga, Don Franklin, Mike Fraser, Paul Funston, John Gerwin, David Gibbs, Phil Gregory, Mike Haramis, Simon Harrap, Phil Hockey, G. N. Hopkinson, Jon Hornbuckle, Roger Jaensch, Alvaro Jaramillo, Ian Jamieson, Dave Johnson, Leo Joseph, Krys Kazmierczak, Greg Kearns, Frank Lambert, Olivier Langrand, Alberto Madrono, Manuel Marin, Alex Masterson, Niven McCrie, Ian McLaren, Clive Mann, Rex Naug, Richard Noske, Fabio Olmos, Fernando Ortiz-Crespo, Jose Fernando Pacheco, Tony Palliser, Hugo Phillipps, Steven Piper, Michael Poulsen, Marcos Garcia Rams, Nigel Redman, Van Remsen, Robin Restall, Gerry Richards, Adam Riley, Chan Robbins, Craig Robson, Danny Rogers, Chris Sharpe, Robson Silva e Silva, Rob Simmons, Tim Stowe, Monica Swartz, Warwick Tarboton, David Westcott, Edwin Willis and Mengistu Wondafrash. In particular, the South-East Asian species accounts benefited greatly from Craig Robson's input, while very extensive and detailed comments on some species, and much unpublished information, were provided by Juan Mazar Barnett, Bill Eddleman, Ian Jamieson, Fabio Olmos and Roger Jaensch. Courtney Conway and Sue McRae very kindly provided much published and unpublished material related to their research projects.

Many people deserve special mention for their contribution to various aspects of the work. John Dunning provided information on rail weights not included in his *CRC Handbook of Avian Body Masses*. Bill Eddleman and Roger Jaensch kindly identified American and Australian marsh plants which were mentioned in the literature only by vernacular names. Both before and during my work on this book Colin Wintle provided much valuable information on the behaviour and breeding of rails, especially flufftails, in captivity, and has made a major contribution through his captive breeding studies of the Striped Crake and the Red-chested and Streaky-breasted Flufftails. At the University of Cape Town's Avian Demography Unit, James Harrison and Peter Martinez provided raw data from the Southern African Bird Atlas Project, while my work with Keith Barnes on parts of the forthcoming Southern African Important Bird Areas book and the South African Red Data Book (Birds) provided very useful input. Alex Masterson, G. N. Hopkinson and Duncan Parkes supplied much valuable information on the eggs, nests and habits of various African rails. Donovan Kotze provided much information on wetland plants, and I am also grateful to him for many stimulating and informative discussions on wetland conservation and management, and for enjoyable time spent in the field.

Information for the introductory sections, and comments on the layout and content of both the introduction and the species accounts, were kindly provided by David Allan, Bob Behrstock, Tom Brooks, Paul Coopmans, Storrs Olson, David Steadman and Steven Piper. Various species accounts were reviewed by David Allan, Juan Mazar Barnett, Peter Boesman, Robin Brace, Tom Brooks, Roselle Chapman, Rob Clay, Guy Dutson, Bill Eddleman, Jon Fjeldså, Phil Gregory, Jon Hornbuckle, Roger Jaensch, Ian Jamieson, Frank Lambert, Olivier Langrand, James Lowen, Fabio Olmos, Tony Palliser and Robin Restall.

I was very fortunate in being allowed to use material from the rail texts of some important forthcoming books. Robert Ridgely provided full details from both volumes of his book on the birds of Ecuador, Jorge Cravino did likewise for his work on the birds of Uruguay, and Nigel Collar provided information from the forthcoming *Asian Red Data Book* (Birds). I am extremely grateful to these authors for the valuable information which they supplied, which has greatly improved my text. I am also grateful to Mauricio Guerrero and Sandra Loor-Vela, who provided information from the CECIA distributional list of Ecuadorian birds, and to Lois Jammes of Armonia (the BirdLife partner in Bolivia) for providing full database information on Bolivian Rallidae. Ian Jamieson very kindly allowed me to quote details of an as yet unpublished study on the breeding of Red-knobbed Coot and Lesser Moorhen in Namibia, while Don Franklin provided copies of important forthcoming publications on the poorly known Chestnut

Rail. I am also grateful to Chris Feare, editor of the *Bulletin of the British Ornithologists' Club*, and Tim Inskipp, editor of *Forktail*, for expediting the publication of Frank Lambert's two new species of rails so that they could be included in this book; Nigel Redman kindly assisted in this process. Last, but not least, I am very grateful to Bradley Livezey for allowing me to include a summary of the results of his important phylogenetic analysis of the Gruiformes (particularly the Rallidae), which is due to be published later in 1998.

Photographic material, both published and unpublished, of living birds and of museum specimens, has helped greatly in the production of plates and in the compilation of descriptions. I would like to thank Jeff Davies, Andrew Dunn, David Eades, Graeme Elliott, David Gibbs, Phil Gregory, Ben Kakebeeke, Ian McLaren, Danny Rogers, Paul Sweet and Warwick Tarboton for providing unpublished material. Information on tape recordings was kindly provided by Greg Budney, Andrea Priori and Richard Ranft.

Work on this book has taken several years to complete and I am particularly grateful to Christopher Helm for his patience and understanding during this period. Nigel Redman did a sterling job as editor and provided much help, encouragement and critical comment. I am also very grateful to Nigel Collar for expertly editing the entire text. Christine Taylor very kindly helped with checking maps, collating references and proof-reading texts, and I am grateful to Pamela Sweet and Jane Flockhart at the Department of Zoology and Entomology, University of Natal, for assistance with computer-related work.

The World Wide Fund For Nature (South Africa), sponsored the project and provided financial support during the preparation of the book. I thank John Hanks, Ian Macdonald and Lisa Padfield for their encouragement and support during this and other rail projects which WWF (SA) have supported.

I would like to thank my wife Janet and my family for putting up with my long working hours and my very limited time at home during the final year of the project, for their help and support, and for reminding me that there is life after writing bird books.

Finally, I must acknowledge the help and encouragement which I received from Stuart Keith when I first began to study rails seriously, and whose work on *Sarothrura* first aroused my interest in the genus. Dylan Aspinwall, whose death in 1995 was such a sad loss to African ornithology, stimulated and encouraged my work on rails and many other birds when I lived in Zambia in the 1970s. Gerald Kaufmann, who died in 1998, also provided early encouragement, and his work on Sora and Virginia Rails stimulated my interest in captive breeding studies and in rail behaviour. I must also express my gratitude to those who have participated in my studies of flufftails and other rails, particularly to Peter Clowes, Erik Forsyth, Edward Smith and Christine Taylor, all of whom have made enormous contributions. To these, and to all the other birders with whom I have shared the excitement of finding and observing rails, I extend my thanks for their help and companionship.

Layout and Scope of the Book

This book is similar in style and layout to other volumes in the series which deal with specific taxonomic groups of birds. As in other volumes, identification is treated in some detail, including the separation of subspecies, while particular attention is given to describing vocalisations, which are frequently of great importance in the location and identification of secretive rail species. However, it is hoped that this volume may also fulfil the role of a handbook in that it attempts to summarise the most significant aspects of the general biology and ecology of all the species described. In particular, attention is given to descriptions of habitat characteristics, which are very important to the understanding of the (often poorly documented) distribution and status of rails. There is also a detailed treatment of status, with particular emphasis on current threats to the birds' survival, and a review, where relevant, of conservation and management measures.

In the text equal prominence is given to the description of those species and races which are known, or are assumed, to have become extinct in the last 400 years (since c. 1600). The number of such taxa is depressingly long, and their inclusion serves a multiple purpose. Firstly, such taxa are of interest in their own right, to illustrate the variety of rails, to improve understanding of rail biogeography and distribution, and to demonstrate the remarkable ability which rails exhibit in adapting to apparently hostile and totally unsuitable environments, for example on remote, small oceanic islands. Secondly, and at least equally importantly, they vividly illustrate the extent of what we have lost, the ease with which such vulnerable species can be exterminated, and often the greed, thoughtlessness and general ignorance with which men treat their environment and the species with which they share it. It is hoped that their inclusion will serve as a warning of the threats which many surviving taxa still face.

Many rail species, especially those of the tropics and including many rare or threatened taxa, are very poorly known and hence the depth of coverage varies enormously from species to species. It is hoped that, by highlighting gaps in our knowledge, interest in discovering more about poorly known species will be stimulated. Although very little has been published on some species, in the last few years there has been a great increase in the amount of information gathered on rails by active birders and by ornithologists, especially in South America and particularly concerning the basic distribution, status and natural history of little-known rails. An attempt has been made to obtain as much unpublished information as possible for inclusion in the book.

A few rail species have been extensively studied in recent years and a large volume of literature exists on these birds. In such cases insufficient space was available to summarise all the existing literature, which has often been comprehensively reviewed in the regional handbooks, notably Cramp & Simmons (1980), Urban *et al.* (1986), Marchant & Higgins (1993) and Poole & Gill (1992–1998). In such cases reference is freely made to the information contained in these works, which often include comprehensive details gleaned from many local publications, and the reader should consult these handbooks for a more detailed list of references.

The first part of the book attempts a brief review of the taxonomy, phylogeny, distribution and general biology of rails, and briefly discusses some of the conservation issues associated with these birds and their forest, wetland and grassland habitats. Most of the information contained within them is fully referenced in the relevant species accounts, and references are only given in the introductory sections when the relevant sources are not given elsewhere in the text. The main part of the volume is devoted to the plates and species accounts. The taxonomy followed, which is discussed in the introductory section, is generally that of Olson (1973b), modified in some cases by more recent morphological studies to which reference is made where appropriate. The order of the genera and species follows that in Sibley & Monroe (1990).

SPECIES NUMBERS

Each species in this book has been given a number. The numbers have no taxonomic significance and merely serve to provide a ready link between species texts, plate captions and illustrations, and to facilitate cross-referencing between species accounts.

PLATES

The plates depict all 133 living species of rails, and 12 of the 17 species (those for which specimens and proper descriptions exist) which have definitely or almost certainly become extinct since c. 1600; two of the remaining five are illustrated by line drawings. In almost all cases the plates were painted from museum specimens, sometimes with the help of colour photographs from published works and photographs and transparencies provided by observers. In the very few cases where no such material was available for study, birds were painted from existing illustrations. This applies, for example, to the Roviana Rail *Gallirallus rovianae*, which is known from one specimen and is illustrated by Diamond (1991), and to the race *peruvianus* of the Bogotá Rail *Rallus semiplumbeus*, known from only one specimen, now lost, and illustrated only by Fjeldså & Krabbe (1990). All living species were illustrated in Taylor (1996b) and that work provided a reference against which to check details of taxa for which few or no specimens were available for study. Regarding extinct species, Sharpe's Rail *Gallirallus sharpei* has not previously been illustrated and is painted from the only specimen in existence; the doubtfully valid Gilbert Rail *Tricholimnas conditicius* is painted from descriptions of the only specimen (originally preserved in spirit); and the strikingly plumaged Tahiti Rail *Gallirallus pacificus*, for which no specimen exists, is illustrated from the painting of the bird which is kept in the British Museum (Natural History).

Subspecific identity in rails is frequently based on relatively minor plumage differences such as variation in the overall tone of the plumage, or simply on measurements. In many species, authors have erected subspecies on the basis of flimsy evidence from an inadequate range of material, sometimes without taking full account of individual and seasonal variation, and of differences which might be attributable to age; the validity of many forms is therefore questionable. Subspecies which differ significantly from each other have been illustrated wherever possible, within the constraints of the availability of specimens and good colour photographs. Plate 14, which shows some of the individual and geographical variation in the Buff-banded Rail *Gallirallus philippensis*, serves to illustrate the complexity of such taxonomic issues.

A few rail species, for example the Weka *Gallirallus australis* and the Clapper Rail *Rallus longirostris*, have distinctive colour morphs and an attempt has also been made to illustrate a representative selection of these forms.

The immature and juvenile plumages of some rail species are still undescribed, while those of others have been poorly described and inadequately or incorrectly illustrated in many field guides. Depending on the availability of specimen and photographic material, distinctive juvenile and immature plumages are illustrated wherever possible. The chicks of some species are also illustrated, both to show some of the variation in the colours of bare parts, plumage and the bare skin of the head (e.g. in moorhens *Gallinula* and coots *Fulica*) and to add interest and variety to the plates.

Species on the plates are arranged as far as possible in taxonomic sequence, but in some cases we have departed from this arrangement in order to juxtapose species similar in appearance and/or from the same geographical region. Plates 16 and 32 are devoted entirely to extinct species. All individuals on a plate are painted to the same scale, but scale obviously differs considerably between plates.

The plate captions provide a brief summary of the range and habitat of each species and briefly list the most important diagnostic features of each age, sex and plumage illustrated. However, it is important to refer to the species accounts for a full treatment of identification criteria. Where space permits, a brief description is also included of races which are not illustrated.

GENUS ACCOUNTS

A brief description of the major features of each genus is given before its species account(s). The genus account gives the number of species in the genus and briefly discusses the taxonomic validity of the genus and any past or current variation in the taxonomic treatment of its constituent species. Mention is also made of the geographical range of the genus, the major characteristics of the species (including the existence of any flightless forms), their habitat requirements, their status and migrations, and the existence of any threatened or endangered taxa.

SPECIES ACCOUNTS

Each species has its own account which is subdivided into the sections listed below. Within most sections of the species accounts references are provided to identify the sources of the information given; these include papers, articles and books which are published or in press, unpublished masters and doctoral theses, unpublished manuscripts obtained from the relevant authors, and information supplied in writing (credited as *in litt.*) or verbally (pers. comm.). The *in litt.* category includes numerous communications which should perhaps be more properly described as 'in e-mail'. References to my own unpublished observations are given as '(P. B. Taylor unpubl.)'. In most cases references are not provided within the following sections: Identification, Description, Geographical Variation, and Breeding and Survival (for exceptions see the section descriptions below). This is because the information in these sections is presented in a condensed form and has been compiled from so many sources that to include all citations would make the text very difficult to read (e.g. breeding data and details of range), or because the information (e.g. plumage descriptions and measurements) has largely been generated by the author.

Abbreviations

Metric measurements are used throughout this book. Abbreviations relating to measurements are as follows: mm = millimetres, cm = centimetres, m = metres, km = kilometres; ha = hectares (1 ha = 10,000m^2 or 100m x 100m); s = seconds, min = minutes, h = hours; kJ = kilojoules; °C = degrees centigrade; SD = standard deviation; SE = standard error; n = sample size; c. = approximately; ppt = parts per thousand; ppm = parts per million.

Points of the compass are usually abbreviated to their initial letters, e.g. N, S, E, W, NE, SW etc. C denotes central (north-central = NC). Note that these abbreviations may represent nouns or adjectives, e.g. 'to the N', 'S (southward) movements occur', or 'N Ghana' (northern Ghana). Abbreviations are not used within proper names so that, for example, country names such as South Africa are always given in full, as are the commonly used names of regions; thus 'East Africa' is the region comprising Kenya, Uganda and Tanzania, while 'E Africa' refers to eastern Africa, i.e. the eastern part of the entire continent.

Other abbreviations used are as follows: a.s.l. = above sea level; I/Is = island/islands; L = Lake; Mt/Mts = Mount/Mountains; R = River; V = Valley; D = District; Dpt = Department; NP = National Park; sp. = species (singular), spp. = species (plural); mtDNA = mitochondrial DNA. Months are represented by the first three letters of the month name. In plumage descriptions, P1-P10/P11 are the numbers of the primary feathers, while secondaries are designated S and tertials T (see Description section below). Two of the museums from which specimen material was obtained are referred to by their initials, namely the Natural History Museum, Tring (BMNH), and the American Museum of Natural History, New York (AMNH); 'Leiden' refers to the Rijksmuseum van Natuurlijke Historie, Leiden; the names of other museums are given in full.

Initial Section (untitled)

Each species account starts with brief information on nomenclature and taxonomy. The original scientific name, author, date and type locality are given, followed by any relevant information on relationships with other species, e.g. whether the species has ever been considered conspecific with any other, and whether it is a member of a superspecies group. The number of races currently recognised is given, plus any relevant information on separation or merging of races and their possible allocation to other species (a more detailed discussion of this topic appears under Geographical Variation). Rail taxonomy is bedevilled by **synonyms**; as well as currently accepted races frequently having been originally described as full species, many species have been given several different generic names since their original description and such names may often be encountered in the ornithological literature and in museum collections; even today, the generic placement of some species varies between authors (where relevant, details are given in the genus accounts). Because of the possibility of confusion over the scientific names of some species, as many synonyms as possible are listed for reference. When appropriate, **alternative names** are also listed, these being commonly or widely used English alternatives to the vernacular name used in this book.

Identification

After mentioning if the species is extinct, this section provides a brief description, including the most important identification features, and gives the major characters involved in the separation of races. Details of distinct colour morphs are also given where appropriate. Overall size is indicated by the inclusion of minimum and maximum body lengths, which are taken from the literature and include the size range of all races; published maximum or minimum lengths outside the given range are included in parentheses with a question mark. For ease of comparison with other forms, the size class is indicated by a descriptive term based on the details in the following table.

Class	Size range (cm)
Very small	12 – 15
Small	14 – 19
Medium-small	18 – 24
Medium	23 – 32
Medium-large	30 – 40
Large	38 – 50
Very large	48 – 63

The wingspan is also given whenever possible, figures being from the literature or from unpublished measurements taken by the author. Other biometric data are provided under Description and Geographical Variation.

The field identification of rails poses many problems, not only because closely related species are often very similar in colour and pattern, but also because many species inhabit dense cover and are very difficult to locate and to see well. Furthermore, details of the juvenile and immature plumages of some species are still unknown. Important field characters to note include the colour and pattern of the head and neck, including the possible presence of a superciliary stripe, an eyestripe, contrastingly coloured bare facial skin, or contrastingly pale colouring on the chin and throat; whether the head and neck differ markedly in colour or pattern from the rest of the body; the overall colour and pattern of the upperparts (plain, spotted, streaked, scalloped, and the presence of contrasting colours on the mantle, back, wings etc); the colour and pattern of the underparts, especially the presence and extent of barring or spotting in the flank region, and the presence (in some moorhens) of a white flank line; and the colour and pattern of the undertail-coverts, which often contrast with the rest of the body plumage and are obvious when the tail is flicked.

Most rails rarely fly when disturbed, preferring to run to dense cover, so the wing pattern is rarely visible well. Many species have a contrasting pattern of bars or spots on the flight feathers, which may sometimes be partially visible in the folded wing, and contrasting colours or patterning on the scapulars, coverts and tertials, which are usually readily visible. The pattern of the underwing-coverts may also be diagnostic but is rarely seen under normal circumstances. Bare part colours are often important and should always be noted if possible, especially the presence of areas of different colours on the bill and (if present) the frontal shield, and on the legs and feet, always bearing in mind that mud or staining may temporarily affect colours, especially of the legs and feet. It is also important to note the size and shape of the bill and (if present) the frontal shield, and the length of the legs and toes.

Many rails show little sexual dimorphism in plumage. All appear to have a distinctive juvenile plumage (acquired by postnatal moult from the downy plumage) which may be retained only for a very short period, and some have a recognisable immature plumage (obtained via the postjuvenile moult). These plumages have previously often been confused, incorrectly described, or not described at all, and their diagnostic characters are mentioned here as well as being described in detail under Description and Geographical Variation. Details of the moult pattern are given in the Moult section.

Rails are vocal birds and their presence is frequently indicated by their calls, which are often given in response to disturbance and usually allow positive identification even if the bird is not seen. A knowledge of rail vocalisations is of great assistance in the successful location and identification of many secretive species. Calls are described in detail under Voice. Knowledge of habitat preferences often gives a good indication of what species should be present in a particular habitat or vegetation type, especially where two or more similar rail species occur in the same area, and habitat characteristics should be noted in detail. A very brief description of the preferred habitat is given at the end of the Identification section; for full details the reader should consult the Habitat section which may also give a comparison with the habitat preferences of other species.

The diagnostic features which assist in distinguishing the bird from other closely related, sympatric or similarly plumaged species are detailed under the subheading **Similar species**. This section may mention the use of vocalisations if particularly appropriate.

Voice

Vocalisations are a very important factor in the location and identification of secretive rail species. Many species have highly characteristic display and territorial calls, which are given frequently during the breeding season but may not be given at other times of the year, when the birds are often correspondingly more difficult to locate. Distinguishing between superficially similar calls of closely related species may be difficult, and relevant differences in calls are described in the text. Information is provided on all types of vocalisation known to be made by each species, including advertising, territorial, aggressive, alarm, contact, courtship and mating calls, plus calls of chicks and juveniles; sexual variation in calls is also summarised.

Calls are described from information in the literature and from sound recordings, but such descriptions are a poor substitute for recordings of the calls themselves, and one observer's description of a call may be very difficult for another observer to interpret. Sonograms are not included because they were felt to be of limited use in describing calls, and the ideal would be to include a CD of calls with the book. Unfortunately this was not possible in the time available, but it is hoped to produce a CD as soon as possible after publication of the book.

Many rails respond well to taped playback of their calls and may be called up and seen at close quarters by this method, while some species also respond well to human imitations of their calls. A cautionary note about the use of playback is given in the Introduction under Voice.

Description

A detailed description is given for each distinct plumage, starting with adult male breeding (otherwise known as definitive alternate), and continuing to adult female breeding, adult non-breeding (definitive basic), immature (first basic), juvenile (juvenal) and downy young (natal); occasionally other terms, such as 'first breeding', are used to define other distinct plumage stages. Terms given above in parentheses refer to the Humphrey & Parkes (1959) terminology, which is described below under Moult. As far as possible all descriptions follow the same sequence: forehead, crown, nape, upperparts, tail, upperwings, underwings, sides of head, chin, throat and underparts. The bare parts are described last. The first plumage to be described is normally treated in considerable detail and all feather tracts are included; for subsequent plumages, only differences from this first plumage are normally described. However when other plumages differ substantially from the first plumage they are described more fully. Similarly, bare part colour and pattern are described for subsequent plumages only if they differ from the previous description. However, when an immature or juvenile plumage has not previously been described, details are usually given.

The effects of feather abrasion and fading on the overall plumage colour and pattern are also discussed in this section whenever appropriate. The stage in development at which juvenile, immature and adult plumage and bare part characters are assumed is also discussed where relevant. For a definition of the plumage and moult terminology used in this section of the book, see below under Moult.

In many species there is little or no sexual difference in plumage, so that the first plumage description is headed 'adult' and refers to the male, with a short additional comment to highlight minor sexual differences in plumage colour and pattern, bare part colours, and body size. Similarly it is often possible to combine descriptions of breeding (alternate) and non-breeding (basic) plumage.

In polytypic species the race described in this section is not selected according to any rule based on geographical distribution or nomenclature, but is usually the race which has been most fully described in the standard works consulted, namely Sharpe (1894), Ridgway & Friedmann (1941), Ripley (1977), Cramp & Simmons (1980), Urban *et al.* (1986) and Marchant & Higgins (1993). Although for many species an initial description was compiled from information given in these standard works, in all cases this has been checked against specimens and/or other available descriptions, and duly modified where appropriate, sometimes being rewritten in the process.

It should be noted that few original plumage descriptions for most rails are available in the literature: those in Ripley (1977) are frequently direct copies of those given by Sharpe (1894) and less frequently those of Ridgway & Friedmann (1941), while species texts in Poole & Gill (1992–1998) either copy the descriptions from Ripley or Ridgway & Friedmann or refer the reader to such standard texts for details.

In the few cases where no material was available for examination and the species concerned is not described in any of the standard texts, recourse was made to the best available description in other scientific publications or to the original description of the type. In all cases where any published description has been copied unmodified, acknowledgement of the source is made in the text.

In an attempt to achieve some consistency in colour names, Ridgway (1912) was used as a colour standard. This work was used in preference to the more recent guide by Smithe (1975) largely because it is also the standard for most of the early plumage descriptions consulted, which were often the only detailed ones in existence for species where no museum material was available for study.

In plumage descriptions the primaries are numbered descendantly, i.e. from the inside of the wing towards the tip, and secondaries are numbered ascendantly, i.e. from the centre of the wing towards the body. Thus the first primary (P1) and the first secondary (S1) are adjacent in the wing. The tertials are also numbered ascendantly.

Measurements

Measurements are all in millimetres, and weights in grams, so the units of measurement are not specified in this section. Wherever possible all measurements were taken by the author, from live birds and museum skins, sometimes supplemented by measurements from the literature. In most cases where no material was available and measurements are all from the literature, the sources are specified.

All measurements made by the author were taken by the following methods. The wing length is the flattened (maximum) chord; the tail length is the distance from the tip of the longest tail feather to the point of insertion of the rectrices into the body; the bill length is the chord of the culmen from the tip to the skull or, in species with a frontal shield, to the rear of the shield; the tarsus is the chord from the intertarsal joint to the distal end of the last undivided scale before the toes diverge.

Unfortunately it is not always possible to make direct comparison between these measurements and those available in the literature, because methods of making standard measurements differ widely. For example in most of the American literature the wing length is the minimum chord, taken with the wing lying in a natural, unstraightened position, and this may be up to 10% less than the maximum wing length (Campbell & Lack 1985). Many authors measure bill length to the start of the feathering at the base of the upper mandible, while others measure it to the gape, or even to the nostril; in species possessing a frontal shield, the shield length may not be included in bill length, or may be measured separately. In some works tarsus length is measured to the point where the toes diverge, or even to the sole of the foot. Wherever possible, measurements from the literature were selected to conform with the standard method of measurement used by the author; where measurements taken by other methods had to be included, the method of measurement is given in the text. Care was taken to ensure that specimens measured by the author did not duplicate supplementary data taken from the literature.

Wherever possible, for each character measured the range, the mean and standard deviation (SD) and the sample size (n) are given in parentheses. In some cases, standard error (SE) is given instead of standard deviation when the latter is not available. For some samples from the literature, fewer data are available, e.g. only the range and mean, or just one of these figures. Measurements of wing and tail are usually given to the nearest mm, and of culmen and tarsus to the nearest 0.5mm; means and standard deviations are rounded to one decimal place wherever possible.

Weights are given to the nearest gram, and means and standard deviations to one decimal place. They comprise unpublished data from the author and other observers, from various national ringing schemes, and from information on the labels of museum skins, as well as data from the literature. The hatching weight of chicks is given if known. In some cases the sources of weights are cited, for example where data are very few and/or are not compiled from unpublished information or from standard reference works.

Geographical Variation

Many rail species are polytypic but, as previously mentioned, the validity of many subspecies is questionable. This section starts with brief comments on the nature of the variation within the species, and the principal characters used in defining the subspecies. It also mentions races which are not recognised here as distinct, and are therefore included with others, and gives sources which the reader may consult for further information and discussion on the relevant taxonomy.

For each race the geographical range is given in detail, with separate descriptions of breeding and wintering areas if relevant. Only references for particularly significant, new or controversial distributional data are cited, to avoid making the text unreadable; all important references are usually cited in the sections on Distribution and Status, and Movements.

The sequence of subspecies attempts to be by geographical range, from North to South and West to East. The description of each subspecies is usually brief, referring only to those characters which serve to distinguish it from other subspecies, and thus often referring only to adult plumages. The reader is referred to the Description section for details of that subspecies which is fully described therein. More comprehensive details are given for subspecies which differ substantially from others, and for some widespread subspecies which were not chosen for the Description section (e.g. *Rallus longirostris crepitans*). Different plumage morphs are also described here. In cases where no description of the immature, juvenile or natal plumage is available for the race included in the Description section, details are included under any subspecies for which these plumages are known. Full details of measurements and weights are given for each subspecies when available.

Moult

In this section all moults are described, together with their pattern, timing and duration, as far as is known. In a few rail species moult details are fairly well known but there is very little information on the moults of most rail species and nothing at all is known for some. Even when a species is fairly well represented in museum collections there may be few or no specimens in active moult. This may be because most species seem to be particularly secretive when undergoing the complete moult, when many are flightless, and they are therefore rarely collected. Furthermore, there is very little juvenile and immature specimen material available for many species, again possibly because of the relative unobtrusiveness of young birds.

The moult terminology used in this book is the slightly modified version of that introduced by Dwight (1900) which is used by Cramp & Simmons (1980) and is in general use throughout the Old World. Because this terminology was based on temperate-zone passerine birds it is by no means satisfactory when applied to tropical birds and to others which show overlap in, or no constant relationship between, the timing of moult and reproduction. An alternative terminology was introduced by Humphrey & Parkes (1959), in which the generations are named in a way which avoids seasonal, reproductive or age criteria. This terminology is in general use in the Americas and the moult of New World rail species is usually described in the literature using Humphrey and Parkes terminology. Therefore in this book mention is often made of this terminology in New World rail species accounts, and also in other species where it is relevant. A comparison of terms used in the two nomenclatures is as follows. Plumages are given in italics, moults in bold print.

Standard terminology	**Humphrey & Parkes**
Natal/Downy chick	*Natal*
Postnatal/Prejuvenile moult	**Prejuvenal moult**
Juvenile	*Juvenal*
Postjuvenile moult	**First pre-basic moult**
Immature (non-breeding)	*First basic*
First prebreeding moult	**First pre-alternate moult**
First breeding/Adult breeding	*First alternate/Definitive alternate*
Postbreeding moult	**Pre-basic moult**
Adult non-breeding	*Definitive basic*
Prebreeding moult	**Pre-alternate moult**
Adult breeding	*Definitive alternate*

Almost all rails appear to have a distinct, often relatively plain, juvenile (juvenal) plumage, acquired via the postnatal (prejuvenal) moult. This plumage may be replaced very quickly, e.g. in the Buff-spotted Flufftail *Sarothrura elegans* postjuvenile (first pre-basic) moult may begin at three weeks of age, before all the natal down has been lost. The postjuvenile moult is usually partial, involving only the head and body plumage, but it may also be complete (e.g. in the American Purple Gallinule *Porphyrio martinica*). This moult often gives rise to a plumage which is distinct from that of both juvenile and adult, and is termed immature (first basic); a subsequent partial first prebreeding (first pre-alternate) moult then produces a first breeding (first alternate) plumage which is often virtually indistinguishable from that of the adult except for the presence of worn and/or relatively pointed juvenile remiges. In some species a further clue to the identification of first alternate plumage is the retention of unmoulted upperwing-coverts or of some juvenile body feathers, which may have a different appearance to those of the adult.

Some species (e.g. the Galapagos Rail *Laterallus spilonotus*, the Little Crake *Porzana parva* and Baillon's Crake *P. pusilla*) miss out the immature plumage, acquiring an adult-type plumage during the postjuvenile moult; this plumage is then retained throughout the first breeding season. In other species, such as the Inaccessible Rail *Atlantisia rogersi* and the Dusky Moorhen *Gallinula tenebrosa*, an immature-type plumage may be retained until the bird is two years old.

Adult postbreeding (pre-basic) moult always appears to be complete and gives rise to an adult non-breeding (definitive basic) plumage which may then be succeeded by the adult breeding (definitive alternate) plumage via a partial prebreeding (pre-alternate) moult. However, in many species the extent of this prebreeding moult is unclear and it may even be omitted. Moult of the remiges is simultaneous in many species, which thus become flightless until the new feathers have grown; in other species the remiges are moulted sequentially or irregularly. The timing of adult moults may be irregular, especially in the tropics and subtropics, and body moult may be prolonged or almost continuous. The Striped Crake *Aenigmatolimnas marginalis* has two complete moults per year, at least in captivity, a pattern which is not known to occur in any other member of the family.

Distribution and Status

For monotypic species this section begins with a detailed description of the geographical range, with separate descriptions of breeding and wintering ranges if relevant. As in the equivalent descriptions for races under Geographical Variation, only references for particularly significant, new or controversial distributional data are cited, to avoid making the text unreadable. For polytypic species this section begins with a very brief summary of the overall distribution, including breeding and wintering ranges if relevant.

Each species has a **distribution map** which shows its range. Note that, although ranges are usually shown as solid blocks of shading, the species is never evenly distributed throughout the area covered (except perhaps on some small oceanic islands) because the distribution of suitable habitat will always be discontinuous and cannot be mapped accurately. For non-migratory species [80%] shading indicates regions in which breeding is known or strongly suspected to occur, while [35%] shading marks areas in which breeding has not yet been recorded or suggested. For migratory species, [55%] shading marks areas where the species is a breeding visitor, [35%] shading shows areas where it is a non-breeding visitor, and [80%] shading shows areas where it occurs throughout the year, either because some birds in the population fail to migrate or because in the non-breeding season some migrants penetrate areas occupied by a discrete resident population, which may belong to a different subspecies. However, many species known to be migratory over at least part of their range do not have breeding and non-breeding areas differentiated on the map, simply because insufficient information is available to allow these ranges to be defined with any degree of accuracy. The maps give no indication of relative abundance in different regions.

Non-migratory species

■ Breeding   ▫ Non-breeding

Migratory species

■ Occurs throughout the year   ▨ Breeding visitor   ▫ Non-breeding visitor

On the maps, question marks denote areas in which occurrence is uncertain, while crosses mark extralimital or vagrant occurrences and also isolated sites at which the species has been introduced. For a very few migratory species migration routes are shown by arrows, or regular passage areas by hatching, but precise information on routes is lacking for most species. However, it can be assumed that a species may occur on migration in unshaded or unmarked regions between its breeding and non-breeding ranges. The ranges of subspecies are not shown on all of the distribution maps, principally because they are frequently imperfectly known or the validity of the subspecies themselves is questionable.

The status of the species is described in detail, if possible giving information on both former and current status, and population figures if available; the sources of information are cited. Particular attention is given to providing status details for globally or regionally threatened taxa. Factors limiting the distribution of the species are given if known. In polytypic species, the status of each race is normally described separately. If a species or race is legally hunted, as are several rails in the USA and Canada, estimates of the annual harvest are included. Any known reasons for recent changes in status are documented, and current threats are discussed. Wherever possible, conservation and management recommendations are included, usually from information in the literature.

For extinct species and subspecies this section describes their status when discovered and the known or supposed reasons for their decline and extinction. For all species information on the predators of fully grown birds is also included here; the reason for this is that predation has been a major cause of most recent extinctions and is thus discussed in detail for extinct species: it therefore seemed appropriate to include it here for all other species as well.

Movements

All types of movement are described in this section, including regular migrations between breeding and wintering areas, moult migrations, all other types of seasonal movement (including altitudinal migrations), and also dispersive movements, nomadism, irruptive behaviour and all types of local and long-distance movement made in response to changing habitat conditions such as rainfall, flooding, desiccation, burning, harvesting and vegetation cutting, and fluctuations in food supply. Adult and juvenile dispersal from nesting areas at the end of the breeding season is also covered. Even flightless species may make some seasonal or dispersive movements, and these are also summarised.

For species or races undergoing regular migrations, passage months and passage routes are given if known. The periods of occurrence of both migrating and wintering/non-breeding birds are given by country if possible, but details of the overall breeding and non-breeding ranges are normally included in the Distribution and Status section. Migration routes, the behaviour of migrating birds, and causes of mortality on migration are also described. Information from recoveries of ringed birds is summarised whenever available and the occurrence of vagrants is also mentioned. Habitat use by migrants is normally covered in the Habitat section.

Habitat

Rails can adapt to a very diverse range of habitat types, both natural and artificial, including terrestrial, semiaquatic, estuarine and littoral wetland habitats, open water, forest, deciduous woodlands, dense thickets, grassland, agricultural lands, and even small oceanic islands, where they may occupy almost any available terrestrial niche. Because they occupy such a variety of habitats, and because a sound knowledge of habitat requirements is crucial to the understanding of the distribution, dispersion, and seasonality of occurrence of rails, a detailed summary of habitat requirements and preferences is given whenever such information is available. For wetland species this may include a description of the characteristic plant species, the height and structure of the vegetation, the nature of the substrate, the degree and seasonality of flooding, the water salinity, the degree of interspersion of open and vegetated areas, and the significance of edge habitats. Less detailed information is usually available for forest and grassland habitats, but as much information as possible is included. The characteristics of breeding and wintering habitat are compared, and habitat use by migrating birds is discussed. Differences in the habitat preferences of sympatric species are also discussed. Finally, the altitudinal range of the species is given.

Attention is also given to the ability of some species to take advantage of artificial or man-modified habitats, because this ability may have a significant effect on the birds' numbers and distribution. It may compensate to a variable extent for loss of natural habitats and may even result in range expansions into previously unsuitable areas, e.g. the colonisation of artificial wetlands in arid regions.

Food and Feeding

In general the most aquatic rail species, such as the gallinules *Porphyrio*, moorhens and coots, are predominantly vegetarian, while those that inhabit terrestrial and palustrine habitats are either omnivorous or predominantly carnivorous, at least seasonally. This section aims to summarise all the types of food which are known to be taken by the species. The diet of many species is very poorly known, and for such species all known foods are listed. For some well studied species a large range of food items, both plant and animal, is often documented: rather than give a full list of the many plant and animal species involved, a summary by genus, family or order may be provided. Good quantitative analyses of diet are lacking for most species, even those which have been well studied, but a summary is given of such figures as are available. Wherever possible, the relative quantities of animal and plant foods eaten in different seasons are documented.

Foraging habitats, foraging methods and feeding behaviour are described, and activity patterns are mentioned if relevant. Any relevant information on drinking and water requirements is also given, and energy requirements are given for the very few species in which this topic has been studied. Information on kleptoparasitism also appears in this section. Finally, available information on the diet of chicks is summarised.

Habits

This section attempts to outline what is known of the daily activities and normal behaviour of rails. The main activity periods are mentioned, and a description is given of those aspects of behaviour which are likely to be relevant to anyone trying to observe the species, i.e. its relative shyness, its use of cover and open areas, its jizz, its normal methods of locomotion and its characteristic movements (such as tail-flicking), its reaction to disturbance (e.g. its willingness to fly) and its normal methods of escape from potential danger. Its flight action is described, as are its propensity for climbing in reeds, bushes, trees and other tall vegetation, and its ability to swim, dive and remain submerged. Reactions to human observers are described, and useful techniques for watching the species may be mentioned. Information is given on roosting behaviour, and on comfort activities such as sunbathing, bathing and preening. Social and sexual behaviour, and behaviour related to breeding, are discussed in other sections of the species account.

Social Organisation

Detailed studies of social organisation have been made in only a few rail species, in some of which it has been found to be complex although the predominant mating system in the family is monogamy. All species for which reliable information is available appear to be territorial during the breeding season, while some hold permanent territories. In this section the nature of the pair bond is described, together with the timing of its formation and its duration (when known). Wherever possible, information is given on the mating system, the degree and persistence of territorial behaviour and the size and nature of the territory both during and outside the breeding season. Details of breeding density are also given if available. Any available information on home ranges (size, degree of seasonal variation, etc) is also given. Information on helping and cooperative breeding is also summarised here.

Social and Sexual Behaviour

This section is entirely concerned with three basic types of behaviour: social, agonistic and sexual. Information given on social behaviour includes that relevant to social hierarchies, interactions with conspecific individuals and with other species, and details of flocking and the seasonal variations in the occurrence of aggregations.

All known types of agonistic behaviour are described, including spacing mechanisms in non-breeding flocks, threat displays, attack, chasing and fighting. Wherever possible brief comparison is made with the displays of closely related species. The nature, intensity and duration of territory defence are described, together with the roles of the sexes (and sometimes also of their offspring) in territory defence. Displays and other methods used in the defence of the nest and young are also described.

Sexual behaviour encompasses that involved with pair formation and maintenance, courtship, copulation, postcopulatory display, etc. It includes activities such as greeting displays, courtship feeding, reverse mounting and allopreening.

Activity which could be defined as play is also described for a very few species.

Breeding and Survival

This long and relatively complex section is usually subdivided into five subsections. If very little breeding information exists for a species, these sections may be dispensed with and details of individual breeding records given. **Season** gives details of the breeding season throughout the bird's range, usually listing breeding months by country. Months are normally those in which egg-laying is known or calculated to occur; however, where laying information is limited, details may also be given of months in which nests, chicks or juveniles have been recorded or in which birds in breeding condition have been collected. Comments on the relationship of breeding dates to rainfall and other proximate factors are given if available.

The **Nest** subsection includes information on nesting habitat, nest site, and the composition, structure, shape, size and position of the nest. The role of the sexes, and of helpers, in nest-building is outlined. If the species builds brood nests or brood/display platforms, this fact is also mentioned here. Under **Eggs**, clutch size is given, factors affecting clutch size are listed, and the production of replacement clutches is discussed. Egg shape, texture, gloss, colour and pattern are described, and egg sizes are given; egg weights (measured or calculated) are also included whenever possible. Data for clutch and egg sizes include range, mean and standard deviation whenever possible; egg measurements are given to the nearest 0.1mm if possible, as are mean and standard deviations. A separate dataset may be provided for each race for which a reasonably sized sample is available. The laying interval is also given.

The timing and period of **Incubation** are given, together with the role of the sexes, and of any helpers, in the incubation process. Information on synchronous or asynchronous hatching is also provided. Depending on the information available, the **Chicks** subsection may contain any or all of the following information about the development and behaviour of the young: the length of time the chicks remain in the nest; the development of their ability to walk, run, climb, swim and dive; the extent and duration of parental care; the chicks' response to predators; their feeding and begging behaviour; stages in the development of self-feeding and other behaviour; the development of the juvenile plumage; the age at which the first flight is made; the age at which the young become independent; and the age at which the young disperse from the natal territory. The condition at hatching is described as either precocial or semi-precocial; definitions of these terms vary but this work follows Nice (1962) in defining precocial as including chicks which leave the nest in the first day or two but follow the parents and are fed by them; chicks which do not leave the nest for at least 3 days are here described as semi-precocial. The plumage and bare parts of chicks and juveniles are given in the Description section. Other topics included here are nest parasitism, hatching and nesting success, egg infertility, survivorship of the young, and the known and suspected predators of eggs and young. Other information reviewed here (if known) includes the age of first breeding, the number of broods raised per season and the interval between clutches. The **Survival** subsection gives details of adult survival and longevity, and the greatest recorded age of individuals of the species. As such information is available for only a few rail species, this subsection is often omitted.

BIBLIOGRAPHY

The Bibliography lists all references consulted during the preparation of the book. Almost all of these references are cited within the text. The few titles which have been listed but not cited anywhere are included to make the bibliography as comprehensive as possible and also to show that the titles have not been overlooked, even though material from them may not have been used. It should be borne in mind that this list does not represent a complete bibliography of material on rails, but the references listed should provide sufficient information and citations to allow the reader to gain a comprehensive introduction to any aspect of rail biology, ecology and conservation.

INDEX OF SCIENTIFIC AND ENGLISH NAMES

Species are listed by their English vernacular name (e.g. Giant Wood-rail), and also by any alternative names, which are cross-referenced to the name which is used in the book (e.g. Florida Gallinule *see* Common Moorhen). Species are also listed by their scientific specific name followed by the generic name as used in the book (e.g. *poeciloperus, Nesoclopeus*). In addition subspecies are listed by their scientific name followed by the species name (e.g. *crepitans, Rallus longirostris*). A number in italic type refers to the first page of the species account and one in bold type to the plate number. Genera used in this work are also listed separately with numbers referring to the genus account.

Many synonyms are given in the species accounts, and these are listed as for the normal scientific names (e.g. *cinereus, Poliolimnas*), but the only number following the synonym is that referring to the systematic entry in which the synonym is listed. Genera, such as *Ortygometra*, which appear only as synonyms do not have their own entry in the index.

TOPOGRAPHY

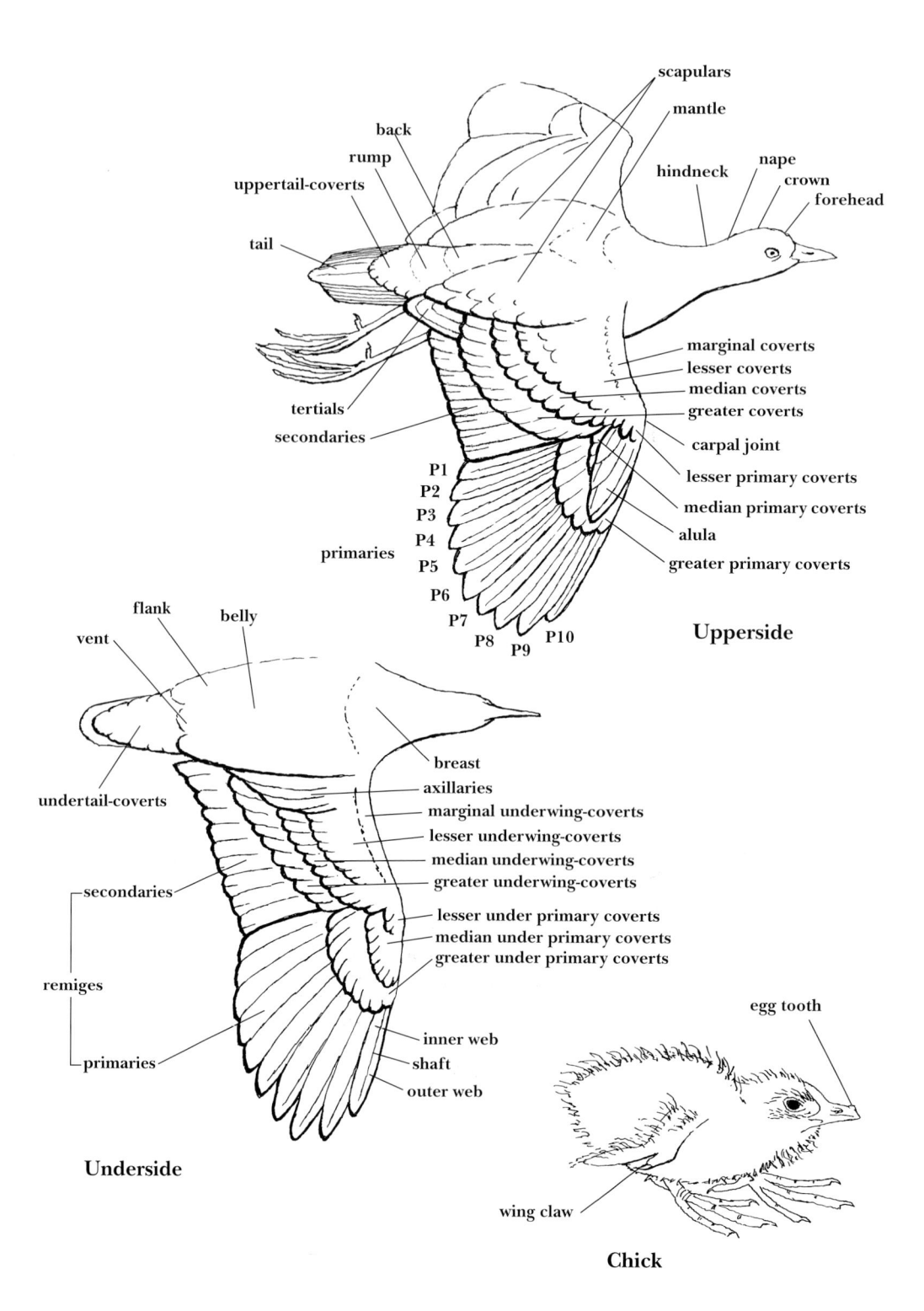

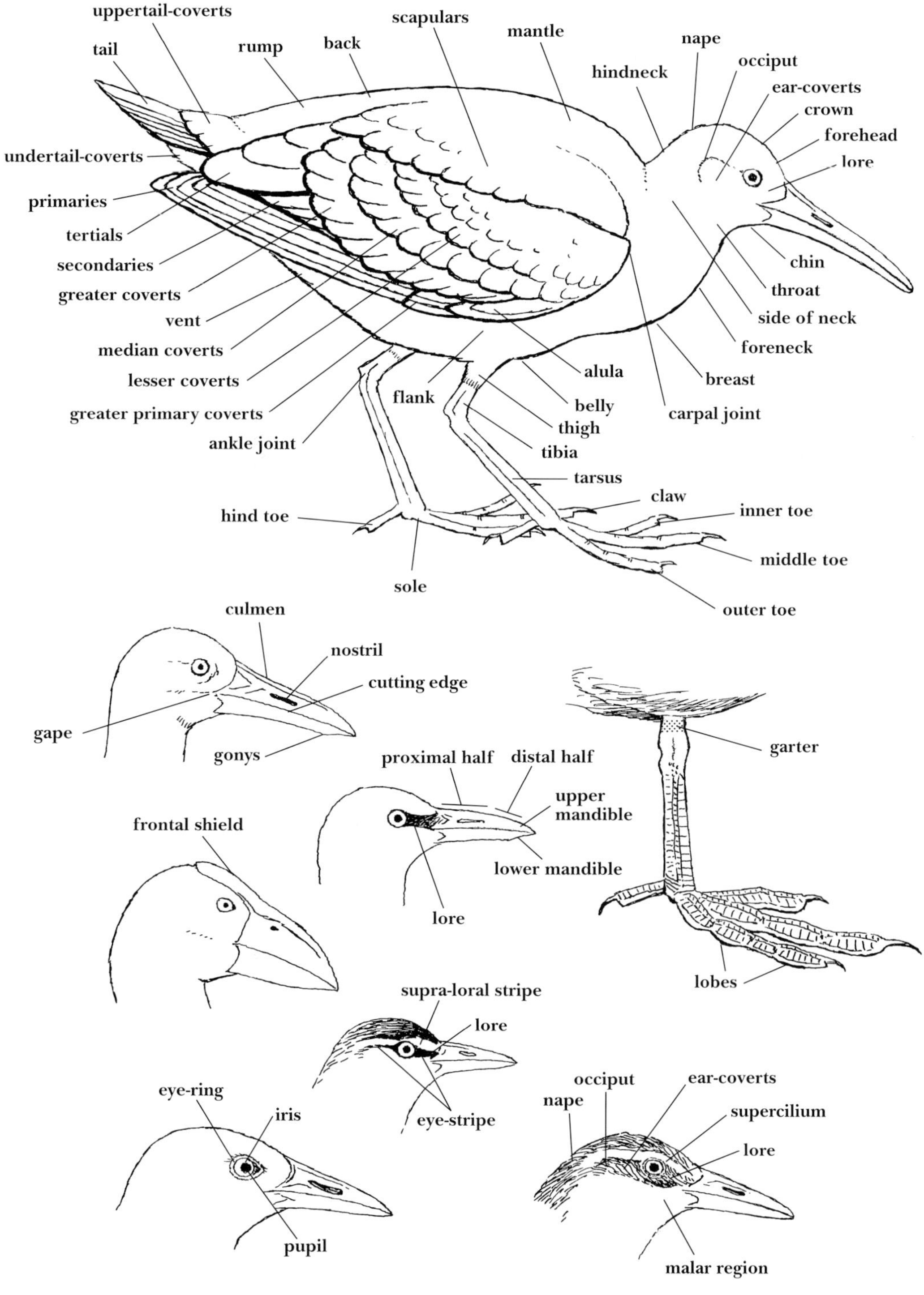
uppertail-coverts
tail
rump
back
scapulars
mantle
nape
hindneck
occiput
ear-coverts
crown
forehead
lore
undertail-coverts
primaries
tertials
secondaries
greater coverts
vent
median coverts
lesser coverts
greater primary coverts
ankle joint
flank
alula
belly
thigh
tibia
breast
carpal joint
chin
throat
side of neck
foreneck
tarsus
claw
inner toe
hind toe
middle toe
sole
outer toe
culmen
nostril
cutting edge
gape
gonys
garter
proximal half
distal half
upper
mandible
frontal shield
lower mandible
lore
lobes
supra-loral stripe
lore
eye-stripe
occiput
ear-coverts
nape
supercilium
eye-ring
iris
lore
pupil
malar region

PHYLOGENY AND CLASSIFICATION

THE GRUIFORMES

The order Gruiformes is a diverse assemblage of small to very large wading and terrestrial birds. Its relationships to other orders are unclear but it is normally considered closest to the Charadriiformes and the Galliformes. It is possibly polyphyletic, but DNA comparisons indicate that it may be monophyletic, composed of morphologically diverse groups that are more closely related to one another than to members of any other order (Sibley & Ahlquist 1990). Its familial composition remains controversial and it is perhaps the least well understood avian order from a phylogenetic perspective (Livezey 1998).

The Rallidae form by far the largest family in the Gruiformes. The family is cosmopolitan and comprises 134 extant species (including two first described in 1998), plus arguably 16 species extinct since 1600 and probably very many flightless species which have become extinct in the last 2,000 years (see Conservation and Extinction). In some classifications the rails have been assigned to an order Ralliformes, while some have allied the Rallidae to the Charadriiformes; DNA-DNA hybridisation evidence indicates that the family shares a common ancestor with both gruiform and charadriiform birds (Sibley & Ahlquist 1985, Sibley *et al.* 1993). It has also been suggested that the charadriiform family Jacanidae may be derived from *Gallinula*-like stock and could be placed within the Rallidae, but resemblances are superficial and are due to convergence: osteological characters, and recent biochemical analyses, place the jacanas alongside the other shorebirds of the Charadriiformes (Jenni 1996).

The relationships of the Rallidae to the other gruiform families have not been fully resolved. The family has been placed in its own superfamily (Ralloidea) within the suborder Grues, together with the Aramidae (Limpkin), Psophiidae (trumpeters) and Gruidae (Cranes) (Sibley & Ahlquist 1972). Skeletal morphology suggests a close alliance of the Rallidae, the Psophiidae and the Heliornithidae (finfoots), while the Aramidae, Eurypygidae (the Sunbittern) and Cariamidae (seriemas) of South America, the Rhynochetidae (the Kagu) of New Caledonia and the extinct Aptornithidae of New Zealand are also closely related, and it has been proposed that some of these families could be included as subfamilies within the Rallidae (Marchant & Higgins 1993). A phylogenetic analysis based on morphological characters (Livezey in press), indicates monophyly of the Grues (Psophiidae, Aramidae, Gruidae, Heliornithidae and Rallidae) and a sister relationship between the Heliornithidae and Rallidae. However, DNA-DNA hybridisation evidence indicates that the Rallidae form a cluster separate from cranes and their allies, which has had a distinct lineage for a long time, and it has been proposed that the rails should be elevated to their own suborder, the Ralli, alongside the Grues (Sibley & Ahlquist 1985, 1990).

THE FOSSIL RECORD

The fossil record tells very little about the origins and relationships of the Rallidae (Olson 1977). The earliest fossils which can be assigned to the family are from the Lower Eocene, about 50 million years ago, and reveal little about how these birds may have resembled modern rails. It is possible that the family existed earlier than the Eocene but there is no solid evidence for this, although recent DNA-DNA hybridisation studies have indicated that the rails may have diverged from the other gruiform groups as long as 86 million years ago, in the Upper Middle Cretaceous (Sibley & Ahlquist 1985). Only in the Upper Oligocene and Lower Miocene, 20-30 million years ago, are three genera of fossil rails (*Rallicrex*, *Palaeoaramides* and *Paraortygometra*) to be found which are based on adequate and diagnostic material, and by this period these birds had achieved a morphology not greatly different from that of modern rails (Olson 1977).

Fossil rails occur fairly regularly in various younger Tertiary deposits from Europe, Asia and North America, and most continental fossil rails from Pliocene and Pleistocene deposits have been assigned to modern genera. As well as the Tertiary evidence in continental deposits, the more recent late Quaternary deposits of oceanic islands have produced numerous well preserved specimens of extinct rails, most of which can be assigned to modern genera, and the discovery of many more fossil species on oceanic islands may be predicted (e.g. Olson 1989). Only five extinct genera of island rails are known, and of these *Hovacrex* may be synonymous with *Gallinula*, while *Capellirallus* and *Diaphorapteryx* of New Zealand, *Aphanapteryx* of the Mascarenes, and the enigmatic *Nesotrochis* from the Greater Antilles, are valid, being morphologically quite distinct from modern genera (Olson 1977).

CLASSIFICATION OF THE RALLIDAE

Arrangements within the Rallidae have varied, and several classifications have been proposed. The first attempt, by Sharpe (1894), listed 50 genera and 165 species of recent rails, while Peters (1934) gave 52 genera and 138 species. The family was traditionally divided into three subfamilies – the Rallinae, the Gallinulinae and the Fulicinae (see Brodkorb 1967), although there was little justification for this treatment and the assignment of many genera to a particular subfamily was often purely arbitrary (Olson 1973b). Subsequently Verheyen (1957) proposed five subfamilies, but with no rationale supporting this classification, and later Olson (1973b) produced a new classification, reducing the number of genera to 35 and the number of subfamilies to two, the Himantornithinae and Rallinae, the former being intermediate between the Rallinae and Psophiidae and containing the single species *Himantornis haematopus*, the Nkulengu Rail of West and west-central Africa. Olson did not provide a full list of species but discussed about 135, while the most recent classification (Sibley & Monroe 1990) lists 34 genera and 142 species (extant or very recently extinct) and follows Olson's taxonomy and sequence except where modern genetic or biochemical data are available.

Olson (1973b) considered *Himantornis* to be the most primitive and distinctive rail. Its downy young are uniquely patterned in the Rallidae and its skull is unlike that of any rail but is very similar to that of the Psophiidae, while several elements of its postcranial skeleton are also distinct from those of other rails and closer to *Psophia*. He found that all other rails were so alike morphologically that further subfamilial separation was unwarranted. As he pointed out, one of the difficulties of rail taxonomy arises from the relative homogeneity of the family, rails for the most part being rather generalised birds with few groups clearly defined by morphological modifications. As a result, some particularly well-marked genera had previously been elevated to subfamilial rank on the basis of characters which, in more diverse families, would not be considered significant.

Although Ripley (1977) subsequently reduced the number of rail genera to 18, Olson's classification has found general acceptance. The most significant subsequent reappraisal of relationships within the Rallidae has been made by Sibley and Ahlquist (1985, 1990) on the basis of DNA-DNA hybridisation. These workers suggested that the flufftails *Sarothrura* of Africa and Madagascar diverged from the rest of the group about 60 million years ago in the Palaeocene and they therefore proposed that the flufftails be placed in a separate family, the Sarothruridae, within its own superfamily, the Sarothruroidea (Sibley & Ahlquist 1985). They subsequently suggested that confirmation of this arrangement should wait until further genetic material becomes available (Sibley & Ahlquist 1990). The flufftails are a distinctive genus: they are strongly sexually dimorphic in plumage and they lay white, unmarked eggs, both of which characters are rare in the family; they also have distinctive hooting or moaning voices. Olson (1973b) had proposed that *Sarothrura* was closely allied to the *Rallicula* (herein included in *Rallina*) forest-rails of New Guinea, mainly on the basis of plumage characters and the fact that *Rallina* species also lay white eggs. He had also suggested a possible link with the genera *Coturnicops* and *Micropygia*, small species which have a similar build and ecological preferences to flufftails; furthermore, the three *Coturnicops* species have white secondaries, a character which they share with the White-winged Flufftail *Sarothrura ayresi* and which is found in no other rail. A lack of genetic material from *Rallina*, *Coturnicops* and *Micropygia* prevented Sibley and Ahlquist from commenting on these possible relationships, and the problem of the affinities of the flufftails remains unresolved. It is worth noting that the voice of *Coturnicops noveboracensis* is totally unlike that of any *Sarothrura* species and that there is no justification for Ripley's (1977) inclusion of the *Sarothrura* species in *Coturnicops* (Keith 1978).

This work follows the taxonomic treatment of Olson (1973b), modified in some cases by more recent studies. It follows Sibley and Monroe (1990) in emphasising the distinctiveness of the genus *Sarothrura* by placing it at the beginning of the family but not assigning it to a separate subfamily. However, this has the disadvantage that the primitive and distinctive genus *Himantornis* cannot be given the subfamilial rank which it deserves.

RELATIONSHIPS, EVOLUTION AND BIOGEOGRAPHY

Most works on the family recognise, for convenience, two broad 'natural groups' within the Rallidae: the crakes, rails and wood-rails, most of which are terrestrial; and the gallinules (including moorhens) and coots, which tend to be more aquatic. The term 'rail' is applied generally to the whole family and also to longer-billed species in many genera, while 'crake' is applied mostly to the smaller, short-billed, species, particularly in the genera *Laterallus* and *Porzana*. Neither term has a precise taxonomic meaning, and different species in different genera are sometimes called by either name, while crake-like birds seem to have evolved independently several times. The term 'gallinule' can cover all the birds in the second group except the coots, though it is often restricted to *Gallinula* and *Porphyrio*.

The geographic origins of the Rallidae have been obscured by the antiquity, cosmopolitan distribution, and inadequate taxonomy of the family (Olson 1973b). However, from Olson's work it is clear that the greatest number of rail species and of peculiar genera, and the most primitive species, are found in the Old World tropics. The New World has fewer groups, most of which are apparently derived from Old World stem groups, although a few genera appear to have specialised and radiated in the New World, some of which (e.g. *Rallus* and *Fulica*) have subsequently re-invaded the Old World. Olson suggested that the most primitive rails are found in forest and that an adaptation to aquatic habitats and grassland may be a more recent development, but admits that such adaptation could have evolved several times in different groups.

The remainder of this section is based on the classification by Olson (1973b), from which most information is taken. In attempting to determine relationships within the Rallidae, *Himantornis* provides some clues as to which species may be the most primitive, useful indications being *Psophia*-like skeletal characters, long slender tarsi, patterned natal down and forest habitat. Flightlessness has evolved many times in rails and therefore has no major phylogenetic significance; the incidence of flightlessness and its associated morphology can be used as a taxonomic character in the family only at the specific or subspecific level.

The primitive nature of the forest-dwelling genus *Canirallus* of Africa and Madagascar is indicated by its sharing two primitive skeletal characters with *Himantornis* and by the possession of patterned natal down in the Madagascar Wood-rail *C. kioloides*. This species was separated into the genus *Mentocrex* by Peters on the basis of its imperforate nostrils (those of *C. oculeus* are perforate) but this character is not generically important. There are significant similarities in the plumage, structure and natal down of *Canirallus* and the four forest-rails endemic to New Guinea (sometimes called chestnut rails and separated into the genus *Rallicula*), and Olson placed these four species in a subgenus of *Canirallus*. However, on the basis of plumage similarities Ripley (1977) included them in the genus *Rallina* with four species of forest and marshland from Asia and Australia, although these latter differ in not being sexually dichromatic. Olson regarded *Rallina* as a separate group, apparently in transition from the woodland habitat of the more primitive rails to the wetland habitat more typical of the family. Opinions still differ as to whether the forest-rails should be included in *Rallicula* or *Rallina*.

This volume follows the sequence of genera adopted by Sibley & Monroe (1990), although the sequence of the first six genera is unsatisfactory in several ways. Interposing the genera *Coturnicops* and *Micropygia* between the genera of the *Canirallus/Rallina* group is somewhat arbitrary and reflects the assumed affinity which these small grassland and marsh crakes have with the flufftails, as described above, while also acknowledging the aforementioned possible affinities between the flufftails and the *Rallina* group and attempting to reflect the primitive status of *Himantornis* and *Canirallus*.

Two New World genera of small crakes, *Anurolimnas* and *Laterallus*, may be derived from Old World *Rallina* stock. Skeletally, *Laterallus* species are more similar to *Rallina* than to the crakes in or near *Porzana*, and two species show striking plumage similarities to South-East Asian *Rallina* (see Olson 1973b for details). The forest- and thicket-dwelling Chestnut-headed Crake *A. castaneiceps* was formerly placed in *Rallina*, to which it bears some resemblance in structure and coloration. The other two *Anurolimnas* species were formerly included in *Laterallus* but may be more properly included with the Chestnut-headed Crake in a distinct genus. However, the Chestnut-headed Crake has a loud, *Aramides*-like call, very distinct from the *Laterallus*-like calls of its congeners (see the relevant species accounts), so separation may well be valid.

The nine species remaining in *Laterallus* form a group of small crakes, predominantly of marsh and wet grassland, the South American representatives of which are relatively poorly known. A recently discovered and very distinctive form (*tuerosi*) of the Black Rail *Laterallus jamaicensis* is confined to marshes fringing Lago de Junín, central Peru (Fjeldså 1983b), and is considered Endangered. It is sometimes treated as a separate species and this may be justified because it differs from the other forms of the Black Rail as much as does the specifically distinct Galapagos Rail *L. spilonotus*; awarding it specific status emphasises the conservation priorities which should be accorded to it (Fjeldså 1983b, J. Fjeldså *in litt*).

The two Old World forest rails in the genus *Nesoclopeus* were both originally placed in *Rallina* before being assigned to *Eulabeornis*, then to *Nesoclopeus* and back again to *Rallina*. Their affinities are not clear: they appear to differ from *Rallina* in some skeletal characters, while some of their plumage characters are suggestive of the *philippensis* group of *Gallirallus*. They may prove to be intermediate between *Rallina* and *Gallirallus*.

The relationships of the group of barred-winged rails currently placed in the genus *Gallirallus* have been greatly obscured by the combination of some species with the more specialised species of *Rallus* and by the creation of several unnecessary genera for flightless forms of the group. The Weka *Gallirallus australis*, a large, flightless species endemic to New Zealand, at first sight appears to be a strange and distinctive rail and was originally the only species placed in the genus. Early classifiers were deceived by the neotenic characters associated with flightlessness and considered it a peculiar, primitive form without close relatives. However it shows marked similarities, in both plumage and skeletal characters, to the volant Banded Rail *G. philippensis*, and the differences in the wings, pectoral girdle, and some plumage features of the adults, are apparently recently derived neotenic characters. It occurs alongside the Banded Rail, this situation reflecting the multiple invasion of New Zealand by *G. philippensis*-like stock. The presence of two fossil relatives, one on North Island and one on an offshore island, indicates that the Weka has a long historical isolation on the New Zealand islands.

Other possible insular derivatives of *G. philippensis* stock are the recently extinct Tahiti Rail *G. pacificus* and the Lord Howe Rail *G. sylvestris*, both flightless species. The flightless New Caledonian Rail *G. lafresnayanus* is also regarded as a possible *G. philippensis* derivative by Fullagar *et al.* (1982), although Olson could see no indication of a relationship between it and the Lord Howe Rail and regarded it as having diverged so far that any external resemblance to a possible common ancestor has been obscured. He accordingly placed it in *Tricholimnas*, moving the only other species in that genus, the Lord Howe Rail, to *Gallirallus*. Other flightless or weak-flying *Gallirallus* species endemic to oceanic islands and possibly derived from *G. philippensis* stock are the Guam Rail *G. owstoni* and the recently described Roviana Rail *G. rovianae*, as well as the recently extinct Chatham Rail *G. modestus*, Wake Rail *G. wakensis* and Dieffenbach's Rail *G. dieffenbachii*, the last of these sometimes being considered conspecific with *G. philippensis* (e.g. Ripley 1977). A recent mitochondrial DNA (mtDNA) study (Trewick 1997) indicates that the flightless taxa *G. sylvestris* and *G. owstoni* are more closely related to the volant *G. philippensis* than to the flightless *G. australis*, which is genetically very distinct from all three. It also confirms that *G. dieffenbachii* merits specific status, and is genetically most similar to the sympatric *G. modestus* although it is morphologically very different to that species.

Another subgroup within *Gallirallus* is formed by the Barred Rail *G. torquatus* of South-East Asia and the New Britain Rail *G. insignis*, to which can be added the recently described flightless Okinawa Rail *G. okinawae*. The presumed extinct volant rail *G. sharpei*, known from a long-overlooked single specimen of unknown origin, is most similar to the Buff-banded Rail but differs from it strikingly in coloration and plumage pattern. The remaining species in the genus, the Slaty-breasted Rail *G. striatus*, is the only member of the group found in continental Asia and is considered by Olson to be an advanced form of the genus that has paralleled the evolution of the *Rallus* group towards a slender marsh-dwelling build.

The species of *Rallus* are much more specialised than those of *Gallirallus* and are highly adapted to a semi-aquatic existence in reedy marshes. Like the specialised small *Porphyrio* gallinules (formerly placed in *Porphyrula*), they have not colonised small, remote oceanic islands (Olson 1973a). The genus appears to have radiated in the New World, only three allopatric species being found in the Old World. The Water Rail *R. aquaticus* of Eurasia is one of the few Palearctic rails which does not migrate to sub-Saharan Africa, where it is replaced by the African Rail *R. caerulescens*. The third species, the Madagascar Rail *R. madagascariensis*, is endemic to Madagascar. All three probably had their origins in a single invasion of *Rallus* from the New World.

The recognition of the King Rail *R. elegans* as a distinct species, rather than a form of the Clapper Rail *R. longirostris*, has long provoked argument, and recent mtDNA studies have yielded inconclusive results (Avise & Zink 1988), but the separation is currently accepted by many authorities. In a recent study involving morphology, distribution and palaeontology, Olson (1997) concludes that *R. longirostris* is restricted to South America, the West Indies and coastal marshes of eastern North America and Yucatán, while all other populations, including those on the Pacific coast of North America (currently regarded as Clapper Rails), plus those in C and W Mexico and in Cuba, are races of *R. elegans* which are derived from an original widespread King Rail stock and which became isolated following desertification of the American west. The earliest record of a large *Rallus* in North America is that of the nominal species *R. phillipsi* (Wetmore 1957) from the Mio-Pliocene of Arizona 5.5 million years ago.

The genus *Lewinia* comprises three species which are often treated as one. Olson regarded the Brown-banded Rail *L. mirificus* of the Philippines as a well-marked form of Lewin's Rail *L. pectoralis*, while the Auckland Rail *L. muelleri* is sometimes also regarded as being conspecific with Lewin's Rail. Furthermore, Lewin's Rail is sometimes placed in the genus *Rallus*, although its skeleton differs in several respects from the true *Rallus* type. However its skeleton is almost identical to that of the much larger White-throated Rail *Dryolimnas cuvieri* of Madagascar and Aldabra, and Olson placed it in *Dryolimnas*, regarding the genus as rather primitive and forming part of a pro-*Rallus* group. Sibley & Monroe (1990) suggest that it is not congeneric with *Dryolimnas* but provide no evidence for this view.

The genus *Crex* is often reserved for the Corncrake *C. crex*, the African Crake *C. egregia* having been retained in the monotypic genus *Crecopsis* (Sharpe 1893) by Olson, a treatment which was endorsed by Dowsett & Dowsett-Lemaire (1980) and Sibley & Monroe (1990). However, Snow (1978) and Urban *et al.* (1986) consider it closely related to the Corncrake, and it is included in *Crex* in the present volume. There is no really convincing evidence in favour of either alternative. It is certainly not referable to *Porzana*, as was suggested by Benson & Winterbottom (1968), and endorsed by Ripley (1977), on the basis of a superficial resemblance to *Porzana albicollis* of South America: its nasal bar is quite unlike that of *Porzana*.

The unpatterned but distinctive Rouget's Rail *Rougetius rougetii* of Ethiopia was placed after *Amaurornis* by Sharpe (1893) and in the "Gallinulini" of Verheyen (1957). However, its tenuous nasal bar is unlike that of *Amaurornis* or of any gallinule, while its inclusion in *Rallus* (Urban & Brown 1971) is unjustified. Pending further study it is probably best retained in its own genus. Similarly, the poorly known Snoring Rail *Aramidopsis plateni* is better retained in its own genus at present. Its resemblance to the Neotropical genus *Aramides* on plumage is superficial, and it differs from that genus in bill shape and in its more robust tarsi; it may have more affinities with a pro-*Rallus* group.

The genus *Atlantisia*, confined to islands in the South Atlantic, is represented by three flightless species, two of which were exterminated soon after the arrival of man while the third, the Inaccessible Rail *A. rogersi*, survives on one island of the Tristan da Cunha group. The genus is remarkable for the extreme size range of its constituent species. The Inaccessible Rail is very small, the Ascension Rail *A. elpenor* was about the size of a Virginia Rail *Rallus limicola*, and the Great St Helena Rail *A. podarces* was the size of a Weka. No other genus of rails varies in size to this extent. Olson considered that these species represent neotenous relict forms of ancestors of a pro-*Rallus* group.

The Neotropical genus *Aramides* appears to be primitive but to have no apparent ties with the primitive Old World genus *Canirallus*. It may be close to the stock that gave rise to the *Amaurornis* assemblage, but its affinities are unclear. The *Gymnocrex* species of forest in New Guinea and Indonesia show plumage similarities to *Aramides* but are separable at the generic level on skeletal characters (see Olson 1973b) although they may be derived from the same stock. The monotypic Neotropical genus *Amaurolimnas* is possibly also derived from *Aramides*; its bill structure is identical to that of the smaller *Aramides* species, it superficially resembles *Aramides* rails in build, posture and bare part colours, and its voice is very reminiscent of *Aramides*.

The proper allocation of species between *Porzana* and *Amaurornis* is difficult. The five *Amaurornis* species *akool, isabellinus, olivaceus, moluccanus* and *phoenicurus* form a rather basic stock from which both the *Porzana* assemblage and the gallinules could have arisen, and some *Amaurornis* species have previously been included in *Porzana* and *vice versa*. There is reasonable evidence to place the Black-tailed Crake *A. bicolor* in *Porzana* (Inskipp & Round 1989), but the Black Crake *A. flavirostris* of Africa and the closely related Sakalava Rail *A. olivieri* of Madagascar are best retained in *Amaurornis*: the skeleton of the Black Crake is virtually identical to that of *A. phoenicurus*.

The genus *Porzana* is possibly polyphyletic and is very difficult to categorise because of differing taxonomic treatments: for example, Ripley (1977) lists 27 species but Olson only 13. A natural group within *Porzana* is formed by the five living species *parva, pusilla, porzana, fluminea* and *carolina*, and the recently extinct Laysan Crake *P. palmeri* (which is probably a very recent derivative of *P. pusilla*). The extinct Kosrae Crake *P. monasa* and the Henderson Crake *P. atra* are very close to the Spotless Crake *P. tabuensis* and could have been derived from it. The affinities of the Ruddy-breasted Crake *P. fusca* are also disputed, but both Ripley and Olson consider it best placed in *Porzana* although it has also been assigned to *Amaurornis*. The Band-bellied Crake *P. paykullii* is placed in *Rallina* by Ripley

(1977) and further study of its relationships may be required (White & Bruce 1986). The White-browed Crake *P. cinerea* was included by Olson in the genus *Poliolimnas* alongside the Yellow-breasted Crake *P. flaviventer*. Although the separation of the latter may be appropriate on the basis of bill structure and plumage features (Olson 1970, 1973b), Mees (1982) has shown that similarities between the two species are superficial while differences are so marked that the idea of a close relationship is untenable. The Striped Crake *Aenigmatolimnas marginalis* of Africa is close to *Porzana* but is separated on the basis of its deeper bill with a very broad, almost vertical, nasal bar and smaller bony nostril.

Amaurornis-like stock could also have given rise to the assemblage of New World rails in the genera *Cyanolimnas*, *Neocrex* and *Pardirallus*. All the species in these genera share a similar drab plumage, with the exception of the Spotted Rail *P. maculatus* which, however, has a similar dark phase of its juvenile plumage; and all but *P. nigricans* have a red spot at the base of the bill. Olson considers that *Cyanolimnas* forms a near-perfect intermediate between the other two genera. The long-billed *Pardirallus* rails are often mistakenly placed in *Rallus*, from which they differ very markedly in skeletal characters.

Olson considers that another line, probably derived from *Amaurornis*, comprises the Australasian monospecific genera *Eulabeornis*, *Habroptila* and *Megacrex*. *Eulabeornis* bears a strong superficial resemblance in plumage to the Neotropical genus *Aramides*, but lacks the barred underwings and slender tarsus of that genus. Ripley (1977) places *Habroptila* in *Rallus* but gives no reason for doing so, and includes *Megacrex* in *Amaurornis*. Olson includes *Megacrex* in *Habroptila* on structural similarities, but Mees (1982) justifies their separation on the basis of differences in the structure of the bill and frontal shield and because the tibiotarsus of *Habroptila* is feathered. Recent investigations using mtDNA indicate that *Megacrex* falls within the *Gallirallus/Rallus* group (Trewick 1997). The distinctive monotypic genus *Gallicrex* tends to bridge the differences between *Amaurornis* and the 'gallinules'.

The purple gallinules, formerly separated into the three genera *Porphyrula*, *Porphyrio* and *Notornis*, are considered by Olson to form an obviously monophyletic group, as was originally suggested by Mayr in 1949. *Porphyrula* differs from *Porphyrio* only in its smaller size, less massive bill and more oval nostril, while the two genera share a number of specialised characters. The three species sometimes placed in *Porphyrula* are closer to each other than to *Porphyrio*, so a case could be made for maintaining them as a subgenus, but they should not be separated at the generic level. The mtDNA study of Trewick (1997) shows that there is no justification for Ripley's (1977) placing of the American Purple Gallinule *Porphyrio martinica* in *Gallinula*. The Takahe of New Zealand, formerly given the name *Notornis mantelli*, is simply a very recent, flightless derivative of *Porphyrio* (Trewick 1997): see the species account for a discussion of its South I and North I forms. The Purple Swamphen *P. porphyrio* shows great geographical variation and at least the races *madagascariensis*, *pulverulentus* and *melanotus* could be redefined as species (Trewick 1997), while there is some justification for also raising *poliocephalus* and *indicus* to specific status (Sangster 1998): further study and evaluation of this complex assemblage of taxa are required.

Despite the distinctive appearance of the extinct Samoan Moorhen *Gallinula pacifica* and the San Cristobal Moorhen *G. nesiotis*, which has previously resulted in their separation into the genera *Pareudiastes* and *Edithornis* respectively, it is now generally accepted that they are correctly placed in *Gallinula*. Similarly, it is no longer regarded as correct to place the Spot-flanked Gallinule *G. melanops* in the monotypic genus *Porphyriops*, or the Tasmanian Native-hen *G. mortierii* and the Black-tailed Native-hen *G. ventralis* in the genus *Tribonyx*: the skeletons of all three species show no differences which can be considered of generic significance when compared to *Gallinula*.

The coots *Fulica* are very similar to *Gallinula* both skeletally and in their adult and juvenile plumages, and are derived from a *Gallinula*-like stock which has become adapted for diving. Their centre of species diversity is South America and it seems likely that they originated there and later spread to the Old World. The genus is well defined and has diverged relatively little from its ancestral stock. The Hawaiian Coot *F. alai* has only recently been treated as a species distinct from the Common Coot *F. atra* and the American Coot *F. americana*, while the Caribbean Coot *F. caribaea* is sometimes regarded as a morph of the American Coot although West Indian populations mate assortatively without evidence of crossing (Sibley & Monroe 1990).

A new phylogenetic study of the Gruiformes, with particular emphasis on the Rallidae and using morphological characters, has recently been completed by B. C. Livezey of the Carnegie Museum of Natural History. Although details cannot be made available before the results are published in full (Livezey in press), the author has kindly supplied a summary of his major findings, and mention has been made of this study in the discussion on the composition of the Gruiformes (see above). Using cladistic techniques, phylogenetic relationships among fossil and modern genera of the Gruiformes were estimated based on 381 primary osteological characters. Relationships among modern species of Grues (Psophiidae, Aramidae, Gruidae, Heliornithidae and Rallidae) were assessed based on these characters augmented by 189 characters of the definitive integument. Autapomorphic divergence (i.e. that involving characters, evolutionarily advanced from the ancestral state, which are unique to the genus under consideration) was comparatively high for 5 genera, including the rail genera *Himantornis* and *Fulica*. Groupings established within the Rallidae included:

- monophyly of the Rallidae exclusive of *Himantornis* and a clade comprising *Porphyrio* (in cluding '*Notornis*' and '*Porphyrula*');

- a poorly resolved basal group of genera, including *Gymnocrex*, *Habroptila*, *Eulabeornis*, *Aramides* and *Canirallus* (including '*Mentocrex*');

- an intermediate clade comprising *Anurolimnas*, *Amaurornis* and *Rougetius*;

- monophyly of two major subdivisions of the remaining rallids, one comprising *Rallina* (including '*Rallicula*') and *Sarothrura*, the other comprising the apparently paraphyletic 'long-billed' rails (e.g. *Cyanolimnas*, *Rallus*, *Gallirallus*) and a variably resolved clade comprising 'crakes' (e.g. *Atlantisia*, *Laterallus*, *Porzana*), and *Amaurornis*, *Gallinula* and *Fulica*.

Relationships among 'crakes' were poorly resolved: *Laterallus* may be paraphyletic and *Porzana* is evidently polyphyletic and poses substantial challenges for reconciliation with current taxonomy. Relationships among the species of *Amaurornis*, *Gallinula* and *Fulica* were comparatively well resolved, while exhaustive, fine-scale analyses of several genera (including *Aramides*, *Porphyrio*, *Rallus*, *Laterallus* and *Fulica*) and species complexes (*Porphyrio porphyrio* group, *Gallirallus philippensis* group, and *Fulica americana* group) revealed additional patterns of relationships. Provisional placements of selected subfossil rallids (e.g. *Diaphorapteryx*, *Aphanapteryx* and *Capellirallus*) were made.

This classification reaffirms the advisability of separating *Himantornis* from the rest of the family, and confirms the affinity between *Sarothrura* and *Rallina*. Its revised groupings of genera differ substantially from those currently in use, including those suggested by the limited amount of DNA studies which have been carried out to date. It is probably wise to defer a comprehensive revision of rallid classification until the results of extensive DNA studies are available for comparison with those from morphological studies of living and extinct forms, and also to take into account other characters, such as voice and behaviour, which might provide clues to affinities at least at the genus level.

NEWLY DESCRIBED SPECIES

Four new living rail species, all from islands and at least two flightless, have been described in the last 20 years. The Okinawa Rail *Gallirallus okinawae* was first seen on Okinawa Island in 1978 and was described in 1981 (Yamashina & Mano 1981). It is considered to have evolved from a progenitor of the Barred Rail, while an unnamed fossil rail from Okinawa may also be ancestral to the Okinawa Rail (Yamashina & Mano 1981). The Roviana Rail *G. rovianae* from the Solomon Islands is known only from the holotype, collected in 1977, and from a number of sight records on several islands in the group (Diamond 1991, G. Dutson verbally). It belongs to the widespread group of Pacific *Gallirallus* species that includes the volant Buff-banded Rail *G. philippensis*, being most similar to that species and to the flightless Guam Rail *G. owstoni* and the extinct flightless Wake Rail *G. wakensis*, and is probably derived from an ancestor similar to the Buff-banded Rail (Diamond 1991). A distinctive form has recently been found on Kolombangara I (Gibbs 1996).

Two new species have just been described by F. R. Lambert from Karakelong I in the Talaud Archipelago, Indonesia (Lambert 1998a, b). The Talaud Rail *Gymnocrex talaudensis* is most closely related to the Bare-eyed Rail *G. plumbeiventris* of the Moluccas, Aru Is and New Guinea. It is a secretive but probably not rare inhabitant of wet grassland and rank vegetation at the edge of forest, and although it may also occur on other islands it is probably endemic to the Talaud Archipelago (Lambert 1998a). The Talaud Bush-hen *Amaurornis magnirostris* is most similar to the Plain Bush-hen *A. olivaceus* of the Philippines and occurs in habitats ranging from rank vegetation near forest edges to dry dense primary forest slopes (Lambert 1998b). The chances are good that other species await discovery on poorly known small islands, and even on larger islands such as New Guinea (e.g. see Mayr's Forest-rail *Rallina mayri* species account).

MORPHOLOGY AND FLIGHTLESSNESS

MORPHOLOGY

The family Rallidae comprises a large but rather homogeneous group of small to large, terrestrial, marsh and aquatic non-passerine birds, many of which live a secretive and skulking existence in dense ground vegetation. The smallest species, the Black Rail *Laterallus jamaicensis* of the Americas, is 12-15cm long and weighs 20-45g, while the Inaccessible Rail *Atlantisia rogersi*, with a length of 13-15cm and a mean mass of c. 40g, is the smallest living flightless bird. The largest living rail is the flightless Takahe *Porphyrio mantelli* of New Zealand, which is 63cm long and stands 50cm tall; males weigh 2.15-4.25kg. In most species the sexes are similar in size, although the male is often slightly larger than the female. Marked sexual dimorphism in size occurs in only a few rails, and in all cases the male is the larger sex; the greatest size difference occurs in the Watercock *Gallicrex cinerea*, in which the male may average 50-60% heavier than the female (few weights are available for comparison) and 17-20% greater in wing and tail measurements. In the Weka *Gallirallus australis*, the male normally averages 30-37% heavier but only 3.5-8% greater in wing and tail measurements, while in the race *hectori* the male may be almost twice as heavy as the female but averages only 4.5-8% longer in wing and tail. The Dusky Moorhen *Gallinula tenebrosa* shows great racial variation in size, nominate Australian birds averaging 58% heavier than *neumanni* of New Guinea, with wing and tail measurements which, for a limited series of *neumanni*, do not overlap.

The bodies of rails are short, and are often laterally compressed for ease of movement through dense low vegetation, whence comes the expression "thin as a rail". The neck is variable in length but can be quite long in some species, and has 14 or 15 cervical vertebrae. The wings are short, broad and rounded. There are normally 10 primaries, fewer in some flightless forms, and a minute 11th primary is present in some of the larger species. There are 10-20 secondaries, and the wing is usually diastataxic (5th secondary absent) but is eutaxic (5th secondary present) in *Himantornis* and *Amaurolimnas*. In some species the alula has a sharp, curved claw which is used by the young, and possibly also by the adults, to assist climbing. Flight over short distances is usually low and weak, and rails usually fly with dangling legs. Despite this apparent inability to fly strongly, some species migrate or disperse over long distances, sometimes making sea and desert crossings, and the family is renowned for its ability to colonise remote oceanic islands. In terms of their morphology and their ability to accumulate food reserves, rails are perfectly capable of apparently sustained and certainly very long flights but their flight performance is relatively poor and they migrate close to the ground. The high degree of vagrancy in the family is indicative of the readiness with which the birds are apparently blown off course by unfavourable winds (see Movements).

The short tail contains 6-16 rectrices (normally 12), is square to rounded, arched and soft, and is sometimes decomposed, especially in flightless species and the genus *Sarothrura*. It is often raised or flicked, showing the undertail-coverts, which in many species are contrastingly coloured or patterned (see Behaviour).

The bill is sometimes quite slender, straight or slightly downcurved, and slightly longer than the head, for example in *Rallus*; it is often quite short and somewhat laterally compressed, as in most gallinules and coots; in the crakes it may be quite small and fine; and it may be massive and deep or laterally compressed, as in some *Porphyrio* species. Long-billed rails probe in soft ground and litter, and species with small, fine bills take primarily insects and small seeds, while species with large, stout bills may graze, dig out roots and tubers, or tear off vegetation. Many species tend to be omnivorous and to have unspecialised bills.

The bill is often brightly coloured and in some species has a patch of different colour at its base. Gallinules and coots have a frontal shield, continuous with the outer covering of the bill and which may be of a similar or contrasting colour to the bill. The colours of the bill and shield often become duller in the non-breeding season, when the shield may also shrink. The male Watercock has the frontal shield continued into a long red horn which almost disappears in the non-breeding season, while the Horned Coot *Fulica cornuta* has a long, extensible and erectile black proboscis with two tufts at its base. The Invisible Rail *Habroptila wallacii* and the New Guinea Flightless Rail *Megacrex inepta* also have a small frontal shield.

The nostrils are usually in a large depression (but not in *Porphyrio*) and are pervious in most species, perforate in some. The olfactory process is well developed in most species, and the ample supply of nasal glands and ducts, with the relative size of the olfactory bulb of the brain, indicates that rails can easily and efficiently discriminate odours, possibly a help in foraging (Bang 1968, Ripley 1977). Rails of littoral habitats have adapted readily to salt water and are able to secrete salt by means of adapted nasal glands (see Food and Foraging).

Rails have well-developed legs which are often long and slender but are usually strong, and may be laterally compressed. The toes are often long, but in species of predominantly dry habitats may be rather short and heavy. The hind toe is slightly raised, and is used as a plantar toe, helping in the balance of the foot structure. In most gallinules, and also in some crakes, notably the Yellow-breasted Crake *Porzana flaviventer* and the White-browed Crake *P. cinerea*, the toes are particularly elongated and the legs modified for walking, jacana-like, on mud and floating or emergent vegetation. In some species the legs and feet are brightly coloured, sometimes becoming duller in the non-breeding season. Leg colour may also change as birds mature, and may be helpful in determining a bird's age, as in the American Coot *Fulica americana*.

All rails can swim; they can also dive and can use the wings under water if necessary. In coots, which are fully aquatic, the pelvis and legs are modified for diving, and most species have well-developed lateral lobes on the toes to aid swimming. The Spot-flanked Gallinule *Gallinula melanops* also has narrowly lobed toes, while those of the Common Moorhen *G. chloropus* are somewhat emarginated.

The plumage of rails is often cryptic, common colours being sombre browns, chestnut, black, blue-grey or grey; however in the genus *Porphyrio* the plumage is predominantly iridescent purple, blue or green. The upperparts are frequently spotted, barred or streaked, but may be completely uniform. The flanks are often strongly barred, for example in most *Rallus*, *Gallirallus* and *Porzana* species, and the undertail-coverts may contrast strongly with the rest of the plumage. The Weka, the Clapper Rail *Rallus longirostris* and the Chestnut Rail *Eulabeornis castaneoventris* have well-defined plumage morphs. In most species the sexes are very similar in appearance, exceptions being the flufftails, the *Rallina* forest-rails endemic to New Guinea, the Striped Crake *Aenigmatolimnas marginalis*, the Little Crake *Porzana parva* and the Watercock. Only the Watercock shows any significant seasonal change in plumage colour or pattern. In other species, most minor differences which exist between breeding and non-breeding plumages are difficult to see in the field, and are often caused by abrasion.

The downy plumage of rails is typically black or dark brown, with few exceptions. In the Andaman Crake *Rallina canningi* the chick is rich chestnut, while in the Madagascar Wood-rail *Canirallus kioloides* and Forbes's Forest-rail *Rallina forbesi* the grizzled chick has rufous colouring on the head and neck and, in the former species, a rufous stripe along the side of the body; the chick of the Chestnut Forest-rail *R. rubra* is black, grizzled with russet or chestnut. However, the Nkulengu Rail *Himantornis haematopus* chick is distinctively patterned with brown, black and white, and this cryptic pattern more closely resembles that of the precocial chicks of other orders; it probably represents a relatively primitive state while the black down of typical rails appears to be a specialised, derived condition (Olson 1973b). The downy young of some rails have distinctively coloured filoplumes or bristles, these being developed most markedly in the coots. *Fulica*, *Porphyrio* and *Gallinula* species also have brightly coloured bare skin on the head. Young chicks of many species have brightly coloured or patterned bills and it has been widely suggested that this prominent feature serves to assist the parents in directing food to the chick's bill in the dim light conditions prevailing in dense vegetation. The bright colours of the bill and throat act as signals for feeding, and in the Common Coot *Fulica atra* this probable signal effect remains conspicuous as long as the young are totally dependent on parental care (Boyd & Alley 1948).

For details of moult patterns and timing, see the appropriate sub-section of the section entitled Layout and Scope of the Book.

FLIGHTLESSNESS

One of the best known features of the family is the incidence of flightlessness in island taxa. Flightlessness is widespread among various avian orders, and numerous flightless forms exist, but the Rallidae show a particular proclivity to flightlessness and all flightless rails occur on islands. Of the 150 rail species described in this book, 56 are known only from islands, including New Guinea, New Zealand and Madagascar. Excluding the doubtfully valid Gilbert Rail *Gallirallus conditicius*, and *Rallus nigra* (for which no information on flying ability is available), 31 (57%) of the remaining 54 species are flightless or nearly so.

Flightlessness has evolved many times within the Rallidae, often and repeatedly on islands without predators and apparently independently each time (e.g. Olson 1973b, Diamond 1991, Feduccia 1996, Trewick 1997). The energetic cost of flight is high, flight muscles rank among the body's most energy-consuming tissues, and the flight muscle and associated flight structure average 20-25% of body weight in typical birds (Feduccia 1996). Thus the energy costs of flight and the maintenance of the flight structure are high and where such costs are not balanced by the benefits of dispersal and escape from predators it is obviously to the bird's advantage to become flightless. When this occurs the muscles and bones of the wing and pectoral girdle are greatly reduced, the feathers become loosely constructed, there is usually an increase in the development of the leg muscles, and the bird may become larger overall. Reduction of the sternal and flight apparatus generally comes about through neoteny, and the modifications may involve only a few genetic changes, for example in genes controlling the relative growth rate of different body parts (Olson 1973b, Feduccia 1996). The frequency with which flightlessness is developed by rails suggests that they are predisposed to it, and they are certainly pre-adapted to coping with some of the restrictions which it imposes. For example, many volant species are behaviourally flightless, preferring to avoid predators by running, many are temporarily flightless during wing moult (a feature shared by several other groups containing flightless forms) when they become particularly secretive and elusive, and the postnatal development of flight in most species is slow.

Flightlessness may evolve in a very short period of time. For example, the White-throated Rail *Dryolimnas cuvieri* has a volant race on Madagascar and a flightless race on Aldabra; these differ markedly in size but very little in other characters, and have obviously diverged very recently. As Olson (1973a) remarked, the span of time needed to evolve flightlessness in rails can probably be measured in generations rather than in millennia. Trewick (1997) found that genetic distances between the volant Purple Swamphen (race *melanotus*) and the flightless Takahe were very small, this being suggestive of the rapidity with which evolution of radically distinct and gigantic flightless forms may occur.

Two interesting manifestations of flightlessness occur in coots. In the Giant Coot *Fulica gigantea*, the adults are normally too heavy to fly but immatures fly readily. Fjeldså (1981) suggests that dispersal in this species is accomplished mainly by immatures and that adults, when paired, remain faithful to one territory for a lifetime and scarcely need to fly. Flightlessness, accompanied by large size, has evolved independently in three grebes during Pleistocene isolation and inbreeding on Neotropical mountain lakes, and a strategy in the Giant Coot of dispersal in the immature period and subsequent site tenacity with virtual flightlessness would be a remarkable avian adaptation which deserves further study. Concerning the Common Coot, Patrikeev (1995) records that on the Caspian Sea coast of Azerbaijan, wintering birds in the Kizil Agach Reserve have a mean weight of 680g after arrival but increase in weight to average 970g by late Nov, with a maximum recorded weight of 1,450g; by Feb they are so fat that 70-80% of the population become flightless and when chased can only run on the water and dive, but when N migration begins, the mean weight is down to only 750g.

HABITAT

The wide distribution of the family is a reflection of the ability of rails to adapt to a very diverse range of habitat types, both natural and artificial. Rails are cosmopolitan and occur almost everywhere except in polar regions, completely waterless deserts and mountainous regions above the permanent snow line. The greatest variety of species is found in wetlands, of which rails occupy virtually every type of terrestrial, semiaquatic, estuarine and littoral habitat, as well as some open water habitats. Some are found at coastal wetlands such as lagoons, bays, saltmarshes, tidal creeks and mudflats, and mangrove swamps. For example, in South America mangroves are an important habitat for four of the *Aramides* wood-rails, although they also occur in forested habitats. In Australia the Buff-banded Rail *Gallirallus philippensis*, which inhabits freshwater wetlands, forest, woodland, heathland, grassland and crops, sometimes also occurs in saltmarshes, tidal mudflats and mangroves, and even occurs on coral cays and other offshore islands: a remarkably broad habitat tolerance for one species.

Rails occupy many types of freshwater wetland including swamps, peat bogs, marshes, dambos (seasonally wet drainage lines in African woodland), floodplains, permanent and seasonal pans and ponds, ditches, paddyfields and wet sugarcane fields, and the marginal and emergent vegetation of streams, rivers, canals, dams and lakes. Some species, such as the Grey-breasted Crake *Laterallus exilis*, the Yellow-breasted and White-browed Crakes, and the *Porphyrio* gallinules, have a preference for floating vegetation, and one factor responsible for the threatened status of the Sakalava Rail *Amaurornis olivieri* of Madagascar is the continued loss of this habitat type.

The most aquatic rails, the coots, occupy waterbodies ranging from freshwater lakes and ponds to brackish water, and bays and arms of the sea. They frequently occur alongside waterfowl (Anatidae) but, unlike some ducks, have not become adapted to a purely saltwater environment. The two largest species, the Giant Coot and the Horned Coot, frequent barren highland lakes of the Andean puna zone, occurring to altitudes of over 5000m, while the Andean Coot *Fulica ardesiaca* and the White-winged Coot *F. leucoptera* occur to almost as high an elevation in similar habitats. The Red-fronted Coot *F. rufifrons*, which occurs at lower altitudes in South America, appears to prefer semi-open waters with much floating and emergent vegetation, and does not have the enlarged toe lobes which are an aid to frequent and prolonged swimming in its congeners.

Other rails inhabit warm forests from lowland to highland areas, including the interior and edge of primary and secondary growth, monsoon forest, swamp forest, sago palm swamps (New Guinea Flightless Rail), gallery forest, dense riverine forest, forested ravines, partially cleared forest and dense scrub, overgrown and abandoned cultivation at forest margins, banana groves, cassava and arrowroot plantations, and dense evergreen or deciduous thickets. Forest habitats may have clear substrates with leaf-litter, moist ground or mud, or may have ground vegetation. Some species, such as the Nkulengu Rail, the White-spotted Flufftail *Sarothrura pulchra*, the Grey-throated Rail *Canirallus oculeus* and Woodford's Rail *Nesoclopeus woodfordi*, are typically associated with wet habitats such as forest streams, swampy areas and muddy patches, and are never far from water, but the four *Rallina* forest-rails endemic to New Guinea are not said to be associated with wet areas. Other forest rails, such as the Rufous-necked *Aramides axillaris* and Grey-necked Wood-rails of the Neotropics extend their range from forest into deciduous woodland. The Buff-spotted Flufftail *Sarothrura elegans* occupies a wide range of habitat types, from stable forest habitats to deciduous thickets which may be occupied only seasonally depending on the presence of suitable cover and a moist substrate, and consequently also the availability of food.

Other species are associated with predominantly grassland habitats, ranging from mosaics of hygrophilous grassland and sedge meadow, and the marginal zones of grassland savanna with pockets of marsh, to predominantly dry grassland including savanna, pampas, hayfields, meadows, rough pasture, alfalfa and crop fields (occasionally including crop stubble), golf courses and airstrips. The grass may be tall or relatively short, but is usually dense. In New Zealand, both the Weka and the Takahe occur in subalpine tussock grassland, the latter exclusively except when forced to move down to subalpine scrub and beech forest when snow covers the grassland in winter. In Africa the Striped Flufftail *Sarothrura affinis* occurs in both upland grassland and, in the south-western Cape, in mesic mountain fynbos – a macchia-type vegetation dominated by heaths Ericaceae, proteas Proteaceae and the grass- and sedge-like restios Restionaceae on dry to seasonally waterlogged ground;

it is the only rail to occur in this habitat. Grassland habitats may be permanently or seasonally occupied; for example, those occupied by the African Crake *Crex egregia* and the Streaky-breasted Flufftail *Sarothrura boehmi* in central and southern Africa are typically seasonally moist to wet and are frequently burned during the dry season, forcing the birds to emigrate after breeding.

The non-breeding habitats of most rails are similar to their breeding habitats, but some wetland and grassland species, especially those which migrate, may occupy different habitats in different seasons, and many species occupy a wider range of habitats in the non-breeding season. During the rainy season in central and southern Africa, sedentary species such as the Red-chested Flufftail *Sarothrura rufa* and the African Rail *Rallus caerulescens* move out from permanently wet reedbeds and sedgebeds to occupy peripheral seasonally wet grassland which provides good nesting habitat but may be burned in the dry season. In North America the winter habitat of the Black Rail includes smaller, more linear and more fragmented wetland patches than does the preferred summer habitat, while in some areas the Yellow Rail *Coturnicops noveboracensis* appears to prefer the drier portions of coastal marshes in winter and the Virginia Rail *Rallus limicola* occupies saltmarshes more widely in winter than in the breeding season. The Little, Baillon's *Porzana pusilla* and Spotted *P. porzana* Crakes, and the Sora *P. carolina*, all occupy a wider range of palustrine wetland habitats outside the breeding season, while Spotted Crakes wintering in central Africa occupy very temporary habitats not used by breeding Afrotropical species. In winter the Water Rail *Rallus aquaticus*, when displaced by frost, may use drainpipes, rubbish dumps, open ditches and other artefacts as refuges, and may occur in gardens. There is even a record of the threatened Austral Rail *Rallus antarcticus* wintering in a garden and subsequently moving back into its normal reedbed habitat. A narrower habitat tolerance in the breeding season may result from restrictions imposed by breeding-habitat requirements, and also from seasonal differences in food requirements.

Any rail is likely to make use of atypical habitats, often artificial and sometimes of very limited extent or with relatively poor cover, when on migration or undergoing dispersal or irruptive movements. Thus wetland species may occur at sewage ponds, flooded patches of grass or forbs, flooded thickets, and predominantly dry grassy areas such as lawns, pastures and hayfields, while species normally inhabiting freshwater sites may temporarily use brackish or saline wetlands. Grassland species may make use of airstrips, golf courses and crop fields. The highly eruptive Black-tailed Native-hen *Gallinula ventralis* of Australia takes advantage of many wetland, grassland and scrub habitats, especially during influxes, and is also known to occur in urban areas.

The ability to capitalise on this great diversity of habitats throughout the world indicates that the family shows great adaptive plasticity. Furthermore, the majority of rails do not have specialised diets, many being omnivorous and opportunistic (see Food and Foraging), and these characteristics allow many species to exploit highly ephemeral or atypical habitats and help them to be successful colonists of oceanic islands. Their ability to make use of open land allows them to colonise islands where, in the absence of competitors, they can radiate to occupy almost any available terrestrial niche. They are capable of adapting to remote oceanic islands of volcanic origin, where conditions are harsh and vegetation is sparse or even almost non-existent, and where no fresh water is available. For example, the recently extinct Ascension Rail *Atlantisia elpenor* lived on Ascension Island, where the terrestrial environment is hostile in the extreme, consisting mainly of bare, waterless tracts of lava and ash. This smallish to medium-sized rail (about the size of the Virginia Rail) apparently obtained its food and water from the eggs and regurgitated prey of the seabirds which formerly nested on the island in great numbers (Olson 1973a). Another *Atlantisia* species, the tiny Inaccessible Rail, apparently occupies the niche of a mouse on its island. It makes tunnels through the vegetation and forages on boulder beaches, making use of natural connecting cavities under boulders when moving around. The Spotless Crake *Porzana tabuensis* is a good example of a species with a wide habitat tolerance which has successfully colonised islands. It normally occurs in a great variety of wetland habitats and also in ferns, heath and scrub, but on some islands it occupies dry, sterile, rocky or stony habitats with no standing water. Similarly, the Madagascar race of the White-throated Rail occurs in forest, wetlands and mangroves, but the Aldabra race is adapted to coral-scrub habitat.

The habitat requirements of few rail species have been studied in great detail, but it has been demonstrated that, for some wetland and forest species which feed chiefly on invertebrates, the structure of the vegetation and the nature of the substrate may be the most important factors influencing habitat suitability. For example, the Virginia Rail is known to avoid marshes with high stem densities or large amounts of residual vegetation, features which are common in older marshes and which impede the birds' movement, whereas vegetation height is not important as long as there is adequate overhead cover. These birds need shallow water and a substrate with a high invertebrate abundance, and are commonest in areas with 40-70% upright emergent vegetation interspersed with open water patches, mudflats and/or matted vegetation. When breeding, they may select those areas of the marsh with a greater abundance of emergent vegetation. Red-chested Flufftails occur in a very wide variety of wetland vegetation types and, like Virginia Rails, avoid areas where the vegetation is very dense and moribund, such vegetation being difficult to penetrate and providing few good nest sites and relatively little invertebrate food. In southern Africa they are adapted to fire-modified wetland habitats, and periodic burning results in improvement in habitat quality expressed in terms of the increased availability of vegetation suitable for occupation and an increase in the density of permanently resident pairs (Taylor 1994). The timing of their return to burned territories is mainly influenced by the development of adequate canopy cover (at least 55% and usually over 65% cover is required) and to the substrate water content (they do not reoccupy seasonally wet areas until the substrate is moist or flooded); the effect of vegetation height is less important but is complementary to that of cover in that birds will return to very tall vegetation (e.g. *Typha*) while cover values are still relatively low (Taylor 1994). The Buff-spotted Flufftail also occupies a wide variety of habitat types encompassing both natural forest and thickets, and also areas dominated by alien vegetation: in southern Africa it has extended its range locally by occupying exotic vegetation around human habitations. Its main food is macroinvertebrates, for which it forages in leaf-litter and damp earth, and which are as abundant on substrates below exotic vegetation as on those below indigenous plants.

Although the majority of rails inhabit wetland habitats it does not necessarily follow that the family has an aquatic or paludicoline origin. Olson (1973b) has pointed out that the most primitive living rail, the Nkulengu Rail, is a forest bird, as are the members of other primitive or unspecialised relict genera such as *Aramides*, *Canirallus* and *Gymnocrex*, while the most specialised and derived genera such as *Rallus*, *Porphyrio* and *Fulica* contain aquatic or marsh-dwelling species. Thus it appears that the progression from generalisation to specialisation in the family is from forest forms to aquatic forms.

FOOD AND FORAGING

Although some species are predominantly vegetarian and some are almost entirely dependent on invertebrate food, rails as a group are characteristically omnivorous and non-specialised feeders, often opportunistic, and well able to adapt to new food sources. Many species appear to feed largely on the most abundant plant or animal foods available at any time, and thus the proportions of different invertebrates in the diet, or of different seeds and other plant material, often reflect the relative availability of these foods. While there is much information on the diet of a few well-studied species, particularly large, easily studied birds such as gallinules and coots, and especially from North America, Europe and Australasia, little or nothing is known about the food and feeding habits of many others. Even for many well-studied species, good quantitative analyses of diet are lacking.

Plant foods include a wide range of vascular material, including seeds and drupes, shoots, stems and leaves, tubers, bulbs, rhizomes and roots, as well as marine and filamentous algae, fungi, lichens and ferns. Some species take cultivated material, such as young vegetables and cereal crops, fodder crops, grain (including rice and maize), fruit (e.g. tomatoes, melons, apples, bananas and drying apricots), and taro. Invertebrates are an important part of the diet of many species, and include worms (particularly Oligochaeta), nematodes, leeches, molluscs, Isopoda, Diplopoda, Chilopoda, Amphipoda, Copepoda, Decapoda (crabs and crayfish), Arachnida (mainly spiders), and a wide variety of insects and their larvae. Some species take small fish and fish spawn, amphibians and their tadpoles, reptiles (lizards and their eggs, small snakes, and turtle eggs and hatchlings), the eggs and young of other birds, the droppings of gulls and ducks, rodents, carrion (ranging from small birds and rodents to the carcass of a cow), and various household items including bread, porridge, stew, dog food, chicken food, biscuits, butter and chocolate. The Lord Howe Rail has even been known to eat rat poison – and presumably to survive!

In general the most aquatic species, such as the gallinules and coots, are largely herbivorous, while those which inhabit terrestrial and palustrine habitats are either omnivorous or predominantly carnivorous, at least seasonally. Forest-dwelling species are relatively poorly known but probably eat fewer plant foods than rails in other habitats. Forest species such as the Andaman Crake, the Grey-throated, New Caledonian *Gallirallus lafresnayanus*, Okinawa *Gallirallus okinawae* and Woodford's Rails, and the *Gymnocrex* species, are known to take very little plant material, if any.

Coots are almost entirely herbivorous but some aquatic insects, molluscs and crustaceans are taken, and they will sometimes eat eggs, fish, carrion, duck-food pellets and even food scraps from campsites. The Common Coot feeds both in water and on land, employing a variety of foraging techniques which are described in the species account.

Rails which forage in mangrove habitats appear to feed largely on crabs, these crustaceans being the main food items recorded for species such as the Chestnut Rail and the Rufous-necked and Grey-necked *Aramides cajanea* Wood-rails. The forest-dwelling Snoring Rail *Aramidopsis plateni* forages for crabs in mountain streams.

Although most rails drink fresh water some species can survive on small oceanic islands where fresh water is scarce or absent; these species may drink salt water or obtain most of their water from their food. Thus the Aldabra race of the White-throated Rail drinks fresh and salt water, and the Buff-banded Rail can exist on islands with no fresh water. Some rails, such as the saltmarsh-dwelling Clapper Rail, possess well developed supra-orbital (nasal) glands, which function in the excretion of salt. The skulls of extinct species such as the Ascension Rail and the Chatham Coot *Fulica chathamensis* show a narrow interorbital bridge and enlarged supraorbital depressions, indicating that their salt-excreting glands were also well developed: the rail is thought to have obtained its water solely from its food (possibly invertebrates such as squid regurgitated by seabirds at their breeding colonies) and the coot is thought to have been adapting to the salt stress in the lagoons of Chatham Island before its extinction (Olson 1973a, 1977).

All rails swallow grit or small stones to help break up food in the gizzard and in some herbivorous species such as the Takahe this material may be ingested in considerable quantities. The Buff-spotted Flufftail, which forages in the leaf-litter of forest and thickets, frequently ingests large quantities of the small, hard seeds of common trees, probably in lieu of grit, sufficient quantities of which may be hard to find.

Black Crakes *Amaurornis flavirostris* are apparently unique among rails in gleaning ectoparasites from the backs of large mammals, while the Nkulengu Rail and the Grey-necked Wood-rail are the only species known to follow army ant columns (presumably to feed on invertebrates which they disturb), although other forest species probably also do so.

Species with long thin bills probe for invertebrate food in shallow water, soft ground and litter, while species with small, fine bills tend to take many small invertebrates and seeds from the substrate, shallow water and low vegetation. Those with relatively unspecialised, straight bills of moderate length and depth take a wide variety of small to large food items, chiefly by probing, gleaning, digging, sifting litter, stabbing at large prey and raking in soft earth and mud. Thick-billed species tear and slice vegetation, and dig or pull up the underground parts of plants. Some gallinules graze, for example the Purple Swamphen *Porphyrio porphyrio* and the Tasmanian Native-hen *Gallinula mortierii*, as also do the coots.

Only coots regularly dive for food; Common Moorhens do so rarely, and Rouget's Rail *Rougetius rougetii* of wetlands in the Ethiopian highlands has been seen to dive into a stream like a dipper *Cinclus* while foraging, while several other marsh-dwelling rails are also known to dive for food occasionally. Coots and *Gallinula* species regularly up-end when feeding and other species such as the King Rail *Rallus elegans* may up-end occasionally when foraging in deep water.

Terrestrial species which forage in leaf-litter and soil use the bill to move or toss aside leaves and debris, and to turn over small stones, while searching for invertebrates; some species also use the feet to move debris aside and scratch in the ground. The Weka and the Lord Howe Rail *Gallirallus sylvestris* pull apart bark and rotten wood to extract food, while the latter chisels harder wood and will also pull aside palm fronds to search for food beneath them. Some species, such as the Rufous-tailed Bush-hen *Amaurornis moluccanus*, Sora and Spotted Crake, run grass heads through the bill to strip seeds from them. The White-browed Crake often feeds while swimming, floating with the neck extended parallel with the surface and suddenly reaching out with its fine-pointed bill to capture water insects.

Large live food items are usually killed and dismembered before being swallowed. The Water Rail first paralyses frogs and fish by making vertical blows with its bill behind the prey's head and then kills the prey by repeatedly striking it with the bill, while the King Rail hacks large prey items, such as crayfish, to pieces before eating them, and the Rufous-tailed Bush-hen crushes frogs in the bill before swallowing them. Large insects, such as grasshoppers, are shaken and battered before being swallowed. The Buff-banded Rail pecks large snail shells to break them, crushes fish in the bill, and holds crabs by the legs in its bill and flicks its head from side to side until the legs break off. Other species knock large prey animals on the ground to immobilise or kill them before swallowing them, while the Spotted Rail *Pardirallus maculatus* hits snails repeatedly on a rock before eating them (Arballo & Cravino in press), and the Chestnut Rail reportedly uses stone anvils to break gastropod mollusc shells, many of which probably contain hermit crabs (Woinarski *et al.* 1998). Several species spear eggs with the bill and either eat the contents *in situ* or carry the egg away to deal with. Large prey may be carried into cover or, if caught while the bird is in water, brought ashore to be dealt with it. Rails often wash food in water.

Using the foot to grasp and manipulate food is recorded in some *Porphyrio* species, such as the Purple Swamphen, Allen's Gallinule *P. alleni*, the Takahe, and the extinct White Gallinule *P. albus*. The Purple Swamphen uses its powerful bill to pull out or nip off grass tillers, and to pull out emergent plants such as *Typha* and *Eleocharis*, and then grasps them in the foot while it chews or slices off the soft fleshy bases. It also holds down *Typha* stems with the foot while stripping off leaves with the bill to eat the leaf bases. Items such as stems, waterlily buds and figs are held in the foot and lifted towards the bill. Plant material is held for eating by the opposition of the hind toe to the three closed fore-toes. The fore-toes are also used to comb through weed dangling from the bill and to strip casing from plants, and the feet are also used to hold down large food items such as fish and carrion while the bird pecks them.

The Takahe holds food items in the feet in a similar way, and also strips seeds from grasses with the bill while holding the stalks in the foot. Allen's Gallinule turns over waterlily leaves with its bill and holds them down with the feet while gleaning food items from their undersides. It is also partial to developing waterlily seedheads, which it breaks off with the bill and holds with one foot while eating, and it sometimes carries other food items in the foot to the bill. The Weka and the Dusky Moorhen also hold down food items with the foot while pecking or hammering at them, but do not grasp or manipulate them in the foot as do the gallinules.

Seasonal variations in the proportions of animal and plant food taken are reported for a number of species, and reflect seasonal changes in the availability of food, the use of different habitats when birds are on migration or in wintering areas, and the possible need for a greater consumption of protein in the breeding season to satisfy the requirements for egg-laying. Many species increase their intake of animal food in the spring and summer, and of plant food in the autumn and winter, and the results of quantitative studies on this subject are summarised in the King Rail, Virginia Rail, Sora and Common Moorhen species accounts.

Very little information is available on the diet of the chicks of most rails, but it seems that the young, even of herbivorous species, are fed primarily on insects and other animal food. Thus, aquatic insects comprise 45-85% of the diet of young chicks of the American Coot, while invertebrates form a large proportion of the diet of Takahe chicks for 4-6 weeks and the chicks become predominantly vegetarian when 6-8 weeks of age. Sora chicks are fed largely on invertebrates, although plant material constitutes a large proportion of the summer diet of the adults. The young of the Dusky Moorhen are fed on annelids, molluscs and insects for the first few weeks, after which plants are taken increasingly. The vegetarian Giant Coot feeds small young on both weed and invertebrates, and the American Coot feeds large quantities of aquatic insects to its chicks during the first week after hatching.

Kleptoparasitism is recorded in coots and gallinules. The Common Coot will snatch food from conspecifics and also from swans and both surface-feeding and diving ducks. The American Coot robs some duck species and is itself kleptoparasitised by others. Allen's Gallinules will chase and rob conspecifics of waterlily seedheads, and will also snatch them from Pygmy Geese *Nettapus auritus*.

VOICE

Most rails are very vocal, as is to be expected in birds which inhabit dense cover where visual contact is often very limited and communication by sound is important. Red-chested and White-spotted Flufftails, for example, are strongly territorial species which inhabit very dense vegetation in forests and wetlands respectively and which may indulge in prolonged bouts of territorial calling with neighbours at a common boundary without being in visual contact, even though the birds may be within 1 m of each other on the ground. In these flufftails, as well as in some other rails, maintenance of territorial integrity may be achieved largely by vocal means, and the territory owners may rarely be in visual contact with each other.

The calls of rails vary from sweet to harsh and include screams, squeals, trills, whistles, whines, hoots, moans, booms, rattles, clicking and ticking notes, snoring noises, humming and buzzing sounds, trumpets, roars, grunts, barks, frog-like croaks and snake-like hisses; calls of some of the smaller species may be very insect-like. The advertising and territorial calls of many species are give in a rather repetitive series and are often loud, but calls of many species in both forest and wetland tend to be masked by other sounds during the day and are normally mostly given during periods when background noise is minimal and signal attenuation from atmospheric disturbance is least (e.g. Wiley & Richards 1982, Taylor 1994). Thus, although calling may occur at any time of the day or night, many species call most commonly in the early morning and the evening, and also during the night.

Most species are mainly silent when they are not breeding and therefore not territorial, but species which are permanently territorial call throughout the year, as do social species, many of which have loud rallying cries. Non-vocal sound signals include wing-clapping, used by Purple Swamphens to distract predators from the young and to attract relief when incubating, beating the water with the wings, used by coots *Fulica* and moorhens *Gallinula* when fighting and by many rails during distraction displays in defence of chicks, and slapping noises made with the feet, these characteristically being made by moorhens and coots in defence displays and also when swimming to splash water at a potential predator; female coots may also slap the ground with one foot prior to copulation.

Authors describe many rail calls as harsh or monotonous, such as *tack*, *kak*, *crek* or *krr*, and many species include a call of this type in their vocal repertoire, sometimes given in a series as an advertising call and sometimes given singly as an aggressive, alarm or warning call. The call of the Yellow Rail may be likened to the noise made by knocking two large pebbles together, and the birds may be induced to respond to sounds produced in this way. Other small crakes (e.g. some *Laterallus* species: see Hardy *et al.* 1996) produce a sequence of notes rather like a trill, such as is made by small grebes. Some gallinules have booming calls, holding the head and neck in such a way as to produce a resonant note, smaller in scale but similar to that produced by rheas Rheidae and cassowaries Casuariidae. Some of the larger rail species can scream loudly when caught.

Characteristic call types are sometimes shared by most or all members of a genus, and some examples will serve to illustrate the range of calls used. Crakes of the genus *Laterallus* have trilling, churring or rattling calls, *Sarothrura* species have hooting advertising calls and *Rallina* crakes often have croaking or nasal calls. The wood-rails *Aramides*, which inhabit mangroves, swamps and lowland forest in Central and South America, have a wide range of remarkably loud and far-carrying yelps, shrieks, cackles, barks, grunts and pops, which are characteristic sounds of the areas in which they occur, and which have given rise to onomatopoeic local names for the birds. The voices of the Grey-necked Wood-rail and the Giant Wood-rail *A. ypecaha* are particularly arresting; both species call in chorus, the former having a varied series of crazed-sounding rollicking, popping and clicking notes, and the latter congregating in groups in the evening to set up a deafening chorus of screams, shrieks and wheezes. The vocabulary of most *Porzana* crakes includes trills, rattles, purrs, chattering, knocking, grating or ticking notes; some also have whistling calls. *Gallinula* species typically emit crowing or clucking calls, and several *Fulica* species have short, explosive, aggressive calls variously rendered *pit*, *hic*, *kik*, *puhlk*, *pssi* etc.

Calling in some species is said to be ventriloquial, and one way in which this effect is produced has been observed in the Buff-spotted Flufftail. The male gives loud, hollow, hooting *ooooooo......* notes, each lasting 3-4s, usually from a completely concealed low perch in dense woody vegetation. The call is often given throughout the night and is one of the most characteristic and evocative sounds of the African forest, and has given rise to many legends and superstitions. While uttering the note the male continually turns his head slowly from side to side, possibly to broadcast the sound effectively in all directions, and this creates the impression that the call varies greatly in volume and makes the source very difficult to locate.

The hoot of this flufftail is a remarkably loud, pure tone and is amplified in the oesophagus, which is enormously distensible for the first 3.5cm of its length; when fully inflated it becomes almost spherical, having a diameter of over 3cm, and in calling birds probably extends around, or even engulfs, the trachea to form a resonating chamber (Taylor 1994).

Duetting is reported in several genera. As normally applied to rails, the term may cover three different forms of vocalisations: true antiphonal duetting, in which a mated pair make precisely timed alternate calls; calling in precise unison by two birds; and two or more birds calling simultaneously or in chorus, but not in unison or antiphonally. A few species are described as making calls in unison, notably the Virginia Rail, the White-throated Rail, the Chestnut Rail, the Slaty-breasted Woodrail *Aramides saracura* and the Tasmanian Native-hen; most of these species are also said to duet antiphonally. The White-throated Rail's calls are not made precisely in unison and in some other species it also seems that true synchrony is not achieved but that individuals may sing according to their own internally derived rhythm (Huxley & Wilkinson 1979). Species claimed to sing antiphonally include the Nkulengu Rail, New Britain Rail *Gallirallus insignis* and Okinawa Rail, Water Rail, Rufous-necked and Grey-necked Wood-rails, and Black Crake, Rufous-tailed Bush-hen, Spotted Crake and Henderson Crake *Porzana atra*; some of these species also call in chorus. Other species recorded as duetting, such as the Red-necked Crake *Rallina tricolor*, Weka, Lord Howe Rail, African Rail and Giant Wood-rail, call either in asynchronous duet or in chorus.

In rails, duetting, including simultaneous song and chorus singing, appears to be principally related to cooperative territorial defence and advertisement, but in some species dual calling probably help to maintain and reinforce the pair bond and may also be part of precopulatory behaviour. In at least the Black Crake, chorus singing may also assist in the maintenance of the extended family group during the breeding season (Taylor 1994). Much remains to be discovered about duetting in rails, especially because the calls of so many species, particularly from the tropics, remain undescribed.

Many rails have an extensive vocal repertoire, using calls to communicate in a wide variety of situations. Distinctive advertising calls are frequently uttered, both in territorial advertisement and maintenance and to attract mates. Advertising calls are frequently far-carrying; for example, the whistle call of the Spotted Crake is audible for at least 2km in calm weather, while the hoot of the Buff-spotted Flufftail is well audible for up to 1km through forest and woodland and up to 2.8km over open ground. In seasonally territorial species the frequency and variety of calling often peak during the establishment of the breeding territory, courtship and the early part of the breeding season, falling off during or after egg-laying and remaining at a relatively low level while the young are reared. In some species there may be a resurgence of territorial calling at the end of the breeding season and this has been shown, in the African Rail and probably in the Clapper Rail, to reflect the inclusion of juvenile calls at this time of the year. In those species which do not hold permanent territories, advertising calling usually decreases markedly, or ceases completely, at the end of the breeding season, but species which hold permanent territories continue to make territorial advertising calls throughout the year, although usually at a reduced level in comparison with the breeding season. The Lord Howe Rail is less vocal during its postbreeding moult, as probably are other permanently territorial species. Some species are known to call in flight when migrating, and many migrant species are described as being silent on their non-breeding grounds.

Most rails respond well to playback of their recorded calls, especially during the breeding season. However, great care must be exercised in the use of playback, as it may have detrimental effects on the birds. Frequent or prolonged playback may result in birds becoming habituated so that they cease to respond, or it could possibly disrupt their normal territorial and breeding activities to an extent which might be detrimental to their breeding success. Studies in Africa indicate that male flufftails seeking to establish a breeding territory may move away from an otherwise suitable site in the face of apparent threats or competition from prolonged playback (P. B. Taylor unpubl.).

BEHAVIOUR

Locomotion and reactions to danger

Rails often walk with strong, precise strides. Being accustomed to cover they often move continuously without long pauses for visual orientation, pausing mainly to feed, and in dense cover they are almost uncannily adept at walking and running without causing any noise or movement of the vegetation. They walk beneath low horizontal stems or projecting roots, or raise the feet high to step over such obstacles, so that there is no noticeable vertical change in the position of the body relative to the substrate. When alarmed, most rails prefer to run rather than to fly, and they can melt quietly and rapidly into cover, often lowering the head and stretching out the neck, and they compress the body to allow easy passage between stems. They often walk with bobbing head and flicking tail, the head movement possibly being connected with acuity of vision, as in other birds such as pigeons and chickens: the head moves forwards relative to the surroundings and, at the end of the forward thrust, is temporarily stationary while the rest of the body catches up, this brief stationary phase being an adaptation to allow a motionless view, with the central fovea, of the bird's surroundings (e.g. Ripley 1977).

The tail is often jerked or flicked, especially when the bird is wary or disturbed. It is usually assumed that this action functions in visual orientation and signalling between individuals, especially when the tail is jerked and spread in alarm before the bird runs to cover. However, it has been shown that in the Common Moorhen and the Purple Swamphen tail flicking is directed towards potential predators rather than towards conspecifics, and is an alertness signal with a pursuit deterrent function rather than a warning to conspecific individuals (Woodland *et al.* 1980, Alvarez 1993), although in both species it also occurs in response to inter- and intra-specific aggression. Ripley (1977) also suggests that the tail jerk, if accompanied by a short pause in forward gait, allows the bird an instantaneous clear view while the head is temporarily stationary.

Terrestrial predators pose a much greater threat to rails which inhabit dense cover than do aerial predators, and numerous observations in Africa (Taylor 1994, P. B. Taylor unpubl.) show that such rails are very sensitive to disturbances at ground level. For example, flufftail species of forest, wetland and grassland are often less active and more wary on windy days – possibly because the noise and the vegetation movement can mask the sound and disturbance made by approaching terrestrial predators. When startled by a sudden and unexpected noise or movement at close quarters, rails in dense vegetation may instantly make a vertical standing jump, often to a height of 20-30cm, before running away. This may well be an effective way of avoiding a sudden strike by a snake or an attack by a terrestrial mammal, and is typically made by rails of the genera *Sarothrura, Rallus, Crex, Porzana* and *Aenigmatolimnas*. When they encounter a real or model snake, forest flufftails will follow it closely, making no sound, until it moves away from the territory or climbs a tree.

Activity patterns

The daily activity patterns of rails have been very inadequately studied, but it is known that most species are active during the day and roost at night. Many species are predominantly crepuscular and most show peak activity, including intensive feeding, very early in the morning and again in the late afternoon and early evening. The middle of the day is often spent sheltering and resting. Some species are said to be mainly nocturnal but what is loosely described as nocturnal activity may often refer only to crepuscular activity, and convincing evidence for regular nocturnal activity exists for only a few species. Flufftail species give advertising calls throughout the night during the breeding season but are not active in any other way at night: they remain in one spot and will not move around in response to taped playback, and they do not forage at night (Taylor 1994). The King Rail calls much at night during the courtship period, and to a lesser extent at other times of the year, but both captive and wild birds move little at night; nothing is known of its roosting habits in the wild (Meanley 1992).

However, the Plumbeous Rail *Pardirallus sanguinolentus*, which gives advertising calls at night, is also active during the night in marshes or adjacent cultivated fields (Johnson 1965). Other terrestrial and marsh species which are known to forage at night as well as by day include the Chestnut Rail, the Grey-necked Wood-rail, the Weka, the Inaccessible Rail, the Clapper Rail, the Buff-banded Rail and the Guam Rail *Gallirallus owstoni*; however, the last three are most active early and late in the day. The Weka is semi-nocturnal, becoming active in the late afternoon and being active in open habitat on clear moonlit nights. Common Coots and Common Moorhens are sometimes active on moonlit nights and on floodlit waters, and it is possible that nocturnal foraging in rails is largely confined to species which occupy open habitats where the visibility at night is relatively good, for example coots and those species which forage in the open on mudflats and along tidal creeks and open channels. Rails which forage in tidal areas must presumably be active nocturnally when low tides occur at night, and the Clapper Rail is known to roost both by day and by night at high tides.

Roosting

Except for species which gather in large flocks, rails normally roost singly, in pairs, or in family groups. They generally roost on the ground in dense cover, but sometimes above the ground in dense vegetation such as bushes and trees. Some forest rails, including some flightless species such as the Okinawa Rail, roost in trees to avoid ground predators, and the Red-necked Crake may use communal roosting platforms. The Plain Bush-hen *Amaurornis olivaceus* is also known to build roosting platforms and other species roost in groups on patches of flattened vegetation. When breeding, many species roost in nests with their offspring, and some species may roost in nests at other times of the year. For example in New Guinea, Forbes's Forest-rail constructs a domed roosting nest in which up to seven adults may sleep (Ripley 1977) and the White-striped Forest-rail *Rallina leucospila* roosts in pairs in a roofed shelter of dried leaves and moss which, according to local people, has nothing to do with nesting and is merely a shelter from the rain (Ripley 1977). In harsh environments shelter at night may be important, as is suggested by the case of a captive Inaccessible Rail which was inadvertently left outside in one of the island's frequent gales; although it dug a hole almost 30 cm deep near the roots of a tussock plant, it died.

Comfort activities

Rails bathe freely, mainly by standing in shallow water, alternately ducking the head in the water and flipping water over the back, and beating the half-open wings in the water; they may also move the body up and down in the water, or dip the tail. Coots may bathe while swimming. After bathing, the birds leave water to oil and preen. In captive Yellow Rails, bathing activity peaked in the evening at around 19:00h and was less frequent in the morning. Captive Red-chested Flufftails bathed daily, always after midday and with peaks in the hottest part of the day and in the early evening, while wild Buff-spotted Flufftails bathed throughout the day, the frequency of this activity increasing during the day to a peak in the two hours before the birds went to roost; chicks also bathed regularly from the age of 10-11 days (Taylor 1994).

After bathing the birds preen vigorously. When conditions permit, they will often sun themselves after bathing and preening. Many species are known to stand in a characteristic sunning posture, with partly open wings spread backwards over the sides of the tail. A similar posture, with shuffling movements, is also used to dry the wings, and is the source of reports of so-called rain-bathing. Purple Swamphens will also sometimes sun by crouching with legs extended forwards, toes splayed and pointed forwards, and wings closed. In Red-chested Flufftails bathing is solitary but sunbathing may be a social activity, involving a pair or family group; the birds either crouch or lie flat on the ground with the tail spread and the wings open, and they frequently preen vigorously while sunning. In captivity, these flufftails were found to sunbathe for the longest periods in the cool of early morning, as soon as warm sunlit patches became available on the ground or on top of dense vegetation; thereafter they sunbathed for shorter periods throughout the day (Taylor 1994). Social sunbathing and preening are also recorded in captive Red-and-white Crakes *Laterallus leucopyrrhus*.

SOCIAL AND SEXUAL BEHAVIOUR

Flocking and gregariousness

Because of the difficulties involved in observing the birds in dense cover, and the often secretive nature of the birds, the social behaviour of most rails is very poorly known, and most detailed studies have been made on the larger, more obvious species such as gallinules, moorhens and coots. In general, rails are solitary or occur only in pairs, family parties or other small groups. The most gregarious species are the coots, with the exception of the Giant Coot, and most species frequently associate in large monospecific flocks outside the breeding season, while non-breeding Common and American Coots also flock near breeding territories during the breeding season. Some flocks may be very large, numbering over 10,000 birds in the Common Coot, in which some flocks may be moulting groups, and coots tend to feed in loose flocks, often with ducks. In the breeding season coots are territorial and often very pugnacious, but the Horned Coot occasionally nests in loose colonies.

Some gallinules, such as the Purple Swamphen and the Common and Dusky Moorhens, also associate in loose flocks when not breeding. In flocks of Common Moorhens, and in all social groups of Purple Swamphens, a social hierarchy may be seen, males being dominant to females and adults dominant to immatures. The Black-tailed Native-hen is usually gregarious and may occur in enormous flocks of up to 20,000 birds during its periodic irruptions. It may also feed gregariously, but it breeds in isolated pairs or in small colonies of up to five nests. Populations of the Takahe are often concentrated into loose colonies but, within these, solitary pairs defend discrete territories.

Most other species are normally solitary, although population densities may be very high in some wetlands. Black Crakes, Lesser Moorhens *Gallinula angulata* and Allen's Gallinules can occur and breed at high densities in the emergent vegetation of irregularly flooded river systems in the Afrotropics.

Territoriality

Available evidence suggests that rails are seasonally or permanently territorial, living either in pairs or in family groups, and that the majority form monogamous pair bonds of at least seasonal duration. Many species defend territories only while breeding. Many rails, whether or not they are defending breeding territories, are markedly aggressive towards conspecifics, and also towards other rail species and other bird species. There are a few published observations of rails feeding in close proximity to other bird species, including other rails, notably involving *Porzana* species.

Winter feeding territories are maintained by the Water Rail, the Spotted Crake and possibly the Corncrake, and also by the African Crake; this phenomenon is probably more widespread than is known. Some sedentary species, such as the Red-chested and White-spotted Flufftails, which defend permanent territories and remain in pairs throughout the year, may allow immatures to remain in the territory throughout the non-breeding period, during which time the young assist with territory defence, and will then eject the offspring at the start of the next breeding season. Even species such as the Black Crake, which are territorial only during the breeding season, may retain a loose association between family members during the non-breeding period.

Mating systems

Monogamy is to be expected as the predominant mating system in the family because, although rail chicks are precocial or semi-precocial, they need intensive parental care at an early age, when they follow the parents and are fed, guarded and brooded by them.

Some form of non-monogamous mating system is known to occur in the wild in only five rails, the Corncrake, the Purple Swamphen, the Common Moorhen, the Dusky Moorhen and the Tasmanian Native-hen, and to occur in captivity in the Yellow Rail and the Striped Crake (and in the latter probably also in the wild). The mating system of these species is described in detail in the species accounts and is summarised here.

The Corncrake was formerly assumed to be monogamous but early observations indicated that it sometimes formed polygynous associations and recent work has shown that, although monogamy does occur, serial polygyny is regular. Males occupy shifting and overlapping home ranges and mate with two or more females, and the female incubates the clutch and rears the young. Serial polygyny has also been observed in captive Yellow Rails, in which one male became dominant over others, claimed the entire enclosure as his territory, sought a second female when his first mate began to incubate, and took no part in incubation or in caring for the young (Stalheim 1974).

The reverse situation has been seen in captive Striped Crakes, which show serial polyandry. The sexes differ markedly in plumage and it is the female which makes advertising calls and establishes a breeding territory, mating with two or more males in succession and taking no part in incubation or in the rearing of the chicks. Evidence from limited observations of wild birds suggests that males normally incubate, indicating that polyandry may also occur in the wild. One case of possible polyandry has also been reported for the Henderson Crake, in which a female made four nesting attempts with one male and one interim attempt with a male from an adjacent territory.

Breeding habitats such as wetlands and lush grasslands are structurally simple and may be highly productive, with food concentrated in a narrow spatial range. In such conditions it may be possible for males to control territories in which two or more females can breed, relegating less successful males to suboptimal territories, or to none at all. This situation may apply to the Yellow Rail, and possibly also to the Corncrake. Polyandry in the Striped Crake, an Afrotropical migrant which often breeds in unpredictably ephemeral wetland habitat, may have evolved in response to the great variability in breeding conditions and the availability of abundant food in the breeding habitat. The situation may be comparable with that seen in some waders which breed in the far north or in montane areas and have mating systems ('rapid multi-clutch' systems) which enable them to increase their reproductive output when conditions suitable for breeding are short-lived.

The social structure and mating systems of the promiscuous *Porphyrio* and *Gallinula* species are more complex. The Common Moorhen normally forms monogamous pair bonds, usually only for the breeding season but sometimes for up to several years. Immatures from earlier broods, and sometimes additional mature birds, often help to tend and feed chicks. Polyandrous trios, where a female forms pair bonds with two males, also occur at a low frequency, while cooperative nesting also occurs when two or more females are paired to the same male and lay in the same nest, subsequently sharing parental care. Brood parasitism also occurs regularly, with females laying eggs in neighbours' nests, often in addition to laying their own clutches.

The Dusky Moorhen is sedentary or dispersive and is territorial only during the breeding season, when birds form breeding groups of 2-7 apparently unrelated birds in which the sex ratio usually favours males. Simultaneous promiscuity occurs, all males copulating with all females, which all lay in one nest, and all members of the group defend the territory and share all breeding tasks; older siblings also sometimes help to care for the young.

The flightless Tasmanian Native-hen defends a permanent territory and is either monogamous or polygamous, usually polyandrous. The breeding unit often consists of an adult pair or a trio of two males, often brothers, and a female; however, there may be up to five members, rarely including two females which are always sisters. Adults normally associate for life. All males copulate with all females in the group, and all females lay in one nest. Nest-building, incubation and parental care are shared by all adults, sometimes assisted by young of an earlier brood or by young of the previous year which have not yet left the group. This strategy is believed to have arisen as a result of the unequal sex ratio in the species and is cited as an example of kin selection; there is no clear reason for the prior existence of an unequal sex ratio (Maynard Smith & Ridpath 1972).

Monogamy prevails in western Palearctic and African races of the Purple Swamphen, but in two other races birds often live in communal groups; in one (*melanotus* of New Zealand) most birds occur in communal groups of up to 12 birds although monogamy also occurs. Stable groups, in which the birds remain together all year, usually consist of kin, while unstable groups are non-kin. Stable groups which maintain permanent territories are polygamous and usually consist of 2-7 breeding males, 1-2 breeding females, and up to seven non-breeding helpers which are offspring from previous matings. Unstable groups are promiscuous, characterised by much aggression and many male members, and usually unsuccessful. Within the stable group mate-sharing is total, incestuous matings are common and homosexual matings also occur. Some males are dominant, but multiple paternity is prevalent. Only the most dominant females breed, laying in a common nest, and all birds in the group care for the young. Most offspring remain as non-breeders before eventually becoming breeders. In this environment habitat saturation, and a shortage of prime breeding territories, appear to be responsible for the communal breeding strategy. In such breeding groups incest-avoidance behaviour was observed relatively infrequently and no examples were observed of deformed young or of anything that could be considered inbreeding depression (Craig & Jamieson 1988, 1990).

Several other rail species are known to live in extended family groups, at least during the breeding season, and to receive assistance from other adults or young in the group in feeding and rearing the chicks. This situation is normal in the Black Crake, which is territorial only in the breeding season and is multibrooded, young of early broods helping to rear chicks of later broods. Cases of young birds helping to rear chicks of later broods are also reported in the Red-chested Flufftail, Rufous-sided Crake *Laterallus melanophaius* (in captivity), Lord Howe Rail, Tristan Moorhen *Gallinula nesiotis* and Common Coot, while in the Weka large young and an extra adult have been seen to help. Some Henderson Rail territories were found to contain one or two extra birds of either sex, and of indeterminate age and origin, which assisted with the defence of the chicks. The Takahe is monogamous, maintains a permanent pair-bond, holds a territory during the breeding season, and breeds in family groups: older young, and sometimes offspring up to two years old, help to rear the chicks. In some of these cases, habitat saturation and a potential shortage of prime breeding territories may be reasons why mature offspring remain in the natal territory as opposed to establishing their own breeding units.

Agonistic and sexual behaviour
This behaviour is often conspicuous. In the Purple Swamphen, postures important in agonistic display include the position of the tail (at rest, raised, expanded or flicked) and of the wings (closed, partly to fully raised, or spread out to the side). Differing upright postures of the body indicate either aggression or anxiety, while horizontal postures indicate greater aggression and often lead to a charging attack and fighting, in which birds peck at each other, grapple with the feet and rip with the claws. Bowing displays are also aggressive, hunched displays can precede either threat or escape behaviour, and submissive birds crouch.

Moorhens and coots share similar agonistic displays, in which the degree of prominence of the frontal shield is often an important component. A low posture, with head and neck held close to the water and the shield prominent, is important in aggression; in this display, coots often erect the neck feathers in a ruff and arch the wings. Birds expel intruders by a charging attack in the low posture, or by a splattering attack when they run across the water with flapping wings; the attacked bird flees in a similar manner but with the head and neck raised. Fighting is preceded by an upright posture and birds strike each other with the bill and feet, often flapping the wings. A water-churning display, when the bird foot-slaps against the water to create a turmoil, is used in some species to terminate a charge and in others as an interspecific defensive display.

The establishment of the breeding territory and the initiation of breeding activity are usually marked by greatly increased calling. Pair-formation and courtship behaviour are poorly known and little studied except in the larger, more conspicuous and less secretive species such as gallinules, moorhens and coots, but it is clear that sexual display has a limited repertoire in the family.

Courtship feeding and allopreening are common, and aggressive-looking courtship chases often lead to copulation. Courtship displays are usually simple and in some species, such as the Buff-banded Rail and Baillon's Crake, copulation occurs with no preceding display. In some other species, the male's courtship display involves bowing, with head down and tail up, and either raising the wings (e.g. in the Water Rail) or spreading them and holding them down (e.g. in the Corncrake). Courtship displays may involve the display of bold flank patterns and/or contrastingly coloured undertail-coverts, and may be accompanied by calling. Male King and African Rails walk around the female with the tail raised and the white undertail-coverts fanned, while the male Virginia Rail runs around the female with raised wings, stretched neck and a high-stepping gait. Male flufftails invariably vibrate their tails rapidly from side to side during courtship displays, which may involve a bowing or an upright stance, and the head and neck feathers are always raised.

More complex courtship and mating behaviour is shown by gallinules, moorhens and coots, and is well documented. In these displays, a bowing posture with the head down serves to make the shield inconspicuous. In the Purple Swamphen, courtship behaviour usually starts with allopreening but the male may also present waterweed in his bill to the female, bowing repeatedly, and the female solicits copulation by adopting the arch-bow posture (see species account). The American Purple Gallinule has billing, bowing, swaying and squat-arch displays in pairing and courtship. Coots and moorhens share similar components in their sexual displays, including a bowing-and-nibbling ceremony, in which one bird is submissive while the other preens it, a greeting and passing ceremony, and a courtship chase; the female solicits copulation by performing an arch-bow display. Although coots and moorhens may display in water, even coots usually copulate out of water, on land or on specially built platforms. Allopreening is common, between pair members, adults and offspring, and siblings; it probably involves functional feather-care to some extent as well as being of social significance.

Most rail species nest solitarily and are generally well separated, but in some wetland habitats, especially where nesting habitat is scarce or territories are small, nests may be very close together. For example, nests of the Little Crake may be as little as 10-15m apart and those of the Common Moorhen as close as 8m. In restricted habitats some species such as the Purple Swamphen may nest in loose colonies, while the nests of the Australian Crake are sometimes clumped, with up to 30 nests reported in a group. Populations of the Takahe are normally concentrated into loose colonies but with spaced-out territories within these, while the Black-tailed Native-hen breeds in isolated pairs or in colonies of 5-500+ nests, 7-10m apart.

BREEDING

Detailed studies have been made of the breeding of relatively few rail species in the wild, notably some of those in the genera *Sarothrura, Gallirallus, Rallus, Crex, Porzana, Porphyrio, Gallinula* and *Fulica*, and in captivity of some of these species plus others such as the Yellow Rail and the Striped Crake. Little is known about the breeding of many other rails and the nest, eggs and young of 23 species remain undescribed.

The factors which govern the timing of breeding are separable into two distinct types: ultimate factors, which are remotely causative and influence the overall time of breeding; and proximate factors, which influence the precise timing of breeding and have a more immediate effect. Thus the availability of food for the chicks may be an overriding ultimate factor, while the environmental cues of daylength, temperature and rainfall are important proximate factors upon which the precise timing of the breeding period may depend. Little or no information on breeding seasonality and timing is available for many species, but most rails appear to breed seasonally, during the spring and summer in temperate regions and in or near the wet seasons in the tropics. There are exceptions to this generalisation, however, most of them involving species of tropical or subtropical regions which may have extended or ill-defined breeding periods. Thus some *Sarothrura* species breed seasonally in central and southern Africa but have an indeterminate breeding season near the equator, while the African Rail breeds in both the wet and dry seasons in the tropics but is mainly restricted to summer breeding in southern Africa.

Many forest species are reported to breed during, or at the start or end of, wet periods, but the Roviana Rail *Gallirallus rovianae* is reported to breed in the dry season, when water is not a problem on the forest floor in its habitat.

Some species may breed throughout the year if conditions remain suitable, examples from Africa being the Black Crake, Common Moorhen and Red-knobbed Coot *Fulica cristata*. Elsewhere, breeding throughout the year is reported in the Guam Rail, Lord Howe Rail, Slaty-breasted Rail *Gallirallus striatus*, Common Coot (in Australia) and Giant Coot. The highly irruptive and dispersive Black-tailed Native-hen of Australia may also breed in any month. Over some of its range the Buff-banded Rail is also known to breed throughout the year, while the Weka of New Zealand has a variable breeding season, the start and duration of which are influenced by climate, food supplies and population size, and which may continue throughout the year in periods when its populations are expanding.

A study of the breeding seasons of waterbirds in south-western Australia (Halse & Jaensch 1989) showed that, for the rails, the laying period was best correlated with peak rainfall (in most species), day length and temperature, while in the Spotless Crake water depth was also important. The timing of vegetation development is often important to the initiation of nesting in rails of marshy habitats, and migratory or seasonally displaced species may not be able to colonise wetlands, or to begin nesting, until the vegetation has developed to a suitable height and density. Thus in Ohio, USA, peak nest initiation of the Common Moorhen occurred when the height of *Typha* was 45-100cm and its growth rate was greatest; also the nesting of American Purple Gallinules is delayed in some habitats until the plant density is adequate. The breeding season of the Tasmanian Native-hen is determined by rainfall, as it depends on fresh young plant growth. Flooding conditions may also influence the start of breeding: for example, the breeding of the Spotless Crake in Australia may be delayed by flooding by up to six weeks (Marchant & Higgins 1993). Breeding timing may also vary with altitude, as a result of the later development of suitable conditions at higher altitudes: in KwaZulu-Natal, South Africa, the start of breeding in the Buff-spotted Flufftail may be 1-2 months earlier in coastal regions than at altitudes above 1,200m.

In many species the nest site is selected by the male, but sometimes may be chosen either by the female or by both members of the pair. Nests are usually concealed in thick vegetation, often near to or in water, but some species nest on the ground far from water or in forest, or up to 10m high in trees when they may use old nests of other birds. Nest materials are often gleaned from any available vegetation – frequently that which is closest to the nest site – and nests are often built by both sexes, but sometimes by one sex only. The nest is usually cup-shaped and well lined, often with a fairly deep bowl; it is domed in some species, notably the White-spotted and Buff-spotted Flufftails, the Chestnut Forest-rail, the Russet-crowned Crake *Anurolimnas viridis*, and in *Micropygia, Anurolimnas, Atlantisia* and most *Laterallus* species. Some nests of the Henderson Crake are also domed, but others are open cups. Domed nests of forest species are often made of leaves but in the Russet-crowned Crake and most *Laterallus* species the nest is a ball of dead grass placed in grass, reeds or bushes.

Open nests in grass, and in emergent vegetation such as reeds, often have the surrounding and overhanging vegetation pulled down or woven into a canopy for concealment from above, while nests in wetlands often have pathways or ramps up to the bowl. Some species build nests that float or are attached to aquatic vegetation; nests on water may be substantial structures to cope with flooding, or may be built up rapidly as the water level rises or as the nest sinks. Gallinules, coots and moorhens, and also other species, build nursery nests for brooding the chicks, and some species also build display platforms.

The Giant Coot builds an enormous permanent nest of aquatic vegetation which forms a raft up to 3m long and projects 50cm above the water. The nest is in water about 1m deep and large nests usually rest on the bottom; the original structure compacts into peat in the centre. The Horned Coot also builds an enormous nest of waterweed but this is usually placed on a conical mound of stones, up to 4m in diameter at its base and about 60cm high, and ending just below the water surface in a platform about 1m across. The weight of the stone structure has been estimated as about 1.5 tonnes; each stone weighs up to 450g and both adults collect stones from the lake bed in shallow water and carry them to the nest site in the bill.

Rail eggs are usually approximately oval in shape (sometimes elliptical or biconical), smooth, and usually fairly glossy. The ground colour is off-white to dark tan, usually more or less blotched or spotted with red-brown, grey, mauve, black, etc. *Rallina* and *Sarothrura* species lay unmarked white eggs, as also do Russet-crowned and Red-and-white Crakes. Clutch size varies from 1-19 (most frequently 5-10), but dumping or laying by more than one female in the same nest may complicate estimates of clutch size laid by an individual. The laying interval is usually 24-48 hours between successive eggs, but is 48-72 hours in the Takahe. Incubation is usually by both sexes but in some species only the female incubates, whereas in the polyandrous Striped Crake only the male incubates. In many species the male incubates by day and the female by night. The incubation period is 13-31 (usually 15-19) days per egg and the start of incubation varies from the time of laying of the first egg to that of the last egg so that hatching may be synchronous or asynchronous according to species. In the Sora, which has a large clutch size, asynchronous hatching may be advantageous in distributing the high energetic cost of feeding the young over a longer period (Kaufmann 1989).

The Chestnut Forest-rail is unusual in that it lays only one large egg, which is c. 27% of the mean adult weight; incubation is at least 34 days, possibly >37 days, the longest of any rail. Incubation is by both sexes and the incubated egg is often left to become cold.

Rail chicks hatch covered in down and are precocial or semi-precocial, staying in or near the nest until all have hatched, and usually leaving the nest within 1-3 days of hatching. They may return to the egg nest to be brooded at night by the parents, but in many species they roost in nursery nests. Chicks are brooded at night and in poor weather while small. They rapidly become chilled if not brooded at night, and suffer from exposure if left out in rain. Newly hatched young may swim and dive in case of need but all, even those of coots, are prone to wetting and rapidly become chilled in water. They are therefore reluctant to stay long in water during their first days of life, except in emergencies.

Chicks are usually fed bill-to-bill at first and they become self-feeding after a period varying from a few days to over eight weeks. The young are normally tended by both parents, sometimes by one parent only and, in some species, also by adult helpers and/or the offspring of previous broods. The parents may split broods for foraging and brooding. Some species, such as the Virginia Rail, may carry young in the bill; both this species and the Buff-banded Rail will also carry eggs in the bill, moving the clutch to an alternate nest if the original site is disturbed. The American Purple Gallinule has been seen to carry young in its bill when in flight. In general, the fledging period of rails is 4-8 weeks but in the Giant Coot it is about four months. The chicks' legs and feet grow rapidly, reaching full size before the rest of the body; in contrast, the growth of the wings is generally much retarded. The first body feathers begin to appear after 6-15 (usually about 7) days, the tail develops quite late in the sequence and the down on the head and neck is often the last to be replaced. A study of the Buff-spotted Flufftail (Taylor 1994) showed that full juvenile plumage was attained in 27-32 days, mean adult mass was reached at about 45 days, the bill attained full length after 16-17 days, the tarsus after about 26 days, the tail after about 40 days and the wings not until about 90 days; the young made their first short flights at 19 days of age (an early age for small rails), when the wings were 79% grown and the mass 74% of the mean adult mass. Sora and Virginia Rails show a comparable rate of development, the tarsus being fully grown, and full juvenile plumage attained, at 28 days. Spotless Crake chicks reached asymptote weight at one month and had fully grown tarsi at 27-28 days (Kaufmann 1988b).

The young usually become independent as soon as they are fully fledged, occasionally before and sometimes after fledging. The time of independence varies from 3-4 weeks (20-21 days in the Buff-spotted Flufftail) to a maximum of 15 weeks in the Weka and the Henderson Rail. They may be ejected from the territory as soon as they are independent, after which the parents may nest again immediately. In other cases the young may stay in the natal territory for the rest of the breeding season, often helping to rear chicks of subsequent broods, before leaving at the end of the season or later. In the Sora, a migratory species, the family group disperses from the home range when the young are 16-22 days old (Melvin & Gibbs 1996).

Usually one or two broods are reared. Clutch losses, especially from predation or flooding, are often high and many species will relay several times after failure. In species with extended breeding seasons several broods may be reared and many species have the capability to extend the breeding season, or to raise more broods per breeding period, when food is plentiful or conditions remain favourable for longer than normal. The adaptive plasticity of rails allows some species to exploit breeding habitats as and when they appear, and to cope with sudden changes in habitat conditions. For example, American Coots in Colorado adapted well to artificial alteration of water level in a marsh, rapidly deserting areas when habitats deteriorated and recolonising them when habitats improved. The birds were not tied to a strict seasonal breeding cycle, so could colonise ephemeral nesting sites, and when conditions changed could quickly relocate and produce a replacement brood. The Black-tailed Native-hen of Australia is widely dispersive and irruptive, and often appears in great numbers after local rain to begin breeding immediately; birds will breed continuously throughout the winter and summer if conditions remain favourable.

The age of first breeding is usually one year, sometimes less. The Lord Howe Rail and Aldabra White-throated Rail attain sexual maturity at nine months, while the Guam Rail can breed at 16 weeks, which has considerably helped the captive breeding programme for this endangered species. However, the Inaccessible Rail retains its immature plumage for 1-2 years, which suggests delayed maturity; this species appears to have reached carrying capacity on the island and regulatory mechanisms such as low fertility and delayed maturity operate to limit its productivity (Ryan *et al.* 1989, Fraser *et al.* 1992).

Movements

A wide variety of movement and dispersal patterns occurs in the family. Some rails are long-distance migrants, many are dispersive, some are strongly irruptive, some nomadic, and others show limited local movements in response to changing environmental conditions. The movement patterns of many species are often unknown or poorly known, and this is due to a combination of factors such as the birds' tendency to move at night and the difficulty involved in locating birds when they are not breeding, when their silence and unobtrusiveness render them easily overlooked and may result in reported absences being more supposed than real. Evidence from studies in Africa suggests that many rail species, even those normally considered entirely sedentary, make at least local movements in response to seasonal variations in habitat conditions, while species normally considered migratory over at least part of their range may not move unless compelled to do so by deteriorating conditions.

Migrating rails fly at night, usually singly or in small groups and at low altitudes. Both Virginia Rails and Soras gather in numbers in marshes before moving, but there is no evidence that they migrate in groups. Virginia Rails often follow the courses of rivers and low, level ground, and may fly less than 1m above the ground. Migrating rails are frequently attracted to lights of many kinds, from lighthouses and floodlights to lighted house windows, and they often strike towers, buildings, overhead power lines, telephone wires and other obstacles, and even become impaled on barbed wire fences. Migrating Clapper Rails may be forced down in considerable numbers by heavy mist, when they enter buildings, fly into vehicles, and perch on buildings and overhead wires. In the former USSR the Water Rail's migrations are nocturnal, crepuscular and auroral, and migrating birds often spend the day resting in sizeable assemblages in bush, in the burrows of mammals, and on manmade constructions.

Most Holarctic species undertake substantial migrations and winter further south in Africa, India, the Oriental region and South America. In the western Palearctic, seven of nine breeding species are wholly or largely migratory, five crossing the Sahara to winter to the south and one, the Water Rail, wintering far into the Sahara; only the Purple Swamphen and the Red-knobbed Coot are mainly resident. The Water Rail migrates on broad fronts and crosses mountains such as the Alps. The Corncrake's main routes into Africa are in the west, between Morocco and Algeria, and in the east via Egypt; few cross the Mediterranean between these two flyways. It overflies North Africa and the Mediterranean region on its return passage, which is more rapid than the southward passage, unlike the migratory *Porzana* species which overfly these regions in the autumn and not the spring. The Common Coot has a specific moult migration, and particularly large flocks of flightless birds are recorded from the Black Sea in August, while many American Coots apparently migrate to large wetlands to moult after breeding although they will also moult in the breeding areas.

All rails which breed in North America are migratory to some extent, at least some populations wintering to the south of the breeding range. The Yellow Rail and the American Coot do not winter as far south as Central or South America. Some species, such as the Black Rail and the Yellow Rail, are known to move over a broad front, while the Sora apparently migrates along the coast and is prone to drift in strong winds, sometimes being recorded far out to sea. Migrant Clapper Rails fly overland and also along the coast.

Relatively little is known about the migrations of those species which breed from India east across Asia to China and Japan and occur south through the Oriental region. Many are known or thought to be at least partially migratory, evidence suggesting that they are predominantly winter visitors to the southern part of their range or summer breeding visitors in the northern areas, or that resident populations in the south are augmented by visitors from the north in the winter. The Red-legged Crake *Rallina fasciata* is both resident and migratory over its normal range but is largely a winter visitor to the southern regions, and is probably also dispersive. The Slaty-legged Crake *R. eurizonoides* is both a resident and migrant species in India, and moves at the beginning and end of the southwest monsoon. The Slaty-breasted Rail is normally regarded as resident but it has local movements in India during extremes of drought or flood, and there is some evidence for it having migratory movements elsewhere in its range. The Band-bellied Crake *Porzana paykullii* is a recognised migrant, but it has an imperfectly known winter distribution.

Less is known about the movements of South American rails than about those of any other continent or subcontinent. The great majority of the species known or suspected to be migratory or dispersive are inhabitants of palustrine wetlands or wet grassland, habitats which have apparently received relatively little attention from ornithologists. For a few of the more obvious species, such as the coots, there is sufficient evidence to show that movements do occur: thus immature Giant Coots are known to disperse widely at night, while the Horned Coot has altitudinal movements or displacements in harsh weather and other coot species are known to have seasonal population movements. Species such as the Ocellated Crake *Micropygia schomburgkii*, the Colombian Crake *Neocrex colombianus* and the Grey-breasted Crake are suspected to have migratory or dispersal movements because individuals have flown into lighted windows at night. The Ash-throated Crake *Porzana albicollis* is suspected of having seasonal movements in Colombia because occurrences in different parts of the country are reported in different periods. Some evidence for nocturnal movement or dispersal is provided by records of Blackish Rail *Pardirallus nigricans*, Rufous-sided Crake and Grey-necked Wood-rail heard flying over the city of Rio de Janeiro at night.

Much more is known about the movements of African rails, but even this knowledge is very incomplete. Some species, such as Striped Crake, African Crake, Streaky-breasted Flufftail and Allen's Gallinule, move away from the equator to breed during the rains. However, such species are not necessarily always migratory: where conditions remain suitable, the African Crake may remain after breeding, while the Streaky-breasted Flufftail is apparently sedentary in Gabon. Other species, such as Black Crake and African Rail, are normally regarded as sedentary but may be locally migratory in some parts of their range, appearing with the rains and disappearing in the dry season, and may occupy temporary wetlands anywhere within their range. The Buff-spotted Flufftail has complex and poorly understood movements which are related to its wide habitat tolerance: it is largely sedentary in stable forest habitat but of seasonal occurrence in habitats, such as deciduous thickets, which are only suitable in the rains. The Striped Flufftail has seasonal altitudinal movements in KwaZulu-Natal, South Africa, moving to lower altitudes after breeding; in some cases the distance moved may be as little as 30-40km (P. B. Taylor unpubl.). In subSaharan Africa, movements of the Corncrake are linked to rainfall and the movement of the Intertropical Convergence Zone, and the Spotted Crake is known to be itinerant in its African winter quarters, appearing very soon after rain, occupying temporarily flooded habitats while water levels are receding, and leaving as soon as such sites become too dry or deeply flooded by subsequent rain. Interestingly, similar behaviour is shown by the closely related Australian Spotted Crake *Porzana fluminea*, which appears after floods and heavy rain, follows receding water levels and leaves when wetlands dry out.

A typical dispersive and adaptable species is the Buff-banded Rail of South-East Asia and Australasia. It has a very wide habitat tolerance and is able to colonise small islands, it eats a wide range of food, and it can breed throughout the year. It may move in response to the availability of water, switching from dry to wet areas depending on rainfall. The Spotless Crake is known to be a highly dispersive species which has spread widely across the Pacific through Polynesia. The Rufous-tailed Bush-hen is possibly nomadic in Australia, where it moves into some regions with the onset of rain and may retire to areas with permanent water in the dry season. Similar behaviour is recorded for the Dusky Moorhen, which is dispersive or nomadic in Australia but does not undergo large-scale irruptive movements. The Black-tailed Native-hen, however, is a classic dispersive and regularly irruptive species which breeds opportunistically whenever and wherever it can. Its irruptions are probably associated with a period of favourable breeding conditions which allows a population build-up and which is followed by harsh conditions which force the birds to move.

Irruptions and movements also occur in Australasian flightless species. In New Zealand the Weka can move for significant distances: subadults may disperse more than 9km from the natal area and can walk more than 4km in a day. The greatest recorded natural movement in this species is 35km by an adult male. Rivers and lakes are apparently no barrier to this bird's movements, and birds on islands may swim or wade at low tide. Mass migrations, associated with the disappearance of many populations on North Island, have occurred for unknown reasons, while rapid increases in populations are also recorded, apparently owing to short-term changes in food supply. The Weka's homing ability is also remarkable, the longest-known return being 130km. Also in New Zealand, the Takahe undergoes altitudinal movements of up to 5-10km, possibly even 30km, when winter snow covers its summer grassland feeding territories, while chicks are recorded moving 400-800m within a few days of hatching. Immature Tasmanian Native-hens disperse widely and are recorded as moving up to 40km. Both adult and juvenile Lord Howe Rails make altitudinal movements over short distances.

A lack of knowledge of the distribution and status of many species sometimes makes it difficult to judge whether isolated occurrences are indicative of long-distance movements, or whether movements are primarily nomadic, eruptive or migratory. For example, data on distribution and occurrence periods suggest that the Speckled Rail *Coturnicops notatus* is unlikely to have regular migrations in South America, but a lack of knowledge of its distribution means that the possibility of its undergoing long-distance random eruptions cannot be confirmed; the best that can be said is that it must undergo postbreeding dispersal. The enigmatic White-winged Flufftail *Sarothrura ayresi* is known primarily from wetlands in Ethiopia and South Africa, and the morphological similarity between these two populations suggests that migration may occur between the centres of distribution. However, the paucity of records from intervening regions, with a very few acceptable records only from Zimbabwe and Zambia, and an overlap in occurrence dates, makes this unlikely, and it is thought that in both regions it may undergo periodic long-distance dispersal when numbers are high, allowing gene exchange between the two populations.

Although most forest habitats are stable and permanent, some species dependent on moist conditions may have to move at least locally if conditions become dry. In New Guinea, both the Red-necked Crake and the Bare-eyed Rail *Gymnocrex plumbeiventris* move out of some forests when the ground dries out seasonally, the former species apparently migrating to NE Australia, where it is a wet-season visitor to the Cape York Peninsula, leaving again when the forest floor dries out. In Africa the White-spotted Flufftail may move locally in response to drought conditions and habitat disturbance.

The widespread occurrence of rails on oceanic islands provides ample evidence of these birds' powers of dispersal and their tendency to vagrancy. The available evidence shows that the phenomenon of dispersal to islands, or to continental habitats far from their home range, is widespread in rails, and a large number of living, recently extinct, or fossil taxa has been found on islands. Many types of rails are involved in long-distance vagrancy, and the high degree of vagrancy in the family is indicative not only of the birds' dispersive ability but also of the readiness with which they are apparently blown off course by unfavourable winds as a result of their relatively poor flight performance. Long-distance vagrancy, especially involving ocean crossings, is often associated with bad weather and prevailing winds. An example is the occurrence of American Purple Gallinules in South Africa, whereas there are no instances of African migrant rails occurring in the Americas because the prevailing winds are against the crossing. The ability of rails to land, rest on the water and take off again would increase their chances of survival over those of many purely terrestrial birds (Olson 1973a, Sick 1993).

Another factor relevant to rails landing on oceanic islands and remaining there is that they are virtually forced to do so – to leave would mean their having to reorientate themselves from a completely unfamiliar location, and to accumulate sufficient food reserves to sustain them in a long return crossing, probably against prevailing winds. In such a situation they are well served by their ability to adapt to marginal habitats and to colonise open areas, and their tendency to omnivory. They may not be the only types of bird to reach remote islands by accident, but they are certainly some of the best survivors in such hazardous situations. The fact that they are ground-dwellers is also an advantage, so long as no mammals are already present. They also have the potential for rapid breeding to establish a viable population.

However, it appears that not all rails are equally adept at surviving on small islands. As Olson (1973a) points out, the Common Moorhen has successfully colonised many oceanic islands (St Helena, the Azores, the Seychelles, the Marianas, Hawaii etc) and has given rise to flightless forms on Tristan da Cunha and Gough Islands. But two of the most notable vagrants in the family, the American Purple Gallinule and Allen's Gallinule, have not established breeding populations on any remote oceanic island, nor have given rise to endemic flightless forms. Olson suggests that one reason for this pronounced difference in success is that the small *Porphyrio* species normally inhabit wetlands with floating vegetation and are consequently relatively specialised, having long legs and toes, and other structural modifications, whereas the moorhen is relatively unspecialised and does not differ significantly from the basic rallid structure. Thus the moorhen may be much better equipped to adapt itself to the drastically different environments encountered on small oceanic islands.

The American Purple Gallinule occurs from the USA south to Argentina, northern populations moving south to winter. It is regularly reported from Tristan da Cunha, 3,800km from the South American mainland, has also occurred on St Helena, about 6,400km from South America, and is of almost annual occurrence in the southwestern Cape, South Africa. Of 21 vagrants to the southwestern Cape, almost all occurred from late April to early July and it appears that birds starting north from Buenos Aires province, Argentina, or from Uruguay, are caught in strong westerly winds and carried across the Atlantic, to appear in South Africa exhausted and emaciated (Silbernagl 1982). The majority of vagrants are immatures, suggesting that adults are not inclined to stray so far off course. Allen's Gallinule of subSaharan Africa and Madagascar has occurred in North Africa, Europe north to Denmark, and offshore islands as far away as the Azores, Ascension Island and St Helena (1,900km from the African continent) in the Atlantic Ocean, and the Comoros and Rodrigues (1,500km east of Madagascar) in the Indian Ocean. It has even reached South Georgia Island in the South Atlantic, c. 4,800km WSW of Cape Town. The Corncrake, which breeds in Eurasia and winters mainly in subSaharan Africa, has a remarkable range as a vagrant, having been recorded as far afield as North America, the Bahamas, India, Sri Lanka, the Seychelles and Australia.

The process of dispersal is continuous and an example of the recent colonisation of islands by a rail species is that of the nominate race of the Paint-billed Crake *Neocrex erythrops* in the Galapagos. This race, which was formerly known only from coastal Peru, was first recorded from the Galapagos in 1953 and is now an abundant breeding bird on Santa Cruz and is also recorded from Santa María. It had not been found by expeditions which had made extensive collections of the Galapagos avifauna in the previous hundred years, and is thus likely to be a very recent colonist, although it has been suggested that the species was possibly overlooked for some time before its discovery. The Paint-billed Crake is not known to have regular migrations, but the nominate race has also been recorded as a vagrant in Texas; there are many instances of the race *olivascens* occurring outside its normal range in South America and even occasionally in Central America, and once (probably *olivascens*) from Virginia. Too little is known about this bird's status and distribution to ascertain to what extent it is migratory or has vagrant habits.

CONSERVATION AND EXTINCTION

Based on the current IUCN criteria (SSC 1994), which are followed throughout this book, available evidence indicates that 33, almost a quarter, of the 135 surviving rail species are globally threatened, while the subspecies *tuerosi* of the Black Rail, sometimes considered a distinct species (e.g. Collar *et al.* 1994), is also endangered. A further seven species are classed as near-threatened, and five as Data Deficient. Thus the survival of 45, nearly a third, of living rail species gives cause for concern. Furthermore, two of the 33 threatened species can almost certainly be added to the list of extinct species: the Bar-winged Rail *Nesoclopeus poecilopterus*, not definitely recorded since 1890, and the forest-dwelling Samoan Moorhen *Gallinula pacifica*, last definitely recorded in 1873. The Kosrae Crake *Porzana monasa*, not recorded since its discovery in 1827/28, is here regarded as definitely extinct.

In addition to these 33 species, in this book it is recommended that two further species, the Chestnut-headed Flufftail *Sarothrura lugens* and the Striped Flufftail, should be considered Vulnerable on the basis of their fragmented distribution, the decline in the quality and extent of their habitats, and their probably very small populations (<10,000 mature individuals of each species remaining). It is also recommended that a further five species be considered Data Deficient on the basis of their restricted distribution, the lack of information on their status and the potential threats to their habitats. These species include three from South and Central America, the Little and Red-winged Wood-rails *Aramides mangle* and *A. calopterus* and the Uniform Crake *Amaurolimnas concolor*, and two from Asia, the Isabelline *Amaurornis isabellinus* and Plain Bush-hens. The inclusion of these seven species would increase to 52 the number of species whose survival gives cause for concern, this being nearly 40% of the 135 possibly extant rail species.

Flightless species are particularly vulnerable, and only two of the 20 surviving species are relatively safe, the New Britain Rail and the Tasmanian Native-hen. However, even the New Britain Rail should not be regarded as completely safe: it has a restricted distribution confined to New Britain, there is no current estimate of its status, and it is known to be trapped and eaten by local people. The Tasmanian Native-hen is apparently still very numerous, despite some drastic local declines in recent years, thought to be a result of disease. It has a history of persecution which continues today: it is traditionally regarded as an agricultural pest, although most claims are unsubstantiated, and thousands were killed in the 1950s when it was declared a pest. It is not protected, is still controlled by shooting and round-ups, and is attacked by feral cats and dogs.

Of the 18 other surviving flightless or almost flightless species, 15 are globally threatened: the Bar-winged Rail and the Samoan Moorhen are probably extinct; the Guam Rail is extinct in the wild (but is being reintroduced); the New Caledonian Rail, Zapata Rail *Cyanolimnas cerverai* and San Cristobal Moorhen *Gallinula silvestris* are Critically Endangered; four species are Endangered; and another five are Vulnerable. The remaining three species are near-threatened or Data Deficient.

The 20 rallid species and races (excluding the doubtfully valid Gilbert Rail) which have definitely or almost certainly become extinct since 1600 (see Table) are all island forms and 18 (90%) of them were flightless. The extinction of these flightless island rails provides a classic example of the particular vulnerability of island endemics, which arises as a result of their long isolation from predators and competitors, and is one of the major factors involved in island extinctions. The worst period for recorded extinctions since 1600 was the late 19th and early 20th centuries, when many oceanic islands were rapidly developed and altered. Some of the factors which led to these extinctions are still applicable to threatened rails today.

The principal causes of extinctions among island rails since 1600 have been introduced mammalian predators such as cats, dogs, rats, mongooses and pigs, indiscriminate hunting by the first people to visit the islands and, to a lesser extent, habitat destruction by introduced goats, rabbits and fire. Hunting wiped out the larger flightless species very quickly, the White Gallinule, the Mauritian Red Rail *Aphanapteryx bonasia*, Leguat's Rail *Aphanapteryx leguati* and the Tristan Moorhen all being destroyed in this way. Introduced predators have probably been responsible for more extinctions than any other single cause, although some species have been wiped out by a combination of this factor and the destruction of habitat. The Laysan Crake *Porzana palmeri* is an example of this, and also of a lost opportunity. Its habitat was destroyed by introduced rabbits but the species survived on two tiny islets on the Midway atoll, to which it had been introduced. However in 1943 rats invaded the islands from a naval craft and by 1944 they had exterminated the rails. Unfortunately no-one had thought to reintroduce them to Laysan, where the rabbits had been eliminated in 1923.

The Guam Rail was brought to extinction in the wild by another type of predator. Although it was taken by local people with dogs and snares, and despite the presence of introduced predators such as cats and pigs, it held its own on Guam until the spread of the accidentally introduced brown tree snake *Boiga irregularis*, which has caused a precipitous decline and extinction among Guam's native forest birds (Savidge 1987). The snake is nocturnal and arboreal, and probably did not prey significantly on adults but took mainly eggs and chicks. This is the first time that a snake has been implicated as an agent of extinction.

Hunting is still a threat to some endangered species with very small populations. The San Cristobal Moorhen, known only from the holotype collected in 1929, and possibly already extinct, has suffered from hunting by local people with dogs, and has also probably been affected by introduced predators such as cats. A similar situation affects the Invisible Rail, but it is also threatened by habitat destruction.

Several island species which are currently holding their own, with stable populations in secure habitats, are still at risk from the possible accidental introduction of mammalian predators to their islands. The Gough Island race *comeri* of the Tristan Moorhen is one such case, although precautions have been taken to minimise the chances of accidental introductions occurring. The Inaccessible Rail is another, and it is also vulnerable to fire and agricultural development as long as the island continues to lack protected status, while the survival of the Auckland Rail *Lewinia muelleri* depends on the continued exclusion of mammalian predators. The Henderson Crake is also in this situation, and was recently threatened by human impact when a millionaire sought to make the island his home. The Okinawa Rail, which is threatened by deforestation, is also vulnerable to the brown tree snake, which has been seen on the island.

Habitat destruction in many forms is certainly a major threat to many continental rail species, although its impact on some threatened island rails may not be critical at present. The Zapata Rail is restricted to the Zapata Swamp in Cuba, where its habitat is under serious threat from dry-season burning, although introduced rats and mongooses are also a serious problem for this almost flightless species. The Plain-flanked Rail *Rallus wetmorei* has a very restricted distribution in coastal Venezuela, where its mangrove and lagoon habitats are being destroyed by housing development, oil exploration and diking, and the one wildlife refuge in which it occurs is under threat. The White-winged Flufftail is under severe threat from wetland habitat destruction and modification in South Africa and Ethiopia, the principal problems being damming, draining, overgrazing and afforestation, although disturbance by birdwatchers is also a problem in South Africa.

Great efforts have been made to save some threatened species, involving captive breeding and reintroduction of birds into the wild, and also habitat management and predator control. The Takahe, the Lord Howe Rail and the Guam Rail are species which have benefited from captive breeding and reintroduction programmes, which are described in their species accounts. The Weka has been introduced to many offshore and outlying islands off New Zealand, sometimes unsuccessfully, and it has been claimed that this species may adversely affect the indigenous fauna and flora on such islands, a factor which must be considered when introductions of any rail are planned to islands on which the species has not previously occurred.

Many rails seem to be undergoing a continuous population decline, largely through loss of their forest, marsh and grassland habitats. The continued wholesale and enormous destruction of indigenous forests may be a severe threat to some species, especially in South-East Asia and South America. The Bald-faced Rail *Gymnocrex rosenbergii* and the Snoring Rail are both threatened by forest destruction, while the very poorly known Red-winged Wood-rail may be threatened by habitat destruction. However, the forest rails of New Guinea, including four endemic *Rallina* species and the New Guinea Flightless Rail, may be relatively secure because no major threats to their habitats are apparent at present.

Palustrine wetlands are under threat throughout the world and are disappearing at an alarming rate, yet it is only in recent years that the threats to these habitats have been fully appreciated and the great importance of these wetlands realised (e.g. Hollis & Bedding 1994). Even now, there is only limited appreciation of the significance to rails and other birds of relatively small palustrine wetland patches. Many rails which inhabit these wetlands are threatened by habitat loss and modification. In Asia, Swinhoe's Rail *Coturnicops exquisitus*, a migratory species which is considered vulnerable because of its apparent rarity, is threatened by the destruction and modification of wetlands in both its breeding and wintering areas. In South America the Rufous-faced Crake *Laterallus xenopterus* is also considered vulnerable, although it may be more widespread than is currently known. Wholesale destruction of wetland habitats in Brazil by drainage and adjacent afforestation with eucalyptus plantations may have serious effects on hitherto undiscovered populations of this poorly known species. All the savanna and paramo marsh habitats of the Bogotá Rail *Rallus semiplumbeus* in Colombia are seriously threatened in many ways, including by drainage, agricultural encroachment, eutrophication and pollution, tourism, hunting, burning, trampling by cattle, and water-level fluctuations. In Madagascar, the endangered Slender-billed Flufftail *Sarothrura watersi* requires small wetland patches adjacent to rain forest; although its habitat may be widely distributed, the totally inadequate protection of wetlands in Madagascar gives no cause for optimism about this bird's long-term chances of survival (O. Langrand *in litt.*). Small crake species, such as the Black Rail, which inhabit the edges of marshes are generally more threatened by habitat destruction, and by disturbance from grazing or agriculture, than are other marsh rails which live in the interiors of palustrine wetlands or alongside open water.

Prime grassland habitats are also disappearing rapidly and their rail populations are under threat. A typical example of the threats facing rails in these habitats is that of the Striped Flufftail of eastern and southern Africa. In South Africa its distribution is fragmented and its grassland habitats are disappearing rapidly as a result of agriculture and, increasingly more importantly, widespread commercial afforestation (Taylor 1997a).

Another grassland species, the Corncrake, is the only globally threatened rail for whose conservation a significant international effort is being made. It breeds in at least 32 European countries (Hagemeijer & Blair 1997) and is threatened mainly on its breeding grounds, where habitat loss is extensive and mechanised hay mowing causes heavy losses of breeding birds, eggs and young. A BirdLife International European Workshop on the Corncrake was held in Poland in October 1994, to discuss the extent of its decline and any possible measures to prevent further population decreases. In Britain, a draft conservation action plan to halt the bird's decline was produced in 1989 and a major conservation effort was then initiated, involving government, conservation and farming bodies in the UK, Ireland and France. As a result of the implementation of conservation management measures, Corncrake numbers in Britain have recently shown a modest increase.

So little is known about many rails that their status and even their distribution cannot be established with any degree of confidence. To address this problem a considerable amount of work needs to be done urgently in many parts of the world. Particularly poorly known genera include the South American *Anurolimnas* and *Laterallus*, and other typical examples are to be found among the *Aramides* species. The Little Wood-rail is a poorly known species of potentially restricted distribution, with habitat preferences which are not properly established. The Slaty-breasted and Red-winged Wood-rails, although not regarded as globally threatened, are potentially at risk because of the destruction of forest habitats, and their status is in urgent need of investigation.

EXTINCTION ON ISLANDS

Extinction on oceanic islands has been far more catastrophic than the above examples indicate, and the magnitude of the catastrophe has been very effectively summarised by Olson (1989) and Steadman (1995), from which accounts most of the following information is taken. Until recently there was no palaeontological record for most oceanic islands and it was easy to assume that the extinctions documented during the historic period, chiefly the result of the depredations of European man on island habitats and their fauna, formed the greater part of the losses which have occurred in the recent past. The extension of palaeontological work to many more islands has shown that we have grossly underestimated the effects of man on insular biotas and has given us some idea of the true magnitude of the losses, the majority of which occurred prehistorically, i.e. before the arrival of Europeans. For example, Olson calculates that in New Zealand and the Chatham Is, out of a total of 96 species of resident land birds, 12 (12.5%) were exterminated in the historic period and a further 32 (33%) prehistorically, and Steadman records that one region of South I, New Zealand, the Punakaiti karst area, has lost 27 of its 50 species of land birds. A study of bird remains from late Holocene cave deposits in New Caledonia shows that at least 25% (possibly up to 40%) of the resident nonpasserine bird species were exterminated prehistorically, almost certainly as a result of human disturbance.

In the Hawaiian Is, prehistoric man-caused extinctions occurred on a massive scale (e.g. Olson & James 1982). Bones recovered from various sites have so far documented the extinction of 60 bird species, including 12 species of flightless rails and comprising 52-54% of the native land bird species of the archipelago (Olson & James 1991, Steadman 1995). The additional 20-25 species which were exterminated in the historic period bring the total extinctions to 80-85, or at least 72% of the known native land bird species. Henderson I in the Pitcairn group was once inhabited by Polynesians, who exterminated at least two species of pigeon from the total land bird fauna of six species (the other four are still extant). At least 27 species of land birds lived on 'Eua I, Tonga, in prehuman times but only six have survived into the past two centuries. On the island of Huahine in the Society group, bones from archaeological sites show that 15 species of land birds were formerly present, whereas only two species nest there today. The mere seven land bird bones from the remote Easter I include those of two endemic species of rails. Many other similar examples are known but it should be noted that these extinct forms may not have all lived simultaneously.

Extinctions of birds following European explorations in the 15th and 16th centuries have been well documented both historically and palaeontologically for islands in the Indian and Atlantic Oceans. The fate of the fauna of the Mascarenes, home of the dodo and the solitaire, is all too familiar and three recent rail extinctions are documented in this book (see Recently Extinct Species). On the small island of St Helena in the South Atlantic only one native species of land bird has survived, whereas fossil remains show that at least four others were probably present when man first arrived in 1502. Drastic changes in the vertebrate composition of the Lesser Antilles in the last 2,000-3,000 years have also been documented through the fossil record; these are almost certainly the result of man's interference.

The scale of extinctions may be illustrated with reference to the rails of oceanic islands. Olson suggests that endemic species of flightless rails occurred on virtually every oceanic island in the world, while Steadman estimates that most or all of the Polynesian islands supported 1-4 endemic species of flightless rails, many of which inhabited the forests which were destroyed by prehistoric peoples. Extrapolating from the number and size of islands in Oceania, Steadman estimates that c. 2,000 species of flightless rails may have been exterminated in Oceania since the arrival of man; the true figure is probably between 1,000 and 3,000 species (D. W. Steadman *in litt.*). He concludes that the loss of birdlife in the tropical Pacific may represent a 20% reduction in the global number of bird species. About one-third of the world's bird species are endemic to islands and from the fossil record it appears that the species diversity of birds on almost all oceanic islands has been reduced by 30-50%, sometimes much more, within the period of man's occupancy – in an instant of geological time. Add to this the thousands of extinctions of other animals and plants which must also have occurred at the same time and, in Olson's words,

> *"we are faced with one of the swiftest and most profound biological catastrophes in the history of the earth".*

RECENTLY EXTINCT SPECIES

In addition to the ten recently extinct, and two almost certainly extinct, species described in the species texts, the following five species have become extinct since 1600. These are species known only from bones or contemporary illustrations for which no specimens or proper descriptions exist. With regard to the relationships of the genus *Aphanapteryx*, to which two of the species belong, Olson (1977) suggests a derivation from the rather closely interrelated group that includes *Gallirallus*, *Dryolimnas*, *Atlantisia* and *Rallus*.

Mauritian Red Rail *Aphanapteryx bonasia*

Apterornis bonasia Sélys-Longchamps, 1848, Mauritius.

This species is known from five contemporary illustrations of varying quality, from a number of descriptions and from numerous bones (see Olson 1977). It was about the size of a domestic fowl and had reddish-brown plumage which was possibly fluffy with the feather structure slightly decomposed. The wings were very short and useless; the tail was very short and apparently decomposed; the bill was long, downcurved and probably dark in colour; the legs were long and powerful, possibly also dark in colour; the iris may have been yellow. The bones show that there was considerable individual variation in the size and curvature of the bill (Greenway 1967).

An extensive and confusing literature has grown up around the extinct rails of Mauritius and hypothetical species have been erected on the basis of real or supposed differences between the original sources, which comprise both illustrations and written accounts of the birds. However Olson (1977), in an exhaustive review of the literature and evidence, concluded that only one species could be upheld.

There are several accounts of the way in which these birds were killed, quoted by Olson (1977). They were strongly attracted by red articles, which they would pursue. They could easily be caught by hand, for example by holding a piece of red cloth in one hand and a stick in the other to strike the birds as they approached. The cries of a wounded bird would attract others, which were then also killed. They made excellent eating and, although quite common in the early 1600s, they had become scarce by the second half of the century; they were very rare by 1693 and probably became extinct soon afterwards (Fuller 1987).

References Greenway (1967), Olson (1977), Fuller (1987), Day (1989), Feduccia (1996).

Figure 1: Two extinct *Aphanapteryx* rails: left *A. bonasia*, right *A. leguati* [After Day 1989].

Leguat's Rail *Aphanapteryx leguati*

Erythromachus leguati Milne-Edwards, 1874, Rodrigues Island, Mascarenes.
This rail is known from the bones of several individuals and from contemporary descriptions; full details are given by Olson (1977). It was apparently similar in size and shape to its congener but was bright grey (the plumage was also described as being flecked with white and grey), with a similarly curved red bill, red legs and feet, and a red orbital ring. The curvature of the bill was variable, and the bill could be almost straight. The call was described as a continual whistling, but birds when pursued would give a call "which one would say came from a person with the hiccup".

They were said to feed on the eggs of tortoises which they found in the ground, and this diet was said to make them so fat that "they often have trouble running", although they normally ran swiftly. Like the Mauritian Red Rail, this species was attracted by red objects and birds could be caught by hand while attacking such lures. They were described as having a most delicate taste (Day 1989) and their orange fat was said to be excellent for ailments. The species became extinct some time after 1730.

References Greenway (1967), Olson (1977), Fuller (1987), Day (1989), Feduccia (1996).

Tahiti Crake *Rallus nigra*

Synonym: *Nesophylax niger*
The identity of this bird is based on illustrations and descriptions, including a painting of a bird from Tahiti by Georg Forster made on Cook's second voyage and a painting by J. F. Miller of a bird named *Rallus nigra* in his *Icon. Animalium* of 1784. It has been assumed that both illustrations refer

to the widespread Spotless Crake, but in a review of the existing information Walters (1988) concludes that neither picture represents the Spotless Crake which, unlike the illustrations, has dark reddish-brown upperparts. The situation is confused by the fact that J. R. Forster wrote a description of a rail, also collected on Cook's second voyage, which in fact disagrees in several respects with the Georg Forster picture and probably represents the Spotless Crake because the upperparts were "dark rusty-fuscous"; Walters (1988) suggests that both the Spotless Crake and the form given the name *Rallus nigra* (which presumably should also be referred to *Porzana*) were collected on Tahiti by the Forsters, who failed to distinguish between them. More confusion has apparently arisen because later authors failed to interpret these differences correctly (Walters 1988).

The bird depicted by Georg Forster is charcoal-grey on the head and underparts, and black on the back, wings and tail; it has a red eye which is surrounded by an area of black skin outlined in orange; the bill is thick and black; and the legs and feet are red. The Miller plate shows a charcoal-black rail with suggestions of dark and less dark areas; the bill is black with a pale yellowish-green line separating the mandibles; the iris and legs are a duller, browner red than in the Forster painting.

Cook's second voyage around the world took place from 1772 to 1775; it is not known how long the rail survived after this.

References Olson & Steadman (1987), Walters (1988).

Ascension Rail *Atlantisia elpenor*

Atlantisia elpenor Olson, 1973a, Ascension I.

All the skeletal elements and a crude contemporary drawing of this rail are known. The discovery of its remains, and an account of its environment and extinction, are documented in fascinating detail by Olson (1973a), from which account the following information is taken.

It was a medium-sized flightless rail, roughly the size of the Virginia Rail and much larger than the Inaccessible Rail; its major bones are c. 25-35% longer than its congener. Its plumage was apparently grey, or dappled white and black; its wings were very small, and its eyes were "red like rubies". It could run very swiftly, flapping the wings while doing so. The birds were taken while running and were described as making good eating.

The climate of Ascension I is mild but the lowland regions are very dry, with near-desert conditions prevailing, and strong trade winds blow with great consistency. The native flora was impoverished and the terrestrial environment hostile in the extreme, consisting mainly of bare, waterless tracts of lava and ash. The rail probably obtained its food and water from the eggs and regurgitated prey of the seabirds that formerly nested on the island in great numbers. It became extinct some time after 1656.

Although large, this species was only half the size of the extinct Great St Helena Rail *Atlantisia podarces*. This species was described by Wetmore (1963) as *Aphanocrex podarces* but was later recognised as a very large flightless *Atlantisia* (Olson 1973a). It was roughly the size of a Weka but was more slender, and had relatively large wings for a flightless rail and extremely long claws; the claws may have been adaptations for climbing and fluttering up the steep valley walls of St Helena (Olson 1977). It probably survived until shortly after the island's discovery in 1502.

References Knox & Walters (1994), Olson (1973a, 1977), Wetmore (1963), Feduccia (1996).

Mascarene Coot *Fulica newtonii*

Fulica newtonii Milne-Edwards, 1867, Mauritius.

There is no contemporary illustration or good description of this species, but bones from at least 24 individuals have been found (Olson 1977). It was a large, flightless coot, the tibiotarsus measuring 77-86mm as against 52-64mm in the Common Coot (Greenway 1967), but it was somewhat smaller than the huge *Fulica chathamensis* of the New Zealand region. It had become very rare by 1693.

There was also apparently a coot on Rèunion, as large as domestic fowl and all black with a "large white crest on the head" (Greenway 1967).

References Greenway (1967), Olson (1977).

RAIL TAXA EXTINCT SINCE 1600

TAXON		DATE OF EXTINCTION	DISTRIBUTION
Extinct species			
Mauritian Red Rail*	*Aphanapteryx bonasia*	c. 1700	Mauritius
Leguat's Rail*	*Aphanapteryx leguati*	post-1730	Rodrigues I (Mascarenes)
Gilbert Rail	*Gallirallus conditicius*	post-1859 (?)	Marshall Is (?)
Wake Rail	*Gallirallus wakensis*	1945	Wake I (N Pacific)
Tahiti Rail	*Gallirallus pacificus*	1930s (?)	Tahiti
Dieffenbach's Rail	*Gallirallus dieffenbachii*	post-1840	Chatham Is
Chatham Rail	*Gallirallus modestus*	1896-1900	Chatham Is
Sharpe's Rail	*Gallirallus sharpei*	date unknown	possibly Indonesia (?)
Tahiti Crake*	*Rallus nigra*	post-1775	Tahiti
Ascension Rail*	*Atlantisia elpenor*	post-1656	Ascension I (S Atlantic)
Laysan Crake	*Porzana palmeri*	1944	Laysan (Hawaiian Is)
Hawaiian Crake	*Porzana sandwichensis*	1884	Hawaii
Kosrae Crake	*Porzana monasa*	1828-1890	Caroline Is
White Gallinule	*Porphyrio albus*	pre-1844	Lord Howe I
Mascarene Coot*	*Fulica newtoni*	post-1693	Mauritius, possibly Réunion
Almost certainly extinct species			
Bar-winged Rail	*Nesoclopeus poecilopterus*	post-1890 (?)	Fiji
Samoan Moorhen	*Gallinula pacifica*	1873-1907 (?)	Savaii (Western Samoa)
Extinct subspecies			
[Buff-banded Rail]	*Gallirallus philippensis macquariensis*	1880-1894	Macquarie I
[White-throated Rail]	*Dryolimnas cuvieri abbotti*	pre-1937	Assumption I (Mascarenes)
[Uniform Crake]	*Amaurolimnas concolor concolor*	post-1881	Jamaica
[Tristan Moorhen]	*Gallinula nesiotis nesiotis*	1873-1900	Tristan da Cunha
Possibly extinct subspecies			
[Lewin's Rail]	*Lewinia pectoralis clelandi*	post-1932	SW Australia

Species not described in the species accounts nor illustrated on the plates (for which no specimens or proper descriptions exist) are indicated with an asterisk.

Note: Gilbert Rail is possibly synonymous with Lord Howe Rail (see text).

PLATES
1-43

PLATE 1: AFRICAN FOREST AND GRASSLAND FLUFFTAILS

1   **White-spotted Flufftail** *Sarothrura pulchra*   Text and map page 151

Gambia and Senegal E to S Sudan and W Kenya and S to N Angola and NW Zambia. Forest and thickets, usually in association with water.

Male bright reddish-chestnut on foreparts and tail; body spotted white. Female has foreparts like male; rest of body, and tail, distinctively barred.

- **1a**   **Adult male** (*centralis*; Congo E to Kenya and S to Zambia and Angola) Head to mantle and breast, and tail, reddish-chestnut; rest of body black, spotted white.
- **1b**   **Adult female** (*centralis*) Body blackish-brown, barred reddish-buff; tail barred blackish.
- **1c**   **Juvenile male** (*centralis*) Juveniles patterned like adults of same sex, but much duller; note traces of black down on head.
- **1d**   **Adult female** (*zenkeri*; SW Nigeria, coastal Cameroon and Gabon) Upperparts blacker.

[*S. p. pulchra* (Senegal and Gambia to Cameroon). Female has broader, more rufous barring than *centralis*. *S. p. batesi* (S Cameroon) Female blacker than other races; chestnut paler.]

2   **Buff-spotted Flufftail** *Sarothrura elegans*   Text and map page 154

Guinea and Sierra Leone E to Ethiopia and Kenya and S to N Angola and E and S South Africa. Forest, dense thickets and old cultivation.

Male has orange-chestnut foreparts, buff-spotted upperparts, and barred tail; female paler brown than other flufftails, finely spotted buff and black on upperparts.

- **2a**   **Adult male** (nominate; Ethiopia S through E and C Africa to South Africa) Head, neck and breast orange-chestnut; tail barred black and chestnut; otherwise sooty-black, closely spotted buff, but flanks and belly spotted white.
- **2b**   **Adult female** (nominate) Almost golden-brown with small black-edged buff spots on upperparts, blackish and buff bars and scallops on face and underparts; flanks and belly barred white; tail red-brown, narrowly barred black and buff.
- **2c**   **Downy chick** 7 days old; remiges, upperwing-coverts and breast feathers growing.

[*S. e. reichenovi* (Sierra Leone E to Uganda and S to N Angola). Male darker above, with larger, coarser spots.]

6   **Striped Flufftail** *Sarothrura affinis*   Text and map page 168

S Sudan to South Africa, mainly in highlands. Grassland and fynbos.

Male has fluffy chestnut tail and yellowish-white streaks on upperparts; female has chestnut and black bars on tail.

- **6a**   **Adult male** (nominate; S and E South Africa) Head to hindneck chestnut; chin and throat white.
- **6b**   **Adult male** (*antonii*; E Zimbabwe to Kenya and S Sudan – see text for description of northernmost populations) Chestnut extends to upper breast.
- **6c**   **Adult female** (nominate) Upperparts blackish, scalloped and barred white to buff; underparts white to buff with black spots and scales, flanks barred; sides of head and neck often tinged rufous.
- **6d**   **Immature female** (*antonii*) Upperparts streaked, washed chestnut from mantle to uppertail-coverts; breast and belly white, with streaks on breast and scattered spots on belly.

100mm

PLATE 2: AFRICAN WETLAND FLUFFTAILS

5 **Streaky-breasted Flufftail** *Sarothrura boehmi*   Text and map page 165

Nigeria E to Kenya and S through NE Zaïre to E Angola and N South Africa. Dry to seasonally flooded grassland.

Short tail. Male streaked white, female scalloped and barred white.

- **5a**   **Adult male** Reddish-chestnut to hindneck and (variably) upper breast; no spots on upperparts.
- **5b**   **Adult female** Throat and breast spotted sooty-black, and sometimes washed faint rufous; rest of underparts barred.
- **5c**   **Adult male in flight** Note white leading edge of wing.

4 **Chestnut-headed Flufftail** *Sarothrura lugens*   Text and map page 163

Cameroon E to W Tanzania and S to C Angola and N Zambia. Dry to wet dense grass and sedges.

Long, fluffy tail. Male has rich reddish-chestnut head and neck, white chin and throat, white streaks on body and wings, and white-spotted tail; female blackish, with white spots and streaks.

- **4a**   **Adult male** Streaks narrow on upperparts, broad on underparts.
- **4b**   **Adult female** Head streaked buff or chestnut; upperparts have short, broad streaks or spots; wings spotted and barred white; underparts streaked and spotted blackish.

[Two races, separated only on size.]

3 **Red-chested Flufftail** *Sarothrura rufa*   Text and map page 159

Sierra Leone E to Ethiopia and Kenya, and S (excluding forested and dry regions) to the Cape. Marshes, reedbeds and grass.

Long, fluffy tail. Male chestnut from head and neck to mantle and lower breast; female brownest of African wetland flufftails, with buff spots, scallops and bars.

- **3a**   **Adult male** (nominate; C Kenya S to South Africa) Short white streaks on upperparts, becoming spots from rump to tail.
- **3b**   **Adult female** (nominate) Upperparts predominantly spotted. A rather pale-marked individual illustrated.
- **3c**   **Immature female** (nominate) Appears darker and less heavily patterned than adult.
- **3d**   **Juvenile male** (nominate) All black; this bird is in postjuvenile moult, with some immature plumage on head and body.
- **3e**   **Adult female** (*bonapartii*; Sierra Leone E to Cameroon, Gabon and Congo) Upperpart markings more crescent-shaped or streaked than in nominate (variable). Browner overall colour compared with 3b shows extent of individual variation.

[*S. r. elizabethae* (Ethiopia; Central African Republic; NE Zaire to W Kenya) Male like *bonapartii* (white streaks longer than in nominate; upperparts less spotted); upperparts of female mainly barred.]

100mm

PLATE 3: WHITE-WINGED AND MALAGASY FLUFFTAILS

8   White-winged Flufftail *Sarothrura ayresi*   Text and map page 172

South Africa, Zimbabwe, Zambia and Ethiopia. Moist to flooded grasses, sedges and reedbeds. White on secondaries diagnostic but not visible in folded wing; birds often easy to flush. Tail short, barred chestnut and black. Sexes similar; female duller.

- **8a** **Adult male** Head blackish-brown, mottled chestnut; vague pale supercilium often present; neck, upper mantle and breast dark chestnut; chin, throat, centre of breast and belly white. Rest of upperparts mixed olive-brown and blackish, streaked white except on some upperwing-coverts; flanks streaked black and white; undertail-coverts barred black and chestnut.
- **8b** **Adult female** Head and neck duller and browner; breast whitish, with rufous tinge and dark scaling. Upperparts and upperwings largely spotted and barred white, flanks spotted white.
- **8c** **Adult male in flight** Note white secondaries, unpatterned outer upperwing-coverts, and relatively bright chestnut foreparts.

9   Slender-billed Flufftail *Sarothrura watersi*   Text and map page 175

E Madagascar. Small, grassy wetlands near forest at higher altitudes.

Tail short, slightly fluffy; bill longish and slender. Adult female brown with paler mottling on head, pale supercilium, whitish chin, throat and centre of breast, and black, chestnut and white barring on undertail-coverts and tail.

- **9a** **Adult male** Almost unpatterned; head to breast and mantle vinous-chestnut, darker on crown and often greyish behind eye; upperparts dark brown with vague darker streaks; posterior underparts ashy-brown; undertail-coverts and tail chestnut with black feather tips.
- **9b** **Immature female** Differs from adult in having small pale spots on back and brown-and-white bars on sides of breast and flanks.

7   Madagascar Flufftail *Sarothrura insularis*   Text and map page 171

E and NE Madagascar. Grassland, secondary bush, forest clearings, cultivation and marshes.

Male has short, fluffy, chestnut tail and yellow-streaked upperparts; female dark, with rufous to ochraceous streaks and bars.

- **7a** **Adult male** Head, neck, breast and tail rich chestnut; body black, streaked yellowish on upperparts and whitish on underparts.
- **7b** **Adult female** Head and upperparts blackish, streaked and barred rufous; face ochraceous, spotted black; chin and throat ochraceous to white; underparts barred ochraceous and black (spotted on breast and sides of neck); tail chestnut, narrowly barred black.

100mm

PLATE 4: PRIMITIVE FOREST RAILS OF AFRICA AND MADAGASCAR

10 Nkulengu Rail *Himantornis haematopus* Text and map page 177

Sierra Leone E to W Uganda and S to C Zaïre. Lowland rain forest.

A large rail, with upright posture, scaly or mottled plumage pattern, heavy black bill, and red legs.

10a/b Adults, showing range of individual variation in plumage colour and pattern, which especially involves colour of pale feather fringes and extent of scaling on underparts.

10c Downy chick Note distinctive pattern, unique among rallids.

---

11 Grey-throated Rail *Canirallus oculeus* Text and map page 179

Sierra Leone E to W Uganda and S to coastal Congo. Wet areas in lowland rain forest.

11 Adult Medium-large, slender rail; forepart of head grey; chin and throat grey to white; neck, breast and tail reddish-chestnut; upperparts olive-brown; upperwing-coverts and flight feathers blackish, spotted and barred white; posterior underparts olive-brown with pale bars; bill black and yellow-green.

---

12 Madagascar Wood-rail *Canirallus kioloides* Text and map page 181

E and NW Madagascar. Forest, and edges of ponds and marshes.

Forehead, eye region and ear-coverts grey; chin and throat white; crown to back olive-brown to olive-green; lower back to tail, and lesser upperwing-coverts, chestnut; rest of upperwing-coverts washed chestnut; primary coverts, primaries and secondaries black, usually barred white only on inner webs; neck to upper belly chestnut; posterior underparts dark brown, barred whitish to rufous. Bill black with blue-grey cutting edge and tip; legs and feet red-brown.

12a/b Adult (nominate; E Madagascar) Shows range of individual variation, **a** being a typically dark-plumaged bird with unusually extensive grey on crown, and **b** being as pale as a typical *berliozi*, with a typical nominate head pattern and sparse barring on outer webs of flight feathers.

[*C. k. berliozi* (Sambirano, NW Madagascar) Paler than nominate; more extensively grey on forehead and white on throat; greenish-olive on mantle; cinnamon from neck to upper belly and from rump to tail.]

10a

10b

10c

11

12b

12a

100mm

PLATE 5: YELLOW RAIL AND ALLIED SPECIES

14 **Yellow Rail** *Coturnicops noveboracensis* Text and map page 184

S Canada and N USA, wintering S and SE USA; also C Mexico. Freshwater and brackish marshes, and wet grassland.

Secondaries white. Upperparts blackish with tawny-buff streaks and narrow white bars; broad blackish-brown eyestripe; supercilium, face, neck and breast tawny-buff; flanks to undertail-coverts blackish-brown, barred tawny-buff and white.

- **14a** **Adult male** (nominate; Canada and USA) Bill corn-yellow in breeding season, otherwise olivaceous to black.
- **14b** **Adult female** (nominate) Slightly smaller; bill as in non-breeding male.
- **14c** **Downy chick** Black, with pink bill and grey-brown legs and feet.

[*C. n. goldmani* (C Mexico at Río Lerma marshes) Blacker crown, black stripes on nape, and predominantly dull cinnamon undertail-coverts.]

13 **Swinhoe's Rail** *Coturnicops exquisitus* Text and map page 183

Extreme SE Russia and NE China, wintering Korea, Japan and SE China. Wet meadows and short grass marshes.

- **13** **Adult non-breeding** Secondaries white. Upperparts blackish-brown, with prominent cinnamon stripes and narrow white bars; hindneck and sides of neck flecked white in non-breeding plumage; face greyish with indistinct dark eyestripe; foreneck, breast, flanks and undertail-coverts barred tawny-ochre, dusky and white; rest of underparts white.

16 **Ocellated Crake** *Micropygia schomburgkii* Text and map page 190

Colombia E to Guyana, Brazil, Peru, Bolivia and Paraguay; also Costa Rica. Dry to wet grassland.

A distinctive crake, having buffy olive-brown upperparts with black-bordered white spots, and tawny-buff underparts with white on chin, throat and belly; legs and feet coral-red.

- **16a** **Adult male** (nominate; probably entire range except Brazil) Forehead and forecrown tawny; spots begin on hindcrown.
- **16b** **Adult female** (nominate) Spots extend further forward on crown.

[*M. s. chapmani* (C to SE Brazil) Paler on upperparts and more russet on crown; no spots from lower back to tail.]

15 **Speckled Rail** *Coturnicops notatus* Text and map page 187

Colombia to Guyana, and C Brazil to N Argentina. Grassland and marshes.

- **15** **Adult** Secondaries white. Predominantly blackish, with dark olive-brown streaks on upperparts, white speckling from head to breast, and white bars on back and upperwing-coverts and from lower breast to undertail-coverts.

14b
14a
14c
13
16b
16a
15

100mm

PLATE 6: NEW GUINEA *RALLINA* FOREST-RAILS

Note: all species on this plate have remiges and underwing-coverts dark brown, barred or spotted white.

17   Chestnut Forest-rail *Rallina rubra*   Text and map page 192

W and C New Guinea. Interior of montane forest.

The smallest *Rallina*. Male reddish-chestnut; lower flanks to undertail-coverts reddish-brown with pale feather tips; female has upperwings and mantle to rump blackish-brown with whitish spots.

  17a   **Adult male** (nominate; Arfak Mts, W New Guinea) Hindneck and nape tinged blackish.
  17b   **Adult male** (*klossi*; Weyland to Oranje Mts, WC New Guinea) The palest race; no black on neck. For variability, see text.
  17c   **Adult female** (*klossi*) Belly to undertail-coverts with faint black and buff barring. For variability, see text.

[*R. r. telefolminensis* (Victor Emanuel and Hindenberg Mts, C New Guinea) Small; like *klossi* but darker].

18   White-striped Forest-rail *Rallina leucospila*   Text and map page 194

Tamrak, Arfu and Wandammen Mts, NW New Guinea. Interior of montane forest.

Black extends further forwards on mantle than in 17, 19 and 20. Uppertail-coverts and tail narrowly barred black, more distinctly than in 19 and 20; lower flanks and belly brown, faintly barred buff.

  18a   **Adult male** Upperwings and mantle to rump black, streaked white.
  18b   **Adult female** Upperparts spotted white.

19   Forbes's Forest-rail *Rallina forbesi*   Text and map page 195

Mountain ranges of C and E New Guinea. Montane and mid-montane forest.

Upperwings and lower mantle to rump dark brown to black, with black-margined buff spots in female. Uppertail-coverts and tail chestnut, finely barred black (sometimes almost unbarred); lower flanks and belly to undertail-coverts blackish-brown, narrowly barred red-brown.

  19a   **Adult male** (*steini*; Weyland Mts E to Bismarck Mts) Dark regions of upperparts dark blackish-brown.
  19b   **Adult female** (*steini*). Upperparts spotted with black-margined buffy spots.

[*R. f. forbesi* (Herzog Mts to Owen Stanley Mts) As *steini*; tail longer. *R. f. parva* (Adelbert Mts) Small; dark brown of upperparts finely vermiculated in both sexes. *R. f. dryas* (Huon Peninsula, E New Guinea) Small; upperparts brownish-black.]

20   Mayr's Forest-rail *Rallina mayri*   Text and map page 196

NC New Guinea. Montane and mid-montane forest.

Predominantly chestnut; upperwings and mid-mantle to rump darker (not as black as in 17, 18 and 19), with black-margined buff spots in female and a few spots or bars in male. Uppertail-coverts and tail indistinctly barred black. Lower flanks to undertail-coverts brown, narrowly barred buff-white.

  20a   **Adult male** (*carmichaeli*; Torricelli and Bewani Mts) Head to mantle and upper flanks darker and duller than in 17, 18 and 19.
  20b   **Adult female** (*carmichaeli*) Black margins of spots widest on coverts and scapulars.

[*R. m. mayri* (Cyclops Mts) Paler and brighter overall.]

17a 17b
17c
18a 18b
19b 19a
20a
20b

100mm

PLATE 7: OTHER *RALLINA* CRAKES

Note: all species on this plate have remiges and underwing-coverts dark brown to black, barred or spotted white.

23 Red-legged Crake *Rallina fasciata*   Text and map page 201

NE India and Burma E to Philippines and Moluccas, and S to Java and Flores. Marshy areas in open country, scrub or forest.

Wings boldly marked: scapulars have buffy-white and black spots and bars; upperwing-coverts and flight feathers barred black and white. Legs and feet red; bill greenish-horn.

- **23a** **Adult male** Ruddy-chestnut on head, neck and breast; chin and throat whitish; flanks to undertail-coverts broadly barred black and white.
- **23b** **Adult female** More cinnamon on head and neck; narrower black bars on underparts.

22 Andaman Crake *Rallina canningi*   Text and map page 200

Andaman Is. Marshland in forest.

- **22** **Adult** The largest *Rallina*; tail long and fluffy. Head, neck, upperparts and breast deep maroon-chestnut; underparts boldly barred black and white; undertail-coverts plain dark chestnut. Pale barring on wings visible only on outer primaries and some greater and median coverts. Bill apple-green; legs and feet olive-green.

21 Red-necked Crake *Rallina tricolor*   Text and map page 197

Molucca and Lesser Sunda Is, New Guinea and offshore islands, Bismarck Archipelago and NE Australia. Moist forest, scrub and thickets, near water.

- **21** **Adult** Head to mantle and breast rich chestnut; chin and throat whitish; upperparts dark olive-brown; flanks to undertail-coverts dark brown with variably distinct fine, buffy barring. Bars on remiges confined to inner webs; not visible in folded wing. Bill green; legs and feet olive.

24 Slaty-legged Crake *Rallina eurizonoides*   Text and map page 203

India E to S China and S to W Java; Ryukyu Is and Taiwan, S through Philippines to Palau Is and Sulawesi. Forests, woods, dense scrub.

Head, neck and breast bright chestnut; chin and throat white; upperparts dark olive-brown; underparts barred black and white.

- **24a** **Adult** (*amauroptera*; India to E Assam, and Sri Lanka) Narrowish white bars on under parts.
- **24b** **Adult** (*formosana*; Taiwan and Lanyu I) Upperparts darker.
- **24c** **Immature** Dark olive-brown, ashier on face and neck; sparse white and black bars on scapulars and coverts; underparts barred dark brown and white.

[*R. e. telmatophila* (Burma to SE China and Java) Large; upperparts darker than in *amauroptera*. ***R. e. sepiaria*** (Ryukyu Is) The largest race. ***R. e. eurizonoides*** (Philippines and Palau Is) Throat pale rufous; white bars on underparts relatively wide. ***R. e. alvarezi*** (Batan I, N Philippines) The darkest race; throat of male chestnut; white bars on underparts narrow. ***R. e. minahasa*** (Sulawesi) Like *alvarezi*, but head to breast paler.]

100mm

PLATE 8: SMALL NEW WORLD *ANUROLIMNAS* CRAKES

25 Chestnut-headed Crake *Anurolimnas castaneiceps*   Text and map page 206

Colombia, Ecuador, Peru and Bolivia. Forest, secondary growth and thickets.

The largest *Anurolimnas*; tail very short. Uniformly olive-brown with contrasting bright chestnut forecrown, face, neck and breast. Eye red, yellow or brown; bill greenish with black tip.

   25a   **Adult** (nominate; S Colombia and NE Ecuador) Olive-brown of plumage tinged rufous; legs and feet dark brown, olive or grey.

   25b   **Adult** (*coccineipes*; E Ecuador, N Peru and NW Bolivia) Olive-brown of plumage tinged greenish; chestnut usually paler and brighter; legs and feet bright red.

26 Russet-crowned Crake *Anurolimnas viridis*   Text and map page 207

Colombia and Ecuador E across Brazil and S to Peru, Bolivia and Paraguay. Thickets and damp grass; sometimes swamps.

A small species; tail relatively long. Upperparts uniformly olive-brown; tail blackish, feathers edged chestnut; forehead to crown, and underparts, chestnut; supraloral or superciliary stripe buff. Eye red or orange; bill dark grey with bluish base; legs and feet rose-red.

   26   **Adult** (nominate; entire range except N Colombia) Lores to ear-coverts grey.

   [***A. v. brunnescens*** (N Colombia) Upperparts browner; lores to ear-coverts yellowish-brown; underparts paler rufous.]

27 Black-banded Crake *Anurolimnas fasciatus*   Text and map page 209

Colombia, Ecuador, Peru and W Brazil. Marshes and wet grass.

   27   **Adult** Head, neck and breast entirely rich chestnut; rest of upperparts olive-brown; flanks to undertail-coverts boldly barred cinnamon and black. Eye dull red; bill blackish; legs and feet bright coral-red.

100mm

PLATE 9: SMALL NEW WORLD *LATERALLUS* CRAKES (1)

28   Rufous-sided Crake *Laterallus melanophaius*   Text and map page 210

Venezuela to Surinam; Colombia, Ecuador and Peru, through Brazil to Bolivia, Paraguay, Uruguay and N Argentina. Marshes and wet grassland.

Upperparts dark olive-brown, lacking any rufous; sides of head, neck and breast rufous; chin to centre of belly white; flanks barred black and white; undertail-coverts plain cinnamon-rufous. Bill blackish-brown, green at sides; legs and feet olive-brown.

    28   **Adult** (nominate; Colombia, E Ecuador, E Peru and W Brazil) Lores and orbital area ashy; upperparts dark olive-brown.

    [***L. m. oenops*** (Venezuela to Surinam, and Brazil S to Uruguay and Argentina) Upperparts paler, more olivaceous; no grey on head; supraloral stripe tinged rufous.]

29   Rusty-flanked Crake *Laterallus levraudi*   Text and map page 213

N Venezuela. Marshes and wet grassland.

    29   **Adult** Relatively unpatterned: upperparts dark olive-brown; sides of head, neck and breast, and flanks to undertail-coverts, cinnamon-rufous; chin to centre of belly white. Bill blackish, tip pale; legs and feet pale brown.

30   Ruddy Crake *Laterallus ruber*   Text and map page 214

E Mexico S to Honduras, N Nicaragua, and NW Costa Rica. Marshes and wet grassland.

Mainly bright chestnut; chin and belly paler; top and sides of head blackish, ear-coverts slaty. Bill black; legs and feet olive-green. Female more chestnut than male from lower back to uppertail-coverts.

    30   **Adult male** Lower back to tail dark brown to blackish-brown.

31   White-throated Crake *Laterallus albigularis*   Text and map page 216

SE Honduras and E Nicaragua to NW Colombia and W Ecuador. Marshes, moist grassland, thickets and forest clearings.

Head pattern variable; neck, mantle and breast rufous; rest of upperparts dark brown; chin and throat white; flanks to undertail-coverts barred black and white. Remiges and coverts may be faintly marked white or cinnamon. Bill dusky greenish; legs and feet brown.

    31a   **Adult** (nominate; Pacific lowlands from SW Costa Rica to N and W Colombia and W Ecuador) No grey on crown or face; narrow white bars on flanks.

    31b   **Adult** (*cinereiceps*; Honduras, and Caribbean lowlands from Nicaragua to Panama) Crown grey-brown; sides of head grey.

    [***L. a. cerdaleus*** (E Colombia) Like *albigularis* but quite uniform red-brown on head and throat, darker on upper breast, and more heavily barred black on belly and vent.]

28

29

30

31b

31a

100mm

PLATE 10: SMALL NEW WORLD *LATERALLUS* CRAKES (2)

32 Grey-breasted Crake *Laterallus exilis* Text and map page 217

Guatemala and Belize to Colombia and Ecuador; Trinidad, Venezuela, the Guyanas, Brazil, Peru, Bolivia and Paraguay. Marshes; dry to wet grassland.

Chestnut patch from hindcrown to upper mantle contrasts with grey from head to upper breast and olive-brown upperparts. Variable white barring on upperwing-coverts, rump and uppertail-coverts (sometimes absent). Black-and-white barring on underparts tinged rufous in posterior regions. Bill largely greenish; legs and feet yellow-brown.

- **32** **Adult female** Paler on breast than male.

33 Black Rail *Laterallus jamaicensis* Text and map page 220

E USA and Central America, wintering S USA to Greater Antilles and Guatemala; W USA, Peru, Chile and W Argentina. Marshes and wet grassland.

Small and dark; head blackish; nape to upper mantle chestnut or rufous; rest of upperparts, and flanks, blackish-brown, spotted or barred white; undertail-coverts variable. Female paler from head to breast.

- **33a** **Adult male** (nominate; E USA and Central America) Head dark slate; nape to upper mantle chestnut-brown; small spots on upperparts; flanks to undertail-coverts narrowly barred or spotted.
- **33b** **Adult male** (*coturniculus*; W USA) Crown chocolate-brown; nape to mantle more rufous; underparts darker.
- **33c** **Immature?** (*tuerosi*; L Junin, Peru) Head to mantle darker; upperparts and flanks heavily barred; undertail-coverts cinnamon.

[*L. j. salinasi* (S Peru, Chile and Argentina) Largest race; nape to mantle brighter; large spots or short bars on upperparts; broad bars on flanks. *L. j. murivagans* (C Peru) Paler than *salinasi*; short bars on upperparts; undertail-coverts pale cinnamon, mottled grey.]

34 Galapagos Rail *Laterallus spilonotus* Text and map page 224

Galapagos Archipelago. Forest, thickets and grass.

- **34** **Adult male** Very dark. Head, neck and breast dark slate-grey; upperparts dark chocolate-brown with scattered small white spots (in breeding season); flanks and belly dark grey-brown, spotted white; undertail-coverts barred blackish and white. Bill black; legs and feet brown. Female may have paler throat. Wings short; flies weakly.

32

33a 33c

33b

34

100mm

PLATE 11: NEW WORLD *LATERALLUS* CRAKES (3), AND SOUTH-WEST PACIFIC *NESOCLOPEUS* RAILS

38 **Bar-winged Rail** *Nesoclopeus poecilopterus* Text and map page 231

Fiji Is (Viti Levu, Ovalau and Taveuni). PROBABLY EXTINCT. Forest and overgrown plantations; possibly swamps.

- **38a** **Adult** Upperparts brown; tail blackish, feathers fringed reddish-brown; lores and ear-coverts pale brown; chin and throat white; rest of face and neck slate-grey; underparts darker slate with narrow grey bars, becoming darker and obscurely barred from rear flanks to undertail-coverts.
- **38b** **Adult** Shows chestnut barring on remiges and white bars on underwing-coverts.

37 **Woodford's Rail** *Nesoclopeus woodfordi* Text and map page 230

Bougainville I, Papua New Guinea; Santa Isabel and Guadalcanal, Solomon Is. Lowland forest, often near water.

Very dark, with blackish-brown upperparts and grey-black underparts; upperwing-coverts variably spotted white; underparts variably and obscurely barred.

- **37a** **Immature** (nominate; Guadalcanal) Head, neck and upperparts rich blackish-brown, supercilium grey; few white spots on primary coverts; underparts sooty-black with obscure greyish bars.
- **37b** **Immature** (nominate) White bars on primaries and some secondaries; restricted white spotting on underwing-coverts.

[*N. w. tertius* (Bougainville) Flanks and belly mottled buff; undertail-coverts vaguely barred white; bill yellowish-horn; legs grey. *N. w. immaculatus* (Santa Isabel) Blacker overall; no white in wings; no pale mottling on underparts; bill horn, grey at base; legs greenish-yellow.]

35 **Red-and-white Crake** *Laterallus leucopyrrhus* Text and map page 226

SE Brazil, Paraguay, Uruguay and NE Argentina. Marshy, grass-dominated areas.

- **35** **Adult** Strikingly patterned: chin to centre of breast pure white, contrasting with bright chestnut head, nape, sides of neck and breast; rest of upperparts dark olive-brown; flanks boldly barred black and white; median undertail-coverts black, laterals white. Bill black and yellow; legs and feet red to pink. Upperwing-coverts sometimes barred rufous or white.

36 **Rufous-faced Crake** *Laterallus xenopterus* Text and map page 228

SC Brazil and C Paraguay. Wet grassy areas in marshes.

- **36** **Adult** Very distinctive, with dark rufous upperparts grading to black on tail, whitish chin and throat, buffy-brown neck and breast, white lower breast and belly, prominent black and white barring on scapulars, upperwing-coverts, tertials and flanks, and black undertail-coverts with white bars at sides. Bill blue-grey, or blackish with pale lower mandible; legs and feet blue-grey to flesh-brown.

100mm

PLATE 12: FLIGHTLESS *GALLIRALLUS* RAILS OF NEW ZEALAND AND THE WESTERN PACIFIC

39   Weka *Gallirallus australis*   Text and map page 233

New Zealand; also offshore and outlying islands. Scrub, forest edges, grassland, wetland margins, shorelines and cultivation.

Large; flightless; longish tail. Plumage very variable; four races recognised; nominate and *scotti* have chestnut, grey and black morphs.

- **39a** **Adult** (nominate; South Island and outlying islands) Chestnut morph. Breast rufous-brown, mottled black; flanks and belly dark brown, tinged rufous, rear flanks sometimes barred; bill pink at base, grey distally.
- **39b** **Adult** (*scotti*; Stewart Island and outlying islands) Black morph. Black, with dark red-brown streaks on upperparts and often across upper breast; bill pinkish.
- **39c** **Adult** (*hectori*; Chatham Island) Palest race: upperparts and breast streaked yellow-brown to buff; lower breast plain grey-brown; flanks and belly olive-brown, barred dark; legs and feet pinkish-red to pinkish-yellow.
- **39d** **Juvenile** (*greyi*; North Island) Dark form: darker than adult; almost plain; breast-band absent.

41   Lord Howe Rail *Gallirallus sylvestris*   Text and map page 238

Lord Howe Island, south-west Pacific. Forest.

- **41** **Adult** Flightless. Predominantly plain olive-brown; chin, throat and supercilium paler; remiges and primary coverts barred russet and dark brown; undertail-coverts barred russet; eye red; bill longish, decurved, pinkish-brown.

42   Gilbert Rail *Gallirallus conditicius*   Text and map page 242

Known from one specimen, now thought to be an immature Lord Howe Rail (see text).

- **42** Differs from Lord Howe Rail only in having paler crown, throat and underparts, and browner head; also slightly smaller.

40   New Caledonian Rail *Gallirallus lafresnayanus*   Text and map page 237

New Caledonia; POSSIBLY EXTINCT. Dense forest.

- **40a** **Adult** Flightless, with longish tail. Upperparts chocolate-brown, blackish from lower back to tail. Lores and ear-coverts brown; supraloral stripe pale, rest of face grey. Underparts dark grey, darker from rear flanks to undertail-coverts.
- **40b** **Juvenile** Almost entirely black; washed chocolate-brown on upperparts and sides of neck and breast.

39a
39b
39c

39d

41
42

40a
40b

100mm

PLATE 13: *GALLIRALLUS* RAILS OF SOUTHERN ASIA AND AUSTRALASIA (1)

44   Barred Rail *Gallirallus torquatus*   **Text and map page 245**

Sulawesi, Philippines and NW New Guinea. Grassland, marsh edges, cultivation and secondary growth.

Upperparts olive-brown to rufous-brown; face, foreneck and throat black; prominent white streak from base of bill, below eye and down side of neck; underparts black, barred white; chestnut or brown breast-band sometimes present; bill black; legs and feet brown.

- **44a**   **Adult** (nominate; Philippines) Upperparts olive-brown; throat unbarred; rich maroon-chestnut breast-band.
- **44b**   **Immature** (nominate) Breast-band olive-brown; chin mottled whitish.
- **44c**   **Adult** (*sulcirostris*; Sula and Peleng Islands) Upperparts deep rufous-brown; throat to upper breast unbarred.
- **44d/e Adults** (*celebensis*; Sulawesi and Muna) Two colour forms, showing range of upperpart colour and variable extent of white barring on throat (in some, unbarred area extends over foreneck).

[**G. t. kuehni** (Tukangbesi Islands) Largest race; upperparts deep olive-brown. **G. t. limarius** (NW New Guinea) Like *celebensis* but throat unbarred; bill long.]

45   New Britain Rail *Gallirallus insignis*   **Text and map page 247**

New Britain. Damp forest and swampy cane grass.

- **45**   **Adult** Flightless; short tail. Head and neck dark rufous-brown; upperparts dark olive-tinged brown, more rufous on greater coverts and tertials; tail blackish; entire under parts black, narrowly barred white from chin to upper belly. Bill black; legs and feet rose-pink.

48   Guam Rail *Gallirallus owstoni*   **Text and map page 258**

Formerly Guam Island; introduced to Rota Island. Forest, woodland, scrub and grassland.

- **48**   **Adult** Flightless; short tail. Upperparts olive-brown; crown, face and nape more rufous. Broad grey supercilium; chin and throat whitish; foreneck and upper breast grey; obscure olive-buff breast-band in fresh plumage only; rest of underparts barred black and white. Black and white barring also visible on remiges in folded wing. Bill black with grey base, or dark brown; legs and feet pale brown or yellow-brown.

100mm

PLATE 14: *GALLIRALLUS* RAILS OF SOUTHERN ASIA AND AUSTRALASIA (2)

47 Roviana Rail *Gallirallus rovianae* Text and map page 256

Solomon Islands. Forest, scrub and overgrown plantations.

- **47** **Holotype; age unknown** (New Georgia form). Flightless; tail longish. Upperparts dark chestnut-brown, more reddish on nape and face; grey eye-stripe does not extend in front of eye; primary coverts spotted white; chin and throat whitish; neck and breast grey; breast-band pinkish-tan; underparts barred dark brown and white; undertail-coverts barred black and buff. Colours of bare parts not known. See text for description of Kolombangara form.

46 Buff-banded Rail *Gallirallus philippensis* Text and map page 248

Malay Archipelago, Australasian region and SW Oceania. Marshes, swamps, mangroves, grassy areas, scrub and woodland.

Plumage distinctive. Rufous facial stripe and hindneck; upperparts brown, mottled blackish and variably spotted or barred white; flight feathers barred rufous and black (but barred white and black on outer 3); chin white; throat grey; rest of underparts variably barred black and white; buff to rufous breast-band present in some races.

- **46a** **Adult** (*mellori*; C to SW New Guinea, Australia) Typical bird: note moderate to sparse spotting and reasonably prominent blackish mottling on upperparts; well-defined breastband; barring above and within breast-band.
- **46b/c Adults** (nominate; Philippines, Sulawesi, Buru, Lesser Sundas) Two colour variations from North Sulawesi: **c** is typical '*chandleri*' with prominently dark-mottled and heavily white-spotted upperparts, heavily barred underparts and no breast-band; **b** is typical nominate with relatively unpatterned pale brown upperparts, narrowly barred under parts and no breast-band – see text.
- **46d** **Adult** (possibly intermediate between nominate and '*yorki*', from Buru) Note relatively sparse spotting on upperparts and narrow black barring on breast.
- **46e** **Adult** (currently assigned to *yorki*; Moluccas; specimen from Buru) Race not properly defined. Bird more like *mellori* (46a) than 46d.
- **46f** **Adult** (*assimilis*; New Zealand) Compared to *mellori* crown paler and more streaked; upperparts paler, more olive and less patterned; tail paler; bill longer; wing shorter. Breast-band may be slightly paler; 46a and 46f show range of breast-band colour: a very pale *assimilis* and a very dark, rich *mellori*.
- **46g** **Immature** (*swindellsi*; New Caledonia and Loyalty Is) Differs from adult in reduced and indistinct facial stripe and nuchal collar; less distinct supercilium; duller, less patterned upperparts; olive-mottled throat and foreneck; less distinct and buff-washed underpart barring; greyish bill; brown iris.

[At least 17 other races, differing in colour and pattern of upperparts and tail-coverts, presence, extent and barring of breast-band, pattern and extent of barring on underparts, and length of wing, bill and tarsus – see text.]

100mm

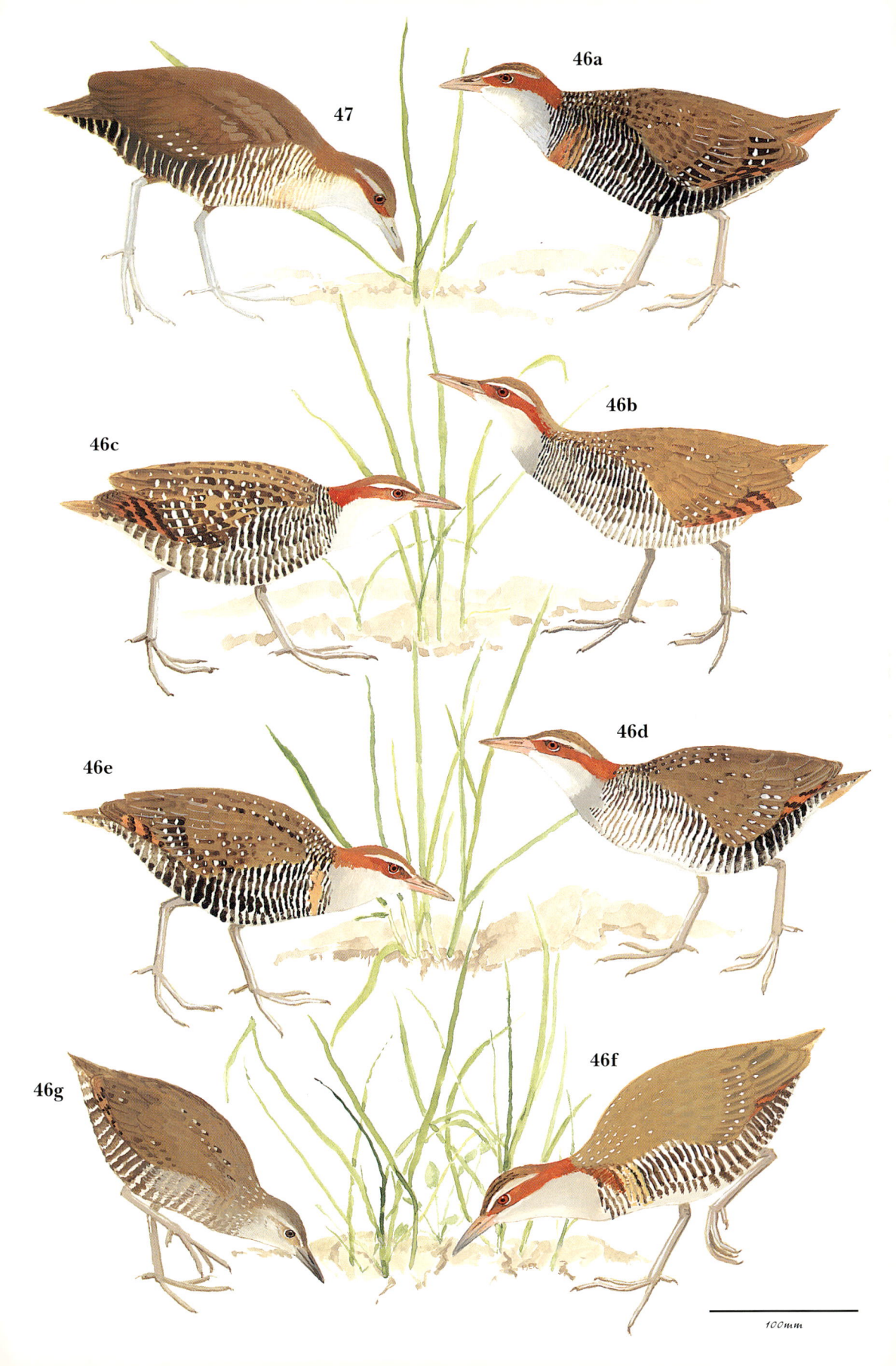

PLATE 15: *GALLIRALLUS* RAILS OF SOUTHERN ASIA AND AUSTRALASIA (3)

54   Slaty-breasted Rail *Gallirallus striatus* Text and map page 265

India to SE China, SE Asia, Greater and Lesser Sunda Islands and Philippines. Freshwater wetlands, mangroves, grassland, scrub and forest.

Forehead to hindneck and sides of neck chestnut; upperparts olive-brown to blackish-brown, barred and spotted white; face to breast grey; rest of underparts barred black and white. Eye red; bill horn with pink-red base; legs and feet olive to grey.

- **54a   Adult** (nominate; Philippines, Sulu Is, N Borneo and Sulawesi) Very dark on upperparts and underparts.
- **54b   Adult** (*gularis*; S China, Vietnam, Cambodia, Malaysia, Sumatra, S Borneo, Java and Bali) A pale race, tending to greyish-olive on upperparts.

[**G. s. albiventer** (India and Sri Lanka, to SE China) Similar to *gularis* but larger and darker; upperparts brownish-olive. **G. s. obscurior** (Andaman and Nicobar Is) Larger and darker than *albiventer*; upperparts blackish; less white on chin and throat. **G. s. jouyi** (Coastal SE China and Hainan) Largest and palest race. **G. s. taiwanus** (Taiwan) Like *jouyi* but smaller. **G. s. paratermus** (Samar Is, Philippines) Upperparts blackish with fewer white markings; underparts dark with narrow bars.]

49   Wake Rail *Gallirallus wakensis* Text and map page 260

Wake Island, Micronesia, Pacific Ocean. EXTINCT. Scrub.

- **49   Adult** Flightless. Lores, ear-coverts and upperparts olive-brown, streaked buffy on upperparts; supercilium pale grey; flight feathers barred cinnamon (white on outer 2 primaries). Chin and throat whitish; rest of underparts brownish to grey, narrowly barred white except on foreneck, centre breast and belly; vinaceous breast-band sometimes indistinct.

43   Okinawa Rail *Gallirallus okinawae* Text and map page 243

Northern Okinawa, Ryukyu Islands (Nansei Shoto), Japan. Forest near water.

- **43a   Adult** Flightless. Upperparts dark olive-brown; sides of head and underparts black, barred white from lower neck to undertail-coverts, whitish loral spot and white line from rear of eye across ear-coverts to side of neck. Bare parts red, bill with ivory tip.
- **43b   Juvenile** Upperparts paler; underparts mottled white; bill and iris brownish; legs and feet yellow-ochre.
- **43c   Downy chick** Black; bill white with blackish base and tip; iris black; legs and feet yellowish.

100mm

PLATE 16: EXTINCT *GALLIRALLUS* RAILS

50   Tahiti Rail *Gallirallus pacificus*   Text and map page 261

Tahiti and Mehetia Islands. EXTINCT. Open areas, marshes and coconut plantations.

50   **Adult** Flightless. Strikingly and unusually coloured: top and sides of head, and entire upperparts, black, with white supercilium, ferruginous nape, white spots from back to tail, and white barring on wings; underparts white except for grey breast and narrow black band across base of throat; iris and bill red; legs pink.

51   Dieffenbach's Rail *Gallirallus dieffenbachii*   Text and map page 262

Chatham, Pitt and Mangere Islands, E of New Zealand. EXTINCT. Scrub and tussock grass.

51   **Adult** Flightless. Predominantly rufous-brown, barred dark brown to blackish, including on wings, neck, breast and undertail-coverts. Supercilium, most of face, and chin and throat bluish-grey; lower throat, flanks and belly black with narrow white bars. Iris reddish-brown; bill quite stout, downcurved towards tip, pale brownish with darker tip; legs and feet pale brown.

53   Sharpe's Rail *Gallirallus sharpei*   Text and map page 264

Distribution unknown (see text). PROBABLY EXTINCT. Known from one specimen.

53   Very distinctive. Predominantly brownish-black, with sides of head, chin and throat grey; upperwings paler and more brownish; mantle, scapulars, upperwing-coverts, undertail-coverts and tail spotted white. Remiges broadly barred white. Flanks marked with small white spots. Bare parts probably red or orange.

52   Chatham Rail *Gallirallus modestus*   Text and map page 263

Chatham, Pitt and Mangere Islands, E of New Zealand. EXTINCT. Bush and tussock grass.

Flightless, long-billed, and darkish brown with buffy barring on underparts.

52a   **Adult male** Predominantly dark brown and grey-brown, with narrow sandy-buff barring on underparts from foreneck to undertail-coverts; primary coverts and outermost primaries also barred. Bill thin, downcurved, dusky brown; legs and feet reddish to brown.

52b   **Adult female** Smaller, bill shorter. Sandy-buff barring extends to mantle, back, rump and upperwing-coverts.

50

51

53

52a
52b

100mm

PLATE 17: CLAPPER AND KING RAILS

55 Clapper Rail *Rallus longirostris* — Text and map page 269

Predominantly coastal in USA, N Mexico, Yucátan Peninsula, and South America S to N Peru and S Brazil; also Bahamas and West Indies. Salt and brackish marshes; locally freshwater marshes.

Large, long-billed; very variable in plumage. Upperparts dark brown to blackish, streaked buffy, olive or greyish; upperwing-coverts contrastingly rufous, brown or olive-brown; pale supraloral streak; sides of head variably grey or vinaceous-brown; foreneck, breast and upper belly buff, cinnamon or rufous; flanks brown to blackish, barred white; undertail-coverts white, medians barred black.

- **55a** **Adult** (*scotti*; Florida) The darkest race. Brown morph. Upperparts streaked grey-brown; breast fawn, washed cinnamon; flanks very dark, with narrow white bars.
- **55b** **Adult** (nominate; Surinam and the Guianas) Relatively pale; upperparts streaked greyish; sides of head pale grey; foreneck and breast buff, washed tawny; white flank bars broad.
- **55c** **Adult** (*crepitans*; coastal Connecticut to N Carolina, wintering S to Florida) Upperparts broadly streaked olive-grey; upperwing-coverts buffy-brown; foreneck and breast grey-tinged cinnamon-buff; legs and feet grey with yellow ankle joint.
- **55d** **Adult** (*obsoletus*; central California) Brown morph (dark individual). Upperparts streaked ashy-brown, tinged olive; upperwing-coverts tinged rufous; sides of head dusky vinaceous-brown; foreneck and breast dull cinnamon, feather edges pale.
- **55e** **Juvenile** (*obsoletus*) Brown morph. Much darker than adult on head and neck; upperparts also darker, streaked colder ashy-brown; breast pale cinnamon, mottled greyish; flanks greyish-olive with cinnamon-tinged bars.

[Seventeen other races described – see text.]

56 King Rail *Rallus elegans* — Text and map page 278

Extreme S Canada, USA, C and E Mexico, and Cuba; winters in S part of range. Freshwater and brackish marshes.

Large, long-billed, and predominantly rusty-coloured; variable amount of grey on ear-coverts; upperwing-coverts russet to tawny; breast and upper belly cinnamon; flanks boldly barred; undertail-coverts white with black spots and bars. Larger and brighter than Clapper Rail.

- **56** **Adult** (nominate; North America and E Mexico) Dark morph shown; this individual has little grey on ear-coverts and a broad, well-defined supercilium (variable characters). Pale morph has paler edges to upperpart feathers and is paler, washed grey, from breast to belly.

[*R. e. tenuirostris* (C Mexico) Duller and paler than nominate; pale-breasted morph has centre of breast and upper belly whitish. *R. e. ramsdeni* (Cuba and Isle of Pines) Like pale morph of nominate but smaller; greyer on sides of head with more white on lower breast and belly.]

55a
55b
55c
55d
55e
56

100mm

PLATE 18: VIRGINIA RAIL AND SOUTH AMERICAN *RALLUS* RAILS

57 Plain-flanked Rail *Rallus wetmorei*   Text and map page 283

Coastal Venezuela. Mangrove swamps. Extremely rare and poorly known.

- **57**   **Adult** Predominantly brown, with streaked upperparts and unbarred underparts; upperwing-coverts fairly uniform; chin and throat pale; belly pinkish-buff; undertail-coverts may appear predominantly buffy-brown or white.

58 Virginia Rail *Rallus limicola*   Text and map page 284

S Canada, USA, Mexico, Guatemala, Colombia and Ecuador. Freshwater and saline marshes.

Smaller than King and Clapper Rails (Plate 17); bill shorter and slightly decurved; side of head extensively blue-grey; upperwing-coverts brighter rufous; throat to breast cinnamon or brownish.

- **58a**   **Adult** (nominate; North America, wintering S to Mexico and Guatemala) Upperparts darker than other races; breast to belly cinnamon to brownish.
- **58b**   **Juvenile** (nominate) Upperparts duller and darker than adult; grey of face tinged olive-brown; throat and foreneck whitish; rest of underparts blackish, mottled and barred whitish on flanks; iris brown; bill blackish; legs and feet dusky-brown.
- **58c**   **Adult (?)** (form *peruvianus*; probably a race of Bogotá Rail [59]; range unknown) Like Bogotá Rail but has more prominent supraloral streak; upperwing-coverts duller and patterned; undertail-coverts lack cinnamon bars: see Bogotá Rail text.

[*R. l. friedmanni* (Mexico, Guatemala) Upperparts streaked paler than in nominate; underparts more pinkish. *R. l. aequatorialis* (Colombia and Ecuador) Smaller and duller; breast to belly greyish-fawn. *R. l. meyerdeschauenseei* (Coastal Peru) Small; brighter and paler in colour than other races; breast to belly dull cinnamon.]

59 Bogotá Rail *Rallus semiplumbeus*   Text and map page 289

E Andes of NC Colombia; possibly also Peru. Freshwater marshes.

- **59**   Larger than Virginia Rail (58); bill longer, slightly downcurved at tip, yellowish-red. Upperwing-coverts uniform chestnut; face, neck, breast and upper belly entirely plumbeous grey; flanks and lower belly blackish, narrowly and irregularly barred white at front, browner towards rear. Legs and feet reddish-brown.

60 Austral Rail *Rallus antarcticus*   Text and map page 291

C Chile and C Argentina S to Tierra del Fuego. Wet grass and marshy areas. Extremely rare and poorly known.

- **60a**   **Adult** Like small Bogotá Rail, but upperparts brighter, streaked blackish and pale sandy-buff; regular black and white barring on flanks. Lores to ear-coverts dark; supraloral streak pale; sides of head, and chin to upper belly plumbeous-grey; undertail-coverts white with cinnamon and black markings. Eye may also be brown, bill horn and legs pale (non-breeding?).
- **60b**   **Adult** Showing olive-brown wash on breast and flanks (in fresh plumage only), white chin and throat (a variable character), and duller light brown streaking at sides of neck and breast.

100mm

PLATE 19: OLD WORLD *RALLUS* RAILS

61   **Water Rail** *Rallus aquaticus*   **Text and map page 293**

Iceland; Europe E across C Russia and S Siberia to N China and Japan, and S to N Africa, Turkey, Iran, NW Himalayas and WC China. Winters in S part of range, S to Sahara, Pakistan, Myanmar and SE Asia. Inhabits dense marsh vegetation at flooded sites with muddy ground.

Upperparts and upperwings olive-brown, streaked blackish. Sides of head and neck, and chin to upper belly, dark slate-blue; flanks barred black and white.

    61a   **Adult** (nominate; Europe S to N Africa and E to C Siberia) Undertail-coverts usually appear white. Eye and bill red; legs and feet flesh-brown.

    61b   **Juvenile** (nominate) Forehead to hindneck blacker; chin and throat whitish; blue-grey replaced by buff or white, with dark barring; flank bars sepia and buffy-white; undertail-coverts buff; bare parts duller.

    61c   **Adult** (*indicus*; N Mongolia and E Siberia to N Japan, S to NE China and Korea; winters S to Assam, and SE Asia) Larger than nominate; brown streak on face; chin whitish; breast and sides of body tinged brown.

    61d   **Downy chick** Bill white with black base and tip.

[***R. a. hibernans*** (Iceland) Slightly warmer brown on upperparts than nominate; grey of head and underparts often tinged brown; flank bars dull sepia and white. ***R. a. korejewi*** (Iran, Turkmenia, Tadzhikistan to Aral Sea and L Balkhash, E to C China and S to NW Himalayas; winters S to Iraq and Pakistan) Larger and paler than nominate.]

62   **African Rail** *Rallus caerulescens*   **Text and map page 299**

Ethiopia S to South Africa and W to E Zaïre, C Angola and NE Namibia. Rare and local West Africa. Inhabits reedbeds, sedges and grass in marshes and beside open waters.

    62a   **Adult** Upperparts and upperwings plain, dark chocolate-brown; chin and throat white; face, neck and breast slaty-grey, tinged blue; flanks and median undertail-coverts barred black and white; lateral undertail-coverts white. Bare parts red.

    62b   **Immature** Duller than adult; upperparts less richly coloured; more white on throat; grey of face and underparts paler, often tinged olive-brown; flanks barred brownish-black and rufous-buff to white. Bare parts brown.

63   **Madagascar Rail** *Rallus madagascariensis*   **Text and map page 302**

E Madagascar. Dense herbaceous vegetation of marshes and wet woodlands.

    63   **Adult** Forepart of head, and chin and throat, grey; upperparts and upperwings olive-brown, streaked blackish; crown to hindneck tinged vinaceous. Ear-coverts, neck, breast and upper flanks dull vinaceous (brighter at sides of breast and flanks), streaked blackish; lower flanks and undertail-coverts dusky olive-brown with faint white bars, outer coverts white. Bill pink-red, slender, decurved; legs and feet grey-brown.

100mm

PLATE 20: *LEWINIA* RAILS OF SOUTHERN ASIA AND AUSTRALASIA

64 Brown-banded Rail *Lewinia mirificus* — Text and map page 304

Luzon and Samar, Philippines. Precise habitat unknown, but has been recorded from wet grassy areas.

- **64 Adult** Forehead to hindneck plain chestnut-brown; rest of upperparts mid-brown with small buffy spots on back, uppertail-coverts and tail, and narrow pinkish-buff bars on scapulars, upperwing-coverts, secondaries and tertials. Pale supraloral line; chin and throat whitish; face to breast grey, washed olive on face and neck; flanks blackish-brown, and undertail-coverts olivaceous-black, barred whitish to pinkish-cinnamon.

65 Lewin's Rail *Lewinia pectoralis* — Text and map page 305

Flores; New Guinea; Australia; Tasmania. Freshwater to saline marshes; wet thickets; cultivated areas; dry upland grassland.

Forehead to hindneck, supercilium and sides of neck rufous, varyingly streaked blackish from forehead to hindneck; upperparts streaked olive-brown and black; scapulars and upperwings prominently barred black, white and olive-brown; chin, and sometimes throat, white; lower face, foreneck and breast grey, tinged buffish-olive; flanks to undertail-coverts barred brown-black and white.

- **65a Adult female** (*brachipus*; Tasmania) Heavily washed olive on face, foreneck and breast; underpart barring cream. More white on chin and throat than in *captus*.
- **65b Adult male** (*captus*; C Highlands of New Guinea) Little olive wash on grey of face, foreneck and breast. Little white on chin and throat.
- **65c Juvenile** (*captus*) Darker, more uniform head and upperparts; brownish wash on foreneck and breast; duller underpart barring; darker bare parts.

[*L. p. exsul* (Flores, Lesser Sundas) Little streaking on crown; throat whitish; breast very grey. *L. p. mayri* (W New Guinea) Crown to hindneck dull; upperparts dark. *L. p. insulsus* (E New Guinea) Crown more olivaceous; throat and neck white; breast relatively pale grey. *L. p. alberti* (SE New Guinea) Small; crown unstreaked; upperparts dark. *L. p. clelandi* (SW Australia, POSSIBLY EXTINCT) Large; breast clearer grey than nominate. *L. p. pectoralis* (E and SE Australia) Crown heavily streaked; throat to breast grey, tinged buffish-olive.]

66 Auckland Rail *Lewinia muelleri* — Text and map page 310

Adams and Disappointment, Auckland Is. Herbfields, tussock grassland and scrubby forest on damp to wet ground.

- **66 Adult** Similar to Lewin's Rail but smaller, bill straighter; no black streaks on forehead and hindneck; upperparts olive-brown with narrow black streaks; upperwings with very restricted black and white barring. Grey of anterior underparts heavily washed buffish-olive.

64

65a

65c

65b

66

100mm

PLATE 21: *DRYOLIMNAS, ROUGETIUS, ARAMIDOPSIS* AND *ATLANTISIA*

67 White-throated Rail *Dryolimnas cuvieri* Text and map page 312

Madagascar and Aldabra. Forest undergrowth, riparian marsh, mangroves and dry coral scrub.

Deep vinous-chestnut head, neck and breast, with prominent white chin and throat; upperparts and upperwings greenish-olive, remiges darker; flanks, lower belly to undertail-coverts dusky brown, barred whitish to buff; lateral undertail-coverts white.

- **67a** **Adult male** (nominate; Madagascar) Black streaks on back, scapulars and tertials.
- **67b** **Adult male** (*aldabranus*; Aldabra) Flightless. Smaller and paler; upperpart streaks absent or much reduced; underpart barring less distinct; base of bill dark red.
- **67c** **Adult female** (*aldabranus*) Base of bill bright pink.

70 Rouget's Rail *Rougetius rougetii* Text and map page 327

Ethiopian and Eritrean highlands. Marshy areas in montane grassland and moorland.

- **70** **Adult** Combination of unstreaked olive-brown upperparts, largely cinnamon-rufous underparts and white undertail-coverts unique in African rallids. Bill and eye red; legs and feet dark red.

71 Snoring Rail *Aramidopsis plateni* Text and map page 329

Sulawesi. Dense secondary growth on forest borders, usually with water.

Large, flightless rail with very short tail and longish, robust bill. Forehead, forecrown and face grey; centre of crown and nape darker; hindneck and sides of neck deep orange-chestnut; mantle, scapulars and upper back slaty grey; rest of upperparts, and visible parts of closed wings, chestnut-brown. Chin and throat white; foreneck and breast grey, lower breast with whitish bars; flanks to undertail-coverts barred black and white.

- **71a** **Adult male** Iris brown to yellowish; bill brownish on upper mandible and tip, yellowish-green on lower mandible; legs and feet black.
- **71b** **Adult female** Brighter rufous on hindneck; less white on chin and throat; iris orange; bill reddish where male's is yellowish; legs and feet slaty blue-grey.

72 Inaccessible Rail *Atlantisia rogersi* Text and map page 331

Inaccessible I, Tristan da Cunha group. Inhabits all island vegetation; also boulder beaches.

Flightless. A very small, dark rail.

- **72a** **Immature male** Blackish overall. This individual is acquiring adult characters: brown wash on upperparts and whitish barring on wings and flanks; adult has more extensive barring and dark grey head and underparts, with blacker lores and ear-coverts.
- **72b** **Adult female** Paler than adult male; underparts faintly washed brown; cheeks and ear-coverts grey, like rest of head; leg colour of both sexes varies from blackish-brown to grey-brown.

67a

67c

67b

70

71a

71b

72a
72b

100mm

PLATE 22: WOOD-RAILS OF SOUTH AND CENTRAL AMERICA (1)

73 **Little Wood-rail** *Aramides mangle*   Text and map page 334

E Brazil. Coastal lagoons and swamps, including mangroves.

- **73** **Adult** The smallest wood-rail. No grey on underparts; grey from head to upper mantle; upperwing-coverts, and lower mantle to upper back, greenish-olive; lower back and rump dark sepia; tail-coverts and tail black. Chin and throat white; rest of underparts tawny, becoming olive on lower flanks, thighs and belly. Greenish bill has red base.

74 **Rufous-necked Wood-rail** *Aramides axillaris*   Text and map page 335

Central and N South America from Mexico S to NW Peru and E to the Guianas. Coastal wetlands, mangroves, swamp forest and cloud forest.

- **74a** **Adult** Head, neck and underparts bright chestnut; triangular grey patch on lower hindneck and mantle; back and upperwing-coverts greenish-olive; lower back to tail, and lower flanks to undertail-coverts, black. Bill green, base yellowish.
- **74b** **Juvenile** Dull grey-brown to olive-brown on head, neck and underparts; upper mantle dull grey; chin and throat extensively whitish; bare parts duller than in adult.

75 **Grey-necked Wood-rail** *Aramides cajanea*   Text and map page 337

Central and South America from Mexico to Peru, Bolivia, N Argentina and Uruguay. Swampy forest, woodland, mangroves and marshes.

A large wood-rail. Head to upper mantle and upper breast grey; occiput blackish to rufescent; mantle, back and lesser upperwing-coverts greenish-olive; rest of upperwing-coverts tawny; breast and upper flanks orange-chestnut. Bill yellow, tip greenish.

- **75a** **Adult** (*albiventris*; S Mexico to NE Guatemala) Broad white patch between black of belly and orange-chestnut of breast; occiput auburn.
- **75b** **Adult** (*plumbeicollis*; NE Costa Rica) Hindcrown and occiput rufous-auburn; mantle rufescent; breast rich dark orange-chestnut.
- **75c** **Adult** (nominate; Costa Rica, Panama and South America) Occiput blackish to brownish; breast and flanks richly toned.

[*A. c. mexicanus* (Caribbean zone of SE Mexico) Occiput rufescent; narrow white line between tawny breast and black belly. *A. c. vanrossemi* (S Mexico to Pacific coast of El Salvador) Paler and larger than *albiventris*. *A. c. pacificus* (Honduras and Nicaragua) Like *plumbeicollis* but mantle more tawny-buff. *A. c. latens* (San Miguel and Viveros, Pearl Is, Panama) The palest race; occiput tinged earth-brown. *A. c. morrisoni* (San José and Pedro Gonzalez, Pearl Is) Upperparts darker than in *latens*. *A. c. avicenniae* (SE Brazil) Like nominate, but occiput grey, nape to back plumbeous, and underparts paler.]

73

74a

74b

75b

75c

75a

PLATE 23: WOOD-RAILS OF SOUTH AND CENTRAL AMERICA (2)

76 **Brown Wood-rail** *Aramides wolfi*   Text and map page 341

SW Colombia, W Ecuador and possibly extreme NW Peru. Forest, woodland, riverine marsh and mangroves.

    76   **Adult** A relatively large wood-rail. Head and nape ashy-grey; neck, mantle and breast vinous-brown; back, scapulars and upperwing-coverts olive-brown; rump, tail, lower flanks and belly to undertail-coverts black. Bare yellow patch on forehead.

77 **Giant Wood-rail** *Aramides ypecaha*   Text and map page 342

E South America in E and SE Brazil, Paraguay, Uruguay and NE Argentina. Marshes, swamps, fields, pastures and gallery forest.

    77   **Adult** The largest wood-rail. Sides of head, foreneck and centre of breast bluish-grey; nape, rest of neck and breast, and flanks to belly, vinous-chestnut; mantle, back, upperwing-coverts and tertials greenish-olive; rump to tail, and undertail-coverts, black.

78 **Slaty-breasted Wood-rail** *Aramides saracura*   Text and map page 344

SE Brazil, extreme NE Argentina and E Paraguay. Tropical forest and woodland.

    78   **Adult** Head greyish-brown, tinged cinnamon at sides; nape to mantle, and sides of upper breast, rufescent to ruddy olive-brown; back and upperwing-coverts greenish-olive; primaries russet; rump sepia; tail-coverts and tail black. Foreneck, sides of neck, breast, flanks and upper belly slate-blue; lower flanks and lower belly blackish. Bill pale green, bluish at base.

79 **Red-winged Wood-rail** *Aramides calopterus*   Text and map page 345

Known distribution fragmented: E Ecuador, NC and SE Peru, and W Amazonian Brazil. Forest, seasonally flooded or near water.

    79   **Adult** A dark wood-rail and the only species with mahogany-red in plumage, on sides of neck and median and greater upperwing-coverts. Upperparts mostly dark greenish-olive; lower back and rump dark rich brown; uppertail-coverts and tail black; chin and throat white; foreneck, breast, upper flanks and upper belly slaty-blue; rest of underparts dusky brown to blackish. Bill greenish.

100mm

PLATE 24: UNIFORM CRAKE AND *GYMNOCREX* RAILS

80 **Uniform Crake** *Amaurolimnas concolor*   Text and map page 346

S Mexico through Central America to Colombia, Ecuador, Venezuela, the Guianas, Peru, Bolivia and Brazil. Wet to dry forest and thickets.

A medium-sized, unpatterned, entirely brown to rufous-brown crake, with red eye, reddish legs, and yellowish-green bill.

- **80a** **Adult** (nominate; Jamaica; POSSIBLY EXTINCT) Upperparts rufous-brown, more cinnamon on scapulars and upperwing-coverts; underparts cinnamon-rufous, chin paler and flanks to undertail-coverts darker. Eye orange to red; legs and feet brown, red or pink.
- **80b** **Adult** (*castaneus*; Venezuela, the Guianas, Peru, Bolivia and Brazil) Upperparts more olivaceous; underparts brighter rufous.

[**A. c. guatemalensis** (S Mexico to Colombia and Ecuador) The smallest and darkest race; upperparts less rufous and underparts more sooty-brown.]

83 **Talaud Rail** *Gymnocrex talaudensis*   Text and map page 352

Karakelong I, Talaud group, Indonesia. Wet grass, scrub and rank vegetation at forest edges.

- **83** **Adult** Head to breast chestnut; upperparts greenish-olive, brighter green on mantle; rump olive-brown; flight feathers tawny; uppertail-coverts, tail and underparts blackish. Bare skin around eye pink, becoming silver-white behind eye; bill bright yellow; legs yellow, pinkish on feet.

81 **Bald-faced Rail** *Gymnocrex rosenbergii*   Text and map page 349

N and NC Sulawesi, and Peleng. Primary and old secondary forest.

- **81** **Adult** A very distinctive species, with conspicuous patch of pale cobalt-blue bare skin around and behind eye, and entirely black face and underparts. Nape to back and scapulars deep maroon-brown; rump and tail black; upperwing-coverts and tertials chestnut, rest of remiges olive-brown. Can fly for short distances.

82 **Bare-eyed Rail** *Gymnocrex plumbeiventris*   Text and map page 350

N Moluccas; New Guinea and nearby islands. Forest, swamps and wet grass.

Distinctive; resembles S American wood-rails. Head to mantle and upper breast vinous-chestnut; throat sometimes grey; lower mantle, back and upperwings olive-brown, but primary coverts and primaries rufous; lower back to tail black; lower breast to upper flanks and upper belly lead-grey; rest of underparts black. Pink bare facial skin visible only at close range.

- **82** **Adult** (nominate; N Moluccas, Misool, N and W New Guinea, New Ireland) Brighter than race *hoeveni*, but great individual variation in plumage.

[**G. p. hoeveni** (S New Guinea and Aru Is) Duller; underparts washed rufous-brown.]

100mm

PLATE 25: *AMAURORNIS* CRAKES OF ASIA AND AUSTRALASIA

84 Brown Crake *Amaurornis akool* Text and map page 353

N Pakistan, India, Bangladesh, W Myanmar, N Vietnam and SE China. N populations winter in S of range. Dense swamp, pandanus groves and riparian vegetation.

Very plain and dark; bill pale greenish, tipped bluish; legs and feet purple to red.

- **84** **Adult** (nominate; N Pakistan and India to W Myanmar) Upperparts dusky olive-brown; sides of head, and underparts, dark slate-grey; indistinct supercilium pale grey; chin and throat white; belly vinaceous and undertail-coverts olive-brown.

  [*A. a. coccineipes* (SE China and N Vietnam) Larger than nominate; legs and feet brighter red.]

85 Isabelline Bush-hen *Amaurornis isabellinus* Text and map page 355

Lowlands of N and SE Sulawesi. Riparian scrub, grass and abandoned cultivation.

- **85** **Adult** The largest *Amaurornis*. Upperparts olive-brown, tinged greyish; lores darker; chin and upper throat whitish; underparts cinnamon-rufous, more vinous on head and neck and browner on lower belly and vent. Legs and feet brownish-green.

87 Rufous-tailed Bush-hen *Amaurornis moluccanus* Text and map page 357

Sulawesi, Moluccas, New Guinea, Bismarck Archipelago, Solomon Is and Australia. Dense vegetation at wetland edges, flooded grass and scrub, forest and cultivation.

Upperparts olive-brown; remiges dark brown; underparts grey, becoming more olive on sides of breast, flanks and sides of belly, and rufous to tawny on undertail-coverts. May have reddish base to upper mandible.

- **87a** **Adult** (*ruficrissus*; S and E New Guinea; Australia) Chin and upper throat grey to whitish; undertail-coverts rufous-buff. Eye red-brown; bill green, yellower basally with swollen red area (breeding only); legs and feet olive, tinged yellow.
- **87b** **Adult** (nominate; Sulawesi, Moluccas, N and W New Guinea) Lores black; darker below than *ruficrissus*, and undertail-coverts sandy-buff; no red on bill.

  [*A. m. nigrifrons* (Bismarck Archipelago to Solomon Is) Darkest race; lower belly rufous-buff; undertail-coverts deep tawny; eye brown-grey; culmen blackish, less red at base. *A. m. ultimus* (E Solomon Is) Smallest and palest race; throat whitish; undertail-coverts deep vinaceous; culmen blackish, no red at base.

88 Talaud Bush-hen *Amaurornis magnirostris* Text and map page 361

Karakelong I, Talaud group, Indonesia. Dry to wet forest, and scrubby habitat at forest edge and along streams.

- **88** **Adult** Very dark, with strikingly robust bill. Head dark brown; upperparts dark rich brown; underparts very dark grey, but flanks and thighs more like upperparts. Bill predominantly pale green; legs olive-brown, yellow at front.

86 Plain Bush-hen *Amaurornis olivaceus* Text and map page 356

Philippines. Wet to dry grass, flooded scrub and forest edge.

- **86** **Adult** Larger and darker than 87. Upperparts very dark olive-brown; tail blackish; underparts very dark slaty-grey, paler on lower belly; vent and undertail-coverts dark rufous-brown. Iris red; bill pale green; legs and feet yellow to yellow-brown.

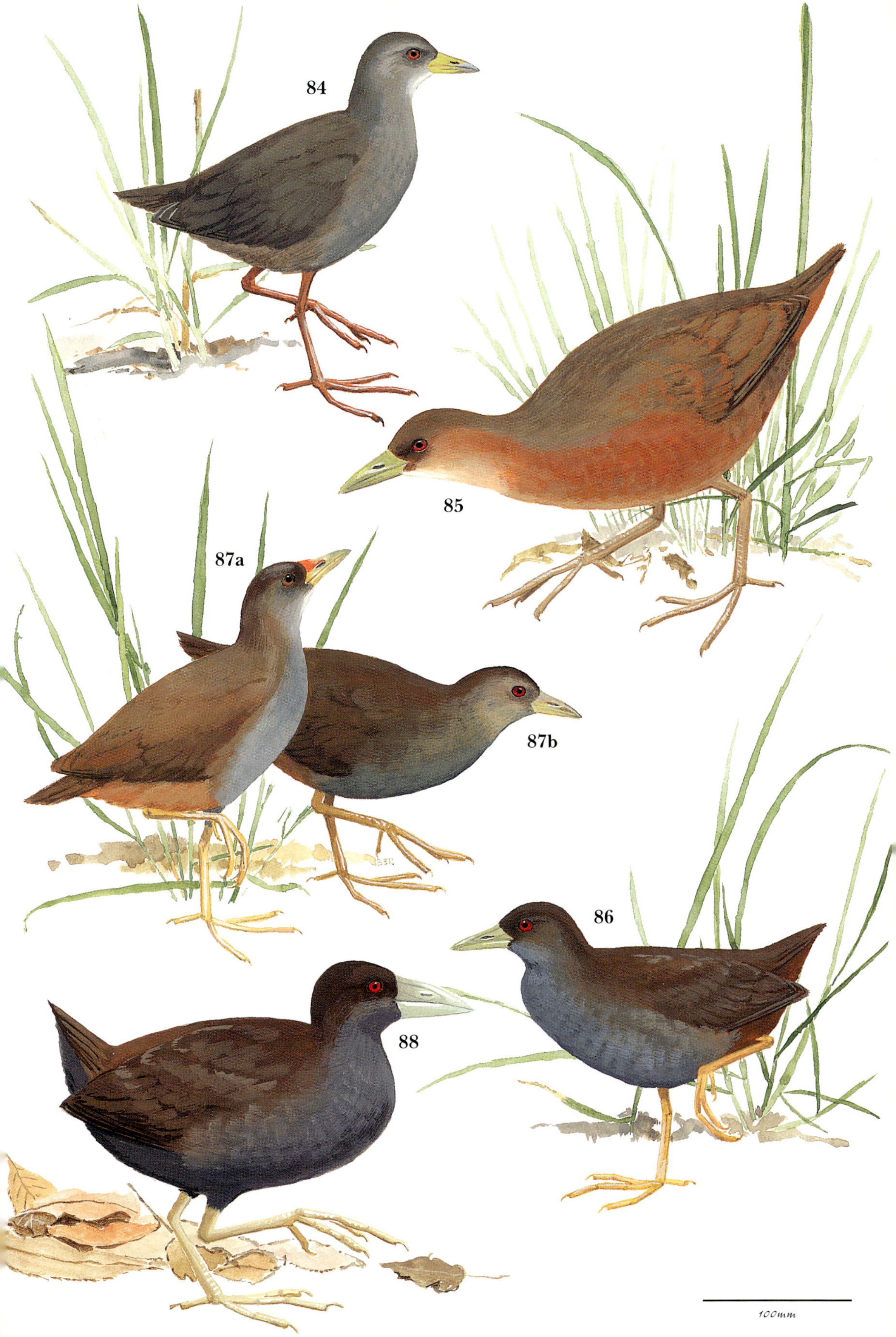
84
85
87a
87b
86
88
100mm

PLATE 26: *AMAURORNIS* CRAKES OF AFRICA AND ASIA

89 White-breasted Waterhen *Amaurornis phoenicurus* Text and map page 363

S and E Asia, and Malay Archipelago; winters mainly in S of range. Marshes, vegetated water margins, grass, mangroves, forest and scrub.

Upperparts and upperwings dark slate-grey, becoming dull brown from lower back to tail; narrow white leading edge to wings; sides of head, and underparts to upper belly, pure white; upper flanks dark slate-grey; lower belly pale rufous; lower flanks and undertail-coverts deep tawny.

- **89a** **Adult male** (nominate; Pakistan and India E to China) Slate-grey on head extends to forecrown.
- **89b** **Adult male** (*leucomelanus*; Sulawesi, Moluccas and Lesser Sundas) Bird from Sulawesi, showing slate-grey extending to forehead.
- **89c** **Adult female** (*leucomelanus*) Bill more olive than in male. Bird from Timor, showing slate-grey extending to ear-coverts and lores, and more extensive on flanks.
- **89d** **Downy chick** Black, including bare parts; bill tip pale.

[*A. p. insularis* (Andaman and Nicobar Is) White on forehead usually extends back beyond eyes; flanks to undertail-coverts darker. *A. p. midnibaricus* (C Nicobar Is) Slate-grey extends only half-way over top of head; underparts more olive.]

90 Black Crake *Amaurornis flavirostris* Text and map page 366

Africa from N edge of Sahel S to Cape. Freshwater wetlands with moderate cover, some flooding and often floating vegetation.

- **90a** **Adult** Appears all-black; olive-brown wash on upperwing-coverts and from mantle to rump often not visible in the field. Iris and eye-ring red; bill greenish-yellow (may appear apple-green or yellow); legs and feet red, duller in non-breeding season.
- **90b** **Immature** Upperparts olive-brown; chin and throat whitish; head, neck and underparts darkish grey; bill has almost changed from black to dull green; iris brownish-red; legs and feet dull red.

91 Sakalava Rail *Amaurornis olivieri* Text and map page 371

W Madagascar. Marshes and riverbanks with dense cover and floating plants.

- **91** **Adult** Blackish-grey; rich dark brown from mantle to rump and on upperwings. Iris red; eye-ring pink; bill greenish-yellow, sometimes reddish at base; legs and feet pinkish-red.

92 Black-tailed Crake *Amaurornis bicolor* Text and map page 372

NE India and Myanmar E to SC China and S to N Thailand, N Vietnam and N Laos. Wet forest, small marshes, grassy streams and wet grassland.

- **92** **Adult** Relatively small. Head, neck and underparts dark ashy-grey; upperparts rufous-brown; uppertail-coverts and tail blackish; chin whitish; undertail-coverts blackish-brown. Iris red; bill pale blue-green with red patch at base; legs and feet brick-red.

89a

89b

89c

89d

90a

90b

91

92

100mm

PLATE 27: CORNCRAKE, AFRICAN CRAKE AND STRIPED CRAKE

69 Corncrake *Crex crex* — Text and map page 320

Eurasia, from Faeroes and British Is E across Europe and C Russia to C Siberia and NW China. Winters in C and S Africa. Dry to moist grassland, crops and marsh edges.

- **69a** **Adult male** Top of head and upperparts, including tertials, streaked brown-black and buff to ashy; upperwing-coverts tawny with variable creamy barring; streak through eye buff; sides of head and neck, foreneck and breast blue-grey; chin and throat white; flanks to undertail-coverts barred rufous-brown and white. Female has warmer buff upperparts and less grey on sides of head, neck and breast. In non-breeding plumage, less ashy on upperparts and less grey on face, neck and breast.
- **69b** **Immature in flight** Shows tawny upperwing-coverts and duller tawny remiges. Blue-grey only on supercilium; foreneck and breast buff-brown; upperwing-coverts less barred than in adult.
- **69c** **Juvenile** Remnants of natal down form black line down centre of throat and breast. Upperpart streaks tinged buffy-yellow; no grey from sides of head to breast; flank bars rufous-brown and buffy-brown; bare parts darker and duller than in adult.
- **69d** **Downy chick** Sooty brown-black, tinged rufous on upperparts; bill pinkish; legs and feet black; small white wing-claw.

68 African Crake *Crex egregia* — Text and map page 316

Sub Saharan Africa, except in arid areas of extreme S and SW. Moist to dry grasslands, marsh edges and cultivation.

- **68a** **Adult** Upperparts, upperwings and tertials blackish, streaked olive-brown almost no dark streaking on nape and hindneck. Narrow white streak above lores and eye; sides of head, sides of neck and breast grey to blue-grey; chin and upper throat white. Flanks to undertail-coverts broadly barred black and white. Iris red or orange; orbital ring reddish; legs and feet grey or brown, often tinged mauve or pink.
- **68b** **Immature** Duller and darker than adult; grey of neck and breast replaced by dark, dull brown; bare parts duller and darker.

109 Striped Crake *Aenigmatolimnas marginalis* — Text and map page 431

Patchily in West Africa, and from East Africa S to N Namibia and N South Africa. Seasonally inundated grassland, pans and marsh edges.

- **109a** **Adult male** Upperparts and upperwings dark brown, streaked white from mantle to rump and on upperwings; sides of head dusky cinnamon; chin and throat white; underparts to breast pale cinnamon, flanks olive-brown, streaked white; lower flanks to undertail-coverts deep cinnamon. Bill heavy, apple-green; legs and feet jade-green; toes long.
- **109b** **Adult female** Forehead to upper mantle dark grey; sides of head to breast and flanks grey, faintly streaked and barred whitish.
- **109c** **Downy chick** Black; bill cream with dark band in front of nostrils.

100mm

PLATE 28: LITTLE, BAILLON'S AND SPOTTED CRAKES

93   **Little Crake** *Porzana parva*   Text and map page 374

Europe and C and S Russia to NW China. Winters Mediterranean S to Africa, Arabia, and E to NW India. Marshes with dense emergent plants and floating/flattened vegetation.

- **93a**   **Adult male** Upperparts olive-brown, streaked blackish; pale lines along scapulars and tertials; white spots in middle of mantle and back; sides of head and underparts blue-grey; lower flanks barred white; undertail-coverts barred black and white; bill green with red base; legs and feet green or yellow-green.
- **93b**   **First-summer female** Some white markings on scapulars retained from juvenile. Grey only on sides of head; chin and throat whitish; underparts buff, bars browner, less distinct, than in male.
- **93c**   **Juvenile** White spots on scapulars, upperwing-coverts and flight feathers; supercilium whitish; no grey in plumage; underparts creamy-buff to whitish, with brownish barring (disappears with wear); undertail-coverts barred brown and whitish; iris brown; bill dull; legs and feet olive-green.
- **93d**   **Downy chick** Black, with green gloss on upperparts, and pink bill.

94   **Baillon's Crake** *Porzana pusilla*   Text and map page 379

Europe, Africa, C and S Russia to N China, Japan, S Asia to Indonesia, Sundas and Australasia; N birds winter to S. Marshes, floodplains and grassland with dense emergents, floating vegetation and shallow water.

Upperparts rich brown with black streaks and black-edged white markings; sides of head and anterior underparts slate-blue to pale grey; posterior underparts barred black and white, bars extending further forward than in Little Crake; bill green; legs and feet greenish, yellowish, brownish or flesh. Primary projection noticeably less than in Little Crake.

- **94a**   **Adult male** (*intermedia*; Europe; Middle East; Africa) Sides of head and anterior under parts slate-blue.
- **94b**   **Juvenile** (*intermedia*) Slate-blue replaced by buff or white; upper flanks and breast barred; eye and bill dull.
- **94c**   **Adult** (nominate; Russia E to China and Japan; winters S and SE Asia, Indonesia and Sundas) Paler grey on underparts; brown streak across ear-coverts.
- **94d**   **Adult** (*palustris*; E New Guinea, Australia, Tasmania) Face and underparts pale; belly white.
- **94e**   **Adult** (*mira*; SE Borneo) Small; face and underparts very pale; throat and breast white; belly barred.
- **94f**   **Downy chick** Black with green gloss; bill yellowish-white with black base; eye and legs blackish.

[*P. p. mayri* (W New Guinea) Like *palustris* but darker; upperparts brighter. *P. p. affinis* (New Zealand) Like *palustris* but darker; bill longer.]

96   **Spotted Crake** *Porzana porzana*   Text and map page 389

Europe and C Russia to SW Siberia and W China; winters Mediterranean, Africa, E to India. Grass/sedge marshes with shallow water and mud; wet meadows.

- **96a**   **Adult** Upperparts olive-brown, streaked black, spotted and streaked white; sides of head, chin and throat blue-grey, spotted white; broad buff-brown streak on ear-coverts and sides of neck; breast olive-brown, tinged blue-grey; flanks barred olive-brown, black and white; undertail-coverts buff. Eye brown; bill yellow with red base; legs and feet olive-green.
- **96b**   **Juvenile** Grey in plumage replaced by grey-brown, brown or cream; flanks less contrastingly barred; bill and legs duller.
- **96c**   **Downy chick** Green gloss on upperparts; bill red, yellow and white.

100mm

PLATE 29: *PORZANA* CRAKES OF AUSTRALIA AND THE AMERICAS

97 Australian Crake *Porzana fluminea* Text and map page 396

Australia and Tasmania. Margins of densely vegetated marshy wetlands.

Upperparts olive-brown, heavily streaked black and streaked and spotted white; forehead, sides of head and underparts to breast dark grey; flanks barred black and white; undertail-coverts white; bill yellowish or greenish, with swollen reddish base to upper mandbile; legs and feet green to olive-yellow.

- **97a** **Adult male** Upperparts washed cinnamon; grey of head and underparts very dark; base of upper mandible usually red.
- **97b** **Adult female** Upperparts duller, no cinnamon wash; grey areas paler; upper lores brown; more white spots on sides of breast; base of upper mandible duller and less swollen.
- **97c** **Juvenile male** White spots of upperparts lack black borders; no white on rump and uppertail-coverts; lores, ear-coverts and sides of breast olive-brown, spotted white; underpart bars dark brown and white; bill grey-olive; iris brown.
- **97d** **Downy chick** Bill black with red base.

98 Sora *Porzana carolina* Text and map page 399

North America; winters USA to South America (Colombia to Guyana and C Peru). Freshwater marshes; also brackish marshes on migration and in winter.

Upperparts olive-brown, streaked black and white; black mask on forehead, base of bill to eye, and chin and throat, often extending as black streak on foreneck; white spot behind eye; sides of head to neck and breast grey; flanks prominently barred; undertail-coverts white; bill bright yellow.

- **98a** **Adult male** Bill with white line at base of upper mandible.
- **98b** **Adult female** Head pattern less prominent; more white markings on upperwing-coverts; bill lacks white line at base, and has olive-green tip.
- **98c** **Juvenile** More white markings on upperparts; grey of head and underparts replaced by buff to olive-brown; flank bars duller; bare parts duller.
- **98d** **Downy chick** Orange throat bristles; red and blue skin on head; bill whitish, with red at base of upper mandible.

99 Dot-winged Crake *Porzana spiloptera* Text and map page 406

S Uruguay and N Argentina. Marshes, grass and riparian scrub.

- **99** **Adult** Very small and dark, with black-streaked olive-brown upperparts, white bars on upperwings, and dark grey underparts becoming black with white bars from lower flanks to undertail-coverts.

100 Ash-throated Crake *Porzana albicollis* Text and map page 407

Colombia to the Guianas; Trinidad; Brazil to SE Peru, NE Bolivia, Paraguay and N Argentina. Marshes, grassland and scrub.

Upperparts blackish-brown, streaked olive-brown; chin and throat white; sides of head and anterior underparts grey; lower flanks to undertail-coverts barred blackish-brown and white.

- **100** **Adult** (nominate; Brazil, Bolivia, Paraguay, N Argentina) Upperparts streaked olivaceous-brown.

[*P. a. olivacea* (Colombia to the Guianas, Trinidad, extreme N Brazil) Smaller and paler than nominate; streaks on upperparts more olive-brown; underparts paler grey. Birds from Peru are as large as nominate but resemble *olivacea* in plumage.]

100mm

PLATE 30: *PORZANA* CRAKES OF ASIA AND AUSTRALASIA

102 Ruddy-breasted Crake *Porzana fusca*   Text and map page 411

Pakistan and India E to S and E China and Japan, and S to Philippines and Sundas. Marshes and wet grassland; also dry bush and cultivation.

Head, and underparts to upper belly, vinous-chestnut; chin and throat whitish; upperparts olive-brown, darker from lower back to tail; lower flanks and belly to undertail-coverts olive-brown, faintly barred whitish, often only from rear flanks and lower belly.

- **102a Adult male** (nominate; Pakistan and N India, E to S China and S through SE Asia to Philippines, Sulawesi and Greater and Lesser Sundas) The darkest race.
- **102b Adult male** (*erythrothorax*; Russian Far East, Japan, E and S China, Korea, Taiwan; winters S to Thailand and Cambodia) Slightly larger and paler than other races.

[*P. f. phaeopyga* (Ryukyu Is) Relatively pale like *erythrothorax*, but smaller. *P. f. zeylonica* (W India and Sri Lanka) The smallest race, slightly paler than nominate.]

103 Band-bellied Crake *Porzana paykullii*   Text and map page 414

Russian Far East, NE China and Korea, wintering SE Asia and Greater Sundas. Marshes, wet to dry meadows with trees and bushes, and crops.

- **103** **Adult** Forehead rufous; rest of upperparts olive-brown; greater and median upperwing-coverts variably marked white; chin and throat white; sides of head to breast cinnamon-rufous; flanks to undertail-coverts boldly barred black and white (dark bars browner in female). Legs and feet salmon to red; bill blue-grey with green base.

104 Spotless Crake *Porzana tabuensis*   Text and map page 416

Philippines and Australasia E across Pacific islands to E Polynesia. Fresh to saline wetlands with dense vegetation; on islands in forest, scrub, and on dry rocky ground.

Head, neck and underparts dark slate-grey; upperparts dark reddish-brown, darker from back to tail; undertail-coverts barred blackish and white. Bill black; legs and feet salmon-pink.

- **104a Adult** (nominate; entire range except W and C New Guinea) Upperparts dark reddish-brown.
- **104b Adult** (*edwardi*; W and C New Guinea) Upperparts darker, less reddish-brown than in nominate; underparts darker grey.

[*P. t. richardsoni* (Oranje Mts, W New Guinea) Relatively pale; upperparts more olive-brown; underparts paler than in nominate.]

100mm

PLATE 31: VARIOUS SMALL *PORZANA* CRAKES

106 Henderson Crake *Porzana atra* Text and map page 422

Henderson I, C Pitcairn group. Forest and dense thickets.

106 **Adult** Flightless. Entirely deep black with slight greyish gloss; iris and eye-ring red; bill black, base and culmen yellowish-green; legs and feet red, legs mottled black. Sexes alike, but female has less yellowish-green on bill (if present, confined to culmen), and legs plain red to orange.

107 Yellow-breasted Crake *Porzana flaviventer* Text and map page 424

Greater Antilles, and S Mexico to Colombia, Trinidad, the Guyanas, Brazil, Bolivia, Paraguay and Argentina. Marshes, swamps, wet fields.

Very small crake; upperparts black with prominent brown streaks and narrow white markings; facial pattern distinctive; anterior underparts cinnamon-buff; sides of lower breast, and flanks to undertail-coverts, boldly barred black and white.

107a/b Adults (nominate; Panama, NW Colombia and rest of range in South America). Showing individual variation in upperpart patterning, colour of sides of head and neck, and extent of white on underparts.

[*P. f. gossii* (Cuba and Jamaica) Smaller and paler than nominate. *P. f. hendersoni* (Hispaniola and Puerto Rico) Small; palest race. *P. f. woodi* (Mexico to Costa Rica) Like *gossii* but smaller. *P. f. bangsi* (N and E Colombia) Upperparts dark like nominate, underparts paler.]

108 White-browed Crake *Porzana cinerea* Text and map page 427

Thailand and Cambodia to Philippines, Sundas, Sulawesi, Moluccas, New Guinea, N Australia, and SW Pacific islands E to Fiji and Samoa. Marshes with floating vegetation, mangroves, grass, thickets, forest.

Slim, with long legs and toes. Top of head grey to black; distinctive face pattern of black eye-stripe with white lines above and below; upperparts blackish-brown with grey-olive to buff-brown streaks; chin and throat white; underparts grey; flanks olive-grey; undertail-coverts buff. Bill olive-yellow to orange-brown, red at base of upper mandible; legs and feet grey-green.

108a/b Adults Showing range of plumage variation: **a** shows brighter upperpart streaking and whiter underparts; **b** shows duller upperpart streaking and greyer underparts.

106

107a
107b

108b
108a

100mm

PLATE 32: EXTINCT *PORZANA* CRAKES OF PACIFIC ISLANDS

95 Laysan Crake *Porzana palmeri* **Text and map page 387**

Laysan, Hawaiian Islands. EXTINCT. Tussock grass and scrub.

- **95** **Adult** Flightless. Upperparts sandy-brown, streaked blackish except on crown and most upperwing-coverts; a few white streaks on scapulars and back; forehead, sides of head, and chin to belly, ashy-grey; sides of neck to flanks and sides of belly, and undertail-coverts, sandy-brown, darker and duller towards rear of body, with irregular white markings on lower flanks.

101 Hawaiian Crake *Porzana sandwichensis* **Text and map page 409**

Hawaii. EXTINCT. Grassland, scrub and forest clearings.

- **101a** **Adult** Flightless. Upperparts dark rich brown, blacker on remiges and rectrices; ear-coverts ashy; underparts cinnamon, becoming russet on flanks and belly, and dull brown on undertail-coverts; faint buffy barring from rear flanks and rear belly. Iris reddish-brown; bill bluish-horn with yellow base; legs and feet orange to red.
- **101b** **Immature** Paler than adult; upperparts warmish brown, paler on forehead, with dark streaks, most prominent on upperwing-coverts and tertials; sides of head pale brown, tinged ruddy, lores and ear-coverts darker; rest of underparts dark vinous-red; iris and legs paler than in adult; bill greenish-yellow.

105 Kosrae Crake *Porzana monasa* **Text and map page 422**

Kosrae I, E Caroline Is. EXTINCT. Swamps, marshes and damp forest interior.

- **105** **Adult** Flightless. Almost entirely black, with bluish-grey reflections; chin and centre of throat paler; remiges and rectrices browner; lesser upperwing-coverts brownish, spotted white; undertail-coverts spotted white.

95

101b

101a

105

100mm

PLATE 33: ZAPATA RAIL AND *NEOCREX* CRAKES OF SOUTH AMERICA

110 Zapata Rail *Cyanolimnas cerverai* — Text and map page 435

Zapata Swamp, Cuba. Dense, tangled bushy swamp.

- **110** **Adult** Almost flightless. Medium-sized; upperparts olive-brown, browner from back to uppertail-coverts; short tail brown-black. Front and sides of head, and underparts to belly, slate-grey; supraloral streak, and chin and upper throat, whitish; flanks greyish-brown, narrowly barred white; lower belly faintly barred whitish; undertail-coverts white. Red at base of bill extends to forehead as narrow, pointed frontal plate.

111 Colombian Crake *Neocrex colombianus* — Text and map page 436

Panama, N and W Colombia and NW Ecuador. Marshes, swamps, wet grass, overgrown forest edges; sometimes on dry ground.

Smallish plain crake, predominantly brown and grey; top of head dark grey; upperparts and upperwings olive-brown; chin and throat white; sides of head to belly slate-grey; flanks and lower belly to undertail-coverts pale cinnamon. Bill greenish to yellowish, with orange to reddish base.

- **111a** **Adult** (nominate; Colombia and Ecuador) Upperparts olive-brown; underparts pale slate-grey.
- **111b** **Adult** (*ripleyi*; Panama) Markedly darker overall than nominate.

112 Paint-billed Crake *Neocrex erythrops* — Text and map page 438

Costa Rica, Panama, Colombia E to the Guianas, Brazil, Peru, Bolivia, Paraguay, N Argentina and the Galapagos Is. Marshes, wet to dry grass, thickets and damp woodland.

Very similar to Colombian Crake but forehead slate-grey; crown olive-brown like rest of upperparts; lower flanks to undertail-coverts barred blackish and white; bill greener, more slender and with bright red base.

- **112** **Adult** (*olivascens*; throughout range except Peru and Galapagos) Darker olive-brown on upperparts than nominate, darker slate-grey on underparts, and with greyer chin and throat.

  [**Nominate** (coastal Peru and Galapagos Is) Paler than *olivascens*, with white chin and throat.]

110

111b 111a

112

100mm

PLATE 34: NEOTROPICAL *PARDIRALLUS* RAILS

113 Spotted Rail *Pardirallus maculatus* Text and map page 440

Greater Antilles; Trinidad; Mexico to Colombia and E to Guianas; SE Peru; E Bolivia; E Brazil S to N Argentina. Marshes, swamps, wet grassland and irrigated fields.

Blackish-brown, with heavy white streaking from head to back and breast, and spots from lower back to uppertail-coverts; upperwings browner, with less white. Flanks and belly barred white; undertail-coverts white. Bill yellowish-green, reddish at base; legs and feet pinkish-red. Sexes alike. Both races have 3 juvenile colour morphs.

- **113a Adult** (nominate; Greater Antilles, Trinidad, and Colombia through South American range) Upperparts streaked white.
- **113b Juvenile** (nominate) Barred morph. Underparts heavily marked.
- **113c Juvenile** (nominate) Dark morph. Almost no white markings.

[*P. m. insolitus* (Mexico to Costa Rica) Upperparts darker and richer, spotted rather than streaked; less white on upperwings. Juvenile pale morph has pale grey-brown breast, weakly barred white.]

114 Blackish Rail *Pardirallus nigricans* Text and map page 443

SW Colombia, Ecuador, Peru, Bolivia, Brazil, Paraguay and NE Argentina. Marshes, swamps and wet grassy areas.

Plain and dark. Upperparts dark olive-brown; chin and throat white to grey; head and underparts dark bluish-slate; lower belly, uppertail- and undertail-coverts and tail black. Bill green, yellower at base.

- **114 Adult** (nominate; entire range except SW Colombia) Chin and throat variably white to slate-grey.

[*P. n. caucae* (SW Colombia) Underparts paler; more white on throat; legs and feet darker red.]

115 Plumbeous Rail *Pardirallus sanguinolentus* Text and map page 445

Ecuador, Peru, Bolivia, Paraguay, Uruguay and SE Brazil, S to Tierra del Fuego. Marshes, waterside thickets, wet grassy areas.

Upperparts olive-brown; underparts slate-grey; lower belly to undertail-coverts brownish; green bill has bright red and blue base.

- **115a Adult** (*luridus*; S Chile and S Argentina) Largest race; no dark feather centres on upperparts; red at base of bill often absent.
- **115b Adult** (*simonsi*; Ecuador, W Peru and N Chile) Relatively pale; dark feather centres on scapulars, tertials, back and rump.
- **115c Immature** (*simonsi*) Paler and browner on head and underparts; chin and throat white; bill duller, lacking blue and red.

[*P. s. tschuddi* (Temperate Peru to SE Bolivia and NW Argentina) Larger than *simonsi*, and upperparts more reddish-brown. *P. s. zelebori* (SE Brazil) Smallest race; heavy black markings on relatively dark and rufescent upperparts; bill yellow-green. *P. s. sanguinolentus* (E Bolivia, Paraguay, extreme SE Brazil, and N Argentina) Larger than *zelebori*; upperparts with dark markings. *P. s. landbecki* (C Chile and SW Argentina) Second largest race; no dark markings on upperparts, which are relatively pale like *simonsi*.]

113a
113c
113b

114

115c
115b
115a

100mm

PLATE 35: LARGE AUSTRALASIAN RAILS

116 Chestnut Rail *Eulabeornis castaneoventris*   Text and map page 448

N Australia and Aru Is. Mangroves.

Very large, with longish tail. Head to hindneck grey; chin white; underparts pink-brown, tinged chestnut; upperpart colour variable, with 3 colour morphs; legs and feet pale yellow or green.

- **116a** **Adult male** (nominate; N Australia) Olive morph (Western Australia), with olive upperparts slightly tinged olive-brown. Bill yellowish-green with greyish-white tip.
- **116b** **Adult female** (nominate) Olive morph, with no olive-brown tinge to upperparts. Smaller than male.
- **116c** **Adult male** (*sharpei*; Aru Is) Like nominate chestnut morph, with chestnut-brown upperparts; bill deeper and heavier, green with yellow tip and red around nostril.

[**Nominate chestnut morph** (Northern Territory) like *sharpei* except for bill. **Nominate olive-brown morph** (Northern Territory and Queensland) has upperparts olive-brown.]

117 Invisible Rail *Habroptila wallacii*   Text and map page 451

Halmahera, N Moluccas. Dense swampy thickets, especially sago swamp, and marsh edges.

- **117** **Adult** A large, flightless rail. Head, neck and forepart of body dark slate-grey; upperwings and rear of body (from lower back) dark rich brown, blacker on uppertail-coverts and tail; lower breast to vent dark slate-grey; undertail-coverts dark brown to blackish. Bare parts bright red.

118 New Guinea Flightless Rail *Megacrex inepta*   Text and map page 452

Lowlands of NC and SC New Guinea. Mangroves, wet thickets and forest, and riparian bamboo.

Flightless, large and powerful; legs long and heavy; tail very short; frontal shield blackish. Forepart of head grey; crown to hindneck dull reddish-brown; mantle, back and scapulars greyish-olive; rump to tail, and upperwings, brownish-olive. Underparts white, often tinged rufous; sides of neck and breast, and flanks, brownish to buffy; sides of belly to undertail-coverts rufous-brown. Sexes alike.

- **118a** **Adult** (nominate; SC New Guinea) Sides of neck and breast, and flanks, brownish to vinaceous.
- **118b** **Juvenile** (nominate) Brownish-black hairy plumes on crown, sides of head and throat; upperparts browner than in adult.

[***M. i. pallida*** (NC New Guinea) Sides of neck and sides of breast paler, buffier; flanks much paler, more olive or pale brown; rump to tail, and remiges, slightly paler brown.]

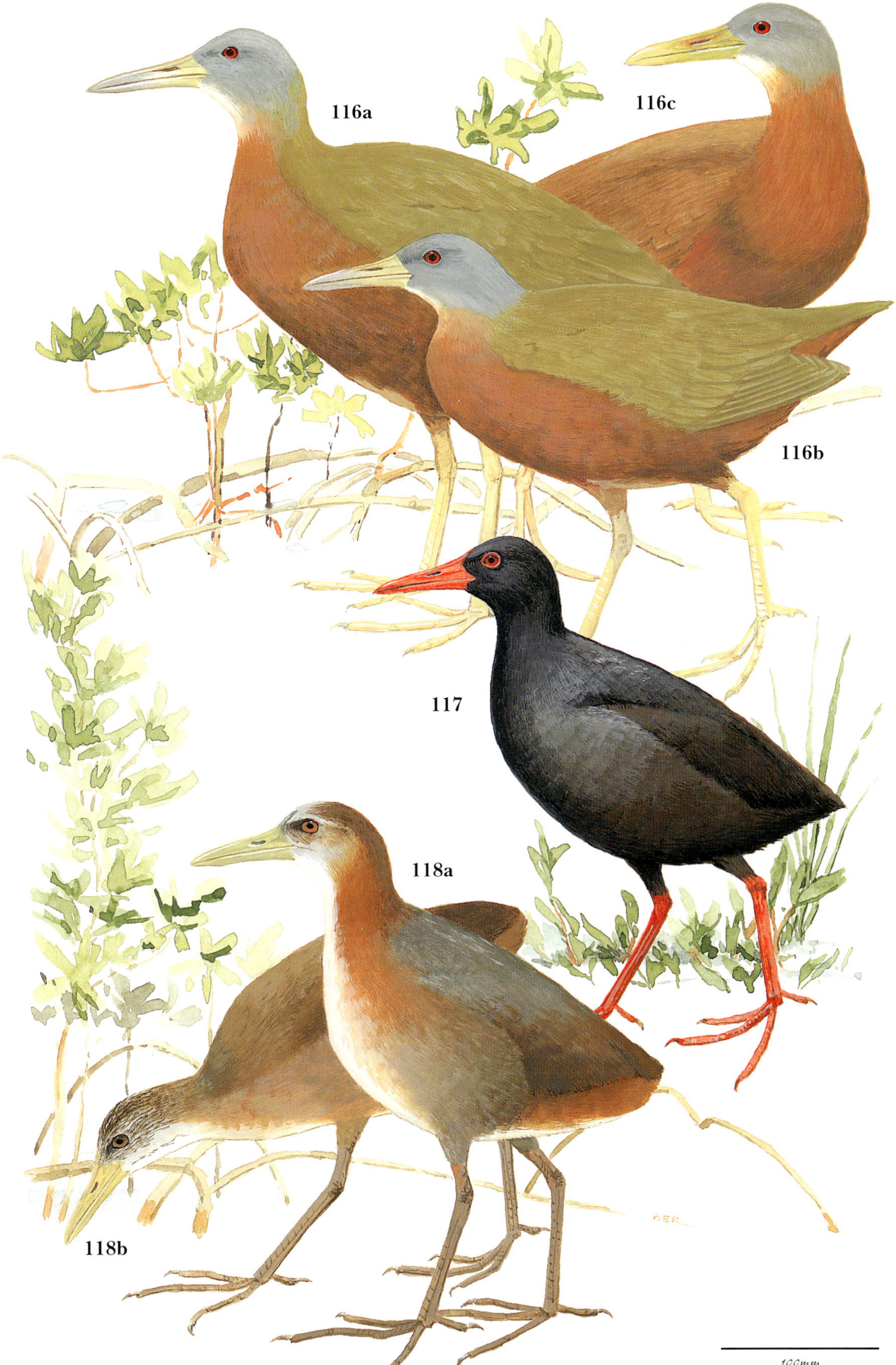
116a
116c
116b
117
118a
118b
100mm

PLATE 36: PURPLE SWAMPHEN

120 Purple Swamphen *Porphyrio porphyrio*   **Text and map page 458**

S Europe, Africa, Madagascar, Persian Gulf and Caspian E across S and SE Asia to S China, Australasia and W Pacific Islands E to Samoa. Densely vegetated marshes and fringes of open water; damp grassy areas.

Large to very large, with massive triangular red bill, red frontal shield, red to pink legs and feet and white undertail-coverts. Very variable in plumage: most races predominantly deep blue to purple or violet on head and body, some with cerulean-blue on head and foreneck; in most races mantle to tail, and upperwings, contrastingly green to black. Representatives of the six subspecies groups are depicted.

     120a Adult (*melanotus*; S and WC New Guinea; Australia (except SW), Tasmania; New Zealand and outlying islands) Head, upperparts and upperwings blackish; neck, upper mantle and underparts purplish-blue; belly and thighs blackish. Legs and feet pinkish-red with dark grey joints.

     120b Adult (*indicus*; Sumatra, Java, Bali, Borneo and Sulawesi) Head and upperparts largely blackish, usually without green tinge. Throat and breast turquoise-green; lesser coverts cerulean-blue; shield large with lateral ridges.

     120c Adult (*pulverulentus*; Philippines) Palest race; head, neck, upper mantle and most of underparts bluish-grey, tinged violet; rest of upperparts, scapulars and tertials olive-chestnut to orange-rufous; upperwings mixed grey-blue, greenish and olive-chestnut; legs and feet pale red.

     120d Adult (nominate; S Europe; N Africa E to Tunisia) Upperparts dark violet-blue, like rest of body; sides of head, and chin to upper breast, cerulean-blue.

     120e Adult (*poliocephalus*; Pakistan, Nepal, India and Sri Lanka E to SC China and N Thailand) Upperparts deep purple-blue; upperwings greenish-blue; sides of head (and sometimes chin to foreneck) hoary grey, tinged cerulean-blue; chin to upper breast cerulean-blue.

     120f Adult (*madagascariensis*; Egypt, sub-Saharan Africa and Madagascar) Hindcrown to upper mantle, and most of underparts, dull purple; rest of upperparts, scapulars and tertials bronze-green to blue-green; upperwing-coverts purple-blue; sides of head, and chin to upper breast, cerulean-blue.

     120g Juvenile (*madagascariensis*) Head brownish, washed pale blue; upperparts olive-brown, washed yellow-green; wings as adult; underparts pale buffy-brown; bare parts dull.

     120h Downy chick Velvety black with whitish filaments on head and upperparts; bare red skin on head and wings; bill white with red base; legs and feet rosy.

[*P. p. caspius* (Caspian Sea, NW Iran, Turkey and Syria) Like *poliocephalus*; larger than *seistanicus*. *P. p. seistanicus* (Azerbaijan (?), Iraq, Iran, SW Afghanistan and Baluchistan) Like *poliocephalus*, but larger. *P. p. viridis* (S Burma, Thailand, Indochina and possibly S China) Head and neck like *poliocephalus*; rest of plumage like *indicus*, having blackish, green-tinged upperparts and green upperwing-coverts; shield large, with lateral ridges. *P. p. pelewensis* (Palau Is) Like *melanopterus* but upperparts more green-glossed, and breast and lesser upperwing-coverts less blue. *P. p. melanopterus* (Moluccas, W Papuan Is and N New Guinea) Very variable; like *melanotus* but lesser coverts and breast blue. *P. p. bellus* (SW Australia) Like *melanotus* but upperparts paler, with variably yellow-olive feather fringes; outer lesser upperwing-coverts blue; chin and throat blue; shield shorter, with squarer base. *P. p. samoensis* (Admiralty and Bismarck Is through W Pacific Is to Samoa) Very variable; like *melanopterus* but with brighter upperparts (green- or blue-tinged black to bright rich brown) and brighter blue to blue-green breast.]

120a

120b

120c

120d

120e

120f

120g

120h

100mm

PLATE 37: ATYPICAL *PORPHYRIO* GALLINULES

122 Takahe *Porphyrio mantelli*   Text and map page 471

SW South I, New Zealand; also introduced to 4 nearshore islands. Alpine tussock grassland; scrub and beech forest; pastures on islands.

Flightless. Very large, thickset species with reduced wings, massive bill and powerful legs and feet; plumage loose with silken sheen.

    122a Adult (*hochstetteri*; South I) Head, neck, outer wings and underparts blue-purple; upperparts and most of inner wings olive; undertail-coverts white; iris red-brown; frontal shield and base of bill red; rest of bill more pinkish; legs and feet pinkish-red.

    122b Juvenile Not fully grown, still largely in second, greyish downy plumage. Head and neck brownish-grey; upperparts becoming olive and underparts becoming dull purple-blue. Iris brown; bill and shield pinkish-brown; legs and feet horn.

[**Nominate** (North I) EXTINCT – see text.]

121 White Gallinule *Porphyrio albus*   Text and map page 470

Lord Howe Island. EXTINCT. Woodland.

    121 Adult Flightless. Entirely white, male possibly with azure-blue tinge to plumage, or some blue feathers in wings; some birds blue and white (possibly hybrids with Purple Swamphen). Bill and shield red; legs and feet red or yellow.

125 Azure Gallinule *Porphyrio flavirostris*   Text and map page 485

Colombia E to Guianas and S through Brazil, E Ecuador and N Peru to Bolivia, S Brazil and N Argentina. Marshy areas with dense grassy vegetation, and rice paddies.

    125 Adult Superficially resembles juveniles of other *Porphyrio* species. Crown to back pale brownish with darker feather centres on back; lower back to tail dark brown; upperwings mostly azure; sides of head, neck and breast pale blue-grey; rest of underparts white. Bill and shield greenish-yellow; legs and feet yellow.

122b

122a

125

121

100mm

PLATE 38: GALLINULES, WATERCOCK AND SAMOAN MOORHEN

123 Allen's Gallinule *Porphyrio alleni*   Text and map page 476

Africa, from Senegal E to Ethiopia and Somalia and S to South Africa; Madagascar. Marshes, floodplains and seasonally inundated grasslands; normally with floating vegetation and dense cover.

- **123a Adult** Head black; neck and anterior underparts purple-blue; upperparts olive-green with blue on forewing and primaries; rump to tail green-black; belly blackish; undertail-coverts white (inverted heart-shaped patch). Bare parts red but shield blue (breeding) or grey (non-breeding), at start of breeding, shield apple-green (female) and turquoise (male). See 124.
- **123b Juvenile** Crown, hindneck and upperparts dark brown, heavily scalloped pale brown except on head and neck; sides of head and underparts buffy, including undertail-coverts; chin, throat and centre of belly white. Iris brown; bill red-brown, shield greyish; legs and feet brownish. See 124.

124 American Purple Gallinule *Porphyrio martinica*   Text and map page 480

E USA, and Central and South America S to N Chile and N Argentina; US birds winter S from Gulf coast. Marshes, flooded fields, and fringes of open waters, often with floating vegetation; also ricefields.

- **124a Adult** Larger than 123. Head, neck and underparts purplish-blue, pale blue on hindneck, sides of mantle and sides of breast; upperparts olive-green, variably tinged brownish; upperwing-coverts largely olive-green (lessers blue); undertail-coverts wholly white. Bill red with prominent yellow tip; shield pale blue; legs and feet yellow. See 123.
- **124b Juvenile** Moulting to adult-type plumage. Upperparts plain brown, tinged greenish; chin and throat cream; sides of head and neck, and underparts, buff-brown; undertail-coverts white; bill yellow-green, paler at tip and pinker at base; shield brownish; legs and feet yellowish. See 123.

119 Watercock *Gallicrex cinerea*   Text and map page 454

Pakistan and India E to E China, and SE Asia to Philippines; winters S to Sunda Is. Reedy marshes; waterside vegetation; rice paddies.

- **119a Adult male, breeding** Large; blackish, upperparts scalloped grey to buffy; undertail-coverts buff, narrowly barred black. Bill yellow, shield and 'horn' red; eye, legs and feet red.
- **119b Adult male, non-breeding** Dark brown cap; upperparts darkish brown, scalloped buffy; head and neck buffy-brown; chin and throat whitish; underparts pale buffy-brown with narrow, wavy dark barring. Eye yellow; bill and shield yellowish; horn absent; legs and feet greenish-brown.
- **119c Adult female** Like non-breeding male, but markedly smaller.

126 Samoan Moorhen *Gallinula pacifica*   Text and map page 488

Savaii I, Samoa. PROBABLY EXTINCT. Montane forest.

- **126 Adult** Small gallinule, almost flightless. Head, neck and breast dark bluish-slate, blacker on sides of head, chin and throat. Rest of upperparts, and upperwings, very dark olive-brown, tinged greenish; rump to tail black. Underparts dark slate, washed olive-green; flanks browner; undertail-coverts black. Bill red; shield yellow; legs and feet red.

100mm

PLATE 39: MOORHENS

128 Tristan Moorhen *Gallinula nesiotis* Text and map page 490

Gough and Tristan da Cunha Is, S Atlantic. Tussock grass, bushes and fern-bush.

- **128** **Adult** Head to mantle, and underparts, black; back to tail, and upperwings, very dark brown; flanks have a few white streaks; lateral undertail-coverts white. Shield and bill red, bill tipped yellow; legs and feet blotched red and greenish-yellow.

130 Dusky Moorhen *Gallinula tenebrosa* Text and map page 503

Borneo, Sulawesi, Sula, S Moluccas, Lesser Sundas, New Guinea, Australia, Tasmania. Open waters with fringing, and often floating, vegetation.

Dark grey-black, upperparts more brownish-olive; some have narrow white flank line. Shield red; bill red and yellow.

- **130a** **Adult breeding** (nominate; Australia, Tasmania) Legs and feet orange-red, dark at rear and on joints and soles; in non-breeding birds bill often olive-black, shield and legs olive-green.
- **130b** **Juvenile** (nominate) Browner on upperparts; paler, more grey-brown, on underparts; undertail like adult; bill and shield dark olive-brown; legs and feet olive-green.

[*G. t. frontata* (Sundas, Moluccas, W and S New Guinea) Slightly smaller; underparts darker; legs and feet red. *G. t. neumanni* (N New Guinea) Smaller and darker.]

131 Lesser Moorhen *Gallinula angulata* Text and map page 509

Sub-Saharan Africa. Wetlands, often seasonal, with emergent grass, sedges etc.

- **131a** **Adult male** Like small Common Moorhen, but greyer on neck; white flank line less distinct; bill more conical; forehead flat; shield and culmen red, bill yellow; legs and feet often yellow-green, also pink to orange. Female paler and greyer on neck and body; throat grey; black on head only at base of bill.
- **131b** **Juvenile** Olive-brown upperparts; buffy-brown sides of head and underparts, greyer on flanks, whiter from chin along central underparts to belly; no white flank streaks; bill brownish-yellow; legs and feet grey-green or yellow-green.

129 Common Moorhen *Gallinula chloropus* Text and map page 492

Eurasia E to Ussuriland and Japan, S to S Asia and Malay Archipelago; Micronesia; Africa, Madagascar and the Americas. N populations winter S from S USA, Mediterranean and S Asia. Freshwater wetlands with open water and fringing/emergent vegetation.

Predominantly grey-black; head and neck black; upperparts dark olive-brown; prominent white flank line; shield and bill red, bill tipped yellow; legs and feet yellow, upper tibia orange.

- **129a** **Adult** (*garmani*; Peru, N Chile, Bolivia, NW Argentina) Large, almost uniformly dark plumbeous; head and neck black.
- **129b** **Adult** (*cachinnans*; N and C America) Upperparts washed rufescent-brown; shield truncated at top.
- **129c** **Immature** (*cachinnans*) Duller and paler than adult; bill and shield greenish-brown, legs and feet olive-greenish (gradually attains adult bare part colours).
- **129d** **Adult** (nominate; Europe, mainland Asia, Africa) Worn plumage: head and neck black; no pale fringes on underparts.
- **129e** **Adult** (nominate) Fresh plumage: head to upper mantle tinged olive-brown; feathers of underparts narrowly tipped white.
- **129f** **Juvenile** (nominate) Head and upperparts dark brown; underparts pale brownish; chin, throat, belly and flank line whitish; bare parts olive-grey, becoming brighter.
- **129g** **Downy chick** White bristles on neck; bare skin pink and blue; legs and feet blackish.

[At least nine other races described – see text.]

100mm

PLATE 40: MOORHENS AND NATIVE-HENS

127 San Cristobal Moorhen *Gallinula silvestris*   Text and map page 489

San Cristobal, Solomon Islands. Dense undergrowth of mountain forest.

- **127**   **Adult** Dark bluish-slate on head, neck and breast; scapulars, mantle and upperwings brown-black, tinged olive; elsewhere dull brown-black; legs, feet and bill scarlet; frontal shield dark grey-blue; bare skin on face yellow.

132 Spot-flanked Gallinule *Gallinula melanops*   Text and map page 512

NC Colombia, Bolivia, Paraguay, E and S Brazil, Uruguay, Argentina and C Chile. Ponds, ditches, marshes, lagoons and lake margins with emergent vegetation.

Forepart of head blackish; rest of head, neck, upper mantle, breast and upper mantle slate-grey; upperparts olive-brown; flanks pale brown-grey, spotted white; undertail-coverts white. Bill and shield green; legs and feet greenish.

- **132**   **Adult** (nominate; Bolivia, Paraguay, Brazil, Uruguay and N and C Argentina) Upperparts washed golden-olive, scapulars and mantle more chestnut.

[*G. m. bogotensis* (Colombia) Axillaries white, unbarred; upperparts often more chestnut-washed. *G. m. crassirostris* (C Chile and S Argentina) Like nominate but larger; bill thicker.]

133 Black-tailed Native-hen *Gallinula ventralis*   Text and map page 515

Australia. Permanent and temporary wetlands in low rainfall areas; dispersive and highly irruptive species.

Large, dark, thickset moorhen with vertically fanned tail and long wings. Female similar to male but duller and paler; less black at base of bill; white flank spots smaller.

- **133**   **Adult male** Upperparts and upperwings olive-brown to olive; blackish feathering round base of bill; tail black; underparts dark grey, becoming blacker on belly and undertail-coverts; large white spots on foreflanks. Iris orange-yellow; bill and shield light green, base of lower mandible orange-red; legs and feet pinkish-red.

134 Tasmanian Native-hen *Gallinula mortierii*   Text and map page 518

Tasmania. Grassland and paddocks near wetlands with abundant cover.

- **134a**   **Adult** Flightless. Very large, thickset moorhen with long, narrow tail often held erect, stout bill and short wings. Conspicuous, noisy and aggressive; runs swiftly. Upperparts brown to olive-brown, more olive on mantle and greyish-olive on scapulars and coverts; tail blackish. Scapulars and coverts with fine white markings. Underparts mostly grey, darkening to black-brown on belly and undertail-coverts; large white blaze on foreflanks. Iris red; bill olive-yellow; legs and feet grey.
- **134b**   **Downy chick** Large, black (becoming browner), with black iris, legs and feet; bill black with large white egg-tooth, pink base to upper mandible and narrow lavender saddle.

127

132

133

134a

134b

100mm

PLATE 41: COMMON, RED-KNOBBED AND HAWAIIAN COOTS

136 Common Coot *Fulica atra*   Text and map page 527

Europe and Asia E to Sakhalin and Japan, S to N Africa, Asia Minor, India, SE Asia, Sundas and Australasia; N populations winter S to W Africa, India, SE Asia, Indonesia and Philippines. Open waters with submerged and (when breeding) fringing/emergent vegetation.

Darkish slate-grey; head and neck black; white tips to secondaries; undertail-coverts black; pointed projection of loral feathering between white bill and shield.

- **136a Adult** (nominate; entire range except Australasia) Bill often tinged yellow or pink; tarsus yellow-green, white at front, grey at rear, tibia orange to red.
- **136b Immature** (nominate) Upperparts more washed olive-brown than in adult; head and underparts variably mottled whitish; bare parts more or less as adult.
- **136c Downy chick** (nominate) Orange-red to yellow round neck and head; pink and blue skin on crown; bill and shield red, bill white and black at tip.
- **136d Adult in flight** (nominate) White tips to secondaries.

[*F. a. lugubris* (NW New Guinea) Smallest race; shield large; underparts relatively pale. *F. a. novaeguineae* (C New Guinea) Shield larger than in *lugubris*; white on secondaries reduced or absent. *F. a. australis* (Australia, Tasmania, New Zealand) White on secondaries reduced or absent; bill tinged blue-grey; legs and feet grey.]

135 Red-knobbed Coot *Fulica cristata*   Text and map page 522

S Spain, N Morocco, and from Ethiopia and Kenya SW to S Angola and S to South Africa. Open waters; breeding sites have fringing or emergent vegetation.

Dark slate-grey (darker than Common Coot); head, neck and lower flanks to tail black. Bill and shield white, often tinged blue; 2 red knobs at top of shield; rounded projection of loral feathering between bill and shield.

- **135a Adult breeding** Red knobs prominent; iris red; legs and feet greenish.
- **135b Adult non-breeding** Red knobs very small, difficult to see; iris red-brown; legs and feet mainly dull slate.
- **135c Downy chick** Grey-black, paler below, tinged silvery; golden-yellow around neck; pink and blue skin on crown; bill red, with white and black at tip.
- **135d Adult in flight** Longer wings than in Common Coot; usually has no white on leading edge of wing or on secondaries.

137 Hawaiian Coot *Fulica alai*   Text and map page 535

Hawaiian Islands. All waterbodies; breeds on open fresh waters with emergent vegetation.

- **137 Adult** Dark slate-grey, paler on underparts, darker on head and neck; tips of secondaries white; white undertail-coverts for inverted heart-shape. Shield swollen, extending to crown; bill and shield usually white, but may be cream, yellow or dark red; some birds have dark subterminal ring on bill. Legs and feet pale grey to bluish, sometimes yellowish or greenish.

100mm

PLATE 42: NEW WORLD COOTS

138 American Coot *Fulica americana* Text and map page 536

N and C America S to West Indies and Costa Rica; winters W to Hawaii and S to Panama; also in Colombia and N Ecuador. Reed-fringed open waters; also estuaries and bays in winter.

Slate-grey with blackish head and neck; secondaries tipped white; undertail-coverts white except basally in centre. Bill white with subterminal broken red-brown band; shield white with red-brown callus at top; legs and feet yellow-green to orange-red, tibiae red, toe lobes grey.

- **138a Adult** (nominate; entire range except Colombia and Ecuador) Secondaries broadly tipped white.
- **138b Juvenile** (nominate) Upperparts olive-brown; rest of plumage pale ash-grey; wings and undertail-coverts similar to adult. Bill and shield ivory-grey, callus small, pale red; legs and feet greyish, soon becoming yellow-green.

[*F. a. columbiana* (Colombia and N Ecuador) Darker; less white on secondaries; bill larger, yellow basally when breeding; shield larger.]

139 Caribbean Coot *Fulica caribaea* Text and map page 543

Bahamas, Greater and Lesser Antilles, Trinidad, Curaçao, and NW Venezuela. Freshwater lakes, ponds and marshes; brackish lagoons.

- **139 Adult** Slate-grey; underparts paler; head and neck blackish; secondaries narrowly tipped whitish; lateral undertail-coverts white. Frontal shield white to yellowish, broad, bulbous and oval, extending to crown; bill white, sometimes with reddish-brown spot or band near tip; legs and feet olive to yellowish.

140 White-winged Coot *Fulica leucoptera* Text and map page 544

Bolivia, Paraguay, SE Brazil, Uruguay, Chile and Argentina. Lagoons, backwaters, ponds and marshes.

- **140a Adult** Slate-grey; underparts paler; head and neck black; secondaries broadly tipped white; lateral undertail-coverts white. Bill and shield commonly yellow (variable); shield small; head appears rounded; legs and feet greenish-grey.
- **140b Immature** Upperparts grey-brown; underparts paler; sometimes whitish from chin to breast; shield pinkish; bill olivaceous or paler.

141 Andean Coot *Fulica ardesiaca* Text and map page 546

S Colombia, Ecuador, Peru, Bolivia, N Chile and NW Argentina. Lakes, ponds, rivers and marshes.

Slate-grey; head and neck black; secondaries usually with small white tips.

- **141a Adult** (nominate; Peru to Bolivia, Chile and Argentina) Lateral undertail-coverts white. White-fronted morph: bill white, shield white or yellow, legs and feet slaty.
- **141b Adult** (nominate) Red-fronted morph, with chestnut shield, yellow bill becoming greener at tip, and green legs and feet.

[*F. a. atrura* (S Colombia, Ecuador and coastal Peru) Lateral undertail-coverts largely black.]

138b
138a
139
140b
140a
141b
141a

100mm

PLATE 43: SOUTH AMERICAN COOTS

143 **Red-fronted Coot** *Fulica rufifrons* — Text and map page 551

S Peru, C Chile, Paraguay, Uruguay, SE Brazil, and Argentina. Marshes, and vegetated lakes, lagoons and ponds, in lowlands.

- **143** **Adult** Medium-large coot. Dark slate-grey; head and neck blacker; prominent white undertail-coverts of inverted heart-shape. Bill yellow, tip green and base reddish; frontal shield dark chestnut-red; bill and forehead make almost straight line; legs and feet olive.

142 **Red-gartered Coot** *Fulica armillata* — Text and map page 548

Chile, SE Brazil, Uruguay and Argentina. Lowland lakes, large ponds, rivers and marshes.

- **142** **Adult** Large coot; dark slate-grey; head and neck black; no white on secondaries; lateral undertail-coverts white. Bill and frontal shield yellow, separated by red along base of culmen; legs and feet yellow, with red 'garter' above ankle joint.

144 **Giant Coot** *Fulica gigantea* — Text and map page 553

C Peru, W and S Bolivia, N Chile and extreme NW Argentina. Ponds and lakes in barren highlands of Andean puna zone.

- **144** **Adult** Very large; heavy-bodied and small-headed, with concave forehead and high knobs above eyes. Dark slate-grey; black on head, neck and undertail-coverts; lateral undertail-coverts with some white streaking. Bill dark red, white and yellow; shield yellow and white; legs and feet dark red.

145 **Horned Coot** *Fulica cornuta* — Text and map page 555

N Chile, SW Bolivia and NW Argentina. High-altitude Andean lakes with dense submerged plants.

- **145a** **Adult** Very large, heavy-bodied and small-headed, like 144, but with long, erectile black proboscis, and two black tufts, above bill. Plumage slaty-grey; blacker on head and neck; undertail-coverts black with lateral white stripes. Bill greenish-yellow, base dull orange, ridge black; legs and feet olive, joints grey.
- **145b** **Downy chick** Black; bill pink or grey, upper mandible partly blackish, base and tip yellow; legs and feet black.

143

142

144

145a

145b

100mm

SAROTHRURA

A distinctive genus of small rails comprising nine species, usually known as flufftails but sometimes called crakes or pygmy crakes, seven of which are endemic to Africa and two to Madagascar. Most species are strongly sexually dimorphic in plumage, a rare character in the Rallidae. The males are chestnut on the head and neck, this colour sometimes extending to the mantle and breast, while the tail is also chestnut in some species; the body is black with white to buff spots or streaks. The females are usually dark brown or black, with pale spots, streaks or bars. In two species, the White-spotted and White-winged Flufftails, the female's plumage pattern is more similar to that of the male, while the White-winged Flufftail has white on the secondaries, a character shown elsewhere in the Rallidae only by members of the genus *Coturnicops*. The tail is fluffy in some species (the generic name translates as 'brush-tail'), and the bill is short and slender. The hooting or moaning songs of the males are very distinctive. Seven species inhabit wetlands and grasslands, and two (White-spotted and Buff-spotted) occur in forest and dense bush. Two species (Streaky-breasted and White-winged) are predominantly migratory, some populations of others also have regular seasonal movements (sometimes altitudinal), and some species are largely sedentary, moving locally only in response to irregular and unpredictable seasonal habitat changes. Six species are known to lay white eggs, completely or almost unmarked, a character unusual in the family. Two species (White-winged and Slender-billed) are globally endangered, one race of Striped is rare and the other data deficient, and Chestnut-headed should be regarded at least as vulnerable.

1 WHITE-SPOTTED FLUFFTAIL
Sarothrura pulchra Plate 1

Crex Pulchra J. E. Gray, 1829, no locality (= Sierra Leone).

Genus sometimes merged into *Coturnicops*, which is more likely to be a derivative of *Sarothrura* stock. Four subspecies normally recognised.

Synonyms: *Coturnicops/Corethrura/Gallinula/Ortygometra pulchra*; *Rallus cinnamomeus*; *Corethrura/Corythura/Ortygometra cinnamomea*; *Corethrura dimidiata*.

Alternative name: White-spotted Crake.

IDENTIFICATION Length 16-17cm. Tail short, not noticeably fluffy. Male unmistakable, with head, neck, breast, mantle, uppertail-coverts, undertail-coverts and tail reddish-chestnut; rest of body black, boldly spotted white. Iris brown; bill blackish; legs and feet brownish-grey to grey. Female also distinctive: foreparts as male; body blackish, barred reddish-buff; tail reddish-chestnut, barred black; iris paler, and bill greyer, than in male. Immature plumage almost identical to adult. Juvenile plumage unique in genus in showing adult pattern; areas chestnut in adults, and underparts, dark rufous-brown; upperparts blackish-brown; male's spots and female's bars pale brown to buff, with some whitish spots on wings of male. Inhabits lowland rain forest, usually in association with water.
Similar species Male easily distinguished by spotted plumage and forest habitat from all other *Sarothrura* species except Buff-spotted Flufftail (2), from which it differs in having darker, more reddish, chestnut foreparts, white spots on upperparts and underparts, and entirely chestnut tail-coverts and tail. Female plumage unlike that of any other flufftail: female Buff-spotted very different, being almost golden-brown with small black-edged buff spots on upperparts, blackish and buff bars and scallops on face, sides of neck, breast, rear flanks and undertail-coverts, and blackish and white bars on flanks and belly.

VOICE All information is taken from Taylor & Taylor (1986) unless otherwise stated. The vocal repertoire is extensive but imperfectly known. The characteristic advertising call of the male is given at any time of day and throughout the year, but most frequently during the breeding season (see Social and Sexual Behaviour). It is a series of 3-14 short, hollow *goong* notes, reminiscent of the call of the Golden-rumped Tinkerbird *Pogoniulus bilineatus*; the series often includes double notes *gui*, which rise in pitch on the second syllable, and is often repeated for several min; a weaker version is given by the female. Males call from the ground inside cover and turn to face in different directions between calls, giving the impression that the calling bird has moved. Often the only thing which can be seen of a calling bird in deep shadow or dense cover is the pale patch which appears and disappears as the throat pulsates with the utterance of each note. Males also give a series of 4-8 shorter, more rapid *goong* notes. Alarm or warning calls include a repeated *kik* or *kek*, given by the male and often followed by the *goong* call; there are also a rapidly repeated *ker* and a high-pitched, rapidly repeated *ki* from both sexes. Male and female also give ringing *klee* notes, and various growls, grunts, gulps, hisses, and quacking and buzzing calls. Unlike other flufftails, this species does not normally call at night. Contact calls made by both sexes to chicks include a low-pitched *gzm* and soft grunts, *duk* and *chuk* notes; downy chicks give squeaks and sharp, metallic cheeps, while older young also give quiet grunts.

DESCRIPTION *S. p. centralis*
Adult male Entire head, neck, mantle and breast bright reddish-chestnut, slightly paler on chin and throat and darker on top of head. Back, scapulars, upperwing-coverts, rump, flanks and upper belly fuscous-black to black, with conspicuous white spots; lower belly, vent and thighs more olive-brown, with smaller, whitish spots and bars. Uppertail-coverts, undertail-coverts and tail reddish-chestnut, tips of rectrices sometimes darker. Remiges blackish-brown with white spots; underwing-coverts and axillaries blackish-brown with narrow white bars. Iris mid-brown; bill blackish, sometimes with whitish base to lower mandible; legs and feet brownish-grey to dark grey.
Adult female Head, neck, mantle and breast as male or slightly duller; back, scapulars, upperwing-coverts, rump, uppertail-coverts, flanks and upper belly black or dark blackish-brown, narrowly barred reddish-buff; lower belly and thighs olive-brown, barred buff. Remiges blackish-

brown with a few small buff bars on inner secondaries and traces of bars on other feathers; underwing-coverts and axillaries blackish-brown with narrow buff bars. Tail reddish-chestnut, barred black. Iris light brown to brownish-grey; bill dark grey; legs and feet dark grey.

Immature male Very similar to adult but slightly duller and with fewer, smaller, spots on upperparts and dark tips to rectrices. Juvenile greater upperwing-coverts retained until birds 1 year old; buff-tinged spots contrast with white spots on rest of upperparts.

Immature female Very similar to adult but with more broken, less regular barring on upperparts.

Juvenile Patterned like adult but chestnut areas are darkish rufous-brown, as are underparts; upperparts are blackish-brown; male's spots, and female's bars, are pale brown to buff. Before becoming fully grown, male shows some whitish spots on back, scapulars and tertials. Eye very dark brown; bill, legs and feet blackish.

Downy young Chick has black down, browner on thighs and belly; bare parts black.

MEASUREMENTS Wing of 232 males 76-88 (81.1), of 78 females 77-89 (79.3); tail of 12 males 36-43 (39.95), of 7 females 36-46 (41.45); culmen to base of 30 males 15-19 (16.6), of 18 females 15-18 (16.7); tarsus of 30 males 28-32 (30), of 18 females 28-32 (30). Weight of 14 males 39-49 (45.2, SD 3.1), of 4 females 41-49 (45.4, SD 4.0).

GEOGRAPHICAL VARIATION Slight, involving size, the relative width of the black and rufous bars in the female and the richness of plumage colours. Birds from the W Cameroon highlands were originally assigned to a separate race, *tibatiensis* (Bannerman 1911) on the basis of larger size when compared with the nominate form, but overlap is extensive and they are now regarded as synonymous with the nominate form (Urban *et al.* 1986). The race *batesi* has sometimes been considered synonymous with either *zenkeri* or *centralis*.

S. p. centralis Neumann, 1908 – Congo, S Central African Republic (Haut Kemo), extreme S Sudan (Benengai), and S through C and S Uganda, Rwanda, Burundi, W Kenya (Kakamega, Malava, Kaimosi, Yala R and Nandi) and Zaïre (except SE) to extreme NW Tanzania (Bukoba), extreme NW Zambia (Salujinga) and N Angola (N of 12°S in Cabinda, Cuanza Norte, Malanje and Lunda). See Description.

S. p. pulchra (J. E. Gray, 1829) – W Gambia (Abuko and Pirang) and SW Senegal (Casamance R), Guinea-Bissau, Guinea, Sierra Leone, Liberia, Ivory Coast (N to Nimba and Comoé NP), Ghana, Togo, Nigeria (N to Zaria and Kano) and Cameroon (E of 11° E and N of 4°30'N). Black barring on upperparts of female narrower, reddish-buff bars broader and more rufous, than in *centralis*. Wing of 56 males 78-92 (83.8), of 26 females 79-90 (83.5); tail of 20 males 32-45 (40.5, SD 3.4), of 16 females 36-44 (39.8, SD 2.9); culmen to base of 32 males 16-20 (17.4), of 19 females 15.5-18 (16.6); tarsus of 32 males 28-32 (30), of 19 females 27-33 (29.3). Weight of 4 males 47.9-53.2 (50, SD 2.3), of 3 females 46, 47.9, 50.6.

S. p. zenkeri Neumann, 1908 – extreme SE Nigeria, coastal Cameroon and Gabon. More richly coloured than other races. Flanks and belly of male tinged chestnut, and white-spotted feathers tipped brown instead of black. Female has dark bars of upperparts blacker than in *centralis* and *pulchra*; width of bars as in *centralis*. Smaller than *pulchra* and *centralis*. 35 males,

8 females: wing of male 70-79 (74.2), of female 75-82 (77.1, SD 2.8); tail of 20 males 29-39 (33.5, SD 2.6), of female 31-35 (33.6, SD 1.7); culmen to base of male 14-17 (15.1), of female 12.5-16.5 (15.1, SD 1.4); tarsus of male 24-28 (26.3), of female 25.5-29.5 (27, SD 1.3). Weight of 1 male 42, of 1 female 40.

S. p. batesi Bannerman, 1922 – Cameroon S of 4°30'N, except coast. Male like *pulchra* and *centralis*; female like *zenkeri* but black barring more intense, chestnut barring on upperparts paler, and chestnut of head paler. Similar to *zenkeri* in size. Wing of 41 males 70-83 (75.3), of 8 females 72-78 (75); tail of 2 males, 37, 38, of 3 females 33, 35, 39; culmen to base of 10 males 15-16.5 (15.6), of 8 females 14-16 (15); tarsus of 10 males 25.5-29.5 (27.5), of 8 females 27-28.5 (27.8).

MOULT Postbreeding moult is apparently complete. Primary moult is often irregularly ascendant, sometimes irregularly descendant, sometimes simultaneously at two foci with other stages or unmoulted feathers in-between, and in one bird (Uganda) the outermost 8 primaries appear to have been moulted simultaneously. The secondaries are moulted with the primaries. Primary moult is recorded from Ghana in Feb and Jul, Nigeria in Aug-Oct and Dec, Cameroon in Jun, Aug and Oct, Uganda in Feb, Mar, May, Jun, Oct and Dec, and Kenya in Jul. Head and body moult is recorded from Liberia in Feb, Ghana in Jun, Nigeria in Jun-Oct and Dec-Jan, Cameroon in Apr-Aug and Oct-Dec, Uganda in Feb-Jun, Oct and Dec, Kenya in Jul, Sep and Dec, and Sudan in Nov. Postjuvenile partial moult begins at 6-7 weeks and is often complete at 14-16 weeks (Taylor & Taylor 1986); it is recorded from Ghana in Jun, Nigeria in Dec and Cameroon in May-Jul and Oct-Dec. The juvenile greater upperwing-coverts are retained until the birds are 1 year old and form a useful ageing character, the buff-tinged spots on those of the male contrasting in the field with the white spots on the rest of the upperparts (P. B. Taylor unpubl.).

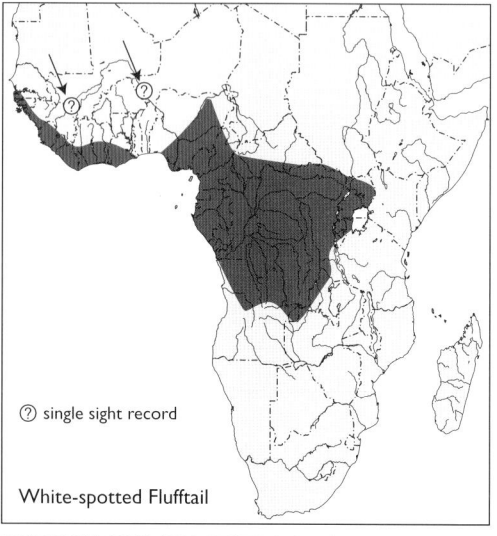

(?) single sight record

White-spotted Flufftail

DISTRIBUTION AND STATUS Gambia and Senegal to Ghana and Togo; Nigeria and Cameroon E to S Sudan and W Kenya and S to N Angola and NW Zambia. There is a questionable sight record from extreme SW Mali (Lamarche 1980) and a single sight record of a female from the Niger R in extreme SW Niger (Giraudoux *et al.*

1990). It is probably locally common to abundant throughout most of its range, including Sierra Leone (Field 1995), but it is apparently uncommon and local, or even rare, in Gambia (e.g. Gore 1990) and rare in Rwanda (Dowsett-Lemaire 1990). It is catholic in the choice of its habitat and is well adapted to forest disturbance, as long as suitable cover remains or develops: it successfully colonises cleared areas and secondary growth in forest and remains along streams after forest clearance (P. B. Taylor unpubl.). However, its numbers must be decreasing in many areas with the continuing large-scale destruction of forest habitat. Predation is not recorded, but possible predators include *Python sebae* (Keith *et al.* 1970) and mongooses (Taylor & Taylor 1986).

MOVEMENTS Normally entirely sedentary, but there is some evidence for local movements in Sierra Leone, where the species appeared near Freetown in an area where it had not occurred in previous years (Field 1995), and in Ghana, where it is found in isolated forest patches on the Accra plains during the wet season (Grimes 1987). At Malava Forest, W Kenya, it occurs irregularly in disturbed and unusually dry habitats (Taylor & Taylor 1986).

HABITAT Mainly lowland rain forest, usually in association with water (forest swamps, streams and pools, and river banks), where it is sometimes found in reeds; it is even recorded from papyrus by Chapin (1939). It also occurs on the forest floor well away from water, not usually deep inside primary forest but most often in edge habitats, disturbed areas and secondary growth, where it may be found in rank herbaceous tangles near paths; a population in disturbed forest at Malava, W Kenya, was 2km from the nearest stream (Taylor & Taylor 1986). In Ivory Coast it also occupies dense shrubbery and cassava plantations in almost completely cleared areas far from water (Demey & Fishpool 1991), while in SW Cameroon it is recorded from village farms and in upland habitats, presumably away from water. It follows rivers and streams out into gallery forest, dense thickets, shrubby growth, neglected cultivation and other rank herbage, and exceptionally occurs in papyrus and other vegetation by lakes (Urban *et al.* 1986). It occurs up to 2,000m in some areas but does not normally frequent montane forest. On streams it is characteristically found where the water is shallow and the banks have good cover of herbaceous or low-growing woody vegetation, and it requires dense cover over foraging areas of leaf-litter, soft earth, mud, sand, gravel or shallow water (Taylor & Taylor 1986).

FOOD AND FEEDING Chiefly invertebrates: earthworms, nematodes, small leeches, small gastropods, myriapods, spiders and many insects, including ants, beetles, bugs (Hemiptera), flies and small moths; the bird also takes terrestrial and aquatic insect larvae, including those of chironomids, mayflies, beetles and Lepidoptera, and small frogs. It occasionally eats a little vegetable matter, including small seeds. The following information is taken from Taylor & Taylor (1986). These flufftails take much food from the surface of humus, mud and shallow water, usually feeding in water only up to 2cm deep, sometimes immerse the head in water to catch prey, and use one foot to stir up muddy stream beds to disturb prey. The bill is used to turn over fallen leaves, to move leaf-litter aside and to probe and dig into soft mould. Birds also forage around the bases of plants and in the dead leaves collected in clusters of stems, search low-growing plants for insects, chase flying insects, and dig fly larvae out of cowpats. In Kenya this species feeds largely on the most abundant suitable invertebrate prey available at any time. The birds forage throughout the day, with peaks in the 2-3h after sunrise and before sunset, are not active during rain, and change their foraging area within the territory every 2-5 days; one pair with 2 chicks foraged over c. 20% of the territory each day.

HABITS All information is taken from Taylor & Taylor (1986) unless otherwise stated. These birds are apparently entirely diurnal, appearing after sunrise and roosting just before sunset, and they do not normally call at night. They are active throughout the day, but less so in the middle of the day. Although this species is rarely seen, its presence is usually easily established by the male's frequently repeated, characteristic song. It is much less difficult to observe than are other flufftails, including the Buff-spotted Flufftail which occurs alongside it in many areas, and is singularly unaffected by the close proximity of a quiet observer. Detailed observations of its behaviour and breeding were made at Kakamega, Kenya, by an observer sitting without any concealment on stream banks; after a time the birds would forage very close to the observer and would even bring young chicks into the open less than 1m away. Like the Buff-spotted Flufftail, birds walk and run over the forest floor and through dense vegetation without causing the slightest noise or movement of vegetation. During normal activity the tail is held horizontally but when the bird is wary or alarmed it may be raised and flicked. White-spotted Flufftails climb well in bushes and low trees and have been found almost 3m above the ground (Friedmann & Williams 1969); they fly into trees when pursued by dogs (Jackson & Sclater 1938). They roost in nests with young chicks and at other times apparently roost above the ground in bushes; captive birds built roosting platforms of crumbled, decayed palm fibre (Yeatland 1952). The male and female regularly bathe together in the middle of the day and in the late afternoon.

SOCIAL ORGANISATION All information is taken from Taylor & Taylor (1986) unless otherwise stated. This species is monogamous and permanently territorial, maintaining a strong permanent pair bond. In W Kenya its distribution along streams was essentially linear, one pair occurring per c. 100-130m of streamside habitat, and 6 territories along streams measured 0.72-1.05 (0.85, SD 0.12) ha. A similar situation prevails in Gabon, where calling birds have been heard every 400-500m, sometimes every 200-300m, along a creek (Brosset & Erard 1986). Eight territories on dry ground in disturbed forest at Malava, W Kenya, measured 0.74-1.25 (0.97, SD 0.17) ha. In the non-breeding season, territory boundaries are more flexible and pairs penetrate adjacent territories more readily, although a core area of at least 60m in diameter is not penetrated by neighbours. Immatures regularly remain in the parental territory until the start of the following breeding season, when they are ejected.

SOCIAL AND SEXUAL BEHAVIOUR All information is taken from studies in W Kenya (Taylor & Taylor 1986). The male and female of a pair normally remain in close contact at all times. Both sexes participate in territory defence but the male plays the more active role, being more vocal and more willing to attack models. In the typical threat display to a model, given throughout the year, the male makes *goong* calls and various growls before walking out of cover with an upright stance, the head and neck plumage roused, approaching the model from the side and pecking downwards at its head. He then walks away, crouches, calls and returns to attack again. During the breeding season the intensity of the attack increases and the bird often makes lunging jumps at the model, delivering vicious pecks at its head and body. Males usually attack female models somewhat less violently, often ignore them in the non-breeding season and frequently attempt to mate with them during the breeding season (but only when their mate is not present). Females attack female models, and occasionally male models, only during the breeding season. Males outside their own territories, especially during the non-breeding season, make a non-attack approach to a model, with compressed plumage, horizontally stretched neck, a crouched stance and a slow, creeping walk.

Immature birds of both sexes make territorial calls while sharing the territory with their parents, and will also attack models much more aggressively than do their parents at this time. When young are present in the territory the adults' frequency of territorial calling, and the level of aggression to models, is reduced; from Jul-Dec calling occurs mainly early and late in the day.

No proper courtship display has been observed. The male approaches the female with small mincing steps, his plumage roused. If she crouches, he mounts, pecks gently at her head and copulates for up to 20s, vibrating his tail rapidly from side to side. Allopreening occurs regularly, and the male occasionally courtship-feeds the female bill-to-bill.

BREEDING AND SURVIVAL Season Gambia, nest-building Aug; Liberia, breeding condition Sept; Ghana, May-Jul; Nigeria, Sept, breeding condition Jun, Aug, Dec; Cameroon, Sept-Oct, breeding condition Apr, Jun; Gabon, Mar; Zaïre, Apr-May, Nov; Uganda Mar-Apr; Kenya, Jan-Mar. In most areas breeds during the rains, probably more seasonally in S areas with a marked rainy season, but in W Kenya lays Jan-Mar, in months of low rainfall. **Nest** Placed in forest in leaf-litter on damp ground, by pool, or on rotten tree root in shallow water in swamp; an oval mound of dry or wet dead leaves, or other damp and rotted vegetation, c. 10cm high x 23cm long, sometimes concealed by covering of dead leaves; lined dry leaves, decayed fibres or grass; entrance a horizontal slit at the side. Nest built by male (Wacher 1993). **Eggs** 2; oval, white with slight gloss; size (n = 6) 30-30.9 x 21.5-22.1 (30.2 x 21.8); calculated weight 7.7. **Incubation** 14 days, by both sexes. **Chicks** Precocial and nidifugous; fed and cared for by both parents but tend to follow parent of same sex; capable of feeding themselves from an early age but remain in close association with parents until fully grown and feathered at 6-7 weeks; feed themselves from 3 weeks of age, after which parents still find food for them; fly at 30-35 days. In W Kenya some young disperse at c. 3 months of age; others may remain in parental territory for up to 9 months, disappearing by the start of the next breeding season (Taylor & Taylor 1986). In W Kenya only 1 brood per season. Age of first breeding 1 year. **Survival** Juvenile and adult mortality probably low.

2 BUFF-SPOTTED FLUFFTAIL
Sarothrura elegans   Plate 1

Gallinula elegans A. Smith, 1839, near Durban, Natal, South Africa.

Genus sometimes merged into *Coturnicops*. Two subspecies recognised.

Synonyms: *Coturnicops/Corethrura elegans*; *Sarothrura buryi/loringi*; *Corethrura pulchra* (part).

Alternative name: Buff-spotted Crake.

IDENTIFICATION Length 15-17cm; wingspan 25-28cm. Tail short, not fluffy. Male unmistakable: head, neck and breast orange-chestnut; tail barred black and orange-chestnut; rest of body sooty-black, closely spotted buff except on flanks and belly, where spots are white. Iris brown; bill blackish; legs and feet blackish or pink-tinged grey. Female has upperparts rich umber-brown with small black-and-buff spots; tail reddish-brown, narrowly barred black and buff; sides of head and neck finely barred brownish-black and buff; chin and throat creamy-white; breast olive-buff, scalloped blackish; rest of underparts barred blackish and white but undertail-coverts barred blackish and rufous-buff. Immature duller than adult; female darker brown on upperparts. Juvenile has upperparts plain dull grey-brown; sides of head whitish; breast and flanks pale grey; dull buff-brown bars on some upperwing-coverts and on remiges and rectrices. Inhabits forest, dense thickets and old cultivation.
Similar species Distinguished by spotted plumage and bushed or forested habitat from all other *Sarothrura* species except White-spotted Flufftail (1), from which male differs in having buff spots on upperparts, black bars on tail and chestnut not extending to mantle. Female very different to female White-spotted, which has reddish-chestnut head, breast and mantle, brownish-black body and upperwings with narrow reddish-buff bars, and chestnut tail with black bars.

VOICE The vocal repertoire is very extensive. A recent study (Taylor 1994), from which the following information is taken unless otherwise stated, identified 119 different vocalisations, of at least 25 basic call types, from adults and young. Of these vocalisations, 44 (37%) have an aggressive or territorial function, 17 (14%) are contact calls, and 24 (20%) are used during courtship and mating. The male's remarkable and unique song is a series of hollow, hooting *oooooooo.......* notes, like a tuning fork, each note lasting 3-4 s and being given 5-8 times per min. Males usually sing from a concealed perch 1-2m up in a bush or tree, and the bird becomes visibly swollen, especially around the neck, when calling. They call most at night, in the morning and evening, and in overcast or rainy weather, and may call continuously for 12h or more; at night the song is audible for 1km through dense forest and for up to 2.8km in more open areas. A ventriloquial effect is often produced by the male turning his head slowly from one side to the other while uttering the note. Recent studies have indicated that the song of individuals may vary little within a season, and that it is possible to distinguish between individual calling birds. Courtship and mating calls of the male include a low-pitched hoot followed by a high-pitched whine: *mooooo-eeeee* (often repeated for long periods), a rapidly repeated short, quiet hoot, quiet growls, and a quiet *gugugu-grooo* to call the female to the nest, while the female has rapid, quiet short hoots, gulps and grunts. Both sexes have a repeated, quiet, low-pitched *ooo* contact call, a rapidly repeated, loud, nasal *né* of aggression, a repeated loud *chek* when attacking potential predators such as rats, and various other hoots, growls, squeaks, and buzzing, ticking, hissing, moaning and whining notes. There is no specifically uttered call equivalent to the *dueh* of the Red-chested Flufftail (3). Although Buff-spotted Flufftails are very vocal during the breeding season, they are normally relatively silent at other times of the year; however, the song is heard throughout the year in Gabon and Liberia (Brosset & Erard 1986; also Rand 1951). Chicks up to 7 days old give cheeping notes, which serve contact and food-soliciting functions, and a squeal of distress. Juveniles have several contact calls, including a trill, a plaintive squeak and a sharp *zik*, and give squeaky, ticking and peeping food-soliciting calls. After independence (21 days) the young develop a limited repertoire of contact, aggressive, warning and alarm calls, including the typical *oop* contact call, *eee* and *klee* alarm calls, the *chek* attack call, and growls and grunts of aggression.

DESCRIPTION *S. e. elegans*
Adult male Head, neck and breast bright orange-chestnut, slightly paler, more orange-rufous, on chin and throat. Upperparts, from mantle to uppertail-coverts, including scapulars, upperwing-coverts and tertials, sooty-black closely spotted buff; tail barred black and pale chestnut. Primaries and secondaries dull brownish-olive, pale greyish-brown on undersides, with buff spots or bars on outer webs; underwing-coverts pale greyish-brown (but lessers dull brownish-olive) with greyish-white fringes and bars; axillaries as lessers, barred greyish-white. Rest of underparts sooty-black to black, closely spotted white, spots becoming buffy on lower flanks and undertail-coverts. Iris dark to mid-brown; bill blackish, with grey cutting edge and often grey to horn base of lower mandible; legs and feet blackish, or dark grey with pink tinge.
Adult female Upperparts from head to rump, including upperwing-coverts and tertials, rich umber-brown to buffy-brown, with small spots which are black on anterior half and buff on posterior half; spots smallest on head and largest on upperwing-coverts, tending to bars on tertials; largest spots have narrow black rear edge; tail more reddish-brown than upperparts, with narrow bars which have black anterior half and buff posterior half. Rest of wing as male. Sides of head and neck finely barred brownish-black and buff; chin and throat creamy-white; breast olive-buff with blackish scalloping; rest of underparts white, barred blackish and washed olive-buff, but bars on rear flanks and undertail-coverts more brownish-black and rufous-buff. Bare parts as male; bill sometimes paler.
Immature male Similar to adult but slightly duller, with brown fringes to crown feathers; spots more irregular in pattern and outline. Distinctive, predominantly unpatterned juvenile greater primary coverts normally retained (in both sexes) through first breeding season.
Immature female Similar to adult but upperparts noticeably darker brown.
Juvenile Predominantly dull grey-brown to sepia-brown, darker on mantle; face whitish; breast and flanks pale grey; centre of belly whitish. Dull buff markings appear progressively on growing remiges; thin dull buff-brown bars visible on some upperwing-coverts and on growing rectrices. Iris dark brown; bill grey-black with pink or horn cutting edge, lower mandible sometimes horn to greyish with dark tip; legs and feet grey-black.
Downy young Black; iris very dark brown, eyelids white; iris fades to dull, dark grey-brown (14 days) and becomes dark brown (brighter in males) at 33-36 days; bill black, gape flanges pink or yellow; bill fades to dark grey-black with paler cutting edge, and gape flanges to grey, at 19-21 days; white egg-tooth present (lost after 2-3 days); legs and feet black, becoming greyish-black with pink tinge (12 days).

MEASUREMENTS Wing of 115 males 83-95 (88.8, SD 3.0), of 60 females 84-96 (89.2, SD 2.8); tail of 81 males 35-47 (40.6, SD 2.7), of 44 females 35-45 (39.6, SD 2.3); culmen to base of 108 males 15-18.5 (16.8, SD 0.7), of 58 females 15-18 (16.3, SD 0.7); tarsus of 110 males 23-27 (25.4, SD 1.0), of 62 females 23-27.5 (25.2, SD 1.1). Weight of 53 males 39.5-58 (46.4, SD 4.1), of 34 females 40-60.5 (45.7, SD 4.1).

GEOGRAPHICAL VARIATION Slight, involving the colour of the upperparts and the size of the spots in the male. The possible races *buryi*, *loringi* and *languens*, known only from single localities in Somalia, Kenya and Tanzania respectively, are normally included with the nominate race (Keith *et al.* 1970).

S. e. elegans (Smith, 1839) – S Ethiopia (Arussi, Kaffa and Sidamo Provinces), N Somalia (Dubar, Wagar Mts), extreme S Sudan (Yei and Torit), W Kenya (Kakamega to Cherangani Hills and Mara R), C and S Kenya (Marsabit to Mt Kenya, Nairobi and Chyulu Hills), Zanzibar, Pemba, Tanzania (East Usambara and Uluguru Mts and Pugu Hills), Zambia, Malawi, NC and E Zimbabwe, Mozambique, Botswana and Namibia (see Movements), and E and S South Africa. See Description.

S. e. reichenovi (Sharpe, 1894) – SE Guinea (Macenta Prefecture), E Sierra Leone (Kono), Liberia (Mt Nimba, Ganta and Firestone Plantation), SW Ivory Coast (Taï NP), S Nigeria (Ubiaja), S Cameroon, Bioko (Fernando Po), Rio Muni, Equatorial Guinea, Gabon, Congo (Sibiti and Mayombe), W, NE and C Zaïre, Rwanda, Burundi, W Uganda (N to Budongo Forest), C Uganda (N to Kifu and Mabira) and N Angola (Cuanza Norte). Averages slightly smaller than nominate; male darker above, with larger, coarser spots. Wing of 30 males 80-91 (86.0), of 14 females 82-92 (86.5, SD 2.5); tail of 6 males 33-40 (35.85, SD 2.7), of 4 females 32-41 (35.25, SD 4); culmen to base of 25 males 14.5-17.5 (15.7), of 10 females 14.5-17 (15.4, SD 0.6); tarsus of 15 males 22-26.5 (24.2), of 10 females 23-28 (24.4, SD 1.0). Weight of 4 males 40-45 (43, SD 2.2), of 3 females 44.5, 44.5, 50.

MOULT Postbreeding moult is complete. Remex moult is usually not synchronised, being in ascendant, descendant or irregular sequence, and secondaries are moulted during or just after primary moult; however 1 South African bird is growing the inner 4 primaries simultaneously, and another shows a simultaneous moult of all primaries and secondaries. Stepwise moult of the primaries is also recorded. Remex moult is recorded from Cameroon in Dec, Sudan in Oct, Kenya in Nov-Dec, Pemba in Aug, and South Africa in Apr-Jul. Head and body moult is recorded from Cameroon in Dec, Apr and Jul, Sudan in Oct, Uganda in Apr, Pemba in May and Aug, and South Africa in Mar-May and Nov-Dec; a partial prebreeding (pre-alternate) moult is indicated by the Nov-Dec South African records. Postjuvenile moult may begin at c. 3 weeks of age, while the rectrices are still growing, and normally involves the replacement of juvenile feathers on the head and body, plus growth of feathers on initially bare tracts, namely the lores, ear-coverts, lesser and marginal upperwing-coverts and all underwing-coverts; it is often complete at c. 10 weeks. In South Africa this moult is recorded in Oct-Jul, and some immatures also show moult in Sep, which may indicate a prebreeding (first pre-alternate) moult. One South African immature in Jul shows primary moult, but this growth of primaries twice in the first year is probably atypical. The juvenile median upperwing-coverts are apparently replaced between the postjuvenile moult and the first breeding season, while the unpatterned greater primary coverts are normally replaced, with their corresponding primaries, after the first breeding season, so that their retention provides a means of ageing birds hatched during the previous breeding season.

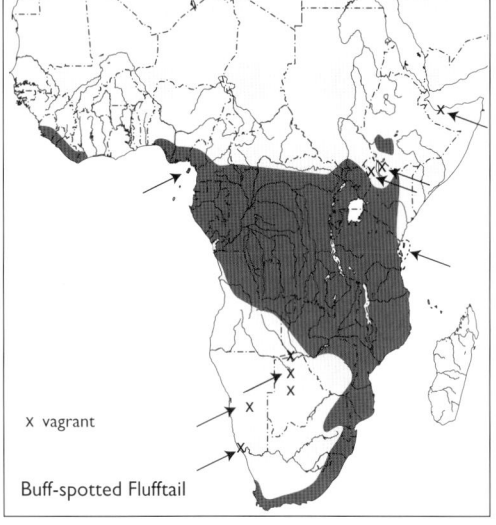

Buff-spotted Flufftail
x vagrant

DISTRIBUTION AND STATUS Guinea and Sierra Leone E to S Sudan and C Kenya and S through C Africa to N Angola, Zambia and E and S South Africa; also in Ethiopia. Its possible occurrence in SW Ghana (Wassaw Region) requires confirmation (Grimes 1987). In W and C Africa it is known only from scattered localities in Guinea, Sierra Leone, Ivory Coast, Nigeria, Congo and Angola, but is more widespread from S Cameroon to Gabon and E through Zaïre to Rwanda (Taylor 1994). It is apparently local and uncommon in Burundi, Sudan, Uganda, Kenya and Tanzania, but is probably widespread and not uncommon in S Ethiopia (Taylor 1994). It is probably widespread over its known range in Zambia, Malawi, Zimbabwe and Mozambique, and is widespread and locally common in E South Africa, especially in KwaZulu-Natal, but is uncommon and local in the Eastern and Western Cape Provinces (Taylor 1994). Although forest destruction must have adversely affected its numbers in some areas, it is probably holding its own by virtue of its ability to colonise degraded forest habitats, overgrown cultivation and exotic vegetation in suburban gardens. In KwaZulu-Natal it has extended its range locally in the recent past, following the creation of habitats associated with human habitation, and its arrival in the SW Cape is probably recent (Hockey *et al.* 1989, Taylor 1994). However, in residential areas it suffers heavy predation from domestic cats, which may have serious local effects on its numbers; other possible predators include genets *Genetta* and various mongoose species (Taylor 1994).

MOVEMENTS These are poorly understood, and the seasonality of the bird's occurrences and the extent of its movements are too complex and imperfectly known to be shown on the distribution map. Although the species is often regarded as resident throughout its range (e.g. Urban *et al.* 1986), it is possibly a migrant in Sudan (Nikolaus 1989), where the only records are from Yei (Oct) and Torit (Jan), and records indicate that it is of irregular or seasonal occurrence in parts of Kenya and Tanzania (e.g. Taylor 1994). For example, at Kakamega Forest, Kenya, it was present in Jul-Sep 1984, then absent until Mar 1985 when a definite influx occurred, and again absent from July 1985 until Apr-May 1986 (Taylor & Taylor 1986). It is recorded from Zambia only from Oct-Apr, but there are Jun-Jul (dry-season) records in neighbouring Malawi (Benson *et al.* 1971, Benson & Benson 1977, Taylor 1979, D. R. Aspinwall pers. comm.). Throughout its range there are isolated records indicative both of vagrancy and of migration, listed in detail by Taylor (1994): the only Nigerian record (Apr) is of a female caught in the evening in a house surrounded by cleared land; in Sudan a male was collected in Jan outside a house at Torit in atypical habitat; the only Somalia record is of a bird collected at the Wagar Mts in May, around hot springs in desolate country; a female collected at Kitgum, Uganda, in Oct was in atypical habitat of long grass and was very fat; on four occasions in Apr birds have been found at or near human habitation in semidesert country at L Turkana, N Kenya; a female was killed at the lighted window of a house near Mbala, N Zambia, in Dec, during bad weather when Angola Pittas *Pitta angolensis* and African Crakes (68), both known migrants, also flew into the windows; in Zimbabwe single birds were found dead in Harare city centre and in Bulawayo in Dec. The only records from Botswana and Namibia are of single birds, in Botswana at Maun in Dec 1991 and the Central Kalahari GR in Apr 1995, and in Namibia at Oranjemund in May 1976, Gobabis in May 1994 and Lianshulu, Caprivi, in Oct 95; all these birds died soon after being found (Winterbottom 1976, Oake & Herremans 1992, Crous & Tebele 1995, Simmons 1995). In some habitats, such as deciduous or exotic thickets, this flufftail's occurrence may be confined to the rains, the only period when sufficient food and cover exist. A recent study in KwaZulu-Natal (Taylor 1994) showed that some individuals remain throughout the year wherever suitable conditions (a moist foraging substrate, and adequate invertebrate food and vegetation cover) persist, but there is strong evidence for regular movements, both altitudinal and coastal, possibly over long distances and involving more first-year birds than adults. The same study found that immatures disperse from the vicinity of the parental territory when c. 7 weeks old.

HABITAT Occupies a wide range of habitats associated with forest or thick bush. It occurs in the interior and at the edges of many types of forest, including *Ficus* forest, juniper-podocarpus forest and bamboo forest, and particularly favours clearings, secondary growth and scrub. It occurs widely outside forest in dense evergreen and deciduous thickets (in South Africa these include *Leucosidea* thickets in upland areas, and the edges of exotic *Lantana* thickets) and also frequents banana groves, arrowroot plantations, neglected cultivation, and occasionally dense scrub within and around conifer and poplar plantations. In South Africa it occurs in suburban and farm gardens, where the predominant dense cover includes clumps or hedges of *Ligustrum lucidum*, *Kerria japonica*, *Chaenomeles lagenaria* and azalea *Rhododendron* sp. (Taylor 1994). Although not typically associated with water, in Gabon it frequents the floors of muddy valleys, in Sudan swampy patches in forest areas, and outside the range of the White-spotted Flufftail it is sometimes associated with forest streams (Taylor 1994). In winter in KwaZulu-Natal it sometimes feeds alongside the Red-chested Flufftail (3) in marshes adjacent to its normal forest habitat (Taylor 1994). It requires dense, low overhead cover such as that provided by secondary growth or thickets, usually some dense ground cover, and clear ground with soft earth or leaf-litter for foraging; it avoids rocky substrates and dry ground (Taylor 1994). Although it occupies vegetation of which grass is a component, references by van Someren (1939) to its occurrence in Kenya in grassland, swamp grass and grass at river margins are probably erroneous. It has a wide altitudinal range, occurring from sea level up to 2,600m (Kenya) and 3,200m (Ethiopia). On migration it may occur in isolated and very small habitat patches or in atypical habitats such as sugarcane plantations (see Movements).

FOOD AND FEEDING Studies in KwaZulu-Natal (Taylor 1994) show that this species eats a wide variety of invertebrate prey, principally earthworms, small gastropods (*Gulella* and *Trachycystis*), Amphipoda, Isopoda, Diplopoda, ants (including *Pheidole*, *Tetramorium*, *Dorylus* and *Pachycondyla*), termite workers and alates, and adult and larval Blattoidea, Coleoptera (Carabidae, Scarabeidae, Lampiridae, Tenebrionidae, Chrysomelidae and Curculionidae), Hemiptera, Lepidoptera and Diptera (Mycetophilidae and Calliphoridae); it also takes Chilopoda, Nematoda, Tricladida, slugs (*Urocyclus*), Collembola, spiders, ticks, grasshoppers and crickets. Grass seeds are eaten occasionally and hard, rough seeds of trees such as bugweed *Solanum mauritianum* and pigeonwood *Trema orientalis* are frequently taken, probably to assist in grinding food in the gizzard. Dog food is also recorded. Breeding birds readily took mealworms (*Tenebrio* larvae) and occasionally ate crushed oat grains, while captive birds ate frog spawn, and grass seeds such as *Pennisetum typhoides* and *Eragrostis tef*. Food items vary from 1-32mm in length (earthworms up to 100mm) and, given a good selection of prey sizes, birds appear to select the larger available items. The daily energy intake of adults is c. 170kJ. The birds feed throughout the day, mostly during the morning and late afternoon, foraging most frequently in leaf-litter but also on open ground and mossy rocks, in short grass, dense ground cover in forest and, in gardens, in soft soil of flower beds and herbaceous borders, at the edges of lawns and in drainage channels. They turn over dead leaves, dig and probe with the bill, scratch with the feet and readily take prey from mud and shallow water. Leaves of low-growing plants are examined in passing for insects, small moths and other low-flying insects are chased, lumps of earth are shaken and dropped to break them or to dislodge prey, and curled-up dead leaves are shaken and prodded, the bird looking into the end of the leaf-tube after each attempt to dislodge prey. Very large prey items are passed sideways through the bill several times before being swallowed. The birds drink daily, often several times per day.

HABITS All information is taken from Taylor (1994) unless otherwise stated. Apart from the prolonged nocturnal singing by the male, this species is diurnal and crepuscular, birds normally appearing just after sunrise and roosting just before sunset. They have relatively larger eyes than other flufftails, an adaptation to living and foraging in the permanently dim light beneath dense cover in forest habitats. They are active throughout the day, less so during the hottest period in the middle of the day. Although normally regarded as shy and very difficult to see, with patience they may be observed closely for long periods within the dense cover of their preferred habitat. During normal activity the tail is held horizontally, but may be raised or flicked if the bird is wary or alarmed. Buff-spotted Flufftails are adept at climbing in low trees and bushes, usually flying up into the lowest branches and then walking or jumping to get higher; they perch readily, crosswise on small twigs and lengthways on larger branches. They normally examine an observer from the ground and they are fond of using rocks as preening and observation posts. They are very unwilling to fly, usually flushing only when surprised outside cover, when they fly low into the nearest cover. When running to escape, their speed, manoeuvrability and silence on the ground are remarkable. Their reactions to movements are extremely rapid and they often appear uncomfortable in windy weather, when the constant noise and movement of leaves must interfere with their perception of possible danger. They usually roost off the ground, either in breeding or roosting nests or on low perches in trees and bushes. Sunbathing has not been observed, but both adults and young bathe daily when water is readily available, the frequency of bathing increasing during the day to peak in the 1-2h prior to roosting. The birds preen frequently, especially in the morning and after bathing.

SOCIAL ORGANISATION Monogamous and territorial, the pair bond being maintained throughout breeding season, and probably permanently in birds resident at the same site throughout the year; non-resident males set up temporary territories during the non-breeding season (Taylor 1994). Two breeding territories in KwaZulu-Natal gardens were 0.34 and 0.5ha (Taylor 1994), while two in forest at Kakamega, Kenya, were 0.86 and 0.94ha (Taylor & Taylor 1986).

SOCIAL AND SEXUAL BEHAVIOUR All information is taken from Taylor (1994). The species is strongly territorial during the breeding season, when both members of a pair take an active part in territory defence. Outside the breeding season sedentary males, and males which set up temporary territories, show much less territorial activity than when breeding but still call in response to taped playback and attack models. At this time of the year most females are very unobtrusive and difficult to observe, but some paired females may show territorial behaviour as strong as that of their mates. Males have been seen to give three types of threat display to models. The commonest is a slow, creeping walk up to a model, the male's tail being fanned horizontally, the body and head plumage roused and the wings drooped; the bird makes repeated *né* calls and often attacks the model with a series of fluttering jumps, pecking violently at the head and neck and striking with the feet. There is also a droop-winged run, in which the bird runs around the model, drooping one wing and sometimes giving *né* calls, and a bow in which the male walks out of cover with roused plumage, raises his head, bows forwards so that his breast touches the ground, and then attacks the model. Two other forms of attack, less intense than the fluttering jump and often made on model females, are a rapid run from cover to deliver several lunging pecks at the model before running back and giving *né* calls from cover, and a silent attack in which the bird runs from cover, delivers a glancing peck without stopping, and veers back into cover. Males show a lessening of aggressive behaviour when the chicks hatch and for c. 16 days afterwards, and during this period will not attack a model male but will attack a model female; at other times during the breeding cycle the male will court and attempt to mate with any live or model females. Females threaten and attack model females but not model males.

The male is bold in the defence of the nest against human intruders, standing in front of the entrance and making loud snake-like hisses. Males with young will threaten any potential predator, from a rat to a human intruder, using a spread-wing threat display in which the wings are opened, held above the back and tilted so that their upper surface faces forwards. The bird walks around giving *chek* calls and then rushes forwards, sometimes striking the intruder with wings or bill. An alternative display involves holding the wings out to the side, drooping the tips to touch the ground, and then calling and attacking as described. In defence of the young, females also perform a drooped-wing display, running around with *chek* calls.

Paired Buff-spotted Flufftails have not been seen to allopreen or to rest for long periods together, and the female frequently seems wary of her mate, whose persistent courtship behaviour throughout the breeding season is often discouraged. In the commonest courtship display the male assumes an upright posture with the neck stretched up, the neck and head plumage greatly roused and the tail wagged or vibrated from side to side (Fig 2). He then struts or runs to the female, who crouches if ready to mate. If she walks away the male follows and the display may develop into a courtship chase, the male upright and the female crouched; if she is willing to mate she stops and crouches to the ground, when the male mounts and copulates for up to 45s, pecking gently at the female's head and neck and wagging his tail from side to side. After copulation the birds may stand close together for a short time. During courtship the male often gives the *mooooo-eeeee* call and other vocalisations (see Voice).

BREEDING AND SURVIVAL Season Few data from W and C Africa: Cameroon, May, Sep, Oct; Fernando Po, undated clutch; Zaïre, Sep; Kenya, breeding condition Feb, Apr, May; Pemba, possibly Jun-Jul. In S Africa breeds during the rains (Sep-May): Zimbabwe, Nov-Feb, Apr; Mozambique, Sep; South Africa, Sep-Mar, May; peak laying month in KwaZulu-Natal, Nov, and birds in high-altitude regions may begin breeding 1-2 months later (Nov) than those at lower altitudes (Sep-Oct). **Nest** Usually built in area shaded by tall trees or within clumps of bushes; placed on ground, often in small excavated depression; usually well hidden in dense ground vegetation such as the forest grass *Oplismenus hirtellus* or exotic *Lamium* or *Vinca* ground creepers in gardens, in tangled cover, under piles of hedge and bramble clippings in gardens, or under the leaves of large plants such as *Isoglossa*, *Plectranthus*, irises or canna lilies; sometimes placed at foot of small tree or close to fallen log. Most nests domed, with an entrance hole at one end; usually built of materials most readily available including dead leaves or grass, moss, twigs, roots and bark; nest lined with fine grass, rootlets, moss or leaf fragments; in dense ground cover, nest approached by tunnel through vegetation. External length of nest 16-19cm, width 13-20cm, height 8.5-10cm; nest chamber length 8-10.5cm, width 8cm, height 6-7.5cm. Some nests are open shallow cups with roofing cover of low vegetation. Nest-building takes 2-3 days, by male only (Taylor 1994). Male may also build roosting nests for use by family. **Eggs** 3-5; oval, white, glossy, sometimes stained; size of 22 eggs 27.3-30.7 x 20.8-22.8 (28.8 [SD 1.0] x 21.6 [SD 0.6]); calculated weight 6.6. Eggs laid at daily intervals. **Incubation** 15-16 days; male incubates during day, female at night. **Chicks** Precocial and nidifugous; leave nest after 1-2 days; cared for by both parents and brooded under wings and body of female. Information on chick behaviour and development taken from Taylor (1994). Chicks can forage at 3 days but are fed by both parents until independent; beg by crouching or standing, sometimes with wing-flapping, and giving begging calls; parent approaches with food and stands still in front of chick, which takes food from parent's bill or picks food up when parent drops it. Chicks 4-15 days old also beg by standing behind parent, pushing head and body forwards between parent's legs, and stretching neck up to take food from parent's bill; when doing this, large chicks may raise parent off ground or cause it to overbalance. Chicks may solicit food from siblings of same brood. Chicks jump well and run strongly from 3-4 days; bathe regularly from 10 days; climb in bushes from 15 days; independent after 19-21 days, when driven off by parents, which often renest immediately. Body feathers and remiges begin to appear at 6-7 days; body almost fully feathered at independence, but many birds (more males than females) have patches of down on head and neck; young can fly at 19 days, when wings 79% of mean adult length; reach mean adult mass at c. 45 days; fully grown at 6 weeks. See Taylor (1994) for full details of plumage development. Clutch losses from predation probably low; chick mortality, from predators or bad weather, sometimes high. Predators of chicks include rats, domestic cats and

Figure 2: Buff-spotted Flufftail male in courtship-strut pose. Note roused plumage of head and neck; tail is in line with body. [After photograph by B. Taylor].

the Southern Boubou *Laniarius ferrugineus* (Taylor 1994). Age of first breeding 1 year. Up to 4 broods per season, KwaZulu-Natal; interval between clutches 30-40 days (mean 36 days). **Survival** Adult mortality, at least in migrants, possibly high.

3 RED-CHESTED FLUFFTAIL
Sarothrura rufa   Plate 2

Rallus rufus Vieillot, 1819, Africa (= Cape Province).

Genus sometimes merged into *Coturnicops*. Three subspecies recognised.

Synonyms: *Coturnicops rufa*; *Crex Jardineii*; *Ortygometra/ Corethrura/Crex ruficollis*; *Porzana/Crex/Gallinula/ Ortygometra/Corethrura/Alecthelia dimidiata*; *Corethrura bonapartii*.

Alternative name: Red-chested Crake.

IDENTIFICATION Length 15-17cm; wingspan c. 25cm. Tail noticeably long, often fluffed out. In male, deep reddish-chestnut of head extends to mantle and lower breast; rest of upperparts black with short white streaks becoming spots on greater coverts and tertials and from rump to tail; underparts blackish with short white streaks becoming spots on undertail-coverts. Iris dark brown; bill blackish, lower mandible blue-grey; legs and feet dark grey, tinged brown. Female paler, more golden-brown, than females of other wetland or grassland flufftails, due to close barring and spotting of ochraceous or rich buff. Immature like adult but in male chestnut duller and white markings fewer, tending to spots on upperparts; female very dark with fewer, paler spots on upperparts and more extensive markings on underparts. Juvenile has upperparts dull black and underparts dull grey-black except for whitish chin, throat and centre belly. Inhabits moist to flooded grass and marsh vegetation; sometimes also dry grass.
Similar species Male distinguished from males of sympatric Chestnut-headed Flufftail (4) and Streaky-breasted Flufftail (5), both of which occur in wetlands, by darker and more extensive reddish-chestnut on head, neck, mantle and lower breast, paler (but not white) on chin and throat, white spots, not streaks, on tertials, rump and uppertail-coverts, and all-dark bill; also differs from male Streaky-breasted in having longer tail with white spots, and blacker upperparts with shorter white streaks. Female distinguished from females of both species by close barring and spotting of rich buff, not white (giving much browner overall appearance), and all-dark bill; has longer tail than female Streaky-breasted. For other differences between females, see Chestnut-headed and Streaky-breasted species accounts. Immature female looks blacker than adult and is thus more easily confused with congeneric females, but usually has buff, not white, markings, at least on upperparts.

VOICE The vocal repertoire is very extensive. Unless otherwise stated, all information is taken from a South African study (Taylor 1994), during which 113 different vocalisations of at least 19 basic call types were recorded from adults and young. Of these vocalisations, 56 (50%) have a territorial or aggressive function, and 20 (17.5%) are contact calls. The male's song is a series of high-pitched short hoots *woooo*, 0.5-0.8 s long, repeated every 0.5-1 s and often continuing for several min; the hoot sometimes rises in pitch *wooaa*. This song is normally given only during the breeding season, by day and by night, but in Gabon it is heard all year (Brosset & Erard 1986). The male's common territorial call is a series of loud notes, each rising in pitch at the end: *dueh* or *doo-a*; the call becomes more strident and lower-pitched as agitation and aggression increase (e.g. in response to taped playback). The female has a similar call, usually higher pitched, sharper, and with the accent on the second syllable *kevic*; this call is given occasionally by males and regularly by immature males and females which share the parental territory. Gulps or grunts may accompany these and other calls. For several weeks at the end of the breeding season, when territorial activity is at its lowest intensity, the male's *dueh* often becomes modified into a quieter, squeaky *squoo-eh* or *squee-a*. Both sexes have sharp *ki* and *ker* calls of alarm, and quiet *oop* courtship calls; contact notes include a soft *oo*, a quiet bubble or rattle, and (to young birds) a subdued version of the *kevic* and a quiet *gruk*. Other calls include growling, chattering, hissing, wailing, moaning, buzzing, humming and ticking notes. Immatures give a small variety of the adults' territorial, alarm and contact calls, while chicks have quiet cheeps, a plaintive repeated *wee-ick* begging and contact call, ticking calls, and *ip* and *up* notes. Recent studies have indicated that there may be little variation in the song of individuals over long periods, and that it is possible to distinguish between individual calling birds using both song and territorial calls.

DESCRIPTION *S. r. rufa*
Adult male Entire head, neck, mantle and breast deep reddish-chestnut to mahogany-red, darkest on top of head and hindneck (dark feather tips) and paler, almost buffy, on chin and throat; lores and cheeks slightly darker, and rest of face paler, than overall colouring. Rest of upperparts, including scapulars, upperwing-coverts, tertials and tail, black with white submarginal streaks (broader than in male Chestnut-headed Flufftail) on each feather; these become elongated spots on greater coverts and tertials, and spots from rump to tail. Primaries and secondaries blackish-brown (almost black when fresh); axillaries and underwing-coverts greyish-brown, narrowly barred white. Rest of underparts blackish-brown, closely marked with short white streaks, becoming spots from rear flanks to undertail-coverts; centre of belly whitish. Iris dark brown; bill blackish with blue-grey lower mandible; legs and feet dark grey with brown to pink tinge.
Adult female Entire upperparts from top of head to tail black to brownish-black, closely marked with spots, bars or broken submarginal streaks of buff to ochraceous-buff; head, body and tail are spotted, scapulars, upperwing-coverts and tertials are barred to streaked. Primaries, secondaries, axillaries and underwings as male. Sides of head and neck buff, finely spotted (on head), scaled and barred blackish. Chin, throat and centre of belly buff to off-white; rest of underparts buff to ochraceous-buff, or even raw-sienna, with close, blackish bars and scales, giving scalloped or chevron effect on sides of breast and flanks; rear flanks to undertail-coverts more grey-brown, with small whitish or buff spots.
Immature male Similar to adult but chestnut duller and darker; white markings fewer, tending to spots on upperparts. Iris very dark, dull brown; bill black with paler cutting edge; legs and feet greyish-black. Both sexes retain

plain greater upperwing-coverts which contrast with patterned medians and lessers.
Immature female Darker than adult; very dark brownish-black with less numerous (and sometimes paler) spots on upperparts and more extensive markings on underparts. Bare parts as immature male.
Juvenile Upperparts, upperwings, underwings and tail dull black; underparts dull grey-black except for whitish chin, throat and centre of belly. Bare parts as immature.
Downy young Black downy chick has pronounced fluffy tail; iris, legs and feet black; bill white with black spot between nostrils, becoming pink with black tip at 1 week, and almost black at 6 weeks (Wintle 1988). White egg tooth present, lost after 2-3 days.

MEASUREMENTS Wing of 68 males 72-82 (76.5, SD 2.6), of 35 females 70-81 (76.6, SD 2.5); tail of 64 males 38-57 (44.5, SD 3.5), of 34 females 39-52 (44.5, SD 2.7); culmen to base of 62 males 13-17 (14.25, SD 0.7), of 31 females 12.5-17 (13.8, SD 0.8); tarsus of 64 males 21-25.5 (23.4, SD 0.9), of 35 females 20.5-26 (22.3, SD 3.6). Weight of 14 males 30-46.5 (38.8, SD 4.1), of 9 females 29.5-42 (36.6, SD 4.8).

GEOGRAPHICAL VARIATION The race *ansorgei* was originally proposed for birds from Angola, but the limited material available does not differ sufficiently from the nominate form to justify separation and more studies are needed (e.g. Keith *et al.* 1970).

S. r. rufa (Vieillot, 1819) – C and S Kenya (Molo E to Karissia Hills and Thika), Zanzibar, Pemba, NE Tanzania (Arusha to Amani, and Uluguru Mts), W and S Tanzania (Ugalla Game Reserve, Isoka, Rukwa, Ufipa Plateau and Matengo highlands), S Zaire, Angola, Zambia (except S and SE), Malawi, Mozambique (except NE), Zimbabwe, Namibia (Caprivi and Omanbonde), N Botswana (Okavango), and N and E South Africa S to the SW Cape. See Description.

S. r. elizabethae van Someren, 1919 – Ethiopia (Shoa, Kaffa, Oromiya and Wollega districts, and including the Jimma area – P. B. Taylor unpubl.); SW Central African Republic (Bangui), N and E Zaïre, Rwanda, Burundi, C Uganda (Mengo and Busoga) and W Kenya (Nandi and Kaimosi). Upperparts of male with longer white streaks and fewer spots; tail is spotted, not streaked as mentioned in error by Taylor (1996b) (but see *bonapartii*); upperparts of female tend to be barred, not spotted, buff. Wing of 128 males 70-81 (76.3), of 74 females 73-81 (77.2); tail of 2 males 39, 44, of 2 females 46, 48; culmen to base of 18 males 13-15 (14), of 11 females 13-15 (13.8); tarsus of 18 males 20-23 (21.5), of 11 females 20-22 (20.9). Weight of 1 juvenile female 33.7.

S. r. bonapartii (Bonaparte, 1856) – scattered records Sierra Leone (mostly in N and W), Liberia, Togo (Kovié, near Lomé) and Nigeria (Ilorin and Gasha-Gumki NP); more continuously S Cameroon, N Gabon and N and S Congo. Smaller than other races; male has upperparts and tail as *elizabethae* but, on feathers of rump to tail, spots sometimes tend to run together, giving more streaked effect; upperparts of female tend to be streaked or crescent-shaped, rather than spotted or barred. Wing of 21 males 66-74 (69.2), of 12 females 66-71 (70.7); tail of 8 males 35-41 (38), of 2 females 40, 42; culmen to base of 20 males 12.5-14.5 (13.4), of 7 females 12-14.5 (13.35); tarsus of 20 males 18-21 (20.1), of 7 females 19-22 (20.5).

MOULT Postbreeding moult is complete and sometimes prolonged. Remiges are moulted sequentially, the pattern being ascendant, descendant or irregular, while stepwise moult is also recorded. Remex moult is recorded from Kenya in Mar, Uganda in Dec, Tanzania in Jul, Malawi in Jun, Zimbabwe in Apr and Aug, and South Africa in Jun and Oct. In C and S Africa (Zambia, Malawi, Zimbabwe and South Africa) some head or body moult is recorded in most months of the year, including the breeding season, but the most heavily moulting individuals are from Jan, Apr, Jun, Sep and Oct, while tail moult is known in Jan, Apr and Oct. Some of these birds are obviously in complete postbreeding (pre-basic) moult, while others may be undergoing a partial prebreeding (pre-alternate) moult, but it is not clear whether significant moult during the breeding season represents the end of a prebreeding moult or the early beginning of a postbreeding moult. Postjuvenile partial moult involves head and body feathers; in captive birds it begins at 5-6 weeks and is complete at 10-11 weeks (Taylor 1994). Records are: Zimbabwe and South Africa, Mar-May, sometimes persisting into Jun-Jul; Pemba, Mar; Malawi, Apr; Zambia, Jul. Until at least the end of their first year, i.e. until the second pre-basic moult, birds retain plain greater upperwing-coverts which contrast with the patterned medians and lessers and provide a useful ageing character.

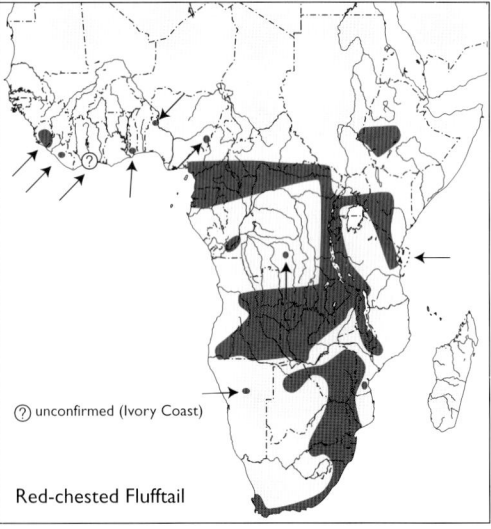

? unconfirmed (Ivory Coast)

Red-chested Flufftail

DISTRIBUTION AND STATUS The most widely distributed flufftail species, occurring in scattered localities from Sierra Leone E to Nigeria and in Ethiopia, and more continuously in Cameroon S to Congo, E to Uganda, W and C Kenya (its occurrence in coastal Kenya at Mombasa is unsubstantiated and probably incorrect) and NE Tanzania, and S through W and S Tanzania, S Zaïre and Angola to E and S South Africa. It also occurs on Zanzibar and Pemba Is. It possibly occurs near Dabou, Ivory Coast, but this requires confirmation (Thiollay 1985). Although probably not uncommon in N and W Sierra Leone (Field 1995), it is apparently rare elsewhere in West Africa and in Uganda (Taylor 1994). It is sparse in Angola, uncommon in Namibia, local but sometimes common in Gabon, Ethiopia, Kenya, Zanzibar, Pemba, Tanzania and Zimbabwe, widespread and locally common in Malawi, Zambia and Zimbabwe, and locally common to abundant in Zaïre and

South Africa (Taylor 1994). Its estimated population in South Africa is more than 33,000 birds and the largest numbers occur in KwaZulu-Natal, where it is ubiquitous in suitable habitat (Taylor 1997a; also see Social Organization). It is often overlooked, although it is relatively easily located by its frequent calling, and its range is probably more extensive than is known. Although it is a successful colonist of artificially created, and sometimes very small, habitat patches, overall numbers must be decreasing with the continual destruction of wetland habitats throughout its range (Taylor 1997a). Predation by a falconry-trained Black Sparrowhawk *Accipiter melanoleucus* is recorded in Gabon (C. Erard *in litt.*), and once by African Marsh Harrier *Circus ranivorus* (R. Simmons *in litt.*). Flufftails may occasionally be killed by fire; as well as escaping by running or flying, they shelter under unburned tussocks and in rodent burrows; individuals deprived of cover by fire, or overcome by heat and smoke, are susceptible to predation by Black-headed Herons *Ardea melanocephala*, which congregate at burned sites very soon after the fire has passed through (Taylor 1994).

MOVEMENTS There is no evidence for regular movements, pairs being entirely sedentary and permanently territorial when conditions permit (Taylor 1994, 1997c). All information is from Taylor (1994) unless otherwise specified. The reduced frequency of calling in the non-breeding season makes the species more difficult to locate during this period. Very local non-breeding-season movements occur when occupied habitat is drastically reduced, e.g. by drying out, burning, or trampling by domestic stock; birds displaced by burns tend to move as short a distance as possible, set up a temporary territory and reoccupy the permanent territory as soon as sufficient regrowth of vegetation has occurred. In South Africa resident pairs were able to tolerate non-breeding-season reductions in territory area of 20-53% for up to four months and of up to 70% for 1-2 months. Immatures disperse widely; in South Africa dispersal occurs at any time between Dec-Oct but mostly between Apr-Sep, and birds are occasionally killed by flying into windows and walls at night. The inexplicable increases in numbers reported from Zimbabwe in Oct-Dec (Hopkinson & Masterson 1984) may represent temporary influxes of birds displaced from other sites by reductions in habitat availability; apparent disappearances during the rains possibly reflect deterioration in habitat quality as a result of some unexplained environmental factor. A female shot at Gafersa, in the Ethiopian highlands, in July, was thought by K. M. Guichard to be a migrant (label on skin in BMNH), presumably because it was in the same marsh as the seasonally occurring White-winged Flufftail (8). In South Africa birds, presumably from a floating population which normally lives a non-territorial existence within or around established territories, occupy marginally suitable areas in seasons when conditions are unusually favourable.

HABITAT Details are from information in Taylor (1994, 1997a) unless otherwise indicated. The species inhabits many types of freshwater palustrine wetland vegetation, from seasonally wet hygrophilous grassland and sedge meadow as short as 55cm to permanently flooded reedbeds up to 3m tall. It occupies swamps, marshes, reedbeds and marshy vegetation fringing rivers, streams, lakes, ponds and drainage lines, and isolated wetland patches. It requires permanent dense cover, usually occupying areas where overall canopy cover is more than 65%, and it generally prefers moist to shallowly flooded ground, although it also occurs in deeply flooded wetlands, including papyrus swamp, as long as there is dense, matted, emergent growth or floating grass to provide a stable substrate. It also requires foraging areas of shallow water, mud, firm ground or short vegetation. Occupied vegetation is often a rich mosaic of plant species, characteristic genera including *Typha*, *Phragmites*, *Juncus*, grasses such as *Leersia*, *Andropogon*, *Arundinella*, *Aristida*, *Echinochloa*, *Agrostis*, *Oryza*, *Eragrostis*, *Paspalum*, *Pennisetum*, *Hemarthria*, *Miscanthus* and *Hyparrhenia*, sedges such as *Cyperus*, *Carex*, *Mariscus*, *Schoenoplectus*, *Eleocharis*, *Bolboschoenus*, *Scirpus*, *Fuirena*, *Pycreus*, *Isolepis*, *Fimbristylis* and *Kyllinga*, forbs such as *Ranunculus*, *Polygonum*, *Colocasia*, *Rubus* and *Epilobium*, and ferns *Cyclosorus*. In South Africa birds will also occupy areas of virtually monospecific plant growth, such as *Typha* or *Cyperus*, but do not occur in pure *Phragmites* or in pure *Carex acutiformis* on very anaerobic substrates, while in the SW Cape region they occur very locally in vegetation dominated by *Juncus kraussii* and *J. acutis* in slightly saline conditions (Taylor 1997d). In forested areas of West Africa the birds occur in dry grass, sometimes near human habitation; in South Africa they sometimes occupy grass-dominated areas which are almost completely dry throughout the non-breeding season, and in the breeding season will forage in dry *Eragrostis* hayfields and lucerne fields adjacent to marshes. Also see the Habitat section of the White-winged Flufftail species account.

The species readily colonises artificially created wetland patches, including those at seepage areas below dam walls and shallowly flooded vegetation at dam intakes and along feeder streams. It will inhabit wetland patches surrounded by cultivated fields or close to human habitation, as long as the habitat is not greatly disturbed or trampled. It is a successful colonist of very small, often isolated wetland patches, even those less than 0.5ha in extent (see Social Organisation). It generally occurs at low to medium elevations; in E South Africa it is widespread from sea level to 2,100m, above which altitude suitable cover does not grow, and it occurs up to 2,700m in Kenya and 2,600m in C Ethiopia (Taylor 1997b). It is quite tolerant of cold weather, in KwaZulu-Natal being permanently resident in upland areas where winter minimum temperatures down to -10°C occur regularly and moderate snowfalls also occur (snow does not usually lie for more than 1-2 days).

FOOD AND FEEDING Information is from Taylor (1994) and Wintle (1988). A generalist feeder, taking a wide variety of invertebrate prey: earthworms, small gastropods, crustaceans (Amphipoda), spiders, the adults and larvae of many insects, including Diptera, aquatic and terrestrial Hemiptera and Coleoptera (including Gyrinidae, Staphylinidae, Curculionidae and Carabidae), and termites and small ants (*Myrmicaria* and *Tetramorium*). Seeds, mainly of grasses (including *Paspalum*), are eaten frequently, but are not fed to chicks, and may form a large part of the diet in the non-breeding season, while some hard seeds are probably ingested to assist with grinding the food. Captive birds ate mealworm *Tenebrio* larvae, pupae and adults, chopped ox heart, cooked egg and milk, and seeds such as millet and munga (sorghum). The birds forage on dry to moist substrates, and in mud and shallow water, sometimes immersing the head in water to seize prey, and they are adept at catching pond skaters (Gerridae) in shallow water; they also probe into the ends of rotten tree branches and stumps. They dig with the bill among plant roots, and in moss and soft earth, to obtain earthworms and arthropods, and they use the bill to move aside moss, leaves and dead plant material. They frequently feed among the roots of emergent plants as these become exposed or flooded by fluctuating water levels. They search low-growing plants for lepidopterous larvae, small moths, flies, Coleoptera and Hemiptera. Occasional bouts of rapid vertical pecking into shallow water are apparently aimed at small crustaceans, mosquito larvae, chironomid larvae and small water beetles. When feeding very intensively captive birds often vibrate the tail rapidly from side to side, as they do during courtship and mating. The daily energy intake of adults is 140-150kJ (Taylor 1994). Birds forage throughout the day, most intensively from 1-3h after daybreak and with a smaller peak 2-3h before dusk; they cease foraging during heavy precipitation and in very hot weather. They drink daily, usually several times per day. Chicks are fed mainly on insects and other arthropods.

HABITS All information is taken from Taylor (1994) unless otherwise stated. The species is normally diurnal and crepuscular, birds appearing at daybreak, roosting at dusk, and being most active in the early morning and late afternoon; nocturnal activities are apparently confined to breeding-season calling by males. The birds are normally very difficult to observe, keeping to the interior of dense cover from which they are almost impossible to flush. When birds are confident the tail is held horizontally, but it is sometimes held erect, especially when the bird is wary or alarmed, and in such circumstances it is often fluffed out, when it looks rather like that of a domestic chicken, and it may also be flicked. The birds frequently climb in vegetation and can perch, crosswise and crouching, on small twigs, like a small passerine; they often climb near the top of dense vegetation to observe a human intruder, and in dense cover flushed birds are usually those which are high in the vegetation (those lower down can easily escape by running). They are said to swim well with tail erect (Sclater 1906) but this is probably misleading, as the birds do not normally swim (Keith *et al.* 1970). The birds usually roost off the ground, often in breeding or roosting nests, and both captive and wild birds are frequently seen climbing around in bushes and tall marsh vegetation at dusk (Pakenham 1943, Taylor 1994). They sunbathe frequently for short periods (see Social and Sexual Behaviour), usually crouching or lying flat on the ground with the tail spread and the wings partly to fully open and raised or drooped. Sunbathing in captive birds occurred throughout the day, and in the early morning captive birds flew to the top of bamboo clumps to bask in the first rays of the morning sun which reached the aviary. Captive birds bathed regularly, mostly in the hottest part of the day and in the late afternoon.

SOCIAL ORGANISATION All information is from Taylor (1994). Monogamous and permanently territorial wherever conditions permit, forming a strong permanent pair bond. In continuous habitat pairs may be found spaced at intervals of c. 50-100m. In KwaZulu-Natal this species occurs very widely in suitable habitat, even in patches of < 0.5ha, at densities of 2-6 pairs/ha; in a recent study, the size of 108 territories in the breeding season varied from c. 0.1-0.45ha, while 74 territories in the non-breeding season were c. 0.05-0.5ha in extent. Most non-breeding territories of <0.1ha were temporary and occupied by birds displaced by fire. Territory size varied with habitat type; the smallest permanent territories tended to be in permanently shallowly flooded reedbeds and sedgebeds (24 territories in summer 0.12-0.175ha, mean 0.14, SD 0.013; 19 territories in winter 0.12-0.14ha, mean 0.14, SD 0.015) and the largest in seasonally wet hygrophilous grassland and sedge meadow (12 territories in summer 0.34-0.45ha, mean 0.4, SD 0.033; 7 territories in winter 0.31-0.49ha, mean 0.41, SD 0.058).

Young of at least the last brood frequently remain in the parental territory during the non-breeding season, and are ejected by the start of the next breeding season. The winter territory area often appears excessive to the foraging requirements of the resident pair, because large temporary reductions in territory area are tolerated (see Movements); when the winter territory size is thus reduced, the territory's resources may not be sufficient to support immatures, which are probably ejected.

SOCIAL AND SEXUAL BEHAVIOUR All information is taken from Taylor (1994) unless otherwise stated. Red-chested Flufftails are territorial throughout the year. Both members of the pair take an active part in territory defence but males usually react more rapidly and frequently to taped playback and to models and mirrors. Both sexes threaten or attack models of their own sex, and their own reflection in mirrors, throughout the year; the intensity of their reaction is least in Mar-Jun, in which period, immediately after breeding, the highest number of territories contain resident immatures. During the breeding season adults avoid territorial Black Crakes (90), which often attack them on sight, but in the non-breeding season the Black Crakes avoid the flufftails, which then attack them: even female flufftails will attack adult Black Crakes in the winter. In captivity the female of a dominant flufftail pair may be more vocal and aggressive than the male (Wintle 1988). The adult's commonest threat display involves the bird approaching a model or mirror with a very slow walk on flexed legs, uttering gulping calls and with the body, neck and head plumage roused and the tail fanned horizontally. The bird attacks with a rush, usually from the side, making fluttering jumps, kicks, pecks at the head, and sometimes buffets with the wings. Females also threaten model males, while males attack model females in the non-breeding season but court and attempt copulation with them during the breeding season, when captive males readily rape other females if the opportunity arises. Immatures of both sexes make territorial calls while sharing the territory with their parents, and immature males occasionally display to, and attack, models. In captivity, an incubating bird will leave the nest to attack a human intruder, running around his feet with open wings and lowered head, making hissing sounds.

The paired male and female usually remain close together and contact notes are frequently uttered. In the courtship display the male stands in front of the female, bobbing up and down so that his breast almost touches the ground, with wings slightly raised and wingtips crossed over the back, head raised and body plumage fluffed out. He often utters quiet booming calls or other subdued notes and usually vibrates the tail rapidly from side to side during both courtship and mating. The female either stands still or walks around during this display and, when ready to mate, crouches with lowered head, giving quiet calls. The male mounts and copulates for 5-40s, pecking at the female's head and neck; the birds then separate and sometimes stand side-by-side briefly before resuming normal activities. Instead of this display there may be a courtship chase, in which the male, sometimes with raised wings, chases the female, sometimes pecking at her head. If she is receptive she stops and crouches, but she may escape by flying up into the surrounding vegetation. The male occasionally gives the courtship display to the female (without mating) in the non-breeding season.

Allopreening is regular, and the male of a pair occasionally courtship-feeds the female bill-to-bill, even outside the breeding season. In captivity sunbathing is often a social activity, involving both adults and young, while two or more individuals of a family may also bathe and preen together. Activity suggested to be play was also recorded in a captive family group comprising a pair and 2 juvenile males. Any bird might start this activity, which often began with a bird standing on tiptoe and flapping the wings vigorously, or making several short, rapid, apparently pointless dashes in various random directions; this bird might then crouch and dash at another flufftail or another might run at it; a vigorous chase would then ensue, the birds often running with wings flapping or raised above the back, and always with raised tails; they sometimes raced at high speed round and round an isolated clump of vegetation. The pursued bird either escaped into dense cover, or both bird stopped, the pursued bird faced the pursuer and both birds crouched, bowed to each other, and then resumed normal activities. Other family members would sometimes join in a chase. Quite frequently, 2-4 birds would suddenly dash at each other, stop when almost touching, crouch, and make a small bow, often raising the wings over the back, before resuming normal activities. These group activities occurred throughout the day but mostly in the early morning, continuously for up to 39 min and periodically for up to 1h.

BREEDING AND SURVIVAL Season Indeterminate in equatorial regions: Sierra Leone, May, Jul; Nigeria, Dec; Cameroon, Mar-Jul, breeding condition Oct, Dec; Gabon, Nov-Dec, Mar; Zaïre, Jan-Feb, May, Jul-Aug; Angola, breeding condition May; Uganda, Oct; Kenya, May; Pemba, Jan-Apr; Tanzania, Feb, Mar, breeding condition Apr, July. Normally breeds during rains in S Africa: Malawi, Feb-Apr; Zambia, Jan-Mar, breeding condition Oct; Zimbabwe, Dec-Mar (confirmed records), Aug-May in captivity; Mozambique, May, Nov; South Africa, Sep-Mar, but Aug-Jan nesting recorded in SW Cape under winter rainfall regime (Apr-Sept); in high-altitude regions of KwaZulu-Natal, birds may begin breeding 1-2 months later (Nov) than those at lower altitudes (Sep-Oct). **Nest** A cup of grass or dead plants, sometimes with slight dome; well hidden in or under a clump of grass or herbs 8-30cm above ground or water surface; often nests in damp to shallowly flooded grass (including *Oryza*) at edge of marshy areas. External diameter 13cm, internal diameter 8-9.5cm, depth of cup 3-4cm. Captive birds also occasionally used domed nests of small passerines, built on ground or in bushes. In captivity males also build one or more roosting nests for use by brood at night. **Eggs** Usually 2-3 (2-5); oval; white with slight gloss; size of 29 eggs confirmed to be of this species 24-29.4 x 17.6-21.5 (27.4 [SD 1.5] x 19.7 [SD 0.7]); calculated weight 6.1. Eggs laid at daily intervals. **Incubation** Usually 16-18 (14-18) days; normally male incubates during day, female at night. **Chicks** Precocial and nidifugous; fed and cared for by both parents; leave nest after 2-3 days and are capable climbers from an early age; normally independent after 3-4 weeks but may solicit food until 8-9 weeks old. Chicks solicit food by standing or crouching, often uttering begging calls and often weaving the head from side to side to attract parent's attention; parents feed chicks bill-to-bill or by dropping food on ground. Chicks held and brooded under wings of female, and sometimes carried thus for short distances (Steyn & Myburgh 1986). Young fully feathered on body at 23 days; remiges and rectrices fully grown at 6 weeks, when young fly. Young may help feed chicks of subsequent broods; young of last brood often remain in parental territory during non-breeding season (Taylor 1994). Clutch losses, from predation (probably mainly rats and mice) or flooding, sometimes high (Taylor 1994). Age of first breeding 1 year. Usually 1-3 (1-4) broods per season; interval between clutches 4-9 weeks; in captivity in Zimbabwe up to 6 clutches laid Aug-May (Taylor 1994). Some southern African breeding records must be treated with caution because of confusion with Streaky-breasted Flufftail: at least 12 of 13 Zimbabwe nest record cards for Red-chested are probably referable to Streaky-breasted. **Survival** Adult mortality probably low; juvenile mortality possibly low.

4 CHESTNUT-HEADED FLUFFTAIL
Sarothrura lugens Plate 2

Crex lugeus [sic] Böhm, 1884, Ugalla District, Tanzania.

Genus sometimes merged into *Coturnicops*. Two subspecies recognised.

Synonyms: *Sarothrura lynesi/modesta/lineata lynesi*; *Coturnicops/Corethrura lugens*.

Alternative names: Long-toed Flufftail; Chestnut-headed Crake.

IDENTIFICATION Length 15cm. Tail slightly longer than in Red-chested Flufftail; often fluffed out. Male is rich reddish-chestnut only on head to hindneck and malar region; chin and throat white; tail black, spotted white; upperparts and upperwings black, finely streaked white; underparts black with broad white streaks. Iris dark grey-brown; bill dusky, most of lower mandible whitish; legs and feet dark brown. Female is streaked black and pale chestnut to buff on head where male is chestnut; body dark brown-black with shorter, broader white streaks than in male; wings predominantly spotted and barred. Both sexes have narrow white line along leading edge of wing. Immature duller than adult, with fewer streaks on upperparts; male dark brown on head and neck, and with fewer streaks on underparts; female more heavily streaked on underparts. Juvenile blackish, with whitish chin, throat and centre belly. Inhabits dense, lush grass and sedges in dry to wet situations.

Similar species Male Red-chested Flufftail (3) darker chestnut on head and neck, this colour extending to chin, throat, mantle and lower breast (but beware moulting juvenile, and immature, Red-chested with incompletely developed chestnut plumage); is not white on chin and throat; rump, uppertail-coverts and tertials have white spots rather than streaks; underparts have narrower white streaking; belly has no pale area in centre; bill all-dark. Streaky-breasted Flufftail (5) has very short tail; male has chestnut extending to hindneck, sides of neck and upper breast, and is sooty-black elsewhere, entire upperparts and tail streaked white; underparts have proportionately slightly more white. Female Chestnut-headed differs from female Red-chested by blackish appearance with white, not buff, markings on body, wings and tail (but note that immature female Red-chested looks blackish-brown because of less numerous buff markings than in adult); streaks on head pale chestnut to buff; sides of head darker and more heavily patterned; sides of lower neck and upper breast have spots or streaks rather than the scallops or chevrons which are characteristic of Red-chested; underparts more heavily streaked blackish, with smaller pale area in centre of belly; bill paler on lower mandible. Female Streaky-breasted has no chestnut or buff wash on head, but sometimes has pale rufous wash on throat and breast; upperparts, upperwings and tail scalloped or barred white; underparts barred black and white, bars becoming spots on breast; for immature, see Streaky-breasted Flufftail species account.

VOICE Most information is from Urban *et al.* (1986). The male's song is a series of moaning, rather guttural hoots *whooo*, each typically lasting c. 1s and often repeated with hardly a pause for up to 1 min; the speed and pitch are variable. Notes sometimes increase in intensity towards the middle of the song, and at the end may die away into short *hoo-boo* notes. Calling is normally noted during the breeding season, but in Gabon the song is heard throughout the year, especially during the rains and notably in October (Brosset & Erard 1986). The territorial call is a series of loud, rapid, far-carrying, pumping *koh* notes, given at a rate of 3 per s for up to 45s and often in crescendo; the call often begins with a few low moaning grunts and dies away to a grunt; both sexes sometimes call together, asynchronously. A low grunt is incorporated with each *koh* note, sounding like an undertone. In response to taped playback, birds often give an irregular series of calls rather like the low *cuk* of the Streaky-breasted Flufftail (Keith *et al.* 1970).

DESCRIPTION *S. l. lugens*
Adult male Head to hindneck and malar region rich reddish-chestnut; chin and throat white. Sides of neck, upperparts from mantle to uppertail-coverts, scapulars, tertials and upperwing-coverts black, each feather with two fine white submarginal streaks; tail black with white spots. Primaries, secondaries, axillaries and underwing-coverts greyish-brown, tipped whitish; marginal coverts, and outer web of outermost primary and alula feathers, white. Underparts black with much broader white streaks, producing paler effect; centre of belly more blackish-grey, mottled white. Iris dark greyish-brown; bill dusky brown to dark slate, underside of lower mandible whitish; distal half of culmen noticeably curved; legs and feet very dark brown or dark slate.
Adult female Chestnut of head paler than in male, varying from quite deep and rich to pale buff, and heavily streaked black (more mottled on side of head); overall tone of body and wings sometimes browner than in male; streaks on mantle broader and shorter, tending to spots; markings on wings are spots and short bars rather than streaks, especially on greater upperwing-coverts and tertials. Sides of neck, foreneck and breast streaked to spotted white; flanks and belly predominantly white with blackish streaks or spots; thighs, vent and undertail-coverts greyish-black with small white spots; centre of belly predominantly white.

Immature male (One *lynesi*) Chestnut of head and neck replaced by dark blackish-brown; upperparts, upperwings and tail blackish-brown (darkest on tail), more lightly and sparsely streaked than in adult; greater upperwing-coverts with small, white, marginal spots to streaks on outer web. Chin and throat whitish, with dusky-drab mottling; rest of underparts sepia to blackish-brown, with white flecks and streaks across breast and down centre of belly, and sparse white flecks on flanks.
Immature female Very similar to adult but duller; head variably streaked pale buff to chestnut, as adult (too little material exists to establish whether any consistent differences occur); upperparts with fewer and smaller white markings which tend more to spots than streaks; greater upperwing-coverts plain; chin and throat more heavily mottled than in adult and underparts more heavily streaked, appearing predominantly dark with white markings which become spots on flanks and undertail-coverts; less white on centre of belly.
Juvenile Head, neck and upperparts black to blackish-brown. Underparts dark grey-brown to brownish-black; chin, throat and centre of belly whitish. No sexual differences noted, but little material available.
Downy young Down black; bill has pale flesh proximal half, is black distally from in front of nostril, and whitish at extreme tip; legs and feet black.

MEASUREMENTS Wing of 31 males 75-82 (78.2), of 15 females 76-82 (79.1); tail of 1 male 47; culmen to base of 27 males 13-15 (14.4), of 13 females 13-15 (14.1); tarsus of 29 males 19-22.5 (20.6), of 13 females 20-22 (20.9).

GEOGRAPHICAL VARIATION Birds from the southernmost part of the range were originally known as a distinct species, *S. lynesi*, but are now regarded as conspecific, differing only in their overall smaller measurements (Keith *et al.* 1970).
S. l. lugens (Böhm, 1884) – Cameroon (Obala, Ngaounyanga), NE Gabon, NW, NE, W and S Zaïre (Bokalakala, Bokilio, Faradje, Kasaji and Kunungu), Rwanda and W Tanzania (Ugalla). See Description.
S. l. lynesi (Grant & Mackworth-Praed, 1934) – C Angola (Bié Province) and N Zambia; see Distribution and Status for a discussion of other claimed occurrences. Smaller than nominate race in all measurements. Wing of 1 male 72, of 6 females 70-76 (72.7); tail of 1 male 46, of 4 females 43-48 (46); culmen to base of 1 male 13, of 6 females 13-14 (13.6); tarsus of 1 male 19.5, of 4 females 19-20 (19.7).

MOULT Nothing recorded.

DISTRIBUTION AND STATUS Known in scattered localities from Cameroon E to W, NW and NE Zaïre and Rwanda, and S to C Angola (Chitau), W Tanzania, S Zaïre and N Zambia. Its occurrence in Nigeria, Congo and the Central African Republic is rejected (Elgood 1982, Dowsett & Dowsett-Lemaire 1989, Dowsett & Forbes-Watson 1993). The only Southern African specimen, from Zimbabwe (Nyanga highlands), is an immature female Red-chested Flufftail (Taylor & Hustler 1993), and a recently claimed sound record from the same locality (Gibbon 1989) is not acceptable. It very possibly occurs in Malawi (Benson & Benson 1977), whence there is a sight and sound record from the Dzalanyama FR, Lilongwe District, in Jan 1988 (Gibbon 1989, 1991), but at least one of the two recorded calls (Gibbon 1989) could be from a Red-chested Flufftail. It has not been found subsequently at this locality, where

Red-chested Flufftail is common (Mallalieu 1995), and further evidence of its occurrence there is required. It may occur more widely than is known but it is generally uncommon, although it was formerly described as locally common in NE Zambia. It has been described as a relict species in unsuccessful competition with the Red-chested Flufftail (Keith *et al.* 1970), although specimen records suggest that in some regions, such as Kasaji (SW Zaïre), it may exist in similar numbers to its congener. Although it is not regarded as globally threatened, it has been listed as a candidate species for treatment as threatened (Collar & Stuart 1985). In view of its fragmented distribution, its generally uncommon status and the continual destruction of its wetland habitats, it should be regarded at least as VULNERABLE.

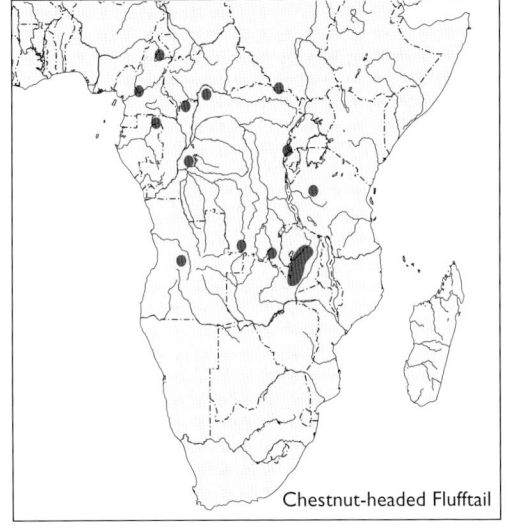

Chestnut-headed Flufftail

MOVEMENTS There is no evidence to suggest that it is anything other than entirely sedentary, with perhaps very local displacement outside the breeding season following habitat reduction by burning, etc (P. B. Taylor unpubl.). A sound record from Malawi of a bird "on migration from C Africa and found in a dry dambo" (Gibbon 1991) requires confirmation (see Distribution and Status) and, even if correct, would not constitute evidence of migration.

HABITAT Inhabits predominantly grassy situations, such as patches of savanna in lowland forest, grass-grown marshes in savanna, lakeside marshes, and rank grass and sedges in dambos (wet drainage lines in woodland). It prefers dense, lush vegetation 0.7-1.5m high, in areas where mud and water are not deep (Urban *et al.* 1986). In the forests of NE Gabon it is recorded alongside the Red-chested Flufftail in "open secondary growth", and it also inhabits moist post-cultivation growth characterised by high densities of shade plants, notably *Aframomum* and arrowroot, in low, damp regions; other flufftail species apparently do not occur in this habitat (Brosset & Erard 1986). It often occurs alongside the Red-chested Flufftail in marshes, and in such situations no obvious ecological segregation between the two species has been established. In N Zambia, both occupy the lusher centres of dambos, while the Streaky-breasted Flufftail (5) occupies shorter grass at the edges, but it is possible that Chestnut-headed Flufftails may prefer the drier regions of such wetlands,

with taller, denser, more tussocky grass (Keith *et al.* 1970, P. B. Taylor unpubl.). It appears not to occupy the very wide range of other palustrine wetland vegetation types in which Red-chested Flufftails are found.

FOOD AND FEEDING Little known. Stomach contents of adults are recorded as insects and hard seeds; those of a chick, small black ants.

HABITS Extremely secretive and difficult to observe or flush in the dense vegetation of its preferred habitats. It often does not emerge from cover in response to taped playback, but it is more easily flushed with the aid of a dog. Flights are short, and the birds tend not to flush a second time.

SOCIAL ORGANIsATION From calling patterns the species appears to be monogamous, and territorial at least during the breeding season and possibly throughout the year.

SOCIAL AND SEXUAL BEHAVIOUR No information available.

BREEDING AND SURVIVAL Season Apparently breeds during the rains in the S part of its range; laying months, derived from chicks, juveniles and collected females containing eggs: Cameroon, Apr, July, Sept; Zaïre, Mar, Apr; Zambia, Mar, Dec. **Nest** The nest and eggs are undescribed.

5 STREAKY-BREASTED FLUFFTAIL
Sarothrura boehmi Plate 2

Sarothrura böhmi Reichenow, 1900, Likulwe, Haut Luapula, SE Zaïre.

Genus sometimes merged into *Coturnicops*. Monotypic.

Synonyms: *Coturnicops boehmi*; *Sarothrura somereni*.

Alternative names: Boehm's Flufftail; Streaky-breasted Crake.

IDENTIFICATION Length 15-17cm. Tail short, not fluffy; entire leading edge of wing white. Male has head, neck, and sometimes upper breast, reddish-chestnut; chin and throat white; upperparts, upperwings and tail sooty-black, streaked white; flight feathers dark grey-brown. Underparts white, broadly streaked black except on centre of belly. Iris blackish; bill black-brown, lower mandible whitish; legs and feet grey to greenish-brown. Female has head, upperparts, upperwings and tail sooty-black, scalloped and barred white; chin, throat and underparts white, barred dull sooty-black (breast spotted); centre of belly white; throat to breast sometimes washed rufous. Immature duller than adult, with fewer, smaller markings on upperparts; male has rump to tail unpatterned sooty-black, thighs and belly almost plain white; female has undertail-coverts brown-black. Juvenile dull sooty-black, underparts paler; chin, throat and centre of belly white; male has faint white markings on upperwing-coverts. Inhabits dry to seasonally wet grassland.

Similar species Similar to sympatric Red-chested Flufftail (3) and Chestnut-headed Flufftail (4) but has much shorter tail, shorter toes and deeper bill. In male, chestnut paler than in Red-chested and extends only to hindneck and upper breast; entire upperparts and tail streaked white (streaks longer than in Red-chested) with no spotting; chin and throat white; more white on lower breast and belly. Female appears blacker overall than adult female Red-chested, with white markings (often difficult to see in flight); tail very short. Male Chestnut-headed Flufftail is chestnut only on head and hindneck; white on throat more extensive; tail long, fluffy, spotted white. Female Chestnut-headed has brownish tinge to head, face and nape; upperparts spotted, rather than scalloped, white; underparts predominantly streaked and spotted black.

VOICE The male's song is a deep, hollow, hooting *hooo*, repeated every 2s. Males also give a short, higher-pitched pumping note rendered *er* or *oe*, repeated every 0.5-0.7s, each note preceded by a low grunt to give a *g'wer* sound; this series sometimes trails off into softer agitated *wu* notes. Males call at any time of the day or night, sometimes for long periods, and may continue calling even when incubation is in progress (Wintle 1988); in Gabon, where the species is resident, the song is heard throughout the year (Brosset & Erard 1986). A call denoting agitation or annoyance is a repeated low *cuk*. Females are almost silent, and make no territorial calls, but both sexes give a snake-like hiss or make wheezy noises when alarmed and when protecting chicks (Wintle 1988). Chicks call *peep*.

DESCRIPTION

Adult male Head, lower throat and neck reddish-chestnut, slightly darker on crown and hindneck; chin and upper throat white; amount of chestnut on underside variable, sometimes extending irregularly to upper breast. Upperparts, including scapulars, upperwing-coverts, tertials and tail, sooty-black, each feather with two fine white submarginal streaks. Primaries and secondaries dark grey-brown; axillaries and underwing-coverts similar, or blackish-grey, coverts with whitish streak on outer webs of lessers and some medians; marginal coverts, leading edge of alula and outer web of outermost primary (P10) white. Underparts white, with broad blackish streaks except on centre of belly. Iris brown to black; upper mandible blackish-brown, lower mandible whitish to pinkish; legs and feet slate-grey, grey-brown or greenish-brown.
Adult female Entire head and upperparts, including scapulars, upperwing-coverts, tertials and tail, dull sooty-black, scalloped and barred white; barring most pronounced on upperwing-coverts and tertials. Rest of wing as male, but white outer web of P10 tinged dusky. Chin, throat and underparts white with dull sooty-black bars which become spots on breast and lower throat; centre of belly white. Some birds have faint rufous wash from throat to breast.
Immature male Differs from adult in having dark tips to duller chestnut feathers of head and neck; streaks on upperparts finer and duller; rump and uppertail-coverts almost unstreaked; tail plain sooty-black; unstreaked juvenile greater upperwing-coverts contrast with streaked medians and lessers; thighs and belly almost unstreaked white.
Immature female Differs from adult in having mantle and scapulars patterned with small, faint streaks and spots; faint spots on tertials and lesser and median upperwing-coverts; greater coverts unpatterned; chin and upper throat off-white; breast and flanks predominantly brown-black with whitish bars; undertail-coverts dusky brown-black.
Juvenile Both sexes sooty-black, with greyish-white chin, throat, breast and centre of belly. Male has fine white lacing on upperwing-coverts (always lacking in female). No further details available.
Downy young Entirely black; upper mandible black, pale at extreme tip and with white egg-tooth (lost after 3 days), lower mandible pale horn, with dark spot towards tip. Small white wing-claw present.

MEASUREMENTS Wing of 33 males 81-91 (85.5, SD 2.9), of 21 females 81-88 (84.8, SD 1.7); tail of 19 males 27-35 (30.4, SD 2.4), of 12 females 27-33 (31.1, SD 1.9); culmen to base of 33 males 12-15.5 (13.5, SD 0.6), of 19 females 13-14.5 (13.6, SD 0.4); tarsus of 22 males 19.5-22 (20.8, SD 0.7), of 13 females 19.5-23 (20.9, SD 1.0). Weight of 4 males 33.5-42 (37.3, SD 3.8), of 2 females 30.5, 35; of 1 juvenile male (59 days old) 26.5, of 1 juvenile female 21.

GEOGRAPHICAL VARIATION Apparently none. The proposed race *danei*, known only from a male taken at sea off the Guinea coast, was described as having the chestnut of the neck extending to the breast, but this character is variable and is also shown by birds from C and S Africa (Keith *et al.* 1970).

MOULT A complete moult occurs after breeding; remiges are moulted sequentially, possibly with a tendency to partial synchrony; remex moult in captive birds (Zimbabwe) occurred Aug-Sep. Migrants in S Tanzania (May) were apparently in suspended remex moult, having a mixture of old and new remiges (Baker *et al.* 1984); this suggests that moult may begin before adults depart from the breeding grounds. Two birds from Zambia, Feb-Mar, were in breeding condition and had body moult. Postjuvenile partial moult begins from the age of 5 weeks. Immature prebreeding moult is partial; in captive birds it occurs 6-7 months after the end of the postjuvenile moult (Wintle 1988).

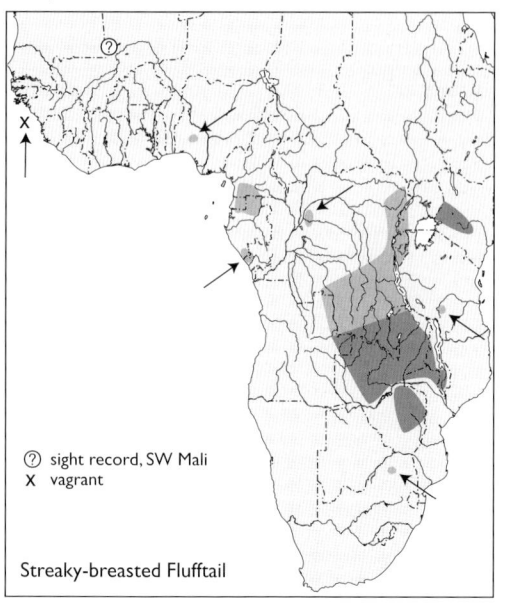

Streaky-breasted Flufftail
? sight record, SW Mali
X vagrant

DISTRIBUTION AND STATUS The distribution of this species is imperfectly known. It is recorded from scattered localities in Nigeria (Ife), S Cameroon (Lolodorf and Bitye), Rwanda, Burundi, W Uganda (Rwenzori NP), W and SC Kenya (Kitale, Kisumu, Mumias District, Nairobi and Machakos), and through Central Africa in CW, SC,

NE, CE and SE Zaïre, Gabon, coastal Congo (Tchissanga and Mpindé), E Angola, S Tanzania (Kilima, Mufindi District), Zambia, Malawi (Kasungu, Mzimba, Lilongwe, Mangochi, Dedza and Blantyre) and Zimbabwe (mainly the central plateau). In years of unusually high rainfall its breeding range may extend to N South Africa, where it was recorded from the Nyl River floodplain (N Transvaal) in 1988, 1991 and 1996, and it may occur on the KwaZulu-Natal coastal plain; in optimum conditions its South African population may be up to 130 birds (Taylor 1997a). There is also an unsubstantiated sight record from Kangaba, SW Mali, Nov (Lamarche 1980). This species is not regarded as globally threatened but its status is difficult to assess because, like other flufftails, it is frequently overlooked. It is often regarded as local and uncommon but may be locally common to numerous in seasons of good rainfall; for example, 100+ calling males were heard on a grass plain in Zambia in Jan 1978 (P. B. Taylor in Urban *et al.* 1986). It is locally very common in NE Gabon (Makokou), where it occurs in similar numbers to the Red-chested Flufftail (Brosset & Erard 1986). Heavy grazing and trampling by cattle may cause premature departure from otherwise suitable breeding habitat, and its numbers in S and E Africa have been adversely affected by habitat loss resulting from overgrazing and the damming, draining and cultivation of seasonal wetlands (e.g. Mallalieu 1995). Predation by a falconry-trained Black Sparrowhawk *Accipiter melanoleucus* is recorded in Gabon (C. Erard *in litt.*) and birds may be taken by other raptors which hunt over short grassland.

MOVEMENTS Imperfectly known; the breeding range shown on the map is hypothetical. Birds in NE Gabon are apparently sedentary, although breeding there is not yet recorded (Brosset & Erard 1986). In E and S Africa it is a migrant, breeding mainly in the S tropics and retreating towards equatorial regions in the dry season when much of its habitat dries out and is grazed and burnt. The timing and period of its occurrences depend on the extent and duration of the rains, and birds may be present for only a very short time in ephemeral habitat. It is definitely recorded only once from Nigeria in Mar, and from coastal Congo in Dec-Jan, and possibly from Mali in Nov (early in the rains) (Farmer 1979, Lamarche 1980, Dowsett-Lemaire *et al.* 1993). Elsewhere it is recorded in Kenya from May-Sep, Zambia from Dec-Apr, Malawi from Jan-Mar, Zimbabwe from late Nov-Apr, and South Africa in Jan 1988, Apr 1991 and Feb-Mar 1996 (Benson & Benson 1977, Taylor 1979, Britton 1980, Irwin 1981, Taylor 1997a, c). It migrates at night. Migrants were attracted to lights at night at Mufindi, S Tanzania, in May-Jun: all were exhausted and were probably moving N after breeding (Baker *et al.* 1984). A presumed off-course migrant was taken at sea in Jun, 150km off the Guinea coast (Urban *et al.* 1986). Birds may migrate when only 5-6 weeks old: 1 in full juvenile plumage flew into a building at night in the Aberdare Mts, Kenya, in Jul (Anon 1992a). Arrivals of birds in breeding habitat at Ndola, Zambia, were frequently noted within 12-24h of heavy rain or storms, sometimes with other wet grassland species such as Spotted Crake (96) and Great Snipe *Gallinago media* (Taylor 1987; P. B. Taylor unpubl.), indicating that the birds follow rain fronts.

HABITAT The breeding habitat in S and E Africa is short grass, temporarily inundated during the rains, such as that at the edges of rivers, drainage lines (dambos) and marshes, and also grass flats and pans. In Zambia it occurs in seasonally moist to flooded grass 30-70cm (often less than 50cm) tall, on ground flooded to a maximum depth of c. 10cm (Taylor 1987). Similar habitat is occupied in Malawi and also in Zimbabwe, where the birds are found in areas characterised by grasses such as *Setaria anceps*, *Sporobolus pyramidalis*, *Eulalia geniculata*, *Eragrostis* sp. and *Bothriochloa insculpta* on ground with numerous puddles of shallow water (Hopkinson & Masterson 1984, Mallalieu 1995). In these regions it often occurs alongside the Striped Crake (109) but it prefers shorter, less dense and often drier vegetation than does the Red-chested Flufftail; in wetlands occupied by both flufftails, Streaky-breasted inhabits the shorter grassland at the edges (Taylor 1997a). However in W Africa it occurs alongside its congener in dry grassland and rank grass, in NE Gabon frequenting areas with large tussocks of tall grass scattered in a continuous cover of shorter grass (Brosset & Erard 1986). It may tolerate light grazing by domestic stock but does not occur in heavily grazed areas (Mallalieu 1995). It generally occurs from 500 to 2,000m; in Kenya it is recorded only above 1,000m.

FOOD AND FEEDING Poorly known. Stomach contents are small seeds, especially grass seeds, and small insects. In captivity it apparently eats much less seed than does the Red-chested Flufftail, and chicks thrive only if fed many small, soft-bodied insects (Wintle 1988).

HABITS Diurnal. The flight is strong for a flufftail, but birds are typically difficult to flush, preferring to escape by running; however, they are more easily flushed by a dog than are other flufftails, presumably because of the relatively short, sparse nature of the preferred vegetation cover. Birds may occasionally be caught by hand, or by a dog, and in Gabon were easily caught by a trained Black Sparrowhawk in airfield grassland (C. Erard *in litt.*). Although able to run rapidly, they normally move slowly in a crouched manner, rather like a buttonquail *Turnix*, and the short tail is normally held depressed, and is raised only when resting or hiding (Wintle 1988). In captivity bathing is frequent but, although the birds can swim, they are reluctant to do so; unlike Red-chested Flufftails, captive birds do not perch in bushes (Wintle 1988).

SOCIAL ORGANISATION Monogamous, and territorial when breeding; the calling pattern in Gabon suggests that resident birds may be territorial throughout the year (Brosset & Erard 1986). In Zimbabwe and Zambia, when the birds are locally abundant during seasons of good rains, calling males may occur as little as 30-50m apart in suitable habitat (Hopkinson & Masterson 1984, P. B. Taylor unpubl.).

SOCIAL AND SEXUAL BEHAVIOUR All observations refer to captive birds (Wintle 1988). Males have a primitive display in which two birds walk side by side with drooping wings, the neck extended forward and the head plumage roused to give the head an almost triangular shape. This display is given when a female is very close. Males also chase each other frequently but, unlike captive Red-chested Flufftails, seldom fight; females are more aggressive to each other than are males. Paired males will rape other females when the opportunity arises. In captivity, Streaky-breasted and Red-chested Flufftails live amicably together, showing no social or territorial interactions.

BREEDING AND SURVIVAL Season Breeds during the rains: S Zaïre, Jan-Mar; Kenya, probably May-Jul; Zambia, Jan-Feb; Malawi, Jan-Mar; Zimbabwe, Nov-Mar (peak Jan);

South Africa, calling recorded Jan-Feb and Apr. In captivity, male selects nest site and builds nest. **Nest** Often in relatively short or sparse grass; a simple pad or shallow bowl of green and/or dry grass placed in a grass or sedge tuft, with growing blades pulled over it in a canopy; internal diameter c. 9-11.5cm; usually placed 2.5-7.5cm above wet to dry (not flooded) ground; often built in crown of burned tuft of perennial grass such as *Sporobolus pyramidalis*, *Setaria sphacelata* or *Aristida* sp., when new blades round perimeter are 20-35cm tall; in captivity, nests freely in *Eragrostis*. Material is added to nest during incubation, resulting in a more substantial structure up to 2.5cm thick. May nest near other rail species, e.g. 1 nest less than 20m from nest of African Crake (68), and nests close to Red-chested Flufftail in aviaries. **Eggs** Usually 3-5 (2-5); ovate; white to creamy-white; slightly glossy; sometimes with sparse small red-brown spots, often concentrated at blunt end; size of 87 eggs (confirmed to be of this species) 24.9-29 x 18.4-20.8 (27 [SD 1.0] x 19.5 [SD 0.6]). Eggs laid at intervals of 1-2 days. **Incubation** 14-18 days; in captivity, male sits during day, female at night. **Chicks** Information from captive birds (Wintle 1988). Chicks precocial; fed by both parents; leave nest after 1-3 days; cheep for food, move head from side to side and point bill slightly upwards towards parent prior to food being offered; begin to feed themselves at 4-5 days; feathers begin to grow at 7 days; young fully grown and able to fly at 5 weeks. Chicks may be rejected at an early age, to allow rapid renesting. Some evidence of desertion and loss of clutches following disturbance by people (Urban *et al.* 1986). Clutch losses from predation often high; eggs possibly often eaten by rats (*Otomys* spp.) and it may be advantageous for birds to nest early in rains before rat population has recovered from reduction by burning in previous dry season (Urban *et al.* 1986). Some breeding records must be treated with caution because of confusion with Red-chested Flufftail, while 12 of 13 Zimbabwe nest record cards for the latter species probably refer to Streaky-breasted Flufftail.

6 STRIPED FLUFFTAIL
Sarothrura affinis   Plate 1

Crex affinis A. Smith, 1828, Cape Province, South Africa.

Genus sometimes merged into *Coturnicops*. Forms a superspecies with *S. insularis*. Two subspecies recognised.

Synonyms: *Sarothrura lineata/antonii*; *Coturnicops affinis*; *Alecthelia lineata*; *Gallinula/Crex/Ortygometra/Corethrura jardinii*; *Corethrura ruficollis*.

Alternative names: Red-tailed Flufftail; Chestnut-tailed Flufftail/Crake.

IDENTIFICATION Length 14-15cm; wingspan 23-24cm. Small flufftail; tail short and fluffy. Male has head to hindneck dull orange-chestnut, this colour extending to upper breast except in South African nominate race; chin and throat whitish; tail dull chestnut; body and upperwings blackish with white streaks, heavy on underparts and washed yellowish on upperparts. Iris blackish; bill dark brown, base of lower mandible pale; legs and feet flesh, tinged grey or brown. Female has upperparts and upperwings blackish, scalloped and barred whitish to buff; tail chestnut, barred black; sides of head and neck buff to rufous, speckled blackish; chin and throat white; underparts whitish to buff with blackish spots and scales, becoming bars on flanks; undertail-coverts rusty, barred blackish. Immature duller than adult; upperparts streaks of male washed buff in nominate race; in race *antonii* male has white streaks on mantle and almost no chestnut on upper breast, while female has chestnut wash to upperparts streaks, fewer spots on belly, and streaks, rather than spots, on breast. Juvenile dull blackish, paler on underparts. Inhabits grassland; also fynbos in South Africa.

Similar species Normally the only flufftail in its habitat. Male distinguished from all sympatric wetland species by chestnut tail and yellowish-white streaking on upperparts; female by chestnut and black bars on tail and rufous wash on face – see White-winged Flufftail (8). Very similar to allopatric Madagascar Flufftail (7), male of which is chestnut from chin to breast and has yellower streaks on back, while female has blackish upperparts closely streaked and barred rufous, and ochraceous face and underparts with blackish spots on breast and bars on rest of undersides.

VOICE The vocal repertoire is extensive but imperfectly known; all information is taken from Taylor (1994) except where otherwise stated. The male's song is a hoot, lower-pitched than that of the Red-chested Flufftail (3), lasting c. 1s and given at intervals of 1.5-2s for up to 10 min; it is audible for up to 2km in still weather conditions. Birds usually start calling after darkness has fallen, and often continue intermittently until just before daybreak, especially on clear, still, moonlit nights; the song is also given during the day, most often in the early morning, at dusk and in cloudy weather. Singing decreases markedly when breeding commences. Recent studies have indicated that the advertising calls of individuals may vary little within a season, and that it is possible to distinguish between individual calling birds. The male's territorial call is a series of loud, sharp, tinny, rapidly repeated *ki* notes, usually followed immediately by a similarly loud and rapid series of lower-pitched, more strident *ker* notes; either series may be repeated or given without the other. Slower, sometimes quieter, versions of this call, and a rapid stuttering *k-k-k*, are given when the bird attacks its image in a mirror. A rapid rattling call, and very rapid, low-pitched short hoots, are given by males in response to taped playback during the breeding season. The female has a quiet, more rapid version of the *ki* and *ker* territorial call. Both sexes have a gruff bark of alarm, and various gulps, rattles, grunts, short hoots, and ticking, buzzing and bubbling notes. An incubating male gave snake-like hisses to human intruders, as well as a mournful repeated *peep* (Porter 1970). Adults give a quiet *grk* to chicks, which have plaintive cheeping notes and a quiet *ip*.

DESCRIPTION *S. a. affinis*
Adult male Top and sides of head, and nape and hindneck, dull orange-chestnut to chestnut; rest of upperparts to rump, including scapulars, upperwing-coverts and tertials, dark blackish-brown, with yellowish-tinged white streaks; uppertail-coverts and tail dull chestnut. Alula, primaries and secondaries dark grey-brown; outer web of outer primary (P10) white; alula feathers streaked white on outer webs; axillaries and underwing-coverts grey-brown with some whitish markings. Chin and throat buffy to white; rest of underparts blackish-brown, streaked white; undertail-coverts dull chestnut. Iris hazel or dark brown to blackish; bill dark brown to greyish-black, lower

mandible flesh to whitish either at base or more extensively; legs and feet flesh-brown or greyish-flesh, occasionally tinged whitish.

Adult female Upperparts from forehead to uppertail-coverts, including scapulars, tertials and upperwing-coverts, dark blackish-brown, scalloped and barred whitish to rich buff; primaries and secondaries grey-brown, with small buff spots on outer webs; tail chestnut with black bars. Sides of head and neck buff, finely speckled black and often with markedly rufous or light chestnut tinge; chin and throat white; underparts whitish to buff with dark blackish-brown spots and scales, becoming bars on flanks; undertail-coverts rusty with dark blackish-brown bars.

Immature Similar to adult but retains some juvenile (unpatterned) greater upperwing-coverts until at least 1 year old. Male has buff tinge to streaks of upperparts; rectrices dusky chestnut. Bill grey-black, lower mandible whitish. Female not described (see *S. a. antonii*).

Juvenile Predominantly plain; dull grey-black on upperparts, grey-brown on underparts. Bill blackish, with pale lower mandible.

Downy young Down black; bare parts black; egg-tooth whitish.

MEASUREMENTS Wing of 13 males 69-77 (73.6, SD 2.8), of 9 females 65-75 (70.8, SD 3.1); tail of 12 males 35-46 (38.4, SD 3.8), of 8 females 35-42 (37.9, SD 2.5); culmen to base of 14 males 13-14 (13.6, SD 0.4), of 9 females 12.5-13.5 (13.1, SD 0.3); tarsus of 15 males 16-19 (17.5, SD 0.8), of 9 females 16.5-18 (17.3, SD 0.6). Weight of 4 males 25-30 (27.8, SD 1.9).

GEOGRAPHICAL VARIATION Males from the South African nominate race differ in plumage to populations further N, all of which are normally included in race *antonii*. Populations from Kenya and S Sudan may be different races but insufficient material is available to substantiate this; they are separated here as "subsp. A" and "subsp. B", following Keith *et al.* (1970), to highlight the differences apparent in existing material. An appreciable degree of subspeciation is to be expected in the isolated populations of this primarily montane species.

S. a. affinis (Smith, 1828) – S and E South Africa, from Cape Peninsula to highlands of KwaZulu-Natal and E Transvaal, and including Lesotho and Swaziland. See Description.

S. a. antonii (Madarasz & Neumann, 1911) – Highlands of E Zimbabwe (Nyanga Highlands and Chimanimani Mts), presumably SW Mozambique (Chimanimani Mts), Malawi (Mulanje, Zomba, Malosa, Viphya and Nyika Plateaux), extreme NE Zambia (Nyika Plateau), S Tanzania (Matengo Highlands) and SW Kenya near Tanzania border (Ndassekera and Nguruman Hills). Larger than nominate; chestnut of male extends to upper breast. Immature male has white streaks on mantle, almost no chestnut on upper breast, and duller chestnut tail with dark feather tips. Immature female has predominantly streaked upperparts, including scapulars and upperwing-coverts; chestnut wash on streaks from mantle to uppertail-coverts; fewer spots on belly; breast streaked rather than spotted. Wing of 7 males 76-85 (80.6), of 3 females 82, 82, 83; tail of 3 males 35-39 (36.6), of 3 females 35, 38, 38; culmen from base of 7 males 12-13 (12.6), of 4 females 12-14 (13); tarsus of 7 males 17.5-19 (18.1), of 3 females 18, 18.5, 19.5. Weight of 1 male c. 30.

Subsp. A – Highlands of Kenya E of Rift Valley (Mt Kenya and Aberdare Mts). Size similar to nominate race. Similar in plumage to *antonii*, but male possibly has chestnut on underside less extensive, and has darker tail than in other races. Wing of 4 males 72-76 (73.8), of 4 females 73-76 (75); tail of 3 males 36, 37, 38, of 3 females 36, 38, 40; culmen to base of 4 males 12.5-13.5 (13), of 4 females 12-14 (13); tarsus of 4 males 15.5-16.5 (16), of 4 females 16-16.5 (16.4).

Subsp. B – Highlands of Kenya W of Rift Valley (Trans Nzoia District), and SE Sudan (Imatong Mts). Similar to subsp. A but larger. Wing of 2 males 82, 84, of 3 females 80, 81, 85; tail of 1 male 34, of 1 female 37; culmen to base of 3 males 13.5, 13.5, 14, of 3 females 13; tarsus of 3 males 18.5, 19, 19.5, of 3 females 18, 19.5, 20.

MOULT Moult of secondaries is ascendant, with a tendency to partial synchrony; it occurs after primary moult and is recorded from South Africa in Nov (S Cape) and Dec (KwaZulu-Natal). Moult of tail and body feathers occurs at the same time; the former can involve the simultaneous replacement of at least half the rectrices.

DISTRIBUTION AND STATUS Under existing climatic conditions its distribution is essentially discontinuous and relict, and it occurs mainly in highland regions from S Sudan through E Africa to the Cape Peninsula (Keith *et al.* 1970, Taylor 1994). Although it was formerly not regarded as globally threatened it was recognised to be uncommon in some parts of its range, and was classed as RARE in the South African Red Data Book (Brooke 1984). This species (subsp. B) has not been recorded in Sudan for over 50 years, and it is now rarely recorded in Kenya, where subsp. A was last recorded in 1974 and subsp. B in 1969, although 70 years ago it was considered locally common (Stoneham 1928, Cave & Macdonald 1955, Britton 1980, Nikolaus 1987, Lewis & Pomeroy 1989, Taylor 1994). In Zambia the race *antonii* may be locally common in its very restricted range, while in Tanzania and Zimbabwe it may occur more widely than is known; it has not been recorded in SW Kenya since 1977 (Benson *et al.* 1971, Fayad & Fayad 1980, Irwin 1981, Taylor 1994). In South Africa the nominate race is less widespread, and its distribution more fragmented, than earlier in the century, and its population at known and predicted sites is estimated at c. 1,730 birds (Taylor 1997a). Its grassland habitats have disappeared at an alarming rate and are now severely restricted in extent; very few birds are found in farming areas, especially where high grazing pressure renders vegetation too short and sparse for occupation, burning is frequent or uncontrolled, and disturbance is too intense (Taylor 1994). However it is still locally common at some protected or well-managed grassland and fynbos sites and it may occur in the eastern Free State, whence it has not previously been recorded (Taylor 1997a). The future of the species is certainly not secure: in view of continuing habitat loss throughout its range the nominate race should be regarded as VULNERABLE, while *antonii* should be regarded as a DATA DEFICIENT taxon and a candidate for treatment as threatened (Taylor 1997a). In the forthcoming book on Important Bird Areas in Southern Africa (Barnes in prep.) the species is classed as VULNERABLE. Predation by Lanner Falcon *Falco biarmicus* and domestic cat is recorded, while the Black Harrier *Circus maurus* and the Slender Mongoose *Galerella sanguinea* are possible predators (Taylor 1994).

1. affinis
2. antonii
3. subspecies A
4. subspecies B

Striped Flufftail

ries included a drainage line. It occurs up to 3,700m in Kenya, 2,500m in Zimbabwe and 2,100m in South Africa.

FOOD AND FEEDING Information is from Taylor (1994). Takes many adult and larval insects, including cockroaches (Blattidae), beetles (including Carabidae and Curculionidae), Hemiptera (including Anthocoridae), Lepidoptera, grasshoppers (Acrididae), crickets (Grillidae) and Diptera, and also ants (including *Tetramorium*) and termites. It also eats earthworms, spiders (including Labidognatha and Salticidae), grass and sedge seeds and some vegetable matter. The birds forage throughout the day, searching clear ground, very short vegetation, low-growing plants and the bases of grass tussocks, and moving aside dead plant material; they catch small ants around nests, and sometimes forage on wet substrates and in shallow water.

HABITS This species is normally very difficult to locate and observe, keeping to the interior of dense vegetation, but may occasionally be seen when it forages in open areas at the edge of, or within, extensive dense cover. Although usually only detected by their distinctive song and territorial call, these birds will sometimes flush when disturbed, when they fly feebly with a fluttering action for a very short distance before dropping into cover and running; they are very difficult to flush a second time. They sometimes freeze when pursued and occasionally may be caught by hand. The species is apparently diurnal, there being no evidence that any activity occurs at night other than the prolonged singing by territorial males. Territorial birds may sometimes be called up at night by taped playback and viewed by torchlight.

SOCIAL ORGANISATION Monogamous; territorial during the breeding season and, when sedentary, throughout the year, probably with a permanent pair bond. The size of 20 breeding territories in *Themeda*-dominated grassland, KwaZulu-Natal, was 1.05-2.3 (1.64, SD 0.36) ha; of 20 home ranges 2.0-3.24 (2.58, SD 0.38) ha; of 1 non-breeding territory 1.5ha (Taylor 1994).

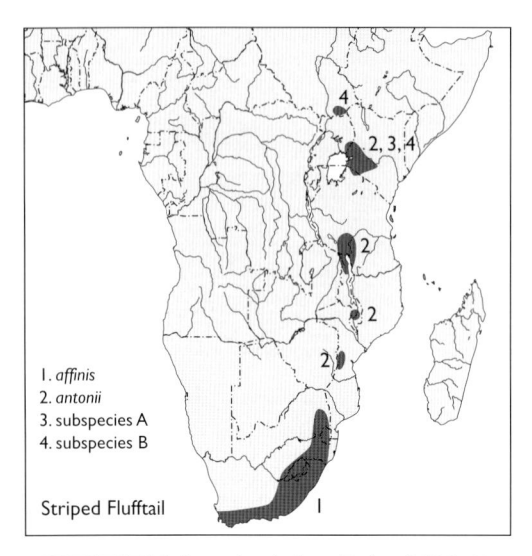

MOVEMENTS Information is from Taylor (1994). Normally regarded as sedentary, with local movements when grassland habitat becomes too dry or is burnt. At Nyanga, Zimbabwe, a bird which flew into a lighted window at night in Nov, at the beginning of the rainy season, was probably undertaking some movement, as were single males which flew into house walls at Stutterheim and Skuitbaai, Cape Province, South Africa, in Jul and Oct. A recent study has shown that in KwaZulu-Natal this species is sedentary at lower altitudes, where cover and food remain at suitable levels all year, but in upland areas (over c. 1,400m a.s.l.) it departs in Apr-Jun and returns in Oct-Jan, departures being associated with a decrease in the availability of invertebrate food and arrivals with increasing invertebrate numbers and the development of vegetation cover. Birds did not reoccupy burned grassland where suitable cover developed later than Jan, after which sufficient time for breeding would not be available. In the study area movements were probably altitudinal, possibly over as little as 35-40km, to altitudes below 1,000-1,200m.

HABITAT Summarised from data in Taylor (1994). Over most of its range the Striped Flufftail typically inhabits dry upland grassland, including that with bracken and/or brambles (such habitat is sometimes described as 'bracken-briar'), or woody vegetation such as *Protea*, *Leucosidea* and *Buddleja*, or near forest edges; it also occurs in crops such as lucerne and millet. In South Africa, where it descends almost to sea level, it is found mainly in sour grassland dominated by *Themeda triandra* with other grasses such as *Hyparrhenia*, *Festuca*, *Tristachya* and *Cymbopogon* species occurring locally. It also occurs in *Psoralea/Osmitopsis* mesic montane fynbos in the extreme SW Cape. Although it is frequently associated with small streams, and with marshy patches in grassland, there is no convincing evidence that it occupies predominantly wetland habitats alongside the Red-chested Flufftail (3) except under severe drought conditions. It requires dense cover with clear ground below for foraging; in a KwaZulu-Natal study it was found in grass 35-100cm tall, did not occur in grassland where mean ground cover was less than 80% and mean vegetation height was less than 35cm, avoided rocky areas and steep or convex slopes, and mainly occupied ground below 1,600 m; most breeding territo-

Figure 3. Striped Flufftail male attacking mirror reflection. [After photograph by B. Taylor].

SOCIAL AND SEXUAL BEHAVIOUR All information is from Taylor (1994). Territorial males attack their image in a mirror, emerging from cover with the plumage roused and the tail raised and fanned out into a prominent 'powder-puff', approaching with a crouching run, drooping and spreading the wings and making jumping attacks at the mirror, striking with the bill and buffeting with the wings. After an attack the male often crouches, faces the mirror with partly open wings (Fig. 3), and then turns to face away from the mirror, thus displaying the raised and fluffed tail, before walking away. In courtship display to a model female the male stands with raised head and drooping wings, then bows until his breast touches the ground. During this display the plumage is roused and the tail is erected and spread.

BREEDING AND SURVIVAL Season Normally breeds during the rains. Sudan, breeding condition May; Kenya, May; Tanzania, Jan; Zambia, breeding condition Jan; Malawi, Apr; Zimbabwe, Jan; South Africa, Dec-Mar but SW Cape Sep (end of winter rains). Of 8 South African clutches in collections, 3 are referable to Buff-spotted Flufftail (2) and 1 to Red-chested Flufftail, while 1 record attributed to Buff-spotted Flufftail is referable to Striped Flufftail (Taylor 1994). **Nest** A bowl of dry grass or rootlets built into a grass or sedge tuft; diameter 9cm; depth of cup 2.5cm; growing grass may be woven into a loose canopy over nest. **Eggs** 4-5 (usually 4); ovate; smooth, sometimes glossy; white or off-white; size 25-27.5 x 18-20.5 (26.3 [SD 0.9] x 19.6 [SD 0.9], n = 8, South Africa), 23.8-25.5 x 19.5-20 (n = 5, Zimbabwe). **Incubation** Probably 15 days, by both sexes; male incubates during day, female in early morning (and thus presumably during night). **Chicks** All hatch within 24 h; precocial and nidifugous; may leave nest within a day of hatching; cared for by both parents. Eggshells eaten by female.

7 MADAGASCAR FLUFFTAIL
Sarothrura insularis Plate 3

Corethrura insularis Sharpe, 1870, Nossi Vola, Madagascar.

Genus sometimes merged into *Coturnicops*. Forms a superspecies with *S. affinis*. Monotypic.

Synonyms: *Coturnicops/Ortygometra insularis*.

Alternative name: Madagascar Crake.

IDENTIFICATION Length 14cm. Small flufftail with short, fluffy tail. Male distinctive, with rich chestnut head, neck, breast and tail; blackish body with yellowish streaks on upperparts and white streaks on wings and underparts. Iris dark brown; bill blackish; legs and feet yellow-brown to pinkish-brown. Female also distinctive, with blackish upperparts closely streaked and barred pale rufous; ochraceous face and underparts with blackish spots on breast and bars on rest of undersides; tail chestnut with blackish bars. Immature male probably resembles adult; immature female like adult but duller, with underparts spots extending to flanks and belly; juvenile male probably almost entirely dark sooty-brown. Inhabits grassland, secondary bush, forest clearings, cultivation and marshes.
Similar species Easily separated from the sympatric Slender-billed Flufftail (9) by the latter's relatively plain plumage: male darker chestnut on head, neck and breast, this colour extending to mantle and anterior flanks; upperparts dark brown; posterior underparts ashy-brown; undertail-coverts and tail barred blackish and chestnut. Female Slender-billed predominantly plain dark brown, white from chin to centre of breast; uppertail-coverts, undertail-coverts and tail barred chestnut, black and white. Very similar to allopatric Striped Flufftail (6), male of which has less extensive chestnut on underparts, and less yellowish streaks on upperparts, while female has a paler, less rufous, pattern of scallops and bars on upperparts, neck and breast, contrastingly rufous markings on head, and spotted underparts with bars only on flanks.

VOICE Information is from Keith (1973) and Wilmé & Langrand (1990) unless otherwise stated. The common call, probably given by both sexes, is a characteristic, loud, high-pitched, ringing series of rapidly repeated single *kee* and double *keekee* notes, given in various combinations but often with several monosyllabic notes at the beginning and end of the series; it normally lasts c. 20s. The volume diminishes towards end of the call, which may be preceded by short trills *drr* and terminated by repeated *kik* and/or rapidly repeated *ki* notes. This call is slightly reminiscent of the territorial call of the Striped Flufftail, although the notes are quite different in quality. On tape recordings of this call (made by C. Carter, R. Stjernstedt and A. Greensmith), notes are repeated at a rate of 3.5-4.5/s; one sequence starts with loud, ringing, rapidly repeated *kli* notes, very similar to an alarm call of the White-spotted Flufftail (2) (P. B. Taylor unpubl.), which slow down, fall in pitch and become typical *kee* notes. Other calls include short versions of, or excerpts from, the full call. The birds call throughout the day and through the entire year.

DESCRIPTION
Adult male Head to hindneck, chin to foreneck, and breast, rich chestnut. Mantle to rump, and scapulars, dark blackish-brown with yellowish streaks; upperwing-coverts dark blackish-brown, streaked white, greaters, and sometimes medians, with more yellowish streaks, and secondary coverts sometimes more spotted than streaked. Remiges dark, dull brown; outer primaries often paler and variably streaked or spotted yellowish on margin of outer web; secondaries and tertials darker, irregularly streaked and spotted yellowish to rufous-white (mostly on outer web); alula dark, dull brown with white spots; axillaries and underwing-coverts grey-brown, spotted whitish. Flanks and belly dark blackish-brown, streaked white; lower belly and vent often more brownish, streaked and spotted white; tail and undertail-coverts rich chestnut. Iris dark brown to almost black; bill black or grey-black, often with small pale patch at tip; legs and feet yellowish-brown to pinkish-brown, occasionally more greyish-pink.
Adult female Entire upperparts, from forehead to uppertail-coverts, blackish-brown; forehead to upper back closely streaked pale rufous; lower back, rump and uppertail-coverts closely barred pale rufous; scapulars and upperwing-coverts barred and mottled pale rufous; remiges brown, primaries and secondaries irregularly spotted to streaked pale rufous, mostly on outer webs; tertials vermiculated pale rufous; tail chestnut, narrowly barred blackish. Lores and sides of head ochraceous-buff, finely spotted blackish on ear-coverts and malar region; chin and throat white to ochraceous; breast and sides of neck reddish-ochre, spotted blackish; spots sometimes in lines giving more streaked effect; rest of underparts blackish-brown, barred ochraceous; bars darker on undertail-coverts and whitish on centre of belly. Bare parts as male.
Immature Male undescribed; probably like adult but duller. Female duller than adult, especially on tail, with spots on underparts extending to flanks and belly. Iris black; bill brown; legs and feet brownish.
Juvenile Male probably almost entirely dark sooty-brown, with some white spotting on breast and upper belly. Female undescribed.
Downy young Chicks have sooty down.

MEASUREMENTS Wing of 20 males 68-79 (71.6), of 11 females 68-75 (71.1); tail of 18 males 43-52 (46.6), of 10 females 41-47 (44.8); culmen to base of 19 males 13-15 (14.1), of 11 females 11.5-14.5 (13.1); tarsus of 17 males 18.5-22 (20.8), of 10 females 19-21.5 (20.2). Weight of 1 female 30.

GEOGRAPHICAL VARIATION None recorded.

MOULT Primary moult begins ascendantly; recorded May.

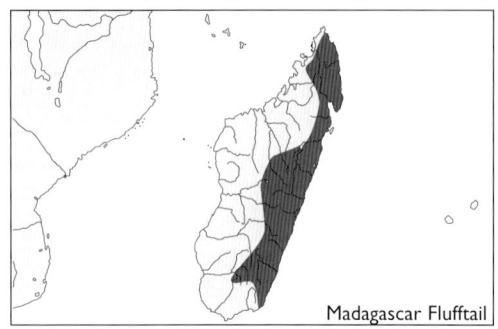

Madagascar Flufftail

DISTRIBUTION AND STATUS Endemic to Madagascar, occurring in the E and on the High Plateau, from Mandena and Midongy du Sud in the SE, N to Mt d'Ambre and NW to Sambirano (Langrand 1990, Wilmé & Langrand 1990). It is incorrectly assumed (Keith *et al.* 1970, Snow 1978) to occur at Manombo in the Southern Domain subdesert region of the SW, and this error was inadvertently included in Taylor (1996b); the correct locality (from Rand 1936) is Manombo (S of Farafangana) on the coast in the SE. It is widespread, evenly distributed and common to abundant over most of its range. It adapts readily to degraded environments and is thus not considered threatened.

MOVEMENTS Unknown; probably none.

HABITAT Tolerant of a wide variety of habitats: grassland of the edges and clearings of both undisturbed and degraded forest; secondary bush, including that comprising large ferns, thick grass, and *Philippia* spp. (Ericaceae); dense cultivation (but it is not recorded from cassava fields); it is also seen on the forest floor, and occurs in marshes with long grass, reeds and sedges, and sometimes rice paddies. It is recorded from sea level to 2,300m.

FOOD AND FEEDING Only insects and seeds recorded.

HABITS Apparently secretive and very difficult to observe; sightings are normally very brief. When alarmed, it runs into dense cover; it is very difficult to flush but sometimes flies for a fewm before running.

SOCIAL ORGANISATION Probably monogamous; apparently territorial at least while breeding and possibly also in the non-breeding season.

SOCIAL AND SEXUAL BEHAVIOUR No information available.

BREEDING AND SURVIVAL Season Lays Oct; probably also Sep and possibly Jul. **Nest** A ball of broad grass blades, thickly lined with finer plant material; external length 17cm, width 12cm; internal length and width 8cm; built on ground in dense grass. **Eggs** 3-4; blunt ovate; white and slightly glossy; size (n = 28) 25-29.2 x 19.5-22.5 (26.8 x 20.7); calculated weight 6. **Incubation** Said to be by female only, but male has been photographed at nest (Taylor 1996b). **Chicks** No information on development. Young may remain with parents after start of postjuvenile moult.

8 WHITE-WINGED FLUFFTAIL
Sarothrura ayresi   Plate 3

Coturnicops ayresi Gurney, 1877, Potchefstroom, SW Transvaal, South Africa.

Sometimes placed in *Coturnicops*, usually with, but sometimes without, other flufftail species. Ethiopian birds first described as *Ortygops macmillani* (Bannerman 1911). Forms a species pair with *S. watersi*. Monotypic.

Synonyms: *Coturnicops ayresi*; *Ortygops macmillani*.

Alternative name: White-winged Crake.

IDENTIFICATION Length 13.5-14.5cm; wingspan 24cm. A small flufftail; easily identified in flight by conspicuously white secondaries; leading edge of wing also white; no white visible in folded wings. Unlike other flufftails, frequently easy to flush and often has strong, direct flight (see Habits). Uniquely in genus, female looks similar to male. Male has blackish-brown head with chestnut mottling and paler, poorly defined supercilium; neck, upper mantle and breast chestnut; chin and throat whitish; upperparts, and median upperwing-coverts, blackish, broadly streaked olive-brown and narrowly white; other upperwing-coverts plain olive-brown; tail barred black and chestnut. Flanks streaked black and white; centre of belly white; undertail-coverts chestnut, black and rufous. Iris and bill blackish; legs and feet grey to purplish-flesh. Female looks darker than male in flight; duller on head and neck; sides of head mottled buff; upper mantle streaked dark brown; white streaks on upperwings replaced by spots; upperwing-coverts more extensively spotted white. Breast whitish, tinged rufous and scaled dark; flanks spotted white. Immature duller than adult; male has white streaks of upperparts replaced by spots or bars, less chestnut on head, mantle and breast, and less white on underparts; female largely paler than adult, with fewer spots on upperwing-coverts, dark scalloping on chin and throat, and no rufous wash on centre of breast. Juvenile male dark grey-brown, mottled chestnut on head and nape, with small white spots on upperparts and upperwings; chin, throat and belly mottled grey-brown and white; tail as adult. Juvenile female blackish-brown, upperparts spotted tawny on mantle and white elsewhere; breast flecked white; throat and centre of belly white; tail as adult. Inhabits moist to flooded marsh vegetation, predominantly of sedges and grasses.

Similar species Easily separable from other *Sarothrura* species, and from all other sympatric marsh-dwelling birds, by white secondaries. On the ground, when this white is not visible, both sexes superficially resemble dull versions of the males of sympatric marsh and grassland *Sarothrura* species. All ages distinguished from both sexes of all these species, except female Striped Flufftail (6), by chestnut and black bars on tail; female Striped differs markedly in having paler upperparts with buff to white scallops and bars, no rufous on hindneck, upper mantle or breast, and different head pattern.

VOICE All information on calls comes from South Africa (Taylor 1994, P. B. Taylor unpubl.) unless otherwise stated. The common call is a low-pitched, short *oop* note, repeated every s and continued for up to 3 min; it is often given in asynchronous duet, the second bird having a higher-pitched note. Calling usually occurs for up to 15 min (occasionally 40 min), at dawn and dusk. This call is very similar to a roosting call of the Crowned Crane *Balearica regulorum* which, however, is louder, usually a double note or a more complex series, and is often given by several birds together. Other calls, rarely heard, include deep mooing notes which are indicative of agitation or aggression in response to taped playback of the *oop* call, and high-pitched short and long hoots. No advertising or territorial calls have been recorded from breeding birds in Ethiopia, but chicks make a loud, plaintive and rather harsh cheeping series and a female, when separated from chicks, called to them with quiet quacking notes and occasionally gave quiet gulps, low grunts, and subdued *crk-crk-crk* calls (P. B. Taylor unpubl.).

DESCRIPTION

Some confusion surrounds description of different age classes of this species because of the paucity of museum material and the incorrect ageing of some specimens. For example, an "adult female" (Durban Museum) from Franklin, Natal (Mendelsohn *et al.* 1983) is in full juvenile plumage, while an "immature female" (BMNH) from Sululta, Ethiopia, dated 22 Sep 1948 (Keith *et al.* 1970) is a male moulting from natal down to juvenile plumage.

Adult male Forehead to nape blackish-brown, proximal half of feathers edged dark chestnut; hindneck and upper mantle dark chestnut, feathers tipped blackish-brown. Lower mantle, back, scapulars and median upperwing-coverts blackish, each feather fringed olive-brown on distal half and with fine white submarginal streak on each web. Lesser and greater coverts olive-brown with vaguely darker centres and no white streaks; some lessers may have small white spots and some greaters vague white submarginal streaks; alula and primary coverts similar, with no white markings except occasionally a few on lessers; marginal coverts white. Primaries dark grey-brown to olive-brown, outer web of outermost primary (P10) whitish on basal three-quarters, outer webs of others sometimes with small buffy spots; inner webs of primaries become progressively paler; P1 brownish-white to white on most of inner web. First secondary olive-brown at base, white on inner web, and mottled brownish and white on outer web; rest of secondaries white, olive-brown basally; first tertial (T1) similar but variably spotted olive-brown on inner web, or patterned as described for T2; T2 olive-brown with variable white spots and bars, or like T3; T3 as scapulars. Axillaries and underwing-coverts greyish-white, olive-brown basally and variably along outer webs. Rump blackish-brown, feathers with narrow white submarginal streaks or spots; uppertail-coverts and rectrices barred black and chestnut. Lores and anterior ear-coverts blackish-brown; rest of sides of head, and sides of neck, dark chestnut, feathers narrowly tipped blackish giving scaly pattern; vague superciliary streak paler chestnut. Chin and upper throat whitish, washed pale chestnut at sides; lower throat, foreneck and sides of breast brighter chestnut; centre of breast pale chestnut; upper flanks like mantle but white streaks broader; lower flanks, thighs and sides of belly broadly streaked black and white; centre of belly white; undertail-coverts chestnut and black, widely tipped pale rufous. Iris dark brown to brownish-black; bill dark brown, or blackish with grey cutting edge; legs and feet grey, or brown to purplish-flesh.

Adult female Similar to male but more blackish-brown, less rufous, on head and neck; sides of head mottled buff; upper mantle dark chestnut with dark brown feather centres, giving streaked effect. White streaking of upperparts and upperwings replaced by spots or short bars, except on back; all secondary coverts, and lesser primary coverts, usually have white markings. Sides of neck and sides of breast paler chestnut than in male; centre of breast whitish, tinged rufous-brown; dark mottling and scaling usually extend all across breast; flanks black, spotted white; undertail-coverts tinged pale rufous. Iris as male, or ashy hazel; legs and feet as male, or dusky pink.

Immature male Less chestnut on head than adult; mantle feathers have broader brownish tips. White streaks on upperwing-coverts of adult replaced by spots; lesser, greater, and primary coverts have variably darker centres, with white spots on secondary coverts (most prominent on greaters) and variable small whitish spots on lesser primary coverts. Face darker than adult, having almost no chestnut; sides of breast less extensively chestnut; underparts less extensively white, centre of belly being washed grey; undertail-coverts more uniformly dull rufous. Legs and feet purplish-flesh.

Immature female Upperparts paler and browner than in adult; head and mantle duller; upperwing-coverts duller, with less distinct pale fringes and fewer spots; chin and throat less contrastingly white, with some dark scalloping; neck and breast duller; no rufous wash across centre of breast. Flanks paler and duller than adult, with less contrasting spots and bars; undertail-coverts tinged whitish. Iris sometimes black; bill dark horn; legs and feet may be dark brown with greenish tinge.

Juvenile male (One specimen, Ethiopia.) Head and nape predominantly dark grey-brown with dull chestnut mottling; hindneck, sides of neck and mantle grey-brown, washed dull chestnut; rest of upperparts, including upperwings, darkish brown with small white spots. Breast and rest of underparts dull grey-brown, mixed grey-brown and white on chin, throat, centre of breast and most of belly; undertail-coverts faintly tinged rufous; tail as adult. Iris grey.

Juvenile female (One specimen, South Africa.) Blackish-brown with small tawny spots on mantle and small white spots on rest of upperparts (except head and neck), whitish flecks on breast, and white throat and centre of belly; tail as adult.

Downy young Down black. Iris blackish-brown, eyelids grey; proximal half of bill pinkish-white, distal half ivory, colours separated by 1mm-wide black band in front of nostril; legs and feet grey-black.

MEASUREMENTS Wing of 14 males 73-80 (76.3), of 11 females 75-80 (76.9); tail of 8 males 35-40 (36.1, SD 1.6), of 6 females 35-43 (38.2, SD 3.3); culmen to base of 13 males 12-13.5 (12.4), of 11 females 12-13.5 (12.5); tarsus of 14 males 17-19.5 (18.5), of 11 females 16-20 (18.5). Weight of 1 male 31.8, of 1 chick, 2-3 days old, 5.4 (P. B. Taylor unpubl.).

GEOGRAPHICAL VARIATION Despite the great distance separating this bird's two centres of occurrence, and the lack of records from most of the intervening regions (see Movements), there appear to be no significant morphological differences between South African and Ethiopian populations.

MOULT An adult female flushed in the former Transvaal, South Africa, in Feb had a gap in the remiges of one wing, indicating moult (Taylor 1994). Adults in Ethiopia, late Jul, are in very fresh plumage (P. B. Taylor unpubl.).

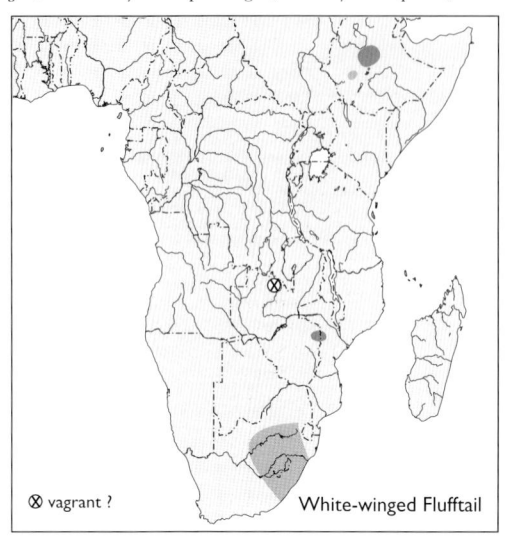

DISTRIBUTION AND STATUS Ethiopia, where formerly known from highlands around Addis Ababa (Sululta Plain, Akaki, Entotto and Gefersa), and at a lower elevation to the SW at Charada, Kaffa; Zimbabwe (Harare area); and South Africa (highlands of KwaZulu-Natal and former Transvaal; also recorded Free State and formerly at coastal localities in E Cape and KwaZulu-Natal). There is one reliable record from Zambia, near Chingola, Solwezi District (Brooke 1964). Sound records from Rwanda (Dowsett-Lemaire 1990) are questionable, sonagraphic analysis indicating that they are calls of the Crowned Crane (Taylor 1994). This globally ENDANGERED and CITES I species is one of the rarest and least known African endemics. From 1939 to 1957 small numbers were recorded occasionally in the Ethiopian highlands; subsequently one bird was seen near Sululta in Aug 1984 and 4 in Aug-Sep 1995, while an estimated 10-15 breeding pairs were present in Aug 1996 (Taylor 1996a). In Aug 1997 a breeding population of at least 200 pairs was found in seasonal and permanent marsh at a new locality near Addis Ababa and it is probable that the species was widespread and locally numerous in the central Ethiopian highlands before intensive human pressure destroyed most of its seasonal marsh habitat (Taylor 1997b). It was recorded in Zimbabwe in Jan-Mar 1977 and 1979 (Hopkinson & Masterson 1984), and possibly bred there in the 1950s (Taylor 1994). In South Africa it was recorded only sporadically after its discovery in 1876, and since the early 1980s 5 highland sites in S KwaZulu-Natal and E Transvaal, South Africa, have held small numbers (maximum overall annual counts 22-29 birds), three of these sites annually in 1990-1992, when regular observations were made. Recent surveys (Taylor 1997a) have identified 5 more sites in the Free State and KwaZulu-Natal where this bird probably occurs annually and the total population at the 10 known sites may be 235 birds. In South Africa the lack of recent records from coastal localities suggests that it may now be confined to the higher-altitude wetlands (Taylor 1994).

This bird's habitats are under severe threat from damming, draining and overgrazing, and its future is precarious (Taylor 1997c). Very large areas of breeding habitat in Ethiopia have been destroyed by overgrazing, trampling and sedge cutting (Taylor 1997b). However an assumed threat to one remaining site from a proposed dam (Atkinson et al. 1995) is unfounded (Taylor 1997b). Much suitable breeding habitat in Ethiopia could be re-established by encouraging local communities to manage and utilise wetland resources more effectively (Taylor 1997b), e.g. by restricting early grazing, delaying the cutting of fodder until Oct-Dec (which would also increasing the yield from seasonal marshes), and (in Jul-Sep) by encouraging the development of alternative sources of freshly cut vegetation used for feeding dairy cattle in Addis Ababa (EWNHS 1996) and for floor covering; some compensation to local communities for loss of revenue early in the season might have to be provided. In KwaZulu-Natal, critically important wetlands are now threatened by desiccation as a result of commercial afforestation in their catchments, as well as by damming, draining, water abstraction, disturbance, and annual burning followed by intensive spring grazing (Taylor 1997a).

MOVEMENTS The apparent lack of subspeciation has been thought to indicate that regular migration occurs between the bird's Ethiopian and South African centres of distribution but the paucity of records from intervening regions, and an overlap in occurrence dates, make this unlikely (Collar & Stuart 1985), while birds may be present throughout the year at one recently discovered marsh near Addis Ababa (Taylor 1997b). However, there may be periodic long-distance dispersal when numbers are high, allowing gene exchange between the N and S populations. Records from Zambia (Nov 1962) and Zimbabwe (Jan-Mar in 1977 and 1979) may reflect such dispersal, and the species is possibly an occasional breeding migrant in Zimbabwe. Much breeding habitat in the C Ethiopian highlands, where most occurrences are recorded from Jun-Sep, is seasonal marsh and is thus unsuitable in the non-breeding season when migration may occur SW to lower-altitude, permanent marshes such as those at Charada, Kaffa (in the Jimma area), whence there is a May specimen (Taylor 1994, 1996a). Guichard (1948) suggested that males arrive in breeding areas before females. In South Africa, where recent records suggest that the species is normally migratory or nomadic, it is recorded from Aug-Mar and in May (Taylor 1994).

HABITAT Most information is from Taylor (1994, 1996a). Ethiopian breeding habitat is seasonal: dense, lush, rapidly growing vegetation, 20-50cm (usually 20-40cm) tall, on firm ground which is flooded to a depth of 20cm (usually to 10cm). Dominant plants include sedges (*Cyperus rigidifolius, C. ?afroalpinus* and *Eleocharis marginulata*), grasses (*Pennisetum schimperi* and *P. thunbergii*) and forbs such as *Uebelinia kigesiensis, Trifolium calancephalum, Ranunculus multifidus, Rumex marginulata, Haplocarpha ?schimperi*, and a *Polygonum* species. Sedges and short grasses tend to dominate in the more shallowly flooded sites, which lie in depressions and at the bases of shallow slopes above seasonal wetlands, as well as within the wetlands themselves. Forbs and taller grasses dominate in the more deeply flooded areas of taller vegetation within the wetlands. In Zimbabwe, birds were recorded from grass 50-100cm tall on dry to moist ground and also from muddy to shallowly flooded marshy ground with grass (*Leersia, Hemarthria* and

Cynodon dactylon) and sedge (including *Cyperus digitatus*) cover (see Hopkinson & Masterson 1984). In Zambia, one bird was found in a pan-like marsh with emergent grass (Brooke 1964). Non-breeding birds in South Africa occur for short periods alongside breeding Red-chested Flufftails (3) in dense hygrophilous grasses (predominantly *Leersia* but also *Andropogon, Paspalum, Eragrostis, Hemarthria, Arundinella* and *Aristida*), sedges (*Pycreus, Kyllinga, Fuirena, Eleocharis, Schoenoplectus, Mariscus, Carex* and *Cyperus*) and rushes *Juncus* spp. averaging 1m tall, on moist to shallowly flooded substrates, and for up to 4 months in dense sedges (principally *Carex acutiformis*, but also *Cyperus fastigiatus*), reeds *Phragmites australis* and reedmace *Typha capensis*, 1-2m tall, on moist to deeply flooded ground not commonly inhabited by breeding Red-chested Flufftails. It has been recorded breeding alongside the Red-chested Flufftail in Ethiopia, occupying typical seasonally flooded vegetation types while the Red-chested occurred in adjacent taller, sedge-dominated, permanently wet areas (P. B. Taylor unpubl.). In Ethiopia it occurs at 2,200-2,600m in the central highlands, and at 1,100m in the SW. It is recorded at 1,300-1,400m in Zambia and Zimbabwe; in South Africa it occurs mostly at 1,100-1,900m and has been recorded rarely at c. 150m in coastal areas.

FOOD AND FEEDING Stomach contents are recorded as water insects, grain seeds and 'vegetable mush'. Recent studies in Ethiopia (Taylor 1996a, 1997b) have provided the following information. Adults take earthworms, small freshwater crustaceans, and the adults and larvae of aquatic and terrestrial insects such as Lepidoptera, Coleoptera (including Chrysomelidae) and Diptera. Small chicks are fed on crustaceans, Coleoptera (including Dytiscidae larvae) and Diptera (including large prey such as Tipulidae and Tabanidae larvae over 2cm long). In the breeding habitat birds forage along muddy cattle tracks, at shallow pools, and at patches of cut vegetation and other small open areas in the dense cover, taking insects and other invertebrates from moist ground, mud and shallow water, and from flattened and low-growing vegetation; both adults and chicks apparently also forage in more deeply flooded vegetation. Foraging has been observed from early to mid-morning and in the late afternoon.

HABITS All information is from Taylor (1994, 1996a). This species is diurnal and crepuscular; most activity is recorded from early to mid-morning and from late afternoon to dusk, but birds may be flushed at any time of the day, and in light to moderate rain. In South Africa no activity has been recorded during the hours of darkness, when calling does not occur and birds do not respond to taped playback. In tall, flooded vegetation the birds are extremely difficult to observe, normally remaining within dense cover, but in short, sparse vegetation they may be glimpsed on the ground just before they take flight. Response to taped playback is often poor, but in tall, dense reedbeds the birds will sometimes approach to within 1-2m of an observer, calling and climbing around c. 1.5m up in the vegetation but remaining invisible and being impossible to flush. However in shorter vegetation this species is often easy to flush, rising up to 30m in front of the observer, remaining airborne for up to 200m and often circling around to fly past the observer. In such circumstances the flight is strong and direct, with legs retracted, neck outstretched, and rapid, shallow wingbeats. However this species also has a weak, fluttering, typical flufftail flight, with dangling legs. The birds normally climb to the top of dense, short vegetation before taking flight. In Ethiopian breeding areas, recently arrived birds, and probably those in the early stages of incubation, flush fairly easily and make low flights of 2-40m. Birds with chicks are extremely difficult to flush and may even be caught by hand. Before taking flight, birds on the ground pause briefly with open wings, possibly as a signal to other individuals.

SOCIAL ORGANISATION Apparently monogamous. In South Africa, observations of calling patterns and reactions to taped playback suggest that birds may be territorial in non-breeding habitat in which residence is prolonged (Taylor 1994). In Ethiopia breeding birds occur at a density of c. 2-4 pairs/ha and no territorial activity has been recorded from birds with nests and young chicks (Taylor 1996a, 1997b).

SOCIAL AND SEXUAL BEHAVIOUR Non-breeding birds in short vegetation appear to move constantly around occupied wetlands, all the birds apparently remaining in loose association with each other (Taylor 1994).

BREEDING AND SURVIVAL Season Nest-building and egg-laying occur Ethiopia during long rains, Jul-Aug; a juvenile, South Africa, Nov, probably from egg laid Aug (Taylor 1994). Despite claims to the contrary, there is as yet no acceptable evidence that species breeds in South Africa (Taylor 1994). **Nest** In Ethiopia, nests probably built in sedge- and grass-dominated vegetation 20-40cm tall, over water up to 10cm deep. Description of nest and eggs from a villager in Ethiopia (Taylor 1996a) agrees in most respects with observations made at Mazowe, Zimbabwe, in 1950s (probably Feb), where 2 unidentified clutches, each of 3 eggs, were found on seasonally flooded ground; nests were shallow cups of grass built in reedy vegetation 10-12cm above water 30cm deep, with growing vegetation pulled over them in a dome. **Eggs** Eggs from these nests were ovate, white, sparsely spotted grey-brown and olive-green, markings most numerous in ring at blunt end; size of 1 egg 27 x 20 (Taylor 1994). Following information is from Taylor (1996a). **Chicks** Both sexes apparently feed and care for chicks. Adults lead observers away from young chicks by running through short vegetation, often across open patches, and hiding briefly in dense cover. Observations in Ethiopia suggest that birds commence nesting immediately after arrival in breeding habitat in late Jul/early Aug, and that entire breeding cycle may occupy as little as 6 weeks, after which all birds may leave breeding habitat, which may have become unsuitable as a result of damage from grazing, trampling and cutting. Some birds may be able to raise a second brood elsewhere, before end of Oct, in late-developing habitat. Natural predation of eggs and young may be low at Ethiopian breeding sites.

9 SLENDER-BILLED FLUFFTAIL
Sarothrura watersi Plate 3

Zapornia watersi Bartlett, 1879, SE Betsileo Country, Madagascar.

Genus sometimes merged into *Coturnicops*. Forms a species pair with *S. ayresi*. Monotypic.

Synonyms: *Lemurolimnas/Ortygometra/Coturnicops watersi*.

Alternative names: Waters's Flufftail; Waters' Crake.

IDENTIFICATION Length 14-17cm. A largely unpatterned flufftail, with short, slightly fluffy tail and quite long, slender bill. Relatively plain plumage of male distinctive: head to mantle darkish chestnut, often with pale greyish patch behind eye; upperparts dull brown with indistinct dark streaks; tail chestnut, with black feather tips. Chin and throat white, breast and upper flanks vinous-chestnut; posterior underparts dark ashy-brown; undertail-coverts as tail. Iris brown; bill black; legs and feet pinkish-grey. Female has upperparts dark olive-brown; sides of head mottled pale; vague pale supercilium or streak behind eye; chin, throat and centre of breast whitish; undertail-coverts and tail barred black, white and chestnut; rest of underparts plain brownish; bill dark brown. Immature duller than adult; immature female has pale spots on back; sides of breast and flanks barred dull brown and whitish. Juvenile female more spotted than immature, spots extending to upperwing-coverts and tertials. Inhabits small grassy wetlands near rainforest.

Similar species Sympatric Madagascar Flufftail (7) is strikingly patterned, with shorter bill, tarsus and toes. Male has rich chestnut head, neck, breast and tail, and is black elsewhere, streaked yellowish and white on upperparts, and white on underparts; legs and feet yellow-brown. Female has pale rufous upperparts, closely streaked and barred black; tail barred chestnut and black; chin and throat whitish; face and underparts ochraceous (more whitish on centre belly), with blackish spots and streaks on breast, and bars elsewhere.

VOICE Details are from Wilmé & Langrand (1990) unless otherwise stated. A muffled, solemn call, probably territorial and uttered during the day and over long periods without a break; described as normally consisting of 3 notes and repeated every 0.3-0.4s. However, A. Riley (*in litt.*) describes the three-note call as a loud *chang-chang-chang*. A 2-note version sometimes precedes the 3-note calls and has the first syllable accentuated *GOO goo*, and longer series of 4-7 notes are occasionally heard; notes vary in pitch. The alarm call is said to be a characteristic *tiec*. Local people in the Périnet Analamazaotra area claim that the species often calls from their fields during the day (see Habitat) and a female flew towards the source of taped playback (A. Riley *in litt.*).

DESCRIPTION
Adult male Head, neck, mantle, lower throat, breast and upper flanks vinous-chestnut, palest on breast; feathers from forehead to mantle tipped blackish-brown to give darker effect, especially noticeable from forehead to hindneck; sometimes with pale greyish lores and patch behind eye; chin and throat white, sometimes with vinous-chestnut wash. Rest of upperparts, including most upperwing-coverts, dark olive-brown with darker, more blackish-brown, feather centres; alula, primary coverts and remiges uniform olive-brown, but tertials more blackish-brown with brighter, more cinnamon-brown fringes; uppertail-coverts, undertail-coverts and rectrices chestnut with black tips, variable in extent on rectrices. Lower flanks and belly ashy-brown. Axillaries and underwing-coverts uniform brown; underside of remiges more grey-brown. Iris brown; bill black; legs and feet pinkish-grey, sometimes with brownish tinge.
Adult female Upperparts plain, dark olive-brown, darker from lower back to uppertail-coverts; crown has small, indistinct tan to rufous-brown streaks or spots; sides of head flecked brownish-olive, with vague pale supercilium or streak behind eye; lores often darker brown. Remiges darkish olive-brown, tertials darker with olive-brown to cinnamon-brown fringes; axillaries and underwing-coverts uniform brown. Tail brownish-black, broadly barred chestnut and narrowly barred white to pale chestnut. Chin, throat and centre of breast whitish, sometimes tinged brownish-olive to tawny-olive, especially across centre of breast; rest of underparts, including sides of neck and sides of breast, almost unpatterned brownish-olive to tawny-olive, but belly to undertail-coverts ashy-brown (Ridgway's Hair Brown); undertail-coverts barred chestnut, black and white. Bill dark brown.
Immature Immature male not described but probably slightly duller than adult, and chestnut on mantle less extensive. Immature female like adult but has scattered small yellowish, buff or tan spots on back; sides of breast and flanks, and belly to undertail-coverts, duller brown than in adult, and barred whitish; white bars very narrow and indistinct on flanks.
Juvenile Possible juvenile from near Antananarivo described as "very dark". Juvenile female like immature, but with more numerous and extensive pale spots on back, rump, scapulars and upperwing-coverts; tertials also spotted or barred; a few narrow short whitish bars on primary coverts; small pale streaks on uppertail-coverts; undertail-coverts possibly dark brown, streaked or barred whitish.
Downy young Not described.

MEASUREMENTS Wing of 7 males 71-79 (73.9, SD 2.8), of 6 females 69-74 (71.3, SD 1.7); tail of 6 males 39-46 (42.8, SD 2.5), of 3 females 41, 44, 45; culmen to base of 7 males 15-18 (16.4, SD 1.1), of 6 females 16-17.5 (16.5, SD 0.5); tarsus of 6 males 22-24 (22.9, SD 0.8), of 6 females 21.5-24 (22.7, SD 1.0), of 1 live male 30 (method of measurement unknown). Weight of 1 male 26.5.

GEOGRAPHICAL VARIATION Apparently none.

MOULT No information available.

DISTRIBUTION AND STATUS This flufftail is a globally ENDANGERED species and is recorded in the literature from only five localities in E Madagascar: "SE Betsileo" in 1875, "near Andapa" in 1930, the Laniera and Ambohibao marshes and the Ikopa River, near Antananarivo, in 1970-1971, Périnet Analamazaotra in 1928 and Ranomafana since 1986 (Collar & Stuart 1985, Wilmé & Langrand 1990). It was considered rare, and possibly a relict species on its way to early extinction, until its uncorroborated discovery in the 1,200km area around Antananarivo in 1970-1971, where it was suspected to occur in all suitable marshes at a density of 1 pair per 2ha, and perhaps to breed in small numbers (Salvan 1972b). Doubt has been cast on these observations because the species has not been recorded from the area subsequently, despite searches being made (Collar & Stuart 1985). However, there is unconfirmed information from R. Thornstrom that it has been rediscovered on the plateau N of Antananarivo, where it is reportedly "quite common" (A. Riley *in litt.*). It was not reliably recorded elsewhere between 1930 and 1986-1987, when it was found at Vohiparara marshes, near Ranomafana (Wilmé & Langrand 1990). More recently, in 1997 it was also found close to the main entrance to Analamazaotra Special Reserve, and it is reportedly not uncommon at Totorofoby Marsh near Périnet (A. Riley *in litt.*). Its total numbers may be considerably less than the maximum of 1,000 suggested by Rose & Scott (1994). Considering only reliable records, its distribution appears coincident with that of the much-pressurised eastern rainforest (Collar *et al.* 1994) (but see Habitat); its particular microhabitat is widely but patchily distributed in small blocks over a broad forest zone and is presumably becoming increasingly rare, but suitable habitat also possibly occurs at nine other localities (Wilmé & Langrand 1990). The Madagascar system of protected areas, which includes wetlands only occasionally, incorporates only two known sites of occurrence, Analamazaotra Special Reserve and Ranomafana National Park. In addition, the Andapa locality may be close to the Anjanaharibe-Sud Special Reserve. In view of the need to expand rice cultivation it is unlikely that wetlands will be included in new protected areas (Wilmé & Langrand 1990). The Laniera marshes near Antananarivo, where the species may have bred (Salvan 1972b), have been turned into rice paddies, but it is not clear whether the possible breeding record was obtained before or after this development. There is an urgent need for distributional and population surveys at all known and possible sites of occurrence, and for an evaluation of sites suitable for protection.

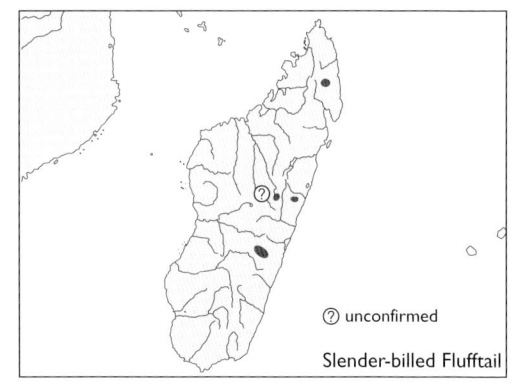

birds occurred at the edge of rainforest in a flooded marsh dominated by *Cyperus rotundus*, with a few specimens of *Pteris* sp., *Sticherus flagellaris* (Gleicheniaceae), *Osmunda* sp. and *Psidia angustifolia* (Compositae); this vegetation covered c. 2ha and the rest of the marsh had been transformed into irrigated rice paddies (Wilmé & Langrand 1990). The Laniera marshes are dominated by *Cyperus articulatus*, *C. papyrus* and *C. latifolius*. It has been suggested that this species may replace the common Madagascar Flufftail at higher altitudes, and that temperature may control its montane distribution; however, its distribution is more probably determined by that of the rainforest (Collar & Stuart 1985). Recent information suggests that its habitat tolerances may be wider than has been supposed: all rainforest had been removed for at least 1km from the site of the 1997 observations at Analamazaotra, and these observations, plus information from a local guide, suggested that the bird also frequents overgrown, weedy, disused agricultural land and the scrubby edges of marshes (A. Riley *in litt.*). The Madagascar Flufftail tolerates a wider variety of habitats and is often recorded from dry areas in grassland, forest, secondary bush and cultivation.

FOOD AND FEEDING No information available.

HABITS Apparently very secretive and difficult to observe, keeping to dense vegetation but occasionally flying short distances. Local people are said to capture this species by beating through the grass and catching the birds by hand or by knocking them down with sticks (Rand 1936).

SOCIAL ORGANISATION The calling pattern indicates that the species is territorial, at least during the breeding season.

SOCIAL AND SEXUAL BEHAVIOUR No information available.

BREEDING AND SURVIVAL Season Unknown; possibly breeds during rains. Male and female in breeding condition Andapa, Sept; male called Ranomafana, Oct and Nov. Adult and juvenile reported at Laniera marshes, near Antananarivo, May (Salvan 1972b), but occurrence there disputed (Collar & Stuart 1985).

MOVEMENTS No evidence for any movements.

HABITAT Small elevated wetlands with *Cyperus* sedges, and adjacent dense, grassy terrain or even rice paddies, near rainforest, at altitudes of 950-1,800m. The Ranomafana

---

HIMANTORNIS

One species, endemic to Africa in lowland tropical rain forest. A large, primitive rail with no close relatives, it has predominantly dull brown plumage, only eight rectrices, a heavy bill, red legs, a patterned natal plumage which is unique in the family, and a very distinctive voice.

10 NKULENGU RAIL
Himantornis haematopus Plate 4

Himantornis haematopus Hartlaub, 1855, Dabocrom, Ghana.

The most primitive of the rails, so distinct from the rest of the family that it has been placed in its own subfamily (Himantornithinae) on primitive skeletal characters, which suggest a link with the trumpeters (Psophiidae) of South America. Monotypic.

Synonyms: *Himantornis whitesidei*; *Psammocrex petiti*.

Alternative name: Nkulenga Rail.

IDENTIFICATION Length c. 43cm. A large rail with upright posture, long slender legs, short toes and a shortish, somewhat heavy bill. Plumage very unlike that of other rails, being mainly dull, darkish brown with pale greyish or brownish feather fringes, giving scaly or mottled effect; chin to foreneck whitish; belly and undertail-coverts often almost unpatterned. Iris, legs and feet red; bill black, green or greyish at base. Plumage very variable, especially in colour of pale feather fringes and extent of scaling on underparts. Sexes alike. Immature has dark upperparts, feathers fringed tawny; underparts pale grey-brown, feathers fringed paler; belly whitish; iris, legs and feet duller than in adult. Juvenile similar but feathers of upperparts with pale grey- or chestnut-tinged fringes; underparts pale brownish, with pale feather fringes. Inhabits lowland rain forest.

Similar species Easily separated from other forest rails by large size, short, relatively deep bill, and red legs and feet. Variably scaly or mottled plumage pattern more reminiscent of a francolin than a rail, but overall shape, leg length and behaviour are obviously rallid.

VOICE Most information is from Urban *et al.* (1986). The song is a rhythmical, antiphonal duet, described as a series of phrases of 6 notes *ko-KAW-zi-KAW-hu-HOOO*, each lasting c. 1.5s and repeated in quick succession, often for long periods. Notes are loud, raucous and far-carrying; reminiscent of monkey calls or even of "men cutting trees". Single notes are sometimes given. During the night calls are given from up to 20m high in a tree, and by day from on or near the ground. The birds call throughout the year, mainly at dusk and during the night (especially at full moon) and most often just before dawn. The contact call between pair members is a short, solemn grunt (Brosset & Erard 1986). Demey & Fishpool (1994) found that this species does not approach playback of its call.

DESCRIPTION
Adult Forehead to nape darkish grey-brown to dark brown, forehead usually paler and greyer; sides of head and neck brownish to greyish, often washed buff; stripe over eye, and lower ear-coverts, usually paler. Rest of upperparts, including scapulars, upperwing-coverts and rectrices, very variable; feathers range from brown with narrow, pale, buff to grey edges or fringes, to dark brown or blackish with subterminal brown or chestnut fringes and whitish to pale grey edges and tips; rectrices usually relatively unpatterned. Remiges dark grey-brown to dark brown; axillaries and underwing-coverts dark brown to blackish, with narrow to broad, whitish to buffy tips. Chin, throat, malar region and foreneck whitish; feathers of underparts vary from grey-brown to blackish, with greyish to ashy-brown fringes varying from indistinct to broad and well-defined; belly to undertail-coverts often almost unpatterned grey-brown to very dark brown, sometimes tinged rufous. Bill black, base of lower mandible light green or bluish-grey; bare skin on lores and around eye black; iris reddish-brown to reddish-orange; legs and feet bright pink to red. Sexes alike; female may be slightly smaller.
Immature Similar to adult; feathers of upperparts with dark centres and tawny fringes; underparts light grey-brown with paler feather fringes; chin, throat and belly whitish. Iris dark brown; tip of bill pale horn; legs and feet dull pale red.
Juvenile A half-grown juvenile in AMNH has natal down on head and neck; feathers of upperparts dark brown with broad pale fringes tinged chestnut subterminally and pale grey-buff terminally; remiges dull brown with pale, dull outer edges tinged yellowish-ochre; feathers of underparts paler, tending to dull yellow-ochre, with greyer fringes (P. Sweet *in litt.*). One possible juvenile in BMNH is similar on upperparts; forehead to upper mantle grey-brown with narrow, indistinct pale chestnut-tinged feather fringes; outer webs of all remiges except outer 3 primaries washed tawny; lower neck to upper belly pale brownish-olive with buffy feather fringes; flanks brown; belly dull brown, feathers tipped buffy; thighs dull brown; undertail-coverts dull brown and buff. Iris very dark olive-brown; bill blackish with pale horn-grey tip and blue-grey base to lower mandible; legs and feet dull light reddish; claws pale grey (P. Sweet *in litt.*).
Downy young Unlike most rails, downy young distinctively patterned: broad blackish-brown stripe from forehead over head and upperparts to tail; lores to ear-patch black; stripe over eye, side of head, chin, throat and underparts creamy-white to brownish-white, palest on chin and throat; brown band across breast; sides of body dull light brown; iris dull brown; legs and feet dull pink; upper mandible blackish; lower mandible grey; 2 small wing-claws.

MEASUREMENTS (6 males, 6 females) Wing of male 211-235 (220), of female 200-222 (212); tail of male 72-94 (85), of female 74-85 (80); bill of male 36-41 (38), of female 33-40 (38); tarsus of male 78-84 (81), of female 70-78 (75). Weight of 3 males 504.5, 509, 595, of 2 females 390, 870 (Colston & Curry-Lindahl 1986, Urban *et al.* 1986).

GEOGRAPHICAL VARIATION Three races have been described: *haematopus* of Liberia to N Zaïre and W Uganda, upperparts brown with buffy feather fringes; *petiti* of Gabon to C Zaïre, upperparts more russet and underparts warmer brown with broader grey feather fringes; *whitesidei* of SC and E Zaïre, generally darker in plumage. In view of the great individual variation in plumage within the ranges of all races, and the overlap of characters between races, it is considered best to treat the species as monotypic until more extensive studies have been made (Urban *et al.* 1986).

MOULT Primaries are moulted in slow ascending order, and primary moult is recorded in Feb (Gabon), Jul (Nigeria), Aug (Tanzania) and Oct-Nov (Zaïre).

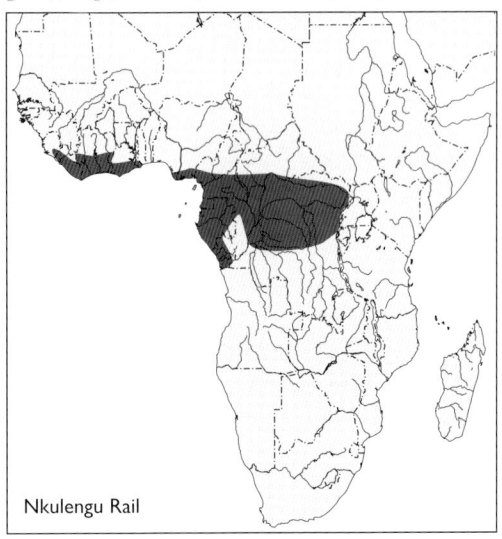

Nkulengu Rail

DISTRIBUTION AND STATUS Sierra Leone, coastal Liberia and Mt Nimba, S Ivory Coast, S Ghana and Togo; also S Nigeria (1 old record from Degema and 2 seen near Lekki Conservation Centre in Sep 1992), S Cameroon, SW Central African Republic, Equatorial Guinea, Gabon, Congo, Angola (Cabinda enclave) and coastal Zaïre (to Mayombe Forest, mouth of Zaïre R), through N and NE Zaïre E to Semliki R and Kivu (Kitutu and Kamituga) and S to C Zaïre (Sankuru R), and in W Uganda (Bwamba Forest) (Louette 1981, Pinto 1983, Thiollay 1985, Urban *et al.* 1986, Grimes 1987, Ash *et al.* 1991, Dowsett & Forbes-Watson 1993, Elgood *et al.* 1994, Field 1995). It is locally frequent to common throughout most of its range; in Gabon, the spacing of calling birds suggested a density of 3 pairs per 200ha (Brosset & Erard 1986). It is probably not uncommon in most forests in Sierra Leone, including secondary forest close to Freetown (Field 1995), while in 1990 it was thought to be widespread in swamps throughout Bwamba Forest, W Uganda, whence there was only one previous record, in 1968 (Ash *et al.* 1991). Widespread destruction of its forest habitat must have reduced its numbers in many areas and it is sometimes trapped for food by local people.

MOVEMENTS Apparently sedentary.

HABITAT Lowland rain forest, usually dense primary or old secondary growth, where it inhabits rank vegetation along streams and rivers, on islands and sometimes in swampy or marshy areas; it occasionally occurs in mangroves (Urban *et al.* 1986). It is also found in forest away from water, and in areas disturbed by logging; it completely avoids the seasonally flooded forests occupied by the sympatric Grey-throated Rail (11) (Dowsett & Dowsett-Lemaire 1991).

FOOD AND FEEDING Snails, millipedes, insects including ants and beetles, small amphibians and hard seeds. Its short toes suggest that it is adapted to walking over relatively firm ground (Chapin 1939) and it commonly forages along small watercourses and among leaf-litter and sticks. It probably takes larger prey, and forages on drier substrates, than does the Grey-throated Rail. In Gabon, it has been seen with large parties of insectivorous birds in the dry season, and also in bird parties following driver ant columns (Brosset & Erard 1986).

HABITS Shy and seldom seen, but its presence is often advertised by its distinctive call. Although most active in the morning and evening, it also forages during the day; it roosts at night in bushes or low trees up to 15m above the ground (Brosset & Erard 1986), and flies into trees when disturbed. Its posture is more upright than that of other rails.

SOCIAL ORGANISATION Probably monogamous and territorial; its frequent calling during the non-breeding season suggests either that it is permanently territorial or that the song plays a role in the maintenance of the pair bond. Foraging birds are usually seen in groups of 2-3, once of 8 (Field 1995).

SOCIAL AND SEXUAL BEHAVIOUR Members of a pair roost together side by side on a horizontal branch (Brosset & Erard 1986).

BREEDING AND SURVIVAL Season Breeding recorded Cameroon, Sep; Gabon, fresh nest Feb; Zaïre, Feb, probably also Mar and Sep (oviduct egg: Bates 1927). **Nest** Only 2 described: 1 in Gabon was a substantial structure of coarse tangled twigs and leaves, 35cm in diameter and depth, placed 1.2m above ground on heap of brushwood in undergrowth of plateau forest far from water; another said to be placed in a tree. **Eggs** Apparently only 1 clutch of 3 found, and 1 oviduct egg (Bates 1927, Chapin 1939, Urban *et al.* 1986): wide oval, not equally pointed at both ends; not glossy; white to creamy-white, with small spots and blotches of red-brown, brown or grey markings, most numerous at small end; size 49-50.5 x 38-39. Schönwetter (1961-62) gives measurements of 7 eggs: 47.9-55.2 x 36.1-38.2 (50.4 x 37.3); calculated weight 38.5.

CANIRALLUS

A primitive genus, closely allied to the *Rallina* forest-rails of New Guinea and comprising two medium-sized, forest-dwelling species, one in Africa and one in Madagascar. The upperparts are predominantly olive-brown, the face grey, the neck, breast and tail chestnut, and the rest of the underparts barred brown and white. The remiges, some upperwing-coverts, the axillaries and the underwings are prominently spotted or barred with white.

11 GREY-THROATED RAIL
Canirallus oculeus   Plate 4

Gallinula oculea Hartlaub, 1855, Rio Butri, Ghana.

Monotypic.

Synonyms: *Canirallus batesi*; *Rallus oculeus*; *Hypotaenidia/Rallina oculea*.

IDENTIFICATION Length c. 30cm. A distinctive, medium-large but quite slender rail. Forepart of head grey; chin and throat grey to white; crown and nape brown; rest of upperparts largely olive-brown, but uppertail-coverts and tail reddish-chestnut. Outer upperwing-coverts, and flight feathers, blackish with large whitish spots and bars. Underparts reddish-chestnut; lower flanks to undertail-coverts olive-brown, narrowly barred whitish to rufous-buff. Iris red; orbital ring yellowish to brownish; bill yellow-green, black on culmen and tip; legs and feet brown. Spots and bars on upperwings obvious in flight but birds unwilling to flush; underwings also boldly spotted white. Sexes alike. Immature has forehead, face and throat brown; upperparts browner than in adult; underparts olive-brown, barring buffier than in adult. Juvenile duller on head and neck but elsewhere brighter, more russet-tinged, than immature; barring as immature. Inhabits wet areas in lowland rain forest.

Similar species Plumage unlike that of any sympatric rail of forest habitat. Smaller allopatric Madagascar Wood-rail (12) differs in having white chin and throat; upperwing-coverts extensively chestnut, with no white barring; spots on flight feathers confined to inner webs; bill blackish, blue-grey on tip and cutting edge.

VOICE The voice is described in some detail by Dowsett & Dowsett-Lemaire (1991). The song lasts 2-3 min; it starts with hollow notes like double drumbeats and after c. 20 s the bird intersperses soft *dou* or *douah* notes, which increase towards the end of the call as the drumbeats fade. It is far-carrying and ventriloquial, and is given both day and night, with a peak at dawn; birds call from a perch 2-

6m above water. There is also a loud, explosive booming of 3-6 notes on a descending scale of half-tones *oue-oue-oue...*, lasting 0.5-1s. Probable alarm calls are rendered *ptik-ptik-ptik...* (Brosset & Erard 1986), *douk-douk-douk...* and a muffled *thouk-thouk....* Captive birds also gave a loud snore, lasting 2s, a soft *coo* and a short *chunk*.

DESCRIPTION
Adult Forehead, forecrown, lores, cheeks and ear-coverts grey; crown and nape brown; hindneck reddish-chestnut, distinctly washed olive-brown in central region; mantle to rump, including scapulars and lesser upperwing-coverts, olive-brown to deep olive; uppertail-coverts and tail reddish-chestnut, occasionally with olive-brown wash. Rest of upperwing-coverts blackish-brown, tipped buff to pale rufous and with white to pale rufous spots or bars, often margined black; a few lesser or marginal coverts may have small white spots. Alula, primaries and secondaries blackish-brown with large, oblong white spots on both webs; outer web and tip of innermost secondaries washed olive-brown; tertials olive-brown, with a few rufous spots or bars at edges of both webs. Axillaries and underwing-coverts blackish-brown with large white spots. Chin and throat grey to whitish; sides of neck, foreneck, breast, upper flanks and upper belly reddish-chestnut, lower belly, lower flanks, thighs and undertail-coverts olive-brown with narrow whitish to rufous-buff bars often bordered dark brown to brownish-black. Iris red; orbital ring yellow-green to pale brown; bill yellow-green with black culmen and tip, to black with sides and base of lower mandible green; legs and feet brown. Sexes alike; female may have less grey on side of head, this colour not extending over ear-coverts.
Immature Similar to adult but forehead, face and throat brown, or grey-tinged brown, not grey; upperparts browner, less olive, sometimes with russet wash; appears to retain juvenile upperwing-coverts, with tawny tips; spots on wings smaller and tinged buffy; anterior underparts olive-brown, washed russet; lower flanks to undertail-coverts olive-brown to bright buffy-brown, with pale tawny-buff to cinnamon-buff bars.
Juvenile Similar to immature but head and neck duller; upperparts often washed brighter russet, especially on fringes of lesser coverts; outer lesser coverts with narrow, tawny to cinnamon, tips; medians and greaters with cinnamon to buffy tips; underparts buffy-brown, tinged russet; lower flanks, and belly to undertail-coverts, with narrow buffy bars.
Downy young Chick has velvety down, blackish-brown when raised but more olive-brown when flattened. One BMNH specimen from Cameroon, 1908, has bill ochraceous-buff, with blackish-brown diagonal band on upper mandible from cutting edge below nostril and running forwards in front of nostril; base of culmen, and proximal half of lower mandible, also blackish-brown; legs and feet brownish. Small white wing-claw present.

MEASUREMENTS (8 males, 5 females) Wing of male 167-179 (173), of female 167-180 (173); tail of male 59-70 (64), of female 58-70 (65); culmen to base of male 37-41 (39), of female 37-38 (37); tarsus of male 49-55 (52), of female 50-57 (52). Weight of 1 male 278, of 2 females 273, 279.

GEOGRAPHICAL VARIATION Birds from Cameroon eastward are sometimes treated as a separate race, *batesi*, on the basis of darker chestnut colouring on the underparts, especially on the throat and breast, but differ too little from other populations to merit subspecific status (e.g. Urban *et al.* 1986).

MOULT Primary moult is partly synchronised, tending to ascendant, and is recorded in Dec (Congo) and Jan (Cameroon). Heavy postjuvenile body moult is recorded in Jun (Ghana).

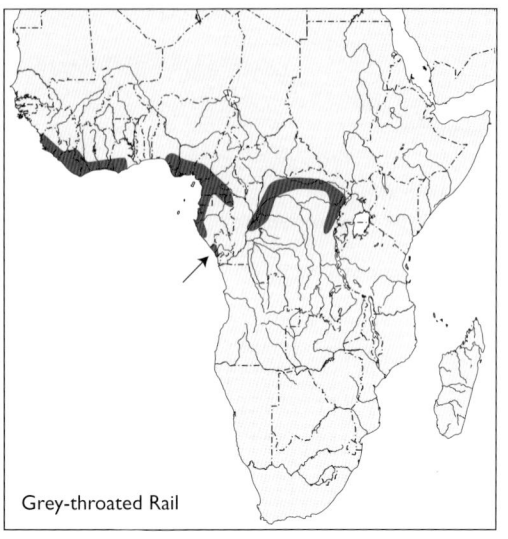

Grey-throated Rail

DISTRIBUTION AND STATUS The known distribution is disjunct: S Sierra Leone (Gola Forest and Tingi Hills), extreme SE Guinea, Liberia, S Ivory Coast (Azagny and Taï to Maraoué) and SW Ghana; S Nigeria, S Cameroon, Equatorial Guinea, N and coastal Gabon and coastal Congo; across N Zaïre E to Ituri district and S Kivu (Bikili and Kibimbi), and to W Uganda (Bwamba Forest) (Louette 1981, Thiollay 1985, Urban *et al.* 1986, Grimes 1987, Gatter 1988, Morel & Morel 1988, Dowsett & Forbes-Watson 1993, Dowsett *et al.* 1993, Elgood *et al.* 1994, Field 1995). Its occurrence in N Central African Republic (Manovo-Gounda-Saint Floris National Park) is doubtful (Carroll 1988). It is regarded as locally numerous only in the Kouilou Basin, Congo, where its density along the Nanga River was estimated at 10-16 calling birds per km^2 (Dowsett & Dowsett-Lemaire 1991). Elsewhere it is apparently uncommon or rare, but its extreme secretiveness, and the unfamiliarity of most observers with its calls, render it very easy to overlook and make its status difficult to assess. In Southwest Province, Cameroon, it is known from two records in the Korup National Park but may be more frequent in seasonally flooded forests along rivers from the S of the park to the coast (Rodewald *et al.* 1994). There are only two old undated records from Nigeria (Benin Province), where its habitat is rapidly disappearing and it may be extinct, and only one old record from Bwamba Forest, W Uganda (Urban *et al.* 1986, Elgood *et al.* 1994). Forest destruction must have affected its numbers adversely over much of its range.

MOVEMENTS None recorded.

HABITAT Primary and secondary lowland rain forest, occurring in ravines and along creeks and forest streams overhung by trees and bordered by rank undergrowth; also swampy or seasonally flooded forest, including areas with mud, tall arrowroot plants and tree ferns (Brosset & Erard 1986), and marshes within forest regions (Ghana, Grimes 1987). The birds normally keep within thickets or patches of fallen branches, avoiding areas of more open water between tall trees (Dowsett & Dowsett-Lemaire 1991). Along the Nanga R, Congo, they occur in low thickets dominated by *Alchornea, Ficus* and *Nauclea* over water, away from the forest proper (Dowsett & Dowsett-Lemaire 1991). In view of its known habitat preferences, its occurrence at "floodplain" in the savanna-park zone of N Central African Republic (Carroll 1988) is very doubtful.

FOOD AND FEEDING Skinks, snails, slugs, small crabs, millipedes and insects including ants, caterpillars and other larvae, and beetles; stomach contents also recorded as pebbles, quartz grains and bits of weed. It has been seen foraging on the half-dry bed of a small forest stream by removing dead leaves with jerking movements of the bill (Demey & Fishpool 1994).

HABITS Extremely shy and very rarely seen; most specimens have been caught in traps. When flushed, the flight is said to be weak and indirect (Mason, P. F. 1940).

SOCIAL ORGANISATION Presumed territorial; in Congo, the territory size was estimated as 6-10ha (Dowsett & Dowsett-Lemaire 1991).

SOCIAL AND SEXUAL BEHAVIOUR No information available.

BREEDING AND SURVIVAL Season Breeds during rains: Nigeria, Apr; Cameroon, Feb, Apr, juvenile Jul; Zaïre, Nov, breeding condition Sep, Dec. **Nest** 2 described; of broad grass leaves, 1 on a stump in swampy bottom of ravine among *Raphia* palms and undergrowth of *Canna*-like plants, another on a low pollarded stump at swampy margin of forest stream; a third (very probably of this species) among roots of uprooted tree over a stream bank (Bates 1927, Serle 1959, Urban *et al.* 1986). **Eggs** 2-3; sometimes glossy; long ovals equally pointed at both ends; creamy-buff with spots of chestnut or mauve-brown and lavender; size of 6 eggs 42.5-44.4 x 30.8-32.5 (43.2 x 31.6); calculated weight 23.5.

12 MADAGASCAR WOOD-RAIL
Canirallus kioloides   Plate 4

Gallinula kioloides Pucheran, 1845, Madagascar.

Formerly placed in the monospecific genus *Mentocrex* on the basis of its imperforate nostrils (those of *C. oculeus* are perforate), but now this character is not considered generically important. Two subspecies recognised.

Synonyms: *Mentocrex/Rallina/Porzana kioloides*; *Corethrura/Ortygometra/Eulabeornis/Gallinula/Rallus/Canirallus griseifrons*.

Alternative names: Grey-throated Wood-rail; Kioloides Rail; Wood Rail.

IDENTIFICATION Length c. 28cm. A medium-sized, predominantly chestnut and olive-brown rail, grey on forehead, around eye and on ear-coverts, and pure white on chin and throat (often bordered by black spots). Crown to back olive-brown (more olive-green in race *berliozi*); lower back to tail chestnut. Lesser upperwing-coverts chestnut; others olive-brown, washed chestnut; primary coverts, primaries and secondaries blackish, usually barred white only on inner webs. Neck, breast and upper belly chestnut; rest of underparts dark brown, barred whitish to rufous (belly barred pale grey-green and brownish in *berliozi*). In the field, lores may appear noticeably white, rather than grey (A. Riley *in litt.*). Iris brown; bill blackish, with blue-grey tip and cutting edge; legs and feet reddish-brown. Sexes alike. Immature duller than adult, with less grey on head; undertail-coverts spotted yellowish. Juvenile has upperparts washed chestnut to tawny; chestnut of underparts mottled dull brown; bars on posterior underparts buff to rufous. Secretive, but readily located by its frequent calling. Inhabits forest.

Similar species May be confused with superficially similar Madagascar race of White-throated Rail (67), which is larger and less stocky, has no grey on head, more extensive white throat patch, vinous-chestnut crown and nape (like rest of head, neck and breast), dark olive-green upperparts with black streaking, no bars on wings, white lateral undertail-coverts, and longer, reddish bill. Grey-throated Rail (11) of Africa larger; has grey to whitish chin and throat with no black spots at border; no rufous on upperwing-coverts; whitish barring on outer upperwing-coverts and flight feathers; and yellow-green to green bill with black on culmen and tip.

VOICE A rather vocal species, its common call being a series of loud, piercing whistles with a rising inflexion. Feeding birds constantly emit muffled, throaty chortles reminiscent of the contact call of the Brown Lemur *Lemur fulvus*, very brief, sharp metallic notes, and also staccato *nak* notes sometimes speeded up to end as a rattle (Langrand 1990). It also utters muffled *bub* notes (difficult to locate) when disturbed, and harsh clucks in response to an imitation of its whistled call (Rand 1936); it responds well to playback of this common call, but is less responsive in areas where taped playback is frequently practised (A. Riley in litt.). When captured, a downy chick gave a high, thin, repeated *tee*, which was answered by an adult with *bub* calls (Rand 1936).

DESCRIPTION *C. k. kioloides*
Adult Forehead, lores and feathers round eye ashy-grey, this colour sometimes extending irregularly over ear-coverts, and even occasionally to hindcrown; grey area of forehead to crown often mottled olive-brown to blackish-brown (dark feather tips); rest of face and ear-coverts, and malar region, bright chestnut, like sides of neck; chin and throat white, often distinctly edged with black dots. Crown to back, including scapulars, bright, dark olive-brown; lower back to tail bright chestnut. Lesser upperwing-coverts bright chestnut (inners more olivaceous); medians and greaters bright olive-brown with variable chestnut wash; primary coverts more blackish-brown, fringed chestnut. Alula, primaries and secondaries blackish-brown, sometimes with small rufous spots on outer webs; inner webs (and rarely basal area of outer webs) with elongated white spots or short bars; outer webs of inner secondaries, and all of tertials, ruddy-olive, tinged rufous. Underwing-coverts and axillaries blackish-brown, broadly barred white; underside of remiges somewhat paler, with elongated spots or bars of white or fulvous. Sides of neck, foreneck, and breast to upper belly, bright chestnut; lower belly, flanks, thighs and undertail-coverts darkish brown, barred whitish to tan; bars of undertail-coverts often rufous. Iris brown; bill blackish, blue-grey towards tip and along cutting edge; legs and feet reddish-brown. Sexes alike.
Immature Duller than adult; less extensively grey on head; has yellowish spots on undertail-coverts.
Juvenile Olive-brown of crown to hindneck replaced by

181

pinkish-cinnamon to vinaceous-tawny, feathers with dark bases; olive-brown of upperparts washed chestnut; remiges as adult but no markings on outer webs (2 specimens); chestnut feathers of underparts extensively dull brown basally, giving irregularly mottled effect; barring on darkish brown posterior underparts is buff to rufous, paler (whitish) on thighs.

Downy young Down velvety. Head and upperparts black, except for light rufous-brown stripe along each side of back and rump, and rufous-brown forehead, supercilium, ear-coverts, throat and sides of neck. Wings black, flecked tawny; upper breast rufous-brown; lower breast black, flecked pale rufous-brown; belly pale rufous-brown, mottled black. Brown down on head, neck and underparts is tipped black, giving mottled appearance to these regions. Iris greyish-black; bill greyish-black, tipped whitish; legs and feet black. Small egg-tooth present.

MEASUREMENTS (10 males, 4 females) Wing of male 127-143 (132.9, SD 4.2), of female 129-135 (132, SD 2.6); tail of male 49-59 (54.5, SD 3.4), of female 45-55 (50.25, SD 4.6); culmen to base of male 29.33 (27.9, SD 9.2), of female 28.5-30.5 (29.1, SD 0.9); tarsus of male 39-47 (42.6, SD 2.5), of female 39-43 (40.75, SD 1.7). Weight of 4 unsexed 258-280 (269).

GEOGRAPHICAL VARIATION Involves size and plumage, and is incompletely understood. Birds from Tsingy de Bemaraha, W Madagascar, are apparently sufficiently different in plumage coloration to be regarded as an undescribed subspecies (O. Langrand *in litt.*).

C. k. kioloides (Pucheran, 1845) – E half of Madagascar, including the High Plateau. See Description.

C. k. berliozi (Salomonsen, 1934) – Sambirano, extreme NW Madagascar. Slightly larger and paler. Grey on forehead extends beyond hind border of eye; white area of throat longer, and pointed at base; occiput yellowish-green; mantle paler, more greenish-olive; rump, and neck to upper belly, cinnamon; belly barred pale grey-green and brownish. One specimen moulting out of juvenile plumage has some pinkish-cinnamon feathers at sides of crown and on upper ear-coverts, and is somewhat duller olive on upperparts than adult. 8 males, 2 females: wing of male 128-141 (134, SD 4.4), of female 134, 135; wing of unsexed 130-141 (Ripley 1977); tail of male 52-58 (54.1, SD 2.1), of female 57, 61; culmen to base of male 31.5-35 (32.9, SD 1.3), of female 32, 32.5; tarsus of male 42.5-46 (43.95, SD 1.1), of female 42, 45.

MOULT Primary moult is partly synchronised, tending to ascendant, and is recorded in May-Jul. Postjuvenile head and body moult is recorded Jan.

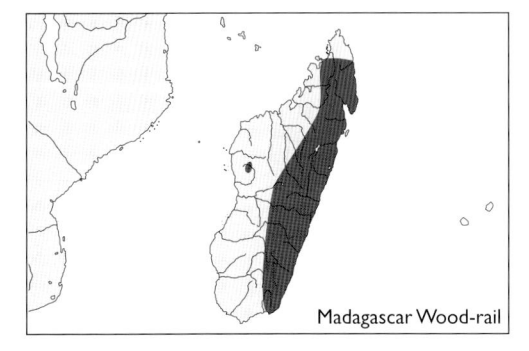

Madagascar Wood-rail

DISTRIBUTION AND STATUS Information comes from O. Langrand (*in litt.*) unless otherwise stated. Endemic to Madagascar, where it is widespread in the E, from the Bemangrily area N to Sambirano in the extreme NW; also recorded in the W, at Tsingy de Bemaraha. The nominate race was regarded as common in 1929-1931 but by the 1970s much of its habitat on the coastal plain had been destroyed (Rand 1936, Keith 1978). At present it is common at middle altitudes but is affected by the general and serious continued destruction of forest habitat, chiefly by slash-and-burn itinerant cultivation and exploitation for firewood. The race *berliozi* is apparently still fairly common in Sambirano Domain (Langrand 1990) but forest habitats in this relatively small area are severely threatened, principally by clearing for rain-fed rice growing and coffee, and have been much degraded outside reserves. It occurs in the Special Reserve of Manongarivo, which is still largely intact (c. 70% of the area is still covered with good forest). It is not known which subspecies occurs on the nearby Tsaratanana massif, which includes large forest blocks and a Strict Nature Reserve; *berliozi* should be regarded as NEAR-THREATENED by habitat destruction unless its occurrence there is confirmed (O. Langrand *in litt.*). The Tsingy de Bemaraha population, first observed in July 1987 and restricted to humid forest in the Tsingy foothills, may be small but is relatively safe because the whole of Tsingy is legally protected as a Strict Nature Reserve.

MOVEMENTS None recorded.

HABITAT Inhabits undisturbed rain forest, slightly degraded contiguous secondary growth with fairly sparse herbaceous ground cover, woodland watercourses and the edges of ponds and marshes with reeds and papyrus. Forest habitat is characterised by an open understorey and a substrate of leaf-litter (A. Riley *in litt.*). It has also been seen in dry deciduous forest on a karstic substrate (Langrand 1990). It occurs from sea level to 1,550m.

FOOD AND FEEDING Only insects, amphibians and seeds are recorded. Birds normally forage in pairs, covering the ground swiftly in the underbrush, stopping suddenly to probe litter and then moving on, sometimes returning several times to the area just searched; as the bird moves, the head is bobbed and the tail flicked (Langrand 1990).

HABITS Secretive and elusive, keeping to the interior of forest and dense cover, but readily detected by its frequent calling. When disturbed the birds normally escape by running, taking wing only in extreme danger. Although these birds are normally seen only on the forest floor, one was once flushed from a perch 4m up in a bush (Rand 1936). In response to taped playback, birds make wide circles around the observer, moving extremely quickly and then freezing (A. Riley *in litt.*).

SOCIAL ORGANISATION Apparently monogamous: usually seen in pairs.

SOCIAL AND SEXUAL BEHAVIOUR No details available.

BREEDING AND SURVIVAL Season Lays May, Jun, Nov; breeding condition Oct; 1 juvenile, almost fully grown, Dec. **Nest** A roughly made bowl of grass, ferns and leaves, placed 2-3m above the ground in bush or tangle of creepers. **Eggs** 2; pinkish-white with sparse rufous and grey speckling at larger end; 2 measure 35 x 27.2 and 37 x 26 (Schönwetter 1961-62); average size 42 x 32 (Ripley 1977).

COTURNICOPS

Three species of small, short-billed rails, one from eastern Asia, one from North America and one from South America. The plumage in the two Holarctic species is superficially somewhat quail-like and all species have white secondaries, a character which they share with the White-winged Flufftail (8). They inhabit wet grassland and marshes. The Holarctic species are migratory, and it has been suggested that the Speckled Rail of South America is either migratory or eruptive, but there is no proper evidence for this and the bird's distribution is imperfectly known. Serial polygyny has been observed in captive Yellow Rails; the breeding habits of the other two species are almost unknown. One species, Swinhoe's Rail of E Asia, is regarded as vulnerable, while the Speckled Rail should be regarded as data deficient.

13 SWINHOE'S RAIL
Coturnicops exquisitus Plate 5

Porzana exquisita Swinhoe, 1873, Yantai (Chefoo), Shandong, China.

Sometimes considered conspecific with *C. noveboracensis*, with which it forms a Holarctic superspecies. Monotypic.

Synonyms: *Coturnicops noveboracensis exquisita*; *Ortygops exquisita*; *Crex/Porzana erythrothorax/undulata*.

Alternative names: Swinhoe's/Asian Yellow Rail; Siberian Rail.

IDENTIFICATION Length 13cm. A very small crake, with white secondaries which are obvious in flight. Top of head and entire upperparts blackish-brown, with prominent cinnamon stripes and narrow white bars from mantle to tail; face greyish-brown with indistinct dark stripe through eye; entire underparts white, with foreneck, upper breast, flanks and undertail-coverts barred tawny-ochre and (often) dusky. Iris brown; bill dark brown above, greenish-yellow below; legs and feet flesh-brown. Sexes alike. Non-breeding adult has small white flecks on nape, hindneck and sides of neck, off-white throat, and some dark spots on breast. Immature like adult non-breeding; juvenile not described. Inhabits wet meadows and short-grass marshes.

Similar species Easily separable from all sympatric rallids by white secondaries and white barring on upperparts. Closely related Yellow Rail (14) of North and Central America very similar in appearance but is larger, and differs in having tawny-buff face with more distinct broad dark stripe through eye, paler and more tawny-buff to tawny-yellow streaks on upperparts, tawny-buff foreneck and breast with indistinct darker feather tips, less white on underparts, white and tawny bars on flanks, and brownish or greenish legs and feet.

VOICE The voice is not recorded, but is presumably similar to that of the Yellow Rail.

DESCRIPTION
Adult breeding Forehead, crown, nape, hindneck, entire upperparts and upperwings dark blackish-brown, with broad tawny-olive to cinnamon feather edges, narrow on crown and broader elsewhere, forming stripes along upperparts; tail blackish-brown, narrowly edged cinnamon; small white spots on hindneck; narrow white bars from mantle to tail and on upperwings, less distinct from rump to tail. Lesser upperwing-coverts tawny-olive to isabelline, narrowly barred or spotted white; medians and greaters as upperparts, but fringes brighter cinnamon; primary coverts plain grey-brown with tawny outer edges; marginal coverts white. Primaries dull buffy-brown to grey-brown with narrow pale edges; outer web of outermost primary (P10) largely whitish; innermost primaries with paler inner webs, whitish shaft-streaks, and white blotches towards tip of inner webs; secondaries dull buffy-brown, distal half of each feather white, first and last secondary often with buffy-brown blotches in white area; tertials as upperparts; axillaries and underwing-coverts white. Lores and indistinct stripe through eye brownish-black with whitish mottling; sides of head greyish-brown with dusky and whitish mottling; chin and upper throat whitish; lower throat, foreneck, sides of neck, breast, flanks and undertail-coverts dull ochre to tawny, barred white (spotted on sides of neck) and often with darker, dusky brown feather centres or bars, especially on flanks and thighs; centre of breast paler, barred white and dull ochre to tawny; lower breast and belly white. Iris brown; upper mandible deep brown, lower mandible greenish-yellow; legs and feet light flesh-brown, darker on joints and claws. Sexes alike.

Adult non-breeding Nape, hindneck and sides of neck have small white flecks; throat off-white; some coarse dark spots extend across breast; possibly less extensive white towards base of secondaries. Upper mandible brown, yellowish at base; lower mandible greenish-brown; legs and feet light brown.

Immature Apparently indistinguishable from adult non-breeding.
Juvenile Not described.
Downy young Not described.

MEASUREMENTS (2 males, 1 female, 4 unsexed) Wing of male 75, 76, of female 80, of unsexed 77-81 (78.75, SD 1.7); tail of male 32, 34, of female 35, of unsexed 29-33 (31.5, SD 1.1); culmen to base of male 13.5, 14.5, of female 13.5, of unsexed 12-14.5 (13, SD 1.1); tarsus of male 22.5, 21.5, of female 22.5, of unsexed 21.5-22.5 (22, SD 0.6).

GEOGRAPHICAL VARIATION None.

MOULT No information available.

DISTRIBUTION AND STATUS This poorly known species breeds in SE Transbaikalia, S Ussuriland and E Heilongjiang (Manchuria) and winters from Japan (Hokkaido, Honshu, Shikoku and Kyushu) and Korea through the Ryukyu Islands (Miyako-jima, and Amami-oshima or Okinawa) to SE China (Fujian and Guangdong). Classed as VULNERABLE, it is regarded as very rare in SE Siberia and rare in Japan (Potapov & Flint 1987, Brazil 1991). It is known from only a few breeding localities, having been recorded recently at Zhalong Nature Reserve, Heilongjiang, and L Khanka, N of Vladivostok (Scott 1989, Collar *et al.* 1994). It is seldom recorded on passage or in winter, recent records being of small numbers on passage at Beidaihe (E China), where it is considered uncommon, and of one bird at Poyang L (S China) in 1989 (Williams *et al.* 1992, Collar *et al.* 1994). The species is presumably threatened by the destruction and modification of wetlands which are occurring in both its breeding and wintering ranges.

MOVEMENTS Poorly known. Potapov & Flint (1987) mention records of migrants at L Khanka in late Apr. Birds are known to winter in SE China (Fukien and Guangdong) and rarely in Japan, and the species is regarded as transient or wintering in N and S Korea and the Ryukyu Islands (Cheng 1987, Brazil 1991, Collar *et al.* 1994). In China, migrants are recorded in Laioning, Shandong, Sichuan and (possibly) Hebei Provinces, and at the Changjiang River (Cheng 1987), while records of 1-2 birds on migration at Beidaihe (280km E of Beijing) are dated 28-30 Sept, 2-14 Oct and 20 and 27 May (Williams *et al.* 1992). Records from Japan fall between 4 Aug and 16 May, most records being in Oct-Apr (Brazil 1991).

HABITAT Breeds in wet meadows and short-grass marshes, its breeding habitat probably being similar to that of the Yellow Rail. On migration and on the wintering grounds it frequents marshes, wet meadows and rice fields.

FOOD AND FEEDING Nothing is recorded, but food is presumably similar to that of the Yellow Rail.

HABITS Difficult to flush; a female collected in SE China was flushed from underfoot and flew for a very short distance with a very feeble and slow flight, like a butterfly (specimen label, BMNH).

SOCIAL ORGANISATION No information available.

SOCIAL AND SEXUAL BEHAVIOUR No information available.

BREEDING AND SURVIVAL Very little information available. Breeding habits probably similar to those of Yellow Rail. Two 19th century records from former USSR are from Transbaikalia (nest with 3 eggs) and L Khanka. Three eggs from "Albacun (or Albasin), Amur-land" (BMNH), dated 15 Jul 1906, are ovate and slightly glossy, creamy-buff, thickly spotted buffy-brown; size 28.85 x 20.71, 29.45 x 22.1 and 29.83 x 20.85. Dement'ev *et al.* (1969) state that eggs resemble small eggs of Corncrake (69), have a pinkish-yellow ground colour with scattered small spots of reddish-brown and lilac-grey, and measure 28-28.3 x 20-20.4; breeding recorded at Darasun (Dauria). Schönwetter (1961-62) and Potapov & Flint (1987) give measurements of 13 eggs as 27-30 x 20-22 (28.5 x 20.7); calculated weight 7; the latter authority states that spots on eggs may run together to form cap at one end.

14 YELLOW RAIL
Coturnicops noveboracensis   Plate 5

Fulica noveboracensis Gmelin, 1789, New York.

Forms a superspecies with *C. exquisitus*, with which it is sometimes considered conspecific. Two subspecies recognised.

Synonyms: *Ortygops/Ortygometra/Porzana/Rallus noveboracensis*; *Porzana goldmani*.

Alternative name: American Yellow Rail.

IDENTIFICATION Length 16-19cm. The largest *Coturnicops* species. A predominantly tawny-yellow and blackish crake; crown and broad stripe through eye brownish-black; broad supercilium and sides of head tawny-buff, mottled brown; upperparts brownish-black (darker in race *goldmani*) with tawny-yellow to tawny-buff streaks and narrow white bars; primaries pale brown but secondaries white (obvious in flight); chin and throat white; foreneck and breast tawny-buff; flanks to undertail-coverts dark brown to blackish, with tawny-buff and narrow white bars (undertail-coverts more cinnamon in race *goldmani*); belly white. Iris reddish to yellowish; legs and feet brownish to greenish. Sexes alike, but male slightly larger than female; bill of breeding male corn-yellow, of female and non-breeding male dark olivaceous to black. Immature darker overall than adult (see Description for details of conflicting plumage descriptions): upperparts black, margined tawny and barred or spotted white, especially on crown and neck (immature *goldmani* lacks white markings on face and upper flanks); sides of neck mottled black, white and buffy. Juvenile not properly described. Inhabits marshes and wet grassland.

Similar species Easily distinguished from all other North American rallids on size and plumage; superficially similar to neotropical Yellow-breasted Crake (105), which is smaller, lacks white on secondaries and has white-streaked upperparts, a broken white supercilium, bold black and white barring from flanks to undertail-coverts, a bluish-black bill and yellowish legs and feet. Closely related Swinhoe's Rail (13) of E Asia has greyer face, with closer dark speckling and less distinct dark streak through eye; darker and more cinnamon stripes on upperparts; foreneck, upper breast, flanks and undertail-coverts barred tawny-ochre, white and dusky; more white on lower breast and belly; legs and feet brownish-flesh. Larger than sympatric Black Rail (33), with proportionally shorter, stouter bill.

VOICE Most information is from Bookhout (1995). The breeding-season song of the male consists of a series of metallic clicks, usually in a five-note pattern *click-click, click-click-click* but sometimes a four-note *click-click, click-click,* easily imitated by tapping two pebbles together, and each series lasting c. 1s. Males respond to such imitations, especially at night, often approaching closely enough to be captured with a long-handled net. This call is given mostly at night, sometimes by day, and sometimes continuously for long periods, up to 17 min being recorded (Fryer 1937); it is audible for up to 1km, and calling birds remain stationary (Robert 1997). Males cease calling as late in the season as Jul-Aug. There is one recorded observation of possible duetting between a captive male and female. Both sexes give cackles, clunks (like distant knocking on a door) and frog-like croaks when breeding, wheezes during hostile encounters and squeaks when retreating from hostile encounters. The female calls chicks with dog-like whines, and gives soft moans when brooding the young. Chicks give *wee* and *peep* calls, while juveniles give peeps and barks.

DESCRIPTION *C. n. noveboracensis*
Two conflicting descriptions of adult and immature plumages exist for nominate race. Some authors (Roberts, T. S. 1932, Walkinshaw 1939, Ripley 1977) describe immature plumage as much darker than adult, with white bars and spots on head and neck, while others (Forbush 1925, Bent 1926, Ridgway & Friedmann 1941) maintain that the darker plumage is adult, and that immature is generally paler. We follow Stalheim (1975), who describes paler plumage as adult and darker, more spotted, plumage as immature; this also agrees with plumage differences given in descriptions of race *goldmani* (Dickerman 1971).
Adult male Forehead to nape brownish-black; hindneck to upper back more rufous-black; lower back to tail, and scapulars and tertials, black. Feathers narrowly edged tawny-buff on crown, more broadly edged elsewhere to form tawny-buff to tawny-yellow stripes along upperparts, and each feather having at least two narrow white bars which are less distinct from rump to tail. Primaries pale brown, inners sometimes tipped white (especially when fresh), outer web of P10 white except at tip; secondaries white, but pale dull brown towards base, on outer web of outermost 3, and on most of inner web of outermost; alula and primary coverts grey-brown to mid-brown, sometimes with small white subterminal spots; outer web of outermost alula feather edged white; marginal coverts white; other upperwing-coverts like back; axillaries and underwing-coverts white, greyish-brown basally. Lores blackish; stripe from rear of eye across ear-coverts to sides of neck brownish-black; broad supercilium, and sides of head, pale tawny-buff, mottled dull brown; chin and upper throat whitish, washed pale tawny-buff; lower foreneck, sides of neck and entire breast pale tawny-buff, feathers with variable narrow dark tips; centre of lower breast and belly white to creamy-white; sides of belly more creamy-buff; upper flanks tawny-buff, feathers variably tipped or centred dull brown to blackish-brown and variably barred subterminally with white; lower flanks to undertail-coverts blackish-brown, barred white and tawny-buff. Iris yellowish-brown to reddish; bill corn-yellow in breeding season, olive-green to olive-black after breeding; legs and feet cinnamon-drab, brownish or greenish.
Adult female Similar to adult male but slightly smaller; bill dark olivaceous to black throughout year (but one had yellow bill in summer).
Immature Darker than adult. Entire upperparts black, feathers margined tawny and barred or spotted throughout with narrow white lines, especially from crown to hindneck; buff line over eye less distinct than in adult; side of neck mottled dull black, white and buffy; lower neck, breast and sides darker buff than in adult; secondaries possibly less extensively white than in adult. Ridgway & Friedmann (1941) also describe a pale morph of both adult and immature plumages, the significance of which is not clear.
Juvenile Not properly described (Bookhout 1995).
Downy young Downy chick all black, sometimes with faint greenish gloss on crown and throat; bill bright pink, becoming dull white (16 days) and then greyish-black (24 days) as juvenile plumage grows; legs and feet greyish-brown; small wing-claw present.

MEASUREMENTS Wing (chord) of 36 males 73-93 (86.2), of 33 females 78-91 (83.3); tail of 27 males 29-38.5 (33), of 25 females 29.5-34.5 (32.1); culmen of 36 males 12-15 (14.1), of 33 females 11.5-15 (13.1); tarsus of 27 males 22-25.5 (24), of 25 females 20-26 (21.8). Weight of 70 males 52-68 (59.2, SD 3.8), of 6 females 41-61 (52.2, SD 7.3).

GEOGRAPHICAL VARIATION Involves size, overall plumage colour of adults, and extent of spotting in immatures. Possible races *richii* and *emersoni* (see Ripley 1977) are included in nominate.
C. n. noveboracensis (Gmelin, 1789) – Breeds in C to SE Canada and NE USA, from NW Alberta, S Mackenzie, C Saskatchewan, N Manitoba, N Ontario, S Quebec, New Brunswick, W Nova Scotia and (possibly) E Maine, S to S Alberta, S Saskatchewan, North Dakota, C Minnesota, S Wisconsin, N Michigan, S Ontario, S New England and Connecticut, with an isolated population in S Oregon. Winters in S and SE USA, from coastal North Carolina S to S Florida, W along Gulf coast to C and SE Texas, and (locally and casually) from Oregon S to S California. See Description.
C. n. goldmani (Nelson, 1904) – C Mexico, in the Rio Lerma marshes at Lerma and San Pedro Techuchulco. Larger; darker on upperparts, with blacker crown and black stripes on nape; undertail-coverts may be predominantly dull cinnamon. Immature similar to that of *noveboracensis* but lacks white spotting on face and upper flanks. Wing of 3 males 90.5, 92, 93, of 1 female 87; tarsus of 3 males 26, 27, 28, of 1 female 24 (Dickerman 1971).

MOULT Information is from Bookhout (1995). Postbreeding moult is complete; remex moult occurs in Aug and the primaries and secondaries are lost simultaneously, resulting in the birds being flightless for c. 2 weeks. Postjuvenile moult includes only the body plumage; it begins in "summer and fall" and is completed in Sep-Oct (nominate race). No information on prebreeding moult is available.

DISTRIBUTION AND STATUS The nominate race is widespread but locally distributed within its breeding range, which extends across C and SE Canada and NE USA; this race winters in S and SE USA, in autumn and winter most birds being found in SW Louisiana and the E Texas rice fields (Cardiff & Smalley 1989). It may be more abundant than existing records indicate, but its breeding range has decreased during the 20th century: it formerly bred in California, and possibly also in NE USA, S to c. 40°N, and in the 19th century it was described as resident in Florida and S Louisiana. A local breeding population in S Oregon, thought to have been extirpated, was rediscovered in 1982, and the species was recorded near historic breeding sites in E California as recently as 1985 (Gaines 1988, Stern *et al.* 1993). The nominate race is listed as THREATENED or ENDANGERED in some US states, and as VULNERABLE in Quebec; it is not legal game anywhere in North America. The race *goldmani* is known only from Mexico, where it was formerly a local resident in the headwater marshes of the upper Río Lerma Valley; it has not been seen since 1964 but may still survive despite the fact that much of the area has been ditched and drained for agricultural development (Dickerman 1971, Howell & Webb 1995b).

Drainage of wetlands is probably responsible for the loss of the bird's southernmost breeding areas during this century, while in Canada many suitable wetlands have been lost since 1950 to drainage and possibly also after building dikes and barriers; in S Quebec it is estimated that at least half the potential Yellow Rail habitat along the St Lawrence R has been lost to diking and road building (Robert *et al.* 1995). Ditching and draining have destroyed several summering sites in Oregon since 1985 (Stern *et al.* 1993). Grazing pressure by cattle probably adversely affects marsh-edge habitat by reducing the height of emergent vegetation (Eddleman *et al.* 1988), but mowing may help perpetuate breeding habitat by preventing the usual vegetative succession (Robert & Laporte 1995), while periodic burning removes invading woody vegetation and is a positive management practice in breeding areas (Stenzel 1982). Mowing seems beneficial for staging and moulting areas in S Quebec (Robert 1997). Manipulation of water levels on wildlife refuges to benefit migratory waterfowl could adversely affect Yellow Rails if the objective is to provide 'hemi-marshes' or deep-water marshes (Bookhout 1995). Predation is almost unrecorded: 2 birds were killed by a raptor and remains have been found in an owl pellet (Bookhout 1995).

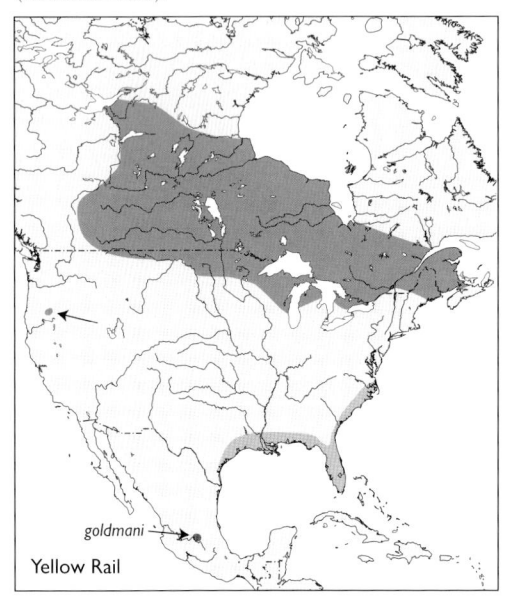

MOVEMENTS The nominate race occupies its breeding areas from late Apr or May to Sep-Oct, arriving in N USA from the last week of Apr and in S Quebec around the middle of May (Robert 1997). Autumn migration takes place from Sep to early Nov, the latest date in S Quebec being 26 Oct and the earliest arrival date in Louisiana 6 Oct (Bookhout 1995); in Michigan and Minnesota, young birds depart from natal sites in late Sep to early Oct (Walkinshaw 1939). Peak migration in Missouri is in mid-Apr, with extreme dates of 27 Mar and 5 May; autumn records fall between early Sep and 21 Oct (Robbins & Easterla 1992). Birds are thought to migrate over a broad front and at least some move in groups; individuals sometimes strike towers, suggesting nocturnal migration (e.g. Pulich 1961). In the summer, this rail is also recorded from SE Alaska, S British Columbia (AOU 1983, Sherrington 1994), Montana and Colorado, and on migration is recorded from Washington, Arizona and New Mexico, and irregularly in most of USA E of the Rocky Mountains; it is casual in Labrador and there is one record of a vagrant in Grand Bahama (Ripley 1977, AOU 1983). Birds from study sites along the St Lawrence R corridor, S Quebec, make a short-distance moult migration to staging and moulting areas at Île aux Grues (Robert & Laporte 1996, Robert 1997).

HABITAT Information is from Bookhout (1995) unless otherwise stated. The breeding habitat of the nominate race comprises the higher and drier margins of fresh- and brackish-water marshes dominated by dense, fairly low sedges and grasses, particularly fine-stemmed emergent species; also swampy meadows, sedge meadows dominated by *Carex lasiocarpa* (and also containing other *Carex* species plus *Scirpus, Juncus, Calamagrostis, Dulichium* and *Eleocharis*), and occasionally wet, cut-over hayfields; birds are rarely found in *Typha* stands. Brackish-water marshland habitats in Canada contain plants such as *Carex, Scirpus, Spartina, Juncus, Eleocharis, Hordeum, Calamagrostis, Lythrum, Sanguisorba, Agrostis, Hierochloa* and *Festuca*. During migration in Missouri birds are found in prairie and wet meadows (Robbins & Easterla 1992). In autumn the species is also found in hay and grain fields; it winters mainly in moist coastal grasslands and marshes, favouring the drier parts of cordgrass *Spartina patens* marshes and also rice fields. Water levels influence bird numbers at breeding sites, fewer birds occurring in drier years; preferred habitat has 0-12cm of standing water and the substrate remains saturated throughout the breeding season (Robert 1997); the maximum water depth recorded at a calling site is 46cm (Stenzel 1982). Breeding marshes are usually large enough to support several pairs of birds: from c. 10ha to several hundred ha (Robert 1997). The encroachment of woody vegetation such as willows *Salix* and birch *Betula pumila* in the absence of fire decreases the quality of breeding habitat. Records of the race *goldmani* are from wet meadows with bunch grass and in sedge and *Typha* marshes in ungrazed areas with vegetation less than 50cm tall, and from up to 2,500m; its habitat is described and illustrated by Hardy & Dickerman (1965).

FOOD AND FEEDING Earthworms, small freshwater snails, crustaceans (including Isopoda), millipedes, spiders, beetles, cockroaches, bugs, grasshoppers, crickets, ants and fly larvae (Bookhout 1995; also Easterla 1962). Snails are often said to be the most important food (e.g. Walkinshaw 1939), but there is no quantitative evidence to support this statement, and a recent study in Quebec showed that nesting birds eat a wide variety of prey, mainly invertebrates (principally arthropods and particularly beetles and spiders) but also a good number of seeds (Robert *et al.* 1997). In the autumn seeds of sedges, and in the winter seeds of *Polygonum*, constitute 5-10% of the diet, and in the winter seeds of rushes *Scleria* and grass *Setaria* each make up 2-5% of the diet (Martin, A. C. *et al.* 1951). This rail forages in areas with shallow water concealed by dense vegetation, taking food from the ground, from vegetation (sometimes climbing into low bushes), and from water, sometimes from 3-4cm below the water surface. It sometimes feeds while swimming.

HABITS Like many other rails and crakes this species is very difficult to see, keeping within dense vegetation through which its progress is mouselike and silent, and it may pass very close to an observer's feet without being even momentarily visible. It is often disinclined to run from danger, moving away slowly or preferring to remain motionless and well concealed in dense vegetation. When flushed it usually makes a low, short, weak flight, when the white secondaries are well visible, and drops suddenly into cover; after landing it may crouch motionless rather than run. Birds probably dive to escape from predators, and will submerge in water after being flushed. Although this species is largely diurnally active (Stalheim 1974), at night during the breeding season males call frequently and the birds are much easier to flush or approach than in the day (Robert 1997). Bathing actions include wing-flapping and tail-dipping; in captive birds bathing was less frequent in the early morning and tended to peak at about 1,900h; wing-sunning occurs after bathing (Stalheim 1974). Bathing, preening and stretching tend to occur together. When roosting, captive birds stand or sit, and either tuck the head under or rest it on the scapulars (Stalheim 1974).

SOCIAL ORGANISATION Presumed territorial and monogamous but the activity areas of breeding males overlap, suggesting gregariousness, and males lack strong fidelity to breeding territories (Bookhout 1995). Territories of 4 breeding males were 5.8-10.5 (7.8, SD 2.3) ha in extent; areas used by females were 1.0-1.7 (1.3, SD 0.4, n = 3) ha during preincubation, 0.1-0.5 (0.3, SD 0.2, n = 4) ha during incubation, and 0.2ha (n = 2) during posthatch (Bookhout 1995). Serial polygyny has been observed in captive birds (Stalheim 1975), the male attempting to pair with a new female as soon as his first mate began incubation, and nests of two females were once located in the territory of one male (Stenzel 1982). In captivity, as the breeding season progressed the territories of subordinate males were reduced until the dominant male claimed the entire pen for his territory (Stalheim 1975). In studies of wild birds, Stenzel (1982) observed young with both parents, as well as a male with chicks and a male and female together at hatching time, so on some occasions the male may share parental duties.

SOCIAL AND SEXUAL BEHAVIOUR Information is taken from Stalheim (1975). When calling, patrolling males may walk upright with the wings lifted and spread, displaying the white secondaries. Vigorous wing-flapping, with the neck extended, is associated with hostile behaviour. Conspecifics intruding into breeding territories may be chased for short distances, and a bird may hop towards an opponent while giving the wheeze call, and then assume a 'swanning' position in which the head is lowered and the wings drooped. Fighting, seldom observed in captive birds, usually consists of pecking the opponent's back or flanks. An attack may be inhibited by the threatened bird crouching and retracting its head, or by turning the head away and making squeak calls while being pecked. Pair formation probably occurs on the breeding grounds.

Allopreening (usually by the male) and courtship feeding have been observed in captive pairs; paired birds sit and sleep together. The male frequently attempts copulation, pursuing the female with a fast, stiff walk; the female often avoids copulation but, when receptive, stands still and may crouch, when the male mounts and treads the female for 30-40s before copulating. Copulation is brief, and the male becomes immobile in a postcopulatory freeze, after which both birds usually shake and preen.

BREEDING AND SURVIVAL Season USA and Canada, lays May-Jul; Mexico, flightless juvenile Sep. All further information is for nominate race; much information from a study of captive birds (Stalheim 1975). **Nest** A cup of fine, dead sedges and grasses; external diameter 7-12cm, depth 3-9cm, thickness 2.5-4cm (2 were 16cm thick); cup diameter 7cm; covered with canopy of dead vegetation which, if disturbed, is rapidly restored by female; nest sometimes possibly has entrance ramp; built in sedges, rushes or grass on damp ground; placed on ground, or up to 15cm above it, beneath dead, procumbent vegetation. Some nests suspended above water 2-10cm deep. Both sexes hollow out crude scrapes in vegetation but female builds nest and may add material to it during incubation; material obtained close to nest site; 1 or more extra nests sometimes used as brood nests. **Eggs** 4-10 (8.03, SD 1.5, n = 34); laid at daily intervals; ovate, often elongate and with tendency to being equally pointed at both ends; creamy-buff, thickly speckled reddish-brown and lilac at large end; small blackish spots rarely occur on remainder of shell; size 28-30.4 x 20.1-21.5 (29.1 x 20.8; n = 79); weight of 9 fresh eggs in one clutch 6.4-7.4 (6.9, SD 0.4). **Incubation** 17-18 days, by female; incubation periods 20-40 min with breaks of 1-5 min (Robert 1997); female remains in nest all night; all eggs hatch within 21-24 h. **Chicks** Precocial; leave nest after 1-2 days; use wing-claw when clambering through vegetation and when climbing into nest; begin to feed themselves from 5 days; in captivity, fed and brooded only by female (but see Social Organisation) for 3 weeks, after which chicks independent. Young largely feathered at 18 days of age; can fly at 35 days, when remiges and rectrices finish growing. Age of first breeding unknown; probably 1 year. Normally one brood per season; may renest after failure.

---

15   SPECKLED RAIL
Coturnicops notatus   Plate 5

Zapornia notata Gould, 1841, Rio de la Plata, Uruguay.

Monotypic.

Synonyms: *Coturnicops/Porzana/Ortygops/Ortygometra notata*.

Alternative names: Darwin's Rail; Speckled Crake.

IDENTIFICATION Length 13-14cm. A predominantly blackish, very small crake, with white secondaries, dark olive-brown streaking on upperparts, fine white speckling on head, neck, mantle and upper breast, and narrow white bars on back and upperwing-coverts and from lower breast to undertail-coverts. Iris red or orange; bill, legs and feet black. Sexes alike. Probable immatures have fewer spots on upperparts (some are barred); white streaks and mottling on lower throat and breast; undertail-coverts sandy-cinnamon. Juvenile undescribed. Inhabits grassland, flooded fields and marshes.

Similar species White secondaries distinguish this species from all sympatric rallids. Black Rail (33) similar in size and overall colour but lacks white markings on head, neck and breast; on underparts has white bars or spots only from flanks and lower belly to undertail-coverts (undertail-coverts are cinnamon in some races); legs and feet olive-brown to blackish-brown. Dot-winged Crake (97) is another very dark small wetland crake but has white spots and bars of upperparts confined to upperwing-coverts, remiges and uppertail-coverts, white barring on underparts only from flanks to undertail-coverts, and brown legs and feet.

VOICE This bird's vocalisations are unobtrusive and easily masked by other marsh sounds. All information is from Teixeira & Puga (1984). It has a *kooweee-cack* call, the first syllable high and brief, the second louder and drier; when giving this call the bird stands upright with the neck vertical. A whistling *keeee* of alarm and a high *kyu* are also given. Although the species is diurnal, a captive bird called frequently from a perch at night.

DESCRIPTION
Adult Head and upperparts dark blackish-brown; feathers from mantle to uppertail-coverts broadly edged dark olive-brown, giving streaked effect; each feather of head, neck and mantle with whitish central spot giving finely speckled appearance; back, scapulars and upperwing-coverts narrowly barred white (2 bars per feather); rectrices blackish-brown, edged dull, dark olive-brown. Primaries olive-brown; secondaries whitish, with dull buffy-brown fringes and darker, dull brown shaft-streak and (in some) subterminal spot; tertials dark blackish-brown, edged dark olive-brown, and with 2 narrow white bars (wing of 1 female examined); underwing-coverts and axillaries white, feathers near edge of wing with dull brown bases. Lores dusky brown; supraloral streak whitish; chin and throat white; throat often mottled or streaked dark blackish-brown. Rest of underparts dark blackish-brown, often paler, more olive-brown, on undertail-coverts, with white spots on neck and upper breast and narrow white barring from lower breast to undertail-coverts; centre of belly more extensively whitish, with greyish or cinnamon tinge. Some birds have sandy-cinnamon bars on undertail-coverts. Iris red or orange, also described as yellow with a wide red ring around pupil; bill, legs and feet black. Sexes alike.
Immature Probable immatures have fewer spots on upperparts (some tending to be barred rather than spotted); white streaks and mottling, rather than roundish white spots, on lower throat and breast; undertail-coverts mostly sandy-cinnamon, rather than olive-brown barred white.
Juvenile Type of putative *duncani*, possibly juvenile, has blacker upperparts with darker, duller feather edges and whiter spots from forehead to mantle; remiges (in sheath) with white central streaks and subterminal crescent; underparts darker, more blackish, with whiter mottling; undertail-coverts tipped sandy-cinnamon. Bird has no traces of down or of juvenile soft, loosely structured feathers, so age uncertain.
Downy young Not described.

MEASUREMENTS Wing (chord) of 7 females 66-78 (69.5), (flattened) of 1 female 69; tail of 6 females 24-31 (28.1); exposed culmen of 8 females 11.5-14 (12.5), culmen to base of 4 females 13-14.5 (13.9, SD 0.7); tarsus of 9 females 19-21 (20.1). Weight of 1 unsexed c. 30.

GEOGRAPHICAL VARIATION None. The one specimen from Guyana was originally described as subspecies *duncani* (Chubb 1916).

MOULT The type of putative *duncani*, from Guyana, which may be juvenile, has all remiges half-grown, and most wing-coverts partly grown, Sep (see Description).

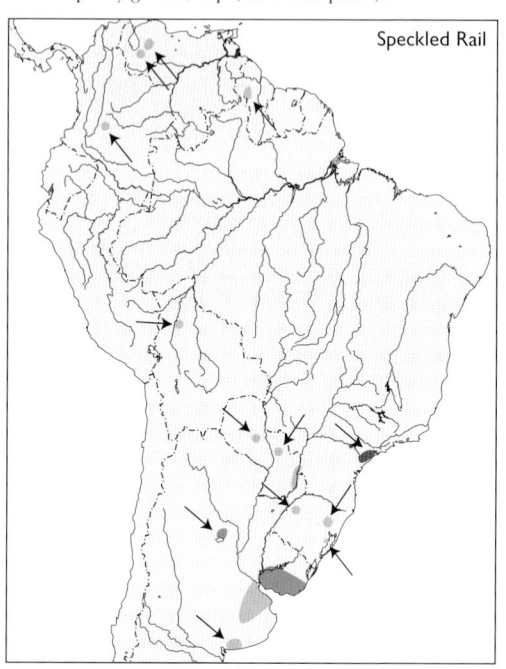

Speckled Rail

DISTRIBUTION AND STATUS South America E of the Andes, very locally in SE Colombia, W Venezuela, Guyana, C and S Brazil, Bolivia, Paraguay, Uruguay and N Argentina (Corrientes, Córdoba, Buenos Aires, Río Negro and possibly La Pampa). This poorly known species is currently regarded as DATA DEFICIENT. Although it occurs over a very large area records are very sparse, with few in any country, the following details being from Collar *et al.* (1992), Wege & Long (1995), Mauricio & Dias (1996) and Arballo & Cravino (in press) unless otherwise stated. Colombia, 1 in 1959 (Río Guayabero, Serranía de la Macarena National Park, S Meta); Venezuela, 1 c. 1915 from the Mérida area, and 1 from Aparición, Portuguesa; Guyana 1 (Abary R); Brazil, 5 localities (Pindamonhangaba, Taubaté and Ipiranga [São Paulo city] in E and NE São Paulo, and Novo Hamburgo and Pontal da Barra marsh, Pelotas, in Rio Grande do Sul); Paraguay, 3 localities (Laguna General Díaz, Presidente Hayes; Horqueta, Concepción; Puerto Bertoni, Alto Paraná); Uruguay, 6 localities (Sarandí and another locality in Durazno; Juan L. Lacaze, Colonia; Parque Lecoqc, Montevideo; Arroyo Maldonado, Maldonado, in Oct 1993; and at sea off Cabo Santa María, Rocha); Argentina, at least 10 records, from Santo Tomé, Corrientes, S to 'Patagonia' (probably near Carmen de Patagones, Río Negro estuary), and it may breed in the Bañados del Río Dulce, Córdoba (Nores & Yzurieta 1980). There are also 4 recent records, all of single birds, from the Reserva Municipal de Biosfera Mar

Chiquita, Buenos Aires, in Jan and May (Martinez *et al.* 1997). The only record from Bolivia is a sighting of 1 bird at the Beni Biological Station in Aug 1997, from the same area in which the Rufous-faced Crake (36) has recently been found (Brace *et al.* in press). It is normally described as very rare, but its apparent occurrence throughout the year at Taubaté, Brazil, where it is not common but not rare, has been taken to indicate that it is difficult to find rather than scarce. This may be so but, until further evidence is forthcoming, it is best to consider the bird genuinely rare. Nine localities from which it is known are listed as key areas for threatened birds; of these, the Colombian and Bolivian sites, 1 Venezuelan site (Mérida) and 2 sites in Argentina (Bañados del Rio Dulce, Córdoba, and Otamendi Strict Nature Reserve, Buenos Aires) are protected (Wege & Long 1995).

At the Reserva Municipal de Biosfera Mar Chiquita, Argentina, Martinez *et al.* (1997) found that uncontrolled fires affected this species: birds did not reappear for up to 1 year after burning; the authors also express concern for unplanned urban development which could cause problems. Martinez *et al.* (1997) record 1 instance of predation: a bird which they flushed was caught in flight by a Cinereous Harrier *Circus cinereus* which was breeding nearby.

MOVEMENTS There has been speculation that this rail undertakes migrations between N and S South America, but this possibility is unlikely in view of the bird's occurrence in Apr-Aug in E Brazil (Taubaté) and Mar-Jun in Paraguay and Uruguay, and the enlarged gonads of the Aug Venezuela specimens (Collar *et al.* 1992). One from Colombia (Mar 1959), and two from ships at sea, have led to the suggestion that birds occasionally erupt large distances randomly from centres of distribution in tropical savannas of N and E South America, but there is no proper evidence that the centres of distribution are in tropical savannas and it is also possible that the bird's distribution is wider and more continuous than is currently known. The two birds taken on ships at sea are the holotype, which was collected on the 'Beagle' in 1831 at the mouth of the Río de la Plata, and an immature which flew aboard a ship off Cabo Santa María, Uruguay, in Nov or Dec 1875. Such records are probably indicative at least of post-breeding dispersal; further evidence for movements comes from S Uruguay (Colonia) where individuals, dead or dying after storms, have been found regularly on the beaches of the upper Río de la Plata near the town of Juan Lacaze (Arballo & Cravino in press), and from Argentina where 1 was found alive in a Buenos Aires suburban garden (date unknown, but during the summer) and was later released elsewhere (J. Mazar Barnett *in litt.*). There is also a possible record from Falkland Islands in Apr 1921, probably of a straggler (the bird was captured alive, died and was not preserved).

HABITAT Recorded from dense marshy vegetation (including rushes, reedbeds and floating vegetation), swamps, flooded rice fields, alfalfa fields, wheat fields, flooded pasture and other wet grasslands, open grassy savanna, crop stubble (e.g. Arballo 1990), brackish coastal *Spartina* spp. grasslands, and humid woodland edges. In Rio Grande do Sul, Brazil, 1 was seen alongside a Plumbeous Rail (113) in marshy habitat characterised by a mixture of sedges (notably *Scirpus giganteus* and *Cladium jamaicensis*) and many grasses, with a few small forbs, on a muddy substrate with little water (Mauricio & Dias 1996). In Argentina, one bird was caught by hand by a tractor driver in a ploughed field (Collar *et al.* 1992), and at the Reserva Municipal de Biosfera Mar Chiquita, Martinez *et al.* (1997) found it in *Scirpus americanus* 60cm tall, *Spartina densiflora* (twice) and *Juncus acutus*. In Beni, Bolivia, it has been seen alongside the Rufous-faced Crake (36) and the Ash-throated Crake (100) in savanna flooded to a depth of 5cm and characterized by tussocky grass c. 1m tall with *Rynchospora globosa*, *Cyperus haspan* and *Tibouchina octopetalia* (Brace *et al.* in press). It inhabits lowlands, to 1,500m.

FOOD AND FEEDING Said to take insects, other invertebrates (including arachnids and crustaceans), and seeds. One stomach contained 80% small grass seeds, 15% arthropod remains and 5% fine gravel (Teixeira & Puga 1984).

HABITS These birds are very difficult to observe, keeping to dense vegetation and usually being seen only when disturbed, for example by a person or a harvesting machine, when they may fly as high as 6m and drop back into cover after c. 70m (Teixeira & Puga 1984). A captive bird was diurnal, and went to roost at dusk on a perch up to 2m above the ground (Teixeira & Puga 1984). Several specimens have been collected at night in open savanna or rice fields with the use of lights (Ripley 1977).

SOCIAL ORGANISATION Solitary or in family groups.

SOCIAL AND SEXUAL BEHAVIOUR A captive bird seemed to show alarm by remaining motionless in a horizontal position with the tail held downwards and fully spread and the wings folded but held obliquely, showing the white tips of some secondaries and exposing the white-patterned underparts (Teixeira & Puga 1984).

BREEDING AND SURVIVAL Season Uruguay, brood of 3 young Dec 1985, in wheat stubble (Arballo 1990); Venezuela, breeding condition Aug; Brazil, breeding condition Dec; Argentina, probably breeds Córdoba province. **Nest and eggs** Undescribed (e.g. Collar *et al.* 1992), although Schönwetter (1961-62) gives measurements of 2 eggs (possibly misidentified) as 32.2 x 24 and 33 x 23.5 and of 8 eggs from "Trinidad" as 23-25.1 x 18.3-19.3 (presumably misidentified, Trinidad being out of bird's known range and the eggs being very small).

MICROPYGIA

One species, closely allied to *Coturnicops* but with distinctively spotted plumage and no white on the secondaries. A poorly known bird, confined to Central and South America where it occurs in dry to wet grasslands, it is considered near-threatened.

16 OCELLATED CRAKE
Micropygia schomburgkii   Plate 5

Crex Schomburgkii 'Cabanis' Schomburgk, 1848, interior of Guyana.

Sometimes placed in *Coturnicops* but differs in vocalisations, nest construction and some anatomical details. Two subspecies recognised.

Synonyms: *Coturnicops/Thyrorhina/Ortygometra schomburgkii*.

Alternative names: Ocellated Rail; Dotted/Schomburgk's Crake.

IDENTIFICATION Length 14-15cm. A very distinctive small crake, having buffy olive-brown upperparts (more tawny on forehead, face and wings) with black-bordered white spots from hindcrown to lower back (race *chapmani*) or tail (nominate) and on sides of breast, and tawny-buff underparts with white chin, throat and centre of belly. Iris reddish; bill horn to blackish, lower mandible greenish to grey; legs and feet red. Tail quite well developed, despite genus name; toes relatively short. Sexes similar but female smaller; usually has spots extending to forecrown, and rich colour of forehead extending less far back on crown. Immature has duller, more grey-brown upperparts with spots only on wing-coverts, scapulars, sides of neck, and sometimes crown; underparts more extensively white or cream, with ochraceous breast-band; sides of breast and flanks dull brown, or barred grey. Juvenile not described. Inhabits dense, dry to wet grassland.

Similar species This crake's distinctive overall colour and pattern easily distinguish it from all other small wetland rallids of the region except perhaps Yellow-breasted Crake (105), which has a distinctive face pattern and heavily barred flanks.

VOICE The song is a sequence of clear, strong *pr-pr-pr* notes, lasting 20-30s. It may be given most frequently at dawn and dusk (Parker *et al.* 1991), but J. C. Lowen (*in litt.*) has noted no variation in call timing and has heard it during the day. The alarm call is a harsh whirring *pjrrr* or *prrrxxxzzz*, likened to the sound of oil sizzling in a frying pan or the rasping of a grasshopper; both sexes (but usually males) also utter a more elaborate sequence, rendered *prrrxxxzzz... crrraaauuu... crrraaauuu...* and uttered as the bird stands upright with the tail held vertical (Negret & Teixeira 1984b). See Hardy *et al.* (1996).

DESCRIPTION *M. s. schomburgkii*
Adult male Forehead, forecrown and sides of head deep tawny to ochraceous-tawny. Rest of upperparts, from hindcrown to tail and including scapulars, upperwing-coverts and tertials, buffy-olive to buffy-brown with small, ovate, black-bordered white spots (one at the tip of each patterned feather); upperpart colour becomes more ochraceous-tawny or tawny on upperwing-coverts; spots become more elongated, almost streaks, on greater coverts, tertials and rectrices; marginal coverts, and adjacent lesser coverts, plain ochraceous-buff. Primaries and secondaries dull olive-brown, outer webs edged tawny-buff (brightest on outer primaries); alula feathers dull olive-brown, outer webs tawny-buff, and blackish-bordered white streak at tip of each feather. Axillaries and underwing-coverts white to cream, tinged tawny-buff; greaters tinged greyish; underside of remiges pale grey-brown. Chin, throat and centre of belly white, often tinged tawny-buff; rest of underparts ochraceous-buff to pale tawny, with black-bordered white spots on sides of breast. Iris coral-red to ruby, orange or reddish-brown; upper mandible horn to black, with blue-grey line from nostril to base, lower mandible white, turquoise, blue-grey or greenish, cutting edges blue-grey; legs and feet salmon, orange-red, coral-red or scarlet.

Adult female Similar to male but smaller; white spotting tends to extend further forwards on crown; rich ochraceous-buff on forehead and crown less extensive.

Immature Upperparts duller than in adult, more grey-brown, with spots only on upperwing-coverts, scapulars, sides of neck and sides of breast, and sometimes on crown; ochraceous-tawny colouring confined to sides of head and neck; underparts more extensively white or cream than in adult, with pale, diffuse ochraceous breast-band; sides of breast and flanks barred grey, or largely dull brown.

Juvenile Not described.
Downy young Not described.

MEASUREMENTS Wing of 11 males 70.5-78 (74.9), of 11 females 71-79 (75.4); tail of 4 unsexed 32-35.5 (34.3); culmen to base of 4 males 14-15 (14.6, SD 0.5), of 3 females 14, 14.5, 15; exposed culmen of 10 males 13-15 (13.8), of 10 females 13-14 (13.3); tarsus of 10 males 19-21 (20), of 10 females 18-20 (19.6). Approximate weight of males 40, of females 24, of unsexed birds 32 (Negret & Teixeira 1984b; Stiles & Skutch 1989); weight of 1 female 23.5 (FMNH, D. E. Willard *in litt.*).

GEOGRAPHICAL VARIATION The nominate race is very variable in the colour of the upperparts and the density of the spotting on the back, but is distinguishable from the southern race on overall plumage colour and the extent of dorsal spotting, and on size.

M. s. schomburgkii (Schomburgk, 1848) – SW Costa Rica (Puntarenas), Colombia in S Meta (Río Guayabero, S Macarena Mts) and NE Meta (Carimagua), patchily in Venezuela and Guianas; birds from Bolivia (Beni, La Paz and Santa Cruz) and extreme SE Peru (Pampas de Heath in Madre de Dios) may be of this form, but their racial affinities are not clear (e.g. Graham *et al.* 1980). The Costa Rican specimen, although assigned to this form, has broader black borders to its profuse white spots and may represent an undescribed population (Dickerman 1968b). See Description.

M. s. chapmani (Naumburg, 1930) – C to SE Brazil (Mato Grosso, Goiás and Bahia to São Paulo); birds from E Paraguay (Canindeyú) presumably also belong here. Larger; crown more russet; upperparts paler, more orange-brown, with spots absent from lower back or rump to tail; spots may also be smaller. One Paraguay bird had white streaks at tips of primaries (J. C. Lowen *in litt.*). Iris bright red; upper mandible black; lower mandible turquoise-green; tarsus bright red (J. C. Lowen *in litt.*). Wing of 7 adults 76-84 (80.5); tail of 4 adults 34-40 (38, SD 2.7); exposed culmen of 4 adults 13-16 (14.5, SD 1.5); tarsus of 3 adults 22, 23, 26.5. Weight of 1 adult 38.8 (J. C. Lowen *in litt.*).

MOULT Primary moult is ascendant and apparently takes place over a long period, during which there is hardly ever more than one feather growing at once; recorded Oct, Dec (Guyana) and Jan-Feb (Venezuela); a bird from Paraguay in Sep had 2 generations of secondaries (J. C. Lowen *in litt.*).

Ocelated Crake

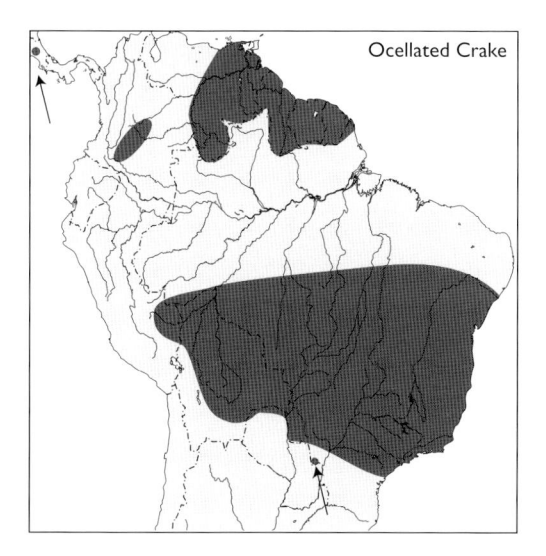

DISTRIBUTION AND STATUS This poorly known crake has an extensive range, occurring in Colombia, Venezuela and the Guianas, locally in Peru, Bolivia and C Paraguay, and through a large area of Brazil. The only record from Costa Rica is of a bird collected on 9 Mar 1967 (Dickerman 1968b). It is classed as NEAR-THREATENED and is regarded as locally distributed and scarce to rare in many areas, probably existing in small populations; this assessment of its status may reflect the difficulty experienced in locating and observing the birds rather than their actual scarcity, but it is threatened by degradation and loss of its tall grassland habitat (J. C. Lowen *in litt.*). It is apparently widespread and numerous in the grasslands of N and C Bolivia (Parker *et al.* 1991, Pearce-Higgins *et al.* 1995, Brace *et al.* 1997), and locally common in C Brazil (Negret & Teixeira 1984b). In Paraguay it is known from two areas (Lagunita and Aguará Ñu) of the Reserva Natural del Bosque Mbaracayú (Dpto Canindeyú), where it was recorded in small numbers in 1994-95, and it is likely to be rare in the country (Lowen *et al.* 1996b); however it may occur in similar habitat in Amambay and Concepción (Lowen *et al.* in press), and J. Mazar Barnett (*in litt.*) has found it to be relatively common in the Paraguayan cerrado. In Brazil, local people catch many for food by setting fire to grasslands and catching (often by hand) birds overcome by the smoke; at such times Aplomado Falcons *Falco femoralis* often take these crakes (Negret & Teixeira 1984b). In Paraguay it is at risk from man-induced fires at Aguará Ñu, where it is possibly also threatened by predation from the Maned Wolf *Chrysocyon brachyurus* (Lowen *et al.* 1996b). In Beni, Bolivia, birds have been known to remain in an area after a fire, when they could be approached very closely (R Brace *in litt.*). In Beni, annual dry-season burning of this bird's savanna habitat is undertaken in many estancias to promote new growth for grazing; this, plus the effect of grazing and trampling by cattle, is adversely affecting the savanna environment (Brace *et al.* in press).

MOVEMENTS No movements are recorded, but birds occasionally fly into lighted open windows at night (Negret & Teixeira 1984b). On 3 Apr 1997, at Ilha Comprida, São Paulo, Brazil, J. Mazar Barnett (in prep.) was given an apparently healthy immature (crown almost unspotted) which had been picked up by a bus passenger (without leaving the bus) from its perch on vegetation 2m above the ground at the side of a road through tall bromeliad- and moss-rich white-sand forest – an obvious vagrant. It is regarded as accidental, and thus presumed migratory, at Emas National Park, Brazil (Forrester 1993). There is a doubtful record of a vagrant from the Galápagos Islands (Meyer de Schauensee & Phelps 1978). Birds must make at least local movements when occupied habitat is destroyed by fire.

HABITAT Occupies a variety of dense grassland habitats, both well drained and seasonally flooded, ranging from almost pure open grasslands (*campo limpo*) to those with an abundance of shrubs and small trees (*campo sujo*) (Parker *et al.* 1991); it sometimes occurs near marshes or forest borders. The substrate may be muddy or even seasonally shallowly flooded. In C Brazil it occurs in dense, dry grassland dominated by *Tristachya leiostachya* (Negret & Teixeira 1984b), and in São Paulo it has been recorded from *campos* recently ploughed to plant introduced trees (Willis & Oniki 1993); in Bolivia it occurs in open *Trachypogon* grassland (Parker *et al.* 1991). The height of occupied grass ranges from c. 1m to very tall grass along dry drainage channels in parched tropical savanna. It occurs in lowlands, up to 1,400m.

FOOD AND FEEDING Beetles (Carabidae, Scarabaeidae), stoneflies, grasshoppers (Acridoidea), cockroaches and many ants. One was caught in a trap baited with oatmeal. The birds forage on the ground, usually in grass or low scrub but occasionally in the open.

HABITS These birds are difficult to observe because they normally remain in dense or tall vegetation, in which they frequently use the tunnels and runways of rats and other rodents such as guinea pigs (*Cavia* sp.). They may freeze when pursued, and may flush from almost underfoot, flying low for a few metres before dropping into the grass, but they normally escape by running. The São Paulo vagrant individual (see Movements), when released in a hotel room for the night, walked with its head down and rear end raised, a posture possibly suited to moving through the thick grass cover of its normal habitat (J. Mazar Barnett *in litt.*).

SOCIAL ORGANISATION Monogamous; territorial at least during the breeding season. In Bolivia, at least eight birds sang in an area of dense moist grassland measuring c. 400 x 200m and six were heard in c. 500m^2 of drier upland grassland with small shrubs, while four territorial pairs counter-called within another c. 500m^2 of open grassland in which their seemingly small territories (perhaps <0.5ha each) appeared to be centred on clumps of shrubby vegetation and tall grass around termite mounds (Parker *et al.* 1991).

SOCIAL AND SEXUAL BEHAVIOUR No information available.

BREEDING AND SURVIVAL Season C Brazil, Oct-Mar; Costa Rica, probably during rainy season; Colombia breeding condition Mar. **Nest** One was a ball of dry grass with a large side entrance, 50cm above the ground, completely concealed in dense wet grassland near a *Mauritia flexuosa* palm grove; it was angled upwards at 45°; external length 20cm, width 14cm and height 17cm; nest cavity length 11cm, width 7cm, height 9cm; entrance 8 x 4.5cm (Negret & Teixeira 1984b). **Eggs** 2; dull white and unmarked, size

24.6 x 19.3 (Negret & Teixeira 1984b); Schönwetter (1961-62) gives size of 6 eggs as 28.3-32 x 23-26 (but means given are incorrect); calculated weight 8.3; eggs also described as buff with light brown spots (Hilty & Brown 1986). **Incubation** By female. No other details available.

RALLINA

This genus comprises eight species of distinctively plumaged rails which inhabit forest or marshland in forest and are confined to Asia and Australasia. The head, neck and breast are chestnut, and the upperparts chestnut, brown or blackish; the pattern of the underparts is variable and ranges from plain reddish-brown with obscure pale feather tips to bold black-and-white barring. The four species endemic to New Guinea are sometimes separated into the genus *Rallicula*; they are medium-small and strongly sexually dimorphic, the females being spotted from mantle to rump and on the upperwings, and the males being either unmarked (three species) or streaked (White-striped Forest-rail) in these regions. The remaining four species are larger and show little or no sexual dimorphism. All species have the remiges and underwing-coverts dark brown or blackish, barred or spotted with white, while the Red-legged Crake also has white bars and spots on the scapulars and upperwing-coverts. The habits and breeding of most species are very poorly known. Six species lay pure white eggs; those of the other two are undescribed. The two species which occur in mainland Asia are both resident and migratory within their normal ranges, while the Red-necked Crake is partially migratory in NE Australia. The species endemic to New Guinea are not globally threatened but two are data deficient, while the Andaman Crake is vulnerable.

17 CHESTNUT FOREST-RAIL
Rallina rubra   Plate 6

Rallicula rubra Schlegel, 1871, northern peninsula of New Guinea.

The four *Rallina* species of New Guinea show marked sexual dimorphism and are therefore sometimes retained in *Rallicula*. Three subspecies recognised.

Synonyms: *Rallicula/Rallina klossi*.

Alternative name: New Guinea Chestnut Rail.

IDENTIFICATION Length 18-23cm. A medium-small rail, the smallest *Rallina* species; relatively common but poorly known. Male almost entirely reddish-chestnut, tinged blackish on nape and hind-neck in two races, and sometimes marked with blackish on scapulars and upperwings (race *klossi*); tail reddish-chestnut; lower flanks and lower belly reddish-brown with paler feather tips (sometimes darker brown, with whitish-buff bars, in *klossi*); remiges dark brown with white bars. Iris brown; bill, legs and feet blackish. Female has mantle to rump, and upperwings, blackish-brown with small white or buffy spots which extend to the dark brown upper mantle; lower flanks and lower belly darker brown with buffy markings in some female *klossi*. Subadult male has grey base to black bill, and dark wine-brown legs and feet. Juvenile male entirely dark blackish-brown. Inhabits the interior of montane forest.
Similar species The four *Rallina* forest-rails endemic to New Guinea are similar in size and plumage, differing mainly in extent of blackish on upperparts and in nature of white patterning on blackish areas. All species appear quite long-legged. Entirely reddish-chestnut plumage of male Chestnut Forest-rail makes confusion possible only with very similar but allopatric male Mayr's Forest-rail (20), which is darker and has indistinct narrow black bars on tail. Female resembles females of Forbes's (19) and Mayr's Forest-rails but has a greater density of spots on blackish-brown regions of upperparts (these areas are dark chestnut with black-bordered buff spots in female Mayr's), spotted upper mantle and no blackish barring on underparts (but some *klossi* have darkish barring); it is very similar to female White-striped Forest-rail (18) but lacks narrow black tail bars of this and the other two species.

VOICE Information is from Frith & Frith (1988, 1990). A shrill, sharp *krill* or *keow*, given every 1-2s and repeated up to c. 130 times; the notes are sometimes squeakier. This call is uttered "with head and bill lifted upwards and gape wide open for each note"; calls are given from the ground and from a perch 1m above the ground. Duetting is recorded, when calls may become faster and sharper, *kee* or *kek*. This species calls frequently. Alarm calls of adults are a sharp *keek* and a soft, sharp *eek*; chicks give soft cheeping or peeping.

DESCRIPTION *R. r. rubra*
Adult male Generally uniform reddish-chestnut, tinged blackish on nape and hindneck (blackish feather tips); shafts of dorsal feathers blackish; alula blackish-brown, barred white; primary coverts and remiges blackish-brown with white bars on inner web; tail uniform reddish-chestnut. Underwing-coverts and axillaries dark brown, narrowly barred or variably spotted white. Lower flanks to undertail-coverts dull reddish-chestnut, with very obscure paler feather tips (sometimes lacking). Iris brown to chestnut; bill blackish with tip and base of lower mandible greyish; legs and feet blackish.
Adult female Not described (see *R. r. klossi*).
Immature Not described (see *R. r. telefolminensis*).
Juvenile Not described (see *R. r. klossi*).
Downy young Down long, silky and black, grizzled with russet or chestnut; iris dark brown; bill blackish with white tip and egg-tooth; legs and feet blackish.

MEASUREMENTS Wing of 6 males 94-103 (97.8, SD 3.3); tail of 6 males 63-70 (65.8, SD 2.5); culmen of 4 males 30-31 (30.5, SD 0.6); tarsus of 4 males 38-39 (38.5, SD 0.6) (Mayr 1938, Diamond 1969).

GEOGRAPHICAL VARIATION Races are separated on size and on differences in overall plumage colour. The possible race *subrubra* is merged with *klossi* because of overlapping measurements and the lack of significant colour differences (Ripley 1977).

R. r. rubra (Schlegel, 1871) – Arfak Mts, W New Guinea. See Description.

R. r. klossi (Ogilvie-Grant, 1913) – Weyland Mts (including Wissel L and Utakwa R) and Nassau Mts to Oranje Mts (including Mt Wilhelmina), WC New Guinea. The following description of this race, from details in Mayr (1938), Rand (1942b), Diamond (1969) and Ripley (1977), apparently refers only to specimens in US collections. The palest race; similar to nominate but paler chestnut overall; lacks blackish tinge on nape and hind-neck, and blackish shafts of dorsal feathers; white bars on underwings much reduced. One male has upperwing-coverts edged blackish; some upperwing-coverts, and inner secondaries, with "obscure spotting" (Rand 1942b). Female slightly smaller than male, with variable upperpart colouring: mantle to rump, and upperwings, blackish-brown; markings on mantle yellowish, white or rust; a few females have streaks instead of some spots on upperparts (especially on anterior mantle); some also have sparse rufous spotting and edging to feathers (especially on posterior mantle). One immature female has distinct dark barring on tail, others do not; a few apparent adults have indistinct tail barring and some dark barring on uppertail-coverts (Rand 1942b). A juvenile male (putative *subrubra*) has upper and undertail-coverts barred "rufous" (presumably chestnut) and black, and tail rufous with some irregular barring (Rand 1942b).

BMNH syntypes (Ogilvie-Grant 1913) differ somewhat from these descriptions, and from those of nominate, as follows. In male, lower mantle darker chestnut than upper; scapulars and tertials chestnut with dark blackish-brown shaft-streaks and fringes, and feathers have basal two-thirds entirely blackish-brown; lesser coverts blackish-brown, barred chestnut and spotted pale tawny; medians blackish-brown, spotted chestnut; greaters largely blackish-brown, edged chestnut; alula, primaries and secondaries blackish-brown with white bars (more spotted on alula); lower flanks, thighs and lower belly darkish olive-brown with whitish to buff bars; undertail-coverts chestnut. In female, extreme upper mantle chestnut; mantle to rump, upperwing-coverts and tertials dark blackish-brown with tawny-buff spots; some mantle feathers edged or washed chestnut; back to rump densely marked with small tawny-buff spots; blackish-brown of upperparts becomes somewhat paler, tending to dark sepia, on all greater upperwing-coverts, primaries and secondaries; inner webs of primaries and secondaries barred whitish, outer webs with small tawny-buff spots. Lower flanks, thighs and lower belly darkish olive-brown with buff spots and short bars; undertail-coverts and tail dark chestnut. The significance of these plumage differences, especially of the dark posterior underparts, is unclear and requires further investigation.

One juvenile male (BMNH) entirely blackish-brown, slightly paler on underparts; primaries and secondaries with whitish spots on inner webs; axillaries and underwing-coverts as adult; tail dark chestnut with narrow blackish-brown bars; bill, legs and feet black. Wing of 15 males 91-102 (95.7), of 13 females 85-100 (95.4); tail of 15 males 50.5-67 (59), of 13 females 46.5-66 (58.1); culmen of male 28-30, exposed culmen of 15 males 22-26 (24.7), exposed culmen of 11 females 23.5-25 (24.4) (Mayr 1938, Diamond 1969; size and mean of Mayr's sample not known). Syntypes in BMNH: wing of male 95, of female 89; tail of male 48, of female 52; culmen to base of male 27.5, of female 26; tarsus of male 37.5, of female 38. Weight of 3 males 84, 88, 91.

R. r. telefolminensis (Gilliard, 1961) – Victor Emanuel and Hindenburg Mts, and Tari Gap, C New Guinea. The smallest race. Similar to *klossi* in overall colour but averages darker, with more maroon and less rufous-brown, particularly on lower back and exposed edges of scapulars, upperwing-coverts and secondaries. Bill more slender than in *rubra*; dark feather tips of nape less prominent; white barring on underwings much reduced. Subadult male has grey base to black bill, and dark wine-brown legs and feet. Wing of 2 males 93, 94; tail of 1 male 56.5; culmen of 2 males 22, 23. Weight of 2 males 71, 76 (Gilliard & LeCroy 1961); of 2 apparently recently hatched chicks 13.4, 14.4 (Frith & Frith 1990).

MOULT As in Forbes's Forest-rail, primary moult tends to be rapid and ascendant, with irregularities commonly occurring, and secondaries are moulted irregularly. An adult male *telefolminensis* was in general moult (heavy on the tail) in May, and a subadult male *telefolminensis* was in general moult, excluding the wings, in Apr.

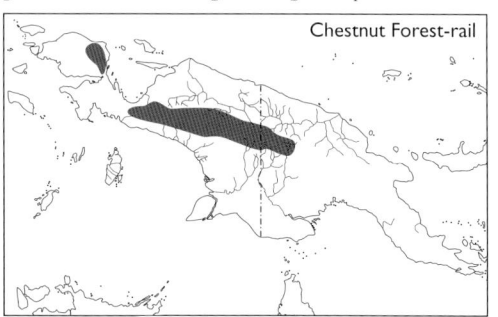

DISTRIBUTION AND STATUS Found in the montane forests of W and C New Guinea, from the Arfak Mts E to the Hindenburg and Victor Emanuel Mts. This rail is not regarded as globally threatened, and is widespread in Irian Jaya where the races *rubra* and *klossi* apparently were formerly abundant (Ripley 1977). The race *telefolminensis* is apparently uncommon in the vicinity of Telefomin in the Victor Emanuel and Hindenburg ranges, but the population reported as common in the Tari Gap forests is presumably also of this race (Gilliard & LeCroy 1961, Frith & Frith 1990).

MOVEMENTS None recorded.

HABITAT Occurs on the ground in the interior of montane forest, between 1,500 and 3,050m. At Tari Gap it is found in mossy mixed lower montane beech forest (Frith & Frith 1988). In areas of sympatry with Forbes's Forest-rail it occurs at higher altitudes.

FOOD AND FEEDING No information available.

HABITS Said to be shy and secretive; no other information available.

SOCIAL ORGANISATION Apparently monogamous. Two adult males responded to one female's alarm call at the nest, suggesting the possibility of cooperative breeding (Frith & Frith 1990).

SOCIAL AND SEXUAL BEHAVIOUR One incubating female stood motionless at the nest entrance, with bill pointing vertically down, when observers approached; she then performed a distraction display by running about the nest area with drooping wings and occasionally flicking the slightly fanned tail; two males and the female ran closely around the observer who had removed the chick from the nest, giving alarm calls (Frith & Frith 1990).

BREEDING AND SURVIVAL Season Lays Oct-Nov, and either Aug or Sep, i.e. at end of period of highest rainfall. **Nest** Large domed structure of moss, grass, leaf skeletons, fibres and fern fronds; lined with fine fibres, rootlets, leaf skeletons and fragments of fern fronds; base extended at one side into ramp leading up to side entrance. Typically placed c. 2m (1.5-2.9 m) up between frond bases in pandanus crown, in undisturbed dense moss forest. Nest dimensions (incm): overall length 35, 42 (ramp 15); height 23, 28; nest chamber length 19, 20, 25 and height 13. **Eggs** One only, large (27% of adult weight); matt white or off-white, sometimes with pinkish hue; size 43.1-45 x 29.3-30.5 (44.3 x 30, n = 5); weight 18.9-21.8 (19.9, n = 5). **Incubation** At least 34 days, possibly >37 days: longest of any rallid known; incubation by both sexes; incubated egg often left to become cold. **Chicks** One young chick possibly killed by torrential rain at night. See Frith & Frith (1990).

18 WHITE-STRIPED FOREST-RAIL
Rallina leucospila   Plate 6

Corethrura? leucospila Salvadori, 1875, Arfak Mountains, New Guinea.

Sometimes placed in *Rallicula*. Forms a superspecies with *R. forbesi* and *R. mayri*. Monotypic.

Synonym: *Rallicula leucospila*.

Alternative name: White-striped Chestnut Rail.

IDENTIFICATION Length 20-23cm. Male has head, neck and breast chestnut, crown and nape darker; mantle to rump, and upperwings, blackish-brown with fine white streaks; uppertail-coverts and tail chestnut, narrowly barred black; lower flanks to belly brown, obscurely barred buffy and blackish; undertail-coverts as tail. Iris olive-brown; bill dark brown; legs and feet black. Female has white spots instead of streaks on upperparts. Immature darker and duller than adult; white markings of upperparts tinged rufous; lower flanks to undertail-coverts plain blackish-brown. Juvenile not described. Inhabits the interior of montane forest.
Similar species Fine white streaks on blackish upperparts easily separate male from the other *Rallina* species of New Guinea. In both sexes, black of upperparts extends further forwards on mantle than in the other forest-rails, and black barring on uppertail-coverts and tail is more distinct; female is blacker on upperparts, and has whiter spots, than any other forest-rail. Female otherwise resembles female of the smaller Chestnut Forest-rail (17) but has narrow black tail bars; females of Forbes's Forest-rail (19) and Mayr's Forest-rail (20) have darker barring on underparts.

VOICE Gives chuckling notes *ko..ko..ko* (Beehler *et al.* 1986). The contact call is a low mewing note (Ripley 1977).

DESCRIPTION
Adult male Head, neck and breast chestnut, paler on chin and throat and washed darker, sooty-brown, on crown and nape; mantle to back, upperwing-coverts, alula and tertials blackish-brown, each feather with two narrow white stripes; lower back and rump paler, with smaller markings; uppertail-coverts and tail darkish chestnut, narrowly barred black. Remiges dark sepia with broad white bars; underwing-coverts and axillaries very dark brown with white bars. Upper flanks as breast; lower flanks to belly darkish olive-brown, obscurely barred pale rufous-buff, bars narrowly edged blackish; undertail-coverts darkish chestnut, narrowly barred black. Iris olive-brown; bill dark brown, paler at base of lower mandible; legs and feet black.
Adult female Slightly smaller than male. Blackish-brown areas of upperparts spotted white.
Immature Similar to adult but darker and duller on chestnut areas; blackish-brown areas of upperparts washed chestnut; white markings of upperparts tinged rufous; streaks of male shorter than in adult; lower flanks to undertail-coverts almost plain blackish-brown (but in 1 BMNH male they are as in adult).
Juvenile Not described; possibly similar to that of Forbes's Forest-rail.
Downy young Not described.

MEASUREMENTS Wing of 4 males 106-111 (108.5, SD 2.1), of female 105-107; tail of 4 males 65-71 (68.3, SD 2.5), of female 58-61; culmen of 3 males 26, 28, 28, of female 23-27; tarsus of 4 males 35-39 (36.5, SD 1.7), of female 31 (Mayr 1938, Ripley 1977; means and sample sizes not given). Weight of 2 males 114, 125 (Diamond 1969).

GEOGRAPHICAL VARIATION None.

MOULT No information available.

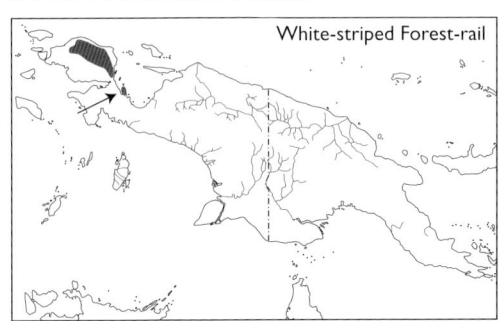

DISTRIBUTION AND STATUS Confined to the Tamrau, Arfak and Wandammen Mountains of the Vogelkop region of Irian Jaya. Very poorly known and classed as DATA DEFICIENT, this shy and secretive forest-rail is apparently uncommon (Ripley 1977). No recent information is available on its status but there is no reason to expect it to be threatened, as its montane forest habitat is likely to be secure, although local hunting with dogs occurs throughout the Vogelkop (Collar *et al.* 1994).

MOVEMENTS None recorded.

HABITAT Occurs on the floor of the interior of montane forest from 1,350 to 1,850m.

FOOD AND FEEDING No information available.

HABITS Shy and secretive. Pairs roost in roofed shelters of dried leaves and moss (Ripley 1977).

SOCIAL ORGANISATION Recorded as occurring in pairs.

SOCIAL AND SEXUAL BEHAVIOUR No information available.

BREEDING AND SURVIVAL Nothing known. Like roosting nest, breeding nest may also be a domed structure, as in Chestnut Forest-rail (Frith & Frith 1990).

19 FORBES'S FOREST-RAIL
Rallina forbesi   Plate 6

Rallicula forbesi Sharpe, 1887, Owen Stanley Range, New Guinea.

Sometimes retained in *Rallicula*. Forms a superspecies with *R. leucospila* and *R. mayri*. Four subspecies recognised.

Alternative names: Red-backed Forest-rail; Forbes' Chestnut Rail.

IDENTIFICATION Length 20-25cm. Male chestnut on head, neck, upper mantle, breast, upper flanks and upper belly; grading to plain dark blackish-brown from lower mantle to rump and on upperwings (dark olive-brown with fine vermiculations in race *parva*); uppertail-coverts and tail chestnut with fine black barring; lower flanks and lower belly to undertail-coverts darkish brown to blackish, with narrow, reddish-brown, black-edged bars. Iris golden-brown; bill dark brown; legs and feet brownish-black. Female has black-margined buff spots on very dark brown or black areas of upperparts. Immature averages browner on blackish areas of upperparts; duller chestnut-brown or greyish-brown below, barred black from lower flanks to undertail-coverts; bill black, tip grey. Juvenile similar to immature; black barring on underparts narrow. Inhabits montane and mid-montane forest.
Similar species Male distinguished from other mountain forest *Rallina* species by unpatterned blackish-brown lower mantle, back, rump and upperwings (finely vermiculated in race *parva*). Red-necked Crake (21), of moist forest at low altitudes, is larger with dark brown upperparts and tail, more extensive underpart barring, green or yellow bill and olive legs. Vermiculations on dark olive-brown upperparts of female *parva* diagnostic; females of all races distinguished from female Chestnut Forest-rail (17) and White-striped Forest-rail (18) by chestnut upper mantle and darker barring on underparts; similar to female of Mayr's Forest-rail (20), which does not occur within its range and which is entirely dark chestnut-brown on upperparts, with black-bordered, darker buff spots from mantle to rump and on upperwings, and less extensive chestnut on mantle.

VOICE A low, frog-like, slowly delivered, repeated *quaaak*; also a rapid *ko..ko..ko* or *bo bo bo bo bo bu bu bu bu uwah uwah uwah uwah...* lasting for more than 1 min (Beehler *et al.* 1986). When alarmed, a low clucking or chuckling is given as birds run off (Beehler *et al.* 1986). The local Kalam name for the species, *kongak* or *kungak*, is onomatopoeic, given for the bird's "noisy call" (Majnep & Bulmer 1977, Schmid 1993).

DESCRIPTION *R. f. steini*
Adult male Head, neck, upper mantle, breast, upper flanks and upper belly chestnut, darker and browner on nape and grading to unpatterned very dark blackish-brown on lower mantle, back, upperwing-coverts and tertials; lower back dark olive-brown; rump chestnut with blackish barring; uppertail-coverts and tail chestnut with fine black barring (sometimes almost lacking). Primaries and secondaries very dark brown with large white bars or spots on inner webs, more conspicuous on undersides; axillaries and underwing-coverts blackish-brown, barred or spotted white. Lower flanks, and lower belly to undertail-coverts, darkish olive-brown to blackish, narrowly barred tawny-buff to reddish-brown, or sometimes greyish-brown, these bars narrowly edged black; narrow bars on belly and thighs sometimes whitish (Rand 1942b). Iris golden-brown; bill dark brown; legs and feet blackish-brown.
Adult female Similar to male but with black-margined buff or buffy-white spots on blackish-brown areas of upperparts. Some females have relatively great contrast between dark brown ground colour and blackish margins to spots, giving a mosaic effect similar to that seen in females of Mayr's Forest-rail (race *carmichaeli*) which, however, shows greater contrast because of its paler upperpart colour.
Immature Duller than adult; blackish-brown to dark olive-brown from mantle to rump and on upperwings, averaging browner than in adult; duller chestnut-brown or greyish-brown below, barred with black from lower flanks to undertail-coverts; bill black with grey tip; some immatures have dark olive-brown back and wings.
Juvenile Not described (see *R. f. forbesi*).
Downy young Chicks have black down with rusty tips and rusty crown, face and throat.

MEASUREMENTS (8 males, 10 females) Wing of male 103-116 (111), of female 105-116 (107.8); tail of male 59-71 (64), of female 60-67 (63); exposed culmen of male 24.5-28 (25.6), of female 22-26 (24); tarsus of 1 male 35, of 4 females 37-38.5 (37.4). Weight of 2 males 88, 91, of 10 females 87-95 (91), of 1 subadult female 81.

GEOGRAPHICAL VARIATION With the exception of the race *parva* from the Adelbert Mts, all populations are very similar in plumage, and other races are separated on biometrics, especially tail length.
  R. f. steini (Rothschild, 1934) – Snow Mts to C Highlands (Weyland, Oranje, Hindenburg, Hagen, Kubor and Bismarck Mts), C New Guinea. See Description.
  R. f. parva Pratt, 1982 – Mt Mengam, Adelbert range, E New Guinea. A small race, comparable in size to the smaller examples of *dryas*. Dark olive-brown to dark chestnut-brown from mantle to rump and on upperwings, finely and obscurely vermiculated in both sexes (Pratt 1982; colour of vermiculations not given), with smallish, black-margined buff spots in female. Wing of 2 males 102, of 3 females 100, 102, 103; tail of 2 males 56, 61, of 3 females 55, 57, 57; exposed culmen of 2 males 26, 27, of 3 females 25, 25, 26.5; tarsus of 2 males 36, 37, of 3 females 34, 34.5, 36.5.
  R. f. dryas (Mayr, 1931) – Saruwaged Mts, Huon Peninsula, E New Guinea. Smaller than other races except *parva*; upperparts generally more blackish-brown than black; tail relatively short. Wing of male 99-113, of female 101-111; tail of male 58-61, of female 61-62; exposed culmen of male 23-25, of 2 females

22.5, 25; tarsus of male 37.5, of female 38; weight of male 78-106 (88), of female 65-96 (82) (Diamond 1969, Pratt 1982, Ripley 1977, some sample sizes not given).
R. f. forbesi (Sharpe, 1887) – Herzog Mts to Owen Stanley Mts, E New Guinea. Similar to *steini* in plumage; tail relatively long. One juvenile male: head pale chestnut-brown, crown feathers tipped darker; supercilium and ear-coverts tawnier. Nape to mantle darker chestnut-brown; back slightly darker, with narrow blackish bars; sides of neck brighter chestnut-brown; rump and uppertail-coverts barred blackish and dark chestnut-brown; tail dark chestnut-brown. Upperwing-coverts, remiges, alula, axillaries and underwing-coverts dark vandyke-brown; upperwing-coverts and alula with narrow chestnut-brown bars at tips and sides of feathers (but lesser coverts largely unbarred); primaries with 3-4 white spots on inner webs, spots brighter on underside of feathers; axillaries and underwing-coverts broadly barred white. Chin and throat cinnamon; upper breast as sides of neck; lower breast chestnut-brown; flanks to belly chestnut-brown with narrow dark bars; thighs barred pale and dark chestnut-brown; undertail-coverts barred blackish and dark chestnut-brown. Iris brown; bill black, extreme tip dirty white; legs and feet black. Wing of 18 males 105-117 (112.8), of 10 females 108-116 (112.7); tail of 5 males 70.5-76 (73), of 5 females 64.5-76 (69.5); exposed culmen of 5 males 22.5-27.5 (25.5), of 5 females 24-27 (25.2); tarsus of 3 males 36.5, 38, 41.5, of 4 females 34, 34, 35, 38.

MOULT Primary moult tends to be ascendant and rapid, with irregularities occurring commonly. The secondaries are moulted irregularly during the primary moult, and several neighbouring feathers may be dropped almost simultaneously. The rectrices are moulted irregularly during the wing moult, all sometimes growing at once. Primary moult of *steini* is recorded in all months from Apr to Jul; moult of body feathers and rectrices is also recorded in Apr. Primary moult of *dryas* is recorded in Feb; in the nominate race, moult is recorded in May, Jul and Sep.

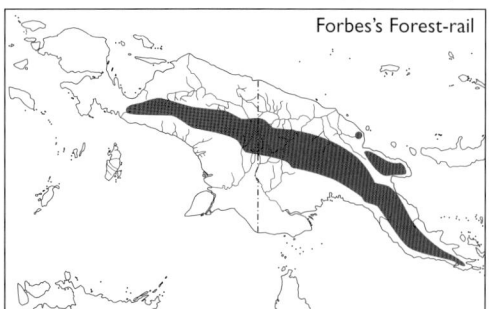

DISTRIBUTION AND STATUS Mountain ranges of C and E New Guinea. Not uncommon locally in E New Guinea but apparently scarce or rare in the W half of its range, where it overlaps with the Chestnut Forest-rail (Coates 1985). This rail is regularly trapped and shot (with bows and arrows) for food (Schmid 1993). The recently described race *parva* is known only from one locality, where it occurs from 1,200 to 1,600m a.s.l (Pratt 1982). A hitherto unnamed chestnut rail from the uninhabited Foja Mts, NC New Guinea, which was frequently seen in October 1979 (Diamond 1985), may be a form of the present species, or of Mayr's Forest-rail, or a new species: see Mayr's Forest-rail species account.

MOVEMENTS None recorded.

HABITAT Inhabits the floor of primary mid-montane and montane forest where ground cover is plentiful, from 1,100 to 3,000m. In areas of sympatry with the Chestnut Forest-rail it occurs largely at lower altitudes.

FOOD AND FEEDING Stomach contents are mostly insects and other animal matter; also some seeds, including those of sedges. Local people say that the rails also feed on various vertebrates, including fledglings of other bird species (Schmid 1993). Birds forage deliberately in dry leaves and tangles, calling quietly.

HABITS Normally very shy and secretive, these birds continually bob the tail and scurry like mice into cover when alarmed. They roost in nests, in pairs or groups (Rand & Gilliard 1967, Ripley 1977). One roosting nest was a football-sized, domed structure of leaf skeletons and moss, placed 2.75m up in a pandanus crown (Mayr & Gilliard 1954).

SOCIAL ORGANISATION The only information is that roosting nests may contain pairs or 3-7 adults of both sexes, and that the birds occur singly, in pairs or in small groups, often foraging in small parties (e.g. Ripley 1977, Coates 1985).

SOCIAL AND SEXUAL BEHAVIOUR No information available.

BREEDING AND SURVIVAL Season Huon Peninsula, Nov. **Nest** Breeding nest thought to be similar to roosting nest, and located on ground or in tree-ferns, *Microsorium* cane or epiphytic ferns, but the only one described was a thick platform of dry vegetable fibre and leaf skeletons, external measurements 18 x 25cm, placed 5-6m up on horizontal fork of small tree in primary forest. **Eggs** Thought to number 4-5; eggs in above-described nest were smooth and glossy white, markedly pointed at one end (Ripley & Beehler 1985).

20 MAYR'S FOREST-RAIL
Rallina mayri   Plate 6

Rallicula rubra mayri Hartert, 1930, Cyclops Mountains, New Guinea.

Sometimes retained in *Rallicula*. Forms a superspecies with *R. leucospila* and *R. forbesi*. Two subspecies recognised.

Synonyms: *Rallicula mayri/leucospila mayri*.

Alternative names: Mayr's Chestnut Rail; Black-tailed Forest-rail.

IDENTIFICATION Length 20-23cm. Male predominantly chestnut, darker on upperparts; a few buffy spots or bars on upperwing-coverts; tail has indistinct narrow black bars; lower flanks to undertail-coverts brown, with narrow, buffy-white, black-edged bars. Iris grey-brown; bill dark grey to black, legs and feet black. Female has dark chestnut back and upperwings with large black-margined buff spots. Immature and juvenile not described. Inhabits montane and mid-montane forest.
Similar species The only *Rallina* species of montane forest

within its range. Predominantly chestnut plumage of male similar to that of brighter male Chestnut Forest-rail (17), which lacks bars on tail. Female similar to female Forbes's Forest-rail (19) and White-striped Forest-rail (18), both of which also have dark bars from lower flanks to undertail-coverts, but differs from these species, and from female Chestnut Forest-rail, in being dark chestnut, with black-bordered buff spots, from mid-mantle to rump and on upperwings. Female Chestnut Forest-rail lacks bars on underparts and tail, and has small white spots which extend to dark brown upper mantle; female Forbes's Forest-rail has more extensive chestnut on mantle and paler buff spots; female White-striped Forest-rail has black extending to upper mantle, and white spots.

VOICE No information available.

DESCRIPTION *R. m. mayri*
Adult male Head, neck and upper mantle darkish chestnut, slightly darker on nape; mid-mantle to rump, and upperwings, slightly darker; a few obscure black and pale buff bars or spots near tips of upperwing-coverts. Remiges blackish-brown, greyer on undersurface, with broad white bars; uppertail-coverts and tail darkish chestnut with indistinct narrow black bars. Underwing-coverts and axillaries very dark brown, barred white. Chin to upper flanks chestnut, somewhat paler than back; lower flanks to undertail-coverts brown with narrow, pale buff to whitish, black-bordered bars (black borders less distinct than in *carmichaeli*). Iris grey-brown; bill dark grey to blackish, legs and feet black.
Adult female Slightly smaller than male; mid-mantle to rump, and upperwings, dark chestnut with black-bordered buff spots.
Immature Not described.
Juvenile Not described.
Downy young Not described.

MEASUREMENTS Wing of male 111-117, of 4 males 111-115.5 (113.6), of female 108-110; tail of male 63-72, of 4 males 65-71.5 (68.2), of female 64-65; culmen of male 23-28.5, exposed culmen of 4 males 23-25.5 (24.5), culmen of female 23-26; tarsus of 1 male 37, of 2 females 35, 36 (Diamond 1969, Ripley 1977; Ripley gives no sample size or mean). Weight of 3 males 123, 123, 129, of 1 female 119.

GEOGRAPHICAL VARIATION There is little geographical variation, the two races being separated on overall colour.
    R. m. mayri (Hartert, 1930) – Cyclops Mts, NE Irian Jaya. See Description.
    R. m. carmichaeli (Diamond, 1969) – Torricelli and Bewani Mts, NW Papua New Guinea. Similar to nominate race but distinctly darker both on upperparts and underparts; pale bars from lower flanks to undertail-coverts with more distinct black borders. In female, extent of black borders to buff spots greatest on upperwing-coverts and scapulars, so that folded wing may appear largely black rather than dark brown. Wing of 4 males 110-118 (114), of 3 females 110, 111, 112; tail of 4 males 62-71 (66), of 3 females 63.5, 65.5, 67; exposed culmen of 4 males 25.5-26.5 (26), of 3 females 23.5, 24, 26. Weight of 2 males 131, 136, of 2 females 112, 123.

UNNAMED FORMS A hitherto unnamed chestnut rail from the uninhabited Foja Mts, NC New Guinea, which was frequently seen in October 1979 (Diamond 1985), may be an undescribed form of this species or of Forbes's Forest-rail, or a new species. Diamond's description runs as follows. "Single individuals and pairs were frequently seen between 1,300 and 1,700m, digging in the ground in bare, poorly lit areas of the forest. The call was a short, quiet, low-pitched, guttural, frog-like grunt. The mantle was dull dark brown, contrasting with the rufous or reddish-chestnut head, neck and tail. Most individuals seen were without dorsal spots or streaks, but one had tiny pale buff spots on the mantle." The birds' weight was estimated as 80-130g. The unspotted birds were probably males and the spotted bird probably a female. The observations suggest that this forest-rail is an unnamed taxon closest to Forbes's Forest-rail of the E Central Range and Mayr's Forest-rail (race *carmichaeli*) of the North Coastal Range.

MOULT No information available.

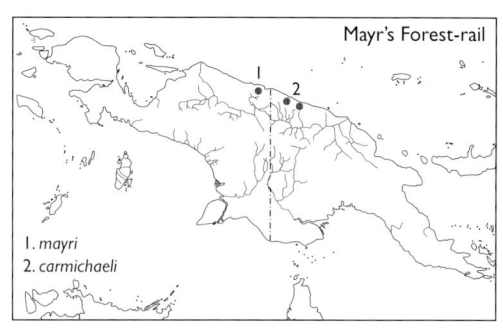

DISTRIBUTION AND STATUS Known only from three isolated mountain ranges in N New Guinea. A very poorly known species, which is classed as DATA DEFICIENT. Its current status is uncertain: it was apparently common in secure habitat in the late 1980s, but it was not found more recently on a visit to the Cyclops Mountains (Collar *et al.* 1994). In view of its localised and highly specific habitat, its total numbers are probably small.

MOVEMENTS None recorded.

HABITAT Floor of montane and mid-montane forest, between 1,100 and 2,200m.

FOOD AND FEEDING No information available.

HABITS No information available.

SOCIAL ORGANISATION No information available.

SOCIAL AND SEXUAL BEHAVIOUR No information available.

BREEDING AND SURVIVAL No information available.

21   RED-NECKED CRAKE
Rallina tricolor   Plate 7

Rallina tricolor G. R. Gray, 1858, Aru Islands.

Monotypic (see Geographical Variation).

Synonyms: *Eulabeornis/Rallus tricolor*.

Alternative name: Red-necked Rail.

IDENTIFICATION Length 23-30cm; wingspan 37-45cm. A medium-sized, chestnut and dark brown rail, with relatively obscure barring on the underparts. Head, neck, mantle and breast rich chestnut; chin and throat whitish; rest of upperparts, including tail, dark olive-brown. Belly and flanks to undertail-coverts dark brown with pale, fine, buffy barring which varies from well-defined to virtually absent on flanks and becomes rufous on undertail-coverts. White bars on inner webs of remiges usually hidden, except when bird runs to cover with partly spread wings; underwing-coverts dark brown-black, boldly barred white. Iris red; bill green or yellow-green, with blue-grey base; legs and feet olive. Sexes alike. Immature not described; probably very similar to adult. Juvenile almost uniform dark olive-brown above and dark brown below; bare part colours similar to adult. Inhabits moist to wet forest, and dense scrub and thickets near water. Loud calls distinctive; birds more often heard than seen.

Similar species Distinguished from sympatric forest rails by combination of large size, unmarked dark brown upperparts and tail, pale fine barring on dark brown underparts, green or yellow bill and olive legs. Undescribed immature may be superficially similar to immature Red-legged Crake (23); likewise, juvenile may be superficially similar to undescribed juvenile Red-legged Crake. Juvenile Rufous-tailed Bush-hen (87) is paler in overall colour, has stouter bill, clean white chin and throat, dull rufous vent and undertail-coverts, and has no barring on underwing.

VOICE Most information is from Marchant & Higgins (1993). The territorial call is given by pairs all the year, mostly at dusk and during the early evening, but also at night and during the day. It is a loud, harsh, penetrating note quickly repeated in a descending scale with the emphasis on the first note *nark-nak-nak...* or *kare-kare-kare...*. Other territorial pairs often respond, and calls are often given in chorus at dusk. In Australia, birds are most vocal in Oct-Dec and are very quiet in Jan; they also call in flight, and may call in response to loud noises. Pairs also utter a monotonous repeated *tock* note, sometimes continuously for hours well into the night, especially during the breeding season. A melodious, rapidly repeated *coot* note has also been noted. Contact calls include low grunts – presumably the repeated *um um um* described by Coates & Bishop (1997), a repeated soft *plop* note like berries dropping into water, and a single, soft *clock* or *click*. When disturbed, the birds utter an abrupt cluck, and a piglet-like squeal is also recorded. Downy chicks give high-pitched, thin whistles, soft peeps (in response to the grunts of adults) when running to cover, and long whistles when lost.

DESCRIPTION
Adult Head, neck, mantle and breast rich chestnut to rufous-chestnut; chin and upper throat paler, often whitish, this colour extending with wear to face below eye, ear-coverts and centre of throat. Rest of upperparts, including upperwing-coverts, tertials and tail, dark olive-brown with a slight greyish-olive tinge (less on tail); primaries and secondaries dark brown, with white bars on inner webs, and sometimes tinged olive on outer webs; some birds have a few orange-buff spots on outer webs of outer primaries (seldom visible in folded wing); tertials unbarred. Underside of remiges dark grey-brown, inner webs barred white; underwing-coverts and axillaries dark brown to blackish, boldly barred white (sometimes with buff tinge). Belly and flanks to undertail-coverts dark brown with pale, fine, buffy barring which varies from bold to indistinct or virtually absent on flanks, and becomes buffy or rufous on undertail-coverts in well-marked birds. Iris red or orange-red, occasionally yellowish; orbital ring orange to orange-yellow; bill green to yellow-green, often with grey culmen and blue-grey base; legs and feet olive to brownish-olive, sometimes tinged yellowish. Bill colour may be brighter during breeding season.

Immature Not properly described, existing descriptions probably being referable to juvenile and aberrant adult plumages (Marchant & Higgins 1993). Some specimens, age unknown, combine duller chestnut foreparts with much reduced barring on both belly and flanks.

Juvenile Neck and top of head brown; hindneck has faint rufous tinge; face, ear-coverts, chin and upper throat mottled cream and grey-brown; rest of upperparts dark olive-brown. Underparts brown; undertail-coverts dark brown with rufous-buff chevrons. Colour of bare parts similar to adult. Ripley (1964) describes a "subadult" (probably juvenile) from Misool as "lighter coloured below than adults, the head with brown bases to the feathers which show through to give a nearly barred effect, the underparts, the throat and belly especially, much paler, the latter with pale creamy bars. The entire throat is cream coloured, the newer feathers having brownish tips".

Downy young Down black, with concealed dark brown bases (possibly exposed with wear when chick older); iris brown; eye-ring pale grey; bill black, sometimes with pale distal third to lower mandible; egg-tooth white; inside of mouth pale flesh-pink; legs and feet grey to brown, later becoming black.

MEASUREMENTS All measurements from Marchant & Higgins (1993). Wing of 6 males 135-150 (141.7, SD 5.4), of 5 females 133-139 (135.8, SD 2.2) (E New Guinea), of 6 males 141-148 (145, SD 2.5), of 3 females 144, 145, 149, of 9 unsexed 139-148 (144.2, SD 2.8) (Cape York Peninsula), of 3 males 144, 151, 155, of 3 females 142, 143, 155, of 11 unsexed 142-155 (147.4, SD 5.4) (Queensland: Cooktown to Ingham); tail of 6 males 65-71 (67.3, SD 2.4), of 3 females 59, 63, 64, of 9 unsexed 59-71 (65.6, SD 3.5) (Cape York), of 3 males 65, 73, 75, of 2 females 70, 70, of 8 unsexed 65-75 (71, SD 3.3) (Queensland: Cooktown to Ingham); exposed culmen of 4 males 26-29 (27.8, SD 1.3), of 4 females 26-28 (27, SD 0.8) (E New Guinea), of 5 males 27.5-30 (28.8, SD 1.1), of 3 females 26, 27.5, 29, of 8 unsexed 26-30 (28.3, SD 1.3) (Cape York Peninsula), of 2 males 27, 27.5, of 2 females 25, 25.5, of 7 unsexed 25-28.5 (27.2) (Queensland: Cooktown to Ingham); tarsus of 4 males 47-52 (48.9, SD 2.4), of 3 females 45, 47, 47 (E New Guinea), of 6 males 45-49 (46.4, SD 1.7), of 3 females 45.5, 46.5, 48, of 9 unsexed 45-49 (46.5, SD 1.5) (Cape York Peninsula), of 3 males 44.5, 46, 48.5, of 3 females 42.5, 46.5, 47, of 10 unsexed 42.5-48.5 (45.6, SD 1.7) (Queensland: Cooktown to Ingham). Weight (Queensland) of 1 male 143, 1 unsexed 166, 1 vagrant 109 (Marchant & Higgins 1993); (New Guinea) of 3 males 169, 214, 231, of 2 females 194, 200 (Ripley 1964); 1 probable migrant 115 (Berlioz & Pfeffer 1966).

GEOGRAPHICAL VARIATION Up to six races have been recognised on the basis of size and plumage differences. Nominate *tricolor* of E Indonesia, Australia, New Guinea and outlying islands, is largest in the W of its range, the increase in size possibly being clinal (Ripley 1977). This form includes *maxima* of New Guinea (larger, with olive-brown wash on mantle and more distinct white or buff underpart barring) and *robinsoni* of Queensland (longer bill, shorter tarsus and less barring on belly). The race *victa* of the Tanimbar Is (smaller than nominate, with ashier upperparts and sparser barring on underparts) sometimes includes *laeta* of St Matthias I (paler orange-tawny than *victa* on crown). The form *convicta* of New Ireland and the New Hanover Is is smaller than nominate and has underparts more heavily barred than in *victa*. However, these differences may be attributable to individual variation, which is considerable, while geographical variation in size is not well understood. On the basis of existing information it is probably better to treat the species as monotypic. Another proposed New Guinea race, *grayi*, was found to be indistinguishable from examples of *tricolor* (Mayr 1949).

MOULT Primary moult is partly asynchronous: the outer 3-5 feathers are dropped at roughly the same time and only when these have grown enough to allow flight are the inner primaries moulted in rapid ascending sequence (Marchant & Higgins 1993). This wing moult is unlikely to cause temporary flightlessness. It is recorded in New Guinea in Apr, Jun-Aug and Dec; Australian birds may moult the primaries in Aug-Nov, and body moult has been recorded in Feb, May and Jul. Postjuvenile moult begins on the face, foreneck and breast, and may involve moult of the primaries.

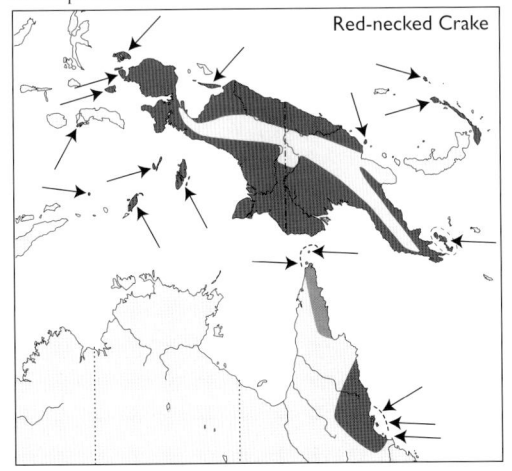

DISTRIBUTION AND STATUS Moluccas (Ambon, and Kur in Tayandu Is) and E Lesser Sundas (Damar, and Larat in Tanimbar Is), Aru Is, New Guinea (including Yapen and Karkar Is) and W offshore islands (Waigeo, Batanta, Salawati and Misool), D'Entrecasteaux Archipelago (Fergusson Is), and Bismarck Archipelago (St Matthias Is, New Hanover and New Ireland); also NE Australia (Queensland), where there are possibly two disjunct populations, from the Torres Strait Is S to the McIlwraith Range, and from Cooktown S to Paluma and inland to Atherton Tableland, while it also occurs on offshore islands (Mason *et al.* 1981). This crake is not regarded as globally threatened, but little information on its status is available because it is shy and seldom seen. In Papua New Guinea it is apparently locally common (Coates 1985), but Gregory (1995) regards it as uncommon in the Ok Tedi area, Tabubil. The Australian population is declining due to the massive and continuing destruction of lowland rainforest habitat for agriculture and residential development, although it is able to exploit some manmade habitats by virtue of its ability to occupy dense vegetation in suburban gardens (Marchant & Higgins 1993). Little is known about its status elsewhere. Predation by the Black Butcherbird *Cracticus quoyi* is recorded in Australia (Joseph & Drummond 1982).

MOVEMENTS Australian data are from Marchant & Higgins (1993). Although regarded as sedentary over much of range, this crake is partly migratory in NE Australia, where it is regarded as a wet-season migrant to the Cape York Peninsula from New Guinea, arriving from Oct to early Jan and leaving in Apr or early May when the forest floor dries out in the winter. It is apparently sedentary further S in Queensland, where rain falls throughout the year and the forest floor is permanently damp. It is recorded from several islands in the Torres Strait, where it is often heard overhead at night and is recorded striking a lighthouse (Booby I) in Nov. It is known from the Moluccas in Jun-Jul and from Tanimbar in Nov-Dec; the only record from Ambon Island (Moluccas) is of one coming to a light at night on 14 June (Berlioz & Pfeffer 1966), possibly a migrant from New Guinea. In Wallacea it may be locally nomadic (Coates & Bishop 1997). In the Port Moresby area, Papua New Guinea, it moves away from forest where the ground becomes hard and surface water disappears in prolonged dry periods, but remains near permanent water (Coates 1985).

HABITAT In Papua New Guinea this crake mostly occurs in rain forest and swamp forest, but it is also found in monsoon forest during the rains and in sago swamps, gallery forest, secondary growth and (once) mangroves (Coates 1985). It occurs mostly in the lowlands, locally up to 1,370m, and normally requires moist habitat, preferably with pools or creeks. In Queensland (Marchant & Higgins 1993) it is recorded from tropical rain forest, usually with a dense understorey of vines, ferns and pandanus, and also from vine-thicket and monsoon scrub; it always occurs near permanent streams or swamps, and is found mostly below 700m but is recorded up to 1,250m. It is also sometimes found in suburban gardens with dense vegetation, and occasionally in *Lantana camara* thickets. The birds apparently roost and nest where the forest and undergrowth are most dense. This species occurs alongside the Bare-eyed Rail (82) around Port Moresby, Papua New Guinea (Coates 1985).

FOOD AND FEEDING Information is from Marchant & Higgins (1993) unless otherwise stated. This species eats mainly invertebrates, including annelids, oligochaetes, molluscs, crustaceans, terrestrial amphipods, spiders, and adult and larval insects; also frogs and tadpoles. Some seeds, including beans, are also recorded. The birds forage methodically, along shallow stream beds and stream margins, in stagnant pools, in leaf-litter and occasionally along the edge of salt water at low tide; one pair foraged on newly bulldozed ground along a roadside. They rake through leaf-litter with the feet and bill, turn over small stones with the bill, and probe pools, moss and debris. At a house in Queensland, birds come to feed on millet grain spread by the back door, and are also attracted to cheese, which they eat with gusto (P. Gregory *in litt.*) – see Social and Sexual Behaviour.

HABITS In Australia, this species has been said to be mainly crepuscular and nocturnal, spending the day in rock crevices, the crowns of palm trees and thick scrub (Blakers *et al.* 1984, Schodde & Tideman 1990) but is also regarded as diurnal, being most active at dusk and in the early evening, although also foraging during the day in forest (Marchant & Higgins 1993). In E New Guinea, activity is recorded from early afternoon to dusk (Coates 1985). It is secretive and wary, more often heard than seen; it often appears nervous, constantly making short, dodging runs on the forest floor; it is usually seen when running to cover with the head extended forwards and the wings partly spread and often fluttering, and it flies rarely, escape flights usually being fluttering and noisy. When alarmed, birds will hide under roots and overhanging banks; they swim readily, and will also dive. Adults and young often drink and bathe at dusk, and may use communal roosting platforms.

SOCIAL ORGANISATION Monogamous, with a strong pair bond; some pairs maintain permanent territories. Birds are normally seen singly, in pairs or in family groups. There is one record of three birds attending two nests, although only one took an active role in both nests (Mason *et al.* 1981).

SOCIAL AND SEXUAL BEHAVIOUR Birds attracted to feed on millet grain (see Food & Feeding) compete with Musky Rat-kangaroo *Hypsiprymnodon moschatus*, which is usually dominant, though the rails will raise their wings to display the black and white underwing pattern as a warning (P. Gregory *in litt.*). Breeding behaviour is summarised by Marchant & Higgins (1993). During the breeding season, pairs strut around with an upright posture, uttering single clicking notes and flicking the tail. Conspecific individuals intruding into the territory are chased away. Incubating birds become increasingly aggressive and difficult to flush as incubation progresses; when the eggs are near to hatching, the bird performs a distraction display, running into water and then squatting, fluffing out the feathers and spreading and fluttering the wings with a shivering motion so that the white bars on the primaries are visible. An adult pecked the leg of a person about to stand on its chicks.

BREEDING AND SURVIVAL Season In Queensland, lays throughout wet season (Nov-Mar), with Dec-Feb peak; in Papua New Guinea lays early in rains (Nov), adults with 3 fully grown young also recorded Nov, and an "active nest" Oct (Coates 1985, Ripley & Beehler 1985); female in breeding condition, Mar (Rand 1942a). **Nest** A shallow cup of dead leaves, twigs and tendrils, diameter 15-20cm, in dense vegetation (ferns, vines etc), between buttresses, in bush or on tree stump 0.6-2m above ground; or depression in ground lined with a few dead leaves. **Eggs** Usually 5 (3-7); oval, slightly pointed at one end; white and lustrous; size of 16 eggs 35.1-39.4 x 25.7-29.2 (37.3 [SD 1.5] x 27.4 [SD 1.3]); calculated weight 17; laid at daily intervals. In Australia, may lay again if first clutch unsuccessful. **Incubation** 18-22 days, by both sexes but probably largely by female; in New Guinea, birds said to cover eggs with leaves, etc, when not incubating (Mason *et al.* 1981); hatching synchronous. **Chicks** Precocial; fed by both parents; leave nest soon after hatching but may return to roost for several days; still fed by adults at 5 weeks; leave parents before they can fly. Young fully feathered at 4-6 weeks. May form pairs when 4-5 months old.

22 ANDAMAN CRAKE
Rallina canningi Plate 7

Euryzona canningi Blyth, 1863, Andaman Islands.

Monotypic.

Synonym: *Castanolimnas canningi*.

Alternative name: Andaman Banded Crake.

IDENTIFICATION Length c. 34cm. A medium-large rail, the largest *Rallina* species and the only one found on the Andamans. Easily distinguished by deep maroon-chestnut head, neck, breast and entire upperparts, bold black and white barring on underparts, unbarred dark chestnut undertail-coverts, red iris, bright apple-green bill with narrow whitish tip, and olive-green legs and feet. Visible pale barring on wings confined to outer primaries and some greater and median coverts. Tail long and fluffy. Sexes alike. Immature is duller overall; barred areas dark grey-brown with chestnut tinge, narrowly banded and streaked dirty white. Juvenile paler than adult; dark bars olive-brown, pale bars dirty white; lower belly buffy-brown, tinged cinnamon. Inhabits marshland in forest.
Similar species No similar rail occurs on the Andamans. Most closely resembles Slaty-legged Crake (24), which is considerably smaller and has chin and throat white, upperparts from mantle to tail (including upperwings) dark olive-brown, undertail-coverts barred black and white, bill green becoming dark brown or grey towards tip, and legs and feet greenish-grey to black.

VOICE A deep croak, *kroop*. The alarm note is a sharp *chick*. When handled, birds sometimes utter a cry "like that of a wounded rabbit" (Ali & Ripley 1980).

DESCRIPTION
Adult Head, neck, breast and entire upperparts, including upperwing-coverts and tertials, deep maroon-chestnut; a few median and greater upperwing-coverts have white and dark brown barring; alula dark brown, feathers edged chestnut and barred reddish-buff on inner webs; primaries and outer secondaries dark brown, edged chestnut, barred whitish on inner webs and also less distinctly on outer webs of outermost 2-3 primaries; axillaries and underwing-coverts black, barred or spotted white. Rest of underparts boldly barred black and white, except undertail-coverts which are dark chestnut. Iris red; bill bright apple-green with narrow whitish tip; legs and feet olive-green. Sexes alike.
Immature Chestnut areas duller; barred areas dark grey-brown with chestnut tinge, narrowly banded and streaked dirty white.
Juvenile Noticeably paler than adult; chestnut of upperparts paler, especially on hindneck; barring of underparts olive-brown, tinged pinkish-cinnamon, and dirty white; lower belly buffy-brown, tinged cinnamon. Iris reddish-brown; bill olive-green with dusky culmen; legs and feet olive-green.
Downy young Downy chick rich chestnut, slightly greyish under the wings; bill dusky olive.

MEASUREMENTS (8 males, 8 females) Wing of male 153-165 (160.3, SD 4.5), of female 150-167 (156.3, SD 5.1); tail of male 72-90 (83, SD 5.6), of female 74-87 (78, SD 4.4); culmen to base of male 32-36.5 (33.9, SD 1.6), of female 30.5-33.5 (31.7, SD 1.0); tarsus of male 51-59 (53.9, SD 2.9), of female 50.5-58.5 (54.3, SD 2.6).

GEOGRAPHICAL VARIATION None.

MOULT Remex moult is ascendant and very rapid, the pattern in two specimens tending to synchronous moult of all primaries and secondaries; this moult is recorded in Oct (Stresemann & Stresemann 1966).

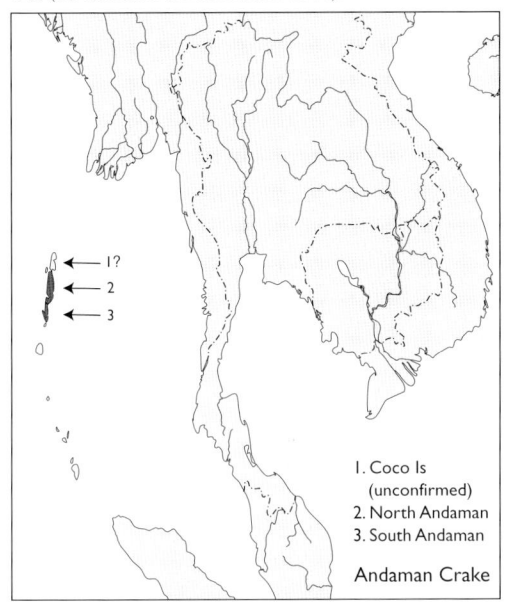

Map legend:
1. Coco Is (unconfirmed)
2. North Andaman
3. South Andaman

Andaman Crake

DISTRIBUTION AND STATUS Endemic to the Andaman Islands. Its precise range is not clear: it is recorded from at least North Andaman and South Andaman Islands, and possibly also occurs on Great Coco or Little Coco Islands, Burma (Collar *et al.* 1994). It was formerly regarded as common, and was apparently snared in some numbers, a total of 80 birds caught in a square mile being recorded by Ali & Ripley (1980). There appear to be only two recent records, of single birds (see Habitat) and the species is classed as VULNERABLE. Although forest cover remains extensive in the Andamans it is suffering slow but continuing loss, while introduced predators may also be a threat to the birds (Collar *et al.* 1994).

MOVEMENTS Apparently sedentary.

HABITAT Marshland in forest; two recent records are from a large, open area of wet marshland and a small marsh on the edge of secondary forest (Collar *et al.* 1994).

FOOD AND FEEDING Largely worms, molluscs and insects (including beetles, grasshoppers and caterpillars); possibly also small fish. Large grasshoppers are shaken and battered before being swallowed (Ali & Ripley 1980).

HABITS Said to be a great skulker in reeds and herbage, seldom seen but readily snared, and to carry itself high on the legs, with the head generally rather drawn in and the feathers ruffled up to produce a rounded outline (Ali & Ripley 1980).

SOCIAL ORGANISATION Presumably monogamous; no further information available.

SOCIAL AND SEXUAL BEHAVIOUR No information available.

BREEDING AND SURVIVAL Season Breeds Jun-Aug, during summer. **Nest** A collection of grass and leaves, placed at foot of large forest tree or under tangled forest undergrowth, not always close to water. **Eggs** Glossy white; 32 eggs measured 37.2-43.1 x 29.7-32 (40.6 x 30.8); calculated weight 21.1; clutch size unknown. **Incubation** By both sexes.

23 RED-LEGGED CRAKE
Rallina fasciata   Plate 7

Rallus fasciatus Raffles, 1822, Bengkulu, W Sumatra.

Monotypic.

Synonyms: *Gallinula/Rallina/Hypotaenidia euryzona*; *Rallus ruficeps*; *Crex/Corethrura/Porzana/Euryzona fasciata*; *Rallina suzuki*.

Alternative name: Malaysian/Malay/Red-legged Banded Crake.

IDENTIFICATION Length 22-25cm. A medium-sized, fairly short-tailed rail with the typically longish legs of its genus, and with barred upperwings. Head, neck and breast ruddy-chestnut; chin and throat paler. Rest of upperparts brownish-chestnut, with buffy-white spots and buffy-white and black bars on scapulars. Upperwing-coverts have buffy-white spots, grading to broad blackish and narrow white bars on greater coverts. Primaries and outer secondaries blackish-brown, barred white. Underparts below breast broadly barred black and white. Iris orange or red; orbital ring red; bill greenish-horn, culmen blackish; legs and feet red. Sexes similar; female more cinnamon on head and neck, with narrower black bars on underparts. Immature browner, less chestnut, on upperparts; face and throat ochraceous-tawny, chin paler; breast and sides of neck as mantle or brighter (more cinnamon); rest of underparts buff to dirty white, obscurely barred brown-black; bill possibly brownish; legs and feet yellowish-brown. Juvenile probably similar to immature but duller, and whiter below. Inhabits predominantly wet, marshy areas in open country, scrub or forest.

Similar species Easily distinguished from Slaty-legged Crake (24) and Red-necked Crake (21) by boldly barred wings and red legs; also from former usually by chestnut wash on chin and throat, and from latter by white barring on underparts. The superficially similar Band-bellied Crake (103) has white bars on wings confined to some upperwing coverts (sometimes missing), dark olive-brown crown, nape and upperparts, and orange-pink legs. Ruddy-breasted Crake (102) is smaller, with unbarred wings, olive-brown from rear crown to tail, and less extensive and narrower white barring on underparts; underwings unbarred.

VOICE Details are from Coates & Bishop (1997) unless otherwise specified. The advertising call is described as a loud, staccato series of *gogogogo* notes, usually given at night but also during the day in rainy weather. Birds also give a series of "devilish-sounding" screams, and very sharp *girrrr R R R R* calls. The territorial call is a loud series of nasal *pek* calls, repeated about every 0.5s and given at dawn and dusk in the breeding season; this is apparently given by the male, and the female sometimes joins in with nasal notes (C. R. Robson *in litt.*). There is also a long, slow descending trill, reminiscent of the Ruddy-breasted Crake

(Lekagul & Round 1991). When two birds meet there is a cacophony of scolding sounds.

DESCRIPTION

Adult male Head, neck and breast ruddy-chestnut, paler to whitish on chin and throat. Rest of upperparts, including scapulars, upperwing-coverts, inner secondaries and tail, brownish-chestnut, darker on rump and uppertail-coverts, with a few buffy-white spots on upper scapulars and buffy-white and black barring on lower scapulars. Lesser secondary coverts have buffy-white spots, grading to broad black and narrow white bars on other secondary coverts and on primary coverts. Inner upperwing-coverts variably washed brownish-chestnut. Primaries and outer secondaries dark brown to blackish, barred white. Underwing-coverts and axillaries barred black and white. Underparts below breast broadly barred black and white (black bars wider); centre of belly white; undertail-coverts sometimes washed rufous; thighs mixed black and white. Iris orange or carmine; orbital ring vermilion; bill greenish-horn, with blackish culmen and carmine gape; legs and feet carmine.

Adult female More cinnamon, less chestnut, on head and neck, with narrower black bars on belly and flanks.

Immature Head, nape and mantle mid-brown, sometimes with a purplish flush; lores, eye region and ear-coverts brighter; malar region paler; lower mantle tinged cinnamon. Back and rump darker brown, faintly barred buff-white; uppertail-coverts and tail dull plain mid-brown. Scapulars mid-brown, tinged cinnamon, each feather with buff central streak; upperwing-coverts darker brown, more cinnamon-brown on greaters, with buff-white spots or bars. Primaries, secondaries and alula dull darkish brown, primaries with 4-5 buff spots on outer webs and 4-5 white spots or bars on inner webs; alula spotted buff; tertials brighter, tinged cinnamon. Underwing-coverts and axillaries dull dark brown, with white spots and bars. Malar region, chin and throat ochraceous-tawny, tending to whitish on chin; breast and sides of neck as mantle or brighter (more cinnamon); rest of underparts buff to dirty white, obscurely barred brown-black. Bill possibly brownish, basal half of lower mandible horn; legs and feet yellowish-brown.

Juvenile Not properly described; probably similar to immature but duller, and more extensively white on underparts.

Downy young Dark smoky-brown.

MEASUREMENTS (10 males, 6 females) Wing of male 118-134 (126.9, SD 4.8), of female 119-127 (123.7, SD 3.8); tail of male 43-51 (47.5, SD 2.4), of female 41-49 (44.2, SD 3.6); culmen to base of male 23.5-25 (24.5, SD 0.5), of female 22.5-23.5 (23.1, SD 0.4); tarsus of male 40-47 (42.7, SD 1.9), of female 41.5-42.5 (41.9, SD 0.4).

GEOGRAPHICAL VARIATION None.

MOULT Primary moult is ascendant, with a tendency to partial synchrony; it is recorded in Borneo, Jul and in Java, May.

DISTRIBUTION AND STATUS Lowlands of SE Asia and the Malay Archipelago, in NE India (Cachar), Burma (except NE), NW and peninsular Thailand, Malaysia, S Vietnam (Cochinchina), Sumatra (on mainland in Utara, Riau, Bengkulu, and Lampung, also Riau, Mentawai and Batu Is, Bangka, Belitung, Enggano and Nias), Borneo, the Philippines (Luzon, Mindoro, Culion, Palawan, Balabac, Mindanao and Basilan) and Flores. It winters (and is also probably resident) S to Java, the Lesser Sunda Is (Lombok, Sumbawa, Alor and Kisar) and the Moluccas (Halmahera, Ambon, Bacan and Buru). Its status is uncertain because the birds are difficult to observe or flush. It is widely distributed but is probably overlooked and is not known to be particularly numerous anywhere except on Flores, where it is locally moderately common (Coates & Bishop 1997); it is apparently rare in S Vietnam (Wildash 1968). It is not regarded as globally threatened.

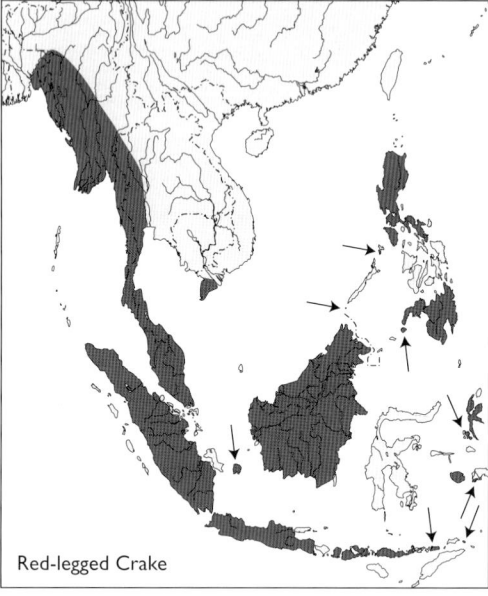

Red-legged Crake

MOVEMENTS These are not properly understood, but the species is both resident and migratory in its normal range, and is also probably dispersive, while it is considered predominantly a winter visitor in the S part of its range. It is resident and migratory in Thailand (where it breeds S at least to Trang) and Malaysia, and probably largely resident in the Philippines where it is recorded in Feb and May-Nov but migrants are also recorded from Dalton Pass, Luzon (McClure & Leelavit 1972, Medway & Wells 1976, Dickinson et al. 1991). Most of the population in Borneo is resident but numbers may be augmented by visitors in winter, and throughout the Greater Sundas it is apparently a local resident, numbers being augmented by winter visitors from mainland Asia; in Sumatra it is recorded between Sep and Jun and breeds sporadically, but it is probably mainly a non-breeding visitor (Smythies 1981, van Marle & Voous 1988, MacKinnon & Phillipps 1993). It is normally considered a migrant to Wallacea but is recorded Apr-Dec, and is apparently resident on Flores; it may breed throughout the Lesser Sundas (White & Bruce 1986). It migrates at night; movements are recorded from peninsular Malaysia in Oct-Dec and Apr-May, and it has been recorded on passage in some numbers at islands and lighthouses in the Malacca Straits (Medway & Wells 1976). Vagrants are recorded from Lanyu I (off Taiwan) in Jun, Palau (W Micronesia) no date available, and Broome, Western Australia, in Jul (a bird found on a pearling lugger) (Serventy 1958, Cheng 1987, Pratt et al. 1987).

HABITAT Reedy swamps and marshes, rice paddies and taro fields, rivers, streams and watercourses, and riparian thickets and wet areas (including the edges of streams and creeks) in forest and secondary growth. It is not often found in open areas away from forest (C. R. Robson *in litt.*). In the Philippines it also occurs on open hillsides and in cogon grasslands (Dickinson *et al.* 1991). It is confined to low altitudes (up to c. 800m) except when on migration, when it is recorded at c. 1,400m; it may occur mainly in the hills in the Lesser Sundas (Coates & Bishop 1997).

FOOD AND FEEDING No information available.

HABITS Shy, retiring and difficult to flush; birds sometimes alight in trees when flushed. On Flores during the breeding season pairs are relatively confiding, and are vocal at night (Coates & Bishop 1997). Claims that the species is nocturnal may refer primarily to calling activity.

SOCIAL ORGANISATION Assumed to be monogamous, and territorial at least during the breeding season.

SOCIAL AND SEXUAL BEHAVIOUR No information available.

BREEDING AND SURVIVAL Season Burma, Jun, Aug-Sep; Thailand, Jul-Aug; Borneo, Apr; Java, Jan-Apr; Sumatra, Jan; Flores (no date). **Nest** Undescribed, but nests have been located in forest undergrowth, one near bank of small stream. **Eggs** 3-6; ovate; chalky white with a definite gloss, sometimes with a few very small, obscure, sepia or darker markings; size 29.8-34.1 x 23.2-25.4 (32.25 x 24.33, n = 24) (Hoogerwerf 1949, Harrison & Parker 1967, Hellbrekkers & Hoogerwerf 1967). **Incubation** By both sexes. No other information available.

24 SLATY-LEGGED CRAKE
Rallina eurizonoides   Plate 7

Gallinula eurizonoides Lafresnaye, 1845, no locality (probably Philippines)

Seven subspecies recognised.

Synonyms: *Rallina telmatophila/formosana/minahasa/ceylonica/superciliaris; Rallus/Ortygometra/Corethrura superciliaris; Zapornia nigrolineata; Porzana amauroptera/ceylonica/superciliaris; Euryzona sepiaria/eurizonoides; Rallus nigrolineatus/telmatophila/minahasa; Ortygometra eurizona.*

Alternative names: (Slaty-legged) Banded/Ryukyu/Philippine Crake.

IDENTIFICATION Length 21-25cm (SE Asian birds 26-28); wingspan 47.5cm. A medium-sized *Rallina*, with bold barring on the underparts but not the upperwings. Head, neck and breast bright chestnut; chin and throat white (but pale rufous or chestnut in races *eurizonoides*, *alvarezi* and *minahasa* of Philippines and Sulawesi). Rest of upperparts, including upperwings and tail, dark olive-brown, with a few black-bordered white bars (usually not visible) on upperwing-coverts. Rest of underparts white, barred dusky black; black bars narrower on undertail-coverts. Iris red to orange; orbital ring greyish-pink; bill greenish, tip dark brown or grey; legs and feet greenish-grey to black. Sexes alike. Immature has chestnut replaced by dark olive-brown, ashier on sides of head and neck; scapulars and wing-coverts with sparse white and black bars; iris brown and legs grey. Juvenile like immature but duller; averages darker above and paler below (but see race *alvarezi*); no bars on scapulars or coverts; iris dark olive-brown; bill and legs black. Inhabits forests, well-wooded areas and dense scrub.

Similar species Distinguished from Red-legged Crake (23) by darker, browner upperparts; white bars on wings confined to a few coverts (not usually visible); whiter chin and throat (usually), contrasting sharply with chestnut of head; and greenish-grey to black legs and feet; in SE Asia is larger than Red-legged, with larger bill (C. R. Robson *in litt.*). Red-necked Crake (21) lacks white barring on underparts and upperwing-coverts, and has green or yellow bill and olive legs. Ruddy-breasted Crake (102) and Band-bellied Crake (103) are brown from crown to tail and lack white bars on remiges; Ruddy-breasted Crake has less extensive barring on underparts, and red legs; Band-bellied Crake has white bars (sometimes missing) on some upperwing-coverts, and orange-pink legs.

VOICE Information is from Ali & Ripley (1980) and Pratt *et al.* (1987). The territorial or advertising call is a repeated *kek-kek* or nasal *ow-ow*, persistently repeated and usually heard in the early morning and at dusk, particularly during the breeding season. It is often given during the night, and even in nocturnal flight, and also during the day in misty, overcast weather. A nasal *kok* is given to other individuals, and is possibly also an alarm call. A subdued *krrrr*, given when alarmed, and a long drumming croak, *krrrrrrrr-ar-kraa-kraa-kraa-kraa*, are also recorded. Birds give a hissing, spitting call to intruders at the nest.

DESCRIPTION *R. e. amauroptera*
Adult Head, neck and breast bright chestnut; chin and throat white. Rest of upperparts, including upperwings and tail, dark olive-brown; a few upperwing-coverts have black-bordered white bars (usually not visible); remiges dark brown with olive-brown outer webs, inner webs sometimes said to have sparse white bars; underwing-coverts and axillaries barred blackish-brown and white. Sides of breast often marked with olive-brown; underparts below breast white, barred dusky-black; black bars narrower on undertail-coverts; belly white with narrow dusky bars. Iris crimson; orbital ring dull greyish-pink; bill green, terminal half of upper mandible and tip of lower mandible dark brown; legs and feet dull greenish-plumbeous to black. Sexes alike; female possibly less barred on upperwing-coverts.
Immature No chestnut in plumage. Entire upperparts, including forehead to hindneck, dark olive-brown, possibly slightly paler than in adult; lores, face, ear-coverts and sides of neck more ashy-brown, mottled warm buff; scapulars and wing-coverts with sparse white and black bars (often absent); white of throat more extensive; breast olive-brown, paler in centre and tinged cinnamon; underparts barred dark brown and whitish. Iris dull brown; bill dull greenish to blackish; legs and feet plumbeous-grey.
Juvenile Similar to immature; upperparts duller than in adult; one has trace of chestnut on forehead; no barring on scapulars and upperwing-coverts; chin and throat whitish; foreneck and breast pale olive-brown; rest of underparts barred olive-brown and whitish.
Downy young Chick has black down.

MEASUREMENTS (7 males, 9 females) Wing of male 127-135 (130, SD 2.9) of female 122-140 (129.9, SD 6.0); tail of male 50-58 (52.5, SD 2.6), of female 47-58 (53.45, SD 3.5); culmen to base of male 26-32.5 (29.35, SD 2.6), of female 27-30 (27.9, SD 1.0); tarsus of male 35.5-45 (41.4, SD 2.9), of female 39-43 (40.7, SD 1.3).

GEOGRAPHICAL VARIATION There is moderate variation in size and plumage colour and pattern, and the populations from India and continental E Asia are paler than those from the islands further E. Races are separated on the colour of the throat, the shade of the upperparts, the width of the white bars on the underparts, and on biometrics. Race *nigrolineatus* is included in *amauroptera* (Ripley 1977). Race *telmatophila* is sometimes merged with *amauroptera* (e.g. Wells & Medway 1976, White & Bruce 1986, Cheng 1987).

R. e. amauroptera (Jerdon, 1864) – India (to E Assam), Nepal and Bhutan, wintering Sri Lanka and possibly Sumatra. See Description. White barring on underparts relatively narrow.

R. e. telmatophila Hume 1878 – S and E Burma (Thayetmyo and S Shan States) and NW Thailand to Laos (see Movements), Vietnam (Tonkin, Annam) and SE China (Hainan, Guanxi and probably SW Yunnan); winters Thailand and Sumatra (including Utara and Lampung), reaching W Java. Birds ascribed to this race have also been collected in Nepal, Bhutan and Bengal (Ripley 1977). Brown of upperparts and chestnut of head, neck and breast somewhat darker than in *amauroptera*; wing longer. Iris bright orange; edges of eyelids light orange; basal area of upper mandible green and blue, base of lower mandible blue. Birds occurring in Hainan may be of this subspecies, but their racial affinities have not been determined. Measurements of 6 males, 3 females: wing of male 129-150 (139.2, SD 7.1), of female 136, 137, 142; tail of male 54-65 (60.15, SD 4.7), of female 58, 60, 64; culmen to base of male 27-31.5 (28.9, SD 1.6), of female 26.5, 30, 30; tarsus of male 39-47 (42.1, SD 3.1), of female 42, 42, 43.5.

R. e. sepiaria (Stejneger, 1887) – Ryukyu Archipelago (Okinawa, Miyako-jima, Ishigaki-jima, Iriomote-jima, Taketomi-jima, Kohama-jima, Kuro-jima and Yonaguni-jima). The largest race. The type has olivaceous-brown head and neck, and may be immature. Wing 152; tail 63; exposed culmen 30; tarsus 46.

R. e. formosana Seebohm, 1894 – Taiwan and Lanyu I. Upperparts darker than in *amauroptera*. Wing 132-133; tail 57-64; exposed culmen 28-29; tarsus 44-48.5 (most data from Ripley 1977, in which sample sizes are not given).

R. e. eurizonoides (Lafresnaye, 1845) – Philippines (Basilan, Biliran, Bohol, Cagayancillo, Catanduanes, Cebu, Fuga, Guimaras, Jolo, Leyte, Luzon, Marinduque, Mindanao, Mindoro, Negros, Panay, Samar and Siquijor) and Palau Is (Angaur, Arakabesan, Koror and Malakal). Throat pale rufous; white bars on underparts relatively wide. Iris reddish-orange; bill olive-green, with blackish culmen; legs and feet dark grey (Rand & Rabor 1960). Measurements of 8 males, 4 females: wing of male 127-137 (134.25, SD 3.5), of female 128-137 (132, SD 3.9); tail of male 53-70 (61.4, SD 5.3), of female 55-68 (61.25, SD 9); culmen to base of male 25.5-28.5 (27.2, SD 1.2), of female 26-28 (26.75, SD 0.8); tarsus of male 41-43.5 (42.4, SD 0.9), of female 40-42.5 (41.4, SD 1.1). Weight of 2 males 112, 118, of 2 females 99, 105, of 2 unsexed 110, 112.

R. e. alvarezi Kennedy & Ross, 1987 – Batan I (possibly also Sabtang), N Philippines. Overall colour darker than in other races; throat of male chestnut, of female pale rufous; inner webs of remiges and rectrices almost black; white barring on underparts narrow. Iris of male dark orange, of female red with orange inner ring; orbital ring orange; upper mandible dark grey with blue-green to green base, lower mandible grey towards tip. Juvenile has upperparts blackish-grey with dark olive-brown wash (absent on lower back and rump); underparts sooty-grey with faint olive-brown wash, and with some white flecks on breast and belly; iris dark olive-brown; bill and legs black. Wing (chord) 123-138 (131.4, SD 6.6, n = 4); tail 65.5-68 (66.9, SD 1.3, n = 4); exposed culmen 20.5-21 (20.6, SD 0.4, n = 4); tarsus 42-44 (43.3, SD 0.8, n = 5). Weight of 1 male 128, of 1 female with oviduct egg 180.

R. e. minahasa Wallace, 1863 – Sulawesi, including Peleng and Sula Is. Olive-brown of upperparts darker than in *amauroptera*, upperpart colour as in *alvarezi* but paler chestnut on head, neck and breast; throat pale rufous to chestnut. White barring on underparts relatively narrow, as in *alvarezi*. Wing 132-152; tail 55-70; exposed culmen 24-25, culmen 26; tarsus 39-42; data from Ripley (1977) and White & Bruce (1986), in which sample sizes are not given, and Riley (1924).

MOULT No information available.

DISTRIBUTION AND STATUS Occurs from India, Nepal and Bhutan E locally to S China, Taiwan and Ryukyu Is, and S through Burma, Thailand, Laos (see Movements) and Vietnam to Sumatra, Java, Sulawesi, Philippines and the Palau Is. Although its distribution was formerly said to include N Pakistan there appears to be no specimen from that country and its occurrence there is not proven although it may occur (Roberts 1991). The races *amauroptera*, *alvarezi* and *telmatophila* are apparently not uncommon over much of their range, but *telmatophila* is uncommon to rare in Thailand, possibly breeding in the N, rarely recorded in Burma (status uncertain), uncommon in Sumatra, and possibly only a vagrant to W Java, whence there are no recent records (Smythies 1986, Lekagul & Round 1991, MacKinnon & Phillipps 1993). Of the two recorded occurrences in Nepal the first was a specimen which was assigned, on wing length, to *telmatophila* (Ripley 1977); the second (in Jun 1957) was a sight record assigned by Inskipp & Inskipp (1991) to *amauroptera*. In Wallacea, *minahasa* is an uncommon to rare resident (Coates & Bishop 1997). In the Ryukyu Is, the resident *sepiaria* decreased seriously in numbers and range on some islands (particularly Iriomote-jima) in the 1980s, coincident with a rapid increase in numbers of the White-breasted Waterhen (89), with which it may be in unsuccessful competition for food and/or nesting sites (Brazil 1991). The status of other 2 races is incompletely known, but *eurizonoides* is uncommon on the Palau Is (Pratt *et al.* 1987) and *formosana* is apparently uncommon to rare throughout its range; however, Perennou *et al.* (1994) maintain that the species is generally fairly common throughout South-East Asia.

MOVEMENTS The races *sepiaria*, *formosana*, *minahasa*, *alvarezi* and *eurizonoides* appear to be largely sedentary. The race *amauroptera* is a resident and a migrant in India, moving at the beginning and end of the SW monsoon (Ali & Ripley 1980). Birds arrive on the W coast of Sri Lanka from India in Oct-Nov, spend a few days in the lowlands and then move up to winter in the hills, leaving in Mar-

Slaty-legged Crake

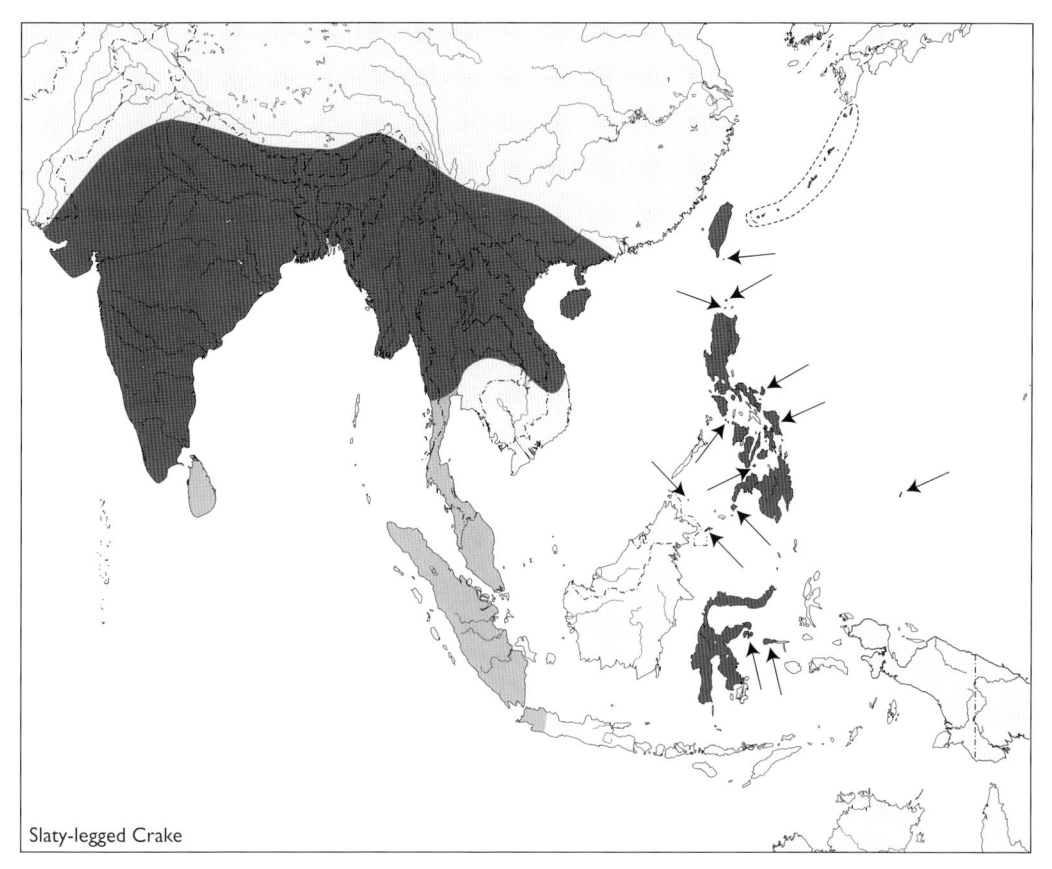

Apr (Ripley 1977). It is probably the best known crake in Sri Lanka owing to its propensity for entering human dwellings along the sea front when it arrives exhausted after the sea crossing (Henry 1978). This species, presumably the race *telmatophila* (Wells & Medway 1976), is a scarce migrant and winter visitor to Thailand and a winter visitor to Malaysia, extreme dates from the Malay Peninsula being 12 Nov and 17 Apr (Medway & Wells 1976); it is also a winter visitor to Sumatra, where it is recorded in Nov and Mar-May, including twice at sea in the Straits of Malacca (van Marle & Voous 1988), and occasionally to W Java (MacKinnon & Phillipps 1993); however, birds assigned here to this race are resident in SE China (Cheng 1987). There is 1 record from Hainan, in Apr (Cheng 1987). There is a Jan specimen from Ok Yam, Laos (Wells & Medway 1976) and it was recorded in Jan 1996 from S Laos (Robson 1996). It is apparently only a vagrant to Singapore, where little habitat remains (C. R. Robson *in litt.*), and there are only 2 records, 1 of questionable provenance (Medway & Wells 1976) and 1 in Apr 1989 (Robson 1989). Some Palau birds are possibly migrant visitors (Pratt *et al.* 1987). Birds move almost entirely at night, often flying into houses when apparently attracted by lights. Long-winged birds, here included in *telmatophila*, straggle to the E Himalayas, whence there are single specimens from Bhutan (Feb), Nepal (undated) and Bengal (Jan) (Wells & Medway 1976).

HABITAT Inhabits forest with a dense understorey, its habitats including forest edges, secondary growth, bamboos, the forest floor inside dense patches of remnant original forest vegetation, and the banks of forest streams. It also occurs in well-wooded areas, dense scrub, long grass with dense bushes, paddy and taro fields, marshes and mangroves. It is recorded from gardens when on passage (Lekagul & Round 1991). Principally a lowland bird, it occurs in well-watered habitats up to 1,600m.

FOOD AND FEEDING Takes earthworms, molluscs, insects including grasshoppers, the shoots and seeds of marsh plants, and also grain.

HABITS A secretive species, difficult to observe and said to be partly nocturnal "like other rails" (Ali & Ripley 1980). It forages at dawn and dusk. During the day the birds hide in trees or tall bushes (e.g. Delacour & Jabouille 1931). The ordinary gait is a jerky, high-stepping walk, with the tail erect and often jerked (Ali & Ripley 1980). When flushed, this species will often fly up into a tree or, if it takes to water, will swim jerkily like a Common Moorhen (129) (Ali & Ripley 1980).

SOCIAL ORGANISATION Monogamous and probably territorial.

SOCIAL AND SEXUAL BEHAVIOUR The incubating bird is very pugnacious; it stands up, fluffs out its plumage and pecks viciously at an approaching hand, repeating the process without leaving the nest (Ali & Ripley 1980).

BREEDING AND SURVIVAL Season India, Jun-Sept (during SW monsoon); Vietnam, Aug; Ryu Kyu Is, Apr-Jul; Sulawesi, Apr; Palau (no date); Luzon (Philippines), juvenile Nov; Batan Is (Philippines), male in breeding condition May, female with oviduct egg and juvenile 3-4 weeks old early Jun. **Nest** Nests in the densest forest as well as in more open scrub. Nest an untidy pad of dead leaves, grass and thin twigs, with slight central depression, built on ground in clumps of bamboo, masses of tangled creepers, or on top of tree stumps up to 1m above ground (2-3m according to Delacour & Jabouille 1931); nest not necessarily located near water; 1 nest in grass 60cm tall beneath *Litchi sinensis* sapling 1m tall (Stusák & Vo Quy 1986). **Eggs** 4-8; creamy-white; size of 90 eggs (*amauroptera/ telmatophila*) 30.9-35.8 x 25-28.1 (30.6 x 26), calculated weight 12.8; 1 clutch, Vietnam, 33-36.5 x 25-27 (34.2 [SD 1.1] x 26.1 [SD 0.6]), weight 12.5-14 (13.3, SD 0.5, n = 8) (Stusák & Vo Quy 1986); size of 4 eggs (*minahasa*) 36.5-38 x 28.2-29 (37 x 28.6), calculated weight 16.5 (Schönwetter 1961-62). **Incubation** By both sexes. **Chicks** Both sexes care for young. A juvenile 3-4 weeks old was almost fully feathered, having down only on forehead, supercilium and malar region.

ANUROLIMNAS

Three species of medium-small South American crakes, all of which have sometimes been included in other genera. All have olive-brown upperparts and are chestnut on the head, neck and breast. The colour of the rest of the underparts differs in each species, in Russet-crowned Crake being chestnut, in Chestnut-headed Crake olive-brown and in Black-banded Crake boldly barred cinnamon and black. The Chestnut-headed Crake is a bird of forest and thickets, the Black-banded Crake inhabits marshes and wet grass, and the Russet-crowned Crake combines the habitat preferences of the other two, occurring in thickets, wet grass and swamps. One species, the Russet-crowned Crake, lays pure white eggs. All three species are apparently sedentary, secretive and among the most poorly known of South American rallids. Their status is thus difficult to assess but none is thought to be threatened although the Chestnut-headed Crake may be uncommon to rare throughout its range.

25 CHESTNUT-HEADED CRAKE
Anurolimnas castaneiceps Plate 8

Porzana castaneiceps P. L. Sclater & Salvin, 1868, Río Napo, eastern Ecuador.

Sometimes placed in *Rallina*, but lacks wing barring and has a very short tail. Two subspecies recognised.

Synonyms: *Rallina/Aramides castaneiceps*; *Micropygia/Ortygometra verrauxi*.

Alternative name: Chestnut-headed Rail.

IDENTIFICATION Length 19-22cm. The largest *Anurolimnas* species, an unpatterned crake which is uniformly olive-brown with a contrastingly bright chestnut forecrown, face, neck and breast. Iris red to yellow, or brown; bill usually greenish with black tip; legs and feet dark brown to olive or grey (but red in race *coccineipes*). Bill rather short and thick; tail very short. Sexes alike. Immature like adult but duller and darker; rufous areas brown or brownish-olive; breast and belly dull brown; throat pale greyish-buff. Juvenile undescribed. Inhabits forest, secondary growth and damp thickets.
Similar species Both other *Anurolimnas* species occur within range of Chestnut-headed Crake; both are slightly smaller, and Black-banded Crake (27) occurs only in wet grass and marsh vegetation; distinguished from Black-banded Crake by unbarred underparts, from Russet-crowned Crake (26) by entirely rufous face and olive-brown lower breast to undertail-coverts, and from both congeners by unpatterned underwing-coverts and predominantly greenish bill; nominate race also differs in leg colour. Uniform Crake (80) very similar but has brown-tinged face, more extensively rufous underparts, different voice, and resembles a small *Aramides* wood-rail in build; like Chestnut-headed Crake it has an upright, wood-rail-like posture. Immature Uniform Crake lacks greyish-buff throat of immature Chestnut-headed Crake and has white shaft-streaks to feathers of throat and breast; may also be darker in overall colour. Smaller allopatric Rusty-flanked Crake (29) and Ruddy Crake (30) of marshes and wet grassland differ in head pattern and in colour of underparts and bare parts.

VOICE Most information is from Parker & Remsen (1987). The song is a loud chorus, like a wood-rail, performed by both adults while standing erect and side by side, and is said to resemble several cuckoo clocks running simultaneously (Hilty & Brown 1986). It is a synchronised duet which starts suddenly and loudly, one bird making a loud, high-pitched *ti-too* call and the other answering immediately with a *ti-turro* sometimes trilled at the end; the two calls sometimes overlap. These calls are repeated without pause for up to 3-4min. When disturbed, the birds utter growls 2-3s long, and soft *tuk* or *tik* notes. See Hardy *et al.* (1996) for a comparison between the very different calls of this species and the Russet-crowned and Black-banded Crakes, which both resemble *Laterallus* spp. in voice.

DESCRIPTION *A. c. castaneiceps*
Adult Forehead, forecrown, side of head, and entire neck and breast rich chestnut; chin and throat paler, orange-rufous to pale buff. Hind-crown to hind-neck, upperwings and rest of upperparts, and lower breast to undertail-coverts, rich olive-brown, tinged rufous; remiges darker (fuscous) above and below; axillaries and underwing-coverts darkish olive-brown. Sexes alike. In some specimens, olive-brown of hind-crown extends to forecrown; in others, chestnut of neck extends to upper mantle; in one female, hind-crown to hind-neck, and upper mantle, almost entirely rich chestnut. Iris variable, red to orange, yellow, yellowish-brown or dark brown; bill entirely greenish, greenish-yellow with black tip, or upper mandible black, greenish below nostril; legs and feet dark fuscous, olive-green or grey. Sexes alike.
Immature Like adult but general appearance darker and

much duller, with rufous areas largely brown or brownish-olive; entire breast and belly dull brown; scattered adult-type chestnut feathers on cheeks, sides of neck and breast; throat light greyish-buff. Bare parts not described.
Juvenile Not described.
Downy young Not described.

MEASUREMENTS (20 males, 22 females) Wing of male 111-126 (118), of female 109-124 (115); culmen from base of male 22-25.5 (24.1), of female 21.5-25 (22.7); tarsus of male 48-57 (52.3), of female 47-53.5 (50.2).

GEOGRAPHICAL VARIATION The two subspecies differ in overall plumage tones and in leg colour.

A. c. castaneiceps (P. L. Sclater & Salvin, 1868) – E Ecuador S of race *coccineipes*, possibly replacing it just S of Río Napo (Ridgely & Greenfield in prep.), and N Peru (including mouth of R Curaray). Presumably it is this race which also occurs in extreme NW Bolivia (Camino Mucden area, Pando) and extreme SW Brazil (R Abunã, Rondônia) (Parker & Remsen 1987, Sick 1997). See Description.

A. c. coccineipes Olson 1973b – S Colombia (Caquetá, Putumayo and SE Nariño; possibly also Amazonas) and NE Ecuador. Very similar to nominate race but browns of plumage more greenish-tinged, brown of belly tends not to extend as far up the breast, and chestnut of underparts usually lighter and tawnier (less brown). Legs and feet bright red. Measurements (6 males, 7 females): wing of male 113-122 (118), of female 112-119 (116); culmen from base of male 22.5-24.5 (23.4), of female 21-24.5 (23.2); tarsus of male 50-56 (52.8), of female 50-53.5 (51.3). Weight of 1 female 126.

MOULT Primary moult is ascendant, with a tendency to partial synchrony; it is recorded in Mar-Apr (Ecuador) and Jun (Colombia). Body moult is recorded Apr (Ecuador).

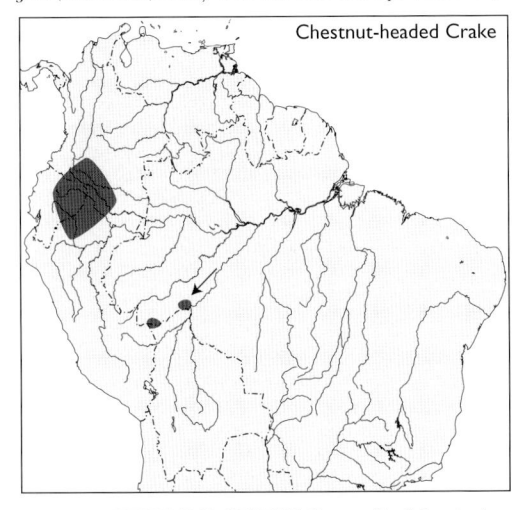

DISTRIBUTION AND STATUS Occurs E of the Andes, from SW Colombia to E Ecuador and N Peru and also in N Bolivia and NW Brazil. This species is not regarded as globally threatened, but is apparently uncommon to rare throughout its range, and possibly of local distribution (e.g. Hilty & Brown 1986), although it may be locally fairly common in E Ecuador (Ridgely & Greenfield in prep.). Its status is difficult to assess because it is very wary and difficult to observe.

MOVEMENTS None recorded.

HABITAT Normally said to inhabit humid forest, either *terra firme* or on seasonally inundated to permanently swampy ground, and the banks of forest streams; it also frequents tall secondary growth and, in NW Bolivia, overgrown gardens surrounded by forest. In such old gardens it is found in dark, damp thickets with an almost impenetrable cover of broad-leaved *Heliconia*-like plants and decaying trunks and branches of trees in areas shaded by stands of trees 4-10m tall and tall banana plants; similar habitat is occupied elsewhere in Amazonia (Parker & Remsen 1987). It occurs from 200 to 1,500m.

FOOD AND FEEDING Food is not recorded; it is presumably insects and other small invertebrates. Birds have been seen foraging on the ground inside dense thickets, flicking aside fallen leaves and probing into debris and rotting wood (Parker & Remsen 1987).

HABITS Pairs with large young were described as extremely wary and very difficult to see, keeping to the interior of dense thickets (Parker & Remsen 1987). This bird walks with a rather upright carriage, like a wood-rail (Hilty & Brown 1986). At the approach of an observer it freezes motionless or slips away (Ridgely & Greenfield in prep.).

SOCIAL ORGANISATION The members of a pair call together, and pairs have been observed with almost fully grown young (Parker & Remsen 1987), so the species is presumably monogamous.

SOCIAL AND SEXUAL BEHAVIOUR No information available.

BREEDING AND SURVIVAL Season Bolivia, pairs with 1-2 nearly grown young Jun, presumably from eggs laid May (Parker & Remsen 1987); Colombia, breeding condition Jun. **Nest and eggs** Undescribed.

26 RUSSET-CROWNED CRAKE
Anurolimnas viridis   Plate 8

Rallus viridis P. L. S. Müller, 1776, Cayenne, French Guiana.

Sometimes placed in *Laterallus*, which it resembles in voice. Two subspecies recognised.

Synonyms: *Laterallus/Porzana/Creciscus/Aramides viridis*; *Rufirallus cayanensis*; *Rallus/Porzana/Creciscus cayanensis/cayennensis*; *Rallus kiolo/poliotis*; *Crex/Ortygometra/Corethrura aurita*; *Crex/Creciscus/Porzana facialis*; *Creciscus pileatus*; *Corethrura/Ortygometra cayennensis*; *Gallinula pileata/ecaudata*; *Corethrura ecaudata*.

Alternative name: Cayenne Crake.

IDENTIFICATION Length 16-18cm. The smallest *Anurolimnas* crake; tail relatively long. Almost unpatterned, having uniformly olive-brown upperparts and chestnut underparts, with chestnut forehead and crown, contrastingly grey (nominate race) or yellow-brown (race *brunnescens*) from lores to ear-coverts, pale supercilium (or supraloral stripe), and blackish tail with chestnut or olive feather edges. Iris red to orange; bill dark grey, bluer at base; legs and feet rose-red. Sexes alike. Immature has entire upperparts olive-brown, tinged chestnut, with some chestnut on forehead; underparts paler chestnut than in adult, washed olive-brown. Juvenile undescribed. A bird of thickets and overgrown areas, damp grass and sometimes swamps.

Similar species Distinguished from Chestnut-headed Crake (25) and Black-banded Crake (27) by grey or yellowish-brown from lores to ear-coverts, and entirely chestnut underparts; also from Chestnut-headed Crake by mottled ashy-brown and rufous underwing-coverts, and dark bill. Like Russet-crowned Crake, Chestnut-headed inhabits thickets and secondary growth, but Black-banded Crake occurs only in marshes and tall wet grass. Larger Uniform Crake (80), of wet forested areas, very similar in appearance but sides of head tinged brown; is more uniform rufous-brown above and below; and has longer, yellowish-green bill; voice also differs.

VOICE A loose, churring rattle, similar to the calls of the Rufous-sided Crake (28) and the White-throated Crake (31) but is slower and louder, with the individual notes more enunciated (Hilty & Brown 1986); it also resembles the call of the Black-banded Crake. The call begins as if "choking" and then flows freely without a pause, falling in pitch and ending with spaced notes; it has been described as having the timbre of a domestic canary (Sick 1993). The alarm is a *kewrr*. This crake is particularly vocal in the evening and early morning, but calls even in the hottest part of the day. It responds well to taped playback (Lowen *et al.* 1997). See Hardy *et al.* (1996) for a comparison between the very different calls of this species and the Chestnut-headed Crake.

DESCRIPTION *A. v. viridis*
Adult Forehead to crown (sometimes to hind-crown) bright chestnut. Rest of upperparts, from hind-crown/nape to uppertail-coverts, and including tertials, uniform olive-brown, sometimes tinged greenish; greater and median upperwing-coverts broadly fringed chestnut; alula, primaries and secondaries darkish brown (Ridgway's Mummy Brown); rectrices blackish-brown, edged olive or chestnut. Narrow supercilium or supraloral stripe buff to grey; lores, side of face, and ear-coverts ashy-grey; cheeks and entire underparts, including undertail-coverts, bright chestnut, with buff throat and centre belly, olive-brown tinge on sides of breast and on flanks, and paler or ashy-tinged thighs. Axillaries olive-brown, fringed chestnut; underwing-coverts pale chestnut with ashy-brown bases. Iris red to orange-yellow; bill dark grey with base of lower mandible bluish-lead colour, or blue-grey tipped dusky; legs and feet rose-red. Sexes alike.
Immature Similar to adult, but forehead to fore-crown, and sides of crown, olive-brown with varying number of pale chestnut feathers; rest of upperparts olive-brown, washed chestnut (less so on sides of neck). Underparts paler chestnut than in adult, washed olive-brown to buffy-brown; centre of lower breast, centre of belly and thighs more pinkish-cinnamon.
Juvenile Not described.
Downy young Not described.

MEASUREMENTS Wing of 7 males 87-92 (89.4), of 10 females 82-97 (90.5); tail of 6 males 28-33 (31.2, SD 1.9), of 3 females 28, 32, 33; culmen to base of 6 males 19.5-21.5 (20.7, SD 0.8), of 3 females 19.5, 20.5, 21.5; tarsus of 7 males 32-37 (34.5), of 10 females 32-37 (34). Weight of 5 males 55-63 (60), of 3 females 63, 71, 73.

GEOGRAPHICAL VARIATION The two subspecies are separated on plumage colours and on size.
A. v. viridis (P. L. S. Müller, 1776) – E of Andes in Colombia (W Meta and E Guiainía), locally in S Venezuela (C Amazonas, Bolívar and Delta Amacuro), through Guianas and Brazil (S to S Amazonas, Mato Grosso, São Paulo and Rio de Janeiro), SW to E Ecuador (Napo and Santiago-Zamora), E Peru and N Bolivia (extreme N Beni, La Paz and Santa Cruz), and in C Paraguay (Canindeyú). See Description.
A. v. brunnescens (Todd, 1932) – N Colombia from lower Cauca Valley (Valdivia) E to middle Magdalena Valley in S Bolívar and S to NE Antioquia (Remedios) and W Santander (El Centro). Browner, less olive, on upperparts; sides of head yellowish-brown rather than grey; underparts paler chestnut; slightly larger. Measurements (7 males, 4 females): wing of male 86-96 (92.1), of female 93-99 (95.3); tail (unsexed) 27-32 (Ripley 1977; sample size and mean not given); exposed culmen of male 17-20 (19), of female 17-19 (18); tarsus of male 35-38 (36.4), of female 35-38 (36.5).

MOULT Primary moult tends to be ascendant, and is recorded in Apr, Peru. Head and body moult recorded Jul, Guyana.

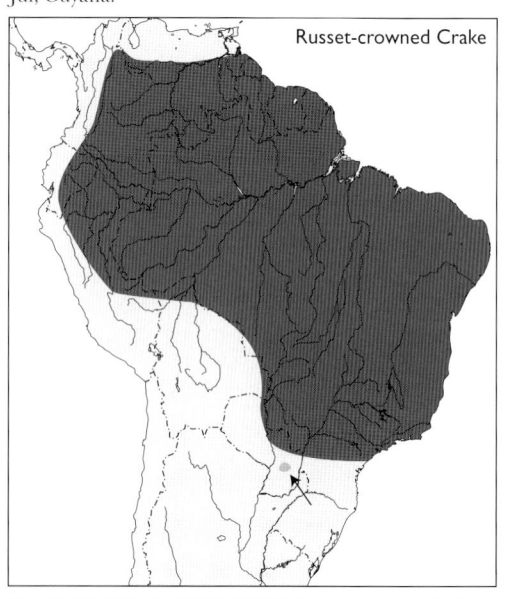

DISTRIBUTION AND STATUS South America E of the Andes, from N and E Colombia, through Venezuela and the Guianas S to E Ecuador, E Peru, N Bolivia, N and C Brazil, and C Paraguay. Although this bird is seldom seen because of its skulking habits, and it is apparently rare to uncommon in Ecuador (Ridgely & Greenfield in prep.), its presence is quite readily detected by its frequent calling and the race *viridis* is thought to be common over much of its extensive range (e.g. Tostain *et al.* 1992, Haverschmidt & Mees 1995); the status of *brunnescens* is not known but is likely to be similar. The recent documentation of this bird's occurrence in N Bolivia at the R Iténez (Beni) and the Parque Nacional Noel Kempff Mercado (Santa Cruz) (Bates *et al.* 1992) and in La Paz (Armonía 1995), and also in C Paraguay at the Reserva Natural del Bosque Mbaracayú, suggests that its range may be expanding with the spread of its favoured secondary growth habitat (Lowen *et al.* 1997).

MOVEMENTS None recorded, although Forrester (1993) regards it as accidental to some localities in C Brazil (Chapada dos Guimarães, Brasília National Park and Araguaia National Park).

HABITAT Inhabits dense thickets of secondary growth saplings at forest edges, thickly overgrown dry wasteland, damp grassy or bushy pastures, brushy manioc pastures, dry fields of thatching grass, overgrown roadsides, and gardens at the edges of towns and villages. Although not normally a marsh bird, in Venezuela it is recorded from swamps as well as from wet marshy meadows (Meyer de Schauensee & Phelps 1978), while in Peru it occurs at the margins of oxbow lakes (Parker *et al.* 1982). In Paraguay it has been found in scrubby secondary growth on a dry substrate within 20m of an extensive marsh adjacent to a river (Lowen *et al.* 1997). On small islands such as Alcatrazes, Buzios and Vitoria, off the coast of São Paulo State, SE Brazil, it is usually associated with dry, open grassy slopes (F. Olmos *in litt.*). It occurs in lowlands up to 1,200m.

FOOD AND FEEDING Takes insects, including ants, and grass seeds. Forages within cover; no other information available.

HABITS Furtive and difficult to see, keeping within cover and being very difficult to flush, but occasionally perching partially in the open in the early morning. It walks daintily on the ground or climbs and runs through grass and thickets, being adept at running over, and climbing in, low branches.

SOCIAL ORGANISATION No information available.

SOCIAL AND SEXUAL BEHAVIOUR No information available.

BREEDING AND SURVIVAL Season Colombia May-Jun; Surinam Jan-Jun, also a pair at empty nest Dec. **Nest** A ball of dead grass with side entrance, sometimes with ladder-like entrance ramp; hidden c. 1m up in branches of shrubs or in coarse herbage, in dense vegetation. **Eggs** 1-3; pure white, or with a very few minute specks of pale yellowish-brown; size of 60 eggs 31.2-36.9 x 22.4-25.5 (32.6 x 24.1); calculated weight 10.2. No other information available.

---

27   BLACK-BANDED CRAKE
Anurolimnas fasciatus   Plate 8

Porzana fasciata P. L. Sclater & Salvin, 1867, near Caracas, Venezuela.

Sometimes placed in *Laterallus*, which it resembles in voice, but its pattern of primary moult (non-synchronised) differs from that of *Laterallus* and is similar to that of *A. castaneiceps*. Species formerly known as *hauxwelli*, as name *fasciatus* erroneously thought to be incompatible with the older species name of *Rallina fasciata*. Monotypic.

Synonyms: *Laterallus fasciatus*; *Ortygometra sclateri*; *Porzana/Anurolimnas/Laterallus/Creciscus/Aramides hauxwelli*.

Alternative name: Hauxwell's Crake.

IDENTIFICATION Length 18-20cm. Easily identified by entirely rich chestnut head, neck and breast, olive-brown upperparts, upperwings and tail, and cinnamon and black barring from flanks to undertail-coverts. Tail short. Iris dull red; bill blackish; legs and feet bright coral-red. Sexes alike. Immature paler than adult; upperparts washed chestnut; underparts obscurely barred olive-brown and dull cinnamon. Juvenile undescribed. Inhabits marshes and tall wet grass.
Similar species Distinguished from Chestnut-headed Crake (25) and Russet-crowned Crake (26) by black barring on cinnamon-rufous underparts, and entirely rufous head and neck; also from Chestnut-headed Crake by mottled cinnamon and dusky underwing-coverts, and blackish bill. Easily separated from Rufous-sided Crake (28) and White-throated Crake (31), which also inhabit marshes, by its red legs and feet, and especially by the other two species' black and white barring on underparts.

VOICE The song is a low churring rattle, similar to that of the Russet-crowned, Rufous-sided and White-throated Crakes, and lasting 9-10s (Hardy *et al.* 1996); it is often given simultaneously by both members of a pair, and is possibly antiphonal (Hilty & Brown 1986). The birds are often vocal, especially in the early morning and late evening.

DESCRIPTION
Adult Entire head, neck and breast rich chestnut. Mantle to tail darkish olive-brown, lightly tinged rufous on remiges; primaries darker, more fuscous; axillaries and underwing-coverts barred olive-brown and pale cinnamon to buffy. Flanks, belly and undertail-coverts barred black and cinnamon to pinkish-cinnamon. Iris dull red; eyelids and gape vermilion; bill blackish (also described as dark horn); legs and feet bright coral-red. Sexes alike.
Immature One bird has head and neck paler chestnut than in adult; olive-brown of upperparts washed chestnut; barring on underparts rather obscure, olive-brown and dull cinnamon.
Juvenile Not described.
Downy young Not described.

MEASUREMENTS Wing of 5 males 94-96 (95), of 8 females 92-99 (95.1); tail of 3 males 25, 26, 27, of 1 female 25, of 1 unsexed 26.5; culmen to base of 3 males 20.5, 21.5, 21.5, of 1 female 20.5; exposed culmen of 5 males 19-20 (19.8), of 8 females 17-19 (18); tarsus of 5 males 34-40 (37.4), of 8 females 34-40 (37.2).

GEOGRAPHICAL VARIATION None.

MOULT Primary moult tends to be ascendant and is recorded in Colombia, Aug. Body moult recorded Colombia, Oct.

DISTRIBUTION AND STATUS Occurs in the tropical zone of upper Amazonia, E of the Andes from E Ecuador (primarily Napo and Sucumbios) and E Peru (including Pebas in NE), N over S and SE Colombia (Caquetá, Putumayo and Amazonas) and E across W Amazonian Brazil (from R Solimões to R Purús). This species is not regarded as threatened but its status is difficult to assess because of its skulking habits. It is thought to be more widespread than the few existing records indicate (Hilty & Brown 1986). In Ecuador it is rare to locally fairly common, and appears to be most numerous close to the base of the Andes (Ridgely & Greenfield in prep.).

Black-banded Crake

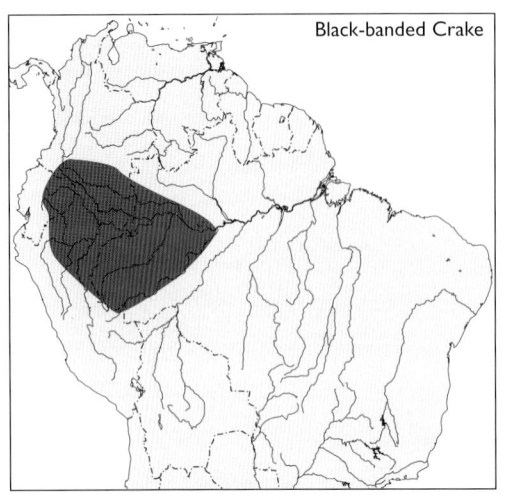

MOVEMENTS None recorded.

HABITAT Inhabits tall wet grass and marsh vegetation, including that fringing streams, pools and oxbow lakes. In Ecuador it occurs in dense shrubbery and grassy vegetation in clearings and at forest and woodland edges, almost always in damp conditions and usually near streams or seepage zones, and also on river islands (Ridgeley & Greenfield in prep.). In the upper Amazon region it occurs on islands with pioneer vegetation, alongside the Grey-breasted Crake (32) (Sick 1997). It occurs in lowlands, up to 500m, rarely up to 1100m in E Ecuador (Ridgeley & Greenfield in prep.).

FOOD AND FEEDING Nothing recorded.

HABITS It is even more secretive than *Laterallus* crakes and skulks in tall wet grass and thick tangled vegetation (Ridgeley & Greenfield in prep.).

SOCIAL ORGANISATION In view of its calling pattern the species is presumably monogamous, forming pairs at least during the breeding season.

SOCIAL AND SEXUAL BEHAVIOUR No information available.

BREEDING AND SURVIVAL Season Colombia, Jun. **Nest** One nest was domed and bulky, of grass with side entrance, situated 1.7m up on vine-covered fallen limb in clearing. **Eggs** Size of 4 eggs 29.8-31.6 x 22-23 (31 x 22.8); calculated weight 8.7.

LATERALLUS

The nine species in this genus comprise a rather heterogeneous group of small to very small crakes confined to the Americas. Two species (the Black Rail and Galapagos Rail) have very dark plumage with variable amounts of white barring and spotting; the other seven are predominantly olive-brown to rufous on the upperparts, most also having the face, neck and breast at least partially rufous (but the Grey-breasted Crake is grey in these regions) and five having bold black and white barring on the flanks. Most species occur in marshes and wet grassland, but the White-throated Crake also inhabits forest clearings and thickets, while the Galapagos Rail normally occurs in forest and thickets. Most species are regarded as sedentary, but most North American populations of the Black Rail are migratory, while the Grey-breasted Crake of South America may have some movements. Some of the South American species are among the least known rallids of the continent. The breeding of only two species, the Black Rail and the Galapagos Rail, has been studied in any detail in the wild, and one species, the Red-and-white Crake, is known to lay pure white eggs. The status of several species is unclear but the Rusty-flanked Crake and Rufous-faced Crake are regarded as vulnerable, the Galapagos Rail is near-threatened, and one race of the Black Rail is endangered.

28   RUFOUS-SIDED CRAKE
Laterallus melanophaius   Plate 9

Rallus melanophaius Vieillot, 1819, Paraguay.

Sometimes considered to include *L. albigularis*. Forms a superspecies with *L. levraudi*. Two subspecies recognised.

Synonyms: *Crex lateralis*; *Porzana oenops/melanophaea*; *Creciscus oenops/melanophaius*; *Gallinula albifrons/lateralis*; *Ortygometra/Corethrura melanophaea/lateralis*; *Laterirallus albifrons*; *Aramides melanophaia*.

Alternative names: Rufous-sided Rail; Rufous-vented Crake; Brazilian Crake.

IDENTIFICATION Length 14-18cm. Slightly larger than other *Laterallus* species. Distinguished from congeners and other small marsh crakes by its combination of dark olive-brown upperparts (darker from rump to tail), pale supercilium, rufous sides of head, neck and breast (greyish on lores and around eye in nominate race), white chin, throat and centre breast, blackish and white barring on flanks and plain cinnamon-rufous undertail-coverts. Some birds have narrow white barring on upperwing-coverts and scapulars, and some have buffy or rufous tinge to white from chin to centre of breast. Iris red to red-brown; bill relatively long and medium-slender, blackish-brown but green at the sides, or greyish but yellower towards base; legs and feet olive-brown or flesh/straw. Sexes alike. Immature similar to adult. Juvenile similar to adult on upperparts; sides of head and underparts duller than in adult; underparts barring indistinct, olive-brown and vinaceous-cinnamon; may appear predominantly brownish-grey. Inhabits marshes and wet grassland.

Similar species Very similar to White-throated Crake (31) of Central America but differs in having no rufous on upperparts, white on underparts extending from chin to belly, and plain rufous undertail-coverts. Also easily confused with sympatric Red-and-white Crake (35) which, however, is rufous from head to mantle, and has sharp demarcation between rufous and white colours at sides of neck and breast, white lateral undertail-coverts, yellow base to bill, and coral-red legs and feet. Sympatric Rufous-faced Crake (36) has boldly barred inner secondaries, upperwing-coverts and scapulars, black tail, partially black undertail-coverts, crimson-red iris, relatively heavy blue-grey (or blackish and turquoise) bill, and blue-grey to flesh-brown legs and feet. Easily separated from sympatric Black-banded Crake (27) by latter's entirely chestnut neck and breast and its cinnamon and black underpart barring.

VOICE An abrupt, loose churring, very similar to the call of the White-throated Crake (31), also described as a loud trill *tsewrrr...* of c. 6s (Sick 1993); it is very similar to the trill of the Red-and-white Crake, which is more strident, slightly higher-pitched and possibly slightly shorter (J. Mazar Barnett *in litt.*). It is also said to have a call rendered *frey-eeé, frey-ee, frey-o-o-o* (Meyer de Schauensee & Phelps 1978). Other calls are described as a high peep, *tsewip-tsip*, similar to a chick, a warning *psieh*, and a short, smooth, repeated *tsewrrr* uttered even at night in flight on migration (Sick 1993); in the Hardy *et al.* (1996) recording, a call apparently similar to the *tsewrrr* is given by a bird while another is churring. It calls frequently and loudly, and is said to duet (Arballo & Cravino in press) and to call with young (Hardy *et al.* 1996).

DESCRIPTION *L. m. melanophaius*
Adult Upperparts, including top of head, nape, scapulars and upperwing-coverts, dark olive-brown, darker and more blackish-brown from lower back or rump to rectrices and on tertials. Alula, primary coverts and primaries dusky olive-brown. Some birds show some narrow to faint white barring on upperwing-coverts and scapulars. Axillaries blackish-brown, barred white; underwing-coverts variable: barred olive-brown and white, or greaters pale grey-brown, medians largely white and lessers olive-brown, or all predominantly pale grey-brown to olive-brown. Lores, feathers around eye, and often supercilium, often tinged grey; ear-coverts, side of neck and side of breast bright ferruginous to cinnamon-rufous (paler and duller on ear-coverts); chin to centre of breast white, sometimes tinged buffy-ochraceous to rufous on chin, foreneck and upper breast. Flanks and thighs barred blackish-brown and white; white barring broadest on flanks; belly extensively pure white. Undertail-coverts dark cinnamon-rufous. Iris bright red or red-brown; bill blackish-brown, bright green at the sides, also described as greyish, yellower at sides and towards base (Brace *et al.* in press); legs and feet olive-brown or flesh/straw. Sexes alike.
Immature Apparently similar to adult, but not properly described; birds probably 6-7 weeks old were slightly reddish on sides of neck and had dark auricular patch (Arballo & Cravino in press). Iris probably pale brown or greyish; bill, legs and feet possibly paler or greyer than in adult.
Juvenile One probable juvenile similar to adult on upperparts, but no grey tinge on head; lores and ear-coverts pale fawn, tinged cinnamon; sides of neck and breast duller than in adult, and feathers tipped dull olive-brown; centre of breast and centre of belly white; flanks and thighs dull olive-brown, vaguely barred dull vinaceous-cinnamon; undertail-coverts dull cinnamon. Young c. 3 weeks old said to be "predominantly brownish-grey" (Arballo & Cravino in press).
Downy young Black; bill orange; legs and feet black.

MEASUREMENTS (10 males, 8 females) Wing of male 76-88 (82.8), of female 75-85 (80.6); tail of male 39-43.5 (Ripley 1977, sample size and mean not given); culmen to base of male 20-22 (21), of female 20-22 (20.6); tarsus of male 26-32 (29.1), of female 28-30 (28.3). Weight of 1 male 60, of 2 females 53, 54, mean weight of 6 unsexed 51.4.

GEOGRAPHICAL VARIATION Putative races *lateralis* and *macconnelli* are included in nominate (Hellmayr & Conover 1942). The two subspecies are separated on overall plumage tones.
L. m. melanophaius (Vieillot, 1819) – coastal Venezuela (Caracas and Sucre) E to Surinam and S through C and E Brazil (E Pará and Bahia S to Rio Grande do Sul; also Alagoas) to Bolivia (Pando, Beni, La Paz and Santa Cruz), Paraguay, Uruguay (San José, Salto and Montevideo) and N Argentina (S to La Rioja and Buenos Aires). See Description.
L. m. oenops (Sclater & Salvin, 1880) – SE Colombia (Caquetá), E Ecuador, E Peru and W Brazil (Amazonas to R Negro). Differs from nominate race in having paler, more olivaceous upperparts, pale rufous-tinged supraloral stripe, no grey tinge on head, and less white on underwing-coverts and axillaries. Measurements (9 males, 1 female): wing of male 71-82 (78.4), of female 76; tail of 1 male 45; culmen from base of male 19-22 (20.3), of female 20; tarsus of male 25-33 (28.6), of female 28. Weight of 6 unsexed 46-57 (51).

MOULT Postbreeding moult is complete and the primaries are moulted simultaneously. Primary moult is recorded in Oct-Nov, Panama. Captive birds attained adult plumage at c. 4 months of age.

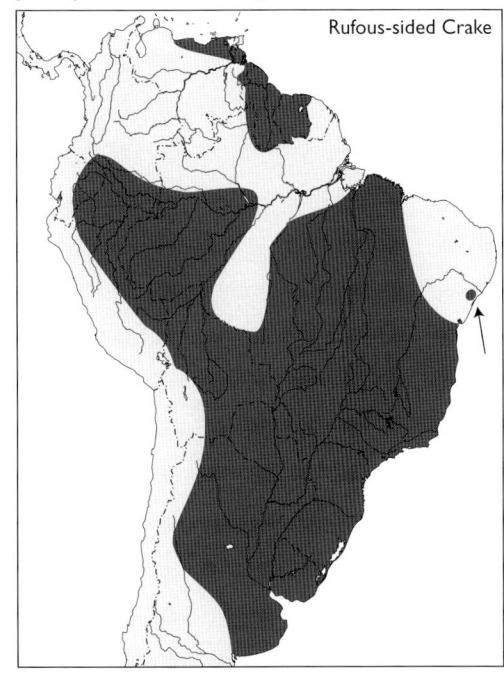

Rufous-sided Crake

DISTRIBUTION AND STATUS This is the most widely distributed South American *Laterallus* species, occurring in tropical and lower subtropical zones E of the Andes, from coastal Venezuela W to E Ecuador and E Peru, E to N Brazil and S through Brazil at least to C Argentina in Buenos Aires Province. It is rare in Guyana (Snyder 1966); in Uruguay it was known only from Montevideo Dpt and there were no records from 1925 to 1986, when it was recorded near Playa Pascual (San José Dpt), and it was also found at the Establecimiento el Espinillar (Salto Dpt) in 1988 (Arballo 1990); it is regarded as a scarce resident and is also recorded from Dpts de Colonia, Canelones, Maldonado, Rocha, Paysandú, Río Negro, Soriano and Treinta y Tres (Arballo & Cravino in press). In Ecuador it is uncommon to locally fairly common (Ridgely & Greenfield in prep.); in Paraguay it is recorded sparsely from the Chaco (W of the Paraguay R) but there are many records from the Orient (E of the river) (Hayes 1995). In Argentina it is apparently generally uncommon, and is rare in Chaco (Contreras et al. 1990); it is also uncommon in Rio Grande do Sul, SE Brazil (Belton 1984), but abundant elsewhere in Brazil (Sick 1993), and is fairly common in E Peru (Parker et al. 1992). Its status is not recorded elsewhere as the species is poorly known and often considered difficult to observe.

MOVEMENTS It is normally regarded as sedentary, but birds have been heard calling while flying over Rio de Janeiro on rainy nights (Sick 1993).

HABITAT Freshwater marshes (especially in dense, sometimes low, vegetation along the edges), swamps, wet meadows and grassland. In Ecuador it occurs in floating mats of vegetation surrounding oxbow lakes, and compared with White-throated Crake is less often in tall grass away from the margins of ponds or marshes (Ridgely & Greenfield in prep.). In Colombia it often occurs some way from water (Hilty & Brown 1986), and in Argentina it frequents scrub and thickets with interspersed trees, and gallery forest, as well as *Typha* beds along rivers (Canevari et al. 1991). In Venezuela it is also recorded from forest, mudflats, lagoons and dry grassland (Meyer de Schauensee & Phelps 1978). In Bolivia, when occurring alongside Grey-breasted Crakes (32) in wet pasture, Rufous-sided Crakes are apparently more restricted to marsh grasses around patches of open water (Parker et al. 1991); in Beni they were recorded alongside Rufous-faced Crake in a belt of *Cyperus giganteus*, up to 20m wide and 2m tall, and partially flooded to a depth of 5cm, which was interspersed with other sedges and some scrub, had narrow 'runs' which were used by the present species, and gave way to fairly open, seasonally (Oct-Apr) inundated savanna dominated by grasses such as *Sorghastrum setosum* (in wet areas) and *Andropogon bicornis* (in drier areas); at the same time the following year, when the *Cyperus* was flooded to a depth of 15-20cm, only Rufous-sided Crakes were found (Brace, et al. in press). In Paraguay they were recorded in dense grass 1m tall near open water at the edge of a marsh (Storer 1981). In the Santos-Cubatão region of São Paulo State, SE Brazil, this species occurs in disturbed mangroves rich in the fern *Acrostichum aureus*, in transitional areas which have some freshwater input and are dominated by tangles of *Hibiscus tiliaceus*, and throughout the coastal plains in freshwater lagoons and ponds rich in floating macrophytes; it also occurs inland, with the Grey-necked Wood-rail (75) and the Blackish Rail (112), along rivers wherever there are swamps rich in *Typha*, sedges etc; further S (in the Peruíbe region) it has been seen alongside the Clapper Rail (55) in stretches of *Spartina* bordering mangrove forest (F. Olmos *in litt.*). At Barra de Icapara, Paraná, SE Brazil, pairs were seen alongside Yellow-breasted Crakes (105) "low in the mangroves when the tide was high" (Bornschein et al. 1997). In Rio Grande do Sul, Brazil, it is found in marshes dominated by *Scirpus giganteus*, where it occurs alongside the much more numerous Red-and-white Crake (Mauricio & Dias 1996). It occurs in lowlands up to 1,000m, reaching 1,350m in W Sucumbios, Ecuador (Ridgely & Greenfield in prep.). In Brazil it occurs near human habitations.

FOOD AND FEEDING Insects, especially Coleoptera (Curculionidae), Homoptera, Orthoptera, Diptera, and spiders; also some seeds and leaves. Captive birds eat mealworms (*Tenebrio* larvae) and blowfly larvae, and occasionally seeds, and are said to take the young of other small birds from nests (Gronow 1969). Birds forage on the ground, in shallow water and on floating vegetation. In captivity, they dip live food in water before eating it.

HABITS Although normally described as shy and secretive, and normally located by its call, which is given from inside dense cover, this species may possibly be less skulking than other small crakes and rails. It sometimes walks in the open at the edge of cover, especially while foraging, when it is very confiding and will approach an observer closely (Arballo & Cravino in press). It is difficult to flush, and normally flies only short distances. Captive birds occasionally perched in trees (Gronow 1969).

SOCIAL ORGANISATION Normally seen singly or in pairs. In captivity it was found to be monogamous, and young of a first brood fed chicks of a second brood (Gronow 1969). In Brazil it was found to occupy territories in marsh alongside the Red-and-white Crake (Mauricio & Dias 1996).

SOCIAL AND SEXUAL BEHAVIOUR Nothing recorded.

BREEDING AND SURVIVAL Season SE Brazil, two young Nov; Uruguay, Oct (small chick 9 Nov), Jan; Argentina, Oct, Dec, Jan. **Nest** Almost globular, with side entrance; external dimensions 17 x 14cm and 22 x 18cm, entrance 6.5 and 7cm; cavity height 9cm. Two nests (Argentina) in cortadera *Scirpus giganteus* on semi-dry to flooded ground (1 was 60cm above water) were made of woven cortadera leaves and lined with shredded cortadera leaves or dry grass and feathers (Contreras 1988, Aguilar & Kowalinski 1996). In Uruguay, 1 nest at edge of water in field of *Zizaniopsis bonariensis* was made of woven dried *Zizaniopsis* stems (J. Cravino *in litt.*); another in flooded *Typha*, 110cm above water surface and 120cm from dry shore, of dry stalks and some leaves, flimsy, height 18cm, breadth 13cm, cavity depth 8cm (Arballo & Cravino in press). A captive pair built 3 nests in 1 season, 2 in bushes 0.6m and >1m above ground, and 1 in nestbox on ground (Gronow 1969). **Eggs** In the wild 4-5, in captivity 2-3; ovate; white to cream, spotted brown or reddish and flecked violet, spots sometimes more concentrated at blunt end; size of 37 eggs 28.9-32.5 x 22-23.7 (30.7 x 23.3), calculated weight 9 (Schönwetter 1961-62, Ripley 1977, Aguilar & Kowalinski 1996). Eggs laid at intervals of 48h (Contreras 1988). **Incubation** 19.5-20 days, starting when clutch complete (Contreras 1988). **Chicks** Semi-precocial; leave nest on 4th day after hatching (Contreras 1988); apparently fledged at c. 3 weeks of age.

29 RUSTY-FLANKED CRAKE
Laterallus levraudi Plate 9

Porzana levraudi P. L. Sclater & Salvin, 1868, Caracas, Venezuela.

Forms a superspecies with *L. melanophaius*. Monotypic.

Synonyms: *Creciscus/Rufirallus/Aramides levraudi*.

Alternative name: Levraud's Crake.

IDENTIFICATION Length 14-18.5cm; male 16.5-18.5 (18), female 14.5-19.5 (17). A relatively unpatterned *Laterallus*, with darkish olive-brown upperparts (darker from lower back and darkest on tail), greyish lores and anterior supercilium (not prominent in some birds), cinnamon-rufous sides of head, neck and breast, and flanks to undertail-coverts, and white from chin to centre of breast and belly. Upperwing-coverts spotted rufous in some birds. Iris red; bill green with black culmen; legs and feet greenish-yellow to pale brown. Sexes alike; female smaller. Immature and juvenile similar to adult but have smaller, thinner bill, paler, drabber underparts and grey-brown legs. Inhabits marshes and wet grassland.

Similar species Resembles the 3 allopatric *Anurolimnas* crakes (25-27) of South America, which are all larger, and lack white from chin to centre of breast and belly; 2 species, Chestnut-headed Crake (25) and Russet-crowned Crake (26), normally inhabit forest and thicket. Chestnut-headed Crake also differs in being olive-brown from lower breast to undertail-coverts, and in having brownish to red legs and feet. Russet-crowned Crake has paler upperparts, is grey or yellow-brown from lores to ear-coverts, and has very short tail with blackish feathers edged olive or rufous. Marsh-dwelling Black-banded Crake (27) has distinctive black barring from lower breast to undertail-coverts. Marginally sympatric Grey-breasted Crake (32) (known from Miranda) and Rufous-sided Crake (28) (from Caracas and Sucre) have barred posterior underparts; similar but larger Uniform Crake (80) of wet forest and thickets has variably rufous-tinged upperparts, more extensively rufous underparts, yellowish-green bill and red to pink legs.

VOICE Information is from Boesman (1997). A descending churring rattle, lasting 3-5s and preceded by gradually accelerating notes audible only at close quarters. When excited the rattle can last several more seconds, rising and falling several times. Although the call is similar to that of several other *Laterallus* species, subtle differences are apparent: for example, it is higher-pitched than that of the Rufous-sided Crake. In Jun-Jul 1996, birds were very vocal, calling from dawn to c. 10:00 and from 16:00 to dusk, with intervals of 10 min to 1h between calling bouts; birds call all year, but mostly May-Aug (P. Boesman *in litt.*). Response to taped playback is good but birds apparently do not respond to playback of Rufous-sided Crake calls. A copy of the call recorded by Boesman is on Hardy *et al.* (1996).

DESCRIPTION
Adult Upperparts, including forehead to hindneck, and upperwings, darkish olive-brown; primary coverts and remiges slightly paler and duller; lower back to rectrices, tertials and secondary coverts darker, grading to dark fuscous on rectrices. Upperwing-coverts variably spotted with rufous or cinnamon-rufous in some birds (occasional entire feathers this colour). Lores ashy-grey; narrow superciliary streak, to just over eye, vinous-white or pale greyish-cinnamon, may be concolorous with lores. Sides of face, ear-coverts, sides of neck and breast, and flanks, thighs, vent and undertail-coverts, dark cinnamon-rufous to orange-rufous; chin, throat, centre of breast and belly white to off-white, with cinnamon tinge on foreneck and buff tinge on belly. Underwing-coverts and axillaries cinnamon-rufous, medians whitish. Iris red to brown; bill green to yellowish-green, with culmen (and sometimes tip) black (P. Boesman and P. Precey *in litt.*); legs and feet greenish-yellow, pale olive or pale brown. Sexes alike but female slightly smaller; bill green or green-grey, with black culmen; legs and feet dark brown, olive-grey or dirty green (P. Precey *in litt.*). Two males have cinnamon-rufous of ear-coverts extending to forehead, lores and supercilium, this character possibly becoming more obvious with age (R. Restall and P. Precey *in litt.*).

Immature Similar to adult. Two young birds, possibly 1 immature and 1 juvenile, distinguishable from adult by paler, drabber underparts and smaller, shallower, more slender bills (R. Restall *in litt.*). Forehead to mantle, scapulars and upperwing-coverts fuscous; back to tail, and tertials, dark greyish-brown; lores, anterior superciliaries, ear-coverts, sides of face, sides of neck and breast, flanks, thighs, vent and undertail-coverts cinnamon; throat, centre of breast and belly off-white; iris dark brown; bill horn (juvenile?) or green, with black culmen; legs and feet grey-brown (P. Precey *in litt.*). Birds having bill rather pale pink-brown with dark culmen, and legs and feet pale pink-brown, may be immatures (P. Boesman *in litt.*).

Juvenile See Immature.
Downy young Not described.

MEASUREMENTS (5 males, 8 females) Wing of male 78-84 (81.2), of female 79-85 (82.2); exposed culmen of male 19-21 (19.4), of female 17-19 (18); tarsus of male 30-32 (30.8), of female 28-30 (29.3) (Blake 1977). Measurements of birds in Colleción Ornitológica Phelps, Caracas, give: male, minimum exposed culmen 18, minimum tarsus length 28; female, minimum exposed culmen 13, tarsus 27-31 (R. Restall and P. Precey *in litt.*).

GEOGRAPHICAL VARIATION None.

MOULT No information available.

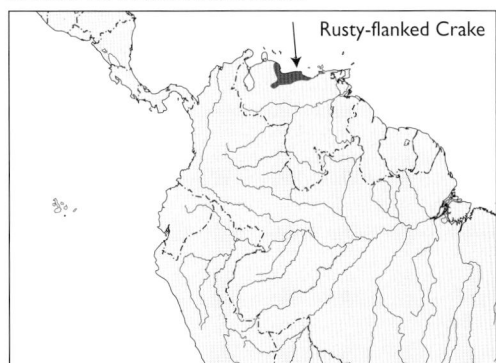

DISTRIBUTION AND STATUS Details are taken from Collar *et al.* (1992), Wege & Long (1995) and Boesman (1997) unless otherwise specified. This species is confined to the Caribbean slope of NW Venezuela, N of the Orinoco River, in Lara, Falcón, Yaracuy, Carabobo, Aragua, Distrito Federal and Miranda. A specimen from NE Brazil

(Paraíba) is a subadult Grey-breasted Crake (Teixeira *et al.* 1989). Regarded as VULNERABLE, before 1995 this crake was recorded infrequently from very few localities, 5 of which are listed as key areas for threatened birds and at most of which the bird's continued presence requires confirmation. The most recent of these records are from only three of these key areas, at two of them in man-made habitats (a pool and a reservoir): in Yacambú National Park (Lara), at Morrocoy National Park and Cuare Faunal Refuge (Falcón) pre-1991, and at Embalse de Taguaiguai (Aragua). In the 1940s it was at least locally common, and both Laguna de Valencia (Carabobo/Aragua) and Laguna de Tacarigua (Miranda) provided a number of specimens. Although it was still considered locally common in the 1980s its population at Yacambú National Park was then small, while only 6-12 birds were seen in c. 3-4ha of habitat at Embalse de Taguaiguai in 1985 but have not been seen since, falling water levels having possibly caused a reduction in available habitat. However in 1995-96 this species was observed at 7 new localities in E Falcón and W Carabobo, and estimates of its population were obtained (Boesman 1997): at Hacienda La Coreanera, W of Sanare (Falcón), 5-10 pairs; Tacarigua Dam (Falcón), 5-20 pairs; Serranía La Mision, S of Sanare (Falcón), 2-5 pairs; Guataparo Lake, W of Valencia (Carabobo), 10-20 pairs; E of Bejuma (Carabobo), 1-3 pairs; Canoabo Dam (Carabobo), 2-6 pairs; and San Pablo marsh (Carabobo) 10-30 pairs; it was still present at Yacambú National Park in Jul 1996.

The discovery of these new sites, at 3 of which healthy populations exist, gives more hope for the bird's survival, and the species probably occurs at other small sites, at which it is able to survive but is easily overlooked; the known population is c. 35-94 pairs and the total population could be several times this figure (Boesman 1997). However, there is no cause for optimism: the bird's numbers must be declining as a result of the general degradation of wetlands caused by industrial waste, pesticides and the lowering of water levels, and its habitats are under severe pressure. For example, the species was seen at Embalse de Taguaiguai in February 1985 but, by January 1986, the reservoir's water level had dropped so much that the birds were not seen there again (Collar & Andrew 1988). The Morrocoy and Cuare areas are threatened by development and pollution, especially from the tourist industry. There are plans to increase the water level of Guataparo Lake, which would inundate the bird's habitat, while agriculture and deforestation are advancing around Canoabo Dam, threatening the small pools at which the crake survives (Boesman 1997). Only the Yacambú site is protected and no site is safe: the bird's survival depends on at least some important wetlands (e.g. Guataparo Lake, Tacarigua Dam and Canoabo Dam) being protected against agricultural development and water level fluctuations, a measure which would benefit the water quality of these important reservoirs (Boesman 1997); San Pablo marsh is also an important site, deserving protection (P. Boesman *in litt.*). A detailed assessment of this bird's ecological requirements is urgently needed so that a conservation strategy may be developed, while more surveys are needed to determine its overall status and to assess threats to its survival. Surveys should be carried in May-Aug, when the birds are most vocal, and it is recommended that areas in Yaracuy should be investigated, this region having seen less human population expansion than Carabobo and Aragua (Boesman 1997).

MOVEMENTS None recorded.

HABITAT Marshes, lakesides, lagoons, and flooded pastures; sometimes dry grassland. 'Swamps' are also recorded, but the species does not occupy wetlands with trees (P. Boesman *in litt.*). It is said to frequent dense fringing aquatic vegetation, and Boesman (1997) noted a preference for two specific habitats: in hilly country, small ponds at least partly bordered by reeds and grasses and with some fringing or adjacent forested slopes; and in more open country, lakes, pools or marshes with rich or very dense aquatic vegetation, not subject to disturbance by humans or cattle. It occurs from sea level to 600m, but at c. 1,400m at Yacambú NP.

FOOD AND FEEDING No information available.

HABITS Described as shy and secretive; even in response to taped playback, birds remained in dense vegetation and went unseen (Boesman 1997).

SOCIAL ORGANISATION Possibly monogamous and apparently territorial: in Jun-Jul 1996 birds responding to taped playback were always seen in pairs, but once 3-4 birds were together (Boesman 1997).

SOCIAL AND SEXUAL BEHAVIOUR No information available.

BREEDING AND SURVIVAL Season A juvenile dated 6 Sep is the only evidence of breeding; this, and observed calling patterns, may indicate that the species breeds May-Jul and that young fledge Aug-Sep (Boesman 1997).

30 RUDDY CRAKE
Laterallus ruber   Plate 9

Corethrura ruber P. L. Sclater & Salvin, 1860, Vera Paz, Guatemala.

Monotypic.

Synonyms: *Creciscus ruberrimus/ruber*, *Porzana/Aramides/Thryocrex rubra*, *Rufirallus rubrus*, *Erythrolimnas ruber*.

Alternative names: Red Crake; Red Rail.

IDENTIFICATION Length 14-16.5cm. A predominantly bright chestnut small crake with blackish top and sides of head, red iris, black bill and olive-green legs and feet. Sexes similar, but male dark brown to blackish-brown from lower back to tail, while female is richer, more chestnut, from lower back to uppertail-coverts; male averages slightly larger than female. Immature like adult but paler on belly and with rump to uppertail-coverts like adult male. Juvenile predominantly sooty-grey or grey-brown, blackish on crown, rump and tail; throat whitish; belly pale grey to cream; iris greyish-white; bill pink to dark horn; legs and feet greenish-grey. Inhabits marshes and wet grassland.
Similar species Similar to allopatric Rusty-flanked Crake (29), but has uniformly rufous underparts, blackish head and no pale tip to bill. Within its own range a distinctive small crake. The only sympatric *Laterallus* species, White-throated Crake (31), has grey face and white throat and is barred from flanks to undertail-coverts. Distinguished from all congeners by dark slate top and sides of head, which contrast with chestnut mantle and sides of neck, and also by unbarred cinnamon to chestnut underparts with paler chin, and by greenish legs and feet.

VOICE Most information is from Howell & Webb (1995b). The territorial or advertising call is an explosive, descending, churring or purring trill, similar to that of the White-throated Crake and typically preceded by quiet piping notes audible at close range. The trill may rise and fall in pitch, and the bird may call continuously for over 30s. Calls are heard throughout the year, most frequently during the breeding season, when birds may call even until midday; calling is normally most frequent at sunrise and about two hours before sunset. A hard, clipped, ticking *chk* or *tek*, often repeated as an insect-like chatter and sometimes followed by quiet low rasps, is apparently a scolding note. A *stchup* is also recorded, and is very similar to a call of the White-throated Crake. Clucks and growls of annoyance are given to taped playback (Hardy *et al.* 1996).

DESCRIPTION
Adult male Forehead to nape, and sides of head, dark grey to blackish-slate, palest on ear-coverts. Sides of neck, mantle to upper back, and lesser upperwing-coverts, bright chestnut, shading to chocolate-brown and very dark brown from lower back to uppertail-coverts; tail blackish-brown. Greater upperwing-coverts and remiges dark brown to dark olive-brown; axillaries and underwing-coverts chestnut with indistinct blackish-brown barring but greaters ashy-brown, fringed rufous; underside of remiges ashy-brown. Underparts cinnamon, heavily tinged deep chestnut on sides of breast, flanks, thighs and undertail-coverts; chin and belly paler. Iris red; bill black; legs and feet olive-green to olive-grey.
Adult female Similar to male but richer, more chestnut, from lower back to uppertail-coverts; tail as in male. Averages slightly smaller than male, especially in culmen length.
Immature Like adult but paler on belly; in both sexes, rump to uppertail-coverts resemble adult male; undertail-coverts dark olive-brown with chestnut fringes.
Juvenile Essentially dark grey-brown to sooty-grey; almost black on crown, rump and tail; interscapular region has brownish cast; throat whitish; breast and flanks dark grey; belly pale grey to cream; undertail-coverts like immature, possibly darker. Iris greyish-white; bill pinkish to dark horn; legs and feet greenish-grey.
Downy young Black downy chick has pink bill with white egg-tooth, and grey legs and feet.

MEASUREMENTS Wing of 6 males 77-86 (82.8, SD 3.2), of 5 females 77-86 (81.4, SD 4.0); tail of 12 males 28-36 (31.2), of 6 females 27-32.5 (30.3); culmen to base of 6 males 21.5-22.5 (21.9, SD 0.5), of 5 females 20.5-22 (20.9, SD 0.6); tarsus of 12 males 30.5-35 (32.6), of 6 females 29-32 (30.8). Weight of 6 unsexed 42-49 (45.3); of 1 male 46 (FMNH; D. E. Willard *in litt.*).

GEOGRAPHICAL VARIATION Up to three subspecies (*ruber*, *ruberrimus* and *tamaulipensis*) were formerly proposed, but the colour characters, and the differences in bill length, used to separate the races are apparently based largely or entirely on variations attributable to sexual dimorphism. There are no plumage or mensural characters of value in distinguishing subspecies (Brodkorb 1943, Paynter 1955, Dickerman 1968a).

MOULT The juvenile quickly attains adult-like (first basic) plumage, and there is apparently no remex moult at this time (Dickerman 1968a).

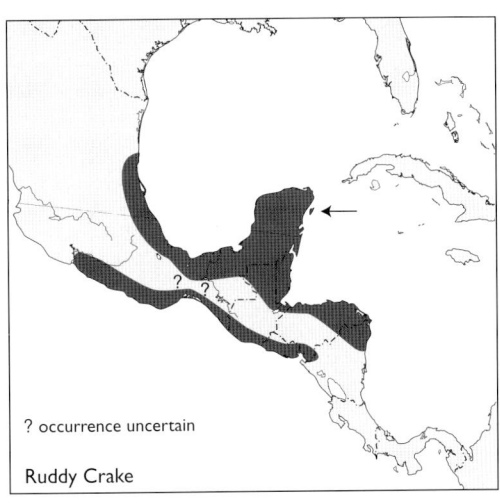

? occurrence uncertain
Ruddy Crake

DISTRIBUTION AND STATUS Tropical and subtropical zones of Central America, from E Mexico (Guerrero and Tamaulipas) along Pacific, Gulf and Caribbean slopes (including Cozumel Island) to Honduras, N Nicaragua and NW Costa Rica (N Guanacaste). This species is difficult to observe, and its current status is unclear, but in the 1960s and 1970s it was regarded as fairly common in some areas, possibly the most abundant crake over much of its range in Mexico, where its calls were heard throughout the Atlantic coastal lowland marshes in spring (Dickerman 1968a). It is apparently rare and local at its two known Costa Rican sites (Miravalles and Orosi) (Stiles & Skutch 1989).

MOVEMENTS It is considered resident throughout its normal range but is possibly a visitor to NW Costa Rica (Stiles & Skutch 1989).

HABITAT Varied freshwater habitats: reedbeds, marshes (including *Scirpus* growth and brushy edges), tall, dense *Eichhornia*, flooded to damp fields, wet meadows and roadside ditches; also tall sawgrass, pastures with tall grass or weeds, and grass in ditches and citrus groves. In Belize, it was recorded from brushy thickets at the edge of seasonally wet savanna in June, at a time when the savanna was quite dry (Russell 1966), and also from abandoned pastures which had not been grazed or mowed for 2-4 years (Saab & Petit 1992). In Costa Rica it was seen in May some way from a marsh, at the edge of thickety low growth bordering a recently charred field covered with stubble (Slud 1964), and during the dry season in Honduras it occurs in grassy areas far from water (Monroe 1968). It inhabits lowlands up to 1,500m.

FOOD AND FEEDING The diet is unknown. Birds feed in open situations at the edges of marshes or on floating mats of vegetation, mainly early and late in the day. Russell (1964) captured a bird in a mousetrap baited with peanut butter and oats.

HABITS Thought to be largely diurnal. Like many crakes it is difficult to flush, normally frequents dense vegetation, and is best detected by its voice; however it does feed in the open. It flies strongly, often with dangling legs.

SOCIAL ORGANISATION Undescribed. It may be monogamous and the pattern of calling throughout year suggests that birds may be permanently territorial.

SOCIAL AND SEXUAL BEHAVIOUR Not described.

BREEDING AND SURVIVAL Season Honduras, Jun-Sep, and incomplete nests found late May; El Salvador, Jul. **Nest** A rather tightly woven ball of dry grass stems, reeds and leaves, lined with finer grass, with side entrance; often placed in grass tussock. One nest, external depth 20cm and diameter 13cm, placed in centre of small tuft of reeds 20cm above water. **Eggs** 3-6; ovate; cream with reddish-brown flecks at large end; size 28.5-29.9 x 23.1-23.6 (29.2 x 23.45, 1 clutch of 4 eggs). No other details of breeding recorded.

31 WHITE-THROATED CRAKE
Laterallus albigularis Plate 9

Corethrura albigularis Lawrence, 1861, Atlantic side of Isthmus of Panama.

Sometimes considered conspecific with *L. melanophaius*, which replaces it E of Andes, but may be closer to *L. exilis*. Three subspecies recognised.

Synonyms: *Creciscus albigularis/cinereiceps/leucogaster/alfara* (*sic*); *Porzana cinereiceps/albigularis/leucogastra/alfari*; *Aramides albigularis*; *Limnocrex albigularis/cinereiceps*.

Alternative name: White-throated Rail.

IDENTIFICATION Length 14-16cm. A small crake with predominantly dark brown upperparts (more rufous on mantle), grey-brown crown and grey face in race *cinereiceps*, white chin and throat (most extensive in nominate race, least in *cerdaleus*), cinnamon-rufous neck and breast, and blackish and white barring from flanks to undertail-coverts; iris orange-red; bill greenish, darker on upper mandible; legs and feet brown, sometimes with yellowish or olive tinge. Sexes alike. Occasionally has remiges and/or upper-wing-coverts faintly mottled or barred white or cinnamon. Immature duller than adult; sides of head to sides of breast grey to grey-brown, flecked cinnamon on breast; flanks to undertail-coverts grey-brown, narrowly barred whitish. Juvenile almost entirely blackish-brown. Inhabits marshes, moist grassland, thickets and forest clearings.

Similar species The only sympatric *Laterallus* species with rufous neck and breast is Ruddy Crake (30), which lacks barred underparts. Contrast between grey head and brown back in race *cinereiceps* is reminiscent of Grey-breasted Crake (32), especially in flushed bird, but Grey-breasted has no rufous on neck or breast. Differs from Rufous-sided Crake (28), which replaces it E of Andes, in having rufous mantle, less extensive white on underparts (only from chin to throat or centre of upper breast), and undertail-coverts barred black and white.

VOICE An abrupt, explosive descending call, described as a whinny or a bubbly, churring trill, which is similar to that of the Ruddy Crake and is apparently a territorial call. It is more prolonged than the trill of the Grey-breasted Crake (32) and is given in a wider variety of situations, more frequently in response to disturbance and taped playback (Stiles & Levey 1988). The alarm note is a sharp, often slurred or metallic *chip* or *chirp*, and in intense alarm a shrill squeal is given (e.g. Slud 1964, Stiles & Levey 1988, Hardy *et al*. 1996). Clear, piercing whistled calls and a *jeer-jeer-jeer* are also recorded (Stiles & Skutch 1989).

DESCRIPTION *L. a. cinereiceps*
Adult Crown dark brownish-grey to grey; nape darker; sides of head grey; mantle variably cinnamon-rufous and dark brown; back to tail, including upperwings, dark brown, darkest from rump to tail. Primaries and secondaries dark brown; tertials darker; underwing-coverts white, mottled dusky; axillaries white, barred black. Chin, centre of throat (and sometimes centre of upper breast) white; sides of throat and breast bright cinnamon-rufous; centre of breast paler. Flanks and thighs to undertail-coverts barred black and white; black bars on belly narrow, sometimes absent. Some birds have cinnamon wash on white bars of undertail-coverts which occasionally extends over belly. Occasional birds have remiges and/or upperwing-coverts faintly mottled or barred with white or cinnamon. Iris orange-red; bill greenish to greenish-grey, with dusky culmen or with black upper mandible and tip of lower mandible; legs and feet dull yellowish-brown, olive or pale brown. Sexes alike.
Immature Similar to adult but duller; head feathers have paler brown tips; sides of head and sides of neck grey, mottled paler; sides of neck and sides of breast grey-brown to dull olive-brown, often mottled cinnamon on sides of breast; chin, throat and centre of breast and belly white; flanks dusky grey-brown to dull olive-brown, narrowly barred whitish; undertail-coverts blackish-brown, narrowly barred white. Iris dark brown; bill blackish, lower mandible pea-green with dusky tip; legs and feet olive-green.
Juvenile First plumage entirely dark blackish-brown, with pale feather bases visible on underparts; belly and thighs vaguely barred cinnamon-buff. One downy chick was growing dull grey, black-tipped feathers on breast and belly (Ridgway & Friedmann 1941).
Downy young Black downy chick has pale brownish-white bill with black band anterior to nostril.

MEASUREMENTS Wing of 9 males 75-81 (77.3, SD 2.0), of 5 females 74-76 (74.8, SD 0.8); tail of 18 males 23-33 (26.5), of 9 females 22.5-28; culmen to base of 9 males 18.5-21.5 (19.9, SD 1.0), of 5 females 18-20 (18.6, SD 0.8); tarsus of 18 males 27.5-32 (29.4), of 5 females 26-31.5 (28.3). Weight of 1 male, with slight amount of fat, 44 (D. Baepler *in litt.*).

GEOGRAPHICAL VARIATION There is variation in the colour and pattern of the head and neck, and in the extent and breadth of black barring on the posterior underparts.
L. a. cinereiceps (Lawrence, 1875) – SE Honduras (R Segovia = R Coco; one specimen, 1954) and Caribbean lowlands of Nicaragua, through Costa Rica to NW Panama (W Veraguas). See Description.
L. a. albigularis (Lawrence, 1861) – Pacific lowlands of SW Costa Rica (Gulf of Nicoya) through Panama (including Coiba I) to N and W Colombia (Pacific coast N to R San Juan) and W Ecuador (S to El Oro). Crown dark brown; side of head brown to pale russet; nape and mantle rufous-brown, concolorous with throat and breast; malar region pale russet. White barring on flanks relatively narrow; broader on undertail-coverts. Wing of 7 males 72-86 (79.15, SD 5.0), of 4 females 75-78 (76.5, SD 1.3); tail of 16 males 25.5-34 (29.1), of 15 females 24-33.5 (29.6); culmen to base of 7 males 17.5-21.5 (19.45, SD 1.2), of 4 females 18-21 (19.4, SD 1.4); tarsus of 16 males 27-32 (29.7), of 15 females 27-30 (28.5). Mean weight of 13 males 49.7 (SD 5.8), of 12 females 45.0 (SD 6.3) (Dunning 1993).

L. a. cerdaleus Wetmore, 1958 – E Colombia (Córdoba to Santa Marta region). Similar to *albigularis* but has fairly uniform reddish-brown head and throat (white of throat much restricted or absent); upper breast deeper reddish-brown; belly and vent conspicuously barred black. Measurements (6 males, 4 females): wing of male 68-74 (71.8), of female 69-77 (72.3); culmen from base of male 17-21 (18.6), of female 17-18 (17); tarsus of male 28-31 (29.3), of female 27-30 (28.3).

MOULT Remex moult is simultaneous, and is recorded in Sep, Nicaragua; the same bird is also in head and body moult.

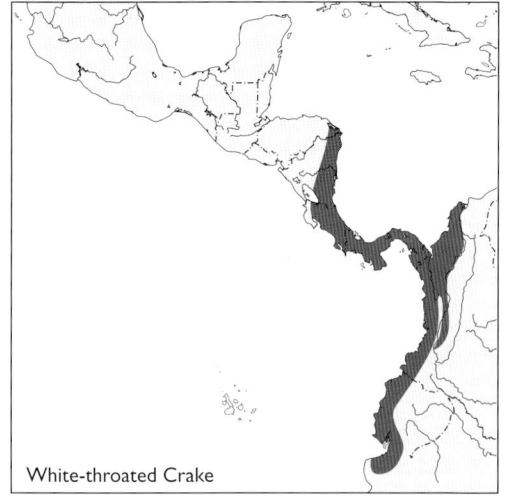

White-throated Crake

DISTRIBUTION AND STATUS Tropical and subtropical zones of Central America and NW South America, from SE Honduras and E Nicaragua to NW Colombia and W Ecuador. Although not regarded as threatened, this species is very difficult to see and its status is unclear. It was considered widespread and common to abundant over much of its range in the 1970s and 1980s (Ridgely 1981, Hilty & Brown 1986, Ridgely & Gwynne 1989, Stiles & Skutch 1989), but in Costa Rica in the 1960s the race *albigularis* was relatively uncommon except in some lowland regions whereas *cinereiceps* was abundant (Slud 1964). In W Ecuador it is fairly common (Ridgely & Greenfield in prep.).

MOVEMENTS Normally resident throughout its range, but in the wet season birds may wander to higher ground with low cover (Wetmore 1965).

HABITAT Marshes (including *Typha* beds), wet fields and damp pastures, grassy or overgrown banks of ditches, canals, ponds, streams and rivers, and drying stream beds; also thickets and forest clearings. Standing water is not required; in Colombia, the species is more associated with grass than with marshes (Hilty & Brown 1986), and during the rains in Panama it ranges widely away from marshes in any low cover (Wetmore 1965). At La Selva, Costa Rica, it occurred alongside the Grey-breasted Crake, but in the wet season (Aug) it occupied the wetter habitats (permanent and temporary marsh, and tall grass on river banks), from which Grey-breasted Crakes were then absent, and occurred only occasionally in tall dry grass where its congener was common (Stiles & Levey 1988). In the dry season (Feb) it was most numerous in permanent marsh and in tall grass on river banks, and occurred uncommonly in tall dry grass, its congener, then being absent from all these habitats, and it was occasional in temporary marsh (which was then dry) and pasture alongside its congener which was more numerous in these habitats (Stiles & Levey 1988). It inhabits lowlands and is recorded up to 1,600m.

FOOD AND FEEDING Takes insects (Orthoptera, Coleoptera, flies and ants), spiders, seeds of grass (*Panicum*), sedges (*Scleria*, *Fimbristylis*), spurge and *Solanum*; also algae. Six stomachs contained c. 90% vegetable matter, comprising fibre and seeds (Wetmore 1965). Although normally confined to dense vegetation, birds venture into more open spots, including tall grass, to feed at dawn and dusk or in dull, rainy weather (Stiles & Skutch 1989). They sometimes wade but apparently are reluctant to swim.

HABITS These birds normally stay within cover and are difficult to see, their presence usually being detected by their churring calls, given at any alarm. Like other small crakes, they may call almost at the observer's feet and yet remain invisible. However they may be approached and observed in *Typha*, while they can sometimes be seen feeding in the open in the early morning and towards dusk, the periods during which they are most active and vocal (Darlington 1931). Normally very difficult to flush, they usually fly across open areas, such as trails, with a weak flight action. On floating vegetation they often run to keep from sinking, and they sometimes perch in low bushes and tangles of reeds (Slud 1964, Wetmore 1965).

SOCIAL ORGANISATION Probably monogamous, and territorial at least during the breeding season. In Costa Rica, it is possibly interspecifically territorial with the smaller Grey-breasted Crake, responding to taped playback of its congener's vocalisations and possibly being the aggressively dominant species (Stiles & Levey 1988).

SOCIAL AND SEXUAL BEHAVIOUR No information available.

BREEDING AND SURVIVAL Season Nicaragua, May, Jul, Aug; breeds Costa Rica in wet season; Panama, Mar; Colombia, breeding condition Dec-Aug. **Nest** A ball with side entrance, of woven stems and leaves of grass, placed up to 60cm above ground or water in grass tussock or bush. **Eggs** 2-5; subelliptical to oval; creamy-white with small spots of cinnamon, reddish-brown or lilac concentrated at larger end; size of 15 eggs (nominate) 27.3-32.5 x 20.6-22 (Wetmore 1965); size of 25 eggs (nominate) 28.3-34 x 22-26 (31.2 x 23.8), calculated weight 9.5 (Schönwetter 1961-62).

32   GREY-BREASTED CRAKE
Laterallus exilis   Plate 10

Rallus exilis Temminck, 1831, Cayenne, French Guiana.

Monotypic.

Synonyms: *Gallinula ruficollis*; *Ortygometra cinerea*; *Porzana exilis/cinerea*; *Creciscus/Laterirallus exilis*; *Rallus cinereus*; *Ortygometra/Aramides cinerea*.

Alternative name: Temminck's Crake.

IDENTIFICATION Length 14-15.5cm. A rather dark small crake, conspicuously bright chestnut from hindcrown to upper mantle; head, neck and breast grey, darker on crown; throat white; upperparts olive-brown with variable narrow white barring on upperwing-coverts, rump and uppertail-coverts; posterior underparts barred blackish and white. Eye red; eyelids yellow to orange; bill dusky to black, with pale green to yellowish-green proximal two-thirds of lower mandible, extending to basal region of upper mandible; legs and feet pale brown, often tinged yellowish. Sexes alike; male darker on breast. Immature grey from head to breast; upperparts brown, lacking any chestnut; barring on underparts duller and more restricted than in adult; iris brown; bill largely horn-brown. Juvenile sooty-black with whitish throat and centre breast; underparts barring indistinct. Inhabits marshes and wet to dry grassland.

Similar species Easily distinguished from congeners with barred flanks by pale grey head to breast, olive-brown upperparts and contrasting chestnut nape and upper mantle. Race *cinereiceps* of White-throated Crake (31) also shows contrast between grey head and brown back, especially when flushed, but has cinnamon-rufous neck and breast. Three other superficially similar grey-breasted crakes with barred flanks are sympatric, but all are larger and lack chestnut on nape and mantle: Paint-billed Crake (112) has red base to bill and red legs; Ash-throated Crake (100) and Sora (98) have patterned upperparts and different head pattern, former having purplish brown legs, latter white undertail-coverts.

VOICE Most information is from Stiles & Levey (1988). The commonest call is an explosive, sharp, whistled note *dit*, *tink* or *keek*, often given in a descending, fairly rapid series of 2-10 notes in which the first is loudest and successive notes are shorter; it is quite musical and at close range a soft introductory note may also be audible (Howell & Webb 1995b, Hardy *et al.* 1996). This call may be repeated with pauses of 3-4s between series. It is given in alarm or excitement, and in response to disturbance and to taped playback. A descending, dry musical rattle or ticking trill, similar to the call of the Rufous-sided Crake (28), is also given, and is probably a territorial song. A quiet, sharp *check*, a low, gruff *chk* and other soft churrs are also recorded (Hilty & Brown 1986, Howell & Webb 1995b).

DESCRIPTION
Adult Head, sides of neck and breast grey, slate-grey on crown and white on chin, throat, foreneck and centre of breast; posterior crown to upper mantle bright chestnut. Rest of upperparts, including upperwings and tertials, olive-brown to brown; upperwing-coverts, and occasionally even outer scapulars, variably barred narrowly with white, barring sometimes absent. Alula, primaries and secondaries dark olive-brown to fuscous, fringed paler; underwing-coverts white with dusky feather bases; axillaries barred blackish-brown and white; underside of remiges ashy. Rump and uppertail-coverts olive-brown to black, narrowly barred white (colour and pattern variable, and white barring may be absent); tail olive-brown to blackish-brown. Belly and flanks to undertail-coverts barred blackish-brown and white, broadest on upper belly; centre of belly white; lower flanks and undertail-coverts tinged buff to rufous. Iris red, eyelids yellow to orange; upper mandible horn-brown to blackish, proximal quarter green to greenish-yellow, lower mandible pale green to greenish-yellow, distal third black; legs and feet pale brown to yellowish-brown or yellowish-buff. Sexes alike, but male has darker grey breast, contrasting more with white throat.
Immature Upperparts plain brown, lacking any chestnut; crown grey; underparts medium grey with duller, more restricted barring than in adult; throat and centre of belly whitish. Iris brown; bill horn-brown to blackish, sometimes with greenish-yellow tinge on lower mandible.
Juvenile Sooty-black (dark blackish-brown); scapulars and upperwing-coverts dark olive-brown, feathers tipped blackish-brown; chin, throat and centre breast whitish, mottled blackish-brown; sides of breast, and upper flanks, blackish-brown; lower flanks, belly and undertail-coverts more dark olive-brown, indistinctly barred cinnamon-buff to vinaceous-buff; iris and bill as immature.
Downy young Chick has black down.

MEASUREMENTS (6 males, 7 females) Wing of male 72-78 (75.2, SD 2.2), of female 72-76 (74, SD 1.6); tail of male 30-36 (32.3, SD 2.6), of female 31-37 (33.9, SD 2.3); culmen to base of male 16.5-18.5 (17.5, SD 0.7), of female 15.5-18 (17, SD 0.9); tarsus of male 21.5-24 (22.8, SD 0.9), of female 21.5-24 (22.7, SD 0.9). Weight of 10 males 26.5-39 (32.4), of 9 females 27-43 (34.7).

GEOGRAPHICAL VARIATION None. The specimen from Honduras was originally described as a distinct race, *L. e. vagans*, but has since been shown not to differ from South American birds (Monroe 1968).

MOULT One non-breeding adult female in simultaneous primary moult (flightless, with all primaries in sheath) was caught in Surinam, Dec, when other birds caught were in breeding condition and not moulting (Haverschmidt 1974). A male in heavy body moult was caught in Bolivia, Jun.

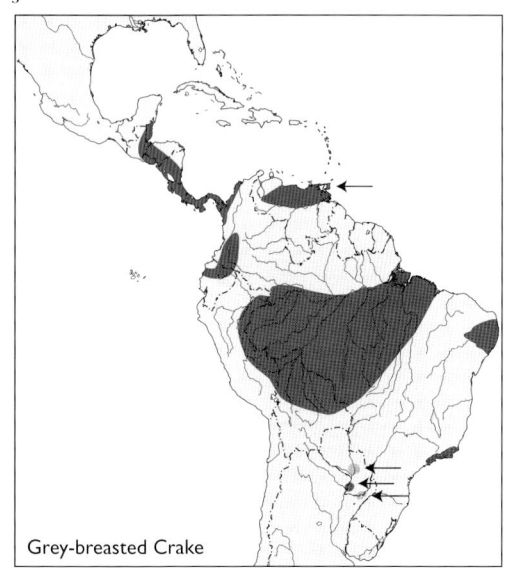

Grey-breasted Crake

DISTRIBUTION AND STATUS Central and South America. From the Caribbean slope of Guatemala and Belize, through SE Honduras (R Segovia = R Coco; one specimen, 1887), Nicaragua, Costa Rica (including Golfo Dulce in SW) and Panama (recorded only from Coiba I, Puerto Obaldía, Fort Sherman and Tocumen) to N and W Colombia (N Chocó N to Barranquilla area at Atlantico,

S to Valle and NW Cundinamarca, and E of the Andes in Meta, Caquetá and Putumayo). In NW Ecuador it occurs from W Esmeraldas S to S Pichincha, and it also occurs E of the Andes (Blake 1977, Ridgely & Greenfield in prep., Ridgely *et al.* in press). Also Trinidad, Venezuela (Miranda E to Monagas) and the Guianas S through Amazonian Brazil (R Negro and R Solimões to E Pará) to E Peru and N and C Bolivia (Pando, Beni, La Paz and Santa Cruz); E Brazil (Paraíba and Pernambuco) and extreme SE Brazil (Rio de Janeiro and São Paulo). There are only 3 Paraguayan records, from Villa Hayes (Dpto Presidente Hayes) in Aug 1979 (Storer 1981), N of Pilar (Dpto Ñeembucú) in Jan 1994 (Hayes 1995), and Monumento Natural Bosque de Arary (Dpto Itapú) where 3 birds were calling in Nov 1995 (Lowen *et al.* 1997); its status is uncertain, but the Nov and Jan records suggest that breeding is possible although the nearest records outside Paraguay are far to the N (Hayes 1995, Lowen *et al.* 1997).

This fragmented distribution suggests that the species may be more widespread than is known, but the birds are usually secretive and very difficult to see. It is reported as locally common in Bolivia, Amazonian Brazil, and the wet lowlands of Costa Rica (Stiles & Levey 1988, Parker *et al.* 1991, Peres & Whittaker 1991), and uncommon to locally fairly common, even numerous, in Ecuador, where it may be increasing in the E as a result of deforestation (Ridgely & Greenfield in prep.). It is locally quite numerous in Surinam; at Paramaribo, Surinam, many birds were recorded in grass fields at a new housing project in Dec 1972 (Haverschmidt 1974). It is apparently local and uncommon to rare elsewhere, especially in Central America where it may be generally less numerous than in South America, a distribution pattern which might be expected in a species retreating from a formerly greater range (Olson 1974a). However, it may be more overlooked than uncommon in Central America, and its numbers in Costa Rica are thought to have increased greatly in recent years as a result of the creation of habitat following the deforestation of most of the country's humid lowlands (Stiles & Levey 1988).

MOVEMENTS Regarded as sedentary, but at Paramaribo, Surinam, two were caught at night at lighted houses in Jul and Feb (Haverschmidt & Mees 1995), while near Puerto Viejo de Sarapiqui in the Caribbean lowlands of Costa Rica one flew into a field station at night in Feb (Stiles & Levey 1988). In Paraguay, it is not clear whether the bird taken at Villa Hayes on 6 Aug 1979 (Storer 1981) was a migrant, a stray or part of a resident population (see Distribution and Status).

HABITAT Marshes, flooded rice fields, riverbanks, lake edges and wet grassy habitats with dense, short to fairly tall vegetation (heights of 50-100cm are recorded); also on floating mats of *Paspalum* grass on a river (Parker *et al.* 1991) and floating grass mats surrounding oxbow lakes (Ridgely & Greenfield in prep.); in Trinidad birds frequent dense *Eleocharis* sedges. The preferred habitat usually has shallow standing water (5-15cm deep) but birds sometimes occur in dry habitat, being recorded from fields of thick grass up to 1m high in a new housing area (most of the birds caught were in breeding condition) (Haverschmidt 1974), and also pasture, coconut groves and airfields. This species occurs alongside the Rufous-sided Crake (28) and the Yellow-breasted Crake (105) in marshes, and is also found in habitat similar to that occupied by the Ruddy Crake (30). At La Selva, Costa Rica, it occurred alongside the White-throated Crake (31) but was largely segregated by habitat preferences (Stiles & Levey 1988); in the wet season (Aug) it occupied the driest habitats (tall grass on high ground, and pasture), from which White-throated Crakes were then absent or occasional. In the dry season (Feb) it was most numerous in pasture and uncommon in temporary marsh (which was presumably then dry), its congener then being recorded only occasionally in these habitats. In the upper Amazon region it occurs on islands with pioneer vegetation, alongside the Black-banded Crake (27) (Sick 1997). It normally inhabits lowlands, but is recorded up to c. 1,700m.

FOOD AND FEEDING Earthworms, spiders, insects (Tettigoniidae and other Orthoptera, Homoptera, Coleoptera, Hemiptera, Heteroptera and larval Lepidoptera), and grass seeds. One was seen foraging by a mud puddle on a dirt road through tall grass on a dark, rainy afternoon, and birds may forage just outside cover in the early morning (Stiles & Levey 1988).

HABITS This species is most active and vocal for c. 1h after sunrise and before sunset. It is regarded as secretive and difficult to observe or to flush, normally remaining in dense cover; when flushed, it flies low, with dangling legs, for a short distance. However in Bolivia, in Jun, birds in wet grassy pasture were fairly common and easily seen, often landing temporarily in small bushes after being flushed, and four were collected (Schulenberg & Remsen 1982). It uses runways under grass; a Paraguayan specimen was trapped in a runway used by swamp rats *Holochilus brasiliensis* (Storer 1981).

SOCIAL ORGANISATION Birds are usually found in pairs and are probably territorial, at least in the breeding season. In the dry season (Feb) in Costa Rica, two pairs occurred in 0.375ha of pasture and 7-8 pairs in an area of c. 2ha (Stiles & Levey 1988). The birds responded to taped playback of White-throated Crake (31) calls and thus the species possibly shows interspecific territoriality with its larger congener. White-throated Crakes respond more vigorously to calls of Grey-breasted Crakes than vice versa, and may thus be the aggressively dominant species (Stiles & Levey 1988).

SOCIAL AND SEXUAL BEHAVIOUR Nothing recorded.

BREEDING AND SURVIVAL Season Belize, male probably in breeding condition Mar; Costa Rica, possible nest Aug; Surinam, Dec (oviduct egg), Feb; Colombia, breeding condition Feb; Panama, possibly Jul but record normally ascribed to Black Rail (33); record from Trinidad, Jul (Belcher & Smooker 1935), possibly not of this species (Ripley 1977), although it is not clear to what other species it may be attributed. **Nest** Spherical with side entrance, typical of genus; of woven grass and weed stems; placed in thick grass; external diameter c. 14cm. **Eggs** 3; long oval; slightly glossy; creamy-white with dark brown to grey spots and smears, most numerous at larger end; size of 5 eggs (including Trinidad clutch) 29.8-32.5 x 22.5-24 (31.2 x 23.2); calculated weight 9.2. No other information available.

33 BLACK RAIL
Laterallus jamaicensis Plate 10

Rallus jamaicensis Gmelin, 1789, Jamaica.

Distinctive race *tuerosi* possibly merits recognition as a separate species. Forms a superspecies with *L. spilonotus*. Five subspecies normally recognised.

Synonyms: *Cresciscus* (sic)/*Creciscus jamaicensis*; *Creciscus murivagans*; *Crex pygmaea*; *Porzana/Ortygometra/Corethrura jamaicensis*; *Creciscus/Gallinula/Rallus salinasi*.

Alternative name: Black Crake.

IDENTIFICATION Length 12-15cm; wingspan 22-28cm. A very small, dark rail; head blackish-grey (crown chocolate-brown in race *coturniculus*); nape to mantle chestnut or rufous; rest of upperparts, and upperwings, blackish-brown, spotted or barred white (*tuerosi* has bold white bars; rump barred pale cinnamon); secondaries and tail spotted white (barred in *tuerosi*). Chin to upper belly slate-grey; flanks, lower belly and undertail-coverts blackish-grey, barred white (bars broadest in *salinasi* and *tuerosi*; undertail-coverts cinnamon in *murivagans* and *tuerosi*). Bill black; iris scarlet; legs and feet flesh to brown. Sexes similar, but female greyer on sides of head (contrasting more with dark crown), and on throat and breast. Immature similar to adult; iris and legs gradually attain adult colours. Juvenile browner on upperparts, with fewer white spots; face and underparts grey, with white chin, throat and centre belly; flanks and undertail-coverts browner than adult; iris brown; legs and feet darker than adult. Inhabits marshes and wet grassland. Advertising call of male distinctive.

Similar species Easily distinguished from most sympatric rallids, and from all congeners except allopatric Galapagos Rail (34), by very dark plumage (but note that young of most rallids are black) with white spots or bars on upperparts and bars on underparts, and contrasting chestnut or rufous patch on nape. Very similar Galapagos Rail, sometimes considered a race of Black Rail, has sparser white spotting and barred undertail-coverts. Dot-winged Crake (97) is a similarly sized, very dark wetland crake; upperparts dark olive-brown, streaked blackish and with white markings restricted to upperwing-coverts, remiges and uppertail-coverts; flanks and undertail-coverts barred white; legs and feet brown. Possibly sympatric Speckled Rail (15) similar in size and colour but has white chin, white spotting on head, neck and upper breast, rest of underparts barred, and black legs and feet.

VOICE Information is from Eddleman *et al.* (1994). In the breeding season the male gives a characteristic *kic-kic-kerr* advertising call ending in a downward slur (the call may have 1-4 introductory notes and 1-3 concluding notes), and the female gives a low *croo-croo-croo* or *who-who-whoo*, rarely heard (an example is given by Hardy *et al.* 1996). When highly agitated, e.g. in response to taped playback, birds utter repeated growls. At the nest, the scolding call of the female is a rapid series of soft, high-pitched, nasal *nk* notes; the male has a similar call, and the female also has a rasping scold call. Single barks, *churt* or *kik*, are also given, described as similar to "a sparrow call note"; also a soft, low-pitched *twirr*, like the sound produced by running a finger along the teeth of a comb. Advertising calls are given both by day and by night, sometimes continuously for long periods; the frequency of calling decreases in high wind, bright moonlight, and cold temperatures.

DESCRIPTION *L. j. jamaicensis*
Adult male Head dark slate-grey, darkest on crown and ear-coverts. Nape to upper mantle chestnut-brown; rest of upperparts, including upperwing-coverts and alula, blackish-brown with small white spots; often darker on rump and uppertail-coverts. Remiges fuscous; secondaries spotted white, forming three or more rows of spots on spread wing; spots reduced, faint or absent on primaries. Rectrices fuscous, spotted white. Axillaries and underwing-coverts dull fuscous with narrow whitish bars. Chin, throat, side of neck, breast and upper belly uniform slate-grey, sometimes paler on chin and throat. Flanks, lower belly, thighs and undertail-coverts blackish-grey with narrow white bars; flanks sometimes spotted white. Iris scarlet; bill sepia or blackish-grey; legs and feet dusky flesh to dull brown.
Adult female Similar to adult male, but paler grey on sides of head and from chin to upper belly. Top of head more contrastingly dark; chin and throat paler grey to white; centre of belly paler grey.
Immature Similar to adult. Some males have white feathers on chin. Bill sepia, darkening as bird matures; legs and feet gradually pale to adult colour by c. 3 months; iris becomes rufous or orange before attaining adult colour at c. 3 months.
Juvenile Predominantly dark brown on upperparts, including top of head, upperwing-coverts and alula; more rufous from nape to upper mantle, with variable number of small white spots from mantle to uppertail-coverts; spots sometimes absent from rump. Sides of head, and underparts to upper belly, grey; chin white to pale grey; throat and centre of upper belly often pale grey; sometimes has pale wash on throat and breast and often has greyish-horn wash on sides of neck and breast. Lower belly white to pale grey; flanks and undertail-coverts browner than in adult, with narrower white bars. Bill sepia, lower mandible paler; iris greenish-olive, becoming amber to hazel; legs and feet darker than adult.
Downy young Black, with oily greenish sheen on upperparts. Bill sepia, with pinkish spot around nostril, gradually becomes lighter, with sepia culmen and buffy lower mandible (but see race *salinasi*); iris fuscous or dark brownish-olive, becoming olive-green at 4-6 weeks; legs and feet darkish dull brown.

MEASUREMENTS Wing of male 74.0 (SD 3.2, n = 60), of female 73.7 (SD 3.2, n = 47); tail of male 33.3 (SD 3.0, n = 37), of female 31.4 (SD 6.0, n = 37); exposed culmen of male 14.3 (SD 1.3, n = 51), of female 13.6 (SD 0.6, n = 46); tarsus of male 22.9 (SD 1.9, n = 59), of female 22.4 (SD 1.0, n = 45). Weight of 33 males 29-43 (34.6, SD 3.1), of 16 females 29-44 (36.2, SD 5.0) (Eddleman *et al.* 1994).

GEOGRAPHICAL VARIATION There is considerable variation in plumage colour and pattern, particularly in the colour of the mantle and back, and of the undertail-coverts, and in the nature of the white patterning on the upperparts and upperwings (see discussion in Fjeldså 1983b). Southern populations average larger in wing and tarsus measurements. In the USA, inland nesting *jamaicensis* were formerly considered a separate race, named either *stoddardi* or *pygmaeus*. Birds from Belize (vicinity of Monkey River) are intermediate in colour, wing and tarsal measurements between *jamaicensis* and *coturniculus*, and are assigned to the former (Russell 1966).

L. j. jamaicensis (Gmelin, 1789) – E USA along Atlantic coast from Connecticut to S Florida, along Gulf coast from S Florida to S Alabama, and in SE Texas; inland sporadically from Colorado, Kansas and Oklahoma to Minnesota, Michigan and E to Connecticut. Central America in E Mexico (locally in SE San Luis Potosí and N Veracruz), Belize, Costa Rica (Bebedero basin at Taboga; Península de Osa at Rancho Quemado; and Río Frío District at Medio Queso) and Panama (1 record); Cuba and Jamaica (wintering; 1 breeding season record, Cuba), and possibly Puerto Rico. Winters from coastal S and E USA (S through Florida, SW to SE Texas, and rarely N to New Jersey) to Guatemala (one record, Dueñas) and Greater Antilles. One record from N Brazil (Belém). See Description.

L. j. coturniculus (Ridgway, 1874) – C California, at 5 coastal localities from Bodega Bay S to Morro Bay, and inland in Yuba County (Aigner *et al.* 1995); irregularly S to NW Baja California; also inland in Salton Trough and on Lower Colorado River in extreme E California and SW Arizona; probably resident. Smaller than nominate; bill smaller and more slender; crown chocolate-brown; nape to upper mantle more rufous, colour extending further down mantle; underparts more deeply coloured. Wing of male (W Coast) 67.3 (SD 2.0, n = 102), (Arizona) 69.6 (SD 1.8, n = 18), of female (W Coast) 66.5 (SD 2.0, n = 150), (Arizona) 69.1 (SD 1.8, n = 15); tail of male (W Coast) 30.6 (SD 1.8, n = 37), (Arizona) 34.4 (SD 1.6, n = 16), of female (W Coast) 31.0 (SD 2.2, n = 64), (Arizona) 34.5 (SD 1.9, n = 15); exposed culmen of male (W Coast) 14.2 (SD 0.6, n = 99), (Arizona) 14.5 (SD 0.8, n = 18), of female (W Coast) 13.6 (SD 0.5, n = 149), (Arizona) 13.2 (SD 1.6 [rather large], n = 16); tarsus of male (W Coast) 21.4 (SD 1.5, n = 101), (Arizona) 21.1 (SD 1.8, n = 18), of female (W Coast) 20.8 (SD 1.4, n = 150), (Arizona) 20.7 (SD 1.0, n = 16). Weight of 36 males 20.5-34 (27.8), of 32 females 23-46 (30.1). Mensural data from Eddleman *et al.* (1994).

L. j. murivagans (Riley, 1916) – a few coastal marshes in C Peru (Lima area). Similar to *salinasi* but head and upperparts paler; white dorsal markings form short transverse bars, which are normally interrupted but may extend across whole width of feather in some upperwing-coverts (see *tuerosi*); remiges spotted; underwing-coverts almost white; undertail-coverts pale cinnamon or light pinkish, with grey mottling. Wing of 1 male 77, of 1 female 73, of 4 unsexed 75-78 (76.75, SD 1.3); tail of 4 unsexed 32-35 (33, SD 1.4); culmen to base of 4 unsexed 14.5-17 (16.25, SD 1.2); tarsus of 1 male 23, of 1 female 22, of 4 unsexed 21.5-22.5 (22, SD 0.4).

L. j. tuerosi Fjeldså, 1983b – Lake Junín, C Peruvian Andes. Two specimens described. Darker head and neck, less chestnut on mantle, and markedly larger and bolder white bars on upperparts, remiges and rectrices than other races, these bars being continuous across both webs of each feather; rump barred pale cinnamon; underwing-coverts white to slate-grey; undertail-coverts cinnamon. Iris dark brown; bill blackish; legs and feet dark olive-brown. Iris colour suggests that specimens are immature. 2 unsexed: wing 74; exposed culmen 14; tarsus 24.5, 25.

L. j. salinasi (Philippi, 1857) – Mejía lagoons, S Peru; Atacama S to Maleco, C Chile, to adjacent Catamarca, Mendoza, San Juan and La Rioja, WC Argentina. Largest race; larger and more russet patch on nape and upper mantle; larger white spots or bars on upperparts and remiges, and broader and more numerous white bars on flanks, than other races except *tuerosi*. Bill may be greyish-brown, and legs and feet greenish (Reed 1941). Bill of chick described as yellow at base, black elsewhere, with small white tip (Reed 1941). Wing of 10 males 75-80 (77.9), of 9 females 75-79 (76.3); tail of 2 females 32, 37, of 4 unsexed 30-37 (33.5, SD 2.9); culmen to base of 1 female 16.5, of 4 unsexed 16.5-18 (17.25, SD 0.6); tarsus of 6 males 19-23 (20.8), of 8 females 20-23 (20.7). Weight of 1 male 30 (Marín 1996); weight of chicks 1 day after hatching 4.5 (Reed 1941).

MOULT In the race *jamaicensis* a postbreeding complete moult occurs Jul-Aug; the remiges and rectrices are moulted simultaneously and birds are flightless for c. 3 weeks; the prebreeding partial moult occurs Feb-Apr, and a postjuvenile partial moult occurs from autumn to early winter (Eddleman *et al.* 1994).

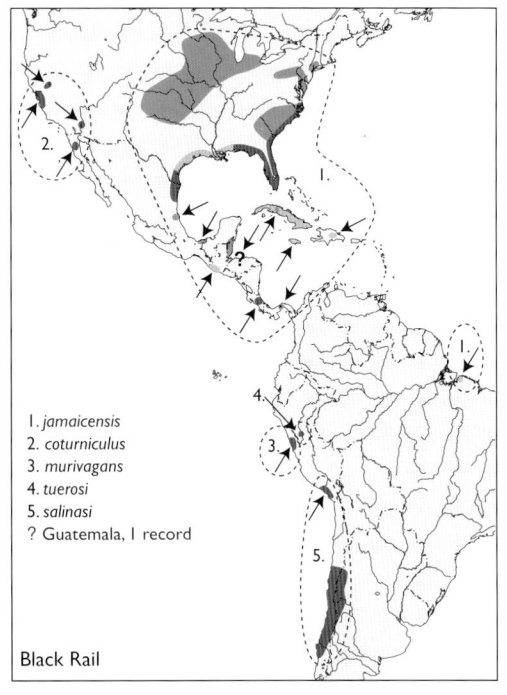

1. *jamaicensis*
2. *coturniculus*
3. *murivagans*
4. *tuerosi*
5. *salinasi*
? Guatemala, 1 record

Black Rail

DISTRIBUTION AND STATUS Locally distributed in the USA, Central America and western South America. In the USA it is recorded sparsely at inland sites in the breeding season and occurs throughout the year in coastal areas (Eddleman *et al.* 1994). The nominate race's Central American distribution is poorly known; it has been recorded very locally in E Mexico and Central America (Belize, Costa Rica and Panama) during the breeding season (e.g. Dickerman & Warner 1961, Harty 1964, Russell 1966, Stiles & Skutch 1989, Howell & Webb 1995b), and it is probably present all year in Costa Rica, where it is possibly widespread but overlooked (Stiles & Skutch 1989); it is also recorded during the breeding season in

Cuba and Jamaica (where it is mainly a winter resident, but may breed very locally in the Greater Antilles) (Bond 1979, AOU 1983, Eddleman *et al.* 1994). It may be only an extremely rare winter visitor to Puerto Rico (Biaggi 1983, Raffaele 1989), and it has only been recorded once from Guatemala, in winter (Salvin & Godman 1903). The sole record from Panama, of a bird flushed from a nest near Tocumen in Jul 1963 (Harty 1964), was long doubted because of possible confusion with the Grey-breasted Crake (32) but the recorded call resembles the common call of the Black Rail (Ridgely & Gwynne 1989). The species may also occur in Honduras, whence there is a record in March 1953 (Monroe 1968). There is 1 recent record from N Brazil (Novaes & Lima 1996; see Movements).

The South American races are sedentary, occurring in W Peru, Chile and W Argentina, with an isolated population (*tuerosi*) in the Peruvian Andes (Fjeldså & Krabbe 1990). The existence of two presumed Colombian specimens, labelled "Bogotá" (doubtful race *pygmaeus*) and "Nouvelle Grenade", suggests that it may occur in the temperate zone of the Colombian E Andes (Hilty & Brown 1986).

Although the species is not globally threatened, the race *tuerosi*, sometimes treated as a full species, is considered ENDANGERED; it is known only from 2 sites on the SW shore of L Junín (near Ondores and Pari), but is likely to occur in large portions of the 15,000ha of wide marshland surrounding the lake; nearby lakes do not have suitable habitat (Fjeldså 1983b). It is at risk from pollution and water level fluctuations which have affected L Junín since at least 1955 (Collar *et al.* 1992); in particular there has been strong desiccation of reedmarshes caused by drought and unsustainable water management by Electro Peru, and also occasional flooding with very acid water from the Cerro de Pasco mines; however the species still survived in Jan 1995 (J. Fjeldså *in litt.*).

All other races are locally distributed and often rare or infrequently recorded. Both races occurring in the USA are regarded as THREATENED or ENDANGERED in most states. In the USA most populations have probably declined drastically this century, and the breeding range has contracted seriously, both of *jamaicensis* at inland sites (from which there have been no confirmed nesting recorded since 1932) and of both races at coastal sites; the situation was worse before the 1970s when laws protecting coastal wetlands were enacted (Eddleman *et al.* 1994). In coastal California, *coturniculus* has almost been eliminated as a breeding species S of Morro Bay, and most birds occur in N San Francisco Bay (San Pablo Bay), but a small population was discovered inland, E of Marysville, Yuba Co, as recently as 1994 (Aigner *et al.* 1995). This race is threatened by marsh subsidence caused by groundwater removal, by the diking of saltmarshes (95% of the tidal marshes present in San Francisco Bay in 1850 were diked or filled by 1979), by water level fluctuation, and by wildfires. There is only one recent record from NW Baja California (Howell & Webb 1995b). The Lower Colorado R population, not discovered until 1969, may be recently established, or may be relict, surviving in reduced habitat; it declined c. 30% from 1973 to 1989, and most of the population has shifted from Imperial Dam and Imperial Co to Mittry Lake (Evens *et al.* 1991).

This species is more susceptible than many other marsh rallids to disturbance from grazing and agriculture, because its habitat is at the edge of marshes. The continued massive degradation and loss of shallow wetland habitats gives cause for greater concern in the future, and management action is urgently needed. In South America *murivagans* is rare and local (Ripley 1977), while *salinasi* is rare (e.g. Reed 1941) and was not recorded from Chile for 25 years until a dead bird was found in Oct 1994 (Howell & Webb 1995a). The race *jamaicensis* is regarded as an extremely rare winter visitor to Puerto Rico, where it was once a breeding resident but was probably extirpated by the introduced mongoose (Raffaele 1989); the situation in Jamaica is apparently similar (deGraaf & Rappole 1995). In North America, avian predators include herons, owls, Ring-billed Gull *Larus delawarensis* and Northern Harrier *Circus cyaneus*; herons may eat large numbers of adults during extremely high tides if adequate cover is unavailable adjacent to tidal marshes (Evens & Page 1986). Domestic cats are also predators.

MOVEMENTS Migrations are poorly known. In the USA, inland populations, and most birds from E coastal populations (all nominate), probably migrate, wintering in coastal E USA and also S to Cuba, Jamaica and Guatemala; there is also an isolated winter record from the Dominican Republic (Eddleman *et al.* 1994). The species is recorded only in the winter in Guatemala, Puerto Rico and Honduras, and its status in Belize appears to be uncertain (see Distribution and Status). Most Cuban records are of winter residents, it is transient in the Bahamas, and it may winter more widely than is known in the West Indies and in its Central American breeding range. The sole record from Brazil is of a female found dead on 3 Nov 1994 after having possibly hit a lighted window in Belém, Pará (Novaes & Lima 1996). The USA spring migration occurs from mid-Mar to early May, the autumn migration from early Sep to early Nov (most mid-Sep to mid-Oct). Birds migrate at night along a broad front and sometimes strike towers and buildings. It is casual or accidental in Canada (Ontario, Quebec, Nova Scotia), Maine, Rhode Island and Bermuda (AOU 1983, Godfrey 1986, Eddleman *et al.* 1994). The Californian race (*coturniculus*) and the South American races are apparently sedentary, but in the USA juveniles disperse widely from breeding areas and may appear in atypical habitats (Eddleman *et al.* 1994).

HABITAT Most information for the USA is from Eddleman *et al.* (1994). Inhabits marshes, both fresh and saline, and wet meadows and savanna; also the borders of ponds (Costa Rica). In North America it occupies breeding sites with shallower water (under 3cm deep in Arizona) than other rallids; most breeding areas have fine-stemmed emergent plants (grasses, sedges and rushes), while the substrate of ideal habitat is usually moist soil with scattered small pools. In the E USA, coastal habitat is characterised by infrequent tidal inundation and is dominated by cordgrass (*Spartina* spp.), pickleweed (*Salicornia* spp.), rushes (*Juncus* spp.), saltgrass *Distichlis spicata* or bulrush *Scirpus olneyi*. The species tolerates more flooding in tidal marshes of the W coast, but requires surrounding higher vegetation for escape during extreme high tides, while occupied sites are characterised by features associated with high tidal elevation and a greater freshwater influence. Sites in coastal California are characterised by taller vegetation and greater coverage of alkali heath *Frankenia grandiflora*. Inland, birds select sites with shallow, stable water levels, gently sloping shorelines and vegetation dominated by fine-stemmed *Scirpus* spp. or grasses.

Structure may be more important than plant species composition in determining habitat suitability (Flores & Eddleman 1995). The habitat of wintering California birds in Arizona differs little from the breeding habitat but in California wintering sites are lower in elevation, smaller, more linear and more fragmented than preferred summer habitat.

In Belize, birds were found in savanna dominated by the grasses *Sporobolus cubensis*, *Paspalum pulchellum* and *Mesosetum filifolium*, and *Rhynchospora* sedges; the vegetation was 25-50cm tall (Russell 1966). At L Junín in the Peruvian Andes the race *tuerosi* inhabits wide rushy (*Juncus*) zones in areas with mosaics of small *Juncus* beds 1m tall and open areas of waterlogged marl sparsely covered with *Chara* and *Myriophyllum*, the *Juncus* beds having partly flooded openings with moss and low herbs (Fjeldså 1983b). In coastal Peru *murivagans* is found in habitat much like that described for North America, being mainly *Distichlis* marsh (J. Fjeldså *in litt.*), while *salinasi* occurs in saltmarsh and inundated fields (Howell & Webb 1995a) as well as in pastures and meadows, including alfalfa fields (Reed 1941). During migration, birds are often found in wet prairie or grassland sites, and sometimes in hayfields. The species occurs in lowlands up to 1,350m, but also at 4,080m at L Junín.

FOOD AND FEEDING Most information is from Eddleman *et al.* (1994). It takes predominantly animal foods, especially in the breeding season: mainly small (under 1cm long) aquatic and terrestrial invertebrates, including Gastropoda, Amphipoda, Isopoda, spiders, ants, aphids, grasshoppers, beetles (including Carabidae, Curculionidae and Hydrophilidae), Hemiptera (including Aphidae, Reduviidae and Kinnaridae), earwigs (Dermaptera) and flies (including Dolichopodidae). Seeds of *Typha* and *Scirpus* are also taken, and birds eat more seeds in winter when less insect food is available; 3 birds in winter ate mostly *Scirpus* seeds, which comprised 47% of the volume of food consumed (Flores & Eddleman 1991). The race *salinasi* takes earthworms, very small fish *Fitzroya*, and larvae of small amphibians *Paludicola* (Reed 1941). Foraging is poorly known; the bill shape suggests generalised feeding methods such as gleaning or pecking at individual items, and thus a reliance on sight to find food; the species is probably an opportunistic forager. Birds probably forage on or near the substrate at edges of stands of emergent vegetation; in tidal marsh they forage above and below the high-tide line.

HABITS Birds are active throughout the day; in the USA, birds fitted with radio transmitters were active from c. 15 min before sunrise to 20 min after sunset, and were occasionally active for short periods (less than 7 min) at night (Flores 1991); the species is active and vocal on moonlit nights in Costa Rica. Secretive and difficult to flush, this crake runs quickly on the ground, probably using the runways of mice (e.g. *Microtus* spp.) in dense vegetation; it may perch in grass tufts in the early morning. It normally flies very little, and rises only when pursued (Reed 1941); short flights are low, and are made with the body held at an angle and the legs dangling, but over long distances the flight is fast and strong. The birds can swim for short distances.

SOCIAL ORGANISATION Probably monogamous, but occasional polygyny is possible because females can rear young unassisted; however, in a study in Arizona the sex ratio was found to be 1:1 (Flores & Eddleman 1991). Breeding birds are territorial, but the nature of the territoriality is poorly known and confusing. Calling birds either shift calling sites frequently over a short period or cease calling entirely when nesting begins. The territory size is estimated at 3-4ha in *jamaicensis* (Maryland) and is probably smaller in *coturniculus* (Eddleman *et al.* 1994). Home ranges in Arizona covered 0.1-0.8 (0.4, SD 0.2)ha and were larger in winter; they overlapped broadly only outside the breeding season (Flores 1991).

SOCIAL AND SEXUAL BEHAVIOUR Poorly known; information is from Eddleman *et al.* (1994). Calling males often charge a tape player with their wings open. The threat display at the nest, given by both sexes late in incubation, consists of the bird raising its wings, walking around the nest and attempting to charge the observer. Calling birds tend to have a clumped distribution, and this has led to speculation that the species may have a lek-type mating system (Kerlinger & Wiedner 1990), but other observations do not support this suggestion.

BREEDING AND SURVIVAL Season USA, May-Aug, peak laying for *jamaicensis* in E USA Jun, for *coturniculus* Apr-May; Belize, breeding condition Jun; Panama, Jul; Peru, Sep-Oct, end of dry season through rainy season; Chile, Nov-Dec, 2-day-old chicks 11 Nov (eggs Oct), oviduct egg 3 Oct (Reed 1941). **Nest** A bowl of dominant local grasses, rushes etc. (e.g. *Typha*, *Eleocharis*, *Spartina*, *Juncus*), with woven canopy of dead or living vegetation and ramp of dead vegetation from substrate to entrance. A nest of *salinasi* was of stems and leaves of marsh plants mixed with damp mud (Reed 1941). External diameter 11-16cm, height 8-10cm; internal diameter 6-8cm; depth of cup 3-7cm (1cm in *salinasi*). Well concealed in dense marsh vegetation (including *Typha*, *Juncus*, *Scirpus*, *Distichlis*, *Salicornia* and *Spartina*), usually in clump-forming plants at sites having both dead and newly grown plant components; usually placed low down in centre of vegetation clump over moist substrate or very shallow water (up to 3cm deep), but may be up to 46cm above substrate; sometimes in depression in ground; nest site often at higher elevation from immediate surroundings. May build nest up in response to high tides. **Eggs** 4-13 (7.6, SD 1.3, n = 49) in *jamaicensis*, 3-8 (6.0, SD 1.3, n = 86) in *coturniculus*, 3-7 (4.9, SD 1.4, n = 8) in *salinasi*, 2 in *tuerosi*; probably laid at daily intervals; subelliptical to elliptical, some tending to spherical; white to buffy-white or pinkish-white, with fine brown spots either evenly distributed or concentrated at large end; size (n = 287, *jamaicensis* and *coturniculus*) 23.7-28.1 x 17.8-20.4 (25.7 x 19.2); weight c. 5; size (n = 4, *murivagans*) 26.5-27.2 x 20.1-20.7 (26.9 x 20.3); size 29 x 19 (*salinasi*); calculated weight c. 6. **Incubation** 17-20 days, by both sexes. **Chicks** Precocial; brooded for first few days; both parents, sometimes only female, care for chicks; brood division may occur for foraging and brooding. Juvenile plumage attained by 6 weeks of age. Nest success high in Arizona, where all eggs in 5 nests hatched; of 44 eggs in 6 Florida clutches, 2 infertile (Eddleman *et al.* 1994). Rats and foxes *Vulpes vulpes* may prey on nests. Age of first breeding probably 1 year. May lay replacement and second clutches. **Survival** Adult survival apparently high in stable habitats, despite predation by herons and other avian predators during extreme high tides, a major source of mortality for populations in tidal marshes (Eddleman *et al.* 1994).

34 GALAPAGOS RAIL
Laterallus spilonotus Plate 10

Zapornia spilonota Gould, 1841. Santiago (James) Island, Galapagos.

Sometimes regarded as a race of the widespread Black Rail *L. jamaicensis*, of which it is an insular derivative, and with which it forms a superspecies. Monotypic.

Synonyms: *Porzana spilonota/galapagoensis/sharpei*; *Ortygometra/Aramides spilonota*; *Creciscus spilonotus/sharpei*.

Alternative name: Galapagos Crake.

IDENTIFICATION Length 15-16cm; wingspan 23-24cm. A small, very dark, short-winged, weak-flying rail, with dark slate-grey head, neck and breast (blacker on sides of head), dark chocolate-brown upperparts, dark greyish-brown flanks and belly, some small white spots from mantle to uppertail-coverts and wings (in breeding plumage), and on flanks and belly, and blackish-brown undertail-coverts with white bars. Eye scarlet; bill black; legs and feet brown. Sexes similar; female may have paler throat. No distinctive immature plumage. Juvenile plain blackish-grey above and sooty-brown below, with black to brown eye. Inhabits forests, thickets and grass.

Similar species Very similar in appearance to allopatric Black Rail (33), but has sparser white spotting, and white barring only on undertail-coverts. Races *murivagans* and *tuerosi* of Black Rail have cinnamon or buff undertail-coverts. Larger (18-20cm) Paint-billed Crake (112) also occurs in the Galapagos; it has no white spotting, is barred from lower flanks to undertail-coverts, and has olive-green bill with red base, and coral-red legs.

VOICE Calls are described in detail by Franklin *et al.* (1989). The vocal repertoire is extensive. The frequently uttered territorial advertising call of adults, similar in form to the advertising call of the Black Rail, is a rapid *chi-chi-chi-chirroo*, the *rroo* being a falling slur; this call is sometimes preceded by several harsh, squeaky notes. The species also has a short (1s) descending trill, sometimes followed by 1-2 *chah* calls, a long (3-9s), low, breathy, wheezy chatter and a long (several s), rapid, coarse rattle; various squeaks, squeals, ticking notes, cackles, warbles and hisses are also recorded (Franklin *et al.* 1989, Hardy *et al.* 1996). Chicks and juveniles cheep constantly, easily betraying their presence, while adults make clucking notes to the chicks. Birds occasionally call at night.

DESCRIPTION
Adult Head, neck and breast dark slate-grey; top of head, lores and ear-coverts darker, more blackish. Upperparts, including wings, uniform chocolate-brown; more blackish-brown from lower back to tail; remiges dusky brown with chocolate outer webs; underwing-coverts dark greyish-brown, narrowly barred whitish. Flanks and belly dark greyish-brown; undertail-coverts blackish-brown with white to buffy bars. Small white spots from mantle to uppertail-coverts (reduced or absent during the non-breeding season), including upperwing-coverts and remiges, and on flanks and belly. Iris scarlet to orange-red; bill black; feet and legs brownish. Sexes similar but female sometimes has paler throat.
Immature See Juvenile.
Juvenile Upperparts blackish-grey, greyer on face and throat; underparts sooty-brown; no white spots. Change from juvenile to adult-type plumage gradual, with no distinctive immature plumage; involves mantle and back becoming increasingly brown, white spotting increasing, iris colour changing from black through brown and orange to crimson, and bill gradually darkening to black.
Downy young Black; iris black; bill white with black tip, black extending from sides and below.

MEASUREMENTS (39 males, 38 females) Wing of male 61-72 (68.3), of female 61-72 (66.6); tail of male 20-26 (24.6), of female 21-27 (23.6); culmen (exposed?) of male 15-17.5 (16.1), of female 14.5-16.5 (15.5); culmen to base of 4 males 17.5-19.5 (18.1, SD 0.9), of 2 females 16.5, 17; tarsus of male 19.5-23.5 (21.5), of female 19.5-22.5 (21.3). Weight 35-45; at hatching 8-9 (Franklin *et al.* 1979).

GEOGRAPHICAL VARIATION None.

MOULT Body moult has been noted in Jul and Nov, once in Jan; most birds have fresh body plumage in Jan.

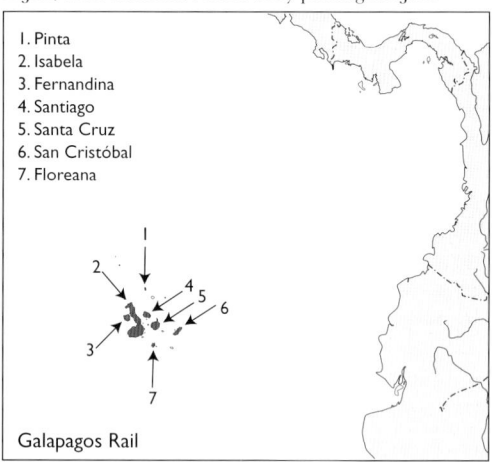

1. Pinta
2. Isabela
3. Fernandina
4. Santiago
5. Santa Cruz
6. San Cristóbal
7. Floreana

Galapagos Rail

DISTRIBUTION AND STATUS Endemic to the Galapagos Archipelago, this rail is known to occur on seven islands: Pinta, Fernandina, Isabela, Santiago, Santa Cruz, Floreana and San Cristóbal. It is regarded as NEAR-THREATENED and is now known only from islands high enough to have extensive humid-zone regions (see Habitat). In the late 1970s it was considered common, especially on Santiago and Santa Cruz, and the population on Santa Cruz appeared stable (Franklin *et al.* 1979). On Pinta no rails were found after goats destroyed the ground vegetation in 1968-1970, but after the introduction of hunting in 1971 to reduce the goat population the undergrowth regenerated and rails were again common in 1973-74 (Franklin *et al.* 1979). A census of 6 occupied islands was conducted in 1986-87 (Rosenberg 1990): the rails were common only in the upper regions of Santa Cruz and of S Isabela; they were present in small numbers on Fernandina; they were not found on Floreana, where they were last seen in 1983; and they probably numbered less than 100 on Santiago, and were very rare on San Cristóbal, both of which islands probably had a very large population before habitat destruction commenced; Pinta was not surveyed but the rails were believed to be common there. Rosenberg (1990) provides population density figures (see Social Organisation) but no overall population estimate.

Because of its weak flight this rail is probably vulnerable to introduced predators such as dogs, cats and pigs, while the Short-eared Owl *Asio flammeus* and Barn Owl *Tyto alba* are possibly natural predators. Most of the archipelago is under national park protection, including much of the bird's remaining natural habitat, and an eradication programme exists to control introduced predators. Of the 7 islands on which the rail occurs, only Fernandina (which has little suitable habitat and thus very few rails) is free of introduced herbivores, which are likely to affect rail populations. For example, the rarity of birds on San Cristóbal, and on Floreana (where the rail may already be extinct), and the restricted area of occurrence on Santiago, are probably due to extensive grazing by introduced herbivores, chiefly goats, cattle and horses (Rosenberg 1990).

Several populations are threatened by loss of habitat, and if the bird's numbers on Santiago and San Cristóbal remain small, early extinction is more likely as a result of natural disturbances, inbreeding, and population changes of predators and herbivores (Rosenberg 1990). Management recommendations (Rosenberg 1990) include the acquisition of plateau areas on San Cristóbal and the exclusion of herbivores from these areas, the restriction of domestic animals on Floreana, and the complete eradication of goats from Pinta. The elimination of introduced herbivores (goats) and predators (pigs) on Santiago, which is the only uninhabited island with a large highland region, would allow the restoration of natural plant communities and might allow this important island to become a reserve for the rail (Rosenberg 1990). The reasons for the rail's recent abandonment of mangrove habitats are not known, but are unlikely to be simply a result of introduced predators or competitors because rails have also disappeared from mangroves on Fernandina, where no such introductions have occurred.

MOVEMENTS None.

HABITAT Grass and forest of mesic regions in the highlands, where it occurs in deep thickets and dense ground cover. Specific habitats recorded are: meadows, *Psidium galapagensis* forest and *Psychotria* thickets on Santiago; *Scalesia* forest, shrubby *Miconia robinsoniana* belt, patches of bracken (*Pteridium aquilinum*), patches of introduced *Psidium guajava* shrubs, open sedge-fern meadows and moist farming regions (including *Pennisetum purpureum* pastures) on Santa Cruz; grass-fern vegetation (including a wet area) on Isabela; and highland fern belt on Pinta (Franklin *et al.* 1979, Rosenberg 1990). Historically it is also recorded from coastal mangroves, from which it is no longer known. It appears tolerant of man-modified habitats, occurring in agricultural land, but does not venture into overgrazed short-grass pasture. Rosenberg (1990) found that rail density increased with elevation on Isabela and Santa Cruz, probably because of the greater moisture in the higher zones, while on Santa Cruz the birds did not occur in vegetation which had <25% cover of herbs >30cm tall and >25% cover of herbs <30cm tall. They appeared to favour areas which contained freshwater pools, and to avoid areas with short herbaceous ground cover; vegetation rendered patchy by grazing may also be avoided because of the increased risk of predation. Precise altitudinal limits are not given, but the rail occurs on all islands higher than 500m and may ascend to 1,700m on Isabela. During census work (Rosenberg 1990) rails were found at 125-695m on Santa Cruz, at 590-1,060m on Isabela, at 800m on Fernandina (where they apparently do not occur at 1,330m, above the cloud belt), at 550-650m on Santiago, and only at 200m on San Cristóbal, where potentially suitable land at 550-600m is heavily grazed.

FOOD AND FEEDING Well described, particularly by Franklin *et al.* (1979). The species eats primarily invertebrates: Gastropoda, Isopoda, Amphipoda, dragonflies, moths, Hemiptera, ants, caterpillars and spiders. Seeds of a solanaceous berry are recorded, and also seeds of the shrub *Miconia robinsoniana* and the grass *Paspalum conjugatum*. Birds feed throughout the day, normally foraging on the ground, walking with the tail depressed, darting the head from side to side, jabbing the bill rapidly into litter, and picking up and tossing or moving aside leaves and twigs. Prey items are taken from stems and overhanging leaves, the birds frequently probe into moss and other epiphytes on *Miconia* trunks and they sometimes climb 0.5m up into vegetation to forage. They also forage around pools, streams and patches of *Sphagnum* bog, wading breast-deep to take prey from the surface of the water. The chicks are fed on insects and spiders.

HABITS Primarily diurnal. The birds are sometimes difficult to see in dense ground cover but their presence is often betrayed by the frequently repeated territorial call. They are exceptionally confident and may forage within a few centimetres of quiet observers; they are also very curious and will circle observers to look closely at them. They normally keep close to cover and when startled they run, rather than fly, moving in a zig-zag manner with body and head low. Like most rails the birds fly with dangling legs; the flight is short and laboured and the birds land clumsily. One bird was seen to hover for a few seconds in front of a mist-net before flying off. Much use is made of small beaten-down runways to move around in dense grass and bracken. This species climbs well and moves adeptly along branches; it has been seen hanging upside down to feed and one walked easily up the wire mesh sides of a cage. It swims well and bathes.

SOCIAL ORGANISATION Apparently monogamous and territorial. Rosenberg (1990) found no seasonal variation in the frequency of territorial calls made in response to taped playback. During a field study on Santa Cruz (Franklin *et al.* 1979) no more than two adults were ever seen together, home ranges appeared not to overlap, and the birds responded to taped playback of calls; one home range measured 132 x 50m. Population densities of 5.45 (SE 0.45) to 8.5 (SE 1.65) rails/ha were recorded by Rosenberg (1990) for occupied census plots, with a mean of 2.1 (SE 0.19) birds/ha over all census plots at higher elevations in the Galapagos National Park (Santa Cruz) and 2.45 (SE 0.5) birds/ha in the Sierra Negra on Isabela.

SOCIAL AND SEXUAL BEHAVIOUR Described in detail by Franklin *et al.* (1979). Two intensities (mild and vigorous) of a foot-stamping display are recorded, both directed at observers and apparently aggressive. Birds often also display an erect tail, apparently an alarm response. Both members of pair possibly defend the territory. Fights with conspecifics (presumably intruders on the territory) involve ramming with the bill and striking with the feet; fights were observed after taped playback of advertising, chatter and rattle calls, and birds gave the last two types of call after fights. Pair members indulge in bill-fencing, allopreening and courtship feeding. Adults defend the nest by pecking at intruders and by performing a spread-wing threat display in which the wings are extended laterally and are tilted forwards so that their upper surface faces forwards (Fig. 4). Introduced black rats *Rattus rattus* appear not to be significant nest predators, probably as a result of this aggressive behaviour by the incubating adult. (The Buff-spotted Flufftail [2] has a very similar spread-wing threat display.) In defence of chicks, both parents perform foot-stamping and spread-wing displays (the latter with loud hissing calls) to human intruders.

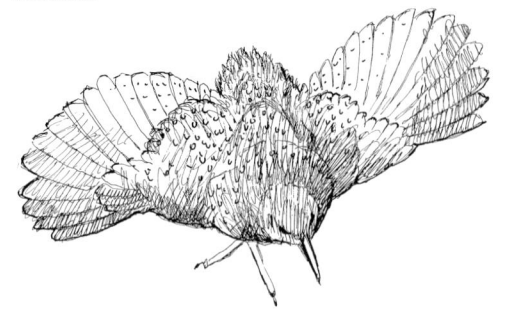

Figure 4: Galapagos Rail threat display. [After Franklin *et al.* 1979].

BREEDING AND SURVIVAL Season Breeds Sep-Apr, middle of cool season to end of hot season. **Nest** A deep, semi-domed cup with side entrance; made of herbaceous stems; placed on ground and covered by dense, low vegetation. **Eggs** 3-6; white to beige with fine red-brown to grey speckling; size of 8 eggs 25-30.2 x 19-22.2 (27.3 [SD 2.0] x 20.4 [SD 1.2]). **Incubation** At least 23-25 days, by both parents; changeovers at one nest occurred about every 50 min. **Chicks** Precocial; leave nest soon after hatching; fed and cared for by both parents. Juveniles feed themselves but stay with parents until almost fully grown; adult-type plumage fully attained by 80-85 days. Age of first breeding 1 year. 1-2 broods per season.

35   RED-AND-WHITE CRAKE
Laterallus leucopyrrhus   Plate 11

Rallus leucopyrrhus Vieillot, 1819, Paraguay.

Monotypic.

Synonyms: *Corethrura/Porzana/Aramides leucopyrrha*; *Corethrura/Laterirallus hypoleucos*; *Creciscus leucopyrrhus*.

Alternative name: White-breasted Crake.

IDENTIFICATION Length 14-17cm. A striking and distinctively patterned small crake. Pure white from chin to centre of breast, this colour being sharply demarcated from adjacent bright chestnut areas of head, neck and breast; rest of upperparts dark olive-brown, darker from rump to tail; sides and flanks boldly barred blackish and white; undertail-covert pattern unique in genus, being black in centre and white laterally. Iris and eyelids red; bill black with yellow base and lower mandible; legs and feet bright coral-red to salmon-pink. Some birds have diffuse rufous barring on upperwing-coverts; a few have more prominent white barring, sometimes bordered with black. Sexes alike. Immature plumage very similar to adult; juvenile has dark brown upperparts, grey to whitish underparts, blackish bill and black legs. Inhabits moist to wet grassland in marshy areas.

Similar species Distinctive plumage characters separate this species from its 3 sympatric congeners, all of which also have barred flanks. Rufous-sided Crake (28) has no rufous from top of head to mantle, lacks sharp demarcation between rufous and white at sides of neck and breast, and has cinnamon-rufous undertail-coverts, dark bill (green at sides) and olive-brown legs and feet. Rufous-faced Crake (36) is buffy-ochraceous on foreneck and breast, and has strong black and white barring on wings, black tail, black undertail-coverts with white bars on outers, relatively heavy blue-grey bill (sometimes blackish with pale lower mandible), and blue-grey to flesh-brown legs and feet; Grey-breasted Crake (32) has grey head, neck and breast, barred undertail-coverts and yellowish-brown legs and feet. Allopatric White-throated Crake (31) has different head pattern, much less extensive white on underparts, entirely barred undertail-coverts, predominantly greenish bill, and brown legs and feet.

VOICE A prolonged throaty chatter; also described as a resonant trill, descending at the end. This call is very similar to the trill of the Rufous-sided Crake, but is more strident, slightly higher-pitched and possibly slightly shorter (J. Mazar Barnett *in litt.*). The contact call is a low squeak or whistle. The male utters a creaking call during courtship display; in another display, both sexes make growling sounds (see Social and Sexual Behaviour).

DESCRIPTION
Adult Head to mantle, side of neck and side of upper breast bright chestnut, slightly darker and duller from crown to hindneck; becoming dark olive-brown, tinged chestnut, on back and upperwing-coverts, including wings; rump to tail, and tertials, darker and duller, more sepia; primaries pale olive-brown. Some individuals have diffuse rufous barring on upperwing-coverts; a few have more prominent white barring, sometimes bordered with black. Underwing-coverts and axillaries white, feathers brownish-black basally. Chin, throat, centre breast and belly immaculate white, sharply demarcated from chestnut sides of neck and breast; sides and flanks strongly barred blackish-brown and white; central undertail-coverts black, lateral coverts white. Iris and eyelids red; bill black with yellow to greenish-yellow base and lower mandible, sometimes predominantly yellowish or greenish, with culmen and tip black to dark blue-grey; legs and feet bright coral-red or salmon-pink. Sexes alike; male may be warmer in plumage tone. Individual variation in plumage characters and bare parts colours is sufficient to allow individual recognition (Mauricio & Dias 1996).
Immature Almost indistinguishable from adult.
Juvenile Upperparts dark brown; underparts grey or whitish; legs blackish; bill black, becoming yellowish-green.
Downy young Black downy chick has black bare parts, with a white tip to the upper mandible; egg-tooth present.

MEASUREMENTS (7 males, 6 females, 18 unsexed) Wing of male 78-84 (81.2), of female 81-85 (83.3), of unsexed 80-86 (82.3); exposed culmen of male 15-17 (16.1), of female 15-19 (16.1), culmen from base of 3 unsexed 19, 19, 22; tarsus of male 30-32 (30.8), of female 30-32 (31.3), of unsexed 30-35 (32.8). Tail of 8 males 41-53 (45.7), of 3 females 48, 49, 55 (Navas 1991). Weight of 1 female 46.5; of 10 unsexed 34-52 (45.4); hatching weight of 1 chick 7.8.

GEOGRAPHICAL VARIATION None.

MOULT Postjuvenile moult begins at 5-7 weeks and is apparently complete at 10-11 weeks of age.

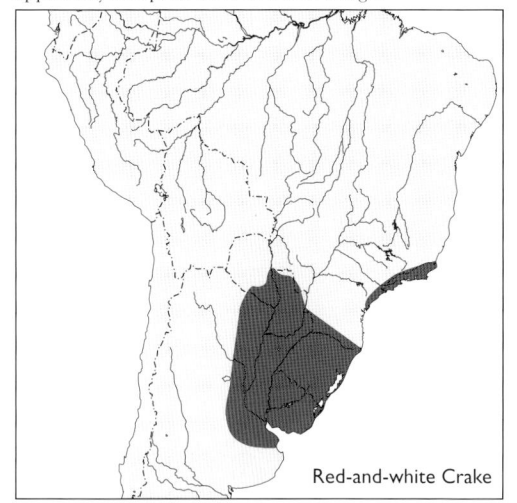

DISTRIBUTION AND STATUS SE Brazil (Rio de Janeiro, E São Paulo, Santa Catarina and Rio Grande do Sul), Paraguay, Uruguay (all departments except Artigas, Rivera, Durazno, Lavalleja and Flores) to NE Argentina (Misiones, Corrientes, E Formosa, E Chaco, E Santa Fe, Entre Ríos and Buenos Aires). A record from Tucumán, N Argentina, is erroneous (Blake 1977). Not globally threatened. It was formerly regarded as locally common in Argentina (Ripley 1977) but scarce to rare in Rio Grande do Sul, Brazil (Belton 1984), and very infrequently recorded in Uruguay (Arballo 1990). Little recent information is available on its status. In Rio Grande do Sul in 1996 Mauricio & Dias (1996) found it locally common to abundant at Pontal da Barra marsh (see Social Organisation), while it also occurred in marshes near Pelotas and at Rio Grande and at Capão Seco. In Paraguay it is recorded only twice from the W (R Pilcomayo and R Confuso) but many times from E of the Paraguay R (Hayes 1995). In Uruguay from 1985 to 1897 it was recorded for the first time in Montevideo and San José Dpts (Arballo 1990); it is now known from 14 of the 19 departments and is considered fairly common (Arballo & Cravino in press). It is said to be abundant in the Punta Lara Reserve, Buenos Aires (Klimaitis & Moschione 1987); however throughout Buenos Aires Province it is endangered as a result of strong pressure from commercial hunting (Narosky & Di Giácomo 1993). It is kept in captivity relatively frequently and, at least in the 1970s, was more available through the bird trade than many other small rails, suggesting that it was fairly common.

MOVEMENTS None recorded.

HABITAT In Paraguay this species occurs in open marshy areas in lowlands, inhabiting dense tussocky or matted grasses (sometimes heavily grazed) 30-50cm tall on moist ground or in water 2-4cm deep, and also coarse grass, 1.5 to over 2m tall, with scattered shrubs or tree-ferns on moist or flooded ground (Storer 1981). In N Argentina it occurs in densely vegetated marshes with rushes and *Scirpus* (Contreras et al. 1990), in dense scrub and thickets adjacent to rivers, and also at forest streams and marshes (Canevari et al. 1991). In Uruguay it inhabits dense emergent vegetation such as *Zizaniopsis bonaeriensis*, fields with submerged sawgrass *Eryngium pandanifolium* and other shrubby wet areas (Arballo & Cravino in press). In Rio Grande do Sul, Brazil, it is found in marshes dominated by *Scirpus giganteus*, and in small residual marshes close to built-up areas (Mauricio & Dias 1996).

FOOD AND FEEDING Takes insects, other invertebrates, and some seeds (Arballo & Cravino in press). The birds forage on floating vegetation, in shallow water, by probing under stones and into mud, and also by searching the ground and occasionally prodding dead leaves or twigs. Food of captive birds included earthworms, insects such as mealworms (*Tenebrio* larvae) and ant pupae, small freshwater crustaceans (including *Daphnia*), tadpoles, millet seed and "germinating seeds". Food items are dipped in water before being swallowed, and captive birds also ate scrambled egg, rice pudding and powdered dog food, and drank "milk nectar" (Gibson 1979).

HABITS This species is difficult to observe and uses runways on the ground beneath dense matted vegetation; in Paraguay birds were trapped in runways made by small mammals and by water runoff (Storer 1981). Captive birds remain close to cover and climb freely in bushes, preferring those with dense foliage; they can walk along narrow branches, using the wings to balance. The birds walk with long strides, the legs raised high, the head bobbed and the body twisted from side to side with each step (Rutgers & Norris 1970); the tail is flicked when the birds are nervous. Captive birds are not as shy and retiring as many other crakes, but are very reluctant to fly and very difficult to flush a second time (Levi 1966); birds in the wild are even more confiding than are Rufous-sided Crakes, and will walk around at the feet of a quiet observer (Arballo & Cravino in press). In captivity the birds bathe freely, often several times a day and commonly in the late evening; they wade to belly depth before immersing the body, ducking the head and flapping the wings and tail; after bathing they may retire to a high perch in foliage before preening. The wings may be held partly opened to dry the feathers. In aviaries birds roost in a nest all the year round, sometimes building a semi-domed roosting nest of grass and a few leaves.

SOCIAL ORGANISATION Monogamous, with a strong pair bond. In Brazil, Mauricio & Dias (1996) found at least 24 pairs in a 16-ha *Scirpus giganteus* marsh, and between Feb and Apr 1995 5 territories c. 20 x 25m (c. 0.05 ha) were delimited, each occupied by a pair and at least 1 other bird (probably juveniles); at the same site, only 5 Rufous-sided Crake territories were found. [These are some of the smallest territory sizes recorded for rails, considerably smaller even than those of Red-chested Flufftails (3) in South Africa.] In captivity, first-brood young, aged c. 8 weeks, helped the parents to feed a second-brood young; in other pairs 1-2 other adults helped feed, guard and brood the young.

SOCIAL AND SEXUAL BEHAVIOUR All information comes from captive birds. Sunbathing is reported to be a social activity and includes much allopreening (Levi 1966), while birds roost in groups in nests. In the courtship display (Kleefisch 1984) the male trots to and fro "in a peculiar fashion" on a determined pathway in the vicinity of the nest, uttering a gentle creaking sound. At the same time the female performs a very rapid take-off and upward flight. Another display, the function of which is not clear, involves both birds standing upright facing each other, fluttering strongly with their short rounded wings and uttering a growling sound. During the breeding season the male feeds the female with animal food.

BREEDING AND SURVIVAL Season Uruguay, Nov; Argentina, Oct-Feb. Information on captive birds is principally from Essenberg (1984), Everitt (1962), Gibson (1979) and Kleefisch (1984). **Nest** Spherical with side entrance; made of grasses and placed 0.9-2.5m up in bushes and other vegetation; 1 nest of dried *Typha* leaves, built at the edge of a field of swordgrass *Cephalanthus glabratus* with a muddy substrate; external diameter 20cm, internal diameter 13cm, circular entrance 7cm (Arballo & Cravino 1987, in press); 1 aviary nest was large pile of wet leaves (dry leaves were left in water until soft and pliable), 18cm deep and 23cm across, placed 2m up in dense honeysuckle (*Lonicera*); another was bowl-shaped and placed 1.5m up in reed clump. Captive birds also use natural holes in tree trunks, half-open wooden nest boxes and baskets, which they line with dry leaves, grass or even strips of paper. Both sexes carry material to the nest, including dried *Typha* leaves up to 1m long (Arballo & Cravino in press). **Eggs** 2-4; sharply tapering; dull, pure white and unmarked; size 29-36 x 23-27 (mean size of 25 eggs 33 x 24.8); weight c. 8.5 (Gibson 1979); calculated weight 11 (Schönwetter 1961-62). **Incubation** 23-24 days, by male and female alternately during day and both parents together at night; female incubates more than male; birds desert very easily after disturbance, even when eggs are about to hatch. **Chicks** Precocial; leave nest as soon as down is dry; leap out of nest and fall to ground. Chicks can feed themselves immediately, but usually do so only after a few days; adults feed and care for them intensively during growing period; chick takes food from adult's bill, or adult shows chick food on ground. Captive chicks took ant pupae from surface of shallow water in dish. Family roosts in nest at night, unless nest difficult of access, when birds roost on ground; 1 chick roosted on the back of an adult until at least 1 week old; chicks preened by adults. Whitish feathers visible on breast and belly, and brown feathers on flanks and back, at c. 10 days; bill begins to assume adult colour from 3 weeks; young fully grown and feathered at 6 weeks, when eye and legs show red tinge; young independent at c. 4 weeks, when second clutch may be laid.

36   RUFOUS-FACED CRAKE
Laterallus xenopterus   Plate 11

Laterallus xenopterus Conover, 1934, Horqueta, Paraguay.

Monotypic.

Alternative name: Horqueta Crake.

IDENTIFICATION Length 14cm. A very distinctive small crake, with dark rufous upperparts grading to black on tail, prominent black and white barring on scapulars, upperwing-coverts and tertials, buffy-ochraceous neck and breast, black and white barring on flanks, and black undertail-coverts with white barring at sides. Tail relatively long, bill short and relatively deep, and tarsus relatively short. Iris red; bill bluish-grey, or blackish with pale lower mandible and turquoise base; legs and feet bluish-grey to flesh-brown. Sexes alike. Immature and juvenile undescribed. Inhabits wet grassy areas in marshes.

Similar species Distinguished from other *Laterallus* species by buffy-ochraceous foreneck to breast, although some Rufous-sided Crakes (28) are tinged buffy ochraceous to rufous from chin to upper breast, usually brighter than in Rufous-faced Crake; from Rufous-sided Crake and Red-and-white Crake (35), its two sympatric congeners with barred underparts, by its boldly barred tertials, upperwing-coverts and scapulars, black tail and partly black undertail-coverts, relatively heavy blue-grey or blue-based bill with markedly curved culmen (Brace *et al.* in press), and often blue-grey legs and feet; some individuals of these other two species have diffuse or narrow rufous or white barring on the upperwing-coverts and scapulars, but never as extensive or prominent as in Rufous-faced Crake. Iris crimson-red, whereas that of Rufous-sided Crake is reddish-brown (Brace *et al.* in press).

VOICE It gives a trill typical of other members of the genus, preceded by several hollow, whistled *tiu* notes; the trill is very similar to that of Rufous-sided Crake but somewhat longer (J. Mazar Barnett *in litt.*).

DESCRIPTION
Adult Top of head, side of face to below ear-coverts, nape, sides of neck and mantle dark rufous, duller and more brown-tinged on mantle and more orange-rufous on sides of neck; lores blackish to brownish-grey. Back and rump sepia; uppertail-coverts and tail black, with brownish sheen; rectrices narrowly edged buffy. Upperwing-coverts, scapulars and tertials black, very prominently barred white; marginal coverts white; alula, primaries and secondaries dark brown, alula feathers with narrow white tips and 1-2 white bars or spots near tips; outermost primary with 1-3 small white subterminal spots or bars on outer web and sometimes 1 on inner web. Underwing-coverts white, slightly mottled dusky; axillaries brown, faintly tipped white. Chin and throat cream to white; foreneck to upper breast buffy-ochraceous, sometimes washed pale bright orange-rufous; lower breast, belly and vent white; undertail-coverts black, some with a trace of whitish pattern, especially on the outers which may be barred black and white (Lowen *et al.* 1996b). Sides of lower breast and flanks white, broadly barred black. Iris bright red; bill bluish-grey with pale horn tip to lower mandible, or upper mandible blackish, lower mandible blackish to pale grey, proximal half of lower mandible and base of upper mandible cobalt or turquoise, cutting edges whitish; legs and feet bluish-grey, pinkish-grey or flesh-brown. Sexes alike.
Immature Not described.
Juvenile Not described.
Downy young Unknown.

MEASUREMENTS (5 unsexed) Wing (chord) 83-91 (86.6), (max) of 1 unsexed 89; tail 42-55 (50.7); tarsus 28-32.5 (30.3); culmen of 1 female 16, of 1 unsexed 17.5. Weight of 5 unsexed 51-68.5 (56.9, SD 7.4).

GEOGRAPHICAL VARIATION None recorded.

MOULT No information available.

DISTRIBUTION AND STATUS This poorly known South American species is regarded as VULNERABLE and is reliably recorded from only SC Brazil (Federal District), C and S Paraguay (Concepción, Canindeyú and Caazapá)

Rufous-faced Crake

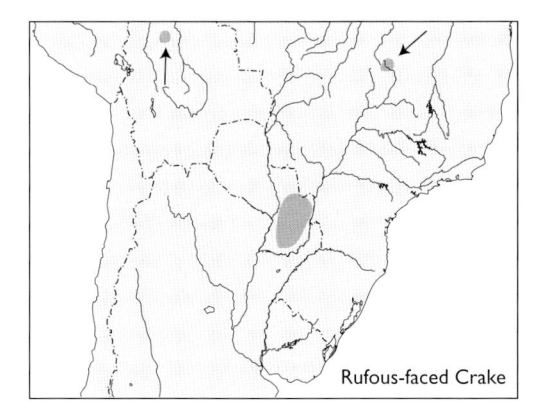

and lowland Bolivia (Beni). In Paraguay the type was collected at Horqueta (Concepción) in Nov 1933, 4 birds were collected in the Curuguaty area (Canindeyú) in 1976-1979 (Collar *et al.* 1992, Hayes 1995), and 1 was mist-netted in the Aguará Ñu area of the Reserva Natural del Bosque Mbaracayú on 17 Sep 1995 (Lowen *et al* 1996b). The species was also found in Oct 1997 near San Juan de Nepumoceno (Caazapá), and was heard N of Curuguaty (probably close to Estancia La Fortuna, the site of the 1970s records) on 30 Nov 1997 (R. P. Clay *in litt.*). Over 1,200km to the NE there were several sight records, and one specimen, from the Brasília area during the period 1978-1989 (Collar *et al.* 1992, Wege & Long 1995). Finally, a single bird was trapped at the Beni Biological Station Reserve in lowland Bolivia in Aug 1976 (Anon 1997b) and there were 3 sightings in Aug 1997 at a site c 2.5km from the 1996 trapping area (Brace *et al.* in press; R. Brace *in litt.*). All known sites of occurrence except San Juan de Nepumoceno are listed as key areas by Wege & Long (1995). The species is doubtless more widespread than existing records suggest, and presumably occurs in wetlands between these three areas. In Paraguay, it has been suggested that it may be found in the wetlands of Concepción, while there are unsubstantiated records (lost specimens) of its occurrence at Lima (Dpto San Pedro) and Pedro Juan Caballero (Dpto Amambay) which should be treated with caution (Collar *et al.* 1992, Lowen *et al.* 1996b). In Bolivia it may be expected to occur widely in the abundance of seasonally wet savanna (the Llanos de Mojos) in Beni, while it may also occur in Santa Cruz (Brace *et al.* in press). The bird's total numbers cannot be assessed; it is probably commoner than is known, but areas where it might occur are poorly studied and the species is difficult to locate. It is judged to be relatively frequent around Brasília (Collar *et al.* 1994). There are no known threats to its survival in Paraguay, but in Brazil the wholesale loss of wet campo habitats from drainage, and from the drying effects of adjacent eucalyptus plantations, may have serious effects on hitherto undiscovered populations; the impact of fire is also a potential threat (Collar *et al.* 1992). In Beni, Bolivia, human population density is low and there is minimal agroforestry development, but extensive annual dry-season burning of savanna is undertaken in many estancias to promote new growth for grazing; this, plus the effect of grazing and trampling by cattle, are adversely affecting the savanna environment (Brace *et al.* in press). Work is urgently needed to relocate birds in Paraguay, to record the bird's voice and to use taped playback to ascertain the range and status of the species.

MOVEMENTS None recorded.

HABITAT Inhabits coarse, tussocky or matted grass on moist to shallowly flooded substrates in marshes. It is reported from marshes with perennial bunch-grass 30-50cm or more tall, completely covering the ground and growing on moist soil or in 3-4cm of water; such marshes are used extensively by small mammals and by Ash-throated Crake (98) and Red-and-white Crake; it is also recorded from coarse grass 1.5-2m tall in 2-3cm of water at the edge of a marsh (Myers & Hanson 1980). Such marshes (cañadones or wet campos) commonly form in low areas of E Paraguay and adjacent parts of Brazil. They range from a few to several hundred metres in width and may form a band on gently sloping valley sides between upland cerrado and riparian forest. The habitat in the Reserva Natural del Bosque Mbaracayú was of this type, being a marshy valley with semi-saturated lower slopes and valley bottom, within undulating cerrado and flanked by gallery forest; there were occasional *Eryngium* plants in the stream bed, scattered low bushes, and grass up to 1.5m tall (Lowen *et al.* 1996b); the Storer (1981) crake was in similar marshy grassland, with *Xuris*, adjacent to forest. In such areas they have been found occasionally on the valley slopes in wet grass c. 30cm tall, possibly when forced to move uphill by flooding at lower levels (J. Mazar Barnett *in litt.*). The Aug 1996 Bolivian bird was trapped in a belt of *Cyperus giganteus*, up to 20m wide and m tall, and partially flooded to a depth of 5cm, which fringed the Laguna Normandia, was interspersed with other sedges and some scrub, and gave way to fairly open, seasonally (Oct-Apr) inundated savanna dominated by grasses such as *Sorghastrum setosum* (in wet areas) and *Andropogon bicornis* (in drier areas); the *Cyperus* belt had narrow 'runs' which were used by Rufous-sided Crakes (Anon 1997b, Brace *et al.* in press). In Aug 1997, only Rufous-sided Crakes were found at this site, which was flooded to a depth of 15-20cm, but 3 sightings of Rufous-faced Crakes were obtained in an area of savanna flooded to a depth of 5cm (surrounding areas were dry) and characterised by tussocky grass c. 1m tall with *Rynchospora globosa*, *Cyperus haspan* and *Tibouchina octopetalia*; Rufous-sided Crakes were not seen here but Ash-throated Crakes and a Speckled Rail (15) were observed (Brace *et al.* in press). This species probably has a more restricted habitat tolerance than either the Red-and-white Crake or the Rufous-sided Crake, and may not tolerate habitat flooded to a depth of more than a few cm. In C Brazil it occurs occasionally in wet palm groves (Negret & Teixeira 1984b).

FOOD AND FEEDING Nothing recorded. Specimens were captured in traps baited with peanut butter, rolled oats, cracked maize and banana (Myers & Hanson 1980).

HABITS The paucity of records indicates that this bird is extremely secretive. It uses runways in grass made by small mammals or water runoff (Myers & Hanson 1980).

SOCIAL ORGANISATION No information available.

SOCIAL AND SEXUAL BEHAVIOUR No information available.

BREEDING AND SURVIVAL No information available.

NESOCLOPEUS

Two medium-large and very dark forest rails, probably flightless, possibly conspecific, and endemic to the Fiji and Solomon Islands in the SW Pacific. One, Woodford's Rail, is close to extinction, the other, the Bar-winged Rail, is almost certainly extinct.

37 WOODFORD'S RAIL
Nesoclopeus woodfordi Plate 11

Rallina Woodfordi Ogilvie-Grant, 1889, Aola, Guadalcanal, Solomon Islands.

Sometimes retained in *Rallina* or even placed in *Rallus*, but *Nesoclopeus* differs in structural characters and may provide a link between *Rallina* and *Gallirallus*. Forms a superspecies with *N. poecilopterus*, with which it may be conspecific. Three subspecies recognised.

Synonyms: *Nesoclopeus/Rallus poecilopterus woodfordi/immaculatus/tertius*; *Rallina/Ardeiralla/Astur/Eulabeornis woodfordi*.

Alternative name: Solomons Rail.

IDENTIFICATION Length c. 30cm. A medium-large, very dark rail with a longish, powerful bill; probably flightless. Head, neck and upperparts (including upperwings) rich blackish-brown; supercilium, chin and throat greyer; a few obscure white spots on upperwing-coverts (none in race *immaculatus*); primaries narrowly barred or spotted white, few or no white markings on secondaries; underwing-coverts spotted white (no white on wings in *immaculatus*). Underparts sooty grey-black, with variable obscure greyish barring on sides of breast and flanks; in race *tertius*, belly and flanks mottled buff, and undertail-coverts vaguely barred white; underparts unmarked in *immaculatus*. Iris red or orange-red; bill black with pale horn tip or entirely ivory/pale cream; legs and feet grey (greenish-yellow in *immaculatus*). Sexes alike. Immature similar to adult but has more white on chin, and browner underparts with pale fringes to breast feathers. Juvenile undescribed. Inhabits lowland forest, grassy swamps and gardens, often near water.

Similar species The closely related Bar-winged Rail (38) of Fiji is paler and more patterned, having brown upperparts, medium to dark slate-grey underparts with faint pale barring, a more marked facial pattern, chestnut bars on remiges, white bars on all underwing-coverts, brown eye, yellow and orange bill and yellow to orange legs.

VOICE Vocal. Alarm call an explosive squealed *ngowh*, more nasal and lower pitched than Buff-banded Rail. Also a series of *kik-kik...* notes (probably a duet) fluctuating in tone and volume – similar to Roviana Rail but less squealing, and to New Britain Rail but less varied (G. Dutson *in litt.*). An unmusical series of metallic shrieks of unvarying pitch, generally delivered in threes (Webb 1992).

DESCRIPTION *N. w. woodfordi*
Known only from the type, which is described as an immature. Head, nape, hindneck and sides of neck rich blackish-brown, paler and greyer on supercilium, chin and throat. Rest of upperparts from mantle to tail, including scapulars, tertials and upperwing-coverts, and axillaries and underwing-coverts, rich blackish-brown; a few obscure white spots on greater primary upperwing-coverts; primaries and secondaries blackish-brown, primaries with 3-4 rows of white spots or narrow bars, spots reduced or absent on secondaries; alula, and all underwing-coverts except lessers and medians on inner wing, spotted white on inner webs; axillaries unspotted. Foreneck and rest of underparts very dark, sooty grey-black, with varying obscure greyish barring (pale feather tips) on sides of breast and flanks. Iris red; bill black with pale horn tip; legs and feet dark grey.
Juvenile Not described.
Downy young Not described.

MEASUREMENTS Wing 173; tail 71; culmen to base 40.5; tarsus 60 (Mayr 1949).

GEOGRAPHICAL VARIATION Races separated on size, extent of white on wings and pattern of underparts.

N. w. tertius Mayr, 1949 – Bougainville I, Papua New Guinea. The smallest race; primaries barred white; upperwing-coverts with a few obscure white spots; belly and flanks mottled buff; undertail-coverts vaguely barred white; iris dark red; bill yellowish-horn at least at base (sometimes completely); legs and feet grey or pale slate-blue. Wing of 5 males 143-157 (151), of 1 female 151; tail of 4 males 68-71 (69.5), of 1 female 70; bill of 5 males 36-38 (37.2); tarsus of 5 males 54-57 (56.3), of 1 female 54 (Mayr 1949). One immature examined by Diamond (1991) differed from adults in having more white on chin and browner underparts with pale fringes to breast feathers.

N. w. immaculatus Mayr, 1949 – Santa Isabel I, Solomon Is. The largest race, and blacker overall; upperparts sooty-black with trace of brown only on upperwing (especially in male); underparts black with no pale mottling; no trace of white on remiges, axillaries or wing-coverts. Iris orange-red; bill pale horn at tip grading to dark grey at base or entirely pale; legs and feet dark greenish-yellow (Webb 1992). One male: wing 179; tail 73; bill 41.5; tarsus 65. One female: wing 170; tail 78; bill 37.5; tarsus 63 (Mayr 1949).

N. w. woodfordi (Ogilvie-Grant, 1889) – Guadalcanal I, Solomon Is. See Description.

MOULT Primary moult is sequential and is recorded from Santa Isabel, Aug.

DISTRIBUTION AND STATUS Bougainville I and the Solomon Is (Santa Isabel and Guadalcanal), E of New Guinea. In the Solomon Is it possibly occurs on Choiseul (Collar *et al.* 1994) and has been reported from Malaita, but its occurrence there is considered unlikely (G. Dutson *in litt.*). There are also records from New Georgia (Sibley 1951, Blaber 1990) but these are most likely to be the recently described Roviana Rail.

This species is globally ENDANGERED and possibly close to extinction. The nominate race from Guadalcanal is known only from the type specimen, an immature collected at Aola in or before 1889, and may be extinct, having fallen victim to introduced predators: cats have wiped out most terrestrial mammals on Guadalcanal (Collar *et al.* 1994). However, a single bird seen in Mar 1998 was thought to be an adult of the nominate race; its plumage was dark cold brown with a plain mid-grey head and heavy dark grey bill. It also differed from *immaculatus* in its calls, and may best be treated as a separate species

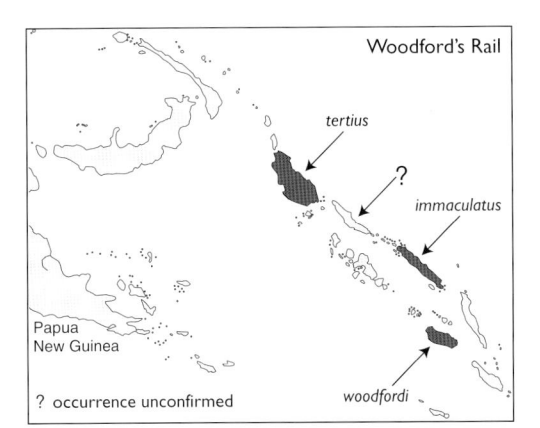

(G. Dutson *in litt.*). Although last collected in 1936, there have been a number of recent observations of *immaculatus* on Santa Isabel (e.g. Webb 1992) where it is considered to be locally common (G. Dutson *In litt.*). This bird occasionally forms part of the local human diet when caught by dogs or in traps set for the Pacific Black Duck *Anas superciliosa*. There were no records of *tertius* from Bougainville between 1936 and 1985, when one bird was seen in forest which has since been logged (Collar *et al.* 1994).

MOVEMENTS None.

HABITAT Lowland forest, secondary growth and swamp forest, occasionally ascending to 1,000m. On Santa Isabel it has been seen in forested river valleys and it also frequents abandoned gardens there (Webb 1992). Also found in grassy swamps, wet grassy coconut plantations and gardens in wet secondary forest. Reportedly absent from closed forest on Santa Isabel and scarce in gardens in hills (G. Dutson *in litt.*).

FOOD AND FEEDING Worms, snails, insects, frogs, lizards and small snakes are recorded; also young shoots of taro plants. The birds occasionally dig in gardens (Webb 1992).

HABITS A timid species which skulks along, stiffly jerking the head back and forth as it walks (Webb 1992). When startled, it stands erect with the tail flicked down from the horizontal position, and then runs rapidly into cover with its head lowered and neck outstretched, sometimes ineffectively flapping its short wings over its back; it is probably flightless (Kaestner 1987, Webb 1992).

SOCIAL ORGANISATION Usually found singly or in pairs.

SOCIAL AND SEXUAL BEHAVIOUR No information available.

BREEDING AND SURVIVAL Season Bougainville, Aug. **Nest** One purported to be of this species, from Santa Isabel in Jul 1988, situated under small bush in lowland forest near a stream; was made of rootlets, leaves and other vegetable matter, measured c. 25cm across and had shallow depression c. 15cm across. **Eggs** In this nest were 6 eggs 30-40mm long and almost identical to known eggs of this species; they were oval, and off-white with dark brown spots, more numerous at large end (Webb 1992). Two eggs collected at Buin, SE Bougainville, are creamy-buff with faint pink tint and large blotches of red-brown and faint purple and lilac, mostly towards large end; size 46.5 x 33.3 and 47.2 x 33.9 (Harrison & Parker 1967).

38 BAR-WINGED RAIL
Nesoclopeus poecilopterus   Plate 11

Rallina poeciloptera Hartlaub, 1866, Viti Levu, Fiji Islands.

Sometimes retained in *Rallina* or even placed in *Rallus*. Forms a superspecies with *N. woodfordi*, with which it may be conspecific. Monotypic.

Synonyms: *Rallus/Eulabeornis poecilopterus*; *Nesoclopeus poeciloptera*.

Alternative name: Fiji Rail.

IDENTIFICATION PROBABLY EXTINCT Length 33-35.5cm. A medium-large, darkish rail with a longish, powerful bill; possibly flightless, or flight very weak. Top of head, and upperparts, brown; more rufescent on upperwings (including tertials); rectrices blackish, fringed reddish-brown. Greater upperwing-coverts, primaries and secondaries barred chestnut and black; axillaries and underwing-coverts black, barred white. Lores and ear-coverts pale brown; sides of head and neck slate-grey; chin and throat whitish; rest of underparts dark slate-grey, narrowly barred paler, but rear flanks to undertail-coverts darker, obscurely barred grey-brown. Iris brown to red; bill yellow, orange towards tip; legs and feet yellow to orange. Sexes alike. Immature and juvenile undescribed. Inhabits forest, overgrown plantations and possibly swamps.

Similar species The closely related Woodford's Rail (37) of Bougainville I and the Solomon Is is darker; head, neck and upperparts blackish-brown; facial pattern almost absent; underparts sooty grey-black with variable greyish to white barring (depending on race); primaries (and sometimes some secondaries) barred or spotted white; a few white spots on upperwing-coverts; limited white barring on underwing-coverts (one race has no white in wings); iris red; bill black to yellowish-horn; legs and feet grey to yellow-green.

VOICE Not described.

DESCRIPTION
Adult Forehead, crown, nape, hindneck, scapulars and mantle to uppertail-coverts brown (tending to raw umber); tertials, and all upperwing-coverts except greaters, slightly more rufescent; rectrices blackish, with reddish-brown fringes. Greater upperwing-coverts, alula, primaries and secondaries pale ferruginous to vinaceous-tawny, tipped and barred black; axillaries and underwing-coverts blackish-brown, barred white. Lores and ear-coverts pale brown; rest of face, and sides of neck, slate-grey; chin and throat whitish, washed grey on throat; foreneck, breast, anterior flanks, belly and thighs dark slate-grey, feathers narrowly tipped pale grey; shafts of breast feathers whitish; rear flanks to undertail-coverts darker, more blackish, irregularly and very narrowly tipped grey-brown. Iris pale brown to red; bill yellow, more orange towards tip; legs and feet yellow to orange. Small black wing-claw present. Sexes alike.
Immature Not described.
Juvenile Not described.
Downy young Not described.

MEASUREMENTS (2 males, 5 females) Wing of male 172, 175, of female 163-177 (166.6, SD 5.9); tail of male 78, 80, of female 66-85 (75.8, SD 7.4); culmen to base of male 46,

48.5, of female 46-51 (47.9, SD 2.1); tarsus of male 64, 66, of female 57-60 (58.8, SD 1.3). Unsexed birds: tail 68-85 (75.8, SD 8.1, n = 4); exposed culmen 42.5-49.5 (46.7, n = 7); tarsus 57.5-64.5 (61, n = 8).

GEOGRAPHICAL VARIATION None.

MOULT Body moult in a first-year bird is recorded in Jun.

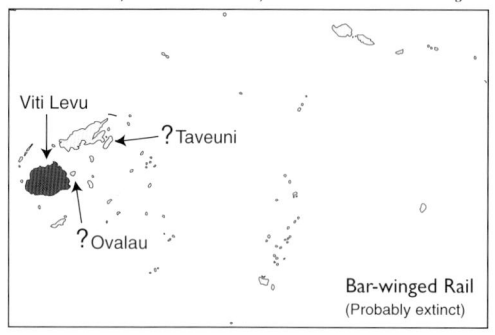

DISTRIBUTION AND STATUS Viti Levu, Taveuni and Ovalau, Fiji. Although this species is considered EXTINCT, there is an unconfirmed record (Holyoak 1979) from N of Waisa, near Vundiwa, Viti Levu, in 1973 (the first since 1890), but the species is thought unlikely to have survived there because a number of other ground-dwelling species, including two non-endemic volant rails, Buff-banded Rail (46) and Purple Swamphen (120), have disappeared as a result of predation by introduced mongooses and feral cats (Hay 1986, Collar *et al.* 1994). It was probably rare even before the introduction of mongooses and was regularly hunted by dogs (Watling 1982b). It is almost certainly extinct on Taveuni (Blackburn 1971, Holyoak 1979). It is known from 12 specimens collected on Viti Levu and Ovalau in the 19th century; no estimate was ever made of its abundance because it was shy and seldom seen.

MOVEMENTS None.

HABITAT Remote forested areas, secondary forest, old taro fields and possibly swamps.

FOOD AND FEEDING Stomach contents are recorded as insects, crustaceans and grass.

HABITS Said to be very shy and rarely seen.

SOCIAL ORGANISATION No information available.

SOCIAL AND SEXUAL BEHAVIOUR No information available.

BREEDING AND SURVIVAL Season Breeds Oct-Dec and possibly Mar. **Nest** One, "of sedges", contained 6 eggs; on another occasion male and 4 eggs obtained by local people. **Eggs** Larger and paler than those of Woodford's Rail, with more profuse and much smaller red-brown spots (Harrison & Parker 1967). Eggs 4-6; ovate; pinkish-cream or brown-cream, speckled light purple and dark red mostly towards blunt end; size 46-51.3 x 33.8-39 (49.8 x 36.6, n = 15); calculated weight 35.4 (Schönwetter 1961-62).

GALLIRALLUS

This genus comprises 15 species (or 16, including the doubtfully valid extinct Gilbert Rail) of mostly medium-sized rails inhabiting Australasia and south-west Pacific islands north to Okinawa in the Ryukyu Islands; only one species, the Slaty-breasted Rail, occurs in continental Asia. The species occupy a wide variety of habitat types, from wetlands and dry grassland to scrub and forest; the widespread Buff-banded and Slaty-breasted Rails occur in all these habitats, while six species are confined to forest and dense woody growth. *Gallirallus* rails are relatively unspecialised and their characteristic features include barred primaries and a relatively short, robust bill (but the extinct Chatham Rail had a long, thin, decurved bill). The relationships of the species in this group have been greatly obscured by the creation of several unnecessary genera for flightless forms, and by the combination of some forms with the more specialised species of *Rallus*. The Slaty-breasted Rail has a more slender skull, bill, pelvis and hindlimb than its congeners and is possibly an advanced form which has paralleled the evolution of the true *Rallus* group towards a slender, marsh-dwelling build (Olson 1973b). Of the 15 accepted species, 11 are flightless or nearly so, having evolved on isolated islands with no significant competitors or mammalian predators. The largest subgroup within the genus, that which contains the Buff-banded Rail and derivatives of its ancestral stock, has given rise to the most flightless forms and includes the Weka, the Guam, Roviana and Lord Howe Rails, the extinct Chatham, Wake, Tahiti and Dieffenbach's Rails, and possibly the New Caledonian Rail. The four extinct flightless species have been lost in the recent past as a result of human activities and the introduction of mammalian predators to their islands. The volant Sharpe's Rail, known from only one specimen of unknown origin, is probably also extinct. Of the seven living flightless forms, the Guam Rail was extinct in the wild but has been bred in captivity and reintroduced to a nearby island, the New Caledonian Rail is critically endangered, the Lord Howe and Okinawa Rails are endangered, and the Weka and the Roviana Rail are near-threatened. Even in the widespread Buff-banded and Slaty-breasted Rails some subspecies are poorly known, rare or threatened. Island forms such as the flightless New Britain Rail and the volant Barred Rail, although not threatened, are very poorly known.

39 WEKA
Gallirallus australis　　　　　　Plate 12

Rallus australis Sparrman, 1786, Dusky Sound, South Island, New Zealand.

Includes *G. troglodytes* (Gmelin, 1789), now regarded as a black morph of *G. australis*. Olson, in Ripley (1977: 366), suggested that the recently extinct *G. minor*, described from bones found on North, South and Stewart Islands, may represent the lower size ranges of the very variable *G. australis* which were eliminated by introduced predators. Considerable geographical variation: historically up to eight subspecies recognised, but currently only four.

Synonyms: *Gallirallus troglodytes/brachypterus/fuscus/townsoni*; *Rallus troglodytes*; *Ocydromus greyi/hectori/scotti/australis/troglodytes/brachypterus/earli/nigricans/finschi*.

Alternative names: Maori/Kelp Hen; Troglodyte Rail; Brown/Black/North Island/South Island Woodhen or Weka; Buff/Eastern/Western Weka.

IDENTIFICATION Length of male 50-60cm (wingspan 50-60cm); of female 46-50cm. Unmistakable: a large (female) to very large (male), thickset, flightless rail with stout bill, legs and feet; tail remarkably long for a flightless species. Marked geographical variation in colour and pattern of underparts. Nominate and race *scotti* have chestnut, grey and black morphs. Chestnut morph has top of head and entire upperparts (including upperwing-coverts) rufous-brown, streaked blackish (barred on tail); supercilium, chin and throat grey to brownish-grey; broad rufous-brown streak from bill to ear-coverts. Breast rufous-brown mottled blackish, forming broad band; flanks, belly and undertail-coverts dark brown to rufous-brown, posterior flanks (sometimes) and undertail-coverts barred blackish. Iris red to brown; bill dull pink to grey at base, cream to grey distally (all pinkish-grey in *scotti*); legs and feet dull brownish-red to orange (pinkish-red to pinkish-yellow in *hectori*). Sexes similar but female smaller; bill paler, with narrower pink base and creamier tip. Grey morph has duller upperparts, and predominantly greyish to olive underparts below a well-marked brown breast-band; black morph is mostly black with dark red-brown streaking on upperparts, and darker brown legs and feet. Race *hectori* palest, with yellow-brown to buff edges to feathers of upperparts; foreneck and breast yellow-brown, streaked blackish; lower breast plain grey-brown; flanks and belly olive-brown; flanks barred black-brown. Race *greyi* resembles grey morph of nominate but facial pattern more distinct; breast-band unspotted and often inconspicuous; flank bars generally absent; lower flanks brown, tinged olive; belly dark grey; undertail-coverts brown, vaguely barred olive to rufous; legs dark. Immature similar to adult of same morph. Juvenile generally darker, though paler in *hectori*; upperparts appear less streaked, and dark feather centres almost absent in some *greyi*; underparts may be more uniform and breast band indistinct; flanks often blotched, not barred; dark forms often have brown spots in plumage; bill dark grey; iris brown; legs brown. Inhabits scrub, forest borders, shorelines, grassland, wetland margins and cultivated land; occurs around human habitations.

Similar species The superficially similar Buff-banded Rail (46) is volant, smaller and slimmer, with more prominent facial pattern, white-spotted upperparts, black-and-white barred underparts and often a buff breast-band. Weka's *coo-eet* spacing call sometimes confused with whistling territorial calls of male Brown Kiwi *Apteryx australis* and Little Spotted Kiwi *A. owenii* but call of both more constant and repetitive; that of former also more guttural, with resonant rolled *r* sound; that of latter more prolonged.

VOICE Information is from Marchant & Higgins (1993). The 'spacing call' of the adult, which is given throughout the year, is a repeated shrill whistle *coo-eet*, rising in pitch. It is given when pair members meet after separation, when one looks for the other, after a territorial encounter or in response to the same call from other birds; it is most often heard at dawn and dusk and in the early evening. A resonating boom call *doon-doon-doon* is given by territory holders during territorial or aggressive encounters. The contact call *ih-ih-ih* is also given before copulation and sometimes to chicks. There is also a courtship growl, an *uhahhuah* dismount call and a lead call *put-put-put* which is used by both sexes in courtship as well as when directing young and when trying to locate a partner. Calls to chicks include a soft *im-im-im* contact call, a repeated *uuurha* food call (also used in courtship feeding), a distress call *uh-uh-uh* and a more agitated distress squeak *uhhreek*. There is also an undescribed call given when a raptor passes overhead. Chicks give a chirping *ieep* and a wavering call.

DESCRIPTION *G. a australis*
Adult chestnut morph Forehead to hindneck black with rufous-brown feather edges giving streaked effect; supercilium pale grey to pale rufous-brown; broad facial streak of rufous-brown with darker markings from bill, through and under eye to ear-coverts; chin and throat grey or brownish-grey; sides of neck rufous-brown with dark brown streaks or spots giving speckled appearance. Mantle and scapulars black with rufous-brown to pale brown feather edges; back and rump darker, rufous-brown with grey-black feather bases giving mottled effect; uppertail-coverts similar but with black shaft-streaks, or black feather centres and rufous notches along feather edges giving almost barred effect; rectrices rufous-brown with black barring. Upperwing-coverts as upperparts: rufous-brown to pale brown, with blackish-brown streaks, becoming more barred towards rear of wing. Remiges variably patterned: usually barred black and rufous-brown, but may be only edged or notched rufous-brown; secondaries may be more mottled than barred; tertials black with broad rufous-brown edges. Underwing-coverts grey-black with rufous-brown mottling; underside of remiges rufous-brown with subdued dark barring. Breast rufous-brown to orange-rufous, mottled black; flanks and belly dark brown with warm rufous tint, posterior flanks sometimes barred black; vent dark brownish-grey; undertail-coverts rufous-brown, barred black. Iris red to brown; bill dull pink to grey at base, cream to grey distally; legs and feet dull brownish-red to orange. Remiges soft and floppy with narrow, stiff shafts; alula has short, sharp, curved claw. Sexes similar but female smaller; bill paler, with narrower dark pink base and creamier tip; legs and feet may be paler pink or orange-brown. **Grey morph** has broader supercilium; chin and throat off-white, grading to pale grey on throat and contrasting more with facial stripe; upperparts less richly coloured than in chestnut morph; breast-band narrower but distinct, lower breast being brownish-grey; anterior flanks, belly and vent grey-brown tinged olive; posterior flanks dark olive-brown, feathers varyingly barred pale brown at tips and sometimes black-brown elsewhere. **Black**

morph is mostly black with dark red-brown speckling on face and streaking on upperparts; chin and throat very dark brown-grey; tail black; remiges black-brown with russet-brown edges (and sometimes notches on inner webs); underparts black to black-brown (palest on lower breast), with variable red-brown markings which are sometimes extensive on upper breast to form vague band; undertail-coverts narrowly barred rufous-brown; legs and feet dark brown.

Immature Very similar to adult. Juvenile remiges and rectrices retained; remiges with worn and frayed tips are a good ageing guide. Bill becomes paler, more grey-pink, along sides, some attaining adult colour by 9 months; iris brown to red-brown at 5-8 months, greyish-red to red at 9-12 months; legs and feet as adult.

Juvenile Highly variable. Usually similar to adult of same morph but darker, with noticeable difference in feather shape (narrower, with more pointed tips). Central streaks of upperpart feathers average narrower, browner and less distinct than in adult; underparts may be more uniform and breast-band indistinct; flanks often blotched rather than barred; darker forms often have brown spots or speckles in plumage, especially on sides of underparts and uppertail-coverts. Bill dark grey by 6th week, dull red base develops by 7th week; iris brown, becoming yellow-brown or greyish-yellow (sometimes olive-green); legs and feet become brown or orange-brown by 6 weeks, then become redder, like adult. Usually smaller than attendant adults.

Downy young Down long, soft and silky; black to black-brown, throat sometimes greyer; in skins, often fades to dark grey-brown. Generally lacks contrast in upperpart colours (see *greyi*). Bill black, with white or pink-white egg-tooth; some have pink round nares; iris dark brown or dark yellow-brown; legs and feet blackish, dark purple, dark brown or pink-grey.

MEASUREMENTS Wing of 20 males 168-205 (186.7), of 9 females 160-183 (171.6, SD 7.6); tail of 18 males 108-146 (128.3), of 8 females 108-136 (122.6, SD 10.4); exposed culmen of 22 males 37-55 (47.5), of 9 females 44-49 (46.5, SD 2.0), culmen to base of 9 unsexed 45-56 (50.5, SD 2.5); tarsus of 22 males 53-71 (61), of 9 females 52-61 (55.7, SD 3.2). Weight of 9 males 612-1250 (978, SD 233), of 6 females 350-1035 (725, SD 270).

GEOGRAPHICAL VARIATION Considerable, involving plumage and, in race *scotti*, size. Partial to almost complete leucism (whiteness in plumage) common; holotype of "*G. townsoni*" is a partly leucistic nominate bird. For the racial distribution of introduced birds, see Distribution and Status. The taxonomic status of birds on Kapiti I, where three races have been introduced, is uncertain and the birds appear intermediate between nominate and *greyi* (Marchant & Higgins 1993).

G. a. greyi (Buller, 1888) – North I. Little variation; most resembles grey morph of nominate but facial pattern more distinct; breast-band of variable width, unspotted and less distinct; rectrices blackish with brown edges; underwing-coverts brown, finely barred grey-black; underparts mostly uniform brown-grey; flank bars generally absent; lower flanks brown, tinged olive at feather edges; belly dark grey; undertail-coverts brown, vaguely barred olive to rufous; legs dark. Juvenile may be darker or paler grey below than adult, with breast-band indistinct or absent; some almost lack dark centres to feathers of upperparts and thus appear very rufous (see Plate 12). One downy chick dark brown with black crown (giving capped effect) and lower back to tail. Wing of 14 males 160-198 (175.8, SD 11.2), of 12 females 155-188 (169.8, SD 11.0); tail of 15 males 99-122 (113.1, SD 6.7), of 10 females 98-118 (107.2, SD 5.9); exposed culmen of 38 males 42-51 (46.7), of 28 females 40-48 (44.3), culmen to base of 4 unsexed 48-54 (50, SD 2.5); tarsus of 49 males 53.5-64 (60.1), of 28 females 52.5-62.5 (56). Weight of male 532-1053 (912), of female 382-1010 (699) (sample sizes not known).

G. a. australis (Sparrman, 1786) – N and W South I. See Description.

G. a. hectori (Hutton, 1874) – formerly E coast and interior of South I but now confined to Chatham I and Pitt I, where introduced. Relatively pale, having yellow-brown or buff edges to feathers of upperparts; tertials tipped and edged olive-buff; supercilium pale grey; chin and throat off-white to pale grey; foreneck and breast yellow-brown, streaked blackish; lower breast plain grey-brown; anterior flanks and belly olive-brown, posterior flanks darker; flanks barred black-brown; legs and feet pinkish-red to pinkish-yellow. Juvenile paler than adult; shows less contrast on upperparts between browner feather centres and olive-buff edges, thus appearing less streaked; underparts more uniform olive than adult, lacking distinct breast-band. Wing of 14 males 155-202 (186.5), of 7 females 156-178 (169.1); tail of 5 males 120-145 (131.6, SD 10.7), of 3 females 107, 122, 125; exposed culmen of 34 males 42.5-53 (50.1), of 20 females 41-50.5 (46.3), culmen to base of 7 unsexed 48.5-55 (52.6, SD 2.4); tarsus of 34 males 53.5-70.5 (61.7), of 20 females 49-60 (54.3). Weight of 8 males 925-1605 (1255, SD 203), of 4 females 525-850 (651, SD 139).

G. a. scotti (Ogilvie-Grant, 1905) – Stewart I. Similar to nominate but smaller; bill entirely pinkish-grey. Downy young generally lacks contrast in upperpart colours (see *greyi*); one had red tinge on underparts – indicative of chestnut morph perhaps. Wing of 95 males 160-226 (182.6), of 46 females 110-199 (164.1); tail of 91 males 74-175 (125.3), of 37 females 105-130 (118); exposed culmen of 96 males 41-54 (49.4), of 48 females 38-48 (44.5), culmen to base of 8 unsexed 45-51 (47.5, SD 2.6); tarsus of 96 males 43.5-64 (58), of 47 females 43-60.5 (48.5). Weight of 70 males 600-1425 (1034, SD 134), of 26 females 500-884 (753, SD 87).

MOULT Information is from Marchant & Higgins (1993). Postbreeding moult is complete, starting with the head, neck and lower breast; it begins from mid-Dec to late Jan and usually takes 65-90 days. Remex moult is simultaneous or partly simultaneous, beginning c. 20 days after the head and being complete after 50-90 days; the secondaries usually moult slightly earlier than the primaries, sometimes simultaneously but more often in groups; occasionally one or more remiges are retained. The greater primary and secondary upperwing-coverts moult with their respective remiges; other wing tracts moult within 10 days of the remiges. The rectrices and tail-coverts moult together, beginning soon after the head. A partial pre-breeding moult, mostly of the head and neck, begins just before the breeding season and possibly continues through breeding; probably not all birds have this moult. Postjuvenile moult is partial, involving the head and body only; it begins at c. 70 days of age and frequently coincides with the end of parental care. The first pre-breeding moult is probably partial but may be suppressed.

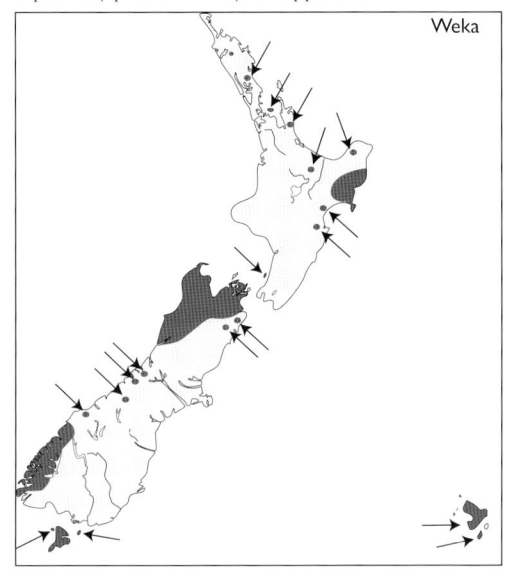

DISTRIBUTION AND STATUS Most information is from Marchant & Higgins (1993). Endemic to New Zealand, the Weka formerly occurred throughout North Island and much of South Island , D'Urville and Stewart Is, and some inshore islands, but it declined dramatically between 1900 and 1940, becoming extinct in E South Island and in much of North Island. The causes of the rapid decline are not clear, but may have included habitat changes such as the conversion of forest and scrubland to farmland, the use of poisoned baits and the introduction of mammalian predators (Heather & Robertson 1997). Its current range is greatly reduced and fragmented as a result of human activity, although in some areas birds are recolonising their previous range. It now occurs mainly in the East Coast region of North Island (Gisborne, and moist areas from Tikitiki SW to L Waikaremoana and S to 39°S), and in N, NW and SW South Island (NW Marlborough, Nelson and N West Coast, and W Southland S from Milford Sound to West Cape and inland to Murchison Mts), with a few scattered populations elsewhere on these islands. It has been introduced to many offshore and outlying islands (sometimes unsuccessfully): the nominate race to Kapiti, Chatham and Macquarie Is (now extinct on last), *greyi* to Kapiti, Rakitu, Mokoia and Kawau Is and Bay of Islands, *hectori* to Chatham, and *scotti* to Solander and other outlying islands off Stewart I, and also to Kapiti and Macquarie Is (now extinct on Macquarie). It is locally rare to common, its total population is estimated at 25,000-100,000 (Rose & Scott 1994), and its global conservation status is NEAR-THREATENED; the race *hectori* has a CITES II classification. The North Island race *greyi* is regarded as threatened because of its restricted and declining distribution, and its failure to re-establish itself in apparently suitable areas; however a successful captive breeding programme aims to release birds into suitable North Island habitats (Heather & Robertson 1997). Weka population sizes can fluctuate markedly, possibly as a result of variations in climate and the availability of food (Beauchamp 1987a), while modification of habitats by introduced mammals, especially browsers, may affect Weka numbers; there is no evidence that disease is responsible for declines. Local irruptions occur and birds readily recolonise areas, but attempts to re-establish birds have sometimes been unsuccessful.

Wekas may sometimes adversely affect the native fauna (such as frogs, lizards and petrels) and flora on offshore islands to which they have been introduced: e.g. on Taumaka I, where they exist at an unusually high density, they have affected the native plants and animals and are significant nest predators of the rare Fiordland Crested Penguin *Eudyptes pachyrhynchus* (St Clair & St Clair 1992). Consequently, in some places removals have been undertaken (sometimes unsuccessfully) although the need for this on some islands is disputed and the problem may have been exaggerated. Most of the impact of Wekas on vertebrate populations is based on analysis of gizzard contents and faeces (which cannot distinguish between freshly killed items and carrion) and on extrapolation of species absences (Beauchamp 1996). A recent study showed that the anti-predator behaviour and gland secretions of two native frogs, *Leiopelma hochstetteri* and *L. archeyi*, were sufficient to allow them to escape unharmed from Wekas, that their habitats are not favoured by Wekas, and that the rocks and logs under which they hid were heavier than those generally moved by Wekas while foraging (Beauchamp 1996).

Once hunted extensively by Maoris and early settlers for food, feathers, oil and medicinal purposes, Wekas are now protected in most places except the Chatham Is, where a legal harvest is permitted. Although sometimes considered a pest because they pull up seedling crops, kill poultry and eat eggs, they also eat pest insects.

Wekas are adept at killing rats, which might otherwise be predators of both eggs and young. Predators of adults and young include skuas *Catharacta*, stoats *Mustela erminea*, weaselsM. *nivalis*, ferrets *Putorius putorius*, dogs and feral cats. Poisoned bait for mammals has reduced populations locally, while recolonisation may be hindered by predators. Many Wekas are killed on roads when population densities are high.

MOVEMENTS Most information is from Marchant & Higgins (1993). Although sedentary and flightless, Wekas can move for significant distances. Adults are normally confined to the territory or the core area of the home range all year but may walk up to 1km from it to good feeding areas. Most subadults disperse up to 6 months after independence, set up their own core areas and either stay at one site, or move between core areas, until they establish a territory or die. Dispersal distances may exceed 9km; subadults can walk 4+km/day and on islands may swim or wade at low tide. The greatest natural recorded movements are 9km (a subadult) and 35km (an adult male), both of which involved crossing major rivers or lakes, and relocated territorial birds were recorded swimming almost 1km to return after release. The Weka's homing ability is well documented and the longest returns include 130km, and 72km in 6 weeks. Mass migrations, associated with the disappearance of many populations in North I, are reported from 1890 to the 1930s; the reasons for these movements are not known. Rapid population increases are recorded, one in South I being due to movements associated with plagues of mice (Moncrieff 1928); others are also possibly due to short-term changes in food supply. Seasonal altitudinal movements, and movements from forest to open ground in summer, are also recorded. Sudden disappearances may be due to disease (McKenzie 1979).

HABITAT Details are from Marchant & Higgins (1993). The Weka frequently inhabits forest (sometimes the edges only), woodland, scrub and grassland (including subalpine tussock grassland). It also occurs on beaches, particularly those with rotting seaweed, and at tidal creeks and bays, and is found at the margins of estuarine and freshwater wetlands. It also occupies modified habitats such as lawns, rough pasture, cultivated land and plantations. It inhabits hills, mountains, cliffs, moraines, sandy and rocky shores, sand-dunes and urban environments. It prefers low vegetation which gives cover but does not hinder movement and it avoids forest with no suitable dense understorey. It occurs from sea level to c. 1,500m a.s.l; it may prefer low altitudes in some areas.

FOOD AND FEEDING Most information is from Marchant & Higgins (1993). The Weka is omnivorous and opportunistic, taking mainly native fruit, vegetation and invertebrates, and sometimes vertebrates; it eats more vegetable matter in winter. It takes foliage such as that of ferns, Poaceae grasses, clover *Trifolium* and many other dicotyledons, and seeds of many plants, particularly Poaceae, Cyperaceae (mainly *Cyperus ustulatus*), Juncaceae, Araliaceae, Liliaceae, Solanaceae, Polygonaceae, Urticaceae and Apiaceae. It also takes roots, tubers (including *Ipomea* and potatoes), fruit (especially that of *Comprosma* and also tomatoes and grapefruit), Bryophyta, fungi and seaweed. Invertebrate prey includes earthworms, *Peripatus*, molluscs (mostly Gastropoda), a wide variety of crustaceans (including Copepoda, Decapoda, Isopoda and Amphipoda), Diplopoda, Chilopoda, Arachnida, Collembola and insects, especially adult and larval Coleoptera, Orthoptera (especially Acrididae and Tettigoniidae), Hymenoptera (including Formicidae), Diptera, Lepidoptera and many eggs. Parts of marine animals such as cephalopods are also taken. Vertebrate prey includes fish, amphibians such as *Litorea auria* and *Leiopelma hochstetteri* (but see Distribution and Status section), lizards (Scincidae), and the unguarded eggs and young of other birds, including kiwis *Apteryx* spp., penguins, ducks, shearwaters, petrels and various small passerines. There are also reports of Wekas killing adult birds such as Mottled Petrel *Pterodroma inexpectata*, House Sparrow *Passer domesticus* and Song Thrush *Turdus philomelos*. It may kill chickens and young ducks and it also eats mammals (Brush-tailed Possum *Trichosaurus vulpecula*, stoats, rats, mice and young rabbits), and scavenges carcases. Cat food, skimmed milk, bread and sandwiches are also recorded. Objects up to 4cm in diameter can be swallowed, and indigestible matter, when eaten in large quantities, is regurgitated.

Foraging takes place mainly on the ground. Birds dig in leaf-litter with the bill to a depth of 8cm, covering up to 50m^2 in dry litter per day, they sometimes pull apart bark and wood to extract food, they lift dead *Rhopalostylis* palm leaves to search below them, they climb in trees and search in tree-hollows, and they follow wild pigs and search where they have rooted. They also search seaweed and debris on rock platforms and beaches (including along the tideline), take petrels from burrows and scavenge at campsites. They spear eggs with the bill and run off with them, and use the feet to hold down large objects which are then hammered with the bill to kill them or break them up. One bird ran off with a tourist's sandwiches while being photographed (D. Coetzee pers. comm.).

HABITS Most information is from Marchant & Higgins (1993). Wekas are inquisitive by disposition but usually secretive and wary in their habits. However, they readily become confident around human habitations: they will visit campsites to glean scraps, they become bold and will take food from the hand, and they may become persistent raiders of rubbish bins and chicken runs. They are fond of investigating bright objects such as spoons. They walk with deliberate strides, flicking the tail when unsure or threatened, and when alarmed they run rapidly into cover, giving a high-pitched shriek. They swim readily and well. They are both diurnal and nocturnal, tending to be particularly active in the late afternoon and (in open habitats) on clear moonlit nights. They roost at night in the open, under logs, among rocks, in burrows or in their own nests, and some territories have specific roosting sites. In the afternoon birds often roost in thickets, occasionally up to 2m above the ground. After bathing, birds may adopt a spread-wing sunning posture.

SOCIAL ORGANISATION Information is from Marchant & Higgins (1993) unless otherwise stated. Monogamous and often permanently territorial, the pair bond usually being permanent and known to last 13+ years in the wild. Non-monogamous associations of two females with one male are recorded twice (Guthrie-Smith 1910, 1914, Beauchamp 1986); in one trio a younger female temporarily joined an established pair and helped to raise young in one season. Males tolerate a second female in the territory, whereas females usually do not, but one pair bond between two females lasted at least 2.5 years when no males were available (Beauchamp 1987a). Disproportionate sex-ratios occur in all populations at some times during the population cycle; some recorded biases may arise because males are more readily seen in the field. Co-operative breeding is unusual and subadults usually leave the territory before the parents breed again, but in captivity young fed chicks of later clutches. Groups of up to 13 subadults and 6 territorial adults have been seen at rich food sources.

Territories are stable, with no size changes during the breeding cycle; territory sizes at Double Cove were 2.6-15.8 (4.5) ha, and on Kapiti I 0.7-4.5 (1.96)ha (Beauchamp 1987a, b). Home ranges of individuals vary greatly, and in non-territorial birds can range from 1.6ha (Kapiti) to 70ha (Double Cove). Intrusion into neighbouring territories seldom exceeds 10m but can be up to 200m, while on Kawau I incursions of over 300m by males seeking food for young are common.

SOCIAL AND SEXUAL BEHAVIOUR Most information is from Marchant & Higgins (1993). Wekas are aggressive, particularly adult males. Face-to-face interactions between territorial birds and intruders generally involve wing-arching, feather-ruffling and posturing, often followed by chases. Both sexes undertake territorial chases and fights, and females chase males as often as males chase females; in long chases the pursuer attempts to seize the tail of the fleeing bird. Fighting, which occurs at territorial boundaries, usually involves ritualised face-to-face movements and parallel walking, sometimes with calls; actual combat, which involves striking with the bill and feet, occurs less often, causes loss of plumage and has been known to result in death. Pecking orders have been reported at food sources. An incubating bird usually refuses to leave the nest when threatened and, if forced off, may peck or eat the eggs. If young are captured, the parents peck vigorously at the captor or look on in agitation, with roused plumage, fanned drooping wings and open probing bill.

Nest-building captive birds had a greeting ceremony in which the pair members faced each other with heads held low and called for c. 30s (Timmis 1972). Either sex may initiate courtship: an initiating female gives the courtship growl and sometimes stands or crouches with the bill almost touching the ground (the pre-copulatory stance); this behaviour leads to allopreening by the male. The male initiates courtship by chasing the female. Courtship usually consists of one bird giving courtship growls and nest-sitting while the other continues normal activities; allopreening, and courtship feeding by the male, also occur during courtship. Copulation is preceded by the birds uttering the booming call and walking stiffly with necks stretched and curved downwards. The female then adopts the pre-copulatory stance and the male mounts from behind and copulates for 3-4s, flapping his wings, moving his tail vigorously from side to side, arching his neck and sometimes pecking the female's nape. Allopreening is recorded between pair members and by adults to young.

BREEDING AND SURVIVAL Season Highly variable, start and duration influenced by climate, food and size of population; start recorded Aug-Nov, end recorded Nov-Mar, but when population expanding may breed throughout year; one semi-tame pair raised 4 broods between Mar and Dec. **Nest** Built on dry ground; situated in or under tussocks, in burrows, under logs, stumps or rocks, in tree-hollows at ground level, also concealed in outbuildings. Up to 7 sites per territory, used for many years. Nest a bowl of sedges (*Carex*, *Uncinia*), grasses and lilies (*Astelia*, *Erycenetia* and *Cordoline*); twigs and moss used if grass not available; lined with finer grasses and sometimes feathers, wool, hair, moss and leaves or leaf-litter. Built by both sexes (in captivity) or by male only; material added during incubation (only in captivity). **Eggs** 1-6, usually 2-4; laid 2 or more days apart; ovate; glossy when fresh; pinkish to creamy-white, blotched light chestnut or pale and dark brown, chiefly at larger end, and streaked purplish and reddish-brown; blotches fade and eggs become creamier before hatching. Size of 39 eggs (nominate) 53.5-62.5 x 37.3-42 (58 x 40.1), of 4 eggs (*greyi*) 57.5-63.4 x 40.5-43.2 (60.4 [SD 2.3] x 41.7 [SD 1.4]), of 2 eggs (*hectori*) 53.5 x 39, 58.8 x 40, of 16 other eggs 53.1-61.2 x 38.1-41.8 (58 x 39.5) (Schönwetter 1961-62, Marchant & Higgins 1993); calculated weight 54 (*greyi*), 53 (nominate), 47 (*hectori*). **Incubation** 26-28 days; female usually incubates during day, male at night, in shifts usually exceeding 4 hrs; hatching asynchronous. **Chicks** Precocial and nidifugous; leave nest after 2-3 days; brooded by parents for 8 days after leaving nest; cared for and fed until independent at 40-108 days. Chicks beg by crouching or by standing upright and flapping wings, pecking at food or at tip of the parent's bill; fed bill to bill or by regurgitation when small; some young feed independently at 21-40 days; late in parental care period male usually more attentive than female. See Social Organisation for examples of cooperative breeding. First body feathers appear at c. 9 days; secondaries begin growing at c. 16 days, followed by rectrices, alula and then primaries; underwing-coverts grow last; body feathered at 4 weeks; young almost fully grown by 6-10 weeks; all feathers grown by c. 100 days, when postjuvenile moult in progress (see Moult); most young disperse by 4 months after independence; attain adult weight at 6-9 months.

Possible egg-dumping recorded (Guthrie-Smith 1910), perhaps as result of polygynous behaviour or long laying period. On Kapiti I over 5 years 44.8% of pairs tried to breed, 27.1% successfully, raising 0.3 young per pair; at Double Cove, South Island, 2.8 young raised per pair, success related to food supply (Beauchamp 1987a, b). Specific nest predators not recorded. Age of first breeding 1 year, but can breed in first year (minimum recorded age 5.9 months). May rear up to 4 broods per year; one captive pair relaid within 5 days of losing newly hatched young. **Survival** One semi-tame pair, protected by people, lived at least 18 years.

40   NEW CALEDONIAN RAIL
Gallirallus lafresnayanus   Plate 12

Gallirallus Lafresnayanus J. Verreaux & Des Murs, 1860, New Caledonia.

Sometimes placed in the genus *Tricholimnas*, or in *Rallus*. Monotypic.

Synonyms: *Tricholimnas/Rallus/Eulabeornis/Ocydromus lafresnayanus*.

Alternative name: New Caledonian Wood-rail.

IDENTIFICATION Length 45-48cm. A large, dark, almost unpatterned rail with strongly developed legs and feet and long, slightly decurved bill. Flightless, but tail noticeably longish. Forehead to fore-crown dark grey; lores, crown and nape dark brown; supraloral stripe pale; ear-coverts brown; rest of face grey, chin and throat paler. Hind-neck to back, and upperwings, rich dark brown; lower back to tail darker, more blackish-brown. Remiges blackish-brown, unbarred. Underparts dark grey, tinged brown on breast; rear flanks to undertail-coverts darker, sometimes blackish. Iris crimson; bill dark horn; legs and feet dark horn to brown. Sexes alike; female somewhat smaller. Immature undescribed. Juvenile largely black, washed rich dark brown on upperparts and sides of neck and breast. Inhabits dense forest.

Similar species The possibly closely allied Lord Howe Rail (41) is paler, predominantly olive-brown, with noticeable rufous and dark barring on upperwing-coverts, remiges and undertail-coverts. The shorter-billed and slightly larger Weka (39) of New Zealand has grey and black morphs, and dark juvenile plumages (see species account and Plate 12) which superficially resemble adult or juvenile New Caledonian Rail, but remiges black with variable rufous-brown barring, edging or notching, and upperparts always streaked, though very vaguely in some juveniles (see Plate 12).

VOICE Unknown.

DESCRIPTION

Adult Forehead to fore-crown dark grey, becoming dark brown from crown to nape; lores dark brown, with pale brown supraloral stripe; ear-coverts brown; rest of face grey, chin and throat slightly paler. Hindneck to back, and upperwings, chocolate-brown, scapulars tinged olive; lower back to tail blackish-brown washed dull rufous-brown. Primary coverts, alula and remiges blackish-brown, unbarred, more rufous-brown on outer webs; axillaries and underwing-coverts barred blackish and off-white; underside of primaries blackish-brown, of secondaries more rufous. Underparts dark grey, tinged brown on foreneck and breast; rear flanks to undertail-coverts darker, sometimes blackish. Iris crimson; bill dark horn; legs and feet dark horn to brown. Sexes alike; male larger, especially in bill. References to fluffy and decomposed plumage probably refer to first-year birds which have retained some juvenile feathers.

Immature Undescribed.

Juvenile Almost entirely black; upper body and sides of neck tinged chocolate-brown; head and throat tinged slate-grey; underparts slaty-black. Foreneck of one specimen has an orange-brown patch. Variable and inconspicuous barring may exist on undertail-coverts. Plumage relatively loose and decomposed. Iris crimson; bill and legs dark horn to brown.

Downy young Unknown.

MEASUREMENTS Wing (chord) of 4 males 187-197 (191.3, SD 4.6), of 7 females 160-188 (179.9, SD 10.5), wing (max) of 1 male 208, of 3 females 172, 184, 190; tail of 3 males 102, 102, 109, of 7 females 94-108 (101.4, SD 4.3); exposed culmen of 5 males 50-63.5 (58.2, SD 5.4), of 7 females 48-54.5 (50.9, SD 2.6), culmen to base of 1 male 68, of 3 females 52, 57, 57; tarsus of 5 males 54-62 (57.8, SD 3.4), of 7 females 51.5-57.5 (53.2, SD 2.1).

GEOGRAPHICAL VARIATION None.

MOULT No information available.

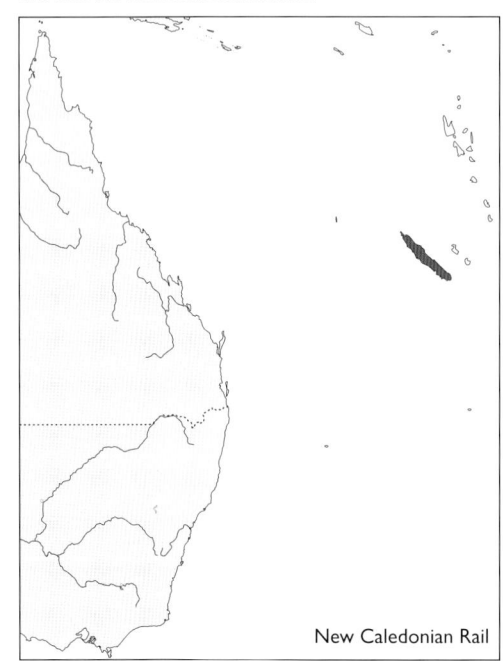

New Caledonian Rail

DISTRIBUTION AND STATUS New Caledonia. CRITICALLY ENDANGERED. Until recently it was thought to be extinct because it had not been recorded by ornithologists this century, the last specimen having been collected in 1890 (Collar *et al.* 1994). However occasional local reports this century suggest that it may still survive in relatively inaccessible areas of forest, albeit in very small numbers (Stokes 1979). It was not found during a search of three areas of the island in 1976, but there was an unsubstantiated sighting from N New Caledonia in 1984 (Hannecart 1988). It is likely to be threatened by introduced species such as dogs, cats, rats, deer and especially pigs; no part of the island is inaccessible to pigs. Hunting is a passion on the island (Stokes 1979) and the bird's habit of crouching when chased makes it vulnerable to dogs (Ripley 1977).

MOVEMENTS None.

HABITAT Dense humid forest, now probably only that in relatively inaccessible areas but formerly also in forested river valleys near the coast. Formerly occurred from almost sea level up to c. 1,000m.

FOOD AND FEEDING Stomach contents are given as worms. Captive birds ate raw meat and vegetable matter, and caught rats. Its diet is thought to be similar to that of the Weka, i.e. almost any invertebrate, including gastropods. Its foraging methods are possibly also similar to those of the Weka.

HABITS This species is said to walk with great rapidity, flicking the tail, to climb and jump "like a cat", to crouch when chased and, when alarmed, to squeeze into the smallest holes and crevices and lie motionless, feigning death (Layard & Layard 1882). It is said to be nocturnal in captivity but in the wild is probably diurnal and crepuscular.

SOCIAL ORGANISATION No information available.

SOCIAL AND SEXUAL BEHAVIOUR Unknown.

BREEDING AND SURVIVAL Season Unknown. **Nest** Unknown. **Eggs** One egg, laid by captive bird in gardens of Zoological Society of London in 1871, is elongated ovate, cream with light brown and lilac spots and scrawls (M. P. Walters *in litt.*). Size of 2 eggs 55.4 x 33.6 and 56.8 x 33.6; Schönwetter (1961-62) gives measurements of what is probably the first egg as 55 x 33.1, and calculated weight 34.3.

41 LORD HOWE RAIL
Gallirallus sylvestris   Plate 12

Ocydromus sylvestris P. L. Sclater, 1869, Lord Howe Island.

Sometimes placed in *Tricholimnas* or *Rallus*. May include *G. (Tricholimnas) conditicius* (42), known from one specimen usually regarded as an immature *G. sylvestris* but sometimes thought to be specifically distinct (see next species account). Monotypic.

Synonyms: *Sylvestrornis/Tricholimnas/Rallus/Cabalus sylvestris*.

Alternative names: Woodhen, Lord Howe Island Rail/Woodhen.

IDENTIFICATION Length of male 34-42cm, wingspan 49-52cm; of female 32-37cm, wingspan 47-49cm. Medium-large to large flightless species, with short tail, short thick legs and rather long bill downcurved at tip. Predominantly plain: olive-brown tinged russet; faint supercilium buffy-olive; chin and throat pale brown-grey; remiges and primary upperwing-coverts barred russet and dark brown; underparts greyish-olive, grading to dark olive-brown towards rear; undertail-coverts narrowly barred dull russet. Old individuals develop white feathers on ear-coverts and sometimes nape, crown or lores. Iris red; bill pink-brown; legs and feet pink-brown or pink-grey. Sexes alike; female smaller. Immature like adult, but pointed juvenile remiges retained; iris orange. Juvenile very similar to adult but upperparts more reddish-brown and underparts browner; bill initially shorter, grey-black becoming browner; iris grey; legs and feet duller. Inhabits lowland forest and montane moss-forest.

Similar species Smaller and less bulky Buff-banded Rail (46) has bold black and white barring on underparts, buff breast-band in some races, spotted upperparts, prominent white supercilium, longer tail, longer, thinner legs, and shorter, straight bill. Like Lord Howe Rail, Buff-banded Rail has rufous and dark brown barring on remiges; this character also shared by closely related Weka (39), but not by possibly closely related New Caledonian Rail (40).

VOICE Information is from Marchant & Higgins (1993). The advertising or territorial call resembles the spacing call of the Weka, being a loud, repeated piercing whistle *coo-eet*, often given in duet by the pair, audible for at least 100m and often answered by neighbours; birds call during the day and occasionally at night. Calling is stimulated by any unusual sound, such as banging on a tin or an aircraft passing overhead. This call may also be given before copulation. The alarm call is a single loud explosive *brr-deeep*, likened to the sound of a sharp pull on a rusty pulley and similar to the common squeak-call of the Buff-banded Rail. Territory holders and young make soft, resonant, drumming or throbbing *booomp* contact notes, audible at close range; this call is also given after copulation. An undescribed greeting call is given by members of a pair when they meet after a short separation, while in captivity the male (and sometimes the female) gives an intimate *weep-weep* call as a prelude to copulation. There is also a soft, repeated purring *prrrup* call, of unknown function, and a distress squeak similar to that of the Weka and given only rarely by birds caught in nets and handled. Chicks chirp before hatching, and in captivity young chicks peep incessantly. Feathered young call persistently and make a wavering call to adults.

DESCRIPTION

Adult Forehead to hind-neck, and broad area from lores over ear-coverts, olive-brown; faint, narrow supercilium buff-olive, merging with obscure buff-olive band from side of neck to foreneck; chin and throat pale brown-grey, sides of face slightly darker; foreneck pale grey-olive. White feathers appear on ear-coverts and sometimes crown, lores and nape of males from c. 7 years of age. Upperparts and tail olive-brown, slightly washed tawny, and often washed more rufous from rump to tail. Primaries, secondaries, greater primary coverts and alula russet to dull cinnamon-rufous, with narrow dark brown bars and tips; tertials have broad, rich olive-brown tips and outer webs, the latter sometimes with narrow faint yellow-brown bars. Greater secondary coverts barred russet and dark brown basally and olive-brown distally; median and lesser coverts like upperparts but with very faint, concealed yellow-brown barring towards feather centres. Underside of remiges russet, narrowly barred dark brown; axillaries and underwing-coverts dark olive-brown, barred russet on axillaries and greaters, and often more yellow-brown elsewhere. Breast and flanks pale grey-olive, washed russet, some flank feathers obscurely barred pale brown; belly and vent more olive-brown; undertail-coverts darker olive-brown with dull rufous barring. Iris bright red to crimson; orbital ring grey; bill pink basally grading to light pink-brown distally, also described as dark horn; legs and feet pink-brown to light pink-grey, sometimes with olive tinge, also described as bluish-green. Sexes alike; female smaller.
Immature Like adult, but pointed juvenile remiges retained; iris orange.
Juvenile Very similar to adult but upperparts slightly more reddish-brown and underparts browner; undertail-coverts may be more barred or suffused rufous-brown. Tips of outer primaries narrow and pointed. Bill initially shorter, grey-black becoming browner; iris dark grey, becoming pale red at 65 days and red at 100 days; legs and feet duller than adult.
Downy young Down sooty-black; iris dark brown; bill dull black with pink or red tinge and white egg-tooth; legs and feet grey, tinged pinkish.

MEASUREMENTS Wing of 30 males 134-151 (140, SD 4.3), of 32 females 126-156 (136.8, SD 6.3); tail of 32 males 53-71 (62, SD 3.9), of 33 females 55-69 (61.6, SD 2.8); exposed culmen of 83 males 47-60.5 (54.8, SD 3.0), of 91 females 42.5-59 (48.6, SD 1.5), culmen to base of 3 males 54, 55, 57.5, of 3 females 48, 49, 50; tarsus of 84 males 44.5-57 (51.3, SD 1.0), of 40 females 41.5-50.5 (46.4, SD 1.6). Weight of 49 males 410-780 (536, SD 11), of 58 females 330-615 (456, SD 9): from Marchant & Higgins (1993); figures for sample size and SD appear erroneous. Hatching weight 17-21.5 (20).

GEOGRAPHICAL VARIATION None.

MOULT Information is from Marchant & Higgins (1993). Postbreeding moult is complete. The contour feathers start to moult just before or with the remiges, which moult more or less together, and the rectrices moult slightly later. The greater coverts are moulted with the remiges and the tail-coverts before the rectrices. Remex moult is recorded in Feb-Mar, moult of contour feathers continuing well after this. There is a partial pre-breeding moult of the head and body feathers. Postjuvenile moult is also partial; in captivity it begins at c. 40 days of age and ends at c. 60 days of age, while in the wild it is noted in Feb-Mar. The first prebreeding moult is probably partial and possibly restricted to the head and neck.

DISTRIBUTION AND STATUS Endemic to the 13km^2 Lord Howe I, E of Australia, this species is classified as globally ENDANGERED and is listed on CITES Appendix I. When the island was first discovered in 1788 this rail was apparently widespread and common, particularly in the lowlands, and many were killed by the first visitors (see Habits). In 1853, 19 years after the first settlement, the rail was restricted to mountainous regions and by 1930 it occurred only in the summit regions of Mt Lidgbird and Mt Gower, where c. 30 birds (at most 10 breeding pairs) survived in 1980. The reason for its ultimate virtual extinction was a combination of habitat degradation by feral pigs and goats, and predation by introduced cats, dogs, pigs and possibly people. Land clearance occurred over only a small part of the island so is unlikely to have had a major effect. Considering the very marginal nature of the habitat on the mountains, it is remarkable that the species survived there for so long (Fullagar 1985). In 1980,

Lord Howe Rail

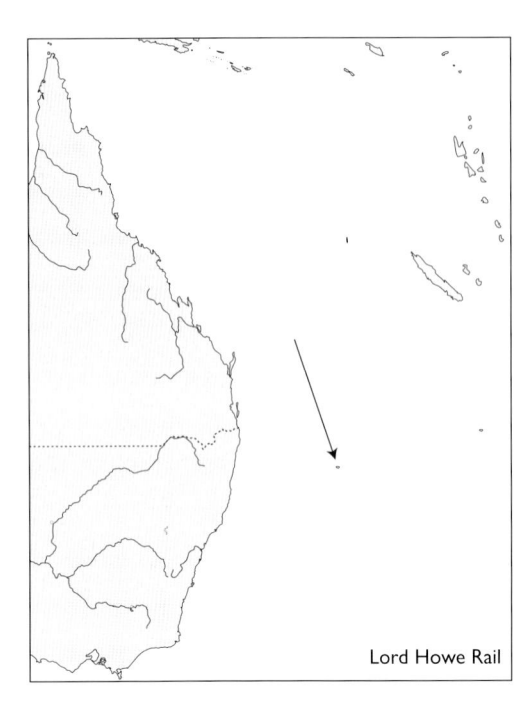

3 pairs were removed for captive breeding at a centre specially built on the island, and over the next 4 years 85 captive-bred birds were released into the wild, where they have become re-established in some lowland areas; measures to control predators such as cats and pigs are effective (Garnett 1993). Recent estimates suggest a population of 170-200 birds, with 40-50 breeding pairs, and some islanders now regard them as pests (Garnett 1993, Marchant & Higgins 1993). The major immediate threat to the population is predation by the Masked Owl *Tyto novaehollandiae castanops*, introduced in the 1920s to control rats, and efforts are being made to eliminate owls from the island (Fullagar 1985, Garnett 1993). Black Rats *Rattus rattus* are not considered a problem, although they sometimes take eggs and young, because the rails can kill them (Garnett 1993); Pied Currawongs *Strepera graculina* also take eggs and young. In settled areas some birds are killed on roads or drown in water tanks (Marchant & Higgins 1993). The rail population is small, vulnerable, largely derived from a limited genetic pool, and may be limited by the extent of available habitat, so constant monitoring is recommended; suitable habitat, particularly *Howea* forest, should be protected from disturbance (Garnett 1993). A recent population viability analysis (Brook *et al.* 1997) shows that the species is acutely sensitive to, and at risk of extinction from, not only catastrophic events such as cyclones or disease epidemics but also minor changes in mortality and fecundity potentially due to inbreeding depression, minor chronic disease, or new exotic predators (e.g. ferrets). A captive zoo population has been recommended but would be limited to 3-5 pairs and is thus not a long-term solution; instead, a remote population needs to be established quickly on another island to minimise the likelihood of the bird's extinction (Brook *et al.* 1997).

MOVEMENTS Largely sedentary. Movements on the island before settlement are unknown, but after the birds were restricted to the southern mountains a few occasionally appeared in lower settled areas, such movements taking place at night, when most road-kills occur (Marchant & Higgins 1993). Ringing has shown that movement occurs between the summit of Mt Gower and palm-covered plateaux 300m below and 1.5km south of the mountain (Miller & Mullette 1985). On Mt Gower, juveniles disperse from the natal territories in Jun-Jul, when 3-5 months old, and then live a skulking existence in occupied territories or inhabit marginal territories, usually disappearing before the next breeding season (Fullagar & Disney 1975).

HABITAT Information is from Marchant & Higgins (1993). Confined to forests of the subtropical oceanic Lord Howe I. Before human settlement, the birds probably occurred over most of the island at low to high altitudes but, as a result of persecution and habitat degradation, they were subsequently restricted to the highest parts of the mountains, up to 825m, in gnarled moss-forest dominated by *Bubbia howeana*, *Dracophyllum fitzgeraldii*, *Negria rhabdothamnoides*, *Leptospermum polygalifolium*, *Hedyscepe canterburyana* and *Lepidorrhachis mooreana*, with an understorey of tree-ferns *Cyathea* and a litter-covered floor. Most vegetation is covered by moss, lichens and small ferns. This is not a preferred habitat, and lowlands birds breed more rapidly, and with far greater success, than mountain birds. The species is now re-established in some lowland forest areas (half the present population occurs in settled lowlands), where it inhabits megaphyllous broad sclerophyll forests, including stands of *Howea* palms (particularly *Howea forsteriana* on igneous soils and *Hedyscepe*), *Ficus columnaris*, *Drypetes australasica* and *Cryptocarya triplinervis*. It occurs in forest bordering pasture and gardens, and its territories may include open areas such as paddocks, pastures and small marshes; a chicken yard and a rubbish dump are also recorded. It rarely occurs in rainforest, which is the most widespread vegetation on the island, preferring other closed forest types on igneous soils, possibly because more food is available near the surface of the soil. It occurs on boulder-covered slopes, steep scree, in valleys, on plateaux, often near cliffs and sometimes near the coast.

FOOD AND FEEDING Most information is from Marchant & Higgins (1993). Mainly earthworms, crustaceans (Amphipoda and Isopoda) and insect larvae (Coleoptera); also Gastropoda, Myriapoda, Arachnida and Hemiptera (Cicadidae and Notonectidae). The eggs and chicks of shearwaters, petrels and domestic fowl are also taken. The birds also eat lichens, fungi, pteridophytes (*Blechnum*), and vascular plants including flowers of *Randia stipulosa* (Rubiaceae) and fruits such as *Cucubis*, tomatoes and strawberries. These birds forage on the forest floor, turning over and raking aside leaf-litter with the long bill but not using the feet, and foraging intensively in small areas (e.g. 0.5m^2 for 20-30 min). They also forage along roads through forest, and in gardens (I. Ferrier *in litt.*). They chisel wood, enlarge holes in soil by gaping with the bill, and dig holes up to 10cm deep for earthworms; they also pull aside large fallen palm leaves to glean underneath, and probe moss and rotten logs. At Mt Gower, 52.2% of feeding sites comprised soil, 30.7% leaf-litter, 4.5% rotting logs, 5.8% moss and 6.8% other sites (Miller & Mullette 1985). To obtain food they frequent areas inhabited or regularly visited by people and will eat meat, stew, butter, porridge, biscuits, bread, chocolate and rat poison, but not muesli bars. Captive birds were also given mealworms (*Tenebrio* larvae), woodgrubs (Cerambycidae), cat food, poultry pellets, cheese and madeira cake (Lourie-Fraser 1983). The rails must have water, which they obtain from streams, pools and droplets on moss.

HABITS Hindwood (1940) describes how occupants of the first ships to arrive at Lord Howe Island found most birds remarkably tame and one report describes "the sport we had in knocking down birds", including "a curious brown bird about the size of the Land Reel [Rail] ... walking totally fearless and unconcerned in all parts around us, so that we had nothing more to do than to stand still a minute or two and knock down as many as we pleased with a short stick if you throwed at them and missed them, or even hit them without killing them, they never made the least attempt to fly away". Another visitor describes knocking down several birds, and "their legs being broken, I placed them near me as I sat under a tree. The pain they suffered caused them to make a doleful cry, which brought five or six dozen of the same kind to them, and by that means I was able to take nearly the whole of them". Today, the rails are alert and wary but not shy (but they are secretive when breeding), and they can readily be observed in their natural environment, despite the thick vegetation which they inhabit. The following information is from Marchant & Higgins (1993) unless otherwise stated. They are curious, and will come to investigate any unusual sound, such as two rocks being tapped together or people talking or whistling. When excited or nervous they flick the tail, and when disturbed they run rapidly and can jump 1.5m vertically using the legs alone. Although flightless, they have been seen to skim horizontally for 3.5m using the wings, with legs dangling. They are diurnal, being most active at dusk and dawn and moving to a roosting site at dusk, but may often move around at night and well before dawn. They may roost in thick ferns, and in small caves and petrel burrows. When sunbathing, they stand with the wings spread sideways (Fig. 5). They will preen in view of observers (I. Ferrier *in litt*.).

Figure 5. Lord Howe Rail – sunning posture. [After Serventy 1985].

SOCIAL ORGANISATION Information is from Marchant & Higgins (1993). Monogamous, apparently with a permanent pair bond, and strongly and permanently territorial. On Mt Gower the territory size may be 2-3ha. Single males may also hold territories. The young of a previous brood often feed and defend young of the next brood and assist in territorial defence.

SOCIAL AND SEXUAL BEHAVIOUR Information is taken from Marchant & Higgins (1993). The male establishes the territory but both sexes defend it. Communal calling by a territorial pair is stimulated by the alarm call or any unusual noise, and occurs for most of the year and at any time of day or night. The birds face each other and call with open, upward-pointing bills. The threat display, made to intruding conspecific individuals and also readily to predators such Pied Currawongs and Black Rats, involves lowering the head and opening and raising the wings above the back; these predators are also chased. Intruders may be chased and attacked with no warning calls being uttered, and a pursuer will jump on the back of an intruder. Birds have been seen to fight by rushing at each other like fighting cocks, with the feet up. The young are fiercely defended, and adults will attack if young are caught and handled. In captivity, at c. 100 days of age, males fight among themselves but females coexist without aggression.

Courtship feeding may occur before copulation, and pair members allopreen, sometimes after copulation. In the pre-copulatory display (seen in captivity) the male gives a beckoning call and walks in step behind the female, who postures with lowered tail, sometimes crouching or standing with lowered head. The female may initiate this display by running in front of the male and calling. An additional display involves the male circling the female with an erect posture; the hunched-up female calls and the male pulls at her nape. To copulate the female crouches down and the male steps on her back, balancing with slightly open or flapping wings, and copulates for c. 15sec. After copulation, both birds stand side by side, the male sometimes with lowered head, and the male gives a low call. Copulation may be followed by allopreening or nest-building. Adults also allopreen young birds.

BREEDING AND SURVIVAL Most information is from Marchant & Higgins (1993). **Season** Normally lays Aug-Jan; 1 lowland pair laid 11 times in 18 months, Jan 1982-Jun 1983. **Nest** A shallow depression in ground, lined with dry grass, ferns, moss, palm fronds and leaves; 10-25cm across and 7.5cm deep; cavity depth 2.5cm; well hidden in thick vegetation, under tree roots or fallen log, or 10-65cm down petrel burrow. In captivity, several trial nests made, and final site chosen by female; both sexes build. Rough brood-nests also constructed. **Eggs** 1-4 in captivity, laid at intervals of 24-36 h; oval and smooth; dull white, minutely dotted and irregularly blotched light chestnut and faint steel-grey, mostly at blunt end; size of 2 eggs 48 x 34.4, 49 x 34.7 (Schönwetter 1961-62); for 33 eggs laid in captivity, mean size 46.9 x 33.4, mean weight c. 29 g, while 2 eggs from semi-tame pair measured 51 x 33.3 and 51.4 x 34.5, weight 30.7, 30.5. **Incubation** 20-23 days, by both sexes; begins when clutch complete; during day, changeover occurs about every 2 h (20 min to 2.5 h). Birds readily desert nests if disturbed. **Chicks** Precocial and nidifugous; move from nest to nursery (brood nest) within 2 days of hatching; fed and cared for by both parents, also by young of previous brood; drink from 4 days old; peck at food from 6 days. Pin-feathers appear from 12 days; young preen from 18+ days; are fully feathered at 28 days and almost fully grown at 65 days, when in heavy postjuvenile moult; leave, or are driven from, natal territory when 3-5 months old. Breeding success greater in lowlands than on Mt Gower, and greatest in Settlement area, where birds receive extra food and water from islanders. On Mt Gower, 1970-79, usually 5 or fewer young fledged from 9 territorial pairs; probably none in some years. Lowlands pair which laid 11 times in 18 months produced 29 eggs and reared 16 young to independence. Nest predators include Pied Currawongs and Black Rats. Can pair and breed when 9 months old. Multiple-brooded pairs in wild and captivity laid clutches 31-66 (45; n = 5) days after previous clutch hatched. **Survival** Survival of immatures apparently depends on establishing new territory or occupying a vacant place in an existing one. Greatest recorded age 13+ years (male).

42   GILBERT RAIL
Gallirallus condiicius   Plate 12

Tricholimnas condiicius Peters & Griscom, 1928, Apaiang Island; later thought to be from the south-west Marshall Islands, probably Ebon (Walters 1987).

Possibly an immature specimen of *G. sylvestris* (41).

Synonyms: Species name sometimes erroneously spelt *conditicus*.

Alternative name: Apaiang Rail.

DESCRIPTION Length 30.5cm. Flightless. Known only from the holotype, a skin prepared from an eviscerated specimen originally preserved in alcohol at the Museum of Comparative Zoology, Harvard. The skeleton has apparently been lost (Olson 1991) but the original description of the holotype (Peters & Griscom 1928) states that examination of the skeleton showed the bird to be an immature. Its plumage is described as follows (no allowance was made for the possible bleaching effects of alcohol). "Top of head buffy-brown, shading into cinnamon-brown on the back, rump and wing-coverts, the latter russet basally with obsolete dusky bars; uppertail-coverts and tail chestnut-brown. Indistinct superciliary stripe, lores, sides of head, throat and foreneck pale greyish-brown; remainder of underparts buffy-brown, becoming more cinnamon on the flanks; undertail-coverts chestnut-brown with indistinct paler barring. Primaries and secondaries dull rufous ('russet'), irregularly barred blackish and indistinctly tipped dusky." It was noted as being similar to the Lord Howe Rail (41) but smaller, with less conspicuously chestnut wing-coverts. **Measurements** Wing 132; tail 58; bill 45; tarsus 47 (Greenway 1952).

Subsequent investigation by Greenway (1952), including a comparison of the holotype with a series of Lord Howe Rail skins, suggested that the specimen is an immature Lord Howe Rail. It differs in plumage from that species only in having a paler crown, throat and underparts, and a browner head and throat. These differences could be explained by the bird's long immersion in alcohol (Greenway 1952). It is also slightly smaller than the smallest female Lord Howe Rail, a difference which may be explained by the fact that it is an immature. However, Walters (1987) maintains that alcohol does not usually affect the brown and grey melanins involved in the colour differences described by Greenway.

Olson (1991) could find no consistent difference in plumage between the holotype and a series of Lord Howe Rail specimens. He pointed out that, although the proportions of its bill differ from those of adult Lord Howe Rails, they are almost exactly duplicated by immatures of that species. He also noted that the feet of the specimen are orangish in colour, as are those of one Lord Howe Rail which had been prepared from a spirit specimen. He found no morphological features by which the holotype could be separated from the Lord Howe Rail at any level and considered it hardly conceivable that two flightless rails could evolve independently on islands c. 4,500km apart and show no morphological differences.

DISTRIBUTION AND STATUS The holotype bears the label "Kingsmill Islands, 1861, Andrew Garrett, Collector" and was originally thought to have been collected by Garrett when he visited Apaiang Atoll in 1859. However, the island is low, rocky and sandy, its primary cover is *Cocos* (Amerson 1969), and it contains no habitats similar to those which are occupied by the bird's two closest relatives, both forest dwellers, the Lord Howe Rail and the New Caledonian Rail (40). Furthermore, Garrett wrote that there was only one species of land bird on Apaiang, which he thought was a hawk but was probably the New Zealand Cuckoo *Urodynamis taitensis* (Amerson 1969). It was suggested (Greenway 1952) that the holotype is a Lord Howe Rail which may have been confused with a consignment of specimens from the Kingsmills (Gilbert Is) and thus mislabelled. It was thought possible that Garrett obtained the specimen from another source (Greenway 1952), who had presumably obtained it from Lord Howe I. No similar birds have been recorded from other islands in groups close to the Kingsmills and, if the holotype did come from Apaiang, the species must have become extinct in these islands very soon after Garrett's visit. These uncertainties, plus the fact that the specimen was considered to be an immature female Lord Howe Rail (see Description), led Greenway (1952) to propose that the bird had been collected on Lord Howe I.

Walters (1987) reviewed the evidence and concluded that the specimen should be regarded as a distinct species. Acknowledging the unlikelihood that Garrett would have obtained a specimen from Lord Howe I, Walters assumed the label to be correct and suggested that the bird was obtained from an island in the Kingsmill (Gilbert) group. Agreeing that Apaiang was a very unlikely source, Walters suggested that Garrett obtained the holotype on Ebon Atoll, at which his boat called immediately after picking him up from Apaiang. Ebon is in the southern Marshall Is, which are close to the northern Gilberts and is one of the lushest, best vegetated and most fertile of the Marshall Is, having 20 species of plants and a jungle-like, dense central area (Amerson 1969). It was also the island from which a dove *Ptilinopus marshallianus*, known only from the holotype, was also described by Peters and Griscom (1928).

Olson (1991) rejected Ebon as the type locality, quoting an overlooked 1861 report on its natural history which describes the few bird species occurring on the island but does not mention a rail. Furthermore, he maintains that the report's description of the atoll suggests that it would be almost as unsuitable a home for a large flightless rail as was Apaiang. Olson maintains that the holotype is a Lord Howe Rail and points out that, if Garrett had obtained the specimen, it should have come into his possession before 1861, but specimens of the Lord Howe Rail were not obtained and described until 1869. Furthermore, it appears that the traffic in natural history specimens from Lord Howe I in that period was through naturalists in London via Sydney and it is unlikely that

Garrett in 1859-61, while he was based in Hawaii, could have obtained a specimen of the Lord Howe Rail about ten years before an example of the same species reached London. Olson therefore concludes that Garrett had no connection with the holotype, and that a curatorial error led to the specimen being wrongly associated with an old label from Garrett's collections.

FOOD AND FEEDING Stomach contents are recorded as "seeds of two species, the chitinous shell of a small beetle, and an entire harvest fly [cicada]". Apparently this material was not retained (Olson 1991).

43 OKINAWA RAIL
Gallirallus okinawae   Plate 15

Rallus okinawae Yamashina & Mano, 1981, near Mt. Fuenchiji, Kunigami-Gun, northern Okinawa.

Forms a superspecies with *G. torquatus* and *G. insignis*. Monotypic.

IDENTIFICATION Length c. 30cm; wingspan c. 50cm. Medium-sized and almost flightless; tail very short and decomposed; legs long and strong. Entire upperparts and upperwings dark olive-brown; chin, throat, sides of head and upper neck black, with whitish loral spot and white line from below rear of eye across ear-coverts to side of upper neck. Some outer primaries narrowly barred white on outer webs; inner primaries broadly barred brownish-yellow. Underparts black with narrow wavy white bars; undertail-coverts blackish-brown, barred buffy-white. Bill scarlet, tip ivory-white; eye-ring, iris, legs and feet red. Sexes alike. Immature like adult but has darker bill tip and culmen. Juvenile has paler, olive-tinged head and upperparts, mottled underparts, brownish bill, brown iris and fleshy yellow-ochre legs and feet. Inhabits broadleaf evergreen forest near water.

Similar species Most similar to races *sulcirostris* and *limarius* of larger, volant Barred Rail (44) of South-east Asia, but differs in having reduced white facial stripe, and red bill and legs. Flightless New Britain Rail (45) larger, with different head pattern, black undertail-coverts, black bill and pink legs and feet.

VOICE A highly vocal species. Information is taken from Brazil (1984, 1991). Calls include a loud *kyo*, a *kwi kwi kwi ki-kwee ki-kwee*, often answered by a *ki-ki-ki*, and a *kyip kyip kyip kyip* given by a pair as a rapidly overlapping chorus in answer to the *ki-ki-ki* call. A rolling *kikirr krarr* is followed by *kweee* notes on a rising scale to become an almost pig-like squeal which is followed by a repeated *ki-kwee-ee* call, rather reminiscent of the neighing of a horse. Deep bubbling *gu-gu-gugugugu* and *gyu-gyu-gyagyagya* calls are given by birds while walking and apparently when alarmed. The birds normally call mostly in the early morning, late afternoon and evening, when their voices carry further in the calm air. Calls are normally given from the forest floor but occasionally from trees, and duetting by roosting pairs has been recorded, both birds giving *kek* calls.

DESCRIPTION
Adult Entire upperparts from forehead to tail, including upperwings, dark olive-brown. Lores, sides of head, sides of upper neck, chin and throat black; whitish spot on lores and white line from below rear of eye across ear-coverts and down side of upper neck. Remiges brown; some outer primaries have narrow white bars on outer webs; one bird also had barring on inner webs, and on alula (Brazil 1984); inner primaries broadly barred brownish-yellow; secondaries unbarred. Axillaries and underwing-coverts blackish-brown with white spots and bars. Entire underparts from lower neck to belly black, with narrow white feather tips forming wavy bars which are reduced or absent on inner thighs and centre of belly; scarcely visible chestnut to olive-brown feathers present on sides of breast. Undertail-coverts blackish-brown, feathers with wide buffish-white terminal band. Eye-ring red; iris bright blood-red; bill scarlet, tip ivory-white; legs and feet red. Sexes alike.
Immature Like adult but has darker bill tip and culmen until c. 1 year old. Brazil (1985a) described a bird with white patch around the eye, poorly defined white line from rear of eye, and pale orange-red bill with yellowish tip; age of this individual is not known.
Juvenile Paler; upperparts and head tinged olive; loral spot tinged brown; white facial stripe shorter, broader and more irregular; underparts mottled rather than barred with white; iris brown; bill mottled brownish; legs and feet fleshy yellow-ochre.
Downy young Black; bill white with blackish base and tip; iris black; legs and feet yellowish (1 photograph shows chick with dull reddish legs).

MEASUREMENTS (holotype, an adult female) Wings (chord; worn) 139, 148; tail 53; exposed culmen 45; tarsus 59. One unsexed live adult: exposed culmen 52.5; tarsus 65. Weight of 1 adult 433; of 1 fully grown 434.5.

GEOGRAPHICAL VARIATION None.

MOULT Wing and tail moult of an adult was recorded in July and of a juvenile (postnatal moult) in June.

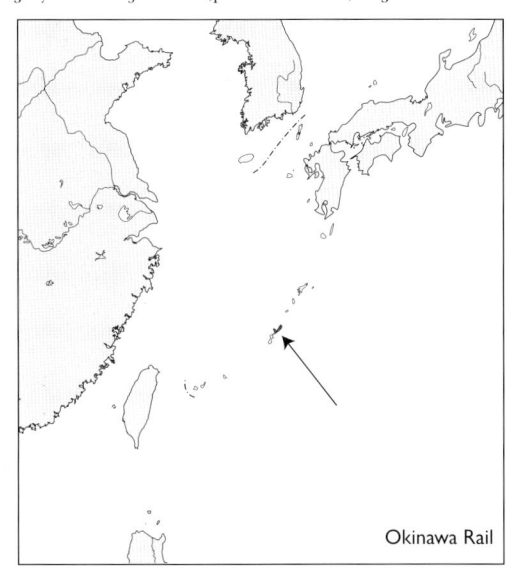

DISTRIBUTION AND STATUS Endemic to Okinawa I in the Ryukyu Archipelago, southern Japan, where it is confined to the largely uninhabited forests of the N quarter (Yambaru), N of a line from the Shioya to Higashi-son. Although there are reports of unidentified rails dating back to 1973 and of unidentified calls up to 20 years earlier, the rail's existence was properly established by T. Mano's observations of one bird near Mt Yonaha in June 1978 and also in 1979 and 1980 (Yamashina & Mano 1981, Brazil 1991). A dead bird was found near Mt Fuenchiji in June 1981 and the species has subsequently been recorded from many localities in Yambaru. It is apparently reasonably common in its very small range (c. 26,000ha), the population being estimated at c. 1,800 birds (Yanagisawa *et al.* 1993), and widespread wherever suitable habitat remains, but is probably declining in numbers. It is a globally ENDANGERED species and is threatened by continuing government-sponsored deforestation (Collar 1987) which has short-term political benefits but is potentially disastrous with respect to erosion, soil infertility, siltation of dams and marine pollution. The preservation of the forest habitat is of prime future importance in relation to local land use (Yamashina & Mano 1981). The Yambaru forests also hold one other globally endangered endemic bird, the Okinawa Woodpecker *Sapheopipo noguchii*, and an endemic frog and beetle, as well as other rare and threatened birds (Collar 1987). Southern Okinawa is highly developed and overcrowded, and in recent years the pace of development has increased in the north, with the building of golf courses, dams and public roads in the hitherto continuous pristine forests; forest habitat on the lower slopes has also been destroyed to create pineapple fields (Brazil 1984, Harato & Ozaki 1993). The rail's southern distributional boundary is gradually retreating north into the Itajii *Castanopsis siebold* forest (Harato & Ozaki 1993). Isolated records on the Motubu peninsula, an area which has suffered great habitat destruction, may refer to a tiny relict population. No birds were found during an intensive search in 1985 and this population must be close to extinction (Brazil 1985b).

The rail is designated a 'Natural Monument' and a special bird for protection; this designation prevents hunting and trapping, but the bird and its habitat receive no protection from disturbance or destruction outside wildlife protection areas, which are themselves not permanently protected (Brazil 1984). Conservation priorities include the preservation and restoration of Itajii forest habitat, the control or elimination of introduced predators, the development of captive breeding techniques, the translocation of rails to unoccupied suitable habitat, the changing of laws so that Okinawa's native flora and fauna may be conserved, and the development of a programme to increase public awareness of the local natural history heritage and the need for effective conservation measures (Harato & Ozaki 1993).

The bird's habit of roosting in trees may be adaptive in avoiding snake predation. However the snake *Trimeresurus flavoviridis* takes eggs and young, while *Dinodon semicarinatus* may also be a predator; furthermore, the arboreal and nocturnal Brown Tree Snake *Boiga irregularis*, which has extirpated the Guam Rail (48) from the wild on Guam (see Guam Rail species account), has been observed on Okinawa and may have the potential to establish itself (Harato & Ozaki 1993). Introduced predators probably include domestic cats and dogs (cats are often observed along forest roads and at dam shorelines) and potentially also the weasel *Mustela itatsi* and the mongoose *Herpestes edwardsi*, which are spreading into the rail's original habitat (Harato & Ozaki 1993). Birds are occasionally killed by vehicles on roads, and dead birds have also been found in grassy areas and pineapple fields where the causes of death are not known (Brazil 1984).

MOVEMENTS Sedentary, but in the winter some may descend to lower altitudes and wander to reach areas just S of the main distribution (Brazil 1991). It is not known what ages of birds are involved in these movements.

HABITAT Its habitat is essentially subtropical broadleaf evergreen forest with dense undergrowth and some water, but it is catholic in its choice of habitats and occurs almost wherever forest and marshy areas still exist (Brazil 1991). It is found in primary and secondary forest, along forest edges and in small forest patches. The dominant tall tree in its forests is Itajii but other trees such as *Quercus miyagii*, *Persea thunbergii*, *Schima wallichii*, *Styrax japonica* and *Distylium racemosum* are also common (Harato & Ozaki 1993). It is also found in Ryukyu Pine *Pinus luchuensis* forest, in meadows and grasslands adjacent to Itajii forest, and in scrub and around cultivated areas close to forest, where the ground is damp or streams, pools or reservoirs occur (Harato & Ozaki 1993). It has also been recorded near highly populated areas and in a lawned garden (Harato & Ozaki 1993). It requires standing water for bathing. It occurs from sea level to the highest hilltops at 498m in mountainous areas.

FOOD AND FEEDING Land snails, insects (particularly locusts), amphibians and lizards, taken on the forest floor; the birds possibly also take some food from shallow water when visiting pools to bathe and drink (Brazil 1991).

HABITS These birds fly very weakly but have strong legs and run very rapidly to escape predators. They avoid open areas and quickly run to cover when disturbed, so are extremely difficult to observe. They are diurnal and crepuscular, being most visibly active in the early morning and in the late afternoon to just after sunset, when they visit standing water to bathe. When near water, birds spend most of their time preening in bouts of 4-20 min between brief bouts of bathing (2-4 min), which involve the bird crouching in shallow water, dipping forwards and downwards and splashing with the wings (Brazil 1984). Birds stand erect to preen, and regularly raise the head to scan around; preening may end with the wings flipped up above the back and with vigorous wing flapping and also jumping; the wings are also spread as if to dry them (Brazil 1991). The birds swim strongly, rather like Water Rails (61), nodding the head back and forth (Brazil 1991). They usually roost in trees, presumably to escape ground predators, although roosting in bushes, and once on the ground, is also recorded (Harato & Ozaki 1993). Roosting birds may be observed by spotlight and a recent study (Harato & Ozaki 1993) found that roost trees were usually Itajii, either alive or dead, and that the birds usually roosted in trees on the hillside slope beside and above forest roads. Roost trees were 6-16m tall (mean 10.6m) and usually grew at an angle, while roosting branches were 2-12m (mean 6.7m) high, 5-30cm (mean 12.7cm) in diameter and relatively horizontal (mean angle 27.7° above the horizontal). The birds usually roosted in a spot free of leaves; if on a dead limb, they usually roosted at the end of the limb. Birds also roosted on the trunks of trees which were leaning or had fallen. One bird was recorded roosting on a telephone pole. Most birds roosted individually but pairs roosted up to 5m apart. When climbing the trees, the rails flap the wings and scratch with the feet. They usually preen and stretch for 10-20 min before leaving the roost at daylight, when most drop directly to the ground.

SOCIAL ORGANISATION Monogamous and territorial, apparently with a strong permanent pair bond. Birds frequently appear in pairs and remain in close proximity to each other throughout observation periods (Brazil 1984).

SOCIAL AND SEXUAL BEHAVIOUR The rails sometimes chase members of their own species, and also (but unpredictably) attack or chase other bird species, including Rufous Turtle Doves *Streptopelia orientalis*, Grey Wagtails *Motacilla cinerea* and Common Moorhens (129); at other times they appear to associate with the Common Moorhen (Ikenaga 1983, Brazil 1984). Pair members often feed and roost together, and allopreening has been observed (Brazil 1991).

Figure 6: Partly fledged Okinawa Rail.

BREEDING AND SURVIVAL Season Lays May-Jul. **Nest** Placed on ground; 1 photograph shows nest, possibly in cavity of rotten tree stump, of dead leaves, grass and fern fronds. **Eggs** 2-4; ovate; white, spotted and blotched brick-red, vinaceous-pink, brownish-olive and light brownish-olive, markings concentrated at the blunt end; no information available on size. **Incubation** Period unknown. **Chicks** No information available. Eggs and young are taken by the indigenous snake *Trimeresurus flavoviridis* (Harato & Ozaki 1993).

44 BARRED RAIL
Gallirallus torquatus   Plate 13

Rallus torquatus Linnaeus, 1766, Philippines.

Sometimes retained in *Rallus* or placed in *Hypotaenidia*. Forms a superspecies with *G. okinawae* (43) and *G. insignis* (45). Five subspecies recognised.

Synonyms: *Habropteryx/Eulabeornis torquatus*; *Hypotaenidia torquata/celebensis/saturata/sulcirostris/Jentinkii(=jentinki)*; *Rallus lineatus/celebensis/ sulcirostris*; *Eulabeornis sulcirostris*.

Alternative name: Philippine Rail.

IDENTIFICATION Length 33-35cm. A medium-large rail; entire upperparts olive-brown (deep rufous-brown in race *sulcirostris*); face, throat and foreneck black with prominent white streak running from base of bill, below eye, across ear-coverts and down side of neck; underparts black, barred white; extent of white barring on throat and foreneck variable, entire area being barred in some *celebensis* while throat to upper breast are unbarred in *sulcirostris*; nominate race of Philippines has maroon-chestnut breast-band. Undescribed form from S Sulawesi has white supercilium, no white cheek stripe, and brown throat. Eye red; bill black; legs and feet brown. Sexes alike.

Immature duller than adult; often washed buffy on underparts; throat whitish; in nominate race breast-band olive-brown and white barring on undertail-coverts tinged reddish-brown. Juvenile like immature but black replaced by dark grey or blackish-brown; white barring tinged vinaceous-cinnamon; one has pale fringes to feathers of crown and upperwing-coverts, barred tertials, and white supraloral streak. Inhabits grassland, marsh edges, cultivation and secondary growth.

Similar species Easily separable from sympatric rallids on plumage. Widespread Buff-banded Rail (46) has spotted upperparts and upperwing-coverts, chestnut facial stripe, rufous and black barring on most remiges, pink bill and grey-pink legs. New Britain Rail (45) and Okinawa Rail (43) are almost flightless and differ in head pattern, while Okinawa Rail has reddish bill and legs, and New Britain Rail has black undertail-coverts and pink legs.

VOICE A loud cacophony of seemingly pulsating, very discordant, harsh croaks and screams, often lasting several seconds, invariably given by more than one bird at a time and usually from dense cover. This call was incorrectly ascribed by Watling (1983) to the Isabelline Bush-hen (85) (Rozendaal & Dekker 1989).

DESCRIPTION *G. t. torquatus*
Adult Entire upperparts, from forehead to tail, olive-brown, often tinged greenish-olive; forehead slightly greyer; lower back to uppertail-coverts duller, sometimes with some indication of blackish barring on rump; rectrices, and sometimes uppertail-coverts, tinged rufous-brown. Lores, feathers round eye, ear-coverts, throat, sides of nape, and sides of neck black; chin sometimes slightly greyer; prominent white stripe extends from base of bill, below eye, across ear-coverts and down sides of neck. Median and lesser upperwing-coverts, and tertials, as upperparts (tertials with darker centres); greater coverts slightly more rufous-brown; primary coverts and primaries rufous-brown, chequered with vague dusky markings on outer webs and barred dusky brown and rufous on inner webs; inner webs of outer primaries often have white barring; secondaries rufous-brown. Alula feathers blackish-brown, olive-brown on outer webs; inner webs barred white, outer webs spotted white. Axillaries and underwing-coverts barred black and white; underside of remiges rufous-brown; inner webs of primaries with blackish-bordered white bars. Foreneck to undertail-coverts black, narrowly and profusely barred white; broad band of rich maroon-chestnut across breast; vent and undertail-coverts tinged rufous-brown. Iris blood-red; bill black; legs and feet brown. Sexes alike; iris of female crimson.
Immature Duller than adult; often washed buffy on underparts; chin, or chin and throat, mottled whitish; breast-band olive-brown; white barring at sides of vent and on undertail-coverts tinged reddish-brown. See Juvenile.
Juvenile One juvenile in BMNH resembles above description of immature, and has all remiges in sheath; black of head replaced by dark grey; chin and throat white with a little dusky mottling; breast-band olive-brown with faint rufous barring; underpart barring blackish-brown and white, some white bars tinged vinaceous-cinnamon; thighs olive-brown washed vinaceous-cinnamon; undertail-coverts barred blackish-brown and vinaceous-cinnamon. See race *celebensis*.
Downy young Not described.

MEASUREMENTS (7 males, 10 females) Wing of male 149-156 (152.6, SD 2.2), of female 135-150 (142, SD 4.9);

tail of male 50-62 (55.7, SD 4.2), of female 45-55 (49.8, SD 2.9); culmen to base of male 41-48 (44.4, SD 2.3), of female 41-47.5 (43.5, SD 2.3); tarsus of male 48-53 (51.1, SD 2), of female 46-52 (48.4, SD 1.7). Weight of 1 female 241.

GEOGRAPHICAL VARIATION Involves upperpart colour, the extent of barring from throat to upper breast, the presence or absence of a rufous breast-band, and biometrics. The nominate race includes the putative forms *maxwelli*, *quisumbingi* and *sanfordi*; race *sulcirostris* includes *jentinkii* and *simillimus*; *Hypotaenidia saturata* is synonymous with race *limarius* (Parkes 1971, Ripley 1977). For a discussion of the proposed form *remigialis*, see race *celebensis*.

  G. t. torquatus (Linnaeus, 1766) – Philippines (Bantayan, Basilan, Biliran, Bohol, Bongao, Boracay, Cagayancillo, Cahayagan, Caluya, Camiguin Norte, Carabao, Catanduanes, Cebu, Dinagat, Fuga, Leyte, Luzon, Marinduque, Masbate, Mindanao, Mindoro, Negros, Panay, Pan de Azucar, Polillo, Romblon, Samar, Semirara, Siargao, Sibay, Sibuyan, Siquijor, Ticao and Verde Is). See Description.

  G. t. celebensis (Quoy & Gaimard, 1830) – Sulawesi and Muna. Upperparts dark olive-brown (feather centres darker and fringes paler); rectrices olive-brown with dusky centres; no breast-band. Extent of white barring on black throat and upper breast variable (see Plate 13): entire area may be barred, upper throat may be unbarred, or unbarred area may extend as far as breast. Immature similar to adult but trace of juvenile's white eyebrow (see below) remains; chin (and sometimes upper throat) white, mottled dusky; breast, belly and undertail-coverts buffier. A juvenile in Leiden is similar but has crown feathers dark olive-brown with olive fringes; irregular white line from base of upper mandible to above eye (a very distinctive character); upperparts paler than in older birds; marginal and lesser upperwing-coverts dusky brown with narrow pale rufous-brown fringes giving barred effect; similar but fainter pale barring extends over the more olive-brown median and greater coverts; tertials dusky brown, narrowly barred rufous-brown; chin and throat white with slight dusky mottling, this pattern extending to side of upper neck; lower throat and foreneck mottled grey-black and white; upper breast blackish, spotted buffy-white; undertail-coverts brownish-black, barred buff. In specimen, bill appears browner than in adult; legs appear paler and more olive (R. Dekker *in litt.*). Measurements from Sharpe (1894) (type specimen), Ripley (1977) and White & Bruce (1986) (sample sizes not given): wing of male 138-168, of female 143-162; tail of male 40-59; culmen of male 35-39; tarsus of male 47-54. Seven unsexed: wing 139-154 (149.1, SD 5.8); tail 51-60 (54.7, SD 2.9); culmen to base 39-42 (40.7, SD 1.1); tarsus 45-53 (49.3, SD 2.9).

  The form *remigialis* has been proposed for birds from SE Sulawesi and Muna, which are larger (wing of 2 males 167, 168, of female 153-162; wing of *celebensis* from N Sulawesi, male 149-164.5, female 143-153), but *remigialis* may be a clinal intermediate and more material is needed to clarify the situation (White & Bruce 1986).

  G. t. kuehni (Rothschild, 1902) – Tukangbesi Is (Binongka and Kaledupa), SE of Sulawesi. Largest race, with deep olive-brown upperparts; no breast-band. Wing of 1 male 174, of 1 female 169; tail of 1 male 75; tarsus of 1 male 54.

  G. t. limarius (J. L. Peters, 1934) – N Irian Jaya (Vogelkop) and Salawati I. Similar to *celebensis*, but always lacks white barring on throat; no breast-band; bill relatively long and strong. Wing of 2 males 150, 151, of 1 female 140; tail of 2 males 58.5, 59; culmen to base of 1 male 47, of 1 female 44; tarsus of 2 males 54.5, 60, of 1 female 57.

  G. t. sulcirostris (Wallace, 1862) – Sula Is (Taliabu, Seho, Mangole and Sanana) and Banggai Is (Peleng). Upperparts deep rufous-brown; sides of neck, and foreneck to upper breast, intensely black with no white barring and no breast-band; at least some birds have white barring of rest of underparts relatively narrow. Deep groove along side of lower mandible extends forwards for at least half of bill length. Wing of male 144-165, of female 140-151, of 1 unsexed 136; tail of 1 male 51, of 1 female 58.5, of 1 unsexed 53; culmen to base of male 35-39.5, of 1 female 43, of 1 unsexed 42; tarsus of 1 female 54.5, of 2 unsexed 49, 49.5 (Sharpe 1894, Ripley 1977, White & Bruce 1986, BMNH).

UNDESCRIBED FORM Baltzer (1990) observed birds at the Tempe lake system, and at Mampie and Maros, in S Sulawesi, which appear to represent an undescribed form. They had plain olive-brown upperparts; the white line on the side of the head passed through and slightly above the eye (not below the eye); and the throat was brown. The breast was faintly and inconspicuously barred black and grey so that the underparts often appeared dark grey, and the barring faded away towards the "rump" (belly?), which was uniform grey. The bill was dark grey to black.

MOULT Two immatures in Leiden, collected at Gorontalo, N Sulawesi, in Jul 1863, have simultaneous moult of all primaries and secondaries, the new feathers being just less than half grown (R. Dekker *in litt.*). One juvenile in BMNH, from Marinduque, May, has all remiges partly grown.

DISTRIBUTION AND STATUS Lowlands of Sulawesi, the Philippines and NW New Guinea. Not globally threatened. This species is very shy and retiring and there is little information on its current status. The nominate race is common in the Philippines, while *limarius* of New Guinea was formerly regarded as quite rare (Rand & Gilliard 1967, Dickinson *et al.* 1991). In the 1980s *celebensis* was reported as common in the lowlands of the Minahasa Peninsula and in north-central Sulawesi (Ripley & Beehler 1985, Rozendaal & Dekker 1989), while Watling (1983) found it far commoner in N Sulawesi than in C areas where the Buff-banded Rail was more abundant; it is currently regarded as widespread and generally common in Sulawesi (Coates & Bishop 1997).

MOVEMENTS None recorded.

HABITAT Occurs in open dense grassland, rice and maize fields (including areas with a scrub component), coastal swamps, at the edges of marshes, at grassland and marshes adjacent to, or within, primary and tall secondary forest, and in dry secondary growth; it is also recorded from mangroves (Coates & Bishop 1997). In S Sulawesi, with other species such as the White-breasted Waterhen (89), Purple Swamphen (120), Common Moorhen (129) and Dusky Moorhen (130), it inhabits mature 'bungkas', large, artificially constructed mats of floating vegetation on lakes; this habitat is apparently not occupied by the Buff-banded Rail (Baltzer 1990). In the Philippines it often prefers areas where grassland, marshy spots and small groves of trees

45 NEW BRITAIN RAIL
Gallirallus insignis Plate 13

Rallus insignis P. L. Sclater, 1880, Kahabadai, New Britain.

Sometimes retained in *Rallus*, or placed in the monospecific genus *Habropteryx* although its affinities are clearly with *G. torquatus* (44). Forms a superspecies with *G. okinawae* (43) and *G. torquatus*. Monotypic.

Synonyms: *Hypotaenidia/Habropteryx insignis*.

Alternative names: Bismarck/Pink-legged/Sclater's Rail.

IDENTIFICATION Length 33cm. A medium-large rail with no distinctive head pattern. Probably almost flightless, although some reports say that it may fly well; tail very short and decomposed. Top and sides of head, hind-neck and sides of neck dark rufous-brown; rest of upperparts dark olive-tinged brown, darker on tail and tinged rufous on tertials and greater coverts; primaries barred and spotted white; entire underparts black, narrowly barred white except from lower belly to undertail-coverts. Eye red; bill black; legs and feet rose-pink. Sexes alike; female averages slightly smaller. Immature and juvenile not described. Inhabits damp forests and swampy canegrass.
Similar species Barred Rail (44) and Okinawa Rail (43) differ in having a black face with prominent white facial stripe, and barred undertail-coverts; Barred Rail volant; has longer tail, brown legs and, in one race, maroon-chestnut breast-band; Okinawa Rail has ivory-tipped red bill and red legs. Sympatric Buff-banded Rail (46) has paler, well-patterned upperparts and upperwings, distinctive head pattern, orange-buff breast-band, rufous and black barring on most of remiges, pink bill and pinkish-grey to olive-grey legs and feet.

VOICE The call is harsh and low-pitched, suggestive of a dog or pig, and birds often duet, calling alternately back and forth to each other for several minutes (Diamond 1972a). Groups of up to ten birds are reported to follow a leader, which regularly utters loud pig-like squeals, often from an elevated calling site, when followers utter low, nasal gulping notes (Bishop 1983). Records of short, loud calls and deep notes, not properly described but said to be suggestive of a dog or pig, may refer to the aforementioned calls and are uttered by day and at night (Ripley 1977).

DESCRIPTION
Adult Forehead to hind-neck, sides of face and sides of neck very dark rich brown, tinged rufous; rest of upperparts, including upperwings and tertials, very dark brown, tinged olive, but tinged more rufous on rump and uppertail-coverts; rectrices blackish-brown. Outer greater upperwing-coverts and tertials tinged rufous; alula and primary coverts blackish-brown, outer webs rufous-brown; primaries blackish-brown, outer webs spotted white and inner webs barred white; axillaries and underwing-coverts dark blackish-brown, narrowly barred white. Malar region, chin, throat and entire underparts almost black, narrowly barred white except on lower belly, vent and undertail-coverts, which are very dark blackish-brown, almost unmarked; lower flanks only faintly barred white; thighs blackish-brown. Bare parts of female: iris blood-red; bill black; legs and feet rose-pink. Sexes alike; female averages slightly smaller.
Immature Not described.

Barred Rail

G. t. torquatus

G. t. celebensis

G. t. limarius

G. t. sulcirostris

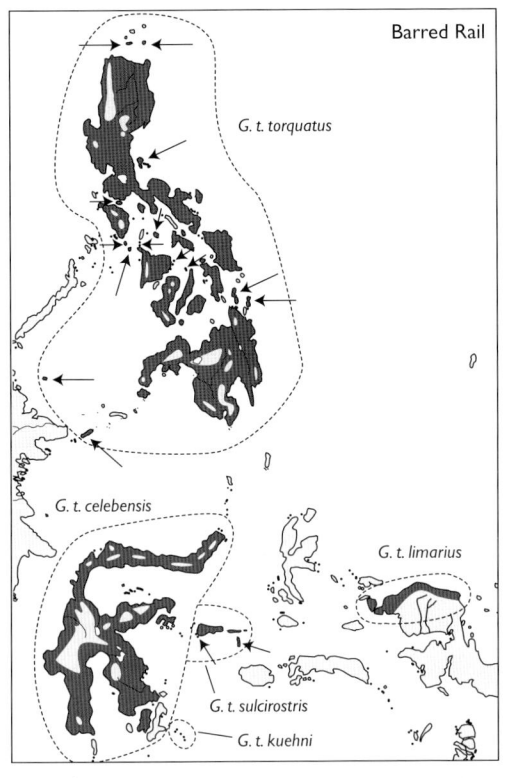

G. t. kuehni

intermingle. It is found on open hillsides and in elephant grass (alang-alang), but it avoids extensive crop fields and closed forest (Stresemann 1941). It is not restricted to sites with water but it favours moist or wet areas, and often occurs near human habitation. It is found from sea level up to 1,000m.

FOOD AND FEEDING Food not recorded. Sometimes forages along roadsides.

HABITS Despite the widespread occurrence of this species, almost nothing is known of its habits and behaviour. It is very shy, in the open it is agile and moves rapidly, and it is difficult to observe. Coates & Bishop (1997) state that its behaviour is "similar to that of the Buff-banded Rail", and that it is regularly seen along the edges of little-used roads through suitable habitat.

SOCIAL ORGANISATION Presumably monogamous.

SOCIAL AND SEXUAL BEHAVIOUR It apparently associates occasionally with the Buff-banded Rail (Coates & Bishop 1997).

BREEDING AND SURVIVAL Season Luzon, Apr, Nov; Sulawesi, Nov-Dec. **Nest** Not described; built on ground. **Eggs** 3-4, usually 3; ovate; creamy-white, with sparse spots and blotches of dark chocolate to reddish-brown, and more numerous spots and blotches of pale lilac; markings most numerous at blunt end; size (nominate) 36.6-39 x 26.9-28.5 (37.5 x 27.6, n = 3), calculated weight 15.8; size (*celebensis*) 40-42 x 28.5-30.3 (40.9 x 29.7, n = 3), calculated weight 20. **Incubation** Incubation, and care of the young, probably undertaken by both sexes. No other information available.

Juvenile Not described.
Downy young Not described.

MEASUREMENTS Wing of 5 males 149-152 (150.4, SD 1.3), of 8 females 138-148 (142.3, SD 3.8); tail of 2 females 44, 51; bill of 8 males 41-46 (42.6, SD 2.0), of female 39-46.5; tarsus of 8 males 61-68 (64.3, SD 2.1), of 8 females 55-63 (58.5, SD 3.0).

GEOGRAPHICAL VARIATION None.

MOULT Remex moult is simultaneous and is recorded in Apr.

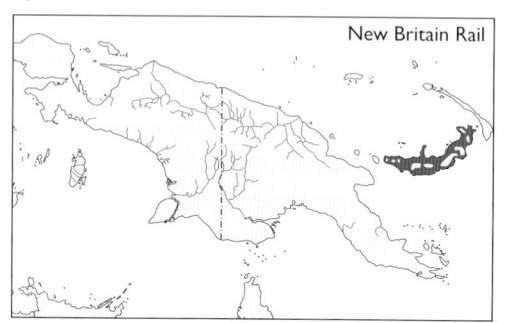

DISTRIBUTION AND STATUS New Britain, Bismarck Archipelago. Not globally threatened. There is no recent assessment of its status but it was regarded as locally common in the 1940s and the 1980s (Mayr 1949, Coates 1985). It is snared and eaten by local people, who consider it very palatable (Mayr 1949).

MOVEMENTS None recorded.

HABITAT Heavy damp forests and mountain valleys; also swampy canegrass (pit-pit); it wanders into gardens when foraging. It is recorded from sea level to at least 1,250m and may occur higher. Bishop (1983) did not find it above 400m in western New Britain but it is reportedly more common at mid-montane altitudes (G. Dutson *in litt.*).

FOOD AND FEEDING Snails, insects (including beetles) and vegetable matter, including seeds. The birds forage on the ground, apparently sometimes in groups, but foraging methods are not recorded.

HABITS Almost unknown. Their presence is often revealed by loud calls (see Voice). Although difficult to see, if the observer remains completely still these rails will pass within a few metres and will occasionally feed within centimetres of the observer's feet (Bishop 1983). Although W. F. Coultas (in Mayr 1949) suggested that it can "cover considerable distances on the wing", this species is probably almost flightless: other observers have never seen it fly and local people emphasise that it always runs to escape (Gilliard & Lecroy 1967, Bishop 1983).

SOCIAL ORGANISATION Social organisation not known often encountered as pairs or singles but may occur in parties.

SOCIAL AND SEXUAL BEHAVIOUR No information available.

BREEDING AND SURVIVAL Season No information available. **Nest** Undescribed; nests on the ground. **Eggs** Clutch size not known; size of 13 eggs 39-43.5 × 29-32.5 (41.5 × 31.5), calculated weight 22.7 (Schönwetter 1961-62). Egg variously described as white and lustreless, flecked at blunt end with light blue-grey and violet spots (Schönwetter 1935), and whitish with brown flecks (Meyer 1936). No further information is recorded.

46 BUFF-BANDED RAIL
Gallirallus philippensis   Plate 14

Rallus philippensis Linnaeus, 1766, Philippines.

Sometimes retained in *Rallus*. Great geographical variation, with up to 26 subspecies recognised; validity of some racial distinctions questionable in view of individual plumage variation, minor sexual size differences, and changes due to age and wear; at least one race (*tounelieri*) of doubtful status (Marchant & Higgins 1993). Extinct *G. dieffenbachii* (51) sometimes included as a race but now generally considered specifically distinct. Twenty extant races currently recognised; 21 if birds from Moluccas are assigned to the doubtful race *yorki* (White & Bruce 1986); one other race (*macquariensis*) recently extinct.

Synonyms: *Hypotaenidia/Porzana/Rallina philippensis*; *Rallus assimilis/pacificus/pectoralis/forsteri/etorques/hypotaenidia/ rufopes/ striatus*(part)*/hypoleucus/pictus/philippinensis/ macquariensis*; *Hypotaenidia pectoralis/etorques/assimilis/ australis/striata*; *Eulabeornis philippensis/assimilis/hypoleucus/ etorques*.

Alternative names: Banded/Land/Pectoral/Painted Rail; Banded Land Rail.

IDENTIFICATION Length 25-33cm; wingspan 40-52cm. A robust, medium-sized rail, with distinctive head pattern and barred underparts. Tail relatively long; when walking, carried erect like a gallinule and flicked incessantly. Crown streaked blackish and brown; supercilium grey-white; rufous-brown stripe from bill through eye to sides of neck and lower hindneck forming distinct nuchal collar; chin white, throat to foreneck greyer. Colour and pattern of upperparts, upperwing-coverts and tail vary with race: olive to rufous-brown with variably prominent blackish-brown feather centres, overall appearance ranging from almost plain olive through well-mottled to very dark; entire region variably spotted or barred white, least in race *swindellsi* and completely barred in *admiralitatis*; mantle almost black, with dense white markings in *lacustris*, *reductus*, *praedo* and *lesouefi*. Remiges dark brown, outer 3 primaries barred white and other remiges barred rufous-brown. Underparts barred blackish-brown and white; barring starts either from sides of neck and lower throat/upper breast, or from breast (when unbarred regions above are grey). Buff to rufous-brown band of varying width and prominence, sometimes barred blackish-brown, across middle of breast in most races: indistinct in *christophori* and some *mellori*, more or less absent in birds from extreme NE of range (*swindellsi*, *sethsmithi*, *ecaudatus*, *goodsoni*), many *wilkinsoni*, some *pelewensis* and *reductus*, and many nominate birds. Relative width of blackish and white bars on underparts varies with race: blackish bars broad in *lacustris*, *lesouefi*, *meyeri* and Sulawesi birds (formerly *chandleri*), white bars broad in *admiralitatis*; extent of barring on centre of belly variable (unbarred in *meyeri*, *anachoretae* and some *reductus*). Undertail-coverts barred blackish-brown and white basally, buff towards tips, thus often appearing buff. Iris red to orange; bill pink, greyish towards tip; legs and feet grey to brown-grey, tinged pink to olive. Sexes similar; female may be slightly smaller. Immature like adult but retains worn juvenile remiges; also has narrow or obscured nuchal collar but uncertain whether some adults may show this feature; iris probably yellow-brown. Juvenile duller than adult; head pattern less distinct and with little chestnut; throat mottled; smaller spots (usually olive-buff) on upperparts; breast-band (if present) less distinct; underparts barred grey, not grey-black; bill grey-black; iris brown. Inhabits dense vegetation of freshwater and coastal wetlands; also mangrove swamps, wet to dry grass, scrub, woodland, forest, cultivated areas and even coral cays.

Similar species Distinguished from congeners by combination of head pattern, variably spotted and barred upperparts and upperwing-coverts, rufous and blackish bars on all remiges except outer 3 (which are barred blackish and white), white chin, grey throat, buff to brown breast-band (in most races), barred underparts (including undertail-coverts), pink bill and pinkish-grey legs. Combination of barred remiges and pattern of head and body distinguish it from other sympatric secretive rails; larger than *Porzana* and *Lewinia* crakes (also see species accounts 64 and 65).

VOICE The commonest call is a harsh *kreek* or *eeeer*, usually repeated 4-5 times with 15-20s between notes, mostly heard in the morning and evening, especially in the breeding season, and probably territorial (Marchant & Higgins 1993). Calling may develop into a chorus if the breeding population is dense. This call is also given after dark and infrequently in the middle of the day. Pair members frequently call back and forth while foraging (Pratt *et al.* 1987). There is also a loud *coo-aw-ooo-aw-ooo-aw-ooo-aw* (Serventy 1985), a low-pitched, aggressive growl, probably equivalent to the angry *coo* mentioned by Dunlop (1970), muffled repeated grunts, given both during threat displays and as contact notes, and explosive hisses with deep growls when the chicks are threatened (Marchant & Higgins 1993). When flushed, birds give a sharp, squeaky *krek*, and birds squeal when retreating from agonistic encounters (Dunlop 1970). Birds at the nest utter a soft, low *kik-kik-kik*. Chicks maintain contact with penetrating peeping calls and give soft, high-pitched squeaks when pursued (Marchant & Higgins 1993). The response of this species to taped playback is unpredictable (Elliott 1989).

DESCRIPTION *G. p. mellori*

Adult Forehead to nape dark blackish-brown, feathers with rufous-brown to brown fringes, giving densely streaked effect; narrow supercilium from base of upper mandible to side of nape pale grey, whiter at front; lores, through and under eye to entire hind-neck, chestnut to rufous-brown, rufous hind-neck forming broad nuchal patch of solid U-shape; narrow white ring around lower half of eye. Mantle olive with dark blackish-brown mottling and fine white spots or bars which become less prominent with wear; scapulars blackish-brown with broad olive to brown feather fringes and irregular white spots, appearing olive with blackish mottling and sparse white spotting; back and rump similar to scapulars but white spots becoming fewer and often absent towards rear; uppertail-coverts olive to brown, with blackish bases; rectrices grey-black with olive or brown fringes. Lesser upperwing-coverts olive to brown with darker centres and sparse, irregular white spots (sometimes absent); marginal coverts white (forming white leading edge); median and greater primary coverts dark brown with olive tips and rufous-brown bars at edges; median and greater secondary coverts similar but barred or spotted white. Alula dark brown, feathers fringed olive and barred white at edges. Remiges dark brown with 4-5 broad rufous-brown bars, but outer 3 primaries dark brown with narrow and widely spaced white bars often washed rufous-brown. Underside of remiges barred dark brown and light rufous-brown, outer 3 primaries narrowly barred white; underwing-coverts and axillaries dark brown, barred white. Upperparts of birds from Mt Wilhelmina region, C Irian Jaya (formerly *randi*), paler olive and more uniform; blackish feather centres much reduced. Chin white; throat, lower ear-coverts, sides of neck and foreneck to upper breast grey, palest (whitish) on throat and darkest posteriorly, often becoming grey-brown from lower neck. Sides of lower neck and entire breast, barred blackish-brown and white, bars becoming broader on lower breast; tips of feathers of middle breast washed pale to rich buff or brown, giving breast-band of varying breadth and colour (sometimes virtually absent) with barring above and below it; dark barring also often visible in breast-band. Flanks and belly blackish-brown with narrow white bars; thighs and vent largely white; undertail-coverts barred grey-black and white basally, buff towards tips (appear buff). Iris bright red to dark orange; bill dirty pink or pink-red at base, grading to grey or brown-grey on culmen and tip; legs and feet grey or brown-grey, tinged reddish, pink, buff or olive. Sexes alike; non-breeding plumage similar.

Immature Very like adult, but pointed juvenile tertials and rectrices, if not worn, indicate age; nuchal collar narrow, often obscured by brown, but not known if some adults also show this feature. Iris probably yellow-brown, becoming red with yellow flecks before attaining adult colour.

Juvenile Forehead to nape duller and browner; supercilium less distinct behind eye; stripe through eye duller, often brown, and indistinct; rufous nuchal collar narrow, obscured by brown feather centres; rest of upperparts duller than adult, with fewer, olive-buff to white, spots; mantle has duller, dark brown feather centres. Wings and tail similar to adult except for pointed remiges and rectrices. Throat and foreneck mottled or barred olive; breast-band variable, as in adults; barring of underparts paler and less distinct, grey and whitish, washed pale buff. Iris dark brown to brown; bill grey-black, tinged pinkish especially at base; legs and feet dull grey, tinged pink.

Downy young Down dense; on upperparts black or sooty-black, on underparts black-brown; hatches with grey eyestripe or cheek patch; down becomes browner at 1 week; iris dark brown; bill at first grey, with grey-black saddle and white egg-tooth, becoming largely black by c. 6 days and all black at 12 days; legs and feet dark grey to black.

MEASUREMENTS Birds from Australian mainland: wing of 30 males 130-156 (145.8, SD 6.2), of 18 females 132-152 (141.2, SD 5.4); tail of 26 males 60-78 (69.3, SD 4.6), of 16 females 54-71 (64.8, SD 4.2); exposed culmen of 25 males 22.5-41.5 (31.9, SD 2.9), of 17 females 27.5-32.5 (29.8, SD 1.4), culmen to base of 7 males 29-39 (34.4, SD 3.3), of 6 females 29-35 (32, SD 2.2); tarsus of 30 males 34.5-45.5 (40.8, SD 2.6), of 18 females 35.5-43 (38.8, SD 2.2). Weight of 16 males 144-234 (181, SD 25.1), of 8 females 123-240 (165.6, SD 36.5). Two males and two females from New Guinea (formerly *randi*): wing of male 150, 151, of female 133, 145; tail of male 65, 72, of female

62, 72; exposed culmen of male 27, 27, of female 24, 25; tarsus of male 43, 44, of female 39.5, 42.

GEOGRAPHICAL VARIATION Considerable, complex and sometimes poorly defined; some trends in plumage previously thought to be geographical are more probably a result of age, wear and individual variation. Principal characters used in racial separation include: overall colour and pattern of crown; colour and prominence of nuchal collar; colour of feather fringes of upperparts and prominence of dark feather centres; extent, size and density of white spotting on upperparts and upperwings; mantle spotted or barred white; width of rufous bars on primaries; presence, extent, colour and barring of breast-band; presence of barring above breast-band; overall tone of underparts and relative width of blackish and white bars; extent of barring on belly; size (including bill length and proportions). See Mayr (1949) for further details. Possible race *wahgiensis* included in *reductus*; *chandleri* in nominate; *randi*, *norfolkensis*, *australis* and some *yorki* in *mellori*; other *yorki* in *tounelieri*. Remaining populations of putative *yorki*, from Moluccas, need further investigation and may prove to be referable to nominate, to be migrant *mellori* from Australia, or to be intermediates. Race *forsteri* is synonymous with *ecaudatus*. For detailed discussion of racial variation, see Mayr (1938, 1944, 1949), Parkes (1949), Junge (1953), Greenway (1973), Ripley (1977), Schodde & de Naurois (1982), White & Bruce (1986) and Marchant & Higgins (1993).

G. p. andrewsi (Mathews, 1911) – Cocos (Keeling) Is. Poorly known but possibly highly distinctive. Upperparts very dark, almost lacking olive-brown feather fringes but with more white spots than any other race, including on rump; nuchal collar rather dull; well-marked breast-band tinged brick-red; well-marked regular barring on underparts; iris bright brown; base of bill pink. Type male has wing 132, tail 59, culmen 30, tarsus 43; 2 unsexed have wing 144, 148; tail 58 (n = 1); culmen 33, 37; tarsus 42, 46 (Mathews 1911, Ripley 1977, BMNH).

G. p. philippensis (Linnaeus, 1766) – Philippines (Batan, Cebu, Luzon, Mindanao, Mindoro and Samar); Sulawesi; Tanahjampea; Butung; Sula Is (Taliabu); Buru; Lesser Sunda Is (Sumba [race?], Alor, Sawu, Roti and Timor). Typical *philippensis* (Philippines): crown feathers fringed olive-brown; nape and hind-neck more olive-brown than rufous; facial stripe from lores to eye may be dusky brown, becoming rufous on ear-coverts; back with broad pale brown feather fringes and small blackish centres, white spots sparse (most numerous on mantle); upperwing-coverts uniform, with white spots and blackish-brown bars almost confined to greaters; primary coverts rufous, banded blackish-brown, and olive-brown at tips; alula chequered blackish-brown; underwing-coverts blackish-brown, broadly edged white; breast-band often absent, but 6 Luzon birds have sandy breast-band (White & Bruce 1986); narrow blackish barring on underparts. Birds similar to this occur on Sulawesi (Plate 14). Putative *chandleri* (Sulawesi) darker olive-brown dorsally with larger blackish-brown feather centres and much more white spotting; nape and hind-neck rufous; breast-band often absent; heavier blackish barring on underparts (Mayr 1944, White & Bruce 1986). See Plate 14. An immature female had red iris, dull brick-red bill with brown tip, and pale brown legs. Wing of male 136-144, of female 129-137; tail of male 58.5-65, of female 53-64; culmen of male 27-33, of female 27.5-30; tarsus of male 40-46, of female 39-43.5 (Riley 1924, Ripley 1977). Measurements of 4 males, 1 female: wing of male 139-146 (142.5, SD 3.1), of female 137; tail of male 65-68 (66.8, SD 1.2), of female 67; culmen to base of male 34-38 (35.75, SD 1.7) of female 36; tarsus of male 40-43 (41, SD 1.4), of female 43.

G. p. pelewensis (Mayr, 1933) – Palau Is (Angaur, Arakabesan, Babelthuap, Garakayo, Koror and Peleliu). Small; like nominate but darker overall; nape more rufous-brown; upperparts with narrower and darker feather fringes, and pronounced whitish spotting which is most obvious on mantle (which is sometimes barred) and uppertail-coverts but lacking on rump; blackish bars on outer webs of primaries broader than rufous bars; breast-band (usually present) more pronounced; underparts lacking any buffy tinge and with darker, more pronounced barring. Measurements of 12 males, 3 females: wing of male 127-143 (134.6), of female 129, 126, 136; tail of male 54-65 (60), of female 56, 57, 58; bill (to lateral feathering) of male 25-29 (27.7), of female 23, 24, 25; tarsus of male 41-46 (43.5), of female 40, 41, 42 (Mayr 1933b).

G. p. xerophilus (van Bemmel & Hoogerwerf, 1941) – Lesser Sunda Is (Gunungapi). Very small; in plumage resembles *mellori* but is brighter; nape deep rust-brown; throat almost white; breast-band orange-brown; blackish and white barring continues for c. 1cm above breast-band; upper mandible deep grey; lower mandible flesh-coloured; legs and feet light brown-grey to light grey. Males and females: wing 123-132; tail 49-56; culmen 26-29 (Ripley 1977). White & Bruce (1986) regard recognition of this form as questionable, because it is not markedly smaller than most birds from Wallacea.

G. p. wilkinsoni (Mathews, 1911) – Lesser Sunda Is (Flores). Larger than nominate. Underparts similar to darkest '*chandleri*' from Timor but less blackish on upperparts; throat slightly darker and greyer; upper breast darker grey (Mayr 1944, Greenway 1973); breast-band often absent. Males and females: wing 145-158, tail 67-73; culmen 29-33; tarsus 45-46 (Rensch 1931, Ripley 1977). Measurements of 3 males, 1 female: wing of male 143, 146, 147, of female 143; tail of male 61, 62, 66, of female 58; culmen to base of male 32, 36, 40, of female 35; tarsus of male 43, 43, 44, of female 44.

G. p. lacustris (Mayr, 1938) – L Sentani region, N lowlands of New Guinea. Large and quite dark: crown feathers with narrow rufous fringes; mantle blackish, heavily marked with small white spots; fewer but larger spots on scapulars and upperwing-coverts; few or no spots on rest of upperparts and tail; feathers of back narrowly fringed greyish-olive (more brownish in *meyeri*); no barring above; well-developed but narrow breast-band; blackish flank bars much broader than white ones; belly completely barred. Measurements of 3 males, 2 females: wing of male 147, 148, 154, of female 132, 147; tail of male 65, 68, 69, of female 62, 66; bill (from lateral feathering) of male 30, 30, 31, of 1 female 27, culmen to base of 1 female 33; tarsus of male 44, 44, 46, of female 39, 41 (Mayr 1938, BMNH). Weight of 3 males 210, 225, 240, of 1 female 185 (Mayr 1938).

G. p. reductus (Mayr, 1938) – New Guinea, in C highlands from low N slopes of Snow Mts E to Mt Giluwe and Wahgi Valley (formerly *wahgiensis*), and on NE coast from China Straits E to Astrolabe Bay, including Long I. Similar to *admiralitatis* but larger; mantle more blackish, barred white; tertials, and back to tail, with little or no white; breast-band variable, absent or inconspicuous and barred blackish (coastal birds) to well-marked and rufous (former *wahgiensis*); centre of belly unbarred apparently only in coastal birds. Iris reddish-brown; legs purplish-grey (*wahgiensis*). See Schodde & de Naurois (1982). Wing of 7 males 136-148 (141.9, SD 4.9), of 8 females 136-144 (138.8, SD 2.9); tail of 6 males 61-66 (63.3, SD 1.8), of 5 females 55-67 (60.7, SD 4.5); culmen of 7 males 27-33 (29.8, SD 2.5), of 8 females 24-32 (27.3, SD 2.6), culmen to base of 2 males 35.5, 37; tarsus of 6 males 43-48 (44.8, SD 1.7), of 6 females 40-43 (41, SD 1.2) (Mayr 1938, Ripley 1964, BMNH). Weight of 1 male 129.

G. p. anachoretae (Mayr, 1949) – Kaniet (Anchorite) I, NE of Ninigo Is. Upperparts rather similar to those of *praedo* but crown rufous to hair-brown, less heavily streaked; nuchal collar purer rufous; mantle and back less blackish but more heavily marked white (often barred) on mantle; almost no white from back to uppertail-coverts. Breast-band very broad, rufous and unbarred; some unbarred grey feathers below breast-band; barring of breast and flanks coarse; large unbarred area in centre belly, as in *reductus*; size large (like *lacustris*). Wing of male 145-150 (147), of 3 females 136, 140, 145; tail of male 65-71 (68.2), of 2 females 64, 65; bill of male 28-31 (29.1), of 2 females 25, 28; tarsus of male 41-45 (43.4), of 3 females 40, 40, 41 (Mayr 1949).

G. p. admiralitatis (Stresemann, 1929) – Admiralty Is, Bismarck Archipelago (Papenbush I, and Pityili, Los Negros group, off N coast of Manus I). Birds from Hermit I and Ninigo Is possibly also belong here (Mayr 1949); also recorded Wuvulu (Coates 1985; race not specified). Crown rufous chestnut, like nape; entire upperparts, including tail, narrowly barred white; white bars on underparts, particularly flanks, wide like the blackish bars; centre of belly unbarred buff. Differs from *pelewensis* in having upperparts barred. One male: wing 137; tail 70; culmen 32; tarsus 46 (Ripley 1977).

G. p. praedo (Mayr, 1949) – Admiralty Is (Skoki I, Sabben group). Very similar to *lacustris* but smaller and more slender, with darker mantle and narrower breast-band. Crown dark rufous, heavily streaked blackish; hind-neck dark rufous, mottled blackish; mantle blackish with numerous white dots, occasionally barred; rest of upperparts very dark with few white dots. Blackish and rufous bars on outer webs of primaries of equal width. Breast-band well defined but usually narrow, occasionally barred blackish, sometimes with a few grey feathers below; breast and flanks evenly and quite narrowly barred; belly completely barred. Wing of male 132-148 (140.4), of female 132-145 (138.1); bill to lateral feathering of male 25.5-30 (28.3), of female 24-28 (26.1); tail of male 56-71; tarsus of male 40-43 (Mayr 1949).

G. p. lesouefi (Mathews, 1911) – Bismarck Archipelago (New Hanover, Tabar and Tanga; New Ireland birds also probably this race). Species also recorded from Watom and Duke of York (Coates 1985); race not given. Small and very dark; almost as dark as *swindellsi*. Crown chestnut-rufous, narrowly streaked blackish (streaks broader in *praedo*, making crown much darker); mantle almost black, finely barred white (spotted in some birds from Tabar and Tanga); fringes of scapulars and back feathers narrow, walnut-brown; back and rump profusely spotted and streaked white; white and rufous bars on outer webs of primaries rather narrow. Breast-band distinct, narrow, often barred blackish; virtually no grey below breast-band; blackish bars on underparts much broader than white; belly well barred (some from Tabar and Tanga have unbarred belly). Wing of 5 males 131-143 (138.2, SD 4.6), of 4 females 132-140 (137, SD 3.6); tail of 4 males 57-64 (60.8, SD 3.3), of 4 females 55-60 (57.5, SD 2.1); bill of 5 males 27-30 (28.8, SD 1.3), of 4 females 25-27 (26.25, SD 0.9), culmen to base of 1 female 32, of 1 unsexed 37; tarsus of 5 males 40-45 (42, SD 1.8), of 4 females 40-41 (40.4, SD 0.5) (Mayr 1949, BMNH).

G. p. meyeri (Hartert, 1930) – Bismarck Archipelago (New Britain and Witu Is). Compared with *lesouefi*, larger and paler; crown more hair-brown, rather than rufous, with heavier streaking; broader olive feather fringes on back, almost hiding blackish-brown centres; upper mantle with small white spots; almost unspotted on tertials and from back to tail. Breast-band broad (Mayr 1949, *contra* Ripley 1977 and Schodde & de Naurois 1982), barred blackish-brown; few or no grey feathers below breast-band; blackish-brown and white bars of breast and flanks narrow, about equal in width on breast, blackish bars wider on flanks; belly and vent unbarred. Wing of 4 males 142-151 (146.5, SD 3.7), of 2 females 137, 141; tail of 3 males 60, 64, 66, of 2 females 58, 62; bill of 4 males 29-31.5 (30.4, SD 1.1), of 2 females 26, 30, culmen to base of 1 male 36, of 3 females 31, 32, 33; tarsus of 4 males 45-46.5 (45.4, SD 0.7) (Mayr 1949, BMNH).

G. p. christophori (Mayr, 1938) – Solomon Is (Bougainville, Choiseul, Florida, Guadalcanal, Ysabel, Uki, San Cristobal and Santa Ana). Similar to *reductus*; blackish streaks on crown narrow and inconspicuous, feather fringes hair-brown; general colour of upperparts rather variable, but mantle less blackish, more brown-tinged; uppertail-coverts spotted white, central rectrices with white bars. Breast-band much reduced; blackish barring on underparts heavier than in *reductus*. Differs from *sethsmithi* in having blacker mantle, blackish bars on breast and flanks much broader than white ones, much darker underparts and reduced barring on belly. Measurements of 2 males, 3 females: wing of male 147, 150, of female 137, 142, 143; tail of male 58, 60, of female 55, 55, 57; bill of 1 male 30, of 2 females 26.5, 27, culmen to base of 1 male 37, of 1 female 38; tarsus of male 44, 47, of 2 females 40, 41 (Mayr 1938, BMNH). Three birds seen on New Georgia had narrow, ragged breast-band; subspecific identity could not be determined (Blaber 1990); however, this observation could refer to Roviana Rail (47) (Diamond 1991).

[***G. p. yorki*** (Mayr, 1938) – Moluccas: Halmahera, Buru, Ambon, Seram, Sawai, Seram Laut Is (Panjang), Tayandu Is (Kur) and Kai Is (Tual); race tentatively recognised by White & Bruce (1986). On available information, not satisfactorily distinguishable from *mellori* or *philippensis* (including putative *chandleri*), and no convincing description exists. Racial identity of birds from this region needs further study; see Mayr (1938), Junge (1953) and White & Bruce (1986).]

G. p. mellori (Mathews, 1912) – EC Irian Jaya (Snow Mts, formerly *randi*), S and SW New Guinea (N to the head of Geelvink Bay), Australia (including Torres Strait and Ashmore Reef [race?]), Tasmania, Lord Howe I and Norfolk I (formerly *norfolkensis*). See Description and Plate 14.

G. p. assimilis (Gray, 1843) – New Zealand, including Stewart and Auckland Is. Bill longer and more slender, and wing shorter, than in *mellori*; also differs in plumage (Marchant & Higgins 1993): crown paler and more streaked; nuchal collar slightly paler; upperparts paler, more uniformly olive, blackish mottling less conspicuous; tail paler, rectrices with broader olive fringes; breast-band possibly slightly paler. Nuchal collar lacking in 1 immature. Measurements of 4 males, 2 females: wing of male 130-156 (145.8, SD 6.2), of female 130, 135; tail of male 62-67 (64, SD 2.4), of female 64, 66; exposed culmen of male 32.5-39 (35.2, SD 3.3), of female 30.5-33.5, culmen to base of 3 males 35, 36, 39, of 1 female 34; tarsus of male 38.5-46 (42.6, SD 3.3), of female 38.5, 42. Elliott (1983) gives means as follows: wing of male 140.4 (n = 41), of female 134 (n = 30); exposed culmen of male 36.4 (n = 41), of female 32.7 (n = 29); tarsus of male 40.8 (n = 40), of female 3.92 (n = 31). Weight of 7 males 126-204 (171, SD 31.3), of 8 females 115-218 (168, SD 36.7); hatching weight 12-13.

G. p. tounelieri Schodde & de Naurois, 1982 – Coral Sea islets, from SE New Guinea archipelagos and Great Barrier Reef (Raine I S to Bunker group) E to Willis and Chesterfield groups, and Surprise group (off N New Caledonia). Doubtfully valid. Compared with *mellori*, wing averages slightly shorter; also said to differ in having facial stripe and nuchal collar duller, darker chestnut; mantle and scapulars darker, heavily mottled blackish on olive and with fewer white spots; middle remiges with narrower chestnut bars, often becoming spots; breast-band narrower and darker (Schodde & de Naurois 1982). However breast-band, though more often reduced in *tounelieri*, shows great individual variation, while all other plumage distinctions are not evident when age and wear considered (Marchant & Higgins 1993). Wing of 54 unsexed 114-138 (128.9, SD 5.9); tail of 48 unsexed 48-62 (55.9, SD 3.0); exposed culmen of 56 unsexed 27-37 (31.4, SD 2.5); tarsus of 53 unsexed 32-54.5 (40.2, SD 4.8). Weight of 52 unsexed 130-290 (188, SD 31.7). Population highly mobile and shows slight cline from E to W, suggesting broad hybrid zone with gene flow from subspecies in adjacent regions (Schodde & de Naurois 1982, Marchant & Higgins 1993).

G. p. swindellsi (Mathews, 1911) – New Caledonia and Loyalty Is. Very dark upperparts with few and faint spots, none on uppertail-coverts; hind-neck has only indications of rufous; underparts very closely barred blackish; only faint wash suggestive of breast-band. Males and females: wing 140-143; culmen 31-33; tarsus 41-43 (Ripley 1977). Measurements of 1 male, 1 female: wing of male 145, of female 130; tail of male 60, of female 51; culmen to base of male 38, of female 35; tail of male 43, of female 39.

G. p. sethsmithi (Mathews, 1911) – Vanuatu (Ambae, Ambrym, Aneityum, Efate, Emae, Emau, Epi, Erromango, Maewo, Malakula, Malo, Nguna, Pentecost, Santo, Tanna, Tongoa and Vanua Lava), Fiji (except Viti Levu and Vanua Levu but including Lau group) and Rotuma. Birds from Santa Cruz Is (Nendo), have poorly developed orange-buff breast-band and are probably closest to *sethsmithi* (Gibbs 1996). In overall colour and markings most similar to *mellori*, but breast-band reduced or absent; uppertail-coverts unpatterned; similar to *goodsoni* but darker overall; larger than *ecaudatus*. Measurements of 5 males, 6 females: wing of male 137-160 (146.7, SD 8.4), of female 136-144 (139.1, SD 2.9); tail of male 55-74 (61.8, SD 7.1), of female 49-59 (55.6, SD 3.5); culmen of male 32.5-40 (35.5, SD 2.8), of female 30.5-38 (34.4, SD 2.5); tarsus of male 43-47 (45.2, SD 1.5), of female 36-45 (41.1, SD 2.9) (Ripley 1977, BMNH).

G. p. ecaudatus (Miller, 1783) – Tonga (Niuafo'ou, Keppel and 'Eua); birds from Futuna and Wallis Is also presumably this form. Variable; similar to *goodsoni* but smaller; paler overall than *sethsmithi* and possibly averages smaller. Faint pectoral wash shows in young but is lost in adults (Mathews 1911); uppertail-coverts barred white. Measurements of 4 males, 3 females and unknown number of unsexed birds: wing of male 137.5-144 (139.4, SD 3.1), of female 129, 131, 134, of unsexed 129-141; culmen of 5 males 30.5-32.5 (31.1, SD 0.8), of 6 females 27.5-32 (29.6, SD 2.0), of unsexed 28-30; tarsus of male 40-43.5 (42.2, SD 1.5), of female 40.5, 42, 43, of unsexed 40-43 (Ripley 1977, Kinsky & Yaldwin 1981, BMNH). Weight of 16 unsexed 142-218 (180) (Rinke 1987).

G. p. goodsoni (Mathews, 1911) – Samoa Is and Niue I. Paler overall than *sethsmithi*; larger than *ecaudatus*. Some adults and immatures lack breast-band; uppertail-coverts barred white (Ripley 1977). Variable; approaches darker forms of Bismarck Archipelago in bright chestnut crown, white-barred mantle and loss of rufous breast-band (Schodde & de Naurois 1982). Males and females: wing 136-155; culmen 32-36; tarsus 43-55 (Ripley 1977). Measurements of 5 males and 5 females from Niue: wing of male 140-158 (149.1, SD 7.4), of female 138-146 (142.7, SD 3.4); tail of male 57.5-63 (60.1, SD 2.4), of female 57-64 (61, SD 2.7); culmen of male 32-35.5 (33.4, SD 1.2), of female 28.5-32 (30.7, n = 4, SD 1.8); tarsus of male 46-49 (47.3, SD 1.3), of female 42.5-47 (45, SD 1.6); weight of 4 males 235-314 (266, SD 38.8), of 4 females 185-225 (205, SD 20.4) (Kinsky & Yaldwin 1981).

G. p. macquariensis (Hutton, 1879) – Macquarie I. Recently EXTINCT. Wing short; possibly flightless. Compared to *assimilis*, plumage darker; upperparts with fewer pale spots; secondaries with olive-brown fringes; breast-band broad, well marked and rufous-brown; flank barring less pronounced; flanks washed olive-brown; bill shorter; iris brick-red (type, possibly immature); tarsus heavy. Unsexed birds: wing 117-131; tail 43-53; culmen 28-35; tarsus 30-41. Formerly said to be dimorphic, with reddish and slightly smaller black forms, but this appears incorrect.

MOULT Postbreeding moult is complete and remex moult simultaneous. In captivity most remiges are lost in 1 day, and all within 5 days, regrowth being complete within 35 days and flight being possible only a few days before this (Elliott 1983). The rectrices moult with the wings or the body plumage. The timing of postbreeding moult probably

Buff-banded Rail

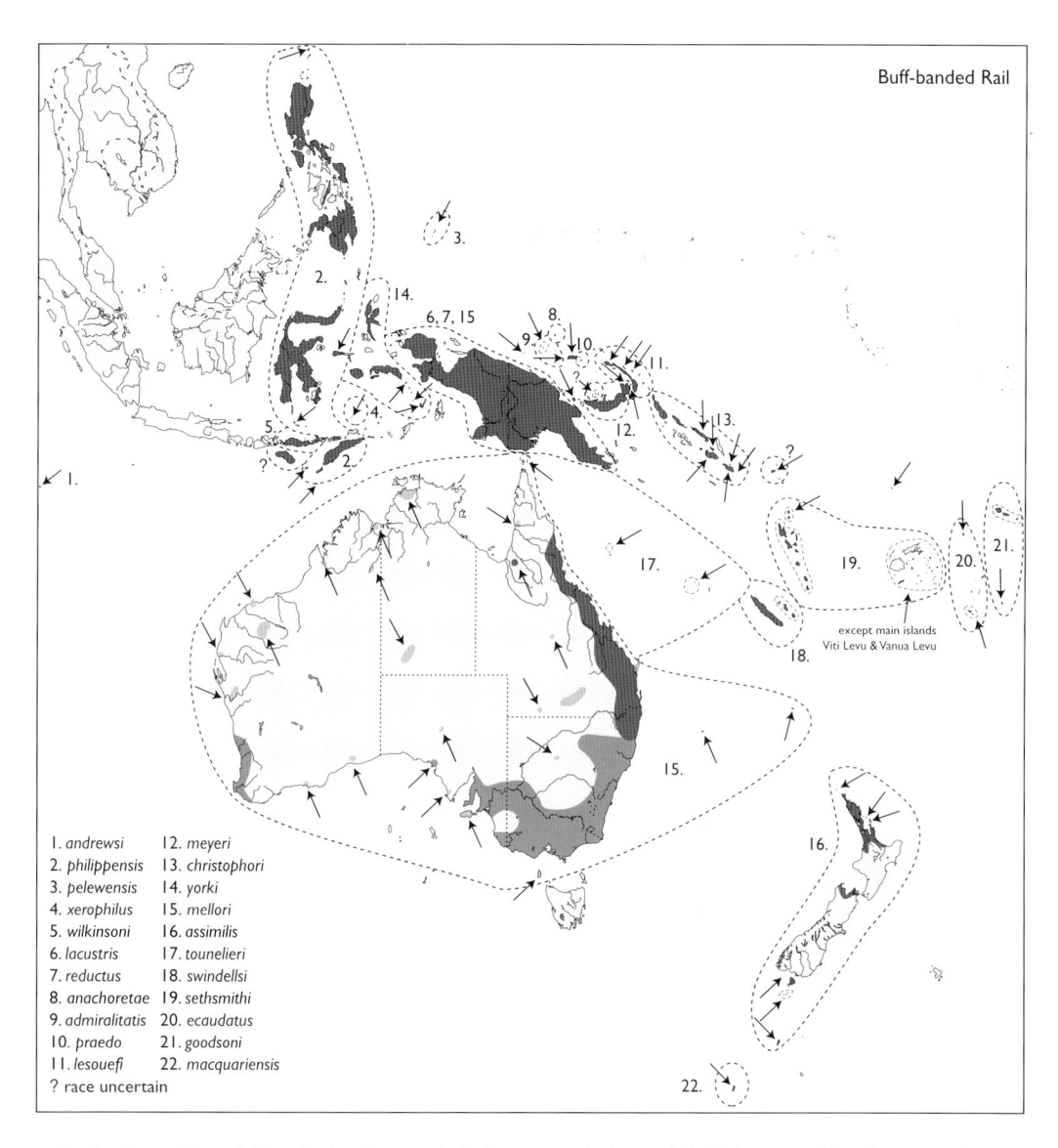

1. andrewsi
2. philippensis
3. pelewensis
4. xerophilus
5. wilkinsoni
6. lacustris
7. reductus
8. anachoretae
9. admiralitatis
10. praedo
11. lesouefi
12. meyeri
13. christophori
14. yorki
15. mellori
16. assimilis
17. tounelieri
18. swindellsi
19. sethsmithi
20. ecaudatus
21. goodsoni
22. macquariensis
? race uncertain

varies in Australia and New Zealand, records being (Marchant & Higgins 1993): SE Australia, Oct-Jan; Coral Sea, Sep-Oct and Dec; New Zealand, Nov and Jan-May. Elsewhere, it is recorded from the Lesser Sundas, Aug, and NW New Guinea, Jul, while a moulting male with enlarged testes was collected in EC New Guinea, May. Moult and breeding may occur throughout the year in Samoa (Robinson 1995). A partial pre-breeding moult of the head and body occurs, probably during Jun-Jul in Australia and New Zealand. Postjuvenile moult begins at 2 months (Marchant & Higgins 1993). The first pre-breeding moult is partial, probably involving only some head and body feathers.

DISTRIBUTION AND STATUS Cocos (Keeling) Is, Lesser Sundas Is, Sulawesi, Philippines, S Moluccas, Kai Is, New Guinea, Bismarck Archipelago, Palau, Solomon Is, Vanuatu, Fiji, Futuna, Wallis, Samoa, Tonga, Niue, Norfolk I, Australia, Tasmania, New Zealand and Auckland I. The species is not globally threatened, but the race *andrewsi* of Cocos (Keeling) Is is ENDANGERED. Whether it was originally introduced to these islands is not clear, but it was once abundant on all islands in the group before clearance of *Pisonia* forest for coconut plantations, hunting, and predation by feral cats and Black Rats *Rattus rattus* eliminated it from the main atoll; it is now confined to North Keeling I (area 1.1km^2), where its total population may exceed 100 individuals, and possibly West Island where it was last recorded in 1991 (Garnett 1993). Even the small populations which may have survived on South and Horsburgh Is until the early 1980s can no longer be found. Although harvesting of birds on North Keeling is illegal, poaching is not effectively controlled, while the introduction of cats or rats to the island is an imminent threat. The bird's population should be monitored more precisely and the island declared a national park, while elimination of cats and rats from other islands could allow re-introductions to be made (Garnett 1993).

Elsewhere, the nominate race is locally common in the Philippines and sparse to locally common in Sulawesi and the Lesser Sundas (Dickinson *et al.* 1991, Coates & Bishop 1997); *pelewensis* was locally abundant on the Palau Islands in the 1980s, but probably less numerous on Babelthuap than in the 1940s (Pratt *et al.* 1980). It was formerly locally common to abundant in New Guinea (*mellori, lacustris, reductus*) (e.g. Coates 1985, Beehler *et al.* 1986). At Tabubil, C New Guinea, a recent great decline in numbers (*reductus*) is attributed to predation by a greatly increased cat and dog population (Gregory 1995). The race *praedo* was regarded as plentiful in its restricted range on the Admiralty Is in the 1940s (Mayr 1949) but more recent information is not available.

In Australia the species is widespread and common to uncommon in coastal, subcoastal and riverine regions, and on many offshore islands (*mellori, tounelieri*) (Marchant & Higgins 1993); it has recently colonised Ashmore Reef (race?), off NW Australia (Coates & Bishop 1997). In New Zealand *assimilis* was formerly widespread but has declined markedly this century and has disappeared from many regions as a result of habitat destruction and predation (Marchant & Higgins 1993); although it is still said to be locally common near the coast of N North I (Heather & Robertson 1997), the coastal population in the Nelson-Marlborough area is vulnerable, being small, scattered, and dependent on unmodified saltmarshes, most of which are unprotected (Elliott 1989). Most habitat loss is caused by wetland drainage, regular mowing of grassy areas, and stock grazing and trampling (e.g. Marchant & Higgins 1993), but this may be somewhat offset locally by the birds' use of artificial wetlands and other man-created habitats. On some islands introduced predators have reduced or eliminated populations, and may prevent colonisation on others. Populations on Norfolk and Lord Howe Is are ephemeral, probably resulting from the occasional arrival of vagrants; it is a vagrant to Tasmania and there is a doubtful nineteenth-century record from the Kermadec Is (Marchant & Higgins 1993).

The race *anachoretae* of Kaniet (Anchorite) I was regarded as very scarce in the 1970s as a result of trapping by introduced labourers (Ripley 1977); its current status is not known. The race *sethsmithi* was probably extirpated by introduced mongooses on Viti Levu and Vanua Levu, Fiji, but was thriving on Taveuni I in the late 1970s and is locally common on Vanuatu (Blackburn 1971, Holyoak 1979, Pratt *et al.* 1987, Bregulla 1992). The species is also common on Nendo, Santa Cruz Is (Gibbs 1996), and *goodsoni* is widespread and common in Western Samoa (Robinson 1995). In the 1980s *ecaudatus* was common to abundant on Tonga, while *goodsoni* of Niue I was very scarce in 1953, increasing in 1956 and very common by 1968 (Wodzicki 1971). In Tonga it is often trapped for food, and the birds are also chased and caught when they become flightless in wet ground cover, especially when their primaries are very worn (Rinke 1987). The race *swindellsi* was formerly present on most of the New Caledonia and Loyalty Is and was apparently not uncommon (Ripley 1977, Hannecart & Letocart 1980). No information is available on the status of races *xerophilus, wilkinsoni, admiralitatis, lesouefi, meyeri, christophori* and the Moluccan populations assigned to *yorki*.

The extinct race *macquariensis* is known from subfossil remains and 3 skins; it was quite common on Macquarie I (area 260km^2) when discovered in 1879 but became extinct between 1880 and 1894. It co-existed with feral cats for at least 70 years but was eventually either displaced by Wekas (39), which were introduced in 1872 and may have taken eggs and young, or disappeared as a result of the introduction of rabbits in 1879; the rabbits reduced the suitable tussock habitat and enabled the population of cats to increase, and the cats may then have preyed on the rails in the winter when rabbits were scarce (Garnett 1993).

Predators of adults and young include cats, dogs, foxes, weasels *Mustela nivalis*, stoats *M. ermina*, rats (*Rattus rattus, R. norvegicus* and possibly *R. exulans*), and probably Swamp Harriers *Circus approximans* and Wekas (Marchant & Higgins 1993). One was taken by a Spotted Harrier *C. assimilis* (Jaensch 1989).

MOVEMENTS Not well known. This rail is regarded as sedentary throughout much of its range but is widely dispersive and possibly also migratory. There are no autumn or winter records in much of S Australia, but it is present all year in the N with a reduced reporting rate in the winter; reporting rates may be affected by the birds being less conspicuous during the non-breeding season; it is of erratic occurrence in the interior of Australia (Blakers *et al.* 1984, Marchant & Higgins 1993). There is evidence for migration or dispersal on Cape York and across the Torres Strait: casualties have been recorded at Booby I lighthouse in all months except Apr and Oct, most in May-Jul and Dec (Stokes 1983, Marchant & Higgins 1993). Birds may move in response to the availability of water, e.g. from drought areas to areas of high rainfall, and may move away from cold areas. Movements are also suggested between islands in the Coral Sea, while observations of periodic dramatic fluctuations in numbers on Bismarck Sea islets indicate that populations there, as in the Coral Sea, may have a nomadic ecology, birds moving to islands on which seabirds are breeding, exploiting temporary food supplies and perhaps also breeding (Schodde & de Naurois 1982). The species is recorded from New Zealand in all months except July, most birds during the summer, and there are records of possible vagrants from several inland sites (Marchant & Higgins 1993). Moving birds fly at night, sometimes hitting telegraph wires, street lights and lighthouses (Marchant & Higgins 1993). The remains of two birds found at a falcon feeding post at Suva on Viti Levu, Fiji, were presumably taken over the sea (Ripley 1977). This species is a vagrant to Mauritius (no details available).

HABITAT This species inhabits many types of permanent and ephemeral wetland, including marshes, swamps, lakes, pans, billabongs, rivers, streams, temporary inundations, estuaries, coastal lagoons, mangrove swamps, tidal mudflats and saltmarshes – in New Zealand, those with *Juncus maritimus, Leptocarpus similis* and *Plagianthus divaricatus*, which provide foraging, nesting and roosting habitat (Elliott 1989); it also inhabits rice paddies, sewage ponds, farm dams and drainage channels. It occurs on beaches, reef flats, sandbanks and coral cays. It is also found in wet to dry grassland, swampy canegrass, dry hill-rice fields, bushed grassland, heathland, bush and scrub (including *Scaevola* bush on Cocos Keeling and low *Pemphis acidula* and *Triumfetta procumbens* bush along shores in Tonga), coconut plantations with *Serato* ground cover (Niue I), taro plantations and paddocks of the legume *Phaseolus atropurpureus* on Niue I, woodland, forest-edge ferns, broadleaf forest, and rocky and grassy uplands (including alpine grassland in New Guinea) (Kinsky & Yaldwin 1981, Stokes *et al.* 1984, Coates 1985, Rinke 1987); it is also recorded in pockets of burnt vegetation on plains (Marchant & Higgins 1993). In New Zealand it avoids *Salicornia*-dominated saltmarshes where cover is low, and *Typha*-dominated wetlands because the tangled nature of this vegetation at ground level hinders foraging movements (Elliott 1987, 1989), but in Queensland it occurs at *Typha*-lined bore-drains (Storr 1973). It often occurs alongside other rail species, and in New Zealand it occupies saltmarsh habitat (*Juncus maritimus* and *Leptocarpus similis*) alongside Baillon's Crake (92) (Owen & Sell 1985). It adapts well to man-modified habitats, occurring in crops, pasture, gardens and parks, on golf courses and airstrips, and in secondary growth on abandoned plantations. It favours dense rank vegetation such as tall grass and herbage, rushes, reeds etc, but also inhabits islands with little cover and no surface water, and may be flushed from petrel burrows; on Macquarie I the extinct race *macquariensis* must have inhabited tussock grass and mosses, while *praedo* inhabits weeds among shrubs and trees on low, sandy islets (Mayr 1949, Day 1989, Marchant & Higgins 1993). Birds range from sea level to 3,600m.

FOOD AND FEEDING Most information is from Marchant & Higgins (1993). Animal food comprises polychaete worms, molluscs (in New Zealand *Ophicardelus costellaris* and *Amphibola crenata*, while the small gastropod *Potamopyrgus estuarinus* is an important element in the winter diet), crustaceans (including crabs, Amphipoda and Isopoda), insects (including Dermaptera larvae, Anisoptera nymphs, Coleoptera, Hemiptera, Hymenoptera [Formicidae], adult and larval Lepidoptera, Diptera), spiders, small fish, frogs and tadpoles (*Litoria raniformis*), lizards, eggs of birds and turtles and young of Sooty Tern *Sterna fuscata*. Plant food includes fruits, seeds, young plants, leaf and flower buds and other vegetable matter. It also eats carrion from road kills, chicken feed, bread and grain. It may feed at any time of the day and night, but is mostly crepuscular. It forages in both dry and wet areas, feeding at the edges of reedbeds and other dense cover, among low scrub, fern, bushes and mangroves, along roadsides, and in gardens, taro plantations, pastures and on airstrips. It probes and pecks in mud and shallow water, pecks at animals encrusting mangrove trunks, and takes seeds from plants and from the ground. It captures prey by stabbing with the bill slightly open, flicks crabs from side to side until the legs break off, crushes fish in the bill, pecks large snail shells until they break, and spears eggs with the bill. It scavenges along strandlines and lagoon shores, taking crustaceans from stranded or rotting vegetation; it also forages at rubbish dumps, and in Tonga it searches for food scraps with domestic chickens around villages, gardens and picnic sites (Rinke 1986, 1987); it was once recorded feeding inside a human habitation. The race *praedo* may take food dropped by nesting Lesser Noddies *Anous tenuirostris*, as well as feeding on flies in the nesting colonies and perhaps noddy eggs and young whenever possible (Mayr 1949), while on the sand cays of the Great Barrier Reef the species may be a significant predator at tern breeding colonies, taking many eggs. It was once recorded hoarding food (Blakers *et al.* 1984). This rail can exist on islands without fresh water but drinks it frequently when it is available.

HABITS Most information is from Marchant & Higgins (1993). This rail is mainly diurnal or crepuscular, and is most frequently seen foraging in the open in the early morning and late afternoon, but it can be active at any time of the day and night; it is particularly active after rain. It is secretive, wary and suspicious, dashing into cover at any hint of danger and reappearing a short time later, but is not shy and, when accustomed to human presence, becomes confident and is frequently visible in the open. When crossing open areas such as roads, it walks with a hunched posture, incessantly flicking the tail. It runs with neck extended, head lowered and tail raised, and walks and runs with the head bobbing rapidly, often pausing to stand erect and look around with the neck stretched and the tail flicking. When chased by a cat, one bird was seen to zigzag. It is very secretive when flightless during the postbreeding moult. Birds roost, loaf and shelter among thick, tall clumps of vegetation, sometimes also in nests not used for laying or brooding, while on small islands they shelter in petrel burrows and under coralline rocks during the heat of the day. They were also found roosting during the day c. 2m above the ground on platform-like *Sesbania* vegetation near water with White-browed (106) and Baillon's (92) Crakes, venturing out to feed at dusk and in the early morning (Mason & Wolfe 1975). They bathe freely, often in roadside puddles after rain, and sunbathe by standing with open, drooping wings.

SOCIAL ORGANISATION Probably monogamous, the pair bond persisting at least for the duration of the breeding season, when birds are territorial; pairs are said to remain on their territories all year (Heather & Robertson 1997). In New Zealand, the species is recorded in saltmarsh at a density of 1 pair per 0.6-3.0ha (mean 1.5 ha) (Elliott 1989) and in mangroves, saltmarsh and grass at 1 pair per 0.7-3.4ha (mean 1.4 ha) (Marchant & Higgins 1993).

SOCIAL AND SEXUAL BEHAVIOUR Most information is from Marchant & Higgins (1993). Although usually solitary, it is occasionally seen foraging in loose groups of up to 15 birds. Buff-banded Rails are aggressive towards conspecifics and other birds, both when feeding and when forming pairs before breeding, and are known to kill other rail species. They often forage amicably and boldly near other bird species, then attack or displace them for no apparent reason. Species involved include the large Masked Lapwing *Vanellus miles*, Peaceful Dove *Geopelia placida*, and Australian Magpie *Gymnorhina tibicen*; bandicoots (Peramelidae) are also attacked. Most interactions with conspecifics occur in cover, but observations suggest a display of unknown significance: a bird flew and ran in front of mangroves, then darted and dived, jumped into the air, displaying the wings and back, and described a rapid figure-of-eight pattern on the ground.

The threat display involves two birds standing facing each other and staring at each other for 15-20 sec before one withdraws; another display involves lowering the head, pointing the bill at the ground, half-spreading the wings and making quiet grunting calls. Most agonistic encounters involve chasing, with neck extended, head lowered and tail raised; birds sometimes fly to pursue a rival. Fighting, less frequently observed, involves pecking, mainly at the head, and birds often face one another, rise together and strike with their bills and claws; between bouts, the birds coo noisily but briefly; retreating birds utter squeals. Unfledged chicks may chase adults, which may either flee or peck the chick. The young often charge at other bird species, and there are many encounters between adults and large offspring. Birds often dash suddenly into groups, scattering the members without making physical contact. The distraction display at the nest involves the adult dashing out of cover, dragging its wings on the ground; the bird may repeatedly charge an intruder with head lowered and back feathers erect.

Courtship feeding and allopreening occur, but there appears to be no courtship display. Chasing is possibly related to sexual behaviour, the male sometimes pursuing the female and pecking at her. Possible mating behaviour involved one bird standing on its toes, the bill pointed upwards, shrugged its shoulders and gave a sharp, repeated squeak, wandering around between calls as though looking for something (Dunlop 1970). In copulation the female does not crouch, but bends forward on stiff legs with the bill almost touching the ground; the male mounts, flaps his wings vigorously, copulates and dismounts after 5-6 sec, when the pair often shake themselves, and then resume feeding. Females sometimes avoid copulation.

BREEDING AND SURVIVAL Season Cocos (Keeling), Jan, May, Jun, Nov; Philippines, May, Nov; Lesser Sundas, Apr, Aug, but all year on Flores; Moluccas, Jul/Aug; New Guinea, mainly in period of most rain, but in Papua New Guinea also in drier months (Jul-Nov); Witu I, Jul; Australia, mainly Aug-Mar, but eggs and chicks recorded all months in tropics (N Queensland); Norfolk I, Dec; New Zealand, Sep-Dec, Mar, dependent chicks Apr; New Caledonia and Loyalty Is, Oct-Jan; Vanuatu, extended breeding season but mainly Oct-Mar; Tonga, Feb, Jul; Samoan Is (in tropical climate), chicks Dec-Apr and Jul-Sep, may breed throughout year on Apia with possible break May-Jul; Niue I, Jul, "probably mid-winter, shortly after solstice" (Wodzicki 1971). **Nest** A cup of short dry grass stems and herbage (fine twigs also recorded), sometimes with bent-over and interwoven rooted stems; or a roughly woven platform of grass and rushes with depression on top; sometimes with flimsy canopy of growing vegetation; nest unlined; external diameter 10-23cm; platform height 17cm; internal cavity depth 2.5-6.5cm. Placed on ground, or up to 15cm above ground or water, in long grass, tussocks, marsh vegetation or crops (oats, lucerne); also in slight depression in ground, in sheltered position under bush, tree, banana leaf, or in hollow; sometimes in debris under trees (*Pisonia*), on tree stump, in fork of tree or shrub (once 1.5m above ground in coconut palm at base of frond), or on sand; 1 nest among pile of boxes. On islands, nest sometimes just a scrape under rocks or logs. Both sexes build. Trial nests, and flimsy brood nests, also constructed. **Eggs** 2-8, usually 4-8 (6-10 in captivity); larger clutches (up to 15) probably from 2 females laying in same nest; oval to rounded oval, smooth and glossy; faint buffy-white to rich cream or pale pinkish-brown, irregularly spotted and blotched (sometimes also streaked) reddish-brown, red, and purplish-red, with underlying markings of violet, liver or grey; spots larger and more numerous at blunt end. Size of eggs: *mellori*, Australia and Norfolk I, 34.2-39.2 x 26.2-28.8 (36.4 x 27.6, n = 37), calculated weight 15.4; *assimilis*, 35.9-43.5 x 27.7-29.8 (40.4 x 28.9, n = 27), weight 13-19 (16, n = 26, SD 1.4); *philippensis*, 36.4-39.2 x 26.8-30.7 (38 x 28.7, n = 9), calculated weight 17.2; *meyeri* & *lesouefi*, 33.5-41 x 27.3-30 (37 x 28.5, n = 33), calculated weight 16.6; *christophori*, 37.3-40.9 x 28.1-29.6 (39.5 x 29, n = 6), calculated weight 18.2; *swindellsi* 33.6 x 27.4, calculated weight 14; *sethsmithi*, 36.1-41.6 x 27.5-29.5 (38.8 x 28.7, n = 4, calculated weight 17.6; *ecaudatus*, 33.5-34.8 x 26.7-27.4 (34 x 26.9, n = 4); *goodsoni*, 37.2-41 x 26.8-29.2 (39 x 28.1, n = 12), calculated weight 17 (Schönwetter 1961-62, Elliott 1983, Rinke 1987). Eggs laid at daily intervals. **Incubation** 18-25 days, by both sexes. If nest disturbed, bird may carry eggs in bill to hastily constructed nest nearby. Eggs hatch within 24 h. **Chicks** Precocial and nidifugous; leave nest soon after hatching; fed and cared for by both parents, possibly largely by female (in wild and in captivity); brood may divide evenly between parents. Adult picks up food, coos softly; chick takes food from adult's bill, sometimes adopting feeding position between adult's legs; chicks forage independently after first week. On islands with little cover, chicks shelter in Wedge-tailed Shearwater *Puffinus pacificus* burrows for shade and/or protection (Dyer 1992). Remiges begin to appear at 5 days; rest of feathers from 7 days; young attain adult weight at 3 weeks; fledge in 5-6 weeks; fly at 2 months, when postjuvenile moult begins. Parents sometimes aggressive to chicks, usually those which are sick or under-developed. Young evicted by parents when 5-9 weeks old. In Western Samoa hatching success over 5 broods 4.2 chicks per brood; survival to fledging averaged 1.4 young per brood; losses apparently due to predation by feral cats (Robinson 1995). In New Zealand, from 8 clutches totalling 38 eggs, only 13 hatched (mean 1.6 young per nest). Some nests destroyed by flooding; those in lucerne paddocks often destroyed by mowing; eggs also trampled by cattle (Marchant & Higgins 1993). Age of first breeding probably 1 year. Breeding recorded up to 3 times (Australia) or 5 times (Western Samoa) per year; probably nests more than once per season in Vanuatu; interval between clutches 2 months (Australia).

47 ROVIANA RAIL
Gallirallus rovianae   Plate 14

Gallirallus rovianae Diamond, 1991, near Munda, New Georgia, Solomon Islands.

Two distinct forms exist, presumed to be the same species but not yet racially separated.

IDENTIFICATION Length c. 30cm. New Georgia form known only from the holotype, sex and age unknown, but apparently fully grown. Medium-sized rail, flightless or nearly so, but with long tail. Unmarked upperparts almost uniform dark chestnut-brown, slightly richer on nape and more reddish-chestnut from lores to ear-coverts; inconspicuous grey stripe from rear of eye almost to nape. Wing barring restricted to small white spots on primary coverts. Chin and throat whitish; malar region to breast grey; breast-band pinkish-tan. Sides of neck and breast, and rest of underparts, charcoal-brown with narrow white bars; undertail-coverts barred pale buff and black. Colours of bare parts in life not known. Kolombangara form darker, especially on underparts which are largely dark grey with very fine black and white barring from sides of neck to flanks and undertail-coverts; lacks buff breast-band; Iris dark red; bill dark grey with yellowish tip and sometimes base; legs and feet dull grey. Inhabits forest, scrub and overgrown plantations.
Similar species New Georgia form closest in plumage to Buff-banded Rail (46) and Guam Rail (48), but appears

less boldly patterned than former, with plain upperparts, and has longer tarsus and charcoal-brown, not black, barring on underparts; has no barring on remiges. Differs from Guam Rail in having longer tail, brown underpart barring which extends to sides of neck, less conspicuous stripe behind eye, white markings on wings confined to primary coverts, and possibly paler bill. Kolombangara form (Fig. 7) is even more distinctive because of restricted barring on underparts; appears a little larger, bulkier and longer-necked than Buff-banded Rail (D. Gibbs *in litt.*) and stands more upright (G. Dutson *in litt.*).

Figure 7. Roviana Rail, Kolombangara form. [After painting by D. Gibbs].

VOICE The call is said to be a rapidly repeated high-pitched note, which has given rise to its Roviana name Kitikete (Diamond 1991). D. Gibbs (*in litt.*) heard the Kolombangara birds make a similar frenetic, duetting *kik-kik-kitikek-kitikek-kitikek*, and also variations in which some notes were shrill and others clicking. Also a sneezed *tchu* or *tchu-ku* similar to Buff-banded Rail and a deep gulping similar to New Britain Rail (G. Dutson *in litt.*).

DESCRIPTION New Georgia form
Holotype; fully grown but age not known; details from Diamond (1991). Upperparts from crown to tail, including upperwings, nearly uniform unmarked dark chestnut-brown; colour slightly richer on nape, becoming richer, slightly reddish-chestnut, from lores to ear-coverts. Inconspicuous grey stripe extends from rear of eye almost to nape; no clear indication that this stripe extends anteriorly towards lores. Upperwing-coverts like back; primary coverts have eight small white spots; underwings unmarked dark brown with two white spots. Chin and throat whitish; malar region to chest grey, sides of neck and breast narrowly barred charcoal-brown and white; interrupted breast-band pinkish-tan. Rest of underparts to upper belly charcoal-brown, narrowly barred white, feather tips washed chestnut; centre of belly pale buff; thighs brown; undertail-coverts barred black and pale buffy, washed pinkish-tan or brick towards feather tips. Bare part colours not known; in skin, bill and legs dull in colour.
Downy young Not described.

MEASUREMENTS (Holotype; from Diamond 1991) Exposed culmen 34.5; tarsus 57. Wing and tail not measured.

GEOGRAPHICAL VARIATION The distinctive form from Kolombangara I is clearly an undescribed race. Details from Gibbs (1996) and D. Gibbs (*in litt.*). Rather darker than New Georgia form, especially on underparts which are darkish grey, finely barred blackish-grey and whitish on sides of neck (very obscure), sides of breast and flanks to undertail-coverts; lacks buff breast-band. Forehead to nape darker brown than rest of upperparts, and hind-neck and sides of neck rather richer chestnut; dark grey supercilium very narrow at base of bill, more prominent behind eye and ending at rear of ear-coverts; lores dark; ear-coverts chestnut, merging with nuchal collar; chin and throat appear darker than foreneck and breast. Folded wing shows a row of whitish spots, probably on primary coverts; primaries may be barred dark brown and chestnut (not clearly seen). Iris dark, dull red; bill dark grey with variable amount of dirty yellowish at tip and sometimes at base; legs and feet dull grey.

MOULT The holotype has all the primaries in sheath (Diamond 1991). This is compatible with the bird being a juvenile with primaries not fully grown, or an adult with simultaneous moult of the primaries. The second possibility is as likely as the first because remex moult is synchronous in all *Gallirallus* species for which it has been described (see species accounts 39-54). The remex moult of the possibly closely allied Woodford's Rail (37) is sequential.

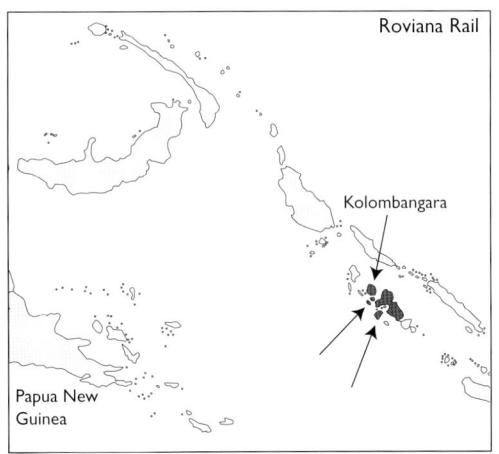

DISTRIBUTION AND STATUS The islands of New Georgia, Kolombangara, Wana Wana, Kohinggo and Rendova in the central Solomon Is. This rail is considered NEAR-THREATENED. Its known distribution is based largely on reports by local people on the islands concerned (Diamond 1991, Gibbs 1996). It may also occur on the nearby islands of Vangunu and Tetipari, where it has not yet been sought, but apparently it does not occur on the other major islands near New Georgia (Gatukai, Simbo, Ganonga, Vella Lavella and Gizo), as local inhabitants do not know it (Diamond 1991). On Kolombangara its local name is Kermete; it is well known to the local people and is common in degraded habitats (Gibbs 1996). It is apparently widespread on New Georgia (Diamond 1991) but is also reported to be scarce there (G. Dutson *in litt.*). Its current status elsewhere is not known, but its extremely restricted distribution gives cause for concern.

An observation by Blaber (1990) of a group of three Buff-banded Rails on New Georgia may refer to Roviana Rail (Diamond 1991), and it is thought that sightings of claimed Woodford's Rails from Ringi Cove, Kolombangara (Finch 1985), may also be referable to Roviana Rail (Gibbs 1996).

MOVEMENTS None recorded.

HABITAT Described as forest, especially second growth, where young trees grow on abandoned garden sites (Diamond 1991). On Kolombangara it is common in scrub, overgrown coconut plantations and other degraded habitats, and ventures into open grassy places such as clearings and airstrips (Gibbs 1996).

FOOD AND FEEDING Said to be omnivorous, taking worms, small crabs, seeds, coconut shoots, and potatoes and taro from gardens (Diamond 1991).

HABITS Flightless, or nearly so, being able to flutter only up to 50cm above the ground; the birds run very fast, zigzag, and can be caught only with dogs (Diamond 1991). More detailed information is given by Gibbs (1996). In the early morning the rails regularly venture out into clearings in scrub but they usually disappear into cover as soon as sunlight reaches the open ground; however, they may venture into the open later in the morning. Once under cover they become impossible to see but can still be heard. Four birds were observed wandering about on a grassy airstrip, one venturing out into the centre of the strip; when disturbed they ran fast to cover, with their heads down, not making the slightest attempt to use their wings.

SOCIAL ORGANISATION The Kolombangara birds appear to live in pairs, often but not always keeping close together (D. Gibbs *in litt.*).

SOCIAL AND SEXUAL BEHAVIOUR Unknown.

BREEDING AND SURVIVAL Season Reported to breed in the dry season (June), when rainwater is not a problem on the forest floor (Diamond 1991). **Nest** Said to be a depression in ground, lined with debris and containing 2-3 eggs. No further information available.

---

48   GUAM RAIL
Gallirallus owstoni   Plate 13

Hypotaenidia owstoni Rothschild, 1895, Guam.

Sometimes retained in *Hypotaenidia* or placed in *Rallus*. Monotypic.

Synonym: *Rallus owstoni*.

Alternative name: Owston's Rail.

IDENTIFICATION Length c. 28cm. Medium-sized and virtually flightless; tail very short and decomposed; toes very short. Head and entire upperparts olive-brown, slightly more rufous on crown, lores and ear-coverts; some birds have pronounced reddish tinge on nape. Broad supercilium pale grey; chin and throat whitish. Flight feathers and adjacent upperwing-coverts barred black and white, forming noticeable patterned area on folded wing. Foreneck to upper breast pale grey; in fresh plumage has indistinct olive-buff breast-band which becomes abraded to leave breast grey. Rest of underparts black with narrow white bars. Iris red; bill black to dark brown; legs and feet pale brown. Sexes alike but female smaller; size difference often noticeable in field. Immature like adult. Juvenile similar to adult but with less extensive areas of grey on neck, breast and supercilium. Inhabits forest, woodland, scrub and grassland.

Similar species Most closely resembles the widespread Buff-banded Rail (46) and the flightless Roviana Rail (47) of the Solomon Is. Larger and longer-billed than Buff-banded Rail, with darker, unpatterned upperparts; facial pattern similar but facial stripe brown, not chestnut; usually lacks chestnut on nape; lacks prominent orange-buff breast-band of most races of Buff-banded Rail; has more extensive black and white barring on remiges, and black bill with grey base. Roviana Rail from New Georgia has browner barring on underparts extending to neck, less prominent supercilium, longer tail and possibly paler bill. Kolombangara form of Roviana Rail has dark grey underparts with very fine black and white barring from sides of neck to flanks and undertail-coverts.

VOICE Information is from Ripley (1977) and Pratt *et al.* (1987). During the breeding season the birds make loud penetrating screeches *keee-yu*. They also give a series of short *kip* notes. They may be largely silent when not breeding. Downy young utter a single light *tsip*. Fully grown birds in a family group with chicks uttered gulping contact calls (Jaffe 1997).

DESCRIPTION
Adult Forehead to nape, hind-neck, sides of hind-neck, stripe from lores through ear-coverts to nape, and entire upperparts including tail, olive-brown; slightly more rufous on crown, lores and ear-coverts; some individuals have pronounced reddish tinge on nape. Broad superciliary stripe pale grey, off-white in front of eye; terminates at nape. Lesser upperwing-coverts olive-brown; medians and greaters blackish-brown, barred white; alula and remiges barred blackish-brown and white, barring decreasing from being extensive and bold on outer primaries to narrow on secondaries as brown of feather centres becomes broader; inner secondaries with broad olive-brown tips; tertials olive-brown; axillaries, underwing-coverts and underside of remiges blackish-brown, barred white. Chin and throat whitish, fading into pale blue-grey or grey on foreneck, sides of foreneck and upper breast. In fresh plumage has indistinct, dull olive-buff breast-band which becomes abraded to leave breast clear grey. Rest of underparts black with narrow white bars; posterior belly slightly paler. Iris red; bill black with grey base, or dark brown; legs and feet pale brown or yellow-brown. Sexes alike but female smaller.
Immature Indistinguishable from adult.
Juvenile Similar to adult but with less extensive areas of grey on neck, breast and supercilium.
Downy young Black, with black bare parts.

MEASUREMENTS Wing of 6 males 120-133 (124.8), of 8 females 112-125 (118.7); tail of 3 males 50-53 (51.7), of 6 females 45-52 (47.2); culmen to base of 1 female 41; tarsus of 1 female 44 (Jenkins 1979, 1 BMHN skin). Ripley (1977) gives tail of male 46-54, of female 38-46; exposed culmen of 8 males 33-43 (39.8), of 9 females 34-42 (38); tarsus of 6 males 49-56 (51.5), of 8 females 43-54 (46.6). Weight of 27 males 174-303 (241.1), of 20 females 170-274 (211.9).

GEOGRAPHICAL VARIATION None.

MOULT No information is available on the moult of adults. The completion of postjuvenile moult is evident at c. 16 weeks, after which age birds are indistinguishable in plumage from adults (Jenkins 1979). The occurrence of a first prebreeding moult is not recorded.

HABITAT This rail formerly occurred in most habitats on Guam, including forest, savanna, secondary grassland, agricultural areas, mown grass (e.g. along roads and telephone lines) bordering scrub communities, mixed woodland and scrub, and fern thickets (Jenkins 1979). It was seldom seen in the interior of mature limestone forest which, like the savanna in S Guam, was a marginal habitat (Jenkins 1979), and it did not occur in freshwater wetland habitats (Jenkins 1979, Witteman *et al.* 1991), although King (1981) states that it occurred in marshlands.

FOOD AND FEEDING Most information is from Jenkins (1979). An opportunistic, omnivorous feeder, preferring animal to vegetable matter. Food includes snails, slugs, insects (Orthoptera, Dermaptera and Lepidoptera), geckoes *Hemidactylus frenatus* and some vegetable matter such as seeds and palm leaves; the birds also take fish, and carrion such as amphibians crushed by cars. The giant African snail *Achatina fulica*, accidentally introduced in 1945, became an important food when it expanded its range into most of the island's habitats. During the dry season the rails were reported to damage crops such as tomatoes, cucumbers and melons, but such damage probably resulted from their obtaining moisture rather than food. The birds ingest coral chips (up to 9mm in diameter) and pieces of snail shell for grit. Captive birds also took ground beef, chicken feed and lettuce leaves, and drank mainly fresh water but also occasionally sea water (Carpenter & Stafford 1970). The birds take food items from the surface of the ground, especially snails and slugs after rain showers. They chase low-flying insects (especially butterflies), and take seeds and flowers from low grasses and shrubs, stretching up to reach items 40cm above the ground. They often forage along field edges and roadsides, but seldom venture far from cover. Their fondness for water, and their inability to maintain weight in captivity on a succulent diet with no water, indicate a high water requirement (Carpenter & Stafford 1970).

HABITS Information is from Carpenter & Stafford (1970) and Jenkins (1979). Guam Rails are secretive and wary, seldom wandering far from cover, and are most active at dawn and dusk, when they venture from cover to forage and bathe at field edges and roadsides. They will also forage at night. They can fly for only 1-3m, and to a height of 1-2m, flights being aided by jumping with the powerful legs, but they usually run from danger, flapping the wings when running rapidly over open ground. They are fond of fresh water and they frequently bathe in rain puddles after early morning showers. In the wild, birds in the open spend much time bathing and preening, while captive birds bathe at least once per day, completely immersing themselves in water. The birds prefer to roost above the ground, in trees and shrubs.

SOCIAL ORGANISATION Monogamous and territorial. One family group observed in the wild consisted of two adults, several immatures (juveniles?) and three chicks (Witteman *et al.* 1991).

SOCIAL AND SEXUAL BEHAVIOUR In captive birds high levels of aggression have caused pairing difficulties (Witteman *et al.* 1991). In the wild, recorded instances of fighting, assumed to be between males, are probably related to territoriality (Jenkins 1979), while chases, accompanied by screeching calls, have been observed in July.

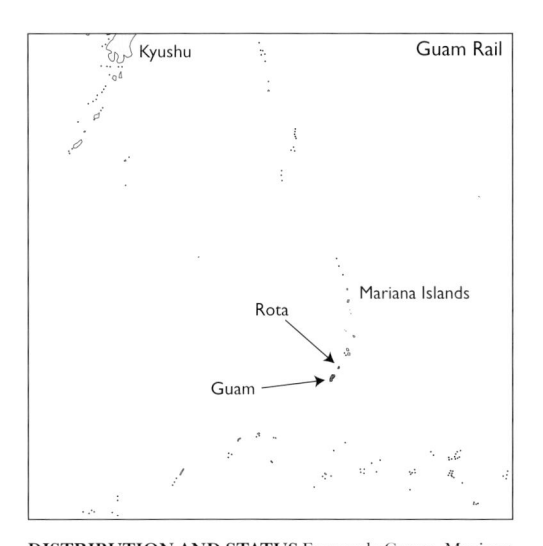

DISTRIBUTION AND STATUS Formerly Guam, Mariana Is. EXTINCT IN THE WILD, but recently reintroduced to the nearby Rota I. It was formerly widely distributed and abundant on Guam (area 541km^2), although taken by local people with dogs and snares and despite the presence of introduced predators such as feral pigs and cats, while its numbers tended to fluctuate with rainfall cycles (Perez 1968). After World War II its population increased and in the 1960s was estimated at 80,000 (Lint 1968). After 1968 it declined rapidly, along with most other indigenous birds, as a result of the spread throughout the island of the accidentally introduced Brown Tree Snake *Boiga irregularis*, a nocturnal and arboreal species which is native to Australia (Savidge 1987). Although adult Guam Rails may be too large to be eaten by most of these snakes, the eggs and young are very vulnerable (Savidge 1987). The rail was listed as a protected species in 1976 but by 1981 its population was reduced to c. 2,000 birds, by 1983 to less than 100, and it became extinct in the wild by 1987 (Collar *et al.* 1994). In 1982 a captive breeding programme was set up and in 1994 c. 180 breeding birds were located on Guam and in 16 zoos in the USA (Collar *et al.* 1994). Captive breeding has been very successful: the captive population has grown rapidly because the birds attain sexual maturity at only four months of age and breed throughout the year, and the founding birds showed surprisingly good genetic diversity (Witteman *et al.* 1991, Haig *et al.* 1993). Efforts are being made to establish a self-sustaining experimental population on the snake-free island of Rota in the N Mariana Is. The first introductions on Rota were unsuccessful, observed causes of mortality being vehicles on roads and predation by cats, but the species has recently bred there for the first time (Derrickson 1996), and more introductions are planned to boost the population. Efforts are being made to eradicate or control the Brown Tree Snake on Guam and it is hoped that Rota will provide a source of wild rails for eventual reintroduction there (Derrickson 1996). Genetic management techniques have been applied in the captive breeding programme to maximise the retention of genetic diversity in the captive population, and to allow the introduction into the wild of the full complement of this genetic diversity (Haig *et al.* 1993).

MOVEMENTS None recorded.

BREEDING AND SURVIVAL Season Breeds throughout year, but possibly peaks during rains (Jul-Nov). **Nest** Located on dry ground in dense grass; shallow cup of interwoven loose and rooted grass; built by both sexes. One nest was 13cm in diameter and 3cm deep. **Eggs** 1-4 (usually 3-4); white to pinkish, with scattered small spots of pink or russet, and pearl-blue, concentrated at blunt end; size (n = 3) 35.7-40.7 x 28-29.1 (38.5 x 28.7). **Incubation** 19 days, by both sexes; hatching asynchronous; eggshells are consumed by an adult (presumably female). **Chicks** Precocial; leave nest within 24 h of hatching; fed and cared for by both parents. Adults locate foraging spots and allow chicks to peck there, often allowing juveniles to forage independently. Adults also catch food, usually insects, which chicks take from their bills; alternatively food is placed on ground for chicks (Jenkins 1979). Chicks begin to attain juvenile contour feathers during fourth week; reach adult weight at 7 weeks; become sexually mature at 16 weeks, when postjuvenile moult is complete. Number of broods per year unknown; in captivity, will start laying again within 3 weeks of hatching chicks (Jaffe 1997). Several introduced species, such as the monitor lizard *Varanus indicus*, three species of rats (*Rattus rattus*, *R. norvegicus* and *R. exulans*), as well as feral dogs, cats and pigs, may have been nest predators (Jenkins 1979).

49 WAKE RAIL
Gallirallus wakensis Plate 15

Hypotaenidia wakensis Rothschild, 1903, Wake Island, Pacific Ocean.

Sometimes placed in *Rallus*. Monotypic.

Synonym: *Rallus wakensis*.

Alternative name: Wake Island Rail.

IDENTIFICATION EXTINCT. Length 22-25cm. Flightless; a small *Gallirallus* species. Lores, ear-coverts and entire upperparts, including upperwings, olive-brown, darker on head, and feathers noticeably fringed buffy except on head and tail; supercilium pale grey. Primaries and secondaries barred and spotted cinnamon; outermost 2 primaries barred white. Chin to throat whitish; rest of undersides ashy-brown or grey, narrowly barred white on sides of neck and breast, and on flanks; undertail-coverts barred olive-brown and white; vinaceous breast-band present but sometimes indistinct. Bill, legs and feet brown (in skins). Sexes alike; female slightly smaller. Immature and juvenile undescribed. Inhabited scrub.
Similar species Closest in appearance to Buff-banded Rail (46), which is brighter in colour, has patterned upperparts, rufous-brown facial stripe and nape, pronounced barring on remiges and barred breast.

VOICE A low chattering call, a clattering noise, and a low cluck audible at close range (Ripley 1977).

DESCRIPTION
Adult Details from 1 skin in BMNH, plus information from published descriptions. Upperparts olive-brown, feathers fringed buffy-brown to buff; forehead to crown darker, with very narrow, indistinct pale feather fringes; rump to uppertail-coverts with slightly darker, duller feather fringes; tail plain olive-brown. Upperwing-coverts and tertials as upperparts, tertials fringed brighter pale buff; alula and primaries buffy-brown, darkening to olive-brown on secondaries; outer edge of primaries pale buff; alula and secondaries fringed duller buff; outer 2 primaries barred white; inners and secondaries barred and spotted vinaceous-cinnamon; underwing-coverts and axillaries olive-brown, barred white; underside of remiges as upperside, but duller. Lores and ear-coverts dark olive-brown; supercilium pale grey. Chin and upper throat whitish; sides of neck and foreneck grey; rest of undersides buffy-brown to medium pale grey, barred white, barring narrow and variable on sides of neck and breast, narrow on flanks; vent to undertail-coverts more prominently barred olive-brown and white; deep vinaceous-buff breast-band present, but indistinct in some specimens; centre of belly whitish. Bill, legs and feet brown (in skins). Some illustrations show supercilium ending at nape, others show it continuing around rear of ear-coverts to join grey at sides of neck (Plate 15). Sexes alike but female slightly smaller.
Immature Not described.
Juvenile Not described.
Downy young Not described.

MEASUREMENTS Wing of unsexed birds 85-100; tail of 1 unsexed 45; culmen of male 25-29, of female 24-27, exposed culmen of 8 unsexed 25-30 (27.4); tarsus of male 32-37, of female 24-27, of 8 unsexed 32.5-35 (33.4) (Rothschild 1903, Ripley 1977, Diamond 1991).

GEOGRAPHICAL VARIATION None.

MOULT No information available.

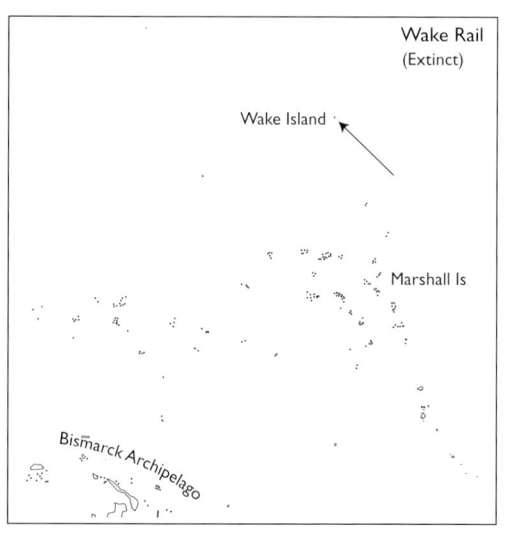

DISTRIBUTION AND STATUS EXTINCT. Formerly confined to Wake I, a small (23km^2) remote island in Micronesia. This rail, the only land bird on the island, survived the potentially dangerous period of early human contact and was quite plentiful before World War II (Fuller 1987). However, from 1942 to 1945 the island was a Japanese garrison and, when the island was repossessed in 1945 the rail was extinct (Fuller 1987). It is fairly certain that the birds were eaten by the Japanese forces, who were cut off from supplies and were living on almost a starvation diet (Greenway 1967). However, during the war rats were very common on Wake and were possibly partly responsible for the demise of the rail (Ripley & Beehler 1985).

MOVEMENTS None.

HABITAT The island is low and largely covered with *Pandanus* scrub. The rails inhabited areas with scrub and low trees *Sesurium* (Ripley 1977).

FOOD AND FEEDING Probably omnivorous, taking most animal foods which were available; worms, gastropods and insects are recorded in the diet. The birds walked deliberately and paused frequently to dig in loose soil with sideways thrusts of the bill to expose food items (A. Wetmore in Ripley 1977).

HABITS These rails were alert and inquisitive, unafraid of people and without any significant enemies. They would venture 3-4m from cover and walk around deliberately and unconcernedly, with head and neck erect and tail occasionally flicked (Day 1989). They were flightless, and ran rapidly into cover at any suspicious movement.

SOCIAL ORGANISATION Nothing recorded.

SOCIAL AND SEXUAL BEHAVIOUR Nothing recorded.

BREEDING Occurred Jul-Aug, possibly also at other times.

50 TAHITI RAIL
Gallirallus pacificus   Plate 16

Rallus pacificus Gmelin, 1789, Tahiti.

Sometimes retained in *Rallus* or placed in *Hypotaenidia*. No specimen exists. Monotypic.

Synonyms: *Hypotaenidia pacifica*; *Rallus ecaudatus*; also *Rallus pacificus ecaudata*, in confusion with *Rallus* (=*Gallirallus*) *philippensis ecaudata*.

Alternative names: Tahitian/Pacific/Red-billed Rail.

IDENTIFICATION EXTINCT. Length 23cm. Flightless; small for a *Gallirallus* species. A striking and unusually coloured rail: top of head, sides of face and entire upperparts black, with white supercilium, ferruginous nape, white spots from back to tail, and white barring on wings; underparts white except for grey breast and narrow black band across base of throat; iris and bill red; legs pink. Immature and juvenile undescribed. Inhabited open areas, marshes and coconut plantations.

Similar species Although features such as rufous nape, white supercilium, spotted upperparts and barred wings suggest that it was probably derived from Buff-banded Rail (46) stock, Tahiti Rail was very different to this and other *Gallirallus* species in general appearance.

VOICE The call is described as "similar to the other rails but with one major difference – the end of the call was terminated with a high-pitched whistle" (Bruner 1972).

DESCRIPTION
Adult Details from Ripley (1977), taken from the original Forster illustration. Top of head and sides of face black; supercilium, from base of bill to occiput, white; hind neck ferruginous; rest of upperparts and upperwings black, sparsely patterned with minute white dots from back to rectrices, and upperwings variegated with broken white bands. Chin and throat white, narrow band across base of throat black; breast grey; rest of underparts white. Wings short. Iris red; bill blood-red; legs and feet fleshy-pink.

Sexes presumably alike. An illustration in a later work (Rothschild 1907) shows an indication of barring on the uppertail-coverts and undertail-coverts, which is not shown in the original Forster painting (Ripley 1977).
Immature Not described.
Juvenile Not described.
Downy young Not described.

MEASUREMENTS Unknown.

GEOGRAPHICAL VARIATION None recorded.

MOULT No information available.

DISTRIBUTION AND STATUS EXTINCT. Formerly confined to Tahiti I and the smaller uninhabited Mehetia I, E of Tahiti in the same archipelago; it may also have occurred on other of the Society Is. It was discovered by naturalists on James Cook's second voyage around the world but no specimen exists and the species is known only from a painting by George Forster kept in BMNH (Fuller 1987). Although King (1981) gives its date of extinction as 1925, Bruner (1972) states that it was last recorded from Mehetia in the 1930s. It became extinct on Tahiti earlier, some time after 1844. It was said to be very common on Tahiti until the end of the 19th century, when it began to decline in numbers, probably as a result of the introduction of cats and rats (Bruner 1972). The reason why it survived longer on Mehetia may be that there were no cats on the island (Greenway 1967, Day 1989).

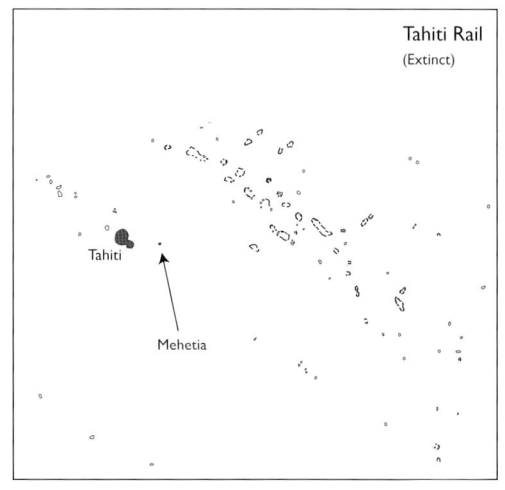

MOVEMENTS Unknown.

HABITAT It was often seen in open areas, sometimes around marshes with other rails, and commonly in coconut plantations (Bruner 1972).

FOOD AND FEEDING It apparently took mainly insects found in the grass, but occasionally ate copra (Bruner 1972).

HABITS Flightless. Described as a rather attractive bird which blended well with its surroundings because of its broken colour pattern, and lacked shyness (Bruner 1972).

SOCIAL ORGANISATION Unknown.

SOCIAL AND SEXUAL BEHAVIOUR Unknown.

BREEDING Nested on the ground.

51 DIEFFENBACH'S RAIL
Gallirallus dieffenbachii   Plate 16

Rallus Dieffenbachii G. R. Gray, 1843, Chatham Islands.

Sometimes regarded as a race of *G. philippensis*, from which it is distinct both in plumage and skeletal characters (Scarlett 1979, Marchant & Higgins 1993); recent investigations using mtDNA sequence data (Trewick 1997) confirm the distinction from *G. philippensis* and indicate that it is close to *G. modestus*. Monotypic.

Synonyms: *Cabalus/Nesolimnas/Ocydromus/Hypotaenidia dieffenbachii*; *Gallirallus/Rallus philippensis dieffenbachii*. *Gallirallus modestus* (52) was described as juvenile *Cabalus dieffenbachii* by Sharpe (1894).

Alternative names: Banded/Chatham Island Banded Rail.

IDENTIFICATION EXTINCT. Length 28-36cm. Flightless. A distinctive, medium-sized rail, predominantly cinnamon to tawny with darker brown to blackish barring, including on wings, neck, breast and undertail-coverts. Head pattern also distinctive: forehead to hind-neck, and stripe through eye, dark rufous-brown; long supercilium, lower ear-coverts to chin and throat bluish-grey. Lower throat, flanks and belly black with narrow white bars; tail dark brown. Iris reddish-brown; bill strong, pale yellow-brown and markedly downcurved towards tip; legs and feet pale brown; legs short. Immature and juvenile undescribed. Inhabited scrub and tussock grass.
Similar species Closest to the Buff-banded Rail (46), of which it is thought to be an insular derivative, but differed most markedly in having darker grey on face, chin and throat, pronounced barring and no spotting on upper-parts, cinnamon and black barring over entire breast and on undertail-coverts, and longer, stronger, downcurved bill.

VOICE The call was said to be shrill and frequently heard (Day 1989).

DESCRIPTION
Adult (Holotype; unsexed) Forehead and crown to hind-neck dark brown; feathers of hind-neck rufous (burnt sienna), vaguely barred dark brown; Upperparts warm dark brown (raw umber), barred cinnamon-buff to ochraceous-tawny, bars edged blackish; pale barring absent from lower back to rump but reappears on uppertail-coverts; rectrices unbarred darkish brown, mottled chestnut near bases. Upperwing-coverts, alula, primaries and secondaries broadly barred dark brown and rufous: lesser coverts with palest, most ochraceous bars, and other feathers having orange-rufous or cinnamon-rufous barring, especially on primaries and outer secondaries; tertials dark brown with rufous to cinnamon spots at edges; underside of remiges as upperside, but slightly duller; shafts pale rufous; axillaries and underwing-coverts blackish-brown, barred white to rufous. Lores, and region through and under eye to upper ear-coverts and nape, darkish rufous-brown; long supercilium, lower ear-coverts, malar region, chin and throat bluish-grey to grey, flecked white on chin and throat. Lower throat feathers blackish, subterminally barred white; foreneck and breast cinnamon, narrowly and irregularly barred blackish, this pattern extending up sides of neck to rear of ear-coverts. Lower breast, flanks, thighs and belly blackish-brown with narrow white bars which are tinged ochraceous on flanks and vent; undertail-coverts broadly barred blackish and cinnamon-rufous. Iris reddish-brown; bill relatively stout and markedly downcurved towards tip, yellowish-brown or pale brown with darker tip; legs short; legs and feet pale brown.
Immature Not described.
Juvenile Not described.
Downy young Not described.

MEASUREMENTS Holotype (adult, unsexed): wing 122; tail 69; culmen to base 37; tarsus 41. Weight, estimated from regressions of femur diameter and body weights of extant rails, 340-400 (Marchant & Higgins 1993).

GEOGRAPHICAL VARIATION No significant variation between subfossil bones from Chatham, Pitt and Mangere Is (Marchant & Higgins 1993).

MOULT No information available.

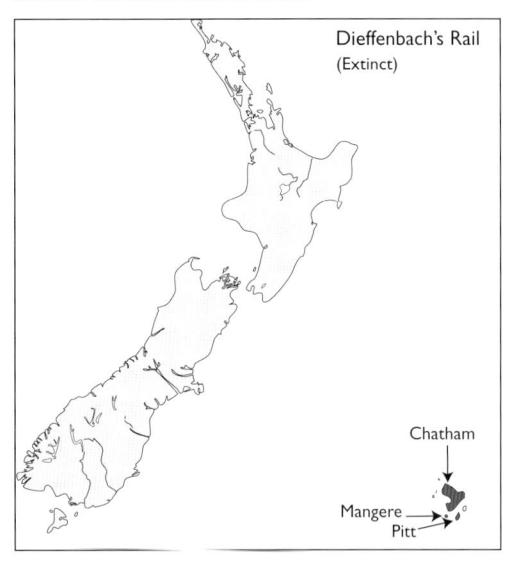

DISTRIBUTION AND STATUS EXTINCT. Known from the type, held in the British Museum, which was collected on Chatham I in 1840 and accessioned in 1842, and from subfossil bones from Chatham, Mangere and Pitt Is. Greenway (1967) also states that there is a specimen in Bremen, not mentioned in most other publications on the species. The abundance of subfossil material suggests that this rail was once common, but it was recorded as scarce by 1840 and was not found from 1872 onwards (Marchant & Higgins 1993), probably having been exterminated soon after the type was collected. It is suggested that the introduction of cats, dogs and rats, and the destruction of habitat by bush fires after settlement by the Polynesians, brought about its extinction (Day 1989, Marchant & Higgins 1993). Local people also caught it with nooses and on Mangere I, where the introduction of cats and the clearance of vegetation did not occur until the 1890s, it is likely to have been exterminated by being taken for human food (Tennyson & Millener 1994). It was sympatric with the smaller Chatham Islands Rail (52) (Tennyson & Millener 1994), which was at first thought to be the juvenile form of Dieffenbach's Rail; a discussion of the confusing situation generated by early distributional and taxonomic reports, including the identification of subfossils, is given by Olson (1973b: 394-395).

MOVEMENTS None.

HABITAT Apparently scrub and tussock grass (Day 1989).

FOOD AND FEEDING Its distinctive bill shape may have been an adaptation to feeding by probing in tussock grass.

HABITS Flightless, with a stunted, rounded wing.

SOCIAL ORGANISATION Unknown.

SOCIAL AND SEXUAL BEHAVIOUR Unknown.

BREEDING Nested on ground (Ripley 1977). No other information available.

52 CHATHAM RAIL
Gallirallus modestus   Plate 16

Rallus modestus Hutton, 1872, Mangere Island, Chatham Islands.

Monotypic.

Synonyms: *Cabalus modestus*; *Ocydromus pygmaeus*. Was also described as juvenile *Cabalus* (=*Gallirallus*) *dieffenbachii* by Sharpe (1894).

Alternative names: Hutton's/Mangere/Modest/Chatham Islands Rail.

IDENTIFICATION EXTINCT. Length 18-21cm. Flightless, with very soft, hair-like body plumage. A medium-small, predominantly dark olive-brown and grey-brown rail, with narrow sandy-buff barring on underparts from foreneck to undertail-coverts; primary coverts and outermost primaries also barred. Iris blue-black(?); bill long, thin, downcurved towards tip, dusky brown; legs and feet reddish to brown. Female smaller than male with shorter bill; one has sandy-buff barring on mantle and sparsely from back to uppertail-coverts, including upperwing-coverts. Immature like adult. Juvenile similar, but underparts plain leaden-grey, with a few bars on upperwing-coverts and primaries. Inhabited bush and tussock grass.
Similar species By far the smallest member of its genus, with distinctive plumage and bill. Shares with Buff-banded Rail (46) the buff barring on outer primaries, and is derived from an ancestral form of Buff-banded Rail which assumed neotenic characters in plumage as well as in the development of flightlessness.

VOICE Unknown.

DESCRIPTION Published descriptions and measurements (Sharpe 1894, Ripley 1977, Marchant & Higgins 1993) all refer to the 3 specimens (adult and juvenile male, and adult female) in BMNH. Neither description (Ripley's is a copy of Sharpe's) mentions that the male and female specimens show differences in upperpart plumage, which are detailed below and illustrated in Plate 16.
Adult male Head, upperparts and tail dark olive-brown, darkest on head, with vague indication of ashy supercilium, and slightly more fulvous-brown on mantle, upper back and upperwing-coverts. Tail much reduced, hidden by coverts. Upperwing-coverts long and fluffy, with vague indications of whitish barring; alula, primary coverts and outer 2-3 primaries olive-brown to dusky brown, barred or notched sandy-buff; other remiges darkish olive-brown, duskier on inner webs. Axillaries and underwing-coverts dark olive-brown to blackish-brown with a few buff spots;

underside of remiges dark brown, outer 2-3 primaries marked as on upper surface; shafts buff. Ear-coverts have faint ashy wash; chin and throat washed ashy-grey. Lower throat and foreneck dull ashy-brown; breast duskier; flanks, belly and vent more grey-brown; foreneck to belly variably barred sandy-buff, barring most distinct on sides of breast and flanks, and slightly tinged whitish on flanks and belly; undertail-coverts sometimes more spotted than barred. Iris described as light brown (Marchant & Higgins 1993) but labels on skins of 1 male and 1 female in British Museum state "eyes bluish-black with reddish-brown ring". Bill narrow, downcurved towards tip, dusky brown; legs and feet reddish-brown or dark brown. Small spur on carpal joint.
Adult female One specimen, BMNH. Similar to male, but bars on underparts brighter and more numerous; these bars also extend (less numerously) over back, rump and upperwing-coverts, and less obviously over mantle. From material examined it cannot be deduced whether upperpart plumage differences are consistent between sexes, are indicative of individual variation, or are age-related. Female smaller than male, with shorter bill.
Immature Apparently like adult (Forbes 1893).
Juvenile Much more uniformly coloured than adult; plain leaden-grey on underparts with only a few sandy-buff bars on outer wing-coverts and primaries. This juvenile, described as missing in 1988 (Knox & Walters 1994), is 1 of the 3 existing specimens: records show that BMNH did not have another in the collection.
Downy young Uniform brownish-black.

MEASUREMENTS Wing of 3 males 85.5, 86.5, 92 (88), of 1 female 76; tail of 3 males 31.5, 32, 34 (32.5), of 1 female 29; culmen to base of 2 males 38, 39, of 1 female 35; tarsus of 3 males 31.5, 34, 35 (33.5), of 1 female 31.5.

GEOGRAPHICAL VARIATION There is no apparent variation between bones from Chatham, Pitt and Mangere Is (Marchant & Higgins 1993).

MOULT No information available.

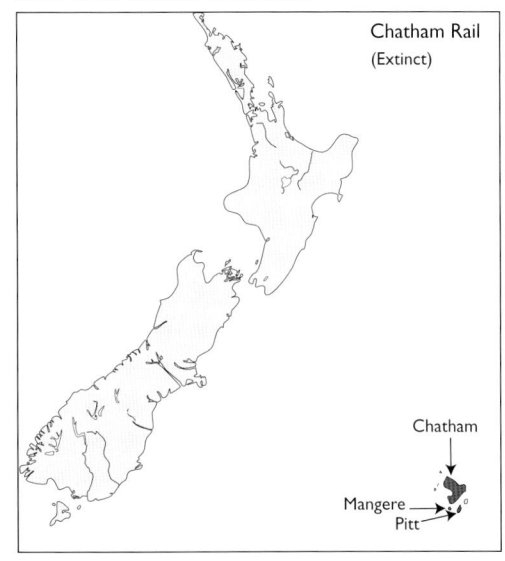

Chatham Rail
(Extinct)

Chatham
Mangere
Pitt

DISTRIBUTION AND STATUS EXTINCT. This rail was endemic to Mangere, Pitt and Chatham Is. It was first discovered in 1871 on Mangere I (area 1.25km^2), where it became extinct between 1896 and 1900 and from where 26 specimens are known in museum collections (Marchant & Higgins 1993). It became extinct on Chatham early in the nineteenth century, and on Pitt later that century; recent skeletal remains have been found on both islands (Marchant & Higgins 1993). It probably arose from an invasion of the islands by a *G. philippensis*-like form, and was sympatric with the larger Dieffenbach's Rail (51), which represented a later invasion of the islands by *G. philippensis*-like stock (Olson 1973b, 1975c). It was at first thought to be the juvenile of Dieffenbach's Rail (Buller 1873); a discussion of the confusing situation generated by early distributional and taxonomic reports is given by Olson (1973b: 394-395).

It has been suggested that the Chatham Rail was gradually swamped off the larger islands of the group as a result of competition from its larger congener, surviving longest on the small, outlying Mangere I (Olson 1975c, Ripley 1977). However recent studies have shown that both species coexisted on Mangere (Tennyson & Millener 1994), and its ultimate extinction there was brought about by the introduction of cats and rats, and the destruction of habitat (Marchant & Higgins 1993). The combined effect of introduced goats and rabbits, and of cutting, burning and replanting to provide pasture for sheep, had completely cleared the original bush on Mangere by c. 1900, while in the 1890s cats had been introduced to kill the rabbits (Greenway 1967). Excessive collecting has also been cited as a factor contributing to the bird's extinction (Oliver 1974), but this appears doubtful because the bird was extinct on Chatham before any were collected, while the Mangere population could not have survived the destruction of its habitat (Greenway 1967).

MOVEMENTS None.

HABITAT Tussock grass and scrub.

FOOD AND FEEDING Stomach contents were the legs and elytra of beetles (Buller 1905). The species was also observed feeding on Amphipoda in bush (Forbes 1893). Its bill appears to be adapted for probing, and it was probably more specialised in its foraging and food requirements than was Dieffenbach's Rail (Ripley 1977).

HABITS Flightless, with a stunted, rounded wing. Apparently nocturnal, or partly so (Greenway 1967).

SOCIAL ORGANISATION Unknown.

SOCIAL AND SEXUAL BEHAVIOUR Unknown.

BREEDING Season Dependent nestling collected 5 Jan. **Nest** Nested in holes in the ground; nest undescribed. **Eggs** One egg known; almost white with faint double spotting of grey and rufous; size 38.5 x 28, calculated weight 16.5 (Schönwetter 1961-62). **Chicks** After hatching, young apparently hid in fallen hollow trees (Forbes 1893).

53　SHARPE'S RAIL
Gallirallus sharpei　　　　　　　Plate 16

Stictolimnas sharpei Büttikofer, 1893, origin unknown.

Originally identified by Schlegel as the young of *Pardirallus maculatus* (see Olson 1986). Known from only one specimen.

Synonym: *Hypotaenidia sharpei*.

IDENTIFICATION PROBABLY EXTINCT. Length 28cm. Volant. A medium-sized, very distinctive species, known from only one specimen, origin and habitat unknown. Upperparts predominantly brownish-black; upperwings paler and more brownish; mantle, scapulars, upperwing-coverts, undertail-coverts and tail spotted white. Remiges dull brown, broadly barred white. Sides of head, chin, throat and neck grey; rest of underparts predominantly brownish-black; flanks with small white spots. Bare parts probably red or orange.

Similar species In its strongly barred wings, overall proportions, size and shape of bill, and all other particulars except pattern and coloration of the plumage, the specimen agrees very closely with the Buff-banded Rail (46), with which it must be congeneric (Olson 1986). The specimen differs from most *Gallirallus* species in lacking strong barring on the underparts, while the white dorsal spotting is shared with only the Buff-banded Rail and Slaty-breasted Rail (54).

VOICE Unknown.

DESCRIPTION
Adult (Holotype; unsexed) Forehead to hind-neck brownish-grey with darker feather centres; rest of upperparts, including scapulars and tail, brownish-black; feathers of mantle and scapulars narrowly margined olive-brown and with 1-2 pairs of irregular white spots; rectrices also spotted white, spots elongated to lateral stripes. Upperwing-coverts paler than rest of upperparts, tending to mummy-brown or dark brownish-grey; broadly fringed pale olive-brown to ochre and irregularly spotted white; spots largest on greater upperwing-coverts. Remiges as coverts, with 4-5 broad white bars on inner webs, continuing as bar-like spots on outer webs and running together to form longitudinal white stripe on outer web of 2-3 outermost primaries. Axillaries black with 3 pairs of white spots; underwing-coverts black, tipped white; underside of remiges heavily barred black and white. Lores and eyelids whitish; streak above and behind eye ashy-grey; sides of head, chin, throat, sides of neck and foreneck darker ashy-grey to slaty-grey. Breast, flanks and belly brownish-black, feathers broadly margined olive-brown; feathers of sides of breast with 1 pair of small white spots, flank feathers with 2 pairs; centre of breast and belly more brownish-grey; thighs ashy-grey; undertail-coverts brownish-black, spotted white. Colour of iris unknown, possibly red; bill, legs and feet probably red or orange.
Immature Unknown.
Juvenile Unknown.
Downy young Unknown.

MEASUREMENTS Holotype (adult, unsexed): wing 140; tail 65; exposed culmen 26; tarsus 41.5.

GEOGRAPHICAL VARIATION Unknown.

MOULT No information available.

DISTRIBUTION AND STATUS PROBABLY EXTINCT. Known only from the holotype in Leiden Museum, for which no locality is given and which was purchased from G. A. Frank, a natural history dealer in Amsterdam with worldwide trade connections (Olson 1986). The specimen was originally said to come from South America (Büttikofer 1893) but this must have been an assumption based on the original misidentification as Spotted Rail (113). Olson (1986) has produced reasoned speculations on the geographical origin of the species, as follows. As it is volant, it is unlikely to have come from an island in Oceania: flightlessness evolves very rapidly in the absence of terrestrial mammalian predators and any endemic species of Rallidae from Oceania are likely to have been flightless. In view of its close similarity in size and morphology to the Buff-banded Rail, it is unlikely to have been sympatric with that species, which is widespread in the Philippines, south-eastern Indonesia, the Australo-Papuan region, and east to Samoa, Tonga and New Zealand. However, the Buff-banded Rail is absent from the islands of the Sunda Shelf and, of all the *Gallirallus* species, only the Slaty-breasted Rail is known from the Asian mainland or any of the islands which were formerly connected to it. Therefore it is suggested that Sharpe's Rail came from an island on the Sunda Shelf, such as Java, Sumatra or Borneo. The fact that this region was long under Dutch influence might have increased the chance of a specimen passing though the hands of a dealer in Amsterdam. Olson also considers it not impossible that Sharpe's Rail could still exist and have been overlooked up to now.

MOVEMENTS Unknown.

HABITAT Unknown.

FOOD AND FEEDING Unknown.

HABITS The wings are not in any way reduced, so the species was not flightless.

SOCIAL ORGANISATION Unknown.

SOCIAL AND SEXUAL BEHAVIOUR Unknown.

BREEDING Unknown.

---

54 SLATY-BREASTED RAIL
Gallirallus striatus Plate 15

Rallus striatus Linnaeus, 1766, Manila, Philippines.

Sometimes retained in *Rallus*, or placed in *Hypotaenidia*. Seven subspecies recognised.

Synonyms: *Rallus gularis/albiventer/indicus/jouyi*; *Lewinia striatus/albiventer*; *Hypotaenidia ferrea/striata/jouyi/obscuriora/abnormis*; *Eulabeornis celebensis/striatus*; *Gallinula gularis*.

Alternative names: Blue-breasted (Banded)/Plumbeous-breasted Rail.

IDENTIFICATION Length 25-30cm. A distinctively plumaged, medium-sized rail with longish straight bill. Male has forehead to hind-neck and sides of neck chestnut, more blackish on centre of crown; rest of upperparts, including wings, olive-brown, barred and spotted white (often most conspicuously on wings) and with blackish-brown feather centres. Chin and throat white (less extensively in race *obscurior*); lores, ear-coverts, foreneck and breast grey; rest of underparts blackish, narrowly barred white. Iris red; bill horn with pink to red base; legs and feet olive-brown to greyish. Female duller on head and neck, paler on upperparts and more whitish on belly. Races separated on size (*jouyi* is largest) and overall colour (*obscurior*, *striatus* and *paratermus* are much blacker on upperparts and darker on breast; *jouyi*, *gularis* and *taiwanus* are palest): see Plate 15. Immature identical to adult but legs slightly darker. Juvenile has no chestnut in plumage; upperparts olive-brown with dark streaks from crown to nape, and fewer white markings and dark feather centres on upperparts; anterior underparts grey-brown; posterior underparts with paler barring than in adult; legs and feet probably grey. Inhabits marshes, flooded fields, mangroves, grasslands, scrub and forest.

Similar species Easily distinguished from most sympatric rallids by its combination of chestnut crown and nape, narrow white bars and spots on rest of upperparts and upperwings, grey foreneck and breast, black-and-white-barred underparts, and fairly long, reddish bill. Superficially very similar to Brown-banded Rail (64), which is sympatric in the Philippines (Luzon), but Slaty-breasted larger and heavier, with stouter bill, and heavier tarsi and toes; remiges usually have bold white barring and spotting (variable, occasionally reduced). Only inner secondaries of Brown-banded Rail are barred, the pinkish-buff bars being most developed on outer margin; scapulars and some upperwing-coverts have black-bordered pinkish-buff bars.

VOICE The species has as a sharp but not loud whistle, and a noisy *ka-ka-ka* (Ripley 1977). The advertising call comprises repeated sharp, metallic *kerrek* or *trrrik* notes which may be run together into a sort of song lasting up to 30s and which are heard in the early evening (Neelakantan 1991). A possible variant of this call is a buzzing *kech*, repeated 10-15 times, starting weakly, becoming stronger and then fading (MacKinnon & Phillipps 1993). A grunting sound, possibly a contact note, is also recorded (Henry 1978). The male gives a low *kuk* call when courting the female, and a low *ka-ka-kaa-kaa* to attract the female during courtship feeding, while chicks give a low whistle when taking food from a parent (Timmis 1974). The race *obscurior* utters a deep croak, apparently reminiscent of the Andaman Crake (22) (Ali & Ripley 1980). This species is said to call infrequently, and for a very short period during the breeding season (Neelakantan 1991).

DESCRIPTION *G. s. albiventer*
Adult male Forehead to hind-neck and sides of neck predominantly chestnut, brightest at sides of neck and becoming blackish in centre of crown. Rest of upperparts, including upperwing-coverts and tertials, olive-brown barred and spotted white; centres of feathers more blackish-brown (variably prominent depending on width of olive-brown fringes); tail predominantly blackish-brown, barred white. Alula, primary coverts, primaries and outer secondaries blackish-brown, variably barred and spotted white, markings sometimes tinged tawny. Underwing-coverts and axillaries blackish-brown, barred white. Chin and upper throat white; lores, ear-coverts, lower throat, foreneck and breast slaty-grey to blue-grey. Flanks, and belly to undertail-coverts, blackish-brown, narrowly barred white. Iris red or orange-brown; bill horn, base rose-pink to scarlet; legs and feet deep olive-brown to olive-grey.
Adult female Chestnut of head duller and streaked black; upperparts duller and paler, more olive (pale fringes more extensive and prominent); bars of underparts tinged fulvous; belly more extensively whitish.
Adult non-breeding According to Sharpe (1894), plumage of non-breeding birds entirely overshadowed with olive-brown; belly and undertail-coverts tinged fulvescent, this colour almost hiding black bars on undertail-coverts.
Immature Published descriptions of young birds refer to juvenile plumage. Birds in juvenile plumage appear to

moult directly into adult-type plumage (chestnut on head and neck, more markedly patterned on upperparts, greyer on anterior underparts and more contrastingly barred on posterior underparts), and captive birds at 5 months old were identical to parents except for slightly darker legs (Timmis 1974).

Juvenile First feathers of upperparts appear darker than in adult, with darker brown fringes, but fully grown birds have head to hind-neck pale olive-brown, sometimes slightly tinged tawny, streaked blackish-brown, and rest of upperparts usually paler than in adult, pale olive-brown with sparser and more obscure blackish-brown feather centres (sometimes hardly visible on mantle); white markings on upperparts fewer, and predominantly spots rather then bars; upperwings more like adult. Chin and throat as adult; sides of head greyish, sometimes tinged tawny; foreneck and breast brownish-grey; underparts paler than in adult, dark bars more olive-brown and pale bars off-white. Iris pale brown; bill black. See nominate and *gularis*.

Downy young Down thick, woolly and black, with faint greenish gloss; iris grey-brown; bill blackish; legs and feet dark brown to black.

MEASUREMENTS (12 males, 12 females) Wing of male 111-126 (120.6, SD 4.2), of female 111-127 (118.5, SD 4.2); tail of male 35-47 (41.7, SD 3.3), of female 35-46 (39.5, SD 2.9); culmen to base of male 36-42 (38.8, SD 1.9), of female 33-41 (37.9, SD 2.8); tarsus of male 36-40 (38.4, SD 1.5), of female 35-38 (36.2, SD 1.3). Weight of adults 100-142.

GEOGRAPHICAL VARIATION Races are separated on size and overall colour, but validity of some requires further investigation (see, for example, comments under *G. s. jouyi*). An aberrant individual of the nominate race, with no white barring on the upperparts, was originally described as a new race, *deignani*, of Lewin's Rail (Ripley & Olson 1973). Race *obscurior* includes *nicobarensis*; race *gularis* includes *Hypotaenidia striata reliqua* (Ripley 1977). Race *paratermus* sometimes included in nominate (e.g. Dickinson *et al.* 1991).

G. s. albiventer (Swainson, 1838) – Sri Lanka, India (except extreme W and N in Punjab, Kashmir, Rajasthan, N Gujarat and Sikkim), Bangladesh (except extreme N) and Myanmar to SC China (SW Yunnan; but see *G. s. gularis*) and Thailand; some may winter in S Thailand and Malaysia. Vagrant to Nepal. See Description.

G. s. obscurior (Hume, 1874) – Andaman and Nicobar Is. Darker and larger than *albiventer*; upperparts much blacker; white on chin and throat much reduced; breast darker grey. Iris dark brown; bill dark horn-brown, base dark dull red; legs and feet dark greenish-horn. Measurements of 5 males, 5 females: wing of male 127-139 (133.8, SD 5.8), of female 118-133 (126.8, SD 5.8); tail of male 40-47 (44.2, SD 2.6), of female 35-48 (41.2, SD 5.6); culmen to base of male 39-45 (42.2, SD 2.1), of female 37-43 (40.6, SD 2.2); tarsus of male 36-40 (38.4, SD 1.5), of female 35-38 (36.2, SD 1.3).

G. s. jouyi (Stejneger, 1886) – Coastal SE China (from Luichow Peninsula, Kwangtung, N to S Kiangsou) and Hainan. The largest race; also the palest, on both upperparts and underparts. Wing of male 138, of female 133; tail of male and female 48; culmen of male 45, of female 41; tarsus of male 44, of female 43 (Ripley 1977). However, Cheng (1987) maintains that *jouyi* is hardly separable from *gularis*, either on plumage or measurements; specimen from Shanghai has wing 121, overlapping with both *gularis* and *albiventer*.

G. s. taiwanus (Yamashina, 1932) – Taiwan. As pale as *jouyi* but smaller; paler than *gularis*, upperparts more greyish-olive: white on belly more extensive. Wing of male 119-128, of female 118-127; culmen of male 38-40, of female 35-38; tarsus of male 39-43, of female 40 (Ripley 1977).

G. s. gularis (Horsfield, 1821) – S China (SE Yunnan, interior of Guangdong and probably Guangxi [Meyer de Schauensee 1984]), Vietnam and Cambodia through Malaysia to Sumatra (Belitung, Simeule and Riau Archipelago), S Borneo, Java and Bali. Note that Cheng (1987) places all Chinese birds in *gularis* (see *G. s. jouyi*) and records this race widely in S Chinese provinces, from Sichuan (Chengdu), W and SE Yunnan, Guangdong, Fujian, and Guizhou (middle and lower Changjiang R). According to Ripley (1977) *gularis* is similar to *albiventer* but paler overall, particularly on upperparts which tend to greyish-olive rather than brownish-olive. One juvenile (Java) still in postnatal moult is predominantly pale cinnamon-brown to bright olive-brown, tinged rufous on nape and greyish on breast; feathers of mantle and back with dark centres; upperwing-coverts, and flanks to undertail-coverts, finely, irregularly and relatively sparsely barred and spotted white, barring more obscure and buff-tinged towards undertail-coverts; bill very dark; legs and feet probably grey. Measurements (3 females): wing 106.5, 110, 112; tail 38, 39, 41; culmen 30, 30, 31; tarsus 32, 34.5, 35.

G. s. striatus (Linnaeus, 1776)– Philippines (Busuanga, Camotes, Cebu, Guimaras [requires confirmation], Jolo, Leyte, Luzon, Marinduque, Mindanao, Mindoro, Negros, Palawan, Panay, Sibuyan, Siquijor and Sulu Is), N Borneo, and also N and NC Sulawesi, where possibly not resident. Also Lesser Sundas at Sawu and Lombok (see Movements). Much darker than *gularis* on upperparts (deep brown), and underparts. A juvenile female had chestnut of head replaced by black; upperparts, including upperwing-coverts and scapulars, black, feathers broadly edged olive-brown; upper back with a few white spots; some spots on scapulars and upperwing-coverts almost bars; primary coverts and remiges black, first primary and some inner secondaries with a few white spots on outer web only (Riley 1924). Measurements of 7 males, 5 females: wing of male 113-122 (117.9, SD 3.2), of female 111-117 (114.8, SD 2.7); tail of male 35-40 (37.9, SD 1.7), of female 32-40 (36.6, SD 2.9); culmen to base of male 36-40 (37.4, SD 1.6), of female 34-38 (36.2, SD 1.8); tarsus of male 32-36 (34.3, SD 1.5), of female 32-33 (32.2, SD 0.4).

G. s. paratermus (Oberholser, 1924) – Samar Island, EC Philippines. A very dark race. Holotype (female) similar to nominate but much darker above, ground colour more extensively and deeply blackish, and feather fringes also darker; white markings from mantle to rump fewer and much smaller; underparts somewhat darker, white barring much narrower and further apart (Ripley 1977). Wing 114; tail 45; culmen 38; tarsus 34.

Slaty-breasted Rail

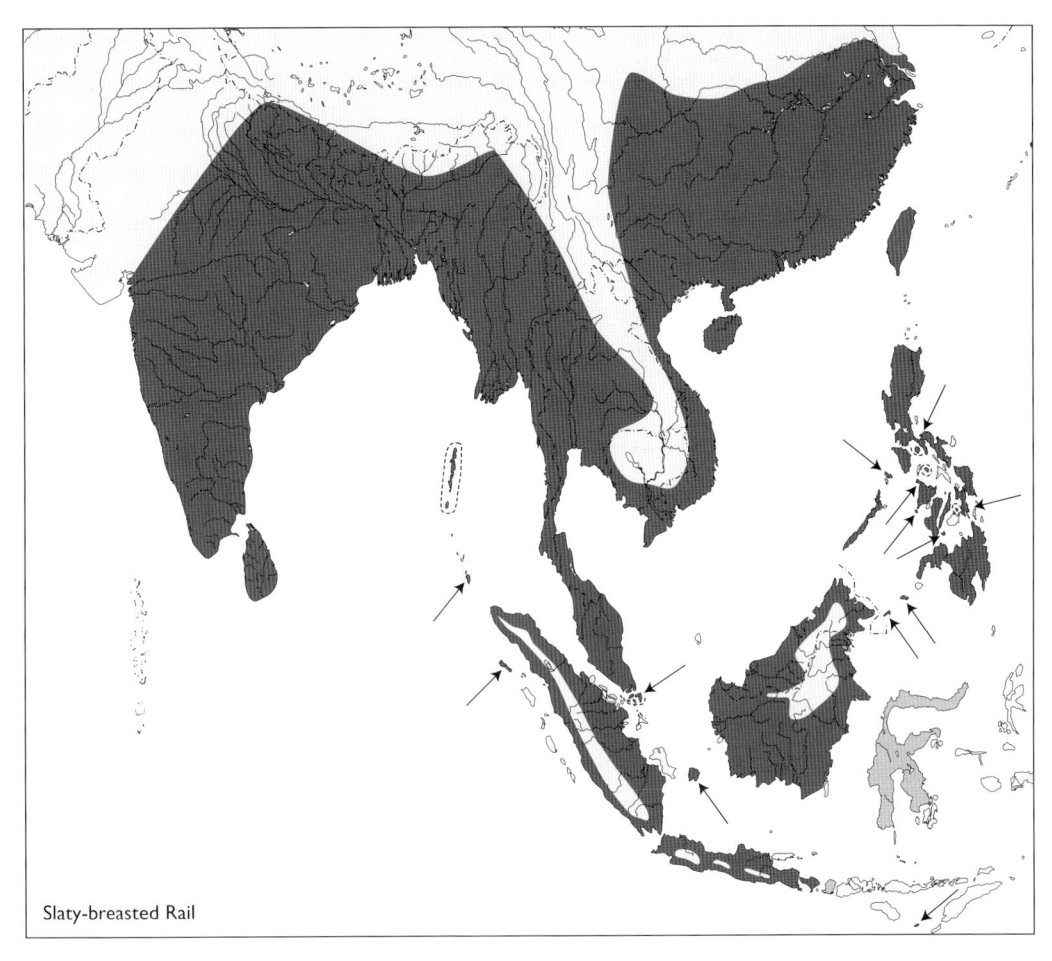

MOULT In captive birds, postjuvenile moult was complete at five months of age (Timmis 1974). No other information is available.

DISTRIBUTION AND STATUS Sri Lanka and India E across Bangladesh and Miyanmar to S China and Taiwan, and S through SE Asia to the Andaman and Nicobar Is, Sumatra, Borneo, Java, Bali, Sulawesi, Sawu, Lombok and the Philippines. On the distribution map the bird's range on the Asian mainland is largely hypothetical, being a combination of the distributions shown on the maps in Ripley (1977) and Cheng (1994), and birds may occur in the northernmost parts of the range only in summer. There is relatively little recent information on the status of this species. The race *albiventer* was formerly regarded as widespread and quite common in India, Sri Lanka and Myanmar (Ali & Ripley 1980, Smythies 1986); its current status in Sri Lanka is given as rare (Kotagama & Fernando 1994), while it is local but not scarce in Bangladesh (Harvey 1990), and common in Thailand (Lekagul & Round 1991). In the Philippines, nominate *striatus* was formerly considered common (Rabor 1977) but is now regarded as uncommon, but very secretive (Dickinson *et al.* 1991); it has been recorded only twice from the Lesser Sundas (see Movements). Formerly *obscurior* was probably common on the Andamans and Nicobars (Ali & Ripley 1980), while *gularis* was uncommon in S Vietnam but common in Malaysia, and is currently regarded as common in Sumatra, Borneo, Java and Bali (Glenister 1951, Wildash 1968, MacKinnon & Phillipps 1993). Formerly, *taiwanus* was apparently fairly common (Ripley 1977). The status of the species (including *jouyi*) throughout its range in S and SE China is given as rare or uncommon (Cheng 1987). The race *paratermus* is known only from the type locality on Samar (Ripley 1977).

MOVEMENTS This species is normally regarded as resident and no regular movements are recorded, but it is speculated that some winter birds in S Thailand and Peninsular Malaysia may be migrant *albiventer*, while the species may be merely a winter migrant to Sulawesi, where it is recorded from Aug-Oct and Mar (Ripley 1977, White & Bruce 1986). It is possibly only a summer breeding visitor to Myanmar and to the northern parts of its range in China (Smythies 1986, Cheng 1987). In India, it moves locally under the stress of drought or flood, and in the Philippines considerable postbreeding dispersal occurs (Ali & Ripley 1980, Dickinson *et al.* 1991). It is a vagrant to Nepal, where the only record is of a bird taken on 16 February 1938 (Inskipp & Inskipp 1991). At least some birds appear to be migratory in Vietnam, as near Hanoi it was seen in crop fields in Apr (see Habitat) and 1 was taken at night at Tam Dao Mt, while it appears in Apr and disappears at the end of Sep at Tranninh, N Laos (Stusák & Vo Quy

1986). In the Lesser Sundas, an old record from Sawu in Aug/Sep has been attributed to vagrancy (White & Bruce 1986) but the species was recorded from Lombok in Oct 1991 and may be a local migrant (Johnstone *et al.* 1993). Two birds from Hong Kong, 23 Oct, were found dead at a floodlit building (BMNH specimens).

HABITAT Inhabits wetlands of many types, including marshy meadows, reedmarsh, paddyfields, mangroves and the margins of ponds. It also occurs in *Imperator* grassland and dry rice fields, in rank grass and bushes, at drainage ditches, gardens and damp areas near villages, in dry scrub and bush near country roads (Sumatra), in forest (Andamans) and even on dry coral islands (Greater Sundas). In Vietnam it was seen in fields among low stands of *Phaseolus*, *Vigna*, soya etc in Apr (???? & ??? 1986), these birds presumably being on passage. It is recorded most commonly up to 1,000m, but also up to 1,500m and rarely (Sri Lanka) to 1,850m.

FOOD AND FEEDING Worms, molluscs, crustaceans (including crabs, and Isopoda in captivity), insects (including beetles, flies and ants) and their larvae, spiders, the seeds and shoots of marsh plants, and grain. Large insects such as locusts are beaten against the ground or a log before being swallowed (Timmis 1974). The birds take much food from the surface of the ground, but also take seeds and insects from vegetation by jumping up, or by perching on plants (Timmis 1974). They probe soft soil with the bill for earthworms and insect larvae, and pick up and sweep aside fallen leaves and grasses with the bill; they do not scratch with the feet (Timmis 1974). They have been observed feeding along the edges of tidal water and paddyfields, in the latter instance apparently taking aquatic insects, and they walk easily over floating vegetation when foraging (Timmis 1974, Rabor 1977). They often wade or swim, and they dive for food (Rabor 1977). One was seen molesting a crab for several minutes on a small mudflat with scattered mangroves (Johnstone *et al.* 1993). Captive birds ate meat, boiled egg, rice, carrot, insect mixture, biscuit meal, fly larvae and mealworms; they occasionally caught mice, which they shook, stabbed and then drowned but did not eat (Timmis 1974).

HABITS This species is described as skulking, rarely flushed and often overlooked, normally frequenting the interior of dense vegetation. However, these rails are sometimes seen in the open, especially during the early morning and late afternoon, when they are most active. They walk with an upright carriage and a high-stepping gait, bobbing the head and jerking the short tail. When alarmed, they crouch and slink into cover. The flight is often described as slow, weak and laboured. It apparently swims and dives well (Delacour & Jabouille 1931). In captive birds (Timmis 1974) preening was a constant occupation to which much time was devoted, while bathing was observed daily; the bird rolled over to one side, extended the other wing and threw water over the back with posteriorly directed head movements; the process was then repeated on the other side. Dust bathing was seen twice in very hot weather, and involved loosening sand with the feet and bill, then crouching down and flicking sand over the body with the bill. Sunning was observed on most hot days, the bird leaning to one side and opening the other wing fully for several minutes. Anting was also observed, when the birds picked up Red Wood Ants *Formica rufa*, swallowed some, placed some among the feathers and rubbed others on various parts of the body. The captive birds always roosted on the highest perches and, like wild birds, they were partly nocturnal, feeding and bathing on moonlit nights.

SOCIAL ORGANISATION Monogamous; the pair bond is probably maintained at least for the duration of the breeding season. Nothing else recorded.

SOCIAL AND SEXUAL BEHAVIOUR No information is available on territorial behaviour, except that Baker (1929) noted territorial disputes. The incubating bird leaves the nest only when the adjacent vegetation is disturbed; one bird ran in circles through the rice paddy, causing rustling noises, and also approached the nest with wings half-open and trailing in an apparent distraction display (Neelakantan 1991). Adults with chicks run away if danger threatens, drawing attention to themselves, while the chicks scatter and hide.

In the courtship display, the male approaches the female with drooping, quivering wings, the tail held erect and constantly flicked forwards over the back, and the head held low and stretched forwards. When close to the female he stops, jerks upright on his toes, spreads his wings and then crouches again. If the female is receptive she crouches, and mating follows (Baker 1935). Timmis (1974) described a courtship display which was performed by either sex, occasionally by both birds together. The birds stood slightly bent forward with the neck outstretched and the wings held at rightangles to the body; they then swayed from side to side, lifting the feet alternately. The male than walked to the rear of the female; she then usually faced him, he circled her with an erect and stiff posture, giving a low *kuk* call, and she continued turning away. When receptive, she crouched and the male mounted, treading her briefly, extended and drooped his wings so that the primaries touched the ground, and fluffed out the body plumage. The female constantly moved her head and neck from side to side jerkily. Mating was frequent during the nest-building period. Courtship feeding was also observed; after passing the food item, the male circled the female and often tried to mount.

BREEDING AND SURVIVAL Season India and Sri Lanka, Jun-Oct, Dec; Andaman and Nicobar Is, throughout year but mostly Jun-Nov; Myanmar, Jul-Aug; Thailand, breeding condition Jun; Sumatra, Jun, Dec; Borneo, Sep, Dec-Feb; Java, Jan-Jul and Sep-Nov but most in Mar-May; Philippines, Aug-Sep. **Nest** Well concealed in thick vegetation at edge of marsh or rice paddy, or in forest (Andamans); a thick pad of matted weeds, reed stems and grass, 20-30cm in diameter and with slight central depression, or a deep saucer 16cm across, 19cm deep and lined with grass and rice leaves. Nest built on ground, on dead reed debris, or up to 30cm above ground in growing vegetation; 1 nest had short entrance tunnel through grass. Both sexes build; male collects most of nest material. **Eggs** 2-9; ovate to broad oval; smooth and fairly glossy; white, cream, pinkish-white, salmon-pink or pinkish-buff, with spots, streaks and blotches of pale rufous to bright reddish-brown and underlying markings of pale purple or lilac; spots sometimes large and merged together, especially at blunt end. Average size of 200 eggs (*albiventer*) 33.7 x 25.8 (Baker 1935); size of 116 eggs (*obscurior*) 32.7-39.2 x 25-29.4 (36 x 27.6), calculated weight 15 (Schönwetter 1961-62); size of 39 eggs (*gularis*) 29.3-36.1 x 24-31.6 (33.5 x 26.2) (Hellebrekers & Hoogerwerf 1967). Eggs laid daily

(Neelakantan 1991); sometimes at more frequent intervals in captivity (Timmis 1974). **Incubation** Normally 19-22 days, sometimes possibly as little as 14 days (Neelakantan 1991); by both sexes, predominantly by female; in captivity started when first egg laid. **Chicks** Precocial; leave nest soon after hatching; fed and cared for by both parents. Parents either hold food in bill for chick to take, or attract chicks to food by moving head up and down or pecking at food. At 8 days, chicks begin to feed themselves. Feathers appear first on breast; young fully grown at 2 months.

RALLUS

Although sometimes expanded to include short-billed rails of the genus *Gallirallus*, and also other genera as diverse as *Aramidopsis*, *Habroptila*, *Nesoclopeus*, *Pardirallus* and *Lewinia*, this genus is best restricted to nine species of slim, long-billed rails with a narrow skull, sternum and pelvis and slender legs, which are much more specialised in form than, for example, the *Gallirallus* rails, and are highly adapted to a semi-aquatic existence in densely vegetated reedy marshes. The genus has its centre of species abundance and diversity in the New World, where six species occur, including the well-known Clapper, King and Virginia Rails. Only three allopatric species occur in the Old World: the Water Rail in Eurasia, the African Rail in subSaharan Africa and the Madagascar Rail in Madagascar, and these form a superspecies, having probably arisen from a single invasion of *Rallus* from the New World (Olson 1973b). *Rallus* species typically have the upperparts streaked brown and blackish; only the African Rail has plain upperparts. All species except the Plain-flanked Rail have a variable amount of grey or grey-blue on the face and anterior underparts, this colour extending to the lower breast in the Bogota, Austral, Water and African Rails and being restricted to the face in the Virginia and Madagascar Rails and in some races of the Clapper and King Rails (it is absent in other races). All species except the Plain-flanked Rail have barred flanks, although the barring is very obscure in the Madagascar Rail. The North American species are some of the best known and most intensively studied members of the Rallidae, but the species endemic to South America and Madagascar are very poorly known. Some species are severely threatened by habitat destruction, particularly the Austral Rail, which is critically endangered, and the Plain-flanked and Bogota Rails, which are globally endangered, while the Madagascar Rail is becoming rare throughout its range. Some North American races of the Clapper and King Rails are endangered, while two South American races of the Virginia Rail are rare and probably threatened by habitat loss.

55 CLAPPER RAIL
Rallus longirostris Plate 17

Rallus longirostris Boddaert, 1783, Cayenne, French Guiana.

This largely saltwater and brackish water species is often considered conspecific with the largely freshwater-marsh *R. elegans*; is provisionally treated here as forming a superspecies. Californian and Pacific coast races *levipes*, *obsoletus*, *yumanensis* and *beldingi* sometimes placed in *R. elegans*, including in a recent morphological, distributional and palaeontological study by Olson (1997); races *tenuirostris* and *ramsdeni* of *R. elegans* sometimes placed in *R. longirostris*. Hybridisation with *R. elegans* occurs occasionally in intermediate habitat in areas of sympatry, and mitochondrial DNA studies of the two species have yielded inconclusive results (Avise & Zink 1988). Twenty-one subspecies currently recognised.

Synonyms: *Rallus levipes/yumanensis/beldingi/pallidus/ crepitans/corrius/crassirostris/cypereti/obsoletus/scottii*; *R. obsoletus rhizophorae*; *R. crepitans waynei/scottii*.

Alternative local names for races of Clapper Rail: California (*obsoletus*), Light-footed (*levipes*), Yuma/ Sonora/San Blas (*yumanensis*), Belding's (*beldingi*), Northern (*crepitans*), Wayne's (*waynei*), Florida (*scotti*), Louisiana (*saturatus*), Brooks's/ Mangrove (*insularum*), Bahama (*coryi*), Yucatan (*pallidus*), Honduras (*belizensis*), Isle of Pines (*leucophaeus*), Caribbean/Cuban/ Hispaniolan/Puerto Rican/Antiguan (*caribaeus*). General names for species include: Salt Marsh-hen; Marsh/Mud Hen; Long-billed/Salt Marsh Clapper Rail.

IDENTIFICATION Length 31-41cm. Medium-large rail with long, slender, slightly decurved bill; great regional variation in plumage colours. Forehead to hind-neck dark brown to blackish; rest of upperparts similar, with buffy-brown, olive or grey feather fringes giving streaked appearance; rump and uppertail-coverts often almost uniform, dark feather centres reduced or absent. Upperparts darkest in race *scotti*; palest in *pallidus* and very pale in nominate. Upperwing-coverts range from warm sepia (*scotti*), cinnamon-tinged (*pallidus*), rufous-tinged (*obsoletus*) or bright mid-brown (*belizensis*) to buffy-brown or olive-brown (*yumanensis*, *crepitans*). Supraloral streak whitish to cinnamon-buff; sides of head vinaceous-brown (*obsoletus*, *levipes*) to grey (very dark in *scotti* and *crassirostris*); chin and throat white to pale buff; foreneck, breast and upper belly very variable, for example ochraceous-buff (*obsoletus*), greyish buffy-brown (*crepitans*), pinkish-cinnamon (*yumanensis*, *belizensis*), vinaceous (*levipes*) or more rufous (*pallidus*), very pale in *leucophaeus*; centre of belly whitish; flanks olive greyish-brown (*cypereti*) to sepia (*scotti*), narrowly (*scotti*, *beldingi*) to broadly (*belizensis*) barred white; undertail-coverts white, medians barred or blotched black. Iris red, orange or brown; bill reddish, orange or yellowish, with dark culmen and tip; bill described as orange to orange-red when breeding and pinkish to pale orange in non-breeding adults and juveniles (Meanley 1985); legs and feet vary from brown-tinged (*obsoletus*, *scotti*) through dull orange (*pallidus*) to grey with yellow ankle joints (*crepitans*). Sexes alike; male averages 20% larger than female and has brighter bill. Immature similar to adult but with duller bare parts. Juvenile similar to adult on upperparts but often darker, and with darker, duller (sometimes greyer) feather edges;

lower back to uppertail-coverts often appear unpatterned (pale feather fringes virtually absent); outer greater and median upperwing-coverts may have narrow white tips and subterminal bars; sides of head and underparts duller, often greyer, than in adult; some may be very dark on underparts, and some *yumanensis* are almost black on flanks and sides, with underparts decidedly grey; belly more extensively white; flank bars absent or indistinct; eye brown; bill paler than in adult.

Normally divided into 3 groups (e.g. Olson 1977, Eddleman & Conway 1998): *obsoletus* group, comprising the 4 races of W North and Central America, which resemble King Rail (56) in large size, relatively bright or buff-tinged feather fringes on upperparts, and relatively bright cinnamon-washed foreneck and breast; *crepitans* group comprising races from E North America, the Caribbean and Yucatan, which are greyer in plumage than W birds and have longer, more slender bills than South American birds; and the South American *longirostris* group with relatively, short, stout bills. Races *obsoletus*, *levipes* and *scotti* have two colour morphs: brown or dark morph, in which upperpart feathers have dark brown to brown-black centres and buff-brown fringes (fringes greyer in *scotti*); and olive or pale morph, in which upperpart feathers have darker, blacker centres and duller, more olive, fringes. Race *caribaeus* has at least two colour morphs, in which upperpart feathers are margined with a variety of olive-greys and browns, while *saturatus*, *waynei* and *leucophaeus* are also said to have two morphs (Ridgway & Friedmann 1941). It is not clear to what extent colour variations may be explained by age-related plumage differences, or even by hybridisation (Eddleman & Conway 1998). Inhabits salt and brackish marshes, and mangroves; locally freshwater marshes.

Similar species Races sympatric with King Rail differ from that species in having more grey on face, variably duller fringes to upperpart feathers, and duller neck, breast, upperwing-coverts (especially lessers) and flank barring. Possibility of hybrids is potential complication. Hardly separable from King Rail on voice, although vocalisations of Clapper often given more rapidly (Meanley 1969); usually separable on habitat, Clapper (except race *yumanensis*) normally occurring in saltmarshes and mangroves, King in freshwater or brackish marshes. Markedly larger than the sympatric Virginia Rail (58), which is darker overall with brighter rufous upperwing-coverts, and has grey sides of head which contrast markedly with rest of head and body plumage. Race *phelpsi* sympatric with Plain-flanked Rail (57) in Venezuelan mangrove swamps, but Plain-flanked Rail smaller; lacks grey on side of head and has plain buffy-brown underparts, lacking all barring; bill brown and iris reddish-brown.

VOICE Very similar to that of the King Rail. Most information comes from Massey & Zembal (1987) and Eddleman & Conway (1998). Most calling occurs in the evening and early morning, but birds are sometimes also heard at night. The territorial call of both sexes (*clapper* call) is a series of loud, rapid *kak*, *chè* or *chack* notes, falling in pitch and becoming slower towards the end, and given with the head and neck stretched vertically; it is also a greeting call between pair members. This call is given most frequently at dusk and dawn and is often answered by other birds; it is also described as ventriloquial (Audubon 1842). The male's mate attraction call is a monotonous series of harsh *kek* notes. There is an agitated, higher-pitched and faster *kek* call, indicating distress and often given by juveniles when chased. During the breeding season the female gives a series of 1-5 *kek* notes followed by a *burr* to attract a mate or to call back a straying mate (Zembal & Massey 1985), and also a low-pitched, soft churring trill which may be a diminutive *burr*.

A call resembling a chicken's squawk is given by birds when startled, and also often preceding a chase (the chase-squeal); a screech is given by birds flushed from nests with young broods and by birds in traps. Birds in the hand give a raucous, repeated *rack-k-k-rack-k-k* (Meanley 1985). There is also a hoarse grunting call, both sexes give a quiet purr like a cat (a contact call) which becomes more agitated in response to disturbance, and a very low, ghostly *hoo* note (possibly from the male only). A snarling note is recorded by ffrench (1973). The contact call to chicks is a subdued, chicken-like cluck; young up to 6 weeks old utter shrill peeps, soft *chitty-chitty* sounds, and also squeals if distressed; they begin to develop adult-type calls from 7 weeks (Adams & Quay 1958). Also see detailed call descriptions in Eddleman & Conway (1998).

DESCRIPTION *R. l. obsoletus*
Adult, brown morph Forehead to hind-neck dark brown, anterior feathers with black shafts and hind-neck feathers edged tawny-olive; mantle to uppertail-coverts, including scapulars, darker (sepia), with very broad olive-tinged ashy buffy-brown fringes; dark centres almost absent on rump and uppertail-coverts; rectrices sepia, fading to olive-tinged ashy-brown at margins. Upperwing-coverts olive-tinged buffy-brown with dark shaft-streaks; outer medians and greaters tinged rufous; primary coverts sepia, edged olive-brown, paler on outer webs; remiges sepia, edged olive-brown, but tertials broadly edged olive-tinged ashy-brown. Alula sepia, edged rufous externally. Underwing-coverts and axillaries dusky brown, barred white. Supraloral streak pinkish-buff, washed rusty; lores and ear-coverts dusky vinaceous-brown; sometimes a vague, paler vinous streak above ear-coverts; eyelids whitish; malar region light ochraceous-buff; chin and upper throat white, sometimes washed buffy. Sides of neck, lower throat, breast and upper belly light ochraceous-buff to dull cinnamon, with pale feather fringes, darker ochraceous on centre of breast and upper belly; flanks and undertail-coverts barred greyish buffy-brown and white; thighs similar brown, unbarred; centre of belly whitish, washed pale ochraceous-buff; lateral undertail-coverts white, medians barred like flanks. Iris reddish-orange; bill brown with orange-yellow base; legs and feet pale brownish-grey to brownish-flesh. Sexes alike; male averages 20% larger (Meanley 1985). **Olive morph** has darker, blacker centres to feathers of upperparts and duller, more olive fringes.
Immature Very similar to adult but duller overall, and paler on underparts.
Juvenile Much darker than adult on upperparts. Forehead to hind-neck, entire upperparts including scapulars, and tail, fuscous-black; feather margins of scapulars, mantle, and upper back dull ashy-tinged brownish-olive; supraloral streak paler than in adult; lores and ear-coverts fuscous-black, washed cinnamon; chin and upper throat white; lower throat, sides of neck, breast and upper belly pale cinnamon, mottled dull greyish-olive to greyish-brown, more uniformly this colour on sides of neck and lower throat; centre belly and lower belly white, washed buffy or pale ochraceous-cinnamon; flanks dull greyish-olive to

greyish-brown, with indistinct bars of ochraceous-cinnamon. Iris brownish, becoming orange-brown (10 weeks); bill and legs becoming more brown- or orange-tinged.

Downy young Down jet-black with greenish gloss on upperparts, black-brown on underparts; patch of white down in each ventral pteryla just below wings, not present in King Rail (Weatherbee & Meanley 1965, Olson 1997); iris dark brown, paling to olive-drab at 6 weeks; bill white or pink with black base and white egg-tooth, becoming black over basal two-thirds (4 weeks), then fading to grey (5-8 weeks); legs and feet grey; vestigial wing-claw present.

MEASUREMENTS (29 males, 24 females) Wing (chord) of male 153.5-170 (161.7), of female 147-161 (151.6); tail of male 68-80 (73.1), of female 60-76 (65.3); exposed culmen of male 55-66 (60.3), of female 49-61 (55); tarsus of male 52-61 (56.1), of female 45-63 (51.1). No published weights.

GEOGRAPHICAL VARIATION Considerable, and rendered more complex by the existence of colour morphs in at least 7 races; for detailed treatment of these, see Oberholser (1937). Principal characters used in racial separation include: darkness of crown and upperparts; colour of paler feather fringes on upperparts; colour and brightness of upperwing-coverts; colour of lores and ear-coverts; overall colour and brightness of underparts from lower throat to upper belly; colour of dark flank bars and width of white flank bars; bare part colours; size. Northernmost and westernmost races generally have longer wing, tail and tarsus; NW races have shorter culmen than those in NE; South American races are smallest, with relatively short, stout bills. Race *yumanensis* includes *rhizophorae* from Sonora, and *nayaritensis* from Nayarit, Mexico (following Ripley 1977; this treatment is regarded as being "not without some justification" by Olson 1997); *beldingi* includes *magdalenae* from Magdalena Bay, Baja California; *insularum* includes *helius* from Florida; *caribaeus* includes *cubanus*, *vafer*, *manglecola* and *limnetis* from Cuba, Hispaniola, Antigua/Guadeloupe and Puerto Rico respectively. See Ripley (1977) for further details.

R. l. obsoletus Ridgway, 1874 – C California, mainly in San Francisco Bay, also San Pablo Bay; formerly also Morro, Tomales, Humboldt and Monterey Bays (Eddleman & Conway 1998). The largest race. See Description.

R. l. levipes Bangs, 1899 – coastal C California (Santa Barbara County) S to N Baja California (Bahia de San Quintin), and S to San Ignatio, where it intergrades with *beldingi* (Ripley 1977). Similar to *obsoletus*, with brown morph and uncommon olive morph, but smaller; upperparts with darker sepia feather centres and more brownish-olive, less greyish, feather fringes; malar area, foreneck, sides of neck, breast and upper belly brighter, more vinaceous-cinnamon; flanks darker and browner. Differs from *beldingi* in having upperparts feather fringes more olivaceous and dark feather centres less conspicuous; breast less reddish, more cinnamon; flanks paler, with wider white bars. Juvenile slightly darker than in *obsoletus*. Measurements of 10 males, 12 females: wing (chord) of male 154.5-167 (161.9), of female 138-155.5 (147.3); tail of male 62.5-69 (66.7), of female 57-67 (62.6); exposed culmen of male 56-61 (58.9), of female 51.5-58 (54.2); tarsus of male 53-60.5 (56.9), of female 47-51 (49.5) (Ridgway & Friedmann 1941).

R. l. yumanensis Dickey, 1923 – SW Arizona and SE California, from lower Colorado R (Needles, California, to river mouth), lower Gila R and Salton Sea, at isolated sites in Arizona, and along Mexican coast S to San Blas, Nayarit. Dickerman (1971) records 1 specimen, similar to '*rhizophorae*', from Laguna San Felipe, Puebla, in Apr 1962, the first record of this coastal form from interior Mexico. Similar to *levipes* but wing shorter; feathers of underparts less brightly coloured. Differs from *saturatus* in having shorter bill and in being brighter, more pinkish (not grey-washed) cinnamon from lower throat to upper belly. Forehead to hind-neck predominantly greyish olive-brown; upperparts dark sepia, feathers fringed dull greyish-olive; upperwing-coverts buffy-brown to dark brown; sides of head dull mouse-grey, lores darker and browner; supraloral stripe buffy-white; malar stripe poorly defined, light pinkish-cinnamon. Juvenile often very dark, some being almost black on flanks and sides, with underparts decidedly grey (W. R. Eddleman *in litt.*) Wing (chord) of 65 males 141-164 (154.1, SD 4.7), of 30 females 133-146 (141.6, SD 3.4); tail of 66 males 54-82 (66.6, SD 5.2), of 29 females 56-69 (60.8, SD 3.7); exposed culmen of 94 males 50-67 (61.0, SD 2.8), of 47 females 50-62 (55.6, SD 2.5); tarsus of 29 males 45-55.5 (49.6, SD 2.4), of 23 females 43-49.5 (45.5, SD 3.0) (Eddleman & Conway 1998). Weight of 109 males 194-347 (269.0, SD 35.8), of 49 females 160-310 (209.5, SD 39.8) (Eddleman & Conway 1998).

R. l. beldingi Ridgway, 1882 – S Baja California, presumably from San Ignatio, where it intergrades with *levipes*, and Magdalena (Ripley 1977) on Pacific coast to La Paz, and on San José and Espírito Santo Is. Smaller and darker than *levipes*; top of head, and feather centres of upper body, fuscous-black, pale fringes buffy-brown; lower throat, sides of neck, breast and upper belly darker and richer, fawn with cinnamon wash; flanks barred white (narrower) and dark brown, often black adjacent to white bars. Slightly paler than *margaritae*. Measurements of 14 males, 12 females: wing (chord) of male 147-160 (155.1), of female 140-150 (144.8); tail of male 55-73 (64.8), of female 54-68 (63.2); exposed culmen of male 53-63 (56.2), of female 49-55.5 (52.7); tarsus of male 48-57 (53.1), of female 45-53 (49.2) (Ridgway & Friedmann 1941).

R. l. crepitans Gmelin, 1789 – coastal Massachusetts, Rhode Island and Connecticut S to NE North Carolina; winters from S part of breeding range S to Florida, principally in South Carolina, Georgia and NE Florida. It is possibly this race which has been recorded from Bahamas in winter (Eddleman & Conway 1988, W. R. Eddleman *in litt.*). Forehead to hind-neck, and rest of upperparts to uppertail-coverts, including upperwing-coverts, buffy olive-brown to light sepia; feathers of upper body, scapulars and some lesser and median coverts conspicuously fringed olive-grey to neutral grey; centres of some scapulars darker (sepia); rest of coverts dark buffy-brown, olive-brown or darker brown, narrowly edged paler; underwing-coverts mid-brown, barred white; rectrices dark olive-brown, fringed buffy grey-brown. Supraloral stripe buffy-white; lores deep grey, tinged brown; ear-coverts and sides of nape grey, sometimes washed brownish; chin and upper throat white; sides of neck, lower throat and breast greyish buffy-brown to grey-tinged cinnamon-buff (less grey on breast); belly whitish; flanks barred dusky brown and white; median undertail-coverts similar. Iris reddish-brown to pale yellow; bill yellow to orange-red, with grey-brown culmen and tip; legs and feet greyish with yellow or orange tibiotarsal joint, or light yellow to pinkish. Juvenile like adult on upperparts but more uniformly olive-brown from lower back to uppertail-coverts; greater and median upperwing-coverts narrowly tipped and subterminally barred whitish; flanks greyer, and belly more extensively white, than in adult. Measurements of 21 males, 17 females: wing (chord) of male 142.5-159.5 (151.1), of female 135.5-160 (146.8); tail of male 55-69 (64.6), of female 55-69.5 (61.9); exposed culmen of male 55-69.5 (63.3), of female 53.5-67 (59.6); tarsus of male 48-56 (51.7), of female 41-56 (48.1) (Ridgway & Friedmann 1941). Weight of 14 males 286.5-394.5 (351.0), of 6 females 236-330.5 (278.1) (Meanley 1985).

R. l. waynei Brewster, 1899 – coastal SE North Carolina, where it intergrades with *crepitans*, to E Florida (S to Martin County at c. 27°N) where it intergrades with *scotti* (Eddleman & Conway 1998). Similar to *crepitans* but slightly smaller; upperparts and flanks darker; underparts more cinnamon, with ashy band across breast (Ripley 1977); said to have dark morph with fuscous feather centres on upperparts and darker lower throat and breast, and pale morph similar to *crepitans* but with darker flanks (Ridgway & Friedmann 1941). Juvenile has white bars on some upperwing-coverts (see *crepitans*). Measurements of 15 males, 16 females: wing (chord) of male 135-152 (145.1), of female 129.5-146.5 (138.4); tail of male 57.5-72 (61.9), of female 56-63.5 (59.9); exposed culmen of male 54-67.5 (62.2), of female 53-62.5 (58.7); tarsus of male 46.5-53.5 (48.2), of female 43.5-50 (46.7) (Ridgway & Friedmann 1941). Weight of 17 males 275-375 (337), of 33 females 200-400 (272) (Eddleman & Conway 1998).

R. l. scotti Sennett, 1888 – coastal Florida, from c. 29°N on E coast to Pensacola on Gulf Coast; intergrades with *saturatus* in W Florida panhandle (Eddleman & Conway 1998). Upperparts darkest of any race, always much darker than *waynei*. Forehead to hind-neck blackish; feather centres of rest of upperparts blackish, fringes either brownish-olive or olive-tinged grey (producing two morphs), becoming dull sepia-brown on rump and uppertail-coverts; upperwing-coverts warm sepia with blackish-brown feather centres; primaries dusky brown; narrow white leading edge to wing. Supraloral streak buffy; lores dusky; ear-coverts and broad post-ocular stripe dark grey; foreneck greyish-olive; breast fawn with cinnamon or tawny wash; flanks sepia-brown, narrowly barred white; undertail-coverts barred blackish and white, some with cinnamon-buff wash. Legs and feet horn-brown. Juvenile has correspondingly dark upperparts, almost black; foreneck to flanks and belly greyish buffy-brown; washed ochraceous-buff on neck and breast; white bars on flanks few and irregular. Measurements of 23 males, 18 females: wing (chord) of male 135-155 (146), of female 128.5-145 (137); tail of male 56.5-72 (63.3), of female 54.5-63.5 (59.2); exposed culmen of male 56-66 (61.6), of female 51.5-60 (55.9); tarsus of male 42-55.5 (49.8), of female 42-51 (45.6) (Ridgway & Friedmann 1941). Those of 15 males, 10 females (BMNH): wing (max) of male 138-151 (145.3, SD 4.2), of female 129-142 (136.1, SD 3.6); culmen to base of male 58-66 (63.5, SD 2.6), of female 53-63 (58, SD 3.3). Weight of 11 males 263-310 (290.2, SD 15.5), of 14 females 199-314 (247.3, SD 30.1) (Eddleman & Conway 1998).

R. l. saturatus Ridgway, 1880 – Gulf coast from Alabama to extreme NE Mexico (Tamaulipas). Paler than *scotti*; darker than *insularum*. Similar to *crepitans* but smaller; much darker and browner; more cinnamon on breast. Dark morph has fuscous centres to feathers of upperparts; flanks dark brown; lower throat to upper belly pale ochraceous-buff, tinged cinnamon; sides of neck and breast washed grey (rarely as grey as in *crepitans*). Pale morph has dorsal feather centres dark buffy-brown to olive-brown, and fringes grey to olive-grey. Iris orange; bill reddish-yellow with dusky culmen; legs and feet pale bluish-horn. Juvenile similar to that of *crepitans*; darker on upperparts than adult, with no pale fringes on feathers of lower back and rump; outer upperwing-coverts narrowly subterminally barred and tipped white. Measurements of 23 males, 16 females: wing (chord) of male 140.5-163 (150.4), of female 131-154 (141.3); tail of male 58-68 (63.6), of female 56-66 (60.9); exposed culmen of male 54-69 (61.7), of female 55.5-64 (59.9); tarsus of male 47-55 (50.9), of female 42-52.5 (47.7) (Ridgway & Friedmann 1941). Weight of 109 males 180-400 (321.7, SD 35.5), of 157 females 180-320 (256.6, SD 34.3) (Eddleman & Conway 1998).

R. l. insularum W. S. Brooks, 1920 – Florida Keys, from Key Largo to Boca Grande, and presumably the Marquesas (Eddleman & Conway 1998). Like pale morph *waynei* but smaller; fringes of upperpart feathers grey; colour of centres varies as in *waynei*; sides of neck more greyish; breast less grey-washed; flanks paler. Upperparts slightly paler than in *leucophaeus*, but underparts markedly more cinnamon. Iris brownish; bill light brown with reddish base and blackish culmen; legs and feet greyish. Juvenile similar to that of *waynei*. See *coryi*. Measurements of 4 males, 4 females: wing (chord) of male 140-148 (144.8), of female 129.5-136.5 (133.8); tail of male 51-64 (57.5), of female 57-60 (58); exposed culmen of male 59-61.5 (60.6), of female 53-59 (55); tarsus of male 47-54 (50.4), of female 44-47 (45.1) (Ridgway & Friedmann 1941).

R. l. coryi Maynard, 1887 – Bahamas. Like *crepitans* but slightly paler on top of head and upperparts, light brownish-olive with broader, paler, neutral grey fringes; sides of neck, and lower throat to upper belly, paler, more pinkish-buff with pale cinnamon wash. Compared with *insularum*, lacks dark feather centres on upperparts; has darker underparts, but paler underparts, than *leucophaeus*. Iris reddish-brown; bill orange-brown with blackish culmen and tip; legs and feet pale brownish-orange or olive-grey. Juvenile like that of *crepitans*, but paler; pinkish wash on underparts confined to lower throat and upper breast. Measurements of 7 males, 10 females: wing (chord) of male 137-150 (146), of female 128.5-141 (134.7); tail of male 53.5-67 (61.3), of female 54-62 (57.5); exposed culmen of male 53-65 (59.5), of female 52-60.5 (55.3); tarsus of male 45-53.5 (49.4), of female 42-50 (46.9) (Ridgway & Friedmann 1941).

R. l. pallidus Nelson, 1905 – Rio Lagartos, coastal Yucatán, SE Mexico. Pale; very similar to *insularum* and *coryi*, especially in colour of feather fringes on upperparts. Crown to hind-neck medium brown; dark feather centres of upperparts olive-brown, pale fringes ashy-grey; upperwing-coverts more cinnamon than primaries and with sparse, narrow white bars; primaries and secondaries isabelline with cinnamon wash; supraloral streak white; lores and sides of head plumbeous; chin and throat white; sides of neck brown, becoming dull cinnamon on foreneck and pale cinnamon-rufous on breast; flanks olive-brown to duller brown, strongly barred white; undertail-coverts white. Iris medium brown; bill dull orange with horn tip and dark horn culmen; legs and feet dull orange (Dickerman 1971). Measurements of 3 males, 2 females: wing (chord) of male 145, 147, 152, of female 138, 141; exposed culmen of male 53.5, 53.5, 54.5, of female 51.5, 52; tarsus of male 54.5, 54.5, 56, of female 48.5, 49 (Warner & Dickerman 1959). Weight of 2 females 237, 301 (Dickerman 1971).

R. l. grossi Paynter, 1950 – Chinchorro Reef and possibly also Cayo Culebra and Holbox I, Quintana Roo, SE Mexico. Bill shortest of all races. Nearest to *pallidus*, but has crown to hind-neck much darker brown; feather centres of upperparts duskier; primaries deeper brown; lores and ear-coverts darker grey; neck and breast more richly coloured, almost vinaceous-buff; flanks darker. Compared to *belizensis*, has shorter wings and paler feather centres on upperparts. One female (type): wing 134.5; tail 52; culmen 48, tarsus 43 (Ripley 1977).

R. l. belizensis Oberholser, 1937 – Ycacos Lagoon, Belize; possibly Half Moon Cay, Belize (Eddleman & Conway 1998). Similar to *saturatus* but upperpart feathers dark brown, fringed pale olive-grey, and breast light pinkish-cinnamon. Top of head to hind-neck sepia, streaked dull buff on hind-neck; lower back to uppertail-coverts sepia, edged olive-grey; remiges sepia; upperwing-coverts brightish mid-brown, greyer on medians and inner greaters; supraloral streak white; sides of head olive-grey; breast to belly light cinnamon with pink tinge; centre of belly dull white; flanks dull brown, broadly barred white. One female (type): wing (chord) 141.5; tail 57; exposed culmen 57; tarsus 48.

R. l. leucophaeus Todd, 1913 – Isle of Pines, Cuba. Upperparts slightly darker than in *insularum*, more like those of *waynei*; underparts much paler than in *waynei* or *coryi*; throat, breast and belly dull white, washed pale cinnamon-buff on upper breast and sides of throat in pale morph, or heavily washed light greyish-olive and pale pinkish-buff in dark morph (Ridgway & Friedmann 1941). Measurements of 12 males, 12 females: wing (chord) of male 135-155 (146), of female 127.5-149 (134.3); tail of male 57.5-67.5 (61.8), of female 53-62.5 (58.6); exposed culmen of male 60.5-66 (63.3), of female 51.5-59.5 (56.3); tarsus of male 50-59 (55.3), of female 45-51.5 (49.3) (Ridgway & Friedmann 1941).

R. l. caribaeus Ridgway, 1880 – Cuba, Jamaica, Hispaniola and Puerto Rico (including Vieques I), Lesser Antilles E to Antigua, and Guadeloupe. Closest in colour to *waynei*. Has at least two colour morphs, in which characteristic dark feathers of upperparts are margined with a variety of olive-greys and browns. Individuals of one population inseparable from another throughout range, although formerly split into 5 races (Ripley 1977). Juvenile darker than adult; underparts greyer (blacker in dark morph); flanks only slightly barred. Measurements of 23 males, 18 females: wing (chord) of male 138.5-159.5 (148.7), of female 128-144.5 (137); tail of male 56.5-67 (61.9), of female 49-60.5 (57.9); exposed culmen of male 58.5-74 (63.9), of female 53.5-65 (58.8); tarsus of male 50-61 (55.3), of female 43.5-59.5 (50.5) (Ridgway & Friedmann 1941).

R. l. cypereti Taczanowski, 1877 – coasts of extreme SW Colombia in Nariño (Tumaco) through Ecuador to extreme NW Peru (Río Tumbes delta). Apparently differs from *crassirostris* and nominate in having margins of dorsal feathers much paler and more greyish, less olivaceous; barring on flanks also paler, olive greyish-brown instead of dusky brown (Blake 1977). Measurements of 2 males, 2 females: wing of male 126, 132, of female 118, 123; exposed culmen of male 52, 53, of female 50, 51; tarsus of male 48, 49, of female 44, 45 (Blake 1977).

R. l. phelpsi Wetmore, 1941 – extreme NE Colombia in Guajira (Ríohacha, Bahía Portete and Puerto López), and extreme NW Venezuela (Falcón and Carabobo). Relatively pale. Similar to nominate but darker on upperparts, especially on crown and dark feather centres of back; breast brighter, more cinnamon. Lower mandible yellowish. Paler on upperparts than *pelodramus*, with brighter breast and whiter belly (Blake 1977). Measurements of 7 males, 6 females: wing of male 133-140 (135.8), of female 120-130 (125.5); exposed culmen of male 47-55 (49.7), of female 44-50 (47.3); tarsus of male 43-49 (46.1), of female 42-45 (43) (Blake 1977).

R. l. margaritae Zimmer & Phelps, 1944 – Margarita I, Venezuela. Darkest of the South American races. Similar to *pelodramus* in size but upperparts blacker, with feather fringes narrower and more brownish, less greyish; breast somewhat redder, less buffy. Lower mandible coral-red with black tip. Measurements of 6 males, 6 females: wing of male 121-139 (129.6), of female 114-126 (120.3); tail of 2 males 47, 49, of 3 females 43, 44, 47; exposed culmen of male 50-57 (53.8), of female 48-53 (50.1); tarsus of male 42-46 (44.2), of female 41-43 (42) (Zimmer & Phelps 1944, Blake 1977).

R. l. pelodramus Oberholser, 1937 – Trinidad. Smaller than *caribaeus*; paler overall than *margaritae*. Upperparts much more heavily streaked blackish than in nominate and *phelpsi*, and thus similar to *crassirostris* from which distinguishable by darker, more rufescent margins to upperpart feathers. Measurements of 5 males, 4 females: wing of male 129-139 (133.8), of female 125-139 (131); exposed culmen of male 46-53 (50.8), of female 47-54 (50.5); tarsus of male 43-48 (46.4), of female 44-49 (45.2) (Blake 1977).

R. l. longirostris Boddaert, 1789 – coastal Guyana, Surinam and French Guiana. Paler above and below than any adjacent races; culmen longer. Iris reddish-brown; bill brownish, lower mandible orange-yellow; legs and feet pale orange-red (Haverschmidt 1968). See Plate 17. Measurements of 4 males, 9 females: wing of male 133-143 (139), of female 126-138 (130.1); exposed culmen of male 50-55 (52.7), of female 46-53 (48.7); tarsus of male 46-48 (47), of female 41-47

(43.8) (Blake 1977). Weight of 2 males 307, 310, of 2 females 259, 262 (Haverschmidt & Mees 1995).

R. l. crassirostris Lawrence, 1871 – coastal Brazil from Amazon estuary (Marajó I), through Maranhão (Mangunca I), Pernambuco (Recife) and Bahia to Espírito Santo, Rio de Janeiro, São Paulo and Santa Catarina. Similar to nominate but upperparts have blacker feather centres and narrower olivaceous margins; sides of head sootier, less brownish; underparts darker cinnamon. Bill stout, as in nominate, but shorter; mainly reddish but darkening towards tip; legs and feet yellowish-brown. Measurements of 1 male, 3 females: wing of male 135, of female 126, 131, 136; exposed culmen of male 55, of female 43, 48, 48; tarsus of male 45, of female 36, 40, 43 (Blake 1977).

A single bird recorded from Panama (Bocas del Toro) (see Distribution and Status) cannot be assigned to any of the described races and presumably represents a hitherto undescribed form. Crown blackish; rest of upperparts quite boldly streaked blackish and greyish-olive; upperwing-coverts dull rufescent; facial area grey; throat white; breast cinnamon-rufous; belly whitish; flanks and undertail-coverts boldly barred blackish and white (Ridgely & Gwynne 1989). Wing 144.5; exposed culmen 49; culmen to base 54; tarsus 49.

MOULT Information is for North American birds. Postbreeding moult is complete, normally occurring in Aug-Sep; remex moult is simultaneous. There is a partial prebreeding (pre-alternate) moult of the contour plumage in early spring; it is poorly known and more study is needed. Juvenile plumage is attained at 7-8 weeks, after which adult-like plumage develops, moult often appearing continuous with acquisition of juvenile plumage; the postjuvenile (first pre-basic) moult is partial, excluding remiges and rectrices (Eddleman & Conway 1998). By 10+ weeks the young are becoming indistinguishable from adults except for their dull bare parts; in the USA, the barred flanks are acquired in Sep and postjuvenile moult is complete in Oct (Bent 1926). The first pre-alternate moult is poorly described, but is probably like adult pre-alternate.

DISTRIBUTION AND STATUS North America, along the Pacific coast from C California S to Baja California, inland in SW Arizona and SE California (lower Colorado R, lower Gila R and Salton Sea), and on the Pacific coast of W Mexico S to Nayarit; also along the Atlantic and Gulf coasts from Massachusetts S to Florida and W to extreme NE Mexico (Tamaulipas). The northernmost populations on the E coast winter from South Carolina to Florida (Eddleman & Conway 1994), small numbers remaining N to Connecticut during mild winters, and the species being uncommon in New Jersey and Delaware in winter (Eddleman & Conway 1998). It also occurs in the Bahamas, in the West Indies S to Antigua and Guadeloupe, on Trinidad and Margarita I, and coastally in SE Mexico (Yucatán and Quintana Roo), Belize, extreme NE and SW Colombia, Ecuador, NW Peru, NW Venezuela, Surinam, Guyana, French Guiana, and Brazil from the Amazon estuary S to Santa Catarina. Its overall distribution in the USA is influenced most strongly by water salinity and ambient temperature. Its normal habitats of saltwater marshes and mangrove swamps have an average salinity of c. 7,480ppm; on the E and Gulf coasts of the USA, in areas where the average salinity is c. 5,670ppm, the species comes in contact with the King Rail and hybridisation is recorded (Ripley 1977, Meanley 1985). The climate of these saltwater marshes is relatively mild in winter, with a frost-free period of 210 days or more and January temperatures which rarely drop below -7°C (Root 1988).

In the USA this species was formerly abundant. Audubon (1842) reported collecting 72 dozen eggs in a day and said that professional eggers collected 100 dozen per day. As a measure of the abundance of *crepitans*, he estimated that, in a series of saltmarshes c. 32km x 1.6km, 1-2 birds could be found every 8-10 steps. Audubon also gave a graphic account of the wholesale slaughter of Clapper Rails in South Carolina marshes, which is quoted by Bent (1926: 281). By the beginning of the 20th century the birds were far less abundant (Bent 1926) and numbers have decreased significantly since then. Although the species is not globally threatened, in the USA 3 races, all from the W coast, are now considered ENDANGERED. The race *levipes* has declined mainly because of habitat loss: in the USA less than 25% of the S Californian wetlands existing in 1900 remain today, and of the c. 10,250ha of habitat originally available to this race, less than 3,500ha remain. Its US population is now estimated at 190 pairs, with c. 240 pairs left in Mexico (Eddleman *et al.* 1988, Ehrlich *et al.* 1992). Its remaining marshes in California are susceptible to violent storms, extreme tides, excessive runoff and siltation (Ehrlich *et al.* 1992). Severe threats are posed by introduced predators, and by habitat destruction in Mexico (Ehrlich *et al.* 1992). Low fertility and hatching success in the N populations may result from contaminants, or from inbreeding: very little genetic variation has been detected in *levipes* populations (Eddleman *et al.* 1988, Fleischer *et al.* 1995). The effects of inbreeding have not been assessed.

The race *obsoletus* was originally threatened by hunting, but industry, agriculture, saltpan construction and urbanisation have drastically reduced its habitat this century (Eddleman *et al.* 1988). About 85% of the original 73,500ha of habitat has been diked or destroyed, and much of that remaining is in a degraded condition (Ehrlich *et al.* 1992); urban development poses a major threat, together with pollution and introduced predators. Most of the remaining population of c. 400 birds in 1991 (down from 4,200-6,000 in the early 1970s) occurs in San Francisco Bay (Eddleman & Conway 1994), where development pressure is greatest. The race *yumanensis* has experienced recent contractions and expansions of its freshwater and brackish water marsh habitats, and its population S to the Colorado R delta is estimated at 1,700-2,000 birds (Ehrlich *et al.* 1992); it is threatened by wetland loss and high water flows, being susceptible to the effects of development projects along the Colorado R. The level of genetic variation in this race does not give cause for concern (Fleischer *et al.* 1995).

Recovery plans for these threatened races include the acquisition, restoration and improvement of habitat (including the restoration of normal tidal flows), the elimination of pollution, the establishment of refuges, and restocking and captive breeding (for *levipes*); effective predator control is also necessary (Ehrlich *et al.* 1992). To avoid possible inbreeding depression in *levipes* populations, translocation of birds from large populations to small ones is recommended (Fleischer *et al.* 1995, Nusser *et al.* 1996).

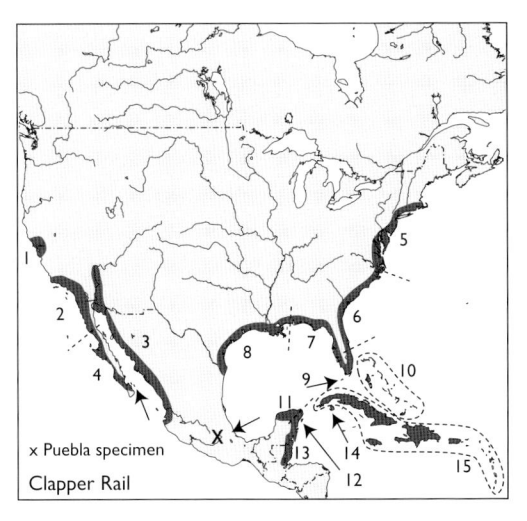

Clapper Rail map legend:
1. obsoletus
2. levipes
3. yumanensis
4. beldingi
5. crepitans
6. waynei
7. scotti
8. saturatus
9. insularum
10. coryi
11. pallidus
12. grossi
13. belizensis
14. leucophaeus
15. caribaeus
16. cypereti
17. phelpsi
18. margaritae
19. pelodramus
20. longirostris
21. crassirostris

x Puebla specimen
? occurrence uncertain
x Panama specimen

Clapper Rail

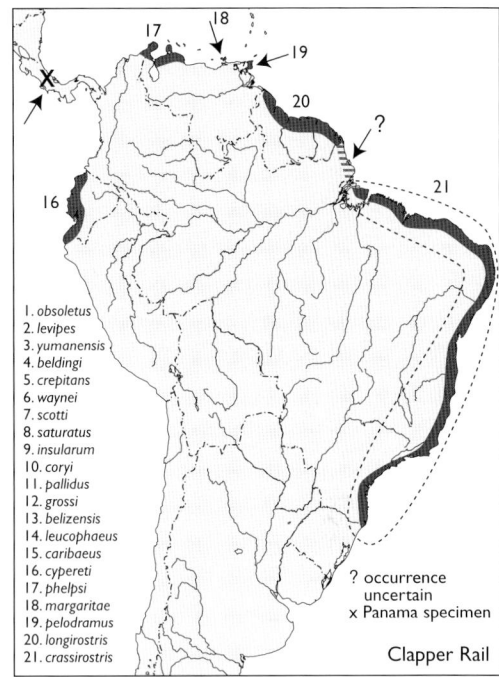

Although most E USA races are relatively numerous, local loss of breeding populations has occurred through habitat loss and degradation (Meanley 1985). All North American races are vulnerable to habitat destruction and the effects of pesticides and contaminants, although the species shows a high tolerance to DDT and DDD and is less susceptible to eggshell thinning caused by chlorinated hydrocarbons than are other birds (van Velzen & Kreitzer 1975, Klaas *et al.* 1980). Birds are also excluded from marshes by the presence of cattle (Bent 1926). Clapper Rails are also hunted extensively in 13 E and Gulf coastal states from Rhode Island to Texas. The estimated mean annual harvest from 1964-1986 was c. 101,000 birds but few reliable estimates are available and the effects of hunting are largely unknown (Eddleman *et al.* 1988, Eddleman & Conway 1994). Hunting pressure is low in the W coast areas where threatened races occur. Analysis of 1966-1991 Breeding Bird Survey data indicates a mean annual increase of 4.3% in NE US populations and no change for the entire USA (Eddleman & Conway 1994). The species is stable in the E USA provided wetland degradation ceases, and continued implementation of wetland protection laws is the most effective conservation technique (Eddleman & Conway 1998).

This species is probably more widespread in Colombia than the few known localities suggest (Hilty & Brown 1986). The nominate race is considered fairly common in French Guiana (Tostain *et al.* 1992) but in Surinam, where it was formerly locally very common (e.g. at Matapica Creek), it appears to have declined in recent years, probably as a result of increased human presence (Haverschmidt & Mees 1995). In Brazil, *crassirostris* is described as scarce in Rio de Janeiro (Forrester 1993) but locally fairly common in the Peruíbe region of São Paulo (F. Olmos *in litt.*; see Habitat). The current status of other races is not known, although *caribaeus* is considered common in Cuba (Garrido & Kirkconnell 1993), but some were formerly considered locally common or fairly common, e.g. *coryi* (Cat I, Bahamas: Buden 1987), *caribaeus* (Ripley 1977, Biaggi 1983, Raffaele 1989), *grossi* (Paynter 1950), *belizensis* (Russell 1964), *margaritae* (Yepez 1964), *pelodramus* (ffrench 1973) and *cypereti* (Parker *et al.* 1982); however, *cypereti* is rare and local in Ecuador, where it should be considered VULNERABLE because of the widespread massive destruction of mangrove forests which has taken place (Ridgely & Greenfield in prep.). The Guadeloupe archipelago holds only c. 40 pairs of *caribaeus* (formerly *manglecola*), and this population is threatened by egg-collecting (for food) and habitat destruction (Benito-Espinal & Hautcastel 1988). The only record from Panama is of a single bird of uncertain racial identity (see Geographical Variation) which was seen at Miramar, Bocas del Toro, Panama, from 25 Jan 1985; it was captured on 19 Mar 1985 and was kept in captivity but died some weeks later (Ridgely & Gwynne 1989).

Natural predators in North America include raccoons *Procyon lotor* and mink *Mustela vison*; introduced predators such as feral cats, red foxes *Vulpes vulpes* and brown rats *Rattus norvegicus* are major factors implicated in the recent decline of the species in California (Eddleman & Conway 1994). Birds are also taken by Great Blue Heron *Ardea herodias*, Bald Eagle *Haliaetus leucocephalus*, Northern Harrier *Circus cyaneus*, hawks (*Buteo*, *Parabuteo*), falcons *Falco*, and owls (*Tyto*, *Bubo*, *Asio*). Audubon (1842) notes that "minks, raccoons, and wild cats destroy a great number of them during the night, and many are devoured by turtles and ravenous fishes; but their worst enemy is man." Natural catastrophes such as storms and hurricanes may cause severe local and regional mortality, from which populations usually recover (Meanley 1985).

MOVEMENTS In North America this species is primarily resident, with dispersal of juveniles and adults after the breeding season, but northernmost populations are partially migratory and there is some evidence from ringing and observation that birds in the SE make movements, sometimes over long distances, which probably involve migration as well as dispersal and wandering (Crawford *et al.* 1983). Spring migration is poorly known; birds begin calling from mid-Mar in New Jersey and Virginia, and arrivals occur through early Apr (Eddleman & Conway 1998). Birds begin to disperse after pairs and family groups break up in Jun-Sep; migration begins from late Aug in Virginia and New Jersey, peaks in Sep-Oct (New Jersey and Carolina), and birds arrive in wintering areas of South Carolina in mid-Sep, with movements on the US Atlantic coast extending through Oct and Nov (Eddleman & Conway 1998). Movements appear to be overland as well as coastal, and migrating birds fly at night, probably at low altitude, and sometimes strike TV towers, wires and tall buildings (Bent 1926, Adams & Quay 1958, Crawford *et al.* 1983, Eddleman & Conway 1998). In North Carolina in Sep-Oct large numbers have been forced down by heavy fog, entering shops, flying into people and cars, and perching on telephone wires and houses, and have been found dead under telephone wires (Adams & Quay 1958). In winter, the race *crepitans* ranges S to Florida, where it occurs alongside at least 3 other races (Crawford *et al.* 1983). The W coast race *yumanensis* was formerly thought to be migratory, but radio telemetry studies indicate that most birds are resident, although some possibly winter on the W coast of Mexico from Sonora to Nayarit (Tomlinson & Todd 1973, Banks & Tomlinson 1974, Eddleman *et al.* 1988, Ehrlich *et al.* 1992). Casual wandering by birds of the *obsoletus* group is recorded on the Pacific coast to the Farallon Is, N to N California (Humboldt Bay) and S to S Baja California (Todos Santos), and by the *longirostris* group on the Atlantic coast N to New Brunswick, Prince Edward I, Nova Scotia and Newfoundland, and inland to C Nebraska, C New York, Vermont, W Pennsylvania, West Virginia and C Virginia (AOU 1983). South American populations are sedentary.

HABITAT Most information is from Eddleman & Conway (1994, 1998). In the USA, optimal habitat is saltmarsh which is flooded at least once daily at high tide, is dominated by *Spartina* of moderate height and has salinity levels exceeding 7,100ppm at low tide and 5,600ppm at high tide (Meanley 1985). Such marshes may be dominated by *Spartina alterniflora* or by *S. foliosa* (often with *Salicornia virginica*), *Spartina patens* or *Juncus roemerianus*. On the E coast, habitat ideally has emergents and scattered shrubs bordering ditches or tidal creeks within 15m of open water, and at least 25% of the wetland area within 15m of a shoreline (Lewis & Garrison 1983). In San Francisco Bay, California, nesting habitat of *obsoletus* is characterised by low marsh vegetation, tidal sloughs, *Salicornia* and *Spartina*, with some *Grindelia cuneifolia*, irregular flooding by the highest tides, and abundant prey; uninterrupted tidal flow is essential to maintain the vegetation communities preferred by this species. Fringing areas of high marsh serve as refugia at very high tides and, although used infrequently, may be very important in reducing mortality during such tides. The species also occurs locally (mostly in the lower Colorado River valley, race *yumanensis*, and also in S California, race *levipes*) in tropical and subtropical freshwater marshes with *Typha domingensis*, *Juncus californicus* and *Scirpus olneyi*, sedges, and sometimes *Phragmites*. Ideal habitat has a mosaic of emergent plants of different ages interspersed with shallow pools. Low stem densities and little residual vegetation are year-round requirements of *yumanensis* in freshwater habitat, while areas of high water coverage and moderate water depth are used for foraging, and areas of shallower water for nesting (Conway *et al.* 1993).

Along tropical and subtropical coasts the species also inhabits mangrove (*Avicennia*, *Rhizophora*, *Laguncularia*) swamps; in Mexico, *yumanensis* occurs in mangroves with an understorey of *Salicornia*, *Mesembryanthemum* and *Distichlis*, and also (Colorado River delta) in estuarine habitat characterised by *Tamarix* with an understorey of *Allenrolfia occidentalis* (Tomlinson & Todd 1973). The Panama bird frequented a freshwater pond with grassy margins just back from the shoreline of Chiriquí Lagoon (Ridgley & Gwynne 1989). In the Peruíbe region of São Paulo State, SE Brazil, the race *crassirostris* is fairly common alongside the Rufous-sided Crake (28) in extensive stretches of *Spartina* bordering closed-canopy mangrove forest, but is not recorded further N in the Santos-Cubatão region, where the mangroves are more disturbed and the canopy is not closed, and extensive stretches of *Spartina* are largely absent (F. Olmos *in litt.*).

Habitat requirements of US populations outside the breeding season are poorly known and rail density may vary as a result of tides, winds, ice coverage and cold rather than vegetation structure or prey abundance (Meanley 1985). Some populations move into different types of vegetation in winter, when *yumanensis* prefers denser cover than in summer, while from Nov-Apr North Carolina birds move from low-level *Spartina alterniflora* marsh into higher mixed *Spartina* and *Juncus roemerianus*, and even into the higher grass-shrub community of *Spartina patens*, *Iva* and *Myrica* (Adams & Quay 1958). Habitat use by migrating birds is unclear.

FOOD AND FEEDING Most information is from Eddleman & Conway (1994, 1998), Ohmart & Tomlinson (1977) and Zembal & Fancher (1988). The food is primarily invertebrates, typically crustaceans (which are important in many areas), and includes: Polychaeta (*Nereis*); molluscs, including slugs, gastropods (*Littorina*, *Cerithidea*, *Helix* and *Melampus*), mussels *Modiolus demissus* and clams (*Corbicula* and *Macoma*); leeches; many crustaceans, including Isopoda, Amphipoda, shrimps (*Orchestia grillus* and the freshwater *Palaeomonetes paludosis*), crayfish (especially *Procambarus* and *Orconectes*) and many crabs (especially fiddler crabs *Uca* and *Sesarma*, and also *Pachygrapsus* and *Hemigrapsus*); aquatic and terrestrial insects, including Coleoptera (Dytiscidae, Hydrophilidae, Carabidae and Curculionidae), Anisoptera and Zygoptera nymphs, Orthoptera larvae and Tipulidae; and Lycosidae spiders. Small fish (e.g. *Gillichthys* and *Fundulus*), frogs and tadpoles, baby terrapins, and mice are also taken, and birds scavenge fish carcases. Vegetable matter taken comprises seeds (including *Spartina*, *Scirpus*, *Polygonum*, sedges Cyperaceae, oak *Quercus*, soyabean *Glycene* and rice), berries of *Sambucus mexicana*, green plant material (e.g. *Salicornia* stem tips) and tubers of woody plants. It will take birds' eggs, and occasionally kills small birds, including ones caught in mist-nets (e.g. Spendelow & Spendelow 1980, Jorgensen & Ferguson 1982). A water snake *Natrix sipedon* 40cm long is recorded once; it is not known whether the snake was killed or found dead (Hoff 1975). This rail feeds opportunistically on the most readily available foods; animal food makes up 90-100% of its diet in spring and summer and up to 89% in winter; more vegetable matter is eaten in the winter when animal food is less abundant. It apparently hunts by sight; it probes exposed and shallowly flooded mud and sand, stirs up soft mud with the bill to bring prey to the surface, takes crabs from burrows by probing, sometimes inserting its head and neck into the burrow, gleans items from the substrate, throws aside surface litter with the bill and seizes small invertebrates hiding under it, and catches fish while wading, swimming and sometimes diving. It mostly forages on substrates adjacent to cover, e.g. at the edges of marshes or on mudflats and grassy areas adjacent to mangroves, and along tidal creeks and in open channels. Sites with at least 50% of the shoreline adjacent to emergent marsh are optimal, and are selected because of prey abundance, ease of movement by foraging birds, and the presence of nearby cover which is especially important for chicks. Birds usually dismember large crabs (by shaking them) before eating them, and eat the limbs and the flesh from the body. Foraging rates of 1,000-2,000 gleans and probes per hour were commonly sustained during peak foraging hours in the late evening; birds took many small prey items in this way and interspersed long intensive foraging bouts with periods of hunting for crabs (Zembal & Fancher 1988). This rail regurgitates pellets consisting largely of fragments of crab exoskeleton and pieces of clam shell. It can drink either fresh or salt water.

HABITS This species has been shown to have peaks of activity for 3h after dawn and 2h before sunset and is assumed to be diurnal and crepuscular (Zembal *et al.* 1989); however, it may show some nocturnal activity in open areas close to cover (e.g. Wetmore & Swales 1931). In coastal habitats it is most active at low tide, and often roosts at high tides (day or night). Like most rails, Clapper Rails are reluctant to flush, preferring to escape danger by running silently, with neck outstretched and tail erect, through dense vegetation; they have many well-defined tracks and paths through such cover. The normal gait is slow and deliberate, and the tail is often flicked when walking, especially if the bird is agitated. They are extremely difficult to flush in the tangled branches and roots of mangroves, but in marsh vegetation they will flush at the feet of an observer and make a slow, laboured flight for a short distance with dangling legs; flights >80m, however, may attain almost the flight speed of a shorebird or duck, suggesting a good capability for migratory flight (Eddleman & Conway 1998). They swim with ease, sitting high in the water with head and neck raised, jerking the head back and forth; when pursued, birds will sink below the surface with the bill above the water, and will remain thus for some time; if hard pressed they will dive and swim under water, using the wings, and if wounded will hold onto submerged vegetation. During extreme high tides, when little cover is available, they may become conspicuous and will swim readily, 'freezing' alongside any patch of emergent vegetation which might afford the slightest cover (Bent 1926). They occasionally fly up onto a tree limb or a fence post in marshland, sometimes standing in full view for several minutes.

Birds often bathe in shallow pools, especially at low tide; they preen for up to 35 min after bathing, and often preen at dusk (Meanley 1985).

SOCIAL ORGANISATION Information is available only for North American populations, and most details are from Eddleman & Conway (1994, 1998). This rail is monogamous and the pair bond is maintained only during the breeding season, when the birds are territorial. Pair formation occurs on the breeding grounds, in Jan-Feb in W and S areas and in Mar in N areas. Breeding density varies with marsh type, ranging from 0.15 birds/ha for *yumanensis* to 10.0 birds/ha for *crepitans* in Virginia; nest density ranges from 1/ha (New Jersey) to 4.2/ha (Virginia). Territory size varies from 0.1-3.59ha (Arizona,

Virginia). Minimum distance between nests at one site was 23m; most were >45m apart (Kozicky & Schmidt 1949). Home range sizes vary greatly with habitat type and season; *yumanensis* has the largest home ranges (mean 7.6ha for males and 10ha for females, ranging in winter up to 43 ha) because of the relatively low productivity of the habitat and the use of different marsh types for foraging and nesting; in coastal saltmarshes, home ranges are smaller, being 0.04-1.66ha in the breeding season and 0.1-2ha in winter. Home ranges are larger in autumn and winter because of increased foraging movements to locate less abundant prey (Conway *et al.* 1993), and home ranges of immatures are larger than those of adults (Zembal *et al.* 1989). Birds are solitary after breeding but there may be considerable overlap in home ranges. Individuals may pair with the same mate in subsequent years, and mated pairs in nonmigratory races will continue to maintain adjacent home ranges after the breeding season. Sex ratios are poorly documented and may be even or male-dominated.

SOCIAL AND SEXUAL BEHAVIOUR Audubon (1842) describes how rails will attack avian predators, such as harriers and owls, which hunt over marshes: "The rail rises a few yards in the air, strikes at the marauder with bill and claws, screaming aloud all the while, and dives again among the grass, to the astonishment of the bird of prey, which usually moves off at full speed. They are not so fortunate in their encounters with such hawks as pounce from on high at their prey, ... against which they have no chance of defending themselves." Birds crowded together at high tides may fight viciously (Stone 1937).

Males are very pugnacious during the breeding season, and intruders on the territory are attacked strongly; vanquished birds are often chased in flight for >100m. The reactions of incubating birds to disturbance vary: unlike King Rails, most leave the nest before the observer approaches but a few remain on the nest until the observer is <1m away, and some feign injury (Kozicky & Schmidt 1949). Rails will attack Laughing Gulls *Larus atricilla* which approach too close to nests; they compete with the gulls for nest sites, and often fight with them, while incubating rails are attacked by gulls (Segre *et al.* 1968). Distraction behaviour includes 'broken wing' displays and aggressive displays (Eddleman & Conway 1998).

Males initiate pair formation by giving the mate attraction call, and paired birds may give the clapper call as a duet within 2 days of the start of male attraction calls (Eddleman & Conway 1994). In a courtship display the female walked slowly, with exaggerated jerking of the body and tail and frequent stops; the male then approached in a series of short runs, "with turns and gestures of body and wings"; the female moved along slowly, finally dropping with her body flat and her head stretched forward; the male then mounted and copulated (Bent 1926). In the courtship display the male fans and exposes the white undertail-coverts, while giving the mate attraction call (Ripley 1977) or pointing the bill downwards and moving it slowly from side to side (Meanley 1985). He also stretches the neck, opens the bill and pursues the female for up to 1m; males courtship-feed females during egg-laying and incubation (Meanley 1985).

BREEDING AND SURVIVAL Season USA, Mar-Aug, mostly Apr-Jun; southernmost races *yumanensis* and *saturatus* from mid-Mar; *obsoletus* from mid-Apr; *crepitans* and *waynei* from early Apr; peak late Apr (*waynei*), mid-May (*yumanensis*), and mid-May to mid-Jun (*crepitans*); late dates (renesting or second broods) early Jun to mid-Aug. Mexico, Mar; Belize, May; Trinidad, Apr-Dec, mostly May-Jun; Venezuela, Apr; Surinam, breeding condition, Jan, May-Jun. **Nest** A platform or cup of dead rushes, sedges and marsh grass, or of sticks and dead leaves; material collected at site; nest lined with finer material; built on ground in clump of vegetation, typically *Spartina* or *Juncus*, 60-120cm tall, or in *Salicornia*; vegetation height is major cue used in selecting nest sites on E coast of USA, and early nests may be in *Juncus roemerianus* because it provides good cover in early spring (Adams & Quay 1958). Race *yumanensis* nests in *Typha* and *Scirpus*; selects sites near high ground in shallow areas dominated by mature marsh vegetation (Conway *et al.* 1993, Eddleman & Conway 1994). Nest may also be situated at base of shrub, on dry hummock or among mangrove roots; is usually on ground not flooded by tide, but sometimes has base in water; often has canopy of dead or growing material; many nests have ramp from ground to rim. Incubating bird may raise eggs above level of encroaching water by tucking plant fragments under them, using material from canopy (Ripley 1977). In some areas, most nests are within 15m of tidal creeks or pool edges: nests averaged 1.7m from open water in Virginia and 7.8m in California. External diameter 15-35cm; cavity diameter 10-18cm, depth 1-10cm; thickness 7-15cm; height of nest mass up to 30cm; rim up to 45cm (average 18-26cm) above substrate; height of dome above nest 7.5-23cm; nests in mangroves relatively high: in Mexico averaged 73.6cm (30-150cm) above substrate (Eddleman & Conway 1998). Male most active in nest-building. Up to 6 brood nests (with no canopy) constructed near breeding nest; built of dead vegetation and may float with rising tide; some brood nests are old nest sites of water birds, or platforms made by muskrats *Ondatra zibethicus* (Eddleman & Conway 1994). **Eggs** 3-16; clutch size 4-16 (9.2) (*crepitans*); 5-15 (9.4) (*waynei*); 7-14 (9.5) (*saturatus*); 4-14 (8.3) (*obsoletus*); 5-11 (7.3) (*levipes*); 2-11 (7.3) (*scotti*); 6-7 (6.5) (*insularum*); California 5-8 (6.7), Mexico 3-7 (5.5) (*yumanensis*); 5-8 (6.4) (*beldingi*); 6.5 (*caribaeus*); clutch size increases with latitude (see Eddleman & Conway 1998). Eggs laid daily; second brood smaller; replacement clutches may be only 4-6 eggs; second nests average 1-3.4 fewer eggs (Eddleman *et al.* 1994). Eggs oval to subelliptical; smooth and slightly glossy; creamy-white, pinkish, or yellowish-cream to greenish- or pale olive-buff, irregularly spotted and blotched dark reddish-brown to brown, with fainter lilac or grey markings; size: *obsoletus* 41-48 x 29.6-33.4 (31.3 x 44.1, n = 60), calculated weight 23; *levipes* 41.5-49.1 x 29.7-32.3 (44.6 x 31, n = 40), calculated weight 22; *yumanensis* 41.8 x 28.8, n = 1, calculated weight 18; *beldingi* 44.6 x 30, 44 x 29, n = 2, calculated weight 20; *crepitans* 37.5-48.5 x 27.5-31.5 (42 x 30.1, n = 75), weight 18-20 (calculated weight 21); *waynei* 36.6-46.2 x 27.8-30.6 (41.5 x 29.1, n = 40), calculated weight 19; *saturatus* 39.5-46 x 28.4-31.6 (42 x 29.3, n = 43), calculated weight 19.9; *scotti* 37-43.5 x 27.7-31.4 (41.1 x 30.1, n = 15), calculated weight 20; *caribaeus* 38.6-44.6 x 28.8-31.1 (41.7 x 29.6, n = 11), calculated weight 20.2; *longirostris* 40-40.1 x 29-29.3 (40 x 29.1, n = 3), calculated weight 18.8. **Incubation** 18-29 days; by both sexes, male during night and sporadically during day, female during most of day; commences before last egg laid, hence large variation in incubation period; hatching asynchronous. **Chicks** Precocial; leave nest as soon as hatched but use brood nests for first few days; one parent usually takes first chicks while other incubates remaining eggs; young fed and cared for by both parents for 5-6 weeks; independent after c. 6 weeks. Parents may carry small young in bill (Skutch 1976). Chicks can swim within 1 day of hatching. Family forages in loosely organised group over an area which may be >50m in diameter; adults break crabs into small pieces, which chicks then pick up; chicks run to eat food which adult has sighted (Adams & Quay 1958, Zembal & Fancher 1988). Feathers grow from 4 weeks, first on breast; fully feathered at 7-8 weeks; fly at 9-10 weeks, when wings fully grown and adult-type plumage and bare part colours developing (Adams & Quay 1958).

Egg-dumping recorded; nests with up to 21 eggs found (e.g. Meanley 1985); also lays in Laughing Gull nests and vice versa, but no parasitic eggs hatch (Segre *et al.* 1968). In USA, nesting success variable, 10-100%; usually high (80%+) in good habitat with good weather and tidal conditions (Eddleman & Conway 1994); egg success also high, often exceeding 90%. Initial losses usually followed by persistent renesting attempts (up to 5; most races average successful completion of >1 nests); most failures due to flooding by extreme high tides but on high ground, and on Gulf Coast, predation is a problem (Eddleman & Conway 1994). Survival of young before fledging often low, usually <80%, sometimes <50% (Eddleman & Conway 1994). Californian birds suffer heavy predation by rats (38% nesting success); Fish Crows *Corvus ossifragus* and Laughing Gulls rob nests (Eddleman & Conway 1998), and crabs are reported to take small young (Bent 1926); other nest predators include raccoons, red foxes, rats and mink. Age of first breeding unknown; probably 1 year. Normally 1 brood per season but 2 recorded in most USA races. **Survival** In USA, observations and hunting returns show autumn populations having an age ratio of 2.0-5.8 young/adult; age ratios as high as 13.6 young/adult observed in New Jersey; annual adult survival (*yumanensis*) 49-67% (Eddleman & Conway 1998).

56 KING RAIL
Rallus elegans   Plate 17

Rallus elegans Audubon, 1834, Charleston, South Carolina.

This largely freshwater-marsh bird is often considered conspecific with the largely saltwater-marsh *R. longirostris* (55), with which it is here provisionally treated as forming a superspecies. Race *tenuirostris* of C Mexico sometimes considered specifically distinct, or transferred to *R. longirostris*; race *ramsdeni* sometimes transferred to *R. longirostris*; California and Pacific coast races of *R. longirostris* (*levipes, obsoletus, yumanensis, beldingi*) sometimes placed in *R. elegans*, e.g. by Peters (1934) and also in a recent morphological, distributional and palaeontological study by Olson (1997). Hybridisation with *R. longirostris* occurs occasionally in intermediate habitats on Atlantic and Gulf coasts, and mitochondrial DNA studies of the two species have yielded inconclusive results (Avise & Zink 1988). Three subspecies currently recognised.

Synonyms: *Rallus tenuirostris/crepitans*; *Aramus/Limnopardalus elegans*.

Alternative names: King Clapper Rail; Marsh/Mud Hen; Rice Chicken; Highland Rail/Mexican Clapper Rail (*tenuirostris*).

IDENTIFICATION Length of race *elegans* 38-48cm; of *tenuirostris* 33-42cm. A large to medium-large, rusty-coloured rail with a long, slender bill, slightly decurved near the tip. Nominate race of North America polymorphic: dark morph has upperparts blackish, mantle to tail with broad olive-brown feather fringes; lower back to uppertail-coverts, and tertials, more uniform olive-brown; most upperwing-coverts russet, sometimes with narrow or ill-defined dark centres and some with narrow whitish bands. Lores and ocular area ashy; supraloral stripe pale pinkish-cinnamon, often continued as indistinct stripe over ear-coverts, which are ashy to dull cinnamon; chin and centre of throat white. Lower throat, breast and upper belly vinous-cinnamon, becoming whitish on lower belly; sides of breast olive with blackish feather centres. Flanks blackish to brown, narrowly barred white, or light brown with blackish-bordered white bars; median undertail-coverts barred black and white; lateral undertail-coverts white with blackish spots. Iris red to reddish-orange; bill brownish, basal half orange-yellow to ochre; legs and feet brown-grey to yellow-brown. Pale morph has forehead to nape dark brown; upperpart feathers fringed buff; upperwing-coverts dark rich tawny; supercilium pale creamy-buff; sides of head more ochraceous; lower throat, breast and upper belly light ochraceous-buff, washed greyish. Sexes alike but female smaller. Immature similar to adult but somewhat duller, with pale brown to yellowish bill. Juvenile has upperparts duller and darker, fuscous-black with brown feather fringes; underparts paler, pale cinnamon to buff or white with indistinct barring on sides and flanks. Race *tenuirostris* of C Mexico duller and paler than nominate; pale flank barring sometimes pinkish-cinnamon; legs and feet brownish-flesh; has dark-breasted morph with cinnamon breast and upper belly, and pale-breasted morph with these areas whitish, washed pinkish-cinnamon. Cuban birds (*ramsdeni*) smaller, with more grey on face and more white on lower breast and belly. Inhabits freshwater marshes, wet fields, and (locally) brackish marshes.

Similar species Much larger than Virginia Rail (58), which is blue-grey on side of head and duller overall, with brighter rufous upperwing-coverts. Differs from sympatric races of Clapper Rail (55) in appearing more rusty-coloured overall: has brighter fringes to upperpart feathers (not so evident in comparison with W and S USA populations of Clapper Rail); little or no grey on the face; richer and more chestnut upperwing-coverts; brighter cinnamon neck and breast; and bolder flank barring. Possibility of hybrids is potential complication in brackish marshes where breeding ranges overlap. See Clapper Rail species account.

VOICE Information is taken from Meanley (1992) unless otherwise stated. The advertising call of both sexes, given throughout the year and often answered by other birds, is a series of *cheup* or *chac* notes, slower and more regular than the advertising call of Clapper Rail, and increasing in tempo towards the end when the notes seem to run together. This call is also given when a bird is startled, when a pair is reunited after separation, and to initiate changeover during incubation. Howell & Webb (1995b) describe a repeated *kek* which is given in flight when a bird is flushed. The male's courtship/mating call, given principally during pair formation and heard both day and night, is a loud, harsh series of *kik* or *kuk* notes. There is also a soft, rapidly repeated *tuk* call, given during courtship and by paired birds, a deep boom given by the male before nesting, and a cat-like purring call given by the female before nesting. As nesting approaches, paired birds begin to use a number of other calls, mostly soft and subdued, which serve as rallying calls; one such call is a soft, repeated *poyeek*. When threatened or disturbed at the nest, birds give a *rak-k-k* or *chur-ur-ur* distress call, and a resonant *gip-gip-* or *kik-kik-*; the *gip* is also a warning call which causes chicks to scatter and hide. Contact calls to the young include a soft, continuously repeated *woof*. Chicks have a begging *chee-up*, a soft, low *we* or *we-up* when brooded, and a constantly uttered *seep* when following the parents. At c. 1 month, the *seep* call is replaced by other calls: a repeated high-pitched *tahee* when about to sleep, a *soo* or *tsoo*, indicating dissatisfaction, and a series of 5-6 rapid, hoarse *keelp* notes, which may express protest. Adult calls develop from 9-10 weeks of age, when young also give a loud, cat-like *meow* when excited or separated from the family group; an adult-like advertising call is heard from 5 months.

DESCRIPTION *R. e. elegans*
Adult, dark morph Forehead to nape uniform blackish-brown with black feather shafts on forehead and crown; hind-neck blackish-brown with slight traces of pale feather edges; rest of upperparts, including scapulars and rectrices, dark blackish-brown to black, with broad olive-brown feather fringes; lower back, rump and uppertail-coverts more uniform, with blackish-brown feather centres less distinct or absent. Upperwing-coverts deep hazel to bright russet, sometimes with narrow or ill-defined dark centres; inners washed olivaceous; some outer medians and greaters with narrow whitish tip and narrow subterminal whitish band; primary coverts sepia. Alula sepia-brown, outer webs rufescent; remiges sepia, edges olive-brown; tertials ruddy-olive with dark blackish-brown centres. Axillaries and underwing-coverts dusky brown to blackish, tipped and narrowly barred white. Lores, circumocular area and ear-coverts dull ashy, tinged cinnamon (ashy colour variable in extent on ear-coverts); supraloral stripe from base of upper mandible to rear of eye, pale pinkish-cinnamon, often continued as indistinct dusky-cinnamon stripe over ear-coverts; whitish spot below eye; lower ear-coverts and sides of throat cinnamon; chin and centre of upper throat white. Lower throat, breast and upper belly vinaceous-cinnamon, becoming paler in centre of upper belly and whitish on lower belly; sides of breast olive with blackish feather centres. Colours of flanks variable: blackish to brown, narrowly barred white, or even light brown with blackish-bordered white bars; thighs whitish, barred brown; vent and median undertail-coverts barred black and white; lateral undertail-coverts white with blackish spots on inner webs. Iris red to reddish-orange; bill brownish; orange-yellow to ochre-yellow on both mandibles from base at least as far as nostrils; tongue and inside of mouth bright orange-red; legs and feet pale brown-grey to olive-tinged yellowish-brown. Sexes alike; female smaller. **Pale morph** has forehead to nape dark brown; feathers of upperparts with paler, buffy margins; upperwing-coverts dark rich tawny; remiges washed olive; underwing-coverts dark brown with white bars; supercilium pale creamy-buff; sides of head more ochraceous-buff; lower throat, breast and upper belly pale ochraceous-buff, washed greyish.
Immature Very similar to adult but somewhat duller, with pale brown to yellowish bill. Full adult colours attained at first prebreeding moult.

279

Juvenile Upperparts duller and darker, feather centres largely fuscous-black; pale feather fringes narrower and greyer than adult on mantle, similar to adult on scapulars, and almost nonexistent from lower back to uppertail-coverts, which are uniform blackish. Lesser and some outer greater upperwing-coverts tipped white and with a narrow white bar behind tip. Sides of head ochraceous-buff, narrowly barred dusky. Underparts much paler than in adult: chin and upper throat white; lower throat to upper breast tinged cinnamon; lower breast and belly whitish; sides of neck and breast ashy; most of neck and breast feathers with dusky or ashy tips; flanks dull ashy-black to dark grey-brown, indistinctly barred white. Iris brownish; tongue and inside of mouth yellow.

Downy young Down entirely black with faint greenish sheen; bill grey-black on basal half, white around nostrils and flesh-coloured on distal half; egg-tooth present for 4-6 days; vestigial wing-claw present.

MEASUREMENTS (18 males, 14 females) Wing (chord) of male 159-177 (163.4), of female 147-162 (154.3); tail of male 56-72.5 (65.9), of female 60-70 (64.4); exposed culmen of male 58-65.5 (62.5), of female 50-63 (61.9); tarsus of male 52-64 (58.4), of female 49.5-58 (54) (Ridgway & Friedmann 1941). Wing (max) of 5 males 160-170 (163.8, SD 3.9), of 4 females 151-166 (159.5, SD 7.7); culmen to base of 5 males 60-64 (62.2, SD 1.7), of 4 females 59-63 (60.75, SD 2.1). Weight of 9 males 340-490 (415), of 9 females 253-325 (306); of newly hatched young 14-16.5.

GEOGRAPHICAL VARIATION Races separated on size, and on plumage colour and pattern.

R. e. elegans Audubon, 1834 – E North America, from E South Dakota, C Minnesota, E Nebraska, Iowa, S Wisconsin, S Michigan, extreme S Ontario, Ohio and W New York, S through E Kansas and C Oklahoma to S Texas and Gulf Coast, E to W Pennsylvania, Kentucky, Tennessee and Georgia, and N through E North and South Carolina and E Virginia to E New York, Connecticut and (rarely) Massachusetts (AOU 1983, Meanley 1992). Also very locally in E Mexico, from Tecolutla S to Tlacotalpan in S Veracruz (Dickerman 1971, Meanley 1992, Howell & Webb 1995b; see Distribution and Status). Winters primarily in S part of breeding range where most abundant: Delaware Valley to SE Georgia, Florida, along Gulf Coast, in rice belts of Louisiana and Texas, and N into Arkansas rice belt (Meanley 1992). The largest race. See Description.

R. e. tenuirostris Ridgway, 1874 – C Mexico, from S Nayarit, Jalisco, Guanajuato and S San Luis Potosí S to Guerrero, Morelos and Puebla; also on Pacific slope in Colima (Dickerman 1971, AOU 1983, Howell & Webb 1995b). **Dark-breasted morph** Duller, paler and less boldly marked on upperparts than nominate; upperwing-coverts slightly duller; underwing-coverts paler brown; malar stripe broad, pale pinkish-cinnamon; bars of flanks dull umber-brown, washed pinkish-cinnamon; pale flank barring sometimes pinkish-cinnamon rather than white; undertail-coverts largely white; legs and feet brownish-flesh. **Pale-breasted morph** has centre of breast and entire upper belly white, washed pale pinkish-cinnamon. Juvenile darker than adult on upperparts; lores and subocular area darker; lower throat and upper breast washed pale pinkish-cinnamon; lower breast and belly white, laterally suffused and mottled grey-brown and washed pale pinkish-cinnamon. Measurements of 4 males and 6 females: wing (chord) of male 151-159.5 (156.5), of female 139-144.5 (142.3); tail of male 62-70.5 (66.5), of female 59-65 (62.9); exposed culmen of male 62-65.5 (63.5), of female 56-60 (58); tarsus of male 56-59 (57.8), of female 47.5-53 (51.3) (Ridgway & Friedmann 1941). Weight of male 271-331, of female 220-268 (Warner & Dickerman 1959).

R. e. ramsdeni Riley, 1913 – Cuba and Isle of Pines. Resembles pale morph nominate, but smaller; ear-coverts and upper cheeks more greyish; more extensively white on lower breast and abdomen. Measurements of 7 males and 9 females: wing (chord) of male 149-153 (151.2), of female 134-144 (141.4); tail of male 53-65 (56.4), of female 52-58.5 (57); exposed culmen of male 53-58 (56.7), of female 46-54 (51.4); tarsus of male 52-56 (54.6), of female 49-55 (51.9) (Ridgway & Friedmann 1941).

MOULT Postbreeding moult is complete, normally occurring in Aug-Sep; remex moult is simultaneous (earliest date 7 Jul) and birds are flightless for almost 1 month after breeding. Some adults renew the contour feathers while still nesting, this possibly representing a prolonged prebreeding (pre-alternate) or an early postbreeding (pre-basic) moult (Meanley 1992); a prebreeding moult of the contour feathers occurs in early spring (Bent 1926). Postjuvenile (first pre-basic) moult is evident at 9-10 weeks, usually occurs Sep-Oct and is apparently complete by Nov (Bent 1926); at 4-6 months, captive young show a partial moult involving the body plumage. One bird from Veracruz was in heavy postjuvenile moult in Jul (Dickerman 1971).

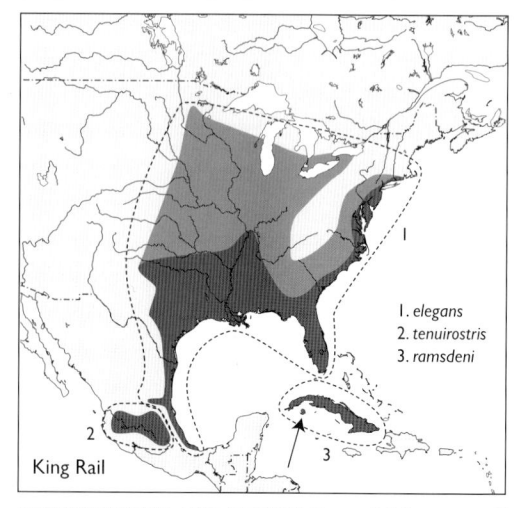

King Rail

1. elegans
2. tenuirostris
3. ramsdeni

DISTRIBUTION AND STATUS From C Minnesota, E South Dakota, E Nebraska and S Ontario S to E Texas and through most of E USA to Atlantic coast as far N as Connecticut; also C and E coastal Mexico, and Cuba and Isle of Pines; northern populations winter in the S parts of this range. Temperature is important in determining the winter distribution, and birds avoid areas where marshes freeze; all wintering sites are in regions which rarely drop below -1°C in Jan (Root 1988). Records from E Mexico may be of migrants, casual birds or scarce winterers (Meanley 1992), but a juvenile was collected at Tecolutla, Veracruz, in Jul 1969, indicating breeding in this region (Dickerman 1971). The King Rail is not globally threatened but was listed as VULNERABLE in Canada in 1985, while in the USA it is listed as ENDANGERED in 6 states and THREATENED or of SPECIAL CONCERN in 6 others. Very little information is available on population densities for this species: winter densities in the 1960s were estimated at 12-15 birds/km using roadside counts (Meanley 1992); for breeding and other density figures, see Social Organisation. Severe declines have been evident in the N part of its range since the 1940s, mainly as a result of the loss, modification and degradation of wetland habitats: c. 54% of the original wetlands in the USA have been lost, and palustrine emergent wetlands declined by 1.9 million ha from the mid-1950s to the mid-1970s (Reid *et al.* 1994). Pesticide use has also adversely affected this species. Alarming declines have been recorded recently from the SW shore of L Erie, the glacial marshes of NW Iowa, the W shore of Maryland, the Smyrna R marshes of Delaware and the Arkansas rice belt (Meanley 1992). At the beginning of the century the species was a common resident in marshes along Missouri's large rivers, but it is now rare and endangered there, being restricted almost entirely to one large wildlife refuge (Reid 1989); a similar pattern is apparent elsewhere in the Midwest (Eddleman *et al.* 1988). Populations in the S USA, especially in Louisiana and Florida, appear more stable. The greatest winter abundance is recorded in S Florida (Everglades) and in the Louisiana bayous around the mouth of the Mississippi R N into Mississippi and W to E Texas (Root 1988); however suitable habitats, especially herbaceous floodplain wetlands, along the Mississippi R corridor, once an important area for the species, have been extensively reduced and degraded, and most good habitats are now found on public refuges (Reid 1989). Birds are often killed in muskrat *Ondatra zibethicus* traps, while both adults and young are frequent road casualties when forced to move to higher ground during floods in the breeding season (Meanley 1992). Historically it was a prized game bird (Bent 1926) but, although it is still listed as a game bird in 13 coastal states, it is now seldom hunted because of the difficulty of finding concentrated numbers and the difficulty of manoeuvring in slough habitats (Meanley 1969). Most hunting probably occurs in the Gulf Coast domestic rice-producing areas of Louisiana and Texas but, even there, <1% of the local population is probably killed (Meanley 1969). The best chance for its long-term survival in North America is considered to be offered by wildlife refuges, where a complex of habitats should be encouraged, including tussocky, shallowly flooded, densely vegetated sites for cover and nesting, beds of perennial vegetation where water depths range from moist to 25cm, and drying patches for brood foraging (Reid *et al.* 1994).

The race *tenuirostris* has a limited range, being confined to the freshwater marshes of C Mexico, and its habitats are under threat from increasing agricultural, industrial and urban development (Williams, S. O. 1989). The current status of the race *ramsdeni* is not clear.

Predators in North America include raccoons *Procyon lotor*, mink *Mustela vison*, foxes *Vulpes*, Hen Harriers *Circus cyaneus*, Great Horned Owls *Bubo virginianus* and alligators *Alligator mississipiensis* (Meanley 1992).

MOVEMENTS North American populations of the King Rail winter primarily in the S parts of the breeding range, but occur casually in winter throughout most of the breeding range. Spring arrivals have been noted in Apr-May, while most breeders at Laurel, Maryland, had departed by late Sep (Meanley 1992, Reid *et al.* 1994). Birds migrate alone and at night; they sometimes strike buildings, telephone lines, lighthouses and towers, become impaled on barbed wire fences, and are killed on roads; they appear in odd places such as city streets during migration periods (Meanley 1992). The Atlantic Coastal Plain and Mississippi Valley are the principal migration corridors (Meanley 1969). In C Mexico (race *tenuirostris*), the occupation of seasonal wetlands in the rainy season (May/Jun-Sep) is indicative of local movements and dispersal (Williams, S. O. 1989). The species is casual or accidental in Manitoba, Quebec, Newfoundland, Prince Edward I, North Dakota, Maine, E Colorado and the Dry Tortugas Is (Florida) (AOU 1983, Meanley 1992). Immatures may disperse widely; in Maryland, most young dispersed from natal area in Aug-Sep (Meanley 1992).

HABITAT Most information is from Meanley (1992). King Rails inhabit freshwater and brackish marshes, both tidal and nontidal; also successional stages of marsh-shrub swamp (including *Cephalanthus occidentalis* swamp), rice fields, flooded farmlands, river margins, roadside ditches, and upland fields near marshes. They do not usually inhabit saltmarsh where Clapper Rails are present, but occur occasionally in saltmarsh during migration. Preferred vegetation types include grasses (Poaceae), sedges (Cyperaceae), rushes (Juncaceae) and *Typha*, while in coastal plain marshes, where the species is most abundant, *Scirpus olneyi*, *Spartina cynosuroides*, *Zizaniopsis miliacea*, *Panicum hemitomon*, *Cladium jamaicense*, *Echinochloa* spp. and *Polygonum* spp. predominate. Vegetation in breeding territories at roadside ditches in Arkansas included *Juncus effusus*, *Carex* spp., *Typha*, *Eleocharis*, *Polygonum*, *Sagittaria* and various Gramineae (Meanley 1957). Nesting occurs in association with perennial vegetation, in water depths from <1 to 25cm, and a selection of shallow-water nest sites is characteristic of this species; however, sites are also recorded in water up to 46cm deep (Reid *et al.* 1994). The habitat requirements are very similar to those of the muskrat (Meanley 1992). This species occurs up to 2,500m in Mexico, where it also uses seasonal freshwater wetlands, having been recorded in grasses, sedges and forbs surrounding shallow pools and seasonal reservoirs, and in muddy, weedy sorghum fields (Williams, S. O. 1989). For foraging habitats, also see Food and Feeding.

FOOD AND FEEDING Omnivorous, but specialising in crustaceans and insects. Crustaceans include crayfish (especially *Cambarus* and *Procambarus*), a critical food item in freshwater marshes, and crabs (especially Red-jointed Fiddler Crabs *Uca minax*), which are dominant food items in brackish tidal marshes; aquatic and terrestrial insects taken include Odonata, Orthoptera (especially *Neoconocephalus* grasshoppers), Coleoptera (especially Dytiscidae, also Carabidae, Scarabeidae and Curculionidae), Diptera larvae (Tabanidae) and Hemiptera (Notonectidae). It also takes molluscs, including slugs and the clam *Macoma balthica*; leeches; spiders; fish, including sunfish Centrarchidae and perch Percidae; and frogs and tadpoles. Many seeds are taken, including those of rice, oats, grasses (*Echinochloa*, *Paspalum*, *Leersia*), *Polygonum*, *Rhynchospora*, the cane *Arundo tecta*, rice-field weeds, and woody plants, while tubers are also recorded. Unusual food items include a small water snake *Natrix*, a mouse, a shrew *Sorex*, acorns, cherries and blackberries; one stomach contained vertebrae of a female Red-winged Blackbird *Agelaius phoeniceus* (Jorgensen & Ferguson 1982). In an Arkansas study, animal food comprised 95% by volume of food in spring, 90% in summer, 74% in autumn and 58% in winter; crayfish 61% in spring; fish 26% in autumn (when many fish became stranded in shallow water on drained rice fields and were easy prey); Dytiscidae 19% in winter; rice seeds made up 16% of the annual diet (Meanley 1992). Birds usually feed either within or close to cover, usually in habitats with shallow water, immersing part or all of the bill; they also forage in deep water, especially when on autumn migration, immersing the head, neck and sometimes the body; they occasionally feed by up-ending like dabbling ducks (Meanley 1992). In tidal areas, most feeding takes place at low tide, and birds may forage with migrant shorebirds on mudflats, taking worms from burrows in the mud (W. L. Dawson in Bent 1926). The birds probe in mud, and also feed on dry land, picking food items from the substrate. When foraging on land they often carry food items to water and immerse them before ingestion. Food items up to the size of crayfish and fiddler crabs are eaten whole; larger prey is usually first carried on to a muskrat lodge or a pile of debris and hacked to pieces (Meanley 1992). Birds (including chicks only 2 days old) regurgitate pellets comprising fragments of mollusc shells and exoskeletons of crustacea. Captive birds ate snow and chunks of ice during a cold spell (Meanley 1992).

There are dramatic shifts in foraging habitat between migration and breeding periods (Reid 1989). Sites used on spring migration and during nesting are in dense vegetation with water 2.5-20cm deep, but broods forage in open mudflats with saturated soil or water up to 7.5cm deep, these sites being close to drier areas of dense, tall vegetation where broods hide during the day. Autumn migration sites have water 1-24.5cm deep and include mosaics of drying swales. Prey densities at brood foraging sites are more predictable than those at sites used by adults at other periods and which contain patchy food resources (Reid 1989).

HABITS This species is mainly diurnal and crepuscular, feeding most actively in the early morning and late afternoon or evening. It normally flies only when flushed or provoked, when crossing a barrier or when migrating. Short flights are made with rapid wingbeats and dangling legs, but in longer flights the legs and neck are stretched out straight and the bird flies rather slowly and close to the ground; on migration it flies at greater heights. It can attain flight speeds of at least 48kph (Meanley 1992). The tail is often held erect when the bird is walking. Birds may swim to cross a creek or pond. Captive birds roost on the ground, either lying down or standing up, and both captive and wild birds move little at night, even on moonlit nights, although wild birds vocalise frequently at night during the courtship period. Birds bathe by squatting in water, ruffling the plumage, moving the body up and down and shuffling the wings; the head is used to flip water over the body (Meanley 1992). Bathing usually lasts 1-2 min. In tidal areas, birds usually bathe at low tide. When foraging among flocks of shorebirds on mudflats, birds may stand still for up to 15 min, resting and sunbathing, and possibly relying on the shorebirds to warn them of danger (W. L. Dawson in Bent 1926).

SOCIAL ORGANISATION Information is from Reid (1989) and Meanley (1992). Monogamous; territorial during the breeding season. Pair formation is thought to occur between early Jan and early Mar. The duration of the pair-bond is not known; present evidence indicates that the birds are solitary after the breeding season. No accurate estimates are available for total populations or even breeding densities over most of the range. Estimates in the early 1960s were 62, 74 and 266 birds/100ha using strip censuses and two-stage sampling, and 2.3 and 3.5/km using roadside counts; more recent estimates range from 7 to 22 adults/100ha in semi-permanent wetlands.

SOCIAL AND SEXUAL BEHAVIOUR Most information is from Meanley (1992). King Rails attack and chase both conspecifics and other rallids which enter breeding territories or approach nest sites. Soras (96) migrating through breeding areas are attacked, as are Virginia Rails (58). The threat display to a conspecific intruder involves the territory holder assuming a partial crouch, drawing in its neck and slowly ruffling or extending its contour feathers. It then chases the intruder out of the territory, on foot or in flight. Fighting between males is likened to "sparring between fighting cocks", and encounters may last for up to 3 min (Meanley 1957). Incubating birds seldom flush until an intruder is within 3m of the nest, when they may fly from the nest and strike the intruder, and may then give a distraction display by spreading the wings and fluttering through the vegetation uttering the distress call (Meanley 1969).

It is unclear if courtship, displays or pairing occur outside the breeding grounds (Reid *et al.* 1994). The courtship display of the male involves walking about with the tail raised and the white undertail-coverts fanned out, sometimes flashing the coverts and flicking the tail slightly. While thus displaying, the male stops occasionally to give the courtship call. When a female enters his territory the male follows her with head and neck outstretched and angled upwards, bill open, tail raised, and undertail-coverts extended. In mated pairs, when the female approaches the male he stands still and bends his neck so that the bill is perpendicular and nearly touching the ground. Courtship feeding has been observed, when males presented crayfish and fiddler crabs to their mates. Copulation takes place near the nest, and continues during egg-laying. The male often gives the advertising call before copulation, which is effected by the female crouching and the male mounting and placing his feet and tarsi on the female's back. 'Symbolic' nest-building has been observed, when a male carried nesting material into a hole in a rice-field dike; two days later the breeding nest was started c. 10m from the symbolic site.

BREEDING AND SURVIVAL Season Lays Florida early Feb to mid-Jul; elsewhere in North America Mar-Apr to Jun-Aug; Mexico, May-Aug, possibly Apr-Sep. **Nest** Nest site usually over shallow water in tidal or nontidal marsh; also in broad roadside ditches with *Typha*, grasses (*Bromus, Echinochloa, Paspalum*) and sedges and rushes (*Carex, Cyperus, Juncus*), and in rice fields, oats (*Avena* var.), swamps, and upland fields near water; one nest was in potato field 198m from saltmarsh edge. Most nests located in fairly uniform stands of perennial vegetation, especially grasses, sedges and rushes. Nest placed in clump or tussock of such vegetation, or between clumps, sometimes on dry ground but usually on moist ground or up to 45cm (usually up to 30cm) above highest water mark; may be raised as water rises after heavy rain. Nest a platform with saucer-shaped depression, or a well-made cup; of dead, dry rice plants, grasses, sedges or rushes; sometimes lined with finer rushes and grasses; at wet sites has base of wet decaying vegetation or flattened stems; usually has canopy of grass, and entrance ramp. Outside diameter 20-28cm, inside depth of cup 1.5cm. Both sexes build but male more active; nest not always completed before first egg laid. Several brood nests, usually without canopies, constructed near breeding nest. **Eggs** 3-15 (most often 10-12; means in 4 US populations 10.5-11.2); late clutches contain fewer eggs; ovate; glossy; creamy-buff to pale olive-buff, sparingly and irregularly spotted pinkish- or purplish-brown, sometimes with a few spots of bright brown; size 38.5-44 x 28-32 (40.95 x 30.1, n = 76); weight of 3 eggs 18.8, 18.9, 20.3; calculated weight 21.1. Eggs laid at daily intervals. **Incubation** 21-24 days, by both sexes; begins when clutch complete; all eggs hatch within 24-48h. **Chicks** Information is from Meanley (1992). Precocial; leave nest soon after hatching; can follow parents for considerable distance at 1 day; fed and cared for by both parents. Chicks take food from adult's bill at 1 day; attempt to pick up food from ground from 2 days; sometimes remain concealed while adults carry food to them; fed almost entirely by adults to 3 weeks; forage independently at 7-9 weeks; begging display still occasional at 9-10 weeks but discontinued soon after, as young become independent. Most broods of 1-3 weeks associate with 2 adults; most older broods with only 1. Tail-flicking occurs from 2 weeks; juvenile plumage evident at 4 weeks; appears first on underparts, then on back, later on head and neck; quills of remiges and rectrices may begin to emerge at 5 weeks; body plumage fully attained at 7-8 weeks; young fly at 9-10 weeks, when postjuvenile moult under way.

Nest success usually high; in Missouri, 54 (81%) of 67 nests were successful (the others being destroyed by mammalian predators) and 39 (58%) hatched complete clutches (Reid 1989); in Arkansas, c. 50% of young survived to 2 weeks. Survival rates during incubation: daily, 0.97-1.0; interval 0.74-1.0; survival from first egg to hatching 0.695 ± 0.071 (SE) (Meanley 1992). Nest predators include raccoons, Fish Crows *Corvus ossifragus* and foxes, and predation is known to limit nesting success in some areas (Meanley 1992). Age of first breeding not clear; presumably 1 year, but 1 nesting female appeared to be in first-year plumage (Meanley 1969). May rear 2 broods per season in S USA, where nesting season is long. Some birds return to same territory in consecutive years.

57 PLAIN-FLANKED RAIL
Rallus wetmorei Plate 18

Rallus wetmorei Zimmer & Phelps, 1944, La Ciénaga, Aragua, Venezuela.

Suggested to be conspecific with *R. longirostris*, but ranges overlap. Monotypic.

Alternative name: Wetmore's Rail.

IDENTIFICATION Length 24-27cm. A medium-sized, predominantly brown rail with streaked upperparts and no barring on underparts. Top of head olive-brown; supercilium pale fawn; lores dark grey; rest of face dull buffy-brown. Upperparts and tail buffy-brown, streaked blackish-brown; upperwing-coverts more uniform buffy-brown with vague darker markings; remiges dark brown. Chin whitish-buff; throat buff; rest of underparts buffy-brown; lower belly more pinkish-buff; undertail-coverts buffy-brown and white, may appear predominantly white or pale brown. Iris reddish-brown; bill relatively straight and slender, dark olive, with paler brown base to lower mandible; legs and feet olive-brown. Sexes alike; female probably smaller. Possibly flies poorly: remiges and rectrices appear soft. Immature and juvenile undescribed. Inhabits mangrove swamps.

Similar species Similar to the sympatric race *phelpsi* of the Clapper Rail (55), but smaller; lacks grey on side of head and has plain buffy-brown underparts, lacking all barring; bill brown and iris reddish-brown. These features also serve to distinguish it from the allopatric Virginia Rail (58), which also differs in being smaller and having rufous upperwing-coverts, a dusky and red bill, and a red iris.

VOICE Undescribed.

DESCRIPTION
Adult Top of head and nape olive-brown with faint suggestion of tawny-olive feather margins; supercilium pale tawny-olive, ending behind eye; lores dark grey-brown; sides of head dull buffy-brown; chin whitish-buff; throat buff, becoming browner at base and grading into buffy-brown of foreneck. Hind-neck, scapulars, mantle to uppertail-coverts and tail dark blackish-brown with broad buffy-brown feather margins, giving streaked effect. Upperwing-coverts almost uniform buffy-brown with vague darker feather centres. Remiges dark brown; primaries almost uniform, secondaries and tertials with buffy-brown margins. Underwing-coverts and axillaries more fawny-brown, with faint suggestions of white at feather tips. Sides of neck dull buffy-brown with vague darker stripes merging into more pronounced streaks of hind-neck; foreneck, breast, flanks and upper belly dull buffy-brown to buffy-brown, with faint suggestion of white margins on rear flank feathers. Lower belly vinaceous-buff; short undertail-coverts buffy-brown with whitish bars, long feathers white with brownish margins; overall, undertail-coverts may appear whitish or brownish. Iris reddish-brown; bill relatively straight and slender, upper mandible dark olive, lower mandible paler brown with dark olive tip; legs and feet pale olive-brown (Zimmer & Phelps 1944). Sexes alike; from available measurements, female probably averages smaller. In Blake (1977) bill colour said to be black, reddish at base, and feet reddish-black.
Immature Not described.
Juvenile Not described.
Downy young Not described.

MEASUREMENTS Wing of 8 males 129-137 (134.6), of 6 females 119-127 (123.1); tail of male 46-57, of female 45-48.5; exposed culmen of 8 males 48-53 (50.5), of 6 females 45-48 (47.1); tarsus of 8 males 48-50 (48.4), of 6 females 40-44 (43.3) (Blake 1977, Ripley 1977; means and sample sizes not given by Ripley); 1 male (BMNH) has culmen to base 56. Zimmer & Phelps (1944) give measurements of the type (female) as follows: wing 115; tail 44; exposed culmen 47, culmen from base 59, tarsus 41.5.

GEOGRAPHICAL VARIATION None.

MOULT No information available.

DISTRIBUTION AND STATUS Summarised comprehensively by Collar *et al.* (1992) and Wege & Long (1995), from which sources most of the following information is taken. This species is globally ENDANGERED, and its population is probably <500 (Rose & Scott 1994). It is extremely rare and very poorly known, being restricted to the coast of N Venezuela, where it is known from at most 9 localities (only 3 designated as key areas for threatened birds) in 3 states: near Chichiriviche (unconfirmed sighting), and at Cuare Faunal Refuge and Tucacas in E Falcón; at Puerto Cabello, Borburata and Patanemo in N Carabobo; and at La Ciénaga, Playa de Cata, and the Henri Pittier National Park in Aragua. The type specimen was collected in April 1943 at a mangrove swamp of 1 square mile (c. 260ha) on the western shore of the then uninhabited bay of La Ciénaga, and many collecting trips were subsequently made to the locality without relocating the species (Zimmer & Phelps 1944). It has not been recorded from Carabobo State since 1945, and the only records since those from Tucacas in 1951 are from Playa de Cata, 15km E of La Ciénaga, in April 1991, from the Henri Pittier National Park, where 1 pair was seen in 1991 and 1993, and from Chichiriviche and Cuare Faunal Refuge (both records possibly referring to the same bird) in the 1980s. It was probably at least locally common when discovered, 11 specimens being collected at Puerto Cabello/San Esteban in 1944-45, and 11 at Tucacas in May 1951. Since these localities are easily accessible, it is likely that the paucity of subsequent records indicates a considerable population decline. Its mangrove habitats are being destroyed by housing development and by expanding oil exploration. The Cuare Faunal Refuge is threatened by squatters, hotels, tourist pressure, illegal hunting, pollution from domestic sewage, pesticides and mercury, a proposed golf course, and the restriction of water flow to and from the sea by the building of a new road. This refuge is one of the most important coastal wetlands in Venezuela and is the country's only Ramsar site; its integrity needs to be ensured. The coastal wetland habitats in the San Esteban and Henri Pittier National Park region are under severe pressure from development projects. The Playa de Cata lagoon has been closed off from the sea by a dike to facilitate the development of the very popular beach for tourism. The 32,000-ha Morrocoy National Park, NE of Tucacas, whence there is 1 recent record, may hold a population of this rail.

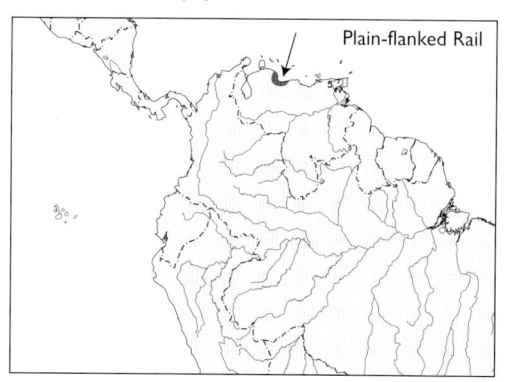

Plain-flanked Rail

MOVEMENTS Apparently sedentary.

HABITAT Coastal mangroves, "mangrove swamp", and shallow saltwater or seasonally flooded brackish lagoons and marshes with emergent and halophytic vegetation dominated by saltwort *Batis maritima* (Scott & Carbonell 1986, Collar *et al.* 1992).

FOOD AND FEEDING Nothing recorded; presumably similar to *R. longirostris*.

HABITS Undescribed.

SOCIAL ORGANISATION Unknown.

SOCIAL AND SEXUAL BEHAVIOUR Unknown.

BREEDING AND SURVIVAL Season Breeding condition: female, Apr; male, May. A Sep specimen is apparently a juvenile. Nothing else known.

58 VIRGINIA RAIL
Rallus limicola Plate 18

Rallus limicola Vieillot, 1819, Pennsylvania.

Forms a superspecies with *R. antarcticus* and *R. semiplumbeus*; former sometimes considered a race of *R. limicola*. Four subspecies recognised.

Synonyms: *Rallus virginianus pacificus*; *R. aequatorialis/ friedmanni*; *Aramus virginianus*.

Alternative name: Lesser Rail.

IDENTIFICATION Length 20-27cm. A smallish-medium-sized rail; upperparts blackish, streaked dull brown (paler in race *friedmanni*, darker in *aequatorialis*); crown, hindneck, and lower back to uppertail-coverts more uniform; upperwing-coverts russet (obvious in flight). Sides of head grey; lores and ear-coverts darker; well-defined supraloral streak buffy; eyelids white. Chin white; throat, foreneck and breast brownish to cinnamon (paler in *friedmanni*, duller in *aequatorialis*); belly and thighs paler; sides of neck and sides of breast browner, streaked dark. Flanks blackish, narrowly barred white (broader in *meyerdeschauensee*); lateral undertail-coverts white; medians black, barred white and tipped cinnamon. Race *meyerdeschauensee* brighter and paler overall. Iris reddish; bill brownish on upper mandible, dull orange-red on lower mandible; legs and feet dusky brownish-red. Sexes alike; female smaller than male. Immature very similar to adult, but darker and duller. Juvenile has duller upperparts with darker, more rufescent, feather fringes; grey on head tinged brown and mottled white; throat and upper foreneck whitish, mottled grey to buff; rest of underparts blackish, mottled white, with poorly defined flank bars and whitish centre to breast and belly. Iris brown; bill blackish, lower mandible becoming reddish-brown; legs and feet dusky to reddish-brown. Inhabits freshwater and brackish marshes, sometimes saltmarshes.

Similar species Much smaller than Clapper Rail (55) and King Rail (56), with slightly decurved bill, extensive blue-grey on side of head, and duller upperparts with brighter rufous upperwing-coverts. Easily separated from other American *Rallus* species by combination of grey on sides of head and cinnamon or brown anterior underparts.

VOICE All information is for North American birds (Kaufmann 1983, Conway 1995) unless otherwise stated. The most frequent call is an antiphonal, duetting grunt, given by pairs during the breeding season and functioning in pair bonding, communication, territory defence and neighbour recognition. The advertising call, known as the *kicker* call and heard infrequently in spring, is a series of loud and rapid *ki ki* notes followed by a fast, descending *krrrrr* or *k-krrrrr* and is thought to be given by unpaired females to attract males (Conway 1995; W. R. Eddleman *in litt.*; *contra* Ripley 1977). A *tik-it* call, probably given by unpaired territorial males seeking a mate, is monotonously repeated throughout the day for a brief period in spring. Other calls include a sharp, piercing or rasping *kiu* and a sharp *kek* of alarm, a short, high-frequency *skew* of alarm from adults disturbed at the nest, and a low clucking call at incubation changeovers. Various low-pitched quiet calls are also given, including soft, nasal peeping contact calls, a repeated, rapid, soft *tipit* contact call, purring calls, and a soft *ka ka ka ka* given by parents to young. A call of unknown function, described by Hardy *et al.* (1996) as the *weka-weka* call, is a series of rapid low quacking notes. Various squeals and squawks are given when birds are attacked. Birds may produce a clacking sound by repeatedly bringing the mandibles together rapidly. South American birds are also said to give single *kuc* notes and a rolling *yrll* (Fjeldså & Krabbe 1990). Chicks give *pee-eep* calls. Migrants seldom call during the first 1-3 weeks after arrival on the breeding grounds. Calling occurs most frequently in the 2-3h around dawn and dusk, but birds may call through the night. Vocalisations decline after Jul and are rarely heard thereafter, including during migration and on the wintering grounds.

DESCRIPTION *R. l. limicola*
Adult breeding Upperparts from forehead to tail, including scapulars, dark blackish-brown to blackish, feathers fringed pale sepia to pale brownish-olive, giving strongly streaked appearance except on crown, hind-neck, and lower back to uppertail-coverts, where feather fringes very narrow, giving more uniformly dark appearance (especially on crown, hind-neck and rump). Upperwing-coverts russet to cinnamon-rufous; inner webs of greater coverts duskier on inner portion; alula, primaries and secondaries fuscous; tertials dark blackish-brown, fringed pale olive-brown, like upperparts; underwing-coverts and axillaries dusky brown with narrow white margins. Claw 1mm long on outer digit. Lores dark slate; ear-coverts slightly paler and greyer; well-defined supraloral streak pinkish-cinnamon to buffy-white; sides of head above, behind and below ear-coverts, and to malar region, ashy to plumbeous-grey. Eyelids and chin white; throat, foreneck and breast brownish to cinnamon or pale vinous, becoming lighter on belly and thighs; sides of neck and sides of breast browner, with obscure darker streaks. Flanks blackish-brown to black, narrowly barred white; lateral undertail-coverts white, medians black, barred white and tipped cinnamon. Iris reddish-brown to russet; bill brownish on tip and upper mandible, dull orange-red on lower mandible (redder in spring); legs and feet dusky brownish-red. Sexes alike; female smaller than male, with duller bill.
Adult non-breeding As adult breeding, but darker and more richly coloured (Oberholser 1974); however, Sharpe (1894) states that upperparts are rather lighter because of broader brown feather edges.
Immature Very similar to adult breeding but somewhat darker; slightly duller than adult non-breeding. First adult (alternate 1) breeding plumage similar to adult breeding, but juvenile remiges retained.
Juvenile Upperparts similar to adult but somewhat duller blackish-brown, with less contrasting and more rufescent feather fringes; upperwing-coverts duller and darker. Lores darkish grey-brown; ear-coverts dull brown; supraloral stripe creamy-white, shading to mouse-grey above ear-coverts and merging with similar colour behind and below ear-coverts to malar region; sides of head mottled whitish and often tinged olivaceous-brown. Chin whitish; throat and upper foreneck whitish, mottled dusky-grey to buff; lower foreneck, breast, flanks and belly predominantly fuscous or fuscous-black, mottled white; flank barring poorly defined; centre of breast and belly largely whitish, sometimes tinged pinky-buff; thighs brown; undertail-coverts black, barred white and tipped cinnamon. Iris olive-brown, becoming brown; bill blackish, lower mandible becoming reddish-brown; legs and feet dusky-brown, becoming reddish-brown.
Downy young Black downy chick has greenish gloss on back; iris black, becoming olive-brown (4 weeks); bill pink or buff with black band anterior to nares, band expanding to cover central third of bill (2 weeks); bill entirely black at 4 weeks; white egg-tooth retained for up to 2 weeks; legs and feet brownish-black.

MEASUREMENTS Wing of 47 males 99-112.5 (106.3, SD 3.2), of 29 females 89-106 (100.1, SD 3.5); tail of 47 males 39-49.5 (43.8, SD 2.4), of 40 females 32-47.5 (40.6, SD 3.0); exposed culmen of 47 males 33.5-58 (40.2, SD 3.4), of 40 females 32-39 (35.7, SD 1.7); culmen to base of 9 males 37-46 (42.6, SD 3.9), of 9 females 34-42 (39.1, SD 2.5); tarsus of 43 males 29-38.5 (35.1, SD 1.9), of 36 females 29.5-35 (32.4, SD 1.3). Weight of 17 males 64-120 (92.3), of 7 females 64-80 (71.9), of 147 unsexed 55-124 (84.1, SD 14.5); weight at hatching 5-8.1 (6.7, n = 7) (Ripley 1977, Dunning 1993, Conway 1995).

GEOGRAPHICAL VARIATION Races separated on size, colours of upperparts and underparts, and width of white flank barring. Race *friedmanni* sometimes included in nominate but variation among fawn-breasted populations is consistent with general ecomorphological trends, and forms are best assigned subspecies rank (Fjeldså 1990). Nominate race includes putative races *zetarius* and *pacificus* (synonyms) of W USA (see Ripley 1977). South American races are smaller than northern races.

R. l. limicola Vieillot, 1819 – Locally in North America, from S British Columbia, C and S Alberta, C Saskatchewan, C Manitoba, S Ontario, NE Minnesota, S Quebec, New Brunswick, Nova Scotia and Prince Edward I S to N Baja California, S Arizona, C New Mexico, Kansas, N Iowa, N Illinois, N Indiana, N Ohio, S Pennsylvania, E Virginia and coastal N Carolina. Very local or casual in other C, S and SE states of USA (Conway 1995). Also occurs SW Newfoundland (AOU 1983). Winters predominantly coastally, from SW British Columbia and Vancouver I S through Baja California and C Mexico (S to Jalisco and Veracruz), E along Gulf Coast, throughout Florida, and NE along coast to New York and Massachusetts; also inland N to Montana, Colorado, Wyoming, Nebraska, Illinois, Michigan and S Ontario (AOU 1983, Conway 1995, Howell & Webb 1995b). Limited numbers winter at Great Lakes and as far N as Maine. Few also winter Bermuda (Amos 1991); also said to winter S to Guatemala (Antigua; Dueñas; Ciudad Vieja) (Ridgway & Friedmann 1941). Upperparts darkest of all races; see Description.

R. l. friedmanni Dickerman, 1966 – C Mexico in Puebla, Tlaxcala and México (probably also C Veracruz and Oaxaca); extreme S Mexico in C Chiapas; C Guatemala. Compared to nominate, upperparts have paler, more olivaceous fringes to mantle feathers and tertials; upperwing-coverts average paler rufous but colour variable; underparts paler, more pinkish, less cinnamon. Mean measurements of 24 males and 15 females: wing (chord) of male 105.4, of female 96.8; exposed culmen of male 39.7, of female 34.8; tarsus of male 35.7, of female 32.5 (Fjeldså 1990). Weight of 1 male 82.

R. l. aequatorialis (Sharpe, 1894) – Extreme SW Colombia, on both slopes of Nariño SW from Túquerres and Sibundoy, and Andes of adjacent N Ecuador from Carchi SW to Loja (San Lucas). Small; duller than other races. Dorsal streaks fuscous, less sharply defined; underwing-coverts darker than in nominate, dark grey to fuscous with darker bars. Anterior underparts greyish-fawn; flanks with rather obscure, thin, off-white bars; thighs drab grey; centre of belly vinaceous-fawn to cinnamon. Juvenile overall sooty-black; some olive-brown feather fringes on upperparts; throat and centre belly mottled white or pale brown; vague indications of pale flank bars; much less white mottling on underparts than juvenile nominate; undertail-coverts black with ochre-brown feather margins separated by white line (Fjeldså 1990). Mean measurements of 10 males and 6 females, and measurements of 68 unsexed birds of all ages: wing (chord) of male 104.1, of female 101.6, of unsexed 93-111 (102.4, SD 4.5); exposed culmen of male 37.1, of female 35.2, of unsexed 31.4-40 (35.9, SD 2.3); tarsus of male 34.3, of female 31.9, of unsexed 30-37 (32.9, SD 4.9) (Fjeldså 1990). Culmen to base of 5 males 36-40 (38.1, SD 1.9).

R. l. meyerdeschauenseei Fjeldså, 1990 – Coastal Peru, from La Libertad (Trujillo) S to Arequipas. A specimen record from Imbabura province, N Ecuador, may suggest a continuous past or present distribution in coastal zone of Ecuador (Fjeldså 1990). Brighter and lighter in colour than other races. Upperparts with paler feather fringes and fuscous to olive-brown feather centres; crown and rump mid-brown. Underwing-coverts slaty with narrow whitish edges. Chin buffy-white; anterior underparts dull cinnamon to orange-cinnamon; flanks with more distinct and wider white barring; undertail-coverts mainly white with semi-concealed dark spots. Smaller than all other races, but measurements overlap. Measurements (3 males, 3 females): wing (chord) of male 96, 98, 101, of female 90.5, 93, 96.5; exposed culmen of male 34, 35, 35.5, of female 32.5, 33, 34; tarsus of 2 males 31, 33.5, of 2 females 28.5, 29 (Fjeldså 1990).

MOULT Information is available for North American birds only (Conway 1995). Postbreeding moult complete; moult of remiges and rectrices simultaneous; occurs before autumn migration, usually Jul-Aug. Pre-breeding moult of body and some wing-coverts occurs Mar. Partial postjuvenile moult begins at 12-14 weeks; occurs Jul-Oct; in Iowa, most had completed moult by mid-Sep.

DISTRIBUTION AND STATUS S Canada, N and W USA, wintering S to CS Mexico and Guatemala; other populations CS and S Mexico, Guatemala, extreme SW Colombia, the Andes of Ecuador, and coastal Peru. In the USA and Canada it is locally rare to common and has suffered considerable habitat reduction as wetlands have been lost to agriculture and to urban, industrial and reservoir development; palustrine and riverine wetlands are currently among the most threatened habitats in the USA. In North America, populations declined 2.2%

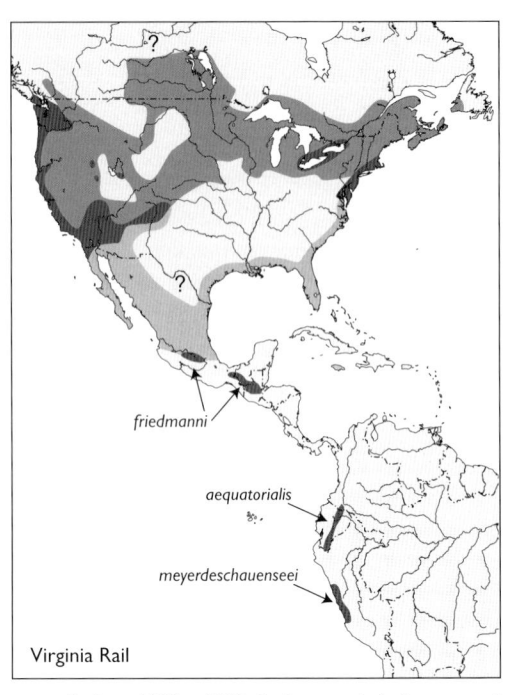

Virginia Rail

annually from 1982 to 1991, during a period when natural droughts also reduced the availability of wetlands; the total population is now considered relatively stable (Conway *et al.* 1994). Winter habitat inland is possibly increased by hot-water discharge from power plants, and the bird's occurrence in other inland areas may be influenced by the presence of wildlife refuges, some of which hold the densest winter populations (Root 1988). Virginia Rails may be legally harvested in 37 states and in Ontario, Canada. Although few hunters take rails and the current annual harvest is probably within sustainable levels (at least on a national scale), harvest surveys are needed. In all but 1 state the rail hunting season is in the autumn, and hunting pressure is greatest on the wintering grounds along the S Atlantic and Gulf coasts. Habitat loss is the greatest threat to the species, and the preservation, restoration and proper management of inland freshwater wetlands are major priorities to safeguard habitats. It is important to maintain or create diverse wetland complexes, and to maintain or emulate natural water fluctuations; such measures benefit a wide range of bird species, including rails (see Sora [98] account, Distribution and Status section, for further details; also Conway & Eddleman 1994 and Conway 1995). Current monitoring programmes do not adequately survey populations, and alternative ones using taped playback should be implemented (Conway 1995). The extent and significance of postbreeding dispersal and emigration (see Movements) deserve investigation, especially to help evaluate the potential impact of the loss of small private wetlands (Johnson & Dinsmore 1985).

The race *friedmanni* is fairly common to uncommon but local in Mexico (Howell & Webb 1995b); its status in Guatemala is not clear. The status of *aequatorialis* in Colombia is also unclear (Hilty & Brown 1986), but in Ecuador it is rare, local and probably declining; agriculture and siltation may have resulted in little habitat remaining in the Andes (Fjeldså 1990, Fjeldså & Krabbe 1990); however it is still locally quite numerous in a few localities

(Ridgely & Greenfield in prep.). The race *meyerdeschauenseei* is one of the rarest sedentary waterbirds in Peru, where it occurs in only a few of the remaining coastal marshes; these are under serious threat from drainage for agriculture and urban development, and from pollution and reed-cutting (Fjeldså 1990).

Predators of adults and large young include mink *Mustela vison*, coyote *Canis latrans*, feral cats, Great Egret *Egretta alba*, Hen Harrier *Circus cyaneus*, and owls; the predation rate is probably high (Conway 1995).

MOVEMENTS Most information is from Conway (1995). Breeding populations in Central and South America are non-migratory, but most North American populations are migratory, although some resident birds occur at least as far N as S New England, especially near the Atlantic coast (W. R. Eddleman *in litt.*; see also Habitat section, and Root 1988). Departures occur mainly in Aug-Oct (Sep-Nov in British Columbia: Campbell *et al.* 1990), but dispersal from N US breeding territories is recorded from late Jul (Iowa) and southern birds may leave as late as Nov; the timing of departures is possibly influenced by weather conditions, and birds congregate at larger marshes prior to autumn migration (Pospichal & Marshall 1954). Birds begin to arrive on the wintering grounds in Sep, and wintering birds are present in Mexico from late Aug to Apr (Howell & Webb 1995b). Migrants leave the wintering grounds by early Apr and arrive on the breeding grounds from Mar to May; arrival occurs in Colorado in early Apr, in Kansas and New York as early as 10-17 Mar, and in Connecticut, Iowa, Michigan, Minnesota, Ohio and Wisconsin between the 3rd week of Apr and the 1st week of May. Arrivals are possibly influenced by weather and vegetation development; males usually arrive 7-10 days before females. Birds migrate at night, singly and at very low altitudes, often following river courses or low, level land. They sometimes strike towers, telephone lines or wire fences, and they are reported from small urban gardens in British Columbia (Campbell *et al.* 1990). Spring migrants are adversely affected by storms. The species is scarce but frequent in Bermuda in Aug-Apr and casual or accidental on Queen Charlotte Is (British Colombia), Greenland and Cuba, and there is one sight record from Puerto Rico (AOU 1983, Campbell *et al.* 1990, Amos 1991). Postbreeding dispersal may initially involve the movement of family groups or individuals away from the vicinity of the brood-rearing home range into other wetlands and also to different habitats such as cornfields; such dispersal was recorded in Iowa in late Jul and early Aug (Johnson & Dinsmore 1985). The stimulus for this emigration may be the maturation and independence of the brood and the male may stimulate the breakdown of the family group by his increasing aggressiveness towards the female and the young (Johnson & Dinsmore 1985). The significance of postbreeding dispersal, which may involve a fairly long-distance movement between wetlands, is unclear: it may simply segregate family members, it may constitute a limited moult migration, or it may involve a shift to an autumn migration staging area (Pospichal & Marshall 1954, Brown & Dinsmore 1985); it deserves further investigation (see Distribution and Status).

HABITAT Most information is from Conway (1995). The Virginia Rail inhabits reshwater marshes with stands of robust emergent vegetation such as *Scirpus*, sedges, *Typha* and tall grasses; swampy grassland, wet meadows and irrigated hayfields; brackish marshes and occasionally coastal saltmarshes; it also occurs in páramo bog in the Andes (Fjeldså & Krabbe 1990). The most important habitat requirements are probably shallow water, emergent cover and moist to muddy substrates with a high abundance of invertebrates; in the USA, freshwater marshes in early stages of succession are favoured, as are moist to shallowly flooded (<15cm) habitats with muddy, unstable substrates; vegetation height is not as important as adequate overhead cover. Birds will also use deep-water habitats, especially those with substantial collapsed or floating vegetation to provide a foraging substrate. They may use interspersion as a proximate cue in selecting habitats rich in macroinvertebrates, being commonest in wetlands with 40-70% upright emergent vegetation interspersed with open water, mudflats and/or matted vegetation. They avoid dry stands of emergents, and also areas with high stem densities or large amounts of moribund vegetation which can impede the birds' movement. Parents with young, and dispersing birds, may temporarily feed in habitats outside wetlands: row crops and weedy cornfields have been recorded. Shallow flooding of manmade impoundments in late summer attracted postbreeding rails; these sites had water up to 27cm deep (mostly <15cm) and robust cover of *Bidens*, *Eupatorium* and *Echinochloa* (Rundle & Fredrickson 1981). On migration, birds use areas of flooded annual grasses or forbs with shallow water, and optimal habitat for migrants includes a variety of water depths, robust vegetation cover, and a diversity of plants including short-stemmed seed-producing species, especially annuals. In winter the birds occupy freshwater, brackish and saltwater marshes, and are also found in moist coastal grasslands.

This rail occurs in both subtropical and temperate zones, from sea level to 3,660m, locally to 3,800m in Ecuador (Ridgely & Greenfield in prep.). In winter it normally frequents regions warm enough to prevent marshes from freezing; its presence in colder areas of the USA in winter possibly results from the use of wetlands artificially warmed by hot-water discharge from electricity-generating plants (Root 1988). Wintering birds are found at sites with flowing water and spring outflow (W. R. Eddleman *in litt.*). Virginia Rails often occur alongside Soras but prefer less deeply flooded to saturated sites, and generally breed in marshes where spring air temperatures are warmer (mean 5.6°C) than in Sora breeding marshes (Griese *et al.* 1980). For a more detailed discussion of the habitat requirements of the two species, see the Sora species account (Habitat section).

FOOD AND FEEDING Animal food includes earthworms, Bryozoa, molluscs (gastropods such as *Helisoma* and *Physa*, and slugs), crayfish, amphipods *Gammarus* spp., many insects and their larvae including Coleoptera (Curculionidae, Dytiscidae and Hydrophilidae), Odonata, Diptera, Hemiptera (Notonectidae), Lepidoptera and Hymenoptera (including Formicidae), spiders, small fish, frogs and small snakes. It also eats a variety of aquatic plants, including *Lemna minor*, and the seeds of many marsh plants, including *Agropyron*, *Carex*, *Cyperus*, *Eleocharis*, *Scirpus*, *Spartina*, *Leersia*, *Zizaniopsis*, *Sparganium*, *Potamogeton*, *Polygonum*, *Nuphar*, *Bidens*, *Cephalanthus*, *Hippuris* and *Chenopodium*. Animal food predominates (85-97% of the diet in summer), and insects comprised almost 62% by volume of the diet of 37 breeding birds (Horak 1970). More plant material is eaten at other times; seeds of marsh plants may constitute 32% of the diet in autumn and 21% in winter as opposed to 12% in spring and 3% in summer (Conway 1995). This species feeds mainly by probing with its long bill in mud and shallow water, but also probes under matted or floating vegetation; it will climb in pursuit of food, occasionally using the wing-claw. Foraging occurs primarily at dawn and dusk; the birds usually remain in or next to cover but may feed in the open at dusk. It consumes a much higher percentage of animal food than does the sympatric Sora; for a comparison of food and foraging habitats with those of the Sora, see the Sora species account (Food and Feeding section).

HABITS Virginia Rails are active throughout the day, with peak activity at dawn and dusk. The birds are secretive, and are more often heard than seen. They walk deliberately; the tail is normally held erect and fanned to expose the black and white undertail-coverts, and is often flicked. Birds seldom fly, except during migration, and normally flush only at close range, making a short flight with dangling legs before dropping into cover. They most often escape danger by running, or by fluttering over floating vegetation, but they may also dive and swim to flee from predators, using the wings to propel themselves under water. After diving, the bird remains submerged while raising its head, or sometimes only the bill and eyes, above the water (Conway 1995). They sometimes climb in marsh vegetation and bushes. Birds roost each night in the same spot, occasionally vocalising during the breeding season but seldom moving (Conway 1995). Sunbathing is recorded, but no details are available.

SOCIAL ORGANISATION Monogamous; the pair bond is maintained only during the breeding season, when the species is territorial, and apparently breaks down before dispersal, shortly after the young fledge (Johnson & Dinsmore 1995): see Movements section. Territory size is difficult to measure: birds defend the immediate area of the nest vigorously but do not defend the territory boundaries as aggressively (Conway 1995). The mean distances between adjacent nests in Minnesota was 46m (Pospichal & Marshall 1954), while in Iowa nests have been recorded only 17.4m apart (Tanner & Hendrickson 1954). The territories of Virginia Rails and Soras often overlap and there may be as little as 1.5m between the nests of the two species (Conway 1995). The density of breeding Virginia Rails has been estimated at 0.1-8.9 pairs/ha (the maximum recorded density is 25 pairs/ha) and 0.2-8.6 birds/ha (Conway 1995). Home range size varies seasonally and with habitat quality; it has been estimated at 0.18 ± 0.02 (SE)ha and 1.64 ± 1.48ha during the breeding season, and 2.41 ± 1.84ha in winter; male and female home range sizes do not differ, and the home ranges of pairs overlap extensively (Johnson & Dinsmore 1985, Conway 1990). Winter territories are loose or nonexistent.

SOCIAL AND SEXUAL BEHAVIOUR There is some evidence of dominance hierarchies, but few empirical data exist (Conway 1995). Birds frequently forage alongside conspecifics and Soras, and the species's tolerance of Soras appears liberal (Pospichal & Marshall 1954). Aggregations of Virginia Rails observed during migration probably result from decreasing habitat and food availability rather than being social gatherings (Conway 1995).

Detailed descriptions of behaviour are given by Kaufmann (1983, 1988, 1989). As pair bonds are formed, pairs engage in vigorous territorial defence, which is probably performed mainly by the male and ceases some time before nesting. Birds adopt an upright stance in hostile encounters, ruffling the neck feathers, or adopt a horizontal stance and stretch the head forward towards the opponent; the latter stance may result in the opponent either facing away to inhibit attack, or fleeing, and a chase may ensue if the opponent does not flee. The wings may be drooped and fanned during chasing, pecking and fighting, or flicked during chasing. A more dramatic spread-wing display, with the remiges pointing upward and the upperwing-coverts turned forward, was performed by birds after chasing an intruder from the territory, in response to a human intruder approaching the nest and by a female to a male which attacked her mate (Kaufmann 1983). Fighting involves jumping into the air and pecking and clawing the opponent's breast. Captive males attack an opponent's back, raking it with the claws, striking with the edges of the wings, and pecking the head, forcing the subordinate under water. When fleeing, birds may jump or turn abruptly and stand motionless to avoid detection. Parents protect young aggressively and approach intruders closely, giving rasping or grunting calls, the wings separated and drooped, the plumage roused and the head and neck bowed and outstretched. Both sexes defend the nest and young, but the female is usually the more aggressive (Conway 1995). Incubating birds may leap at or peck intruders, or strike severely at them. At the nest, adults perform a distraction display to a human intruder, giving alarm calls, spreading and drooping the wings, lowering the head, raising the tail and either running in tight circles or moving forward unevenly, dodging between clumps of vegetation (Wiens 1966).

The courtship period is brief and is identified by the short duration of the male's *tick-it* call (see Voice). Either male or female may initiate the formation of the pair bond; birds stand quietly side by side for periods of up to 30 min during 1-2 weeks. In the courtship display, the male runs around the female with wings raised above his body and flitting his tail, bowing in front of her at each pass (Audubon 1842). Before a precopulatory chase, the male walks stiffly towards the female with his bill pointed upwards, giving low growls; the head and neck move backwards and forwards, the throat is puffed out and the undertail-coverts are fanned. If not receptive, the female flees, otherwise she stand still, lowers her tail and stretches her head and neck forward. The male then mounts from behind and the female lowers her head and raises her tail, and the male treads her briefly before copulating for a few seconds, arching his wings. After copulation, the male makes a head-flick and a body shake, the female makes a wing shuffle or a body shake, and both sexes may peck at the substrate several times. Copulation has been observed up to 20 days before the first egg is laid. Allopreening between pair members, and preening of young by adults, are common during the breeding season, while allopreening is also common during the winter, when it is not restricted to pair members; bill nibbling is also common in paired birds (Kaufmann 1988).

BREEDING AND SURVIVAL Season North America, Apr-Jul, occasionally Mar, laying mid-Mar to mid-Jul, hatching 1 Apr-3 Aug; Mexico, Apr-May, Aug (Dickerman 1966, Binford 1972); Ecuador, chick/juvenile Aug. **Nest** Almost always in robust emergent vegetation (e.g. *Typha*, *Scirpus*) of freshwater marsh or in rank vegetation near fresh water; a platform, loosely woven cup or substantial cup of coarse grass, reeds and *Typha*, well concealed in sedge tussocks, at base of *Typha* clumps or on piles of broken reeds or driftwood; often over, touching, or slightly submerged in water up to 70cm deep (usually <30cm); loose canopy of growing vegetation often constructed. Placed up to 60cm above substrate or water, sometimes with entrance runway, or built up from substrate to a height of 20cm; external diameter 14-20cm, height 7-8cm; internal diameter 10-13cm; depth of cup 1-6cm. Female probably selects nest site but both sexes build; construction usually begins at start of egg-laying or shortly before; takes 3-7 days but material may be added during laying and incubation, especially if water level rises. Parents (or male?) may build 1-5 brood nests near breeding nest. **Eggs** 4-13 (8.5, n = 115 North American clutches); oval to short oval; very slightly glossy; creamy-white to buff, sparsely and irregularly spotted brown, lilac or grey; often more spots at large end; paler, less heavily marked, and less glossy than Sora eggs; size (nominate) 29.4-34.3 x 22.45-25.5 (32.05 x 23.75, n = 175), weight 7.15-11.3 (9.19, n = 74), calculated weight 10.5; size (*aequatorialis*) 27-31 x 21-23 (29.4 x 22.1, n = 6), calculated weight 7.8. Eggs laid daily. **Incubation** By both sexes, with changeovers every 1-2 h; begins up to 5 (usually 1-2) days before last egg laid; period 18-20 (normally 19) days; all eggs usually hatch within 48 h; parents occasionally carry eggs in bill, moving clutch to another nest; nest desertion during incubation not uncommon. **Chicks** Precocial; preen within 1-4 h of hatching; use wing-claws to grasp vegetation and aid initial movements; are fed within 1-2 h; can swim at 2-4 h; run and drink at 1 day; bathe at 2-7 days; leave nest at 3-4 days; parents seen to carry young chicks in bill and one picked up chick that fell into footprint and could not escape (Skutch 1976). Chicks cared for by both parents; can feed themselves at 3-7 days, but fed until 2-3 weeks old (>1 month in captivity); beg by peeping loudly and pecking at parent's bill; probe at 14-16 days; exhibit aggression at 16-22 days; wash food at 16-28 days; brooded until independent, at 3-4 weeks, after which adults shift home range out of territory. Feathers appear on breast at 2-2.5 weeks; fully feathered at 4 weeks, when first flights made; full juvenile plumage, and adult body mass, attained at 6 weeks; postjuvenile moult begins at 12-14 weeks.

Brood parasitism by Brown-headed Cowbird *Molothrus ater* recorded once, and rails reportedly will reject cowbird eggs; single cases of Virginia Rail laying in Sora nest, and vice versa, also recorded. Overall nesting success (North America) 53%; chick mortality probably high, most broods being small (2-5) (Conway 1995). Nest predators include snakes, muskrat *Ondatra zibethica*, weasels *Mustela* spp., raccoon *Procyon lotor*, hawks, blackbirds and wrens; young chicks taken by fish (*Esox* and *Micropterus* spp.), Sandhill Cranes *Grus canadensis* and frogs; many nests lost to flooding in some areas (Conway 1995). Both sexes can breed in first year. In USA may produce 2 broods per season, especially in S. Birds recorded returning to same breeding location for up to 4 successive years. **Survival** Annual survival rate (36 radio-tagged birds) 0.526 ± 0.195 (SE); annual survival probability (88 ringed birds; capture-recapture data) 0.532 ± 0.128 (SE) (Conway *et al.* 1994).

59 BOGOTA RAIL
Rallus semiplumbeus   Plate 18

Rallus semiplumbeus P. L. Sclater, 1856, Bogotá, Colombia.

Sometimes considered conspecific with *R. limicola*; forms superspecies with *R. limicola* and *R. antarcticus*. Race *peruvianus*, known only from type specimen and sometimes regarded as race of *R. limicola*, was originally described as a species and may merit specific status. Two subspecies normally recognised.

Synonyms: *Aramus/Limnopardalis semiplumbeus*.

Alternative name: Peruvian Rail (*peruvianus*).

IDENTIFICATION Length 25-30cm. Medium-sized; upperparts, from top of head to tail, streaked dark blackish-brown and olivaceous-brown; upperwing-coverts uniformly chestnut (possibly duller in race *peruvianus*); remiges blackish, secondaries and tertials fringed olivaceous-brown. Sides of head, sides of neck, foreneck, breast and upper belly plumbeous-grey; lores and ear-coverts darker; narrow supraloral stripe (if present) whitish (more prominent in *peruvianus*); chin and throat whitish to grey. Flanks and lower belly dull blackish, narrowly and irregularly barred white to brownish-white, brownest at rear; undertail-coverts largely white, with some black and buff barring. Iris dark brown; bill long, slightly downcurved and yellowish-red, culmen and tip blackish; legs and feet reddish-brown. Sexes alike; female smaller. Immature not described. Juvenile like adult but with slightly sooty feather tips on breast, whitish throat, and black bill with red base. Inhabits freshwater marshes.

Similar species Structurally similar to Virginia Rail (58) but larger, with uniform deep chestnut upperwing-coverts, entirely plumbeous-grey face, lower throat, foreneck, breast and upper belly, irregular white barring on the flanks (tinged brownish towards rear flanks), yellowish-red bill and dull red legs. Differs from much smaller Austral Rail (60) in olive-brown feather fringes to dorsal plumage and less regularly spaced and narrower white barring on flanks.

VOICE A variety of squeaks, grunts, and whistling and piping notes; the most frequently heard call (from breeding birds) was a high-pitched rattle or trilling whistle which started low, grew in intensity and frequency, then wavered and trailed off rapidly; this call was similar to the 'sharming' announcement call of the Water Rail (61) and, as in that species, the calling bird opened its bill slightly and vibrated its throat (Varty *et al.* 1986). Another call is described as a clear, piercing, high-pitched *peeep*, like the signal whistle of a ground squirrel *Spermophila* (Hilty & Brown 1986). Birds also give a brief, rapid, clattering *titititirr* when disturbed (Fjeldså & Krabbe 1990). Breeding birds are highly vocal, calling throughout the day from just after dawn to dusk; calling does not occur at night (Varty *et al.* 1986). Chicks were generally silent but made occasional soft *cheep* and *peep* calls; parents summoned young with a soft call (Varty *et al.* 1986).

DESCRIPTION *R. s. semiplumbeus*
Adult Upperparts, from forehead to tail, dark blackish-brown, feathers with olivaceous-brown fringes; crown darker, with less pale streaking; sides of hind-neck and sides of upper breast almost plain, slightly brighter brown; rectrices with darker brown fringes. Upperwing-coverts

uniform burnt-sienna to chestnut; marginal coverts white; remiges fuscous to fuscous-black, secondaries and tertials fringed olivaceous-brown; axillaries and underwing-coverts blackish-brown with narrow white terminal bars. Sides of head, sides of neck, foreneck, breast and upper belly almost uniform plumbeous-grey; lores and ear-coverts darker; narrow supraloral stripe whitish (apparently not present in some birds, according to Fjeldså 1990); chin and throat whitish (but sometimes grey, uniform with face and foreneck, according to Fjeldså 1990). Flanks and lower belly dull blackish-brown, narrowly and irregularly barred white to brownish-white, brownest at rear; lateral undertail-coverts white, with subterminal black spots; central feathers blackish-brown, barred buff or white. Iris dark brown; bill long, slightly downcurved, yellowish-red, culmen and tip blackish; legs and feet reddish-brown. Sexes alike but female smaller.

Immature Undescribed. Extent of white on chin and throat may vary with age (see Juvenile).

Juvenile Like adult but with whitish throat and slightly sooty feather tips on breast, giving indistinct narrowly scalloped or barred effect; bill blackish, base of lower mandible red (Fjeldså & Krabbe 1990).

Downy young Black; upper mandible white or ivory, with central black bar, lower mandible black; small egg-tooth present; legs and feet dark blue-grey. At 2 weeks, chicks appeared slate-grey on body, black on head with paler areas below and behind eye; bills had white base, black central area and small white tip (Varty *et al.* 1986).

MEASUREMENTS Wing of 10 males 105-117 (112.7), of 10 females 98-110 (103.9); tail of 3 males 44, 47, 49, of 2 females 45, 48; exposed culmen of 10 males 43-45 (44.1), of 10 females 37-41 (38.1), culmen to base of 1 male 47.5; tarsus of 10 males 38-44 (41.4), of 10 females 35-44 (38.2) (Blake 1977, Ripley 1977, BMNH).

GEOGRAPHICAL VARIATION Race *peruvianus* is known only from one specimen, originally in the Raimondi collection, University of Lima, Peru, but now apparently lost (Fjeldså 1990). The original description is thus the only information available, and the bird was illustrated from this by Fjeldså & Krabbe (1990). Originally described as a species, this bird was listed as a subspecies of Virginia Rail by Peters (1934) and Blake (1977), but Meyer de Schauensee (1966) suggested it may be a race of Bogotá Rail, which it appears to resemble more closely. The lack of a specimen makes it unwise to maintain the species rank proposed in the original description and it is best regarded as a race of Bogotá Rail (Fjeldså 1990).

R. s. semiplumbeus P. L. Sclater, 1856 – Bogotá, Colombia. Cundinamarca and Boyacá departments, E Andes of Colombia. See Description.

R. s. peruvianus (Taczanowski, 1877) – locality unknown, possibly Peruvian highlands. From details in original description and in Fjeldså & Krabbe (1990), resembles nominate but supraloral stripe more prominent; upperwing-coverts possibly duller chestnut, with either dark feather centres (see Plate 18) or vague whitish bars as in Austral Rail; throat more extensively pale (an immature character?); undertail-coverts possibly lacked cinnamon bars, being white laterally and having black and white barring centrally. Wing 112; bill 40; tarsus 37 (Sharpe 1894).

There is also an unconfirmed record from Ecuador (Laguna Kingora, Sigsig), subspecies unknown (Blake 1977), which is apparently a misidentified juvenile Virginia Rail (Ridgely & Greenfield in prep.).

MOULT No information available.

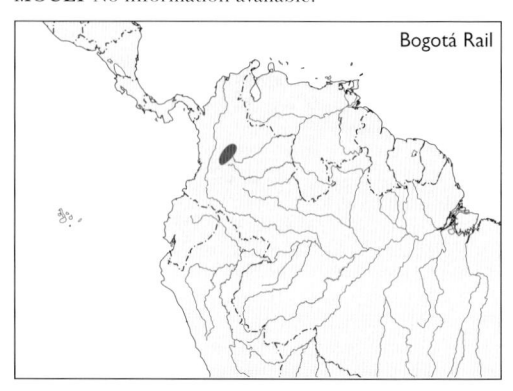

DISTRIBUTION AND STATUS ENDANGERED. This species is restricted to relatively few lake or marsh areas of the Bogotá-Ubaté savannas and some surrounding higher-altitude areas in Colombia. In Boyacá department it occurs at Laguna de Tota, while in Cundinamarca department it occurs more widely, being recorded from at least 21 sites (Collar *et al.* 1992): Laguna de Fúquene; Laguna de Cucunubá; Subachoque; Torca; Cota; Laguna de Pedropalo; "Suba Marshes" and Laguna de Juan Amarillo; El Prado; a marsh near Funza; Laguna de la Florida (whence come many recent records); a roadside pool near La Florida; La Holanda; Usaquén (a suburb of Bogotá); Laguna de la Herrera; a marsh 15km ENE of Bogotá; Techo; Embalse del Muña; Páramo de Chingaza and the adjacent Carpanta Biological Reserve; and Laguna Chisacá. It is uncommon to locally fairly common and, despite habitat destruction, it may occur in numerous localities where suitable habitat (even in small patches) remains. Healthy populations remain in a few areas, notably L de Tota with c. 400 birds in 1991, L de la Herrera with c. 50 territories, L de la Florida with c. 55 pairs, and probably L de Cucunubá; birds at L de Fúquene are possibly subjected to excessive hunting pressure (Collar *et al.* 1994, Wege & Long 1995). Rose & Scott's (1994) population estimate of 600-700 birds should probably be regarded as a minimum figure.

Only a few lakes with high plant productivity exist in the Colombian Andes, but until recent disturbance by man the Ubaté and Bogotá plateaux held enormous marshes and swamps (Fjeldså & Krabbe 1990). Throughout this region this species has suffered enormous habitat loss caused by drainage, and few suitable characteristically vegetated marshes remain because of pollution and siltation. All major savanna wetland localities are now seriously threatened with final destruction, mainly from drainage but also from agricultural encroachment, erosion, diking, eutrophication from untreated sewage effluent and agrochemicals, pollution from insecticides, tourism, hunting, burning, trampling by cattle, harvesting of reeds, fluctuating water levels, and increased water demand, especially for irrigation (Collar *et al.* 1992). For example, L de Tota is in urgent need of environmental management and its fringing wetland habitat is reduced to <175 ha; many of its reedbeds are decaying and provide shelter but little food, and encouraging new growth of reeds and maintaining a diversity of emergent vegetation types are urgent requirements to improve the habitat for rails and other reedbed birds (Varty *et al.* 1986). The marsh at L de la Herrera is threatened by hunting, cattle trampling and irrigation schemes; L de la Florida is threatened by development projects and illegal settlement (Wege & Long 1995). However, the species can utilise areas where submerged vegetation has been replaced by floating plants (see Habitat) and it is apparently able to survive in small patches of habitat.

Some páramo populations of this rail occur in protected areas such as Chingaza National Park, the Carpanta Biological Reserve and possibly Sumapaz National Park, but savanna wetlands enjoy no legal protection although they are important in providing water for domestic and industrial use (Collar *et al.* 1992). The major conservation priority must be to secure the long-term future of the larger remaining wetlands, while further searches for the species should perhaps concentrate on páramo wetlands, which have been less disturbed than savanna wetlands and where this rail may occur widely and in significant numbers. The savanna wetland populations, especially those of the smaller marshes away from the main lake areas, should also be assessed (Collar *et al.* 1992).

MOVEMENTS None recorded.

HABITAT Information is from Collar *et al.* (1992). Restricted to the savanna and páramo marshes of the temperate zone of the Colombian E Andes, characteristically occurring in areas of dense tall fringing reeds (e.g. *Scirpus californicus*, *Typha latifolia*, *Juncus* and some *Cortadera*) and some *Alnus acuminata* swamp, with vegetation-rich shallows containing *Elodea*, *Myriophyllum* and *Potamogeton*. It is found in rushy fields, reedbeds (often with open, regenerating burnt areas), reed-filled ditches and fens fringed with dwarf bamboo *Swallenochloa*; it is said to be closely associated with *Typha* and *Scirpus* reedbeds (Varty *et al.* 1986). It may also tolerate modified habitats where floating mats of *Azolla*, *Ludwigia* and *Limnobium* replace submergent vegetation as a result of pollution or siltation; however, it avoids *Eichhornia crassies*. It has also been seen at a small roadside pool. Its altitudinal range is 2,500-4,000m, occasionally down to 2,100m.

FOOD AND FEEDING Information is from Varty *et al.* (1986) and Collar *et al.* (1992). Food comprises primarily aquatic invertebrates and insect larvae, but also worms, molluscs, dead fish, frogs, tadpoles and plant material. Birds forage in areas with a thin carpet of floating plants such as *Azolla* and *Limnobium*, at the edges of reedbeds, on marshy shorelines, in flooded grass and marsh, in areas of regenerating burned *Scirpus*, on patches of dead waterlogged *Typha*, and at the water's edge. They take insects trapped on the water (including a hesperidid butterfly), and they probe mud and floating vegetation, usually not completely immersing the bill but sometimes immersing the whole head, and they investigate floating dead reeds. They either make quick darting runs, followed by intense probing at one spot, or walk quickly among vegetation probing to both sides as they do so. They repeatedly travel along definite paths to foraging areas, and they generally forage alone.

HABITS Most information is from Varty *et al.* (1986). A skulking species of dense cover which is active from dawn to dusk and visits more open areas, including the edges of reedbeds, in the early morning. It is not particularly shy, and moves in a typical rail fashion, with the tail often flicked (particularly when the bird is alarmed); it sometimes darts through vegetation with the neck extended and low. It is not infrequently flushed from immediately in front of an observer's feet, when it flies only a short distance. It will also jump, and it swims easily for short distances.

SOCIAL ORGANISATION Monogamous and territorial. Two breeding territories covered between 0.2 and 0.45ha (Varty *et al.* 1986).

SOCIAL AND SEXUAL BEHAVIOUR All information is taken from Varty *et al.* (1986). Breeding adults drove off other species which approached the chicks, including Spot-flanked Gallinules (130), American Coots (136), Rufous-collared Sparrows *Zonotrichia capensis* and Great Thrushes *Turdus fuscater*. The adults made charging attacks with the body low, the neck outstretched and the bill pointing forwards, and they stabbed at the other bird, usually around the head. Once a Spot-flanked Gallinule had to dive to avoid serious injury from a particularly aggressive rail. Rails did not attack Least Bitterns *Ixobrychus exilis* and both species were often seen in close proximity.

BREEDING AND SURVIVAL Season L de Tota, chicks and dependent juveniles Jul-Aug; new nest Sep; these and other observations suggest breeding Jul-Sep; also empty nest Oct. The following information is from Varty *et al.* (1986). **Nest** Two nesting territories were in a combination of vegetation types, but the 3 nests found were in *Typha*. No details of nest or eggs available; brood of 4 chicks (2 days old) and 2 chicks (13-14 days old) seen. **Chicks** Precocial and nidifugous. Fed and cared for initially by both parents, after 2 weeks by 1 parent (sex unknown); chicks begin to peck at objects at 2 days but still dependent on adults for food 2 weeks later. Parent presents food bill-to-bill; breaks up large food items such as fish before feeding them to chicks. Chicks swim well and run quickly, but when young tend not to follow foraging adults. Apparent nest-building in Aug, by bird with young, suggests possibility of birds breeding twice per season.

60 AUSTRAL RAIL
Rallus antarcticus   Plate 18

Rallus antarcticus King, 1828, Straits of Magellan.

Sometimes regarded as conspecific with *R. limicola* or *R. semiplumbeus*, with both of which it forms a superspecies. Monotypic.

Synonyms: *Rallus rufipennis/f. uliginosus*; *Ortygometra antarctica*; *Aramus antarcticus*.

IDENTIFICATION Length 20cm. The smallest *Rallus* species. Entire upperparts, and sides of upper breast, streaked blackish and pale sandy-buff; streaks small from crown to hind-neck and broadest on scapulars; upperwing-coverts russet, variably barred whitish (bars with faint dusky margins). Lores blackish-brown; ear-coverts paler; supercilium plumbeous-grey, whitish in front of eye; chin and throat white to grey; sides of head, foreneck, breast and upper belly plumbeous-grey; in fresh plumage often has strong olive-brown wash from breast downwards. Flanks and belly with broad blackish and narrow white bars; undertail-coverts mostly white, marked cinnamon and

291

black. Iris reddish-brown; bill red, brighter on lower mandible; legs and feet dark purple (breeding?) or pink-red. Iris can also be brown; bill dusky horn, lower mandible pinkish with dusky tip. Sexes alike. Immature undescribed; juvenile like adult but throat white and breast feathers tipped sooty-grey. Inhabits wet grass fields and marshy areas.
Similar species Similar to Virginia Rail (58) and Bogotá Rail (59) but smaller, with sandy-buff (not brown) fringes to dorsal feathers and regular black and white barring on flanks. Also differs from Virginia Rail in having plumbeous-grey throat, breast and upper belly.

VOICE The call, recorded in Jan 1998, is a series of 5-10 high-pitched, strident, slightly metallic *pic* or *pi-ric* notes, quite similar to the *tik-it* vocalisation of the Virginia Rail but more rhythmical and higher pitched; birds did not respond to playback of Virginia Rail calls (J. Mazar Barnett *in litt.*).

DESCRIPTION
Adult Entire upperparts, including top of head and tail, dark blackish-brown; feathers fringed pale sandy-buff to tawny-olive, giving well-streaked appearance; streaks small and narrow from crown to hind-neck, broadest on scapulars; narrowly streaked area extends down sides of breast and forwards in a wedge at base of sides of neck, in which regions pale streaks are more olive-tinged or darker, duller brown, and surrounding feathers are extensively brownish. Upperwing-coverts russet to hazel, variably (none to many, vaguely to prominently) barred whitish with blackish-brown lower margins; alula, primaries and secondaries dusky brown with sandy-buff edges and whitish tips; tertials blackish-brown, broadly fringed sandy-buff; axillaries and underwing-coverts barred blackish-brown and white. Lores blackish-brown; ear-coverts somewhat paler; supercilium plumbeous-grey, whitish on supraloral area; rest of sides of head plumbeous-grey; chin and throat white, grading into darkish plumbeous-grey foreneck, breast and upper belly, or concolorous with grey foreneck; in fresh plumage, often has strong olive-brown wash from breast downwards. Flanks and belly with regular barring, blackish-brown bars broad and white bars narrow; thighs and centre of belly plain grey, tinged olive; undertail-coverts mostly white, tipped cinnamon and subterminally barred black. Sexes alike. One female (Buenos Aires, Jun) had iris reddish-brown; upper mandible dark red, lower mandible bright red; legs and feet dark purple. Another female (Tierra del Fuego, Feb) had iris brown; bill dusky horn, lower mandible pinkish-drab with dusky tip; legs and feet pale. A Jan bird had pink-red legs (J. Mazar Barnett *in litt.*). Difference in bare part colours possibly related to breeding condition.
Immature Undescribed; must be very similar to adult.
Juvenile Like adult but with white throat and sooty-grey feather tips on breast (Fjeldså 1990).
Downy young A possible chick of this species has whitish bill with distal half of upper mandible black, and black streak on the middle of the lower mandible (Fjeldså 1990).

MEASUREMENTS Wing of 3 males 93, 96, 98, of 6 females 92-96 (93.5, SD 1.3), of unsexed 85-106; tail of 1 male 47, of 4 females 39-45 (42, SD 3.4), of unsexed 38-47; culmen of 1 male 34, exposed culmen of 2 males 32, of unsexed 27-34, culmen to base of 4 females 32-36 (33.6, SD 1.7), of 6 unsexed 31-35 (32.5, SD 1.7); tarsus of 3 males 27, 30, 30, of 6 females 27-30 (28.5, SD 1.4), of unsexed 24-34 (Blake 1977, Navas 1991, BMNH). Fjeldså (1990) gives measurements of 16 unsexed birds as follows: wing (chord) 90-104 (93.6, SD 4.6); culmen 29-32 (30.8, SD 1.8); tarsus 26.8-32 (28.9, SD 2.7). Weight of 1 female 60 (Humphrey *et al.* 1970).

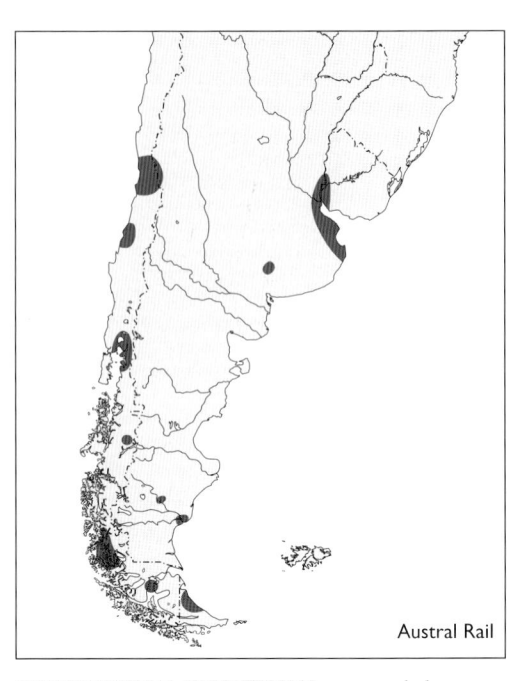

Austral Rail

GEOGRAPHICAL VARIATION None recorded.

MOULT No information available.

DISTRIBUTION AND STATUS C Chile and C Argentina patchily S to Tierra del Fuego. This bird's status is **CRITICALLY ENDANGERED**. Most information is from Collar *et al.* (1992). Until 1998 it was known from 19 localities S of 33°S, only 2 of which are listed as key areas for threatened birds by Wege & Long (1995), with additional records from unspecified sites. In Argentina it was recorded as follows: there were possible 19th century sightings near Concepción del Uruguay, Entre Ríos Province; in Buenos Aires Province it is recorded from Partido de Lomas de Zamora, from Cabo San Antonio in Jul 1899, Carhué in Apr 1881, and possibly from Barracas al Sud (2 nests and eggs in Nov 1900, records doubtful); in Río Negro from El Bolsón in Oct 1959, and in the 1980s "with some regularity" at the same site, where there are still extensive marshes (Wege & Long 1995); in Chubut from Valle de Lago Blanco in Nov 1901; in Santa Cruz from lower Río Chico in Mar 1897; in Tierra del Fuego from Río Grande Norte, and from Viamonte in Feb 1931. In Chile: a record of a nest from "C Chile" (Oates 1901); in Valparaiso from "Valparaiso"; in Santiago from Fundo, San Ignacio, where a nest with 6 eggs was found in Nov 1940 but identity is uncertain, and from Viluco; in Colchagua from Cauquenes; in Llanquihué from Puntiagudo, Pella and the Río Cayutué (all at Lago Todos los Santos); in Magellanes from Bahía Tom, NE Isla Madre de Dios, in Apr 1879, from Puerto Mayne, Isla Evans, in Mar 1880, and from Punta Arenas in Jan 1876 and Feb 1883. For further details, see Hellmayr & Conover (1942) and Collar *et al.* (1992). The species was rather common at Carhué, Buenos Aires, in Apr 1881, but it seems to have become rare everywhere by the turn of the century and to have virtually disappeared since then: only two specimens have been collected since 1901 and the only records since 1959, when the last was collected, have been the El Bolsón sightings in the 1980s, despite its occurring in a region where ornithological activity has not been insignificant. However, in Jan 1998 4 calling birds were found in an extensive reedmarsh on the Río Chico at Estancia La Angostura, near Gobernador Gregores, Santa Cruz (J. Mazar Barnett *in litt.*).

The reasons for its decline are unclear although the draining, grazing and cutting for hay of its wet grassland habitats, and the overgrazing and disappearance of practically all the tall grass habitat in Patagonia, have been suggested. Lago Todos los Santos, in SC Chile, is threatened by proposed development, despite its being within the contiguous Puyehue and Vicente Perez National Parks (Scott & Carbonel 1986); searches there in the late 1980s produced no records.

Regarding the conservation of this almost unknown species, the immediate priority is to locate any remaining populations by surveying known sites of occurrence and other localities where potential habitat still exists. In SC Argentina, it is reportedly seen occasionally near El Bolsón, Río Negro, and it should be looked for in Nahuel Haupi and Los Alerces National Parks (Wege & Long 1995). Now that its calls have been tape-recorded, searches should be less difficult and it may prove to be more widespread and numerous than is currently known (J. Mazar Barnett *in litt.*)

MOVEMENTS It is regarded as possibly a summer breeding visitor to Tierra del Fuego (Humphrey *et al.* 1970). Northward post-breeding migration may occur, at least in the southernmost populations, in Mar-Apr (Navas 1962).

HABITAT Wet fields, rushy parts of meadows, rushy lake shores and reedbeds, mostly adjacent to coasts. In Chile some birds spent the winter months in the garden of a house before returning to adjacent extensive dense reedbeds (Johnson 1965b). The 1899 Cabo San Antonio birds were in flooded *Spartina densiflora*, while the 1998 Río Chico birds were in extensive marsh, flooded c. 50cm deep, with areas of *Myriophyllum* and lush grass, and large patches of dense, tall, partly dead *Scirpus californicus*, at 570m a.s.l. (J. Mazar Barnett *in litt.*).

FOOD AND FEEDING The diet is probably similar to that of the Virginia Rail. One stomach contained caddisflies *Limnophilus meridionalis* (Trichoptera). Birds wade to search for insect larvae (Canevari *et al.* 1991). Birds wintering in a garden fed on grubs found among decomposing leaves under raspberry bushes, and even ate leftover dog food (Johnson 1965b).

HABITS Almost nothing is known. Birds wintering in a garden appeared very tame (Johnson 1965b). The 1998 Río Chico birds kept within dense reed cover, even when approaching observers closely in response to taped playback; 1 bird swam, almost submerged, across water between patches of cover (J. Mazar Barnett *in litt.*).

SOCIAL ORGANISATION In Jan 1998 the 4 Río Chico birds called from a 300 x 200m area of marsh and reedbeds (J. Mazar Barnett *in litt.*).

SOCIAL AND SEXUAL BEHAVIOUR Not described.

BREEDING AND SURVIVAL Season Chile, possibly Nov; Argentina, possibly Nov; probably breeds throughout range (Navas 1962), but data lacking. **Nest** A possible nest of this species (San Ignacio, Santiago, Nov 1940) was placed on ground, under a thick bramble *Rubus* bordering an irrigation canal; eggs in a depression scantily lined with grass stems and rushes (Collar *et al.* 1992); 2 other possible nests (Barracas al Sud, Buenos Aires, Nov 1900) in grass tussocks 20cm above water in a lagoon (Hartert & Venturi 1909). **Eggs** 8, possibly also 4 and 6; oval; pinkish-cream to pinkish-buff, spotted red-brown and purplish; size 29.8-34.5 x 23.1-25.5 (31.8 x 23.8, n = 9), calculated weight 9.9 (Schönwetter 1961-62). Note that Oates (1901) describes 8 eggs from C Chile (Berkeley James Collection) which are too large for this species and have been re-identified as Plumbeous Rail (113); while the 4 eggs from Chile assigned by Oates to Plumbeous Rail have been re-identified as Austral Rail (M. P. Walters *in litt.*). Oates describes these 4 eggs as resembling those of Water Rail (61) and gives size: 31.8-34.8 x 23.6-24.1.

61 WATER RAIL
Rallus aquaticus Plate 19

Rallus aquaticus Linnaeus, 1758, Great Britain.

Forms a superspecies with *R. caerulescens* and *R. madagascariensis*. Four subspecies recognised.

Synonyms: *Rallus minor/indicus/germanicus/sericeus/fuscilateralis/japonicus*; *Scolopax obscura*; *Aramus aquaticus*.

Alternative names: European/Indian Water Rail.

IDENTIFICATION Length 23-28 (30?)cm; wingspan 38-45cm. A medium-sized rail. Forehead to hindneck, and entire upperparts, including upperwings and tail, olive-brown (olive-buff in race *korejewi*) streaked black; streaks broadest on scapulars and tertials. Lores blackish; sides of head and neck, chin, throat, breast and upper belly dark slate-blue (paler in *korejewi*, often tinged brown in race *hibernans*). Sides of upper breast olive-brown with some narrow black streaks; flank feathers black (sepia in *hibernans*) with white bars and pale buff tips. Undertail-coverts usually appear white, but may have variable amount of black or buff markings depending on feather wear and age of bird, and overall appearance may be barred black and white or entirely buff (see Similar Species). In flight, outer wing and secondaries blackish-brown, more uniform than inner wing; narrow white leading edge to inner wing. Iris red; bill red with blackish culmen; legs and feet flesh-brown, sometimes tinged green or red. Female smaller, with more slender bill. Race *indicus* slightly larger than nominate; has brown streak from lores below eye to ear-coverts; chin white; breast and sides of body extensively tinged brown. In many nominate birds, a variable number of greater and median upperwing-coverts have white bars, black-bordered or on black feathers; white markings may also occur on alula, primary coverts and even remiges, while some have white speckles on mantle, back or rump; race *indicus* has more boldly and extensively barred wing-coverts. Immature very like adult but duller; chin mottled grey and white; small pale supraloral streak; ear-coverts olive-brown; breast tinged olive-brown; bare parts duller than in adult. Juvenile similar on upperparts but forehead to hindneck darker, less streaked; has off-white supraloral streak; grey of head and underparts replaced by buff or white with brown to black bars and feather tips; lower breast to belly white to cream-yellow; flanks barred buff, whitish and sepia; undertail-coverts often appear entirely buff. Iris brown, tinged yellow or red; bill black, basal half of lower mandible dull red; legs and feet dark horn-brown, tinged pink, olive or yellow. Inhabits dense marsh vegetation at still or slow-moving water with muddy ground.

Similar species Differs from all American congeners except Plain-flanked Rail (57) in lacking contrasting rufous or chestnut on upperwing-coverts. African Rail (62) of sub-Saharan Africa is larger, with unstreaked, deep vinous-brown upperparts and brighter red legs and feet. Easily distinguishable on plumage characters from rather long-billed Slaty-breasted Rail (54); long red bill and (usually) predominantly white undertail-coverts separate it from sympatric *Rallina* and *Porzana* crakes. However juveniles, and birds in fresh plumage with extensive and unabraded buff tips to undertail-coverts, may appear to have entirely buff undertail-coverts like Spotted Crake (96); birds with unusually heavy black and white markings on inner undertail-coverts may appear to have entirely barred undertail-coverts, especially when buff tips have worn away, and may thus resemble Little (93) and Baillon's (94) Crakes in this character (Becker 1995).

VOICE Most information is from Cramp & Simmons (1980). A highly vocal species. The announcement call ('sharming') comprises grunting notes rising in the middle of the call to high-pitched, trilling whistles, likened to piglets squealing, and usually dying away into more grunts. This is a territorial and display call, used throughout the year and serving for advertisement, alarm, warning and location. Pair members often call antiphonally in duet, both when close together and far apart, the male giving the lower and slower notes. The courtship-song, given by both sexes, is rendered *tyick-tyick-tyick*, often ending with a trill *tyüirr* which is probably given only by the female; the male calls more loudly and may continue for hours; in W Europe this call is given in Feb-Aug. The female's version could be confused with the advertising call of the female Little Crake (91) but is higher-pitched and less musical. There is also a series of 3+ ticks followed by a thin, agonised, wheezy scream, *tic-tic-tic-wheee-ooo*. The normal flight call is a clear, sharp whistle; birds also sometimes give the announcement call or the courtship song in flight. The female gives soft crooning notes to the male, who may utter little grunts at intervals. When showing the nest site to the female, the male gives a loud, repeated creaking call reminiscent of the song of the Corncrake (69) (Andreas 1996). Both sexes give an incessant purring note when at the nest with chicks, while the female gives *dug-dug* notes to call young together; there is also a warning shriek to the young (Buxton 1948). The birds are vocal by day throughout the year but calling is generally most frequent during territory establishment and early in the breeding season, when birds may call all night; calls are usually given from cover. Newly hatched young cheep faintly, while the begging call of young chicks is rendered *tyk-tyk-trik*; contented chicks are said to give a *yyiy-yyiy* call.

DESCRIPTION *R. a. aquaticus*

Adult Forehead to hindneck, and entire upperparts and tail, including scapulars, secondary coverts and tertials, olive-brown with black feather centres, giving streaked appearance; olive-brown fringes broader on secondary coverts and tertials; rump and uppertail more uniformly olive-brown. Alula, primary coverts, primaries and secondaries almost uniform blackish-brown, with white markings on inner edge of alula and olive-brown tinge to outer webs of secondaries; marginal coverts white with black bases, giving narrow white leading edge to inner wing and bend of wing. Axillaries and underwing-coverts barred black and white; underside of remiges ashy-black. Lores dark grey to blackish; supraloral area, broad stripe over eye, sides of head and neck, chin, throat, breast, upper belly and sides of belly dark slate-blue, rarely with small olive-brown spot on ear-coverts. Sides of upper breast olive-brown with some narrow black feather centres showing as streaks; feathers of flanks elongated, black with white bars and pale buff tips. Centre of lower belly and thighs slate, feathers broadly tipped pale buff or cream-yellow; shorter undertail-coverts black, broadly tipped buff and subterminally marked white; longer undertail-coverts white, basally grey-black, and tipped warm buff, but feathers may be variably barred black and white, with buff tips (Becker 1995); variable pattern has been used for individual recognition (King 1980). Iris bright red or orange-red; bill bright red, culmen and tip black with red tinge; legs and feet flesh-brown, joints darker brown, sometimes with green, orange or red tinge. Sexes alike; female smaller, with thinner bill and duller bare parts. In c. 60% of nominate adults, a variable number of greater and median upperwing-coverts have white bars, black-bordered or on black feathers, but c. 20% have white faint or absent; white markings may also occur on alula, primary coverts and even remiges, while some have white speckles on mantle, back or rump. In fresh plumage, feathers of underparts often narrowly and contrastingly edged white, front part of chin appearing white. In worn plumage, olive-brown of upperparts paler; dark feather centres more obvious because of wear of pale fringes; buff feather-tips of flanks, lower belly and undertail partially disappear; remiges tinged brown. In Aug-Nov, bare parts often duller: iris orange-brown or hazel; bill pale red, culmen and tip dark brown.

Immature Very like adult but duller, with mottled grey and white chin, small pale supraloral streak, olive-brown patch on ear-coverts and olive-brown tinge on breast. Fresh feathers on flanks often have broader and less contrastingly white tips than adult, often subterminally bordered brown; underparts usually less blue; fresh grey feathers have narrow bright brown fringe, sometimes visible in the field. Bare parts like adult, slightly duller in some. First-breeding birds retain variable amount of immature non-breeding or juvenile feathers, some remaining mainly immature non-breeding; wings as juvenile.

Juvenile Upperparts like adult but olive-brown fringes narrower, sometimes duller; appears more blackish from forehead to hindneck. Lores dark brown; ear-coverts slightly paler; narrow off-white streak from base of upper mandible to above eye; chin and throat off-white to pale buff. Grey of head and underparts replaced by buff or off-white, with brown to black bars and feather tips; flanks variably barred whitish to buff, and dark brown to black (Becker 1995); lower breast, belly and vent white, pale grey, buff or cream-yellow, variably mottled or barred blackish on breast and belly; undertail-coverts often appear entirely buff. Wings like adult, but white bars on outer upperwing-coverts absent or much reduced in c. 60% of birds, but as many as in many adults in c. 20%. Iris becomes olive-green and then yellow-brown to orange-red; bill blackish to horn, cutting edges of upper and basal half of lower mandible dull red; legs and feet dark horn-brown with pink, olive-green or yellow-brown tinge (brighter in males).

Downy young Down black, with metallic blue-green gloss on upperparts; has visible patch of red skin on hindcrown; iris grey to brown-black, gradually changing to olive-green; bill ivory-white with black base and tip, blackening later; egg-tooth white; legs and feet grey-brown to blackish with pale red tinge. Wing stump and clawed alula whitish.

MEASUREMENTS Wing of 126 males 119-132 (125, SD 2.8), of 124 females 110-121 (116, SD 2.4); tail of 60 males 47-59 (53, SD 2.6), of 56 females 45-55 (48.4, SD 2.2); exposed culmen of 55 males 39-45 (41.4, SD 1.75), of 58 females 34-40 (37, SD 1.3); tarsus of 63 males 39-46 (42.6, SD 1.5), of 58 females 36-41 (38.5, SD 1.4). Weight: monthly measurements, Sep-Apr, of 170 males 88-190 (114-144), of 145 females 74-138 (98-107); heaviest Dec-Feb (Cramp & Simmons 1980). Hatching weight 8-16 (Glutz von Blotzheim *et al.* 1973).

GEOGRAPHICAL VARIATION Races are separated on colour and size, all except E Asiatic *indicus* differing only slightly in these characters. The race *korejewi* includes *deserticolor*, *tsaidamensis* and *arjanicus*; *indicus* includes *japonicus* (Sharpe 1894, Ripley 1977).

R. a. hibernans Salomonsen, 1931 – Iceland; probably also Faeroes in winter. Upperparts slightly warmer brown than nominate, with more restricted black centres to feathers; often has brown tinge to underparts and sides of head; grey of underparts less slate-blue; flanks barred dark sepia, not black; bill slightly shorter; feathering denser, especially in winter. Wing of 7 males 119-129 (123, SD 3.6), of 6 females 113-120 (117, SD 3.1); tail of 1 male 56, of 1 female 58; exposed culmen of 7 males 38-42 (40.3, SD 1.4), of 6 females 33-39 (35.9, SD 2.0); tarsus of 1 male 41, of 1 female 39.

R. a. aquaticus Linnaeus, 1758 – Europe from British Isles, S Norway, S Sweden and S Finland and Russia (regions of Leningrad, N Kalinin, S Yaroslavl, Gorki, S Kirov, and Ufa); SW to Portugal and S to Mediterranean (including Balearics, Corsica, Sardinia and Sicily), N Morocco, N Algeria (mostly within 100km of coast), Tunisia (S to Sfax), Libya (Cyrenaica), lower Egypt (Nile delta and Wadi El Natrun), Saudi Arabian Gulf wetlands (presumably this race), Turkey, Black Sea, Caucasus, Azerbaijan, N Caspian Sea area and W Kazakhstan (Ilek and Khobda Rivers); and E in narrow belt (range not well known) across W Siberia N to Omsk and Baraba Steppe, and N Kazakhstan to Semipalatinsk and Zaisan Nor. Winters S to N Africa, where extends far into Sahara in Algeria (S to Daïet Tiour, El Goléa and Aïn Amenas) and Libya (Brak, Sebha), in Egypt along Nile S to near Luxor (also 3 records from W desert oases), and E to Azerbaijan and N Iran (S Caspian Sea area). See Description.

R. a. korejewi Zarudny, 1905 – S and E Iran (Luristan to Kirman, and E Khorasan from Tabas N along Hari Rud and Tedjen Rivers to Qarri Band), the S Aral Sea region, NE Aral Sea E along Syr Valley to (former) Turkestan, SE Turkmenistan (Kelif) to Tajikistan (Kirovabad), NW Tajikistan N to L Balkhash and the Sasyk and Ala Ku; E to N and NE Tibet, China in Xinjiang (=Sinkiang) (from W at Kashi, Tien Shan Mt, to E at Lop Lake), NW Gansu (= Kansu), E Qinghai (=Tsinghai) to Qaidam Basin, and SW Sichuan (= Szechwan) to Xichang; Cheng (1987) regards it as present all year in these regions of China; also S to NW Himalayas (Kashmir and Ladakh). Winters in Iraq, Iran (including S Caspian Sea area), Afghanistan and Pakistan (Punjab and Baluchistan, occasionally Sind); Vaurie (1965) regards it as wintering E to C China (Qinghai and C Gansu) and possibly to coastal E China; wanders to NW India S to Sehore in Madhya Pradesh and is occasional in E Arabia, including Yemen. Note that Perennou *et al.* (1994) record this species (possibly this race?) from scattered localities in S and W India (not marked on the distribution map); as it is not otherwise known from these regions, the authors acknowledge that these observations need to be verified. Slightly larger than nominate; upperparts have paler, olive-buff, fringes and more restricted black on feather centres; underparts slightly paler slate-blue. Measurements of 10 males, 10 females: wing of male 125-129 (126.6, SD 1.2), of female 108-121 (115.5, SD 3.9); tail of male 46-57 (52.8, SD 3.0), of female 41-52 (46.8, SD 3.9); culmen to base of male 44-51 (47.4, SD 1.9), of female 40-47 (43, SD 2.2); tarsus of male 38-43 (41.2, SD 1.5), of female 34-42 (37.4, SD 2.3).

R. a. indicus Blyth, 1849 - N Mongolia and E Siberia (upper R Yenisey and middle R Lena) to Sea of Okhotsk, Amurland, Sakhalin, Ussuriland and S Kuril Is; NE China in Heilongjiang (= Heilunkiang), Kirin, Liaoning and N Hebei (= Hopeh); also Korea and N Japan (Hokkaido and N Honshu). Winters Bangladesh (?), Assam, Burma (Arakan, Inle L and Myitkyina District), Thailand, N Vietnam and N Laos, E to SE China in Guangdong, Fujian and Guizkou, and Hainan, Taiwan and S Japan (Ryukyu Is N to Honshu); also occasionally in N Borneo. Perennou *et al.* (1994) also record it from the Philippines (Luzon) without comment, presumably in error. Slightly larger than nominate; has brown streak from lores below eye to ear-coverts; chin whiter; breast and sides of body extensively tinged brown; adults and juveniles have more boldly and extensively barred wing-coverts. Wing of 10 males 120-138 (128.5), of 6 females 122-133 (124.8); tail of 8 males 50-55 (52.8), of 6 females 48-56 (51.3); culmen of 8 males 37-45 (41.5), of 6 females 35-40 (37); tarsus of 8 males 38-44 (41.7), of 6 females 38-44 (40.3).

MOULT Information is for the nominate race only (Cramp & Simmons 1980). The postbreeding moult is complete; remiges and rectrices are moulted simultaneously, rendering the birds flightless for c. 3 weeks between early Jul and early Sep. Body moult starts soon after breeding (Jul, sometimes Jun) and is mainly finished before late Aug/early Sep. Prebreeding moult is partial, involving the head and, at least sometimes, part of the body; it occurs Feb-May. Postjuvenile moult is partial, involving the entire body; it starts soon after fledging (at c. 10 weeks) and is often complete before migration, ending mid-Jul to early Oct, rarely later. The first prebreeding moult is highly variable in extent and timing, but appears to be gradual, occurring Nov-Apr; in some birds no moult occurs. Second calendar-year birds moult into adult body plumage while raising young (Becker 1995). Birds do not moult during migrations.

DISTRIBUTION AND STATUS Iceland and Eurasia, from British Isles and S Scandinavia E discontinuously across C Russia and S Siberia to N China and N Japan, and S to Mediterranean region, N Africa, Saudi Arabia, Turkey,

Water Rail
? occurrence uncertain
x vagrant

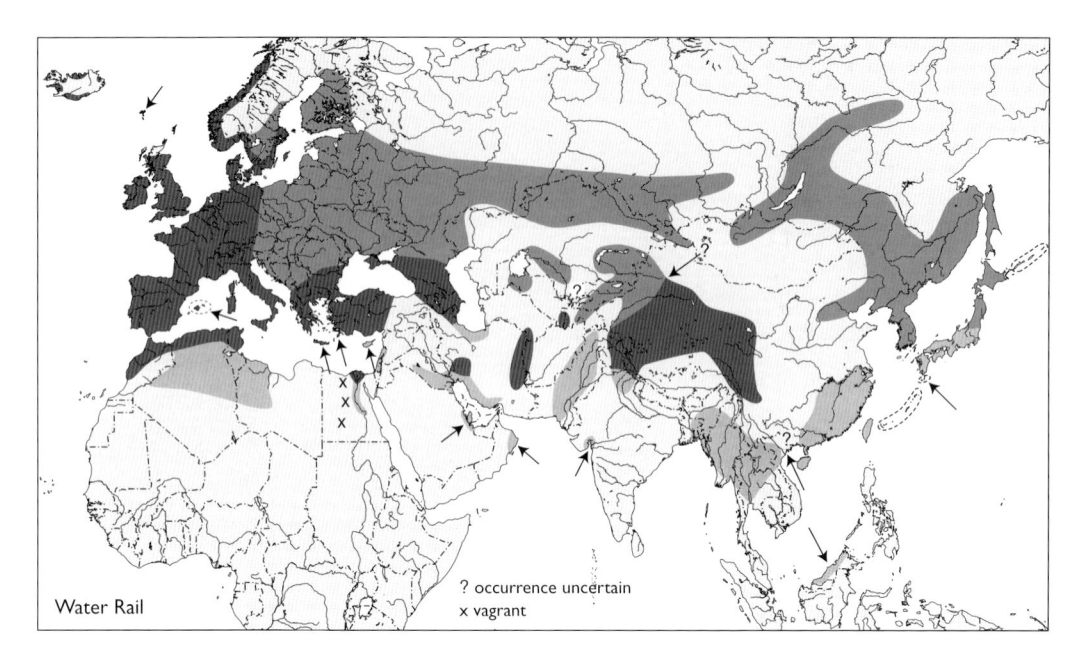

Iran, N Himalayas and WC China (Gansu and Sichuan). S and W populations are mainly resident, but N and NE populations are mostly migratory (see Distribution map), wintering S to N Africa and Arabia, and E to Pakistan and SE Asia. There is no evidence of significant recent changes in the status of the nominate race, which is regarded as locally common over much of its range, including N Africa (Morocco) and Azerbaijan, although wetland habitat loss must have affected its numbers adversely (Patrikeev 1995, Hagemeijer & Blair 1997). There has been some expansion in Denmark, Sweden, Finland and Spain during the 20th century, but also some recent decrease in Sweden, and it has benefited locally in the UK from increased habitat provided by gravel-pits and disused canals (Cramp & Simmons 1980). Populations fluctuate in some areas due to harsh winters. It bred in Cyprus in 1958 and in Libya in 1978; it may breed in Syria (Cramp & Simmons 1980). Population estimates in the W Palearctic (Cramp & Simmons 1980, Hagemeijer & Blair 1997) are (in breeding pairs): Denmark, 500-1,000; Sweden, 10,000-20,000; Finland 350; Britain and Ireland, 2,000-4,000; Spain 10,000-70,000; France, 10,000-100,000; Belgium, 250-400; Italy, c. 5,000; Netherlands, 1,500-4,500; Germany 5,000-15,000; Austria 8,000-12,000; Poland 10,000-20,000; Belarus 8,000-16,000; Croatia c. 10,000; Ukraine c. 7,000. The total population in Europe is estimated at 130,000-239,700 pairs (geometric mean 158,600), in Russia 10,000-100,000 (31,600) pairs and in Turkey 10,000-30,000 (17,300) pairs; overall, the population is believed to be declining or stable, declines being strongest in E and SE Europe where economic development and agricultural intensification are rapidly changing the landscape (Hagemeijer & Blair 1997). In Israel it is a quite common autumn and less common spring passage migrant, daily counts of 15-75 birds being recorded at suitable sites during peak movements in autumn but usually <10 in spring, while the wintering population is 1,000-2,000 birds and it is a casual breeder (Shirihai 1996). In Saudi Arabia this species (presumably this race) is a breeding resident in small numbers at Gulf wetlands (Jenkins 1981). In the former USSR, enormous numbers were said to winter in Transcaucasia and even at L Sevan, at altitudes of c. 2,000m, some even remaining in severe weather; however, high losses occur during winter snows throughout the former USSR wintering grounds, caused partly by various kinds of predator (Dement'ev et al. 1969).

The race *hibernans* is most numerous in the S of Iceland; its current status is not clear but, as formerly (Cramp & Simmons 1980), it is probably scarce in some areas. The current status of the races *korejewi* and *indicus* is unclear in many areas but they were formerly regarded as at least locally common. However *indicus* is uncommon throughout Japan and is a rare winter visitor to Thailand, Burma and N Laos, while it was recorded once in the late 19th century from Sri Lanka, where it must be a mere straggler (Delacour & Jabouille 1931, Henry 1978, Smythies 1986, Brazil 1991, Lekagul & Round 1991). In Pakistan, wintering birds (*korejewi*) can be locally frequent to plentiful and the species may breed in the country occasionally (Roberts 1991). This bird's range in Asia is probably more extensive than is known.

Little is recorded on predation, which is apparently a minor cause of mortality compared with losses during hard winters and on migration; birds are sometimes killed by various avian and mammalian predators (e.g. Potapov et al. 1987).

MOVEMENTS Most information for the W Palearctic is taken from Cramp & Simmons (1980). The race *hibernans* probably migrates to the Faeroes in winter, occurs on passage at Scottish islands and in winter in Ireland, and wanders to Greenland and Jan Mayen. The nominate race is mainly resident in North Sea countries, SW Europe, the Mediterranean basin, Turkey and S Russia, but migratory or partially migratory elsewhere in Europe, where a continental climate prevails; birds from N and C Europe move S and SW to winter in the Mediterranean basin, North Africa (extending far into the Sahara in Algeria and Libya) and E to the S Caspian Sea area. In Europe and Russia juveniles from early broods can disperse from

Jul but true autumn migration begins in Aug, peaks in Sep-Oct, and ends in Dec; spring migration begins in Feb, peaks in Mar to mid-Apr and ends by early May (Dement'ev *et al.* 1969, Cramp & Simmons 1980). Birds pass through Morocco in Sep, Algeria Sep-Oct, Egypt Sep-Nov and Israel end-Jul to mid-Dec (mainly mid-Sep to end Nov); it occurs in coastal Libya Oct-Apr and at Saharan sites in Dec-Apr, and winters in Israel mainly Nov-Mar; wintering birds leave Egypt by mid-Apr. It is a scarce passage migrant in Oman, occurring in Oct-May and occasionally overwintering (Gallagher & Woodcock 1989). Passage is recorded from Azerbaijan in Sep-Nov and Mar-May (Patrikeev 1995); in autumn in the former USSR the bulk of the European migrants fly to W Turkmenia between Uzboi and the Caspian coast (Dement'ev *et al.* 1969). It wanders to Spitsbergen, Madeira, the Canary Is and the Azores (Cramp & Simmons 1980).

The race *korejewi* is partly migratory; it winters discontinuously from Iraq E to Pakistan and C China, and wanders to E Arabia and NW India; it occurs in Pakistan from Oct-Mar (Roberts 1991); it is a vagrant to Nepal (3 records, Oct, Jan and Feb) (Inskipp & Inskipp 1991). It may occasionally occur in Israel on passage (Hovel 1987). The race *indicus* is largely migratory, occurring in winter S to N Borneo (3 records, Brunei and Sarawak); it occurs in N Japan May-Oct and winters in Japan S to the Yaeyama Is in Oct-Mar/Apr; it is recorded once from Nepal (Jan 1937), when it occurred alongside *korejewi* (Inskipp & Inskipp 1991); it is rarely recorded on passage in NE and S Bangladesh, Feb-Apr (one recent record); and it winters rarely in NW, C and coastal W Thailand (Lekagul & Round 1991), N Laos (Delacour & Jabouille 1931) and E Tonkin (David-Beaulieu 1939). Birds migrate in loose groups, at night, on broad fronts, and even cross mountain systems such as the Alps; casualties are recorded at lighthouses and powerlines. In Azerbaijan in late Sep, weakened migrating birds appear in the city of Baku and can be caught (Patrikeev 1995), while Baker (1929) records that on their first appearance in India they are often so exhausted as to allow capture by hand without attempting to move. In the former USSR migrations are in the evening, by night and at dawn; birds migrate singly but frequently spend the day in sizeable assemblages, resting wherever the rays of the morning sun catch them; in Turkmenistan and C Asia daylight haunts include mammal burrows, saxaul brush, and man-made constructions; high losses occur during migration and birds are commonly killed at telephone lines and other overhead wires, television masts and lighthouses (Dement'ev *et al.* 1969, Potapov *et al.* 1987). There are reports of birds making at least local movements on foot (e.g. Dement'ev *et al.* 1969).

HABITAT Occupies almost any type of dense, fairly tall, riparian and aquatic vegetation at water which is still or slow-moving, and fresh, brackish or waste; substrates are usually predominantly permanently waterlogged to shallowly flooded, the preferred water depth being 5-30cm; it prefers base-rich and species-rich, eutrophic to polytrophic habitats (Hagemeijer & Blair 1997). It usually requires muddy ground for foraging and occurs in the densest vegetation. Breeding habitat includes reedbeds and other emergent vegetation of swamps, marshes and fens, and at the fringes of open waters. Occupied vegetation includes *Phragmites*, *Typha*, *Iris*, *Sparganum* and *Carex*; in larger wetlands the close proximity of trees (e.g. willow *Salix*) or other fringing scrub, and of drier patches forming a habitat mosaic, is preferred to large uniform wet tracts. Suitable habitat may form small pockets or narrow strips in other habitat types. It inhabits claypits or gravel pits, peat workings and other small wetlands with muddy margins and good cover; it also occurs on floating islands, rice paddies and lotus ponds; on Japanese offshore islands it is found amongst grasses and dwarf bamboo stands (Brazil 1991). In Kashmir it occurs at inundated sugarcane fields (Ali & Ripley 1980), as does the African Rail in South Africa. It inhabits both fresh and saline lakes and marshes, and exceptionally (Scilly Is, UK) occurs on tidal sandy beaches with seaweed among rock pools and rocks for shelter; in Borneo it is apparently recorded from mangroves (MacKinnon & Phillipps 1993). On migration and in winter birds exploit a wider range of habitats, such as farm sewage outfalls, island bracken, flooded thickets of blackberry *Rubus*; even, when displaced by frost, drainpipes, rubbish dumps, open ditches and gardens (Cramp & Simmons 1980). In Iceland in winter this rail is largely dependent on marshy areas with warm water from volcanic springs and it sometimes enters tunnels in the snow over such streams; when not feeding it spends much time in holes and channels in the lava (Cramp & Simmons 1980). It is vulnerable to extreme conditions such as ice or severe floods. Although occurring mainly in lowland areas, it breeds to 1,240m in the Alps and is resident up to 2,000m at L Sevan, Armenia (Cramp & Simmons 1980).

FOOD AND FEEDING Information is primarily from Cramp & Simmons (1980), and also Potapov *et al.* (1987). Takes predominantly animal matter, but also regularly eats plant material. Animal food includes worms, leeches, Gastropoda (including *Bulinus*, *Zonites*, *Helix*, *Lymnaea*, *Planorbis*, *Pisidium* and *Bithynia*), shrimps, small crayfish, spiders, and many terrestrial and aquatic insects and their larvae, including Dermaptera, Odonata, Trichoptera, Heteroptera, Homoptera (including Coccidae), Coleoptera (Carabidae, Hydrophilidae, Chrysomelidae, Dytiscidae, Staphylinidae and Curculionidae), Diptera (including Tipulidae), Formicidae (occasionally), Lepidoptera larvae and Orthoptera. Small vertebrates, either killed or eaten as carrion, include amphibians (frogs, toads and newts), fish, birds and mammals. Plant foods include duckweed *Lemna*, leaf buds (e.g. *Betula*), flowers (e.g. *Juncus*), shoots, roots (especially of *Rorippa nasturtium-aquaticum*), seeds (including *Phragmites*, *Carex*, *Setaria*, *Glyceria*, *Robinia*, *Ranunculus*, *Menyanthes*, *Polygonum*, *Potamogeton*, Compositae and legumes), berries and fruits (e.g. in winter, fruit of *Hippophaë*); birds eat more plant material in the autumn and winter. Fish and frogs are first paralysed by vertical blows on the spinal cord behind the head and are then killed; strikes are made with the bird's head, body and neck relatively stiff, most movement occurring at the ankle joints (Glutz von Blotzheim *et al.* 1973). Food obtained on land or in mud is normally washed in water before it is swallowed. Birds forage on dry ground or mud near water, and also wade in shallow water, taking items on and below the surface, and from emergent vegetation; they sometimes feed while swimming, and rarely by diving. They leap up to take insects from vegetation, climb in vegetation to take berries of *Solanum dulcamara*, and fly into apple trees to remove fruit which is then eaten on the ground (Cramp & Simmons 1980). In both breeding and winter territories, birds use definite paths, the courses of which are fixed by the position of favourable food sources; successful hunting spots are repeatedly visited (Koenig 1943). The young are fed mainly on insects and spiders.

HABITS Although normally shy and retiring, this rail may sometimes be seen in quite open habitat, especially in winter, and may become relatively bold in the presence of people; see the photograph in Taylor (1996b: 121). The birds have a slow, high-stepping walk, a faster loping half-run, and a sudden, crouching run for cover. The tail is carried raised or horizontal; it is constantly flicked when the bird is alarmed but held still during mere suspicion of danger. When disturbed by people, the birds run quickly to cover, often then freezing in a crouched position with the bill pointing down; a similar response has also been observed in chicks. It normally flies reluctantly when disturbed; the flight action is usually rather weak and fluttering, with dangling legs and rather obviously long wings; however, it frequently flies between feeding areas and is a strong but low flier when on migration. It climbs well, and swims readily over short distances with a moorhen-like action. Birds generally roost at dusk but nocturnal migrants rest during the day (for behaviour on migration, see Movements), and birds call throughout the night early in the breeding season. In captivity, chicks are active at night and roost during the day; in the wild, the roosting behaviour of parents and young is not clear.

SOCIAL ORGANISATION Monogamous; highly territorial when breeding. The pair bond is maintained for the duration of the breeding season and, in migratory populations, is apparently formed soon after the birds' arrival on the breeding grounds; however, Dement'ev *et al.* (1969) suggest that pair bonds are formed before spring migration, "since pairing off is already noted in winter quarters". It often breeds singly or in small numbers at very small wetland patches, but large concentrations of breeding birds occur in extensive wetlands. At high population densities birds may nest 20-50m apart; in England and Europe, breeding density averages c. 0.25-2 pairs/ha (Percy 1951, Glutz von Blotzheim *et al.* 1973); in Germany, densities were 7.5 pairs/ha (6 pairs in 0.8ha), up to 3.6 pairs/ha (for sites of 1-3ha), and up to 1.34 pairs/ha (for sites of 33-100ha) (Flade 1994). Territory size is variable: in C Europe 200-450 (300) m^2 (Berg & Stiefel 1968); in S Sweden 160-590 (320) m^2 (Bengtson 1967). Individuals may defend a winter feeding territory or show aggression when approached too closely (e.g. Dement'ev *et al.* 1969). In England, 8-9 birds wintering along a stretch of stream c. 64m long and 4m wide were fairly evenly distributed in loosely defined feeding territories (King 1980).

SOCIAL AND SEXUAL BEHAVIOUR Information is from Cramp & Simmons (1980) unless otherwise specified. Although normally solitary outside the breeding season, when they are generally well spaced because of their aggressive behaviour, at times in winter birds may congregate in groups of up to 30. They drive away other species such as Common Moorhen (127) and *Porzana* crakes, but will also feed alongside moorhens. In a group of rails wintering along a stream in England the dominant birds were males and were the larger, more clearly patterned individuals with deeper hued bills and feet (King 1980). Disputes between territorial birds often involve a charging attack with the body lowered and the neck stretched forwards; sometimes the pair attack the intruder together, when the intruder usually flees. The victor may stand erect on its toes and make a long announcement call. Fighting and stabbing with the bill are also recorded. Wintering birds bickered a lot, and when contesting feeding places had encounters reminiscent of cockerels fighting (King 1980): on meeting, both birds 'sharmed', with bill slightly open and throat vibrating rapidly, standing on the toes with body, head and neck stretched up and breasts sometimes touching; then each began to head-jerk, with the bill thrusting at the opponent, stepping threateningly forward or momentarily retreating. These encounters lasted up to c. 20 sec, ending when 1 bird retreated a little with head turned half away and fully stretched head and neck almost touching the ground; it then quietly retreated to its own feeding area while the victor, retaining the upright aggressive stance and with wings partly outstretched, 'sharmed' while moving its head to and fro before also resuming feeding.

An incubating bird may remain on the nest even when touched, or may attempt defence, even attacking the intruder; it frequently runs around the intruder's feet, with ruffled feathers, uttering distress or aggressive calls (Percy 1951), or attempts to lure the intruder away by injury-feigning (Turner 1924); it is also known to carry the young away in its bill (Percy 1951).

When showing the selected nest site to the female, the male raises the back feathers, opens the remiges in an arch over the back, spreads the undertail-coverts, points the bill vertically downwards and makes a loud creaking call (Andreas 1996). The male's precopulatory display consists of head-bowing with the bill pointing down and touching the breast, the wings raised so that the striped flanks are prominent, and the tail raised high with the undertail-coverts alternately fanned and closed. Allopreening occurs; the male courtship-feeds the female; and the incubating female leaves the nest to display to the male, walking round him and uttering soft crooning notes, rubbing her bill against his and taking short runs to and from him (Turner 1924).

BREEDING AND SURVIVAL Season Iceland, Jun; W and C Europe, late Mar-Aug (most mid-May to end Jul); Morocco and Algeria, May-Jun; Tunisia, Jun; Egypt, late Mar-late May; Israel, chicks 25 Apr and 10 May; former USSR, May-Jul; Kashmir, late May-early Aug. **Nest** A substantial (but often loosely made) cup of dead grass stems, sedges, rushes, 'aquatic weeds' and leaves; material usually from plants at nest site; usually built in thick reeds or rushes on ground near or in water but also recorded in vegetation on dry land; rarely on tree stump or in open; surrounding vegetation often pulled down into loose canopy. External diameter 13-16cm; height c. 7cm; may be built up if water level rises. Both sexes build; in captivity, male chose nest site and showed it to female, and male did most of building, which usually took only 1 day (Andreas 1996). **Eggs** 5-16 (usually 6-11); blunt oval, sometimes elliptical; smooth and variably glossy; off-white to pinkish-cream or pale pinkish-buff, more or less spotted and blotched red-brown and pale purple, markings often concentrated at large end. Size of eggs: *aquaticus* 32-40 x 24.1-27.2 (36 x 25.8, n = 120), calculated weight 13; *korejewi* 33.2-40.7 x 23-27 (36.9 x 25.3, n = 48), calculated weight 12; *indicus* 32-38 x 25.1-27 (35 x 26, n = 38), calculated weight 13. Eggs laid at daily intervals. **Incubation** 19-22 days, by both sexes but female normally does larger share; begins when clutch complete; hatching usually synchronous; if asynchronous, male tends first chicks while female continues to incubate. Parent may aid hatching by enlarging hole in eggshell; when most eggs hatched, remaining eggs occasionally deliberately destroyed by parent spearing with bill (Buxton 1948, Percy 1951). **Chicks** Precocial; fed and cared for by both parents, apparently only until 20-30 days old (Cramp & Simmons 1980; Andreas 1996); loose bond keeps young together for some time after this, despite some intolerance. Young brooded mainly on nest for first few days, while one parent brings food and presents it bill-to-bill to brooding bird which feeds chicks bill-to-bill; chicks often fed on partially regurgitated matter; young beg by raising wings slightly, stretching or jumping at parent's bill and calling (see Voice); begin to feed themselves at c. 5 days but dependent on parental feeding for at least 14 days. Young use wing-claw when climbing in vegetation. Adults brood young under wings; also regularly carry chicks into or out of nest, or across open water, grasping chick by head, neck or shoulder in bill (Cramp & Simmons 1980). First contour feathers visible in field at 14-18 days (Becker 1995); young fully independent and capable of flight at 7-9 weeks. Age of first breeding 1 year. 2 broods normal; replacements laid after egg loss. **Survival** Oldest recorded bird 5 years 7 months (Potapov *et al.* 1989).

62 AFRICAN RAIL
Rallus caerulescens   Plate 19

Rallus caerulescens Gmelin, 1779, Cape of Good Hope.

Forms a superspecies with *R. aquaticus* and *R. madagascariensis*. Monotypic.

Synonyms: *Rallus caffer/caeruleus*; *Aramus caerulescens*.

Alternative names: Cape/Kaffir/African Water Rail.

IDENTIFICATION Length 27-28cm. A distinctive, medium-sized *Rallus*, having upperparts and upperwings plain dark chocolate-brown, darker from crown to hindneck. Chin to upper throat white; forehead, sides of head and neck, foreneck, and breast dark slaty-grey, often tinged bluish; flanks, belly and median undertail-coverts black, narrowly barred white; outer undertail-coverts white. Bare parts red, bill with dusky culmen. In flight shows narrow white leading edge to inner wing only. Sexes similar but female smaller; face to breast may be less slaty-blue, with some pale markings; may have some brown on sides of neck and breast and less regular barring on flanks. Immature duller than adult; more white on throat; sides of neck and breast brown; flanks brownish-black, barred rufous-buff to white; outer undertail-coverts tipped rufous-buff; bare parts brown, becoming redder with age. Juvenile has darker upperparts; face, sides of neck and breast dark brown; breast brown, mottled pale; flanks to undertail coverts blackish-brown, barred rufous; centre of breast and belly buff; bare parts brown. Inhabits dense reeds, sedges and grass in flooded marshes and swamps and beside open water.
Similar species Unpatterned, deep chocolate-brown upperparts unique character within genus, while long red bill easily distinguishes it from sympatric rails. Water Rail (61), which occurs in North Africa, is smaller, with streaked upperparts and flesh-brown legs and feet; similar-sized Rouget's Rail (70) of highland marshes in Eritrea and Ethiopia has largely cinnamon-rufous underparts, shorter bill and purplish-red legs and feet.

VOICE Information is from Urban *et al.* (1986) unless otherwise specified. The common territorial call, given by both sexes, often together, is a loud high trill winding down into spaced single notes and falling in pitch and volume; the notes have a low, 'pumping' undertone like those of the Water Rail. The call is often taken up by other pairs or individuals, while nearby birds may also join in with a series of low clucking or grunting notes. One bird flushed from a nest gave a slower and very plaintive version of the territorial call (P. B. Taylor unpubl.). Other calls include a wheezy, high-pitched *kreee* or *kreeea* which is given during territorial and aggressive responses to taped playback; a loud rattle, given by both sexes and apparently a warning call, which is uttered most commonly for 2-3 months at the end of the breeding season, when it normally replaces the territorial call; a repeated breathy *aah*, given by both sexes before and after copulation; deep, resonant gulps given by the male when courtship-chasing; a prolonged purring call given by both sexes and used to maintain contact with the mate, to attract the mate and the young, and also possibly to indicate mild agitation (it may develop into the rattle call); and a sharp *chip* given by birds in the hand (Taylor 1994 and unpubl.). In southern Africa the rattle call is virtually indistinguishable from the advertising rattle of Baillon's Crake (94) (Taylor 1994). There is also a shrill *ri-ri-ri* when fighting; a deep, pumping *krock* given during the threat display; and a low growl of warning; the warning call to chicks is a squeaky *zii-zii*, while adults leading chicks give short snores (Schmitt 1976). Other calls to chicks are muted *krr* and *k-k* notes, while adults with young give a low hissing growl to an intruder (P. B. Taylor unpubl.). The young make squeaky calls, which develop into a feeble version of the territorial call at 6-7 weeks; calling of immatures peaks in Mar-Jun (Taylor 1994).

DESCRIPTION
Adult Crown to hindneck blackish-brown; rest of upperparts, including scapulars and upperwing-coverts, dark chocolate or vinous-brown; rump to uppertail-coverts often slightly darker; tail blackish. Alula, primary coverts and remiges blackish-brown; marginal coverts predominantly white; occasional individuals have white spots or bars on secondary and median coverts; axillaries and lesser underwing-coverts blackish, narrowly barred white; rest of underwing-coverts dark grey or dark brownish-grey, narrowly barred white; underside of remiges dark brownish-grey. Chin white; throat white to pale grey (variable); forehead, entire sides of head, sides of neck, foreneck, and breast dark slaty-grey, often with bluish tinge in good light; lores and upper ear-coverts darker. Flanks, thighs and belly black, narrowly barred white; centre of belly brownish-grey, feathers tipped buff. Median undertail-coverts barred black and white like flanks, sometimes with rufous-buff tips; outers pure white. Iris blood-red to red-brown; bill bright red, culmen dusky; legs and feet red; in non-breeding season, upper mandible, legs and feet darker. Small black wing-claw present. Sexes similar but female smaller; face to breast may be greyer, less slaty-blue, with some pale markings; may have a little brown on sides of neck and breast and less regular barring on flanks. Adults may be sexed by external examination of pubic bones, points being adjacent in males but

separated in females which have laid eggs; birds with bill >50 and weight >170 should be males (Schmitt 1976).

Immature Duller than adult; upperparts less richly coloured, may be olive- or grey-tinged; chin and throat more extensively white; sides of neck and sides of breast brown; grey of underparts paler, duller and sometimes tinged olive-brown; flanks brownish-black barred rufous-buff to white; outer undertail-coverts tipped rufous-buff; iris brown; bill dusky brown; legs and feet dull brown; bare parts become redder with age.

Juvenile Upperparts and upperwings darker; face, sides of neck and breast dark brown; breast brown, mottled pale; flanks to undertail-coverts blackish-brown, faintly barred rufous to whitish (sometimes almost spotted on rear flanks); centre of breast and belly buff; iris brown to blackish; bill becomes black, with pink nostril spot, at 8-9 weeks; legs and feet dark brown.

Downy young Down black, with greenish sheen at 1 week; iris greyish, becoming olive (2-3 weeks); bill pink, distal half black, becoming black except for proximal one-third of upper mandible (4-5 weeks); legs and feet slate.

MEASUREMENTS Wing of 54 males 109-135 (122), of 38 females 105-126 (115); tail of 14 males 38-48 (44.9), of 12 females 38-44 (40.8); bill of 55 males 42-59, of 38 females 40-56 (46); tarsus of 14 males 38-46 (41.6), of 12 females 33-42 (38.4). Weight of 66 males 146-205 (179.6), of 50 females 120-170 (145.6); weight varies greatly during year and birds may lose up to 20% of weight during severe winter (Schmitt 1976). Hatching weight c 9.5.

GEOGRAPHICAL VARIATION None.

MOULT All information is from Schmitt (1976) for South African birds, unless otherwise stated. There is a complete moult once a year, when the remiges are moulted simultaneously and the birds are flightless for just over 3 weeks; the rectrices are usually dropped before the remiges, and body moult usually takes at least 6 weeks to complete. Remex moult normally occurs in Aug-Nov but is also recorded in Dec-Jan, and takes place either before incubation or after the young become independent; birds do not moult the remiges while incubating. Body moult occurs in Aug-Nov and Feb-May; some individuals moult during both periods. At 6-7 months of age, immatures show at least a partial moult into adult-type plumage, involving the head and body, which is complete at 8-9 months, but 3 birds also moulted the remiges at 8-10 months. Some birds retain the juvenile undertail-coverts until 10 months old. There is no information on the timing of the postjuvenile moult. Body moult is also recorded in Kenya, Mar.

DISTRIBUTION AND STATUS Confined to sub-Saharan Africa. In West Africa, where its status is uncertain but it is apparently rare and erratic (see Movements), it is recorded from W Sierra Leone (Ribi R area) and Cameroon (Ndop, Yaoundé) (Parrott 1979, Fotso 1990); NE Gabon (Makokou) (Brosset & Erard 1986); and Congo (Odziba) (Dowsett & Dowsett-Lemaire 1989); its occurrence in NE Central African Republic (Carroll 1988) requires confirmation. It has apparently also been recorded recently from Gambia (Anon 1995d; no details available). In NE Africa it is uncommon to rare in S Sudan, where it is known only from Gilo/Itibol in the Imatong Mts (Nikolaus 1987), and in the Ethiopian highlands (Urban *et al.* 1986), but is locally common in permanent marshes in the Jimma area of SW Ethiopia (Taylor 1996a). Its principal range includes E Zaïre (Kivu, Katanga); Rwanda; Burundi (common on lower Ruzuzi R); East Africa (uncommon and local) in W and NE Uganda (Pakwatch, Rwenzori NP, Mbarara, Kidepo Valley NP), WC Kenya (N to Siaya and Nandi, E to Thika, SE to L Jipe) and throughout Tanzania; Zambia (widespread except in Luangwa and Zambezi Valleys); Malawi (probably common below 1,500m); Mozambique (probably widespread and locally common in the S, especially along the coast); Zimbabwe (local and generally uncommon); the interior highlands of Angola (widespread); N Botswana (sparse and uncommon in Okavango, Linyanti and Mpandamatenga areas) to NE Namibia (Caprivi Strip); and through most of E South Africa (locally common), including Swaziland, to the SW Cape (Benson *et al.* 1971, Clancey 1971, Benson & Benson 1977, Britton 1980, Irwin 1981, Urban *et al.* 1986, Dowsett & Forbes-Watson 1993, Penry 1994, Taylor 1997a). There are scattered records from the arid W regions of South Africa and isolated records from N Namibia (Etosha Pan, Bushmanland pans), S and W Namibia (Hardap Dam, Windhoek) and seasonally flooded pans in Hwange NP, Zimbabwe (Taylor 1997a). Its occurrence in coastal Kenya (Lewis & Pomeroy 1987) requires confirmation. It is undoubtedly under-recorded in many areas, but its overall numbers must be decreasing as a result of the continual degradation and destruction of wetland habitats throughout its range. However, it is apparently under no immediate threat and is able to colonise relatively small, artificially created, wetland patches (Taylor 1997c). In South Africa its total population is estimated at c. 15,800 birds, 87% (13,700) of these being in KwaZulu-Natal, the former Transvaal and the SW Cape (Taylor 1997a); it is more widespread in the SW Cape than was previously thought (Hockey *et al.* 1989).

Known predators are the African Marsh Harrier *Circus ranivorus* and the Slender Mongoose *Herpestes sanguineus* (Jackson & Sclater 1936, Schmitt 1976); the species is also recorded as being killed by fires (Schmitt 1976).

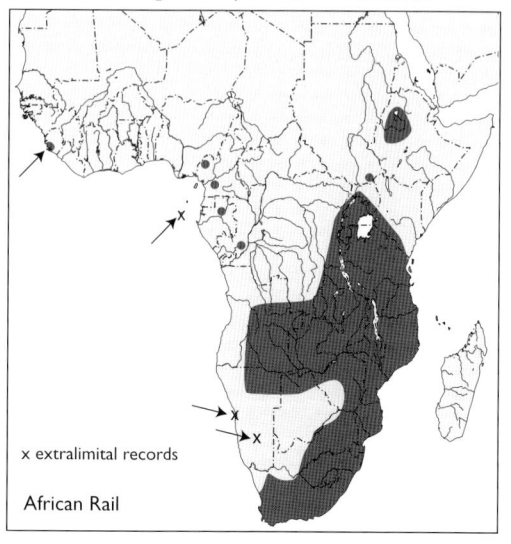

African Rail

MOVEMENTS There is no evidence for regular migrations, and most seasonal variations in numbers may be explained by the dispersal of immatures or by local (and sometimes longer-distance) movements to and from habitat rendered temporarily unsuitable by burning, drying out or the reduction of cover (Taylor 1997c). In South Africa it may be common on the Nyl R floodplain in years of high rainfall but is absent in drought years (Tarboton *et al.* 1987), while during a 6-year study in the S Transvaal Schmitt (1976) recorded notable influxes of birds in adult-type plumage every Apr-May. Increases in reporting rates at the end of the breeding season (May-Jun) in South Africa probably reflect the peak in calling by immatures which occurs at this time (Taylor 1994, 1997c). There may be some seasonal movements in N Namibia and N Botswana (e.g. Penry 1994), and its known occurrences in West Africa are erratic: W Sierra Leone, May-Oct; Cameroon, Jan, Feb, Oct; NE Gabon, Oct-Nov; Congo, Feb (Parrott 1979, Brosset & Erard 1986, Dowsett & Dowsett-Lemaire 1989, Fotso 1990, Field 1995). One came to lights at night in Tsavo West NP, SE Kenya, in Dec 1985 (1987, *Scopus* 9: 149). It is a vagrant to São Tomé (undated skin: Parrott 1979).

HABITAT Information is from Taylor (1997a) unless otherwise stated. This rail occupies reedbeds and dense rank growth in permanent, seasonal and temporary swamps and marshes, and beside lakes, pools, rivers and streams. It normally prefers sites at least parts of which are permanently flooded, but it will also occupy temporary habitats and during the breeding season it moves out into seasonally inundated dense hygrophilous grassland and sedge meadow adjoining permanent cover, and into seasonally wet sugarcane adjacent to marshes (South Africa). It sometimes occurs at mature pans, in paddyfields (West Africa), and in thick secondary growth along rivers (Sudan) (Urban *et al.* 1986). It prefers shallowly flooded areas with mud and/or floating vegetation for foraging. Occupied vegetation types include beds of *Phragmites*, *Typha*, *Carex*, *Cyperus*, *Bolboschoenus*, *Juncus* and *Eleocharis*, as well as mixed sedgebeds, grass (including *Leersia*, *Oryza*, *Echinochloa* and *Hemarthria*), *Polygonum*, *Epilobium* and other plants (Hopkinson & Masterson 1984, Tarboton *et al.* 1987, Taylor 1997a). Occupied sedges and grasses are usually robust and c. 50cm or more in height (P. B. Taylor unpubl.), and dense vegetation is favoured, especially where the plant species composition is diverse and there are channels and runways linking scattered patches of more open growth (Hopkinson & Masterson 1984). In South Africa this species is tolerant of brackish conditions – more tolerant in the SW Cape than is the Black Crake (88) – and occurs at brackish sites dominated by *Phragmites australis*, *Juncus kraussii*, *J. acutis*, *Scirpus littoralis*, *S. nodosus* or *Cladium mariscus*, the sedges sometimes being interspersed with glasswort *Sarcocornia pillansiae*, or with mats of *Isolepis* in shallow water (Taylor 1997a, d). It is also recorded from brackish *Remeiria maritima* sedge marshes in Sierra Leone (Field 1995).

Its altitudinal range is imperfectly described. In C Ethiopia it is recorded at 2,600m (Taylor 1997b); in Kenya it is recorded up to 3,000m but is scarce at the coast; in South Africa it extends to at least 1,870m (considerably higher than the Black Crake) and, although it occurs widely in coastal regions of the S and SW Cape, in KwaZulu-Natal and the former Transvaal it may be less widespread and numerous at low altitudes and in coastal areas. However, although it is infrequently recorded on the littoral and coastal plains of KwaZulu-Natal, where it has been suggested to be an uncommon winter visitor (Robson & Horner 1996), it may be more widespread, regular and numerous there than is known.

FOOD AND FEEDING Information is from Urban *et al.* (1986) with supplementary observations by P. B. Taylor (unpubl.). Food includes earthworms, crabs, aquatic and terrestrial insects and their larvae, spiders, small fish and small frogs; also some vegetable matter, including seeds; occasionally carrion (crayfish, crabs and small mammals). This species forages mainly on moist to shallowly flooded ground, but sometimes on dry ground; it probes mud, clumps of wet grass and mats of floating vegetation; while wading it takes food from the water surface and also by immersing the head and neck; and it takes prey from low vegetation. It climbs readily in tall reedbed vegetation, sometimes possibly to forage. It feeds both within and outside cover, and frequently forages along the edges of reedbeds.

HABITS Diurnal and crepuscular, being particularly active in the early morning and from late afternoon to dusk. In South Africa, Schmitt (1976) found it active throughout the day in the winter months but inactive for up to 3h in the middle of the day during the summer. This rail is normally skulking, keeping within or close to dense cover, but inside cover it is not shy, keeping c. 3m from an intruder and uttering territorial calls (Schmitt 1976). When accustomed to the presence of man, and not molested, it will emerge from cover to feed at the edges of roads and tracks through wetlands, and will even walk boldly across wide roads (P. B. Taylor unpubl.). When foraging, it walks with long, even strides on flexed legs, but at other times it moves more jerkily, flicking the tail when agitated. When alarmed it stands upright, raises the tail and flattens the plumage. When ground cover is good, these rails normally escape danger by running, but in more open situations, and when surprised, they will fly: flights are normally short, the birds flying low and with dangling legs. However, when a predator such as a harrier *Circus* approaches, the birds 'freeze', dropping on to the tarsi and remaining motionless until the danger has passed (Schmitt 1976). When its habitat is burned this species, like the Black Crake, will remain in much less dense and smaller patches of cover than will the smaller Red-chested Flufftail (3), and will forage on completely open burned ground up to 10m from such cover, flying back to cover if disturbed (Taylor 1994). It swims well when undisturbed, and climbs well; birds have been seen 4m above water in dense *Phragmites* (Schmitt 1976). African Rails bathe in shallow water, or on submerged stems in deeper water, and sunbathe with slightly open wings (Schmitt 1976). Pair members allopreen (P. B. Taylor unpubl.).

SOCIAL ORGANISATION Monogamous; strongly territorial when breeding, but the territory is apparently relinquished after breeding. In South Africa, territories are established in Jun-Jul in the Transvaal (Schmitt 1976); in KwaZulu-Natal they are established in Jul-Aug, when calling increases greatly, and relinquished in Apr, when territorial calling decreases dramatically, and the rattle call (see Voice) predominates in Mar-Jun (Taylor 1994). Adults attacked model African Rails most frequently during the first 3 months of the breeding season, from Sep to Dec (Taylor 1994). Territories covered 150m of river frontage before hatching and 50m 2 weeks after hatching (Schmitt 1976). In KwaZulu-Natal, 7 territories were 1,250-3,550 (2,754, SD = 793)m^2 (P. B. Taylor unpubl.). The overall density of breeding birds in extensive optimal habitat may be up to 4 pairs/ha, and in moderate to variable and patchy habitat may average c. 1 pair/ha or less (Taylor 1997a).

SOCIAL AND SEXUAL BEHAVIOUR Nesting Black Crakes and Common Moorhens (127) are tolerated within the breeding territory (Schmitt 1976). However, during the breeding season African Rails will attack Black Crakes and Red-chested Flufftails (Taylor 1994); in captivity they are very aggressive towards other rail species in the breeding season and will attack and kill species as large as themselves (C. C. Wintle pers. comm.). When stimulated by taped playback of African Rail territorial and *kreee* calls (see Voice) during the breeding season, mated pairs make loud territorial calls very frequently; the birds often run to each other and stand upright, side by side or facing each other, stretch the neck up and call with the bill slightly open and horizontal; during such calling, the birds' flank feathers are often somewhat roused and spread, and this accentuates the bold flank barring (P. B. Taylor unpubl.). In response to taped playback, territorial birds will attack model African Rails, and sometimes model Black Crakes and Red-chested Flufftails, by crouching, extending the neck horizontally to point the bill at the model, rousing the plumage, drooping the wings, spreading the lateral undertail-coverts and making a silent rush at the model to stab it with the bill; the male is apparently more prone to show such aggressive behaviour than the female (P. B. Taylor unpubl.). When fighting, 2 birds face each other with an elongated stance, then jump into the air towards each other, attacking with their bills for short periods; one bird is either driven away or the two eventually resume feeding. An incubating bird remained standing on the nest and gave a threat display to a human intruder; when chicks are trapped the adults become aggressive, bending the legs, lowering the head, stretching the neck, rousing the plumage and pecking the hand which is removing the chick from the trap (Schmitt 1976).

The following information on courtship and copulation is from unpublished observations (P. B. Taylor) unless otherwise stated. This species has a typical rail courtship chase, sometimes preceded by purring calls from the male to attract the female, and preceded and followed by deep, resonant gulps from the male. During the chase, both birds are low and stretched with the head and tail in line with the body. Before copulation the male sometimes crouches, holds his neck vertically and his head horizontal, and makes repeated, low, breathy *aah* calls with the bill slightly open; he then follows the female around until she crouches with her head lowered, when he approaches from behind, mounts and copulates for up to 20s; during copulation his outer undertail-coverts are fanned, his tail is wagged from side to side and he gently pecks the female's head and neck. In the postcopulatory display the male walks slowly away from the female with a crouching gait, his head lowered, his undertail-coverts still spread and his inner remiges spread and raised so that their tips meet over his back (the primaries are apparently closed in the usual position at the sides of the body). The female raises her head and utters breathy *aah* calls before walking off. Apparent reverse mounting has been observed (Brooke 1992).

BREEDING AND SURVIVAL Season Ethiopia, Aug; SW Zaïre (Upemba NP), 'dry season'; Kenya, May-Jun; Malawi, Feb-Mar; Zambia, Jan, breeding condition May; Zimbabwe, Jan-May; Mozambique, Jan; South Africa, Sep-Mar, with single records Jul and Aug but in suitable habitat winter breeding possibly occurs more frequently than is known. **Nest** A cup of aquatic plants (*Carex, Cyperus, Eleocharis,* *Juncus, Typha, Rorippa*); well concealed in aquatic vegetation, usually over water but occasionally on almost dry ground; external diameter 15-20cm; depth 5-10cm; depth of cup 2-5cm; nest rim 10-40cm above water. Both sexes build, and also construct brood platforms 20-50cm above water. **Eggs** 2-6 (4.1, n = 16 clutches); broad oval; creamy-white to pinkish-cream, profusely spotted and blotched light brown, red-brown, purple and grey, especially at large end; size (n = 56) 32.5-40.3 x 25.8-29.1 (38.1 x 27.5), calculated weight c. 16. **Incubation** About 20 days, by both sexes. **Chicks** Precocial; leave nest soon after hatching; fed and cared for by both parents until 6-8 weeks old; parent holds food items near tip of bill, young bird crouches and reaches up to take food. Body feathers appear at 2-3 weeks, remiges at 4-5 weeks; well feathered by 6-7 weeks; remiges fully grown at 8-9 weeks; young reach adult weight at 3 months. Age of first breeding 1 year. Not known to be multiple-brooded, but will relay if clutch fails. Nest predators include African Marsh Harrier.

63 MADAGASCAR RAIL
Rallus madagascariensis   Plate 19

Rallus madagascariensis Verreaux, 1833, Madagascar.

Forms a superspecies with *R. aquaticus* and *R. caerulescens*. Monotypic.

Synonyms: *Rallus poliocephalus*; *Biensis typus/madagascariensis*; *Eulabeornis madagascariensis/bernieri*.

IDENTIFICATION Length 25cm. A medium-sized, predominantly darkish brown rail. Forehead, forecrown, face, chin and throat grey; rest of upperparts, including upperwings, olive-brown with blackish streaking; crown to hindneck washed vinaceous. Ear-coverts, sides of neck, foreneck, breast and upper flanks dull vinaceous, brighter on sides of breast and flanks; with blackish streaks which are most conspicuous on sides of breast. Lower flanks, belly and median undertail-coverts dusky olive-brown, with narrow faint whitish bars; lateral undertail-coverts white. Iris red; bill markedly long, slender and decurved, rosy with black culmen; legs and feet grey-brown. Sexes alike, but female smaller, with markedly shorter bill. Immature duller than adult, with brownish bill and brown eyes. Juvenile dull; grey replaced by pale olive-brown; chin and throat whitish; anterior underparts dull buff-brown, not vinaceous. Inhabits dense herbaceous vegetation of marshes, river banks and wet woodlands.

Similar species Distinguished from other medium-sized Madagascar rails by long, slender, decurved bill, lack of white on throat, almost unbarred flanks, and olive-brown, dark-streaked upperparts; from Madagascar Wood-rail (12) also by white lateral undertail-coverts. Differs from Old World congeners in having grey only on face and throat, vinaceous underparts, narrow and faint whitish bars on dull olive-brown lower flanks, and greyish-brown legs and feet.

VOICE A sharp *tsi-kia* when flushed; also a *kik-kik* (Rand 1936). Available tape recordings (made by C. Carter and I. Sinclair) contain 6 distinct calls, functions unknown: a sharp, high-pitched *kia* or *kio*; a low-pitched, rather frog-like rattle *crrrr*; a rapidly repeated series of fairly high-pitched *ké* notes, falling a little in pitch and slowing down slightly (presumably the advertising call); a sharp, squeaky *tsip* or *tsi*; a clicking *crik*; and a sharp, high-pitched *queea*. It calls frequently during the day, from dense cover. It is responsive to playback (A. Riley *in litt.*).

DESCRIPTION
Adult Forehead, forecrown, lores and circumocular region to malar region dark slaty-grey, darker on lores and around eye, and narrowly streaked blackish, most prominently on forecrown; chin and throat somewhat paler, ashy-grey, and unstreaked. Upperparts olive-brown, tinged greenish, with blackish feather centres giving streaked effect; crown to hindneck with slight vinaceous wash; lower back to uppertail-coverts slightly duskier than mantle and upper back; rectrices blackish, edged olive. Scapulars and upperwing-coverts like mantle and upper back; greater upperwing-coverts, alula, primary coverts and remiges predominantly dusky brown, with olive-brown fringes. Axillaries and underwing-coverts olive-brown, with narrow, faint white fringes; underside of remiges dark olive-brown. Ear-coverts, sides of neck, foreneck, breast and upper flanks dull brownish- to russet-vinaceous, slightly brighter on sides of breast and upper flanks than elsewhere; centres of feathers blackish, forming streaks which are heaviest on sides of breast. Lower flanks, thighs, belly and median undertail-coverts dusky olive-brown, with narrow and obscure whitish bars (pale feather tips) except on thighs, white bars sometimes vaguely edged dusky brown; lateral undertail-coverts white. Iris scarlet or brick-red; bill rosy with black ridge to upper mandible and dark, or dark ivory, tip; legs and feet greyish-brown. Sexes alike, but female smaller, with markedly shorter bill and shorter tarsus.
Immature Similar to adult but duller; iris brown; bill brownish.
Juvenile Upperparts like adult but feather fringes duller, and feather centres slightly paler. Grey on head replaced by pale olive-brown, often tinged greyish; chin and throat off-white; foreneck, breast and upper flanks pale, dull buffy-brown, streaked blackish-brown; lower flanks, thighs, belly and median undertail-coverts dusky olive-brown, with obscure paler barring; centre of belly creamier, washed vinaceous-cinnamon; undertail-coverts probably as in adult.
Downy young Not properly described: down black; bill of specimens horn with blackish tinge to basal area and centre of upper mandible; legs and feet darkish brown.
Measurements (10 males, 11 females) Wing of male 108-118 (112.1, SD 3.3), of female 100-112 (105.6, SD 3.4); tail of male 40-51 (43.8, SD 3.9), of female 36-45 (38.9, SD 2.4); culmen to base of male 55-70 (62.3, SD 5.0), of female 42-54 (49.2, SD 3.9); tarsus of male 38-44 (40.7, SD 1.7), of female 34-39 (36.6, SD 1.3). Weight of 1 male 148 (Benson *et al.* 1976).

GEOGRAPHICAL VARIATION None.

MOULT No information available.

DISTRIBUTION AND STATUS Madagascar, in the E and the E part of the high plateau, from Ivohibé in SE to Marojejy in NE; between these localities it is recorded from Manombo, Iampasika, SE Betsileo, Périnet, L Alaotra, Foulepointe, Nossi-Vola, and Andapa; there is also a record from "Boiboahazo", locality untraceable (Dee 1986). In Nov 1997 it was recorded from Moramanga Marsh, E of Antananarivo (A. Riley *in litt.*). Although not regarded as globally threatened, it is becoming rare throughout its range as a result of pressure on wetland habitats in E Madagascar, where it was formerly (1929-1931) common, especially at high altitudes; it was regarded as quite abundant near Antananarivo in 1969-1971 (Salvan 1972b). Direct destruction of wetland habitats is greatly affecting this rail and is continuing, especially to create rice paddies for feeding Madagascar's rapidly growing population – see comments under Slender-billed Flufftail (9) on wetland conservation – and no Madagascar wetland has full legal protection (O. Langrand *in litt.*). Another threat is habitat alteration through the effects of deforestation, such as variations in the quantity, quality and frequency of water available from wetland catchments, and siltation as a result of extensive soil erosion; fertilisers and pesticides from rice paddies are also affecting wetlands and their invertebrate communities (O. Langrand *in litt.*). It has been listed in the Manombo and Analamazaotra Special Reserves, and in the Marojejy Strict Nature Reserve, and may occur in other reserves, but it is easily overlooked because of its shyness (Langrand 1990). It is still relatively common in some pristine wetlands (e.g. Torotofotsy marsh, near Périnet), but only small populations exist in the few areas which are protected; its long-term survival depends on the implementation of a wetland protection programme and the creation of additional protected areas (O. Langrand *in litt.*).

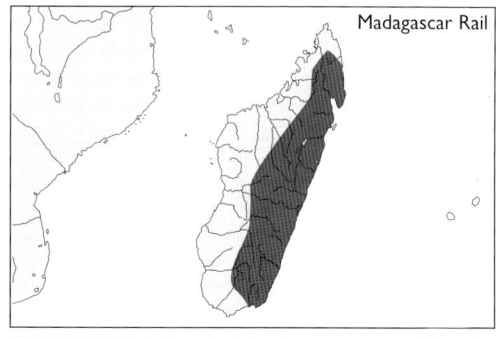

MOVEMENTS Although it has been suggested that this rail is possibly migratory (Salvan 1972b), no evidence for this has been found.

HABITAT Small to large marshes with dense long grass, reeds and sedges; river margins; also grassland at forest edges and dense herbaceous vegetation of wet woodland and forest (Rand 1936, Langrand 1990). It occurs from sea level to 1,800m, more frequently at high altitudes.

FOOD AND FEEDING Only invertebrates are recorded. This rail moves about slowly in dense aquatic vegetation, searching for food and probing the mud with its long bill (Langrand 1990). Sexual differences in bill length may have some effect on the comparative foraging strategy of males and females.

HABITS This species is very secretive and shy, and is rarely flushed or seen. However, it responds well to taped playback (A. Riley *in litt.*).

SOCIAL ORGANISATION Not described; it occurs singly or in pairs (Langrand 1990).

SOCIAL AND SEXUAL BEHAVIOUR No information available.

BREEDING AND SURVIVAL Season Lays Aug-Oct. **Nest** Not described; built on ground in aquatic vegetation. **Eggs** Reddish-brown, spotted darker brown at large end; size of 8 eggs 42-43.7 x 31.8-34 (42.7 x 33.1), calculated weight 25.5 (Schönwetter 1961-62). No record of clutch size.

LEWINIA

Three species of medium-sized rails which are sometimes considered conspecific. They are often placed in *Rallus* and bear a strong superficial similarity to the Slaty-breasted Rail (54), which is now included in *Gallirallus*. The skeleton of *Lewinia* rails does not agree with that of true *Rallus* and Olson (1973a, b) considers them generalised, and more primitive, than Slaty-breasted Rail, forming part of a pro-*Rallus* stock; he includes them in the genus *Dryolimnas* because skeletally they are virtually identical to the much larger White-throated Rail *D. cuvieri* (67) of Madagascar and Aldabra. However Sibley & Monroe (1990), citing only M. Bruce (pers. comm.), maintain that the *Lewinia* superspecies is not congeneric with *D. cuvieri*. Until further studies have been made, I prefer to retain the genus *Lewinia* on the basis of the fundamental differences in plumage between it and *Dryolimnas*, and the geographical isolation of the two genera. One species, the enigmatic Brown-banded Rail of Luzon in the Philippines, is known mainly from records of migrating birds and is considered globally endangered. Its habitat requirements and breeding areas are unknown. The Auckland Rail was thought to be extinct but has been found to survive on two islands in the Aucklands group. It is vulnerable, and its survival depends on the continued exclusion of mammalian predators from its islands. The more widespread Lewin's Rail occurs in New Guinea, Australia and Tasmania. One race, from SW Australia, is probably extinct; the others are locally common to uncommon.

64 BROWN-BANDED RAIL
Lewinia mirificus Plate 20

Rallus mirificus Parkes & Amadon, 1959, Aritao, Luzon, Philippines.

Forms a superspecies with *L. pectoralis* and *L. muelleri*. Sometimes placed in the genus *Dryolimnas*; sometimes regarded as a race of *L. pectoralis* (e.g. Paynter 1963) but separation on criteria given by Parkes & Amadon (1959) appears valid. Monotypic.

Synonyms: *Dryolimnas mirificus*; *Lewinia pectoralis mirificus*.

Alternative name: Luzon Rail.

IDENTIFICATION Length c. 23cm. Forehead to hindneck plain chestnut-brown; rest of upperparts almost unpatterned mid-brown, with some small buffy spots on back; uppertail-coverts and tail darker brown, with small black-bordered pinkish-buff spots; scapulars and upperwing-coverts with black-bordered bars of pinkish-buff; marginal coverts white (presumably showing as white leading edge to inner wing in flight); remiges blackish-brown, secondaries and tertials barred or spotted pinkish-buff. Narrow supraloral streak pale tawny; chin and throat whitish; lores, sides of face, foreneck and sides of neck deep olive-grey, washed chestnut-brown; breast mid-grey, washed brown at sides. Flanks and belly blackish-brown, with narrow, black-bordered, whitish to pinkish-cinnamon bars; undertail-coverts olivaceous-black, barred pinkish-cinnamon. Iris probably brown; bill mostly pinkish-red; legs and feet probably dark brown. Sexes alike. Immature duller: dark greyish-brown from forehead to hindneck; greyer on cheeks and upper breast; little barring on underparts; bill darker than adult. Juvenile undescribed. May inhabit wet grassy areas.

Similar species Differs from adult Lewin's Rail (65) in much duller overall colouring; lacks Lewin's Rail's prominent olive-brown streaks from mantle to tail and on upperwings, and black streaking from forehead to hindneck and on rest of upperparts; upperwings lack prominent barring of Lewin's Rail; chin and throat duller white; flank barring duller; bill shorter. Upperparts more similar to those of juvenile Lewin's Rail, but pattern of upperwings and colour of underparts markedly different to those of Lewin's. See also Auckland Rail (66). Superficially very similar to sympatric Slaty-breasted Rail (54), but Slaty-breasted larger and heavier, with stouter bill, and heavier tarsi and toes; remiges usually have bold white barring and spotting (variable, occasionally reduced). Only inner secondaries of Brown-banded Rail are barred, bars being pinkish-buff and best developed on outer margin. Distinguished from other sympatric *Gallirallus* species by smaller size, head pattern, extent and pattern of pale markings on upperparts, and less extensive barring on underparts; from sympatric *Porzana* species by longer bill, rufous crown and nape, and pattern of upperparts.

VOICE Unknown.

DESCRIPTION
Adult Description based on that of type, an adult female, from Parkes & Amadon (1959). Forehead to nape chestnut-brown, duller (auburn) on forehead; hindneck brighter (chestnut). Mantle mid-brown, feathers with vaguely defined blackish centres; back to rump slightly darker brown, feathers of back with small black-bordered spots of pale pinkish-buff along edges, decreasing in size and abundance anteriorly and posteriorly; scapulars as mantle, feather edges with black-bordered bars of pinkish-buff; uppertail-coverts and rectrices dark dull brown, with small black-bordered pinkish-buff spots. "Larger upperwing-coverts" patterned like scapulars, "smaller" chiefly mid-brown; marginal coverts white; primaries and secondaries blackish-brown, secondaries with a few small pinkish-buff spots along outer margin; tertials mid-brown, blacker towards centres, with prominent black-bordered marginal bars of pale pinkish-buff or pinkish-buff, best developed on outer webs. Axillaries and underwing-coverts olivaceous-black mixed with white. Narrow supraloral line pale tawny-brown; lores and ear-coverts deep olive-grey, washed chestnut-brown, paling to white on chin and throat; foreneck and sides of neck as ear-coverts, darkening to mid-grey on breast; sides of breast washed brown; centre of breast faintly washed brown. Flanks blackish-brown, narrowly barred whitish or pale pinkish-buff to pale pinkish-cinnamon, bars narrowly bordered black; belly pale grey in centre, darker and browner at sides, barred as flanks; undertail-coverts olivaceous-black, broadly tipped light pinkish-cinnamon. Iris probably brown; bill pinkish-red, brownish on culmen and towards tip; legs and

feet possibly dark brown to blackish-brown. Sexes alike. In "spring males" (presumably a May specimen), base of bill brighter red (Amadon & duPont 1970), possibly indicating breeding condition.

Immature Duller than adult: dark greyish-brown from forehead to hindneck; greyer on cheeks and upper breast; little barring on underparts; bill darker than adult.

Juvenile Not described.

Downy young Unknown.

Measurements 1 adult and 3 immature males, 4 adult females. Wing of immature male 103, 106, 107; of female 103-108 (105.75, SD 2.1); culmen from feathers of male 24-26 (25, SD 0.8), of female 25, 26, 27, culmen from base of 1 female 27; tarsus of male 28-30 (28.75, SD 0.9), of female 29-30 (29.25, SD 0.5).

GEOGRAPHICAL VARIATION None recorded.

MOULT Unknown.

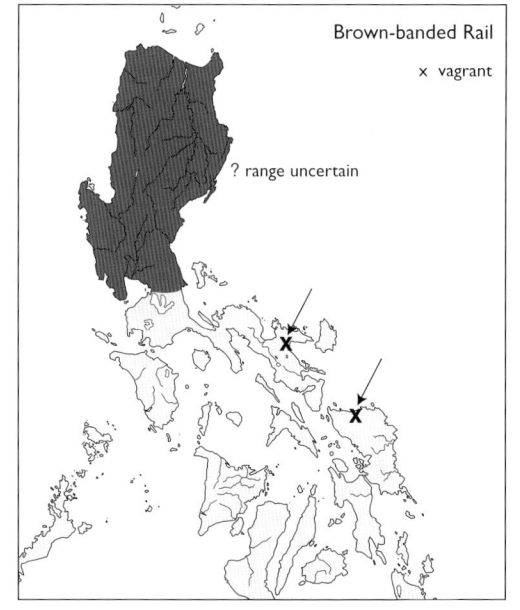

DISTRIBUTION AND STATUS Luzon and Samar, Philippines. ENDANGERED. This uncommon and secretive rail is known largely from records of migrating birds at Dalton Pass in the Sierra Madre Mountains, Nueva Vizcaya Province (see Movements). Recorded localities are as follows: in W Luzon, Nueva Ecija (Papaya and elsewhere, and between Bamban and Camp O'Donnell, Tarlac), Nueva Vizcaya (Imugan, Santa Fe and Dalton Pass), Benguet (Baguio), Mountain Province (Liwan, Kalinga Apayao); in S Luzon, Camarines Norte (Labo area) (McClure & Leelavit 1972, Dickinson et al. 1991, Collar et al. in press). There is also a museum specimen from Catarman, Samar I (Collar et al. in press). Although this species is described as uncommon and secretive (Dickinson et al. 1991) there appear to have been no observations made of it in the field. Its population is unknown, but over 30 birds were collected, and 191 ringed, at Dalton Pass in 1965-70; the paucity of known localities, and the paucity of records since 1979, suggest that the species is probably becoming increasingly rare (Collar et al. 1994, in press). Rails, snipe and other birds that can be caught in rice fields are regularly sold from roadsides, but the Brown-banded Rail seems not to be among them (Poulsen 1995). However, bird-catchers at Dalton Pass are likely to take this species in some numbers, although trappers questioned in 1989-90 did not report catching it; the impact of trapping on its numbers cannot be assessed (Collar et al. in press).

MOVEMENTS In 1965-70 191 birds were trapped at Dalton Pass in May-Jun and Oct-Dec, apparently while undertaking post-breeding dispersal or intra-island migration, the latter possibly between the Cagayan Valley and the Central Plain or between the Sierra Madre and the Cordillera Central (Poulsen 1995). Almost all records are from the foothills, and its apparent movements agree with that recorded during the post-breeding dispersal of other rail species which breed in Luzon in the rainy season (Dickinson et al. 1991). The specimen from Samar I was collected at the Samar Institute of Technology, Catarman, in Apr 1959; this, and the undated specimen from Labo, S Luzon, may be overshooting migrants (Collar et al. in press).

HABITAT The precise habitat is unknown, but birds have been recorded from wet grassy areas; unlike its congeners, it may have strict ecological requirements (Collar et al. 1994) although this is probably not as likely as its sharing their diverse habitat preferences (e.g. see Lewin's Rail). A female was found dead at Baguio on a road through forest of mature pine mixed with sapling pine and many shrubs, while the bird from Tarlac was in a "small (10m x 20 m) undisturbed riverside swamp, primarily composed of *Ipomea aquatica* and surrounded by cane grass *Saccharum spontaneum* on an infertile substrate of volcanic debris" (Collar et al. in press). All records are from mountain foothills, migrating birds being recorded at an altitude of c. 1,000m.

FOOD AND FEEDING Unknown; food and feeding habits are presumably similar to those of Lewin's Rail.

HABITS Unknown. Possibly similar to those of Lewin's Rail.

SOCIAL ORGANISATION Unknown.

SOCIAL AND SEXUAL BEHAVIOUR Unknown.

BREEDING AND SURVIVAL Unknown; it probably breeds during the rainy season. The female from Baguio, Dec, was immature, skull 10% ossified (Collar et al. in press); 3 males from Papaya, Nov, were immature (Amadon & duPont 1970); 2 from Dalton Pass, Jun, had slightly enlarged gonads, while 1 from Sep and 5 from Nov had undeveloped gonads (Collar et al. in press). Brighter bill colour in May (Amadon & duPont 1970) could be indicative of breeding condition (see Description).

65 LEWIN'S RAIL
Lewinia pectoralis   Plate 20

Rallus pectoralis Temminck, 1831, Oceania (= New South Wales).

Forms a superspecies with *L. mirificus* and *L. muelleri*, both of which are sometimes considered races of *L. pectoralis*. Sometimes placed in *Dryolimnas*, or retained in *Rallus*.

Eight subspecies recognised, one (*clelandi*) possibly extinct.

Synonyms: *Hypotaenidia/Lewinia/Eulabeornis brachypus*; *Rallus brachipus/lewini/muelleri/striatus insulsus*; *Dryolimnas pectoralis*.

Alternative names: Pectoral/Short-toed/Slate-breasted/ Water Rail; Lewin('s) (Water) Rail.

IDENTIFICATION Length 23-27cm; wingspan 31-35cm. Medium-sized, rather stout rail with short tail, small head and moderately long, very slightly decurved bill. Forehead to hindneck, broad supercilium and sides of neck rufous (duller in races *mayri*, *captus* and *insulsus*), streaked black on crown and nape (least in *exsul*, *alberti*), sometimes also on forehead and hindneck; rufous may extend to upper mantle (nominate, *captus* and *alberti*). Upperparts and tail streaked black and olive-brown (tinged rufous in nominate and *brachipus*). Lesser upperwing-coverts, usually hidden when wings folded, streaked black and olive-brown; scapulars and most other coverts black-brown, with olive-brown tips and white bars; marginal coverts white. Primaries, primary coverts and secondaries dark brown; inner secondaries fringed olive-brown, with small white spots which become more bars on tertials. Chin sometimes white; throat, ear-coverts and foreneck to central breast grey (throat white in *insulsus*, *alberti*), tinged buffish-olive (most in *mayri* and *brachipus*; least in *exsul* and *clelandi*); lower breast and flanks to undertail-coverts brownish-black, narrowly barred white, or cream to buff (*brachipus*). Iris brown or red; bill pink to pinkish-red, dark brown or grey on distal third and culmen; legs and feet grey, tinged pink. Female slightly smaller; has duller rufous head and hindneck with heavier black streaking. Immature like adult. Juvenile much darker overall; lacks rufous on head and hindneck; upperparts may almost lack paler streaks (*exsul*, *captus*, *mayri*); face, foreneck and breast washed dusky brown; white on throat more extensive; underpart barring dark brown and cream or white, sometimes with little pale barring (*exsul*); undertail-coverts washed cinnamon (*mayri*) or isabelline (*captus*); bare parts duller and darker than in adult. Inhabits grass and marsh vegetation at freshwater to saline wetlands; also dry thickets and upland grassland.

Similar species Differs from Brown-banded (64) and Auckland (66) Rails in having noticeable streaks on head and hindneck, prominently dark-streaked upperparts and prominently barred upperwings. Distinguished from sympatric *Gallirallus* species by smaller size, and by pattern of upperparts, upperwings and underparts; from sympatric *Porzana* species by longer bill, rufous from forehead to hindneck, and pattern of upperparts and upperwings. Allopatric Slaty-breasted Rail (54) resembles Lewin's Rail but is larger and heavier, with stouter bill, and heavier tarsi and toes; upper body and upperwings olive-brown to blackish-brown, with darker feather centres, and with generally distributed white bars and spots; remiges and underwing-coverts also have white bars; lacks Lewin's Rail's mixture of black and white barring and olive-brown markings on scapulars, upperwing-coverts and tertials.

VOICE Poorly known. The commonest call is a repeated loud *crek* or *kek*, which is sometimes preceded by low rasping grunts, and which may be given for up to 30 s, becoming louder and faster and then slower and softer (Marchant & Higgins 1993). This call is apparently given only in the breeding season, when calling is heard most in the afternoons (Leicester 1960). Fletcher (1913) and Serventy (1985) give the characteristic breeding-season call as a metallic, repeated *jik* or *tick*, like the sound of two large coins being struck together, the notes often being run together into a chattering series; this presumably refers to the same call. An incubating female responds to the calls of the male by giving a deep purr like a cat (Fletcher 1913). The female also gives snuffling grunts, like "a little pig with a cold" (Skemp 1955), when with young, and this is also apparently a warning call to chicks in the nest. There are also a loud, repeated *tick* and a loud, harsh *crick*, both apparently given in response to an intruder; a whining moan *oo-er* is recorded (Leicester 1960); a "half booming, half cooing sound" is also recorded from an adult with young (L. Holden in North 1913). This rail may also have a whistle call like the Auckland Rail. Chicks give peeping or squeaking calls.

DESCRIPTION *L. p. pectoralis*
Adult male Forehead to hindneck, broad supercilium and sides of neck bright rufous, black feather centres on crown and nape giving heavily streaked effect; dark shaft-streaks on forehead concealed in most birds (c. 30% have visible streaking); those on hindneck concealed in fresh plumage (c. 60% of birds show streaking); rufous may extend to feather edges of extreme upper mantle. Scapulars black with narrow white bars, and varyingly tipped olive-brown; rest of upperparts, including most lesser upperwing-coverts, black, with olive-brown (often slightly rufous-tinged) feather edges giving heavily streaked effect; in some birds feathers have small white marginal spots; uppertail-coverts in some birds have white bars at edge of each web; rectrices brownish-black, edged brownish-olive. Median coverts, greater secondary coverts and longest lesser coverts black-brown, with olive-brown tips and white bars not meeting at shafts; marginal coverts white. Primaries, primary coverts, alula and outer 3-5 secondaries dark brown; P6-9 usually with small white spots on outer edge; inner secondaries dark brown with olive-brown fringes which broaden on inner feathers; outer edges have small white spots which become more elongated on inner feathers and form incomplete white bars on tertials. Underside of remiges brownish-grey; axillaries and underwing-coverts brownish-grey, tipped white. Chin sometimes white; feathers immediately below eye (lower eyelid area) white; throat, ear-coverts and foreneck to central breast grey, strongly tinged buffish-olive; upper throat paler grey; some feathers of central breast barred white. Lower breast, flanks, thighs, belly and undertail-coverts brownish-black, feathers barred and tipped white to cream; bars sometimes tinged buff on lower breast, belly and vent; extent of barring variable and reduced by wear. Iris brown or red; variation poorly understood; bill pink to pinkish-red, grading to dark brown or dark grey on distal third and culmen; legs and feet pale grey to slate-grey, with varying pink tinge, sometimes largely pinkish.
Adult female Duller than male on head and neck; supercilium and stripe down side of neck brownish-rufous and narrower than in male, grading to rufous-brown on forehead; dark streaking on crown and nape more extensive than in male; c. 87% of females have dark streaks on forehead, and c. 94% on hindneck. Bare parts as male, but brown-yellow iris also recorded.
Immature As adult, but some retain a few juvenile feathers on flanks and belly.
Juvenile Forehead to hindneck black-brown with narrow

brown feather edges; rest of upperparts, and tail, as adult but feather edges olive-brown with slightly less rufous tinge; rump largely dark brown, especially when olive-brown feather tips worn off. Upperwing as adult; white bars on tertials, tertial coverts and greater secondary coverts tend to be narrower and shorter; underwing as adult. Chin and upper throat white, throat feathers varyingly tipped light brown; ear-coverts and lower throat grey with olive-brown feather tips; feathers at sides of throat varyingly tipped dark brown to give mottled to overall brown effect; upper breast dark brownish-grey, with faint olive tinge and paler patch in centre; lower breast, flanks, belly and undertail-coverts dark brown; flanks varyingly tinged olive and with sparse white to cream barring, other regions with wider bars. Iris dark brown; bill grey-black; legs and feet grey to black; assume adult colour before juvenile plumage lost.
Downy young Down black, glossy; iris black-brown; bill black, at first with buff tip to lower mandible; legs and feet dark grey.

MEASUREMENTS Wing of 16 males 97-110 (101.7, SD 3.2), of 14 females 96-103 (98.9, SD 1.8); tail of 12 males 42-52 (45.5, SD 2.6), of 14 females 39-46 (42.6, SD 1.6); exposed culmen of 17 males 30.5-39.5 (33.6, SD 2.2), of 14 females 28-33 (30.6, SD 1.6); tarsus of 17 males 28.5-35 (31.2, SD 1.5), of 14 females 26-32 (29.7, SD 1.3). Weight of 9 males 71-100 (80.8, SD 9.3), of 8 females 63-111 (75.4, SD 21), of 5 juveniles 42-76 (65.1, SD 14).

GEOGRAPHICAL VARIATION Races separated on prominence of crown streaks; colour of crown, sides of head and neck, upperparts, foreneck and breast; extent of white on throat; and biometrics. Race *mayri* includes *connectens* (Ripley 1964). An aberrant individual of the Slaty-breasted Rail, with no white barring on the primaries, was originally described as a race, *deignani*, of Lewin's Rail (Ripley & Olson 1973).

L. p. exsul (Hartert, 1878) – Flores, Lesser Sundas. Differs from nominate in having little black streaking on rufous crown; back more olive, less rufous-tinged; feathers of lower back and rump with small white spots along edge of inner web; chin and throat whitish; foreneck and breast very grey; bill short. Juvenile lacks rufous on head and olive margins to feathers of upperparts; has little barring on underparts (Ripley 1977). Measurements of 3 birds: wing 101-108; tail 44-46; culmen 27-29.5; tarsus 28-30 (Ripley 1977).

L. p. mayri (Hartert, 1930) – Arfak and Weyland Mts, W New Guinea; altitude 1200 2200m. Like *alberti* but slightly larger. Differs from nominate in having crown dull chestnut with dull blackish-brown streaking; sides of head and neck less bright chestnut; feather edges of upperparts darker. Iris brown; bill reddish, culmen and tip blackish; legs and feet dark grey. Juvenile (one specimen examined) similar to that of *captus* on upperparts; upperwings barred whitish to buffy or greyish-white; chin and throat white, feathers of lower throat and sides of throat tipped dusky brown; sides of head, sides of neck, foreneck and upper breast washed dull sepia; lower breast pale, tinged greyish, and whitest in centre; flanks barred dark brown and creamy to buffy-white, with overall cinnamon wash; undertail-coverts washed cinnamon. Wing of male 98-109, of female 96-108, exposed culmen of male 34-41, of female 32-37 (Mayr & Gilliard 1951, Gilliard & LeCroy 1961). Weight 91-103.

L. p. captus (Mayr & Gilliard, 1951) – Snow Mts and C highlands of New Guinea (Victor Emanuel Mts, Mt Giluwe, Mt Hagen and Wahgi Valley), E to Eastern Highlands Province; altitude 1,450-2,600m. Nearest to *alberti* but wing and bill longer; forehead to hindneck, sides of head and neck, and upper mantle, darker and more maroon, less reddish-brown. Compared with *mayri*, crown and hindneck brighter; has much less white on throat; breast and foreneck less heavily washed olive; small patch of white below eye; bill and wing shorter. Iris golden-brown in male, more "rusty brown" in female; bill greyish-flesh with blackish culmen; legs and feet pale grey, tinged pink. A partly grown juvenile (remiges in pin) was dull smoky-black above, with blacker crown; underparts sooty-black with large whitish areas on throat and abdomen; sides of face grey (Gilliard & LeCroy 1961). Fully grown juvenile blackish-brown above, forehead to hindneck more blackish; mantle to rump with indistinct blackish streaks; underparts similar to adult but no barring on belly and no olive wash on breast; lower breast and centre of belly whitish; wings and flanks barred as adult, but pale bars creamy; undertail-coverts have ill-defined isabelline wash (Ripley 1977). Juvenile male has iris blackish-brown; bill greyish-brown with blackish tip; legs and feet grey to pinkish-grey; juvenile female has brownish-grey iris and dark pink-grey bill (Gilliard & LeCroy 1961). Wing of 13 males 95-107 (101, SD 3.7), of 11 females 95-104 (98.7, SD 2.5); tail of 3 males 35, 37, 37, of 1 female 31; exposed culmen of 13 males 27-37 (33.3, SD 2.3), of 15 females 27-34 (31, SD 2.0), culmen to base of 3 males 35, 36.5, 36.5, of 1 female 33; tarsus of 3 males 31, 32, 32.5, of 1 female 28. Weight of 5 males 65-92 (78.6, SD 10.5), of 7 females 58-90 (74.15, SD 9.8), of 3 juvenile males 53, 73, 75, of 3 juvenile females 69, 77, 78.

L. p. insulsus (Greenway, 1935) – Herzog Mts, E New Guinea. Differs from *alberti* in having top of head more olivaceous, not clear reddish-brown; upperparts darker brown; white bars on secondaries and upperwing-coverts narrower, tinged brownish; subocular stripe white rather than reddish-brown; lores and cheeks lighter grey; entire throat and neck white; breast lighter grey. 1 female: wing 103; culmen 29; tarsus 30 (Ripley 1977).

L. p. alberti (Rothschild & Hartert, 1907)– mountains of SE New Guinea, altitude 1,800-2,500m. Differs from *exsul* in having chestnut of head and neck reaching anterior mantle; wing shorter. Differs from nominate in more uniform crown (no black streaks); rufous-chestnut anterior mantle; upperparts darker; throat whiter; size smaller. Unlike *insulsus*, bars on secondaries and upperwing-coverts broader and clear white; chin and upper throat white. Iris chocolate-brown; bill dark brownish-horn with reddish base; legs and feet fleshy-grey. "Young" more olive above; breast lighter grey with rusty tinge (Ripley 1977). Wing of 3 males 87, 93, 94, of 5 females 93-97 (94.3, SD 1.5); tail of 1 male 33, of 1 female 40; exposed culmen of 3 males 27.5, 29, 30, of 3 females 27, 28, 29, culmen to base of 1 female 32.5; tarsus of 1 male 28.5, of 1 female 31.

L. p. clelandi (Mathews, 1911) – Extreme SW Australia, from Margaret R E to King George's Sound. Larger than nominate; bill longer and deeper; breast clearer grey, with only small olive-buff feather tips in fresh plumage. Wing of 1 male 115, of 4 fully grown 109-114; tail of 1 male 45, of 1 female 52; bill of 1 female 46, of 4 fully grown 39-45; tarsus of 1 female 36.5, of 4 fully grown 35-37 (Harrison 1975 [no means given], Marchant & Higgins 1993, BMNH).

L. p. pectoralis (Temminck, 1831) – E and SE Australia, from NE Queensland (Julatten and Atherton tablelands) patchily in coastal region S and W to SE South Australia (Adelaide region, Kangaroo I and [once] Eyre Peninsula). See Description.

L. p. brachipus (Swainson, 1838) – Tasmania. Averages larger than nominate; grey feathers of neck and breast have broader olive tips, giving much stronger olive tinge; pale barring on underparts cream or buff, cream bars often extending to lower breast and flanks. Wing of 11 males 102-106 (104.3, SD 1.5), of 19 females 97-105 (101.7, SD 2.4); tail of 12 males 40-50 (45.4, SD 2.8), of 17 females 41-48 (44.8, SD 2.2); exposed culmen of 12 males 31.5-39 (35.5, SD 1.9), of 17 females 30-35.5 (32.5, SD 1.2); tarsus of 12 males 31-34.5 (33.4, SD 1.1), of 19 females 30-33.5 (31.4, SD 0.9). Weight of 9 males 73-112 (97.5, SD 11.5); of 14 females 71-102 (87.3, SD 8).

MOULT Most information is from Australian populations (nominate and *brachypus*) (Marchant & Higgins 1993). Postbreeding (pre-basic) moult is complete, and moult of the remiges and rectrices is simultaneous, rendering the birds flightless for a period of unknown duration. Feather wear suggests that most birds moult the wing between late Feb and Apr but that some may moult as early as Nov and others as late as May. Body moult is more protracted and mostly occurs in summer and early autumn, but a few moulting birds are known from all months except Jun-Aug. Postjuvenile moult is complete in at least some birds, but it is not clear whether all birds undergo wing moult. It occurs in Sep-Apr and moult of the flanks and belly is not always complete, the unmoulted juvenile feathers allowing some birds in immature (first basic) plumage to be aged. In race *captus* of New Guinea, moult of wings, tail and body is recorded in both sexes in May, while one female with a much enlarged ovary had medium general body moult (Mayr & Gilliard 1954, Gilliard & LeCroy 1961).

DISTRIBUTION AND STATUS Lesser Sundas (Flores), New Guinea at mid-montane altitudes, E and SE Australia, and Tasmania. Although the species is not globally threatened, the race *exsul* is known only from 4 specimens from S and W Flores and was last recorded in 1958-59 (Paynter 1963), while the race *clelandi* of SW Australia is known from four specimens collected at 3 localities (Margaret R, King George's Sound and Bridgetown) and has not been seen since 1932 (Whittell 1933), despite being regarded as reasonably common at that time; it was not recorded during survey and atlas work in the 1980s and it is probably extinct as a result of the draining and burning of its wetland habitats (Garnett 1993). The nominate race and *brachipus* are widespread and uncommon to common throughout their ranges, with no recent decline noted; the birds are able to occupy disturbed and artificially created habitats, including sewage farms and saltworks, and to occur close to human habitation and even in industrial areas (Marchant & Higgins 1993); in Queensland and New South Wales, A. Palliser (*in litt.*) has found the nominate race quite common in rain forest habitats in the tropical and subtropical areas from Cairns to Wollongong. However, local extinctions have probably been caused by wetland drainage and by burning for grazing, while cover is trampled and destroyed by domestic stock (Marchant & Higgins 1993). In New Guinea the race *captus* may have been fairly common in recent years (e.g. Diamond 1972b), but other races may be uncommon to rare (Rand & Gilliard 1967); however, Coates (1985) and Beehler *et al.* (1986) regard the species as locally common throughout New Guinea. In Australia, predators of young and adults are cats, dogs, Grey Goshawks *Accipiter novaehollandiae* and Kelp Gulls *Larus dominicanus*, while foxes may also be predators; birds are also killed on roads and during grass-mowing, and are caught in traps for rats and other mammals (Marchant & Higgins 1993).

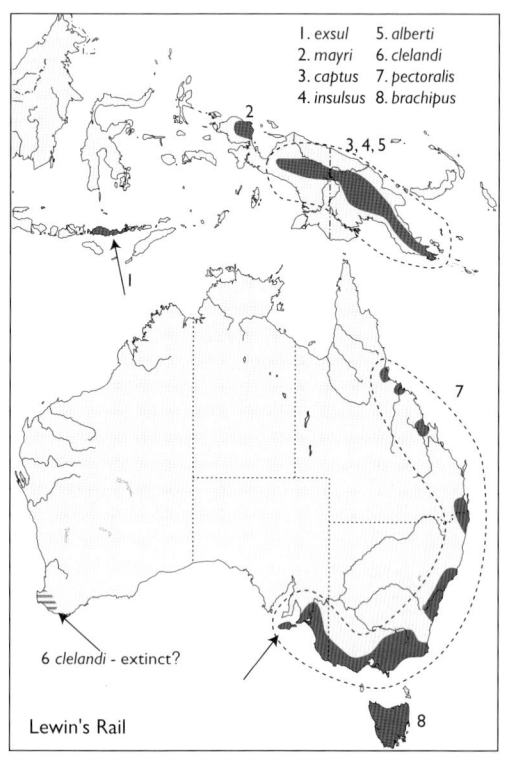

MOVEMENTS Information is from Marchant & Higgins (1993). Movements are virtually unknown, as sightings are rare and observation difficult; this species is seldom seen to fly. In Australia it may be partially migratory: it is present at some sites all year but in the S the lack of records and the lower reporting rates in winter may reflect some seasonal absence (or may reflect the difficulty of observing the birds in winter). It may move to coastal and subcoastal New South Wales to breed; there are no records from S South Australia in Jun-Aug. In Tasmania birds may move locally in response to changing conditions. Movements appear to be nocturnal and one bird struck a window at night; dead birds have also been found under power lines and caught in a wire fence. Birds probably fly c. 12km to and from Maatsuyker Island and mainland Tasmania. No movements are recorded in New Guinea or Flores.

HABITAT In different parts of its range this rail occupies very different habitat types. In Australia (Marchant &

Higgins 1993) it most often occurs at freshwater to saline wetlands, either permanent or ephemeral and usually with standing water: swamps, marshes, inundations, creeks, lakes, small to large pools, farm dams, marshy streams, saltmarshes and estuaries. It requires dense fringing or emergent cover of long or tussocky grass, reeds, rushes, sedges or bracken *Pteridium*; it occurs occasionally in thickets of wetland shrubs such as *Melaleuca*; and it sometimes uses wetlands in rainforest, woodland, riverine forest or wet heathland. It occasionally occurs away from water in parks, gardens, pasture, hayfields, lucerne, tangles of bramble *Rubus* or *Lantana*, and even in dry areas of grass *Gahnia* and ferns under a canopy of scrub. In tropical and subtropical regions (Queensland) it occurs in many rainforest and wet sclerophyll areas, particularly where introduced *Lantana* is present, and not necessarily always near creeks; here it is a bird of low, dense undergrowth (A. Palliser *in litt.*). In contrast, in New Guinea it occurs in medium-height, dense, dry upland grassland at 1,400-2,600m, as well as in thick cover near water (Rand & Gilliard 1967, Coates 1985).

FOOD AND FEEDING Details are from Marchant & Higgins (1993). Eats largely invertebrates: Polychaeta, Lumbricidae, molluscs, crustaceans (Isopoda and Chilopoda), arachnids, and insects including Orthoptera (Gryllidae and Acrididae), Coleoptera (Coccinellidae, Tenebrionidae, Curculionidae and "water-beetles"), Diptera (Stationmyidae larvae), Trichoptera larvae, Lepidoptera larvae, and Formicidae adults. Birds also take small fish, frogs and tadpoles, and occasionally eggs of birds; they take some vegetable matter, such as leaves of aquatic plants. A captive bird also ate slugs and cooked rabbit meat. This species forages on dry ground, dry peat, soft soil and mud, in mangroves, and often in shallow water (<5cm deep) at wetland edges, small pools and channels; it usually forages within, or close to, dense vegetation, emerging from cover only briefly. Occasionally it feeds on garden lawns (Leicester 1960) and in short pasture. It pecks and probes (the probe rate is c. 1 per s), it often immerses the head in water, and it drills vigorously with the long bill. Foraging bouts are interrupted by sudden dashes into cover, especially with large prey items.

HABITS This species is diurnal and crepuscular, probably having peaks of activity in the early morning and evening; it is particularly active on overcast days (Gilbert 1936). It is secretive, wary and very difficult to see or flush, even when hunted with dogs; it makes runways beneath dense vegetation and uses them to escape when disturbed (Leicester 1960), and it may call from very close range but remain invisible. However, in the evening birds emerge from cover to feed, and may be seen flying, with dangling legs, for short distances across thick grass or swimming easily across small stretches of open water, while birds may not be shy in cover, and may occasionally be seen at close quarters in the open (North 1913, Ripley 1977). Birds flushed in hayfields may land and run under new-mown hay, when they can be caught without great difficulty (Leicester 1960). They are probably particularly secretive when flightless during the postbreeding moult. When alarmed, they flick the tail and bob the head up and down. Young are apparently more approachable than adults (Marchant & Higgins 1993). Birds bathe near cover and preen in cover, and sometimes bathe and preen with Spotless Crakes (102) (Watson 1955). A juvenile was seen sunning itself in the open in the early afternoon, while another bird climbed vegetation to sun itself on a cool day; birds sometimes loaf in open areas such as the sandy margins of creeks (Marchant & Higgins 1993).

SOCIAL ORGANISATION Little known; probably monogamous and territorial. Birds are usually seen singly, in pairs or in family groups. Nesting is solitary.

SOCIAL AND SEXUAL BEHAVIOUR Almost nothing is known about its agonistic behaviour, although one bird drove a Spotless Crake and a rabbit from a clump of vegetation (Leicester 1960). Birds are cautious when approaching the nest. When disturbed, the incubating bird readily slips away and hides, sometimes at the base of the nest and once even against an intruder's boot (Fletcher 1913); the disturbed bird may dive. In late incubation the bird is less willing to leave the nest (Leicester 1960); it will usually do so when the overhead cover is parted but it may remain and allow itself to be touched (Skemp 1955). The adult may defend the nest by chasing away threatening intruders (Marchant & Higgins 1993). Sexual behaviour is unknown.

BREEDING AND SURVIVAL Season In Australia usually Aug-Jan; in New Guinea breeds Sep-Mar, and possibly also Jun-Aug, during both rainy and dry seasons. All following information on Australian populations, from Marchant & Higgins (1993). **Nest** Built at edge of marsh or other flooded area; usually above water but also on or just above dry ground near water. Nest a shallow cup or saucer-shaped structure, well hidden in clumps or tussocks of grass or sedge (e.g. *Carex, Gahnia, Juncus, Cladium*), or in samphire *Halosarcia* spp.(?); sometimes covered by blackberries *Rubus fruticosus*; also in tangled grass under fern *Gleichenia circinata*; sometimes a solid platform in denser cover; often has approach runway or ladder. Some nests may be destroyed by high tides. Nests often at high density along drainage areas and river flats. Average height above ground 64cm; one nest was 1.25m above ground; external diameter 18(10-24)cm and depth 8 (5-11)cm; cavity diameter 12-13cm and depth 4(2-5)cm; cup 10-60cm above mud or water. Nest woven from rushes or dead grass stems; lined with fine green stems; often with canopy woven from overhanging vegetation; construction takes 4 days to over 1 week; replacement nests built more rapidly. May also build nursery nests or platforms. **Eggs** 3-8; colour differences and largest clutch size suggest occasional laying by 2 females; oval or elliptical; smooth and more or less glossy; dull white to creamy-pink, warm brown or reddish-buff, with freckles, short streaks, spots and blotches of purplish to chestnut-brown or black, with underlying violet-grey; markings evenly distributed, or concentrated at blunt end. Size (n = 15) 33.8-38.6 x 25.4-27.2 (34.9 [SD 1.2] x 26.3 [SD 0.6]); calculated weight 12-13. Eggs laid at daily intervals. **Incubation** 19-21 days, probably by female only; one pair re-laid within 20 days of clutch failure. Eggs usually hatch within 24h. **Chicks** Precocial and nidifugous; remain in nest for up to 24h but will leave nest if warned by female; young fed and cared for by both parents. May raise 2 broods per season; last brood may remain with parents during autumn. Hatching success of 18 eggs 72% (13 hatched); eggs taken by water rat *Hydromys chrysogaster*.

66 AUCKLAND RAIL
Lewinia muelleri   Plate 20

Rallus muelleri Rothschild, 1893, "Auckland Island".

Sometimes placed in the genus *Dryolimnas*, or regarded as a race of *L. pectoralis*. Forms a superspecies with *L. mirificus* and *L. pectoralis*. Monotypic.

Synonyms: *Dryolimnas/Hypotaenidia/Hyporallus muelleri*; *Lewinia pectoralis muelleri*.

Alternative name: Auckland Island Rail.

IDENTIFICATION Length 18-21cm. Forehead to hindneck darkish rufous, narrowly streaked blackish only on crown; rufous brightest on broad supercilium, across ear-coverts and down sides of neck. Rest of upperparts, and upperwings, olive-brown, narrowly streaked black; black and white barring on some outer and median upperwing-coverts, and on tertial and median coverts at base of wing; marginal coverts and outer edge of alula white. Lores dark brown; chin whitish, grading to grey on malar region, throat, sides of neck, foreneck and breast, whole area strongly washed buffish-olive; centre of breast paler. Flanks, belly and undertail-coverts black with narrow white bars. Iris brown; bill pinkish-red with darkish grey or brown-grey culmen and tip; legs and feet pink, tinged grey. Sexes alike but female may be duller. Immature duller than adult; juvenile apparently similar to that of Lewin's Rail but with darker lores and ear-coverts, and with pale streaking occupying 30% of crown and 60% of upperparts. Inhabits grassland or herbs on damp to wet ground; also scrubby forest.
Similar species Much smaller than Lewin's Rail; bill slightly straighter; plumage somewhat softer and denser; no dark streaking on forehead and hardly any on hindneck; upperparts olive-brown with sparser, narrower dark streaks; barring on upperwing-coverts much restricted; primaries unspotted. Brown-banded Rail (64) of the Philippines is also larger; upperparts darker brown and almost unpatterned except for narrow pinkish-buff bars on wings and small buff or pinkish spots on back, uppertail-coverts and tail; pale supraloral line present; breast greyer; dark flank barring perhaps slightly browner and pale barring buffier.

VOICE Most information is from Elliott *et al.* (1991). The commonest calls are a loud descending *crek* given at c. 1 per s and repeated c. 10 times, and a loud, sharp, whistle-like call at c. 4 per s and repeated c. 50 times; both calls are audible for several hundred metres. The birds also make a variety of grunts and clicks. The distress call is a sharp single note; a captive bird made a two-note call *kek kuk* when agitated, and also gave muffled sounds like faint thumps (Falla 1967). Chicks make squeaking calls. Some calls are similar to those of Lewin's Rail.

DESCRIPTION The original description (Rothschild 1893) appears not to refer to this species and the type is now apparently lost (Falla 1967). A skin, supposedly from Auckland I, and obtained by von Hügel in New Zealand in 1874, is a Lewin's Rail and was probably not collected on Auckland I (Greenway 1967).
Adult Forehead to hindneck darkish rufous, brightest in stripe from broad supercilium across ear-coverts and down sides of neck, and with narrow blackish feather centres only on crown (hardly any on hindneck). Rest of upperparts, including scapulars and tail, olive-brown, feathers with narrow black centres giving relatively sparsely streaked effect; fringes of uppertail-coverts and rectrices possibly slightly rufous-tinged. Upperwing-coverts largely as upperparts, black-and-white barring restricted to some outer and median coverts and to tertial and median coverts at base of wing; marginal coverts and outer edge of alula white; primaries dark greyish-brown, unspotted; inner 7 primaries have very narrow dull olive-brown outer edges, as do outer 5 secondaries; inner secondaries and tertials dark greyish-brown with olive-brown fringes. Axillaries and underwing-coverts grey with white tips; underside of remiges brownish-grey. Lores dark brown to blackish-brown; chin, and sometimes upper throat, whitish; malar region, throat, sides of neck, foreneck and breast grey, feathers with broad buffish-olive tips giving entire area strong buffish-olive tone which remains even when plumage worn; centre of breast paler grey, less strongly washed buffish-olive. Vague dusky barring and some white barring may appear on lower breast, grading into black and white barring of flanks, belly, thighs and undertail-coverts; barred feathers are tipped buffy. Iris brown; bill pinkish-red, grading to darkish grey on distal third and culmen; legs and feet pink, tinged grey. Some variation in extent of rufous on head and neck but not known if this is sex-related. Adult female may be duller than male, with less extensive rufous on head and neck, as in Lewin's Rail.
Immature Not properly described. Apparently duller than adult (Falla 1967).
Juvenile Apparently similar to that of Lewin's Rail but with darker lores and ear-coverts, and with pale streaking occupying 30% of crown and 60% of upperparts.
Downy young Not described.

MEASUREMENTS (6 adults) Wing (chord) 76-85 (80.7, SD 3.7); tail 29-42 (36.3, SD 5.5); exposed culmen 24-32.5 (29.4, SD 3.2); tarsus 27-29 (28, SD 0.8) (Elliott *et al.* 1991). Weight of 1 female 63, of 4 adults 89-100 (93, SD 4.8), of 1 juvenile 63.

GEOGRAPHICAL VARIATION None.

MOULT No information available.

DISTRIBUTION AND STATUS Known only from Adams and Disappointment Is, Aucklands group (Elliott *et al.* 1991, Marchant & Higgins 1993). This species is VULNERABLE. It was probably eliminated on Auckland I by feral cats and/or pigs in the 1860s, and was thought to be extinct on other islands until 1 was caught on Adams I in 1966 and a population of several hundred was found on there (area 10,000ha) in 1989 and another of c. 500 on Disappointment I (400ha) in 1993 (Collar *et al.* 1994); suitable habitat probably occupies c. 10% of Adams I. The population on Adams I is said to be c. 1,500 birds (Heather & Robertson 1997). These islands are part of the Auckland Is Nature Reserve and access is strictly controlled; the survival of the species depends on the continued exclusion of mammalian predators, especially cats, pigs and rats *Rattus rattus*, which are present on Auckland I only a few hundred away (Collar *et al.* 1994). Adams I has no introduced mammals but has New Zealand Falcons *Falco novaeseelandiae* and Brown Skuas *Catharacta skua*, which are possible predators of the rails. The species possibly also occurs on Ewing I (57 ha), whence there are 3 unsubstantiated reports since 1942 (Falla 1967), but there is now little, if any, suitable habitat on the island; formerly there was potentially suitable *Poa litorosa* tussock-grassland habitat but this was replaced by *Olearia lyalli* forest during this century, which may have resulted in the decline or disappearance of the rail (Elliott *et al.* 1991). Rose I may have suitable *Poa litorosa* habitat, but rabbits may prevent the vegetation being suitable for rails, while habitat on Enderby I (688 ha) could become suitable for the introduction of the rail if cattle and rabbits were removed (Elliott *et al.* 1991).

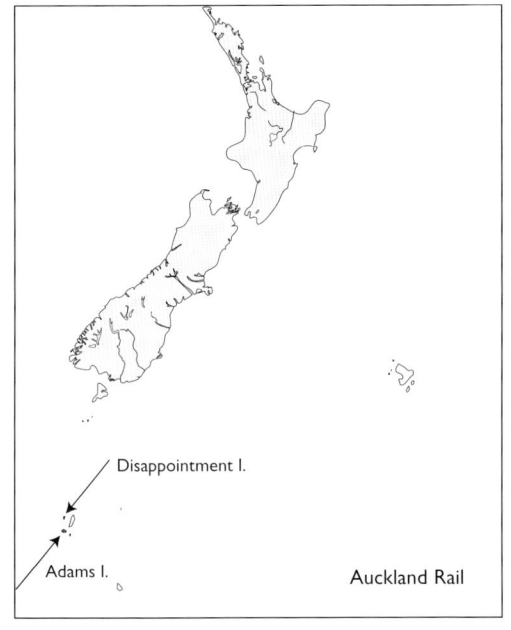

Auckland Rail

MOVEMENTS Probably none, but birds are very difficult to observe.

HABITAT Information is from Elliott *et al.* (1991). On Adams I this rail has been found in four different vegetation types: coastal herbfield just above the high water mark and dominated by the megaherbs *Pleurophyllum criniferum, P. speciosum, Anisotome latifolia* and *Stilbocarpa polaris* and the sedge *Carex trifida*; herbfield, on steep slopes below coastal cliffs, with the same herbs but no *Carex*, and with the grass *Chionochloa antarctica* and some *Bulbinella rossii*; grasslands, up to 250m a.s.l., dominated by *Carex appressa* and with *Blechnum procerum, Chionochloa antarctica* and *Pleurophyllum crinium*, with low-growing shrubs of *Myrsine divaricata, Coprosma ciliata* and *C. cuneata*; and tussock-herbfield at 300-500m, of *Chionochloa antarctica, Pleurophyllum* spp., *Anisotome antipoda, Bulbinella rossii* and, at low altitudes, the 3 woody shrub species already mentioned. Birds also ventured into scrubby forest of *Metrosideros umbellata, Pseudopanax simplex, Myrsine divaricata* and *Dracophyllum longifolium* with a 2.5m high canopy and a 1m subcanopy over dense patches of *Carex appressa* and *Blechnum procerum*. All vegetation types were on damp to wet ground and had a dense canopy (a subcanopy in the forest) c. 1m off the ground and open areas beneath.

FOOD AND FEEDING Little information is available, but the food and foraging methods are probably very similar to those of Lewin's Rail. It was once seen foraging at a rubbish dump. A captive bird ate insects and other invertebrates, and was also fed on mealworms (*Tenebrio* larvae) and "mash" (Falla 1967).

HABITS Most information is from Elliott *et al.* (1991). This species is very secretive and inhabits dense vegetation in which observation is very difficult. Birds were heard calling between dawn and c. 30 min after dark, and are presumed to be active during this period and also possibly for longer after dark. Two birds fitted with radio transmitters were active between 08:00 and 22:00hrs. This rail can fly quite strongly, but apparently does so only infrequently (Falla 1967).

SOCIAL ORGANISATION Probably monogamous and territorial, as is Lewin's Rail.

SOCIAL AND SEXUAL BEHAVIOUR Unknown; probably similar to Lewin's Rail.

BREEDING AND SURVIVAL Season Recorded Nov; season probably Oct-Dec. Two nests described (Elliott *et al.* 1991, Marchant & Higgins 1993): both in wetlands, in tussocks of *Carex appressa* and *Chionochloa antarctica*, sheltered by overhanging or interlaced vegetation and with well-defined entrance runway. **Nest** A shallow cup of the above grass and/or sedge; external diameter 14-23cm and depth 7-16cm; internal diameter 7cm and depth 2-7cm; one nest 14cm above ground, the other had material down to ground. One clutch of 2 eggs recorded. **Eggs** Oval to elongated oval; cream, with brown, red-brown and pale grey spots and blotches concentrated at larger end; size of 2 eggs 35.1 x 26.5, 32.8 x 24.2; weight 12.0, 8.5 (eggs incubated for c. 2 days). **Incubation** Period probably similar to that of Lewin's Rail. A wild-caught immature lived in captivity for 9 years (Heather & Robertson 1997).

DRYOLIMNAS

This genus is sometimes expanded to include the rails in the genus *Lewinia* (for a discussion of this, see the *Lewinia* genus account). As currently defined, however, it includes only one species, the White-throated Rail, which is endemic to the Malagasy region. This rail bears a superficial resemblance to the *Canirallus* rails of Africa and Madagascar but differs markedly in skeletal characters (Olson 1973b). Of its three races, one is recently extinct, one is flightless and restricted to islands in the Aldabra atoll, where it inhabits dry coral scrub, and one is widespread and fairly common in Madagascar, occurring mainly in forest and wetland habitats. The Aldabra form is the last surviving flightless rail in the Indian Ocean region.

67 WHITE-THROATED RAIL
Dryolimnas cuvieri   Plate 21

Rallus Cuvieri Pucheran, 1845, (Mauritius).

Sometimes included in the genus *Canirallus*, but skeletally very close to *Lewinia*. Race *abbotti* of Assumption I recently extinct; two extant subspecies recognised.

Synonyms: *Rallus/Eulabeornis gularis*; *Dryolimnas aldabranus*; *Rougetius abbotti/bernieri/gularis/aldabranus*; *Rallus/Calamodromus bernieri*; *Canirallus kioloides/cuvieri*.

Alternative names: Cuvier's/Aldabra (White-throated) Rail.

IDENTIFICATION Length 30-33cm. A medium-large, dark, predominantly rufous and brown rail, with prominent white chin and throat. Head, neck, and breast to upper belly deep vinous-chestnut; rest of upperparts, including scapulars, tertials and tail, greenish-olive, with black streaks on back, scapulars and tertials (usually much reduced or absent in race *aldabranus* and extinct race *abbotti*). Upperwing-coverts unstreaked; primary coverts, primaries and outer secondaries sooty-brown, olive-green towards tips; axillaries and underwing-coverts black, broadly barred white. Flanks dark olive-brown; lower belly to undertail-coverts dusky, barred buffy; lateral undertail-coverts white. Iris reddish (chestnut or orange in *aldabranus*); bill pinkish to red, tipped black; legs and feet dark olive. Sexes alike; female averages slightly smaller than male; female *aldabranus* has bright pink base to upper mandible, male dull or dark red base. Immature duller than adult, with brownish bill and brown eyes. Juvenile brown where adult vinous-chestnut; upperparts dingy olive; underparts duller than in adult; dark streaks on upperparts, and white barring on underparts, reduced and less distinct. Inhabits rank forest undergrowth, riparian marsh, mangroves, and dry coral scrub.

Similar species Distinguished from other rails on Madagascar by larger size, dusky lower belly and thighs, white on throat and undertail coverts, and straight bill of medium length, pinkish or reddish with dark tip. Superficially similar Madagascar Wood-rail (12) smaller and stockier; has grey on forehead and face; white throat patch smaller; crown to hindneck olive-brown to olive-green; lower back to tail, and lesser upperwing-coverts, chestnut; some white barring on remiges; no white lateral undertail-coverts; bill shorter, deeper and blackish, with blue-grey tip and cutting edge.

VOICE Rather vocal. The following information is from Huxley & Wilkinson (1977, 1979) and Wilkinson & Huxley (1978) unless otherwise stated. The normal contact call between mated birds is a muffled staccato grunt *mp* followed by a *yeah*; the grunt is relatively infrequently given on its own, but may be used by foraging birds. When agitated, or disturbed by an intruder, or during or after a mild territorial encounter, birds utter the grunt followed by a click; there is also a loud warning click *mptok* which is sometimes followed by a whistle to give a call rendered *mptiuu*. There is a low, rasping purr, given when mated birds meet, during allopreening and courtship feeding, before copulation, after song duets and during nest site selection, nest building and incubation. A high-pitched squeal, which rapidly attracts other rails, is given in situations of extreme distress and by birds in the hand, while a more raucous squeal, often preceded by a *tok* call, is given during nest defence. The song, usually uttered towards the evening, and sometimes during the night, often begins with a series of *mp yeah* notes and develops into a series of loud, piercing whistles, rising in pitch and intensity and developing frequency changes; each whistle is preceded by a *mp* note. Birds often duet, initially antiphonally and then gradually becoming co-ordinated, eventually calling in unison; the calls die away after up to c. 1 min. Rand (1936) describes a sharp squeak *tsikeu* which is given when the bird is surprised, and Penny & Diamond (1971) note a similar call, a high-pitched, two-note whistle, given when a cat approached two captive birds; Rand also mentions a low, plaintive call only audible at close range. Van Someren (1947) describes a harsh two-part call: a whistle followed by a short rattle.

Chicks emit a contented peep (a whistle of rising pitch) until c. 1 week old, usually when brooded. They also have a *tiuu* whistle of descending pitch, which is a location call given when they become separated from the family party; a twitter is given by chicks following the parents and, more excitedly, when about to be fed. The *tiuu* call develops into the adult *mptiuu* after independence, while the twitter is not given after independence. Chicks also give a version of the adult distress and alarm calls. The first adult-type calls are given from c. 12 weeks of age.

DESCRIPTION *D. c. cuvieri*
Adult Head, neck, and breast to upper belly deep vinous-chestnut; washed dull olive on crown; malar region, chin and throat white. Rest of upperparts greenish-olive, with black central streaks to feathers of back and scapulars; lower back to uppertail-coverts uniform greenish-olive; rectrices dark greenish-olive, more blackish basally. Upperwing-coverts largely unstreaked, but some birds have narrow dusky shaft-streaks; alula, primary coverts, primaries and outer secondaries sooty-brown, outer webs paler, some with olive-green outer edges; tertials olive-green, with black central streak; axillaries and underwing-coverts brownish-black, broadly barred white; underside of remiges dusky, washed olive. Flanks dark olive-brown; lower belly and thighs dusky, barred fulvous or ashy-buff; median undertail-coverts blacker, feathers barred fulvescent to buff; lateral undertail-coverts white, sometimes barred dusky brown or dull chestnut towards tips. Iris reddish; bill dark pinkish-olive to dark red, with black tip; legs and feet dark olive. Wing-claw c. 10mm long present. Sexes alike, but female may have different bill colour (see *aldabranus*).
Immature Somewhat duller than adult, with brownish bill and brown eyes.
Juvenile Duller than adult. Head, neck and breast dull olive-brown, neck and breast becoming mixed with dull chestnut; rest of upperparts dingy olive, less green-tinged, and streaks less distinct (but more distinct in 1 bird); chin and throat whitish, spotted dull chestnut; posterior underparts paler and duller, barring on belly more diffuse, reduced or absent. Iris greyish-brown; bill black; legs and feet sepia.
Downy young Down black, shiny, with olive tinge; iris olive-green; bill, legs and feet black or dark brown; down becomes dull brown with strong olive tinge (race *aldabranus*).

MEASUREMENTS (7 males, 12 females) Wing of male 147-160 (154.3, SD 4.5), of female 142-162 (147.8, SD 5.3), excluding largest female, which may be wrongly sexed, wing 142-151 (146.5, SD 3.0); culmen of male 41-46 (44.4, SD 1.8), of female 38-45 (40.3, SD 1.9); tarsus of male 41-46 (43.6, SD 1.7), of female 40-46 (42.5, SD 2.1) (Benson 1967). Weight of male 276, of female 258, of downy young 22.5 (Ripley 1977).

GEOGRAPHICAL VARIATION Races separated on overall colour, extent of white on throat, presence or absence of black streaks on upperparts, width and prominence of white barring on belly, iris colour, and size.

D. c. cuvieri (Pucheran, 1845) – Madagascar; formerly possibly also Mauritius. See Description.

D. c. abbotti (Ridgway, 1894) – Assumption I, Indian Ocean; possibly this form also on Astove and Cosmoledo (Benson 1967, Stoddart 1971). Recently EXTINCT. Differed from nominate in being smaller; upperparts paler, more greyish-olive, with dark streaks narrower, fainter or absent (very variable); white on throat more extensive, tending to reach upper breast; white barring on underparts broader and coarser. Iris red; bill black, base of lower mandible to gonys bright red; legs and feet dark brown. Wings shorter than in nominate; was in process of losing power of flight. Immature has iris brownish-grey; bill, legs and feet brown. Wing of 3 males 133, 136, 141, of 1 female 136; culmen of 4 males 40-45 (43, SD 2.4), of 1 female 41; tarsus of 4 males 39-41 (39.5, SD 1.0), of 1 female 39 (Benson 1967).

D. c. aldabranus (Günther, 1879) – Aldabra Is (Malabar, Ile aux Cèdres and Polymnie). Differs from nominate in being smaller and paler, especially on flanks and lower belly; black streaks on upperparts and upperwings absent or much reduced (but very variable); white bars on belly less distinct; iris chestnut to orange; bill black, base dull or dark red in male, bright pink in female; legs and feet blackish-brown. Immature has unstreaked upperparts and more distinct barring on belly (Sharpe 1894). Juvenile has breast tinged reddish on predominantly greenish-olive base; iris changes from olive to yellowish-olive (Penny & Diamond 1971). Wing of 7 males 111-123 (118.1, SD 3.7), of 5 females 110-120 (115, SD 3.7); tail of 5 males 28-42 (38.4, SD 5.9), of 5 females 32-40 (36, SD 3.4); culmen of 17 males 39.5-47 (44.1, SD 1.9), of 14 females 39-43 (40.2, SD 1.2); tarsus of 17 males 39-44.5 (41.9, SD 1.6), of 14 females 36.5-43 (39.9, SD 1.7). Weight of 32 males 145-218 (188.6, SD 17.8), of 22 females 138-223 (176, SD 21.9); weight of individuals varies widely, showing no correlation with time of day, season or sex; greatest individual variation 155-223 (female) (Penny & Diamond 1971).

MOULT Almost all birds handled by Penny & Diamond (1971) showed extreme abrasion of the remiges, possibly due to a combination of the degeneration of the feather structure concomitant with flightlessness and of wear caused by movement through dense undergrowth. Remex moult is simultaneous or partly simultaneous; 4 *aldabranus* were in advanced remex moult, Jan-Feb, with primaries and secondaries moulted concurrently but in irregular sequence, groups of 2-4 feathers being dropped simultaneously from different areas (Penny & Diamond 1971); 1 adult male *abbotti* had all remiges 0.5-2.5cm long, Mar. In *aldabranus*, moult of rectrices is recorded in Jan-Feb, and head and body moult in Oct-Feb.

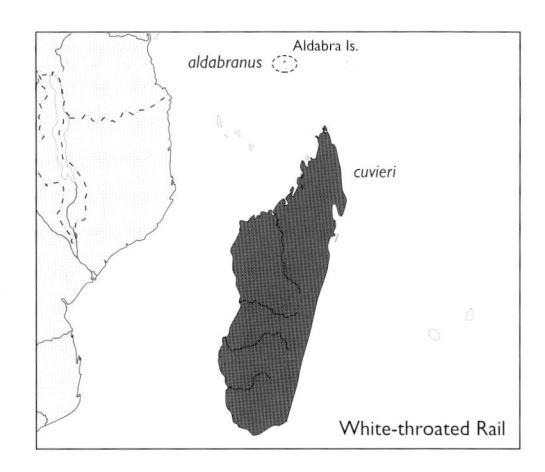

White-throated Rail

DISTRIBUTION AND STATUS Madagascar and Aldabra Is; formerly Assumption I and possibly Mauritius (e.g. Benson 1967). This rail is not globally threatened, although the race *aldabranus* has been considered RARE (Collar & Stuart 1985). The race *cuvieri* is common over much of its range in Madagascar, the greatest concentration of records coming from EC and SC areas (Dee 1986). In 1997 it was very common in the NE around Maroansetra, where c. 10 birds were recorded in 5ha (A. Riley *in litt.*). It may have occurred at a density of c. 1 pair per 10ha near Antananarivo in the 1970s (Salvan 1972b). However it is rarer on the central plateaux, where it may have almost disappeared, probably as a result of hunting (Salvan 1972b, Milon *et al.* 1973). It is obviously very adaptable, occupying a wide range of habitat types. On Aldabra, the race *aldabranus* occurs on 2 major and 1 smaller islands, the largest population being on Malabar (Hambler *et al.* 1993); it was apparently wiped out on most other islands shortly after the introduction of cats in about 1890. It disappeared from Ile Michel as recently as the 1970s (Collar 1993). Its current population is c. 8,000 birds, at an estimated overall density of 1-2 birds/ha, and has remained relatively constant for 20 years, but its distribution continues to contract (Hambler *et al.* 1993); in 1997 the population appeared to be stable at a similar density to that found by Hambler *et al.* (1993), while it was confirmed that rails still exist on Ile aux Cedres, where their survival had been in doubt (R. Chapman *in litt.*). There are no known significant predators of adults, although birds are occasionally eaten by humans; the threat posed by feral cats is currently not serious but colonisation by cats may constitute a major long-term threat, so the establishment of captive breeding populations has been recommended: the species will probably thrive well in captivity (Penny & Diamond 1971, Hambler *et al.* 1993). The birds apparently can tolerate large populations of Black Rats *Rattus rattus* (Penny & Diamond 1971) but these may take eggs and chicks, as do land crabs *Cardisoma carnifex* (King 1981). Aldabra is a World Heritage Site and the rail could presumably be reintroduced to other islands in the group, including the largest island, Grand Terre, if cats could be eliminated; this proposal has recently been reiterated by Hambler *et al.* (1993). In view of the fact that this race is confined to the Aldabra Atoll, urgent consideration should be given to establishing a remote population, either in the Seychelles or on islands elsewhere in the Indian Ocean (e.g. Ile aux Aigrettes, off Mauritius), to safeguard the taxon against extinction by catastrophic events such as cyclones, tidal waves or disease epidemics (R. Chapman *in litt.*): see Brook *et al.* (1997) for a discussion of this topic in relation to the Lord Howe Rail (41). The introduction of diseases carried by domestic chickens is a potential threat, but chickens are currently confined to Picard, where only 1 rail currently exists; because of this threat, rails from Picard should never be moved to any other island (R. Chapman *in litt.*). The race *abbotti* was very common on Assumption I in 1906 (Nicoll 1906, Meade-Waldo 1908), but Nicoll (1908) predicted its extinction due to imported rats which were very abundant and probably ate many eggs; it was still very common in 1908 (Fryer 1911) but was extinct by 1937 (Vesey-Fitzgerald 1940).

MOVEMENTS There is no evidence for movements in Madagascar. On Aldabra, ringing has shown that populations may be transient and fluid to some degree, even during the breeding season, and large numbers of birds may be attracted to long-term camps, dispersing when the people leave; they are not attracted to temporary camps (Penny & Diamond 1971). Widespread or rapid dispersal, relatively short-term site fidelity (<5 years) and nomadic movement by the rails are all possible (Hambler *et al.* 1993).

HABITAT The nominate race occupies diverse habitats on Madagascar: forest with luxuriant herbaceous ground cover; woodland watercourses; stream edges with tall, dense herbaceous growth; wetlands, including marshes with long grass, reeds and sedges, and rice paddies; mangroves and coral islet beaches; it ranges from sea level to 1,800m but is less numerous above 1,100m (Rand 1936, Langrand 1990). On Aldabra, *aldabranus* occurs mainly on extremely rough, heavily dissected coral limestone covered with dense mixed-species bush or very dense scrub dominated by *Pemphis acidula*; it also occurs in less dense vegetation and mangroves, and forages on pebble and sand beaches (Penny & Diamond 1971, Huxley & Wilkinson 1977). Its densest populations apparently exist in dense scrub habitats, it is relatively sparse in the less dense vegetation types, and is probably common in mangroves (Penny & Diamond 1971). The extinct *abbotti* occupied marshy to dry coral sites on Assumption I (Ripley 1977).

FOOD AND FEEDING Most information is from Penny & Diamond (1991). This rail eats predominantly animal food and also scavenges. Its main food is invertebrates: small molluscs, including Gastropoda such as *Melanoides* and *Littorina*; small ghost crabs, including *Ocypode cordimana*; and many types of insect, including Diptera, Coleoptera, termites, and the eggs and larvae of ants *Camponotus maculatus* which it obtains by pecking at the nest; it scavenges dead crabs. It takes tortoise droppings, possibly to extract insects; it takes Tabanidae flies and gleans parasites from the carapaces of tortoises (Honegger 1966, Penny & Diamond 1977); and it takes insects (Coleoptera and Diptera) and their eggs from decaying tortoise carcases. It eats eggs and hatchlings of Green Turtle *Chelonis mydas* at nests on the beach, extracting buried hatchlings with the bill after probing (Frith 1977); turtle eggs are usually speared with the bill and carried off to be eaten in cover (R. Chapman *in litt.*). On Aldabra, birds scavenge for scraps at campsites, and birds were attracted to traps baited with cheese, biscuits, fruit loaf, cake crumbs, broken hermit crabs, flesh of *Conus* molluscs, and chopped fish. It forages in leaf-litter, including *Casuarina* needles on Aldabra where it also feeds at litter disturbed by passing tortoises, in shallow water, in mud among mangroves, and on rocky or sandy shores. When feeding in water, it submerges the head and rakes vigorously into the mud with its bill. It drinks both fresh and salt water, but prefers fresh water for drinking and bathing. When dealing with hard food items such as small crabs, the bird delivers a series of pecks vertically downwards at the item, pausing frequently to examine the object and to reorientate it if necessary. Aldabra Giant Tortoises *Geochelone gigantea* respond to the visual stimulus of a rail within 1m by adopting an immobile erect stance, and also make this response to light tactile stimulation of their soft posterior regions; it is suggested (Huxley 1979) that this is a co-operative posture assisting a symbiont to clean ectoparasites from the soft skin thus exposed.

HABITS Apart from instances of nocturnal calling, there is no evidence that this species is active during the night. It is apparently not as secretive as many other rails, even in Madagascar (Rand 1936), and it may be bold in the presence of man; birds near Maroansetra fed in the open near a busy path frequented by many pedestrians (A. Riley *in litt.*). On Aldabra, if an observer stays still birds will approach him and give circumspect pecks "apparently out of curiosity" (Penny & Diamond 1971). Birds seen in open situations jerk the tail while walking; when flushed they may fly up and perch in low bushes. The flight appears feeble in the Madagascar race; Aldabra birds are flightless but use the wings to assist in jumping, and the Assumption I race apparently had some flying ability but normally ran when pursued. Bathing is reported but no details are available except that, after bathing, captive birds adopted a spread-wing posture (Penny & Diamond 1971). On Aldabra the birds keep in the shade and are very rarely seen feeding in sunlit areas (Penny & Diamond 1971), but in Madagascar they are not averse to frequenting sunlit areas (A. Riley *in litt.*). A male was once seen climbing to the top of a small shrub and perching under the canopy; the function of this behaviour is unclear, but may be connected with choosing a nest-site (Penny & Diamond 1971).

SOCIAL ORGANISATION Monogamous and permanently territorial, forming permanent pair bond; unmated females also hold territories. One instance of two males copulating with the same female is recorded (Penny & Diamond 1971). On Ile Malabar, Aldabra, C.R. Huxley (in Collar 1993) found a territory every 70m along a transect through dense mixed scrub, 1 per 87m in less dense *Pemphis* scrub and 1 per 120m in open mixed scrub. Site persistence appears to be high in some cases, but few recaptures of ringed birds are recorded.

SOCIAL AND SEXUAL BEHAVIOUR This rail is aggressive towards introduced black rats. Frith (1977) observed a large rat approaching camp scraps on which an adult and immature rail were feeding; the adult rail pecked the rat severely on the nose and the rat retreated after its repeated approaches had been met by similar attacks. It is also aggressive to crabs *Birgus latro*: when both species are competing for food, the rails will peck the crabs and chase them back (R. Chapman *in litt.*).

In territorial attacks the defending male ruffles the back feathers, holds the head down and runs at the intruder, which flees; only once was the pursuer seen to strike the fleeing intruder, using the bill only (Penny & Diamond 1971). Immatures from the previous breeding season are tolerated within territories, wander widely between territories and do not respond to taped playback

(Hambler *et al.* 1993). In a dispute over a feeding territory witnessed by Frith (1977), two birds faced each other 30cm apart, with bills lowered, giving an occasional grunt accompanied by an upward flick of the bill. They then leaped at each other, feet first, beating the wings and kicking vigorously; if one bird secured a grip, both fell to the ground and pecked each other. Such fights could result in significant injury, and the conspicuous wing buffeting suggests that the wing-claw may be used as a weapon (Frith 1977). Similarly violent fighting was also reported by Abbott (in Ridgway 1895).

A greeting display between paired birds which meet during foraging involves both birds standing quite still facing each other for a few seconds with necks extended, displaying the white throat patch, and then giving purring notes when they separate to feed (Penny & Diamond 1971). During the song, the bird adopts an upright pose, displaying the white throat patch, and fans and displays the white outer undertail-coverts (Huxley & Wilkinson 1979); when duetting (Fig. 8), birds stand close together, progressively opening and raising their bills until they are wide open and pointing up at an angle of c. 60°; duetting is sometimes followed by copulation (Penny & Diamond 1971).

The female solicits copulation by spreading and lowering the white outer undertail-coverts, and lowering the bill until the tip touches the ground; the male mounts, droops his wings and copulates briefly. Copulations between paired birds are not followed by any post-copulatory display, the birds simply separating to feed (Penny & Diamond 1971), but after mating with a female with which he is not paired, the male walks around the female with neck extended, head held low, and wings spread and slightly raised; this may be an appeasement posture (Penny & Diamond 1971, Frith 1977). The female has been seen to allopreen the male, on one occasion after taking food from his bill, and may solicit copulation after allopreening (Penny & Diamond 1971).

Incubating birds, particularly females, defend the nest fiercely and will attack a human intruder. In the defence display the female raises and spreads the wings to form almost a ruff behind the head, holds the head low with the bill pointing at the intruder, and repeatedly rushes forwards to deliver several pecks, giving the *mptok* call; the female repeatedly gives this display while circling the intruder, and the male usually stays in the background, giving purring calls and adopting a less intense threat posture (Penny & Diamond 1971).

BREEDING AND SURVIVAL Season In Madagascar, possibly breeds mostly during rains, laying Oct, Feb, Mar, breeding condition Nov, Jan, also chicks aged 2 weeks and 8-10 weeks with adults in late Nov (A. Riley *in litt.*); on Aldabra lays Dec, chicks hatch Dec-Jan; Assumption, Feb (black chicks 12 Mar). Female normally selects nest site, to which male is led. **Nest** Flimsy, a few twigs and leaves placed in depression among rocks; or a more substantial deep bowl of interwoven leaves and grass, placed on or near ground in dense vegetation such as grass and *Euphorbia* in Madagascar and among stems of small shrubs on Aldabra. **Eggs** Usually 3-4 (3-6); broad oval, sometimes almost spheroidal (nominate race); smooth and glossy; white to creamy-white, sprinkled and dotted various shades of browns, reddish-browns and greys, with underlying pale purple spots; markings sometimes more numerous around large end; size: nominate 38-42.5 x 28.8-32.5 (40.8 x 31.4, n = 18), calculated weight 22; *aldabranus* 40.6-47.4 x 26.6-31.5 (43.2 x 29.6, n = 10), calculated weight 21. **Incubation** Period unknown. Following information, all relating to *aldabranus*, is principally from Penny & Diamond (1971). **Chicks** Both parents tend young. Chicks follow adults very closely; adult places food items in end of chick's bill. Feathers appear first on upperparts. At 6 weeks young are c. 75% grown and can be largely self-feeding, but young often fed until full independence at 12-15 weeks, after which parents become aggressive to them. Immatures may stay near parents until onset of next breeding season. In *aldabranus* typically only 1 young per brood survives, probably due to predation. Age of first breeding 9 months or more. **Survival** Longevity, from recoveries of ringed birds, 5-8.5 years (Frith 1977, Hambler *et al.* 1993).

Figure 8: Aldabra White-throated Rails duetting (female on left). [After Huxley and Wilkinson 1979].

CREX

Two species of medium-sized rails which typically inhabit grassland rather than marsh. Both species are streaked on the upperparts and have grey from the sides of the head to the breast; in the Corncrake the grey is more extensive in males than females and in adults than immatures. The Corncrake has tawny upperwing-coverts which are distinctive in flight. Both species have barred underparts, the bars being black and white in the African Crake and rufous-brown and white in the Corncrake. In plumage, the African Crake bears a strong superficial resemblance to some *Porzana* crakes and has been included in that genus, but the tenuous nasal bar is unlike that of these and most other crakes. The African Crake is confined to sub-Saharan Africa, where it is widespread and locally common; it is both resident and migratory, many birds moving away from the equator to breed during the rains. The Corncrake is a long-distance migrant which breeds in the Palearctic and winters in central and southern Africa; in its breeding areas it has suffered extensive loss of habitat, while mechanised grass cutting has caused heavy losses of breeding birds, eggs and young, and its numbers have declined drastically; it is now regarded as globally vulnerable.

68 AFRICAN CRAKE
Crex egregia Plate 27

Ortygometra (Crex) egregia W. K. H. Peters, 1854, Tete, Zambesia (Mozambique).

Sometimes placed in genus *Porzana* on the basis of superficial resemblance to *P. albicollis*, but plumage and skeletal differences make the relationship untenable. Sometimes placed in the monospecific genus *Crecopsis* but probably more appropriately included in *Crex* on the basis of similarities to *C. crex*. Monotypic.

Synonyms: *Ortygometra fasciata/angolensis*; *Porzana/Crecopsis egregia*.

IDENTIFICATION Length 20-23cm; wingspan 40-42cm. A medium-small crake, distinctive when seen well. Male has upperparts, upperwing-coverts and tertials blackish, streaked olive-brown; nape and hindneck almost plain olive-brown; primaries and secondaries dark brown. Narrow white supraloral streak continues to above eye. Sides of head and neck, and lower throat to breast, grey to bluish-grey, this colour often extending behind ear-coverts and almost entirely across nape to isolate dark crown as a distinct cap; chin and upper throat white; sides of breast washed olive-brown. Flanks and belly to undertail-coverts broadly barred black and white; centre of belly white. Iris red or orange; orbital ring salmon-pink to red; bill pink, reddish or purplish, with grey tip and dusky culmen; legs and feet pale brown or grey, often with mauve, red or pink tinge. Sexes similar but female normally slightly smaller, with less contrasting head pattern, duller grey face to breast, less regular underpart barring with narrower black bars, and more extensively unbarred centre of belly. Immature darker and duller than adult on upperparts; grey of neck and breast replaced by dark dull brown or grey-brown; underpart barring brown-black and off-white; iris hazel; orbital ring yellow-brown; bill grey to blackish; legs and feet darkish grey. Juvenile has duller, darker upperparts, face, foreneck and underparts dark grey-brown with no barring; belly whitish; bill very dark; iris grey; legs and feet grey to brown. Inhabits moist to dry grasslands, marsh edges and sometimes cultivation.

Similar species Easily separable from Corncrake (69), which also has streaked upperparts, by smaller size, darker upperparts, lack of tawny on upperwing-coverts, entirely grey face, foreneck and breast, and black and white bars from flanks to undertail-coverts. In flight, Corncrake has longer, less blunt and rounded wings, shallower wingbeats, longer tail and white leading edge to inner wing; white bars on blackish marginal coverts of African Crake give less prominent pale leading edge in flight. In profile, head of African Crake appears longer and flatter because of longer, heavier-looking bill, flatter forehead and less rounded crown; bill of Corncrake often looks short and conical, with sharp gonydeal angle. Other sympatric crakes (*Porzana, Aenigmatolimnas*) are smaller, with white markings on upperparts, different underparts pattern and short bill; juvenile African Crake not fully grown is closer in size to Striped (109) and Spotted (96) Crakes but upperparts duller, darker, less patterned, lacking any white, and underparts plain, becoming barred during post-juvenile moult. African Rail (62) has dark, unpatterned chocolate-brown upperparts, long red bill and red legs and feet.

VOICE Information is taken from Taylor (1985a and unpublished observations) unless otherwise stated. The male's advertising call is a series of rapid, quite high-pitched, grating *krrr* notes (*contra* Mackworth-Praed & Grant 1957, Snow 1978, Maclean 1993, Zimmerman *et al.* 1996 and others), repeated, at a rate of c. 2-3 notes/s, for up to several min. This call is given most frequently early in the breeding season and is normally uttered from the ground, early and late in the day but sometimes up to 3h after dark and from 1h before dawn. Calling birds adopt an upright posture with neck extended and bill horizontal; at close range the throat can be seen to vibrate. Territorial males also give this call when chasing intruders on the ground or in flight.

Both sexes give a sharp, loud *kip* call, often repeated and sometimes modified to a descending double note *kiu*; territorial males give a harsher version, a sharp *kyip* or *ky-rr* which may grade into the advertising *krrr*. The *kip* call has aggressive, territorial and alarm functions; it is given during territorial encounters, including immediately before or after a bird flies from long grass, and in flight; it is also given at the approach of a human intruder, and sometimes by birds when flushed. A quiet version was given by one pair prior to copulation. Birds flying over Harare, Zimbabwe, at night in Dec-Jan give both *kip* and *krrr* calls, and may be either on migration or circling overhead (Masterson 1991). When calling on the ground the bird adopts a pose similar to that described for the *krrr* call. When breeding territories are established, and nesting is under way, birds make *krrr* and *kip* calls very infrequently and may respond poorly to taped playback, but territorial birds give *kip* calls throughout the non-breeding season, especially at times when more African Crakes arrive in the area.

There is also a wheezy *kraaa*, given during the threat display and the postcopulatory display, and occasionally before copulation; imitation of this call brings crakes out to within 10m of an observer standing in the open. A growling version *graaa* of this call is given by territorial adults in response to taped playback. A squeaky *queeaaa* or *kaa-quee-kaa* is occasionally given by the female after copulation. There is also a loud *tsuck* of alarm, and birds in the hand make gurgling noises. Newly hatched chicks make a soft *wheeeez* call, and older chicks cheep (Masterson 1991). An agitated adult at a nest made repeated soft *koorr* calls; when chicks run to cover the adult makes a soft, low, resonant boom *hmmm*, repeated infrequently, and also a soft equine snort (Masterson 1991). The alarm call of larger chicks and juveniles is a loud, chittering *chi-chi-chi....*

DESCRIPTION
Adult Forehead and crown blackish, feathers narrowly edged olive-brown; forehead and forecrown may be tinged darkish grey or blue-grey; nape, hindneck, mantle to uppertail-coverts, scapulars and upperwing-coverts blackish, feathers with broad olive-brown fringes which are broadest on nape and hindneck (dark feather centres hardly visible) and become narrower on back and rump; rectrices blackish, edged olive-brown. Primaries and secondaries dark brown; tertials blackish, broadly edged olive-brown like upperwing-coverts; marginal coverts, axillaries and underwing-coverts blackish, narrowly barred white; underside of remiges dusky brown. Narrow supraloral streak white, continuing to above eye and very narrowly bordered blackish along lower margin. Lores, sides of head and neck, and lower throat to breast, grey to

bluish-grey; grey may extend behind ear-coverts and almost entirely across nape to isolate dark crown as a distinct cap; chin and upper throat white; sides of breast washed olive-brown. Flanks and belly to undertail-coverts broadly barred black and white; centre of belly white; thighs white to brownish; sides of vent tinged vinous-buff. Iris red or orange; orbital ring salmon-pink to red, brightest in breeding season; bill pink, reddish or purplish, becoming grey towards tip and with dusky culmen and blue-grey cutting edge; legs and feet pale brown or grey, often with mauve, reddish or pinkish tinge. Sexes alike, but female normally looks slightly smaller, with less contrasting head pattern, duller grey face to breast, no grey tinge on forehead and forecrown, less regular underpart barring with narrower black bars, and more extensively unbarred centre of belly (Taylor 1996b).

Immature Similar to adult but darker and duller; forehead and crown blackish, with very narrow pale streaks; feather fringes on upperparts darker than in adult; foreneck, sides of neck and breast dark dull brown to grey-brown; primaries and secondaries blackish; more white on centre of belly; underpart barring brown-black and off-white; iris hazel; orbital ring yellow-brown; bill grey to blackish; legs and feet darkish grey, sometimes with pinkish tinge.

Juvenile Duller and darker than adult. Feather fringes of upperparts darker and duller, although feather centres sometimes paler, more blackish-brown, than in adult; face, foreneck and underparts dark grey-brown with no barring; belly whitish; bill blackish to dark brown; iris grey; legs and feet grey to brown.

Downy young Down black; iris and bill blackish, white egg-tooth present; legs and feet blackish-brown; small wing-claw present.

MEASUREMENTS (14 males, 15 females) Wing of male 117-133 (122.8), of female 117-131 (123.9); tail of male 36-48 (41.7), of female 36-42 (39.2); culmen of male 23-31 (25.8), of female 21-26.5 (23.9); tarsus of male 38-43 (40.8), of female 39-43 (40.4). Weight of 3 males 121, 127, 141, of 16 unsexed 92-137 (117.5).

GEOGRAPHICAL VARIATION None.

MOULT A complete postbreeding moult occurs, and remex moult is synchronised. A specimen from Togo has lost all primaries in one wing and 6 in the other, while all other remiges are old and all rectrices are missing; 1 from Gambia has the 5 outer primaries growing (all the same length); 1 from S Nigeria, Aug, has 9 outer primaries in pin, the innermost, and all the primaries in the other wing, being old. Two migrating adults killed at buildings on the KwaZulu-Natal coast in Apr (fresh specimens: P. B. Taylor) had fresh remiges and body plumage, and 1 had also replaced the rectrices; this indicates that much or all postbreeding moult may occur before migration. When birds start to breed, the remiges are often moderately worn. Body moult is recorded from S Nigeria in Dec, W Sudan in Aug, S Sudan and W Uganda in Jun, S Tanzania in Feb, Malawi in Apr, Zaïre in May, and South Africa in Feb and Aug.

DISTRIBUTION AND STATUS This crake occurs almost continuously through sub-Saharan Africa from Senegal E to Kenya and S to KwaZulu-Natal, South Africa, except in arid areas of S and SW Africa, where its W limits correspond roughly with the 300-mm summer rainfall isohyet (Urban *et al.* 1986, Avery *et al.* 1988). Its N limits are: Senegal (only 3 records, from Croc, Casamance and

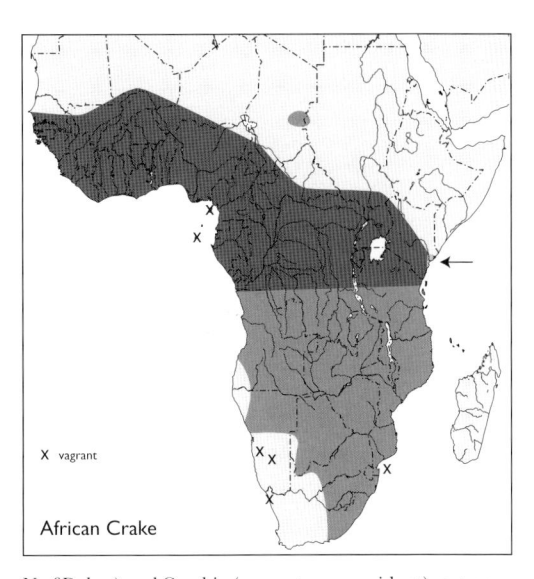

African Crake

N of Dakar) and Gambia (scarce to rare resident), extreme SE Mauritania (probably a vagrant), and S Mali (Banifing-Baoulé and Sankarani Rs), SW Niger (1 record Niamey, possibly a vagrant), extreme NE Nigeria and Cameroon, SW Chad, WC Sudan (Zalingei swamp, Darfur), SW Sudan (Juba and Yei) and C Kenya (N to L Turkana, where it is presumably a vagrant) (Lamarche 1980, Urban *et al.* 1986; Nikolaus 1987, 1989, Lewis & Pomeroy 1989, Giradoux *et al.* 1990, Gore 1990, Dowsett & Forbes-Watson 1993, Elgood *et al.* 1994, Rodwell *et al.* 1996). At the S limits of its range, in Angola it extends SW to Quilengues and Vila da Ponte, Huila, it occurs sporadically S to C Namibia and in N and E Botswana, straggling to S and SW Namibia, and in South Africa it is confined to the E regions (including Swaziland), being only a vagrant to the Northern and Eastern Cape Provinces and Lesotho (Pinto 1983, Urban *et al.* 1986, Bonde 1993, Taylor 1997a, c).

Its distribution and status are relatively well documented, as it is less secretive than other crakes. It is widespread and locally common throughout most of its range except in rain forest and arid regions, and has been recorded as particularly numerous around Bétérou, C Benin (Claffey 1995), in NE Gabon at the Ivindo basin (Brosset & Erard 1986), in Burundi on the plain of the lower Ruzuzi R and the shores of L Tanganyika (Urban *et al.* 1986), and at Balovale, NW Zambia (White 1945). In South Africa its total population is estimated at c. 8,000 birds, 97% of these occurring in KwaZulu-Natal and the former Transvaal, and considerable areas of suitable habitat are protected in the Kruger National Park and Greater St Lucia Wetland Park (Taylor 1997a). Overgrazing, cultivation and wetland destruction must have reduced its habitats in many areas: for example, most prime grassland habitat on the S KwaZulu-Natal coast has disappeared as a result of human settlement and the planting of sugarcane, while moist wetland habitats have also been drastically reduced in this region (Clancey 1964, Taylor 1997a). Cott & Benson (1971) give it a very high palatability rating, and numbers are killed for food in some regions (e.g. Brosset & Erard 1986). Notwithstanding these two factors it appears to be under no immediate threat. As for the Corncrake, some of its grassland habitats may have increased locally in recent years as woodland is felled and some agricultural areas are no longer farmed (Taylor 1997a).

Predators include Serval *Felis serval*, domestic cats, Black-headed Heron *Ardea melanocephala*, Dark Chanting Goshawk *Melierax metabates*, African Hawk Eagle *Hieraaetus spilogaster*, and possibly Wahlberg's Eagle *Aquila wahlbergi* (Aspinwall 1978, Urban *et al.* 1986).

MOVEMENTS The seasonality of its occurrences is too complex and poorly documented to be mapped accurately, and the distribution map is largely hypothetical. It moves away from the equator, both N and S, to breed during the rains, being primarily a wet-season breeding visitor S of c. 5°S, although it also breeds in equatorial regions (Taylor 1997a). In C and S Africa the main periods of occurrence are: S Tanzania, Dec-Apr, with nocturnal movements recorded from late Apr to late May (Britton 1980, A. J. Beakbane *in litt.*); Zambia, Zimbabwe and Botswana, Nov-May, in Zambia most arrivals being in Dec and departures in late Apr-early May (Benson *et al.* 1971, Taylor 1979, 1985a); Malawi, Dec-Apr (Benson & Benson 1977); Mozambique, Oct-Mar (Clancey 1971); Namibia and South Africa, Nov-May (Taylor 1997a, c), with movements on the KwaZulu-Natal coast in Apr (see Moult). Birds normally arrive at breeding sites as soon as there is sufficient grass cover (Irwin 1981, Taylor 1985a). Scattered dry-season records from SE Tanzania, Zambia, Zimbabwe, Namibia and South Africa show that some birds remain after breeding when suitable habitat persists but, throughout these regions, most of the bird's habitat becomes unsuitable during the dry season because of desiccation, burning or disturbance. Stragglers to the Namib Desert and the Namibian W coast are attributed to the effect of prolonged E winds (Avery *et al.* 1988). In East Africa, most occurrences in Kenya fall in Apr-Dec and are possibly of migrants which breed S of 5°S, but it is known to breed in C and W Kenya and N Tanzania in May-Jun, while it occurs in WC Sudan only in the rains and probably breeds in Aug (Britton 1980, Brown & Britton 1980, Taylor 1985a, Nikolaus 1987). It is present throughout the year in some West African countries, and in equatorial regions (e.g. Zaïre and Burundi, Urban *et al.* 1980), but it is commoner at certain times, suggesting the influx of migrants, and it shows marked N-S seasonal movements in Nigeria, where many move N in the rains to breed, and most records from the S are in the northern dry season (Elgood *et al.* 1973, 1994). Similar N-S movements have been noted in Senegal, Gambia, Ivory Coast and Cameroon (Urban *et al.* 1986); May-Jul passage occurs in the SW Central African Republic (Green & Carroll 1991); local movements occur in Sierra Leone (Field 1995).

Seasonally occurring non-breeding populations are recorded in coastal Kenya, May-Dec, presumably from a S African breeding population (Taylor 1985a), and in NE Gabon, Oct-Mar (not mapped) (Brosset & Erard 1986); at both sites the birds are rare or absent for the rest of the year. At the Bamburi site, Kenya, the first birds arrived in early May and, as habitat became more extensive, numbers built up until late Oct as a result of irregularly occurring influxes; in Nov-Dec numbers fell as habitat dried out. Precisely dated arrivals of nonbreeding birds at Bamburi frequently occurred during or just after heavy rainfall, especially in the May-Aug period (Taylor 1985a), indicating that such birds often accompanied or followed rain fronts.

This crake migrates at night, when birds hit lighted windows and are found in city centres; it has been caught at lights in mist in S Tanzania and SE Kenya in May and Nov (A. J. Beakbane *in litt.*). One undated specimen (South African Museum) was taken at sea, 32km off the mouth of the Limpopo R. It is a vagrant to São Tomé and Bioko Is (Urban *et al.* 1986).

HABITAT The habitat is predominantly grasslands, from those (often seasonally moist or inundated) at the edges of freshwater swamps, marshes and open waters (including farm dams) and along drainage lines such as dambos, to tall grass savanna, dry grassland in lightly wooded country, and grassy areas in forest clearings and around exotic plantations. It also occurs in rice, maize and cotton fields, neglected cultivation and rank herbage, and at the edges of irrigated sugarcane fields, in moist sugarcane adjacent to marshy areas, and on airfields; it may occur close to human habitation (Irwin 1981, Urban *et al.* 1986, Taylor 1996b). Dominant grasses in moist breeding habitat in Zimbabwe include *Setaria*, *Sporobolus*, *Bothriochloa*, *Hyparrhenia* and *Eragrostis* spp. (Hopkinson & Masterson 1984); in the Kruger NP, South Africa, it occurs in grassland dominated by *Ischaemum afrum*, *Dinebra retroflexa* and *Sporobolus*, *Diheteropogon*, *Andropogon* and *Eragrostis* spp., and in KwaZulu-Natal it is recorded from tall stands of *Paspalum urvillei* and *Andropogon appendiculatus* (Taylor 1997a). It breeds at temporary wetlands in the moist grasslands of the Okavango, especially at pans with dominant grasses such as *Oryzidium*, *Eragrostis* and *Echinochloa* (Hines 1993). In tussocky grassland habitat at Ndola, Zambia, it preferred cover 0.3-1m tall but occurred in grass up to 2m tall, and the substrate was normally moist, only rarely being flooded to a maximum depth of 10cm; its breeding territories often included or abutted thickets or large termite mounds with dense vegetation, which were apparently used as refuges (Taylor 1985a). At a non-breeding area on the Kenya coast it occupied similar, but often drier, grassland habitat and was also found at the edges of *Typha* and *Phragmites* reedbeds around small dams (Taylor 1985a). It normally occupies moister grassland habitats than does the Corncrake, which is also found more frequently in taller grass (Taylor 1985a). It occurs from sea level up to at least 2,000m but is rare in high-altitude grasslands (Taylor 1996a). Its grassland habitat is frequently burned in the dry season, forcing the birds to emigrate.

FOOD AND FEEDING Information is from Taylor (1985a) and P. B. Taylor (unpubl.). Earthworms; gastropod molluscs; insects and their larvae, especially termites, ants (including *Pheidole*, *Tetramorium* and *Pachycondyla*), Coleoptera (including Curculionidae and Carabidae), Hemiptera, and grasshoppers; and small frogs; small fishes may also be taken. It also takes seeds, especially grass seeds, and vegetable matter including grass blades, green shoots and leaves. It forages within cover, and in the open on roads and tracks, picking up insects and seeds from the ground, turning over dead grass and litter, probing the bases of grass tussocks, and digging with the bill in wet to dry ground. It also makes rapid runs to catch fast-moving prey, stretches up to take prey from low-growing plants, and wades in shallow water to seize prey from below the water surface.

HABITS Information is taken from Taylor (1985a) and Urban *et al.* (1986) unless otherwise stated. This crake is diurnal and is active throughout the day, particularly at dawn and dusk, during light rain and after heavy rain; peak visible activity in Zambia and Kenya is recorded for c. 75-90 min starting c. 30 min after sunrise, and for c. 60 min before dark. It is less shy and more easily flushed than are other crakes, either by a human observer or a dog; birds are not infrequently seen at the edges of roads and tracks through grassland, and newly arrived migrants are often seen in the open. When flushed, birds usually fly low over the grass, resident birds usually for < 50m but others, particularly new arrivals, sometimes for over 100 m; they often land in a flooded area, or in or behind a thicket, and crouch immediately on landing, stretching up afterwards to look around if in the open; they often give *kip* calls when disturbed. In dense grass birds sometimes stretch up to look around over the vegetation, or stand on a low eminence for the same purpose. In very short, tussocky grass birds can elude a dog by their speed and manoeuvrability on the ground, and in grass c. 10cm tall they can run, in a crouched position with the neck stretched horizontally, so that they are hardly visible. Birds foraging in the open allowed the quiet approach of an observer with a dog to c. 5-10m; when suspicious they stretched the body and neck up to look at the observer, and then crouched, resuming feeding if the observer remained motionless; if alarmed they sometimes made a vertical jump followed by a short run before pausing to watch the intruder. This species can be approached to within 1m by an observer in a vehicle, or can be called up to a vehicle by taped playback; as a car stopped alongside a bird on a track, the crake remained crouched, alternately rousing and flattening the plumage.

In Kenya and South Africa birds have been found roosting in depressions at the base of, or within, grass tussocks. Birds bathe at roadside puddles in grassland, subsequently preening either within or adjacent to cover, and they sunbathe for brief periods, standing with the wings open and drooping and the neck stretched forward (P. B. Taylor unpubl.).

SOCIAL ORGANISATION Probably largely monogamous, with a pair bond of unknown duration, and pair formation may occur before leaving the wintering area, but single males have been seen to give courtship chases to two females within the territory (Taylor 1985a, P. B. Taylor unpubl.). It is territorial on both the breeding and non-breeding grounds; in 2 seasons at Ndola, Zambia, 8-9 pairs bred at a density of 1 pair/5-5.2ha of variably suitable habitat, while at Bamburi, Kenya, 14 territories were each occupied by 1-2 birds and the overall population density of resident birds in variably suitable habitat was 1 bird/2.73ha, and 1 bird/2ha for the maximum population, (Taylor 1985a). Breeding territories at Ndola were occupied for 21-144 (81.6, SD 37.2, n = 17) days; non-breeding territories at Bamburi for 25-235 (109.9, SD 58.9, n = 17) days. Breeding density at Harare, Zimbabwe, is estimated at 1 pair/ha in uniformly good habitat (Masterson 1991).

SOCIAL AND SEXUAL BEHAVIOUR All information is taken from Taylor (1985a and unpublished observations) unless otherwise stated. Birds are normally seen walking or foraging singly, in pairs or in family parties, but newly arrived migrants on breeding and non-breeding grounds have been recorded in groups of 3-8. In Zambia the crakes occasionally forage alongside Great Snipe *Gallinago media* and Blue Quail *Coturnix chinensis*, and are sometimes flushed alongside Corncrakes. Intraspecific territorial behaviour is frequently observed, and the common threat display given by males early in the breeding season, both to intruding males and in response to taped playback of *krrr* calls, involves the bird adopting an upright stance and raising and spreading the long feathers of the flanks and belly to produce two large, rounded ventro-lateral fan-like projections of boldly barred underparts feathers (Fig. 9). In this pose the bird walks or runs towards the intruder, or two displaying males walk side by side. During the display the female may walk alongside the male with her feathers also fanned, but less widely. The display is often accompanied by *kraaa* calls.

Figure 9: African Crake threat display. [After photograph by B. Taylor].

The aggressive chase differs from the courtship chase (see below) in that the birds' plumage is not roused and that both pursued and pursuer have the neck stretched forward; if the pursued bird takes flight, the pursuer may also fly. Paired females will vigorously attack other females in the territory, including any female which the male has also courted. Fighting at territorial boundaries involves two birds making flying jumps at each other, with vigorous pecking.

Incubating birds tend to sit until an intruder is almost at the nest, when they slip away into cover or occasionally fly; no nest defence display is recorded, although at hatching an adult scurried around the nest and between the observer's feet, giving *koorr* calls (Masterson 1991).

Pair formation, with increased calling and courtship chasing, was observed at Bamburi, Kenya, for c. 4-6 weeks prior to the birds' departure for unknown breeding grounds, while at breeding areas in Zambia and South Africa, courtship chases were also observed soon after the birds' arrival and before nesting commenced. In the courtship chase both birds often rouse the body plumage and the female runs with crouched stance and retracted neck, pursued by the male who adopts a more upright stance with neck stretched or head moving backwards and forwards. If receptive, the female stops, crouches, and lowers her head and tail, when the male approaches and mounts; during or after the chase, to encourage her to crouch he may peck at her neck and neck. When the male approaches at a walk, the female may walk ahead, with roused plumage and a rolling gait, before stopping and crouching. Copulation is brief, lasting only 2-5s, but may be frequent: one pair copulated 9 times in 90 min; the male may spread and vibrate his tail during copulation. Postcopulatory behaviour varies: in the Bamburi non-breeding population, pair members stood side by side in a normal pose before moving off to feed, but in breeding areas in South Africa there is often a graceful postcopulatory display, in which the two birds stand side by side in an upright posture with the body plumage roused, the head held up and the neck curved in a swan-like S-shape, and give quiet *kraaa* calls with the bill slightly open; when calling, the white chin and throat are prominent. Reverse mounting has been observed once. Although paired birds frequently preen together, especially after copulation, allopreening has not been observed.

BREEDING AND SURVIVAL Season Breeds during rains. Senegal Jul; Sierra Leone, Jun-Aug; Ivory Coast, chicks Apr, May, Oct; Ghana, Apr; Nigeria, Jul-Sep in N, Jun-Sep in C, Apr-Nov in S; WC Zaïre (Kwilu), Dec-Mar; C and W Kenya, May-Jun, breeding condition Jul; N Tanzania, Jun; S Tanzania Jan-Mar; Angola, Apr, breeding condition Jan; Malawi, Jan-Mar; Zambia and Zimbabwe, Dec-Mar, peak Jan-Feb; Botswana, Jan; South Africa, Oct-Mar. **Nest** A shallow cup or bowl of grass blades, placed in scrape or depression on dry to moist ground, hidden under grass tussock or small bush; sometimes 2-15cm above dry ground or standing water in dense grass or other plants; occasionally floating. Loose canopy of grass blades often present. First egg often laid when nest flimsy or just a pad of grass, and building continues during egg-laying (Masterson 1991). External diameter 21-23cm, depth 4-9cm; cup diameter 11-12cm, depth 2-5cm; walls 1-5cm thick, and base of cup 2-4cm above ground (Masterson 1991). **Eggs** 3-9 (5.5, n = 45 clutches); ovate; polished and glossy; white to cream or pinkish-cream, spotted and blotched reddish-brown and lilac, mainly at larger end; size (n = 112) 31.5-38.9 x 24-27.9 (34.5 x 25.5); weight 9.6-13.1 (11.6, n = 8). Eggs laid at daily intervals. **Incubation** By both sexes; period normally estimated at 14 days but for some eggs may be 19-24 days because, although incubation commences with laying of first egg, all eggs hatch within 48h (Masterson 1991). **Chicks** Precocial; leave nest soon after hatching; fed and cared for by one or both parents; chicks escape danger by running into dense cover. Fledging period 4-5 weeks; young can fly when two-thirds grown. Newly hatched chicks were taken by a boomslang *Dispholidus typus* (Haagner & Reynolds 1988). Length of breeding season suggests that more than 1 brood may be reared, but no evidence for this is available. Suggestions (e.g. Elgood *et al.* 1973) that birds may breed twice a year, once N of the equator and again after migrating S to C Africa, are not supported by available evidence.

69 CORNCRAKE
Crex crex   Plate 27

Rallus Crex Linnaeus, 1758, Sweden.

Great variation in colour within all populations renders impractical the recognition of race *similis* of Kazakhstan. Monotypic.

Synonyms: *Gallinula/Ortygometra crex*; *Crex pratensis/ herbarum/alticeps*.

Alternative names: Corn Crake, Landrail, Land Rail.

IDENTIFICATION Length 27-30cm; wingspan 42-53cm. Medium-sized crake with distinctive tawny upperwing-coverts. Adult male (breeding) has top of head and entire upperparts, including tertials and tail, brown-black with buff-brown to ashy streaks; upperwing-coverts bright tawny, with variable number of creamy bars, edged dark brown, on medians and greaters; primaries and secondaries duller tawny; leading edge of wing cream. Lores and central ear-coverts buff; rest of sides of head, sides of neck, foreneck and breast blue-grey; breast sometimes buff-tinged, or predominantly buff-brown with blue-grey tinge; flanks and undertail-coverts barred rufous-brown and white, white bars usually narrowly bordered dusky brown; belly and vent creamy-white. Iris pale brown or hazel; bill pale flesh-brown, pinker on lower mandible; legs and feet pale greyish-flesh to red-grey. Female slightly smaller; has slightly warmer buff upperparts, narrower and duller grey streak over eye, and sometimes less grey on face, neck or breast. Adult non-breeding plumage similar to breeding, but upperparts more rufous-brown, less ashy; male has less grey on sides of head, and little or none on neck and breast; female has faint grey tinge in streak over eye, and often none on face, neck and breast. Immature very like adult but has less barring on upperwing-coverts, less grey on head, grey tinge on sides of foreneck and often no grey on breast. Juvenile like adult but has narrower, more buff-yellow, feather fringes on upperparts; less barring on upperwing-coverts; grey of sides of head, foreneck and breast replaced with buff-brown, sometimes with white dots on breast; flanks less contrastingly barred; iris duller; legs and feet dark grey. Inhabits dry to moist grasslands, young cereal crops, marsh edges and cleared forest areas.

Similar species Separable from African Crake (68), which is sympatric in wintering areas, by larger size, paler upperparts, tawny upperwing-coverts, less grey on face, neck and breast, red-brown and white bars on underparts, and bare part colours. For further details, see African Crake. Easily distinguishable on plumage from other sympatric crakes in breeding and wintering ranges; could be confused with gamebirds (Galliformes) in flight, but tawny upperwing-coverts and dangling legs diagnostic.

VOICE Information is taken from Cramp & Simmons (1980) unless otherwise specified. The male's advertising call is a loud, monotonous, rasping double call *krek-krek*, which is given almost continuously at night from the ground or a low perch such as a wall or bush, exceptionally in flight at night, and is audible for up to 1.5km. The bird stretches the neck upwards, opens the bill wide and calls with head and neck nearly vertical. A calling bout may be introduced by a series of growling sounds. As well as announcing the establishment of the territory and acting as a challenge to conspecific intruders, this call also serves to attract females. In the first 2-4 weeks after arrival in the breeding area males may call for long periods, both day and night; when closely associated with a female, the male calls infrequently but calls frequently again from the latter part of the laying period (RSPB 1995); later in the season the call becomes shorter and weaker. A recent study (Peake *et al.* 1998) shows that the advertising calls of individuals varied little over short and long periods, and that it is possible to distinguish between individual calling birds. Although the advertising call is occasionally given on migration, in the autumn and in Feb, the species is silent on the wintering grounds; in South Africa, birds are attracted to taped playback of the advertising call, especially in Mar-Apr, but do not respond (P. B. Taylor unpubl.). The male also has a growling-mew call, like the grunting squeal of young pigs, which is given with the bill closed and is used during aggressive encounters and in sexual display; variations include a growling squeal, and also a grunt and whistle which may be made as one call or uttered separately. The female has a high-pitched cheep (function unknown); an alarm call given by wintering birds is a short, loud *tsuck* (Taylor 1984).

The female calls the young with *oo-oo-oo* sounds; the distress call to chicks is a series of short hard quacking notes; when driven away from the chicks the female utters a call like the alarm note of the Common Moorhen (129) interspersed with series of short croaks; there is also a repeated, hoarse *peek-peek*. In defence of chicks the female makes a clicking noise with the mandibles and, at times, a high-pitched creaking note. The contact call of the young is a soft, whispering *peeick-peeick*; the begging call is vaguely reminiscent of the chirping of sparrows *Passer* spp.; and the agitation call is a trill which develops into a rasping sound in fully feathered young.

DESCRIPTION

Adult male, breeding Forehead, crown, upperparts of body, tertials and tail brownish-black with buff-brown feather fringes which are variably tinged ashy except at feather tips; in fresh plumage, upperparts appear buff-brown to rufous-brown; in worn plumage upperparts appear ashier, and blackish feather centres more prominent. Upperwing-coverts rich tawny; medians and greaters with faint dark brown bars and variable number of off-white to cream tips and bars and some spots; greater primary-coverts, alula and remiges dull tawny, primaries tinged dull black near tips and on inner webs, and with black shafts; outer web of P10, and marginal coverts, cream; underside of remiges grey, tinged vinaceous-cinnamon; axillaries and underwing-coverts tawny but greaters vinaceous-pink to vinaceous-cinnamon, greyer basally. Lores and central ear-coverts buff, with broad blue-grey streak extending from above lores to sides of nape; chin and throat white, often tinged buff or grey; rest of ear-coverts, malar region, sides of neck and foreneck blue-grey; breast blue-grey, buff-brown feather bases variably visible so that some birds appear pure blue-grey, others buff-brown with grey tinge; Flanks and undertail-coverts barred rufous-brown and white, white bars usually narrowly bordered dusky brown; belly and vent white, tinged cream to buff. Tertials reach to wingtip in folded wing. Males tend to develop more barring on upperwing-coverts until third calendar year (Cramp & Simmons 1980). Iris pale brown or hazel; bill straight and rather heavy, tapering to point; pale flesh-brown with darker brown tip; cutting edges and base of lower mandible flesh; bill deeper flesh-pink early in breeding season; in spring and summer, small knob-like swelling at base of culmen; legs and feet pale greyish-flesh to red-grey, sometimes tinged brown in spring, exceptionally whitish.

Adult female, breeding Very similar to male but averages slightly smaller; upperparts slightly warmer buff; blue-grey streak over eye narrower and duller, sometimes partially buff-brown; grey tinge on sides of head, neck and breast sometimes more restricted and less sharply demarcated from surrounding buff-brown areas. Adults may apparently be sexed on the ratio of wing length to total head length (Niemann 1995).

Adult, non-breeding Similar to breeding, but upperparts more rufous-brown, less grey; male has less grey on side of head, and little or none on neck or breast; female has faint grey tinge in streak over eye, neck and breast, which are buff-brown.

Immature Very like adult but probably with more limited barring on upperwing-coverts. Grey limited to streak above and behind eye and to faint grey tinge on ear-coverts, malar region and sides of foreneck (sometimes also on breast of male). First breeding plumage like adult but primaries more worn; in some birds, grey feathers on sides of head or breast partly mixed with unmoulted, worn brown ones; upper mandible apparently never all pink, having dark tip or dark line along culmen (Niemann 1995).

Juvenile Similar to adult, but brown feather fringes on upperparts narrower, tinged more buff-yellow and less grey; upperwing-coverts uniform rufous-chestnut, pale buff bars usually less well defined (and often shorter) than in adult and restricted to inner greaters and medians (but some may be as heavily barred as adult); grey of sides of head, neck and breast replaced with buff-brown to buff, sometimes with white dots on breast; chin, throat, belly and vent pale buff; flanks less contrastingly barred, pale bars sometimes not bordered dusky. Iris light brown to grey-brown; bill pale horn-brown with darker tip, cutting edges and lower mandible more flesh-pink; legs and feet dark grey, changing in late summer to brown-grey or olive-grey and pale flesh.

Downy young Chick has sooty brown-black down, tinged rufous-brown on upperparts; iris grey-brown; bill pink, becoming black-brown after a few days; legs and feet black, becoming buffy-brown; small white wing-claw present.

MEASUREMENTS Wing of 33 males 132-150 (140.8, SD 4.04), of 16 females 128-145 (135.5, SD 4.7); tail of 55 males 41-53 (46.8, SD 2.9), of 33 females 40-49 (44.7, SD 2.4); bill to skull of 18 males 23-26 (24.4, SD 1.0), of 10 females 22-25 (23.6, SD 1.1); tarsus of 54 males 34.5-43 (39.1, SD 1.7), of 33 females 35-40 (37.6, SD 1.3). Weight of 81 males 129-210 (165.5), of 16 females 138-158 (145). Hatching weight 9-14.

GEOGRAPHICAL VARIATION Slight; a cline of decreasing colour saturation probably runs eastwards and populations from Kazakhstan are slightly paler than those from W Europe – more grey, less brown above, and less buff below; intermediates occur in the Ural basin; however, there is great individual variation in colour within all populations (Vaurie 1965, Cramp & Simmons 1980).

MOULT Most information is from Cramp & Simmons (1980). Postbreeding moult is complete and remex moult simultaneous. Moult starts shortly after nesting, in mid-Jul to late Aug, occasionally in late Jun or Sep, and ends in late Aug to mid-Sep, body moult sometimes continuing into Oct. The wings and wing-coverts moult with the tail and body, the flanks and hind parts moulting first and the sides of the head last. Most birds moult in nesting areas, but some may postpone the moult until after postbreeding dispersal or until they reach the winter quarters. Exceptionally, a female from KwaZulu-Natal on 11 Mar had all remiges, rectrices and greater upperwing-coverts almost fully grown and in sheath, and heavy head and body moult (P. B. Taylor unpubl.). Pre-breeding moult is partial, involving some or all of the head and body plumage, the tail and sometimes some wing-coverts; it occurs in Dec-Mar. Postjuvenile moult, involving only the head and body, begins at c. 34-38 days, when birds are capable of flight

Corncrake

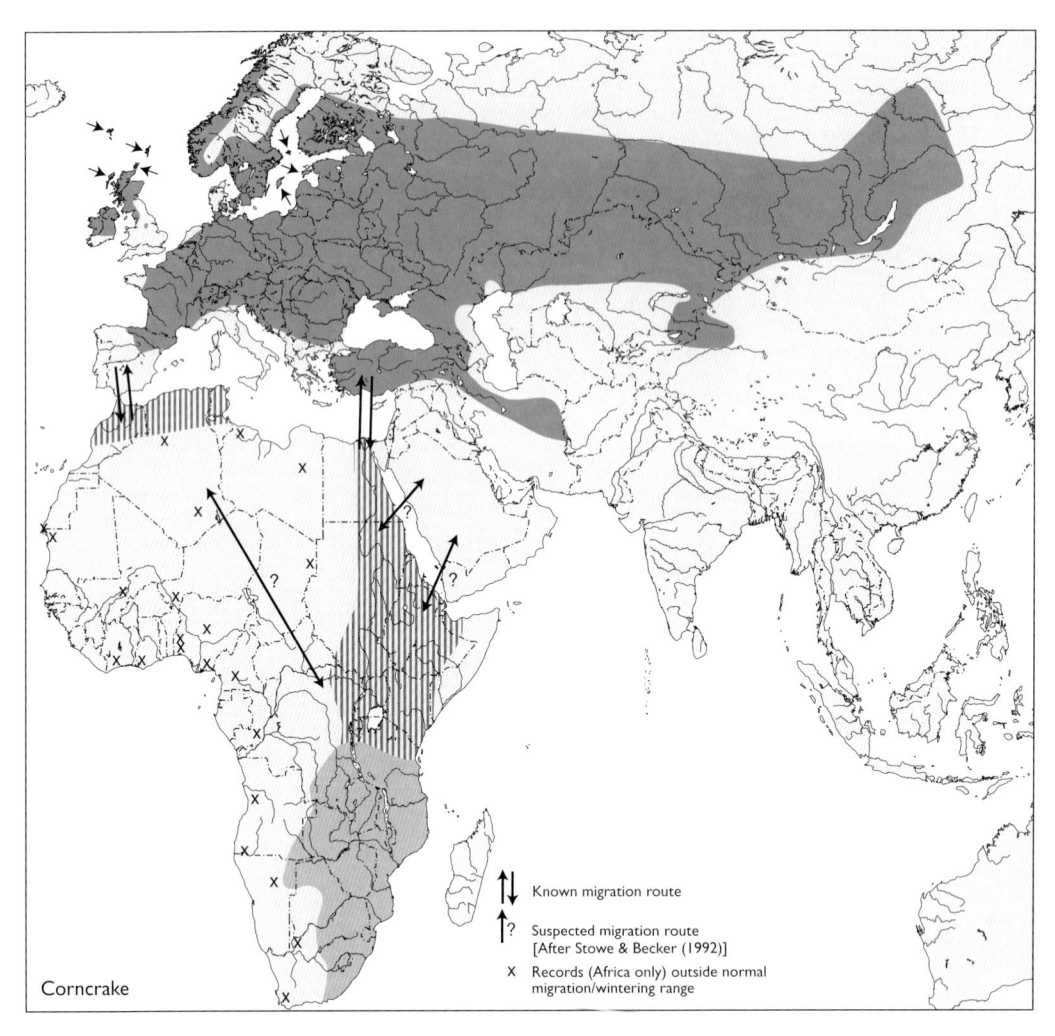

but remiges are not quite fully grown; it starts from mid-Jul to mid-Sep and is complete after c. 1 month. The first prebreeding moult is partial, involving the head, body and some wing-coverts (sometimes fewer feathers than in the adult) and occurs Feb-Apr.

DISTRIBUTION AND STATUS W and NW Europe, from the Faeroes, Britain and Ireland, coastal Norway and S Sweden, E across Europe and C Russia to almost 120°E in C Siberia (W Yakutia and L Baikal) and S to France, N Italy, Balkans, N Greece, N Iran, Transcaucasus, Kazakhstan, Altai and W China in W Xinjiang (=Sinkiang) (Vaurie 1965, Cramp & Simmons 1980, Cheng 1987, Sibley & Monroe 1990). It also breeds rarely in N Spain and in Turkey (Cramp & Simmons 1980, Tomialojc 1994), and it is said to breed in Afghanistan, presumably in the N (Vaurie 1965). Outside this range it is also recorded on passage in Portugal, Syria, Iraq, Lebanon, Israel, Jordan, Arabia (Saudi Arabia, Kuwait, Yemen and Oman), Uzbekistan, Turkmenistan, Armenia, Azerbaijan (where it bred locally in the past), and Xizang Region, W China (Vaurie 1965, Cramp & Simmons 1980, Cheng 1987, Stowe & Becker 1992, Collar et al. 1994, Patrikeev 1995, Shirihai 1996).

The European population of 'pairs' (= singing males) is estimated at c. 87,500-96,900 (geometric mean 91,300) and the world population at c. 122,900 (97,500-196,900) (Tomialojc 1994, Hagemeijer & Blair 1997). The species is classed as globally VULNERABLE. In N Europe and W Siberia the spread of cultivation favoured range extensions (Voous 1960), but more recently there has been a marked decrease in range. Although it is apparently still common and widespread in parts of its range, in Europe there has been a very steep population decline of >20% (17 countries) or >50% (10 countries) in the period 1970-1990; this decline continues and the bird's status is critical in some W European countries. Some breeding population estimates, which generally refer to 'pairs' (this equates to calling males), are as follows (data from Cramp & Simmons 1980, Sheppard & Green 1994, Tomialojc 1994, Niemann 1995, RSPB 1995 and Hagemeijer & Blair 1997). It was formerly a scarce breeder in the S Faeroes but there are very few records this century; its population in Britain and Ireland fell from c. 5,000 pairs in the late 1960s to c. 700 calling males in Britain in 1978 and 480 in 1993; in Ireland only 174 calling males were recorded in 1993, a decline of 81% since 1988. In Scandinavia its range has decreased in Norway since the late 19th century, most markedly from 1910-40; range and population decreases were noted in Sweden early this century but the population may now be stable at 250-1,000 pairs; from being common in S Finland in 1900, numbers fell to 500-1,000 pairs in 1992. From being widespread in France in the 1930s it is now absent or local in many areas, with an estimated population of up to 2,200 calling males in 1984 but only 1,050-1,150 in 1992. Similar declines elsewhere in Europe include: in Belgium and Luxembourg, only 10-55 pairs in 1990; in the Netherlands, from being very common and widespread early this century down to 450+ calling males in 1969 and currently 150-200 in dry seasons to 400-600 in wet seasons; in Germany the population in 1991 was probably c. 1,150 calling males; in the former Yugoslavia, numbers had fallen drastically to 1,000-1,500 calling males in 1991; in Hungary it was common up to the 1960s but then decreased markedly, with only 220-235 calling males reported in the 1980s. There are 6,600-7,800 pairs in Poland, it is widespread but rare in Czechoslovakia, and has declined in Austria (400-600 pairs), Bulgaria (now rare) and in the former USSR where estimates for the European region (Belarus, Estonia, Latvia, Lithuania, Russia and Ukraine) total 77,000-180,500 pairs. It is still locally common in C Siberia (Rogacheva 1992).

The Corncrake occurs chiefly on passage in North Africa but winters occasionally in Morocco, Algeria, Tunisia (where it is rare or irregular, and is recorded mainly in spring) and Egypt (Beni Hassan, Feb) (Urban *et al.* 1986). It winters from Zaïre and C Tanzania S to E South Africa, the centre of its winter distribution including E Zaïre, S Tanzania, Zambia, Malawi, Zimbabwe, Mozambique, N and E Botswana and E South Africa (Stowe & Becker 1992); it also occurs, albeit rarely, in Swaziland and Lesotho (Bonde 1993, Parker 1994), and uncommonly in the Free State, South Africa (Taylor 1997a). Its status in Africa is summarised by Stowe & Becker (1992), as follows. It occurs widely but very sparsely in West Africa (see Movements) but is more widespread in Zaïre; it may winter occasionally in Sudan, but not in Ethiopia, where it occurs only on passage; and in East Africa it is mainly a passage migrant, with a few winter records from Kenya, Uganda and NW Tanzania; it is scarce to rare in Uganda. Its overall status in the winter range is difficult to assess, the scarcity of records from some areas possibly reflecting poor coverage and the difficulties involved in finding the birds, but it is apparently widely but thinly distributed (Stowe & Becker 1992); it has been regarded as locally common only in S Tanzania and in Zimbabwe (Irwin 1981, Stowe & Becker 1992), and may be common on the coastal plain of S Mozambique (V. Parker in Taylor 1997a). Its South African population is estimated at c. 2,000 birds, c. 95% of these in KwaZulu-Natal and the former Transvaal (Taylor 1997a).

In breeding areas, where the birds need tall cover throughout their period of occurrence, the drainage of sites, the loss of hay meadows to the production of maize, poplars and sheep, the intensification of agriculture, and changes in grassland management are the main causes of habitat loss and of heavy losses of breeding birds, eggs and young; for example, mowing destroys virtually all nests (but rarely kills incubating females) and may kill up to 95% of young chicks (Sutherland 1994, Tomialojc 1994). In particular, the rapid introduction of mechanised grass cutting, earlier hay harvesting, and repeated cutting for silage throughout spring and summer, have seriously restricted the available habitat and have reversed gains made in recent centuries through forest clearance and traditional grassland management. Other factors, such as unfavourable climatic trends early in the breeding season, may account for the failure of birds to occupy apparently still suitable habitat (Cramp & Simmons 1980). Eutrophication of wet meadows may also be a factor in the long-term decline of Corncrake populations (Reichholf 1991). The effects of other factors, such as pesticides, and of the loss of nesting habitats along field margins, are poorly understood (Stowe & Becker 1992). A strong attachment to traditional localities is balanced by the persisting tendency to colonise previously uninhabited areas.

Threats to migrating Corncrakes include netting on the E Mediterranean coasts, especially in Egypt where c. 4,600 birds were caught on S passage in 1991, c. 9,000 in 1993, and c. 14,000 in 1994 (an exceptional year); this total includes many juveniles (Anon 1996, Baha el Din *et al.* 1996): W European birds may be involved as well as those from E populations (see Movements). Baha el Din *et al.* (1996) calculate that the 1993-94 estimates represent 0.5-2.7% of the European autumn population, which is given as 7x the number of singing males recorded by Tomialojc (1994) [but their calculation is based on world population figures (92,000-200,000): the percentages should be 0.95-2.85%, based on European population figures (87,000-97,000)]. In the last 30 years, there has been a marked decrease in the number of birds migrating through North Africa (Stowe & Becker 1992). In Malta, where it was once a common migrant, few are now recorded each year, and sometimes none. In Israel, where it was formerly regarded as a common migrant, only c. 20 are now recorded on autumn passage and c. 20-200 in spring (Shirihai 1996).

Corncrakes do not appear to be threatened on the African wintering grounds, where grassland habitats may be increasing as woods are felled and some agricultural areas are no longer farmed, although grassland habitats are suffering locally as a result of overgrazing and cultivation (Stowe & Becker 1992). However, lack of information on population sizes in Africa, and local variations in numbers with rainfall, make it difficult to assess to what extent major changes have occurred. A limited threat may be posed by hay cutting in Zimbabwe in Mar-Apr (Stowe & Becker 1992) and in South Africa throughout the birds' period of occurrence (P. B. Taylor unpubl.). In Zambia and South Africa extensive areas of suitable habitat are protected in national parks (Stowe & Becker 1992, Taylor 1997a).

Conservation action in Europe has been discussed and planned in some detail, and a BirdLife International workshop on the subject was held in Poland in Oct 1994 (Sutherland 1994). Establishing reserves in important breeding areas may have significant results: for example, in S Bavaria, establishing a reserve of 2,355ha in 1980 in an area where 10-20 calling males occurred resulted in the presence of 58 calling males in 1989 (Bezzel & Schöpf 1991). Countries with low agricultural intensity and high Corncrake populations are in E Europe, and the persistence of Corncrakes in this region depends on the management of marshlands to prevent succession to scrub, and the adoption of agricultural policies which do not encourage further intensification of grassland management (Tomialojc 1994, Green & Rayment 1996). It is significant that all large populations of Corncrakes occur in countries where hand-mowing of hay remains a common practice (Sutherland 1994), and important conservation measures include management to protect nests, adults and young from destruction when mechanically cutting meadows for hay or silage, supporting traditional land management practices which benefit Corncrakes, and encouraging the continued production of hay. The most promising conservation approaches include excluding grazing and mowing from strips of tall vegetation (e.g. at the edges of wetlands) and changes in the timing and method of mechanised mowing. Mowing from the centre of a field outwards, instead of from the periphery inwards, can reduce chick mortality from 38-95% to 8-19% (Sutherland 1994, Niemann 1995). Although traditional conservation schemes aim to delay mowing dates to enhance the survival of young, Schäffer & Weisser (1996) show that early mowing of areas within grassland habitat provides refuges for young when the rest of the grass is mown, and that strip-mowing should enhance breeding success. Leaving areas unmown to provide late cover is also advisable, to allow second broods to fledge and to provide areas where adults can moult; such unmown patches can also provide valuable early cover in the following season. Mortality on migration should be investigated, and ways sought to reduce such mortality, notably in North Africa. Because of the tendency of Corncrakes to move between breeding grounds each year, site-specific conservation measures may not be effective, and an integrated national and international conservation approach is required (Stowe & Hudson 1991b). In Britain, where a conservation action plan was drafted in 1989 (see RSPB 1995), Corncrake numbers have recently increased in areas where conservation management measures have been implemented successfully; the Corncrake population of the Isle of Coll reserve has increased from 6 to 17 calling males, 111 calling males were recorded on nearby Tiree, and population decline has been slowed or reversed in areas where farmers have been paid to modify mowing practices (Sutherland 1994, Williams, G. *et al.* 1994). Conservation measures which result in an increase in the production of juveniles may not result in an overall increase in the breeding population unless accompanied by measures to increase suitable breeding habitat in subsequent seasons.

The recent confirmation that the advertising calls of individuals vary little over short and long periods, and that it is possible to distinguish between individual calling birds, should be of considerable benefit to census work and studies of survival (Peake *et al.* 1998).

Little information is available on predation. In Scotland, predation by mammals and raptors is the main cause of adult mortality (Niemann 1995). Birds are killed by feral and domestic cats, feral ferrets, otters, and mink *Mustela vison*, while foxes *Vulpes vulpes*, buzzards, gulls, and Hooded Crows *Corvus corone* are also cited as predators. In Gabon, one was caught by a falconry-trained Black Sparrowhawk *Accipiter melanoleucus* on an airfield (Brosset & Erard 1986).

MOVEMENTS Most information comes from Cramp & Simmons (1980), Urban *et al.* (1986) and Stowe & Becker (1992). This crake is almost wholly a long-distance migrant, although there are numerous Dec-Feb records from W Europe, especially in the 19th century when the breeding population was much larger. The main flyways into Africa are a W route via Morocco and Algeria, and a more important E route via Egypt; few birds cross the Mediterranean between these flyways, but some W European birds may use the E flyway. Some, probably from the E Palearctic, enter Egypt and Sudan via Arabia, the Red Sea and the Gulf of Aden. As the bird's status in West Africa appears to be that of an occasional migrant, most birds using the W flyway probably make a SE crossing of the Sahara to winter in C and S Africa; such migrants may fly via the mountains of SE Algeria. Autumn movements occur from Aug to Nov (peak Sep) in Europe and the former USSR, and the main passage across the Mediterranean is from early Sep to mid-Nov; the species passes through Morocco, Egypt and Israel in Aug-Oct (mainly mid-Aug to mid-Sep in Israel: Shirihai 1996), and Cyprus and Arabia in Aug-Nov (mainly Sep-Oct in Arabia). It occasionally winters in the Mediterranean basin and N Africa. It arrives in Sudan in Sep-Oct and most pass through Kenya in Oct-Dec; few are recorded S of the equator before mid-Nov. Few birds winter in Kenya and most reach C and S Africa in late Nov-Dec; arrivals are recorded in Zambia from late Oct to late Dec. The birds spend a relatively short period in the winter quarters and departures occur mainly from late Feb-Apr; the last birds leave South Africa and Zambia in early Apr and Tanzania and Sudan in Apr, and passage is recorded in Kenya Mar-May, with a very late bird on 2 Jun; movements on the East African coast are recorded in Apr and early May. Return passage is more rapid, birds crossing the Mediterranean from late Mar to mid-May, with a peak in the second half of Apr; passage is recorded in Morocco Feb-May (mainly Mar-Apr), Egypt Mar-Apr (with birds less numerous than in autumn), Israel Mar to mid-Jun (mainly mid-Apr to mid-May: Shirihai 1996), the Gulf of Aden in May, and Arabia and Cyprus Mar-May, while it is recorded in spring in the oases of S Algeria. Breeding grounds in Europe and the former USSR are occupied from mid-Apr to late May, but in C Siberia birds normally arrive in Jun (Rogacheva 1992); males arrive shortly before females (Niemann 1995). Passage is also recorded from Azerbaijan in Sep-Nov and Apr-May (Patrikeev 1995). Movements and distribution S of the Sahara are linked to rainfall and the movements of the Intertropical Convergence Zone: birds pass through Kenya in Oct-Dec, during one of the two rainy seasons, reach wintering areas just after the rains start there, remain throughout the rainy season and then return across the equator during the second rains there in Mar-Apr. Observations in Zambia (Taylor 1984) suggest that local movements may take place in wintering areas in response to changing local conditions.

Corncrakes migrate at night, at a low altitude, and often strike lighthouses and overhead wires; in South Africa birds have struck telephone wires and barbed wire fences (P. B. Taylor unpubl.). In Africa ringed birds have been recovered in Zaïre (Kasai), Nov and Dec, from Sweden and Germany; Angola, Mar, from France; and Congo, Jan, from the UK. Recoveries from Europe and the Middle East include one which shows a marked W shift (Hungary to the Netherlands) and two a shift to the E (Netherlands to Syria and Sweden to Ukraine). Extensive movements during the breeding season can also occur: a calling male, which was ringed in the Netherlands on 23 May 1972 and disappeared soon afterwards, was found dead on 1 August (9 weeks later) in Latvia, not far from Leningrad (van den Bergh 1991).

There are many instances of the Corncrake occurring casually outside its normal range. It is a vagrant W to the Canary Is, the Azores, Madeira, Iceland, Greenland (c. 20

records), North America (c. 17 records) from Baffin Island along the Atlantic coast (Newfoundland, Nova Scotia, Maine, Rhode I, Connecticut, New York, New Jersey, E Pennsylvania and Maryland: AOU 1983) and also Bermuda in Oct 1847 and five times since Oct 1978. To the SE and E it has occurred in Tibet (1 record, Sep), India/Pakistan (2 records, 1 in Oct), Sri Lanka (1 in Sep, 1 in Oct and 1 on a ship offshore), Vietnam (Cochinchina, Jan 1996), Seychelles (2 records, Oct and Nov), and Australia (New South Wales, Jun 1893, and on a ship off W Australia, Dec 1944); a 19th century New Zealand record is now not accepted (Vaurie 1972, Ali & Ripley 1980, Marchant & Higgins 1993, Robson 1996, Skerrett 1996). In Africa it is very rare or a vagrant to Libya (2 records, Apr and Sep), W Mauritania (2, Oct and Apr), Mali (1, Nov), SW Niger (3, Apr), Chad (2, Sep), Ivory Coast (1, Sep), Ghana (1, Feb), Nigeria (4, Apr), Cameroon (1, Dec), Gabon (3, Feb), Congo (2, Jan and 1, Mar), Rwanda (1, Feb), W Somalia (4 old records, 1 in Sep, 2 in Oct and 1 in Dec), Angola (2, Mar), N Namibia (1, Dec) and the Western Cape Province, South Africa (1 in 1864). The high degree of vagrancy is indicative of the bird's dispersive ability and also of the readiness with which individuals are blown off course by opposing winds; however, it may winter more widely in West Africa than is known, especially in view of the fact that its midwinter distribution seems inadequate for the putative numbers of migrants that may be involved. A specimen "from Natal dated 3 July 1985" (Stowe & Becker 1992) is, in fact, dated 3 January.

HABITAT Breeding habitat is essentially dry to moist meadows and other grasslands, including alpine meadows, the fringes of moorland (peatland, mires), marsh fringes and cleared forest areas. In C Siberia the Corncrake occurs within the taiga zone in meadows and river valleys (Rogacheva 1992). Corncrakes prefer cool, moist grass or herbage from 20-50cm tall to head height, and not so dense that it is difficult for them to walk through. They avoid very marshy areas, standing water, river and lake margins, forests, woodland, stands of bushes, reedbeds and other dense vegetation above c. 50cm tall, and open ground with rocks, gravel or sand. Dispersed solitary bushes, hedgerows or reed-margins within the habitat are used as calling sites. Common herbaceous vegetation at breeding sites includes *Anthriscus, Campanula, Chaerophyllum, Cirsium, Crepis, Equisetum, Filipendula, Heracleum, Iris, Lychnis, Lysimachia, Myosotis, Oenanthe, Ononis, Ranunculus, Rorippa, Rumex, Salvia, Scabiosa, Thalictrum, Tragopogon* and *Urtica*, and grasses such as *Alopecurus, Agrostis, Festuca, Holcus, Poa* and *Phalaris*, as well as "Leucojo-Caricetum" sedge meadow and habitats dominated by robust sedges *Carex* spp. (van den Bergh 1991, Flade 1991, Stowe *et al.* 1993, Sheppard & Green 1994, Niemann 1995). Unmown, ungrazed grass, which is difficult to penetrate because of matted vegetation from previous years' growth, is unsuitable (Niemann 1995). Locally, Corncrakes inhabit abandoned land as well as fields of cereals, rapeseed, peas, clover, potatoes or fodder plants, sometimes when breeding and especially after breeding, when adults may also move into tall vegetation along ditches, roads and dams (e.g. Flade 1991); they have also bred in unmown young conifer plantations (RSPB 1995).

In Germany, Croatia and Poland, Flade (1991) found that newly arrived birds occupy, and breed immediately in, *Carex/Iris/Festuca* meadows; 4-6 weeks later most calling males occupy hay meadows dominated by Poaceae, while adults move to high herbage along ditches to moult after breeding: embankments or fallow areas adjacent to the breeding habitat are very important as moulting sites, and as temporary habitat during harvest. In Scotland, first nests (in May-Jun) tend to be in *Anthriscus, Phragmites, Iris, Urtica, Filipendula, Phalaris* and other rough, tall vegetation at field margins or in marshy areas; in Jun-Jul early cover may become too dense and most birds move to hay and silage meadows for the second brood, the meadow vegetation by then having grown to a suitable height (Niemann 1995, RSPB 1995). The bird's original breeding habitat probably comprised riverine meadows of the *Carex-Iris-Typhoides* type, and other treeless wet grasslands in planar, foothill and montane zones (Hagemeijer & Blair 1997).

On the wintering grounds the Corncrake occupies similar habitats, occurring predominantly in dry grassland and savanna, including upland *Themeda triandra* grassland with cover as low as 30cm and grassland with *Hyparrhenia* spp., *Andropogon appendiculatus, Paspalum urvillei* and other robust grasses up to 2m tall, and it often occurs in areas which are burnt during the dry season (Taylor 1984, 1997a); it also occurs in rank grass (especially near rivers, sewage ponds and pools), and in relatively short grass in wetter areas, including that along drainage lines; it has been recorded in tall grass within young conifer plantations, and in South Africa it occasionally occurs in moist sedgebeds and reedbeds (Taylor 1997c, P. B. Taylor unpubl.). In the Kruger National Park, South Africa, it is found in grassland dominated by *Urochloa mosambicensis, Cenchrus ciliaris, Themeda triandra, Panicum coloratum* and *P. maximum* (Taylor 1997a). It also occupies *Eragrostis* hayfields, old lands and pastures, the edges of maize fields bordered by grassy areas, fallow fields and neglected cultivation, lucerne fields, uncut grass on airfields, the edges of sugarcane plantations bordering grassland and marsh, and even suburban gardens (Taylor 1997a). On migration it also occurs in wheatfields and at golf courses. It ranges from sea level to 3,000m (Russia), in South Africa up to at least 1,750m.

FOOD AND FEEDING Most information is from Cramp & Simmons (1980). Omnivorous, but the main food items are small invertebrates: earthworms (Annelida), molluscs (snails such as *Ena* and *Laciniaria*, and slugs such as *Agriolimax*); Isopoda; Diplopoda; Arachnida (spiders and harvestmen); Coleoptera adults and larvae (Curculionidae, Carabidae, Staphylinidae, Elateridae, Cantharidae and Dytiscidae, and including the pest weevil *Sitona*); Hemiptera (Heteroptera and Jassidae); Diptera (Tipulidae and Syrphidae); Dermaptera (e.g. *Forficula*); Orthoptera (Acrididae, Tettigoniidae and Gryllidae); Odonata; Lepidoptera adults and larvae; Mantodea (Dement'ev *et al.* 1989); Dictyoptera; and Hymenoptera (Formicidae). It also eats small frogs, small mammals and (in captivity) small birds. Green parts of plants, seeds of *Spergula* (Caryophyllaceae) and *Anchusa* (Boraginaceae), and seeds of grasses and grain, including barley and wheat, are also taken. Its diet is similar on the wintering grounds; in South Africa it takes many Coleoptera (Curculionidae, Carabidae and especially dung beetles Scarabaeinae), ants (e.g. *Tetramorium* and *Bothnoponera* spp.) and termites, plus cockroaches (Blattoidea), Orthoptera (Acrididae and Gryllidae), Hemiptera and Diptera, and also grass blades and seeds of grasses and legumes (P. B. Taylor unpubl.). In wintering areas it normally forages within cover, occasionally on open grassy tracks or dirt roads, taking food from the ground, low-growing plants and the interior of grass tussocks; it shifts and probes litter with the bill, and runs to catch active prey (P. B. Taylor unpubl.).

HABITS Information comes mainly from Cramp & Simmons (1980) and Taylor (1984). On the breeding grounds it is skulking but apparently quite tame, and is often out of cover. On the wintering grounds it is more secretive than African Crake and, unlike that species, is rarely seen in the open, although it occasionally feeds on open tracks in grassland and at the edges of dirt roads, when a slow and quiet approach may be made by an observer to within 10m of a foraging bird. In Zambia it was active from c. 10 min after daybreak until dusk, and it is most active early and late in the day, after heavy rain and during light rain. When suspicious it may flick the tail and when surprised in the open it retracts the neck and runs rapidly for a fewm before pausing and standing erect, with neck stretched up, to watch the intruder; if not greatly alarmed it may appear more inquisitive than nervous. It is not difficult to flush on the wintering grounds with the aid of a dog, and it usually flies for < 50m, sometimes landing in or behind a patch of bushes or a dense thicket; in the tallest cover birds sometimes escape by running. Its flight action is often weak and fluttering but it normally flies more strongly than species such as African Crake, and in extended flights the action is stronger and steadier and the feet are drawn up; the migratory flight is also strong and sustained. It normally has a high-stepping gait and when escaping danger it runs swiftly, easily negotiating grass tussocks or rows of crops and moving with the body horizontal and slimmed; in short grass it runs in a crouched position with the neck stretched horizontally. Like the African Crake, in dense cover it stretches up to look around over the vegetation, or stands on a low eminence for the same purpose. It will swim in an emergency. When captured it may feign death, recovering instantly if escape is possible.

In breeding areas, calling males are active at night but parents and young apparently roost at dusk (Gilmour 1972). During nocturnal migration birds rest for at least part of the day, settling at dawn. In wintering areas, birds roost at night (P. B. Taylor unpubl.).

SOCIAL ORGANISATION A monogamous pair-bond of seasonal duration was formerly assumed, but serial polygyny regularly occurs, males occupying shifting and overlapping home ranges, and mating with two or more females, remaining with a female only until the second half of the laying period (RSPB 1995, Taylor 1996b, T. Stowe *in litt.*). Nests are often well separated but may be as little as 20-55m apart and the male's territory may cover severalha (Mason 1940, 1941). In the Outer Hebrides the male's home range size varied from 3 to 51ha (median 15.7 ha), and from May to Jul did not exceed 28ha (median 8 ha); female home ranges were 0.4-28ha (median 5.5 ha), the median sizes of core home range areas during pre-incubation being <3 ha, and at other times <1ha (Stowe & Hudson 1991a).

Corncrakes are solitary on the wintering grounds: at Ndola, Zambia, single birds occupied discrete areas of 4-9ha for up to 4 months; some variations occurred in the extent of habitat utilised because patches were rendered temporarily unsuitable as a result of flooding, late growth of vegetation, grass cutting, and cultivation; the amount of suitable habitat occupied by each bird at any time was c. 4.2-4.9ha (Taylor 1984).

SOCIAL AND SEXUAL BEHAVIOUR Most information is from Cramp & Simmons (1980), who extensively quote the studies by Mason (see References). This species is normally solitary, but local concentrations of breeding birds may occur and pairs, or groups of c. 20-40, are occasionally formed on migration, when diurnally resting flocks may contain several hundred birds. In S areas of the former USSR, such as the S Crimea coast, it was formerly said to form "huge flocks" on autumn migration (Dement'ev *et al.* 1969). Corncrakes may join flocks of Common Quail *Coturnix coturnix* when crossing the Mediterranean.

Females apparently take no part in territory defence. Intrusion by a male giving the craking-call is regarded as a challenge by the territory holder, which approaches, crake-calling vigorously with the wings partly open and drooping and the head and neck inclined forward; the intruder then usually moves away or stops calling. If the intruder does not leave as a result of this 'song combat', the birds may meet, walking with head and neck raised and wings drooped to touch the ground, the primaries touching behind and over the tail. Both may then run round in seemingly aimless curves, sometimes giving the growling-mew call. Each bird then makes growling-mew calls and lunges towards its opponent with the wings raised and twisted over the back so that the upperwings face forwards, the head retracted and held low, and the body feathers (especially on the breast) roused. A fierce fight lasting several minutes may then ensue, the birds making flying jumps at each other with vigorous pecking, and sometimes striking with the feet.

The incubating bird sometimes allows close approach before running from the nest, or may remain on the nest; this, together with the crouching behaviour of non-incubating birds, leads to many deaths during hay-cutting and corn-harvesting operations. A bird with chicks may face an intruder and make clicking noises and creaking calls.

A male may attempt copulation with a stuffed decoy without any prior display, but may also give a brief courtship display in which the neck is extended and the head held low, the wings are opened and directed backwards with the tips touching the ground, the uppertail-coverts are erected and fanned, the rectrices are spread, and the flank and neck feathers are sometimes puffed out; in this posture the male may circle the decoy and, approaching from the rear, make a flying jump on to its back, where it may tread the decoy, seize it by the back of the neck and attempt copulation (Mason 1941, 1945, 1950). In live encounters, the male usually approaches the female from behind; if the female turns to face him, he may circle her until she stops moving. Males make many attempts to copulate with decoys (27 attempts in 35 min are recorded), and may try to feed the decoy (Mason 1945); courtship-feeding is also reported in captivity (Heinroth & Heinroth 1928).

BREEDING AND SURVIVAL Season Apr-Aug. **Nest** Site in grassland; sometimes in safer sites along hedgerows, near isolated trees, or in bushy or weedgrown areas. In areas where grass grows relatively late, first nest often in tall herbs or marsh vegetation, second in grassland. Nest built on ground, in dense continuous vegetation or in tussock; sometimes only a scrape but usually a shallow cup of grass, weeds and brambles, lined dead leaves; often with surrounding stems pulled over top in loose canopy;

average external diameter 12-15cm; depth 3-4cm; probably built by female only. **Eggs** 6-14, usually 8-12; Sweden 6-12 (8.9, n = 45 clutches); clutches of up to 19 by 2 females; short oval; smooth and slightly glossy; grey-green, sometimes milky-blue or tinted reddish; spotted and blotched purple, red-brown or grey; size 33-41 x 24-29 (37 x 26, n = 200), weight 13-16. First-clutch eggs laid at daily intervals; second clutch laid more rapidly, sometimes 2 eggs per day (RSPB 1995). **Incubation** By female only; period 19-20 days (first clutch), 16-18 days (second clutch); incubation on second nest begins on average 12 days after first brood abandoned and 42 days after start of incubation of first clutch (RSPB 1995); incubating female forages within a fewm of nest; hatching synchronous. **Chicks** Precocial; usually leave nest 24h after hatching but may remain in nest for up to 2 days; cared for by female alone, sometimes by 2 adults (not clear if 2nd is male or female); fed bill-to-bill at first; self-feeding after 3-4 days. Young independent at fledging or earlier; parent-young bond strong, usually only of short duration but may last up to 9 weeks or to time of departure; early dispersal of brood, at 10-15 days, associated with second nesting attempt; second brood chicks attended by female for 15-20 days (RSPB 1995). Brood may move to neighbouring area after hatching, particularly after mowing or flooding; female moves young progressively farther from nest each day (Stowe & Hudson 1991a); parent and fully grown young reported to fly daily to additional feeding grounds up to 6.4km from nest site, returning at dusk to roost (Gilmour 1972). Fledged at 34-38 days, when capable of flight; postjuvenile moult then begins. Age of first breeding 1 year. 1-2 broods; in W Europe 2 broods normal, one hatching mid-Jun, the second late Jul; replacements laid after egg loss.

Breeding success apparently low in fertilised meadows and arable land, and reproductive output probably declining with mechanisation and fast farming practices (Tomialojc 1994, Hagemeijer & Blair 1997). Nest success in undisturbed habitat remarkably high at 80-90% (RSPB 1995), but considerable destruction of clutches and mortality of young caused by timing and methods of mowing; in Outer Hebrides, 19 chicks killed by mowing represented c. 25% of chicks found, and 12 nests (17% of those found) were destroyed (Stowe & Hudson 1991a); machine mowing can kill 38-95% of chicks in a meadow (Sutherland 1994). Before independence, 10-20% of whole broods are lost; partial losses average c. 50% of first brood chicks and <40% of second brood chicks (RSPB 1995). In Scotland, predation of nests and broods is infrequent (Niemann 1995). KwaZulu-Natal breeding records (Sclater 1906) are erroneous and two "Corncrake" eggs (Durban Museum) are of African Rail.

ROUGETIUS

One species, a medium-sized rail endemic to marshy areas in the highlands of Ethiopia and Eritrea. Its affinities are uncertain: it has a tenuous nasal bar and is thus probably closer to the *Gallirallus-Rallus* group than to the gallinules or *Amaurornis* as was formerly proposed (Olson 1973b). It has a medium-long bill, uniform brown upperparts without the patterning typical of most *Rallus* species, largely cinnamon-rufous underparts, and white undertail-coverts. It may have suffered considerable loss of habitat in recent years, and it is currently considered near-threatened, but it is still locally common in some parts of its range.

70 ROUGET'S RAIL
Rougetius rougetii   Plate 21

Rallus Rougetii Guérin-Méneville, 1843, Ethiopia.

Monotypic.

Synonyms: *Rallus/Rougetius abyssinicus*; *Eulabeornis/Rallina abyssinica*; *Eulabeornis/Calamodromus rougeti*.

Alternative name: Abyssinian Rail.

IDENTIFICATION Length c. 30cm. A distinctive, medium-sized, brown and cinnamon-rufous rail. Top of head to ear-coverts, entire upperparts, upperwings and tail dark olive-brown; line from base of upper mandible above lores to eye pale buff (may be absent); chin whitish, throat pale cinnamon; sides of neck, and breast, belly and upper flanks, cinnamon-rufous; lower flanks and vent dark olive-brown; undertail-coverts white. Iris reddish-chestnut; bill red and medium long; legs and feet purplish-red. Sexes alike. Immature paler and duller than adult; dark brown of crown does not extend below eye; face, chin and throat grey; iris brown; bill blackish, base dull red; legs and feet brownish-red. Juvenile undescribed. Inhabits marshy situations, usually near pools and streams, in highland grassland and moorland.

Similar species Combination of unstreaked olive-brown upperparts, largely cinnamon-rufous underparts, and white undertail-coverts unique among African rails. Similar-sized African Rail (62) has upperparts dark chocolate-brown; sides of head, and underparts to breast, dark slate-grey; posterior underparts barred black and white, except for all-white outer undertail-coverts; bill longer; legs and feet red.

VOICE Information is taken from unpublished observations (P. B. Taylor) unless otherwise stated. The common territorial and advertising call is a loud, ringing, repeated *wreeeee-creeeeuw*, given throughout the year during the day and on moonlit nights, mainly in the morning (sometimes until midday) and evening (Urban *et al.* 1986). Two or more birds (adults or immatures) may call together. Pair members stand upright and side by side when calling; one bird makes repeated *wreeeee-creeeeuw* calls while at the end of each call the other bird gives an almost synchronised, shorter *cr-creeeeuw*. Calls indicative of aggression include a sharp, low-pitched *kak*, a low-pitched, repeated *ke-ke* of extreme aggression, a rapid, high-pitched *kikikikikiki* (almost a rattle) followed by a repeated *krrr*, and various grunts and growls. The alarm call is a shrill, clear, piercing *dideet* or *di-dii* (Urban *et al.* 1986). Before and during courtship chases, birds may utter a bubbling call, a gulping *gug-gug-gug* call and various quiet squeaky, rattling and growling notes.

DESCRIPTION
Adult Top of head, lores, sides of face, ear-coverts, entire upperparts, upperwings and tail dark olive-brown, darkest on lower back, rump and uppertail-coverts, and palest, more olive-tinged on hindneck and mantle; sides of head paler than crown, and tinged pale cinnamon. Indistinct line from base of upper mandible above lores to eye pale cinnamon-buff, occasionally absent. Upperwing-coverts and tertials slightly darker than mantle and upper back; marginal coverts cinnamon-rufous; primaries and secondaries olive-brown with slight bronzy lustre; axillaries and underwing-coverts bronzy-brown. Chin whitish, becoming pale cinnamon on throat; sides of neck, breast, belly and upper flanks cinnamon-rufous; lower flanks and vent dark olive-brown; undertail-coverts white. Iris reddish-chestnut; bill red, sometimes duller and browner towards tip; legs and feet purplish-red. Sexes alike. Iris and leg colour may become slightly duller outside breeding season.
Immature Paler and duller brown on upperparts; brown of crown does not extend below eye; face, chin and throat grey; underparts washed-out orange. Iris brown; bill blackish, dull red at base; legs and feet brownish-red.
Juvenile Not described.
Downy young Described by Urban (1980) as yellow-brown with black crown, black along sides of face, russet neck and black bill and legs.

MEASUREMENTS (8 males, 8 females) Wing of male 124-136 (131), of female 124-137 (131); tail of male 45-54 (49), of female 41-51 (46); bill of male 30-35 (33), of female 28-34 (31); tarsus of male 46-57 (51), of female 44-51 (47) (Urban *et al.* 1986). Weight of 1 male 220, of 1 female 170.

GEOGRAPHICAL VARIATION None.

MOULT No information available.

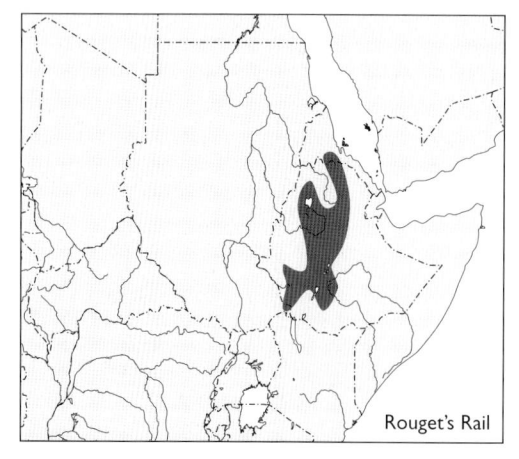

DISTRIBUTION AND STATUS Confined to the highland areas from Eritrea south through Ethiopia to the Kaffa (Maji Plateau), Gamu-Gofa and Sidamo regions and the Bale Mts, where it was formerly regarded as widespread and common to locally abundant (Urban *et al.* 1986). It is currently regarded as NEAR-THREATENED; in 1989, visits to areas where the species had not been uncommon in 1975 yielded extremely few sightings and the observers concluded that the greatly increased grazing pressure in marshlands and along streams had depleted vegetation cover to the extent that much habitat had become unsuitable (Ash & Gullick 1989), while grassland habitats were also being ploughed up for cereal growing (J. S. Ash *in litt.*). Such changes are often extensive and significant, and have presumably affected populations in some areas, while recent droughts have possibly also caused reductions in numbers. However in 1996-97 the species was still widespread and locally common in highland wetlands within 100km of Addis Ababa and the birds occur close to human habitation, even occupying artificially created and man-modified habitats in the centre of Addis Ababa (Taylor 1996a, 1997b). This species is not molested by local people, whose religious beliefs encourage the protection of many birds, but traditional beliefs are declining and such protection cannot be guaranteed for the indefinite future; formal conservation plans for wetlands and their birds are urgently needed (Negere 1980, Taylor 1996a).

MOVEMENTS None recorded.

HABITAT Restricted to the Ethiopian and Eritrean highlands at 1,500-4,100m a.s.l., where it normally frequents marshy situations in grasslands and moorlands. It is found in grass, sedges, reeds, forbs and bushes at the margins of pools and streams and in seasonal and permanent wetlands, and in marshy meadows, patches of tussocky marsh grass and lobelias in wet hollows; it also occurs in *Alchemilla* bogs and among *Alchemilla* or heaths on dry ground (Urban *et al.* 1986). It is also recorded from forested regions (e.g. Sailom Forest) in open areas at pools with fringing vegetation and emergent grass (J. Atkins pers. comm.). It adapts readily to artificial and man-modified habitats, such as meadows, lawns, overgrown paths and flowerbeds, shrubberies, hedges and thickets in parks and gardens, and occurs very close to human habitation in Addis Ababa, even in the centre of the city (Taylor 1996a, 1997b). In such areas it often frequents lush grass 30-100cm tall and is attracted to feed at areas where grass has been recently mown; it often occurs at ditches and drainage channels in grass (P. B. Taylor unpubl.). It occurs in relatively sparse cover along polluted streams (J. Atkins pers. comm.). Its habitat preferences are more catholic than those of the African Rail (62), alongside which it has definitely been recorded only at one site (at 2,600m a.s.l.), where it may principally occupy streambank vegetation while the African Rail occurs in adjacent sedge marsh and sedge meadow (Taylor 1997b); it apparently replaces the Black Crake (88) in many freshwater habitats of the highland regions (Taylor 1996a).

FOOD AND FEEDING Information is from unpublished observations (P. B. Taylor) unless otherwise stated. These birds take earthworms, crustaceans, snails, aquatic and terrestrial insects (especially Coleoptera) and seeds (Urban *et al.* 1986). They forage in dry to flooded grass, sedges and other marsh vegetation, in ditches, drainage channels, and moist flowerbeds, on bare mud, and in shallow water. When foraging, they often walk slowly and deliberately, with long strides. Invertebrate prey is taken from bare ground, short grass, low vegetation, mud and water. Under trees and bushes the birds probe the ground and the leaf-litter, flick aside dead leaves with the bill and stretch up to examine cracks in the bark of tree trunks, from which they extract insects. They probe and dig with the bill into short grass, grass cuttings and soft soil, and move aside growing grass blades close to the ground. They are also attracted to recently mown grass, where they forage extensively. In streams and rivers, this species is recorded as hopping from stone to stone and has been seen to dive into the water at a waterfall like a dipper *Cinclus*, reappearing in calmer water (Urban *et al.* 1986). Its local name translates as "Devil's Chicken", which was apparently bestowed because of the bird's habit of walking into rivers and streams and disappearing from view; this is presumably a manifestation of the above-described foraging behaviour, which therefore may not be uncommon. Birds feed in the open in full sunlight; in rain they continue to feed but tend to keep under the cover of trees or bushes if possible.

HABITS Information is from unpublished observations (P. B. Taylor). Although it has been said to be nocturnal (Mackworth-Praed & Grant 1957), the only published evidence of nocturnal activity involves calling. The birds are active throughout the day, especially in cool or overcast weather, but most foraging takes place in the early morning and late afternoon. In the middle of the day birds rest at the edge of, or within, cover, when much preening and wing-stretching take place; birds also preen for long periods, sometimes continuously for 20 min or more, in the late afternoon. One bathed for several minutes in a grass-grown ditch just before sunset. The birds walk with long strides, jump well, climb actively among tall grass tussocks and slide down sloping matted vegetation. They stand upright when looking around, flick the tail frequently only when disturbed, and when alarmed in the open they run with the head and tail raised. When moving into cover they stretch the neck horizontally and hold the tail down. Often confiding, they may stand in the open in full view and allow close approach, but they become wary when frequently disturbed by people. When feeding close to an observer, birds often hold the tail still but fluff out the white undertail-coverts into a 'powder-puff'. Birds may forage in the open, even alongside roads and tracks and around buildings in the centre of Addis Ababa. They make use of ditches in grassland for concealment, and readily walk into thickets.

SOCIAL ORGANISATION Monogamous, apparently with a permanent pair bond (Taylor 1996a). Permanent territoriality is indicated by the frequent occurrence of territorial calling in the non-breeding season, when pairs also respond strongly to taped playback of territorial and aggressive calls (P. B. Taylor unpubl.). In optimum habitat the territory may be very small; 12 pairs were recorded in 1km (Dorst & Roux 1973), while territories as small as c. 0.18ha were recorded in an Addis Ababa park (P. B. Taylor unpubl.). One chick was accompanied by 10 adults, suggesting some degree of co-operative breeding (J. S. Ash *in litt.*). Immatures remain in the parental territory for an indefinite period after breeding, and join their parents in making territorial calls in response to calls of other territory holders and to taped playback (Taylor 1996a).

SOCIAL AND SEXUAL BEHAVIOUR All information is from unpublished observations (P. B. Taylor). In response to taped playback of territorial and aggressive calls, an adult will sometimes aggressively chase immatures which are normally tolerated within the territory, or will even chase another adult. Such chases between pair members sometimes develop into courtship chases, in which both birds move with a crouching run and with head stretched out, and tail and neck horizontal. The male runs with slightly drooping wings and pecks at the female's head and neck; if receptive the female stops and crouches to allow mating, observed instances of which have been brief and always carried out in partial cover. Bill-fencing between adults has also been observed in response to taped playback of territorial calls. Pair members and immatures often forage close together, and rest and preen standing side by side; allopreening between pair members occurs both during and outside the breeding season.

BREEDING AND SURVIVAL Season Lays Mar-Oct. **Nest** A pad or shallow cup of dead rushes or grass placed on wet ground among high rushes, in rushes over water or in grass tussocks. **Eggs** 4-8; oval; cream to ivory or white, very faintly tinged buff, with fine specks and irregular spots of red-brown, lilac and grey concentrated at larger end; size of 7 eggs 42.4-46.8 x 30.6-31.6 (44.1 x 31.1) (Harrison & Parker 1967). Mackworth-Praed & Grant (1957) and Schönwetter (1962) refer to a much smaller egg, suggesting confusion with another rallid species (Harrison & Parker 1967). Eggs laid at daily intervals. **Incubation** By female, probably also by male. **Chicks** Both parents tend chicks, which apparently remain with adults at least until fully grown (Urban *et al.* 1986, Taylor 1996a).

ARAMIDOPSIS

The only species in this genus is a medium-sized, flightless rail of secondary forest, confined to Sulawesi. It is locally distributed, rare, and very poorly known. It has been seen on only four occasions in recent years and is considered vulnerable.

71 SNORING RAIL
Aramidopsis plateni   Plate 21

Rallus plateni A.W.H. Blasius, 1886, Rurukan, Sulawesi.

Sometimes retained in *Rallus*, but bill shape and plumage are somewhat similar to a pro-*Rallus* group (*Dryolimnas-Atlantisia*), near to which this species is provisionally placed (Olson 1973b). Monotypic.

Alternative names: Platen's/Celebes/Platen's Celebes Rail.

IDENTIFICATION Length c. 30cm. Flightless, with very short tail, short rounded wings and heavy legs and feet; in the field appears tailless, with longish, very slightly downcurved bill. Forehead, forecrown, sides of head, foreneck and breast grey; hindneck and sides of neck deep orange-chestnut; mantle, scapulars and upper back grey; rest of upperparts, including median upperwing-coverts, and tertials, chestnut-brown; rest of upperwings darkish brown. Chin and upper throat white; lower breast has whitish bars; flanks, belly and undertail-coverts blackish, with narrow white bars. Iris red, brown or yellowish; bill brownish, most of lower mandible yellowish-green; legs and feet black. Sexes similar but female has brighter rufous hindneck and often less white on chin; iris orange; bill horn-coloured, with most of lower mandible reddish; legs and feet blue-grey. Immature and juvenile undescribed. Inhabits dense secondary growth on the borders of primary forest, usually with water.

Similar species Occurs alongside similarly sized but quite different Bald-faced Rail (81), which may have similar snoring call; the only other sympatric rallids with which it may be confused are similar-sized Buff-banded Rail (46), which can occur near forest and has patterned head, upperparts and breast, and smaller, marsh-dwelling Slaty-breasted Rail (54), which has similarly coloured underparts and chestnut nape, but has distinctively barred upperparts and no grey on mantle and scapulars.

VOICE A distinctive but rather quiet call, lasting 1-2s: a very brief *wheez* followed by a longer snoring noise *ee-orrrr* (Lambert 1989). The call is given throughout the day and may be uttered frequently; it is possibly similar to the call of the Bald-faced Rail. Birds also give a brief, quiet, deep sigh *hmmmm* (Lambert 1989).

DESCRIPTION
Adult male (Sharpe 1894, Ripley 1977) Forehead and crown ashy-grey, washed olive; centre of crown and nape darker brown, becoming deep orange-chestnut on hindneck, this colour extending to rear edge of ear-coverts and down sides of neck and being brightest at edges, darker in central regions. Mantle, scapulars and upper back slaty-grey, upper mantle tinged cinnamon; lower back to tail chestnut-brown. Upperwing-coverts and tertials chestnut-brown; outer upperwing-coverts dusky, washed olive-brown; primary coverts, primaries and outer secondaries uniform sepia-brown; primaries barely longer than secondaries. Underwing-coverts and axillaries black, barred white; underside of remiges shining brownish-dusky. Lores, sides of crown and sides of head ashy-grey; chin and upper throat white; foreneck and breast lead-grey with vague sandy margins to feathers of throat and breast; lower breast has whitish bars; flanks, belly and undertail-coverts blackish-brown, narrowly barred white, bars less distinct on belly; olive-brown patch on each side of vent; thighs dusky brownish with faint white bars. Iris brown to yellowish; upper mandible and tip brownish, rest of lower mandible yellowish-green; legs and feet black.
Adult female Similar to male but hindneck brighter chestnut-rufous; white on chin less extensive; iris orange; upper mandible horn-coloured, sides sometimes reddish, bill tip dark brown, rest of lower mandible reddish; legs and feet slaty blue-grey.
Immature Undescribed.
Juvenile Undescribed.
Downy young Undescribed.

MEASUREMENTS Wing of 6 males 143-160 (152.3, SD 6.1), of 7 females 145-157 (150, SD 4.4); tail of male 33; culmen of 5 males 52-58 (54.8, SD 2.1), of 6 females 50-57 (53, SD 2.3); tarsus of male 62, of female 60 (Coomans de Ruiter 1947a, Ripley 1977).

GEOGRAPHICAL VARIATION None.

MOULT No information available.

DISTRIBUTION AND STATUS N, NC and SE Sulawesi. This rail is locally distributed, uncommon to rare and elusive and is regarded as VULNERABLE. It is known from 14 specimens (one is probably lost), all but one collected before 1940 (Coomans de Ruiter 1947a), the last being collected in Apr 1980, from Torro Itulawi, Donggala, in Lore Lindu National Park (Watling 1981, 1983). It has been seen on four occasions in Lore Lindu since 1980: near Sidaunta in Aug 1983 (Andrew & Holmes 1990; see Breeding), twice in the Sopu R valley in Jul 1987 (Lambert 1989), and in hill or lower mountain forest in 1992 (Collar *et al.* 1994); Coates & Bishop (1997), who regard it as uncommon, also record it as occurring at Gunung Rurukan and Tomohon (in the N), Kantewoe (NC), and Mekonga Mts (SE). It is vulnerable to deforestation, which is already quite widespread in parts of Sulawesi, and perhaps also to introduced or feral predators. On the Minahasa Peninsula it was apparently locally quite common in the past but, even as long ago as 1941, was described as threatened by increasing deforestation along the Menado R and was said to have been almost extirpated in some areas by large-scale snaring (Stresemann 1941).

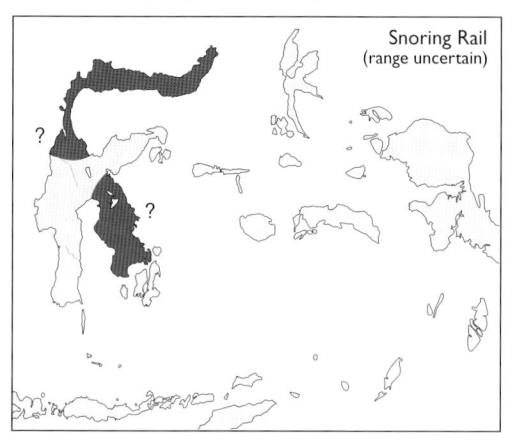

Snoring Rail (range uncertain)

MOVEMENTS None known.

HABITAT Dense liana and bamboo secondary growth on the borders of lowland or highland forest; it occasionally enters adjacent primary forest. On the Minahasa Peninsula it is recorded from secondary growth on mountainsides skirted at the base by plains of elephant grass and small bushes. In Lore Lindu National Park it is recorded from lightly disturbed hill forest (Andrew & Holmes 1990; see Breeding) but is apparently commonest in wet, impenetrable secondary vegetation (Watling 1983). Lambert (1989) observed this species in old secondary growth, little different in structure from primary forest, with a dense understorey dominated by rattans, on very wet and peaty ground bordering a fast stream; it occurred in wetter, more impenetrable habitat than that normally frequented by the Bald-faced Rail. A reference to its entering rice fields (Watling 1983) could result from confusion with the Buff-banded Rail (Lambert 1989). Its habitat is said to be similar to that occupied by the Invisible Rail (115) on Halmahera (Stresemann 1941). It occurs from sea level up to 1,300m.

FOOD AND FEEDING Described as possibly eating largely crabs, with lizards also recorded, and a captive bird survived on crabs for some time (Stresemann 1941); stomach contents of 2 museum specimens are given as crabs (BirdLife files). It forages for crabs in mountain streams (Coomans de Ruiter 1947a), and has also been seen feeding in a shallow muddy gully (Lambert 1989). It normally forages in dense cover (Lambert 1989), but it occasionally forages along the damp margins of little-used dirt roads through forest, when it feeds unhurriedly, pecking at the ground from side to side (Coates & Bishop 1997) – this foraging method suggests that prey types other than crabs and lizards are taken.

HABITS Apparently not especially shy and sometimes not difficult to observe, but it is inconspicuous, elusive and easily overlooked (Coates & Bishop 1997).

SOCIAL ORGANISATION Normally seen singly or in pairs, occasionally in small family groups (Coates & Bishop 1997). Two birds seen foraging together seemed to keep in contact by calling (F. Lambert *in litt.*).

SOCIAL AND SEXUAL BEHAVIOUR No information available.

BREEDING AND SURVIVAL The only breeding information is an unsubstantiated observation of an adult foraging with 2 chicks on 18 Aug 1983 (Andrew & Holmes 1990).

ATLANTISIA

This genus contains one extant and two recently extinct flightless rails of the remote South Atlantic islands of Inaccessible, St Helena and Ascension. All are probably descended from a common mainland ancestor of pro-*Rallus* stock which independently evolved flightlessness and divergent body sizes in response to widely different environmental conditions; no other genus of rails varies so much in size. The only surviving species, the Inaccessible Rail, is a diminutive, very dark bird with decomposed and hair-like plumage. It occupies all habitats on Inaccessible I and is considered Vulnerable because of its restricted distribution and the threats posed by possible habitat destruction and the introduction of predators.

72 INACCESSIBLE RAIL
Atlantisia rogersi   Plate 21

Atlantisia rogersi Lowe, 1923, Inaccessible Island, Tristan da Cunha.

Monotypic.

Alternative names: Inaccessible Island Rail; Tristan Rail; Island Hen/Cock.

IDENTIFICATION Length 13-15.5cm. Smallest flightless bird in the world. Predominantly very dark: male has dark grey head; lores and ear-coverts blacker; upperparts dark chestnut-brown; scapulars and upperwing-coverts narrowly barred whitish; underparts dark grey; belly to undertail-coverts darker; flanks to undertail-coverts barred whitish to buff. Iris red to orange; bill blackish-grey; legs and feet blackish-brown or grey-brown. Female smaller; paler grey, especially on head, throat and breast, with faint brown wash on underparts; lores and ear-coverts grey. Immature male blackish-grey with no white barring, and orange-red iris; immature female slate-grey with brown wash, buffy or faint white barring, and brown or orange-red iris. Juvenile black; iris brown. Inhabits all available vegetation types, from tussock grass and ferns to tree thickets; also boulder beaches.

Similar species No similar species occurs on Inaccessible I. Adults easily identifiable by very dark plumage, red or orange-red eye, and whitish barring on scapulars, upperwing-coverts, flanks and belly.

VOICE Highly vocal, and calls are often loud. Most information is from Fraser *et al.* (1992). Pair members give a loud trill when meeting and in territorial interactions with other adults, this call being similar to the trill of the Little Grebe *Tachybaptus ruficollis*. Confrontations between adults are accompanied by a loud squealing followed by a prolonged twitter, or one bird squeals while the other utters a long, excited *keekeekeekee...* twitter, varying in length and ending in a *chitrrrr*; the victorious bird in territorial encounters may utter a *weechup* call. A hard *tchick* of aggression is also recorded. Foraging adults give a slow, wooden, rather monotonous, repeated *chunk, tchick* or *tchock*, a soft, enquiring *choptchaptchick* and a soft *tchick-tchuck* or *chip chip*. The *chip* call is also given by incubating birds or birds disturbed at the nest, while a hard, repeated *chip* of alarm is given when Brown Skuas *Catharacta antarctica* fly overhead. When alarmed, birds utter a sharp, repeated *pseep* as they scuttle off (Elliott 1957). Vocal contact between members of pairs and family groups is maintained almost continuously.

Calls of unknown function include a soft *tik tik*, a harsh rattling trill, a brief *schkreek*, a high-pitched, enquiring *tip*, a rising *queechick*, a *squick*, a twanging *chong chick*, a short *t-chip* and a low, guttural *queechock*. Chicks have a repeated *tchwip* contact call, and a give resonant *tchwee* if visual contact with the adults is lost.

DESCRIPTION
Adult male Head very dark grey-black, black from lores to ear-coverts; rest of upperparts, and entire upperwings, very dark chestnut-tinged brown; scapulars and upperwing-coverts very long (normally hide remiges), each feather washed hair-brown to chestnut, and with single narrow indistinct whitish terminal bar (sometimes absent), often irregularly bordered black; remiges with slight whitish barring towards tips (often only visible on outermost primary); axillaries and underwing-coverts as upperwings. Underparts very dark grey; flank feathers tipped whitish (more prominent in older birds); belly to undertail-coverts very dark grey, tinged dull brown and with indistinct white to buff tips forming narrow bars. Iris fiery-red to orange-red; bill blackish-grey, lower mandible sometimes tinged reddish; legs and feet blackish-brown or dark greyish-brown.
Adult female Slightly smaller and paler grey, most noticeably on head, throat and breast, with faint brown wash on underparts; lores and ear-coverts grey.
Immature male Blackish-grey with no brown wash, no white barring, and orange-red iris.
Immature female Medium to pale slate-grey with brown

wash, buffy or faint white barring, and brown or orange-red iris.
Juvenile Sooty-black, with no white barring; iris brown.
Downy young Chicks have soft black down; iris dark brown; bill, legs and feet black; gape silvery-white; white egg-tooth present; wing-claw well developed.

MEASUREMENTS (8 males, 6 females) Wing of male 51-59 (54.25, SD 2.9), of female 51-56 (53.7, SD 1.7); tail of male 27-32 (29.25, SD 1.7), of female 25-30 (27.8, SD 2.1); culmen to base of male 21-24.5 (22.6, SD 1.1), of female 20-21.5 (21, SD 0.5); tarsus of male 20-24.5 (22.3, SD 1.4), of female 19-23 (21.3, SD 1.3). Weight of 6 males 38-49 (41.8, SD 3.8), of 6 females 34-42 (36.9, SD 2.8); hatching weight 7.8.

GEOGRAPHICAL VARIATION None.

MOULT Remex moult is synchronised and is recorded in 2 individuals, Feb, at which time the rectrices were fully grown in both birds (Stresemann & Stresemann 1966). The immature plumage may be retained for 1-2 years, indicating delayed maturity.

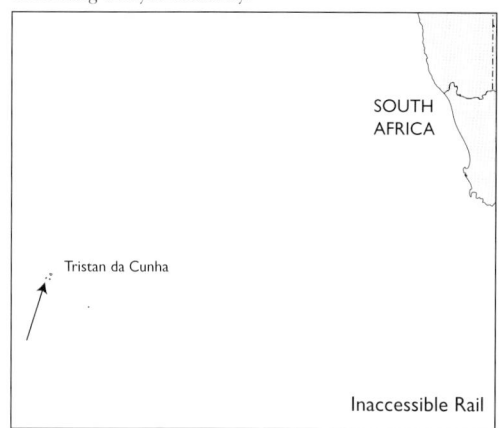

Inaccessible Rail

DISTRIBUTION AND STATUS Confined to the uninhabited Inaccessible I (area 16km²), Tristan da Cunha group. This rail is abundant within its restricted range and is in no immediate danger, but it is classed as VULNERABLE because of its restricted distribution and the permanent threats posed by the possible introduction of predators and the destruction of habitat. Its total population is estimated at 8,400-10,000 birds (Fraser *et al.* 1992), living at a high density: in 1982-83, the estimated density in mixed fern and tussock at sea level was 15 birds/ha; in pure tussock grass 10 birds/ha; in mixed ferns and tussock on the upland plateau 5 birds/ha; and in plateau tree-fern and tree habitat 2 birds/ha; young birds increase the overall numbers substantially with, for example, 12 chicks recorded in 1ha of tussock grass (Collar & Stuart 1985, Fraser *et al.* 1992). The population is probably at the carrying capacity of the environment, with regulatory mechanisms such low fertility and delayed maturity operating to limit productivity (Ryan *et al.* 1989). The most important potential threat is the risk of mammalian predators (especially domestic cats, and rats *Rattus rattus*) becoming established on the island as a result of the infrequent visits by Tristan islanders (Collar & Stuart 1985). The two other *Atlantisia* rails, *A. elpenor* of Ascension Island and *A. podarces* of St Helena, have become extinct since the advent of man to the islands (Olson 1973a). Last century, goats and pigs were present on Inaccessible and did much damage to the vegetation, but were eradicated by 1950, to be replaced by a few cattle and sheep which were removed before 1960 without having caused any damage (Collar & Stuart 1985). Concern that the invasive exotic New Zealand flax *Phormium tenax* might overrun the island's natural vegetation has been allayed by recent observations (Collar & Stuart 1985).

The only significant natural predator of adults is the Brown Skua, which apparently does not pose a threat. In 1872 and 1909 the island's tussock grass was burned, and the 1909 fire burned for at least a month; these events would probably have killed many rails but no details of mortality exist (Collar & Stuart 1985). Alien plants and invertebrates occur on the island but apparently have no adverse effect, while introduced centipedes may be an important food item of the rails (Collar & Stuart 1985). Proposed conservation measure include the introduction of the rails to Gough or Nightingale Is as a safety measure, but this may adversely affect the endemic invertebrates of these islands (Collar & Stuart 1985). Captive breeding has also been recommended to build up reserve stocks, and would also facilitate research and promote public awareness (Fraser *et al.* 1992). Scientific collection of this rail requires the strictest control (Wace & Holdgate 1976). Inaccessible has no protected area status, and agricultural development is still advocated; it is one of the least man-modified islands in the world, and the granting of full protection to the island, and the prohibition of agriculture, are strongly recommended to help ensure the survival of the island's birds (Fraser *et al.* 1992).

MOVEMENTS None.

HABITAT Occurs in all vegetation types at all altitudes, and even on the steepest slopes. The population density is highest in coastal tussock-grass *Spartina arundinacea*, especially where this vegetation is more open and mixed with ferns *Blechnum penna-marina* which form luxuriant undergrowth or mats of vegetation (Hagen 1952), and in patches of sedges *Carex thouarsii* and *Scirpus sulcatus* on the peaty plateau (Fraser *et al.* 1989); the species also occurs in fern-bush and island-tree thickets with *Phylica arborea* and *B. palmiforme* on the plateau, and in *Empetrum* and *Scirpus* heath at the highest altitudes. It also forages on exposed boulder beaches on and above the high-tide line. It is apparently absent from short dry tussocks on cinder cones.

FOOD AND FEEDING Most information is from Collar & Stuart (1985) and Fraser *et al.* (1992). Earthworms, amphipods, isopods, mites, centipedes, insects (weevils and other beetles, flies, small Homoptera, moths and caterpillars); also berries of *Empetrum* and *Nertera*, and seeds of dock *Rumex*. It does not eat fish or bird carrion, which are both readily available on the island, and birds ignored cheese and bread in traps. Chicks are fed on centipedes, earthworms, caterpillars, moths and amphipods. This rail forages in every available habitat, including very short vegetation, boulder beaches and marshy seepage zones. It jabs rapidly at its insect prey with its neck extended. Foraging actions are slow and deliberate, rather mouse-like, and the species appears to occupy the ecological niche of a mouse or shrew. The estimated standard metabolic rate of adults (mean mass 39.4g) is 19.5 ± 5.4 (range 12.5-25.5) kJ/day, "only 60-68% of the basal metabolic rate predicted for a 39.4g bird from allometric equations for resting nonpasserine birds" (Ryan *et al.* 1989). These authors suggest that, as flightless birds typically do not have a lower metabolic capacity than volant birds (Brackenbury 1984), it is likely that this rail's low metabolic rate is a manifestation of its life history strategy. Its high population density, delayed maturity and small clutch size suggest that carrying capacity has been reached on the island in the absence of major predators or competitors, and such conditions probably favour energy conservation features such as low metabolic rate, small size and the evolution of flightlessness.

HABITS Most information is from Collar & Stuart (1985) and Fraser *et al.* (1992). These tiny rails are retiring and partly subterranean, frequently making use of natural connecting cavities under boulder beaches, and small tunnels (often formed by regular use of favourite paths) through *Blechnum* and sedge mats, for access and concealment; they also often enter seabird burrows, which may be occupied. When undisturbed, their movements are deliberate but when alarmed they retract the head and run very rapidly among the tussock grass in almost mouse-like fashion. They occur most commonly where low small ferns provide rodent-type tunnels of shelter up to 1m long. Shelter appears to be important, as a captured bird which was accidentally left outside on the night of a gale was found to have died, but not before it had dug itself a hole 30cm deep by the roots of a tussock. When foraging out of cover, these rails are extremely wary. All ages of birds climb freely in *Phylica* and tussock grass up to a height of 1.5m, flapping the wings to aid balance and probably using the well-developed wing-claw. They are active mainly during the day but they also forage at night; captive birds were active at times throughout the night and free-living birds called between 04:00 and 22:00 (Ryan *et al.* 1989). They may roost in old nests (Elliott 1957), presumably built for breeding.

SOCIAL ORGANISATION Monogamous; the pair bond is permanent and the species apparently has a loose, flexible territorial system, generally keeping to a discrete area but occasionally going outside it (Collar & Stuart 1985). Adults with chicks are occasionally joined by a bird in subadult plumage (Fraser 1989), but no evidence of cooperative breeding has been found. The population density in optimal habitat is estimated at 10-15 birds/ha (Collar & Stuart 1985), and territories are likely to be small, c. 0.01-0.04ha in extent (Fraser *et al.* 1992).

SOCIAL AND SEXUAL BEHAVIOUR Information is taken from Fraser *et al.* (1992). Birds occur in pairs or in family groups of 2 adults, 1-2 immatures and occasionally also 1 subadult. Territorial behaviour is frequently evident, birds meeting at common boundaries calling and displaying before each bird retreats into its own territory. The birds face each other, standing a fewcm apart with lowered heads and bills pointed to the ground; they sometimes slowly circle about a point midway between them. After a short time one bird slowly retreats, or 2-3 skirmishes occur, involving loud trilling and buffeting, followed by a fast chase in which one bird is driven off; the victorious bird may then utter the *weechup* call. The courtship display involves similar vocalisations and movements but ends when the birds move off silently together. The male courtship-feeds the female, silently offering earthworms or centipedes, which the female accepts after bobbing her head. The incubating female is sometimes fed on the nest by the male, and will also accept food from a human observer; when accustomed to human observers the incubating bird sits tight and has to be removed to allow the eggs to be examined.

Adults vigorously defend the nest against Tristan Thrushes *Nesocichla eremita*, repeatedly running up and pecking this predator, and retreating into cover between attacks; a female rail also rammed a thrush, which flew off. When confronted by a thrush at other times, adult rails draw in the head and rouse the plumage before moving rapidly into cover.

BREEDING AND SURVIVAL Season Lays Oct-Jan. **Nest** Built on ground, at base of tussock-grass clump, in sedges or in ferns under dense cover of fallen tussock-grass; oval or pear-shaped, with entrance at narrower end and access track or tunnel extending for up to 0.5m through vegetation. Nest carefully woven; thinly domed or sturdily roofed, with side entrance; constructed of vegetation in which sited (either dead tussock leaves or dead sedges); tussock nests constructed of broad curled leaves, cup lined with finer strips; 1 nest lined with dead leaves, possibly from *Malus* or *Salix* tree. If disturbed, incubating bird may escape through thin dome (Fraser 1989). Material may be added after clutch complete, possibly in response to wet, cold weather (Fraser *et al.* 1992). One nest had external length 17cm, external width and height 13cm, cavity diameter 9cm and entrance diameter 4cm. **Eggs** 2 (n = 8 clutches); 1 set of 3 possibly from 2 clutches (Fraser *et al.* 1992); oval; cream, buffy-white or greyish-white, sparingly marked with smallish chocolate-rufous spots and a concentration of brown speckles over underlying slate or lavender-mauve spots at broad end; size 31.6-34 x 21.7-24 (33.4 x 23, n = 11); calculated weight 9.4; mean weight of 2 eggs 2-4 days before hatching 8.7 (SD 0.1), c. 25% of female's mass. The following information is from Fraser (1989). **Incubation** By both sexes (period unknown), and either sex may take larger share; nest change over preceded by *chip* calls, or sometimes trills; eggs hatch 12.5-32h apart; chicks may call in eggs for up to 45h before hatching. **Chicks** Precocial; walk within 2h of hatching; leave nest c. 1 day after hatching; roost in nest only for first 2 nights; fed and cared for by both parents, 1 adult staying with chicks while other forages; chicks peck at tip of adult's bill and are fed bill-to-bill; chicks bask in warm sun. At 4 days, chicks move up to 5m from nest and climb up to 1m in sloping tussocks, extending wings for balance; attempt self-feeding at c. 7 days. Fledging period unknown; at least some young are still with parents when subadult. Immature plumage may be retained for 1-2 years, suggesting delayed maturity. In 1982, of 8 eggs found, 5 (63%) hatched, indicating low fertility and low breeding success; chick mortality high, chiefly from predation by Tristan Thrushes; wet weather also believed to cause losses (Fraser 1989).

ARAMIDES

A genus comprising seven species of medium to large South American rails, commonly known as wood-rails, two of which, the Rufous-necked and Grey-necked Wood-rails, have ranges extending through Central America to Mexico. They are relatively primitive, and inhabit mangroves and other coastal wetlands, swamp forest, marshes, and dry forest and woodland. Their loud and striking calls are often well known and have given rise to onomatopoeic local names. The plumage is distinctive. All species have black upper- and undertail-coverts and tail, the black in some extending to the rump, lower flanks and lower belly. This contrasts strongly with the predominantly olive to olive-green upperparts and upperwings, and with the rufous or chestnut underparts of most species. The primaries are chestnut to tawny in all species except the Red-winged Wood-rail, which has dusky brown primaries and mahogany-red upperwing-coverts; all species have the underwing-coverts barred, either black and white or black and rufous. All have some grey in the plumage, this colour being restricted to the head in the Brown Wood-rail and to the mantle in Rufous-necked Wood-rail, while in the Slaty-breasted Wood-rail the anterior underparts are entirely slaty-blue. The eye and orbital ring are usually red, the moderately long and strong bill predominantly yellow or greenish, and the long legs and toes red. Most species are poorly known and are in urgent need of further study. The Brown Wood-rail, which is classed as Vulnerable, has a restricted range and appears to have become rare because of destruction of its mangrove habitats. Two other rarely observed species, the Little and Red-winged Wood-rails, should possibly be classed as Data Deficient pending further investigation of their range and status.

73 LITTLE WOOD-RAIL
Aramides mangle Plate 22

Gallinula mangle Spix, 1825, coast of Bahia, Brazil.

Genus sometimes merged into *Eulabeornis*. Monotypic.

Synonyms: *Rallus/Eulabeornis mangle*; *Aramides chiricote/ruficollis*.

IDENTIFICATION Length 27-32cm. Smallest member of genus. Forehead to upper mantle grey to slaty-blue, ashier on top of head; mantle, upper back and scapulars greenish-olive; lower back and rump dark sepia; uppertail-coverts and tail blackish. Upperwing-coverts and tertials buffy-olive; primary coverts, primaries and secondaries tawny; axillaries and underwing-coverts blackish-brown, barred white to cinnamon. Sides of head and sides of neck ashy-grey; chin and throat white. Foreneck, breast and flanks tawny; belly isabelline; thighs and lower flanks more olive; undertail-coverts black. Iris and eye-ring red; bill greenish with red base; legs and feet red. Sexes alike. Immature and juvenile not described. Inhabits coastal swamps and lagoons, including mangrove swamps.
Similar species Sympatric with Grey-necked Wood-rail (75) and Slaty-breasted Wood-rail (78), the former having a wider habitat tolerance (see Habitat) and the latter inhabiting forest. Differs from all congeners in having red base to greenish bill, and from all except Red-winged Wood-rail (79) and Rufous-necked Wood-rail (74) in having underwing-coverts brownish-black, narrowly barred white or cinnamon, instead of rufous with black bars. Lacks grey underparts of Slaty-breasted Wood-rail and Red-winged Wood-rail, and extensive grey on head and foreneck of Grey-necked Wood-rail and Giant Wood-rail (77); Rufous-necked Wood-rail chestnut on head, neck and underparts; Brown Wood-rail (76) darker, especially on underparts, and has yellow patch on forehead.

VOICE Not described.

DESCRIPTION
Adult Forehead to crown ashy-grey; nape to upper mantle similar, or more slaty-blue; mantle, scapulars and upper back greenish-olive; lower back and rump dark sepia; uppertail-coverts and tail blackish, tinged rich dark brown. Upperwing-coverts and tertials buffy-olive, but marginal coverts pale cinnamon-rufous, and primary coverts dull tawny, edged cinnamon-rufous; alula, primaries and secondaries tawny, becoming more cinnamon-brown towards tips; axillaries and underwing-coverts blackish-brown, narrowly barred white to cinnamon. Lores and feathers over eye ashy; sides of face and ear-coverts pale ashy-grey; malar region, chin and throat white; sides of neck ashy-grey. Foreneck, breast and flanks tawny to cinnamon; belly isabelline; thighs and lower flanks greenish-olive to grey-brown; undertail-coverts blackish. Iris and eye-ring red; bill greenish with red base; legs and feet red. Sexes alike.
Immature Not described.
Juvenile Not described.
Downy young Not described.

MEASUREMENTS (Sexes combined) Wing 157-170; tail 52-59; exposed culmen 36-41, culmen to base 42-44.5; tarsus 49-53 (Blake 1977, Ripley 1977, BMNH). Weight of 2 males 230, 245, of 6 females 164-265 (227.5, SD 38.4) (Teixeira *et al.* 1989, D. C. Oren *in litt.*).

GEOGRAPHICAL VARIATION None.

MOULT No information available.

DISTRIBUTION AND STATUS E Brazil, from NE Pará (W to Belém [Forrester 1993]), Maranhão, Ceará, Pernambuco, Alagoas and Bahia States S to São Paulo. Between these limits it is probably less widespread than indicated on the distribution map, as it is apparently confined largely to coastal localities in NE and E Brazil, with only a few inland localities known (Ripley 1977). It is not considered globally threatened, and is regarded as very common in the mangroves of Alagoas and S Pernambuco (Teixeira *et al.* 1989), but its status in other parts of its range is unclear. In view of its potentially restricted distribution, the almost complete lack of knowledge about its natural history, and the possibility of threats to its wetland habitats (similar to those facing the Brown Wood-rail), it should at least be considered DATA DEFICIENT and worthy of urgent investigation.

MOVEMENTS A bird once flew against a hotel in Rio de

Janeiro (Sick 1993). No other evidence of movements is recorded.

HABITAT Mainly coastal swamps and lagoons, including mangrove swamps; details of inland habitats are not recorded, but it penetrates forest adjacent to mangroves (Sick 1993). The larger sympatric Grey-necked Wood-rail occurs locally in mangroves (the race *avicenniae* is apparently almost confined to mangroves, where it occurs alongside the present species), but occurs more widely in swampy forests and woodland, on river banks, at pools and on dry ground with grass and bushes (see Grey-necked Wood-rail species account, Habitat section).

FOOD AND FEEDING At low tide in the estuary of the R Traripe, Bahia, birds were seen emerging from mangroves to forage on the river banks (Taylor 1996b). Foraging birds flick dead wet leaves aside with the bill (Willis & Oniki 1993).

HABITS Birds fly readily to perch high in bushes (Willis & Oniki 1993), presumably when disturbed.

SOCIAL ORGANISATION Unknown.

SOCIAL AND SEXUAL BEHAVIOUR Unknown.

BREEDING AND SURVIVAL Schönwetter (1961-62) gives measurements of 7 eggs as 46-54 x 34-36.2 (51 x 35.2), calculated weight 34.6. No other information available.

74 RUFOUS-NECKED WOOD-RAIL
Aramides axillaris   Plate 22

Aramides axillaris Lawrence, 1863, Barranquilla, Colombia.

Genus sometimes merged into *Eulabeornis*. Monotypic.

Synonyms: *Eulabeornis axillaris*; *Ortygarchus manglè/ ruficollis/axillaris*; *Aramides ruficollis*.

Alternative name: Rufous-crowned Wood-rail.

IDENTIFICATION Length 28-30cm. A relatively small wood-rail. Head, neck and underparts to lower flanks and belly bright chestnut; face paler; chin and throat white. Lower hindneck and mantle bluish-grey, forming distinct triangular patch; scapulars, back and upperwing-coverts greenish-olive; lower back to tail blackish-brown to black. Primaries cinnamon-rufous (obvious in flight); secondaries greenish-olive; axillaries and underwing-coverts blackish, barred white to rufous. Lower belly ashy-grey; lower flanks to undertail-coverts black. Iris orange-brown; eye-ring red; bill green with yellowish base; legs and feet red. Sexes alike. Immature probably similar to adult. Presumed juvenile duller and paler; forehead to hindneck buffy-brown; rest of upperparts, and upperwing-coverts, olive-brown, with dull grey patch on upper mantle; chin and throat grey-buff to white; rest of underparts greyish olive-brown, lower flanks to undertail-coverts darker; bare parts duller than in adult. Inhabits mangroves and other coastal wetlands, swamp forest, forest undergrowth and cloud forest.

Similar species Only slightly larger than Little Wood-rail (73), like which it has underwing-coverts black, barred white or rufous. Differs from all congeners in lacking grey on head, neck, breast, flanks and upper belly, these regions being bright chestnut; also usually has orange-brown rather than red eye. Smaller and shorter-legged than the commoner Grey-necked Wood-rail (75). Uniform Crake (80) is smaller with shorter bill and legs, and is almost uniform rufous-brown, lacking grey and greenish-olive on upperparts and black on hindbody and tail; it frequents mangroves only very locally.

VOICE An incisive, loud *pik-pik-pik*, *pyok-pyok-pyok* or *tuktuktuk*, also described as a yelp, repeated c. 8 times; often calls antiphonally (ffrench 1973); there is also a gruff *kik* of alarm. The *pyok* call is also given, more continuously, in response to taped playback (Hardy *et al.* 1996). These birds call most frequently at dawn and dusk, but also often at night. They may respond to imitations of the calls of the chicks, such as the peeping of young chickens (Meyer de Schauensee & Phelps 1978). Adults with chicks make an irregular clicking noise, likened to the pecking of a large woodpecker, which appears to be a contact and rallying call. When threatening an intruder, an adult gave grunts and squeaks; when the intruder retreated, a second adult gave a two-part call consisting of a high-pitched *queeng*, similar in quality to the song of the Grey-necked Wood-rail, followed by a much deeper, less audible booming sound (Parker *et al.* 1995).

DESCRIPTION
Adult Head and neck chestnut to burnt-sienna, paler on lores, ear-coverts and malar region; lower hindneck and mantle bluish-grey, forming distinct triangular patch; back, scapulars, upperwing-coverts and tertials greenish-olive; lower back and rump blackish-brown; uppertail-coverts and tail black. Alula cinnamon-rufous, feathers more olive at tips; primary coverts greenish-olive, washed cinnamon-rufous at base; primaries cinnamon-rufous, tending to greenish-olive at tips; secondaries more greenish-olive; underside of remiges slightly darker and duller than upperside; axillaries and underwing-coverts blackish-brown, barred white to pinkish-cinnamon. Chin and throat white, tinged pinkish-cinnamon (possibly only in female); breast, flanks and belly bright chestnut to burnt-sienna; lower belly ashy-grey; thighs dark slate; lower flanks to undertail-coverts blackish. Iris orange-brown, possibly sometimes red; eye-ring red; bill green with yellowish base; legs and feet pinkish-red to red. Sexes alike.
Immature Probably very similar to adult, as in other *Aramides* species.

Juvenile Much duller than adult. Forehead to hindneck buffy-brown, tinged cinnamon; rest of upperparts, and upperwings, paler and duller than in adult, olive-brown with dull grey patch on upper mantle; sides of head, and chin to foreneck, pale buffy-brown to dirty white, strongly washed vinaceous-cinnamon; dirty white; rest of underparts buffy-brown, washed orange-cinnamon; lower flanks to undertail-coverts deep brownish-olive; iris brownish to pale red; bill duller than in adult; legs and feet dull red-brown.

Downy young Down is black (Parker *et al.* 1995): no other information recorded.

MEASUREMENTS (18 males, 13 females) Wing (chord) of male 163-174 (169), of female 145.5-170 (163.6); tail of male 53-63 (58.3), of female 47-62.5 (57.3); exposed culmen of male 39.5-46 (43.7), of female 37.5-46 (42.2); tarsus of male 52.5-63 (59.5), of female 50-60.5 (57.6) (Ridgway & Friedmann 1941). Measurements of 4 unsexed: wing (max) 165-179 (172, SD 6.2); culmen to base 39-42 (40.75, SD 1.3). Weight of male 262-327, unsexed mean 275.

GEOGRAPHICAL VARIATION None recorded.

MOULT The juvenile attains adult plumage "over a few months" (Howell & Webb 1995b).

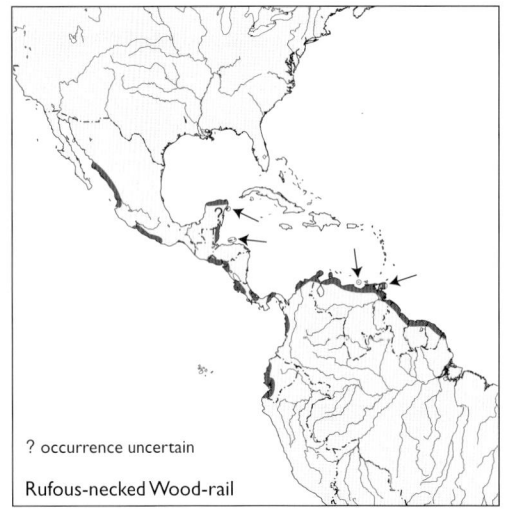

? occurrence uncertain
Rufous-necked Wood-rail

DISTRIBUTION AND STATUS Pacific slope of C Mexico (Sinaloa, Nayarit, Guerrero); Yucatán (Isla Mujeres and Las Bocas de Silán); Belize; El Salvador; Honduras (Isla Guanaja and Roatán in Bay Is, and Pacific coast of Bay of Fonseca); W Nicaragua (San Cristóbal and Volcán Mombacho); Costa Rica (Golfo de Nicoya and Caribbean foothills of Cordillera de Guanacaste, also possibly Sarapiquí lowlands); Panama (on Caribbean coast in NW Bocas del Toro and Canal Zone, and on Pacific coast in S Coclé); NW coastal Colombia (Cartagena E to NE end of Santa Marta Mts, and on Pacific coast at Chocó); S to W Ecuador (Guayas, El Oro and S Loja, also once in Esmeraldas), and extreme NW Peru in Tumbes (Puerto Pizarro and El Caucho [Best *et al.* 1993, Parker *et al.* 1995]), and E through coastal Venezuela (Zulia, Falcón, Lara, and Carabobo to Monagas, also Patos and Los Roques Is), Trinidad and Guyana (Demerara R and Bartica) to Surinam and French Guiana. Not globally threatened. This rail is apparently fairly common to common in a few areas, such as at Puerto Pizarro in extreme NW Peru (Parker *et al.* 1995), and locally in Ecuador (Ridgely & Greenfield in prep.), but generally seems to be local and uncommon to rare throughout its range, although it must be overlooked to some extent.

MOVEMENTS None recorded.

HABITAT Mangrove swamps, coastal marshes, lagoons and mudflats, and swamp forest; in South America, also the floor of wooded foothill forest, dense forest undergrowth, forest edge, forest creeks and deciduous woodland. It mainly occurs in coastal and low-altitude areas in Central America, but in South America it also occurs in primary and secondary cloud forest, ascending to 1,400m (Ecuador) and 1,800m (Venezuela) in this habitat (Meyer de Schauensee & Phelps 1978, Best *et al.* 1993).

FOOD AND FEEDING This species eats mainly crabs. It forages in dense cover and along creeks, but also emerges onto open mudflats at low tide to catch crabs, normally keeping close to cover but occasionally venturing up to 15m from cover.

HABITS These rails are most active at dawn and dusk. They are secretive and difficult to see, apparently normally remaining within dense cover such as mangroves, but at low tide both adults and immatures may forage on open mudflats (though rarely far from cover). Birds have been seen walking on the branches of mangroves *Avicennia* (Haverschmidt & Mees 1995). Wetmore (1965) notes that their mannerisms in walking and jerking the tail, and their alert though furtive posture are like those of other *Aramides* species. He reports that "as I sat at the landing place [at a mangrove swamp] ... to skin out a heron that threatened to spoil, one of these rails came quietly across the mud of the swamp to watch me". This bird's confiding behaviour was apparently its undoing, as Wetmore continues "In the hand this proved to be an immature female".

SOCIAL ORGANISATION Probably monogamous. Three family parties have been reported: 2 adults with 4+ chicks, 2 adults with 1 chick, and 3 adults with 1 chick (Parker *et al.* 1995); the last observation suggests that some form of cooperative breeding may occur.

SOCIAL AND SEXUAL BEHAVIOUR Information is from Parker *et al.* (1995). An adult with chicks dashed towards a human intruder, displaying the rufous primaries by spreading the wings sideways, and approached to within 1.5m, giving grunts and squeaks. This attracted a second adult, which gave similar displays while the first bird, uttering clicking notes, led the chicks away. As the intruder retreated, the second bird followed for c. 50m.

BREEDING AND SURVIVAL Season Trinidad, Jul, Oct; Peru, Feb (unconfirmed). **Nest** A bowl or platform of small twigs, lined with weed stems and dead and green leaves; placed 1.8-3m up in small tree or vine tangle, or on dead stump; sometimes over water. **Eggs** 5; pointed oval, inclining to conical at large end; slightly glossy; cream to creamy-white, with small spots of cinnamon, pale to rich brown and purple, and subsurface markings of faint lilac-grey; very similar in appearance to the eggs of Grey-necked Wood-rail; size (n = 10) 43-47 x 31-36.1 (45.6 x 32.7), calculated weight 26.7. No further information available.

75 GREY-NECKED WOOD-RAIL
Aramides cajanea Plate 22

Fulica Cajanea P. L. S. Müller, 1776, Cayenne, French Guiana.

Genus sometimes merged into *Eulabeornis*. '*A. gutturalis*', known from a single specimen supposedly from Lima, Peru, is a poorly prepared specimen of *A. cajanea* (Meyer de Schauensee 1966). Nine subspecies recognised.

Synonyms: *Aramides cayanea/cajanus/mexicanus/ruficollis/ cayennensis/albiventris/vanrossemi/plumbeicollis*; *Ortygarchus/ Crex cayennensis*; *Fulica major/ruficollis/cayennensis*; *Rallus/ Aramides ruficeps/chiricote/maximus*; *Gallinula ruficollis/ ruficeps/cayennensis/cayanensis*; *Rallus hydrogallina*; *Eulabeornis cajaneus*.

Alternative name: Cayenne Wood-rail.

IDENTIFICATION Length 33-42cm. A widespread, medium-large wood-rail, with a distinctive and well-known call. Head, and neck to upper mantle and upper breast, grey, darker on forehead and crown and with back of head blackish or brown (S races *latens*, *morrisoni* and nominate), grey tinged brown (*avicenniae* and some nominate), or rufescent (all other races); chin and throat white; mantle, back, scapulars, tertials and lesser upperwing-coverts greenish-olive, overall colour varying individually and racially (race *plumbeicollis* has markedly rufous mantle, *avicenniae* has nape to back plumbeous); rump to tail black. Rest of upperwing-coverts, and primaries and secondaries, tawny; axillaries and underwing-coverts rufous, barred black. Breast and upper flanks orange-chestnut; lower flanks and belly to undertail-coverts black; race *mexicanus* has narrow white line between black of belly and orange of breast; *albiventris* has obvious large patch of white at base of breast. Iris, eyelids, legs and feet red to pink-red; bill yellow, green towards tip, sometimes orange at base. Sexes alike; female averages slightly smaller than male. Immature similar to adult. Juvenile duller than adult: belly sooty-black, flecked buff; bill and legs dusky; eyes brownish. Inhabits swampy forest, woodland, mangroves and marshes.

Similar species Only member of genus with entirely grey head and neck (with rufous or dark patch on occiput and grey extending to upper mantle) and, in two races, white on belly. Brown Wood-rail (76), which also occurs from Colombia to Peru but is rare, is almost as large but has grey only on head; upperparts and underparts much darker, predominantly brown, with no orange-chestnut on breast; bill entirely yellowish-green; bare yellow patch on forehead.

VOICE In Brazil this bird is called "[saracura] tres-potes", and in Argentina "chiricote", both names being onomatopoeic and referring to the common call which is a prolonged and varied series of loud notes, variable in form and variously described as *chirin co chirin co-co-co*, *trAY po trAY po trAY po po po po*, *kook-kooky*, *kook-kooky*, *ko-ko-ko*, *koo koo whiwho wee-wee-oo*, *kook-kak*, *kré-ko* etc. At a distance the call is melodious and pleasing but at close range the effect is spoiled by low, rattling or scratchy notes (Skutch 1994). The species is very vocal, calling most often at dawn, dusk, and at night, and also after rain; birds often call in duet, sometimes synchronized and often lasting for several minutes (Sick 1993), or in a chorus which gradually winds down until only 1 bird is calling *kré-ko* at the end. Calling activity peaks in the pre-nesting period, and birds hardly ever call during incubation (Skutch 1994). There is a legend from Brazil that, when calling, the bird also makes a "sonorous, intense, loud, but not very smelly tone with its rear end, which can be frightening"; this could refer to a ventriloquial, repeated *bo* which accompanies the *trAY po* call and may be made by the female (Sick 1993). Birds normally call from the ground, but one called while walking along a tree branch severalm up (Darlington 1931). The alarm call is a loud, harsh cackle or clucking shriek, and a bird driven from its nest gave queer throaty grunts. A bird watching two others fight made "smacking sounds" (Skutch 1994). Low grunting clucks *puk* or *kuk*, and similar notes rendered *ook*, *k-luk* or *kruk*, often reveal the nearby presence of a bird in cover. Foraging birds give a low series of 3-4 upslurred *umm* notes; the rails are attracted by an imitation of this contact note (Munn in Ripley & Beehler 1985). Newly hatched chicks gave weak peeps, and a slight, low-pitched whistle when handled; adults call chicks with clucks like a domestic hen (Skutch 1994).

DESCRIPTION *A. c. mexicanus*
Adult Forehead and crown dusky-slate; occiput russet or auburn; hindneck to upper mantle grey; lower mantle, back, scapulars, lesser upperwing-coverts and tertials greenish-olive; lower back more brownish-olive; rump to tail black. Rest of upperwing-coverts, primaries and secondaries rich tawny to ferruginous; primary coverts and primaries with dusky tips; greater coverts and inner secondaries washed greenish-olive towards tips; innermost secondaries greenish-olive with tawny central regions. Axillaries greenish-olive; underwing-coverts barred tawny and dark blackish-brown. Chin and throat white; sides of head ashy-grey, becoming paler and purer grey towards rear and on sides of neck and foreneck, extending to upper breast. Breast and upper flanks ochraceous-tawny; lower flanks and belly to undertail-coverts black; usually has narrow white line between black of belly and tawny of breast; thighs slaty-grey to dusky-brownish. Sexes alike. Iris red; eyelids, gape, legs and feet rose to red; bill yellow, becoming green on distal half, often orange at base.
Immature Similar to adult.
Juvenile Duller and more uniform than adult: belly sooty-black to very dark grey, flecked with buff to tawny-cinnamon (feather tips); bill and legs dusky to black; eyes brownish. See race *plumbeicollis*.
Downy young Down of upperparts dull black; deep auburn or reddish-cinnamon on head, foreneck and upper breast (sometimes only on cheeks and rear of head), and sometimes has dark band running along centre of crown; rest of underparts blackish-brown; indefinite line of deep mouse-grey extends from upper breast to foreneck; iris dark, surrounded by dull reddish bare skin; bill black with small white egg-tooth, base flesh-coloured; legs and feet black.

MEASUREMENTS (6 males, 3 females) Wing of male 186-200 (193.5, SD 3.8), (chord) of female 173-175 (173.7); tail of male 51-61 (56.3, SD 3.8), of female 54.5-59 (56.5); culmen to base of male 63-70 (66.5, SD 3.4), exposed culmen of female 60-63 (62); tarsus of male 75-86 (79.5, SD 4.0), of female 74-79 (76.6) (Ridgway & Friedmann 1941, BMNH). Wing of male 180-197, of female 173-185; tail of male 67-70, of female 55-62; culmen of male and female 50-56; tarsus of male 74-78, of female 68-77 (Ripley 1977, sample sizes and means not given).

GEOGRAPHICAL VARIATION Races separated on colour of occiput, presence of white patch between black belly and orange breast, intensity of rufous on mantle, overall colour, and on size. Within the very large range of the nominate race there is considerable variation in plumage which has resulted in new taxa (e.g. race *chiricote*) being proposed; however, most such variation appears to be poorly correlated with geographical distribution, and the various named taxa do not appear to be valid (Hellmayr 1929, Hellmayr & Conover 1942, Ripley 1977). The exception is the recently described race *avicenniae* from coastal SE Brazil (Stotz 1992).

A. c. mexicanus Bangs, 1907 – Caribbean slope of SE Mexico (S Tamaulipas, México, Distrito Federal, Veracruz, Hidalgo, Oaxaca, Tabasco and NW Chiapas). See Description.

A. c. albiventris Lawrence, 1867 – S Mexico (E Chiapas), Yucatán Peninsula, Cozumel I, Belize and NE Guatemala (Petén, Alta Verapez and Izabel). Like *mexicanus* but mantle more suffused rufous; broad, conspicuous white patch between black of belly and orange-chestnut breast; occiput auburn. Measurements of 5 males, 5 females: wing of male 180-189 (184.4), of female 172-196 (183.5); exposed culmen of male 57-65 (62.3), of female 58-65 (60.6); tarsus of male 77-82 (79.3), of female 73-81 (76.9) (Blake 1977). Tail of 3 males 60-66 (62.8), of 4 females 60-66 (62.7) (Ridgway & Friedmann 1941). Weight of 1 male 466 (Ripley 1977).

A. c. vanrossemi Dickey, 1929 – S Mexico (Oaxaca, SW Chiapas), Pacific coast of Guatemala and El Salvador. Intergrades with *mexicanus* in S Oaxaca (Blake 1977). Very similar to *albiventris* but slightly paler (Land 1970, Blake 1977, *contra* Ripley 1977), larger and with longer, heavier bill. Measurements of 8 males, 9 females: wing of male 188-203 (194.3), of female 188-197 (192.7); exposed culmen of male 65-71 (68.6), of female 63-66 (64.4); tarsus of male 80-86 (83.1) of female 78-84 (80.8) (Blake 1977). Tail of 2 males 60.5, 70, of 1 female 60 (Ridgway & Friedmann 1941).

A. c. pacificus A. H. Miller & Griscom, 1921 – Caribbean slope of Honduras, and throughout Nicaragua. Smaller than *vanrossemi*; similar to *plumbeicollis* but mantle less rufescent, more tawny-buff, and less distinct from greyish-olive of lower mantle, back, scapulars and upperwing-coverts; primaries average slightly darker. A few specimens from Honduras have a small amount of white at base of breast (Monroe 1968). Measurements of 8 males, 9 females: wing (chord) of male 171-186 (178.2), of female 166-183 (174.6); tail of male 47-58 (54.4), of female 45-57 (52.8); tarsus of male 77-85.5 (82.1), of female 74-81 (78.3).

A. c. plumbeicollis Zeledon, 1888 – Caribbean lowlands of NE Costa Rica. Differs from all other races in having posterior crown and occiput extensively rufescent-auburn, and mantle markedly rufescent, contrasting sharply with olive of back, scapulars and upperwing-coverts; breast rich, dark orange-chestnut. Juvenile male has mantle less extensively rufous; feathers of lower back to uppertail-coverts extensively margined and washed umber; anterior feathers of centre of belly broadly tipped cinnamon-buff (Ridgway & Friedmann 1941). Measurements of 5 males, 8 females: wing (chord) of male 172-186 (178.2), of female 168-179 (172); tail of male 51-59 (56.3), of female 50.5-56 (52.3); exposed culmen of male 53.5-63 (59.6), of female 54-60 (55.2); tarsus of male 74-87 (83.6), of female 69.5-82 (76.4).

A. c. latens Bangs & Penard, 1918 – Pearl Is, Panama (San Miguel and Viveros). Somewhat paler overall than all other races, especially on crown, neck and underparts; chin and throat faintly tinged grey; breast and flanks more cinnamon-buff, less rufous; hindcrown and occiput tinged dull earth-brown. Similar in size to *morrisoni*; smaller than nominate. Wing (chord) of 2 males 168, 173, of 2 females 161, 172; tail of 2 males 57.5, 59.5, of 2 females 57.5, 64; culmen of 2 males 50, 52, of 2 females 51, 51.5; tarsus of 2 males 68, 69, of 2 females 67, 60.5.

A. c. morrisoni Wetmore, 1946 – Pearl Is, Panama (San José & Pedro Gonzalez). Similar to *latens* in size; grey of hindneck darker, clearer, less brownish; back, scapulars and upperwing-coverts darker, more olivaceous-green. Measurements of 6 males, 6 females: wing (chord) of male 165-179 (174.3), of female 161-173 (166.5); tail of male 54-64 (58.3), of female 54.5-57 (54.9); culmen of male 47-59 (52.5), of female 48-52 (49.8); tarsus of male 64-71 (67.6), of female 63-70 (66.5).

A. c. cajanea (P.L.S. Müller, 1776) – Costa Rica (except NE), Panama (including Coiba and Cébaco Is), Colombia (on Pacific coast S only to Baudó Mts, and not in E Guajira), Venezuela, Trinidad, Surinam, the Guianas, E Ecuador, E Peru, Bolivia except Oruro and Potosí (Armonía 1995), Brazil (excluding the range of race *avicenniae*), Paraguay, Uruguay, and N Argentina (S to Tucumán, Santa Fe and Buenos Aires). Differs from N races in having occiput blackish to brownish, but some have occiput grey, as in *avicenniae* (Bornschein & Reinert 1996); upperparts clearer olive (but shows great individual variation, ranging from golden-olive to dull green); breast and flanks generally darker and richer rufous (but *plumbeicollis* may be darker). Bill as in other races, but also said to be yellow, tinged green on basal half, base red (Meyer de Schauensee 1962). Measurements of 12 males, 6 females: wing (chord) of male 171.5-201 (187.6), of female 166.5-189 (177.2); tail of male 53-80.5 (66.8), of female 55-68 (61.9); exposed culmen of male 46.5-58 (54.1), of female 46.5-56.5 (52.3); tarsus of male 68-74 (71.4), of female 66.5-75.5 (70.1). Note that Ripley (1977) incorrectly duplicates measurements of *mexicanus* for this race. Weight of 8 males 365-460 (412, SD 30.1), of 7 females 340-462 (388.3) (Ripley 1977, Belton 1984, Haverschmidt & Mees 1995, D. C. Oren *in litt.*); mean of 18 unsexed 397 (Dunning 1993); 1 male (FMNH) only 210 (D. E. Willard *in litt.*); 1 male (Barick Mus., Univ. Nevada: D. Baepler) only 240.

The specimen originally designated *A. gutturalis* by Sharpe (1894) is, as stated by Meyer de Schauensee (1966), an example of *A. cajanea*. The collecting locality is uncertain, being given as "Lima", and the bird's racial affinities are unclear. It most closely resembles the nominate form, which occurs E of the Andes in Peru, but differs in the following respects: forehead to occiput blackish-brown, tinged chestnut; grey of hindneck tinged olive-brown and does not extend as far on to upper mantle; grey of foreneck does not extend to upper breast; upperparts markedly tinged rich brown rather than greenish, especially on upperwing-coverts, which are washed burnt-sienna; primaries and secondaries darker than in nominate, chestnut with dusky tips; inner secondaries and alula predominantly very dark brown; breast and flanks deeper and richer in colour (more burnt-sienna) than most nominate; thighs and centre of belly dark ashy-brown. Wing 175; tail 63; culmen 50; tarsus 68.

A. c. avicenniae Stotz, 1992 – Brazil, in coastal São Paulo (S to Ilha do Cardoso, and probably N to Peruábe), Paraní (Guaratuba, and probably R de Borrachudo), and probably S to S limit of mangroves along Santa Catarina coast. Resembles nominate, but nape to back plumbeous, lower back with slight olive tone; occiput essentially grey, with brown wash reduced or absent (but see nominate); underparts paler rufous (cinnamon). Wing of male (type) 186; tail 70; tarsus 69. Apparent intergrades with nominate race occur on N São Paulo coast and its offshore islands (Alcatrazes, Buzios, Vitoria, Ubatuba), these birds being intermediate in colour of mantle and occiput, and darker rufous below than *avicenniae*, more like typical nominate (Stotz 1992).

MOULT Remex moult is simultaneous and recorded from Ecuador, Jun, Brazil (Pará), Apr, and Venezuela, Mar; the Venezuela bird had enlarged gonads and very worn rectrices with no sign of moult (Friedmann & Smith 1950). Young at 7 weeks resembles adults in plumage (Skutch 1994).

Grey-necked Wood-rail

DISTRIBUTION AND STATUS SE Mexico through Central America and South America, SW to E Ecuador and E Peru, E across Brazil and S to N Argentina and Uruguay. This species is not globally threatened, and is apparently still fairly common to common over much of its extensive range, although it must be adversely affected by habitat destruction, at least locally (see the comment in the Habitat section on its local extinction on São Sebastião I). In view of its ability to occupy forest edge and secondary growth habitats it may even have benefited locally from forest destruction (J. Mazar Barnett *in litt.*).

Its total population is estimated at 100,000-1,000,000 (Rose & Scott 1994). The nominate race is locally distributed in French Guiana (Tostain *et al.* 1992). In the 1970s the race *latens* was known only from four specimens; its present status is unknown. The recently described race *avicenniae* is fairly common in the mangrove-lined seaways and lagoons of Iguape-Cananéia, including Ilha do Cardoso, and the nominate race is locally common in the mangrove swamps of Santos-Cubatão, 90km N of Peruíbe, and on small offshore islands such as Alcatrazes, Buzios and Vitoria (see Habitat) (F. Olmos *in litt.*). This wood-rail is able to adapt to habitat modification: in Costa Rica and Brazil it persists in remnant streamside woods or forest patches amid pasture or agricultural lands, it may occupy cultivated areas such as ricefields and sugarcane plantations, and it can also exist close to cities (Slud 1964, Stiles & Skutch 1989, F. Olmos *in litt.*). In Brazil it is sometimes regarded as a pest, being accused of destroying rice when the plants sprout (Sick 1993).

MOVEMENTS None described, but records at 2,300m on the E slope of Colombia's Central Andes are probably of wanderers from lower elevations (Ridgely & Gaulin 1980). In Brazil the species overflies cities such as Rio de Janeiro on rainy nights, when its calls can be heard (Sick 1993), while Forrester (1993) regards it as accidental, and thus having some migratory movements, in Iguaçu National Park, Paraná, and Itatiaia National Park, Rio de Janeiro.

HABITAT Humid forest, both permanently swampy and seasonally inundated (such as *igapó* forest in Brazil); gallery forest; forest streams and rivers; oxbow lakes; seasonal pools near gallery woodland in the *llanos* tropical savanna (Colombia); palm groves; secondary growth and deciduous woodland; coastal swamps and mangroves; marshes and marshy thickets, particularly in forested areas; ricefields, and even sugarcane fields and scrubby pastures as long as they include wet areas; also seasonal pools near woodland. In the Pantanal de Mato Grosso, SW Brazil, it occurs around seasonal lagoons and swamps covered with emergent and floating macrophytes and with surrounding patches of bare ground caused by trampling cattle; the whole area is seasonally flooded except for semideciduous forest belts on higher ground (F. Olmos *in litt.*). The race *avicenniae* is apparently almost completely restricted to mangroves, where it is found alongside the smaller Little Wood-rail (73), but it occasionally occurs along streams through low forest on white sands (Stotz 1992). This species normally frequents dense cover and in Colombia it sometimes occurs some distance from water. It occurs in lowlands and up to 2,000m, rarely 2,300m (Ridgely & Gaulin 1980).

F. Olmos (*in litt.*) provides the following details of its habitat preferences in São Paulo state, SE Brazil. In the Santos-Cubatão region, this species is the most commonly recorded rail in the coastal mangrove forests, and also in the nearby flooded *restinga* (sandplain) forest dominated by *Tabebuia cassinoides*. In neighbouring dry-ground and hill forest it is replaced by the Slaty-breasted Wood-rail (78), which also occupies isolated mesophytic forest fragments and riverine forests in the hinterland W of the Serra do Mar, a region in which the Grey-necked Wood-rail prefers the rather sparsely distributed *Typha*- and sedge-rich swamps along the rivers. In the mangroves, the Grey-necked Wood-rail occurs where the canopy is usually closed, and also in transitional areas which have some freshwater input and are dominated by tangles of *Hibiscus tiliaceus*. However, on small offshore islands such as Alcatrazes, Buzios and Vitoria, it is locally common in dry forest of the type inhabited elsewhere by Slaty-breasted Wood-rail; it is one of the largest birds on these islands, where the only other rail is the small Russet-crowned Crake (26), and apparently has been able to expand its niche in the absence of other ground-dwelling arthropod eaters. On the larger São Sebastião I, which has a diverse bird community, it was restricted to wetlands and it apparently became extinct as these were drained, despite the fact that extensive areas of hill forest remained.

FOOD AND FEEDING In mangroves it preys heavily on crabs; elsewhere it takes molluscs (including the large water snail *Pomacea flagellata*), arthropods (including Diplopoda, cockroaches and locusts), frogs, seeds, berries and palm fruits. It also takes maize, rice (dry or cooked), bananas and fruits of the pejibaye palm *Bactris gaseipes* and the African oil palm *Elaeis guineensis* (Skutch 1994). To extract fragments of oil palm pericarp, the birds hammer the fruits with the bill, pivoting the body about the legs like a woodpecker hammering at a tree (Skutch 1994). The birds hammer large *Pomacea* snails with the bill for 1-4 min, making a hole in the side of the shell to extract the snail, and one bird killed a water snake 30cm long by giving the snake many blows with the bill, particularly around the head, as well as picking it up and shaking it; after 45 min it managed to swallow the snake, which still writhed feebly (Kilham 1979). It jumps high to break off clusters of berries. In Santos-Cubatão, São Paulo, Brazil, pairs and family groups forage under breeding colonies of Little Blue Heron *Egretta caerulea* and Scarlet Ibis *Eudocimus ruber*, apparently feeding on fallen eggs and small nestlings as well as taking insects (F. Olmos *in litt.*); it also raids the nests of waterbirds for eggs (Teixeira 1981). It often forages in cover but sometimes openly in short grass close to thickets; it wades in streams and forages on muddy tracks; it forages extensively on muddy shorelines between mangrove forest and water. It digs and probes with the bill in soft ground and mud, it moves aside leaf-litter and other debris with the partly open bill or picks up dead leaves and tosses them aside, and it digs into cattle dung for insects and horse droppings for undigested grains of maize or perhaps intestinal parasites (Teixeira 1981, Skutch 1994). It sometimes accompanies other birds following army ants (Slud 1964).

HABITS Although active during the day, particularly in the early morning and late afternoon, this species is also said to feed regularly at night (Ridgely & Gwynne 1989); however, according to Wetmore (1965) and Skutch (1994), although it may be active in the hour or two after sunset and also in periods of bright moonlight, it usually roosts at night in trees and bushes on branches 2-3m up or on a platform of weed stalks, dead leaves and other coarse vegetation. In the Pantanal de Mato Grosso it is active all day, even at midday in temperatures > 35°C (F. Olmos *in litt.*). It is normally wary, secretive and suspicious, seldom allowing the observer more than an occasional glimpse as it slips quietly through heavy ground cover, but it sometimes forages in the open although never far from cover. In areas where it has become accustomed to the proximity of man it may be bold and confident (Kilham 1979). It does not always flee from an approaching observer, and may permit a fairly close approach before walking into dense cover. It flies rarely; when disturbed it usually runs swiftly, flushing only if chased by dogs when it flies up to a low branch. It frequently perches in bushes and trees. It moves stealthily, with the head held forward and the tail at an angle, but it also walks with a fairly upright posture and, when suspicious, it flicks the tail, jerks the head and moves with halting steps. In Magdalena, Colombia, birds are seen more frequently in the dry season, when they are forced more into the open to find food (Darlington 1931). These rails sun themselves by standing with their backs towards the sun and spreading the wings widely (Skutch 1994).

SOCIAL ORGANISATION Monogamous; the pair bond is apparently permanent (Skutch 1994), and birds forage in pairs along beaches adjacent to mangrove forest (F. Olmos *in litt.*).

SOCIAL AND SEXUAL BEHAVIOUR In Colombia, birds may congregate in numbers near pools in the *llanos* dry season (Hilty & Brown 1986), while in Venezuela they are often seen "in bands of 10-12 individuals" (Friedmann & Smith 1950), and in dry forest on small islands off the São Paulo coast they are commonly observed in groups of 4-6 birds (F. Olmos *in litt.*).

These rails fight by confronting one another like fighting roosters and attacking with the bill; they also strike with the feet and grip each other's neck with the bill (Shore-Bailey 1926). The following information is from Skutch (1994), unless otherwise noted.

When approached or disturbed, an incubating bird sits until the intruder reaches the nest, when it leaves. A bird disturbed from a nest with newly-hatched young walked around in nearby cover giving throaty clucks, and the pair also gave a brief loud duet. A bird with eggs or chicks led an opossum *Didelphis marsupialis* away by walking and running in front of it, while the bird's mate walked behind the opossum.

A semi-tame pair remained together all year but showed some aggression to each other when taking food provided for them; however, at times when breeding was thought to be imminent one bird courtship-fed the other, either bill-to-bill or by placing food at its partner's feet. Aviary mating between a male Grey-necked and a female Giant Wood-rail (77) is recorded, but the clutch was not hatched because the female was injured (Shore-Bailey 1926).

BREEDING AND SURVIVAL Season Mexico, breeding condition Jan, May; Costa Rica, Apr-Sep; Panama, Apr; Colombia, breeding condition Mar-Jun; Venezuela, Jul-Sep, female with enlarged gonads, Mar; Trinidad, May-Aug; Surinam, Mar-Jul; Uruguay, Dec (from young two-thirds grown, 6 Feb), breeding commences Oct; Brazil, in Rio de Janeiro territorial songs Nov-Dec, courtship and nests Feb, eggs hatch Mar, in Bahia 2 young 30 days old Dec (Teixeira 1981), in São Paulo pair with 5 one-third grown chicks, early Oct (F. Olmos *in litt.*); recently described race *avicenniae* found breeding at Cananéia (2 eggs and 3 newly hatched young) 7 Jan 1998, while other nests being built or abandoned indicated protracted breeding season (R. Silva e Silva & F. Olmos *in litt.*). **Nest** A deep bowl of twigs, dried weed stems and similar materials, or a compact, bulky mass of dead leaves and twigs with a shallow central depression; sometimes of leaves and wet aquatic plants (Teixeira 1981); usually placed 50cm to 7m up in bush, vine, low tree or on tree stump, often over or near water, sometimes on ground; 1 nest of *avicenniae* an ellipsoidal platform of twigs and leaves, 3m above water in top of tilted *Laguncularia* tree in mangrove forest (R. Silva e Silva & F. Olmos *in litt.*); external diameter 30-40cm; internal diameter c. 15cm; height c. 16cm; depth of central depression <4 to 9cm. One nest was on pollarded tree stump, well concealed by growing shoots, ferns, moss etc (Skutch 1994). **Eggs** 3-7; oval, sometimes biconical oval, with slight gloss; creamy-buff to whitish, spotted cinnamon to brown, with underlying pale purple to grey; markings may be large and confluent at large end. Size of eggs: *albiventris* 47.5-51.8 x 34.5-38.6 (50.3 x 35.6, n = 8), calculated weight 35; *plumbeicollis* 50-54 x 34.9-37.3 (51.8 x 36, n = 6); nominate 43.3-52.2 x 31.5-35.6 (48.2 x 34.2, n = 50), calculated weight 31; weight of 4 fresh eggs (nominate) 25.1-27.1 (Haverschmidt & Mees 1995). **Incubation** 20+ days, by both parents; at 1 nest change overs occurred in forenoon and afternoon, at least once, at dusk (Skutch 1994); in captivity, male incubated by day and female by night (Shore-Bailey 1926). **Chicks** Precocial; leave nest after 1-2 days; young up to 40 days old sometimes use brood nest (Teixeira 1981), in captivity built by male. Chicks fed and cared for by both parents; fed while still in nest; fed bill-to-bill and by adult placing food on ground for chick to pick up. One young two-thirds grown still had down on crown and foreneck (Wetmore 1926); in captivity, young independent at 29 days of age, when 75% grown. At 25 days, one downy young, c. half as tall as adult, appeared greyish; it was almost fully grown at 7 weeks and disappeared c. 1 week later. Adult fed almost fully grown juveniles on maize grains which it placed on ground for juveniles to pick up (Skutch 1994). Chicks hatched Mar stay with parents until May, when 1 adult leaves; remaining parent stays with young until Jul (Teixeira 1981).

76 BROWN WOOD-RAIL
Aramides wolfi   Plate 23

Aramides wolfi Berlepsch & Taczanowski, 1883, Chimbo, Ecuador.

Genus sometimes merged into *Eulabeornis*. Monotypic.

Synonym: *Eulabeornis wolfi*.

Alternative names: Brown-backed/Wolf's Wood-rail.

IDENTIFICATION Length 33-36cm. A medium-large wood-rail, predominantly brown, with grey head and black hindbody and tail. Head and nape ashy-grey; chin and throat white; neck, mantle and breast vinous-brown, warmest in tone on mantle and darkest on breast; back, scapulars, upperwing-coverts, inner secondaries and tertials olive-brown; rump to tail black; primaries and outer secondaries chestnut. Flanks and upper belly tawny; lower flanks and belly to undertail-coverts black. Underwing-coverts chestnut with black bars. Iris and eye-ring red; bill yellowish-green; forehead has bare yellow patch (sometimes absent?); legs and feet red. Sexes alike. Immature and juvenile undescribed. Inhabits forest, woodland, riverine marsh and mangroves.

Similar species Differs from Grey-necked (75), Giant (77), Slaty-breasted (78) and Red-winged (79) Wood-rails in having no grey on foreneck or underparts, from Rufous-necked Wood-rail (74) in lack of rufous or chestnut from head to breast, and from Little Wood-rail (73) in size, overall darker colouring, blackish lower flanks and belly, and bill colour.

VOICE A frequently repeated *kui-co-mui* or *kui-co*, much like the call of the Grey-necked Wood-rail (Olivares 1957, Hardy *et al.* 1996). Two or more birds may sing in chorus, and in response to taped playback a single individual gave repeated *kui-co* and *kui-co-co* calls, 1 sequence also preceded by *co* calls (Hardy *et al.* 1996).

DESCRIPTION
Adult Crown, nape, and sides of head ashy-grey, darker on crown and variably tinged pinkish to vinous elsewhere; hindneck vinous-brown, sometimes tinged grey; mantle richer vinous-brown, more chestnut-tinged; back, scapulars and upperwing-coverts bright olive-brown, sometimes tinged greenish; rump to tail black. Alula and primary coverts dull chestnut, becoming more dusky-olive towards tips; primaries and outer secondaries chestnut, dusky brown at tips; inner secondaries and tertials predominantly olive-brown, slightly duskier than upperwing-coverts; axillaries and underwing-coverts chestnut with black bars (indistinct on axillaries); underside of remiges chestnut with dusky tips. Chin and throat white or ashy-white; sides of neck, foreneck and breast vinous-brown to warm sepia, darkest on breast; flanks and upper belly ochraceous-tawny to tawny, washed olive; lower flanks black, mixed with brown; thighs vinous-brown; belly to undertail-coverts black. Iris red; eye-ring brick-red; bill yellowish-green with bare yellow patch on forehead (see Ripley 1977; this character not evident in some specimens); legs and feet pink to pale red, but Sharpe (1894) noted that in one specimen the thighs are blackish. Sexes alike.
Immature Not described.
Juvenile Not described.
Downy young Down sooty-black; iris brown; eyelids red; surrounding skin shows red through down; legs and feet brown to grey-black; bill reddish-brown with yellow tip.

MEASUREMENTS (9 males, 6 females) Wing of male 165-187 (175.7), of female 164-176 (171.2); exposed culmen of male 50-60 (54.8), of female 51-55 (53.2); tarsus of male 67-77 (71.3), of female 66-74 (70) (Blake 1977, BMNH); tail (unsexed) 48-65 (Ripley 1977, BMNH). Weight of 1 male 454 (FMNH, D. E. Willard *in litt.*).

GEOGRAPHICAL VARIATION None recorded.

MOULT No information available.

DISTRIBUTION AND STATUS SW Colombia (N to Serranía de Baudó in Chocó) to W Ecuador (Esmeraldas, Imbabura, Pichincha, Los Ríos, Guayas and El Oro) and possibly extreme NW Peru (Tumbes). Classed as VULNERABLE, this very poorly known species has a restricted range and appears to have become extremely rare, at least in Ecuador, owing to extensive destruction of its mangrove and forest habitats (Collar *et al.* 1994, Ridgely & Greenfield in prep.). Its inability to survive in isolated patches of humid forest is predicted from observations of one such habitat island in Ecuador (Leck 1979). In Colombia, where it was unreported for years, a possibly large population was discovered in mangroves at the Ensenada de Utría NP, Chocó, in Mar 1996 (Porteous & Acevedo 1996). It is recorded once from Peru, at Puerto Pizarro, Department of Tumbes, Sep 1977 (Graves 1982), although there is some doubt about the identity of this bird and further confirmation of its occurrence in Peru is desirable. There are a few recent records from Ecuador (Quinindé, Jatun Sacha Bilsa, Río Palenque and Manglares-Churute Ecological Reserve) (Clay *et al.* 1994, Ridgely & Greenfield in prep.). Apart from Río Palenque in Ecuador, none of the key areas for threatened birds listed by Wege & Long (1995) is important for this species, and there is an urgent need for further investigation of its distribution, status and natural history.

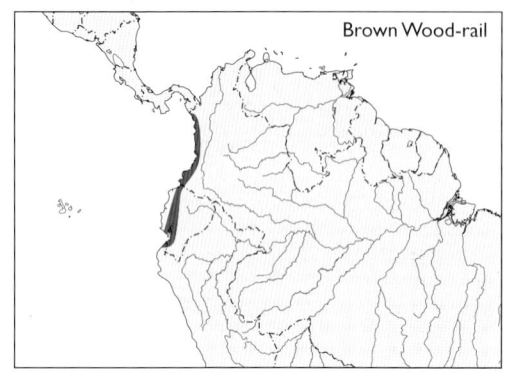

MOVEMENTS None recorded.

HABITAT Forest, secondary growth, forested rivers and mangroves; also riverine marsh and swampy woodland. Recorded up to 1,300m.

FOOD AND FEEDING Stomach contents of 1 bird (FMNH) were "beetles, larvae and small snails" (D. E. Willard *in litt.*).

HABITS Virtually unknown. It is apparently shy and secretive, emerging from cover less often than the Grey-necked Wood-rail, but it may emerge to feed at mudflats exposed at low tide (Ridgely & Greenfield in prep.); it will clamber over mangrove roots; when surprised in the open it runs and may then fly to cover.

SOCIAL ORGANISATION No information available.

SOCIAL AND SEXUAL BEHAVIOUR No information available.

BREEDING AND SURVIVAL Not described.

77 GIANT WOOD-RAIL
Aramides ypecaha Plate 23

Rallus ypecaha Vieillot, 1819, Paraguay.

Genus sometimes merged into *Eulabeornis*. Monotypic.

Synonyms: *Rallus/Aramides ypacaha*; *Crex melampyga/melanopyga*; *Aramides ipacaha/ipecaha*; *Gallinula/Aramides/Aramus gigas*; *Ortygarchus melampygus*; *Eulabeornis ypecaha*.

Alternative names: Ypecaha (Wood)/Great Rail.

IDENTIFICATION Length 41-49 (possibly up to 53)cm. The largest *Aramides*; an impressive and handsome bird, with upright stance and elegant gait; calls are loud, harsh and distinctive. Forehead and forecrown ashy-grey; sides of head, foreneck and centre of breast pale bluish-grey; chin and throat white. Nape, hindneck, sides of neck, sides of breast, flanks and belly vinous-chestnut. Mantle to back, scapulars, upperwing-coverts, inner secondaries and tertials greenish-olive; primaries and outer secondaries tawny; rump, uppertail-coverts, tail, undertail-coverts and vent black; centre of belly white; thighs ashy-grey. Iris, legs and feet red; bill green or greenish-yellow, with yellow base. Sexes alike; female slightly smaller. Immature apparently indistinguishable from adult. Juvenile slightly duller and paler in plumage; legs brownish-red; bill tinged brownish. Inhabits reedy and grassy marshes and swamps, fields and pastures near water, and gallery forests.

Similar species Of the 3 other *Aramides* species which are grey on face and foreneck, Grey-necked Wood-rail (75) has grey hindneck and rufous or brown-black patch on occiput, while Slaty-breasted (78) and Red-winged (79) Wood-rails are darker and extensively grey below, latter also having striking mahogany-red on upperwing-coverts and sides of neck.

VOICE In the evening, and occasionally during the day, especially in overcast weather, these birds congregate in groups and set up an astonishingly powerful, even deafening, chorus of screams, shrieks and wheezes (see Social and Sexual Behaviour). The common call of the evening chorus has been likened to the sound of a rusty windmill pump (Wetmore 1926). The calls are described as reminiscent of a hysterical human saying *all..wacky....all..wacky*, a piercing, high, single *eeeeeeok*, a bark, a low *keaaw* and an explosive *puk* (Belton 1984). Sick (1993) describes the calls as a trisyllabic, repeated *keyo-BYke*, a repeated *BY-kare* followed by *KOa*, and a sequence initiated with a series of simple *keyuh* notes reminiscent of the cries of an *Amazona* parrot; some multisyllabic motifs are sometimes followed by a low moan. There is also a high-pitched *weeee* (Hardy *et al.* 1996). The local name *ipacaa* is onomatopoeic, describing the common *WAI ca-A* call (J. Mazar Barnett *in litt.*). On catching sight of an intruder, birds utter a powerful cry not unlike that of a Blue Peafowl *Pavo cristatus*; this alarm call is taken up by other individuals (Hudson 1920).

DESCRIPTION
Adult Forehead and forecrown ashy-grey; hindcrown reddish-brown; nape and hindneck vinous-chestnut to dull vinaceous-tawny; sides of neck brighter vinous-chestnut. Mantle, back, scapulars, upperwing-coverts, inner secondaries and tertials greenish-olive; lower back browner; rump to tail black. Alula and primary coverts greenish-olive, with slight tawny tinge; primaries and outer secondaries tawny-russet, buffy-olive towards tips and on outer webs of outermost primaries; axillaries ferruginous; underwing-coverts ferruginous, barred black; underside of remiges tawny. Lores, supercilium, ear-coverts and malar region light bluish-grey, ashier on supercilium; chin and upper throat whitish; lower throat, foreneck and centre of breast light bluish-grey; sides of breast, flanks and belly pale vinous or delicate brownish-pink; centre of belly white; vent and undertail-coverts black; thighs ashy-grey. Iris reddish-brown to carmine-pink; bill greenish-yellow to apple-green or lime-green, yellow to orange at base and grey-green at tip; legs and feet red to carmine-pink. Sexes alike.

Immature Apparently indistinguishable from adult (see Moult).

Juvenile Similar to adult, but paler or duller overall; black areas of adult are dull brownish. Captive birds 3 months old were slightly duller than adults in plumage; legs brownish-red; bill with brownish tinge (Brown 1974).

Downy young Down is described as very dark mahogany-brown, or rusty-brown with darker upperparts, but a chick in BMNH has down buff-brown to olive-brown; bill dark brown with paler tip and gonys; legs and feet dark brown; small wing-claw present.

MEASUREMENTS (12 males, 10 females) Wing of male 216-237 (223.7), of female 212-226 (218.5); exposed culmen of male 68-77 (73), of female 65-73 (70.5); tarsus of male 77-87 (82.3), of female 73-83 (79.3) (Blake 1977). Tail of 6 males 80-92.5, of 2 females 81-86; culmen of 6 males 73.5-83, of 2 females 66, 78 (Ripley 1977, BMNH). Weight of 3 males 655, 695, 860, of 2 females 565, 765 (Dunning 1993, D. Oren *in litt.*).

GEOGRAPHICAL VARIATION None recorded.

MOULT A male from Corrientes, Argentina, 18 Oct, is moulting all secondaries and has slight body moult (D. Baepler *in litt.*). Captive-bred birds were indistinguishable from adults at 5 months of age, but had slightly duller plumage and bare parts than adults at 3 months (Brown 1974).

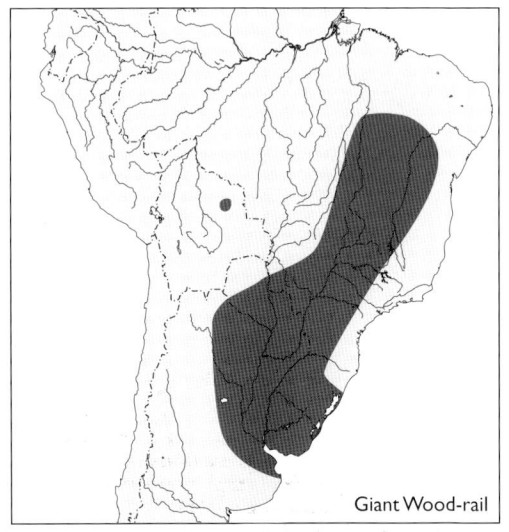

Giant Wood-rail

DISTRIBUTION AND STATUS E and SE Brazil, in S Piauí, Bahia, Minas Gerais, E Mato Grosso, Goiás (Araguaia NP and Ilha do Bananal [Forrester 1993]), Santa Catarina (no specific locality known) and Rio Grande do Sul, and also recorded from Santana do Araguaia, extreme SE Pará (D. Oren *in litt.*); Bolivia at Rivers Alto-Negro (Santa Cruz) and Area Protegida Corvalcn (Tarija) (Armonía 1997); Paraguay; Uruguay; NE Argentina (E Formosa, E Chaco, Corrientes, Misiones, N Santa Fe, Entre Ríos and N Buenos Aires, occasional Córdoba and Salta). This rail is not globally threatened and it was formerly common to abundant over much of its range. In the 1980s it was common in Rio Grande do Sul, Brazil, where it is a frequent victim of steel traps set in marshes for fur-bearing animals (Belton 1984). It is regarded as local in NE and C Brazil, but more abundant in the S (Rio Grande do Sul) and in Paraguay, Uruguay and Argentina (e.g. Klimaitis & Moschione 1987, Contreras *et al.* 1990, Sick 1993, Hayes 1995, Arballo & Cravino in press). In view of its preference for more open habitats and light woodland it may not have suffered as much from habitat destruction as some of its forest-dwelling congeners, and its total population is estimated at 100,000-1,000,000 (Rose & Scott 1994). It is often kept in captivity in South America, and is sold commonly in local village shops (Ripley 1977). In Argentina, although it is subjected to some hunting pressure, it continues to exist in good numbers (Contreras *et al.* 1990).

MOVEMENTS None recorded.

HABITAT Open marshes, lagoons, estuaries, sawgrass swamps, lightly wooded swampy areas, fields and pastures near water and cover, and gallery forests. It apparently prefers dense marsh vegetation, especially when interspersed with trees and bushes; it particularly favours riparian marshes and reedbeds, and is sometimes found far from water. On the R de la Plata it frequents extensive beds of *Erythrina crista-galli* (Hudson 1920). In Uruguay it is usually found in transitional zones such as reedbeds and shrubby areas adjacent to waterways, as well as in open areas and sometimes in forests far from water (Arballo & Cravino in press). It is restricted to lowland areas.

FOOD AND FEEDING Very little information is available. It has a varied diet, including arthropods and some vegetable matter (including seeds and fruit), although it often takes predominantly molluscs (*Planorbis*, *Ampullaria*) (Arballo & Cravino in press). Stomach contents include a small mussel shell, a crab claw, small seeds and decomposed vegetable matter. In captivity it eats insect larvae, eggs, minced meat, apple and meal. The birds are solitary feeders, and emerge from reedbed cover in the early morning and evening to forage (Durnford 1877).

HABITS When unmolested by man this rail is apparently bold, and even when persecuted it retains a characteristic inquisitive boldness (Hudson 1920). It moves with a deliberate yet easy, elegant and somewhat stately gait, lifting the feet fairly high, drooping the toes as the foot rises and straightening them out as the foot descends; the head is bobbed back and forth, and the tail jerked at every stride. It is vigilant and usually keeps well concealed, but is more likely than Grey-necked and Slaty-breasted Wood-rails to appear in the open and less likely to take cover if danger threatens (Belton 1984). On misty days it will wander some distance from cover, and it occasionally runs out of cover to pause, with twitching tail, to look from the crest of some low bank before moving off into cover again (Hudson 1920); it will also use fence posts or termite mounds as observation points (Arballo & Cravino in press). It runs faster than a man, but when surprised on open ground it lies close, like a tinamou (Tinamidae), not flying unless almost trodden on, when it rises with a loud whir and flies swiftly to cover, gliding the last few metres before dropping into the vegetation (Hudson 1920). In wooded areas it usually flies into a tree when disturbed and it roosts in trees, having been found as high as 15m from the ground. It struts around on branches and, when perched, retracts the head. When alarmed, or before flight, the neck is stretched out and the head held high.

SOCIAL ORGANISATION Normally seen singly or in pairs; monogamous in captivity.

SOCIAL AND SEXUAL BEHAVIOUR This species has a reputation for being bold and pugnacious and, in areas where they are not persecuted, birds come into the open by day and attack domestic poultry around houses and in village streets adjacent to their marshes (Hudson 1920).

Captive birds, if allowed to run about in the hen-yard, soon dominate the flock and become a nuisance because of their egg-stealing propensities (Ripley 1977). Birds hastily converge upon a spot where one raises an alarm call on detecting a human intruder, and will follow the intruder for a considerable distance, watching closely from cover (Hudson 1920).

Although normally solitary by day, in the evening these birds congregate in groups in selected spots and indulge in a remarkable communal display, rushing frenziedly about with the wings spread and the bill raised vertically, and giving an astonishingly powerful chorus of screams, shrieks and wheezes. This display apparently does not have a sexual function, and birds have not been observed to fight or to show aggressive behaviour during the performance (Hudson 1920). If a human observer is visible the birds will ignore him while giving this wild chorus but as soon as they have finished they will flee into cover.

A dance, which is assumed to have a courtship function, is described by Friedmann (1927). The bird prances around, repeatedly bringing the legs up very high, the toes almost touching the body, and opening and closing the wings more or less in time with the motion of the legs; the wings are not more than half opened, the neck is stretched and the head bobbed back and forth. While dancing, the bird occasionally utters a harsh, wheezy call.

Aviary mating between a male Grey-necked and a female Giant Wood-rail is recorded, but the clutch was not hatched because the female was injured; when nesting, the female attacked human intruders, pecking viciously at their legs and feet (Shore-Bailey 1926). A downy chick in BMNH is a hybrid Grey-necked Wood-rail x Giant Wood-rail.

BREEDING AND SURVIVAL Season Uruguay, mainly Sep-Feb; no information available from other countries. **Nest** Elaborately constructed of grasses and weed stems; built in a variety of situations, both on ground and in trees, shrubs, tangles of vines or broken-down vegetation; nests in dense clumps of strawgrass *Scirpus giganteus*, at base of thistles *Cynara cardunculus*, and in *Eryngium* bushes and other dense vegetation (Navas 1991, Arballo & Cravino in press); usually near water, and may be built 1+m above water. External diameter 28-54cm; height 13-15cm; internal depth 6-7cm. **Eggs** 4-7; broad oval; glossy; variable in colour, usually white but may be cream, yellow, green, orange or pink, with small mauve, brown or reddish-brown spots; size of 30 eggs 45.6-56 x 34.2-38.6 (50.8 x 36.2), calculated weight 45. Other information from captive breeding (Brown 1974). Eggs laid at daily intervals. **Incubation** 24 days, by both sexes. Eggs hatched within 24h. **Chicks** Left nest after 3-4 days; fed and cared for by both parents for 8-9 weeks; began to feed themselves at 3-4 weeks; last brood stayed with parents for over 3 months. Chicks half grown at 5 weeks; renesting occurred when young of first brood 6 weeks old; 3 clutches were laid in the season.

78 SLATY-BREASTED WOOD-RAIL
Aramides saracura   Plate 23

Gallinula saracura Spix, 1825, Brazil.

Genus sometimes merged into *Eulabeornis*. Monotypic.

Synonyms: *Rallus nigricans/melanurus/hydrogallina*; *Gallinula plumbea*; *Aramides plumbeus*; *Ortygarchus surucura*; *Eulabeornis saracura*.

Alternative name: Saracura Wood-rail.

IDENTIFICATION Length 34-37cm. A medium-large wood-rail. Top and sides of head greyish-brown, tinged pale cinnamon from lores to ear-coverts, becoming rufescent on nape and hindneck and ruddy olive-brown on mantle and sides of upper breast. Back, scapulars, upperwing-coverts and inner remiges greenish-olive; lower back brown; rump sepia; uppertail-coverts and tail black. Primaries russet; axillaries and underwing-coverts chestnut, barred black. Chin and throat white; foreneck, sides of neck, breast, flanks and upper belly slate-blue; lower flanks and lower belly blackish; undertail-coverts black. Iris and eye-ring carmine; bill light green, bluish at base; legs and feet brownish-red or yellowish-brown. Sexes alike. Immature and juvenile undescribed. Inhabits tropical forest and woodland patches.

Similar species The only *Aramides* species with grey of head and neck extending as far as the belly apart from Red-winged Wood-rail (79), which has mahogany-red sides of neck, and median and greater upperwing-coverts. Differs from all congeners in having green bill with bluish base, and duller reddish or yellowish-brown legs and feet. May look like a large Blackish Rail (114), which sometimes occurs alongside it in Brazilian marshes, but the latter has no cinnamon or rufous in the plumage and is darker overall.

VOICE All information is from Belton (1984), unless otherwise indicated. This rail has a variety of loud, resounding, noisy cries, including *po-quit kwaa kwaa kwaa*, *po-peek*, and *po-pereek*, which are often given for several minutes continuously, usually in duet but not in unison. It also has an irritated, rapidly repeated series of *quir* notes, sometimes with other sounds such as repeated *pik* or *koke* notes leading up to the *quir* series. The birds are heard most frequently during the day. Sick (1993) mentions a repeated, loud *kee* of alarm, and a *bAHik* note, presumably equivalent to the *po-peek*, which is accompanied by a weak, hoarse sound apparently produced in the trachea; he also maintains that this species duets in such perfect synchrony that it is difficult to tell that two birds are calling.

DESCRIPTION
Adult Forehead and crown greyish-brown; hindcrown browner; nape and hindneck more rufescent, almost dull chestnut; mantle ruddy olive-brown, this colour extending variably to sides of upper breast. Back, scapulars and upperwing-coverts dark greenish-olive; lower back brown; rump sepia; uppertail-coverts and tail black. Alula and primary coverts dusky brown, slightly tinged rufous, and greenish-olive towards tips; primaries russet, greenish-olive towards tips and on outer webs of outermost primaries; secondaries greenish-olive; axillaries and underwing-coverts barred blackish-brown and tawny, lessers ruddy olive-brown. Lores and circumocular region greyish, variably washed pale cinnamon to brownish; ear-coverts similar but washed rufescent at rear; malar region pale

greyish with slight cinnamon tinge; chin and throat white, shading into slate-blue on foreneck, sides of neck, breast, flanks and upper belly; lower flanks and lower belly blackish, thighs slate-grey; undertail-coverts black. Iris and eye-ring carmine to pale red; bill light green, usually bluish at base, but Belton (1984) records a brownish-yellow base (indicative of seasonal variation?); legs and feet pale red, or dark red tinged brownish-grey, or yellowish-brown. Sexes alike.

Immature Not described.
Juvenile Not properly described.
Downy young Not properly described; at least some down is rusty-coloured (Navas 1991).

MEASUREMENTS Wing of 7 males 188-199 (192.5), of 7 females 174-196 (186.8); exposed culmen of 7 males 50-64 (58.4), of 7 females 52-60 (55.7); culmen to base of 2 males 57, 62, of 4 females 51-63 (59.25, SD 5.6); tarsus of 7 males 71-78 (74.2), of 7 females 69-76 (71); tail of 2 males 61, 62, of 4 females 60-68 (64.25, SD 3.9). Weight of 1 male 540.

GEOGRAPHICAL VARIATION None recorded.

MOULT No information available.

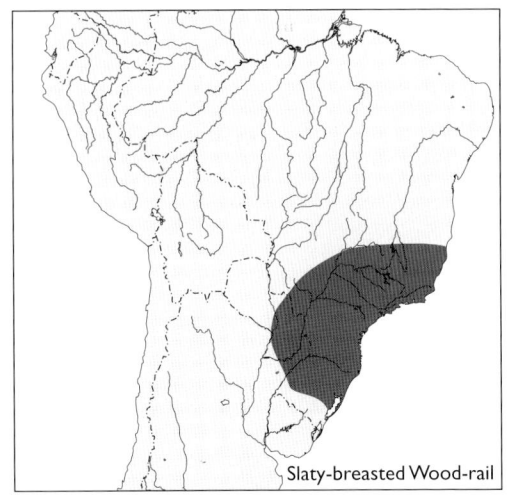

Slaty-breasted Wood-rail

DISTRIBUTION AND STATUS SE Brazil, from S Minas Gerais E to Espírito Santo and Rio de Janeiro, and S to Rio Grande do Sul; extreme NE Argentina (Misiones) and E Paraguay. This rail is not regarded as globally threatened, but it is infrequently observed and its status is unknown over much of its range. Olrog (1984) regarded it as locally common in NE Argentina. In Brazil, it was common in N and E Rio Grande do Sul in the 1980s (Belton 1984), and Forrester (1993) records it from 9 national parks and reserves in SE Brazil, while in the São Paulo hinterland W of the Serra do Mar it is the common rail in isolated mesophytic forest fragments and riverine forest (F. Olmos *in litt.*). It is described as uncommon to rare in E Paraguay (Hayes 1995) and Lowen *et al.* (1995) described it as rare at Estancia Itabo (Dpto Canindeyu) in the Atlantic forest of Paraguay in 1992. Its status elsewhere is in need of investigation, but in some areas it may be more overlooked or disregarded than genuinely uncommon; in view of its preference for forest edge and secondary growth habitats it may even have benefited to some extent from forest destruction (J. Mazar Barnett *in litt.*).

MOVEMENTS None recorded.

HABITAT Forests and woodland patches. It prefers, but is not dependent on, swampy or boggy areas or stream edges and it is not usually seen in open marshes; birds from nearby woods occasionally entered a garden (Belton 1984). In the Santos-Cubatão region of São Paulo State, SE Brazil, this species prefers forested situations near water, such as impoundments made by road cuts and rivers, but also occurs in dry-ground and hill forest near flooded *restinga* (sandplain) forest which is occupied by Grey-necked Wood-rail (75); in the hinterland W of the Serra do Mar it is found in isolated mesophytic forest fragments and in riverine forests, while in the same area the Grey-necked prefers the rather sparsely-distributed *Typha*-rich swamps (F. Olmos *in litt.*). It also inhabits forest edges and secondary growth (J. Mazar Barnett *in litt.*). See the Habitat section of the Grey-necked Wood-rail species account.

FOOD AND FEEDING No information on food is available. In Misiones, Argentina, this species occurs near human habitations and comes out to feed alongside chickens and other domestic animals (J. Mazar Barnett *in litt.*).

HABITS The only information is from Belton (1984). This rail is often seen crossing or running along roads in moist forest. It is shy, normally disappearing immediately it knows that it is being observed.

SOCIAL ORGANISATION No information available.

SOCIAL AND SEXUAL BEHAVIOUR Not recorded.

BREEDING AND SURVIVAL One nest recorded, at Novo Hamburgo (Rio Grande do Sul), spring 1961; situated 2m above ground in thick bush; eggs 4 or 5 (Voss in Belton 1984). Builds nest in trees and bushes; eggs dusky, with spots (Canevari *et al.* 1991), also described as yellowish-pink with many small spots of dark red and purplish, markings more numerous at larger end (Navas 1991). Schönwetter (1961-62) gives measurements of 20 eggs from Brazil and Paraguay as 45.6-56 x 34.2-38.6 (49.8 x 36.6), calculated weight 35.4.

79 RED-WINGED WOOD-RAIL
Aramides calopterus   Plate 23

Aramides calopterus P. L. Sclater & Salvin, 1878, Sarayacu, Ecuador.

Genus sometimes merged into *Eulabeornis*. Monotypic.

Synonyms: *Aramides callopterus/calloptera*; *Eulabeornis calopterus*.

Alternative name: Ecuadorian Wood-rail.

IDENTIFICATION Length 31-33cm. One of the smaller wood-rails, with rather dark plumage. Upperparts mostly dark greenish-olive, darker and browner from forehead to nape; lower back and rump dark rich brown; uppertail-coverts and tail black. Median and greater upperwing-coverts mahogany-red; primary coverts, primaries and secondaries dusky brown; tertials dark greenish-olive. Axillaries and underwing-coverts barred black and white. Lores, circumocular area and cheeks ashy; ear-coverts pale vinous; chin and throat white; sides of neck mahogany-red; foreneck and rest of underparts slaty-blue; lower flanks dusky brown; lower belly, thighs and undertail-coverts blackish. Iris colour not described; bill greenish; legs and feet reddish or pink. Sexes alike. Immature similar to adult but with blackish bill, legs and feet. Juvenile undescribed. Inhabits seasonally flooded forest, and forest near water.

Similar species The only *Aramides* species to have mahogany-red ear-coverts, sides of neck, and median and greater upperwing-coverts. Separated from all congeners except Slaty-breasted Wood-rail (78) in having grey of head and neck extending to belly. Underwing-coverts black with white bars, a character shared only with Little Wood-rail (73) and Rufous-necked Wood-rail (74).

VOICE Not described.

DESCRIPTION
Adult Forehead dusky-slate; crown to nape dusky olive-brown, becoming somewhat paler, dark greenish-olive, from hindneck to back and on scapulars; lower back and rump dark rich brown; uppertail-coverts and tail black. Lesser upperwing-coverts dark greenish-olive, like scapulars; median and greater upperwing-coverts mahogany-red, greenish-olive basally; alula, primary coverts, primaries and secondaries dusky-brown, slightly olivaceous on outer webs; tertials dark greenish-olive, browner towards tips. Axillaries and underwing-coverts blackish-brown, barred white. Lores, circumocular area and cheeks ashy; ear-coverts pale vinous; chin and throat white, tinged pearly grey at sides and on lower throat; sides of neck mahogany-red; foreneck and rest of underparts slaty-blue, grading into dusky brown on lower flanks and dusky blackish on lower belly, thighs and undertail-coverts. Iris colour not described; bill greenish; legs and feet reddish or pink. Sexes alike.
Immature Similar to adult but with blackish bill, legs and feet.
Juvenile Not described.
Downy young Not described.

MEASUREMENTS Wing of 6 males 164-171 (166.1), of 7 females 156-173 (165.2); tail of 1 male 48, of 4 females 45-52 (48.25, SD 3.0); exposed culmen of 6 males 41-46 (43.6), of 7 females 44-46 (45); culmen to base of 1 male 51, of 4 females 46-50 (47.75, SD 1.7); tarsus of 6 males 57-62 (60.5), of 7 females 57-62 (59.2).

GEOGRAPHICAL VARIATION None described.

MOULT No information available.

DISTRIBUTION AND STATUS E and NE Ecuador (specimens from Concepción, Guamayacu, Bobonaza and Tigre Rivers; recorded Río Suno), NC and SE Peru (Río Tigre and SE Loreto) and Brazil in Amazonas (Upper R Juruá in SW, upper R Urucu and Tefé in NC, and Manaus in E). This species is not considered globally threatened, but it is very rarely observed and its status is unknown over most of its range. In Peru it is described as uncommon (Parker *et al.* 1982); in Amazonas it was described as rare earlier this century and uncommon on the upper R Urucu in the late 1980s (Peres & Whittaker 1991), while it is scarce at Tefé and accidental at Manaus (Forrester 1993); in Ecuador it is very rare and seemingly local, but presumably much overlooked (Ridgely & Greenfield in prep.). Its status elsewhere is in urgent need of investigation. In view of its scarcity in some areas, its restricted distribution in threatened habitats and the lack of knowledge of its natural history, it should be classified as a DATA DEFICIENT species.

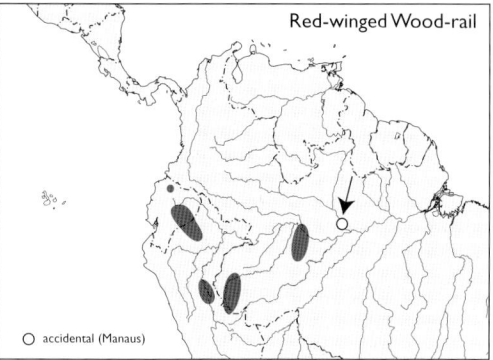

Red-winged Wood-rail

○ accidental (Manaus)

MOVEMENTS None recorded. Forrester (1993) regards it as "accidental" (1-2 records since 1960) in Manaus, W Brazil, and thus assumes it to be a migrant there.

HABITAT Seasonally flooded *igapó* forest, *terra firme* forest, and forest in the vicinity of streams and creeks. In Ecuador it occurs at 300-900m a.s.l (Ridgely & Greenfield in prep.).

FOOD AND FEEDING Not described.

HABITS Not described.

SOCIAL ORGANISATION Solitary or in pairs (Peres & Whittaker 1991).

SOCIAL AND SEXUAL BEHAVIOUR Not described.

BREEDING AND SURVIVAL Unknown.

AMAUROLIMNAS

One species, a plain, entirely brown to rufous crake with yellowish-green bill and reddish legs. It occurs in forests and thickets of Central and South America, and is probably closely related to the *Aramides* wood-rails of the region. Its status is unclear and it should be considered Data Deficient; the nominate race from Jamaica has been presumed to have become extinct.

80	**UNIFORM CRAKE**	
	Amaurolimnas concolor	Plate 24

Rallus concolor Gosse, 1847, Jamaica.

Sometimes included, with *Aramides*, in *Eulabeornis*. Possibly derived from *Aramides* stock, and has identical bill structure to the smaller *Aramides* species. Three subspecies recognised, one (*concolor* of Jamaica) generally believed recently extinct.

Synonyms: *Eulabeornis/Aramides/Porzana/Rufirallus concolor*; *Corethrura Gautemalensis* (sic)/*cayennensis*; *Porzana guatemalensis*; *Rallus castaneus*; *Rallina castanea*; *Rufirallus boecki/castaneus*; *Erythrolimnas/Aramides boecki*.

Alternative names: Red Rail, Wood Rail.

IDENTIFICATION Length 20-23cm. A medium-small, entirely brown to rufous-brown crake. Birds from Central America, Colombia and NW Ecuador (race *guatemalensis*) have plain olive-brown upperparts, more rufous on scapulars, upperwing-coverts and tertials, and contrastingly dark on outer remiges; underparts rufous-brown, paler on throat and belly and darker from flanks to undertail-coverts. Birds from the rest of South America (race *castaneus*) are larger, with more rufous upperparts and brighter rufous underparts. Iris orange to red; short, rather thick bill yellowish-green; legs and feet dark brownish to red, orange or pink. Sexes alike. Immature similar to adult but duskier on upperparts and underparts, with white shaft-streaks to feathers of throat and breast, and yellow eyes. Juvenile greyish, with brownish eyes and dull legs. Inhabits wet to dry forest and thickets.

Similar species Superficially resembles a diminutive *Aramides* wood-rail in build, posture and bare part colours, but has no grey in plumage and lacks the barred underwing-coverts, and the black hindbody and tail, characteristic of *Aramides*. Similar to Chestnut-headed Crake (25) and Russet-crowned Crake (26) but differs from the former in having browner forecrown and more extensively rufous underparts, from the latter in having a longer, yellowish-green bill, and from both in having brown ear-coverts, more rufous upperparts, and different voice. Immature Chestnut-headed Crake has greyish-buff throat; it lacks immature Uniform Crake's white shaft-streaks on throat and breast, and may be paler in overall colour.

VOICE Gives loud, arresting whistled calls, reminiscent of *Aramides* wood-rails. The following information is taken from Stiles (1981) and Howell & Webb (1995). The song is a series of 6-20 upslurred *tooee* whistles, loudest during the middle of the sequence; this call is apparently used for territorial advertisement and was heard in Costa Rica chiefly in Aug-Dec, which is probably the breeding season. It is sometimes given at night. At high intensities (e.g. when birds are responding to imitations of the call), the loudest notes often have a flutelike break in the middle *toourleee*. Pair members give clear, but not loud, whistled *tooo* notes to each other. The alarm call is a sharp, nasal *kek* or *plik*, while a soft, low-pitched, pigeon-like *cuuuuhuuuu* is possibly an aggressive call. The extinct nominate race was said to utter a cluck like a hen (Greenway 1967).

DESCRIPTION *A. c. concolor*
Adult Crown and nape bright brown; forehead slightly tawny; lores and face tawny, more ochraceous on ear-coverts. Rest of upperparts bright brown to cinnamon-brown, with hazel to cinnamon-rufous wash on scapulars and upperwing-coverts; alula, primaries and secondaries pale sepia but tertials like coverts, as also are rectrices. Sides of neck and underparts cinnamon-rufous, paler, more ochraceous-buff on chin, more cinnamon on centre of breast and belly, and darker from flanks to undertail-coverts. Underwing-coverts and axillaries uniform buffy-brown with cinnamon-rufous tinge, lessers tipped cinnamon-rufous; underside of remiges buffy-brown. Iris orange to vermilion; bill yellowish-green with dusky culmen; legs and feet dull purplish-crimson to reddish, orange or pink. Sexes alike. Leg colour may become duller during the non-breeding season.
Immature Similar to adult but duskier overall, with less rufous wash on upperparts; underparts dingy vinous-brown, with white shaft-streaks to feathers of throat and breast; centre of belly and thighs dull yellow-brown; iris yellow.
Juvenile Not properly described; plumage apparently greyish. Has brownish to yellow eyes and dull legs and feet.
Downy young Not described.

MEASUREMENTS (Unsexed birds) Wing (chord) 115-125.5 (120.9, SD 3.9, n = 9), wing (max) 126, 126, 129 (n = 3); tail 45-53 (48.8, SD 2.2, n = 12), exposed culmen 25-29.5 (27.3, SD 2.0, n = 9), culmen to base 29 (n = 3); tarsus 39-44 (41.4, SD 2.0, n = 12) (Ridgway & Friedmann 1941, BMNH).

GEOGRAPHICAL VARIATION Birds from the N and NW part of the range are smaller and darker than other forms.

A. c. concolor (Gosse, 1847) – presumed extinct; formerly occurred on Jamaica. See Description.

A. c. guatemalensis (Lawrence, 1863) – Locally from S Mexico (Veracruz, Oaxaca, Tabasco and Chiapas) through Central America (except El Salvador) to Panama (W Chiriquí, W Bocas del Toro, Veraguas, Caribbean slope of the Canal area, San Blas, Darién, and Pearl Is), and S from W Colombia (Chocó, Cauca, Nariño, and a specimen labelled "Bogotá") to NW Ecuador (Esmeraldas at San Javier and San Mateo). The smallest race; strikingly darker than *concolor*, more olive, less rufescent, on upperparts and browner on underparts. Wing of 7 males 117-125 (119.7), of 3 females 121, 123, 127; tail of 5 males 39-42 (41.8), of 5 females 41-46.5 (44.6, SD 1.4); culmen of 13 males 25-28.5 (26.5), of 8 females 23-29 (26.6); tarsus of 15 males 38-43 (40.6), of 8 females 38-43 (40.8). Approximate weight 95 (Stiles & Skutch 1989); weight of 1 male 133 (Teixeira *et al.* 1986).

Ripley (1977) records that this species, possibly this race, occurs in W Peru, but gives no further information. A single unsexed bird in BMNH, dated 1894 and collected "10 miles S of Lima", approaches *castaneus* in colour of underparts and has slightly darker upperparts, possibly closer in tone to *guatemalensis*; wing 122, tail 48, culmen to base 27, tarsus 37.

A. c. castaneus (Pucheran, 1851) – N Venezuela (Carabobo) and the Guianas, E Peru (Loreto and Pasco) to E Bolivia (Santa Cruz) and Amazonian and E Brazil (R Solimões, R Negro and R Guaporé regions, Pará, Bahia, Alagoas, Espírito Santo, Rio de Janeiro, São Paulo and Paraná). Birds from E Ecuador in Napo (near Loreto) and Pastaza (Sarayacu) are presumably also of this race. Birds from E Ecuador, in Napo (near Loreto) and Pastaza (Sarayacu) are presumably also of this race. Similar in size to *concolor* but more olivaceous on upperparts and brighter rufous on underparts. Wing of 14 males 120-138 (126.1), of 7 females 118-130 (124); tail of 1 male 46, of 5 unsexed 44-52 (48.4, SD 3.0); exposed culmen of 12 males 23-31 (26.4), of 7 females 23-29 (26.3), culmen to base of 5 unsexed 28-32 (29.5, SD 1.6); tarsus of 14 males 38-46 (41.1), of 7 females 38-43 (40.7). Weight of 1 male 133.

MOULT Primary moult is sequential, and is recorded N Brazil (Rio Acara, Pará), Nov.

DISTRIBUTION AND STATUS Central America from S Mexico to Panama, and South America in Colombia, Ecuador, E Peru, Bolivia and Brazil. This species was formerly considered locally distributed and by the 1970s the race *guatemalensis* was regarded as very rare except in NE Costa Rica, where it was locally fairly common, while the race *castaneus*, known from only a few specimens, was also regarded as rare (Ripley 1977, Stiles 1981). More recently, *guatemalensis* has been described as rare in Panama and in Colombia (where it is definitely known from only three localities), while *castaneus* was described as apparently rather common in coastal Alagoas, Brazil, but rare in E Peru (Hilty & Brown 1986, Teixeira *et al.* 1986, Ridgely & Gwynne 1989, Parker *et al.* 1992. In both NW and E Ecuador the species is very rare and local, being known only from old specimens (Ridgely & Greenfield in prep.). Because of its secretive habits this species is undoubtedly overlooked, and is possibly more widely distributed than is known, but it is certainly being adversely affected by the destruction of its forest habitats. It should be considered DATA DEFICIENT and worthy of investigation, and possibly rare and threatened in some parts of its range. The nominate race was last reported from Jamaica in 1881, its extirpation presumably being due to the introduction of a mongoose species in 1872 although it also had cats and rats to contend with (Bond 1979, Day 1989). The only record of *castaneus* from French Guiana is a specimen from Cayenne, collected in 1833, and there is one nineteenth century specimen of *guatemalensis* from Guatemala (Land 1970, Tostain *et al.* 1992).

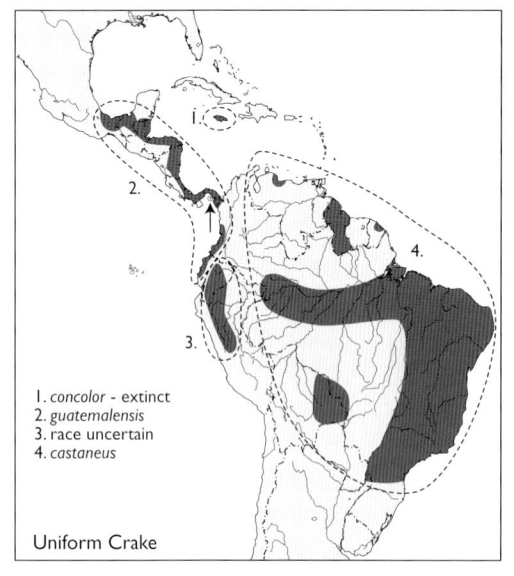

1. *concolor* - extinct
2. *guatemalensis*
3. race uncertain
4. *castaneus*

Uniform Crake

MOVEMENTS None recorded.

HABITAT Forested swamps, moist and flooded forests and forest remnants, shallow clear creeks bordered with palms in tall forest (Brazil), damp ravines, and heavy vine-tangled thickets along the edges of streams. It occurs locally at the edges of mangroves and is also often found away from water in dense secondary growth adjoining forest (especially favouring areas with hanging dead and decaying leaves in *Heliconia* thickets) and in dense thickets bordering cultivation (e.g. Kiff 1975, Stiles 1981). In E Peru it is said to occur at the margins of oxbow lakes (Parker *et al.* 1982). It is essentially confined to lowlands from sea level to c. 1,000m.

FOOD AND FEEDING Takes earthworms, small invertebrates including insects and spiders (Lycosidae), and small frogs (e.g. *Eleutherodactylus*) and lizards; also eats some seeds and berries. It normally remains in cover but sometimes forages in the semi-open on shady forest floor. It forages deliberately, walking slowly and pecking into leaf-litter, detritus and dead leaves hanging from plants; it also probes in wet earth to the depth of the bill and digs with the bill in soft mud (Stiles & Skutch 1989). A small lizard *Anolis* sp. was killed with a few pecks before being swallowed (Stiles 1981).

HABITS Race *concolor* of Jamaica was apparently very sluggish, preferred to run rather than fly, and flew heavily only for a short distance when flushed; it sometimes perched on low trees by the roadside, when it allowed close approach (Greenway 1967). Although difficult to locate in dense, tangled vegetation (except by its loud calls), this species is not particularly shy in cover and may approach a motionless observer closely. The birds usually have an erect stance and walk with head high and tail cocked, having the aspect of a small, short-billed wood-rail *Aramides*. The tail is jerked in agitation; when alarmed the bird scurries off with the body horizontal and the tail low. Birds may perch low down in thickets.

SOCIAL ORGANISATION Probably monogamous, and territorial during the breeding season.

SOCIAL AND SEXUAL BEHAVIOUR No information available.

BREEDING AND SURVIVAL Season Belize, male in breeding condition Apr; Costa Rica, female in breeding condition, Jul, most song Aug-Dec, one probable nest of this species, Nov (Stiles 1981); Panama (Pearl Islands), eggs, possibly of this species, Sep (Wetmore 1965). **Nest** Costa Rica nest was loose cup of leaves filling a hollow in top of a vine-covered stump beside a seldom-used trail in a treefall clearing in swamp forest, c. 5m from stream and 1m from dense thicket. **Eggs** This nest contained 4 eggs which were pale buffy, with blotches and spots of reddish-brown, purplish-brown and grey concentrated near larger end; size 33.3-34.4 x 25.75-26.2 (33.7 x 26) (Stiles 1981). Size of other eggs assumed to be of this species: 2 from Panama 32.7 x 27.3, 32.6 x 26.5; 1 from Guatemala 34 x 31.5 (Wetmore 1965); Panama eggs are pointed short subelliptical. A putative egg of this species from Brazil is described as reddish-grey with very sparse violet and rust-brown flecks; size 33 x 26.5; this egg has probably been lost (Schönwetter 1961-62, Stiles 1981).

GYMNOCREX

Three species of very distinctive, long-legged forest rails of Indonesia and New Guinea. All have bare facial skin, which is cobalt-blue in the Bald-faced Rail and contrasts strikingly with the black and deep maroon-brown plumage. The Bare-eyed and Talaud Rails strongly resemble the *Aramides* rails of Central and South America in their plumage. All species are very poorly known and the Bald-faced Rail is considered Vulnerable. Despite its restricted distribution the Talaud Rail, which was discovered only in 1996, is probably not threatened but should be considered Data Deficient.

81 BALD-FACED RAIL
Gymnocrex rosenbergii   Plate 24

Rallina rosenbergii Schlegel, 1866, Kema, Sulawesi.

Sometimes merged, with *Aramides*, in *Eulabeornis*. Genus shows some similarities to *Aramides*, possibly being derived from the same stock. Monotypic.

Synonyms: *Gymnocrex rosenbergi*; *Eulabeornis/Schizoptila rosenbergii*.

Alternative names: Bare-faced/Rosenberg's (Bare-eyed) Rail; Blue-faced Rail.

IDENTIFICATION Length c. 30cm. Can fly for short distances. Very distinctive appearance: a medium-sized, very dark rail with very conspicuous patch of pale cobalt-blue bare skin around and behind eye; bill quite thin and sharply tapering; legs long and strong-looking; toes quite short. Top of head greyish-black; nape to back and scapulars deep maroon-brown; rump to tail black; upperwings (including tertials) chestnut; primaries and secondaries olive-brown. Sides of head, and entire underparts, greyish-black. Iris brown to red; eyelids reddish; bill pale greenish-yellow, tinged dusky on upper mandible; legs and feet brownish-grey to green. Sexes similar. Immature and juvenile undescribed. Inhabits primary and old secondary forest; also bushy abandoned cultivation.

Similar species Conspicuous patch of pale cobalt-blue bare skin behind eye, and black underparts, make this species impossible to confuse with any sympatric rallids. Occurs alongside the similarly sized Snoring Rail (71), which is very different in plumage but may have a similar call.

VOICE A snoring sound, which could possibly be confused with the call of the Snoring Rail; also a quiet clucking call when disturbed.

DESCRIPTION
Adult Forehead and crown greyish-black, tinged reddish-brown; nape and hindneck deep maroon-brown; mantle, back and scapulars similar but slightly paler; rump, uppertail-coverts and tail black; upperwing-coverts and tertials chestnut; alula, primaries and secondaries bright olive-brown, outer edges washed chestnut. Axillaries, underwing-coverts and marginal coverts blackish-brown, each feather broadly tipped whitish (sometimes buff-tinged). Lores, sides of head, chin, throat and entire underparts greyish-black. Bare skin around and behind eye pale cobalt-blue; eyelids red to pink. Iris brown to red; bill pale greenish-yellow, tinged dusky on upper mandible; legs and feet variously described as brownish-grey, yellowish-green or green (Riley 1924, Eck 1976). Sexes similar, but see comment under Geographical Variation.
Immature Not described.
Juvenile Not described. A specimen at Dresden, formerly assumed to be a juvenile because it has all remiges in sheath, is probably an adult (Eck 1976).
Downy young Not described.

MEASUREMENTS (Sexes combined) Wing 194-208; tail 70-74; culmen 36-39.5, culmen to base 44.5, 50; tarsus 67.5-73 (Ripley 1977, BMNH).

GEOGRAPHICAL VARIATION Eck (1976) notes that a comparison of skins from Sulawesi and Peleng showed that the Peleng birds are somewhat paler, less intensively mahogany-brown on the upperparts, and more slate-grey than grey-black on the underparts. This may represent geographical variation, but Blasius (1897) found in birds from Sulawesi that the male was paler and greyer below than the female. In Sulawesi birds, the iris is said to be red-brown to red and the eye-ring grey-blue, while in Peleng birds the iris is light brown to brown and the eye-ring red to pink. Birds from both Sulawesi and Peleng are described as having green legs and feet, but Blasius described the legs and feet as grey-brown. It is not clear whether such differences reflect geographical, seasonal, sexual or individual variation.

MOULT One specimen has all remiges in sheath, indicating simultaneous moult (Eck 1976).

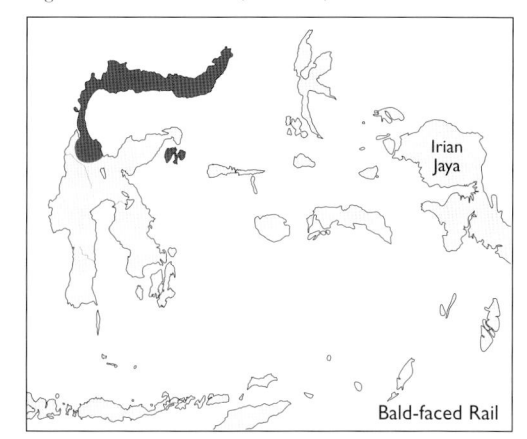

DISTRIBUTION AND STATUS N and NC Sulawesi, and Banggai Is (Peleng). This rail is difficult to observe and is uncommon to rare, but is almost certainly greatly under-recorded (Collar *et al.* 1994, Coates & Bishop 1997); it is classed as VULNERABLE. It is known from 1 Sulawesi specimen and 3 from Peleng, other specimens having been lost, and there are also recent Sulawesi sight records from the Lore Lindu National Park (Watling 1983, Lambert 1989), the Dumoga-Bone National Park (Rozendaal & Dekker 1989), and Gunung Rurukan, Tomohon, Likupang (in the N) and Laboea Sore (NC) (Coates & Bishop 1997). Although Stresemann (1941) regarded it as one of the rarest birds in Sulawesi, Coomans de Ruiter (1947a) found it more numerous there than the Snoring Rail. It is presumably vulnerable to forest destruction and degradation, which are already quite widespread in parts of Sulawesi (Collar *et al.* 1994). It may be trapped for food in some areas.

MOVEMENTS None.

HABITAT Primary and old secondary forest; also bushy, abandoned rice fields. In Lore Lindu National Park it was recorded with the Snoring Rail in old secondary forest, but it appeared to differ in its preference for drier areas with a thick understorey of small saplings, some bamboo clumps, and less dense rattan; when disturbed, one bird ran into the wetter, more impenetrable habitat occupied by the other rail species (Lambert 1989). It is recorded from sea level up to 1,500m.

FOOD AND FEEDING It apparently feeds principally on insects (including Coleoptera) and snails (Coomans de Ruiter 1947a; data on museum skins from BirdLife files).

HABITS This rail is shy, elusive, inconspicuous and difficult to observe. One bird observed by Lambert (1989) was very wary and, on spotting the observer (who was hidden behind a fallen log), ran off with its tail cocked over its back, making a quiet clucking call; it was not seen or heard again. This species looks strong-legged, walks with an upright posture, and takes long strides when moving rapidly (F. R. Lambert *in litt.*).

SOCIAL ORGANISATION It is normally seen singly, occasionally in pairs (Coates & Bishop 1997).

SOCIAL AND SEXUAL BEHAVIOUR Unknown.

BREEDING AND SURVIVAL No information available.

82 BARE-EYED RAIL
Gymnocrex plumbeiventris   Plate 24

Rallus plumbeiventris G. R. Gray, 1862, Misool Island, off north-west New Guinea.

Sometimes placed in *Eulabeornis*, but has more in common with *Aramides*. May form superspecies with *G. talaudensis*. Two subspecies normally recognised, but their validity is questionable; a third (probably extinct) may be valid.

Synonyms: *Rallus hoeveni/intactus*; *Rallina plumbeiventris/intacta*; *Gymnocrex intactus*; *Eulabeornis plumbeiventris/griseoventris*.

IDENTIFICATION Length 30-33cm. A medium-large, heavily built rail with long legs and short toes; pink bare skin around eye visible only at close range. Head, neck, mantle and upper breast vinous-chestnut; throat may be largely grey (possibly commoner in race *hoeveni*); lower mantle, back, scapulars and upperwings olive-brown, with golden tinge in bright light, but primary coverts and primaries vinous-chestnut; lower back to tail black. Lower breast, upper flanks and upper belly lead-grey; lower flanks to undertail-coverts black. Bare skin on lores and behind eye flesh-pink; eyelids orange-red (male) or dull red (female); iris red to brown; bill brownish-olive with greenish-yellow base; legs and feet coral-red. Sexes alike. Non-breeding plumage may be duller. Immature apparently like adult but mixed black and rufous on belly; juvenile undescribed. Questionable race *hoeveni* duller, and underparts washed rufous-brown; form *intactus* darker on upperparts and paler on belly. Inhabits dry to marshy forest, swamps and wet grass.

Similar species A distinctively plumaged rail, similar in overall colouring and pattern to *Aramides* rails of C and S America, especially Red-winged Wood-rail (79); also somewhat similar to much larger Chestnut Rail (116), which, however, has grey head, entirely pinkish-brown underparts, greenish bill and yellow or green legs. The allopatric Talaud Rail (83), recently discovered on Karakelong (Talaud Is), is superficially similar in overall plumage pattern but has head, neck and breast deeper chestnut; chin dark brown; mantle, back and upperwing-coverts bright greenish-olive with iridescent green patch on mantle; rump olive-brown; underparts sepia; flight feathers with greenish-olive outer webs; bare skin behind eye silver-white; also has longer, all-yellow bill, longer yellow legs, and pinkish feet. All similar-sized sympatric rails of wet habitats have barring on underparts, except Rufous-tailed Bush-hen (87) which is olive-brown above and grey below, with rufous-buff belly and undertail-coverts, greenish-yellow legs and a short bill. The sympatric Red-necked Crake (21) is also chestnut from head to mantle and breast but is somewhat smaller, with dark olive-brown upperparts, dark brown underparts with fine, pale buffy to rufous barring, a shorter bill, and olive legs and feet.

VOICE A loud gulping *wow-wow-wow-wow*, only heard at the start of the wet season and probably a territorial call; when foraging, a repeated pig-like grunt (Beehler *et al.* 1986). Coomans de Ruiter (1947a) describes the call as a powerful, repeated *too-oop*, rather like the sound produced by striking a hollow bamboo with a piece of wood, which is followed by a grunting sound; the *too-oop* call is audible from kilometres away. Coates (1985) describes a continual gulping or grunting *uw uw uw...*, virtually identical to a call of the Red-necked Crake, uttered by birds running over the forest floor; this call, described as a frequently repeated low grunting *ung ung ung...*, is given by foraging birds (I. Burrows *in litt.*).

DESCRIPTION *G. p. plumbeiventris*
Adult Entire head and neck vinous-chestnut to burnt-sienna, slightly darker from forehead to hindneck and usually slightly paler on chin and throat; some have chin and upper throat whitish, feathers tipped chestnut; see race *hoeveni* for comments on birds with greyish throat. Upper mantle and upper breast vinous-chestnut to burnt-sienna; rest of mantle ochraceous olive-brown, this colour extending over back, scapulars and upperwing-coverts; lower back to tail black; feathers of lower back and rump sometimes faintly tipped ochraceous olive-brown. One bird had feathers of rump plumbeous, broadly edged chestnut (P. Gregory *in litt.*); not clear if this variation is age-related. Alula, primary coverts and primaries vinous-chestnut with olive-brown tips; marginal coverts predominantly white, mixed with blackish; secondaries and tertials ochraceous olive-brown, like upperwing-coverts; secondaries slightly rufescent on outer webs and tinged dusky on inner webs. Axillaries and underwing-coverts blackish, each feather with broad white tip giving boldly spotted effect; underside of primaries vinous-chestnut; underside of outer secondaries raw-umber, tinged chestnut, often becoming more fuscous to blackish on inner secondaries and tertials, and shading to ochraceous olive-brown on distal part of outer webs. Lower breast, upper flanks, upper belly and thighs leaden-grey, shading to black from lower flanks to undertail-coverts; some birds have entire belly blackish. Bare skin of lores and behind eye flesh, flushing to scarlet when bird handled (P. Gregory *in litt.*); eyelids orange-red (male) or dull red (female); iris red, red-brown or brown (non-breeding or immature?); bill colour variable; often brownish-olive, with greenish-yellow cutting edge and base; bill of 1 bird was dark, with greenish-yellow only towards base (P. Gregory *in litt.*); legs and feet coral-red. Photographs, supplied by P. Gregory, show birds with iris almost burnt-sienna. Sexes alike. Breeding plumage may be brighter than non-breeding plumage (F. Lambert *in litt.*).
Immature Not properly described: apparently similar to adult but with belly mixed rufous and black.
Juvenile Not described.
Downy young Not described.

MEASUREMENTS Wing of male 184-189, of 4 females 184-188 (186, SD 1.8); tail of male 67, 77, of female 71.5, of unsexed 66; culmen of male 50, exposed culmen of male 47, culmen of female 48, of unsexed 52; tarsus of male 57, 58, of female 58, of unsexed 52 (Rand & Gilliard 1967, Ripley 1977, BMNH). Weight of 2 unsexed 255, 320, the former bird thin but not emaciated (P. Gregory *in litt.*).

GEOGRAPHICAL VARIATION The two races are separated only on the relative brightness of the plumage, but much individual variation occurs and the validity of *hoeveni* is questionable (Mees 1965). Knox & Walters (1994) accept a third form, *intactus*, for which they tentatively retain specific status. This form is known only from the type collected by L. Brazier in the Solomon Is. It has been consistently ignored, mainly because a number of birds allegedly collected by Brazier in the Solomons were actually collected on New Ireland, and has been considered to be a synonym of *G. plumbeiventris* (e.g. Mayr 1933b). However it differs in having the brown of the neck darker and less chestnut, with a more maroon bloom, while the wings and back are slightly darker, and the belly very slightly paler, than in *G. plumbeiventris*; wing 175; tail 47; culmen to base 50; tarsus 57. Knox and Walters (1994) suggest that the locality is probably correct and that the specimen represents an extinct (or overlooked?) form of Bare-eyed Rail from the Solomons. If so, it is probably only subspecifically distinct.

G. p. plumbeiventris (G. R. Gray, 1862) – N Moluccas (Morotai, Halmahera and Bacan), Misool, New Guinea (except S), Karkar I and New Ireland. See Description.

G. p. hoeveni (von Rosenberg, 1866) – Aru Is and S New Guinea (Setekwa R to Fly R). Three specimens in BMNH, labelled "Queensland", one previously described as *Eulabeornis griseoventris*, and all collected by J. T. Cockerell, are probably from the Aru Is (see Mathews 1915). Similar to nominate but crown to hindneck duller, washed dark olive; primaries duller rufous; rump and tail less blackish; grey of underparts duller and more washed rufous-brown; breast duller reddish-brown; head and neck duller. Some birds have throat largely grey, as illustrated in Ripley (1977); Rand (1933a) gives this as a character of *hoeveni* but some nominate birds also have grey to whitish chin and upper throat (e.g. Rand & Gilliard 1967; photographs supplied by P Gregory). Wing of 1 male 192, of 4 females 187-198 (194, SD 4.8), of 3 unsexed 195, 197, 200; tail of 1 female 66, of 3 unsexed 63, 71, 74; culmen to base of 1 female 52, of 3 unsexed 46, 47, 50; tarsus of 1 female 52, of 3 unsexed 54, 55, 56 (Rand 1938, Ripley 1977, BMNH).

MOULT Remex moult is synchronised, and is recorded from Halmahera, May, when tail and upperparts moult was also in progress (Stresemann & Stresemann 1966).

DISTRIBUTION AND STATUS N Moluccas, New Guinea, Aru Is, Karkar I, New Ireland; formerly possibly also Solomon Is. This species is not globally threatened, but very little information is available on its status because it is easily overlooked, being very shy and difficult to observe. It was formerly regarded as not uncommon throughout its range, although the nominate race was regarded as rather thinly spread and is currently regarded as probably uncommon in Wallacea (Coates & Bishop 1997). The provenance of the form *intactus* is discussed under Geographical Variation; if the type locality of "Solomon Is" is correct, P. Gregory (*in litt.*) suggests that the form may not be extinct because good habitat could still be intact, unless the bird came from only a small island.

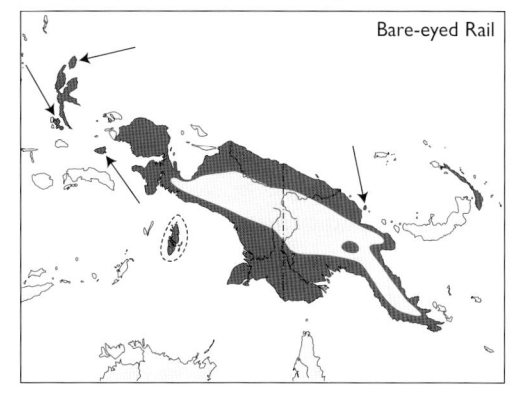

MOVEMENTS Sedentary; locally nomadic in New Guinea. In the Port Moresby area it is mostly seen in Dec-May, but there are also records for Aug-Sept; it vacates lowland rainforest near the Brown R during long periods of dry weather, when the ground becomes very hard, while it also visits monsoon forest in the Port Moresby area during the rains (Coates 1985). One found perched in a tree in urban Port Moresby, 15km from the nearest forest, in Jan, is indicative of movements (Coates 1985), and 1 flew into buildings at night at Ambua Lodge, Mt Hagen (Tano 1996). During a severe drought 1 arrived, apparently overnight, at a school in Tabubil on 8 Sep 1997, and roosted during the day on a rafter; it appeared thin but not emaciated and was released in secondary scrub (P. Gregory *in litt.*).

HABITAT This rail occurs on the floor of primary forest and also in marshy forests, swamps (including sago swamps) and wet grassy areas near rivers and lakes; in SE New Guinea it also occurs in monsoon forest during the rainy season. It prefers forest with a damp substrate. It is found in lowlands up to 1,200m, but in E New Guinea it is also recorded from the Mt Hagen area at 1,600m.

FOOD AND FEEDING Eats insects, including large Coleoptera, and also molluscs (Gastropoda); it probably also takes a variety of other small animals. Birds foraging on the forest floor have been observed scratching dead leaves away with the feet, and also shovelling leaves aside with sweeping movements of the closed bill (I. Burrows *in litt.*).

HABITS This species is said to be very active by day. It is shy, secretive, elusive and infrequently seen, but is inquisitive if the observer remains motionless. It runs swiftly to and fro over the forest floor, carrying the black tail erect and continually uttering a gulping or grunting sound (Coates 1985). It often holds itself horizontally, without raising the head (Beehler *et al.* 1986). A foraging bird frequently bobbed its head forwards and downwards, and it flicked its wings when crossing a fallen log, the underwing-coverts flashing pale in the poor light (I. Burrows *in litt.*). A captive bird stood for long periods in a shallow pool, while another hid in thick ground vegetation, squatting down without moving (P. Gregory *in litt.*).

SOCIAL ORGANISATION Territorial, at least when breeding; it occurs in pairs or family parties.

SOCIAL AND SEXUAL BEHAVIOUR Seen to attack a Common Paradise Kingfisher *Tanysiptera galatea* which was digging for food (Coates 1977). A bird foraging close to a Red-necked Crake chased the smaller rail away 3 times, each time running at it with head down, bill pointed forwards, wings held tightly closed and tail cocked (I Burrows *in litt.*). It is said to be pugnacious and to have a raised-wing threat display (Ripley 1964), presumably showing the black and white underwing-coverts. The bare facial skin flushes scarlet when the bird is stressed, and P Gregory (*in litt.*) suggests that this feature, together with the underwing pattern, may have some significance in display.

BREEDING AND SURVIVAL Season Misool, Nov; Karkar, Feb; breeds in wet season. **Nest and eggs** In Sepik Region, New Guinea, nest said to be of grass, built on ground when water level low, and eggs said to be white (Gilliard & LeCroy 1966); however, this does not agree with a nest at Karkar, which was situated between roots of a large tree and contained 8 eggs, smooth and glossy, light pinkish-cream with rufous-brown spots and underlying purplish-grey patches; size 41.1-42 x 30.4-31.5 (41.5 [SD 0.4] x 31.2 [SD 0.4]) (Rothschild & Hartert 1915). A possible nest near Port Moresby in Feb was a few fragments of dead leaves on bare ground between the roots of a small tree in foothill monsoon forest; it contained 1 egg, whitish-buff with grey spots and blotches and larger chestnut spots and blotches, size 39 x 30 (Coates 1985). Schönwetter (1961-62) gives size of 15 eggs as 38.2-42.2 x 28.3-32 (40.4 x 30.7), calculated weight 20.9.

83 TALAUD RAIL
Gymnocrex talaudensis   Plate 24

Gymnocrex talaudensis Lambert, 1998, Karakelong I, Talaud Archipelago.

May form superspecies with *G. plumbeiventris*. Monotypic.

IDENTIFICATION Described only from the holotype (unsexed). Head, neck and breast chestnut; chin dark brown; mantle, back and upperwing-coverts bright greenish-olive, more iridescent green on mantle; rump olive-brown; uppertail-coverts, flanks, and belly to undertail-coverts, blackish; flight feathers tawny with greenish-olive outer webs. Iris red; bare skin round eye pink anteriorly and silver-white (obvious in field) behind eye; bill bright yellow; legs yellowish; feet pinkish. Immature and juvenile unknown. Inhabits wet grass, scrub and rank vegetation at forest edges.
Similar species Allopatric Bare-eyed Rail (82) superficially similar in overall plumage pattern but has head to mantle and upper breast vinous-chestnut; lower mantle, back, scapulars and upperwings olive-brown, tinged gold; primary coverts and primaries vinous-chestnut; lower back to tail black; lower breast, upper flanks and upper belly lead-grey; lower flanks to undertail-coverts black. Bare skin around and behind eye flesh-pink; iris red to brown; bill shorter, brownish-olive with greenish-yellow base; legs and feet coral-red.

VOICE One individual, as it ran to cover, was thought to give a series of at least 15 rapid, high-pitched *peet-peet-peet* calls (Lambert 1998a).

DESCRIPTION Details from text and illustrations in Lambert (1998a).
Adult Described only from the holotype, presumed adult on plumage colour and texture, and on bright colours of bare parts. Head, neck and breast chestnut, with slight maroon bloom, becoming sepia-tinged from forehead to forecrown and around bill, and dark sepia on chin. Mantle, back and upperwing-coverts bright greenish-olive, mantle feathers with distal third iridescent, thus appearing bright shining green; rump olive-brown; uppertail-coverts blackish-sepia; rectrices missing, presumed same colour as tail-coverts. Axillaries and underwing-coverts sepia, many feathers tipped white to give large white spots. Flanks and belly to undertail-coverts blackish. Bare skin around eye pink, with prominent silvery-white patch behind eye; iris bright red; bill bright yellow, slightly duskier on distal third and darker around large nares; legs yellow, becoming more pinkish on feet.
Immature Unknown.
Juvenile Unknown.
Downy young Unknown.

MEASUREMENTS Wing and tail of holotype unmeasurable; bill (gape to tip) 58; tarsus 68 (Lambert 1998a).

GEOGRAPHICAL VARIATION None.

MOULT The holotype, collected in Sep, is in heavy moult, presumably postbreeding, with remiges, rectrices and body feathers in sheath (Lambert 1998a).

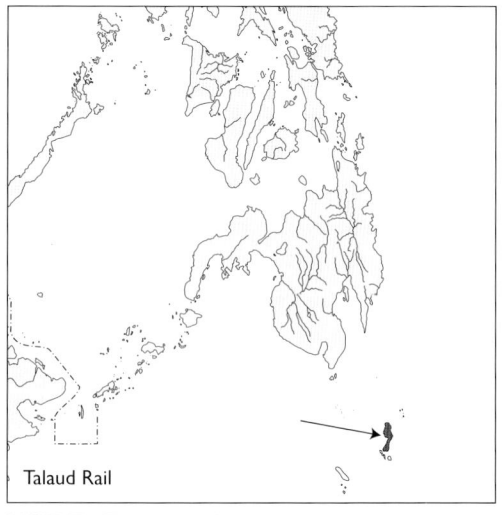

DISTRIBUTION AND STATUS All information is from Lambert (1998a). This species is currently known only from Karakelong I in the Talaud Archipelago. One bird was seen briefly near Tarohan on 15 Aug 1996, another was seen near Beo in Mar 1997 and the holotype was obtained on 6 Sep 1996 from a man in Beo who was selling birds caught near Rainis village. According to local people it is relatively widespread on Karakelong and is not uncommon in suitable habitats. It is occasionally eaten, being caught in traps for terrestrial birds or by using dogs, but it is apparently difficult to catch and pressure from trapping must be very low. At present it is probably not threatened because Karakelong still has a diversity and abundance of wetland habitats, particularly rank grasslands bordering forest, and the forests are still relatively intact. However, Lambert recommends that the species be classed as DATA DEFICIENT and suggests that, since its full ecological requirements are unknown, future changes in land use could place it under threat: although the largest island in the group, the island is relatively small, covering only c. 600km². The forests are still relatively intact, and there are 2 protected areas totalling c. 21,800ha but there is virtually no provision to enforce this protection, and both areas are under pressure from smallholder encroachment, while there may be a long-term threat of forest clearance for transmigration. It may also be present on the other larger islands in the archipelago, namely Salebabu and Kabaruang, although these have little natural forest.

MOVEMENTS Unknown; presumably sedentary.

HABITAT Frequents long wet grassland and scrub, and swampy secondary growth and rank herbage, at the edge of forest which may itself be swampy; it also reportedly occurs in the interior of tall primary forest with a dense understorey of saplings and herbs along a shallow stream, but such records could be misidentifications (Lambert 1998a). The site of the Aug 1996 sighting had wet roadside ditches with lush grass, several streams, permanent marshy areas and small remnant forest patches; littoral swamp forest was present c. 150m from the locality, which was <100m from the sea. The holotype was obtained at an altitude of <100m.

FOOD AND FEEDING The stomach of the holotype contained small fragments of snail shell and a very small beetle (Lambert 1998a).

HABITS Described as extremely shy; just after dawn 1 bird walked casually across a tarmac road in front of a vehicle, paused in short grass under small cocoa trees, then ran into cover at the forest edge (Lambert 1998a).

SOCIAL ORGANISATION Unknown.

SOCIAL AND SEXUAL BEHAVIOUR Unknown.

BREEDING AND SURVIVAL Unknown; apparent post-breeding moult recorded Sep.

AMAURORNIS

As treated here, this genus includes nine species of mostly medium to medium-large rails from Africa, Asia and Australasia. Olson (1973b) recognised that one of the most difficult problems in rail taxonomy concerns the proper allocation of the species included in this genus and *Porzana*. Distinctions between the two genera are inadequately defined and incompletely understood, and some *Amaurornis* species have previously been included in *Porzana*, and vice versa. The species *akool, isabellinus, olivaceus, moluccanus, magnirostris* and *phoenicurus* are generally accepted as belonging in *Amaurornis* and represent a rather basic stock from which both the *Porzana* assemblage and the gallinules could have arisen. The other species, *flavirostris, olivieri* and *bicolor*, are sometimes included in *Porzana*, and although this may well be appropriate in the case of *bicolor* (e.g. Olson 1973b, Inskipp & Round 1989), the other two species are probably best retained in *Amaurornis* (Olson 1973b).

Amaurornis rails are typically plain and dark in colour, being either predominantly black or various shades of olive-brown, dark rich brown, slate-grey and rufous, the exception being the strikingly marked White-breasted Waterhen which has white and chestnut underparts. All species have yellow or greenish bills, and most have brightly coloured legs and feet. They inhabit marshes, swamps, dense riparian vegetation, wet grass, wet forest, and sometimes cultivation such as rice paddies, maize and sugarcane. The species commonly known as bush-hens also occur in dry grass, bush and forest, sometimes far from water. Species such as the Rufous-tailed Bush-hen and the White-breasted Waterhen are particularly catholic in their choice of habitats. Most species are typically lowland birds but the Black-tailed Crake is primarily a highland species. The Brown Crake, Plain Bush-hen and Black-tailed Crake are infrequently seen and poorly known but the Black Crake and White-breasted Waterhen are widespread, locally common and easily observed. The very poorly known Isabelline Bush-hen should be regarded as Data Deficient, while the Sakalava Rail of Madagascar is critically endangered as a result of habitat loss and persecution. The recently described Talaud Bush-hen is apparently widespread on the one island from which it is known, but should possibly be regarded as Near-threatened.

84 BROWN CRAKE
Amaurornis akool Plate 25

Rallus Akool Sykes, 1832, Deccan.

Two subspecies recognised.

Synonyms: *Rallus griseopectus/niger; Porzana akool; Gallinula coccineipes/modesta; Ortygometra akool/griseopectus; Corethrura akool/modesta; Hypotaenidia akool; Amaurornis coccineipes*.

Alternative name: Crimson-legged Crake.

IDENTIFICATION Length 24-29cm. A medium-sized crake, similar in size to the Water Rail (61), this species has a typical *Amaurornis* shape, a rather short, heavy bill, robust legs and long toes. A very plain crake, with olive-brown upperparts, white chin and throat, dark slate-grey underparts shading to vinaceous and olive-brown from belly to undertail-coverts, and indistinct pale grey supercilium. Iris red; bill pale greenish with bluish tip; legs and feet purple or red, becoming brown in non-breeding season. Sexes alike; female averages smaller. Race *coccineipes* larger, with brighter, redder legs. Immature similar to adult but with brown iris. Juvenile has darker upperparts, and brown-tinged underparts; bare parts darker and duller than in adult. Inhabits dense swamps, pandanus groves and riparian vegetation.

Similar species The smaller sympatric Black-tailed Crake

(92) has dark rufous-brown upperparts contrasting with dark grey head, neck and underparts; tail black; bill blue-green with a red patch near base; legs and feet dull to bright red. Easily distinguished from immatures of sympatric Red-legged (23), Slaty-legged (24), Ruddy-breasted (102) and Band-bellied (103) Crakes by lack of white barring on underparts.

VOICE Ali & Ripley (1980) describe 3 calls: a short plaintive note, heard at daybreak and just before sunset; a shrill rattle, attributed to this species but not proved; and a long vibrating whistle, gradually falling slightly in pitch. Other descriptions, probably referring to the vibrating whistle, are a call similar to that of the Dabchick *Tachybaptus ruficollis* and a high, rippling trill lasting 3-4 s (Roberts 1991).

DESCRIPTION *A. a. akool*
Adult Forehead, crown and entire upperparts, including tail, tertials and most upperwing-coverts, uniform darkish olive with brown tinge. Alula, primary coverts, primaries and secondaries fuscous, outer edges more olive-brown; underwing-coverts and axillaries dark olive-brown. Chin and throat white; supraloral area to eye dusky grey-brown; lores, supercilium, sides of head, sides of neck, foreneck, breast and upper belly dark slate-grey, palest on supercilium (showing as vague pale stripe), ear-coverts and malar region; lower belly and vent slightly more vinaceous; flanks and undertail-coverts dark olive-brown. Iris blood-red; bill horny-green to pale green, with bluish tip; legs and feet purple or red in breeding season, otherwise duller, more fleshy-brown. Sexes alike but female averages smaller; bill green, culmen dark brown, tip of lower mandible lavender.
Immature Similar to adult but iris brown.
Juvenile Upperparts darker and browner than adult; grey of underparts washed pale olive-brown, heavier and more vinaceous on lower flanks, lower belly and vent. Tip and upper mandible dark, possibly blackish; lower mandible pale, possibly greenish; legs and feet duller than in adult.
Downy young Chick has black down, with bottle-green sheen on upperparts and brown-tinged on underparts; anterior half of bill pale (ivory in 1 specimen), rest of lower mandible ivory, upper mandible with black band down centre and pale ochre base; small pale wing-claw and small white egg-tooth present.

MEASUREMENTS (12 males, 12 females) Wing of male 121-130 (125.8, SD 2.7), of female 117-126 (121.4, SD 3.2); tail of male 49-58 (53.4, SD 3.0), of female 48-59 (53.1, SD 4.2); culmen to base of male 30-35 (33.1, SD 1.4), of female 28-32.5 (30.95, SD 1.2); tarsus of male 46-53 (49.5, SD 2.5), of female 42-49 (46.2, SD 2.2). Weight of male 114-170, of female 110-140.

GEOGRAPHICAL VARIATION Races are separated on size and leg colour.
A. a. akool (Sykes, 1832) – Kashmir and N Pakistan (Rawal Lake), E through lowlands and duns of Nepal to W Assam; S through India (except Kerala, Tamil Nadu and S Andhra Pradesh); Bangladesh and W Myanmar (Arakan). See Description.
A. a. coccineipes (Slater, 1891) – SE China in Jiangsu (= Kiangsu), S Anhui (= Anhwei), Zhejiang (= Chekiang), Hunan, Guizhou (= Kweichow), Fujian (= Fukien), Guangdong (= Kwangtung) and Guangxi Zhuang (= Kwangsi); Hong Kong and NE Vietnam (Tonkin). Larger than nominate, with brighter, redder legs. Iris red; bill dark olive-green; legs and feet carmine. Measurements of 5 males, 3 females: wing of male 130-134 (132.4, SD 1.5), of female 128, 133, 134; tail of male 59-63 (61.2, SD 2.0), of 2 females 58, 59; culmen to base of male 32.5-35 (34, SD 1.0), of female 31, 35, 35; tarsus of male 50-53 (51.8, SD 1.6), of female 49, 50, 52. Tail of male 63-67, of female 62-64; tarsus of male 53-55, of female 48-50 (Ripley 1977).

MOULT No information available.

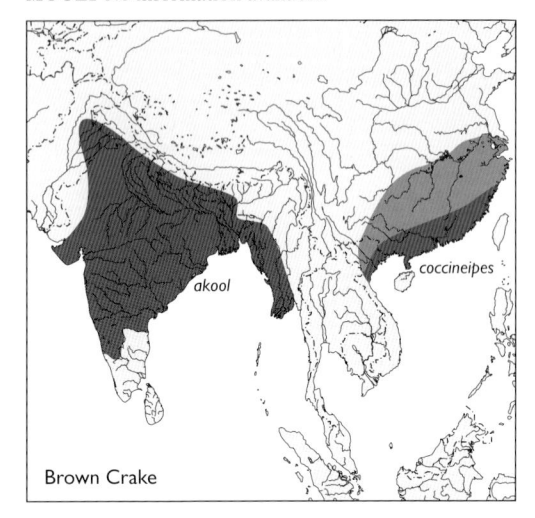

Brown Crake

DISTRIBUTION AND STATUS Kashmir, India, Nepal, Assam, Bangladesh and W Burma; and SE China to NE Vietnam. It was also recorded from Rawal Lake, N Pakistan, in Jul-Aug 1987, a W extension of range (Roberts 1991). This crake is not regarded as globally threatened. It is infrequently seen because of its shy and secretive nature, but it was formerly regarded as fairly common throughout its range although it is apparently rare in Kashmir (Roberts 1991). No precise information is available on its current status, but only a few records of the nominate race were obtained in India and Nepal during the 1987-91 Asian Waterfowl Census (Perennou *et al.* 1994) and there are no recent records from Bangladesh (Harvey 1990). The race *coccineipes* is regarded as uncommon in China (Cheng 1987).

MOVEMENTS Resident, but possibly also a migrant over relatively short distances. N populations winter in the S parts of the range but the extent of seasonal occurrences is often not clear, and is shown on the distribution map only in China, where the race *coccineipes* is regarded as only a summer visitor to Guizhou, Hunan, Jiangsu, Zhejiang and S Anhui (Cheng 1987).

HABITAT Dense swamps, *Pandanus* palm groves, sugarcane fields, reedbeds and other dense herbage along rivers and other watercourses, and irrigation channels. Lowlands to 800m, possibly higher.

FOOD AND FEEDING Worms (Annelida), molluscs, adult and larval insects, and seeds of marsh plants. Birds forage outside cover on short grass and at the drying margins of pools, pecking at the bases of water weeds and grass clumps and apparently gleaning gastropods and insects (Roberts 1991); they forage in open meadows during the morning and evening (Delacour & Jabouille 1931).

HABITS Information is taken from Ali & Ripley (1980) unless otherwise specified. Often regarded as shy and secretive, this crake is seldom seen and is said to be largely crepuscular, emerging cautiously at edges of reedbeds and bushes to feed in the early morning and evening, when it walks around with constant jerks of the short tail and runs rapidly into cover with lowered head when suspicious or alarmed. However in Nepal and N India it is less secretive and less crepuscular than *Porzana* species and may sometimes be seen foraging for long periods in the open close to cover (K. Kazmierczak pers. comm.). It will sometimes hide under a large stone or in a hole in a bank and remain there for many minutes, looking out cautiously before emerging. It climbs in reeds, bushes and *Pandanus* as agilely as does the White-breasted Waterhen (89). It is apparently less difficult to flush than its Asian congeners but has a similarly feeble flight action with dangling legs. It will fly up into trees to roost (Roberts 1991).

SOCIAL ORGANISATION Monogamous.

SOCIAL AND SEXUAL BEHAVIOUR The incubating bird is reluctant to leave the nest and often betrays its site by running off it like a rat when almost trodden upon (Ali & Ripley 1980).

BREEDING AND SURVIVAL Season India, Mar-Oct, chiefly May-Aug; varies with locality but mainly during monsoon. **Nest** A pad of grasses, reeds, rush blades and sticks, with slight central depression, well concealed in raised grass tussocks or bush in or near edge of swamp; 1 nest was c. 1.5m above ground, in flood-deposited debris on small acacia tree on riverbank. **Eggs** 5-6; broad oval; little gloss; creamy-white or buff to pale salmon-pink, sparsely flecked, streaked and blotched pale reddish-brown, purple-brown, or pale brick-red; markings often more concentrated at larger end; size (n = 70, nominate race) 33-39.5 x 25.4-29 (36 x 27), calculated weight 14.5. **Incubation** By both sexes; period unrecorded.

---

85 ISABELLINE BUSH-HEN
Amaurornis isabellinus   Plate 25

Rallina isabellina Schlegel, 1865, Gorontalo, Sulawesi.

Monotypic.

Synonyms: *Rallus isabellinus*; *Euryzona/Oenolimnas/ Amaurornis/Erythra/Gallinula isabellina*.

Alternative names: Isabelline Waterhen; Celebes Bush-hen.

IDENTIFICATION Length 35-40cm. The largest *Amaurornis* species. Upperparts olive-brown, tinged greyish; chin and upper throat whitish; sides of head and entire underparts, including underwing-coverts and axillaries, cinnamon-rufous, more vinous on ear-coverts and sides of neck, paler around white of chin and throat, and browner on lower belly and vent. Iris red; bill pale green; legs and feet brownish-green. Sexes alike. Immature like adult but olive of head replaced by brown; bill brownish. Juvenile undescribed. Inhabits riparian scrub, grass and abandoned cultivation.
Similar species Paler than congeners, and easily distinguished by grey-tinged, olive-brown upperparts and entirely cinnamon-rufous underparts.

VOICE A very striking loud cacophony of sharp, slightly pulsating screeches and chattering notes, apparently resembling the call of the Barred Rail (44) but usually ending with a loud, clear *tak-tak-tak-tak*, from which its local name "Taktak" is presumably derived (Ripley 1977). The call is also described as being similar to that of the White-breasted Waterhen, but drier and more strident (Coates & Bishop 1997). Several birds usually call together, and calling may last for several minutes.

DESCRIPTION
Adult Forehead, crown and upperparts, including upperwing-coverts and tertials, olive-brown, washed greyish-olive; lower back to tail brighter, more cinnamon-brown. Upperwing-coverts variably edged pale cinnamon-rufous; tertials brighter, more cinnamon-brown, than upperparts; alula, primaries and secondaries cinnamon-brown or brighter, darker towards tips; axillaries and underwing-coverts pale cinnamon-rufous; underside of remiges bronzy-brown. Lores and feathers below eye dusky-brown, faintly tinged cinnamon; chin and upper throat whitish; ear-coverts and sides of neck ochraceous-tawny to cinnamon-rufous; malar region and throat paler; rest of underparts cinnamon-rufous, darker and browner on lower belly and vent. Iris red; bill pale green; legs and feet brownish-green. Sexes alike.
Immature Like adult, but olive of head replaced by brown; bill brownish.
Juvenile Not described.
Downy young Not described.

MEASUREMENTS Wing of male 154-175, of female 154-167; tail of male 63-70, of female 63-70; culmen of male 36-37, of female 35; tarsus of male 63-67, of female 62-65 (Ripley 1977, sample sizes and means not given; BMNH). Stresemann (1941) gives wing of 7 males 168-175 (172), of 8 females 154-164 (161.2).

GEOGRAPHICAL VARIATION None.

MOULT No information available.

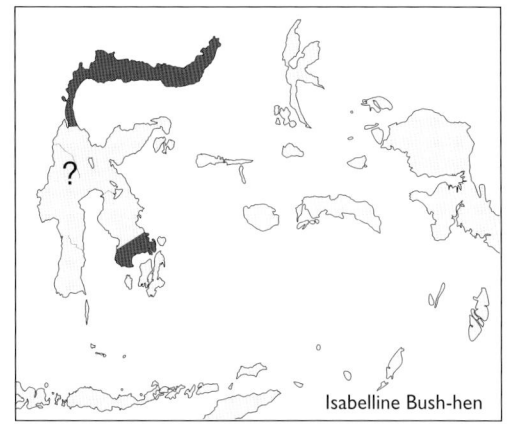

Isabelline Bush-hen

DISTRIBUTION AND STATUS Endemic to the lowlands of Sulawesi. It is definitely known only from the N peninsula of Sulawesi S to Tawaya, and from the SE; its occurrence elsewhere in Sulawesi requires investigation. It is not regarded as globally threatened. It was regarded as relatively common, at least locally, in the 1980s (Watling 1983), and it is currently regarded as widespread and generally common (Coates & Bishop 1997); it was not recorded during the Asian Waterfowl Census in 1987-91 (Perennou *et al.* 1994). In view of its restricted range, and the lack of information on its ecology, it should probably be regarded as a DATA DEFICIENT species in urgent need of investigation.

MOVEMENTS None recorded.

HABITAT Grass with low bushes near streams and rivers with forested edges, and at the edges of rice and maize fields; dense rank grassland (especially alang-alang *Imperata cylindrica*), including tall grass bordering forest; and secondary vegetation, often at forest edges. It is relatively common in dry, rank vegetation, especially old, fallow gardens in the first stages of reversion to forest (Watling 1983). It may occur far from water, and it frequents drier habitat than does the sympatric White-breasted Waterhen (89). It occurs from sea level up to 1,400m.

FOOD AND FEEDING The food is not recorded. Birds have been noted foraging along a vehicle track in secondary coastal woodland and open areas of alang-alang (Andrew & Holmes 1990); birds pick up food from the surface of the ground (Coates & Bishop 1997).

HABITS It is generally shy, though noisy, but is occasionally conspicuous and confiding. It is nervous, constantly flicking the tail, and walks slowly with the head erect, then runs quickly for a short way with head and neck horizontal; this mode of progression is maintained while foraging (Coates & Bishop 1997).

SOCIAL ORGANISATION Unknown. Both 1 and 2 adults have been seen with chicks (Rozendaal and Dekker 1989, Andrew & Holmes 1990).

SOCIAL AND SEXUAL BEHAVIOUR It normally occurs singly or in pairs, occasionally in small family groups.

BREEDING AND SURVIVAL Season Young found in May and Aug: 2 adults with 5 chicks in late May 1986 (Andrew & Holmes 1990), and 1 adult with chicks on 10 Aug 1985 (Rozendaal & Dekker 1989). **Nest** undescribed. **Eggs** 2 eggs measured 39.2 x 31.1 and 42 x 32, calculated weight 22 (Schönwetter 1961-62).

---

86 PLAIN BUSH-HEN
Amaurornis olivaceus   Plate 25

Gallinula olivacea Meyen, 1834, Manila, Philippines.

Formerly considered conspecific with *A. moluccanus*, but sufficiently distinct to merit separation (McAllan & Bruce 1988), probably best considered as forming superspecies (see Geographical Variation). Monotypic. Synonyms: *Tribonyx/Amaurornis olivacea*.

Alternative name: (Common) Bush-hen.

IDENTIFICATION Length 24-31cm. A medium-sized, dark, unpatterned rail with long legs and toes, short stout bill and short tail. Top of head, upperparts and upperwings very dark olive-brown, slightly paler on ear-coverts and tinged blackish from top of head to hindneck and on lores; tail blackish. Malar region, chin, throat and underparts very dark slaty-grey; thighs and lower belly paler; vent and undertail-coverts very dark rufous-brown. Iris red; bill pale green; legs and feet yellowish-brown to dark yellow. Sexes alike; female somewhat smaller. Immature probably like adult; juvenile darker on upperparts, and washed olive-brown on underparts. Inhabits wet to dry grass, flooded scrub and forest edge.

Similar species Similar to Rufous-tailed Bush-hen (87), but larger and darker, with more olive-tinged upperparts, dark slaty-grey underparts (including grey thighs) and dark rufous-brown undertail-coverts; bill has no red spot at base; iris red, not brown; legs and feet normally less pure yellow. See also the recently discovered Talaud Bush-hen (88).

VOICE No proper description is available, but the voice is presumably similar to that of the Rufous-tailed Bush-hen, although the shrieking call is considerably more harsh, lower pitched and less melodious than that of Australian Rufous-tailed Bush-hens (Lambert 1998b).

DESCRIPTION
Adult Forehead, crown, upper sides of head and entire upperparts very dark olive-brown, slightly paler on upper ear-coverts and often tinged blackish from top of head to hindneck, on lores and on rectrices; ear-coverts sometimes entirely slaty-grey. Upperwings as rest of upperparts; primaries and secondaries slightly darker brown with less olive tinge; underwing-coverts and axillaries dark slaty-grey, washed olive-brown; underside of remiges dark grey. Lower ear-coverts, malar region, chin, throat and underparts very dark slaty-grey; sides of neck, sides of breast, and flanks washed olive-brown, most prominently on flanks; thighs and lower belly paler, washed vinaceous-cinnamon; vent and undertail-coverts very dark rufous-brown, but paler, more rufous-buff, at tips in some birds (Lambert 1998b). Iris blood-red; bill pale green; legs and feet yellowish-brown to dark yellow. Sexes alike but female averages smaller. Chin sometimes described as light grey (e.g. duPont 1971) – see Immature.
Immature Not described, but must be almost identical to adult. Some birds have whitish bases or tips to feathers of chin and throat but it is not clear whether this character is age-related.
Juvenile Very similar to adult but upperparts darker brown, more sepia in tone, and grey of underparts more heavily washed olive-brown.
Downy young Not described.

MEASUREMENTS (8 males, 6 females) Wing of male 155-176 (164.6, SD 6.7), of female 151-157 (153.7, SD 2.5); tail of male 49-68 (55.5, SD 5.9), of female 44-55 (50.5, SD 4.3); culmen to base of male 35-39 (37.8, SD 1.2), of female 32.5-37.5 (35.75, SD 1.8) ; tarsus of male 58-64 (62.1, SD 1.9), of female 53-61 (57.2, SD 2.7). Weight of 2 males 292.5, 312.5, of 1 female 250 (Rand & Rabor 1960).

GEOGRAPHICAL VARIATION Formerly considered conspecific with the Rufous-tailed Bush-hen, although originally retained as a separate species (e.g. Sharpe 1894), but considered by McAllan & Bruce (1988) to be sufficiently distinct to merit separation, a treatment followed by Sibley & Monroe (1990). Compared with all races of the Rufous-tailed Bush-hen it is larger and darker; upperparts darker and more olive-tinged; underparts dark slaty-grey; thighs dark grey; undertail-coverts dark rufous-brown; bill pale green with no red spot at base; iris red, not brown; legs and feet normally less pure yellow.

Parkes (1971) discusses a possible "small island effect" in the population from Siquijor, particularly the females;

this population is longer-billed, and also longer-winged, than populations from other Philippine islands. Thus in 4 Siquijor females, exposed culmen is 35.5-38 (37) and wing 160-166 (164.5); in 10 from other islands, exposed culmen is 31-35 (34.3) and wing is 148-162 (155.2). 1 Siquijor male has exposed culmen 39, wing 171; 10 from elsewhere have exposed culmen 36-39 (37.25), wing 166-178 (168.4).

MOULT No information available.

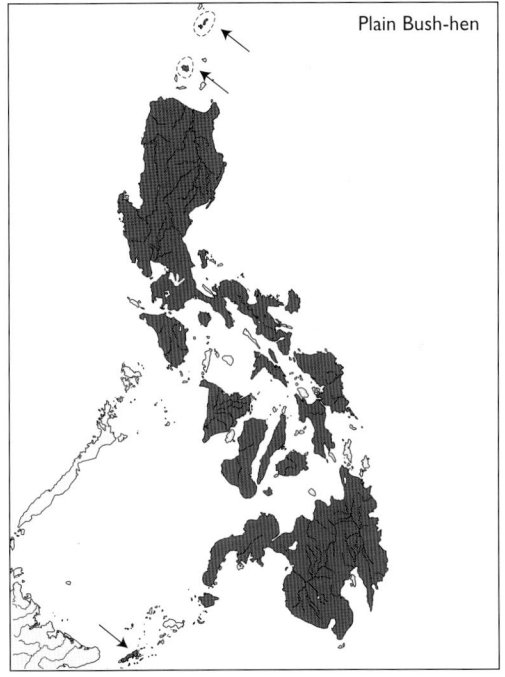

Plain Bush-hen

DISTRIBUTION AND STATUS Philippines, from Basilan, Batan, Bohol, Calayan, Cataduanes, Cebu, Leyte, Luzon, Marinduque, Masbate, Mindanao, Mindoro, Negros, Panay, Polillo, Sabtang (sight records), Samar, Siquijor and Ticao; it is not recorded from the Sulu Archipelago except for a sight record from Sibutu; records from Palawan and Siasi are not acceptable (Dickinson *et al.* 1991). This crake is not globally threatened and was formerly considered common, but it is regarded as a rare resident by Dickinson *et al.* (1991). In NE Luzon it was recently found to be common in the Palanan area, Siagot and Cayapa, and uncommon in N Maconacon, Minanga, Baliuag and near Dimapnat (Danielsen *et al.* 1994). In view of the possible change in its status, and the general lack of information about its biology and ecology, it should be regarded as DATA DEFICIENT and its current status should be investigated in detail. Recent observations (Danielsen *et al.* 1994) show that it is able to colonize man-modified habitats (see Habitat). F. Lambert (*in litt.*) doubts whether it is threatened, because its preferred habitat has probably increased in areas close to human activity, but hunting could be a problem.

MOVEMENTS None recorded.

HABITAT Waterside grass clumps, flooded scrub and swampy grassland. It nests in swampy conditions but also ranges through neighbouring grassland and forest edge (Dickinson *et al.* 1991). It also occurs in cultivated areas, scrub, and degraded and selectively logged forest (Danielsen *et al.* 1994), and is usually encountered in secondary growth and adjacent damp grasslands (Lambert 1998b).

FOOD AND FEEDING Not recorded: likely to be similar to Rufous-tailed Bush-hen.

HABITS Shy and secretive. See Rufous-tailed Bush-hen.

SOCIAL ORGANISATION Apparently monogamous: 1 downy chick was accompanied by 2 parents (Rand & Rabor 1960).

SOCIAL AND SEXUAL BEHAVIOUR Nothing recorded; presumably similar to that of Rufous-tailed Bush-hen.

BREEDING AND SURVIVAL Season Siquijor, Feb; Bohol, 1 downy chick a few days old, 3 Apr (egg laid Mar); Mindoro, May; Luzon, Sep. **Nest** Nests in swampy conditions. **Eggs** Broad oval, with little gloss; creamy-white, spotted, streaked and blotched reddish-brown and underlying pale purple; markings more concentrated at larger end; 5 eggs measure 39.6-46 x 29.6-33.1 (42.2 x 31.6), calculated weight 23.3. **Incubation** Unrecorded. **Chicks** Both parents apparently care for the young. No further details available, but presumably similar to Rufous-tailed Bush-hen.

87 RUFOUS-TAILED BUSH-HEN
Amaurornis moluccanus   Plate 25

Porzana moluccana Wallace, 1865, Ambon and Ternate, Moluccas.

Frequently considered conspecific with *A. olivaceus*, but sufficiently distinctive to merit separation; probably best considered as forming superspecies (see *A. olivaceus* species account). Four subspecies recognised.

Synonyms: *Amaurornis/Porzana/Rallina/Erythra moluccana*; *Gallinula olivacea/Frankii/moluccana/yorki*; *Amaurornis olivacea/olivaceus/ruficrissa/ruficrissus*.

Alternative names: Rufous-tailed Moorhen/Crake/Rail; Brown/Rufous-vented Rail/Moorhen; Common Bush-hen.

IDENTIFICATION Length 22-26cm. A medium-sized, rather dark rail with long legs and toes, short stout bill, short tail and upright posture. Very noisy; presence often revealed by distinctive duetting call. Entire upperparts, including top of head to hindneck, upperwing-coverts and tail, olive-brown (paler and less olive in race *ultimus*, darker and browner in *nigrifrons*; rump and uppertail-coverts browner in *ruficrissus*); primary coverts and remiges dark brown with olive-brown outer edges. Lores olive-brown (black in nominate, with whitish mark over lores); ear-coverts grey (tinged brown in nominate); chin and upper throat variably dark grey (nominate) to whitish (*ultimus*, female *nigrifrons* and some *ruficrissus*); lower throat to belly grey (darkest in nominate and *nigrifrons*, the latter having lower belly rufous-buff, not grey, while *ultimus* has lower belly pale vinaceous and nominate has lower belly vinous-isabelline); sides of breast, flanks and sides of belly more olive (browner in nominate); thighs, vent and undertail-coverts rufous-buff (*ruficrissus*), vinous-isabelline with undertail-coverts sandy-buff (nominate), deep vinaceous (*ultimus*), or rufous-buff with undertail-coverts deep tawny (*nigrifrons*). Iris brown-grey (*nigrifrons*), or reddish-brown to dark brown or red. Bill green, becoming paler or yellower basally; culmen dusky; in *ruficrissus*, slightly swollen area at base of culmen reddish to orange-yellow; after breeding, bill colour fades and red patch shrinks and fades to grey-brown; this feature poorly developed in *nigrifrons* and absent in other races. Legs and feet olive, tinged brown or yellow, or golden-yellow (some nominate and *nigrifrons*). Sexes alike; female slightly smaller. Immature very similar to adult, but throat often whitish. Juvenile also similar to adult but throat white; in worn plumage has slightly browner upperparts and olive-brown or pale brown breast and belly; iris dark brown; bill blackish-brown, changing to adult colour; legs and feet olive-grey. Inhabits overgrown margins of freshwater wetlands, swampy grassland, flooded scrub, forest and cultivation.

Similar species Similar to Plain Bush-hen (86) but smaller and paler, with browner (less olive) upperparts, paler grey underparts, thighs vinaceous to rufous, like undertail-coverts, and olive-yellow legs and feet. See also the sympatric, recently discovered, Talaud Bush-hen (88).

VOICE The common territorial call, made by either sex, is a series of 8-12 loud shrieks or cat-like wails, decreasing in loudness, and (in Australian birds) followed by 5-6 muted notes in quick succession; the call is often given as an antiphonal duet, with notes sometimes overlapping. Each series may last 10-12s, with a brief pause before the next, and may be introduced by hoarse, chicken-like clucks which accelerate and become louder before changing to wails or becoming a downslurred querulous chatter (Diamond & Terborgh 1968, Coates & Bishop 1997). In New Guinea, the call commonly starts as 5-15+ short, high-pitched clucks, the first being lower than the rest *hoo hu-hu-hu...* and a second bird may call at the same time, giving *ho* notes of a lower pitch or short pigeon-like *buk-buk-coo-buk...* notes, which suddenly change to upslurred wails *whee*, when the first bird immediately begins to make rasping, downslurred notes to form a duetted *whee-kyou* or *whee-kroow* series, often interspersed with chicken-like notes; sometimes the first bird gives the downslurred note so that the call becomes *koo-kwee* (Coates 1985). Calling birds may walk along in dense, tall grass out of sight of each other. The call is given during the day, especially in overcast weather, and also at night, often for long periods of time and sometimes all night; the birds are vocal all year, particularly in the breeding season. It was regularly heard from 19:00-22:00h in May-Jun on Santa Isabel, and the overlap between the pair's notes resulted in a yodelling sound *woodle-woodle-woodle...*; each duet lasted up to 20s, with an interval of 15-30s before the next duet (Webb 1992).

A persistent and monotonous piping, given while feeding, may last for up to several hours (Marchant & Higgins 1993). Loud to soft clicking calls are given near the nest and as contact calls to the young; this may be the repeated clucking *bk* given by an adult whenever it found food for an accompanying juvenile (Diamond & Terborgh 1968). A soft but far-carrying clucking is sometimes given by several birds in unison but also by single birds, usually starting just after dark, and is also given by incubating birds on the approach of an intruder. Gilliard & LeCroy (1967) describe the bird as very noisy, calling *ka-ka-ka*, pair members calling to each other at dawn. A low, harsh, croaking call is also recorded (Rand 1942b). An adult gave *chuc* calls to a human intruder who caught the chicks and a bird gave hissing and grunting notes when its eggs were being taken (Marchant & Higgins 1993); a continuous *chat chat chat* call is possibly an alarm (Coates 1985). Chicks give a soft cheeping whistle or a piping note, and may also squeak.

DESCRIPTION *A. m. ruficrissus*
Adult Top of head to hindneck, and upperparts to back, olive-brown; rump and uppertail-coverts often browner; rectrices olive-brown, inner webs darker brown. Most upperwing-coverts olive-brown; greater secondary coverts have dark brown inner webs and olive-brown tips; alula and primary coverts dark brown with narrow dark olive tips and outer edges; remiges dark brown, with olive-brown outer edges, outer webs of tertials olive-brown. Axillaries olive; underwing-coverts dark brownish-grey with pale tips (whitish on greater primary coverts, brown to olive-brown on others); underside of remiges dark brownish-grey. Lores olive-brown; ear-coverts grey; chin and upper throat variably grey to whitish; Lower throat, foreneck, sides of neck, breast and belly grey, paler when fresh; sides of breast, flanks and sides of belly more olive; thighs, vent and undertail-coverts rufous-buff, this colour sometimes extending along mid-belly and grading to pinkish-buff. Iris dark reddish-brown to dark brown or red. Bill green, becoming pale green or olive-yellow at base; smooth, slightly swollen area at base of culmen reddish, orange or orange-yellow; after breeding, green of bill fades, and red patch on culmen shrinks in size and fades to grey-brown. Legs and feet olive-yellow to olive, sometimes tinged brown. Sexes alike but female slightly smaller.
Immature Very similar to adult, but some retain juvenile white feathers on chin and throat; relatively narrow, pointed and worn juvenile primaries also diagnostic.
Juvenile In Australian birds, similar to adult but with white chin and upper throat; in fresh plumage foreneck and breast have stronger olive-brown tinge; in worn plumage, top of head, hindneck, upperparts and secondary coverts olive-brown, slightly browner than adult, and entire breast and belly olive-brown or light brown. Iris dark brown; bill at first blackish-brown, later developing adult colour in patches; legs and feet olive-grey.
Downy young Downy chick black; possibly fades later to black-brown on body and wings; iris dark brown; bill grey-black, tip grey-buff to pink-buff, on lower mandible buff extending over gonys, small white egg-tooth present; legs and feet dark grey to blackish.

MEASUREMENTS (Australian birds) Wing of 12 males 145-159 (152, SD 3.7), of 6 females 133-150 (143.8, SD 5.6); tail of 11 males 61-71 (64.5, SD 2.7), of 6 females 52-63 (58, SD 3.7); exposed culmen of 12 males 30-35 (31.5, SD 1.4), of 6 females 27.5-29.5 (28.9, SD 0.8); tarsus of 14 males 49.5-55 (51.7, SD 1.5), of 6 females 44.5-50 (48.3, SD 1.7). Weight of 2 New Guinea females 175, 180 (Diamond 1972b); of 9 Australian males 129-208 (177, SD 26.4), of 1 probable female 161, of 3 immature females 131, 143, 163 (Marchant & Higgins 1993).

GEOGRAPHICAL VARIATION Races separated on bill colour, colour of supraloral area and undertail-coverts, and shade of upperparts and underparts. Nominate race includes *frankii* of N and W New Guinea (see Mayr 1938, Marchant & Higgins 1993); *ruficrissus* includes putative *yorki* (Ripley 1977).

A. m. moluccanus (Wallace, 1865) – Sulawesi (Sangihe and Siau); Talaud Is (Karakelong); Moluccas (Ambon, Bacan, Buru, Halmahera, Mangole, Obi, Seram, Taliabu, Ternate and Tidore), coastal N New Guinea, and Misool, Biak, Karkar and Long Is. Also recently recorded from Kai Is (Kai Kecil) and Tanimbar Is (Yamdena), subspecies uncertain (Coates & Bishop 1997). Differs from *ruficrissus* in darker underparts, duller, less rufous undertail-coverts, and no red at base of culmen. Lores black with whitish supraloral streak; ear-coverts grey, tinged brown; malar region, chin and throat to upper belly slaty-grey; sides of breast washed olive-brown; flanks clearer brown; lower belly, vent and thighs vinous-isabelline; undertail-coverts and sides of vent more sandy-buff. Iris reddish-brown to rusty-brown; bill greenish-yellow (male) or green (female), culmen more brownish; legs and feet yellowish-tan to golden-yellow. Females generally much paler on lower belly and undertail-coverts (Mayr 1938). Birds from E New Guinea darker than those from Arfak Mts in NW, and those from Moluccas; forehead and lores blackish; upperparts more olive-grey, less brownish (Mayr 1949). Immature duller than adult; throat whitish. Wing of 9 males 129-157 (144.7, SD 10.3), of 6 females 135-143 (139.2, SD 3.5); tail of 4 males 54-60 (56, SD 2.7), of 3 females 45, 51, 54; culmen to base of 7 males 31-36 (33.7, SD 1.8), of 7 females 29-33 (31.3, SD 0.6); tarsus of 7 males 48-58 (51.4, SD 3.3), of 7 females 47-51 (48.6, SD 1.4). Weight of 2 males 182, 205, of 2 females 200, 205.

A. m. nigrifrons (Hartert, 1926) – Bismarck Archipelago (Tolokiwa, Umboi, Sakar, New Britain, Watom, Witu, Duke of York, New Ireland, New Hanover, Lihir, Tabar and Tanga) (Coates 1985) E to Solomon Is (Bougainville and New Georgia). Smaller than *ruficrissus*; culmen narrow and blackish, with less developed swollen base; very dark on upperparts and underparts; upperparts suffused with brown; breast deep bluish-grey; throat grey in males, sometimes whitish in female; lower belly rufous-buff with no grey; undertail-coverts deep tawny or vinaceous. Iris brown to brown-grey; legs and feet dull gold to dull mustard-yellow. Birds from Bougainville are intermediate in colour between those from Witu and Gower Is (Mayr 1949). Wing of 20 males 130-146 (139.2, SD 4.5), of female 126-142 (mean for New Britain 132.3, elsewhere 133); tail of 1 male 47, of 2 females 48, 53 (BMNH); bill of male 29-33 (mean for Witu 31.7, New Britain 30.2, elsewhere 31.5), of female 25-30 (mean for New Britain 28, elsewhere 28.8); tarsus of male 54-58 (mean for Witu 56.2, New Britain 55.9, elsewhere 57), of female 49-54 (mean for New Britain 51.5, elsewhere 51.2) (Mayr 1949). Weight of 1 male 179, 1 female 177 (Gilliard & LeCroy 1967).

A. m. ultimus Mayr, 1949 – E Solomon Is, from Malaita (Gower), San Cristobal and Santa Ana. Smaller than *nigrifrons*; tarsus relatively long; overall colour paler. Top of head and upperparts rather brownish, without olive tint; grey of underparts pale and ashy; throat whitish in both sexes; middle of lower belly pale vinaceous without grey; vent and undertail-coverts deep vinaceous, not rufous. For inter-island plumage variation, see Mayr (1949). Bill green, culmen blackish. Wing of 11 males 124-134 (129.5, SD 3.0), of female 121-132 (125.5); bill of male 28-32 (30.4), of female 25-30 (28.3); tarsus of male 50-55 (53), of female 47-52.5 (50.5). Weight of 6 males 189-248 (212, SD 21.8), of 3 females 135, 162, 192.

A. m. ruficrissus (Gould, 1869) – S and E New Guinea, and NE and E Australia in Western Australia (Kimberley Division), Northern Territory (coastal areas from Daly R E to Cape Arnhem), and Queensland (Cape York) to extreme NE New South Wales (Clarence R). Birds recently discovered on Misima I, Louisiade Archipelago (Tolhurst 1996), may be of this race. See Description.

MOULT Postbreeding moult is complete, and remex moult is simultaneous. Remex moult is recorded in New Britain, Sep; in Australia, many adults moult in late summer but some perhaps in autumn. A partial pre-alternate (prebreeding) moult may occur. Partial postjuvenile moult is recorded in Australia, Mar-Apr; juvenile remiges are replaced in the complete moult of the second summer (second pre-basic).

DISTRIBUTION AND STATUS Sulawesi, Moluccas, W Papuan Is, New Guinea, Louisiade Archipelago (Misima I), Bismarck Archipelago, Solomon Is and NE Australia. This species was formerly regarded as common and widespread in New Guinea, and Gregory (1995) found it common in the Tabubil area in 1991-94; it is moderately common in Wallacea (Coates & Bishop 1997), and common on Santa Isabel, Solomon Is (Webb 1992). Only a few records were obtained during the Asian Waterfowl Census, 1987-91, all from Papua New Guinea (Perennou *et al.* 1994). There are no current accurate estimates of its abundance in Australia, where it is apparently local and uncommon to reasonably common (Marchant & Higgins 1993). It may be widespread along the coast of Northern Territory and Queensland, ranging along the E coast to Rockhampton, from where there may be a gap in its distribution S to 26°S before it occurs again S to Brisbane. It may have undergone a range expansion in Northern Territory from 1977, but it is just as likely to have been overlooked previously. A recent study (Muranyi & Baverstock 1996) has shown it to be more widespread in NE New South Wales than was previously thought; it occurs from the Queensland border S to Evans Head, and may range S to Grafton, where it was found breeding in the 19th century. Its occurrence in Western Australia was confirmed in 1985, after 10 unconfirmed sightings from 1979 to 1983, but it was not recorded during the Australian Atlas (Marchant & Higgins 1993). In Australia it is obviously well able to exploit man-modified habitats, especially on farmland, at artificial wetlands and in gardens, but it can survive clearing only where small patches of scrubby vegetation remain on swampy ground. No information is available on predation but birds are occasionally killed on roads in Australia.

MOVEMENTS Information comes from Marchant & Higgins (1993) unless otherwise stated. It is recorded throughout its range (including in Australia) in all seasons and is considered resident over most of its range. However, it is possibly nomadic in Australia, moving into seasonally wet habitats in some regions with the onset of the rains and retiring to areas with permanent water in the dry season. Its arrival is reported in some areas after rain and in one area it left when rain fell further N. It is thought to cross the Torres Strait between Australia and New Guinea regularly, records from Booby I being for Nov-Jan and May-Jun (Stokes 1983); arrivals are reported at Cape York in

Rufous-tailed Bush-hen

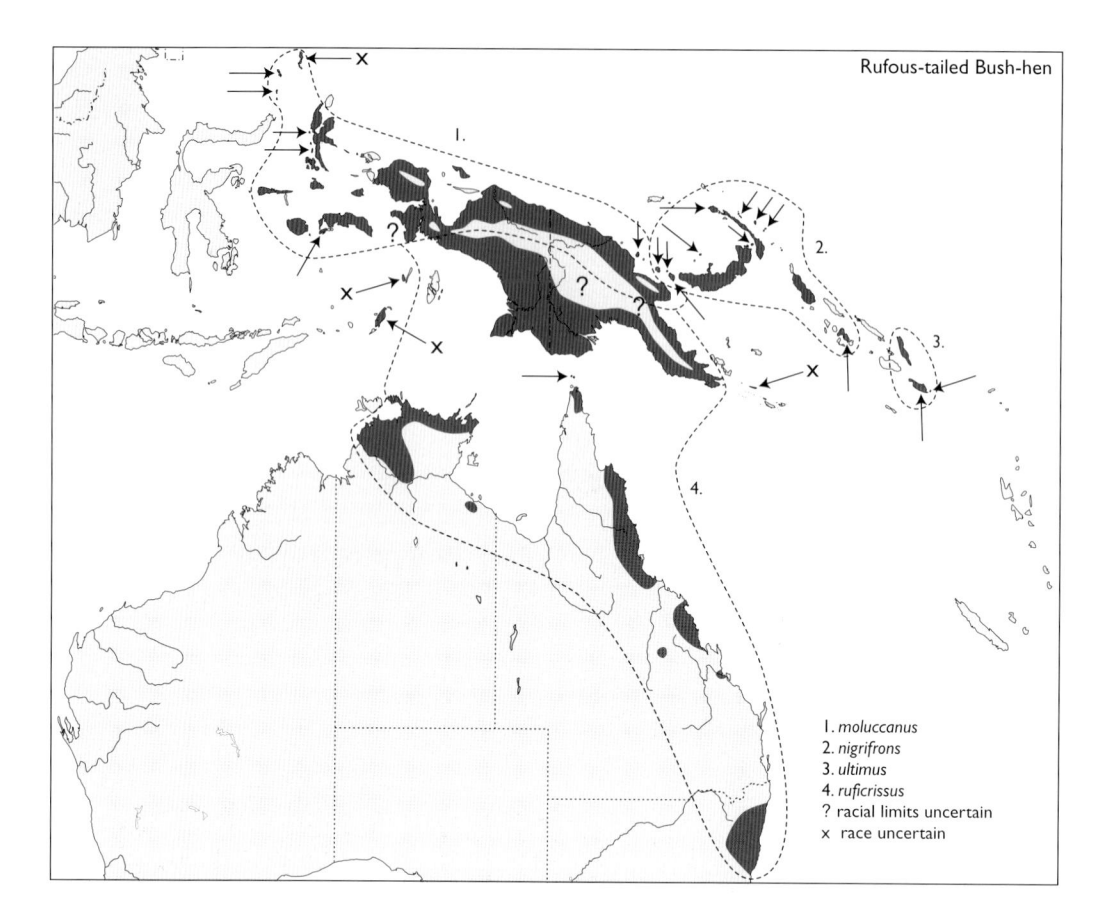

1. *moluccanus*
2. *nigrifrons*
3. *ultimus*
4. *ruficrissus*
? racial limits uncertain
x race uncertain

late Jan, and birds have been noted flying at night S from the Cape York region at the onset of the wet season. It may make local movements when habitat becomes temporarily unsuitable through damage. Changes in calling patterns may explain lower reporting rates in some parts of Australia during the winter. It is a vagrant to W Micronesia (Palau), where an immature was found dead in the sea, May 1979 (Pratt *et al.* 1987).

HABITAT Dense stands of grass, reeds and forbs, sometimes bamboo, with pockets of dense bushes, *Pandanus* palms or stands of bananas, near or bordering still or flowing fresh water including rivers, streams, lakes, ponds, dams, billabongs, waterholes and roadside ditches. It also occurs in flooded scrub and swampy grasslands, grass bordering swamp forest, dry to swampy rainforest (often near marshes and lagoons or in remnant patches or secondary growth), and sometimes in open forest or woodland, farmland adjacent to forest, grass bordering roads through rainforest, *Lantana camara* thickets, sugarcane, and pasture and cleared land far from water. It is even recorded near human habitations, such as villages in forest clearings, and suburban gardens. On islands in the Torres Strait it occurs in areas bordering mangroves. On Karakelong (Talaud Is) it occurs alongside the Talaud Bush-hen (88) in open, scrubby habitats (Lambert 1998b). It usually inhabits the tallest and densest available vegetation, up to 2-4m tall. It is found primarily in lowlands, but occurs up to 1,500m in New Guinea.

FOOD AND FEEDING Oligochaeta, Orthoptera (Acrididae and Tettigoniidae), Coleoptera, Lepidoptera (adults and larvae), and occasionally frogs; also vegetable matter, including seeds of grasses (Poaceae) and green shoots. It forages within or at the edges of wetland vegetation, in cane fields and short pasture, on ground in forest, at roadsides, and in shallow water, and occasionally while swimming. It gleans from the substrate and from low vegetation, probes into leaf-litter, and reaches or leaps up to take seeds or to capture insects on vegetation or in the air; it runs grass heads through the bill to remove seeds. It crushes items such as frogs in the bill before swallowing them.

HABITS This crake is diurnal and crepuscular, but is also claimed to be active mainly at night (Marchant & Higgins 1993), when it calls for long periods. It is shy and secretive, normally keeping to dense cover, and is frequently heard but seldom seen. However it often frequents roadside vegetation and is sometimes seen crossing roads and tracks or walking in the open at roadsides very early in the day (Coates 1985), and sometimes it may be observed easily during wet, overcast weather (Czechura 1983). It is extremely difficult to flush and seldom flies if it can run for cover. Its normal gait is a slow, stalking walk and when alert or suspicious it adopts a very upright posture, often with the short tail held erect and sometimes flicked upwards in a moorhen-like manner. It often wades, and it swims occasionally but seldom walks on floating vegetation despite having long toes; when in the open along streams it keeps within the shade at the stream edges. It climbs readily in bushes and low trees, and pairs usually roost in bushes but sometimes among dense grass (Ripley 1964, Clarke 1975, Beruldsen 1976). Birds sunbathe on a perch with the wings partly spread. Bathing occurs daily; one pair bathed most days after 17:00h, using a bird bath in full view of a house and tolerating the presence of observers at a distance of 2.5m (Clarke 1975); birds stand thigh-deep in water, plunge the head and shoulders into the water, throw the head back, fluff the feathers and shake; the wings are then extended and slowly moved up and down (Marchant & Higgins 1993).

SOCIAL ORGANISATION These birds are normally seen in pairs or family groups. It is probably monogamous, and is territorial at least during the breeding season. Of 9 territories near streams, birds usually remained within 25m of the stream but wandered up to 100m (Beruldsen 1976); in NE New South Wales, birds were recorded up to 500m from water, usually within 100m (Muranyi & Baverstock 1996).

SOCIAL AND SEXUAL BEHAVIOUR All information is taken from Marchant & Higgins (1993). The shrieking call is given in response to loud noises, while breeding birds may call and search for conspecifics which utter shrieking calls. The incubating bird will attempt to distract an intruder with a display which includes repeated clucking calls and loud fluttering actions (Clarke 1975). Alternatively it may sit tight and peck at the intruder; as incubation progresses, the adults become more aggressive and may fly at an intruder. When defending chicks in the nest, the adult raises the wings, gives hisses and cat-like and guttural growling sounds, stamps the feet and makes short lunges with flapping wings. A broken-wing display is also recorded, the bird dragging the wings and spreading and depressing the tail. An adult with young moved slowly backwards and forwards, repeatedly flicking the wings, while the young and the other parent ran into cover.

BREEDING AND SURVIVAL Season Moluccas (Bacan), Sep; New Guinea, in dry season in E highlands with chick early Jul and adults with enlarged gonads Jul-Aug (Diamond 1972), in wet season (chicks Feb and May) and dry season (adults with 2 juveniles, Sep, after unseasonal rain in Jun and Sep) in Port Moresby lowlands (Coates 1985, Tolhurst 1992); Misima I, Aug (Tolhurst 1996); New Britain, female almost ready to lay, early Sep; Australia, Oct-Apr. Almost all following information from Australian birds. **Nest** Nests usually near or over water; often in bushy vegetation; also near forest edge or in secondary growth; often adjacent to tree trunk, vines, stump or fence post. Nest bowl-shaped and fairly substantial, of grass, sometimes with dry leaves, lined with dry grass or twigs; material added during incubation; built in tall, dense grass, sometimes in reeds or weeds, or on branches of *Lantana*. Grass stems bent down to form base and sides and loosely interwoven into rough canopy; ramp of trampled vegetation often present. External diameter 20-30cm, depth 10-13cm; cavity diameter 10-15cm, depth 4-6.5cm; built 10-200 (76, SD 51, n = 21)cm above ground or water. Roosting nests also constructed; few have canopy. **Eggs** 4-7 (5.8, SD 1.3, n = 12 clutches, Australia); 1 clutch of 13 probably from 2 females; first clutch usually 6-7 eggs, second usually 4-5 (Beruldsen 1976). Eggs broad oval to elliptical; moderately lustrous; white or creamy-white to pale pinkish-cinnamon, irregularly spotted and blotched cinnamon, purplish-red or purplish-grey, with underlying markings of pale purple to violet-grey; markings often concentrated at larger end. Size of eggs: *ruficrissus* 36.3-41.9 x 26.7-29.7 (39.5 x 28.3, SD 1.5, n = 23), calculated weight 16.9; *nigrifrons* 35-43 x 27-31 (38.7 x 29, n = 34), calculated weight 18. Laying interval 1-2 days. Birds desert readily if nest found before clutch complete. **Incubation** At 1 nest 25 days (18 days from completion of clutch); both parents incubate, possibly mostly female until last few days; hatching asynchronous. **Chicks** Semi-precocial(?); fed and cared for by both parents; usually fed in cover. No information on growth and development. Hatching success (Australia) 32% (17 eggs hatched from 52 laid), causes of loss including predation, flooding and desertion; 3 other clutches of 7 all hatched successfully (Marchant & Higgins 1993). May nest twice per season (Australia): re-lays 14-21 days after clutch loss; 1 pair made 4 nesting attempts in 10 weeks (Beruldsen 1976).

88 TALAUD BUSH-HEN
Amaurornis magnirostris Plate 25

Amaurornis magnirostris Lambert, 1998, Karakelong I, Talaud Archipelago.

Monotypic.

IDENTIFICATION Length 30.5cm. A relatively large, robust and very dark bush-hen, with large head and strikingly robust bill. Head dark brown; upperparts and upperwings very dark rich brown, slightly more rufous-tinged on lower back and rump; entire underparts very dark grey, becoming richer dark brown (like rump) on flanks and thighs. Iris red; bill long and broad, with arched culmen, pale green with pale blue-green tip to lower mandible and dusky basal part of culmen; legs dark olive-brown, yellow at front. Immature and juvenile unknown. Inhabits dry to swampy primary forest; also scrubby habitat at forest edge and along streams.
Similar species Differs from other bush-hens in having darker underparts and lacking contrastingly pale undertail-coverts, and in its larger head and longer, broader bill with distinctly arched culmen. It is primarily a forest species, whereas other species typically inhabit scrub, grass, forest edges and cultivation. Rufous-tailed Bush-hen (87) is noticeably smaller and paler, especially on underparts, with undertail-coverts (in nominate race, Talaud Is) sandy-buff; bill green with yellow base and dusky culmen; legs and feet olive. Smaller allopatric Plain Bush-hen (86) from Philippines also dark, but undertail-coverts contrastingly dark rufous-brown; bill pale green, shorter and more tapered; legs and feet yellow-brown and less heavy. Allopatric Isabelline Bush-hen (85) somewhat larger, but with smaller head and bill, and longer, more slender legs; plumage noticeably paler and brighter, with entire underparts cinnamon-rufous.

VOICE The only call definitely attributed to this species by Lambert (1998b) is a monotonous series of very loud, rather frog-like notes, best described as low-pitched barks. Although significantly different from the monotonous piping call of the Rufous-tailed Bush-hen, it is similar in pattern. Playback of this call elicited an immediate response from calling birds, which investigated the source of the sound. It is not clear whether this species also makes the distinctive growling shrieks typical of other bush-hens; however, quiet shrieks, best described as croaking with rasping, on a tape-recording of its barking call may have been made by this species and are quite similar to the shrieking call of the Plain Bush-hen (Lambert 1998b).

DESCRIPTION From details in Lambert (1998b), who uses colour names (starting in uppercase) from Smithe (1975). Tracts for which two colours are given apparently contain a mixture of fresh and worn feathers; although the relationship between colour and feather age is sometimes not specified, presumably the paler colour refers to the old, abraded feathers.
Adult Described from the holotype (unsexed). Forehead to nape Dusky Brown, some feathers tinged paler; mantle Fuscous, Cinnamon-brown, becoming slightly more olive on back; lower back and rump Burnt Umber; upperwing-coverts as mantle but with patches of Cinnamon-Brown (mostly on worn feathers) more pronounced on greaters and medians; primaries and secondaries Fuscous with diffuse Cinnamon-Brown leading edges; underwing-coverts dark brown; rectrices missing. Lores and orbital region sooty Dusky Brown; chin to neck Dark Neutral Gray; breast and belly patchy Dark Neutral Gray and Fuscous, becoming more Burnt Umber on flanks and thighs; undertail-coverts Dark Neutral Gray, as belly. Iris bright red; bill long and broad, with arched culmen; lower mandible pale green but distal third pale blue-green; upper mandible pale green but culmen dark olive from base to beyond nares; legs dark olive-brown, front of tibiotarsus bright yellow; feet not described, presumably dark olive-brown.
Immature Unknown.
Juvenile Unknown.
Downy young Unknown.

MEASUREMENTS From holotype (unsexed): wing 168; bill (gape to tip) 45; tarsus 62 (Lambert 1998b).

GEOGRAPHICAL VARIATION None.

MOULT The holotype, collected in Sep, has very worn plumage and is in heavy moult (presumed postbreeding) of body, wings and tail (Lambert 1998b).

DISTRIBUTION AND STATUS All information is from Lambert (1998b). This species is recorded only from Karakelong I in the Talaud Archipelago. It is likely that it must also have occurred on neighbouring islands but whether it survives on them (particularly on Salebabu and Kabaruang) is not known. Evidence suggests that it is widespread on Karakelong and it can probably tolerate limited habitat modification. Bush-hens are occasionally trapped for food using snares, or are caught by dogs, but the present levels of hunting and trapping are unlikely to constitute a significant threat. At present this species is probably not threatened by habitat loss because Karakelong is the largest island (c. 600km²) in the group, has a diversity and abundance of wetland habitats, including forested swamps and rank scrub and grass bordering forest, and still has good areas of forest from near sea level to the highest peak. The forests are still relatively intact, and there are 2 protected areas totalling c. 21,800ha but there is virtually no provision to enforce this protection, and both areas are under pressure from smallholder encroachment, while there may be a long-term threat of forest clearance for transmigration. Because its exact ecological requirements are unknown, and because of its potential vulnerability to introduced mammalian predators, Lambert suggests that the species be considered NEAR-THREATENED.

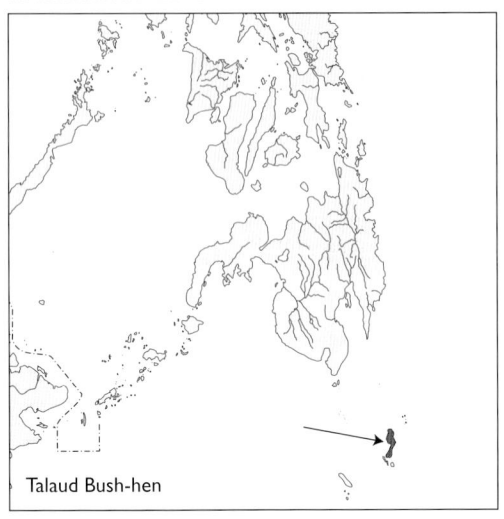

Talaud Bush-hen

MOVEMENTS Unknown; presumably sedentary.

HABITAT All information is from Lambert (1998b). This species has been recorded from the following habitats: undisturbed tall primary forest on steeply sloping ground, with an understorey dominated by rattan; similar forest on flat ground with scattered very swampy patches and a relatively sparse understorey; wet scrubby vegetation with permanent water at forest edge; a narrow strip of littoral swamp forest bordered by coconut plantation; and at the edge of a small stream covered with scrub and patches of trees, and bordering more open dense herbage and grassland. Sightings in various habitats at different altitudes suggest that it is an adaptable species and can probably tolerate limited habitat modification. It occurs alongside the Rufous-tailed Bush-hen in open, scrubby habitats but is probably more of a forest bird which wanders into adjacent scrub, rank vegetation and overgrown plantations. It ranges from near sea level up to at least 400m altitude.

FOOD AND FEEDING Unknown.

HABITS Said to be exceedingly shy but very inquisitive, approaching an observer closely in cover and running away at the slightest alarm (Lambert 1998b).

SOCIAL ORGANISATION Sightings have been of single birds, and once of a pair (Lambert 1998b).

SOCIAL AND SEXUAL BEHAVIOUR Unknown.

BREEDING AND SURVIVAL Unknown; apparent post-breeding moult recorded Aug.

89 WHITE-BREASTED WATERHEN
Amaurornis phoenicurus   Plate 26

Gallinula phoenicurus Pennant, 1769, Sri Lanka.

Races *javanica* and *chinensis* included within nominate due to overlapping measurements (Ripley 1977), but may well be valid (e.g. Medway & Wells 1976). Four subspecies recognised.

Synonyms: *Fulica chinensis*; *Gallinula Javanica/leucomelana/ leucomelaena/erythrura/phoenicura*; *Amaurornis phoenicura/ insularis/leucomelaena*; *Rallus sumatranus/phoenicurus*; *Zapornia thermophila*; *Porzana/Porphyrio/Erythra/Rallina/ Erythrura/Pisynolimnas phoenicura*; *Tribonyx erythrura*; *Erythra leucomelana/major/chinensis/ javanica*.

Alternative name: White-breasted Swamphen.

IDENTIFICATION Length 28-33cm; wingspan 45-54cm. A medium-sized species, plumage distinctive. Top of head, and upperparts to upper back, dark slate-grey, ashier on head, neck and upperwing-coverts; lower back to uppertail-coverts dull brown; primaries, secondaries and tail dark brown; wings have narrow white leading edge. Sides of head, and underparts to upper belly, white, separated from slaty-grey upperparts by blackish band (underparts more olive in *midnicobaricus*); upper flanks dark slaty-grey (more extensive in *leucomelanus*, blackish in *insularis*); lower belly and thighs pale cinnamon; lower flanks and undertail-coverts deep tawny (darker in *insularis*), colour obvious because bird often raises and flicks tail. Head pattern variable: in nominate, slaty-grey extends forwards to forecrown, while forehead and entire face, including supercilium, are white; in *leucomelanus*, grey usually extends to forehead and, in birds from Lesser Sundas, to ear-coverts and lores; in *midnicobaricus*, grey reaches only half-way over top of head; in *insularis*, grey does not extend forward to level of eyes; mixed characters occur in birds from Sulawesi, Greater Sundas and Lesser Sundas (see Geographical Variation). Iris red; bill sage-green, lower mandible yellower, base of upper mandible swollen and reddish; legs and feet greenish-yellow to orange-yellow. In non-breeding season, male's bill becomes olive. Sexes similar but female smaller, bill like non-breeding male. Some birds have blackish sides of breast strongly barred olive. Partial albinism reported in some areas, e.g. Maldives, Andamans, Nicobars, Moluccas and Sulawesi. Immature duller than adult: upperparts more olive-brown; white sides of head mottled brown; underparts tinged brownish; rufous areas duller; bill darker and duller. Juvenile similar but white of underparts, especially chin, throat and sides of breast, more heavily tinged dull brown; bill dark; legs and feet brownish-yellow. Inhabits marshes, swamps, moist to dry grass, vegetated margins of flowing and still water, mangroves, forest and scrub.

Similar species Unmistakable by virtue of the combination of dark slate-grey upperparts and flanks, prominent white facial mask and underparts, and rufous rear flanks, vent and undertail-coverts.

VOICE The characteristic and very loud advertising call comprises a variety of roars, grunts, cackles, croaks and chuckles, followed by a monotonous *kru-ak, kru-ak, kru-ak-a-wak-wak*. The birds call throughout the night, in the early morning and in the evening, and also on cloudy overcast days, and calls may continue for 15 min or more without a break; male and female often call together. Calls are often given while the bird is partially hidden on a perch in the tops of reeds, shrubs or bamboo clumps. Birds are very vocal during the breeding season but silent at other times; in Japan, calls are heard from mid-Mar to the end of May (Brazil 1991), and in Pakistan from early Feb to Oct (Roberts 1991). The volume of sound is astonishing for the size of the bird, and the roars and cackles are said to be "more appropriate to a bear being roasted over a slow fire" (Ali & Ripley 1980). However in Sulawesi the call is described as incorporating bubbles, nasal screams and squeals, but not roars, and is said to be more musical and less harsh than that of the Rufous-tailed Bush-hen (87) (Coates & Bishop 1997). A single note *kuk* is sometimes repeated monotonously for several minutes, even after dusk (Roberts 1991). Captive birds gave a sharp, squeaky note, apparently a contact call (Glenister 1951).

DESCRIPTION *A. p. phoenicurus*
Adult Forecrown to hindneck dark ashy; upperparts to upper back, including scapulars and tertials, dark slaty-grey, with slight olive tinge in museum skins which may not occur in live birds (Abdulali 1978); lower back to uppertail-coverts dull dark brown; tail blackish-brown. Upperwing-coverts dark ashy-grey. Primaries and secondaries dusky brown, ashier on outer edges and tips; marginal coverts and outer web of outermost primary white, this colour extending a little to edge of primary coverts; axillaries and underwing-coverts dark slate with narrow whitish tips. Forehead, sides of head (including supercilium), sides of neck, and underside from chin to upper belly, white, separated from slaty-grey of upperparts by blackish band running from rear of ear-coverts to sides of body; sides of breast black, feathers white with inner webs; upper flanks dark slaty-grey; sides of lower back, lower flanks and undertail-coverts deep tawny; lower belly and thighs pale cinnamon. Some birds have blackish sides of breast strongly barred olive. Iris crimson to reddish-brown or raw-umber; bill sage-green, tip and lower mandible paler and yellower, base of upper mandible swollen and red; legs and feet dull chrome-yellow to yellowish-green. Sexes alike but female slightly smaller, bill olive, washed brown on upper mandible. Bill of non-breeding male as female.

Immature Duller than adult: upperparts more olive-brown, white of face obscured by brown feather tips; underparts tinged brownish; rufous areas duller; iris brown; bill darker and duller than in adult.

Juvenile Similar to immature but white of underparts, especially chin, throat and sides of breast, more heavily tinged dull brown; bill dark; legs and feet brownish-yellow.

Downy young Downy chick black, underparts more blackish-brown; iris pale grey; bill dusky or blackish, bill tip paler, greyish-white or horn-coloured; legs and feet dusky brown.

MEASUREMENTS (12 males, 12 females) Wing of male 155-179 (168.4, SD 6.5), of female 145-165 (156.7, SD 6.6); tail of male 60-73 (66.15, SD 3.9), of female 55-68 (60.8, SD 4.1); culmen (including shield) of male 34-44 (39.3, SD 2.9), of female 34-39 (35.8, SD 1.7); tarsus of male 49-57 (53.3, SD 2.7), of female 47-57 (51.7, SD 2.6). (Java) Wing of male 139-158, of female 136-144; (Sulawesi) wing of male 143-159, of female 137-144 (White & Bruce 1986). Weight of male 203-328 (228.5), of female 166-225 (196.5), of 10 unsexed 152-230 (188.7).

GEOGRAPHICAL VARIATION Races separated on extent of slate-grey on head, extent and shade of grey on flanks, and colour of underparts. Partial albinism seems to occur in some areas, such as the Maldives, Andamans, Nicobars, Moluccas and Sulawesi (van Bemmel & Voous 1951, Ripley 1982). Much individual variation occurs in both size and plumage pattern, and unstable mixed populations exist (see nominate and *leucomelanus*). Because of this variation, race *javanica* is included within nominate and *variabilis* in *leucomelanus* (White & Bruce 1986). Races *chinensis*, *cleptea* and *maldivus* are also included in nominate, and *leucocephalus* in *insularis* (Ripley 1977).

A. p. phoenicurus (Pennant, 1769) – Pakistan, India, Nepal, Bangladesh and Burma; C and E China along the Changjiang R and tributaries from Sichuan and Guizhou E to Shanghai, N to S Henan and S Shaanxi, and S through Fujian, Guangdong, Guangzi Zhuang, and to SE and W Yunnan. Also Hainan, Xisha, Taiwan and Ryukyu Is, and S to Maldives, Laccadives (probably), Sri Lanka, Thailand, Malaysia (including islands offshore of Malay Peninsula), Vietnam, Laos, Cambodia, Philippines (Basilan, Batan, Biliran, Bohol, Bongao, Cagayan, Calamianes, Catanduanes, Cebu, Culion, Dinagat, Leyte, Luzon, Marinduque, Mindanao, Mindoro, Negros, Palawan, Panay, Siargao, Siquijor, and Sulu and Tawitawi Is) to N Sulawesi (Talaud Is, Sangihe and Tahulandang), Sumatra (throughout the mainland, and also Riau and Lingga Is, Bangka, Belitung, Simeulue, Nias, Batu and Mentawai Is, and Enggano), Borneo, Sarawak, Brunei, Sabah, Natuna Is, Java (including Tinjil and Deli Is), Kangean Is and Bali. N populations winter to S, including to S Thailand, peninsular Malaysia, Singapore, N Laos and Sumatra and ranging W to Arabia. Currently expanding N into Japan (see Distribution and Status). See Description. Unstable mixed populations exist in the Greater Sundas and these populations have been separated as race *javanica*, perhaps unnecessarily (White & Bruce 1986). Siebers (1930) found that birds from Java and Sumatra had P1 = P9 whereas birds from Wallacea had P1 = P7/P8; however Ripley (1977) could find no constant difference in wing formula. See *A. p. leucomelanus* for discussion of racial affinities of N Sulawesi birds.

A. p. insularis Sharpe, 1894 – Andaman and Nicobar Is. Larger than nominate; legs and bill stouter; more white on forehead (usually extending back beyond eyes); undertail-coverts darker, more chestnut; flanks blackish. Wing of male 165-176 (168.7), of 10 females 152-170 (159, SD 4.9); tail of male 70-75, of 10 females 53-72 (61.6, SD 5.3); bill (including shield) of male 40-43 (41), of 10 females 34-43 (37.9, SD 2.4); tarsus of 2 males 50, 59, of 10 females 49-60 (52.8, SD 3.8) (Ripley 1977, sample sizes and means not given; BMNH).

A. p. midnicobaricus Abdulali, 1978 – C Nicobar Is. Slate-grey extends forwards only half-way over top of head; underparts more olive, and bill heavier, than in *leucophalus* (=*insularis*). Validity of this race questionable in view of partial albinism known to occur in region (White & Bruce 1986). Both sexes: wing 157-171 (Abdulali 1978).

A. p. leucomelanus (S. Müller, 1842) – Sulawesi (Sulawesi, Muna [?], Butung, Tukangbesi, Salayar, Tanahjampea and Kalaotoa), Moluccas on Sula Is (Taliabu and Mangole), Buru and Seram, and Lesser Sundas (Lombok, Sumbawa, Flores, Palu (=Paloe), Sumba, Alor, Roti, Timor, Wetar, Roma (=Romang) and Damar). Slate-grey on head usually extends to forehead (base of shield) and, in birds from Lesser Sundas, over ear-coverts and lores; grey on flanks more extensive than in other races, white being reduced to a median area. Birds from Sulawesi, Butung, Sumbawa and Flores very variable: some have face white with slate-grey mottling; forehead sometimes slate-grey, sometimes white (White & Bruce 1986). Stresemann (1936) assigned these birds to race *variabilis*, but this is untenable for unstable populations affected by introgression. Birds from N Sulawesi (Talaud and Sangihe) are included in nominate by Ripley (1977); close affinity of many Talaud and S Philippine bird species supports this arrangement, but Sangihe population may be closer to other Sulawesi forms and further study of Talaud and Sangihe material needed (White & Bruce 1986). Measurements of 5 males, 8 females: wing of male 150-167 (156, SD 6.6), of female 141-150 (146.4, SD 3.2); tail of male 60-65 (62.8, SD 2.3), of female 49-61 (55.9, SD 3.9); bill (including shield) of male 37-39 (37.8, SD 1.1), of female 33-36 (34.2, SD 1.1); tarsus of male 49-57 (53.6, SD 3.4), of female 47-52 (48.6, SD 1.7). Weight of 3 males 212, 214, 276, of 1 female 168.

MOULT No information available.

DISTRIBUTION AND STATUS S and E Asia, from Pakistan and India E to E China and Japan, S to Sri Lanka, the Maldive, Andaman and Nicobar Is and Malaysia, and through the Philippines, Sulawesi, the Moluccas and the Greater and Lesser Sundas. It has been known from Seram (Moluccas) only since 1987 (Bowler & Taylor 1989). N populations winter to the S, ranging W to Arabia and S to Sumatra (van Marle & Voous 1988). This species was formerly regarded as common throughout its range. More recently, it has been regarded as locally frequent to common in Pakistan (Roberts 1991), local in Bangladesh (Harvey 1990), very common in Sri Lanka (Kotagama & Fernando 1994) and Thailand (Lekagul & Round 1991), common in Burma, Vietnam, Wallacea, Borneo, Sumatra, Java and Bali (Smythies 1981, 1986, Watling 1983, Stusák & Vo Quy 1986, van Marle & Voous 1988, Rozendaal & Dekker 1989, MacKinnon & Phillipps 1993, Coates & Bishop 1997), common over its range in China in summer (Cheng 1987), and fairly common but local in the Philippines (Dickinson *et al.* 1991). Many records of this species were obtained during the Asian Waterfowl Census, 1987-91, from Pakistan, India, Nepal and Sri Lanka E to China and Taiwan, and SE through Burma and Thailand to Vietnam and Malaysia (including Brunei and Sabah); however, no population estimates could be made and the only large concentration found was at Kaziranga National Park, India (330 birds in 1 year) (Perennou *et al.* 1994).

It has possibly recently colonised Seram (Coates & Bishop 1997). In Japan (Brazil 1991) it was accidental before 1970, after which it underwent a major range extension, becoming a common breeding resident in the Ryukyu Is (Okinawa and Yaeyama); it is now so common on Iriomote-jima that it is thought to be ousting the Slaty-legged Crake (24). Its expansion N is continuing, and it now also occurs fairly regularly on the main Japanese islands, having bred on Kyushu; it is recorded

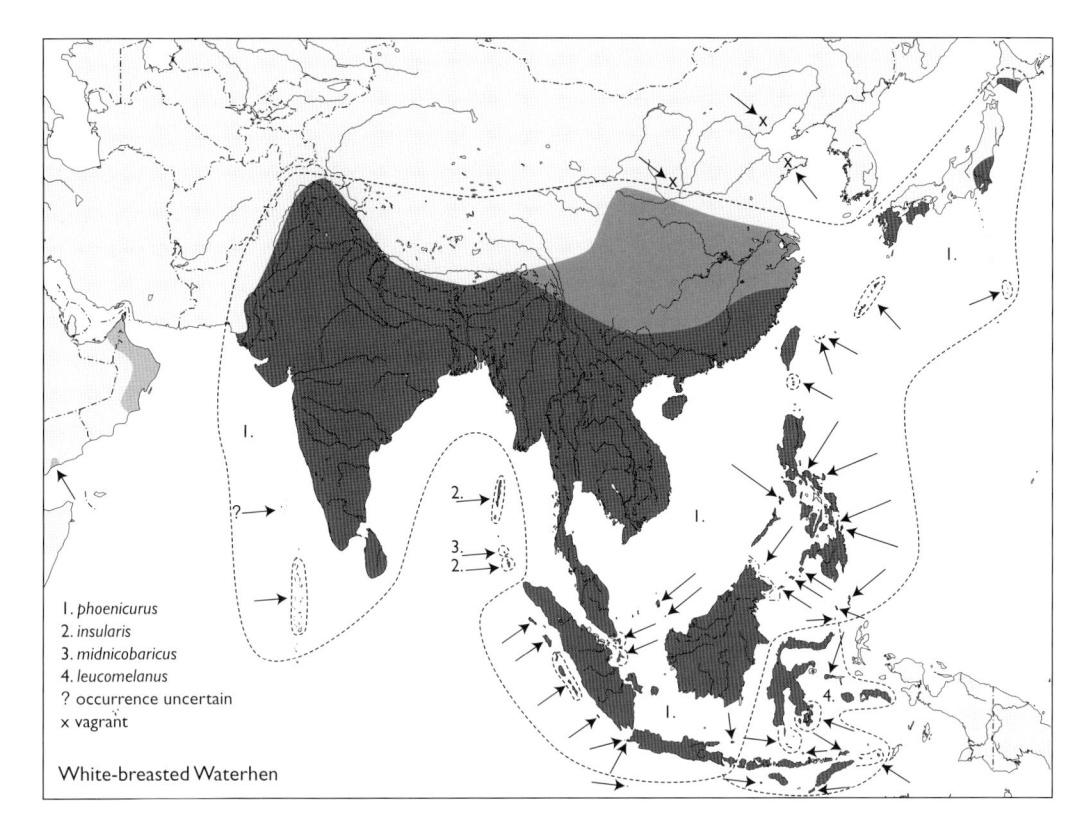

1. *phoenicurus*
2. *insularis*
3. *midnicobaricus*
4. *leucomelanus*
? occurrence uncertain
x vagrant

White-breasted Waterhen

E to the Ogasawara-gunto (Bonin Is). Most records occur from Mar-Aug, peaking in May. It has recently colonised Christmas I in the Indian Ocean (Coates & Bishop 1997).

It is capable of utilising manmade or man-modified habitats: in Malaysia it occurs at overgrown mining pools (Medway & Wells 1976) and in Pakistan it appears to have benefited from the construction of canals (Roberts 1991). It may occur near to human habitation: in Hanoi, Vietnam, it occurred in the campus area of the University and several pairs lived and bred in *Eichhornia* at a hotel in the city (Stusák & Vo Quy 1986). No information on predation is available; the species is often killed on roads in Borneo (Smythies 1981).

MOVEMENTS Resident throughout most of its range, but N populations winter S to Sumatra, occurring W to Arabia, where 13 are recorded from Oman (Oct-Jan), 1 from the United Arab Emirates (Nov) and 1 from Yemen (Mar) (Kirwan 1994). Small numbers, possibly northern migrants, have been ringed at Dalton Pass, Luzon, Philippines (McClure & Leelavit 1972, Mees 1986). Migrants have been recorded from the Malay Peninsula and from islands in the Straits of Malacca in Nov-Dec and Mar, while wintering birds occur in Sumatra from Nov to mid-Apr, most in Utara but 1 from Barat (Medway & Wells 1976, van Marle & Voous 1988). Over most of its range the extent of its migration from the northernmost breeding areas is not clear and the seasonality of its occurrence is shown on the distribution map only for China (from Cheng 1987). Its current expansion N into Japan is indicative of the mobility of the species. Birds migrate at night, sometimes coming to lighthouses. In China it is accidental in Hebei, S Shanxi and Shandong (Quingdao) (Cheng 1987), and it is a vagrant to the Seychelles (1 record, Dec, from Mahé) (Skerrett 1996).

HABITAT Reedy or grassy swamps, marshes, sago palm swamps, tall grass with reeds and shrubs, grassland, bamboo stands, wet scrub, rice fields and sugarcane, sewage ponds, the shores of rivers, ponds, waterholes, ditches and lakes, and the borders of forest and mangroves. In S Sulawesi, with other species such as the Barred Rail (44), Purple Swamphen (120), Common Moorhen (129) and Dusky Moorhen (130), it inhabits mature 'bungkas': large, artificially constructed mats of floating vegetation on lakes (Baltzer 1990). In Vietnam it occurs on almost all plant-overgrown waters including 'plantations' of *Ipomoea aquatica*, *Eichhornia crassipes* and *Nelumbo nucifera* (Stusák & Vo Quy 1986). It also occurs in thick forest (Andaman and Nicobar Is), swamp forest (Tinjil and Deli Is, W Java), at forest edges and clearings, in mangrove swamps, and in scrub and bushes far from water. It often occurs close to human habitation, e.g. at village ponds, and enters compounds and public parks; it runs about under roadside hedges and often feeds in short grass at roadsides and on lawns (Malaysia). In Pakistan it favours seepage zones along irrigation canals, in forest plantations as well as in other cultivation (Roberts 1991). In Thailand, Deignan (1945) found it highly tolerant of dry conditions, its numbers decreasing very little in the dry months, when birds were seen in dry fields some way from cover. On islands off W Java it is recorded once from dry forest and sometimes from relatively dry edge habitats along beaches (Holmes & van Balen 1996). It ranges up to c. 1,500m, but to 2,000m in the Nilgiri Hills, SW India.

FOOD AND FEEDING Worms (Annelida); molluscs (Gastropoda such as *Melania*, *Corbicula* and *Unio*); aquatic Crustacea; aquatic and terrestrial insects and their larvae, including Blattidae, Coleoptera (Tenebrionidae, Carabidae and Curculionidae such as *Myllocerus*), Orthoptera (Ensifera and Acridoidea), Formicidae (*Pheidole*) and Apidae (*Apis florea*); spiders; and small fish. A relatively small amount of plant material is also taken, including grain, young rice, grass and weed seeds, and the shoots and roots of marsh plants. It gleans from the ground and pecks seeds from standing grass. It forages in the open along water margins and at the edges of dense cover; it wades in shallow water, walks on floating vegetation, and feeds on beaches at low tide (Maldives). Foraging takes place throughout the day but peaks in the early morning and the evening.

HABITS It is inclined to be crepuscular and will forage in the open at dusk. It is often not particularly shy and is frequently seen out of cover, and it will even enter villages, compounds and public parks, but it runs rapidly to cover at any sign of danger. Its normal walk is slow, almost sedate, with the tail carried erect and constantly jerked, but it will also run while foraging. It climbs well in reeds, bushes and trees; it swims buoyantly, like a moorhen, and occasionally dives. If suddenly disturbed it may open and shut the wings before running to cover, or it may rise almost vertically and fly in a laboured fashion for a short distance, with dangling legs, before dropping into cover; when flushed in forest it may perch briefly in trees. During the hotter part of the day it retreats into dense cover. Two birds kept in an aviary by Glenister (1951) for 2 weeks became absurdly tame, followed him around the garden and back into the aviary and, when released, refused to depart until driven away.

SOCIAL ORGANISATION Despite this bird's widespread occurrence and relative abundance, remarkably little information is available on its social organisation and behaviour. It is normally seen singly or in pairs, sometimes in groups of up to 5. It is presumed to be monogamous, with a pair bond at least for the duration of the breeding season, when the birds are territorial.

SOCIAL AND SEXUAL BEHAVIOUR Territorial birds are intolerant of conspecifics, and individuals have been seen driving off approaching birds (Roberts 1991). Breeding males are pugnacious, but their fights are rather formalised and usually innocuous (Ali & Ripley 1980).

BREEDING AND SURVIVAL Season India and Pakistan, Jun-Oct (mainly in SW monsoon); Bangladesh, Jul-Aug; Andaman and Nicobar Is, Jun-Jul; Sri Lanka, principally during SW Monsoon but may breed in any month; China, May-Jun; Vietnam, May; Japan and Ryukyu Is, Apr-Oct; Myanmar, May to at least Jul; Malay Peninsula, Feb-Oct, and newly hatched chicks Dec; Sumatra, Jan, May, Jul, Sep, Nov; Java, all months; Lesser Sundas, downy chicks Apr-May; in Borneo, and on other tropical islands, may breed in almost any month. **Nest** A shallow cup-shaped pad of twigs, creeper and other plant stems, dried reeds, rushes and *Typha* leaves; placed on or close to ground in grass or tangled undergrowth at margin of pond, ditch or flooded rice field or between bushes (e.g. *Tamarix*, *Prosopis*); or concealed in shrub, bamboo clump or pandanus thicket up to 2-3m above ground, sometimes far from water. Nest sometimes bulkier, resembling collection of rubbish caught up in thick tangles of vegetation. Both sexes may build. **Eggs** 3-9; rather long oval; only slightly glossy; cream to greyish-cream or pinkish-white, boldly streaked and blotched chocolate-brown or red-brown, and greyish-purple; markings often more concentrated at blunt end. Size of eggs: nominate, S India and Sri Lanka, 37-42 x 28-31.5 (39.5 x 30.0, n = 46), calculated weight 19.5; *chinensis* (= nominate), India E to China and S to Malay Peninsula, 37-45 x 28-31.9 (40.5 x 29.7, n = 100), calculated weight 19.6; *insularis* (presumably includes *midnicobaricus*), Andamans and Nicobars, 37.2-43 x 29-32.2 (40.8 x 31.0, n = 50), calculated weight 21.5; *javanica* (= nominate), Java, 35.2-41 x 27.1-29.9 (38 x 28.6, n = 42), calculated weight c. 17. **Incubation** 20 days, by both sexes. **Chicks** Precocial; leave nest soon after hatching but remain in vicinity for 5-6 days (Roberts 1991); fed and cared for by both parents; one parent leads young, the other follows at rear. Young chicks climb well in reeds, and dive into water to hide when nest approached. One adult built floating, unattached platform of parts of *Eichhornia* plants, on which it roosted with 7-day-old young under its wings (Stusák & Vo Quy 1986). Probably multi-brooded in Japan and Ryukyu Is, which has aided rapid expansion there in recent years (Brazil 1991).

90 BLACK CRAKE
Amaurornis flavirostris   Plate 26

Gallinula flavirostra Swainson, 1837, Senegal.

Sometimes included in *Porzana*, or placed in the monospecific genus *Limnocorax*, but is inseparable from *Amaurornis* on skeletal characters. Forms a superspecies with *A. olivieri* of Madagascar. Monotypic.

Synonyms: *Rallus niger/aethiops*; *Corethrura/Porzana nigra*; *Limnocorax aethiops/niger/capensis/senegalensis/mosambicus/erythropus/flavirostra*; *Gallinula aterrima/nigra/carinata*; *Ortygometra erythropus/flavirostra/nigra*; *Porzana flavirostra*.

Alternative name: Black Rail.

IDENTIFICATION Length 19-23cm. Medium-small; plumage entirely black to dark slate, palest from mantle to back and on breast and upper belly, with olive-brown wash from lower mantle to rump and on upperwing-coverts (often not obvious in the field); iris and orbital ring red; bill relatively short, robust and greenish-yellow; legs and feet bright red, duller in non-breeding season; legs appear quite long. Sexes alike; female averages slightly smaller. Immature paler; upperparts olivaceous-brown; chin and throat whitish; head, neck and underparts grey; bare parts dull. Juvenile dull greyish-brown with whitish chin and throat; bill blackish, with pink around nostrils in youngest birds; bill gradually becomes darkish green; iris grey to brown; legs and feet dark brown. Advertising call unmistakable. Inhabits many types of freshwater wetland with moderate cover of emergent or fringing vegetation and some permanent flooding, and often with floating vegetation.

Similar species All-black plumage, with greenish-yellow bill, and red eye and legs, make adult easily distinguishable from all other African rails. Paler immature and brown juvenile lack any barring, spotting or streaking, have dark undertail-coverts, and are developing adult bare part colours; these characters make them easily distinguishable from sympatric rails. Very similar to Sakalava Rail (91) but upperparts washed olive-brown (in the field often appearing slaty-black, like rest of plumage) as opposed to rich dark brown, and has bright red orbital ring, legs and feet.

VOICE Information is taken from unpublished data (P. B. Taylor) unless otherwise specified. The very characteristic advertising call is a duet: one bird gives a harsh repeated chatter *krrrok-krraaa*, which increases in volume and often ends in a crowing *krraaa*; the other joins in with a softer, almost dove-like purring or crooning call, which may comprise single notes *crrooo* or often a phrase such as *coo-crr-OOO*; birds may call apparently independently, but sometimes the notes at the end of the sequence are precisely timed and antiphonal (Urban *et al.* 1986). Birds usually crouch, with raised heads, to give the duet; either sex may give either call, and duetting birds may be adults, immatures or juveniles. Other family members often join in, and all calling birds may crouch in a circle, heads facing inwards. An immature once crouched alongside a loudspeaker and called with the playback duet. Calling may be used in territorial advertisement and also in reinforcing the bond between pair members and between members of the family group. Adults also give very soft *croo* calls, function unknown. Captive birds gave an annoyance call rendered *shakk*, and a *twoop* call to chicks (Sclater & Moreau 1932); alarm calls are a repeated, sharp *chip* or *tyuk*, and a harsh *chack*; immatures also make harsh *kyaar* and *chaa* calls, possibly signifying alarm, and a trapped bird gave a plaintive, loud *cheeeerp*. Contact calls, made to birds of all ages, are soft *pu* or *bup* notes, subdued clucks *buk*, and also quiet *grr* calls which are also given by birds being allopreened. When soliciting food chicks make quiet squeaks, *tsip* calls and, in common with juveniles, quiet *chee* and *cheea* calls. One immature gave quiet, plaintive *teek* and *teekeeteek* notes when adopting a submissive pose to an adult male. Trapped chicks make loud squeaks and squeals of alarm, to which adults respond with loud cat-like hissing and spitting noises. This species is much less vocal outside the breeding season.

DESCRIPTION

Adult Entire plumage black or nearly so; head and neck black; mantle dark slate, grading to blackish-brown on rump and uppertail-coverts; lower mantle and back washed olive-brown, not obvious in the field; tail black; upperwing-coverts dark slate, washed olive-brown; remiges brownish-black; axillaries and underwing-coverts blackish. Breast and upper belly dark slate; flanks, lower belly and undertail-coverts black. Iris and orbital ring red; bill greenish-yellow; legs and feet bright pinkish-red to red. Small white wing-claw present. In non-breeding season, legs and feet dull red. Sexes alike; female averages slightly smaller. Bill of most males has hook at tip of upper mandible; this is present in only c. 10% of females; any bird with bill length >25 and hooked upper mandible should be male; adults may be sexed by external examination of pubic bones, points being adjacent in males but separated in females which have laid eggs (Schmitt 1975).

Immature Upperparts olivaceous-brown; chin and throat whitish; head, neck, underparts, axillaries and underwing-coverts darkish grey; bare parts duller than adult. Older immatures often appear richer brown on upperparts and darker, blackish-grey on underparts.

Juvenile Dull greyish-brown with whitish chin and throat; greyer on breast, lesser upperwing-coverts, alula and remiges; some upperwing-coverts and scapulars sometimes with variably paler, brighter brownish fringes; marginal coverts with broad whitish fringes. Iris brown or grey-brown, becoming dark red (2-4 months); bill blackish, changing to darkish green (2-4 months); legs and feet dark brown to blackish, becoming dull red.

Downy young Down black; has slight greenish sheen at 1 week; iris grey, becoming olive (1 week) and then brown (4-6 weeks); bill pink with narrow black band round centre, often becoming black with pink at tip and around nostrils (1 week) and all blackish (8 weeks) but sometimes retaining some pink at base at 8 weeks, when some birds may have greenish nostril patches; legs and feet slate, becoming black (1 week) and then dark brown (2-3 weeks); brownish-black wing-claw well developed.

MEASUREMENTS (All from South Africa) Wing of 37 males 97-114 (106.4, SE 0.65), of 31 females 94-109 (101.6, SE 0.83); tail of 16 males 39-49 (43.1), of 14 females 37-48 (41.1); culmen of 37 males 24-28 (25.5, SE 0.16), of 31 females 22-25 (23.4, SE = 0.15); tarsus of 16 males 38-45 (42), of 14 females 35-42 (38.6) (Schmitt 1975, Urban *et al.* 1986, P. B. Taylor). Weight of 45 males 78-118, of 34 females 70-110 (Schmitt 1975, no means given), mean of 8 males 96.5, of 3 females 81; weight of 167 unsexed 69-118 (89.1). Weight varies greatly during the year and birds may lose up to 30% of weight during a severe winter (Schmitt 1975). Hatching weight 8-9.

GEOGRAPHICAL VARIATION Sharpe (1894) found much variation in overall plumage tone throughout range, the most intensely black birds coming from Gabon; no other authority has commented on this.

MOULT All information is from South Africa unless otherwise stated. Postbreeding moult is complete and remex moult is simultaneous, occurring in Dec-Mar and rendering the birds flightless for up to 3 weeks. Rectrices are usually moulted with the remiges, but sometimes earlier. One female moulting the rectrices in Dec contained a fully developed egg; this bird did not moult the remiges until the following Mar (Schmitt 1975). Body moult occurs in Sep-Apr and normally takes 5-8 weeks, with peaks in Oct and Mar, indicating 2 body moults per year (prebreeding and postbreeding). A flightless female was recorded in Zaïre, Dec; in SE Zaïre there are 2 complete moults per year (Apr-Jun and Oct-Nov), suggesting that birds may breed twice a year (Verheyen 1953). Postjuvenile partial moult occurs between 2 and 4 months of age. It is not clear whether the darker, richer plumage of many immatures older than 6 months is a result of continued moult or of feather abrasion, or both.

DISTRIBUTION AND STATUS Africa, from the N edge of the Sahel to the Western Cape Province, South Africa, except in the desert regions of the extreme NE and SW of its range (Urban *et al.* 1986). Its N limits are: extreme S Mauritania; Senegal; S Mali; S Niger, and also N Niger at Lac d'Arrigui, where it was apparently quite common, and breeding, in 1970; C Chad; C and S Sudan; Ethiopia; Djibouti; Somalia S of c. 2°N; and Kenya (absent from inland N areas away from L Turkana except for isolated records at Moyale and Daua R) (Fairon 1971, Urban & Brown 1971, Morel 1972, Lamarche 1980, Ash & Miskell 1983, Urban *et al.* 1986, Nikolaus 1987, Giraudoux *et al.* 1988, Lewis & Pomeroy 1989, Dowsett & Forbes-Watson

1993). It extends S almost continuously, to Namibia (sporadically to c. 25°S and also at 27°S), N and E Botswana, and most of South Africa, including scattered localities in the dry C regions, especially along the Orange R (Taylor 1997c). It occurs on islands off East Africa (Zanzibar, Pemba, Mafia) but is absent from the Gulf of Guinea Is, West Africa (Urban *et al.* 1986). The commonest and most widespread crake in Africa, it is less secretive than other terrestrial wetland rails of the region and has a well-known advertising call, so its occurrences are relatively well documented. It is common to abundant over most of its range, but is local in forested regions of West and Central Africa and is very localised and rare in the drier W and C regions of southern Africa (Urban *et al.* 1986, Taylor 1997c). Under favourable conditions its population density may be very high in some areas, and the total population estimate of >1,000,000 birds (Rose & Scott 1994) may well be of the correct order of magnitude. It must have been adversely affected by loss of wetland habitats but it is under no immediate threat anywhere. It readily occupies artificially created wetlands and temporary habitats, and in the Ivindo basin, NE Gabon, it has benefited from human activity, being found only where forest has been destroyed and marsh vegetation has been cleared along the river (Brosset & Erard 1986). In South Africa its total population is estimated at c. 51,500 birds, 50% (25,700) of these being in KwaZulu-Natal and 29% (15,000) in the former Transvaal (Taylor 1997a); it is the most widespread terrestrial rallid species in the country and may well be considerably more numerous than is currently known.

Predation by man may not be heavy because of the relative unpalatability of its flesh (Cott & Benson 1970); avian predators include African Marsh Harrier *Circus ranivorus* and Purple Swamphen (120) (Urban *et al.* 1986).

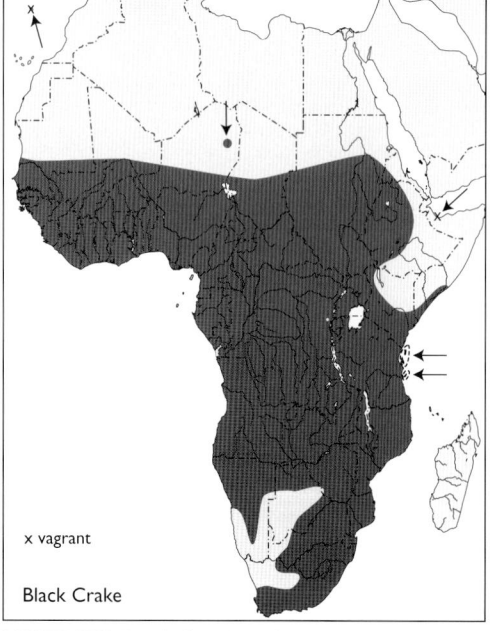

MOVEMENTS It is largely sedentary but also locally migratory, while young birds may undergo extensive dispersal. In the drier N parts of its range (N Ghana, N Nigeria and the Sahel zone of Sudan), it appears with the rains and disappears in the dry season (Urban *et al.* 1986); it moves in Zimbabwe when habitats dry out (Irwin 1981); it occupies temporary waters in East Africa (e.g. Jackson & Sclater 1938) and NE Namibia (Hines 1993); and a presumed migrant was captured at night, at lights in Tsavo West NP, SE Kenya, in Dec (Lewis & Pomeroy 1989). Fluctuating reporting rates in other areas may be explained by the birds being more visible during the non-breeding season, when vegetation cover may be reduced and birds feed more often in the open, and/or by variations in calling, which decreases markedly outside the breeding season when territories are relinquished (Taylor 1994, 1997c). This species is a vagrant to Madeira, Jan 1895 (Bannerman & Bannerman 1965), and NW Somalia in the Guban (at Daimoli), Mar 1919 (Archer & Godman 1937).

HABITAT The Black Crake occupies many types of freshwater wetland, its normal requirements including moderate vegetation cover and some permanent flooding. It inhabits rank grass, sedges, reeds, papyrus, swampy thickets, bushes and other vegetation beside flowing and still waters; it occurs in the interior of dense or extensive reedbeds and in the dense undergrowth of boggy clearings in forest; in open areas it may occupy grassy marshes (Urban *et al.* 1986, Taylor 1997c). Occupied vegetation includes *Phragmites, Typha, Scirpus, Cyperus, Mariscus, Eleocharis, Carex, Schoenoplectus*, and grasses (*Leersia hexandra, Echinochloa, Oryzidium, Panicum repens*) (Hopkinson & Masterson 1984, Taylor 1994, P. B. Taylor unpubl.). It is often found on ponds covered with *Nymphaea* and other floating vegetation such as *Aponogeton* and *Potamogeton*, and on waters with floating and emergent *Polygonum*; in South Africa it frequents stands of elephant's ear *Colocasia antiquorum* var. *esculenta* at the edges of sedgebeds (Taylor 1997a, P. B. Taylor unpubl.). It occurs in thickets of *Nuxia appositifolia* overhanging river banks in the Okavango; it likes tangled vegetation, in which it climbs, roosts and sometimes nests, and (presumably for this reason) it appears to be more tolerant of habitat with much dead vegetation than are other rallids (Hopkinson & Masterson 1984). In the SW Cape, South Africa, it is apparently the only rail to occupy the dense beds of palmiet sedge *Prionium serratum* which fringe and cover streams and rivers; it climbs freely in this tangled, spiny vegetation and forages on clear ground, mud and in shallow water beneath the dense, low palmiet canopy (Taylor 1997d). In the SW Cape it also appears less tolerant of saline conditions than is the African Rail (62) (Taylor 1997a) but it apparently occurs at the margins of coastal lagoons in Ghana (Grimes 1987) and was once seen at the edge of mangroves in coastal Kenya (P. B. Taylor unpubl.). It may occupy seasonal and ephemeral pans, interdune pans, and temporarily flooded areas along rivers (e.g. Jackson & Sclater 1938, Hines 1993). It is very adaptable: in dry regions it will occupy tiny streams with thin cover, while it is quite bold and tolerant of disturbance and occurs close to human habitations (Urban *et al.* 1986, Taylor 1997c). It occurs from sea level to 3,000m. It is widely sympatric with the African Rail but tolerates a wider range of habitat types and in KwaZulu-Natal, South Africa, it is most widespread and numerous below 1,200m, rarely being recorded above 1,800m, in contrast to the African Rail which occurs commonly up to at least 1,800m but may be less widespread and numerous at low altitudes and in coastal areas (Taylor 1997a).

FOOD AND FEEDING Information is taken from Urban *et al.* (1986) and from unpublished observations (P. B. Taylor) unless otherwise stated. This species takes earthworms, molluscs, crustaceans, adult and larval insects including Diptera, Lepidoptera, Coleoptera, Hemiptera and larval Odonata, small fish, frog tadpoles and small frogs; it also takes seeds and other parts of water plants, including duckweed *Lemna* and seeds of waterlilies *Nymphaea*, and will eat crushed oats. It takes eggs and nestlings, including those of weavers *Ploceus* spp. and of herons such as the Rufous-bellied Heron *Ardeola rufiventris*, and it will kill small birds caught in mist-nets. It also scavenges carcasses of crabs, crayfish, frogs, fish and small birds. Captive birds eat mealworms (*Tenebrio* larvae), bread, minced raw meat, rice, chopped maize, and occasionally seed (Rutgers & Norris 1970, Brosset & Erard 1986). Black Crakes often feed in the open along shorelines, on floating vegetation, in short grass or at cultivated plots adjacent to wetland habitats, and on dry or burned ground, and may venture some distance from cover to do so. They take food from the surface of the ground and water, and from growing vegetation; they catch insect prey in shallow water by immersing the head and neck; they make rapid darting pecks at small fish; they probe into mud, dead vegetation, and the exposed ends of broken and cut reed stems; they shift fallen vegetation to search beneath it; they pull down dead and living leaves (especially of *Typha*) and examine them for prey; and they climb in reeds to search for food and to catch flying insects. Birds readily perch and feed on the backs of hippopotamuses *Hippopotamus amphibius*, and also perch on warthogs *Phacochoerus aethiopicus*, apparently gleaning ectoparasites (Dean & MacDonald 1981). One bird repeatedly struck vertically downwards at the shell of a gastropod c. 1.5cm in diameter and then washed the prey in water, presumably to remove shell fragments. When fed dry items such as mealworms (*Tenebrio* larvae) the crakes frequently wash them in water before swallowing them or feeding them to the young.

HABITS The Black Crake is predominantly diurnal. In South Africa, Schmitt (1975) found it active throughout the day in winter but inactive for up to 3h in the middle of the day during the summer, and it was markedly active after rain. It occasionally calls at night and 1 was dazzled by torchlight and caught by hand when it walked onto a flooded path through tall reeds at night (P. B. Taylor unpubl.); in Sudan it is reported to be active on moonlit nights (Archer & Godman 1937). It bobs the head and jerks the tail as it walks; it treads slowly and deliberately over floating vegetation like a jacana, sometimes spreading its wings to keep its balance; it swims well, and may escape danger by diving (Urban *et al.* 1986). When agitated it adopts a very upright pose, stretching the neck up, holding the tail vertical and raising the crown feathers. When alarmed it flicks the wings, raises the tail vertically, lowers the head and rushes to cover, often fluttering the wings, but it flies to cover if any appreciable distance of open ground has to be covered; it may also 'freeze' in cover and feign death when handled (Sclater & Moreau 1932). Flights are normally low and may be weak, with dangling legs, or strong and direct. Birds often climb in bushes, reeds and papyrus, ascending severalm above the water, sometimes possibly to roost (Urban *et al.* 1986). This species is bolder and easier to see than other crakes, and in populated areas it may forage on short grass and in cultivated plots some distance from cover, even when people are nearby (P. B. Taylor unpubl.). When water levels recede at lakes in Kenya and Zimbabwe these crakes continue to inhabit reedbeds from which they must cross 10m or more of open ground to reach foraging habitat at the water's edge (Hopkinson & Masterson 1984, P. B. Taylor unpubl.). In South Africa when its habitat is burned this species, like the African Rail, will remain in much less dense and smaller patches of cover than will the smaller Red-chested Flufftail (3), and will forage on completely open burned ground up to 10m from such cover (Taylor 1994). Within cover the birds are confident and will walk or climb to within 1m of an observer. Immatures of a family group studied in South Africa accepted mealworms from the hand of an observer in the hide (P. B. Taylor unpubl.). In contrast, flightless birds in remex moult are very secretive, remain in cover and are very difficult to observe (Schmitt 1975).

Bathing occurs at any time of the day though possibly most frequently in the morning, while sunbathing has been observed very occasionally: a preening bird pauses, stands motionless for c. 10-20s with the body and neck stretched upwards and the wings open, and then preens again before repeating the sunbathing action (P. B. Taylor unpubl.).

SOCIAL ORGANISATION Monogamous; territorial when breeding. In NE Gabon Brosset & Erard (1986) imply permanent territoriality: "par couples territoriaux, vus ou entendus quotidiennement dans les mêmes sites". In South Africa, males establish territories at the beginning of the breeding season, and relinquish them after breeding (Urban *et al.* 1986). In KwaZulu-Natal, where breeding is recorded in Sep-Apr, the frequency of adult calls increased greatly from mid-Aug and decreased again in Apr; adults responded to taped playback most strongly in Sep-Apr; they attacked, or showed strong interest in, model Black Crakes and their own reflection in mirrors, in Sep-May; and they ignored or avoided models in Mar-Aug and mirrors in Mar-Sep (Taylor 1994). In the former Transvaal, territories covered c. 100m of river frontage, with little depth (Schmitt 1975); in KwaZulu-Natal, 6 territories were 1,775-3,115 (2,028, SD = 534.8) m^2 (P. B. Taylor unpubl.). Observations by Pitman (1929) suggest that breeding density may sometimes be very high, and in good and extensive habitat the overall breeding density may be 3 family groups/ha (Taylor 1997a). Members of the family group remain in association during the non-breeding season, and paired captive birds always remain together (Taylor 1994).

Cooperative breeding is normal, the offspring remaining with the family group until the end of the breeding season and helping to incubate clutches and rear young; helpers roost around the nest (Schmitt 1975, Taylor 1997c). Most immatures disappear from the breeding area before the start of the next breeding season, but in one territory a single bird remained to help rear young in the following season (P. B. Taylor unpubl.).

SOCIAL AND SEXUAL BEHAVIOUR During the breeding season this crake is very aggressive and will chase or attack not only conspecific individuals but also birds of many other species, including kingfishers *Alcedo*, doves *Streptopelia*, bulbuls *Pycnonotus*, weavers *Ploceus*, widows and bishops *Euplectes* and mannikins *Lonchura* (Taylor 1994, P. B. Taylor unpubl.). When a Burchell's Coucal *Centropus superciliosus* attempted to approach a feeding station to eat mealworms, it was prevented from doing so by an immature Black Crake which walked around near the food with its head down and tail raised, staring at the coucal until the latter moved away (P. B. Taylor unpubl.).

Territorial males chase conspecific males, immature Common Moorhens and African Rails; however, nests of all 3 species may be as little as 10m apart (Schmitt 1975). In captivity Black Crakes are very aggressive to other rail species during the breeding season, and will attack and kill species as large as themselves (C. C. Wintle pers. comm.). In KwaZulu-Natal during the breeding season Black Crakes attack Red-chested Flufftails which share their breeding area, but in the winter they are attacked by the flufftails, which are permanently territorial; there is no evidence that either species is excluded from the territory of the other (Taylor 1994). In a South African study of a small wetland with 1 territory, 2 adult males and 2 adult females were resident at the site and the breeding pair comprised a different combination of these individuals in each of the 3 seasons of the study; the territorial pair did not exclude the other adults, which occasionally visited feeding stations at the same time as members of the current breeding group (P. B. Taylor unpubl.; see also Brooke 1975).

In studies involving the use of taped playback, models and mirrors, the following levels of aggressive behaviour have been observed in adults and immatures during the breeding season (P. B. Taylor unpubl.). When making an uncertain or mildly aggressive approach to a model or mirror, the bird adopts an upright pose similar to the stance of a nervous bird (see Habits), with the neck stretched up and the crown feathers raised; the bird flicks the tail, jerks the head, sometimes flicks the wings open and closed above the back, and walks up to the model or mirror and pecks it repeatedly. One immature repeatedly raised one wing while pecking at its own reflection. A more aggressive bird retracts the neck and often adopts a crouched stance before making an attack, and the crown and nape feathers are raised. In strong attacks the bird may make fluttering jumps at the model or mirror, striking with the feet and/or pecking at the head and neck, or may make lunging pecks to the head and body, or may jump onto the model and peck downwards at its head and neck. A less strong attack involves the bird pushing at the mirror with its breast, shoulder or side, while pecking downwards at the head and neck.

An incubating bird will defend the nest from an intruder by opening the wings, rousing the plumage and uttering a deep growl (Watson 1969). Adults and fully grown young in family groups followed a marsh mongoose *Atilax paludinosus* through reedbeds, remaining in cover close to the predator and uttering subdued alarm calls (P. B. Taylor unpubl.).

Courtship and mating normally take place in dense cover and have not been described, although apparent courtship chases have been seen (P. B. Taylor unpubl.). Allopreening is frequent and takes places between pair members, and between adults, helpers and young (Brooke 1975, P. B. Taylor unpubl.). Fully grown young in the family group are normally submissive to the breeding pair; when approached by adults, they frequently crouch in a pose reminiscent of the food-begging pose of the young (see Breeding), often shuffle the wings, and are sometimes allopreened (P. B. Taylor unpubl.).

BREEDING AND SURVIVAL Season Under suitable conditions breeding may occur throughout year, with seasonal peaks in most regions during or following rains. Mali, May-Oct; Senegal and Gambia, Dec-Mar, May-Jun, Aug-Sep; Sierra Leone, Jan, Mar, Oct, Dec; Liberia, mating from Apr, young in Nov; Burkina Faso, juvenile Apr; Ghana, Jun, Apr and Aug (half-grown young May and Sep), chicks Mar; Niger, Jun, large juveniles Jan; Nigeria, Aug-Nov, chicks Jun, Aug, juveniles Feb; Cameroon, Oct, young Mar-Apr; Gabon, Mar, Jul; Zaïre, almost all months, mostly Jul-Dec in NE and Feb-Mar in S; Sudan (Darfur), Oct; Ethiopia, Apr-Oct; Somalia, adults with young Dec; Burundi, Mar; Uganda and W Kenya, all months (peaks Apr, Nov); rest of Kenya and NE Tanzania, all months (peaks Apr-Jun, Oct); rest of Tanzania, Apr, Oct-Nov; Zanzibar and Pemba, chick Feb, breeding condition Oct; Zambia, Nov-Aug (peak Dec-Mar); Malawi, Mar-May, Jul-Sept, Dec; Mozambique, May-Jun; Zimbabwe, all months except May and Jul (peak Jan-Feb); Botswana, Aug-Oct; South Africa, Aug-May (peaks Nov, Jan-Mar). **Nest** A deep, bulky bowl of reeds, rushes, sedges, creepers, grasses or other water plants; dry materials usually predominate; nest often lined with fresh green leaves; placed in vegetation such as *Typha, Carex, Cyperus, Sium, Phragmites, Pontederia* or grasses over water, with rim 20-50cm above water, but sometimes floating among stiff grass stems or other retaining vegetation; also on ground or in grass tussock near water; sometimes up to 3m high in bushes, possibly as protection against sudden flooding; one nest in forest on bed of fern stalks above mud. External diameter 10-30cm, depth 8-18cm; cup depth 5-9cm. Both sexes build, sometimes helped by young of previous broods which also help collect nesting material; adults may build extra nests or platforms for roosting and preening; platforms built by other species also used. **Eggs** 2-6 (usually 3); laid at daily intervals; 2 females once laid 6 and 4 eggs in same nest. Eggs elliptical oval or elongate ovate; white to cream or pinkish-buff, finely spotted all over or with larger spots concentrated at blunt end; spots light brown, reddish-brown or chestnut over pale purple or slate. Size (n = 27, Uganda) 29.5-33.4 x 22-24.4 (31.4 x 23.2); (n = 21, South Africa) 31-35.7 x 23-26 (33.8 x 24.2); (n = 70, entire range) 28-35 x 22.5-25.2 (31.7 x 23.7), calculated weight 9.7. **Incubation** 13-19 days, by both parents, sometimes assisted by young of previous broods; begins with laying of 4th egg (Ripley 1977); hatching asynchronous. **Chicks** Precocial; leave nest 1-3 days after hatching; use wing-claw in climbing; fed and cared for by parents, helped by young of previous broods, for at least 3-6 weeks; chicks and juveniles solicit food by crouching, with head and bill raised, often making quiet calls (see Voice); chicks wave wings, fledged birds sometimes shuffle wings; older bird usually offers food, which young bird takes from its bill; food items sometimes dropped for young to pick up; young may still beg for food when 3 months old (Schmitt 1975). At 2-3 weeks body feathered, head and neck still downy, and remiges developing (half grown at 3-4 weeks); juvenile plumage attained by 4-6 weeks; young fly at 5-6 weeks; apparently independent from 6-12 weeks, when some may leave territory (Schmitt 1975). Nest predators include monitor lizard *Varanus niloticus*, and possibly slender mongoose *Herpestes sanguineus* and vlei rat *Otomys irroratus*; young are eaten by African Marsh Harrier and Purple Swamphen (Schmitt 1975). Able to breed within 1 year of hatching. Up to 4 broods recorded per season in the wild and in captivity; when helpers present, clutch may be laid when young of previous brood are as little as 3 weeks old (Schmitt 1975, Taylor 1994, P. B. Taylor unpubl.).

91 SAKALAVA RAIL
Amaurornis olivieri   Plate 26

Porzana Olivieri Grandidier & Berlioz, 1929, Antsalova, Province of Maintirano, west Madagascar.

Although sometimes placed in *Porzana*, has close affinities with the common *A. flavirostris* of Africa, with which it forms a superspecies. Monotypic.

Alternative names: Olivier's Crake/Rail.

IDENTIFICATION Length c. 19cm. The smallest *Amaurornis*. Appearance distinctive: predominantly dark slate-grey, but rich dark brown from mantle to rump and on scapulars, upperwing-coverts and tertials; bill greenish-yellow; eye red; eyelids, legs and feet pinkish-red. Sexes alike. Immature and juvenile not properly described, but probably browner overall than adult. Inhabits streams and marshes with emergent cover and floating vegetation.
Similar species Distinctive plumage and bare part colours easily distinguish it from other Madagascar rails. Similar to Black Crake (90) of Africa in size (averages slightly smaller), predominantly black plumage, and colour of bare parts, but Black Crake is merely washed with olive from mantle to rump and on upperwing-coverts, these areas normally appearing slaty-black like rest of plumage, whereas Sakalava Rail is bright chestnut in these regions. Bill slightly stouter than in Black Crake; washed reddish at base in some birds.

VOICE One bird, apparently disturbed or alarmed, gave a repeated *kwook*, like the sound of two sticks being knocked together (Ramanampamonjy 1995).

DESCRIPTION
Adult Head, neck and underparts dark slate-grey, washed olive-brown on crown, nape and flanks. Mantle to rump, scapulars, upperwing-coverts and tertials, dark rich brown, washed hazel, contrasting with head and neck; uppertail-coverts blackish-grey, tinged dark brown; tail black. Greater and median upperwing-coverts very dark brown except at edges; alula, primaries and secondaries very dark brown, washed hazel on outer edges; narrowly tipped pale cinnamon-rufous when fresh; axillaries as upperwing-coverts, or dark grey-brown; underwing-coverts dark grey-brown. Chin and throat like rest of underparts (see Plate 26, and Plate 15 in Taylor 1996), or throat slightly paler (possibly an immature character, as in Black Crake) and chin whitish; lower belly and thighs to vent sometimes washed olive-brown; undertail-coverts blackish-grey. Iris red; eye-ring pink to pale red; bill greenish-yellow, in one specimen washed reddish at base; legs and feet pinkish-red. Sexes alike.
Immature Not described (see Adult).
Juvenile Not described; may be browner than adult (O. Langrand *in litt.*).
Downy young Downy chick black; long down of head, neck and upperparts with bottle-green gloss; iris brownish; bill black, tip horn, base of upper mandible to nostril pinkish-white. Small blackish wing-claw present.

MEASUREMENTS 4 males, 3 females: wing of male 102-105 (103, SD 1.4), of female 104-112; tail of male 47-61 (54, SD 5.7), of female 47, 52, 54; bill of male 25-29.5 (27.75, SD 1.9), of female 24-27; tarsus of male 38-41 (39.25, SD 1.5), of female 35-38 (Benson & Wagstaffe 1972, Ripley 1977, BMNH).

GEOGRAPHICAL VARIATION None recorded.
MOULT Specimens from Mar are in very worn plumage.
DISTRIBUTION AND STATUS WC Madagascar, where it is rare and localised, with very few records. CRITICALLY ENDANGERED. It is known from three widely separated areas in the Sakalava country of lowland W Madagascar: Antsalova, near Maintirano, and the region of Lakes Masama and Bemamba; Ambararatabe near Soalala (6 of 7 specimens were collected along the Tsiribahina R), and L Kinkony; and Nosy-Ambositra on the Mangoky R (Collar & Stuart 1985). It was not seen between 1962 and 1985, despite searches from the mid-1980s in suitable habitat, but in May 1995 a single bird was seen near the small island of Antsalovakely, on the W shore of Lake Bemamba (Ramanampamonjy 1995). Like the Madagascar Little Grebe *Tachybaptus pelzelnii*, it has possibly suffered from the loss of lilypad habitat owing to rice cultivation and the impact of introduced fish (Collar *et al.* 1994). It also suffers from systematic exploitation for food: at Lake Bemamba, Ramanampamonjy (1995) was told that its flesh was good and that it had become scarce in the area because it was trapped and its nests were robbed early in the mornings in Sep-Oct; the informant indicated that this species had not been seen or caught in the area for 6 years, and only a few people in the area recognised the bird from pictures. The eggs from the only recorded nest were eaten by local people (Benson & Wagstaffe 1972). As these birds are not very shy, it is thought unlikely that the species has been overlooked in accessible habitat, but some potential wetland habitats are not severely threatened and populations may survive between Antsalova and Mahajanga (O. Langrand *in litt.*).

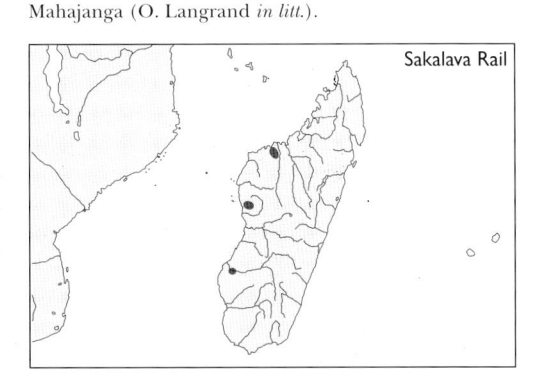

MOVEMENTS None described.
HABITAT Streams (sometimes narrow and deep) and marshes offering stretches of open water, patches of floating vegetation (especially waterlily *Nymphaea* leaves) for foraging, and adjacent dense cover of reedbeds such as *Phragmites* for shelter; also among grass and bushes emerging from the water in a clearing of a flooded palm-covered valley (Rand 1936, Collar & Stuart 1985). One at Lake Bemamba was seen in an open area of *Nymphaea* and *Polygonum senegalense* fringed by large tufts of *Cyperus dives* (= *immensus*) (Ramanampamonjy 1995).

FOOD AND FEEDING Food unknown. Apparently forages on floating vegetation near cover, like the Black Crake (Rand 1936).

HABITS Not particularly shy, but less active and bold than the Black Crake. Like that species it walks, jacana-like, on floating vegetation away from cover (Rand 1936). According to local people it is very active in the early morning and was formerly (when commoner) found foraging close to villages (Ramanampamonjy 1995).

SOCIAL ORGANISATION Nothing known.

SOCIAL AND SEXUAL BEHAVIOUR Nothing known.

BREEDING AND SURVIVAL Season Lays Mar; well-grown young also seen late Mar; apparently also nests Sep-Oct (see Distribution and Status). The following information is from Benson & Wagstaffe (1972). **Nest** One nest placed 50cm above ground level in *Typha* near water, in marshy area dominated by *Nymphaea* and *Phragmites* and with stretches of open water. This nest held 2 eggs, probably a complete clutch; the female was taken at the nest. **Eggs** Creamy-white; slightly glossy; heavily freckled all over with reddish-brown on underlying greyish-lilac. Also said to nest in *Phragmites mauritianus* (Ramanampamonjy 1995).

92 BLACK-TAILED CRAKE
Amaurornis bicolor   Plate 26

Taxonomy. *Porzana bicolor* Walden, 1872, Rungbee, Darjeeling.

Sometimes retained, possibly more appropriately, in *Porzana* (e.g. Olson 1973b, Inskipp & Round 1989); skeletal or DNA study needed. Monotypic.

Synonym: *Porzana elwesi*.

Alternative names: Elwes's/Rufous-backed Crake.

IDENTIFICATION Length 20-22cm. A medium-small, very dark rail. Dark rufous-brown upperparts and upperwings contrast with dark ashy-grey head, neck and underparts; top of head slightly darker grey, and sides of head and throat slightly paler grey; chin whitish; uppertail-coverts and tail blackish; undertail-coverts dusky brown to blackish. Iris blood-red; eye-ring pink; bill pale bluish-green with pale tip and red patch near base (red duller in non-breeding season, and absent in some birds); legs and feet brick-red. Sexes alike; female has duller bill. Immature has darker, duller upperparts; underparts washed vinaceous; iris brown. Juvenile predominantly dark sepia on upperparts and dull brownish-olive on underparts; sides of head paler; breast to belly mottled whitish. Inhabits wet forest, small marshes, grassy overgrown streams and wet grassland.
Similar species Sympatric Brown Crake (84) larger and duller, with uniformly olive-brown upperparts, including head and neck, dark slate-grey underparts shading to vinaceous and olive-brown from belly to undertail-coverts, flesh-brown, purple or red legs and feet, and green bill with bluish tip. Paler and larger than rather similar Sakalava Rail (91) of Madagascar, which has greenish-yellow bill tinged red at base.

VOICE Details are from Inskipp & Round (1989). The song comprises an initial rasping *waak-waak* call, which is often audible only at close range, followed by a long, descending trill lasting c. 12-13s, with an interval of 5-15 min before repetition. The trill is similar to that of the Ruddy-breasted Crake (102) (C. R. Robson *in litt.*). It is given mainly at dusk, but is also heard in the early morning. A foraging bird uttered high-pitched *keck* or *kick* notes.

DESCRIPTION
Adult Head and neck dark ashy-grey, slightly darker on top of head and nape, and paler on sides of head and throat; rest of upperparts to rump, including upperwing-coverts and tertials, dark rufous-brown (very dark cinnamon-brown); uppertail-coverts and tail blackish. Alula, primary coverts, primaries and secondaries dark brown with rufescent-brown outer webs; marginal coverts in carpal region white; axillaries and underwing-coverts dark olive-brown, slightly washed rufous-brown. Chin whitish; rest of underparts dark ashy-grey, breast slightly darker; sides of vent and undertail-coverts dusky-brown to blackish. Iris blood-red; orbital skin pink; bill pale bluish-green, tipped paler and greyer (or sometimes black); red patch near base of both mandibles, which is brighter in breeding season, and which may be entirely absent in many birds (Inskipp & Round 1989); in male, bill brighter green than in female (C. Robson *in litt.*); legs and feet dull red to bright brick-red or deep pinkish-red, fleshy-grey also recorded.
Immature Similar to adult, but brown of upperparts darker and duller; grey of underparts washed pale vinaceous; centre of belly pale; iris brown.
Juvenile Crown to hindneck, and entire upperparts and upperwings, including remiges, dark sepia, palest on hindneck and more olive-brown on remiges; greater coverts contrasting paler. Lores, forepart of head and orbital region sepia; rest of head and neck pale greyish-brown, becoming whitish on chin and throat. Rest of underparts dull brownish-olive, mottled and barred whitish on centre breast, lower flanks and centre belly. Bill possibly dark brownish, with pale tip and blackish culmen; legs and feet possibly reddish-brown.
Downy young Down black, more brownish-black on underparts and glossed bottle-green on head and upperparts; iris hazel; bill pale pinkish-mauve with paler tip; legs dusky, faintly washed mauve-plumbeous; small pale wing-claw present.

MEASUREMENTS Wing of 4 males 113-118 (115.5, SD 2.1), of 10 females 106-117 (112.2, SD 3.6); tail of 2 males 51, 52, of 8 females 41-58 (50.3, SD 6.1); culmen to base of 2 males 23.5, 26, of 4 females 25-26 (25.75, SD 0.5); tarsus of 2 males 37, 39, of 8 females 34-39 (36.75, SD 1.9).

GEOGRAPHICAL VARIATION None recorded.

MOULT Remex moult is simultaneous and is recorded in Nov, Sikkim.

DISTRIBUTION AND STATUS E Nepal (possibly), NE India (Sikkim, Bhutan to Arunachal Pradesh [= E Assam], Khasi and Cachar Hills, and Manipur) and N, NW and E Myanmar, E to SC China (S Sichuan, E Guizhou, Yunnan and Xizang [= Tibet]) and S to N Thailand, N Vietnam (W Tonkin and C Annam), and N Laos. Although not considered globally threatened it is infrequently recorded because of its skulking habits. It was formerly regarded as locally common (e.g. Ludlow & Kinnear 1944, Ripley 1977, Smythies 1986), but was regarded as scarce in Vietnam (Delacour & Jabouille 1931). Several recent records (since the mid-1980s) have come from Doi Inthanon National Park, NW Thailand, where it has been recorded throughout the year and breeding has been confirmed (Inskipp

& Round 1989). Cheng (1987) regards it as rare in SC China, and it is apparently rare (but possibly much overlooked) in Thailand (Lekagul & Round 1991). There is no other information on its current status, but habitat destruction has probably affected its numbers adversely. It is regarded as a vagrant to E Nepal, from which country there are 19th century specimens of doubtful origin (Ludlow & Kinnear 1937, Inskipp & Round 1989) and 3 specimens from the extreme E, also of doubtful origin, taken in May 1912 (Stevens 1925), while there is a possible record from EC Nepal in Mar 1981; it may be overlooked due to confusion with the Brown Crake (83) (Inskipp & Inskipp 1991). It possibly occurs in Bangladesh (Khan 1982). A skin purporting to come from Malaysia is probably mislabelled (Inskipp & Round 1989).

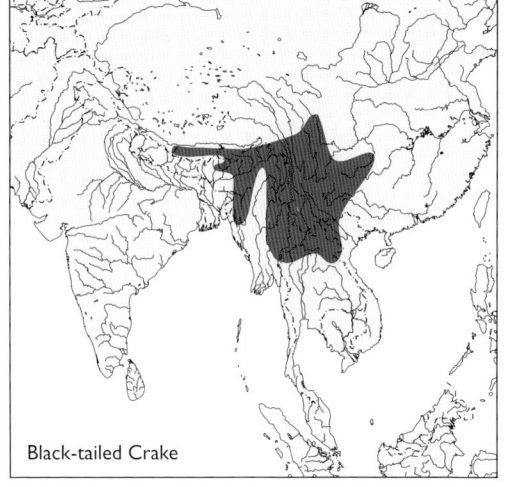

secondary-growth trees (up to 12m tall) along the banks (Inskipp & Round 1989). Although this crake is recorded from lowlands, it is primarily a highland species, occurring mainly from 1,000 to 1,830 m, but locally down to c. 200m in N Burma (Smythies 1953) and possibly at 2,600-3,600m in Bhutan and Sikkim (Ripley 1977).

FOOD AND FEEDING Worms, molluscs, insects (including small grasshoppers), and seeds of marsh plants. It is recorded feeding with a rapid pecking action, the wings drooped and the tail cocked and frequently jerked (Inskipp & Round 1989); in the morning and evening it sometimes appears in open meadows looking for grasshoppers and small worms (Delacour & Jabouille 1931).

HABITS It often remains within very dense cover, but it will feed in the open. It is a great skulker, emerging at the edge of cover to feed in the early morning and evening and progressing with the typical slow, jerky movement of many rails; if alarmed, it rapidly returns to cover, running with lowered head, neck and tail, if necessary flying or swimming. It was recently recorded feeding away from cover near people in the mid-afternoon (Inskipp & Round 1989).

SOCIAL ORGANISATION Monogamous. In Thailand, 3-4 calling birds were heard along 200m of streamside (Inskipp & Round 1989). Baker (1929) records "nearly a dozen" of these birds breeding in an area of c. 0.5ha in India.

SOCIAL AND SEXUAL BEHAVIOUR No information available.

Black-tailed Crake

MOVEMENTS Resident, but it has been recorded as occurring in N Vietnam and Laos only in winter (Meyer de Schauensee 1984); the extent of its seasonal occurrence is not clear, and is not shown on the distribution map.

HABITAT Forest patches, scrub and rushes in and around rice cultivation; swampy patches, ponds and streams in forest; grass-bordered streams and pools, often shaded by trees; small marshes in grass and cultivation; sometimes wet grassland. Recent records from Thailand are mainly from slow-flowing permanent streams, 1-5m wide, the beds of which are overgrown with lush, dense, short (30cm) grass; the streams are shady, with dense herbage and small

BREEDING AND SURVIVAL Season NE India (Khasi Hills, SC Meghalaya), mid-May to Aug; Myanmar, May and Jul; NW Thailand (Doi Inthanon), 2 adults with 3 young, 6 Aug; Vietnam, chick 21 Jun. **Nest** A rough pad of loosely assembled twigs and grass, with slight central depression; built on wet ground in forest undergrowth; sometimes placed 1-2m (once c. 6m) above ground in bushes or trees; 2 nests in China were on small, reedy islands in mountain streams (Thayer & Bangs 1912). Both sexes build. **Eggs** 5-8; pale cream to pale salmon-pink, boldly blotched deep red-brown, purplish-brown or brick-red, with underlying markings of grey or lavender; markings more concentrated at blunt end; some eggs rather feebly marked, resembling those of White-breasted Waterhen (89); size of 87 eggs 31.3-36.3 x 24.5-27 (33.9 x 26.1), calculated weight 12.6. **Incubation** By both sexes. No other details available.

PORZANA

As mentioned under *Amaurornis*, there is considerable uncertainty about the proper allocation of species included in that genus and in *Porzana*. The genus *Porzana* is possibly polyphyletic and is very difficult to categorise because of differing taxonomic treatments. As defined here it includes 16 species, following Sibley & Monroe (1990), although the Neotropical Yellow-breasted Crake may be more appropriately placed in a separate genus *Poliolimnas* (Olson 1970), while there is reasonable evidence to remove the Black-tailed Crake from *Amaurornis* and place it in *Porzana* (e.g. Inskipp & Round 1989). The White-browed Crake is included by Olson (1973b) in *Poliolimnas* because of similarities to the Yellow-breasted Crake, but Mees (1982) has shown that the similarities are superficial while differences between the two species are very marked.

The species included in *Porzana* vary very widely in plumage. A natural group within the genus is formed by five living and one recently extinct species, the Little, Baillon's, Spotted, Australian Spotted

Crakes, the Sora and the Laysan Crake. All have grey underparts, and are streaked olive-brown and black on the upperparts, with variable white markings. Other species are predominantly black, e.g. the Spotless and Kosrae Crakes, while two species of eastern Asia, the Ruddy-breasted and Band-bellied Crakes, have olive-brown upperparts, and rufous-chestnut underparts with variable dark and pale barring in the posterior region. *Porzana* crakes are of worldwide distribution in temperate and tropical zones, and inhabit many types of palustrine wetland, though the habitat of living and extinct island forms includes forest, thickets and grass. Five species are long-distance migrants. Three flightless species of oceanic islands, the Laysan, Hawaiian and Kosrae Crakes, are recently extinct, having fallen victim to introduced predators and, in the case of the Laysan Crake, also to habitat destruction by introduced mammals. Of the 13 living species, two are considered Vulnerable: the very poorly known Dot-winged Crake of South America, which is threatened by habitat loss and disturbance, and the flightless Henderson Crake, an entirely black species very closely related to the Spotless Crake, which is confined to Henderson Island in the Pitcairn group.

93 LITTLE CRAKE
Porzana parva   Plate 28

Rallus parvus Scopoli, 1769, probably from Carniola.

Monotypic.

Synonyms: *Zapornia parva/pusilla*; *Rallus mixtus/pusillus/ peyrousii/minutus*; *Zapornia/Porzana/Gallinula/Crex pusilla/ minuta*; *Ortygometra olivacea/pusilla/minuta/parva/pygmaea*; *Crex parva*; *Gallinula minutissima*.

IDENTIFICATION Length 18-20cm; wingspan 34-39cm. Small, rather slim crake with longish neck, wings, tail and legs. Upperparts dull olive-brown with blackish feather centres, pale olive-brown to buffy feather fringes giving noticeable stripes along scapulars and tertials, and some white markings down centre of mantle and back. Marked sexual dimorphism in plumage: male has blue-grey face and underparts, with narrow white bars on rear flanks and black and white barring on undertail-coverts; in female blue-grey replaced by buff except for pale ash-grey supercilium, lores and cheeks, brownish ear-coverts, and whitish chin and throat, while barring from rear flanks to undertail-coverts more brownish and less distinct. Iris and eye-ring red; bill green with red at gape, duller in female; legs and feet green to yellow-green. From postjuvenile moult through first breeding season male and female retain some juvenile whitish markings on scapulars and upperwing-coverts (absent in full adult); male has olive wash on belly; female has paler buff underparts. First-winter birds have red-brown iris; gape has less red than in adult. Juvenile resembles female but has more white spots on upperwing-coverts, flight feathers and scapulars, almost white supercilium, darker brown ear-coverts, faint brownish barring or mottling on sides of head, and whitish to cream-buff underparts with brownish barring (broader on flanks) which may largely disappear with wear; undertail-coverts barred brown and dirty white; iris olive-green, becoming brown; bill olive-green with darker olive culmen and pinkish gape; legs and feet olive-green. Inhabits dense emergent vegetation of freshwater marshes and the edges of open waters, especially in areas with floating vegetation or flattened/dead vegetation at the water surface.

Similar species Slim shape a useful distinction from sympatric *Porzana* species. Unlike Spotted Crake (96) and Striped Crake (109), has barred undertail-coverts. Confusion most likely with Baillon's Crake (94) but has paler, duller brown, more uniform upperparts and upperwing-coverts, lacking contrasting white markings except on centre of mantle and back (but before complete moult after first breeding season birds also show some white markings on scapulars and upperwings), and with obvious pale olive-brown stripes along scapulars which continue along edges of tertials (these characters noticeable at rest and in flight), red spot at base of green bill, green legs and feet, much greater primary projection in folded wing (wings reach almost to tip of tail), longer tail (often not obvious in the field) and stronger, less fluttering flight. In Baillon's, white spots on upperparts smaller, more extensive and numerous, and more distinct, many being centred and bordered black; usually has more white on leading edge of outer wing but this character very variable in both species and not a consistent field character in Europe (Becker 1995). Male Little also separable by fainter underpart barring, starting further back, on rear flanks and lower belly. All ages separable from juvenile Baillon's on structure, and on upperparts colour and pattern, including pale lines along scapulars and tertials, and fewer white markings; adult female also differs from juvenile Baillon's in having pale salmon-buff underparts. Juvenile less strongly barred than Baillon's on underparts and has fewer white markings on upperparts and wings, but appears brighter than Baillon's because of paler fringes to upperpart feathers and less heavily patterned underparts. However, note that some juvenile Baillon's of nominate race (sympatric with Little) have white underparts with brown breast-band and flanks, and faint white barring on lower flanks only, thus closely resembling some juvenile Little on underparts: such birds are separable on pattern of upperparts.

VOICE Information is from Cramp & Simmons (1980). This crake is vocal in the breeding areas, although birds are often silent for long periods; some calls resemble those of Baillon's Crake and even amphibians. The advertising call of the male is a loud, rather nasal *quack* or *quek*, resembling the barking of a small dog, which is sometimes audible from 1.5km or more, and is repeated at 1-2s intervals for up to several min before accelerating and descending in pitch and ending on softer, more guttural notes (e.g. *quek-quak-ak-uk-u-u-u*); the next series begins almost immediately. Males sing mainly at night, sometimes continuing well after dawn; they may also call at any time during the day, when the series are usually shorter and with the terminal series of guttural notes omitted; calling in C Europe starts a few weeks after arrival in the breeding area and lasts from mid-May to early Jul. The female's advertising call is a fast, hard trill, often preceded by a few sharp *quek* notes, e.g. *quek-quek-quar-r-r-r-r-r*; this call was formerly mistakenly attributed to Baillon's Crake, and may be confused with a call of the Water Rail (61). The female also has a sharp *kik*, often given in duet with the advertising call of her mate. The alarm call is a sharp *tyiuck*, similar to that of Baillon's Crake. Contact calls include subdued trills, low *quek* notes and long sequences of soft, musical, rhythmic tapping sounds *dug-dug-dug*.... The nest-defence call is a snake-like hiss, and the warning call is rendered *rä(e)-trä(e)-trä(e)*...; when greatly agitated, the male utters a rapid series of 30-40 *tyutt* sounds. The contact note of chicks is a penetrating, long *i* or *tseeh*, rising in pitch.

DESCRIPTION

Adult male Centre of forehead, crown, hindneck, upper sides of breast, and entire upperparts, including scapulars, most upperwing-coverts, tertials and tail (but excluding lower mantle and back) olive-brown (variably more brown to pale rusty-brown), with dull blackish centres to scapulars, inner upperwing-coverts, tertials, rectrices, feathers of rump and sometimes feathers of crown; lower mantle and back black with olive-brown feather fringes, and double row of oval to elongated white dots which become less obvious with abrasion and may disappear by spring (Becker 1995). Olive-brown on edges of inner webs of tertials paler than on other upperpart feathers; this, with olive-brown region between dark scapulars and dark feathers of lower mantle and back, gives prominent pale lines along upperparts, visible at rest and in flight. Inner fringes of tertials fade to pale yellowish, thus enhancing prominence of pale lines on upperparts. Primary coverts, primaries, secondaries and alula dark sepia-grey, outer webs of outer primaries tinged olive-brown; outer web of outermost primary sometimes edged or mottled white, but this colour not conspicuous; axillaries and underwing-coverts uniform slate-grey, tinged olive-brown. Upperparts fade to drab brown or pale yellow-brown when worn (Cramp & Simmons 1980). Sides of forehead, and entire sides of head, pale slate-grey; chin paler ash-grey, whiter in fresh plumage; rest of underparts slate-grey; breast and belly feathers narrowly fringed whitish when fresh, this pattern disappearing with wear to give uniform slate-grey appearance; rear flanks and vent slightly darker and barred white; undertail-coverts black, broadly tipped and barred white; lateral lower flank feathers and undertail-coverts often tinged rufous-brown to brown. Iris and narrow orbital ring scarlet; bill grass-green, sometimes greyer or more olive at tip; gape red, colour sometimes extending forwards at base of both mandibles, and in older birds reaching nostrils and almost reaching base of culmen (Becker 1995); some second calendar year birds may have salmon-pink at base of bill (Cramp & Simmons 1980); legs yellow-green to green, more blue-green at joints; feet green or olive-green. Outside breeding season bill colours duller, red at gape paler or almost absent.

Adult female Upperparts, tail and wings like adult male but olive-brown extends slightly further on sides of forehead, crown and hindneck. Lores, streak over eye, and area around and below eye, pale ashy-grey; ear-coverts contrastingly brownish-buff; chin and throat off-white; sides of neck and body, and all underparts, salmon-buff to pale cream-buff, often paler on centre of breast and belly; rear flanks brownish, with dirty dark grey and whitish washed-out barring; short (outer) undertail-coverts barred buff and white, longer (inner) coverts barred black and white, rather brown-tinged; axillaries and underwing-coverts grey-brown. Bill duller than in male. In fresh plumage, feathers of underparts narrowly fringed off-white; when worn, upperparts duller and paler as in male, and pale grey areas on head purer, not buffy-tinged.

First adult No distinct immature plumage. After postjuvenile moult resembles adult, but upperparts not completely moulted from juvenile so that some white-spotted feathers retained on scapulars, upperwing-coverts and tertials, but white marks tend to abrade and become less visible than in juvenile. Primaries and rectrices fade from blackish-brown to brown by spring. Male has grey belly slightly washed olive-brown; female has belly paler buff, less salmon, than in adult. Bare parts as adult, but male may show less red at gape; iris duller red in first winter.

Juvenile Upperparts like adult, but feathers of crown and mantle more extensively black, appearing darker; mantle, scapulars and upperwing-coverts have white dots, often resulting in irregular streaks along scapulars; greater upperwing-coverts, primary coverts, primaries, secondaries, alula and tertials with variable number of white dots and marks at tips or subterminal short streaks on outer webs; in some birds only a few inner secondaries and tertials marked white towards tips, in others, all feathers have white markings; outer edge of outermost primary variably white, as in adult. Sides of head cream-white with white to buff supercilium and brown ear-coverts, and indistinct brownish barring or mottling; forehead, chin and throat more whitish. Underparts cream-white to cream-buff, with narrow relatively indistinct and wavy brownish barring which is broader on rear flanks; tips of feathers of breast and flanks variably grey; barring disappears with wear so that forehead, sides of head and neck, chin, throat, breast and parts of belly may appear almost uniformly whitish; undertail-coverts barred brown and dirty white. Iris becomes grey-brown, then olive-green, bright green and then progressively light brown or red-brown; bill pale green with darker olive culmen and yellow-tinged tip; gape dark flesh-pink or dull red; legs and feet dull olive-green, front of tarsus tinged yellow; some have legs and feet pale grey-brown or flesh-pink. Juvenile plumage retained longer than in Baillon's and Spotted Crakes.

Downy young Black downy chick has green gloss on upperparts, head and throat; iris grey-brown to brown-black; bill white with pink or yellow tinge, and grey to blackish at base especially on lower mandible; legs and feet grey-brown to dull black. Small patch of red skin visible on hindcrown; small white egg-tooth and wing-claw present. At 4 weeks, bill grey, with black tip and lower mandible; legs and feet black.

MEASUREMENTS Wing of 24 males 99-111 (106, SD 2.4), of 22 females 99-109 (103, SD 3.3); tail of 21 males 49-60 (51.4, SD 2.6), of 15 females 50-58 (52.1, SD 2.6); culmen of 25 males 17-20 (18.5, SD 0.6), of 22 females 16-19 (17.4, SD 0.6); tarsus of 25 males 30-34 (31.9, SD 0.9), of 20 females 29-32 (30.8, SD 1.1). Weight of 10 males 30-72 (50.0, SD 13.0), of 11 females 36-65 (49.5, SD 9.9) (Cramp & Simmons 1980). Hatching weight 5.5-6 (Glutz von Blotzheim *et al.* 1973).

GEOGRAPHICAL VARIATION None. Includes the form *illustris* from Turkestan (Ripley 1977).

MOULT Information is from Cramp & Simmons (1980). Postbreeding moult is complete, and moult of flight feathers is simultaneous. The underparts and head start to moult in May-Jun but stop when the flight feathers and

? occurrence uncertain
↘ passage route
x vagrant

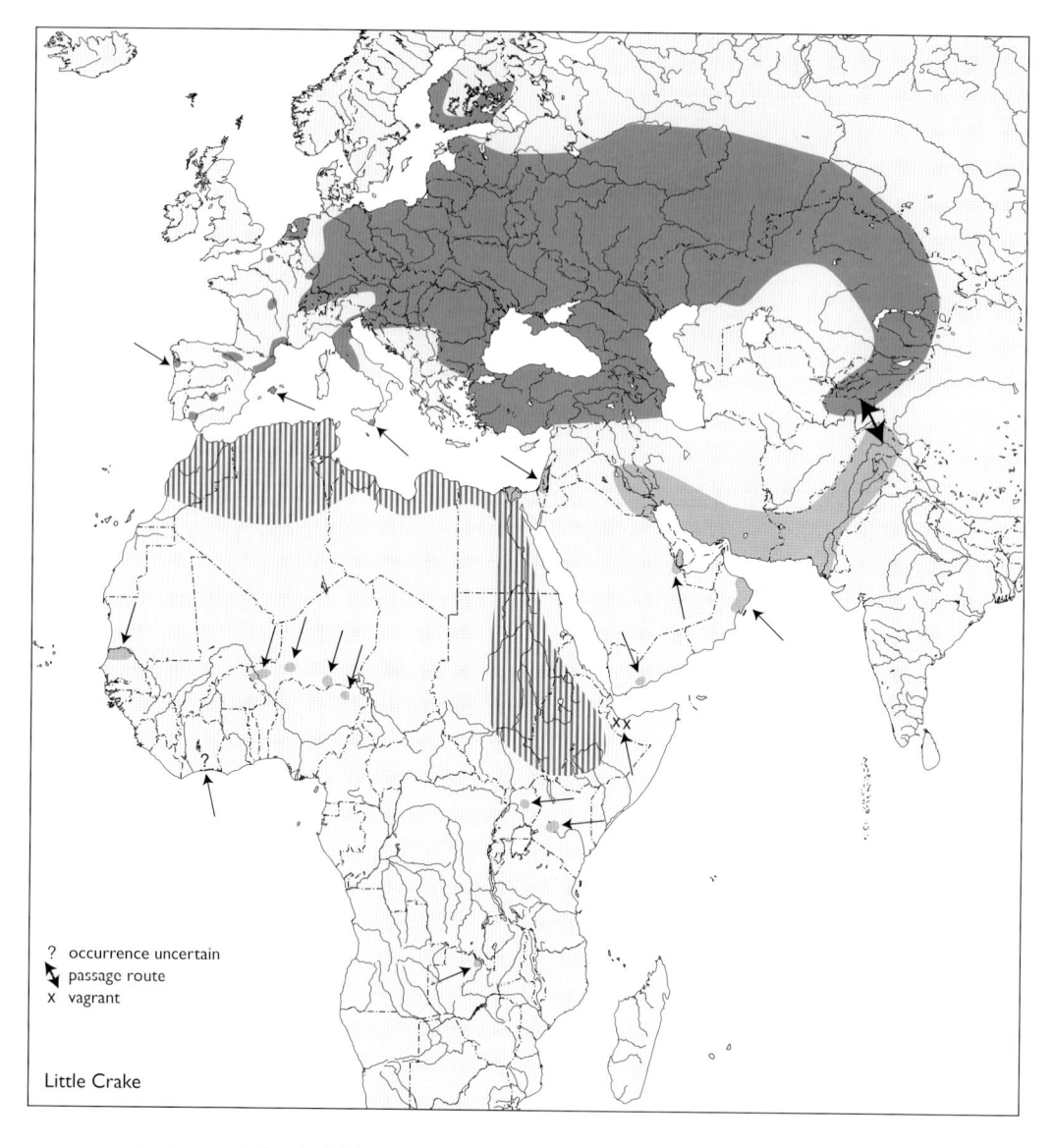

Little Crake

coverts are simultaneously lost in Jul-Aug, usually with many rectrices. When the wing feathers have regrown, the head and underparts are moulted, followed by the upperparts. Some birds are fully moulted by Oct but in others moult is interrupted by migration, and upperparts and sometimes underparts may moult in the winter quarters; all African wintering birds are in fresh plumage by Jan. Prebreeding moult is partial, but its extent is unknown and differences shown by bright spring plumage may be due mainly to abrasion. Postjuvenile moult is usually partial but sometimes complete, and starts in Sep; by Oct some birds have complete adult-type plumage below and some on sides of head, but many have not moulted significantly by then and moult often starts after arrival in winter quarters. From Oct/Nov feathers of entire head, body and tail are replaced, plus a variable number of wing-coverts and tertials; remex moult may also occur in some birds.

DISTRIBUTION AND STATUS Locally from Portugal and Spain E across Europe NE to Finland and Estonia, S to Italy, Sicily and Bulgaria, and E to Ukraine; E across the Transcaucasus, SE Transcaspia, W Tajikistan, Kazakhstan, to NW China in W Xinjiang (Tien Shan and Tarim basin E to Lop Nor). Its winter distribution is imperfectly known due to its secretive behaviour, and few wintering sites are known, so the distribution map is very tentative; it winters from the Mediterranean S to W Africa (Mauritania, Senegal, Nigeria, Niger) and E Africa (Uganda, Kenya), and E in Egypt, Israel, Oman and possibly elsewhere in Arabia, birds having been collected in Aden, Saudi Arabia, Bahrain and Kuwait. To the E, it winters in Iraq, S Iran (Khuzistan, Seistan), S Afghanistan, Pakistan (Baluchistan, Sind, Punjab) and NW India (Gilgit, and 1 record Bombay). W Palearctic breeders probably winter from the Mediterranean S to Africa and E possibly to Arabia, Iraq and Iran. It winters occasionally in Europe

(Netherlands and Hungary), W to Britain whence there are 2 Jan records (Cramp & Simmons 1980, Bradshaw 1993). Dowsett & Forbes-Watson (1993) also record it from Mauritania and Yemen.

It is not globally threatened. Its current status is poorly known; it was formerly thought to be less severely affected than its congeners by wetland drainage and reclamation, and in some areas possibly favoured by habitat changes (Glutz von Blotzheim *et al.* 1973). In Europe it bred regularly in 24 countries during the European Breeding Bird Atlas survey (Hagemeijer & Blair 1997) but its strongholds are in the S and SE. The atlas gives its European population as c. 16,300-20,200 (geometric mean 17,950) breeding pairs, with an additional 10,000-100,000 (31,600) pairs in Russia and 1-100 (10) in Turkey. The largest populations are in Austria, 4,000-6,000 (4,900) pairs (at Neusiedlersee); Romania, 3,000-6,000 (4,200); Ukraine, 3,500-4,000 (3,700); Belarus, 2,000-3,000 (2,500); Poland, 750-1,100 (900); and Hungary, c. 600 (Hagemeijer & Blair 1977). Its total population is estimated to be between 25,000 and 1,000,000 birds, and its numbers are decreasing (Rose & Scott 1994); declines are reported in 12 European countries, which together hold probably >25% of the European population (Hagemeijer & Blair 1997). It was formerly regarded as an uncommon to scarce breeder in Europe, but increases were reported in the Netherlands after 1942, and in Hungary with the introduction of rice cultivation and the expansion of fish ponds (Cramp & Simmons 1980); in well protected wetlands, populations have remained stable or increased locally at least since the 1980s and, despite declines in some regions since the mid-1970s the species has extended its range N, colonising Estonia in the 1930s, Sweden in the 1940s and Finland since the 1970s (Hagemeijer & Blair 1997).

In the former USSR, in the past it was regarded as rare in the N but common to fairly abundant in the S, with total numbers fairly large (Dement'ev *et al.* 1969); in Azerbaijan it is currently regarded as uncommon in summer, probably breeding (Patrikeev 1995). It is rare NW in China (W Xinjiang), where it breeds in the Tarim basin and occurs on passage in Tien Shan (Cheng 1987); it is scarce to rare in Pakistan and uncommon in NW India (Ali & Ripley 1980, Roberts 1991). It is quite common on passage in Israel but rare in winter (Shirihai 1996), scarce in Egypt (Goodman *et al.* 1989), and rare in Oman (Gallagher & Woodcock 1980). Apart from the considerable numbers recorded on passage through N Africa, including many recent records from Algeria and Libya, records from Africa are very few. A total of 12 birds has been recorded in N Nigeria (Dec, Feb, Mar), 4+ in S Niger (Sep, Oct, Jan), 7 in W Uganda (Dec 1901), 4 at Thika in C Kenya (Jan-Feb 1983-85) and 1 in Zambia (1-13 Mar 1980) (Jackson & Sclater 1938, Taylor 1980 and unpubl., Wilkinson *et al.* 1982, Urban *et al.* 1986, Giradoux *et al.* 1988, Ash 1990, Rodwell *et al.* 1996). However it is not rare in N Senegal, where Morel & Roux (1966) record 15 birds for the Sep-Nov period, and 25 records were obtained at Djoudj NP in Dec 1991 and Jan 1992 (Rodwell *et al.* 1996). It is uncommon in Sudan, where it is recorded only in the autumn (Nikolaus 1987), and rare in Ethiopia and Eritrea (Urban & Brown 1971). There are 2 old records from NW Somalia, in Sep and Oct (Ash & Miskell 1983, Clarke 1985). However, apparently large numbers breeding in the W of the former USSR, the considerable passage through N Africa, and its occurrence in the deserts of Algeria and Libya, suggest a wider sub-Saharan distribution and greater numbers than are currently known. It bred in Algeria (Halloula and Fetzara Lakes) in the mid-19th century, and more recent breeding is suspected at Biskra and Djelfa (Cramp & Simmons 1980); it also possibly bred once in the Nile Delta, Egypt, where a clutch of eggs taken in April 1916 is from either this species or Baillon's Crake (Goodman *et al.* 1989). Ash (1990) maintains that the occurrence of this species in NE Nigeria, where he recorded it alongside the Spotted Crake in Feb-Mar 1987, suggests the possibility of "a breeding area in NW Africa or even within Nigeria": this speculation is entirely without foundation, and existing Nigerian records fit the bird's known pattern of occurrence as a Palearctic migrant wintering in the sub-Saharan region.

In the breeding areas loss of habitat through wetland destruction and degradation, and intensive reed-harvesting, gives cause for concern (Hagemeijer & Blair 1997). It is not clear whether habitat loss in non-breeding areas is also significant, but the species is able to occupy artificially created wetland patches. Predators of adults and young include the European Marsh Harrier *Circus aeruginosus* (Dement'ev *et al.* 1969).

MOVEMENTS Information is from Cramp & Simmons (1980) and Urban *et al.* (1986) unless otherwise specified. S movements occur from late Aug-Nov, probably peaking Sep-Oct; it occurs as a migrant on all the larger Mediterranean islands, and it passes through N Africa (Morocco to Egypt) from late Aug-Oct; however, movements through Israel occur from mid-Jul to end-Nov, peaking mid-Aug to end-Sep, and wintering birds are present from Oct to Mar (Shirihai 1996). It occurs in Sudan from Sep-Nov and Somalia in Sep-Oct (Ash & Miskell 1983, Nikolaus 1987); it is present in Oman from Aug-Apr (Gallagher & Woodcock 1980). In Britain and Ireland, 29 of 31 records are from early spring and autumn, with a peak in early Nov (Bradshaw 1992). Presumed wintering birds are recorded in Senegal, N Nigeria, S Niger, W Uganda, C Kenya and Zambia (see Distribution and Status). It probably winters regularly S to the equator in Kenya and Uganda and, like the Spotted Crake, it is probably itinerant in its African wintering areas in response to habitat changes (Taylor 1987). Return passage is noted in N Africa from Mar-May (sometimes in Jun) and in the Mediterranean from late Feb-May; passage in Israel occurs from Feb to mid-Jun, mainly early Mar to mid-Apr, and there are a very few records of summering birds in May-Jul (Shirihai 1996). A presumed migrant (very fat) is recorded from Ethiopia in Apr (Cheesman & Sclater 1935). Most birds arrive on the breeding grounds in late Mar-Apr. Most N African records are in spring, suggesting that autumn migrants may overfly the region. It is a vagrant or an uncommon (overlooked?) winter visitor to Liberia (Gatter 1988, no dates) and possibly Ivory Coast (1 Jun record, questionable on date and habitat – see Thiollay 1985); it is accidental in Ireland, Denmark, Norway, Finland, Portugal, Lebanon, Syria, Azores, the Canary Is and Madeira. The longest distances for ringing recoveries are Switzerland to SW France (6-17 Oct), 780km SW, and Czechoslovakia (Aug) to S Italy (Mar), 950km S.

In the former USSR, S passage begins in mid-Sep and continues until Nov, while arrivals occur from the first half of Mar to the second half of May, predominantly in April (Dement'ev *et al.* 1969). In Azerbaijan passage has been noted in Nov and Mar-Apr, and it formerly remained during warm winters but there is no recent information (Patrikeev 1995); it occurs in Pakistan from Sep to Mar and is recorded on passage through Gilgit in Oct-Nov (Roberts 1991). Birds normally migrate singly; concentrations are rarely recorded, possibly because the species is secretive. In W Uganda, this species and Jack Snipe *Lymnocryptes minimus* arrived together at Butiaba on 6 Dec (Jackson & Sclater 1938).

HABITAT Most information is taken from Cramp & Simmons (1980). In the breeding season it occurs typically in temperate and steppe zones, usually in lowlands but in favourable seasons extending into boreal regions. It frequents dense emergent vegetation of freshwater marshes and other wetlands (usually natural or semi-natural, and eutrophic rather than oligotrophic), including the margins of lakes and rivers, floodplains (where it may occupy the smaller ponds and oxbows in regularly inundated areas), and flooded woodland such as alder *Alnus* coppices. It is fond of floating debris or plants along the edges of reedbeds, and prefers still or slow-moving water. In the breeding season it differs from sympatric congeners in favouring monospecific or mixed stands of tall plants such as *Scirpus*, *Typha*, *Carex*, *Sparganium*, and also *Phragmites* where freedom from regular cutting or burning allows a mixture of dead and living stems; such stands may be in fairly deep water (>50cm) and important requirements are small stretches of open water and a layer of broken or horizontal stems at ground or water level (Cramp & Simmons 1980, Hagemeijer & Blair 1997). In C Europe, deltas of the larger rivers are coastal strongholds (Hagemeijer & Blair 1997). Outside the breeding season it occupies a wider range of habitats and may occur alongside Spotted and Baillon's Crakes at small ponds and other wetland patches, and it is also recorded from seasonally inundated grassland 30-50cm tall in 1-20cm of water (Taylor 1987). At Thika in C Kenya it foraged alongside Baillon's Crakes in tangled stems of robust *Polygonum* spp growing over shallow water, and took refuge in adjacent tall *Cyperus* and *Typha* beds (P. B. Taylor unpubl.). It frequents rice fields in both the breeding and wintering areas. It prefers more deeply flooded habitats than does Baillon's Crake, and it appears to be more tolerant of water level fluctuations, including rises in the water level during the breeding season, than are Baillon's or Spotted Crakes. It typically breeds in lowlands below 200m but in Armenia it nests at c. 2,000 m, and on passage it occurs to 2,000m in N Pakistan and 2,100m in Armenia (Dement'ev *et al.* 1969, Roberts 1991). On migration it occurs in atypical habitats.

FOOD AND FEEDING Mostly insects, especially water beetles (Hydrophilidae) and including Scarabeidae, and also Hemiptera (including Cicadidae), Neuroptera, and adult and larval Diptera, and seeds of aquatic plants (*Carex*, *Sparganium*, *Polygonum*, *Potamogeton* and *Nymphaea*); it also takes worms (Annelida), gastropods (*Valvata*, *Bithynia*, *Ancyclus*, *Planorbis*), spiders and water mites (Arachnida), and aquatic vegetation such as young shoots. This crake is quite aquatic in its feeding habits; it forages while swimming, wading or walking over stems and leaves, particularly floating leaves; it takes food from the water surface, from emergent vegetation, from the surface of mud (P. B. Taylor unpubl.) and by probing in mud (Roberts 1991). Birds energetically hunt spiders inhabiting old *Typha* stands (Dement'ev *et al.* 1969).

HABITS It is normally difficult to observe because of the densely vegetated nature of its habitat, but it will show itself fairly freely in the open (Cramp & Simmons 1980); foraging birds may approach very close to an observer in the open, sometimes moving as much as 10m from cover, and may become fairly bold when accustomed to human presence although the tail is usually flicked constantly when the bird is close to the observer (Taylor 1980 and unpubl.). Birds often run to escape danger, even sometimes when chased by a dog (Taylor 1980), but will also fly laboriously for a short distance before dropping into cover (Cramp & Simmons 1980); however in some circumstances, the escape flight is longer (up to 200 m), faster and more powerful, with shallow wingbeats (Taylor 1980). Birds will dive to escape pursuit, and will either remain submerged or will move away underwater (Roberts 1991, P. B. Taylor unpubl.). This species moves freely over floating vegetation and climbs easily up stems and among growing emergent plants (sedges, reed etc) and bushes overhanging the water. When moving through dense, flattened vegetation at ground level it assumes an elongated shape, with the neck stretched horizontally, and moves smoothly and rapidly, causing no disturbance of the vegetation (P. B. Taylor unpubl.). Birds possibly roost at night since their activities are predominantly diurnal.

SOCIAL ORGANISATION Monogamous and territorial; the pair bond is maintained only during the breeding season. In favourable habitat nests are frequently only 30-35m apart (Dement'ev *et al.* 1969). At Neusiedlersee, Austria, the breeding density in the vast reedbeds averages 1-2 pairs/ha of suitable habitat, with up to 5 pairs/ha in optimum habitat; at the N end of the breeding range densities are much lower: 2.5 pairs/km^2 in W Estonia and 0.3-2 territories/km^2 in SE Finland (Hagemeijer & Blair 1997).

SOCIAL AND SEXUAL BEHAVIOUR Birds may occur in groups on migration, and in autumn sometimes associate with Spotted Crakes (Bauer 1960b). In experiments with wild birds, stuffed specimens were attacked anywhere in the territory, but stuffed Spotted Crake and Water Rail were attacked only when <0.6m from the nest; the attacking bird aimed blows persistently at the back of the decoy's head (Koenig 1943).

Both sexes give advertising calls before pair formation, and the male may continue to do so afterwards, probably only to indicate occupation of the territory in response to potential threats from new arrivals or when population density is high and territorial conflict greater (Glutz von Blotzheim *et al.* 1973). Courtship and copulation are not described. At change over during incubation the sitting bird leaves the nest of its own accord, moves a fewm away and calls to its mate, which replies without delay and arrives to incubate (Glayre & Magnenat 1977).

The incubating bird will often remain on the nest in the presence of a human observer, may attempt to close gaps in the vegetation around the nest, and may peck an approaching hand, ruffling the feathers and stretching the wings forwards in an intimidation display, as well as making snake-like hissing sounds (Koenig 1943).

BREEDING AND SURVIVAL Season C Europe and former USSR, May-Aug; bred once in Egypt, Apr. **Nest** A shallow to deep cup of plant stems and leaves gathered from close to nest site; placed in thick vegetation (sedges, *Typha*, *Phragmites*, willows etc) over or near water; will choose nest sites approachable only by swimming; nest sometimes raised on tussock or platform of dead material; nest base often on surface of water, but when water drops nest may be 20-30cm above surface. External diameter 11-20cm, internal diameter 10-16cm, height 2-9cm, depth of cup 2-7cm. Nest building takes 3-7 days; both sexes probably build; male also builds brood platforms around breeding nest. **Eggs** 4-11 (usually 7-9); short oval, sometimes biconical; smooth and glossy; buff to yellow-buff, densely marked with brown to rufous-brown spots and blotches. Size of 145 eggs 27.5-33.5 x 19-23 (30.4 x 21.7), fresh weight 6.3-8.7 (8.0). Eggs laid at daily intervals; replacements laid after egg loss. **Incubation** 21-23 days (15-17 days per egg), by both sexes; hatching asynchronous; female may incubate during hatching while male cares for first-hatched young. **Chicks** Precocial and nidifugous; fed and cared for by both parents while small (in 1 pair, female initially brooded young and male brought food); become self-feeding after a few days; probably brooded in nest for much of breeding period. Parents may divide brood for short periods while foraging. Young can fly at 45-50 days, when fully fledged, but are independent and probably deserted by parents before this; 2 broods quite frequent. Age of first breeding 1 year. Nest predators include rats *Rattus norvegicus* and mice *Arvicola terrestris* (Glutz von Blotzheim *et al.* 1973). **Survival** The oldest ringed bird was at least 6 years (Glutz von Blotzheim *et al.* 1973).

94 BAILLON'S CRAKE
Porzana pusilla   Plate 28

Rallus pusillus Pallas, 1776, Dauria.

Forms superspecies with recently extinct *P. palmeri* of Laysan I (Hawaii). Possible race *obscura*, of sub-Saharan Africa and Madagascar, of doubtful validity. Six subspecies recognised.

Synonyms: *Gallinula bailloni/stellaris/pygmaea/pusilla/minuta*; *Zapornia pygmaea/pusilla*; *Rallus intermedius/bailloni/punctatus/pusillus*; *Phalaridium bailloni*; *Ortygometra auricularis/affinis/bailloni/pygmaea/minuta/pusilla/palustris*; *Porzana obscura/palustris/affinis/intermedia/pygmaea/bailloni*; *Crex pygmaea/bailloni/pusilla*.

Alternative names: Marsh/Tiny Crake.

IDENTIFICATION Length 17-19cm; wingspan 33-37cm (*intermedia*). Small, rather dumpy crake, short-tailed and short-winged, with short primary projection in folded wing; often appears short-necked; in flight, looks especially round-bodied and usually appears dark. Crown and upperparts, including upperwings, rich brown (shade and brightness varying racially), with black streaks and numerous white markings (spots, streaks, wavy lines etc), most edged black and many with black in centre. White usually visible along leading edge of wing in flight but this character variable; white marginal coverts sometimes do not show on inner wing (Europe). Sides of head, and entire underparts to upper belly and foreflanks, slate-blue in race *intermedia* (Europe and Africa), paler in nominate (Asia) and *affinis* (New Zealand), and pale grey in other races (throat and breast white in *mira* of Borneo); chin paler in female and in non-breeding birds (may not be a consistent character), and more extensively pale, often white, in *mira*, *mayri* (New Guinea) and *palustris* (Australia). Brown streak from rear of eye across ear-coverts in nominate, some *intermedia*, and female and some male *palustris*. Rest of flanks, and lower belly to undertail-coverts, barred blackish and white, belly unbarred in *palustris* and *affinis*. Eye red; bill green; legs and feet greenish, olive, yellowish or flesh-coloured. Sexes similar; immature like adult. Juvenile has grey of underparts replaced by rich buff to white, often mottled from face to breast, and (in *intermedia* and some nominate) dark barring extending forwards to upper flanks and breast; iris and bill initially brownish, and legs and feet brown to grey-flesh, but bare parts rapidly attain adult colours. Inhabits relatively short emergent and fringing vegetation of marshes, flooded grassland, floodplains, open waters and irrigated crops; often forages on floating vegetation.

Similar species Separable on plumage, size and structure from sympatric congeners. Most easily confused with Little Crake (93): see that species for discussion of relevant field characters. Legs and feet may be flesh-coloured but are frequently greenish-grey, olive or yellowish and thus may resemble those of Little Crake. White on marginal coverts said not to show on leading edge of wing in Europe (Becker 1995), but birds in Africa usually show narrow white leading edge to entire wing in flight (P. B. Taylor unpubl.), as they also apparently do in Asia (e.g. Lekagul & Round 1991). Australian Spotted Crake (97) larger and darker, with dark brownish-olive upperparts; more numerous and regular white markings on upperparts (including from crown to hindneck); darker grey sides of head and anterior underparts; white spots on sides of neck and breast; white undertail-coverts; and red or orange spot at base of pale green to yellowish bill. Juvenile Australian Spotted Crake darker and duller on upperparts than juvenile *palustris*, greyer on underparts, with darker barring extending to belly, and pale scalloping on foreneck and breast (scalloping also present in immature). See also Lewin's Rail (65), Spotted Crake (96) and Striped Crake (109).

VOICE Most information is from Cramp & Simmons (1980), supplemented by observations from P. B. Taylor (unpubl.). The male's advertising call is a hard, dry rattle, in African birds identical to the rattle call of the African Rail (62), lasting 1-3s and repeated every 1-2s. It is sometimes preceded by a number of dry *t* sounds. The call is audible for at least 250m, and birds call mainly at night, but in Africa they often call early and late in the day, and may respond well to taped playback. Brief versions of the call are also given, e.g. during courtship flights. The normal alarm call of both sexes is a sharp *tac* or *tyiuk*, similar to that of the Little Crake; a rather explosive *krrrik* at the nest is also an alarm call. Another call, apparently with a territorial or alarm function, is a series of loud, energetic, rapidly repeated grating notes, often in a phrase such as *kraa-kraa-kraa-chachachachacha*; African breeding birds normally give a higher-pitched, less sonorous and faster version of the call than do W Palearctic birds. This call is said to be given infrequently in the W Palearctic, but is often heard in Africa, especially in response to taped playback and most frequently in the breeding season. There is little information from Australia (Marchant & Higgins 1993), where many observations may be made without hearing the species call (R. Jaensch *in litt.*).

In Pakistan and India the territorial or advertising call

is described as a high-pitched *crake* or *crek*, repeated in an accelerating series until it becomes a trill rather like a that of Little Grebe *Tachybaptus ruficollis* (Bates & Lowther 1952, Ali & Ripley 1980, Roberts 1991). This call appears to resemble the advertising call of Little Crake and to differ markedly from the normal calls in Europe, Africa and Australia, and merits further investigation. Playback of the call caused a curious and possibly aggressive male Baillon's Crake to emerge from cover (Roberts 1991).

The nest-defence call is a fierce-sounding *kree-HEET*; the warning call is *tye(ä)k-tye(ä)k...*; a possible warning call near the nest is described as *oo*; and very agitated birds give a repeated *tyutt* sound, as does the Little Crake. Intimidation calls include soft, low growls. Other calls include soft grunts, clicking calls given during the distraction display, soft whining and sharp querulous calls when harassed, and a high-pitched call in flight (Cramp & Simmons 1980, Marchant & Higgins 1993).

DESCRIPTION *P. p. intermedia*
Adult male breeding Central forehead and crown to hindneck rufous-brown, indistinctly spotted black on crown; upperparts, including scapulars, upperwing-coverts, tertials and tail, rufous-brown; inner scapulars, feathers from mantle to tail, tertials and inner coverts with black centres, and small white spots or streaks outlined by black on mantle, back, scapulars, outer tertials, inner and greater wing-coverts, and some rectrices; white marks usually with small black speck or streak in centre, except on wing-coverts. Primary coverts, primaries, secondaries and alula dark grey-brown, slightly rufous-brown on outer webs, some with small white dots or streaks near or at tips of outer webs (less than in juvenile); marginal coverts, and outer edge of outermost primary (P10) and alula, white. Underwing-coverts and axillaries dark grey, tinged brown, irregularly barred or spotted white. Broad streak over eye, sides of head and neck, chin to breast, upper belly and foreflanks uniform slate-blue; chin sometimes slightly paler grey. Rest of flanks, lower belly, vent, thighs and undertail-coverts boldly barred black and white. In fresh plumage, slate-blue feathers of underparts have narrow white margins; when worn, hindneck partly streaked black, and white marks on upperparts and upperwing-coverts partly abraded. Iris scarlet, sometimes bright red-brown or orange-red; narrow eye-ring yellow, more orange-yellow when nesting; bill deep green, slightly darker olive-horn on culmen and tip; legs and feet flesh-brown, olive-brown, olive-grey, yellow-grey or olive-green. Becker (1995) records that in European birds the legs are pale olive-green to olive-brown, and considers brown or flesh colours to be either artefacts or temporary changes as a result of 'sun-tan'. The legs and feet of adults and juveniles from E, C and S Africa are often flesh with a yellow tinge, yellowish or yellow-brown (sometimes tinged greenish), but some adults are pinkish-grey or greenish-grey (P. B. Taylor unpubl.).
Adult female breeding Like adult male breeding, but chin in 60% of females paler, more ash-grey, than rest of underparts; some have indistinct brown streak over ear (Cramp & Simmons 1980). Becker (1995) disputes that the brown on the ear-coverts is a sex-related character; it does occur in some males.
Adult non-breeding Like breeding, but chin of male ash-grey, of female off-white.
First adult No distinct immature plumage; after post-juvenile moult, birds are identical to adults in colour and pattern of plumage and bare parts, but wings are relatively more abraded.
Juvenile Upperparts, wings and tail like adult but crown duller and more heavily streaked; white markings mostly lacking black specks in centres. Primaries and secondaries variably spotted white along outer webs; primary coverts and alula also variably spotted white towards tips of outer webs. Sides of head and neck rufous-brown, mottled off-white, pale buff, or pale grey on feather bases; ear-coverts slightly darker; chin and central belly white to pale buff; breast and foreflanks white, pale buff or cinnamon-buff, closely barred dull brown; rest of flanks, and belly to undertail-coverts, white, heavily barred dark grey to dull black, bars browner on belly. Iris olive-green or brown with olive ring round pupil; bill dark horn, base of upper mandible and most of lower mandible pale horn tinged olive, or brownish-yellow; legs and feet flesh-grey, flesh-brown or olive-brown.
Downy young Black downy chick has bottle-green gloss; iris bluish-brown to black; bill ivory or bone-white to straw-yellow, with some black at base; legs and feet grey-brown to black. Some red skin shows on hindcrown and some blue skin over eye; small white wing-claw and egg-tooth present.

MEASUREMENTS Birds from Europe and N Africa (Cramp & Simmons 1980): wing of 19 males 89-97 (92.9, SD 2.6), of 11 females 87-96 (91.0, SD 3.2); tail of 20 males 39-48 (43.2, SD 2.3), of 12 females 40-46 (43.2, SD 2.0); exposed culmen of 21 males 16-18 (17.2, SD 0.8), of 12 females 15-17 (16.2, SD 0.7); tarsus of 24 males 26-30 (28.3, SD 1.0), of 15 females 25-29 (27.0, SD 1.5). Birds from S Africa (20 males, 20 females): wing of male 81-89 (84.9, SD 2.1), of female 79-91 (83.7, SD 3.0); tail of male 37-44 (39.8, SD 1.9), of female 37-45 (40.9, SD 2.2); culmen to base of male 18-20.5 (19.5, SD 0.6), of female 17-20 (18.5, SD 0.8); tarsus of male 23.5-27 (26.2, SD 1.0), of female 22.5-27 (25.2, SD 1.2). Benson (1964): 45 unsexed from Europe, wing 84-95 (90.0), culmen to base 17-20.5 (18.7); wing of 84 unsexed from Africa 76-93 (84.2), of 26 unsexed from Madagascar 80-94 (84.8); culmen to base of 84 unsexed from Africa 17-21 (18.4), of 26 unsexed from Madagascar 18-21 (19.5). Weight of 1 male 42, of 1 female 46 (Europe); of 7 unsexed 30-38 (35.1, SD 2.9) (Africa). Hatching weight 5-5.5 (Glutz von Blotzheim *et al.* 1973, Fjeldså 1977).

GEOGRAPHICAL VARIATION Variation distinct over much of range, races being separated on size and colour, but the boundary between *intermedia* and nominate in E Europe is indistinct (Cramp & Simmons 1980). The possible race *obscura*, of sub-Saharan Africa and Madagascar, is said to differ from *intermedia* in being generally darker, especially darker grey below, and in having a shorter wing and bill (Mackworth-Praed & Grant 1957, 1962, Ripley 1977). However Benson (1964) found no difference between European and African birds in the colour of the upperparts, and noted a considerable overlap in the colour of the underparts and in bill length. Although on average European birds have longer wings there is considerable overlap in wing length, and Benson considered the race *obscura* not distinct; it was recognised by Cramp & Simmons (1980) on the basis of wing length, but not by Urban *et al.* (1986). The race *mayri* is possibly synonymous with *palustris* (e.g. Marchant & Higgins 1993), and the putative form *fitzroyi* is included in *palustris* (Peters 1934).

P. p. intermedia (Hermann, 1804) – (includes *obscura*) Continental Europe (Portugal, Spain, France, Belgium, Netherlands, Germany, Switzerland, N Italy, Sardinia, Austria, Hungary, former Yugoslavia, Romania, Moldova, Bulgaria, Greece); Turkey (race?); Jordan (race?); Africa, locally in N Morocco, Tunisia, Egypt, possibly N Algeria, and Ethiopia, and from NE Zaïre (E Kasai, Kivu, L Edward), Rwanda, W Uganda, Kenya, C and S Tanzania, Zanzibar, S and SE Angola, Zambia, Malawi, Mozambique, Zimbabwe, N Namibia, Botswana and E and S South Africa; also Madagascar. European birds winter in Africa, including Senegal and possibly S to equator; also in Egypt and Iraq; occurs on migration in Mauritania, Sudan, S Iran (Fars) and Arabia. The largest race; richly coloured and boldly patterned. See Description.

P. p. pusilla (Pallas, 1776) – C and S Russia and S Siberia, from Ukraine (R Dnestr) to mouth of Volga R and N to Smolensk, Gorky and Bashkiria (Ufa) regions, S to Orsk and Semipalatinsk, L Balkhash, S Altai, and E across S Siberia via Tyumen, Tomsk and N along the Yenisey R to Mirnoye at 62°15'N (Rogacheva 1992), to NW of L Baikal (Kirensk), Transbaikalia, N Mongolia, S Amurland, Ussuriland and S Sakhalin; China (extreme W Xinjiang and in NE Nei Mongol, Heilongjiang, Liaoning, Hebei, Henan and S Shaanxi); and Japan (Hokkaido, N Honshu). Also Turkestan (R Syr Darya), R Amu Darya NW to Aral Sea and E to Tajikistan and Pamir Mts; E and SW Iran (Luristan, Khorasan, Baluchistan); and N India (Kashmir and lower Himalayas). Also irregularly in Sumatra (including Belitung I) and N Sulawesi; Moluccas (Seram) and Lesser Sundas (Flores). Winters S Iran, Pakistan, Nepal, India, Sri Lanka, Andaman Is, NW Bangladesh, Burma, Thailand, Malay Peninsula, Vietnam, N Laos, Cambodia, SE China (Guangdong); also S Japan (S Honshu, Shikoku, Kyushu), Ryukyu Is (Amami-oshima, Okinawa, Ie-jima, Miyako-jima, Ishigaki-jima) and occasionally Izu Is (Hachijo-jima) and Kita-iwo-jima. Also winters Sumatra, W Java, Bali, and the Philippines (Dinagat, Luzon, Marinduque, Negros, Palawan). Transient in Tibet, C China (Xinjiang, E Nei Mongol, Qinghai, Gansu, N Shaanxi, Shandong, Hubei, Hunan, Jiangxi, Zhejiang, Fujian, Guangxi Zhuang and SE Yunnan) and Korea. Large; usually has well-marked brown streak from rear of eye across ear-coverts; upperparts slightly paler than in *intermedia*, with white marks smaller, more numerous and more heavily speckled and outlined black; rufous-brown often extends to sides of lower neck and sides of upper breast; chin off-white; underparts paler, darker ash-grey (even paler in female, which has chin and throat white, while some are largely white, grey being restricted to sides of head); belly tinged white; barring extending over belly, as in *intermedia*. Juvenile often less heavily barred below than in *intermedia*; some have white underparts with brown breast-band and flanks, and faint white barring on lower flanks only, thus closely resembling some juvenile Little Crakes on underparts. Wing of 43 males 87-96 (91.1, SD 1.4), of 36 females 84-94 (88.8, SD 2.1); exposed culmen of 42 males 16-18 (17.0, SD 0.7), of 37 females 15-17 (16.2, SD 0.8); tarsus of 51 males 26-30 (28.5, SD 1.1), of 46 females 26-29 (27.8, SD 1.1) (Cramp & Simmons 1980). Weight of 2 females 32, 36; of unsexed 28.5-52; of 11 juveniles 29-44 (33.3) (Cramp & Simmons 1980).

P. p. mira Riley, 1938 – SE Borneo. Known only from the holotype. The smallest race; similar to nominate from N China but crown and upperparts more rufous-tinged; sides of head paler grey; throat and breast white; foreneck very pale grey; belly barred; bill of specimen olive-yellow, base and culmen bronze. Wing 86; tail 39; culmen 16; tarsus 27.

P. p. mayri Junge, 1952 – New Guinea (Wissel L area in Weyland Mts). Possibly synonymous with *palustris*. Smaller than *palustris*, with upperwings and sides of neck warmer brown, less tawny; underparts paler grey than in nominate and *intermedia*. Measurements of 2 males, 2 females: wing of male 76, 79, of female 78; tail of male 39, 42, of female 45; culmen of male and female 15; tarsus of male and female 24 (Ripley 1977).

P. p. palustris Gould, 1843 – E New Guinea (Port Moresby area and C Highlands), Australia (mainly in E, SE and SW) and E Tasmania. Small; similar to nominate but much paler, and underparts paler than in *affinis*. Up to 15% of males have rufous-brown eyestripe from lores to ear-coverts; male has sides of head, most of throat, and foreneck, light grey; chin and upper throat varyingly paler grey to white, possibly paler in breeding plumage; throat and breast pale grey, whiter in centre; belly white, unbarred or faintly barred brown-grey; flanks, thighs and undertail-coverts barred grey-black and white. Iris red-brown to red, sometimes light hazel or orange-red, and grey inner ring also reported; narrow eye-ring yellow to orange-buff in nesting birds; bill green, culmen and tip dark grey or olive-grey; in darkest, bill mostly dark grey, washed olive at sides and yellow at base; legs and feet olive-yellow to olive. All females have some trace of rufous-brown eyestripe: in 65% it is continuous from lores to rear of ear-coverts; Upperparts browner than in male, with stronger olive tinge (male predominantly cinnamon-brown). Juvenile has iris, bill and legs brown; adult colour attained before postjuvenile moult. Downy young has bill yellowish-white, with black gape; some have black at base of upper mantle and extending over 20% of lower mandible from base; legs and feet dark grey. Wing of 16 males 76-93 (82.3, SD 3.8), of 12 females 77-85 (81.1, SD 2.2); tail of 15 males 35-47 (43.1, SD 3.0), of 9 females 36-47 (42.7, SD 3.2); exposed culmen of 16 males 14-16.5 (15.5, SD 0.7), of 10 females 14-15.5 (14.8, SD 0.5); tarsus of 16 males 23.5-27.5 (25.3, SD 1.2), of 12 females 23-26 (24.9, SD 0.8) (Marchant & Higgins 1993). Weight of 15 adults (including 2 known vagrants weighing 17 and 23.8) 14-42 (29.2, SD 6.7); of 7 fully grown juveniles 20-38.7 (28.2, SD 6.6) (Marchant & Higgins 1993).

P. p. affinis (J. E. Gray, 1846) – New Zealand, including Stewart I, Little Barrier I, and Chatham Is. Slightly larger than *palustris*, with longer bill; grey of sides of head and anterior underparts paler than in nominate and *intermedia*, but darker than in *palustris*; lores dark grey-brown to black-brown; belly unbarred. Bare parts of all ages as in *palustris*, but bill larger. Wing of 22 males 78-89 (84.0, SD 2.6), of 22 females 69-88 (83.8, SD 3.6); tail of 8 males 34-50 (45.6, SD 5.3), of 9 females 39-50 (45.7, SD 3.5); exposed culmen of 21 males 16-19.5 (17.9, SD 0.9), of 22 females 15-21.5 (17.1, SD 1.3); tarsus of 21 males 25.5-31 (27.5, SD

Baillon's Crake

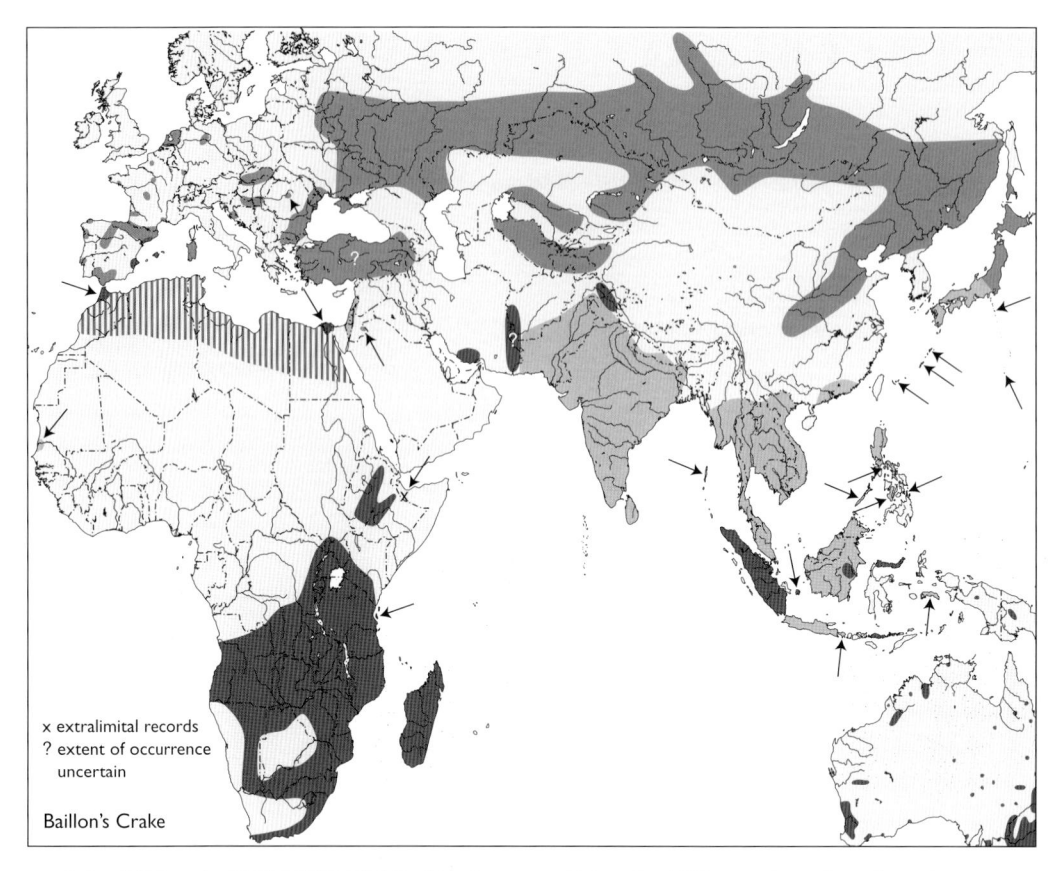

x extralimital records
? extent of occurrence uncertain

1.4), of 21 females 20-32.5 (26.8, SD 2.2) (Marchant & Higgins 1993). Weight of 5 males 31-45 (36.2, SD 4.3), of 1 male with no fat 22.9, of 7 females 32-43 (35.7, SD 3.3), of 3 light females 25.5, 22.4, 20, of fat female with oviduct egg 55.3; of 8 unsexed 40-46 (41.7) (Kaufmann & Lavers 1987, Marchant & Higgins 1993).

MOULT Postbreeding moult is complete, and remex moult is simultaneous. Rectrix moult may also normally be simultaneous (recorded once in Australia). In the W Palearctic some adults moult in the breeding areas in Jul-Sep but others apparently migrate in worn breeding plumage and moult in the winter quarters. The underparts are moulted first, from the foreneck downwards. The prebreeding moult is partial; it is poorly known and occurs in spring. Postjuvenile moult is partial, involving the entire body and tail; some European birds have not moulted by late Oct and thus apparently delay moult until the winter quarters are reached, but others are nearly fully moulted by late Oct, while some have renewed head and most of underpart feathers by early Sep; most are adult by early spring. In C Kenya (P. B. Taylor unpubl.) non-breeding birds arriving from late Dec to Feb consisted largely of adults in worn plumage and juveniles in very fresh plumage; simultaneous remex and covert moult of adults was recorded in Jan-Feb, with varying amounts of body moult in the same period, and 1 adult had almost completed its postbreeding moult on 11 Feb, while postjuvenile partial moult was recorded in Feb-Apr and some birds were still in full juvenile plumage as late as mid-Apr. A mid-Nov specimen from Butembo, extreme NE Zaïre, has all remiges in pin.

In Australia, adult postbreeding body moult is recorded in Oct-Jan and wear of primaries indicates that some adults moult the remiges in Jan-Feb, before the N migration, but some birds have worn primaries in Mar (Marchant & Higgins 1993). In at least some Australian birds postjuvenile moult (recorded Aug-Apr) is partial, involving only the body feathers; it begins on the upperparts and the last feathers moulted are on the sides of the breast and the flanks, a few juvenile feathers sometimes being retained in these areas (Marchant & Higgins 1993).

DISTRIBUTION AND STATUS Europe E through C and S Russia and S Siberia to N China and Japan, and S to Jordan, Africa, Madagascar, SW Asia and Kashmir; also Indonesia, New Guinea, Australia, Tasmania and New Zealand. Palearctic races are migratory, wintering in Africa, Middle East, Indian subcontinent, SE Asia, SE China and S Japan to the Greater Sundas, Borneo, the Philippines and Seram. Not globally threatened, but some subspecies are rare or poorly known. In the W Palearctic *intermedia* is now regarded as rare and very local, having declined considerably since the 19th century. In Europe its numbers are difficult to estimate because of the erratic nature of its occurrence in response to large annual variations in flooding and thus the availability of suitable breeding habitat, and because of the difficulty involved in censusing the birds, but the European Breeding Bird Atlas (Hagemeijer & Blair 1997) records the species

Baillon's Crake

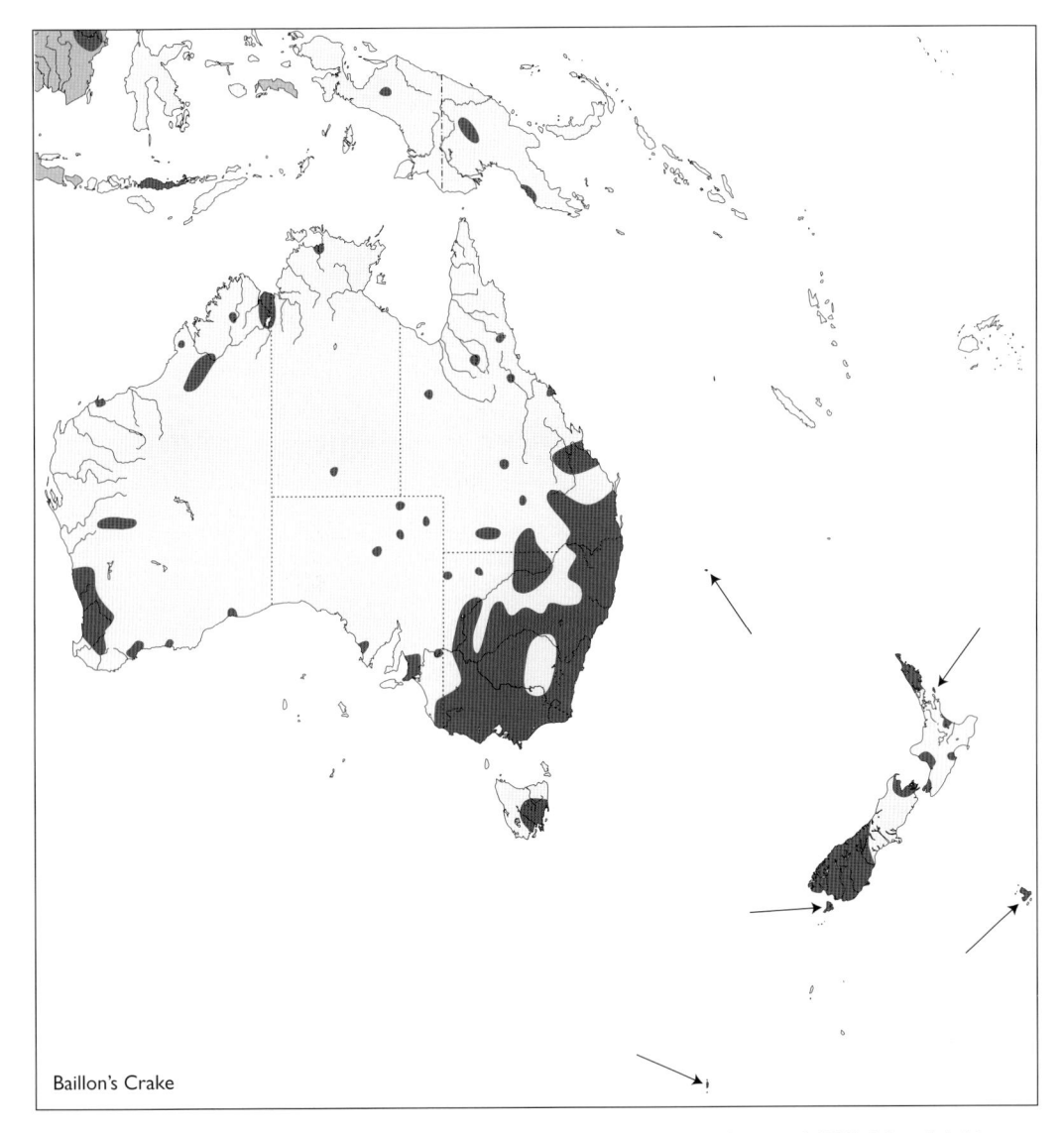

recently from 14 European countries. Its stronghold is in Spain, where 3,000-5,000 (geometric mean 3,900) breeding pairs occur and the Coto Doñana probably holds thousands of nesting birds in good years but few if any in dry years (Llandres & Urdiales 1991); it may be under-recorded in Portugal and Spain, where extensive potential breeding areas such as rice fields are rarely studied (Koshelev 1994). Romania has 100-1,000 (300) pairs and other countries very few; the total European population is estimated at 3,550-5,800 (4,450) pairs, but may be up to 10,000 pairs. European Russia also holds a substantial breeding population (500-2,000 pairs), as may Ukraine (no estimate available), and the Turkish population is 1-100 (10) pairs. It has not been recorded breeding in Poland or Britain this century (Cramp & Simmons 1980).

It formerly bred in N Algeria (L Zana) and may still do so, and in Egypt it is a rare breeding resident in the Nile Delta and Suez Canal area and a rare passage and winter visitor from early Oct to mid-Apr (Cramp & Simmons 1980, Goodman *et al.* 1989); it breeds in Morocco (Larache and perhaps other areas) and possibly Tunisia (L Ichkeul) (Urban *et al.* 1986); it is rarely seen in Oman (Gallagher & Woodcock 1980); it is a scarce migrant and a rare winter visitor to Israel (Shirihai 1996); birds of unknown origin winter in Iraq (Cramp & Simmons 1980); and it is an uncommon to rare breeding species in highland Ethiopia (Urban & Brown 1971). It is widely distributed and scarce to locally common in East, Central and Southern Africa but its status is unclear in some areas. In Zaïre it is known from only 3 records in the NE (Lippens & Wille 1976), and it is probably overlooked in Kenya, Uganda and N Tanzania. It is normally regarded as resident in C and S Africa, and in Zambia it has been recorded in all months except Apr and Oct (Benson *et al* 1971, P. B. Taylor unpubl.) but most records from the region fall during the rains and it is recorded from Malawi only in Dec-Jun and from Zimbabwe mainly in Jan-Apr with isolated records in Jun-Jul (Benson & Benson 1977,

Irwin 1981, Hopkinson & Masterson 1984). The only record from Zanzibar, in Aug 1903, is assigned to "*obscura*" rather than to Palearctic *intermedia* (Brooke & Parker 1984). It is patchily distributed throughout Madagascar (Langrand 1990). Very few estimates of abundance are available for Africa, but non-breeding birds in C Kenya occupied flooded grassland at a density of 4 birds/ha and muddy shorelines with dense reeds and shallow pools at the very high density of 41 birds/ha (P. B. Taylor unpubl.). In South Africa, where the species was formerly regarded as probably rare (e.g. Brooke 1984), recent studies have shown it to be widespread and locally numerous, with an overall estimated population of over 5,400 birds; at 1 site in KwaZulu-Natal in Jan 1992, part of a 120ha wetland held an estimated 120-140 breeding pairs, with unsurveyed potential habitat for many more (Taylor 1997a, c).

The status of the nominate race is largely unclear but it is currently very rare in Azerbaijan, there being no recent records (Patrikeev 1995). It formerly bred abundantly in Kashmir (Ali & Ripley 1980) and it is a relatively common migrant and winter visitor in Pakistan (>100 birds at 1 site in SW Baluchistan in Jan) in Sep-May, probably also breeding in small numbers (Roberts 1991). It is a scarce winter visitor and passage migrant (possibly breeding) in Nepal (Inskipp & Inskipp 1991). It is also rare in Bangladesh (1 recent record, Feb) (Harvey 1990) and in Sri Lanka (Kotagama & Fernando 1994), but is fairly common in China (Cheng 1987). In Thailand it is common in winter (Lekagul & Round 1991), and in the Malay Peninsula it is recorded between Oct and May (Medway & Wells 1976). It may breed in N Korea (Austin 1948). It is uncommon to rare in Japan, where it is a summer visitor (May-Sep) in the N and a migrant and winter visitor (Sep-Nov and Mar) to the S and the Ryukyu Is (Brazil 1991); it is normally regarded as a winter visitor to the Philippines but is recorded in Aug, Oct, Jan and Mar-Jun (Dickinson *et al.* 1991); it possibly breeds in N Sulawesi, where an immature moulting remiges in Jun was presumably bred locally (White & Bruce 1986); in the Moluccas it is known from Seram (1 record, Oct), and in the Lesser Sundas it is known from Flores in Apr-May, possibly a breeding population differing slightly from Palearctic birds (Paynter 1963). In Borneo the nominate race is known from only 4 records but may be a regular winter visitor (Jan-Apr) which has been overlooked (Smythies 1981). In Sumatra it is probably mainly a non-breeding visitor (recorded Jan-Apr) from E Asia or Australia, and breeds exceptionally or irregularly (van Marle & Voous 1988); it is rare in winter in Java (MacKinnon & Phillipps 1993), and it was recorded from Bali in Mar 1995 (Mason 1996).

The race *mira* is known from the holotype collected from the R Mahakam in SE Borneo in 1912 but may be widespread in the extensive lake area of the Mahakam R (Smythies 1981) and perhaps elsewhere in Borneo; its occurrence in the Malay Peninsula, N Sumatra and Java has been suggested (Chasen & Hoogerwerf 1941, Voous 1948) but existing evidence makes this unlikely (White & Bruce 1986). The race *mayri*, endemic to WC New Guinea, is probably very rare and local, being known only from the Wissel Ls region, Weyland Mts (Rand & Gilliard 1967).

In Australia, *palustris* is widespread and apparently locally common in the E, SE and SW, breeding mainly in the SE (but also in the SW) and being most widespread in New South Wales and Victoria (see Marchant & Higgins 1993 for details); maximum densities recorded are 6 birds/ha, but more normal densities are 1-2 birds/ha (Marchant & Higgins 1993); an exceptional density of 17-27 birds/ha was recorded at L Bindegolly, SW Queensland (Woodall 1993). It occurs regularly in the tropics, e.g. in the far N Northern Territory, and is also widespread but possibly not so regular in the central/arid zone (R. Jaensch *in litt.*). It is apparently rare and local in Tasmania (Blakers *et al.* 1984), and in New Guinea, where it is known only from Tari and Mt Giluwe in the C Highlands and from the Port Moresby area but is undoubtedly widely overlooked (Diamond 1972b, Coates 1985); it may occur in the floating grass swamps of the vast Fly and Sepik floodplains, New Guinea (R. Jaensch *in litt.*). The race *affinis* is locally common in New Zealand and is probably widespread, though not often reported; there are scattered records from Stewart I but only 1 from Little Barrier I (in 1922); it formerly occurred on Chatham Is in moderate numbers but its present status is uncertain (Marchant & Higgins 1993, Heather & Robertson 1997); its total population is estimated at <10,000 birds (Rose & Scott 1994).

Like most other wetland rails this species is locally threatened throughout its range by habitat destruction and modification, and its overall numbers are declining (e.g. Owen & Sell 1985, Hagemeijer & Blair 1997, Taylor 1996b, 1997c). It frequently utilizes ephemeral or seasonal wetlands, and such habitats are often drained, overgrazed or cultivated; however it does occupy artificially created wetlands and other vegetation, such as rice fields, dam margins, sewage ponds, saltworks, parks, gardens and airfields (Marchant & Higgins 1993). In Europe many principal breeding areas in the Don, Kuban, Dnieper, Dnestr and Danube R floodplains are threatened by drainage, reed-cutting, reed-burning and heavy grazing, and especially by sudden large changes in water level, caused by discharges from dams, which result in breeding failure (Koshelev 1994). Changes of water levels in the nesting period should be avoided, cutting and burning of vegetation near the water's edge should be controlled, and the maintenance of natural vegetation around fish ponds and rice fields should be encouraged (Koshelev 1994). In SC Siberia this bird's wetland habitats are threatened by agricultural activities (Rogacheva 1992). In South Africa its breeding localities are probably few and should be afforded special protection, especially those sites which are of international importance in terms of the total breeding population in S and C Africa (Taylor 1997c). In S Australia, the loss of winter/spring wet, shallow open palustrine wetlands (which were easily drained) must have affected this species more than Australian Spotted or Spotless (104) Crakes (R. Jaensch *in litt.*). Predators include domestic cats and dogs, and also the Australian Black-shouldered Kite *Elanus axillaris* (Marchant & Higgins 1993), while mortality from collision with powerlines during migration is high (Koshelev 1994).

MOVEMENTS The extent of its migrations is unclear, and the distribution map is conjectural. In Europe most *intermedia* move S from late Aug to Oct, although exceptional Dec-Jan records exist for Europe N to Britain, where it is a vagrant. Migrants occur all round the Mediterranean and probably overfly N Africa to winter in sub-Saharan Africa, where its distribution is poorly known because of its close resemblance to resident birds; presumed Palearctic birds have been obtained in NW Somalia in Sep (Archer & Godman 1937), N Sudan in Apr and Sep-Oct (Nikolaus 1981, 1987), and Senegal in

Nov and Jan (Cramp & Simmons 1980), while it is recorded as a passage migrant and winter visitor to Mauritania (Dowsett & Forbes-Watson 1993). Although Palearctic birds may reach the equator in East Africa, recent speculations (Lewis & Pomeroy 1989, Zimmerman *et al.* 1997) that Kenyan birds are of Palearctic origin are unfounded: observations and trapping in C Kenya in the 1980s, involving the same populations which gave rise to these speculations, established the presence of fresh-plumaged juveniles and of flightless adults in heavy postbreeding moult (see Moult), the dates of which are incompatible with known breeding and moult patterns in Palearctic birds (P. B. Taylor unpubl.). Return passage occurs in Mar-May, most returning migrants in N Africa being recorded in Mar-Apr, when trans-Saharan migration is indicated by records from oases in Libya (Jaghbub, Sebha), Algeria (Beni-Abbès, Laghouat, El Goléa) and Morocco (R Dra) (Cramp & Simmons 1980, Urban *et al.* 1986). Migrants pass through Asia Minor, and it occurs in Oman from Aug-Dec (Cramp & Simmons 1980, Goodman *et al.* 1989). In Israel, autumn passage of very small numbers of birds (usually 1-4 per site) is recorded from mid-Aug to mid-Nov (mainly in Sep), and a more substantial return passage (5-10 birds per site) occurs from mid-Feb to early Jun, mainly mid-Mar to mid-Apr (Shirihai 1996). It is a vagrant to the Faeroes, Ireland, Sweden, Denmark, Malta, Syria, Kuwait, Libya, the Azores, Madeira, and the Canary Is (Cramp & Simmons 1980), and to N Yemen (Hollom *et al.* 1988) and NW Somalia (Archer & Godman 1937).

In sub-Saharan Africa, populations are usually considered resident but occurrences are often seasonal (see Distribution and Status) and at least local movements are recorded in E, C and S Africa in response to seasonal and man-induced habitat changes (Brooke 1984, Urban *et al.* 1986, Tarboton *et al.* 1987, Taylor 1997a, c, P. B. Taylor unpubl. obs. from Kenya, Zambia and South Africa). Populations are certainly itinerant, if not properly migratory, probably moving only when forced to do so.

Nominate birds winter to the S of the breeding range. There is a possible record from Israel (Shirihai 1996), but 2 specimens from Egypt (Meinertzhagen 1930) are referable to *intermedia* (Goodman *et al.* 1989). In Asian regions of the former USSR, southward movements begin in late Jul or early Aug and end in late Oct; return is recorded from Apr to late May (Dement'ev *et al.* 1969); movements in India and Pakistan occur in Sep-Oct and from Mar to early Jun, and birds often fly into houses at night (Ali & Ripley 1980, Roberts 1991). It is a passage bird in N and NE Tibet, recorded in Aug-Sep and May (Vaurie 1972) and in Korea, occurring Sep-Oct and Apr-May (Austin 1948). In SE China it occurs on passage at Beidaihe from Aug to Oct and in May-Jun (Williams 1986, Williams *et al.* 1992). In Vietnam, large numbers have been recorded passing through Saigon in Mar, C Annam in Apr, and Tonkin in winter (Delacour & Jabouille 1931); local seasonal movements are recorded in S Vietnam (Wildash 1986). It is a vagrant to Hong Kong and Taiwan (King *et al.* 1975, Cheng 1987).

The race *palustris* is possibly migratory in Australia (Marchant & Higgins 1993), where many disappear from the S part of the range in Apr-Sep; probably migratory immatures, with much fat, are recorded in Northern Territory in Jul, and some birds probably cross the Torres Strait, where there are May-Jun records from the Booby I lighthouse; return to N Victoria is recorded during the spring. Erratic movements are also recorded in response to vegetation changes (e.g. dieback), drought and flooding; birds sometimes hit bridges and buildings. It is a vagrant to Macquarie and Lord Howe Is.

The migratory status of the races *mira*, *mayri* and *affinis* is unknown; 1 *affinis* on South Island, New Zealand, struck a lighthouse in Feb, 18km from the nearest suitable habitat (Marchant & Higgins 1993).

HABITAT Freshwater to saline, permanent to ephemeral palustrine wetlands with dense vegetation and often floating plants, including marshes, swamps, floodplains, flooded meadows, seasonally inundated grass, the margins of open water, peat bogs, well-vegetated pans, irrigated crops, temporarily inundated depressions and marshy artificial wetlands. It occurs occasionally in saltmarsh (e.g. Elliott 1989, Heather & Robertson 1997). In S Australia it is the typical crake of floating plants and grass-dominated marshes, whereas in the N the White-browed Crake (108) dominates (R. Jaensch *in litt.*). Although suitable habitat is often very shallowly flooded, breeding birds will occur in grassland and sedges flooded to a depth of 30cm (Taylor 1987, Hagemeijer & Blair 1997) and birds also occur on floating vegetation, or in flooded tall shrubs, in water up to 2m deep (Marchant & Higgins 1993). Its breeding habitat is often seasonally or irregularly flooded and in many regions is usually characterised by relatively fine-stemmed and lightly foliaged, low, dense, tussocky or continuous vegetation such as sedges and grasses (*Eleocharis*, *Carex*, *Cyperus*, *Juncus*, *Scirpus*, *Phalaris*, *Leersia* etc), sometimes intermingled with *Typha* (in Europe and C and S Africa); in drier areas forbs such as *Lythrum* and *Althaea* are occupied (Europe). Breeding habitat in Zambia was grass-dominated, with a vegetation height of 0.3-2m (Taylor 1987).

For a comparison of its habitat preferences with those of its congeners in the W Palearctic and Africa, see the relevant sections of the Spotted and Little Crake species accounts, and also Cramp & Simmons (1980). In Australia and New Zealand its ecological segregation from the Spotless Crake is not entirely clear, but although the two species may occur in the same wetlands, even in the same reedbed or lignum stand, they are often segregated by Baillon's apparent preference for floating vegetation at the deeper edges while Spotless prefers to remain inside tall vegetation or will use muddy edges (R. Jaensch *in litt.*); the two species also differ in their choice of breeding sites (Kaufmann & Lavers 1987). In New Zealand it nested in pure stands of tussock sedge *Carex secta* whereas Spotless Crakes at the same site nested in sedges with an overstorey of *Typha orientalis* (Kaufmann 1987). In Queensland, at a lakeshore site where it occurred alongside Australian Spotted Crake, Woodall (1993) found that the occurrence of both species correlated with good cover of lignum *Muehlenbeckia cunninghamii* (used as a refuge) in shallow water and *Cyperus gymnocaulus* on land, a relatively narrow belt of water between the shore and the lignum, and the % cover of *Myriophyllum verrucosum* on land (of significance as cover and/or as a source of food). In New Zealand it occurs in saltmarsh habitat (*Juncus maritimus* and *Leptocarpus similis*) alongside the Buff-banded Rail (46) (Owen & Sell 1985).

Outside the breeding season it occurs in a wider variety of habitats, including tussocky cover interspersed with mud patches around ponds and lakes; the interior and edges of reedbeds 2-3m tall (*Phragmites* and *Typha*, sometimes with *Sesbania* bushes) with extensive mud and shallow puddles (Kenya); flooded *Polygonum* beds; damp grassland; sewage ponds; and dense grassy vegetation of parks, golf courses, airfields etc (Australia). Its habitats may be subjected to fluctuations in water level and, in C and E Africa, non-breeding birds are attracted to feed at muddy lakeshores and dam margins where rapidly falling water levels create shallow pools and expose good foraging substrates (Taylor 1987, P. B. Taylor unpubl.). It normally occurs up to 1,500m, but breeds up to 1,800m in Kashmir (Ali & Ripley 1980) and occurs to at least 1,900m in Africa and 2,450m in New Guinea (Coates 1985, Taylor 1997a).

FOOD AND FEEDING Mostly adult and larval aquatic insects: Plecoptera, Coleoptera (including Hydrophilidae, Carabidae), Hemiptera (including *Nepa*), Odonata (*Xanthocnemis*), Diptera (including Drosophilidae and Culicidae, and larvae of Stratiomyidae and Chironomidae), Trichoptera (Phryganeidae); also Annelida up to 10cm long, molluscs, small crustaceans, Lepidoptera (small moths), spiders, fish up to 2cm long, amphibians, green plant material, and seeds including rushes, sedges (e.g. *Carex*), *Oryza* and 'water weeds'. Invertebrates may form 80-95% of the diet (Koshelev 1994) but it may be less exclusively insectivorous than is the Little Crake (Mason & Lefroy 1912). One bird caught and ate a frog 2.5cm long, and birds also scavenged the squashed carcase of a large frog (P. B. Taylor unpubl.). In captivity it also takes small lizards and geckoes. It often forages while walking on floating vegetation and broken reeds; it also forages on mud, probing and taking prey from the surface, and in shallow water (occasionally to belly depth), immersing the bill to seize prey; it probes into the bases of decaying plants, into dense grass tussocks and into or beneath matted waterside vegetation, detritus, dead stems and roots; it also feeds while swimming and is reported to dive for food. In Crete it fed along the water's edge on debris stirred up by Water Rail (61) foraging 30-80cm ahead (Cramp & Simmons 1980). It normally forages close to, or within, dense cover. Non-breeding birds in C Kenya fed largely on the most abundant small aquatic invertebrates available, including Ostracoda and Copepoda larger than 1.5mm, showing a preference for feeding in shallow drying puddles, where prey was most concentrated; in such conditions individuals foraged intensively for up to 2h in small areas, e.g. 1 adult and 1 juvenile fed for 70 min at a 12 x 8m stretch of mud and puddles, revisiting previously searched patches many times during this period (P. B. Taylor unpubl.). Kenya birds foraged alongside Spotted Crakes, which in this situation took larger prey, sometimes fed in deeper water, took more prey from beneath the surface of water and mud, and were more catholic in their choice of foraging habitats (Taylor 1987, P. B. Taylor unpubl.). In Australia and New Zealand it often forages on floating vegetation, a niche not greatly exploited by Australian Spotted and Spotless Crakes, but it sometimes forages on mud alongside Spotless Crake (R. Jaensch *in litt.*) and it also sometimes forages alongside Australian Spotted Crake, which may be able to wade in deeper water than Baillon's (Woodall 1993).

HABITS It is normally most active from dawn to mid-morning and from late afternoon until at least dusk, sometimes at night. Normally secretive, this crake will forage in the open but usually remains near dense vegetation; however it is not particularly shy and may forage very close to an observer standing quietly in the open. It moves with a slow walk, often with the tail erect and flicked frequently; it walks and runs on floating vegetation, sometimes fluttering the wings to aid balance; it often climbs in robust emergent vegetation. It often appears nervous, dashing to cover for no apparent reason but soon re-emerging, and it will hide under lotus leaves. When alarmed, it stands erect, then takes to cover, sometimes flying; it will run along low branches and will also dive (Marchant & Higgins 1993). It climbs readily in vegetation such as *Cyperus*, *Typha*, *Phragmites*, *Sesbania* and *Muehlenbeckia*, and may rest and preen in such vegetation, and also on vantage points such as branches and stumps above water (Bryant 1942, Woodall 1993, P. B. Taylor unpubl.). The normal flight over short distances is weak and fluttering, with dangling legs, when the short rounded wings are very noticeable, but longer flights are stronger, with the feet trailing and with rapid wingbeats. Birds have been observed roosting during the day on platform-like *Sesbania* vegetation with White-browed Crakes and Buff-banded Rails, venturing out to feed at dusk and in the early morning (Mason & Wolfe 1975). On nocturnal migration, birds pause shortly before sunrise to roost during the day, generally in open stretches of marsh and small ponds (Dement'ev *et al.* 1969). Birds bathe in the morning, and after bathing 1 bird climbed 40cm up into a lignum clump to preen in the sun for 9 min before returning to the water (Woodall 1993).

SOCIAL ORGANISATION Usually solitary, in pairs or in family groups, but forages in loose groups of up to 10 individuals in non-breeding areas in Kenya (P. B. Taylor unpubl.). Monogamous and territorial; in the W Palearctic, and in Africa, the pair bond is maintained only while breeding. The existence of 2 occupied nests only 3m apart may imply a departure from strict monogamy (Cramp & Simmons 1980). In Australia, birds returning to the breeding grounds appear to concentrate and then disperse, possibly moving to the nest-sites as soon as pairs have been formed (Marchant & Higgins 1993); in New Zealand pairs are said to remain on the territory throughout the year (Heather & Robertson 1997). In KwaZulu-Natal, South Africa, breeding birds occurred at a density of 4-5 pairs/ha in parts of a 120ha wetland (P. B. Taylor unpubl.).

SOCIAL AND SEXUAL BEHAVIOUR In non-breeding quarters in Kenya up to 10 adults and juveniles foraged close to each other; adults often foraged within 15-30cm of juveniles, but any foraging bird which was approached to within 1m by another of the same age class (adult or juvenile) would often give a jump with open wings, making a splash in the water, and would chase the other bird for severalm (P. B. Taylor unpubl.). In Kenya this species also foraged alongside non-breeding African Rails (62), Black Crakes (90), Spotted Crakes and various shorebird species, including Painted Snipe *Rostratula benghalensis*; in South Africa, breeding Black Crakes and African Rails sometimes chased Baillon's Crakes (P. B. Taylor unpubl.). In Australia Baillon's Crakes seem to avoid Spotless Crakes (Marchant & Higgins 1993), but differences in feeding may mitigate against these two species coming into frequent contact with each other, even in the same marsh (R. Jaensch *in litt.*). In New Zealand (where the same caveat may apply) Spotless Crakes were observed chasing Baillon's Crakes at a marsh where both species were breeding, but Baillon's Crakes answered playback of Spotless Crake calls and came out to investigate the sounds, while Spotless Crakes responded poorly to playback of Baillon's Crake calls (Kaufmann & Lavers 1987).

Pair members chase intruding conspecific individuals from the territory, pursuer and pursued running rapidly, splashing the water with the feet and fluttering the wings (P. B. Taylor unpubl.). The incubating adult may remain on the nest even when touched, and may defend the nest by lunging and pecking at the intruder or by performing a 'rodent-run' distraction display, scurrying away through the reeds like a rat and splashing the water to attract the intruder's attention, or by injury-feigning with wing-drooping and fanning (Hobbs 1967, Kaufmann & Lavers 1987); in another distraction display the bird parades with feathers raised, head lowered, and wings drooping to show the white outer edge of P10 (Cramp & Simmons 1980), while an intimidation display involves raising and ruffling the back feathers, extending the wings outwards like fans, and walking around the observer uttering grumbling calls (Glayre & Magnenat 1977).

The male makes a courtship flight, circling 6-8m up and uttering short rattles, and is largely silent after pair formation. The following information is from South Africa (P. B. Taylor unpubl.), where pair formation occurred 1-4 days after the arrival of birds in the breeding area. During taped playback of advertising and territorial calls, pair members meeting after a short separation usually crouched face to face, their bills almost touching and the head and neck plumage roused, and called together, giving rapid, grating *kaa* or *kraa* notes (a component of the territorial call) before separating to forage. Copulation is brief: the male approaches the female from behind, with his head raised and his body plumage slightly roused; the female crouches, her head and neck close to the ground and her tail raised, and the male mounts and copulates for 2-3s; the birds then stand side by side briefly before separating to forage. Members of newly established pairs forage close together and frequently allopreen on the head and neck for periods of up to 10 min. In a display of unknown function a female climbed on to a grass tussock and assumed a full-bow posture, with head, neck, body and tail in an almost vertical straight line, the bill touching the toes; she remained in this posture for c. 2 min while the male, with an elongated horizontal posture, silently circled her in the cover of the tussock, with a very slow creeping walk and frequent pauses.

BREEDING AND SURVIVAL Season Europe, May-Jul; Morocco, May, breeding condition Apr; Egypt, Apr; Israel, May-Jun; Ethiopia, Jul; C Kenya, juveniles Jan-Apr, probably from eggs laid at early as Nov-Dec, breeding area unknown (P. B. Taylor unpubl.); Tanzania, breeding condition Apr; Malawi, Jun, breeding condition Mar (there is still no proper breeding record from Africa between 9°N and 15°S); Zimbabwe, Jan-Mar; Botswana, breeding condition Jan; South Africa, Sep, Nov-Mar; former USSR, May-Jun; Kashmir, May-Aug (mostly Jun-Jul); Sumatra, Feb; Japan, May-Aug; Australia, Sep-Jan, exceptionally also Feb-Mar; New Zealand, Oct-Nov; nests during or just after wet season. **Nest** A cup or platform of material available within c. 3m of nest site: dead leaves, dry rush stems, grass, or water weed such as *Triglochin procera*; lined smaller and softer portions of same material; often with vegetation pulled over to form canopy; on ground in thick vegetation close to water, in soft grass or on tussock in water; often in *Carex, Bolboschoenus, Eleocharis, Juncus, Typha, Polypogon*; built 4-60cm above water level; sometimes anchored in water in growing rice crops or on floating vegetation; occasionally in or under low bush or samphire *Halosarcia*; in Australia will also nest in low shrubs with ingrowing grass or sedge (R. Jaensch *in litt.*). Nest rather flimsy; material added during incubation, especially if water rises or if nest exposed for viewing. External diameter 9-19cm, height 8-15cm, internal diameter 6-11cm, cup depth 1-3cm. Nest may take 10-12 days to complete; both sexes build. Some nests have adjoining platform where young sit after hatching. **Eggs** 4-11 (mean 7.4, n = 18 clutches, Hungary) Europe; 4-7 (mean 5.9, n = 8 clutches) Australia; 2-7 (4, n = 7 clutches) Africa; short oval to elongate oval or elliptical; smooth and glossy; buff, yellow-buff or pale olive, or pale brown tinged olive to dark olive-brown; occasionally unmarked but usually heavily marked with short flecks and streaks of darker brown, sometimes also with yellow or with underlying lilac. Size of eggs: *intermedia* (Europe to Iran) 25.2-31 x 19-22 (28.6 x 20.7, n = 95), calculated weight 6.3; *intermedia* (Africa and Madagascar) 26.1-31 x 19.4-24.9 (28.5 x 20.7, n = 41); *pusilla* 26-30.2 x 19-22 (28.2 x 20.2, n = 100), calculated weight 6.3; *palustris* 25.4-29.2 x 18.8-21.6 (27.6 [SD 1.1] x 19.9 [SD 0.7], n = 16); *affinis* 26.5-28.6 x 18.9-20.1 (27.8 [SD 0.9] x 19.6 [SD 0.4], n = 6). Eggs laid at daily intervals. **Incubation** Period 16-21 days, by both sexes; in W Palearctic, 14-16 days per egg, beginning before clutch complete, hatching asynchronous; in Australia at least 16-18 days, eggs hatch within 24h. **Chicks** Precocial and nidifugous; remain in nest for at least 24h after hatching; fed and cared for by both parents, and brooded by both when small; become self-feeding after a few days. When danger threatens, chicks may drop into water and swim away or hide under nest rim (Marchant & Higgins 1993). First feathers appear by c. 10 days; fledging period c. 35-45 days; becomes independent before fledging, sometimes when only half-grown (Cramp & Simmons 1980). In Australia, from 24 eggs laid, 12 hatched; also 5 newly hatched young died after heavy storms, and 1 nest with eggs deserted probably because habitat dried out (Marchant & Higgins 1993). Age of first breeding 1 year. In W Palearctic and Kashmir 2 broods suspected, and recorded in former USSR. **Survival** No data.

95 LAYSAN CRAKE
Porzana palmeri   Plate 32

Porzanula Palmeri Frohawk, 1892, Laysan Island.

Forms superspecies with the widely distributed *Porzana pusilla*. Monotypic.

Synonym: *Pennula palmeri*.

Alternative names: Laysan/Laysan Island Rail.

IDENTIFICATION EXTINCT. Length 15 cm. A small, flightless crake with very short tail. Top of head, upperparts and upperwings sandy-brown, streaked blackish; crown and greater upperwing-coverts almost unstreaked; a few white streaks on scapulars and centre of back. Forehead, sides of head, and chin to belly ashy-grey, darker from throat to upper breast; light brown streak over lores and along top of ear-coverts; sides of neck and breast, flanks, sides of belly, and undertail-coverts sandy-brown; lower flanks with a few irregular, black-edged, white markings.

Iris red; bill, legs and feet green. Sexes similar; female said to be slightly smaller and paler. Young birds said to have underparts pale buff. Inhabited grass patches and *Scaevola* thickets.

Similar species Superficially similar to small congeners with grey anterior underparts, particularly Baillon's Crake (94), which it resembles closely in overall plumage pattern and colour and from which it probably arose (Olson 1973a); compared with Baillon's it is slightly smaller, with very few white markings on upperparts, very reduced barring on predominantly sandy-brown posterior underparts, heavier legs and longer bill.

VOICE The voice was variously described as warbling, rattling and chirping. The evening communal song, given soon after dusk, was described as being like "a handful or two of marbles being thrown on a glass roof and then descending in a succession of bounds"; the birds all started calling together and continued for only a few seconds before falling silent (Frohawk 1892). Two birds when meeting sometimes produced a rattling noise, their throats becoming swollen and their bills slightly open (Fuller 1987). A chirp of 1-3 short, soft notes, presumably a contact call, was given intermittently all day by captive and wild birds (Greenway 1967). The chicks were very vocal, but the calls are not described.

DESCRIPTION
Adult (Sharpe 1894). Forehead sandy-brown, washed ashy-grey; top of head to hindneck, and upperparts, including scapulars, sandy-brown or cinnamon-buff, feathers of all but crown with blackish-brown centres giving streaked effect; streaks narrowest and least prominent on hindneck and upper mantle, well-defined but narrow on hindcrown and nape; scapulars slightly brighter; a few streaks of white sometimes visible on outer webs of scapulars and central feathers of back (these regions also have some concealed white streaks); rectrices short, decomposed, blackish-brown, edged sandy-brown. Upperwing-coverts sandy-brown, with blackish-brown feather centres only on greaters and outer medians (others have ill-defined slightly dusky centres); primary coverts and primaries dark brown, narrowly fringed sandy-brown; secondaries and tertials sandy-brown with blackish-brown centres; remiges soft; only 8 primaries. Underwing-coverts cinnamon-buff, greaters and primary coverts often with white tips; underside of remiges pale grey-brown. Sides of head, including supercilium, ashy-grey; brown streak over lores and along top of ear-coverts; chin, throat, foreneck, breast and belly ashy-grey, often slightly paler on chin and throat, and often duskier from throat to upper breast; sides of neck, sides of breast, flanks, sides of belly, thighs, sides of vent and undertail-coverts sandy-brown, duller and darker from lower flanks to undertail-coverts; lower flanks with a few irregular spots, streaks and bars of white narrowly edged, or with an anterior spot, of black. Iris ruby-red; bill pale olive-green; legs and feet green. Sexes similar; female said to be slightly smaller and paler in colour (Munro 1944).

Immature/Juvenile Young birds (presumably juveniles) said to have underparts pale buffy-tan or pale buff (e.g. Olson 1973a).

Downy young Down black; bill yellow; legs and feet black.

MEASUREMENTS (8 unsexed) Wing 59-63 (61, SD 1.4); tail 19-27 (24.4, SD 2.6); culmen to base 19.5-23 (21.5, SD 1.0); tarsus 22-26 (24.25, SD 1.6).

GEOGRAPHICAL VARIATION None.

MOULT No information available.

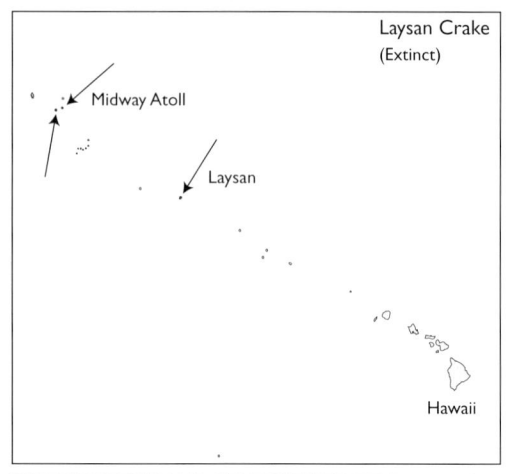

DISTRIBUTION AND STATUS EXTINCT. Laysan I, but also introduced to islands of the Midway Atoll. The rail was first recorded on Laysan in 1828, by Russian sailors who also reported the same or a very similar species on Lisianski I, 184km to the W, but there were no subsequent records from Lisianski until the species was introduced there many years later (Berger 1981, Fuller 1987). At first the species managed to survive on Laysan, despite guano-digging operations which lasted until c. 1906 (Baldwin 1947). However its habitat was destroyed by rabbits *Oryctolagus cuniculus*, which were introduced by the guano diggers in 1903, and by guinea pigs *Cavia porcellus* which were introduced at around the same time (Fuller 1987). Although the crake's numbers were estimated at 2,000 in 1911, by 1923 the rodents had totally destroyed the grass patches and *Scaevola* thickets inhabited by the birds, which finally disappeared from Laysan between 1923 and 1936 (Fuller 1987, Day 1989). However, in 1891 birds were introduced to Eastern I in the Midway Atoll, 480 km to the NW, and more were relocated there in 1913, when birds were also released on Lisianski I (Fuller 1987). An introduction to Pearl and Hermes Reef was unsuccessful, the colony being apparently destroyed by storms in 1930 (Baldwin 1945). Apparently other reintroductions, to Laysan in 1923 and to a large island of the main group (Greenway 1967), and to "cane fields on the main islands" (Munro 1944) also failed, although Ripley (1977) and Day (1989) maintain that no reintroductions to Laysan were made. The Lisianski habitat was destroyed by rabbits and the crakes did not survive (Fuller 1987), but on Eastern I the birds thrived and some were transported to nearby Sand I in 1910 (Berger 1981). The crakes flourished on these 2 islands, being common at the start of World War II (Fuller 1987). However rats *Rattus rattus* were accidentally introduced to both Eastern and Sand Is from a US Navy landing-craft which drifted ashore in 1943; the rats soon wiped out the crakes, the last of which were seen on Sand I on 15 Nov 1943, and on Eastern I probably in Jun 1944 (Berger 1981, Fuller 1987). After the extermination of rabbits on Laysan in 1923 the vegetation recovered well and suitable habitat for endemic bird species was again available, but tragically no reintroduction of the crake was attempted (Ripley 1977). Frigate birds *Fregata* possibly were natural predators (Munro 1947).

MOVEMENTS None.

HABITAT Predominantly *Scaevola* thickets and the coarse, tussocky grass *Ammophila arenaria*, but also thickets of *Juncus*; when plentiful the birds also occurred near houses and on beaches, or generally in the open (Baldwin 1947, Greenway 1967).

FOOD AND FEEDING Most information is from Berger (1981). Practically omnivorous. Its principal food was insects such as beetles (especially Dermestidae), moths, caterpillars, earwigs *Dermaptera*, and flies (including Sarcophagidae larvae from bird and animal carcasses). Other foods were: spiders; the eggs of petrels and small terns (including noddies *Anous*), which were usually opened by other animals (see Habits, and Social and Sexual Behaviour) but apparently occasionally by the crakes themselves, when the birds would jump high off the ground to give more force to their strikes with the bill (Munro 1944); the flesh of decaying seabird carcasses; and some seeds and plant material. The birds also ate scraps, including pieces of meat, and crumbs from the table. They foraged in open areas; they dug in sand for insect larvae and pupae (from carcasses) by using the bill as a shovel and flipping the sand sideways with it; they searched vegetation for insect larvae; and they came round houses in the evenings to catch moths, and entered houses to look for other food items. They were easily caught in traps baited with a chicken egg, which they could not break (Bailey 1956). Their reactions were rapid enough to allow them to snatch flies from mid-air (Fuller 1987). Fresh water was taken if provided.

HABITS Most information is from Baldwin (1945, 1947) and Berger (1981). Diurnal, being most active in the morning and from late afternoon until dark, but sometimes remaining active throughout the day. Calls were heard during the night, but there was no other evidence of nocturnal activity. Although flightless, this crake was very active, restless and quick-moving, using its wings in running, jumping and intimidating smaller birds. It ran swiftly over the sand between grass tussocks, and it crept, mouse-like, in and out of petrel burrows and through the grass; when pursued it would take refuge in petrel burrows. It would stop in the shade of plants to peer at objects, with one foot raised and the tail drooped, before continuing on its way with many stops and starts. It was alert and highly inquisitive, showing little fear of man. On Midway, during World War II, the birds would spring onto tables, and hop onto the laps of servicemen or search under their feet for scraps and crumbs from the table (Day 1989). One observer noted that the birds would bathe in a water pan which he set out, even when he sat 1 m away; another claimed that they were bold enough to hunt for flies between his feet, provided he remained still, and birds entered buildings in search of food; the crakes also clambered over the legs of 2 men, intent on reaching the yolk of albatross eggs which the men were blowing (Bailey 1956). Fresh water, although not normally available to the birds, was readily and repeatedly visited for bathing and drinking when provided. Birds were easily caught in traps (see Food and Feeding), and in hand-nets: if a net was placed edgewise on the ground a rail would presently appear to examine the new phenomenon at close range, and would sometimes almost walk into it (Fisher 1906).

SOCIAL ORGANISATION Possibly monogamous, but proper evidence is lacking. Details are from Baldwin (1947). The maximum density of birds on Laysan, in c. 240ha of suitable habitat, was probably 1 pair per c. 0.25ha (calculations based on Baldwin's figures). The density on Midway was apparently only 25% of that on Laysan. In captivity, 2 birds existed amicably in an enclosure 2m x 4m but fights occurred when more than 2 were present.

SOCIAL AND SEXUAL BEHAVIOUR These crakes were inclined to be pugnacious, and would sometimes chase away Laysan Finches *Psittirostra cantans* from eggs of noddies and terns which the passerines had broken open and were about to eat (Fuller 1987). Incubating birds were not disturbed by the presence of an observer, and would return immediately if lifted off the nest (Baldwin 1947).

Two birds, possibly males giving a threat display, approached each other with feathers erect and, when close together, made rattling calls; they then suddenly ceased and slunk away in opposite directions (Fisher 1906). Two birds, said to be male and female, sat for several minutes close together, each in turn holding its head close to the ground while the other bird allopreened the feathers of the crown and the back of the neck (Baldwin 1947).

BREEDING Laysan, mostly Apr-Jul, but downy young seen as early as Mar 17; eggs found May-Jun. **Nest** Built on or close to ground; usually sheltered by grass tussocks or *Juncus* on Laysan, and under *Scaevola* and other plants on Midway; woven of matted grasses with some feathers and other materials; nest also often hollowed out of dry vegetation in grass tussock or an accumulation of dried *Juncus* leaves; sometimes naturally roofed by matted grass, sometimes ball-shaped with side entrance; often lined with softer materials such as down or soft dry stems; approach runway often constructed. **Eggs** 2-3, occasionally up to 5; blunt oval or elliptical ovate; pale olive-buff to creamy-buff, spotted greyish-lilac and raw sienna, markings sometimes mainly at larger end; size of eggs, 28.6 x 20.9, 28.6 x 21.7, and 31 x 21, calculated weight c. 7 (Schönwetter 1961-62, Ripley 1977). **Incubation** Possibly by female only, which apparently stayed closer to nest than male (Fisher 1906, Greenway 1967). **Chicks** Soon became self-feeding; when c. 5 days old, could run as fast as adults; 1 chick 3 days old was described as like "a black velvet marble rolling along the ground. Its little feet and legs are so small and move so fast they can hardly be seen" (Hadden 1941).

96 SPOTTED CRAKE
Porzana porzana   Plate 28

Rallus Porzana Linnaeus, 1766, Europe = France.

Monotypic.

Synonyms: *Gallinula/Crex/Ortygometra/Zapornia porzana*; *Ortygometra/Porzana maruetta*; *Gallinula maculata/punctata/leucothorax/gracilis*; *Porzana fulicula*.

Alternative name: Eurasian Spotted Crake.

IDENTIFICATION Length 22-24cm; wingspan 37-42cm. Medium-small, rather plump crake with shortish neck and legs, and markedly spotted appearance. Male has upperparts olive-brown with black feather centres and with fine white spots and streaks; upperwing-coverts more uniform olive-brown with white spots and streaks; broad supercilium, cheeks and throat blue-grey, variably spotted white; broad streak across ear-coverts and down sides of neck buff-brown, variably speckled white; dense white spots above dark lores and dark feathers at base of bill may give impression of white supraloral line; chin to foreneck slate-blue, spotted white; breast olive-brown, tinged grey (greyer when worn), and variably patterned white; flanks broadly barred olive-brown, black and white; lower breast, belly and vent white or cream; undertail-coverts bright buff. Iris bright brown; bill yellow, greener towards tip and orange-red at base of upper mandible; legs and feet olive-green. In flight, white leading edge of wing usually well visible, and bill colours may also be obvious. Sexes similar but female averages smaller. Male in non-breeding plumage has less grey and more spotting on face and underparts; adult female has less grey and more spots than male on these areas in both plumages, but sexes may be inseparable in field except on size in paired birds. Rare variations in colour of undertail-coverts range from dark feather tips to black and white barring. Immature like non-breeding adult but has more white spots on side of head; bill changes from yellow-brown with yellow base to olive-green with orange base. Juvenile similar to immature but has streak over eye brown or cream with tiny white spots, neck mottled grey-brown and off-white, breast olive-brown to bright brown with white or buff markings, and flanks less contrastingly barred; eye greenish; bill olive-brown to greenish-horn, darker at tip and more orange at base; legs dull green. Inhabits short-vegetated marshes, bogs, lake shores etc, with shallow water and mud; also moist to wet meadows and (in winter) ephemeral wetland patches.

Similar species Readily distinguished from sympatric Rallidae on size, structure and plumage, major characters being spotted appearance, lack of extensive pure grey on underparts, buff undertail-coverts (usually obvious on bird retreating to cover when disturbed) and yellow bill with greenish tip and orange-red spot at base of upper mandible. No sympatric crake has white on leading edge of wing so broad and noticeable in flight. Striped Crake (109) slightly smaller and more slender, with shorter wings, shallower wingbeats, deeper bill and longer, more projecting legs and feet in flight; also has striped upperparts, reddish-cinnamon lower flanks and undertail-coverts (well visible as bird flies away from observer), and no dark barring on underparts. Worn adults in late summer, with largely unpatterned grey anterior underparts, may conceivably be confused with male Little Crake (93) (Becker 1995) but this unlikely except in poorest and briefest of views. Sora (98) differs most obviously in colour and pattern of head, neck and breast, and lack of red on bill (see Sora species account).

VOICE Most information is from Cramp & Simmons (1980). The male's advertising call is a short, sharp ascending whistle *whitt*, suggesting a whiplash, repeated about once per second for up to several minutes. It is given in spring and summer, mainly from dusk through the night, and is often audible for 0·5-1km, or for at least 2km in calm weather; calling is inhibited by rain and cold weather. Breeding birds are apparently almost silent, and calling late in the breeding season is attributed to unpaired males seeking mates. Wintering birds in Africa are apparently generally silent, but in N Senegal several hundred were calling at 1 locality in Jan-Feb 1987, and it was heard in E Zaïre in Apr (Lippens & Wille 1976; Rodwell *et al.* 1996). The female has a softer version of this call, sometimes given in duet with the male. After territory establishment a series of quiet, musical *hui* notes is given with the advertising call. Males attracted by imitations of the advertising call give sharp, croaking *qwe* calls and a deep, cooing *grrr grrr*. A soft version of the advertising call is used at the nest-site, probably as a pair-contact call. During the precopulatory display the male gives a ventriloquial *brrr-brrr*. The distress call, given when pursued or captured, is *krek* or *kreek*, sometimes increasing to a real scream; in less extreme circumstances, a hard *eh* is given. Adults call chicks (and sometimes each other) by means of repeated *dug* or *gug* notes, and a gurgling *kyörk* is also used; the warning call is a sharp *tshick*, which makes chicks immediately silent. Other calls, of uncertain function, include possible variations of the above calls and also a loud, throbbing, repetitive ticking. In an aggressive encounter with a Pied Kingfisher *Ceryle rudis*, a crake gave a sharp *tck* call (Taylor 1987). Young have a *tyI(e)* food-call, which becomes a plaintive scream when the chicks are fed; after several days the young utter a *sch-IB*; juveniles, when frightened, give a loud hoarse *keck* or *kyeehk*.

DESCRIPTION
Adult male breeding Centre of forehead and crown, hindneck and sides of mantle olive-brown, with black feather centres and finely dotted white; sides of hindneck tinged grey. Mantle to uppertail-coverts, and scapulars and upperwing-coverts, olive-brown; scapulars and mantle feathers with broad black centres; feathers of back to uppertail-coverts with narrower olive-brown fringes; upperwing-coverts plainer and browner than rest of upperparts; mantle to uppertail-coverts, and upperwing-coverts (excluding primary coverts), with narrow white dots or short white streaks, often margined black; white markings less numerous on lesser coverts, and often V-shaped on median and greater coverts; scapulars have longer streaks; tertials like scapulars but inner webs widely margined pale olive-brown to yellow-buff (bleaching during breeding season), outer webs largely blackish with long white bars or incomplete V-marks. Primary coverts, alula, primaries and secondaries dark grey, broadly margined olive-brown on outer webs; primary tips blackish; innermost secondaries may have white spots on inner and outer webs; marginal coverts (and often adjacent lesser coverts on upperwing and underwing) white, and outer webs of alula feathers, P10, and sometimes P9, also white, giving prominent white leading edge to wing. Axillaries and underwing-coverts black-brown, axillaries barred white; underwing-coverts barred and spotted white. Tail feathers black, fringed olive-brown and variably streaked or dotted white at sides. Broad streak over eye, from base of upper mandible to rear of ear-coverts, slate-blue, spotted white; white disappears with wear, leaving only some tiny white spots along sides of crown and above eye. Feathers at base of bill, and on lores, darkish brown to blackish; broad streak from behind eye, across ear-coverts and down sides of neck buff-brown, partly speckled white (including some visible white feather bases), white markings abrading with wear to give more uniform buffy-brown colour; dense white spots above dark lores may give impression of white supraloral line; sides of head below eye, and chin, throat and foreneck, slate-blue, darker in malar region and at base of bill, with small white spots (most numerous adjacent to buff-brown streak) which abrade to give more uniform grey appearance; old males are noticeably darker, more blackish, on malar region, lores and chin when in worn plumage. Breast slate-grey, washed olive-brown

(feather tips), feathers narrowly margined white when fresh, and each feather has black-margined white dot on each web near tip; flanks broadly barred olive-brown, black and white; lower breast, belly and vent white or cream, lower breast with grey feather bases; undertail-coverts cream-yellow to buff. With wear, upperparts become darker and more uniform as white markings and pale feather edges abrade; however, pale fringes to tertials still prominent; breast becomes purer grey as olive-brown and whitish tips abrade, and adults before postbreeding moult may appear uniform dark grey-brown on breast. According to Becker (1995), at end of breeding season breast may become so worn that subterminal white spots become visible; as these are elongated and black-edged, breast may show narrow white wave-like bars continuous with broader barring on flanks. Great individual variation in plumage pattern and colour occurs; plumage characters can be used for individual identification (e.g. Taylor 1987). Rare variations in colour of undertail-coverts range from dark feather tips to black and white barring (see Becker 1995). Iris yellow-brown to brown (bright red-brown in older males); bill yellow to yellow-green, shading to olive-green at tip and with orange-yellow to orange-red base to upper mandible (brightest in spring, duller during moult, and extending to base of lower mandible in older males) and yellowish-white gape; legs and feet pale olive-green to pale green, tarsus often tinged yellow or orange in spring.

Adult female breeding Sometimes separable from male on plumage. Cheeks and sides of throat average more heavily speckled white; grey of sides of head less pure slate; spots on breast often larger, tending to barring.

Adult male non-breeding Like adult male breeding, but sides of head more extensively mottled white, as in adult female breeding; grey of head paler, and pure only on streak over eye, lower sides of head, chin and centre of throat; white markings on breast often less distinctly outlined black.

Adult female non-breeding Like adult female breeding, but more heavily speckled white on head; pure grey restricted to anterior supercilium and central chin.

Immature Similar to non-breeding adult but has more extensive white spots on side of head, and narrower black margins to white markings on breast; bill changes from yellow-brown with yellow base to olive-green with orange base.

Juvenile Similar to immature but upperparts may be more strongly spotted white (but degree of spotting variable in all ages and not diagnostic); no grey on head; streak over eye olive-grey, brown or cream with many tiny white spots; ear-coverts washed-out brownish; lores, lower sides of head, and sides of neck, mottled grey-brown and off-white; chin and throat white, variably speckled dull grey; breast dull olive-brown to bright brown, each feather with 2 white or buff dots or bars (often not outlined black) near tip; flanks less contrastingly barred olive-brown and white (with little black), white bars narrower and more wave-like; some birds have underparts washed buff. Iris olive to deep green, becoming brown from pupil outward during autumn; bill olive-brown or greenish-horn, becoming yellow-brown with yellow base, and then olive-green with orange-yellow base; legs and feet grey-green to olive-green.

Downy young Black downy chick has green gloss on head, throat and upperparts; iris grey to brown-black; upper mandible red at base, yellow in middle and white at tip, yellow and white areas separated by thin black band; lower mandible black with red-brown base and narrow white band in middle; legs and feet grey-brown to black. Skin of face and crown blue, of nape, neck and body flesh-pink. Small white egg-tooth and wing-claw present. Iris becomes dark brown at c. 3 weeks, and bill becomes progressively blacker, with yellow-green base and narrow band in middle, and slightly red base to upper mandible; legs and feet grey-green. Fully grown young have black bill with yellow-green base; at fledging, iris grey-green; bill black.

MEASUREMENTS (122 males, 118 females) Wing of male 117-128 (122, SD 2.7), of female 111-123 (118, SD 3.1); tail of male 42-54 (47.5, SD 2.8). of female 44-52 (47.3, SD 2.6); exposed culmen of male 18-22 (19.7, SD 1.2), of female 17-20 (18.4, SD 0.9); tarsus of male 32-37 (34.1, SD 1.2), of female 30-35 (33.0, SD 1.3). Becker (1995) gives wing lengths of 109-135. Weight of 149 unsexed (Europe and Africa, all months, including migrants) 47-147 (87.1); monthly means (Europe, Mar-Oct) 71.2-96.0; weight heavy prior to migration but low on arrival, e.g. male, Morocco (Apr) 47, female, Algeria (May) 60 (for further details, see Cramp & Simmons 1980). Hatching weight 5-7.5 (Fjeldså 1977).

GEOGRAPHICAL VARIATION None.

MOULT Information is from Cramp & Simmons (1980) unless otherwise stated. Postbreeding moult is complete, with simultaneous remex moult; it usually occurs in or near the breeding area in Jun-Oct, during the last stages of rearing the young; birds in the first breeding season (single-brooded) start as early as Jun and older birds (double-brooded) in Aug-Sep (Becker 1995); birds migrating in Sep are in fresh plumage. Body plumage starts to moult first and continues during the period when the remiges and coverts are lost, with almost simultaneous loss of the rectrices; new wing and tail feathers are fully grown after c. 3 weeks. Pre-breeding moult is poorly known, but apparently involves at least the head and breast, and occurs in Dec-Apr, mainly in Feb-Mar. Juvenile plumage is retained for only 3-4 weeks (Becker 1995) and postjuvenile moult is partial, involving the head and body but not the remiges or (apparently) the greater coverts; it starts in Jul and ends from mid-Sep to mid-Oct, most birds moulting late Aug to early Sep; birds often moult during stops on migration. The first prebreeding moult is apparently like the adult prebreeding, but some birds partially retain the immature non-breeding plumage.

DISTRIBUTION AND STATUS W and C Eurasia, from British Isles and Spain E across S Scandinavia, N Mediterranean and the Balkans to W & C Russia, Caucasus and Iran, continuing to Kazakhstan, SW Siberia (Suva) and NW China (W Xinjiang). It may breed in NW Kashmir, where there is a 19th century specimen from early Jul (Roberts 1991). It winters from the Mediterranean and the Middle East S to W Africa, and from Ethiopia S to S Africa and W to Angola and Namibia, and also in Pakistan, India (S to Mysore), and C Thailand (very rarely), and irregularly in the SW Caspian Sea area. It is not globally threatened. Numbers fluctuate widely in most areas, due to the nature of its preferred habitat, but decreases have been evident over most of its European range this century due to wetland drainage; it is now local and uncommon to rare in most regions (Cramp & Simmons 1980). Its population in Europe and Africa is estimated as 100,000-1,000,000 birds and is thought to be decreasing (Rose & Scott 1994). In Britain it was local but fairly common before the mid-19th century, but disappeared due to

Spotted Crake

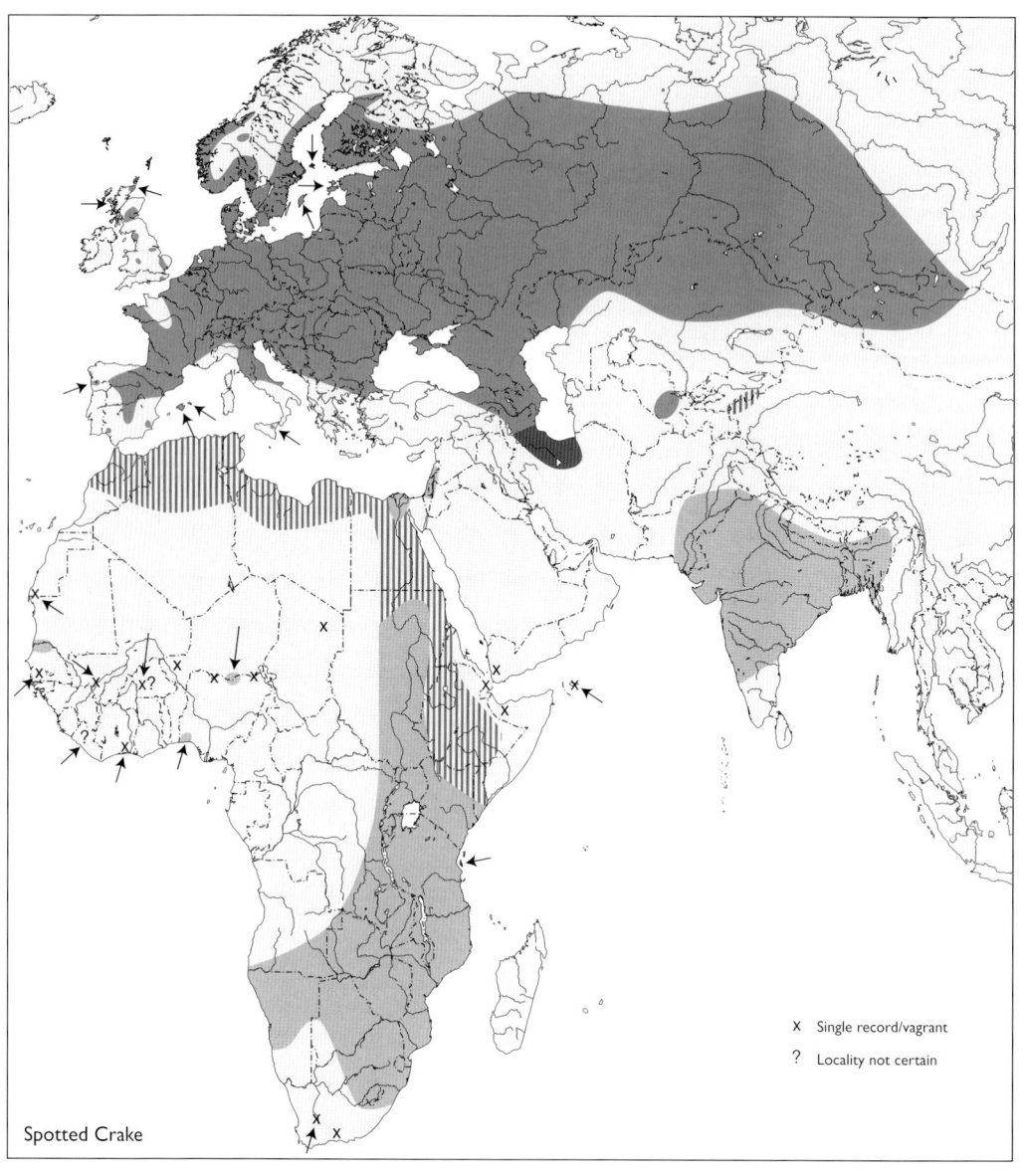

X Single record/vagrant
? Locality not certain

wetland drainage and is now a rare breeder; it may have nested annually in Ireland until the same period (Cramp & Simmons 1980). In Europe (Hagemeijer & Blair 1997), its strongholds are in the E, where it remains relatively common because low-intensity agriculture is still practised, leaving large areas of wet meadows. About 85% of the European population of 50,000-180,000 breeding pairs (43,200-152,800 pairs) inhabits Russia, 10,000-100,000 (geometric mean 31,600); Belarus, 24,000-28,000 (26,000); Romania, 5,000-20,000 (10,000); and Ukraine 4,200-4,800 (4,500); Poland, Estonia and Finland, and probably Lithuania and Latvia, each has a population of a few thousand pairs, but in W and C Europe only France, Spain, Germany and Hungary seem to have >500 pairs each; Turkey has 1-100 (10) pairs (Hagemeijer & Blair 1997). Only in Finland have both the range and population increased since 1970, a N expansion as the lowering of lake water levels (probably beginning in the 19th century) and eutrophication provided more habitats; declines are reported in almost half the 29 European countries in which the species breeds, most strongly in W and C Europe where habitat loss from wetland drainage and agricultural intensification has been most severe (Hagemeijer & Blair 1997). In the former USSR it is numerous to locally abundant, being commonest in the N and in Transcaucasia (Cramp & Simmons 1980).

In Israel it is a scarce autumn and uncommon spring passage migrant, and a quite rare winter visitor from Nov to Feb-Mar in low-lying areas, the total winter population being up to several tens (Shirihai 1996). In Saudi Arabia it is an uncommon migrant in all areas, and also possibly a winter visitor (Jenkins 1981). In Africa, it is regarded as locally common on spring passage from Morocco to Libya, and in winter in Egypt, Senegal and Burundi, while in E

Africa it is much commoner on passage than in winter but it is a scarce (possibly under-recorded) winterer in Kenya and Uganda, and it may be a fairly common passage migrant and winter visitor to the S Sudan even though records are few (Urban et al. 1986, Nikolaus 1989, Short et al. 1990, P. B. Taylor unpubl.). Recent observations at the Djoudj NP in N Senegal showed that this species was exceptionally common there in Jan-Feb 1987, with "several hundred calling..., 1 every c. 10m in flooded areas" (Rodwell et al. 1996). It is generally scarce elsewhere in Africa except in the major known wintering region, which comprises Zambia, Malawi, Zimbabwe and probably Mozambique, within which region it is possibly not uncommon but overall numbers are difficult to assess because of erratic occurrences and annual fluctuations; it is very uncommon in South Africa, but is apparently common at some temporary pans in Bushmanland and Kavango, extreme NE Namibia, and it is a vagrant to the S Cape, whence there are 2 records (Clancey 1971, Urban et al. 1986, Hines 1993, Taylor 1997c). On the basis of suitable habitat, its potential South African population is estimated at c. 160 birds, 150 of these in KwaZulu-Natal and the former Transvaal (Taylor 1997c). There is an unconfirmed oversummering record from Lesotho, Jun 1912 (Bonde 1993), probably incorrectly dated. This crake's wintering habitat in Africa is probably decreasing as a result of wetland destruction.

It was formerly numerous in S and C USSR, and was locally abundant further N and in Transcaucasia; it is still common in Azerbaijan and elsewhere, but is decreasing in some areas (Dement'ev et al. 1969, Patrikeev 1995). Its status elsewhere is difficult to assess: it was formerly said to be commoner than supposed in India and Pakistan (Ali & Ripley 1980) but is currently regarded as scarce in Pakistan (Roberts 1991); it is rare in China (Cheng 1987). In the long term, this species is vulnerable to changes in water levels, whether caused by wetland modification and drainage or by climatic changes, but it can occupy artificially created habitats, especially in its wintering areas (Taylor 1997c). Predators include the Marsh Harrier *Circus aeruginosus* (Potapov & Flint 1987).

MOVEMENTS Most information is from Cramp & Simmons (1980). European birds move S to SW in the autumn; birds from Germany, the Netherlands, Switzerland and Poland have been recovered in S France and Iberia, and birds from Germany and the Netherlands in Italy. A few winter in temperate Europe, mostly in North Sea countries and only exceptionally in regions influenced by the continental climate; more winter in S Europe and N Africa (Egypt and Morocco), and many in W, E and SE Africa. The relative paucity of wintering records from Africa is attributed to the bird's secretiveness and the inaccessibility of its habitat. Autumn dispersal begins in Jul; some birds halt in Aug to moult and are flightless for 2-3 weeks; in Germany, first-brood young remain in the breeding area until Sep, and second-brood young until Oct (Becker 1995). There is a marked S movement from late Aug to early Sep; most reach or pass the Mediterranean by mid-Nov; marked passage is recorded in Israel from mid-Aug to mid-Nov (most mid-Sep to mid-Oct), in Egypt and Sudan Sep-Oct, and in Kenya Nov-Jan (Cramp & Simmons 1980, Lewis & Pomeroy 1989, Shirihai 1996).

This species is recorded in Sudan and Ethiopia/Eritrea from Sep-Jun (Archer & Godman 1937, Cave & Macdonald 1955, Nikolaus 1987); it occurs in East Africa from Nov to early May, and in C and S Africa (E Zaïre and S Angola to N and NE South Africa) in Nov-Apr with peak numbers in Jan-Mar. In W Africa it occurs in Senegal from Sep-Mar, and there are isolated Sep-Mar records from Mauritania (Port Etienne = Nouadhibou), Gambia (Pirang, Oct), Mali (2 records, Dec-Jan), Chad (Abéché), Niger (1 found dying, Apr), Liberia, Ivory Coast (1 record, Apr), and Nigeria (4 localities, Nov-Mar) (Lamarche 1980, Thiollay 1985, Urban et al. 1986, Gatter 1988, Gore 1990, Sauvage 1993, Elgood et al. 1994, Rodwell et al. 1996). There is a possible record from Burkina Faso in Mar (Thonnerieux et al. 1989). The earliest Zambian arrival is 11 Dec but by mid-Dec birds have reached Zimbabwe, Botswana, Namibia and South Africa (Urban et al. 1986; Taylor 1997c). In Africa it is probably more common and widespread than records indicate. It is itinerant in the African wintering areas (Brooke 1974); at Ndola, Zambia, all arrivals occurred in Dec-Apr, all precisely dated arrivals (n = 22) were within 12-24h of heavy local rain, indicating that the birds follow rain fronts, and birds departed when conditions become too dry, too flooded or too disturbed, the maximum period of residence being 23 days and mean periods in different habitats being 1.0-6.8 days; most arrivals occurred overnight (Taylor 1987). By the end of April the birds have left S Africa, and most East African records involve return movements in Apr-May; coastal movements are indicated by 1 very fat bird from Zanzibar on 22 Mar (Pakenham 1979), and 1 with Corncrake (69) at Mombasa, Kenya, on 6 May (P. B. Taylor unpubl.). Passage through N Africa and the Mediterranean is more marked in spring, suggesting that more overfly these regions in autumn; return movements are recorded from Morocco to Libya and Sudan in Mar-May, Israel from end-Feb to end-May (mainly mid-Mar to mid-Apr), and Europe in Mar-Apr; European breeding grounds are reoccupied in April (Cramp & Simmons 1980, Hogg et al. 1984, Lewis & Pomeroy 1989). Birds of unknown origin pass through Iraq and the Near East; passage occurs in Saudi Arabia and Aden in Apr and Nov, and in Oman Sep-Dec; it may overwinter regularly in these regions (Cramp & Simmons 1980, Gallagher & Woodcock 1980).

In the former USSR birds move from breeding marshes to the vegetated muddy shores of large lakes from late Jul to Aug, and migratory movements are recorded in Aug-Oct and Mar-May (Dement'ev et al. 1969). Movements are noted in Azerbaijan in Oct-Dec and Mar-Apr, and it sometimes winters, mostly in benign seasons (Patrikeev 1995). It arrives in Pakistan and N India in Sep-Oct, and return movements are noted in NW Pakistan in Mar-Apr (Ali & Ripley 1980; Roberts 1991). Birds migrate at night, fly low when migrating, usually within 3m of the ground, and sometimes strike power lines (Dement'ev et al. 1969, Patrikeev 1995).

It is accidental in Iceland, the Faeroes, Ireland, Albania, the Azores, the Canary Is and Madeira, and is a vagrant to Greenland (9 records), the Lesser Antilles (St Martin) and the Seychelles (Nov 1982) (Ripley 1977, Cramp & Simmons 1980, Phillips 1984). It is also a vagrant to Djibouti, Somalia (1 record, Oct), Socotra, Yemen, W Myanmar and Tibet (Archer & Godman 1937, Vaurie 1972, Smythies 1986, Dowsett & Forbes-Watson 1993). One individual was found at the same wintering site in the former Transvaal, South Africa, in 2 successive years (Tarboton et al. 1987).

HABITAT Freshwater wetlands (not oligotrophic) with dense cover of sedges and rushes (*Carex, Eleocharis, Cyperus, Juncus* etc), grass (e.g. *Panicum, Poa, Deschampsia*), *Polygonum, Iris, Equisetum* and other emergents; sometimes with trees such as *Acacia, Sesbania, Betula, Salix, Alnus* and pines. It normally requires very shallow water interspersed with ample stands of low cover and rich in invertebrate food; it prefers shorter vegetation than does *Rallus*, and frequents areas where the substrate is moist, muddy or shallowly flooded (<30cm, usually <15cm). Optimum conditions are found in wetlands with a range of water depths or where suitable foraging areas are produced by variations in water level, and in the breeding areas suitable conditions are usually found only in fairly extensive wetlands (Cramp & Simmons 1980). It occurs in permanent or seasonal marshes, in fens, bogs and damp meadows, at sewage ponds, pools in flooded grassland, and at the margins of dams, lakes and sluggish rivers. In the former USSR it sometimes nests in marshes so densely covered with birches and pines that they resemble young woods, and also breeds in damp meadows and on the grass-covered shores of lakes (Dement'ev et al. 1969). On migration it may occur in atypical habitats, and in winter it exploits a wider range of habitats. In Africa it often occupies ephemeral habitats, with rapidly changing water levels, not normally inhabited by breeding Afrotropical rallids, and typically occurs in dense grassland or sedges 0.3-1.5m tall, moist to shallowly flooded (7.5cm), with muddy patches, shallow puddles and shallow pools in the vegetation; it is also recorded from grass-fringed sludge-drying ponds with mud and shallow water (<10cm deep), the edges of drainage ditches lined with dense reeds and grass, and at a dam with dense reeds, long grass and much floating vegetation; it was even recorded inside a temporarily flooded *Acacia/Lantana* thicket (Taylor 1987). Dominant vegetation in wintering habitats includes grasses such as *Panicum, Leersia, Acroceras, Hemarthria* and *Oryza*, sedges such as *Cyperus, Schoenoplectus, Eleocharis* and *Mariscus*, and also *Typha* and *Juncus* (Hopkinson & Masterson 1984, Taylor 1997c). In Zambia, birds resident for more than 1-2 days occurred only in habitat with patches of mud and/or shallow pools, at which they could forage, and in such habitat it often occurs alongside Painted Snipe *Rostratula benghalensis* and Great Snipe *Gallinago media*, and occasionally breeding Striped Crake (109) and Baillon's Crake (Taylor 1987). In Kenya it occurred at L Naivasha, at shallowly flooded (up to 7.5cm) and muddy areas with clumps of *Cyperus* sedges up to 1.5m tall, and *Sesbania* bushes; non-breeding Baillon's Crakes were also present but the Spotted Crakes remained longer as the water level fell, feeding at drying pools where cover was insufficiently dense to attract Baillon's (Taylor 1987). It prefers lowlands but in the former USSR ascends locally to 1,800m and exceptionally to 2,420m (Dement'ev et al. 1969), and it is recorded at 2,800m in Kenya (Jackson & Sclater 1938).

FOOD AND FEEDING Most information is taken from Cramp & Simmons (1980) and Taylor (1987). It is omnivorous, taking mainly small aquatic invertebrates and parts of aquatic plants. It takes earthworms (Annelida), molluscs (*Bithynia, Lymnaea, Physa* and slugs), Arachnida (including spiders and water mites), insects and their larvae, including Trichoptera, Odonata, Diptera (including Tipulidae, Athomyiidae, Athericidae), Coleoptera (especially Hydrophilidae, Dytiscidae, and also Scarabeidae, Curculionidae, Spercheidae, Chrysomelidae), Hemiptera (including *Corixa* and *Naucoris*), Lepidoptera and ants (Formicidae), and small fish (1-2cm long) stranded in drying pools. Plant material includes algae, shoots, leaves and roots, and seeds (*Panicum, Oryza, Carex, Schoenoplectus, Scirpus, Potentilla*). In captivity it apparently steals eggs and kills fledglings (Rutgers & Norris 1970). It forages in water up to 7cm deep, and on wet to dry mud; it picks food from the surface of the substrate and immerses its head in water; it stretches up to strip seeds from grass inflorescences; it also walks on floating vegetation, turns over waterlily *Nymphaea* leaves and takes prey from their undersides, and swims occasionally (Brooke 1974, Taylor 1987). It usually keeps close to cover but in the wintering areas it sometimes feeds 10-15m from cover on open mud and in shallow water with groups of shorebirds such as Wood Sandpipers *Tringa glareola* (see Habits), and it also forages alongside other bird species (see Social and Sexual Behaviour). Although it is said to take only small prey (Koenig 1943), in Kenya, when foraging alongside Baillon's Crake, it appeared to take larger prey than did its congener, which takes prey as small as 1mm in length (see Baillon's Crake species account), and it also fed in deeper water and took more prey from beneath the surface of water and mud.

HABITS Diurnal or crepuscular; in the winter quarters birds are active from dawn to dusk, retreating to cover as darkness falls; although visible at most times of the day, the most intensive foraging activity takes place around sunset (Parnell 1967, Taylor 1987). On the breeding grounds birds normally roost at night, although breeding-season calling may continue all night. On nocturnal migration birds roost during the day, from dawn onwards, frequently on dry land covered with herbage and shrubbery (Dement'ev et al. 1969). Wintering birds in Zambia roosted in dense grass at the edges of sewage ponds (Taylor 1987). Captive birds roost on a perch off the ground, usually standing on 1 leg, the pair roosting close together; in cold weather it squats with the feet buried in the feathers, and may roost in the thickest available cover (Cramp & Simmons 1980).

Detailed observations of behaviour come from the winter quarters (Taylor 1987), from which most of the following information is taken. In Zambia this species is one of the easiest rails to flush with a dog from wet grassland, and may be flushed repeatedly. The normal gait is deliberate, with long strides, forward and backward jerks of the head, and vertical flicks of the tail. Birds more than 1-2m from cover usually appear less confident, flicking the tail and jerking the head more often, and sometimes moving with a more unobtrusive creeping gait with lowered head. When moving between patches of cover, birds often make a crouching run with the neck stretched forward, the plumage compressed and the tail cocked, and this is also the commonest way of retreating to cover when disturbed, although birds will sometimes fly. They swim and dive readily, normally over short distances in cover and across channels and streams, but will also dive when alarmed, swimming submerged, and will crouch if escape is not possible (Cramp & Simmons 1980). Birds wintering at a sewage works were not very shy, and foraged on open mud and in shallow water among shorebirds for up to 45 min, sometimes 10-15m from cover; when emerging to feed they sometimes flew directly to open areas 5-10m from cover. These birds tolerated an observer standing quietly in full view at ranges down to 15m; after a disturbance they were often the first species to return to foraging areas, sometimes emerging within 30s of the disturbance. They ignore observers in cars (Parnell 1967, Taylor 1987). The appearance overhead of raptors such as Black Kite *Milvus migrans* or Gymnogene *Polyboroides typus* elicited the typical escape-dash to cover.

In Zambia, bathing and prolonged preening were observed in the open at all times of the day (Taylor 1987).

SOCIAL ORGANISATION Monogamous, the pair-bond being formed very soon after arrival in the breeding area and maintained only during the breeding season. It is territorial when breeding and also in the winter quarters (Brooke 1974, Taylor 1987). The breeding territory is very small, sometimes as little as 400-800 m² being defended, though the area used during the season may be 2-3 times larger; thus nests are sometimes 45-70m apart, or even as close as 10-15m, though the density is more often extremely low (Glutz von Blotzheim *et al.* 1973, Cramp & Simmons 1980). Other figures are up to 10 calling birds in 30ha (Austria), 60-70 males in 5-21km² of habitat (SE Finland) and 0.5-4 pairs/km² (W Estonia) (Hagemeijer & Blair 1997); USSR figures range from 11 nests in a 4ha meadow to 0.4-33 individuals/km² depending on habitat (Potapov & Flint 1987). Birds wintering at sludge drying ponds in Zambia sometimes occurred in 'pairs', i.e. two birds which arrived together and occupied the same small pond; such birds held individual territories of 210-315(mean 248)m² which did not normally overlap; 3 of 4 such pairs were thought to consist of a male and a female (on plumage characters and size), which suggests that birds may associate in pairs during the wintering period and that temporary territoriality may be intersexual (Brooke 1974, Taylor 1987).

SOCIAL AND SEXUAL BEHAVIOUR Usually seen singly, in pairs or in family groups, but small groups of 2-4 birds may feed together on migration. In Zambia, wintering birds normally foraged alongside shorebirds such as Painted Snipe, Great Snipe, and Wood and Green Sandpipers *Tringa ochropus*, and also Yellow Wagtails *Motacilla flava*, with no aggression noted except once, just before the crakes left in Apr, when a crake repeatedly chased Wood Sandpipers from its immediate vicinity (Taylor 1987). A Pied Kingfisher once flew in to hover low over deep water 2m from where a crake was feeding; the crake immediately gave a sharp *tck* call, flew low over the water and landed immediately below the kingfisher, which flew off; the crake then swam and waded back to the water's edge to continue feeding (Taylor 1987). A Lesser Moorhen (131) which shared the crakes' habitat in Zambia appeared to be dominant to them: although it was never aggressive, whenever it approached closer than 35-40cm to a crake, the crake moved a short way away before continuing to forage (Taylor 1987); in Zimbabwe a crake fled into cover when approached by a juvenile Common Moorhen (129) (Brooke 1974). In Kenya, Baillon's and Spotted Crakes foraged in close proximity to each other (P. B. Taylor unpubl.).

Wintering birds in Zambia each occupied a discrete foraging territory and chased any Spotted Crake which entered this area, while on 'neutral' ground an approach to <3m also often elicited a brief chase, 1 bird retreating to its normal area; chases involved the aggressor running (sometimes with fluttering wings) or flying at the other bird, which immediately ran or flew off, and no fights were observed (Taylor 1987). Two birds fighting in the winter quarters sparred and fluttered up into the air, both eventually taking to cover (Brooke 1974). Aggressive pursuit of other individuals, on foot, by swimming or in flight, and sometimes with fluttering wings, is also recorded on the breeding grounds, where it may be confused with courtship-chases (Cramp & Simmons 1980).

In W Siberia, birds may nest in colonies of Black-headed and Little Gulls *Larus ridibundus* and *L. minutus*, and terns such as Common *Sterna hirundo*, Black *Chlidonias niger*, and White-winged Black *C. leucoptera* (Potapov & Flint 1987). In courtship (Cramp & Simmons 1980), the male steps deliberately towards the female, raising his half-open wings, then stretches his neck to its full extent while giving the advertising call; the female turns away suddenly and initiates a courtship-chase, the birds running, swimming or flying until the female allows the male to catch up. Before copulation, the male flutters up to the female with roused plumage, the female stops, then walks slowly with her head bent forward, and the male runs up and jumps on her back. Copulation is probably brief (3.8 s is recorded), and afterwards the pair self-preen and seek food (Bengtson 1962).

The incubating bird usually sits tight when approached, even pecking a human intruder's hand or, like the Little Crake (91), making a pecking attack, with the feathers roused and the wings spread and pushed forwards rather like a shield (Glutz von Blotzheim *et al.* 1973).

BREEDING AND SURVIVAL Season Europe, Apr-July, laying as late as Jun in N; former USSR, late Apr-July. **Nest** A thick-walled cup of dead leaves and stems of available vegetation, usually lined with finer grasses, roots and leaves; placed in thick vegetation close to or over standing water, often on tussock or built up well above water level, but also on tussocks among damp meadow grasses, and on ground, even bare earth; concealing vegetation often pulled over to form canopy; external diameter 12-17cm, height 5-15cm, cup diameter 10-12cm, cup depth 4.5-7cm. Both sexes build. **Eggs** 6-14, usually 8-12 (mean 10.3); larger clutches may be from 2 females; broad oval; smooth and fairly glossy; buff to olive-buff, well marked with red-brown and grey to pale purple spots and blotches; size of 100 eggs 29.1-37.5 x 22.2-26.8 (33.6 x 24.5); weight 10.4-12.8 (Potapov & Flint 1987). Eggs laid at rate of 1 per 1.5 days; replacements laid after clutch loss. **Incubation** 18-19 days per egg; up to 24 days for clutch; by both sexes; hatching asynchronous, taking 2-3 days, exceptionally 8 days for very large clutches – see Potapov & Flint 1987; in captive pair, male fed female and first-hatched chicks in nest (Pauler 1968). **Chicks** Precocial and nidifugous; remain in nest until all clutch hatched, then leave in 8-10h. Chicks fed and cared for by both parents; in captivity, chicks tended mainly by male, female's attentiveness steadily declining (Pauler 1968), conversely, female alone said to rear chicks (Rutgers & Norris 1970). Chicks beg by crouching with spread wings and moving head to and fro; swim and dive well; roost in specially constructed brood-nest, sometimes in breeding nest. Chicks begin independent feeding at c. 3 days, and may disperse before fully feathered, sometimes being chased away by parents at c. 20 days (Cramp & Simmons 1980); early dispersal possibly linked with potential second nesting attempt. First contour feathers appear after 2-3 weeks; fledging period up to 45 days; young fly at 7-8 weeks. Of 48 clutches,

Hungary, 25 (52%) hatched, 19 (40%) predated or disappeared (predation mainly by vole *Arvicola terrestris*), 4 (8%) infertile; of 180 eggs in 25 nests, 150 (83%) hatched (Szabó 1970). Other nest predators include red foxes *Vulpes vulpes*, Hooded Crows *Corvus corone* and Magpies *Pica pica*, and losses also occur to raccoon dogs *Nyctereutes procyonoides*, badgers *Meles meles*, polecats *Mustela eversmanni*, stoats *M. erminea*, pine martens *M. sibirica* and weasels *M. nivalis* (Potapov & Flint 1987). Age of first breeding 1 year. Usually double-brooded, but single-brooded in first breeding season (Becker 1995). **Survival** The oldest ringed bird is 7 years 2 months (Glutz von Blotzheim *et al.* 1973).

97 AUSTRALIAN CRAKE
Porzana fluminea   Plate 29

Porzana fluminea Gould, 1843, New South Wales.

Monotypic.

Synonyms: *Rallus/Porzana/Ortygometra novae hollandiae*; *Ortygometra fluminea*.

Alternative names: Spotted/Australian Spotted/Water Crake.

IDENTIFICATION Length 19-23cm; wingspan 27-33cm. Medium-small crake, with short bill and longish tail. Male has upperparts from crown to tail, and upperwings, brownish-olive, heavily streaked black (least on hindneck) and streaked and spotted white; Head (except crown), and underparts to breast, dark grey, spotted white on sides of neck and breast; flanks and belly strongly barred black and white; undertail-coverts white. White leading edge to outermost primary often visible in flight. Iris red; bill pale green to yellowish, base of upper mandible usually red and swollen (may be less swollen, or may fade to brownish-yellow, perhaps seasonally); legs and feet green to olive-yellow. Female smaller than male; has duller olive upperparts, brown stripe across upper lores, grey of face to breast much paler, more extensive white spotting on sides of breast and neck, and duller, less swollen, spot at base of culmen. Immature very similar to adult, but has narrow white fringes to grey breast feathers in fresh plumage; upperwing-coverts patterned like juvenile. In juvenile, white spots on upperparts lack black borders and are less striking, rump and uppertail-coverts lack white marks, and secondary coverts all have white tips; upper sides of head and sides of breast olive-brown, spotted white; underparts barring dark brown and white; bill grey-olive with darker culmen and tip; iris dark brown. Juvenile female has less cinnamon, more olive, ground colour to upperparts and breast. Inhabits dense vegetation at margins of permanent or ephemeral, freshwater to saline, wetlands.
Similar species Sympatric Baillon's Crake (94) smaller and paler, with richer cinnamon-brown upperparts; fewer and less regular white markings on upperparts (and no white spots from crown to hindneck); paler grey underbody; no spots on sides of neck and breast; barred undertail-coverts; and no red or orange spot at base of bill. Juvenile Baillon's is normally paler, with cleaner white belly; no grey in plumage; no patterning on foreneck and indistinct patterning on breast. Smaller Spotless Crake (104) darker and more uniform, with no patterning on upperparts and flanks, barred undertail-coverts, blackish bill and red legs and feet. Larger Lewin's Rail (65) has longer bill, shorter tail, rufous forehead to hindneck, distinctive pattern on upperparts and upperwings, and barred undertail-coverts.

VOICE Poorly known; most information is from Marchant & Higgins (1993). The common call is a sharp, metallic double note; it also has a sharp single-note call, a querulous version of which is given when the bird is forced off the nest by disturbance. There is also a chattering call similar to the 'ratchet call' (rattle) of Baillon's Crake; a prolonged wheezing note; and a yelping call. It is also said to utter a single *kek* (Cox & Pedler 1977). Chicks give a squeaking call when handled, which attracts the adults.

DESCRIPTION
Adult male Crown and entire upperparts brownish-olive to olive, tinged cinnamon; feathers with black centres and with white edges (crown to hindneck) or bold, black-bordered white spots (rest of upperparts) giving well-patterned effect, least prominent on hindneck; rump has small white spots; uppertail-coverts have less prominent black markings; rectrices black, fringed brownish-olive and with black-bordered white spots along edges; upper scapulars lack black central streaks; longest feathers edged, not spotted, white. Upper lesser upperwing-coverts brownish-olive with small white spots; other coverts brownish-olive with a few black-bordered white spots; alula and primary coverts dark brown, narrowly fringed olive; primaries and secondaries dark brown, with pale brown tips which become broader with wear, and narrow pale brown outer edges; P10 has white outer edge; tertials brownish-olive, with black centres and white edges. Axillaries and underwing-coverts barred black and white, but greaters grey, broadly tipped white; underside of remiges grey. Forehead and sides of head dark grey, becoming blacker on chin and lores; lores sometimes tinged brown; ear-coverts streaked or mottled white in c. 60% of birds; throat, sides of neck, foreneck and breast dark grey, feathers at sides of neck and breast with small black-bordered white spots; flanks and sides of belly barred black and white; centre of belly and thighs dark grey, barred white; undertail-coverts white. Iris red, sometimes with narrow yellowish or yellow-olive inner ring; iris may also be brown or orange-red; bill pale green or olive-yellow, base of culmen usually swollen and red or orange-red; legs and feet green to olive-yellow. May show seasonal variation in appearance of spot at base of culmen: swelling may disappear and colour may fade to brownish yellow.
Adult female Somewhat smaller than male; also differs in having duller, more olive, upperparts; brown stripe across upper lores; ear-coverts streaked or mottled white in c. 90% of birds; grey of head to breast paler; more extensive white spotting on sides of neck and sides of breast; base of culmen sometimes as male but often duller, less swollen, and rufous-brown, brownish-yellow or orange-rufous.
Immature Very similar to adult, but in fresh plumage grey breast feathers have narrow white fringes which may wear away. Upperwing-coverts patterned like juvenile; narrow white tips to coverts are lost with wear. Bare parts similar to adult.
Juvenile Similar to adult on upperparts but white spots lack black borders and are thus less striking; rump and uppertail-coverts lack white marks; secondary coverts all have white tips; supercilium pale brown to dark olive-brown, heavily flecked white; ear-coverts olive-brown,

washed buff and varyingly streaked whitish; grey areas of rest of head, and of underparts, have white spotting and whitish feather tips; sides of breast broadly olive-brown, spotted white; belly off-white; underpart bars dark brown (grading to olive-brown) and white. Iris brown to dark brown; bill grey-olive with darker culmen and tip; legs and feet as adult. Juvenile female has less cinnamon, more olive, ground colour to upperparts and breast.

Downy young Black downy chicks have faint greenish sheen on head and upperparts; later, body colour fades to dark brown; iris black; bill black with red base and white egg-tooth; bill becomes olive from centre; legs and feet dark olive-green or blue-black.

MEASUREMENTS Data from Marchant & Higgins (1993). Wing of 34 males 88-108 (101.3, SD 3.9), of 17 females 94-103 (97.9, SD 2.1); tail of 31 males 46-58 (51.7, SD 2.8), of 15 females 45-53 (50.3, SD 2.4); exposed culmen of 31 males 18.7-22.1 (20.6, SD 0.8), of 17 females 15.9-21.2 (18.7, SD 1.1); tarsus of 33 males 27.5-31.5 (30.1, SD 1.1) of 16 females 27.5-30 (28.6, SD 0.7). Weight of male 50-75 (65.6, SD 6.9, n = 10), of female 50-61 (57.3, SD 5.2, n = 3), of all ages unsexed 57-81 (68.1, SD 6.8, n = 12).

GEOGRAPHICAL VARIATION None. The putative form *whitei* is not distinct (Greenway 1973).

MOULT Primary moult is probably simultaneous, but no moulting specimens appear to exist. Postbreeding moult is complete, and feather wear suggests that primary moult can occur as early as Nov and as late as Apr; body moult is recorded throughout the Dec-Apr period (Marchant & Higgins 1993). Postjuvenile moult is apparently partial, involving body feathers only, and is recorded Jan-Apr; the last juvenile feathers to be lost are from the flanks and the sides of the breast and throat; at least some birds retain juvenile wing feathering through the first winter (Marchant & Higgins 1993).

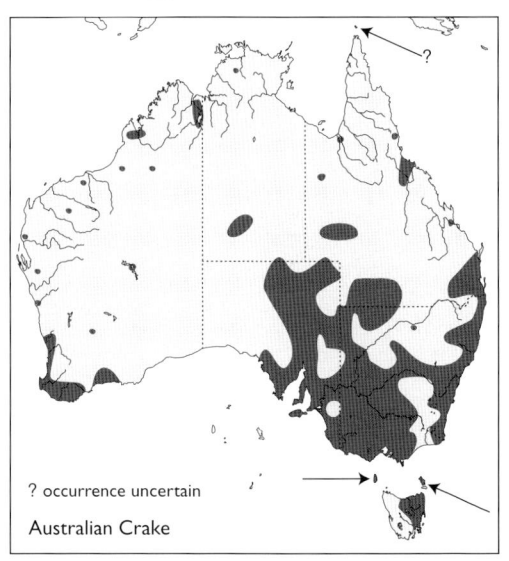

? occurrence uncertain
Australian Crake

DISTRIBUTION AND STATUS Australia and E Tasmania. Information is taken from Marchant & Higgins (1993) unless otherwise specified. As for several other Australian crakes there is a location bias of observer numbers and effort in the south, as well as poor access to many inland and tropical areas where suitable habitat is typically seasonal or ephemeral (and (though locally extensive) is overall sparse and scattered; such considerations make it difficult to interpret distributional data (R. Jaensch *in litt.*). In Australia it is recorded mainly in the SE and SW, being widespread in New South Wales and Victoria, locally common in E South Australia, and local in SW Western Australia. It is sparsely recorded elsewhere, being generally rare in Queensland (but commoner in the SW), and very local in Northern Territory where it was first recorded in 1967 and is locally common in the Barkly region in wetter years when larger lignum swamps are inundated (Jaensch & Bellchambers 1997). It is now probably moderately common around Alice Springs, presumably mainly at artificial sites. It is uncommon in E Tasmania. It is not globally threatened. Apart from records of large numbers during irruptions (see Movements section, and also Marchant & Higgins 1993), measures of abundance are not available, but a high density of 12-19 birds/ha was recorded at L Bindegolly, SW Queensland (Woodall 1993); such high densities are probably not unusual in SE Australia (R. Jaensch *in litt.*). Its range was reduced in Tasmania as early as 1910 by wetland drainage (Littler 1910), but whereas habitat loss through wetland drainage and modification in mainland Australia (especially in the south) has probably reduced its local occurrences, its overall range is not significantly altered (R. Jaensch *in litt.*). Grazing cattle may trample its habitat at wetland edges (Badman 1979). It readily occupies artificial wetlands (see Habitat), the construction of which provides some localised compensation for wetland losses and has also allowed it to expand its range into previously uninhabitable areas (e.g. Alice Springs and Mt Isa). Overall its status is probably less affected by habitat loss than other Australian crakes because of its common occurrence inland (R. Jaensch *in litt.*). Dogs and feral cats kill adults and young, and other predators include the tiger snake *Notechis scutanus* and raptors, while birds nesting in paddocks may be killed by lawnmowers.

MOVEMENTS It is possibly dispersive and irruptive rather than regularly migratory (see Distribution and Status for the caveat regarding observer bias). In some areas (e.g. S Murray-Darling and W L Eyre region) it is present all year, but in others it apparently occurs only in the autumn and winter, while at Port Phillip Bay (S Victoria) it normally occurs only in late summer and autumn (Watson 1955, Marchant & Higgins 1993). No seasonality in the overall pattern of movements is suggested, despite higher reporting rates in the summer: these are possibly related to increased calling or to drying of inland swamps (Blakers *et al.* 1984, Emison *et al.* 1987). Birds may move across the Torres Strait, where they have been recorded on Booby I in Jan and May (but these records may be erroneous), and the Bass Strait, where they have been recorded on King I and Flinders I, while the species is also recorded from Pelsart I (Western Australia) (Marchant & Higgins 1993). Numbers fluctuate with changing habitat conditions: the birds may become abundant after floods and heavy rain, but they seem to follow receding water rather than deeper, flooded conditions, and are known to congregate at the edges of drying bore drains or swamps (Blakers *et al.* 1984, Marchant & Higgins 1993); the occurrences of this species may often be related to its apparent preference for muddy substrates (R. Jaensch *in litt.*). It may be irruptive, as large numbers sometimes

397

appear and depart suddenly; examples are 500 birds round Berri (South Australia) from Sep 1972 to Apr 1973, many records from New South Wales in Oct 1977, an influx into Victoria in spring and early summer 1982, and 40+ birds in NE Kimberley in early May 1988 (Marchant & Higgins 1993).

HABITAT A highly opportunistic species, usually inhabiting the well-vegetated margins of permanent or ephemeral freshwater, brackish or saline wetlands, and also being the typical crake of inland arid areas (R. Jaensch *in litt.*). It occurs at estuaries, tidal creeks, mangroves, saltmarshes, swamps, marshes, lakes, ponds, lagoons, claypans and floodplains; also among waterlilies *Nymphaea*. It prefers the edges of drying wetlands rather than more deeply flooded areas (it is the typical crake of muddy substrates), and it is more tolerant of saline conditions than are other crakes and small rails (Blakers *et al.* 1984, R. Jaensch *in litt.*). It usually occurs among dense vegetation such as samphire (Chenopodiaceae, presumably *Halosarcia* spp.), lignum *Muehlenbeckia cunninghamii*, canegrass *Eragrostis australasica*, sawgrass *Cladium/Baumea*(?), bluebush (presumably *Chenopodium auricomum* and/or *Maireana* spp), saltbush, mangroves, reeds and rushes, rank grass or dense thickets of *Calistemon*, *Melaleuca* or other shrubs, or even floating vegetation such as *Triglochin* and waterlilies (Marchant & Higgins 1993). However, occupied vegetation (e.g. canegrass) is not always dense and this species will probably tolerate sparser cover more readily than will Spotless and Baillon's Crakes (Bryant & Amos 1949; R. Jaensch *in litt.*). It sometimes occupies scrubby cover round saltmarshes and saltworks, and it may occur some distance from water. It is often at artificial wetlands, such as saltworks, sewage ponds, flooded gravel pits, bores and drains, and occasionally in grassy areas e.g. lawns, pastures or golf courses; it once used puddles formed by water from washing cattle trucks, and it has also been recorded at a rubbish dump (Marchant & Higgins 1993). In South Australia it favours dense reeds at artesian bores and mound springs (Badman 1979). It breeds in clumps of dense vegetation such as samphire, lignum, sedges, rushes, long grass or *Sesbania* or other shrubs in swamps and other inundated wetlands (Marchant & Higgins 1993); like Spotless and Baillon's Crakes, it usually nests over water (R. Jaensch *in litt.*). In Queensland, at a lakeshore site where it occurred alongside Baillon's Crake, Woodall (1993) found that the occurrence of both species correlated with good cover of lignum (which was used as a refuge) in shallow water and *Cyperus gymnocaulus* on land, a relatively narrow belt of water between the shore and the lignum, and the % cover of *Myriophyllum verrucosum* on land (of significance as cover and/or as a source of food). Although it does occur in lignum in water too deep for wading, such habitat is more typical of Baillon's Crake (R. Jaensch *in litt.*).

FOOD AND FEEDING Most information is from Marchant & Higgins (1993). Molluscs (freshwater gastropods); crustaceans (Ostracoda); adult and larval insects including Dermaptera, Orthoptera (Acrididae), Hemiptera (Notonectidae, Cicadellidae), Coleoptera (Carabidae, Chrysomelidae, Curculionidae), Diptera (including Chironomidae), Lepidoptera and Hymenoptera (including Formicidae); spiders; and frog tadpoles. It also takes algae, plant material, and seeds of Fabaceae (*Medicago* and *Trifolium*) and Cyperaceae. It forages mainly at wetland margins, gleaning and probing on mud, peat, or in shallow water (up to 5cm deep), and also near or among grass, marsh vegetation or shrubs, and among floating plants such as *Triglochin*. It wades and swims, probing and lunging under the water and at emergent vegetation, often submerging the entire head; a bird also swam in a rapidly drying pool, feeding on flies (Watson 1955). Large food items are brought ashore, knocked on the ground and swallowed in several gulps (Bryant & Amos 1949).

HABITS This crake is diurnal and is particularly active early and late in the day. It is usually unobtrusive, keeping to the thickest stands of cover during most of the day, and running to cover when disturbed. However it is generally bolder than other crakes when feeding, although it never ventures far from cover. It moves with a slow, stalking walk, with the tail constantly flicked, and when alarmed it makes a sudden crouching run to cover with the tail raised; it will also fly if pressed, or take to water if forced from cover. It swims and dives readily, but usually only for short distances in cover or across channels or streams. It is seldom seen in flight but it will fly over open water between stands of emergent vegetation; short flights are laboured and fluttering, with dangling legs. A bird chased by a Black Falcon *Falco subniger* took refuge in a house (McGilp 1923). Birds roost in dense vegetation, and it has been recorded resting and preening on a platform of criss-crossed *Triglochin* (Bryant & Amos 1949).

SOCIAL ORGANISATION Normally seen singly, in pairs or family groups. It is monogamous, and the pair bond may be maintained outside the breeding season, pairs having been seen together in winter (Watson 1955). It is probably territorial when breeding, and appears to nest singly; however, nests are sometimes clumped, with up to 30 in a group, and 2 nests were found 50-60m apart (Marchant & Higgins 1993).

SOCIAL AND SEXUAL BEHAVIOUR Birds may be seen in relatively large gatherings at favourable habitat, and in loose groups of up to 100, probably where food is abundant (Kingsford 1991). Feeding birds occasionally rush at each other when close but usually make no attempt to drive other birds from the area; an Australian Crake was once seen to chase a Baillon's Crake but on other occasions both species foraged together with no agonistic interactions (Woodall 1993). When disturbed, the incubating bird is known to stand at the entrance with raised wings and to call continuously; 1 incubating bird pecked the intruder before leaving the nest and feeding nearby, and then returned to attack the intruder's hand in the nest by jumping onto the hand and digging its bill into the fingers (Bryant & Amos 1949).

BREEDING AND SURVIVAL Season Breeds Aug-Apr, most Aug-Jan; all atlas records were S of 27°S (Blakers *et al.* 1984). **Nest** Varies from flat, flimsy structure to cup of fine woven material; of wet or dry rushes, or dry grass, lined with soft grass; sometimes has dome or canopy of interlaced rushes; material added to nest during incubation. External diameter 7-9cm, depth 5cm; some have ramp from substrate to rim. Usually built 2-50cm above water, in rushes, sedges, grass, low bushes (e.g. lignum, *Glycyrrhiza*), among overhanging tree branches, or in waterlilies; often in centre of clump or tussock; classic nest site has some form of rigid support, such as small shrub, with fine grass or sedge growing through it (R. Jaensch *in litt.*); 1 nest built on ground, partly shaded by samphire. **Eggs** 3-7; rounded or elongate oval; smooth and glossy; brownish-olive to light greenish-olive, freckled, spotted and blotched purplish-brown, purplish-grey and red, sometimes more so around larger end; obvious mottling easily distinguishes eggs from those of Baillon's and Spotless Crakes (R. Jaensch *in litt.*). Size of eggs: 29.3-34 x 22-24.4 (30.9 x 22.9, n = 28), calculated weight 8.9 (Schönwetter 1961-62) and 29.5-34.5 x 20.8-24.4 (31 [SD 1.2] x 22.9 [SD 0.6], n = 28) (Marchant & Higgins 1993). Eggs probably laid at daily intervals. **Incubation** By both sexes, period not known; may start before clutch complete, as hatching asynchronous. **Chicks** Precocial; tended by both parents, possibly until after fledging. No information on growth or development. May sometimes rear 2 broods per year. Nests may be abandoned if water level drops.

98 SORA
Porzana carolina Plate 29

Rallus carolinus Linnaeus, 1758, North America = Hudson's Bay.

Monotypic.

Synonyms: *Rallus porzana/virginianus/stolidus/olivaceus; Gallinula/Ortygometra/Aramides/Crex/Galeolimnas carolina.*

Alternative names: Sora/Carolina Rail/Crake; Meadow Chicken; Ortolan.

IDENTIFICATION Length 19-25cm; wingspan 35-40cm. A medium-small, rather plump rail, with short, thick yellow bill contrasting with black face mask. Crown and upperparts olive-brown, with black feather centres and narrow white streaks from mantle to tail and on scapulars; streaks absent or reduced to a few spots from back to tail; centre of crown black; upperwing-coverts olive-brown with variable white markings. Distinctive head pattern of black mask (forehead, lores, face to eyes, and chin and throat, and sometimes extending in line down foreneck), white spot behind eye, and grey sides of head (including broad supercilium). Underparts to breast pale slate-grey; upper flanks barred grey and white; lower flanks to vent barred olive-brown, white and black; central belly and undertail-coverts white. Iris reddish; bill bright yellow, with white line at base of upper mandible; legs and feet yellowish-green. Sexes similar but female slightly smaller, with less prominent head pattern and more white dots on upperparts and upperwing-coverts; bill has olive-green tip and lacks white at base. Immature like adult but black patch on chin and throat less extensive and mottled grey; grey of head and neck paler; grey of breast often olive-tinged. Juvenile has upperparts and wings like adult but with more white streaks; predominantly buff to olive-brown on head, neck and breast in areas where adult is grey; is blackish only on lores and at eye; chin and throat white; flank barring duller than adult; iris brown; bill greenish-yellow; legs olive-green to greenish-yellow. Inhabits freshwater marshes dominated by emergent vegetation; also brackish and coastal marshes on migration and in winter.

Similar species Only likely to be confused with Spotted Crake (92), each species being a vagrant within the other's range; similar in size and bulk, or slightly smaller, but longer-necked and with striking head pattern (black mask, black streak sometimes extending to upper breast, and triangular white spot behind eye); no white spots from crown to hindneck; less obvious white streaks on mantle and scapulars; fewer white spots from back to tail; grey foreneck and breast with no spots; in some birds, largely unpatterned upperwing-coverts (white markings only on inner greaters); tertials with duller olive-brown fringes and narrow white streaks (no incomplete V-bars as in Spotted Crake); narrower and less conspicuous white leading edge to wing; whitish undertail-coverts; and yellow bill, with no red at base but often with greenish tip. Bill looks proportionally longer than in Spotted Crake, and heavier at base; has obvious nasal groove. Juvenile appears similar to juvenile Spotted Crake but differs in having crown and sides of head buff, with browner ear-coverts and with no spots and streaks, central crown-streak and lores black, neck and breast unspotted and buff, and bill greenish-yellow.

VOICE Information is from Melvin & Gibbs (1996) and Kaufmann (1983). The characteristic breeding-season territorial and contact call, given by both sexes, is a high-pitched descending whinny *whee-hee-hee-hee-hee...*, of 10-30 notes and usually lasting 2-3s; it is also given in response to loud noises, is given frequently during spring migration, and is occasionally heard in the non-breeding season. The female's version is shorter, more variable, and higher-pitched. Birds may call less often in the first week after arrival on the breeding grounds, and calling may increase between first and second nesting. This call is given with the bill pointing down, except in territorial disputes, when the bird faces its opponent. Another common call is a plaintive, ascending, whistled *kerwee*, possibly used to attract mates, and given on spring migration and immediately after arrival on the breeding grounds; it may be shortened to 1 syllable, *kee* or *weep*, similar to the call of the Spring Peeper frog *Pseudacris crucifer*. Males give soft, low cooing notes *gwoo* during courtship chases and in 'swanning' displays. Incubating birds give a gargling call at nest-relief, and a short *tug* when the mate brings food; a sharp, repeated *quink* or *kuk* is given by birds disturbed at the nest; and a nasal coot-like call by adults when feeding. Kaufmann also describes a sharp *kiu* of alarm, like an alarm call of the Virginia Rail (58), a short, high-pitched variant *keep* from adults when the nest is approached, and soft peeping contact calls. Various squeals and squawks are given when birds are attacked. Chicks give soft, wheezy *peep* or *queea* calls, louder when begging for food and becoming a shrill *queee* when alarmed.

DESCRIPTION
Adult male Forehead, lores to above and below eye, sides of head at gape, and chin and throat, black, forming prominent mask; centre of crown black; sides of crown and hindneck dark olive-brown; feathers of mantle, scapulars and tertials black with broad olive-brown fringes and narrow but distinct white streak along each edge (streaks occasionally irregular or broken); streaks short on outer scapulars and sometimes on lower mantle; back to tail black, feathers narrowly fringed olive-brown, with a few white spots on some. Upperwing-coverts dark olive-brown, variably spotted or barred white – sometimes only a few inner greaters with black-edged white markings, occasionally many greaters and most inner medians and lessers barred, spotted or streaked white; primary coverts, alula and remiges dark sooty-grey, tips of primaries black

and outer webs tinged olive; outer web of P10 and of outer alula feather narrowly edged white; marginal coverts white. Axillaries sepia, barred white; underwing-coverts sepia, broadly barred and fringed white. Sides of head (including broad streak over eye) and sides of neck ashy blue-grey; grey streak over eye sometimes extends across upper forehead above black mask; tiny triangular black-edged white spot behind eye; indistinct olive-brown patch or streak on ear-coverts (may be absent). Breast pale slate-grey; sides sometimes tinged olive-brown and barred or spotted white; upper flanks barred ash-grey and white; lower flanks and sides of belly barred olive-brown, white and narrowly black; centre of belly and vent white, sometimes tinged buff, and grey feather bases sometimes visible; undertail-coverts white, variably yellow-buff at edges. In fresh plumage black feathers of throat narrowly tipped pale grey, this possibly being responsible for black throat-stripe appearing narrower and broken in autumn and winter, as described by Melvin & Gibbs (1996) – see also Immature; grey feathers of breast broadly bordered white at tip. When worn, white feather bases may show on chin and throat. Iris red, reddish-brown or amber; bill chrome-yellow, with narrow white band at base of upper mandible; culmen with small knob-like swelling at base during nesting; legs and feet yellowish-green.

Adult female Like male breeding, and often indistinguishable, but shorter scapulars, lower mantle and wing-coverts on average with more and narrower white dots; black on sides of head and throat slightly less extensive; grey streak over eye more often continues across forehead. Bill chrome-yellow, shading to olive-green at tip; lacks white band at base of upper mandible.

Immature Like adult but black area on chin and throat narrower and shorter, with much grey mottling especially in female; black on sides of head less intense; lower sides of head and sides of neck paler grey; grey of breast often tinged olive; white feather edges on breast wider than in fresh adult and partly retained into summer. Juvenile remiges and rectrices retained through first breeding season (Bent 1926); juvenile tertials and greater upperwing-coverts also often retained (Cramp & Simmons 1980). Paler tips to black and grey areas of head and underparts abrade, giving first breeding plumage similar to adult.

Juvenile Forehead olive-brown; sides of crown to hindneck somewhat brighter brown than adult; rest of upperparts, and upperwings, like adult breeding but white streaks slightly narrower, more numerous and with edges less smooth, often becoming dots on longer scapulars and tertials. Tertials narrower and more pointed than in adult, with finer, longer white streaks; sides of tail more often speckled white. Sides of head whitish to buff; blackish on lores and around eye, and with ear-coverts darkish olive-brown; sides of neck dark olive-brown; chin and throat white, often spotted buff; foreneck buffy; breast buff in centre, olive-brown at sides with variable small white dots or bars (sometimes absent); flanks barred olive-brown (sometimes darker brown), white, and narrowly blackish-brown, barring duller and less contrasting than in adult; pale grey bases of breast and belly feathers may show when plumage worn. Iris pale brown to brown; bill greenish-yellow, brightest at base of lower mandible, darkest at tip; legs and feet olive-green to greenish-yellow. Note that Becker (1995) gives iris olive-green, bill bright brownish, and legs bright brownish to flesh, but these features are not borne out by available photographs or descriptions.

Downy young Chick has glossy black down with orange bristles at base of lower mandible; iris black; bill whitish or bluish-grey, with fleshy red cere and white egg-tooth at hatching, later darkening to greenish; legs and feet pink, changing to grey and then yellow-green; bare skin on crown reddish, above eyes bluish (see photograph in Taylor 1996b); small wing-claw present. Throat bristles and egg-tooth lost at 1-2 weeks; cere at 5th week.

MEASUREMENTS Wing of 16 males 109-120 (112.4, SD 2.9), of 8 females 103-116 (110.9, SD 4.1), of 57 unsexed 92-123 (109.4, SD 5.1); tail of 16 males 46-60 (52.0, SD 3.8), of 9 females 40-54 (48.4, SD 4.5), of 55 unsexed 38-60 (51.4, SD 4.2); exposed culmen of 16 males 18.5-24.5 (20.8, SD 1.7), of 8 females 19-22 (20.5, SD 0.9), of 63 unsexed 17-24 (20.8, SD 1.5); tarsus of 16 males 36-44 (40.9, SD 2.0), of 9 females 36-44.5 (39.9, SD 2.4), of 57 unsexed 36-45.5 (40.8, SD 2.2) (Melvin & Gibbs 1996). Weight of 23 males 66-112 (85.1, SD 11.8), of 13 females 65-100 (77.1, SD 9.5), of 86 unsexed 49-126 (74.8, SD 13.1) (Cramp & Simmons 1980, Melvin & Gibbs 1996). Hatching weight 5.0-7.5 (6.35, n = 21) (Walkinshaw 1940).

GEOGRAPHICAL VARIATION None.

MOULT Postbreeding moult is complete, with simultaneous moult of remiges and rectrices, and occurs in Jul-Sep, occasionally starting in Jun; body moult occurs at the same time (Cramp & Simmons 1980); some may not moult before arrival in the wintering area (Melvin & Gibbs 1996). Prebreeding (definitive pre-alternate) moult is partial, involving the head and body, occasionally with some rectrices, and occurs from Dec or Jan to Mar (Cramp & Simmons 1980); first-year birds undergo this moult (first pre-alternate) in Mar or earlier (Melvin & Gibbs 1996). Plumage changes from non-breeding to breeding are mainly due to abrasion of feather tips (see Description). Postjuvenile moult is partial, involving head, body, rectrices, most wing-coverts (usually excluding greaters) but not flight feathers and usually not longest tertials (Cramp & Simmons 1980); however, it is also said to exclude the rectrices (Melvin & Gibbs 1996). It starts with the forehead, sides of head, chin and breast, gradually followed by the mantle, scapulars, flanks and throat, and later by the rest of the head and body, and the median and lesser wing-coverts (Cramp & Simmons 1980). It often begins at 12-14 weeks but its timing is highly variable – it may begin in Aug, in the hatching area, or not until the winter quarters have been reached in Nov-Dec; there is usually no active moult during migration (Cramp & Simmons 1980). Most birds complete this moult between Nov and Feb, but a few retain some juvenile feathers until spring (Cramp & Simmons 1980), or remain in decidedly juvenile plumage throughout the first winter (Melvin & Gibbs 1996).

DISTRIBUTION AND STATUS Details for North America are from Melvin & Gibbs (1996). It breeds locally in North America, from SE Alaska and Canada (S from SW Northwest Territories, NW and NC British Columbia, WC Mackenzie, N Saskatchewan, N Manitoba, N Ontario, SC Quebec, S Newfoundland, New Brunswick, Prince Edward I and Nova Scotia) to N, WC and W USA (along the E seaboard S to N New Jersey; and S to S Pennsylvania, SC Ohio and Indiana, C Illinois, Iowa and Nebraska, E Colorado, E and S New Mexico, C and NW Arizona and S Nevada). In the W, it occurs to W and NE California (possibly also SW but present status uncertain), Oregon,

Washington and C British Columbia; it also breeds on SE Victoria I and in SW British Columbia. It also breeds locally in C Kansas, NE West Virginia, NW Virginia, and very locally or irregularly at other sites S of the regular range. It is more likely to breed near the N edge of its range when drought conditions exist in S breeding areas (Trapp *et al.* 1981). It winters in the USA (along the Atlantic Coastal Plain from the Delmarva Pen S to Georgia and Florida, W along the Gulf Coast to SW Texas, S New Mexico, S and NW Arizona, extreme SE Nevada and extreme SW Utah; also on the Pacific coast from S Oregon S through SW California and Baja California. The densest wintering populations occur in S Florida, S Louisiana, S Texas and the lower Colorado R valley of SW Arizona. It also winters in Mexico, the West Indies (including Bermuda, the Netherlands Antilles, Trinidad and Tobago), Belize, Guatemala, Honduras, Costa Rica, Panama (including Coiba I), Colombia (Cartagena S locally to Santa Marta Mts, middle and upper Cauca V, temperate zone of E and W Nariño, E Andes in Boyacá and Cundinamarca, and lowlands E of Andes in W Caquetá), Ecuador (S to Guayas, Chimborazo, and once in Azuay and Pastaza; Ridgely & Greenfield in prep.), and N and C Peru (S to Junín) and E to Venezuela (Zulia, Falcón, Mérida, Portuguesa to Anzoátegui, S Amazonas, Margarita I, transient to Los Roques and La Orchila Is) and Guyana (Bartica). Small numbers also winter N of the normal wintering range, e.g. to N USA, and to S Canada (see Campbell *et al.* 1990). Belcher & Smooker (1935) attributed to this species a nest with 6 eggs found in Trinidad in May 1927 and hypothesised a local race; this has never been discovered and the egg measurements match those given by Belcher & Smooker for a clutch of the Grey-breasted Crake (32) (Gochfeld 1972); however, there is some doubt that the latter clutch is of Grey-breasted Crake, although it is not clear to what other species it may be attributed (Ripley 1977).

Not globally threatened; it is the most abundant and widely distributed North American rail. Population size has not been estimated. In Trinidad 40-80 birds occurred in 26ha of wetland (1.5-3 birds/ha) (Gochfeld 1972); for breeding density, see Social Organisation. It is estimated that populations declined 3.3% (SD 1.7) annually in 1966-91 in North America, were stable in 1982-92 in Canada but declined 8.5% (SD 2.1) annually in the US during the same period; declines were most severe in C North America, where wetland loss has been greatest (Conway *et al.* 1994, Melvin & Gibbs 1996). The species continues to decline throughout much of its range, at least in North America, where habitat loss and degradation probably limit the population size. Many of the wetlands most important for this species are amongst the most threatened in the USA, including coastal marshes in California, Florida, Louisiana, New Jersey and Texas, as well as palustrine emergent wetlands in S Florida and the Prairie Pothole region, and W riparian wetlands (Melvin & Gibbs 1996). In SC Colorado, decreasing water levels reduced suitable habitat by 95% in Jul-Aug 1975-76, and irrigation contributed to midsummer drying of wetlands which caused premature dispersal of rails to concentrate in wetlands with more water (Griese *et al.* 1980). Furthermore, harvesting of wild rice may disturb or reduce food available to resident or migrant Soras, causing numbers to decline in harvested areas (Fannucchi *et al.* 1986).

The presence of power plants discharging hot water into rivers may allow the species to extend its wintering range N into colder regions (Root 1988). It apparently occurs quite frequently close to human settlements, even on the outskirts of cities (e.g. Bent 1926, Melvin & Gibbs 1996).

Sora migrations are relatively conspicuous, and it occurs in concentrations at beds of wild rice, giving rise to large bags by shooters. It is traditionally a game species, and is hunted in 31 USA states and 2 Canadian provinces, with bag limits of 10-25 birds per day (Melvin & Gibbs 1994). An estimated 26,800-94,400 Soras were shot annually in 1964-76, and 37,900-97,700 annually in 1989-90 (49,000 ± 29,300 of these in Louisiana) (Melvin & Gibbs 1996). Little information is available on population trends or the effect of hunting on the populations, but interest and participation in hunting are currently low and the annual kill may be within sustainable levels, although the effects of incidental killing by waterfowl hunters and other bird hunters are unknown (Eddleman *et al.* 1988, Melvin & Gibbs 1994, 1996). Some are killed in traps for fur-bearers, while pesticides may be a threat in commercial rice fields and in wetlands of California and Arizona, and ingested lead shot may also pose a threat (Eddleman *et al.* 1988, Melvin & Gibbs 1996). Conway *et al.* (1994) have also suggested that radio-tagging Soras may result in increased mortality.

Melvin & Gibbs (1994, 1996) provide a summary of management needs in the USA, which are generally applicable to palustrine wetland habitats for rails in other parts of the world. The preservation and proper management of emergent wetlands are major priorities to safeguard and improve breeding, migration and wintering habitats. Efforts should be made to minimise the effects of wetland drainage, filling, siltation, eutrophication and other pollution, and invasion by exotic plants. Management should encourage diverse stands of both fine-leaved and robust emergents, including *Scirpus*, *Carex*, *Cyperus* and especially *Typha*, as well as moist-soil annuals at wetland margins (Rundle & Sayre 1983, Johnson & Dinsmore 1986a). Soras, like other rails, seem most abundant near the edges of cover types and at vegetation/open water interfaces, so periodic gradual drawdowns to encourage horizontal zonation of vegetation, and management to achieve the maximum interspersion of emergent vegetation and open water areas, should improve habitat quality; constructing wetland impoundments with irregular or sloping bottoms helps achieve these objectives. After summer drawdowns, late-summer flooding will provide habitat for post-breeding and migrant Soras (Rundle & Fredrickson 1981), while spring habitat in wetlands with annual grasses and *Polygonum* is encouraged by reducing water levels over winter to protect vegetation from ice damage and waterfowl, and to encourage the growth of perennial vegetation.

Surveys to monitor harvest and population trends have been lacking but improved estimates of harvest are now possible, and standardised taped-playback surveys could be used to monitor population trends (Melvin & Gibbs 1994). The extent and significance of postbreeding dispersal and emigration (see Movements) deserve investigation, especially to help evaluate the potential impact of the loss of small private wetlands (Johnson & Dinsmore 1985).

Elsewhere, this species is common to frequent throughout Mexico, where it possibly breeds locally in N Baja California), generally uncommon but probably overlooked in Puerto Rico and the Virgin Is, locally common to numerous in Panama, widespread but local in Costa Rica, locally common in Venezuela, locally uncommon to common up to 2,600-3,000m in Colombia, and rare in Peru (Parker *et al.* 1982, Hilty & Brown 1986, Raffaele 1989, Stiles & Skutch 1989, Fjeldså & Krabbe 1990); in Ecuador it is rare to locally uncommon, possibly now occurring less often than formerly (Ridgely & Greenfield in prep., Ridgely *et al.* in press). An undocumented winter record from Brazil (Bonito, Pernambuco: Ridgway & Friedmann 1941) is not accepted (Blake 1977).

Predators in the USA include Coyote *Canis latrans*, feral domestic cats, Bobcat *Felis rufus*, Hen Harrier *Circus cyaneus*, Cooper's Hawk *Accipiter cooperii*, Red-shouldered Hawk *Buteo lineatus*, Peregrine *Falco peregrinus*, Barn Owl *Tyto alba*, Great Horned Owl *Bubo virginianus* and Short-eared Owl *Asio flammeus* (Melvin & Gibbs 1996).

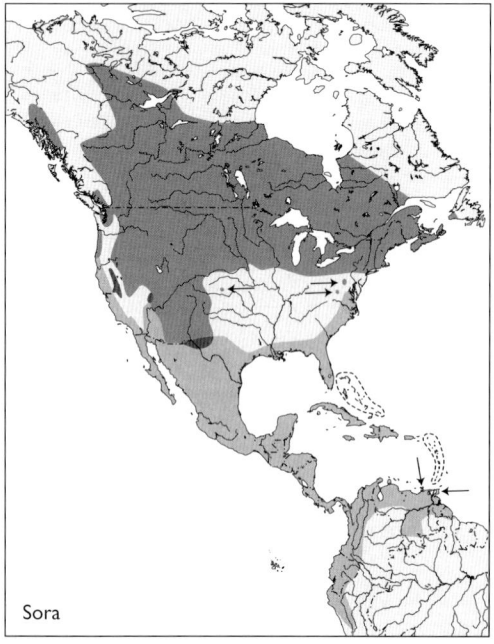

Sora

MOVEMENTS Migratory. Initial dispersal from breeding sites to adjacent wetlands or upland fields occurs from late Jul to early Aug (Johnson & Dinsmore 1985); in Minnesota, birds leave natal marshes in late Aug to early Sep to concentrate at larger wetlands prior to migration (Pospichal & Marshall 1954). Movement begins in late summer and early autumn, when birds gather in numbers around lakes and freshwater and brackish marshes. The birds are cold-sensitive and depart with the advent of frosts in Sep-Oct (Melvin & Gibbs 1996). Peak autumn migration is widely reported in Sep and the first half of Oct, but in Missouri large numbers appear from late Aug and in Arkansas passage is significant throughout Oct, most birds having gone by early Nov (Melvin & Gibbs 1996). Birds migrate on a broad front, over land and sea, fly low, and are prone to drift in strong winds, occasionally being taken far out to sea (Melvin & Gibbs 1996). Migrating birds are commonly killed at tall, lighted towers and probably by overhead wires; at 1 site they were killed in greater numbers than other rails (Stoddard & Norris 1967).

Soras arrive in Bermuda from late Jul to mid-Nov (Amos 1991). Birds wintering in SW Arizona arrive as early as 22 Jul and depart as late as 7 May (Melvin & Gibbs 1996). They are present in Mexico from mid-Aug to May, Panama late Sep to early Apr, Costa Rica Oct-Mar, Colombia Oct-Apr (rarely May), Ecuador Dec-Mar, and the West Indies and Venezuela Sep-May (Dickerman 1971, Meyer de Schauensee 1978, Bond 1979, Hilty & Brown 1986, Ridgely & Greenfield in prep., Ridgely & Gwynne 1989, Stiles & Skutch 1989, Howell & Webb 1995b). Return is noted from late Mar, when the first arrivals occur in N Ohio, British Columbia and Massachusetts; it arrives in S Michigan and Colorado in early Apr and thereafter is recorded widely (Melvin & Gibbs 1996). The main passages are along the Atlantic coast (Maryland) from mid-Apr to mid-May and mid-Aug to late Oct. A few birds may oversummer in Colombia (Hilty & Brown 1986). It is accidental in EC Alaska, Queen Charlotte Is, S Labrador, Bermuda, Greenland, Britain and Ireland, Spain, France and Sweden (Cramp & Simmons 1980, AOU 1983, Campbell *et al.* 1990). Ringed birds occurred at the same site in Arizona for 2-3 consecutive winters (Melvin & Gibbs 1996). Migrants appear occasionally on roads, airport runways and sandy beaches (Melvin & Gibbs 1996).

Postbreeding dispersal may initially involve the movement of family groups or individuals away from the vicinity of the brood-rearing home range into other wetlands and also to different habitats such as cornfields; such dispersal was recorded in Iowa in late Jul and early Aug (Johnson & Dinsmore 1985). The stimulus for this emigration may be the maturation and independence of the brood, and the male may stimulate the breakdown of the family group by his increasing aggressiveness towards the female and the young (Johnson & Dinsmore 1985). The significance of postbreeding dispersal, which may involve a fairly long-distance movement between wetlands, is unclear: it may simply segregate family members, it may constitute a limited moult migration, or it may involve a shift to an autumn migration staging area (Pospichal & Marshall 1954, Brown & Dinsmore 1985); it deserves further investigation (see Distribution and Status).

HABITAT Most information is from Melvin & Gibbs (1996). Preferred habitats throughout the year are freshwater emergent wetlands but it often occurs in brackish and saltwater marshes during migration, and occasionally when breeding; in some areas, e.g. Mexico and the Caribbean, it occurs in mangrove swamps (Raffaele 1989, Howell & Webb 1995b), and in Canada occasionally in willow swamps (Campbell *et al.* 1990). Optimal breeding habitat in North America is marshes with shallow and intermediate water depths, and a high interspersion of fine-leaved and robust emergents, flooded annuals, and patches of open water, such areas probably being attractive because they provide a good supply of seeds (especially of sedges) which are important foods during the breeding season. It is also attracted to areas with abundant floating and submerged vegetation, probably because such areas provide good substrates for invertebrate prey near the water surface. Dominant plants in freshwater breeding habitat include *Typha*, *Carex*, *Cyperus*, *Sparganium* and sometimes *Scirpus*; in brackish or salt marshes it nests occasionally in *Spartina* or *Phragmites*. The water depth in 71 breeding territories was 0-92cm (mean 38.4, SD 16.1), the mean vegetation height 128cm (SD 42.5) and the mean number of stems/m^2 121.9 (SD 80.9); birds did not use areas with flattened emergent vegetation until new sedge growth was 20-30cm tall and provided 80-100 stems/m^2 (Johnson & Dinsmore 1986a). It also feeds in upland fields and row crops during brood-rearing and postbreeding dispersal in late summer, and it visits grass fields, corn- and stubblefields, rank weed growths and brushy hillsides. Migrants occur in freshwater, brackish and salt marshes; stands of wild and cultivated rice, and of flooded annual grasses or forbs, provide important feeding habitat, and birds apparently prefer sites which offer tall dense cover as well as shorter seed-producing plants. Migrants also appear away from water, in fields, pastures and gardens, and on lawns.

Wintering birds occur in freshwater, brackish and salt marshes, particularly those with good interspersion of shallow water and emergent vegetation; they also occur in mixed shrubs near wetland-upland edges, in mangroves, and at mudflats, drainage ditches and the edges of ponds and rivers (Melvin & Gibbs 1996). At the N limits of its wintering range it sometimes uses wetlands kept unfrozen by sewer outlets, freshwater springs and hot-water discharges from power plants (Root 1988, Veit & Petersen 1993). This species breeds up to an altitude of 3,500m in the USA (Colorado), and it is recorded to 4,080m in C Peru (L. Junín).

Soras nest in the same marshes as Virginia Rails and appear to select identical nest sites (Kaufmann 1989). No strong niche-segregating mechanism was found in habitat use by breeding birds in Iowa (Johnson & Dinsmore 1986a); however, compared with Virginia Rails they are said to select areas with shallower water in winter and deeper water in spring (Melvin & Gibbs 1996), to occur in wetlands with larger areas of *Typha* and greater interspersion of vegetation and water (Crowley 1994), and to nest in wetter habitats, more often in *Typha* stands, in Canada (Peck & James 1983, Campbell *et al.* 1990). Soras will breed in very small marshes (<0.5 ha), but may occur less frequently than Virginia Rails in wetlands <5ha in area (Melvin & Gibbs 1996). In Colorado, Virginia Rails often occur alongside Soras but prefer less deeply flooded to saturated sites, and generally breed in marshes where spring air temperatures are warmer (mean 5.6°C) than in Sora breeding marshes (Griese *et al.* 1980). Because of their foraging behaviour and omnivorous diet, migrant Soras may be able to exploit a wider range of water depths than migrant Virginia Rails, which seem to occur more frequently in water <15cm deep and on saturated soils (Sayre & Rundle 1984).

FOOD AND FEEDING Most information is from Melvin & Gibbs (1996), and also Webster (1964). Omnivorous, taking mainly seeds of wetland plants, and aquatic invertebrates. Common plant foods include seeds of 'wild rice' *Zizania* or cultivated rice *Oryza*, and *Polygonum*, *Ludwigia*, *Lippia*, *Bidens*, *Juncus*, *Carex*, *Cyperus*, *Scirpus*, *Eleocharis* and grasses (including *Panicum*, *Digitaria*, *Echinochloa*, *Leersia* and *Setaria*); birds also sometimes eat the vegetative parts of duckweed *Lemna* and pondweed *Potamogeton*. Invertebrate foods include: molluscs (Physidae); crustaceans; and insects including adults, larvae and pupae of Diptera (including Calliphoridae), Odonata (including *Nemobius* and other Gryllidae, Acrididae, *Conocephalus* and other Tettigoniidae), Coleoptera (Curculionidae, Carabidae, Hydrophilidae, Dytiscidae), Heteroptera (*Mesovelia*), Orthoptera (including Gryllidae, Acridoidea) and Lepidoptera (especially larvae of *Coleophora* moths). Soras consume more plant and less animal material throughout the year than do Virginia Rails.

Plant material, especially seeds and particularly wild rice, forms a large part of the diet in late summer, autumn and winter, while the proportion of invertebrates in the diet increases in spring, e.g. from 17.5% of the total food volume in autumn (n = 20 birds) to 37% in spring (n = 18 birds) in migrants from Missouri (Rundle & Sayre 1983); *Carex* seeds made up 31% of the total volume in spring and grass seeds 66% in autumn. In Arkansas rice fields, rice made up 74.6% of the volume of food eaten by 56 Soras in Sep-Oct (Meanley 1960). In Connecticut, 51 Soras collected from a slightly brackish riverine tidal marsh in Sep had eaten 94% *Zizania* by volume, whereas 7 from a highly brackish wet meadow downstream had eaten 91% animal material, including 66% Odonata and 18% moth larvae (Webster 1964).

This species forages on mud and in shallow water, and on floating vegetation by raking mats of duckweed with the feet and pulling aside other vegetation with the bill; it strips seeds from grass panicles, picks seeds from the substrate, and takes food items from the water surface. It sometimes forages while swimming in open water like a coot. It does not raise the bill while drinking. Stands of robust emergents interspersed with shorter species or floating and submerged vegetation and debris provide good habitat for invertebrate prey, while during brood-rearing and premigration periods birds may use shallower portions of wetlands where seed-producing plants provide food. In late summer it may forage for short periods in upland fields and row crops. Migrants feed in wetlands with tall emergent vegetation, where rice, or flooded annual grasses and forbs, are preferred feeding habitats. In Trinidad, wintering birds foraged on exposed mudflats among stands of emergent vegetation (Gochfeld 1972).

HABITS It is diurnal, emerging from cover to feed in the early morning and evening, when it appears quite furtive, keeping close to cover. Its gait is similar to that of Spotted Crake but its stance is more upright (Cramp & Simmons 1980). On the breeding grounds it is often difficult to flush, and it flies reluctantly, but in Colombia and Mexico it is readily flushed (Hilty & Brown 1986, Howell & Webb 1995). In Central America it is said to be a little less terrestrial than most other rails and to climb in shoreline and marsh vegetation (Wetmore 1965, Land 1970). It swims and dives readily, and sometimes submerges with only the bill and eyes above the water. In Costa Rica it roosts communally in *Typha* or other thick vegetation at night (Stiles & Skutch 1989). This species appears to be cold-sensitive (see Movements) and becomes lethargic at temperatures near freezing (Kaufmann 1989).

SOCIAL ORGANISATION Monogamous and territorial. The pair bond is maintained only during the breeding season, when the species is territorial, and apparently breaks down before dispersal, shortly after the young fledge (Johnson & Dinsmore 1995): see Movements section. Breeding density is estimated at 0.1-1.6 birds/ha and 0.6 pairs/ha (Melvin & Gibbs 1996), while brood-rearing home ranges averaged 0.91ha (SD 0.02, n = 8) and wintering home ranges averaged 0.78ha (SE 0.5, n = 27) in SW Arizona (Johnson & Dinsmore 1985, Conway 1990). The distance between 46 nests in Minnesota was

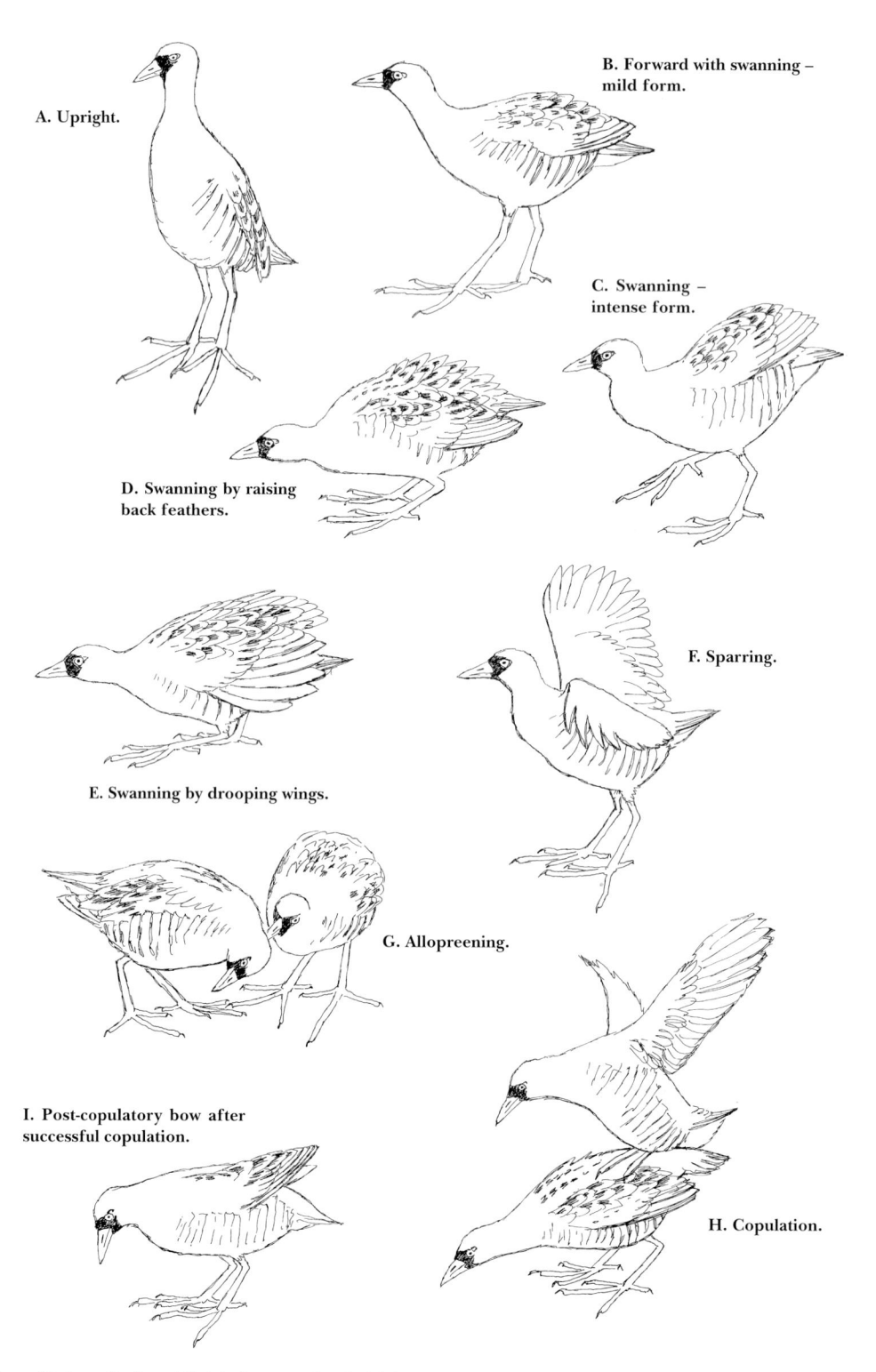

Figure 10: Sora displays. A-F: Displays associated with hostile behaviour. G-I: Displays associated with sexual behaviour. [After Kaufmann 1983].

1.2-30 (mean 9) m, the closest active nests being 3m apart (Pospichal & Marshall 1954); minimum distances between nests in other areas are 12-25m (Melvin & Gibbs 1996). The mean distance between Sora and Virginia Rail nests ranges from 4.3-25m; active nests of Sora and King Rail (56) were as close as 31m (Melvin & Gibbs 1996). In captivity, 3 pairs defended territories of 45-55m² around the nests, probably smaller than territories in the wild (Kaufmann 1989). See also Virginia Rail species account, Social Organisation section.

SOCIAL AND SEXUAL BEHAVIOUR On the wintering grounds Soras are recorded alongside White-throated Crakes (31) (Wetmore 1965). Birds frequently forage alongside Virginia Rails, and the latter species's tolerance of Soras appears liberal (Pospichal & Marshall 1954). Soras vigorously defend the breeding territory against conspecifics and against Virginia Rails (Kaufmann 1983, 1989). Threat displays include assuming an extended upright posture, stretching the head towards the opponent, pecking the substrate and 'swanning' – bending down, fluffing the contour feathers, raising the scapulars and back feathers, spreading the undertail-coverts, and sometimes drooping the wing feathers. Swanning displays are usually given by opposing males at territory boundaries; the contrast between the black facial mask and the bright yellow bill appears to reinforce frontal threat displays, and the facial coloration is heightened during the breeding season (Kaufmann 1983). Chasing by males is the most important method of establishing and maintaining territories; it is used on its own against Virginia Rails, and is combined with swanning displays against conspecifics; birds also splash water while treading rapidly with the feet ('churning'), usually performing this action at the end of a chase (Kaufmann 1983, 1989). Sparring consists of 2 birds facing each other and jumping up and down; this becomes fighting when the birds simultaneously peck and claw each other; birds also fight while lying backward, supported on the wings; fighting most frequently occurs between males, occasionally between females (Kaufmann 1983).

Most pair formation is recorded in the last week of Apr and the first 2 weeks of May (Melvin & Gibbs 1996). Sexual behaviour in captive birds is described by Kaufmann (1983, 1987, 1989). The first stage of pair formation involves 2 birds standing immobile in sight of each other for up to 30 min; after 2-4 weeks they begin to bathe, feed and preen near each other. Pairs frequently allopreen for 1-5 min as part of courtship and of pair-bond maintenance, bowing and facing towards or away from the other bird, and preening usually precedes copulation. In the courtship chase, the male walks with a slightly stiff and erect posture with head held rigid and elevated, and undertail-coverts spread, making soft cooing sounds; when the female crouches the male mounts from behind, the female lowers her head, and copulation occurs for a few seconds, after which the male dismounts or the female runs out from below the male. Postcopulatory displays include head-bowing, and elevating of the wing and tail feathers in the male, and body-shaking in the female.

BREEDING AND SURVIVAL Season USA and Canada, May-Jul or early Aug. **Nest** A loosely to substantially woven basket of available vegetation, lined with finer sedges and grasses, and leaves, and either supported by surrounding stems or built within vegetation clump; usually built over shallow standing water c. 18-22cm deep (range 5-41.5, n = 52 nests) near borders of vegetation types or patches of open water, sometimes with base at or below water surface. Dominant plants at nest sites include *Typha*, *Sparganium*, *Scirpus*, *Carex*, *Cyperus* and grasses; prefers sites in *Typha* or sedges, or in mixture of robust and fine emergents (e.g. *Typha* with an understorey of *Carex*); nest often near border of vegetation types or of vegetation and open water; occasionally nests in *Spartina alterniflora* or *Phragmites communis* in saltmarshes. At first, nest may be crude pile of vegetation onto which first egg is laid; male then gathers material which female builds into nest as laying proceeds. Site selected by female in captivity (Kaufmann 1989). External diameter of nest 12-20 (15)cm; internal diameter 9-13 (11)cm; depth 6cm; height of rim above water 5-23 (12)cm; sometimes raised 1-2cm with additional material if water rises, but may be built up to a height of 48cm to overcome flood conditions; internal diameter 7-8cm and depth 5cm. Often with canopy of surrounding vegetation bent over and tucked into rim on opposite side; ramp of vegetation often constructed to lip of nest. Several brood nests or resting platforms often constructed by male nearby. **Eggs** 5-18 (usually 8-11; means 9.4-11.7); ovate; smooth and glossy; cream, rich buff, cinnamon or pale olive, irregularly spotted brown and russet; size of eggs 29.5-33.6 x 22.2-23.8 (32.0 x 22.8, n = 261); mass 6.5-10.2 (8.4, n = 128). Eggs laid at daily intervals, occasionally every 2 days. **Incubation** Period 16-20 days; 11-22 days (mean 18.7, n = 22 clutches) also recorded; begins any time from laying of 1st to 9th egg; both sexes incubate during day and night; at first incubation may be part-time for several days; hatching asynchronous, period 2-17 days, averages of 7.2-10.5 days recorded; first eggs hatch synchronously, remainder asynchronously; parents may neglect last 1-2 eggs, causing delayed hatching; one adult tends first-hatched chicks while other continues incubation. **Chicks** Apparently semi-precocial; can run and swim at 1 day, but leave nest 2-4 days after hatching; cared for by both parents; beg for food with loud peeping calls, gaping and wing-flapping, from 1 day, also peck at tip of adult's bill; become self-feeding after a few days but are also fed for 2-3 weeks and brooded for up to 1 month; chicks preen and allopreen when >1 day old, and adults frequently preen chicks. Feathers begin to emerge at 2-3 weeks; able to fly at 4-5 weeks, when independent; adult proportions and full juvenile plumage attained in 6th week (Jul-Aug). Dispersal of family group from home range occurs when young 16-32 days old.

Brood parasitism: King Rail flushed from nest containing 9 King Rail eggs, 7 Virginia Rail eggs and 1 Sora egg (G. Swales in Meanley 1992); 1 Virginia Rail nest parasitised by Sora, and 1 Sora by Virginia Rail (Miller 1928, Tanner & Hendrickson 1956). Sorensen (1995) gives evidence for conspecific nest parasitism and egg discrimination in 1 nest: 3 eggs different in appearance to remaining 12 of clutch were repeatedly buried in nest-lining by incubating bird, and 1 egg different in size, shape and colour to all other eggs was laid 5+ days after last of clutch. Nesting success: proportion of nests hatching at least 1 egg 0.60-0.83 (n = 162 nests, 5 states) and 0.53 (n = 203 nests, North America) (Conway et al. 1994, Melvin & Gibbs 1996). Overall hatching success in 3 states 0.67 (531 eggs hatched of 789 laid) (Melvin & Gibbs 1996). Chicks are vulnerable to exposure or drowning (Pospichal & Marshall 1954). Confirmed or suspected egg predators include Striped Skunk *Mephitis mephitis*, Coyote, Raccoon *Procyon lotor*, Marsh Wren *Cistothorus palustris*, Common Grackle

Quiscalus quiscula and Common Crow *Corvus brachyrhynchos* (Melvin & Gibbs 1996). Age of first breeding not known, but probably 1 year (W. R. Eddleman *in litt.*). 1-2 broods recorded; second sometimes possibly renesting after failure of first attempt (Pospichal & Marshall 1954).
Survival Survival probability during non-breeding season (Aug-Apr) of 23 radio-tagged birds in Arizona only 0.308 (SE 0.256) for all ages and sex classes; survival probability possibly lowered by effects of radio transmitters or underestimated because of emigration (Conway *et al.* 1994).

99   DOT-WINGED CRAKE
Porzana spiloptera   Plate 29

Porzana spiloptera Durnford, 1877, Belgrano, Buenos Aires, Argentina.

Sometimes included in the genus *Laterallus*. Has been confused with *Laterallus jamaicensis salinasi* (*Ibis* 1888, Ser. 5(6): 285). Monotypic.

Synonyms: *Porzana salinazi/salinasi*; *Zapornia spiloptera*; *Laterallus spilopterus*.

Alternative names: Dot-winged Rail; Durnford's Crake.

IDENTIFICATION Length 14-15cm. Very small dark crake, with top of head blackish; upperparts dark olive-brown, streaked blackish; white markings on upperparts confined to narrow bars on upperwing-coverts and some spots on remiges and uppertail-coverts; sides of head, and underparts, dark slate-grey; lower flanks to undertail-coverts black, barred white. Iris red; bill blackish; legs and feet dark brownish. Sexes alike. Immature possibly duller on upperparts. Juvenile blackish-brown above, feathers tipped vinaceous-buff to cinnamon; upperwing-coverts barred white; sides of head to throat greyish-white; breast grey; flanks vinaceous-grey, barred cinnamon-buff; undertail-coverts blackish, marked vinaceous, white and buff. Inhabits marshes, grass and riparian scrub.
Similar species Superficially very similar to Black Rail (33) in size and overall colour but latter has nape and upper back chestnut to rufous, rest of upperparts blackish-brown, with white markings from back to tail; races *murivagans* and *tuerosi* have cinnamon undertail-coverts. Speckled Rail (15) also similar in size and colour but has white chin and throat, white speckling on head, neck, mantle and breast, white barring on back, upperwing-coverts and lower breast to undertail-coverts, and black legs and feet. Larger Paint-billed Crake (112) has unmarked olive-brown upperparts, paler grey underparts, olive-green bill with red base, and coral-red legs and feet.

VOICE Not described.

DESCRIPTION
Adult Forehead and forecrown almost black; crown to hindneck very dark blackish-brown, feathers edged dark olivaceous-brown; mantle to uppertail-coverts, and scapulars, dark olivaceous-brown to tawny-olive, feathers with broad blackish-brown centres; uppertail-coverts with a few white spots; rectrices blackish-brown, edged olive-brown. Upperwing-coverts dark olivaceous-brown, with broad blackish-brown centres and a narrow white bar across centre of each feather; greater coverts and alula brown, alula with a few white spots along feather edges; primary coverts blackish-brown; primaries and secondaries dull brown, outermost primary with whitish spots at edge of outer web, and some secondaries with small white wedge-shaped mark near tips; tertials dark olivaceous-brown with broad blackish-brown centres, like upperparts; axillaries blackish-brown, barred white; underwing-coverts white, mottled blackish-brown basally, but primary coverts ashy, like underside of remiges. Lores, and feathers around eye, almost black, like forehead; rest of sides of head, chin, throat, sides of upper neck, foreneck, breast, upper flanks, and belly, dark slate-grey, feathers tipped buff when fresh; sides of lower neck, and sides of breast, dull ochraceous-brown with vague darker feather centres; centre of lower belly whitish; thighs grey-brown with some white; lower flanks, vent and undertail-coverts blackish-brown with narrow white bars. Iris crimson to scarlet; bill very dark horn, almost black; legs and feet dark brownish.
Immature Not properly described; apparently similar to adult but colour of upperparts less intense (Nores & Yzurieta 1980).
Juvenile A "presumed immature" (probably juvenile because of growing remiges) is described as blackish-brown above, feathers tipped vinaceous-buff to vinaceous-cinnamon; many upperwing-coverts with subterminal white bar; growing remiges dark greyish-brown; sides of head, chin and throat whitish, washed greyish vinaceous-buff; breast and belly dull grey; flanks dark vinaceous-grey with buff to cinnamon bars; undertail-coverts blackish with vinaceous tips and dull white to buff markings (Ripley & Beehler 1985).
Downy young Not described.

MEASUREMENTS Wing of 3 males 70, 73, 75, of 1 female 76, of unsexed 74-77; tail of 1 male 35; exposed culmen of 1 male 16, of 1 female 16.5, culmen to base of 2 males 17, 18; tarsus of 3 males 22, 23, 23.5, of 1 female 26 (Hellmayr & Conover 1942, Blake 1977, BMNH).

GEOGRAPHICAL VARIATION None.

MOULT No information available.

DISTRIBUTION AND STATUS S Uruguay and N Argentina. Its status is VULNERABLE. Unless otherwise stated, all information is taken from Collar *et al.* (1992), Wege & Long (1995) and Arballo & Cravino (in press), which should be consulted for further details. In Uruguay it has been recorded only from Arroyo Pando (Canelones Dpt) before 1926, from Montevideo Dpt (no details available), and Arroyo Solís Grande (Maldonado Dpt) in 1973. In Argentina, it is recorded from only 6 provinces, 3 of them doubtfully: the single records from La Rioja and San Juan possibly refer to Black Rail, while the bird's occurrence in San Luis is unconfirmed. In Córdoba, up to 10 birds were observed at Bañados del Río Dulce in 1973-1974, while in Santa Fé 1 was collected in 1906. In Buenos Aires Province the bird is known from 16 localities, records from up to 8 of them falling within the last 10 years; records are all of 1-2 birds. Four of the Argentinean sites are listed as key areas for threatened birds: Bañados del Río Dulce & L Mar Chiquita Natural Park (Córdoba), and Otamendi Strict Nature Reserve, Estación Biológa Punta Rasa and the Bahia Blanca area (Buenos Aires); Bahia Blanca is unprotected.

  The status of this poorly known and secretive crake is unclear. It is regarded as rare and local in Uruguay. In Argentina it was formerly regarded as locally frequent to abundant in Buenos Aires Province, where it is currently regarded as locally rare (or very difficult to locate) to fairly common; the lack of sightings could be attributed to the paucity of observers. At the Reserva Municipal de Biosfera Mar Chiquita, Buenos Aires, Martinez *et al.* (1997) recorded 1-3 individuals of this species on 41 occasions between Feb 1983 and Jan 1997. It was formerly considered rare in La Rioja but common at the Bañados del Río Dulce site in Córdoba; its status in other provinces is unclear. Reclamation and burning of marsh areas, flooding (e.g. at the Bañados del Río Dulce site), intrusion by cattle, overgrazing, burning, disturbance by visitors, and projected development are cited as current risks to the bird's survival (see also Martinez *et al.* 1997, who found that birds did not reappear for up to 1 year after fires). Cattle are absent from the R Luján marshes and the Estación Biológia Punta Rasa, two Argentinean areas which are apparently healthy for the species, but the latter area is at risk from a recreational development project and the bird's population is declining presumably because of an increase in visitors. Distributional surveys are urgently needed within the bird's limited range, the voice should be recorded, and the possibly beneficial effect of better management of cattle should be studied.

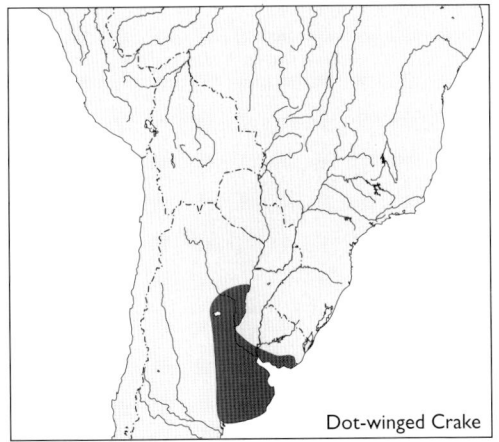

MOVEMENTS None recorded. One male (BMNH specimen) from Isla Ella, Delta del Parana, Argentina, 4 Feb 1917, could possibly be indicative of movements. Between 1983 and 1997 at the Reserva Municipal de Biosfera Mar Chiquita, Buenos Aires, Martinez *et al.* (1997) recorded this species in all months except Jul and Aug.

HABITAT Freshwater and brackish wetlands, including tidal and temporary marshes, swamps, wet marshy meadows, and wet to dry grassland; also cord grass, riparian scrub and flooded scrub. The holotype, and another specimen, were taken in gardens almost in the city of Buenos Aires (Durnford 1877). In the R Luján marshes, the habitat is dominated by *Spartina densiflora* and some *Eryngium*, and birds have occurred in dense *Spartina* up to 70cm tall, with permanent brackish surface water; elsewhere, it is found in *Paspalum* grass (M. Pearman in Collar *et al.* 1992). At the Reserva Municipal de Biosfera Mar Chiquita, Buenos Aires, where it occurs in periodically inundated halophytic vegetation, 39 of 41 records were from *Spartina densiflora*, the other 2 being from *Juncus* (Martinez *et al.* 1997). In Uruguay, 1 was seen in *Typha* at the margin of a small pool which contained diverse floating plants and was surrounded by scrub (Arballo & Cravino in press).

FOOD AND FEEDING It is recorded as feeding on aquatic insects and other invertebrates, and also seeds and marsh weeds (Canevari *et al.* 1991). There is no information on feeding habits.

HABITS It is generally solitary and is very difficult to observe, keeping permanently in cover; occasionally it may be flushed, when it flies for a short distance before dropping into vegetation and disappearing rapidly; it is very difficult to find again (Canevari *et al.* 1991).

SOCIAL ORGANISATION Generally solitary; occasionally 2-3 birds are seen together.

SOCIAL AND SEXUAL BEHAVIOUR No information available.

BREEDING AND SURVIVAL Season The only recorded nest was found near Buenos Aires, but no other details were given (Anon 1888), while a juvenile was seen with an adult at Punta Norte del Cabo San Antonio, NE Argentina, some time in 1987/88 (Collar 1992). An immature found dead at the mouth of the Arroyo Solis Grande in Feb 1973 is indicative of breeding in Uruguay (Arballo & Cravino in press).

100 ASH-THROATED CRAKE
Porzana albicollis   Plate 29

Rallus albicollis Vieillot, 1819, Paraguay.

Two subspecies recognised.

Synonyms: *Rallus olivaceus*; *Crex mustelina/gularis/olivaceus*; *Ortygometra olivacea/albicollis*; *Corethrura mustelina/olivacea*; *Aramides/Galeolimnas/Mustellirallus albicollis*.

Alternative names: White-necked/White-throated Crake.

IDENTIFICATION Length 21-24cm. Medium-small, short-billed crake. Top of head and entire upperparts, including upperwings and tail, blackish-brown, streaked olivaceous-brown. Chin and throat white; sides of head, foreneck, breast, upper flanks and belly grey, upper flanks tinged olive-brown; lower flanks and undertail-coverts blackish-brown, narrowly barred white. Iris reddish; upper mandible brownish to blackish, lower mandible green; legs and feet purplish-brown to brownish. Sexes alike. Immature duller, with brownish wash on grey of underparts. Juvenile has whitish chin, and predominantly olive-brown face, throat and underparts, with no barring from lower flanks to undertail-coverts but deeper brown wash in these regions. Inhabits marshes, savanna, pastures, scrub and secondary growth.
Similar species Sora (98) is slightly smaller and bulkier; has white streaks on upperparts, distinctive face pattern (sometimes with black streak from chin to centre of breast), more extensive and broader white flank bars, white and buff undertail-coverts, a bright yellow bill, and greenish legs and feet. Paint-billed Crake (112), the only other sympatric, short-billed crake of similar size with grey breast and barred flanks, has unmarked olive-brown upperparts, olive-green bill with red base, and coral-red legs and feet. Larger Zapata Rail (110) of Cuba has unpatterned brownish-olive upperparts, dark slate flanks with faint white bars, white undertail-coverts, yellowish-green bill with red base, and red legs and feet.

VOICE Canevari *et al.* (1991) describe the male's call as reminiscent of that of the Limpkin *Aramus guarauna*, a repeated *carrrreau-carrrreau*, while the female replies (in antiphonal duet) with a sharp *kere-kere-ker-kere kreuuu*. Narosky & Yzurieta (1989) also mention a duetting call. Sick (1993) describes the male's song as a melodious hum ending in a prolonged *grrrrrrehhhyo*, while the female constantly barks a *kehrre*; the call is given as *bewrewt* and the warning note *keAH*. The song is described by Hilty & Brown (1986) as a repeated, loud, fast series of vibrating notes, sounding like a machine-gun *d'd'd'd'-ou*, and the call as a sharp *tuk*. This species is most vocal in the early morning and evening.

DESCRIPTION *P. a. olivacea*
Adult Forehead to hindneck, and entire upperparts, including scapulars and upperwing-coverts, very dark blackish-brown with broad olivaceous-brown feather fringes (sometimes tending to pale greyish-olive) giving streaked effect; sides of neck similar but blackish-brown feather centres tend to be less prominent; rectrices fuscous-black, fringed olivaceous-brown; alula, primary coverts and remiges dusky-brown, narrowly edged olivaceous-brown; axillaries and underwing-coverts pale olive-brown, feathers edged whitish; underside of remiges pale olive-brown. Lores, supercilium and sides of head pale ashy-grey or blue-grey; chin and throat white, shading to ashy-grey or blue-grey on foreneck (sometimes also on throat); breast, belly, vent and upper flanks ashy-grey or blue-grey, paler on central breast; upper flanks tinged olive-brown, and feathers of lower belly and vent edged whitish; thighs grey, faintly barred white; lower flanks and undertail-coverts blackish-brown, narrowly barred white, undertail-coverts sometimes barred pale brown. Iris reddish-brown to red; upper mandible brownish to blackish, lower mandible green; legs and feet purplish-brown to brownish. Sexes alike.
Immature Similar to adult but duller; with buffy-brown to pale olive-brown wash on grey of underparts, especially on sides of breast, flanks and thighs; lower flanks, lower belly and vent strongly washed dark vinaceous-buff, with greyer feather bases giving vaguely barred effect; undertail-coverts barred dull brown and dark vinaceous-buff.
Juvenile One presumed juvenile has off-white chin, predominantly olive-brown face, throat, foreneck and underparts, and no barring from flanks to undertail-coverts but deeper brown wash in these regions.
Downy young Not described.

MEASUREMENTS (10 males, 5 females) Wing of male 93-109 (104.3), of female 98-103 (100.8); exposed culmen of male 24-31 (27.3), of female 22-24 (23.3); tarsus of male 31-38 (35), of female 33-36 (34.2) (Blake 1977). 15 unsexed: wing 99-112 (105.4, SD 4.5); tail 35-47 (40.4, SD 2.9); culmen to base 24-30 (27.5, SD 1.5); tarsus 31-36 (33.2, SD 1.5). Weight of 6 males 90-112 (99), of 1 female 105 (Haverschmidt & Mees 1995).

GEOGRAPHICAL VARIATION Races separated on overall colour and size. Race *olivacea* includes *typhoeca* (Ripley 1977).

P. a. olivacea (Vieillot, 1819) – Colombia (upper Cauca V, lower Magdalena V, and E of Andes from Arauca to S Meta and extreme NE Meta), Venezuela (Zulia, Portuguesa, Carabobo, Aragua, NW and SE Bolívar), extreme N Brazil (N Roraima), Trinidad and the Guianas. See Description.

P. a. albicollis (Vieillot, 1819) – Brazil (NW Pará, Rondônia, S Goiás, S Mato Grosso and Mato Grosso do Sul, and Paraíba, Pernambuco, Bahia and Minas Gerais S to N Rio Grande do Sul), N and E Bolivia (Beni, La Paz, Santa Cruz and Chuquisaca), Paraguay (Orient region, E of R Paraguay) and N Argentina (Tucumán, Misiones, Corrientes and Buenos Aires). Slightly larger than *olivacea*, with darker edges to upperpart feathers (brownish rather than olivaceous), and darker grey underparts. Specimen from Corrientes has upperpart feathers fringed more olivaceous, possibly more similar to *olivacea* (D. Baepler *in litt.*). Measurements of 10 males, 6 females: wing of male 108-119 (114.4), of female 105-118 (109.3); exposed culmen of male 27-32 (29.5), of female 25-29 (26.6); tarsus of male 35-37 (36.3), of female 34-37 (35.8) (Blake 1977). 15 unsexed: wing 100-120 (109, SD 4.8); tail 35-47 (41.5, SD 3.8); culmen to base 26-30 (28.3, SD 1.3); tarsus 33-40 (36.1, SD 1.8). Weight of 1 male 100 (Marjorie Barrick Mus, Univ. Nevada: D. Baepler *in litt.*).

Two males from extreme SE Peru (Pampas de Heath in Madre de Dios) have wings as long as *albicollis* but have paler grey underparts and olivaceous margins to dorsal feathers, characters typical of *olivacea* (Graham *et al.* 1980). Weights of these 2 birds: 112, 114 (Dunning 1993).

MOULT No information available.

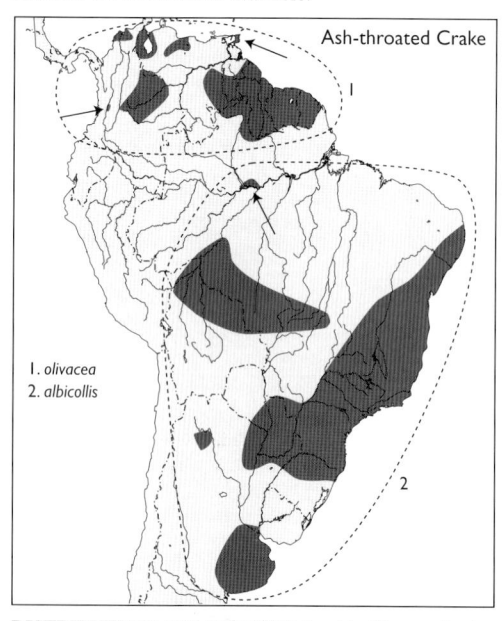

Ash-throated Crake

1. *olivacea*
2. *albicollis*

DISTRIBUTION AND STATUS Colombia, Venezuela, the Guianas, Trinidad, Brazil, N and E Bolivia, SE Peru, Paraguay and N Argentina. Not globally threatened. The race *olivacea* was formerly regarded as quite common to abundant in Guyana, local in Colombia and rare in Trinidad (Beebe *et al.* 1917, Snyder 1966, ffrench 1973, Hilty & Brown 1986). The race *albicollis* was regarded as local in Argentina (Olrog 1984); apparently the only published record from Corrientes is a specimen, collected in Oct 1995, in the Marjorie Barrick Museum, University of Nevada (J. Gerwin *in litt.*). The species was apparently not uncommon in extreme SE Peru (Pampas de Heath) in

1977 (Graham *et al.* 1980). Little information is available on the current status of this species: it is common along the littoral in French Guiana, where it was formerly also described as common inland along rivers (Snyder 1966, Tostain *et al.* 1992); it is regarded as scarce in some parts of Brazil but more frequent in others, including in the E (Forrester 1993, Sick 1993, do Rosário 1996), uncommon to rare in E Paraguay (Hayes 1995), and probably of local occurrence in many other regions. In Beni, Bolivia, it is apparently widespread, and locally fairly common, in the wetter regions of the savanna (Brace *et al.* in press, R. Brace *in litt.*). From Buenos Aires Province, Argentina, there are diverse old records of doubtful origin and the only recent sighting is from Otamendi (Narosky & Di Giácomo 1993). Its distribution is probably more continuous than is known.

MOVEMENTS Nothing is definitely recorded, but some seasonal movements are possible in Colombia, records from NE Meta being for Dec-Mar and from W Meta Mar-Sep (Hilty & Brown 1986), while birds are reported to occur at the edges of *Typha* marsh in Aug at Ciénaga Grande, none being present there in the dry season (Darlington 1931). In French Guiana an exhausted bird was found on a bridge in Jul (Tostain *et al.* 1992).

HABITAT Freshwater marshes, marshy lakes, Moriche swamps, wet grasslands, rice fields, drainage ditches, savanna, dry to damp taller grass of grazing lands, tall reeds along roadsides, and secondary growth. It likes areas thickly overgrown with grass, and in marshy habitats it appears to prefer the drier areas (e.g. low reeds and grass at the edge of *Typha* marsh). In Argentina it occurs in scrubland and dense thickets adjacent to marshes in forest, and it may also occur some way from water (Canevari *et al.* 1991); the Corrientes specimen was taken in grazed bunchgrass pasture with tall grass and pools of water at the perimeter (D. Baepler *in litt.*). In Paraguay it occurs alongside the Rufous-faced (36) and Red-and-white (35) Crakes in marshes with perennial bunch-grass 30-50cm or more tall, completely covering the ground and growing on moist soil or in 3-4cm of water; the small mammals which occur extensively in such marshes make runways through the dense vegetation, and the crakes use these runways (Myers & Hanson 1980). Such marshes (cañadones or wet campos) commonly form in low areas of E Paraguay and adjacent parts of Brazil (see Rufous-faced Crake, Habitat section). In Beni, Bolivia, it has been seen alongside the Rufous-faced Crake and the Speckled Rail (15) in savanna flooded to a depth of 5cm and characterised by tussocky grass c. 1m tall with *Rynchospora globosa*, *Cyperus haspan* and *Tibouchina octopetalia* (Brace et al. in press). In Surinam it is regarded as less of a marsh bird than other small rails (Haverschmidt & Mees 1995). It inhabits lowlands up to 1,200m.

FOOD AND FEEDING Insects and their larvae (Lepidoptera, Formicidae, Coleoptera), other invertebrates, and grass seeds. It occasionally skulks partially in the open near cover, presumably feeding.

HABITS Often an inquisitive and noisy bird. It flushes readily (often from near the observer's feet), flies for a short distance with dangling legs and then drops into cover; it will land or climb in dense vegetation and bushes. Before hiding it may stop to investigate the source of a disturbance by looking out from cover and stretching the neck to look around.

SOCIAL ORGANISATION Normally seen singly or in pairs.

SOCIAL AND SEXUAL BEHAVIOUR Apart from the duetting behaviour nothing is recorded.

BREEDING AND SURVIVAL Season Trinidad, Jul-Sep, probably also Oct; Guyana, Feb-Jul, peaking May, but possibly nests throughout year. **Nest** A large open bowl of roughly woven coarse dry grass, sometimes with some weed stems or a few leaves; placed on or just above ground between, or at the base of, clumps of savanna grass, or in reeds; often near base of tree stump, sheltered by roots. External diameter c. 20cm, depth c. 10cm; cup diameter c. 10cm, depth c. 5cm. **Eggs** Usually 2-3, possibly up to 6; long ovate to almost round; smooth and almost without gloss; pale pinkish-cream to white, heavily and finely spotted chocolate-brown and lilac-grey at larger end, paler eggs often with fewer spots sparsely scattered throughout, some almost unmarked. Size of eggs: *olivacea* 29.2-36.2 x 22.7-27.8 (33 x 25.6, n = 57), calculated weight 11.8; *albicollis* 32.7-36.2 x 25.2-27.8 (35.2 x 26.8, n = 22), calculated weight 13.9. No other information available.

101 HAWAIIAN CRAKE
Porzana sandwichensis   Plate 32

Rallus Sandwichensis Gmelin, 1789. Hawaiian Islands = Hawaii.

Porzana millsi is now regarded as synonymous with the present species, probably representing the immature form, or possibly a colour morph. Monotypic.

Synonyms: *Pennula millei/millsi/wilsoni/sandwichensis/ecaudata*; *Rallus ecaudatus/obscurus*; *Porzanula sandwichensis*.

Alternative names: Sandwich Crake/Rail; Moho; Mills's Crake; Dusky Rail.

IDENTIFICATION EXTINCT. Length 14-15cm. Small; flightless; tail very short, hidden by coverts. Upperparts, including upperwing-coverts, dark rich brown. Ear-coverts dark ashy; malar region, and chin to breast, cinnamon; chin sometimes greyish-white; sides of neck and sides of upper breast washed dark brown; flanks and belly russet; undertail-coverts dull brown; faint buffy barring from rear flanks and rear belly to undertail-coverts. Iris reddish-brown; bill bluish-horn, yellow at base; legs and feet orange to red. Sexes probably alike. Immature paler overall; top of head to hindneck brown, tinged ruddy; upperparts and upperwings ruddy-brown, streaked blackish-brown. Sides of head, and chin, pale brown with ruddy tinge, darker on lores and ear-coverts. Throat and entire underparts dark vinous-red, washed darkish brown on sides of neck; centre of throat may be greyish-brown. Iris and legs paler than in adult; bill greenish-yellow. Juvenile undescribed. Inhabited grassland and scrub adjacent to forest, and forest clearings.

Similar species Probably derived from an ancestral form of the larger, volant Ruddy-breasted Crake (102) of S and E Asia and the Malay Archipelago (Olson 1973b), which it resembles in overall plumage colour and pattern. Ruddy-breasted Crake has crown and entire upperparts olive-brown, darker on remiges and rectrices; forehead, sides of head and underparts vinous-chestnut; flanks and lower belly olive-brown, barred white; undertail-coverts sepia, barred white; iris, legs and feet red; bill greenish-brown.

VOICE Undescribed. Its local name, moho, means "a bird which crows in the grass" (Munro 1944).

DESCRIPTION Two colour forms have been described. The paler form originally received the name *Rallus sandwichensis*, and the darker the name *Pennula millei* (a misprint for *millsi*). Greenway (1967) suggested that it was improbable that two similar species inhabited Hawaii at the same time, and that the paler form is an immature plumage of the darker. Ripley (1977) agreed with this view, which is now generally accepted.

Adult *Pennula millei* (sic); presumed adult (Ripley 1977). Upperparts, including upperwing-coverts, dark rich brown, almost burnt-sienna, slightly darker and duller from lower back to uppertail-coverts, and markedly duller from forehead to hindneck; uppertail-coverts with blackish bases, giving somewhat mottled effect (Sharpe 1894); rectrices reduced and concealed, blackish, narrowly edged rich brown. Remiges darker brown, washed blackish in centres and edged dark, rich brown; axillaries and underwing-coverts greyish-brown. Lores olive-brown; ear-coverts dark ashy; chin greyish-white or dull cinnamon (sexual or age-related difference?); malar region, throat and foreneck dull cinnamon, darkest on malar region; lower throat and centre of breast orange-cinnamon, becoming darker and browner at sides of neck, and on upper flanks; lower flanks, belly and thighs duller, more russet; vent to undertail-coverts dull brown; faint buffy-white black-margined barring from rear flanks and rear belly to undertail-coverts. Iris reddish-brown; bill bluish-horn, yellow at base; legs and feet orange to red.

Immature *Rallus sandwichensis*; presumed immature (Ripley 1977). Top of head to hindneck brown with ruddy tinge, slightly paler from forehead to forecrown; upperparts and upperwings ruddy-brown, centres of feathers dull dark brown (on mantle) to blackish, producing broadly but rather diffusely striped appearance (darkest and most prominent on upperwing-coverts and tertials); upperwing-coverts much elongated; remiges blackish, edged rusty-brown; rectrices blackish. Sides of head, and chin, pale brown with ruddy tinge, slightly darker on lores and ear-coverts. Throat and entire underparts dark vinous-red, darker on sides of neck, and on thighs, where washed darkish brown; Ripley (1977) states that centre of throat is paler brown, almost greyish. Iris and legs paler than in adult; bill dull greenish-yellow.

Juvenile Not described.

Downy young Not described.

MEASUREMENTS (2 adults) Wing 66, 70; tail 14, 20; culmen 18, 20; tarsus 28, 29. 1 immature: wing 73; tail 18; culmen 19; tarsus 30.

GEOGRAPHICAL VARIATION None.

MOULT No information available.

DISTRIBUTION AND STATUS EXTINCT. In historic times it was confined to Hawaii, where it certainly occurred on the windward (E) side of Kilauea, perhaps also in an area extending c. 65km along the coast in this region, and in the Olaa district (Greenway 1967); it is also recorded as originally being very common over most of the island, where Hawaiian chiefs hunted it for sport with bow and arrow and had it served as a delicacy at their tables (Munro 1944, Fuller 1987). It apparently also occurred on Molokai, where it was well known to the older inhabitants in the 1890s, and it may also have occurred on other islands (Perkins 1903, Munro 1944). The first example of the paler form (*sandwichensis*) was collected on Captain Cook's final voyage in 1780, and only 2 specimens of this form exist in collections, plus an illustration in the British Museum. The last specimen of the darker form (*millsi*) was taken in 1864 and the species was last reported in 1884 (Fuller 1987). The reasons for its disappearance have not been positively determined but dogs and cats have been common on the islands since the earliest days of human settlement, and presumably played some part in its extirpation, while introduced rats (*Rattus rattus* and *R. norvegicus*) were probably an important factor (Fuller 1987). The mongoose *Herpestes auropunctatus* has also been blamed, but the rail must have been very close to extinction when the mongoose was introduced in 1883 (Greenway 1967). It is said that the rail "seems to have lived on friendly terms" with the Hawaiian Rat *Rattus hawaiiensis*, and to have been in the habit of hiding in the rat's burrows (Perkins 1903).

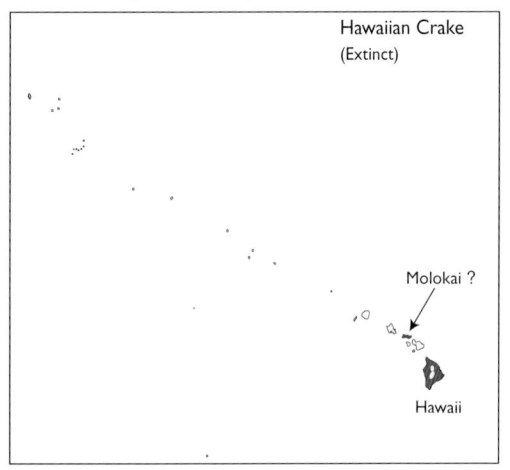

Hawaiian Crake (Extinct)

Molokai ?

Hawaii

MOVEMENTS None.

HABITAT It apparently inhabited open grassy country or low scrub just below the heavy rain forest; within the forest belt it occurred in forest clearings, and in open country covered with scrub (Perkins 1903, Fuller 1987).

FOOD AND FEEDING Not described.

HABITS Although flightless, this small rail was very active and fast-moving. It apparently took refuge in rat burrows in times of danger (Perkins 1903). Nothing else is recorded of its habits.

SOCIAL ORGANISATION Nothing recorded.

SOCIAL AND SEXUAL BEHAVIOUR Nothing recorded.

BREEDING Although the nest was seen by several people there is no description of nest, eggs or young (Munro 1944).

102 RUDDY-BREASTED CRAKE
Porzana fusca   Plate 30

Rallus fuscus Linnaeus, 1766, Philippines.

Sometimes placed in *Amaurornis*. Four subspecies recognised.

Synonyms: *Gallinula/Corethrura/Porzana erythrothorax*; *Porzana phaeopygia*; *Limnobaenus fuscus/rubiginosus/phaeopygus*; *Rallus rubiginosus*; *Zapornia flammiceps*; *Ortygometra fusca/flammiceps*; *Corethrura fusca/rubiginosa*; *Euryzona rubiginosa*; *Rallina fusca/fuliginosa*; *Amaurornis fuscus*; *Crex/Limnobaena fusca*.

Alternative name: Ruddy Crake.

IDENTIFICATION Length 21-23 (26.5?)cm. A medium-small, relatively plain, red-brown and brown crake. Forehead, forecrown and sides of head vinous-chestnut, contrasting with rather cold olive-brown from hindcrown to hindneck and on upperparts (slightly paler in races *phaeopyga* and *erythrothorax*), including tertials and most upperwing-coverts; darker from lower back to tail. Chin and centre of throat whitish, becoming more chestnut with age; rest of underparts to upper belly chestnut (paler in *phaeopyga* and *erythrothorax*); flanks and lower belly olive-brown with variably extensive, narrow, faint off-white barring; undertail-coverts blackish-brown, narrowly barred white. Iris and narrow orbital ring red; bill greenish-brown, or blackish with greenish base; legs and feet reddish-orange to brick-red. Sexes similar but female usually paler, with whiter throat. Immature darker than adult; underparts duller; chin, throat and centre of belly white; iris, legs and feet red-brown. Juvenile darker on upperparts; crown as back; sides of head and neck dull brownish-olive; underparts mostly dull brownish-olive, barred and spotted white; flanks and thighs dull grey-brown; undertail-coverts as adult; iris brown. Inhabits marshes, reedbeds, wet grassland alongside open water, and rice paddies; sometimes also dry bush and cultivation.

Similar species Band-bellied Crake (103) is similar in size but has white bars (sometimes missing) on some upperwing coverts, bold black and white barring on lower underparts, and orange-pink to red legs and feet. Red-legged Crake (23) is larger and brighter, with bold black and white barring on remiges, upperwing-coverts and underparts, and more prominent red orbital ring. Slaty-legged Crake (24) has white bars on inner webs of remiges, bold underpart barring and grey to black legs and feet. Neither *Rallina* species shows Ruddy-breasted's contrasting combination of chestnut forehead and forecrown, and olive-brown hindcrown and hindneck.

VOICE Usually silent, but in the early morning and evening during the breeding season it utters what is presumably a territorial and advertising call: hard *tewk* or *kyot* notes (likened to knocking on a door) every 2-3s, often speeding up and usually followed by a bubbling call similar to that of a dabchick *Tachybaptus* and descending slightly in pitch; this trill resembles that of the Black-tailed Crake (92) (C. R. Robson *in litt.*). It gives a short low *chuck* when foraging. Some authors also describe a soft *crake*, uttered singly at intervals, or a *keek-keek-keek* (Ali & Ripley 1980). Calling males may keep exclusively to hillocks, slopes and other dry areas in breeding marshes (Potapov & Flint 1987).

DESCRIPTION *P. f. fusca*
Adult Forehead, forecrown, supercilium and sides of head, dull vinous-tinged chestnut. Hindcrown, nape, hindneck and rest of upperparts, including most upperwing-coverts and tertials, olive-brown, usually tinged greenish; lower back to tail somewhat darker and browner; alula, primary coverts, primaries and secondaries dark olive-brown, washed olive on outer edges; lessers sometimes partially white, or tinged or blotched chestnut; chestnut spots occasionally occur on medians or even greaters; marginal coverts whitish. Axillaries and underwing-coverts olive-brown, narrowly tipped or barred whitish to give inconspicuous white spots; underside of remiges dusky olive-brown. Chin and centre of throat whitish, becoming more chestnut with age; rest of throat, foreneck, breast and upper belly chestnut, darker and more vinous-tinged at sides of neck and sides of breast; flanks, thighs and lower belly greenish-tinged olive-brown with narrow faint off-white barring, often only on hind flanks and lowermost belly, and sometimes tinged cinnamon; vent and undertail-coverts blackish-brown, narrowly barred white. Iris crimson; orbital skin plumbeous-grey, thin orbital ring red; bill blackish with greenish base, or horn-green, horn-brown to dark greenish-brown with tip of lower mandible yellowish; legs and feet reddish-orange to brick-red, or red-brown, also described as yellow-olive, reddish on anterior surfaces. Sexes similar but female usually paler, with whiter throat.
Immature Similar to adult but darker; duller and browner (tawnier) on underparts; chin, throat and centre of belly white to buffy-white, belly faintly and narrowly barred olive-brown; iris, legs and feet red-brown.
Juvenile Has darker upperparts than adult, often more blackish-brown; forehead, supercilium and ear-coverts dull brownish-olive, tinged vinous; crown same colour as rest of upperparts; malar region, chin, throat and foreneck dull white, foreneck speckled brownish-olive; rest of underparts mostly dull brownish-olive variably barred and spotted white; flanks and thighs duskier grey-brown; undertail-coverts as adult. Iris brown to grey-brown; bill dull brown with paler lower mandible; legs and feet dull brown.
Downy young Glossy black downy chick has bottle-green sheen on upperparts; underparts browner, especially on belly and thighs. Iris glaucous blue-brown; bill black with pink base to upper mandible (behind nostril) and tip to both mandibles; egg-tooth white; legs and feet reddish-black; small white wing-claw present. Upperparts become browner as chick grows.

MEASUREMENTS (15 males, 17 females) Wing of male 95-110 (101.9, SD 4.6), of female 92-107 (100.9, SD 4.4); tail of male 39-48 (43.5, SD 3.4), of female 38-47 (43, SD 2.7); culmen to base of male 22-25 (23.8, SD 0.8), of female 20.5-24 (22.45, SD 0.9); tarsus of male 32-39 (35.2, SD 1.7), of female 30-36 (33.8, SD 1.9). Weight: unsexed mean 60 (Dunning 1993).

GEOGRAPHICAL VARIATION Races separated on size and on shade of upperparts and (especially) underparts. Possible races *bakeri* and *rubiginosa* included in nominate (see Mayr 1944, Deignan 1945, White & Bruce 1986). There is an increase in size from S to N in E Asia (Medway & Wells 1976).

P. f. fusca (Linnaeus, 1766) – N Pakistan, N India (Kashmir E to Assam and W Bengal), Nepal and Bangladesh to S China (SW Sichuan and SE, W and SW Yunnan) and Vietnam; also Thailand, Myanmar,

? status uncertain
x vagrant

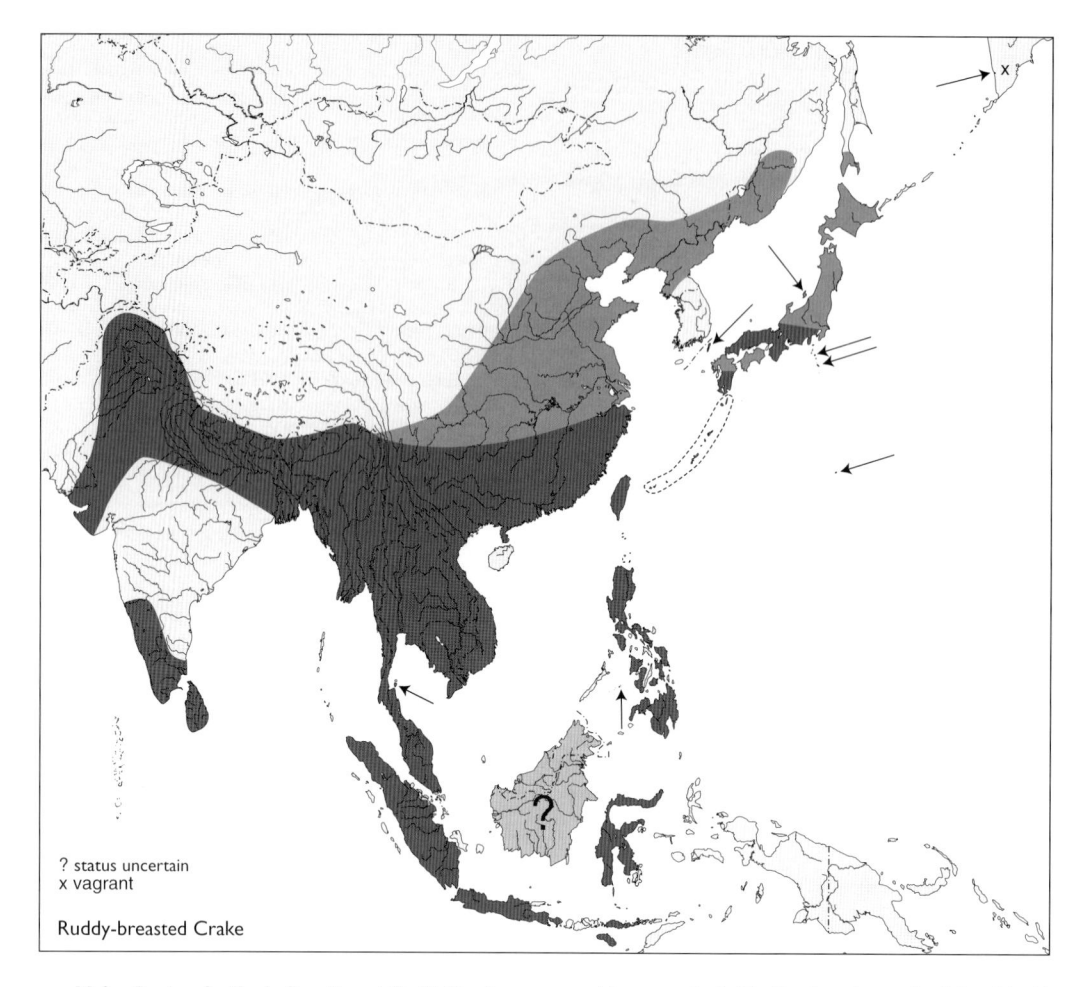

Ruddy-breasted Crake

Malay Peninsula (including Samui I), Philippines (Bohol, Cagayancillo, Leyte, Luzon, Mindanao, Mindoro, Negros, Panay and Samar), Sulawesi, Lesser Sundas (Flores and Sumba) and Greater Sundas in Borneo, Java, and Sumatra (including Utara and Barat); some N populations may winter to S. Birds from SE Yunnan are sometimes assigned to *erythrothorax* (Cheng 1987). See Description.

P. f. zeylonica (Stuart Baker, 1927) – W and SW India (N at least to Bombay), and Sri Lanka. Smaller than nominate and tends to be paler, though variable in colour. Measurements of 7 males, 4 females: wing of male 90-107 (97, SD 5.7), of female 95-103 (98.3, SD 3.4); tail of male 42-48 (45, SD 2.0), of female 42-46 (44.5, SD 1.7); culmen to base of male 22.5-23.5 (23.2, SD 0.4), of female 22-23 (22.4, SD 0.5); tarsus of male 33-36 (34.4, SD 1.1), of female 30-33.5 (31.9, SD 1.7).

P. f. erythrothorax (Temminck & Schlegel, 1849) – Japan, including Is of Daito, Hachijo, Izu, Mikura, Sado, Tanegashima, Tsushima and Yakushima, and Ryukyu Is (Ishigaki, Iriomote and Okinawa); E & S China (Liaoning, Hebei, Shaanxi, Jiangsu, Zhejiang, Fujian, Guangdong, Hunan, Guizhou and Guangxi Zhuang) S to Hong Kong, Korea and Taiwan (but see *P. f. phaeopyga*); also recorded in Russian Far East (W Kamchatka, Primorskiy Kray in S Ussuriland, and Moneron I, Sakhalin I and nearby islands). N populations winter to S, from S Japan and S China to Myanmar (possibly), Thailand, N Laos, Vietnam and Cambodia. The largest race, averaging slightly paler than others on upperparts, and paler vinous-chestnut on underparts. Legs and feet coral-red to brick-red. Juvenile has sepia upperparts, tinted olive on scapulars, upperwing-coverts and tertials (Ripley 1977). Measurements of 17 males, 14 females: wing of male 105-117 (109.7, SD 4.1), of female 98-118 (108.5, SD 5.8); tail of male 43-52 (47.6, SD 2.4), of female 42-51 (46.9, SD 2.9); culmen to base of male 21-26 (23.8, SD 1.3), of female 22-25 (23.2, SD 0.9); tarsus of male 30-37.5 (35, SD 1.8), of female 31-37 (33.7, SD 1.4).

P. f. phaeopyga Stejneger, 1887 – Yakushima I (1 specimen) and Ryukyu Is (Amami-oshima, Ishigaki, Kikai, Minami-daito, Okinawa, Okinoaeabu, Yonaguni and Yoron). Birds from Taiwan are also assigned to this race by some authors (e.g. Cheng 1987, Potapov & Flint 1987). Similar to *erythrothorax* but smaller, with darker breast, flanks, lower belly and undertail-coverts. Intermediate in colour and size between *erythrothorax* and darkest males of nominate; averages slightly larger than nominate. Wing of 7 males 97-114 (104) (Vaurie 1965).

MOULT In Malaysia (Selangor) birds in immature plumage have been netted only from May to mid-Nov, implying that adult plumage is assumed by the age of c. 6 months (Medway & Wells 1976).

DISTRIBUTION AND STATUS S and E Asia and the Malay Archipelago, from Pakistan, India and Sri Lanka E through Nepal, Bangladesh and SE Asia to S and E China and Japan, and S to the Philippines, Greater Sundas and Lesser Sundas; also locally in the Russian Far East. An out-of-range record from W China (W Sinkiang), from the 1987-91 Asian Waterfowl Census, appears on the species map (Fig. 114) in Perennou et al. (1994) but is not mentioned in the text and is presumably an error. This crake is shy and difficult to flush, and is poorly known in many regions, but it was formerly considered common to abundant over much of its range (e.g. Ripley 1977). It is not regarded as globally threatened and has recently been estimated as locally uncommon to fairly common in Nepal (Inskipp & Inskipp 1991), rare to scarce in Pakistan (Roberts 1991), common in Thailand (Lekagul & Round 1991) and in the summer in China (Cheng 1987), locally common to abundant in Sumatra, Java and Bali (MacKinnon & Phillipps 1993), uncommon in the Philippines (Dickinson et al. 1991), and apparently rare in Wallacea (Coates & Bishop 1997). It is rare in the Russian Far East, at the N extremity of its range (Potapov & Flint 1987).

The status of the nominate race in Bangladesh is uncertain: Harvey (1990) regards it as a winter visitor (Dec-Mar) to wetlands in the NW and SE, but Ali & Ripley (1980) state that it breeds in the S (Sundarbans). According to Vaurie (1965) it winters S to the Sundarbans in adjacent S Bengal, where it may also breed. It may be largely a summer resident in Myanmar, and there are few winter records (Smythies 1986). It is probably resident throughout its Wallacean range (White & Bruce 1986), but it may be only a scarce winter visitor to Borneo (MacKinnon & Phillipps 1993), whence there are only 4 published records (Smythies 1981). The race *erythrothorax* is the commonest rail in Japan, but it may be only a straggler to Korea, whence there are only 2 specimen records, in Jul and Oct (Austin 1948, Vaurie 1965). The race *phaeopyga* is resident in the Ryukyu Is (Brazil 1991).

Although this species was inadequately covered by the Asian Waterfowl Census, 1987-91, scattered records were obtained over much of its range and in E China the Yancheng shore had an unusually large number of birds (5-year mean 58 birds; counts made in 2 years) and is potentially an important site (Perennou et al. 1994).

MOVEMENTS Probably largely resident, but migratory in some regions. The status of some populations is unclear, especially on the Asian mainland, and the demarcation on the distribution map between resident and seasonal occurrences is largely hypothetical. N populations of *erythrothorax* move S to winter in S China (S of Yangtze R) and Indochina, and migrants are recorded in Shandong (Yantai) and Taiwan (Cheng 1987), and in Hebei (at Beidaihe) in May (Williams 1986); this race is a summer-breeding visitor to Japan (May-Oct), small numbers also wintering in S areas of Honshu and Kyushu (Brazil 1991). In Sri Lanka numbers of *zeylonica* are augmented by migrants which arrive in Oct-Nov and leave in Mar-Apr. This crake is a vagrant to Christmas I (Indian Ocean), with 2 specimens, Aug 1897 and Sep 1940 (Marchant & Higgins 1993).

HABITAT Reedy swamps, oxbows, marshes, water meadows and stream banks; reedbeds; wet grassland at the edges of lakes, ditches and canals; the edges of rice paddies; dry bushland and forest paths, possibly during postbreeding dispersal (Delacour & Mayr 1946, Dickinson et al. 1991); also mangroves (Coates & Bishop 1997), especially on passage (C. R. Robson *in litt.*). During the summer in Japan it is sometimes found in dry cultivated vegetable fields, and also along the shores of mountain lakes, while in the winter it frequents marshes and field drainage ditches (Brazil 1991). In Wallacea it occupies the same habitat as White-breasted Waterhen (89) (White & Bruce 1986). It usually avoids deep water and seems to prefer rice fields and the sedge-choked depressions on the fringes of permanently inundated swamps (Roberts 1991). It normally occurs in lowlands, but breeds up to c. 2,000m in the Himalayas (Ripley 1982).

FOOD AND FEEDING Molluscs; aquatic insects and their larvae, including Coleoptera (Carabidae, Chrysomelidae and others) and mosquitoes; spiders; and the seeds, succulent roots and shoots of sedges and other marsh plants. It probably feeds mainly in cover, occasionally venturing out to the edges of reedbeds, where it forages on mud at the margin of still or flowing water; it also forages among the roots of mangroves.

HABITS A very poorly known species, it is said to be extremely shy, disappearing into cover at the slightest alarm; it is difficult to flush, and flies reluctantly. However, in Sri Lanka it is said to be less shy than some other crakes (Henry 1978), while Coates & Bishop (1997) regard it as moderately shy and elusive but less so than most other *Porzana* species. The escape flight is feeble, the dangling red legs being conspicuous, and birds normally fly no higher than 2m for a distance of 10-15m before dropping into dense cover. This species is apparently much less vocal than many other crakes and so is more easily overlooked. It is active mainly at dusk and in the early morning (Roberts 1991), when it will forage at the edges of reedbeds or short distances from dense cover, and is seldom seen during the day; it roosts in bushes at night. It flicks the tail and jerks the head while walking, and it swims well (Roberts 1991).

SOCIAL ORGANISATION Apparently monogamous.

SOCIAL AND SEXUAL BEHAVIOUR It is normally seen singly or in pairs, and occasionally in small groups. Several calling males may be found in a small area (Potapov & Flint 1987).

BREEDING AND SURVIVAL Season Sakhalin, May; W Kamchatka, female with developed ovary May; Kashmir and N Pakistan, Jun-Aug; Bengal, SW India and Sri Lanka, Jun-Sep; Japan (including Ryukyu Is), Mar-Sep; Malay Peninsula, Feb, Apr and May; Philippines, Aug-Sep; Sumatra, Dec, and a chick in mid-Mar; Java, Mar-Jun, Sep-Oct, Dec; Singapore, Nov. **Nest** A pad of dry grass, shredded sedge leaves, rush leaves, rice straw, roots and leaves of aquatic plants; lined dry sedge leaves; placed on swampy ground among grass, reeds or rice; sometimes among low bushes; growing plants sometimes bent over to form canopy; framework may be loose and somewhat untidy, but is strong; 1 nest built 20cm above water in sedges 1m tall, another found c. 1m above ground in a bush. Nest diameter 17-20cm; cup diameter 8-10cm, depth 2.5cm. Both sexes build; may also build roosting nest near breeding nest. **Eggs** 3-9; elliptical to ovate, with variable gloss; creamy-white to pale cafe-au-lait or pinkish-cream, spotted and blotched pinkish-brown to rufous-brown, or olive-sepia where markings cover underlying spots of greyish-lavender to purplish-grey or mouse-grey; markings often more numerous at larger end. Size of eggs: nominate (includes *bakeri*), 29-34.2 x 21.8-25.2 (32.3 x 22.85, n = 110), calculated weight 9.25; *zeylonica*, 27.4-33 x 21.1-23.2 (30 x 22.5, n = 36), calculated weight 8.3; *erythrothorax*, 29-35 x 21.4-25.2 (31.5 x 22.8, n = 50), calculated weight 9; *phaeopyga*, 30.5-34.7 x 21-25.2 (32.8 x 23.4, n = 22), calculated weight 9.8. **Incubation** 20 days, by both parents; hatching asynchronous. **Chicks** Precocial; both parents feed and care for young, which leave nest after 1-2 days and are dependent on parents for food during first few days of life (Potapov & Flint 1987, Roberts 1991).

103 BAND-BELLIED CRAKE
Porzana paykullii   Plate 30

Rallus Paykullii Ljungh, 1813, Bandjarmasin, Borneo, and Jakarta, Java.

Sometimes placed in the genus *Rallina*; further study of relationships required. Monotypic.

Synonyms: *Limnobaenus/Porzana/Rallus/Rallina paykulli(i)*; *Porzana/Rallina mandarina/rufigenis*; *Crex/Gallinula/Rallina erythrothorax*.

Alternative names: Large/Chestnut-breasted/Siberian Ruddy/Chinese Banded Crake.

IDENTIFICATION Length 20-23 (27?)cm; wingspan 42cm. A medium-small crake. Forehead rufous; crown to nape, and entire upperparts, including upperwings and tail, olive-brown; greater and median upperwing-coverts variably tipped and barred white (white sometimes absent). Chin and throat white; sides of head, neck and breast cinnamon-rufous but mid-regions, and lower breast, often paler; flanks and belly to undertail-coverts boldly barred black and white. Whitish leading edge to wing visible in flight. Iris and orbital ring red; bill blue-grey, pea-green at base; legs and feet salmon to red. Sexes similar; in female, dark bars of underparts brown, and white markings on coverts may be less conspicuous. Immature has cinnamon-rufous areas duller; legs purplish-brown. In first breeding plumage may have foreneck and centre of breast pure white. Juvenile darker on upperparts than adult; has more white markings on upperwing-coverts; sides of head, and neck and breast, buffy-brown with obscure dark brown to blackish bars; underpart barring less regular than in adult, white bars tinged buff; belly off-white. Bill horn, tip dark; legs and feet darker than in adult. Inhabits marshes, and wet to damp, richly vegetated meadows with bushes and trees; often breeds in drier situations, including hayfields and cereal crops.
Similar species Similar to Red-legged (23) and Slaty-legged (24) Crakes, which are larger, with chestnut crown and nape, and more extensive underpart barring. Red-legged Crake also has richer brown upperparts, boldly barred primaries and outer secondaries (remiges of Band-bellied Crake are unbarred), more extensively barred upperwing-coverts, more extensive underpart barring with broader white bars, a more obvious red orbital ring, and red legs. Slaty-legged Crake usually has no visible bars on upperwing-coverts, and dull greyish-pink orbital ring and greenish-grey to black legs and feet. Sympatric Ruddy-breasted Crake (102) has unbarred wings, and narrow and inconspicuous white barring on flanks, belly and undertail-coverts.

VOICE The advertising call, assumed to be uttered by the male, is distinctive. It is described as a loud, metallic clangour tailing into brief trills like the sound of a wooden rattle. Calls are uttered at brief intervals throughout the night, and at dusk and dawn, but rarely during the day. In Ussuriland they are given from the birds' arrival in mid-May until mid- to late Jul (Dement'ev *et al.* 1969, Knystautas 1993). Birds often call from up to 2m above the ground in bushes and trees (Potapov & Flint 1987). The call is also described as resembling intermittent drumbeats (Ripley 1977).

DESCRIPTION
Adult Forehead vinaceous-rufous; crown to nape, and rest of upperparts, including upperwings and tail, olive-brown, remiges slightly paler; greater and median upperwing-coverts variably tipped and barred white (white sometimes inconspicuous or absent), white bars often edged darker brown; marginal coverts barred white or almost entirely white, forming narrow white leading edge to inner wing; outer web of outermost alula feather and outermost primary mottled white, sometimes entirely white; axillaries and underwing-coverts white, barred dark olive-brown to blackish. Chin and throat white, sometimes tinged warm buff; sides of head, neck and breast cinnamon-rufous; middle of foreneck, centre of breast and lower breast often paler; thighs, flanks and belly to undertail-coverts boldly barred blackish-brown and white; centre of belly often white. Iris crimson; orbital ring red; bill blue-grey, blackish on culmen and extreme tip, pea-green at base; legs and feet salmon-pink, orange-pink or red, also orange, tinged sooty on tarsus (Sato *et al.* 1997). Pea-green at base of bill may be more prominent in males than in females (Sato *et al.* 1997). Sexes similar; in female, dark bars of underparts browner, and white markings on coverts may be less conspicuous.
Immature Similar to adult but cinnamon-rufous of face, neck and breast duller; legs purplish-brown. Dement'ev *et al.* (1969) describe a first alternate (first breeding) plumage which differs from full adult in having foreneck and centre of breast pure white, and only narrow rufous band on forehead.
Juvenile Top of head, entire upperparts and most upperwing-coverts dark olive-brown, darker than adult; more extensive white bars on upperwing-coverts, edged darker olive-brown; outer web of outermost primary white or barred olive-brown and white; next primary (P9) sometimes spotted white on outer web; alula feathers olive-brown, tipped white and barred white on outer webs; all underwing-coverts, including marginals, barred olive-brown and white. Sides of head, and neck and breast, variously described as dingy grey with slight ochraceous tinge and scattered small whitish patches, and (more typically) buffy-brown to pale cinnamon, with obscure dark brown to blackish-brown bars; chin and throat whitish, throat often with numerous small grey streaks (feather bases?); barring on flanks, belly and undertail-coverts less regular than in adult, and white bars tinged buff; belly off-white. Iris colour not described, probably brown; bill horn, tip dark; legs and feet darker than in adult.

Downy young Upperparts black; underparts dull black; iris dark brown; bill ivory with white egg-tooth; upper mandible with blackish tip and broad black band in front of nostrils; lower mandible black at base, ivory at tip; legs and feet black, tinged reddish.

MEASUREMENTS Wing of 13 males 122-135 (128.7, SD 3.9), of 9 females 117-130 (124); tail of 13 males 48-56 (51.7, SD 2.7), of 3 females 46, 49, 49; culmen to base of 13 males 25-30.5 (26,7, SD 1.3), of 3 females 25, 25, 27; tarsus of 13 males 36-42 (38.8, SD 1.7), of 3 females 35, 36, 38. Weight of 4 males 114-132 (119.75, SD 8.3), of 1 female 102, of 1 unsexed migrant 96 (Potapov *et al.* 1987, Sato *et al.* 1997).

GEOGRAPHICAL VARIATION None.

MOULT Information is from Dement'ev *et al.* (1969). Postbreeding moult is apparently complete; it occurs on the breeding grounds, beginning in early Aug and ending in mid-Sep. Remex moult is simultaneous and the rectrices are also moulted simultaneously, before, with or after the remiges. Postjuvenile moult and first prebreeding moult (if any) apparently occur in the winter quarters.

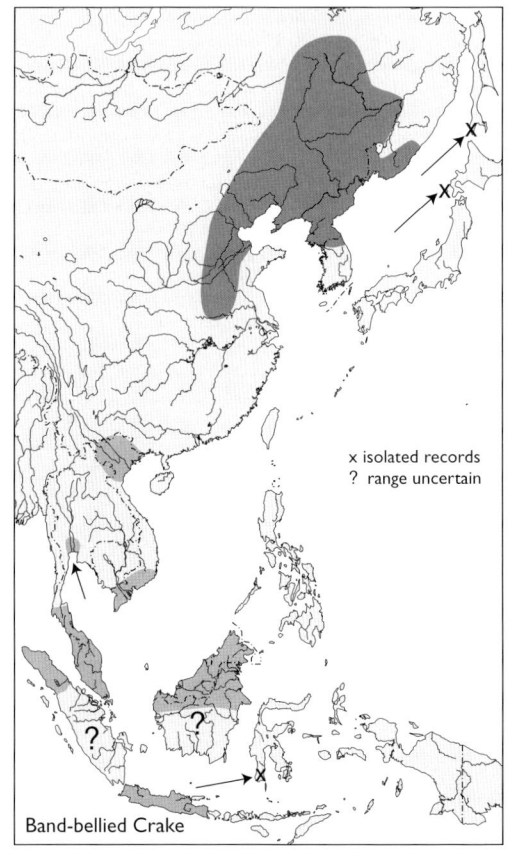

DISTRIBUTION AND STATUS Russian Far East in SW Amurland (middle and lower Amur V) and S Ussuriland; NE China (Heilongjiang, Jilin and Liaoning to Hebei and N Henan); and Korea (N from Kyonggi Do). This crake's winter distribution is imperfectly known. It winters in Vietnam (Tonkin and Cochinchina), peninsular Malaysia, N Sumatra, Java and N Borneo. It also winters very rarely in C Thailand (Lekagul & Round 1991). One specimen from Basilan, Philippines, is a Red-legged Crake (Dickinson 1984). This species is regarded as NEAR-THREATENED, and little information is available on its current status and distribution. In the past it was considered locally common to abundant at lower altitudes in its restricted range in the USSR (Dement'ev *et al.* 1969), present in small numbers in peninsular Malaysia, and occasional in Borneo (Medway & Wells 1976, Smythies 1981). It is regarded as locally fairly common in Ussuriland (Knystautas 1993), uncommon in China (Cheng 1987) and scarce in the Greater Sundas, where it is recorded from N Sumatra, N Borneo and Java (MacKinnon & Phillipps 1993). There are only 2 records from Sumatra, both specimens, the second in Jan 1918 (van Marle & Voous 1988), and 1 from S Sulawesi in Apr 1979 (Escott & Holmes 1980). During the Asian Waterfowl Census, 1987-91, only a few records were obtained, from China, and it is regarded as generally very scarce (Perennou *et al.* 1994). Its status and distribution urgently require investigation, especially in its wintering range.

Potapov & Flint (1987) make the following comments on its status and conservation. It has a restricted distribution and, although it is still abundant in some parts of its breeding range, in most areas it appears only sporadically. In the Russian Far East it is threatened by intensive agriculture and industrial development, as well as by other forms of habitat destruction, and it has locally become a threatened or extinct species although it is not included on any red data list. Many birds are killed by mowing machines, and others are killed by hunters, while the increasing application of weedkillers, insecticides and artificial fertilisers also has a negative effect. All sites with a high density of breeding birds should be conserved, and agricultural activities should be controlled. For example, the mowing of hay, etc, should take place from the centre of the field outwards, to give birds the chance to escape, and methods of flushing birds from the vicinity of mowing operations should be introduced. Spring hunting should be banned.

MOVEMENTS It is transient in Inner Mongolia and Shandong (Yantai and Qingdao), southwards through S China (Jiangsu, Hunan, Fujian and Guangdong), and in late spring and early autumn in Korea (specimens taken in the N from 16 Sep to 26 Oct), where it is also a summer resident (Austin 1948, Potapov & Flint 1987), and it has been recorded in Hong Kong (Dickinson 1984). In the Russian Far East departure times are unclear, but Sep specimens exist and departures are recorded in Oct, while birds do not arrive in Ussuriland until mid-May and N movements continue until mid-Jun (Dement'ev *et al.* 1969, Knystautas 1993, Sato *et al.* 1997). Night-flying migrants have been taken in peninsular Malaysia at Kuala Selangor lighthouse and Fraser's Hill in Oct, Nov, late Apr and early May (Medway & Wells 1976). In the Russian Far East it has also been recorded on Moneron I, SW of Sakhalin I (Potapov & Flint 1987). The first record for Hong Kong was of a bird which flew into a wall, 11 Oct 1977 (BMNH). A migrant, probably an adult male, was mist-netted on Oshima I, off SW Hokkaido, Japan, on 25 May 1993, having probably been blown across the Sea of Japan by strong WSW winds while flying N over the adjacent Russian mainland (Sato *et al.* 1997).

HABITAT The breeding habitat is lowland marshes and meadows with tussocks, rich herbaceous vegetation, and thickets or small trees, and also hayfields and cereal crops. It also nests in marshy or damp situations (including tussocky bogs and hilly moors) in mountains, woodlands and forests, characteristic trees of such sites including birch *Betula*, willow *Salix*, alder *Alnus* and larch *Larix*. It usually avoids open water, large marshes and extensive meadows, preferring damp to dry situations with thick grass. In the Russian Far East it is often found along field edges and fences, and in wet areas and thickets around gardens, sometimes even occurring among buildings (Dement'ev *et al.* 1969). On passage in peninsular Malaysia it frequents open inland swamps (Medway & Wells 1976). The wintering habitat in the Greater Sundas is described as wet grass and paddyfields; however, in Vietnam it is grassy hummocks with bushes or small trees in meadows and swamps, birds usually avoiding very damp sites (Wildash 1968). It is apparently mainly a lowland species, but is recorded up to c. 1,200m on migration.

FOOD AND FEEDING Mostly molluscs, crustaceans and insects, including Coleoptera (Hydrophilidae, Curculionidae, Carabidae and Silphidae); also seeds, including those of 'rushes' and 'Amur vetch' (*Vicia*?) (Potapov & Flint 1987). It forages in damp situations, and also in drier areas in fields and gardens, and near dwellings.

HABITS Information is from Dement'ev *et al.* (1969). This crake is rarely seen but in the breeding season its presence is easily established by its characteristic call. On the breeding grounds its habits apparently resemble those of the Corncrake (89) rather than wetland crakes; it avoids even shallow water and prefers moist or dry soil, and it runs through thick herbage and bushes with extreme rapidity and agility. When chased it runs through the grass from point to point, giving calls all around the pursuer (even to the rear). It is extremely difficult to flush, and resorts to flight only when disturbed in areas lacking cover. The escape flight is feeble, with dangling legs, and the bird soon drops into the closest thicket, but will sometimes fly up into the top of a tree, from which it then gives its grating call. It is said to be primarily nocturnal (Flint *et al.* 1984), apparently only because of its calling behaviour.

SOCIAL ORGANISATION Usually solitary and presumed monogamous. It usually nests in discrete pairs, and on the lower Iman up to 10 calling males are recorded perkm^2 (Dement'ev *et al.* 1969). It sometimes occurs in densely populated colonies. On migration it is encountered singly but several may sometimes be flushed close to each other (Potapov & Flint 1987).

SOCIAL AND SEXUAL BEHAVIOUR Not described.

BREEDING AND SURVIVAL Season In Russian Far East, breeds from late May. **Nest** Nests in hummocky, not too damp, meadows, hay and cereal fields, and swamps with thickets. Nest usually built in depression in ground; sited on tussock or in grass, tall cereal crops or shrubs; lined with fine twigs and grass stalks; size 15 x 17cm. **Eggs** 5-9 (usually 7); ovate; dirty white with small pale ochre spots and dark violet flecks, markings most dense at blunt end; size 33.4-36.2 x 26-26.4. **Chicks** According to Dement'ev *et al.* (1969), juvenile flies well when almost fully grown; age of first breeding 1 year; probably 2 broods per season. Nothing else recorded.

104 SPOTLESS CRAKE
Porzana tabuensis   Plate 30

Rallus tabuensis Gmelin, 1789, Tongatapu Group, Tonga.

Forms superspecies with *P. atra* and *P. monasa*. Validity of subspecies requires confirmation, but three subspecies currently recognised.

Synonyms: *Rallus/Aramus/Corethrura/Zapornia tahitiensis*; *Porzana plumbea/vitiensis*; *Crex/Porzanoidea/Ortygometra plumbea*; *Rallus minor/tenebrosus/minutus*; *Gallinula/Porzana immaculata*; *Corethrura/Ortygometra tabuensis*; *Porzana/Zapornia spilonota*; *Zapornia umbrina*; *Phalaridium tabuense*.

Alternative names: Sooty Crake/Rail; Black Rail.

IDENTIFICATION Length 15-18 (20?)cm; wingspan 26-29cm. Small; head, neck and underparts dark slate-grey (paler in race *richardsoni*, darker in *edwardi*), paler to whitish on chin; upperparts dark reddish-brown (paler in *richardsoni*, darker and less rufous in *edwardi*), darker brown from lower back to uppertail-coverts and blackish-brown on tail; undertail-coverts grey-black, barred white. Flight feathers dark brown; white on outer edge of outermost primary and on leading edge of wing. Iris and orbital ring red; bill black; legs and feet salmon-pink. Sexes alike. Immature like adult. Juvenile paler and browner; head and underparts duller, dark grey-brown; chin and throat white; upperparts duller, less red-brown, than in adult; some birds have narrow white supercilium extending to above eye; bill black with pink base; iris brownish-orange; legs and feet grey-brown to brownish-flesh. Inhabits dense vegetation of freshwater to saline palustrine wetlands; also occasionally in bushes or scrub and (on islands) in forest or rocky areas.

Similar species Easily distinguished from all sympatric small crakes by distinctive appearance; calls also distinctive. Two other species are probably insular derivatives of Spotless Crake stock (Olson 1973b, Ripley 1977): the larger, flightless Henderson Island Crake (106) is entirely black, with red legs and feet, and black bill with yellowish-green base and culmen; the probably extinct Kosrae Crake (105) of Kosrae I is also black, with pale chin and throat, white spots on some upperwing-coverts and on undertail-coverts, and red legs and feet.

VOICE Information is taken from Marchant & Higgins (1993), and from details provided by R. Jaensch (*in litt.*). The calls are distinctive. The commonest calls, given during the breeding season, are a loud trilling purr lasting 1-3s, which resembles a motor or a sewing machine and is preceded by a soft quarrelling sequence, and a single or repeated loud, sharp, high-pitched *pit* or *poot* (which is probably the harsh *crack* call mentioned by Buddle 1941a) often interspersed with a harsh nasal *harr*. There is also a trilling whistle. It also has soft bubbling calls, while a murmuring which is frequently given by two birds (presumably a pair) is also described but may be the soft quarrelling sequence referred to above. Other calls are a *mook*, a *mint-mint*, and a clicking which is used in the distraction display. It has been recorded duetting with the Fernbird *Bowdleria punctata* (Skinner 1979). Chicks first give a weak twittering, and subsequently make a single, thin, rising call which is constantly repeated.

DESCRIPTION *P. t. tabuensis*
Adult Forehead to nape blackish-grey, crown sometimes

developing brown tinge with wear; mantle, back and scapulars dark red-brown, becoming dark rufous-brown to brown with wear; lower back and rump dark brown; uppertail-coverts dark grey-brown; tail blackish-brown. Upperwing-coverts dark red-brown; primary coverts, alula, primaries and secondaries blackish-grey, fading to dark grey-brown with wear; tertials dark brown; outermost primary (P10) and outermost alula feather have white outer edge; marginal coverts white; axillaries and underwing-coverts dark grey, tipped or barred white. Sides of head, sides of neck and foreneck, dark grey, grading to varyingly lighter grey (even whitish) on chin and sometimes down centre of throat; rest of underparts lead-grey, becoming darker or more brownish when worn; undertail-coverts grey-black, barred white. Brightness of upperpart colour variable. Iris red to crimson; orbital ring pinkish-red to orange-red; bill grey-black to black; legs and feet dark pinkish-orange to salmon. Sexes alike.

Immature Apparently identical to adult, but some can be aged on traces of remnant juvenile plumage, which remains longest on chin, throat and centre of breast and belly and may form noticeably pale line (Marchant & Higgins 1993). Bare parts not described.

Juvenile Forehead to hindneck dark brown; mantle, back and scapulars dark reddish-brown, browner and less red than adult. Upperwings as adult, but lesser, median and secondary coverts dark reddish-brown, like rest of upperparts. Lores dark brown; some birds have narrow white supercilium not extending behind eye; ear-coverts and sides of throat dark grey-brown; chin and centre of throat white; breast to belly dark grey-brown (lacking leaden wash of adult), feathers in centre of breast and belly with whitish tips which fade and become larger with wear; rest of underparts like adult. Iris brownish-orange; bill black with pink basal third and pale grey tip; legs and feet pale grey-brown to brownish-flesh, sometimes developing adult colour before postjuvenile moult complete.

Downy young Black down of chick has greenish sheen; bill black with pink or horn saddle between nostrils, and white egg-tooth; later has pink base to both mandibles; iris black or blue-grey; legs and feet dark brown-grey, pink at rear of tarsus. Iris becomes dark olive; after 4 weeks develops orange tinge.

MEASUREMENTS (Australian birds) Wing of 27 males 82-92 (86.4, SD 2.5), of 17 females 78-91 (84.5, SD 3.0); tail of 23 males 45-57 (50.7, SD 2.7) of 13 females 44-54 (49.2, SD 3.5); exposed culmen of 27 males 17.3-22.2 (19.2, SD 1.2), of 14 females 16.2-20.2 (18.1, SD 1.1); tarsus of 27 males 26-32.5 (29.6, SD 1.6), of 17 females 25-32.5 (28.2, SD 1.5) (Marchant & Higgins 1993). Australia and New Zealand, weight of 11 males 34-56 (45.3, SD 6.4), of 8 females 21-58 (37.55, SD 6.4), mean of 21 unsexed 46.95 (Marchant & Higgins 1993). New Caledonia, weight of 1 male 41.5; New Guinea, weight of male 42-54, of female 38.5-43 (Ripley 1977, sample sizes and means not given).

GEOGRAPHICAL VARIATION Slight, involving size, and the colour of the upperparts. The validity of races is questionable: see discussion in White & Bruce (1986). The form *plumbea* of S Australia, Tasmania, New Zealand and Chatham Is is included in nominate *tabuensis*; it supposedly has brighter, redder upperparts and longer wing (Ripley 1977), but colour of upperparts too variable, and differences in wing length too slight, to support separation (Marchant & Higgins 1993, *contra* Banks 1984); the forms *immaculata*, *oliveri*, *vitiensis*, *caledonica* and *filipina* are also included in nominate (Ripley 1977). There is a slight clinal increase in wing length with latitude (Onley 1982b).

P. t. tabuensis (Gmelin, 1789) – Philippines (Luzon, Mindoro, possibly Palawan); E Moluccas (Tiur, Watubela Is); New Guinea (Anggi Lakes, Arfak Mts, and Astrolabe Bay region - also see note after *richardsoni*); Bismarck Archipelago (Wuvulu); Vuatom I; Caroline Is (Kusaie = Kosrae); Solomon Is (Guadalcanal); Australia (mainly SW, W, E, SE and NE, largely absent from the interior, also occurs N and E Tasmania); New Zealand (including Three Kings, Poor Knights, Mayor and Stewart Is); Norfolk I; Chatham Is; New Caledonia; Santa Cruz Is (Tinakula); Vanuatu (Erromango, Efate, Tanna, Aneiteum); Kermadec Is (Raoul); Fiji Is (Viti Levu, Ovalau, Ngau, Kandavu); Tonga Is (Tongatapu, Late, Honga Hapai, Fanua Lai, possibly Niuafo'o); Samoa Is (Savaii, Tau); Niue I; Cook Is (Mangaia, Mitiari, Atiu); Leeward Society Is (Raiatea); Society Is (Tahiti, Moorea); Tuamotu Is (Manui, Manihi, Raraka, Mangareva, Rangiroa, Ahe, Takapoto, Tikei, Faaite, Tuanake, Hiti, Apataki, Kaukuru, Aratika, Toau, Kauhei, Tikahau, Makemo, Napuka [recently extinct], Marutea Atoll); Pitcairn Is (Oeno); Marquesas Is (Nukuhiva, Hatutu, Fatu Hiva, Hiva Oa, Ua Pou, Tahuata); Tubuai Is (Tubuai, Rapa); Gambier Is (recently extinct); and Ducie I (unconfirmed). For further details of Polynesian islands inhabited, see Ripley (1977), Holyoak & Thibault (1984) and Pratt *et al.* (1987). Fossil remains of *P. tabuensis* from caves on Mangaia are of uncertain subspecies, and this island has also yielded remains of the extinct *P. rua*; it is possible that the rails surviving on the island are not *P. tabuensis* (as assumed by Holyoak & Thibault 1984) but are *P. rua*; as there are no modern specimens, this theory cannot be tested (Steadman 1986). The smallest race. See Description.

P. t. edwardi Gyldenstolpe, 1955 – W and C New Guinea (Ilaga Valley, C Irian Jaya; Waghi Valley [Mt Giluwe area] E to Okapa area, Papua New Guinea). See Note after *richardsoni*. Upperparts deep chocolate, darker and less reddish-brown than nominate; underparts darker grey; bill longer. Measurements of 4 males, 4 females: wing of male 79-87 (82.8, SD 3.3), of female 80-84 (81.8, SD 1.7); tail of male 40-47 (44, SD 3.2), of female 38-43 (41, SD 2.2); culmen to base of male 21.5-22 (21.75, SD 0.3), of female 19-20.5 (19.5, SD 0.7); tail of male 28.5-30.5 (29.3, SD 0.9), of female 26-28 (27.25, SD 1.0). Weight of 3 males 47, 50, 52, of female 44-51 (Ripley 1964; Diamond 1972b).

P. t. richardsoni Rand, 1940 – W New Guinea (Oranje Mts, C Irian Jaya); occurs between disjunct populations of *edwardi*. See Note below. Upperparts paler than nominate, more olive-brown; underparts paler grey. Wing of 8 males 83-88 (84.6, SD 1.8), of 8 females 80-89 (83.4, SD 3.1) (Rand 1942b); culmen of male 16-17, of female 15-16 (Ripley 1977).

Note: Coates (1985) also records this species from the following regions of New Guinea, races unspecified. Irian Jaya: Tamrau Mts (*tabuensis*?); Wissel Lakes (*edwardi*?); Baliem Valley (*edwardi*?); and from Merauke W to Frederik-Hendrik I. Papua New Guinea: Tari; Goraka and Awande areas; Wau area; Popondetta; near Bereina; and Port Moresby region. It was also recorded from the Ok Tedi area, Western Province, at 550m, in Oct 1977 (P. Gregory *in litt.*)

? occurrence uncertain

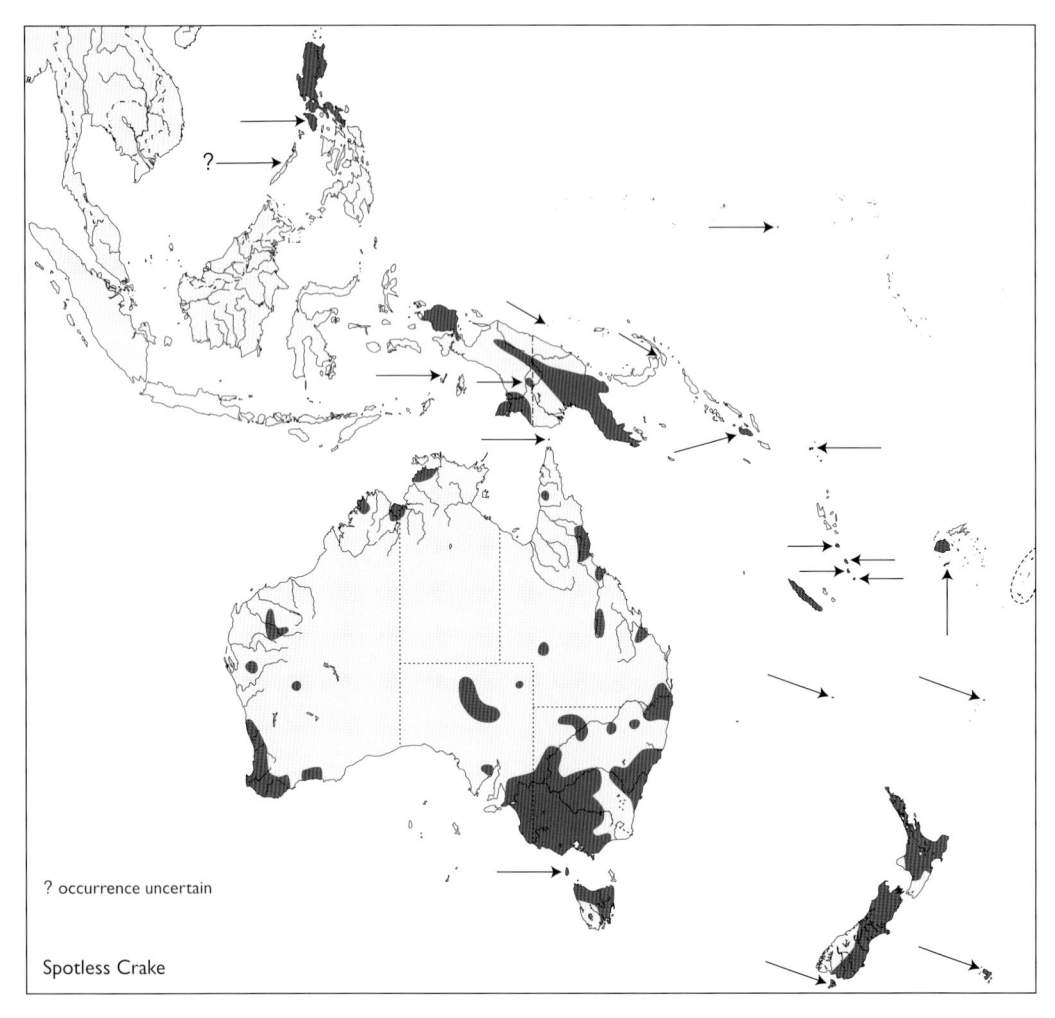

Spotless Crake

MOULT Postbreeding moult is complete; remex and rectrix moult are simultaneous, the tail being moulted with, or before, the remiges. Body moult usually begins on the head, then the upperparts (mantle first) and flanks; the throat and scapulars moult last. In Australia/New Zealand, the remiges are moulted in late summer/autumn, some birds completing moult in late Feb, others not until May; body moult can begin in Sep, and is usually completed by Mar-Apr (Marchant & Higgins 1993). Postjuvenile moult is partial, and is recorded in Sep-May in Australia and New Zealand (Marchant & Higgins 1993).

DISTRIBUTION AND STATUS Very widespread in the Pacific region, from the Philippines, New Guinea, Australia and New Zealand E across the Pacific to Micronesia and Polynesia. Not globally threatened. There is only 1 record from the Moluccas, where it may be a vagrant (White & Bruce 1986) although resident populations may exist on small islands in E Wallacea (Coates & Bishop 1997). It was probably overlooked in the Philippines (Delacour & Mayr 1946) and in New Guinea, except that in New Guinea *edwardi* was thought fairly common within its restricted range (Rand & Gilliard 1967, Diamond 1972b); the race *richardsoni* was "occasionally seen" (Rand & Gilliard 1967).

It is currently regarded as rare in the Philippines, where it was recently reported from Mindoro and possibly Palawan (Dickinson *et al.* 1991). In the W part of its range it may be more widespread than is known (White & Bruce 1986). Its distribution and status in New Guinea are imperfectly known; apparently it is locally fairly common in upland regions but rare or very local elsewhere (Coates 1985); it could have escaped detection in lowland areas such as the Sepik and Fly floodplains (R. Jaensch *in litt.*). In the Bismarck Archipelago there is only 1 record from Vuatom I (off New Britain) in 1936 (Mayr 1949); it is also present on Wuvulu I off the N New Guinea coast (Coates 1985).

In Australia suitable habitat has been lost through the drainage of wetlands, but the species is probably still plentiful in many areas (though overlooked), though only a vagrant or a very localised regular migrant in small numbers in Northern Territory (Marchant & Higgins 1993). However, R. Jaensch (*in litt.*) has recorded 20+ birds in drying *Typha* at a semi-permanent swamp near Darwin and considers that this could be an annual occurrence, possibly of birds from local permanent swamps; it is common and resident in permanent but artificial *Typha* swamp at Kununurra, in extreme NE Western Australia,

418

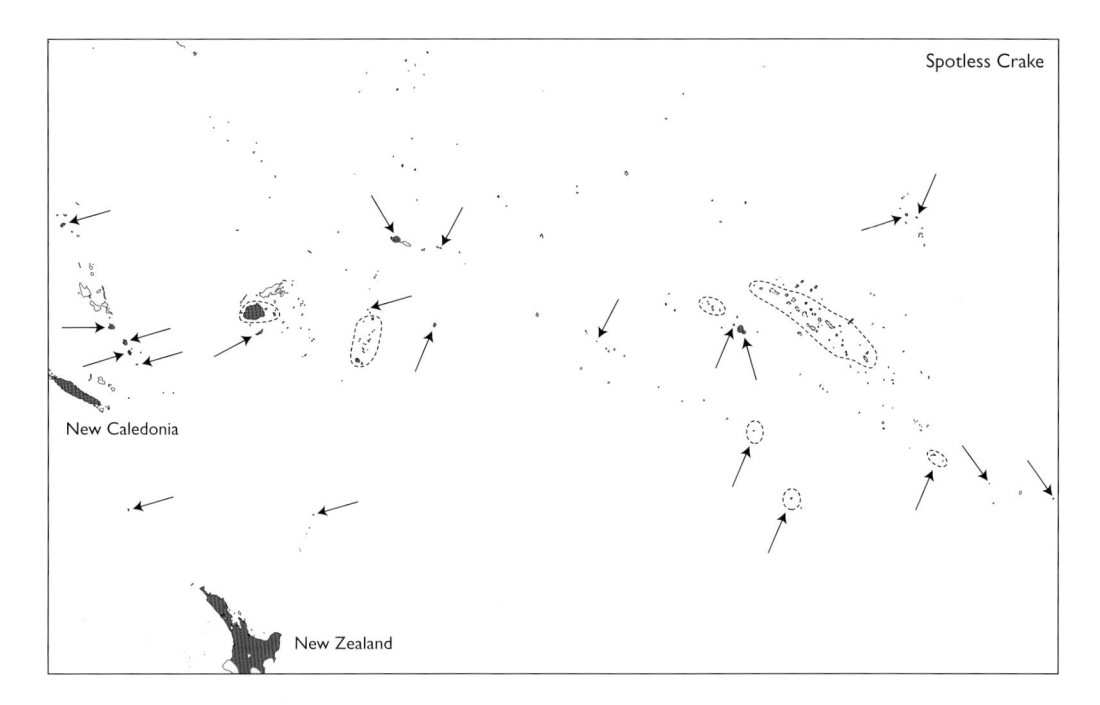

and has bred there (Jaensch 1989). Post-breeding densities of >10 birds/ha (even possibly several times this density) are not uncommon at drying sedge swamps in SW Australia and in some parts of S South Australia; but it is rarer than the Australian Crake (97) and Baillon's Crake (94) in the L Eyre basin, EC Australia (R. Jaensch *in litt.*). Birds readily occupy artificial wetlands, farmland and sometimes garden habitats, and in some areas the rise in the water table due to irrigation has possibly enhanced habitat locally (Marchant & Higgins 1993). In New Zealand, where its wetland habitats have been reduced, it is more widespread in North I, where it may be locally common, and it is recorded more sparsely throughout South I (Marchant & Higgins 1993, Heather & Robertson 1997); on Aorangi, Poor Knights Is, it increased from the 1930s with the development of low mixed forest after human occupation ceased and pigs were removed (pigs adversely affect habitat by restricting regeneration of low forest, destroying nesting cover and disturbing leaf-litter); however it declined from the 1950s as low mixed forest was replaced by tall forest, and the decline is expected to continue as habitat is reduced (Onley 1982a). It has become rare on Norfolk I, where its population has been reduced by rats, and extinct on Raoul I (Kermadecs) (Marchant & Higgins 1993).

Elsewhere the species was formerly regarded as either genuinely uncommon, e.g. on Niue I (Wodzicki 1971), rare, e.g. on New Caledonia (Hannecart & Létocart 1983), or overlooked, e.g. in E Polynesia (Holyoak & Thibault 1984). It was reported from Guadalcanal, Solomon Is, by Stevens & Tedder (1973). It appears to have suffered population reductions and local extinction throughout the Pacific islands, often where it has encountered man and his introduced commensals (e.g. Holyoak & Thibault 1984). It is local and uncommon on Vanuatu, where it could face extinction as a result of swamp pollution or destruction (Bregulla 1992). In E Polynesia it may still be widespread but it disappeared from the Gambier Is and Napuka in the recent past, and it has not been recorded for over 70 years from many other islands (Tikehau, Ahe, Manihi, Tikei, Apataki, Kaukuru, Aratika, Toau, Kauehi, Raraka, Makemo, Faaite, Tuanake, Hiti, S Marutea, Ducie), and on Oeno (Pitcairn Group) the population is very small (Holyoak & Thibault 1984); it may survive on Niue I (Kinsky & Yaldwin 1981). In the Fiji Is it may survive in isolated swamps on Viti Levu, and may occur on Vanua Levu, despite the presence of an introduced mongoose (Watling 1982b). In 1985 it was rediscovered on Tau I, American Samoa, where its population is small and probably decreasing as habitat diminishes with the reduction in subsistence agriculture (Engbring & Engilis 1988). Predators of adults and young include cats, and possibly rats.

MOVEMENTS It is generally regarded as resident over much of its range. Most of the following information is from Marchant & Higgins (1993). Lower reporting rates in the non-breeding season are possibly due to the birds being less conspicuous when not calling. In Australia, relatively few records in the S during the winter may suggest movement, although it is not clear what factors might be responsible for departures (R. Jaensch *in litt.*), and there are no winter records from Tasmania. It is possibly regularly migratory in Queensland, where it is recorded in the N only from Oct-May, and on islands in the Torres Strait, where casualties are recorded from Dec-Feb and in May. In N Queensland, 1 came to house lights at night in fog, in May 1991 (Crouther 1994). Records from N Australia, islands of Western Australia, and elsewhere, suggest that birds may move well beyond the normal range. Sudden or seasonal changes in numbers may occur in response to rainfall or receding water, and irruptions, probably induced by good rainfall, are recorded (see Marchant & Higgins for details). On Poor Knights Is (New Zealand) young disperse to drier habitats (Marchant & Higgins 1996), and Heather & Robertson

(1997) mention "occasional records of birds found in towns, far from their usual habitats". One from Tiur (SE Moluccas) in Jan 1899 may have been a vagrant (see Distribution and Status). This species is known to be highly dispersive, having spread widely E across the Pacific through Polynesia, and vagrancy is to be expected.

HABITAT Usually dense vegetation in, or at the margins of, freshwater to saline wetlands, either permanent or ephemeral. It occurs in marshes, swamps, wet grassland, saltmarsh, mangroves and peat bogs; also at the margins of rivers, streams, tidal creeks, ponds and lakes. It may prefer wetlands with shallow, slow-flowing water. Breeding habitat is often characterised by large, unbroken stands of dense, tall emergents (rushes, reeds, sedges such as *Carex secta* and *C. lessonia*, grass, and shrubs such as *Melaleuca*), and an overstorey of willows *Salix*, or (more usually in Australia, R. Jaensch *in litt.*) tall reedbed vegetation such as *Typha orientalis*, may be an important requirement of breeding habitat in New Zealand, where the sympatric Baillon's Crake appears to require pure stands of tussock sedge *Carex secta* (Kaufmann & Lavers 1987, Kaufmann 1987). It is also recorded from pure *Typha* or *Scirpus pungens* reedbeds, and in mixed emergents including *Juncus canadensis* (O'Donnell 1994), and in the interior of E Australia it is occasionally recorded in inundated lignum *Muehlenbeckia cunninghamii* shrubs (R. Jaensch *in litt.*). It frequently inhabits artificial wetlands such as saltworks, sewage farms, rice paddies, taro ponds (Wuvulu I), irrigation channels, farmland, golf courses, lawns and gardens. In Australia, it seems to exploit isolated and sometimes quite small habitats very successfully, and it is not usually found in saline conditions whereas the Australian Spotted Crake readily occupies saline habitats (R. Jaensch *in litt.*). It also occurs on fern-covered hillsides, and in heathy flats and coastal scrub. In far SW Western Australia it is apparently the only crake normally occurring in acidic (pH <6), heavily vegetated swamps (R. Jaensch in Marchant & Higgins 1993). In Papua New Guinea, 1 bird was found in dry secondary growth with a few emergent tall trees, at 550m a.s.l. (P. Gregory *in litt.*), and in the E highlands of New Guinea it occurs in grassland (Diamond 1972b).

On oceanic islands it occupies many habitat types, from humid mountain forest to wet or dry vegetated slopes (grass, ferns etc) on volcanic islands, and thickets, small pools and the edges of coconut groves on atolls (Holyoak & Thibault 1984). On Aorangi (Poor Knights Is) it prefers mixed low *Coprosma-Myrsine* forest with a 3-4m canopy, sparse understorey, thick leaf-litter, and dense low cover where gaps occur in the canopy; it also occurs in other types of forest (Onley 1982a), and on islands it will also occupy rocky or stony habitats without standing water (Serventy 1985, Marchant & Higgins 1993). In New Caledonia it is found more in dry areas than in wetland but this may reflect observer effort (Y. Letocart per R. Jaensch *in litt.*). It ranges from sea level up to c. 3,150m in New Guinea, where it is most widespread in the highlands.

FOOD AND FEEDING Information is from Marchant & Higgins (1993) unless otherwise indicated. Annelids (Oligochaeta); molluscs (Gastropoda); crustaceans (Ostracoda and Amphipoda); Collembola; adult and larval insects, including Ephemeroptera, Coleoptera (Chrysomelidae), Diptera (Calliphoridae, Tipulidae) and Orthoptera (Acrididae, Tettigoniidae); spiders; eggs of Wedge-tailed Shearwater *Puffinus pacificus*, Black-winged and Kermadec Petrels *Pterodroma nigripennis* and *P. neglecta*, and Black Noddy *Anous minutus*; also seeds (including grass seeds), ripe fruit (*Rubus*), and shoots of grasses (Poaceae) and aquatic plants. It is also recorded feeding on the carcass of a cow (Marchant & Higgins 1993). It usually feeds on the ground, but also in trees (Kermadec Is), foraging among foliage and nests of noddies (Soper 1969). It forages on mud and in shallow water, in or near marsh vegetation, along tidelines adjacent to dense vegetation, at the margins of still and flowing water, and in short grass; it also feeds in and around petrel burrows (Merton 1970). On Rapa Iti (Tubuai Is) it has been observed foraging on pebble beaches in the middle of the day (Holyoak & Thibault 1984). It searches leaf-litter in forests, scratching with the feet and moving or turning over litter with the bill, clearing areas up to 23cm in diameter (Buddle 1941a), and it sometimes feeds in blackberry *Rubus* and other thickets. It has been recorded feeding with domestic fowl. Both adults and chicks wash food. On islands it is able to survive without fresh water.

HABITS Most information is from Marchant & Higgins (1993). It is crepuscular and diurnal. Birds are secretive and normally inhabit dense cover, moving furtively and darting between patches of cover, and peering out from cover before running across open areas. Unlike other *Porzana* crakes, in wetlands it rarely ventures into the open to feed unless forced to do so by the drying up of habitat (R. Jaensch *in litt.*), and when in the open, birds stop and look round every few m; however in New Guinea the species is said to be visible occasionally in the early morning and late afternoon when it forages in the open close to cover (Coates 1985). They walk along branches and fallen logs, and may climb in vegetation up to 2m tall. The normal gait is a slow, stalking walk, with the tail constantly flicked. They wade and swim readily, and often dive to escape enemies (Ripley 1977), and they also run to cover when alarmed, or make a short low flight often alighting with the wings spread. Birds roost at night, sometimes using roosting nests especially when adults and young roost together; sunbathing has been recorded on a stick protruding from the water, and on a clod of earth at the base of reeds.

SOCIAL ORGANISATION Monogamous, possibly pairing for life. Strongly territorial, possibly throughout the year as birds rarely leave the territory and also respond to taped playback throughout the year, although they apparently call less frequently outside the breeding season (R. Jaensch *in litt.*); an observed increase in calling and chasing before the breeding season is possibly associated with pair formation and the establishment of territories (Marchant & Higgins 1993). Seven nests were found along c. 800-1,000m of a small stream near Canberra, Australia, but none had eggs and some could have been from previous years; although it is rare to find >1-2 active nests close to each other, birds may nest in close proximity (R. Jaensch *in litt.*). There were possibly 5 pairs in an 800 x 25m strip of rushes adjoining a watercourse (Morton 1953), a mean territory size of 0.4ha; 13-19 occupied territories were in 86ha of marsh in New Zealand (Kaufmann & Lavers 1987); 22 territories on Aorangi, Poor Knights Is, occupied 33.5ha, a mean size of 1.5ha, but in the richest habitat (mixed low forest with thick leaf-litter) 14 territories occupied 14.5ha, a mean size of c. 1ha, and the smallest territory was 0.225ha (Onley 1982a, Jones et al. 1995). Some territories are said to be as small as 5-120m^2, and home ranges may be 0.04ha (Bryant & Amos 1949, Hadden 1970); such small areas are comparable to areas of thick cover, within territories at Aorangi, which were used by adults with small young, the family groups often not moving more than 10m (Onley 1982a). Solitary birds may also hold territories (Onley 1982a). There is a record of three adults accompanying young (Onley 1982a).

SOCIAL AND SEXUAL BEHAVIOUR Most information is taken from that summarised by Marchant & Higgins (1993). Some agonistic behaviour was recorded towards Baillon's Crakes, which were chased, but Baillon's Crakes answered playback of Spotless Crake calls and came out to investigate the sounds, while Spotless Crakes responded poorly to playback of Baillon's Crake calls (Kaufmann & Lavers 1987). This species will feed in the same wetlands as Baillon's and Australian Spotted Crakes but may use a different feeding strategy, such as keeping more within cover, which reduces contact with the other species (R. Jaensch *in litt.*). Few boundary disputes have been seen, and mostly occur where birds are densely packed. A territory holder approaches an intruder, sometimes giving warning trills, and the intruder usually retreats, normally not being pursued. Aggression within pairs has been noted, and taped playback sometimes causes birds to give loud short calls interrupted by splashing and fighting. In Papua New Guinea, a bird was attracted to playback of Red-necked Crake (21) calls (P. Gregory *in litt.*).

Copulation has been recorded once: the male followed the female round a *Carex* tussock several times; the female then stood on the tussock with body arched and bill pointed downwards, and the male mounted, balancing with outstretched wings, and copulated for several seconds before dismounting and walking off; the female stretched her head up and followed the male (Kaufmann & Lavers 1987).

At change over the incubating bird shakes as its calling partner approaches; the relieving bird may bring and incorporate nest material. If disturbed, the incubating bird will stand on or beside the nest, or may give the distraction display. This is given to predators and to human intruders; the bird splashes water and rustles vegetation, sometimes dashing across open water with the head held down, the body hunched, the tail depressed and the wings slightly opened and drooped. Another display involves the bird flitting from the ground into vegetation, with the back horizontal, the tail raised, and the wings extended with the edges close to the ground and the remiges pointed upwards; bird will also squat in mud with repeatedly quivering wings. An incubating bird did not attack a mounted Spotless Crake specimen placed near the nest, but was very wary of it and left the nest, only resuming incubation over 1h later (in the presence of the specimen) (Hadden 1970).

BREEDING AND SURVIVAL Season Australia, Sep-Jan, in SW Western Australia laying time correlating best with maximum day length and up to 2 months after, and also with the period 2-3 months after maximum water depths (Halse & Jaensch 1989); New Zealand, Aug-Jan; New Caledonia, Oct; Vanuatu, probably Sep-Feb; Moorea (Society Is), Jul; E Polynesia, small young Mar (Raraka), large young Dec-Apr, possibly breeds throughout year. **Nest** Shallowly to deeply cupped, woven of dry grass, fine sedge and rushes, lined with finer material or pieces of soft dry *Typha* (R. Jaensch *in litt.*); lower part may be made of rush stalks, with cupped platform of finer leaves; material obtained from surrounding vegetation. Nest placed in clump of sedges or rushes (e.g. *Carex, Baumea, Lepidosperma, Typha*), centre of grass tussock, or small bush (e.g. *Melaleuca* sp.) in reedbed; often placed over water but may be up to 90m from water; also against stump of tree-fern, in fork of fallen branch in dense *Microleana*, or under tangled ferns, brambles or dense bushes; some nests on ground at base of sedges (e.g. *Carex*) or grass; occasionally in paddocks. External diameter 10-23cm, depth 5-20cm; cup diameter c. 8cm, depth c. 4cm; height above water 3-150cm. Role of male in building uncertain; female may bend reeds to form canopy over nest; one or more ramps of nesting material may be constructed from water to rim. Material may be added to nest during incubation. May build several non-functional nests before building breeding nest; these used for resting and brooding, sometimes for second nesting. **Eggs** 2-7 (usually 3-4); elliptical; slightly glossy; dull creamy-brown (*contra* Ripley 1977) with uniformly distributed pale chestnut-brown flecks, or with few darker spots and streaks concentrated at one end. First clutch of season may be greener and paler (to match new green growth?), 1 clutch in older reeds darker brown (Fletcher 1914, 1916b). Size of eggs: nominate (Poor Knights Is, New Zealand) 29.5-34.5 x 22.5-24 (31.5 [SD 1.7] x 23.5 [SD 0.5], n = 16); weight of 3 eggs, New Zealand, 9.0, 9.1, 9.5, of 11 others 8-9 ± 0.5 (Kaufmann & Lavers 1987). Eggs laid at daily intervals, most eggs laid between 09:00 and 12:00; later clutches larger than earlier ones (Kaufmann & Lavers 1987); island populations may lay smaller clutches than mainland birds (e.g. Onley 1982a). **Incubation** 19-22 days, by both sexes; at 1 nest 40% of incubation during daylight done by male; hatching synchronous, or asynchronous in large clutches (Kaufmann & Lavers 1987). **Chicks** Precocial; leave nest after 1-2 days; can clamber out of nest within 4h of hatching; begin to feed themselves at 3 days (in captivity); bathe from 5 days. Fed and cared for at first by both parents, later usually accompanied by only 1 parent; initially fed at nest (period unknown); captive chicks fed without assistance at 17 days; chicks beg until c. 6 weeks old, running to adult and then sitting on tarsi and toes, waving wings asynchronously; young probably attended until 4-5 months old. First feathers appear at c. 15 days; captive birds were almost fully feathered at 40 days but remiges were not fully grown at 66 days; 2 captive chicks reached asymptote weight at 1 month (Kaufmann 1988b). May have 2 broods in Australia and New Zealand; if nesting delayed, may raise 1 large brood rather than 2 smaller broods; relays after clutch loss. For 16 clutches (Australia and New Zealand), 61 eggs laid, 33 (54%) hatched (Marchant & Higgins 1993). **Survival** No information.

105 KOSRAE CRAKE
Porzana monasa Plate 32

Rallus monasa Kittlitz, 1858, Kosrae (Kusaie) Island, Caroline Islands.

Formerly sometimes placed in the monotypic genus *Aphanolimnas*. Forms a superspecies with *P. atra* and *P. tabuensis*. Monotypic.

Synonyms: *Kittlitzia/Aphanolimnas monasa*.

Alternative names: Kusaie/Ponape/Kittlitz's Crake; Kusaie Rail.

IDENTIFICATION EXTINCT. Length 18cm. A small, flightless crake. Plumage black, with bluish-grey reflections; chin and centre of throat paler; lesser upperwing-coverts brownish with white spots; undertail-coverts spotted white; iris, legs and feet red; bill black. Sexes presumably alike. Immature and juvenile not described. Inhabited swamps, marshes, and wet dark places in forest.
Similar species Very similar in size and plumage to the widespread Spotless Crake (104), of which it is probably an insular derivative (Olson 1973b), and the Henderson Crake (106). Bill proportionately slightly larger and heavier than in congeners, both of which lack white spotting on upperwings. Volant Spotless Crake has head, neck and underparts dark slate-grey; mantle, back, scapulars and inner wing-coverts dark reddish-brown; rump to tail dark brown; marginal coverts and leading edge of outermost primary white; undertail-coverts narrowly barred white; tail longer. Flightless Henderson Crake entirely deep black with slight greyish gloss; lacks white on undertail-coverts; bill blackish, with yellowish-green base and culmen.

VOICE No precise description is available; there is only the often-quoted translation of Kittlitz's remark in Hartlaub (1893): "one hears in these places from time to time its alluring voice resounding".

DESCRIPTION
Adult Plumage largely black, with bluish-grey reflections; chin and centre of throat paler; remiges and rectrices somewhat browner; outer edge of outermost primary dull brownish; lesser upperwing-coverts brownish with white spots; undertail-coverts spotted white. "Underwing with a few inconspicuous white bars" (Mayr 1945). Iris, orbital ring, legs and feet red; bill blackish. Sexes presumably alike.
Immature Not described.
Juvenile Not described.
Downy young Not described.

MEASUREMENTS No measurements available.

GEOGRAPHICAL VARIATION None.

MOULT No information available.

DISTRIBUTION AND STATUS Confined to Kosrae (Kusaie) I in the E Caroline Is. Almost certainly EXTINCT. Following its discovery, and the collection of 2 specimens, by Kittlitz in 1827-28, this crake has not been recorded, despite searches. Kittlitz regarded it as uncommon even in 1828. It was formerly considered sacred by the islanders, and thus apparently remained unmolested, but within 50 years of the arrival of Christian missionaries, and of rats from whaling ships in the 1830s and 40s, it had probably become extinct (Greenway 1967). The only 2 specimens are in Leningrad.

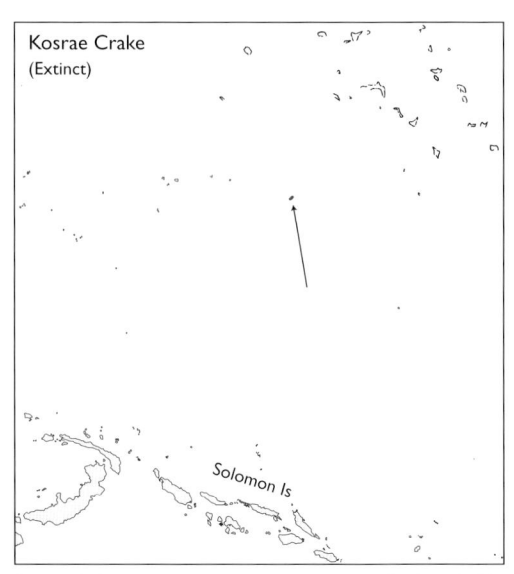

MOVEMENTS None.

HABITAT It occupied a variety of habitats, including swamps and marshes near sea level as well as taro patches and "continually wet, shadowy places in the forest" (Kittlitz, in Hartlaub 1893).

FOOD AND FEEDING Nothing recorded.

HABITS This small crake apparently lived singly. Although it has become a legend among the local inhabitants, nothing else is recorded of its habits (Ripley 1977). Its local names were "satamanot" and "nay-tay-mai-not", meaning "to land in the taro garden", which suggest that the bird may still have had some powers of flight (Fuller 1987). However, measurements of the carpometacarpus from x-rays of the two surviving specimens show that the species must have been flightless (Steadman 1986).

SOCIAL ORGANISATION No information available.

SOCIAL AND SEXUAL BEHAVIOUR No information available.

BREEDING AND SURVIVAL No information available.

106 HENDERSON CRAKE
Porzana atra Plate 31

Porzana atra North, 1908, Henderson Island.

Formerly sometimes assigned to the monotypic genus *Nesophylax*. Forms a superspecies with *P. monasa* and *P. tabuensis*. Monotypic.

Synonym: *Nesophylax ater*.

Alternative names: North's/Henderson Island Crake; Chicken Bird.

IDENTIFICATION Length 18cm. A small, flightless crake; legs well developed for running. Entire plumage deep black; iris and eye-ring red; bill black, base and culmen yellowish-green; legs and feet red, legs mottled black. Sexes alike, but female has yellowish-green on bill (if present) only on culmen; legs plain red to orange. Immature has orbital ring dark; legs and feet brownish-orange. Juvenile greyer than adult on throat and underparts; iris brown; bill, legs and feet black. Inhabits dense forest and thickets; also coconut groves.

Similar species Very similar in size and plumage to extinct Kosrae Crake (105), and also to Spotless Crake (104), which occurs as close as neighbouring island of Oeno and is probably derived from same ancestral stock (Ripley 1977); differs from both congeners in having entirely deep black plumage with no white markings on wings or undertail-coverts, no brown on upperparts, and greenish base and culmen to black bill.

VOICE Information is from Jones *et al.* (1995) unless otherwise stated. The contact call is a nasal *kak*, uttered frequently by paired adults foraging out of sight of each other. This note is repeated in a rapid series as an alarm call to warn chicks or to attract another adult. There is also a loud, continuous churring call, uttered simultaneously as a duet by pair members which are sometimes joined by other family members; the churring is preceded by short, nasal buzzing clucks (see Social and Sexual Behaviour). The birds are most vocal at dawn and dusk, and this species is the first bird to call in the mornings (Bourne & David 1983). Outside the breeding season birds respond less well to taped playback. Adults also give a quiet murmuring call while brooding chicks and as a contact note to chicks, while the soft, nasal buzzing is also given after both copulation and allopreening. Juveniles give soft, almost continuous peeping notes while foraging; in early Apr, some begin to attempt the churring duet with adults.

DESCRIPTION

Adult Entire plumage deep black with slight greyish gloss, or sometimes brownish cast; underwing-coverts and underside of remiges slightly greyer; iris and eye-ring red, eye-ring often blotched black; bill black, yellowish-green at base and along culmen; legs and feet red, legs mottled black. Sexes alike, but female has less yellowish-green on bill (if present, this colour is confined to culmen ridge); eye-ring plain red; legs plain red to orange, darker on anterior surface of toes and tarsus. Only 9 primaries.
Immature Similar to adult but orbital ring dark; legs and feet dusky, brownish-orange.
Juvenile Greyer than adult on the throat and underparts; iris brown; eye-ring, bill, legs and feet black. By early Apr, legs of some birds begin to turn reddish at tarsal joint and bills develop yellow markings (Jones *et al.* 1995).
Downy young Down short; deep velvety black; bill black with paler tip, and paler base of upper mandible (from rear of nostril); legs and feet black.

MEASUREMENTS Wing of 20 males 75-88 (83.2), of 17 females 79-87 (83.4); tail of 11 males 35-44 (39.5), of 12 females 36-43 (40.25); exposed culmen of 16 males 22-24 (22.9), of 15 females 19-22 (20.9); tarsus of 17 males 32-38.5 (35.4), of 16 females 30-36 (33.1). Weight of 10 males 69-87 (80.1, SD 5.3), of 12 females 66-88 (74.9, SD 4.1).

GEOGRAPHICAL VARIATION None.

MOULT Information is from Jones *et al.* (1995). Remex moult is simultaneous and rapid, taking less than 1 month and occurring in Feb-Apr, when a complete postnuptial moult takes place. Postjuvenile moult is partial, involving only body feathers, and is recorded in Feb.

DISTRIBUTION AND STATUS Henderson Island, in the C Pitcairn group. Regarded as VULNERABLE, this crake is confined to a 37-km² raised-reef island with no permanent fresh water, which has been uninhabited by man since at least 1600. The crake population was estimated at c. 6,200 birds in 1992, probably the carrying capacity of the island, as most territories held more than 2 adults (Jones *et al.* 1995). The main predator is the Polynesian rat *Rattus exulans*, introduced by man centuries ago; rats take eggs and chicks but apparently do not pose a threat to the bird's survival and the crakes are very aggressive towards them (Jones *et al.* 1995). However, any inadvertent introduction of a more aggressive predator (e.g. another *Rattus* species) to the island could well result in the crake's decline and extinction, as has happened with many other island rails (Jones *et al.* 1995). The bird is still vulnerable to possible human impacts: in 1982-83 an American millionaire sought to make the island his home, and the proposal was considered seriously by the British authorities before being rejected on environmental and technical grounds (Fosberg *et al.* 1983, Serpell *et al.* 1983, Hay 1986).

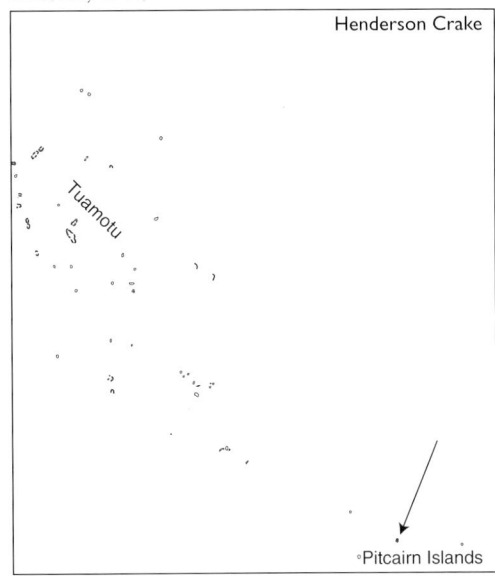

MOVEMENTS Juveniles disperse from natal territories in Mar-Apr (Jones *et al.* 1995).

HABITAT Information is from Jones *et al.* (1995) unless otherwise stated. Thick, species-rich *Pisonia* forest to open, species-poor *Pisonia/Xylosma* forest, and *Timonius* thickets, of the island plateau, and *Pandanus-Thespesia-Argusia* embayment forest on some beaches; also in coconut groves. It is absent only from low vegetation and pinnacled limestone at the exposed S end of the island. It prefers shady understorey of low dense forest with a ground cover of thick leaf-litter (Graves 1992).

FOOD AND FEEDING The diet includes large nematodes, terrestrial molluscs, insects (Coleoptera, adult and larval Lepidoptera, and small unidentified insects), spiders, and eggs of the skink *Emoia cyanura*. It is an opportunistic feeder, taking advantage of seasonal increases in prey species (Jones *et al.* 1995). It forages by turning over leaf-litter with the bill, head-tossing the litter aside, and scratching with the feet; it gleans prey items (including skink eggs) from the underside of leaves.

HABITS This crake is active throughout the day. Bold and curious, it will approach visitors and is liable to sneak up on them from behind; if disturbed, it runs for a short distance before stopping to look back (Bourne & David 1983). During a recent study birds were captured using a tape lure and a handnet, but soon became wary of people who carried a net (Jones *et al.* 1995).

SOCIAL ORGANISATION All information is from the study by Jones *et al.* (1995). Monogamous, with a permanent pair bond, and presumed permanently territorial. Territory sizes in beach forest ranged from 0.28 to 0.42ha; mean territory size in plateau forest was estimated as 0.81-1.13ha; and the overall mean territory size was 1.02ha. Birds live in pairs, or cooperative groups of 3-4 adults (mean group size 2.3 adults): young of the previous brood, and other adults in the family group, help to feed chicks and to protect eggs and chicks from predators. It is not clear whether such additional adults are related to the breeding pair. The presence of a helper had no significant effect on the number of breeding attempts in the territory.

Polyandry was once recorded, a female making 4 nesting attempts with one male and 1 with a male from an adjacent territory. The attempt with the second male was made while the 2 chicks from the 3rd attempt with the first male were 1 month old. Due to the female's double responsibilities, the 3 eggs laid for the second male were incubated less often than normal; 2 were taken by a predator, and the third hatched but the chick soon disappeared. The female returned to her original mate for the final nesting attempt.

SOCIAL AND SEXUAL BEHAVIOUR Pairs often forage close together, usually within 10m of one another (Graves 1992). In the duetting display one bird begins making *kak* calls and runs rapidly, with slightly raised wings and lowered head, through the undergrowth towards its mate; the birds then make buzzing clucks which become the duetting churr (lasting 3.7-15s, mean 9.3s) as soon as they are in physical contact; duetting birds normally face each other with necks extended upwards and bodies in contact, but sometimes stand side by side or at rightangles to each other; juveniles also join in duets (Graves 1992, Jones *et al.* 1995). During the breeding season, duetting is often followed by copulation, which is described by Jones *et al.* (1995). The female walks rapidly in circles c. 0.5m in diameter, her head stretched downwards, her bill pointing at the ground and her wings half open and drooping; the male follows immediately behind for 6-13 circuits (n = 3) before the female stops and holds her tail to one side; the male then mounts and copulates for c. 5s. The birds then stand side by side, preening and allopreening, particularly around the neck and head, and again make the soft, nasal buzzing call before moving off. Allopreening (duration 0.25-2.0 min) often occurs after duetting (Graves 1992). Between breeding seasons pairs remain on the territory, maintaining the pair bond with duetting, allopreening and occasional copulation.

BREEDING AND SURVIVAL Information is from Jones *et al.* (1995) unless otherwise stated. **Season** Breeds Jul-Feb. **Nest** Some nests spherical, c. 20cm in diameter with an opening c. 10cm wide, of shredded *Pandanus* palm leaves, placed up to 30cm above ground in *Pandanus* leaf clump or at base of *Pandanus* trunk; others open-topped, built in low vegetation or up to 60cm above ground in *Asplenium nidus* ferns. Male builds; female works on nest after laying, up to last week of incubation. Also builds 1-5 more lightly constructed roosting nests for use after hatching. **Eggs** 2-3 (mean 2.3, SD 0.5, n = 6); dull olive-green, or beige tinged greenish; closely speckled brown or reddish; size of 8 eggs 33.6-36.5 x 25-25.6 (35.1 [SD 1.0] x 25.2 [SD 0.4]). Eggs laid at daily intervals. **Incubation** 21 days, by both sexes; both adults roost in nest at night, often with young of previous brood; hatching synchronous. **Chicks** Precocial; leave nest soon after hatching; fed and cared for by both parents; female initially spends more time foraging and feeding chicks, male more time brooding; in 3rd week no brooding occurred and both parents spent same amounts of time foraging and feeding; parents helped by young of previous brood and by other adults in the family group; helpers also assist in protecting eggs and chicks from crabs and rats. Young forage actively from 2-3 weeks but are fed by parents throughout development; brood fed on average 13.2 times/h; young fully feathered and capable of independent foraging at c. 1 month, but remain associated with parents for much longer, sometimes being fed even when 15 weeks old. From late Mar to early Apr many juveniles disperse from natal territories, when fights between juveniles observed; but some remain with parents even when adult. Some pairs lay second clutch when young of first clutch are c. 1 month old; also lays replacement clutches after loss of eggs or chicks; in 4 cases laying commenced 2-3 weeks after death of chicks; may make up to 5 consecutive nesting attempts. Assuming mean clutch-size of 2.3, hatching success of observed breeding attempts was 93% (60 chicks from 28 breeding attempts in which young hatched); chick survival not significantly affected by presence of helpers. Of 60 chicks hatched, 19 survived to 1 month old (0.95 chicks per pair). Polynesian rat *Rattus exulans* is a nest predator, as also probably is Pacific rat *R. kiore*, but adult crakes very aggressive to rats and often chase them; chicks vulnerable to predation, and to heavy rainfall, in first week of life. Hermit crabs (species not given) also take chicks. **Survival** Based on small sample, annual adult survival at least 43%; recruitment rate not known, but probably compensates easily for annual losses, so that population is stable; many juveniles apparently present during Apr-May dispersal phase.

107 YELLOW-BREASTED CRAKE
Porzana flaviventer Plate 31

Rallus flaviventer Boddaert, 1783, Cayenne, French Guiana.

Does not closely resemble any other *Porzana* crake; sometimes placed in genus *Poliolimnas* (Olson 1970), a separation which may be appropriate. Also see *Porzana cinerea* (108). Five subspecies recognised.

Synonyms: *Aramides/Poliolimnas flaviventer; Porzana/Ortygometra/Hapalocrex flaviventris; Laterirallus/Crybastes gossii/gossei; Rallus superciliaris/minutus; Ortygometra/Crex/Corethrura/Erythra/Porzana minuta.*

Alternative names: Yellow-bellied/Yellow-breasted Rail; Twopenny Chick.

IDENTIFICATION Length 12.5-14cm. Very small crake, with overall tawny-buff appearance, boldly barred flanks, and long toes which enable it to walk on floating vegetation. Crown to hindneck blackish-brown; upperparts black with prominent brown streaks (varying racially from tawny-brown to sandy-buff or ochraceous) and narrow white streaks; upperwing-coverts plainer and browner, with less black and with white spots and bars; white to buff leading edge to wing. Distinctive facial pattern of dark line through eye, and broken white line above and behind eye; sides of head ashy-buff to cinnamon-buff; chin and throat white; underparts whitish, foreneck and breast washed cinnamon-buff to buff (racially variable); sides of lower breast, and flanks to undertail-coverts, boldly barred black and white; white on undertail-coverts sometimes washed rufous-buff. Iris brown; bill black; legs and feet yellowish. Sexes alike. Immature similar to adult but flank barring continues to sides of breast. Juvenile has indistinct dusky barring on neck and breast. Inhabits marshes, swamps and rice fields.

Similar species Distinctive facial pattern diagnostic, and unique among New World rails. Only other sympatric crake of similar size and overall colour is Ocellated Crake (16), which has more extensive white markings (with no dark streaking) on upperparts, unbarred flanks and undertail-coverts, and red legs and feet. Slightly larger Yellow Rail (14), which occurs in C Mexico, has white secondaries, lacks white streaks on upperparts; breeding male has yellow bill.

VOICE A low, harsh, rolled or churring *k'kuk kurr-kurr* or *je-je-je-jrrr*; a plaintive, squealing *kreeihr*, a single or repeated *kreer* or *krreh* reminiscent of the Sora (98) (Stiles & Skutch 1989, Howell & Webb 1995b); a *tuck* of medium pitch and strength; and a high-pitched, softly whistled *peep* or clear *kleeer*. Hilty & Brown (1986) describe the call as a hoarse, slightly downscale *zeee-eee-eee-eee* (presumably equivalent to the *kreer* mentioned above), given at c. 1 note/s, and a loud, ringing, almost scraping *clureéoo* (presumably equivalent to the *kreeihr* described above); Bornschein *et al.* (1997) describe a high-pitched *fi fi fuu* or *fi fuu* call and a probable alarm call *fii*. Hardy *et al.* (1996) give 6 calls: a high-pitched, harsh, repeated *krrr* (which they term the *shurr-shurr* call); a *kreer* or *kreeo* (presumably the Sora-like call referred to above); a high-pitched, scratchy *keek*; a plaintive *kleer* or *keer*; a high-pitched *kweep* or *weep*; and a short *ki* or *ke*.

DESCRIPTION P. f. gossii
Adult Forehead to nape very dark blackish-brown; hindneck sometimes paler, more fuscous; mantle to rump, and scapulars, dark brown, feathers fringed tawny-brown, more cinnamon-buff on scapulars and more rufescent on rump; feathers of scapulars, mantle and back with white shaft-stripes broadly bordered blackish-brown to black; feathers of back and rump with relatively indistinct dusky centres and some with short white terminal shaft-streaks or median spots; uppertail-coverts blackish with scattered small white bars; rectrices dark brown-black, fringed brown. Upperwing-coverts like scapulars but plainer and paler, medians and greaters with blackish and white areas much reduced, and white markings mostly spots and bars; primary coverts sepia-brown; alula, primaries and secondaries olive-brown, edged paler brown; outer edge of first primary white to buff; tertials blackish, fringed brown to cinnamon-buff; marginal coverts white to buff, barred dusky olive-brown; axillaries white, barred dusky blackish; underwing-coverts white to buff with a few dusky olive-brown bars. Lores to eye blackish, this colour extending in a variably distinct streak behind eye across upper ear-coverts; prominent white supraloral streak extends to just behind and above rear of eye, separated by small, diagonal black line from shorter, buff-tinged white streak running from top rear edge of eye across top of ear-coverts. Ear-coverts, sides of head and sides of neck tawny-buff to buffy hair-brown (ashier), merging into white chin and upper throat. Underparts whitish, foreneck and breast washed pale to rich tawny-buff; sides of lower breast, and flanks to undertail-coverts, regularly barred dark blackish-brown and white; white bars on undertail-coverts sometimes washed rufous-buff; thighs white, less prominently barred blackish. Iris brown to reddish-brown; bill pale bluish-black to black; legs and feet yellowish. Sexes alike.

Immature Similar to adult but flank barring continues to sides of breast (*woodi*).

Juvenile Has indistinct dusky barring on neck and breast.

Downy young Down is black (R. M. Fraga *in litt.*).

MEASUREMENTS A large race; almost as large as nominate. 16 males, 11 females: wing (chord) of male 63.5-71 (67.6), of female 65-74 (68.5); tail of male 26.5-33.5 (30.6), of female 27.5-34 (31); exposed culmen of male 16-18 (17.2), of female 15.5-17.5 (16.3); tarsus of male 20-24 (22.7), of female 20-24.5 (22.5) (Ridgway & Friedmann 1941).

GEOGRAPHICAL VARIATION Races separated on size and colour shades.

P. f. gossii (Bonaparte, 1856) – Cuba and Jamaica. See Description.

P. f. hendersoni Bartsch, 1917 – Hispaniola and Puerto Rico. Smaller than *gossii*; feather fringes of upperparts paler, more sandy-buff, than in other races. Measurements of 3 males, 5 females: wing (chord) of male 62, 62.5, 63, of female 61-64 (62.5, SD 1.4); tail of 1 male 30, of 2 females 29; exposed culmen of male 15.5, 16, 16.5, of female 15-16 (15.5, SD 0.5); tarsus of male 21, 21, 22.5, of female 22-22.5 (22.4, SD 0.2) (Ridgway & Friedmann 1941, Dickerman & Warner 1961).

P. f. woodi van Rossem, 1934 – S Mexico (Michoacán, Guerrero, Puebla, Veracruz and Chiapas), S Guatemala (La Avellana), El Salvador (L Olomega), Nicaragua (R San Juan) and NW Costa Rica (Guanacaste). Smaller than *gossii* but similar in overall colour; feather fringes of upperparts darker, more ochraceous and less sandy-buff than *hendersoni*. Two males had dark red iris, dark olive to blackish-olive bill, and pale, dull yellow legs and feet (Dickey & van Rossem 1938). Measurements of 7 males, 6 females: wing (chord) of male 61-65.5 (62.5), of female 63-66 (63.9); tail of 2 males 26.5, 29; exposed culmen of male 15-17 (16.3), of female 14.5-16 (15.25); tarsus of male 21-24 (22.3), of female 21.5-22.5 (22.15) (Ridgway & Friedmann 1941, Dickerman & Warner 1961). Weight of 6 males 24-28 (26.1), of 10 females 20.2-28.3 (23.85) (Dickerman & Warner 1961, Orians & Paulson 1969, Dickerman 1971, 1975).

P. f. bangsi Darlington, 1931 – N and E Colombia (Santa Marta and lower and middle Magdalena V, and E of Andes in W Meta). Upperparts darker than in other races. Similar to nominate on upperparts, but sides of neck and breast much paler buff, resembling *gossii* but having upperwing-coverts, back and rump markedly blacker and more variegated with white. Measurements of 6 males, 3 females: wing (flat) of male 66-67 (66.6), of female, 60, 66, 70; exposed culmen of male 16-18 (16.6), of female 15, 17, 18; tarsus of male 20-22 (21.3), of female 20, 21, 22 (Blake 1977). ***P. f. flaviventer*** (Boddaert, 1783) – Panama (E to E Panamá Prov. and Coiba I), NW Colombia (Sinú and middle Cauca V), Venezuela (Mérida, Portuguesa, Carabobo, Distrito Federal, Miranda and NE Bolívar), Trinidad and the Guianas, and S to N, EC and S Brazil (Amazonas at the R Negro, and from Amapá and Pará E to Minas Gerais and S to Rio de Janeiro and Rio Grande do Sul), E Bolivia (Santa Cruz), Paraguay, Uruguay (Rocha, Colonia and Maldonado) and N Argentina (Tucumán, Corrientes, Córdoba, Santa Fe, Entre Ríos and N Buenos Aires). Recent records from NW Ecuador (N Manabí; Ridgely & Greenfield in prep.) possibly this race. The largest race; sides of neck and breast deep buff to bright, deep cinnamon-buff, darker than in other races. Plate 31 shows individual variation in plumage. Bill described as dark neutral grey; sides and base of lower mandible dull greenish-olive (Wetmore 1965). Measurements of 14 males, 13 females: wing (chord) of male 64-73 (69), of female 66-79 (70.7); exposed culmen of male 15-17 (16.35), of female 14.5-15.5 (14.9); tarsus of male 22-25.5 (23.7), of female 20.5-25 (22.8) (Dickerman & Warner 1961). 11 unsexed: wing 66-71 (68.2, SD 1.7); tail 28-33 (30.6, SD 1.7); culmen to base 16-18.5 (17.5, SD 0.8); tarsus 22-24 (22.7, SD 0.8). Weight of 9 males 22-30 (26), of 2 females 24, 28 (Haverschmidt & Mees 1995).

MOULT No information available.

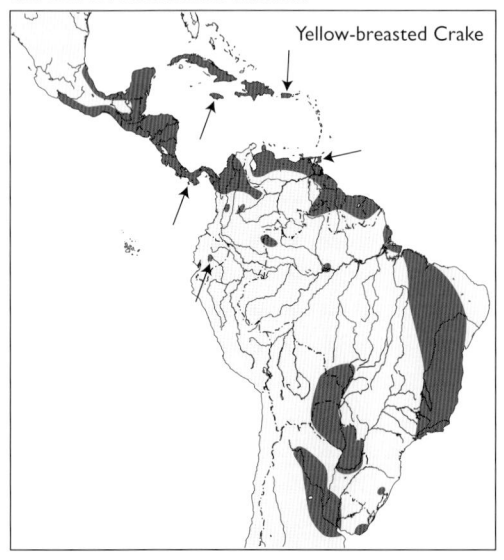

DISTRIBUTION AND STATUS Greater Antilles, S Mexico to Panama, Colombia and Ecuador to the Guianas, Trinidad, and Brazil to Bolivia, Paraguay, Uruguay and N Argentina. It is undoubtedly more widespread than is known: in Central America it possibly occurs in Honduras and Belize (see distribution map), although reports from Belize require confirmation (Howell & Webb 1995b), while it probably occurs in SE Guainía, Colombia, and in Chaco, Argentina (Hilty & Brown 1986; Contreras *et al.* 1990). The first records from Ecuador come from W of Chone, N Manabí, where a few sightings were made from Nov 1996 to Feb 1997 (Ridgely & Greenfield in prep.). In N Argentina it has recently been recorded from Estancia San Juan Poriahu and also from near Loreto (Corrientes) (R. M. Fraga *in litt.*: see Breeding). It is not globally threatened, and was formerly regarded as locally abundant in Cuba (*gossii*), very local but sometimes common in Mexico (*woodi*), rare in Haiti (*hendersoni*), and locally common in Colombia, Trinidad, El Salvador and Panama (nominate) (Wetmore & Swales 1931, Dickey & van Rossem 1938, Dickerman & Warner 1961, Wetmore 1965, Ripley 1977, Ridgely 1981, Hilty & Brown 1986). The race *woodi* is currently regarded as frequent to uncommon but local in Mexico (Howell & Webb 1995b) and locally common in Costa Rica (Stiles & Skutch 1989), while *hendersoni* is uncommon in Puerto Rico (Raffaele 1989). The nominate race is local on the littoral plain in French Guiana (Tostain *et al.* 1992) and rare in Argentina (Canevari *et al.* 1991, Narosky & Di Giácomo 1993); it is apparently rare in Paraguay, where there are many records W of the R Paraguay but only 3 to the E (Hayes 1995), and is occasional in Uruguay, where it has been recorded only twice, in 1960 (Rocha) and 1995 (Maldonado) (Arballo & Cravino in press).

MOVEMENTS It is normally regarded as sedentary, and no definite movements are recorded, but it is seen only in Mar-Jul at Hacienda Corocora, W Meta, Colombia (Hilty & Brown 1986). In Costa Rica it probably makes local movements associated with changing water levels (Stiles & Skutch 1989). In the Ciénaga swamps, Colombia, it probably retreats into permanently flooded *Typha* in the dry season when the area of floating grass and weeds is much reduced (Darlington 1931); however, in an apparently similar situation in Trinidad it did not move into taller vegetation, unlike its larger congener, the Sora (Gochfield 1972). One was found at night under a village street light in Veracruz (Mexico), in May 1964 (Dickerman 1971).

HABITAT Dense vegetation of freshwater marshes and the grassy edges of marshes, lakes and ponds; also flooded fields (including rice fields), and areas of floating vegetation such as water hyacinth *Eichhornia* (Dickey & van Rossem 1938) in choked pools and lagoons; it occurs less frequently in swamps. In SE Brazil, Bornschein *et al.* (1997) found it in a 2.8-ha marsh in an abandoned rice plantation with areas of low and sparse wet vegetation, shrubs and grass, as well as in *Typha* reedmarsh and in marshes with dense closed *Cladium mariscus* and *Fuirena robusta*. In large wetlands it usually occurs in shallowly flooded edge vegetation. Typically occurring in flooded areas and on floating or semisubmerged aquatic vegetation, it occupies more of an aquatic niche than most other crakes and rails. It rarely occurs in saltwater habitats; at Barra de Icapara, Paraná, SE Brazil, 1 specimen in the São Paulo Museum was taken from a pair seen "low in the mangroves with pairs of Rufous-sided Crakes [28] when the tide was high" (Bornschein *et al.* 1997). It occurs from sea level up to 2,500m.

FOOD AND FEEDING Small gastropods, Arachnoidea, and insects including Diptera larvae, Coleoptera, Hymenoptera (Formicidae: Myrmicinae) and Orthoptera. It also takes some seeds, including grass seeds. It forages among emergent plants, on muddy shores, on floating plants and in small open areas of the marsh, takes prey from the water, and emerges from cover in the early morning and evening to feed at marsh edges.

HABITS Often regarded as essentially crepuscular. Although they normally remain in cover during the day, these birds will feed in the open in the early morning and, more briefly, towards sunset. They may not be particularly furtive at any time, possibly just difficult to see because of their small size and the dense nature of the vegetation which they inhabit (e.g. Hilty & Brown 1986). They walk and climb through emergent grass and reeds, perching frequently, and they run across floating plants, sometimes flapping the wings to prevent themselves from sinking; they climb easily in tangled growth. They flush easily (unless with chicks – see Social and Sexual Behaviour); in flight the legs dangle and the head droops, and flights are usually short, with a feeble flight action; the birds can seldom be flushed a second time. In matted grass they appear to fly from a perch high in the vegetation (Dickerman & Warner 1961), as does the Red-chested Flufftail (3). Soon after dawn they often climb in the top of flooded grass clumps to rest briefly in the early morning sun (Wetmore 1965).

SOCIAL ORGANISATION These birds appear to congregate in small groups in limited areas (Wetmore 1965). Presumably monogamous, as 2 adults have been seen accompanying chicks (R. M. Fraga *in litt.*).

SOCIAL AND SEXUAL BEHAVIOUR The larger White-throated Crake (31) is dominant and will drive away Yellow-breasted Crakes which approach too closely (Wetmore 1965). Adults with chicks allowed rather close approach (R. M. Fraga *in litt.*).

BREEDING AND SURVIVAL Season El Salvador, male in breeding condition Aug; Costa Rica, Jul; Puerto Rico, Mar; Trinidad, male in breeding condition Feb; Colombia, 'immature' Feb; Argentina (Corrientes), pair with 2 small chicks and 1 adult with 4 small chicks, 4 Oct 1997 (R. M. Fraga *in litt.*). **Nest** Cup-shaped and loosely built, of grass or other aquatic vegetation, situated near the ground among water plants or marsh grass, or on floating vegetation such as *Pistia*. **Eggs** 3-5; pale cream to whitish, lightly spotted brown or lavender. **Chicks** Apparently tended by both parents (R. M. Fraga *in litt.*).

108 WHITE-BROWED CRAKE
Porzana cinerea Plate 31

Porphyrio cinereus Vieillot, 1819, no locality (designated Java).

On the basis of bill structure and face pattern, which render it distinct from *Porzana* crakes, sometimes placed in the genus *Poliolimnas* with *Porzana flaviventer*, which shows similar characters (Olson 1973b), but similarities are superficial, and differences between the two species are very marked (Mees 1982). Until recently several races commonly recognised, but differences slight, and in most characters subject to overlap (see Geographical Variation). Monotypic.

Synonyms: *Porzana/Poliolimnas/Porphyrio cinereus*; *Erythra/Porzana/Ortygometra quadristrigata/leucophrys/cinerea*; *Rallus quadristrigatus*; *Corethrura sandwichensis/quadristrigata/mystacina*; *Gallinula/Ortygometra leucosoma*; *Rallus/Corethrura tannensis*; *Crex quadristrigata*; *Ortygometra ocularis/superciliaris*; *Zapornia sandwichensis*; *Gallinula superciliosa/mystacina/superciliaris*.

Alternative names: Ashy/Grey-bellied Crake; White-browed (Water) Crake/Rail.

IDENTIFICATION Length 17(15?)-21.5cm; wingspan c. 27cm. Small, slim-bodied crake with relatively long legs and toes. In fresh plumage forehead and crown grey, becoming black with wear; hindneck grey-olive; upperparts and upperwings blackish-brown with grey-olive to buff-brown streaks (feather fringes), least from rump to tail. Prominent face pattern: broad, diagonal blackish stripe from base of bill through eye to rear crown, bordered by triangular white stripe from base of upper mandible to above eye, and narrow suborbital stripe extending over ear-coverts; rest of sides of head, neck, and sides of breast and belly grey (extent and depth variable – sometimes predominantly white), grading to white on chin, throat, and (variably) centre of breast and belly. Flanks olive-grey to darkish brown, sometimes tinged rufous; undertail-coverts buff to cinnamon-buff, or browner. Narrow white leading edge to wing visible in flight. Iris and eye-ring red; bill olive-yellow to orange-brown, upper mandible red at base; legs and feet pale green to grey-olive. Sexes alike. Immature like adult. Juvenile shows adult plumage pattern but less striking: grey and black on head replaced by brown, and white facial stripes tinged yellow-brown, making facial pattern less distinct; grey of neck, breast and flanks replaced by buff to pale brown; bill has thin orange band at base; legs and feet olive to blue-grey. Inhabits fresh to saline palustrine wetlands with floating vegetation; also mangroves, grass, thickets and forest.
Similar species Similar in size to Australian Crake (97), but all ages are easily distinguished from sympatric crake species by striking diagnostic face pattern; also has pale, unpatterned underparts, and relatively long legs and toes.

VOICE Most information is taken from Marchant & Higgins (1993). The common call is a loud, nasal, chattering *chika*, rapidly repeated 10-12 times, made by both members of a pair as a recognition call (e.g. when pair members meet after separation), and sometimes by several pairs in response to disturbance. It is given frequently, and is audible at several hundred metres. Spontaneous calling by a pair is not known to stimulate other pairs to call. Six birds were reported to feed and chatter continuously, and birds also give an occasional sharp, loud *kek-kro* while feeding. Other calls include a quiet, repeated *charr-r* of alarm, plaintive cries given during the distraction display, squeaky cries, and nasal grunts, and undescribed conversational calls audible for only a fewm. Coates & Bishop (1997) also record a repeated, nasal *hee* note, often given simultaneously by several birds; this may be equivalent to the "high-pitched thin reedy piping which has a ventriloquial quality" of King *et al.* (1975). Young utter incessant thin nasal squeaking notes (Ripley 1977).

DESCRIPTION
Adult Feathers of forehead and crown blackish with broad grey fringes, this region appearing predominantly grey when fresh and blackish with variable grey streaking when

worn; nape similar, but grey fringes broader, seldom fully lost with wear. Upper hindneck grey, or mixed grey and greyish-olive; lower hindneck greyish-olive; mantle feathers greyish-olive, with variable darker olive-brown central wedges often exposed with wear, when remaining fringes discolour to olive-brown; scapulars and back blackish-brown with broad grey-buff to pale brown feather fringes giving streaked effect (brownest in worn birds); rump, uppertail-coverts and tail dark brown, feathers narrowly fringed brown. Alula, primary coverts and remiges dark brown; outer webs of outermost primary (P10) and longest alula feather white; primaries often narrowly fringed greyish-olive to pale brown; tertials and their coverts as scapulars; other coverts brownish-olive, fringed grey-olive, and with basal two-thirds grey-brown, this colour exposed towards carpal joint; marginal coverts white. Underside of remiges pale grey; axillaries pale grey-brown to grey, tipped whitish; underwing-coverts grey, greaters and outer medians tipped white, others tipped pale brown. Broad diagonal blackish-brown stripe from gape and base of lower mandible to eye, continuing to join dark hindcrown; contrasts strongly with short triangular white stripe, broad at base of upper mandible and tapering to narrow above centre of eye, and long, narrower, white suborbital stripe which continues around top of ear-coverts. Ear-coverts grey to pale grey, grading ventrally to white on chin and throat and posteriorly to greyish-olive of lower hindneck; sides of neck pale grey; foreneck whitish, often washed pale grey. Breast and belly white, grading to pale grey at sides (extent and depth of grey variable, colour sometimes extending to centre of breast, but some birds predominantly white); flanks olive-grey to darkish brown, sometimes tinged rufous; thighs white and grey-brown; vent and undertail-coverts buff to cinnamon-buff, or even darkish brown. Iris red; orbital ring dark pink to red; lower mandible, and tip and cutting edge of upper mandible, olive-yellow to orange-brown; rest of upper mandible dark olive-brown with pale red band at base (band may also be orange); legs and feet pale green to greyish-olive, tinged yellow-olive at front. Sexes alike; female may average slightly smaller.

Immature Apparently like adult.

Juvenile Has adult plumage pattern but forehead, crown, nape and hindneck look brown, feathers being dark brown with broad paler brown fringes which grade to greyish-olive on lower hindneck; eyestripe grey-brown, and white facial stripes tinged yellow-brown, making facial pattern less distinct; ear-coverts and sides of neck pale brown, grading through buff to white on chin and centre of throat; grey of breast and flanks replaced by buff to pale brown. Iris red, orange or brown; bill as adult but with thin orange band at base of upper mandible; legs and feet olive, blue-green or light blue-grey, lacking yellow tinge of adult.

Downy young Downy chick coal-black; also described as grey (Ripley 1977); a large chick, with emergent body feathers and remiges, had black down with glossy pale grey tips; bill black, tinged brown at base of lower mandible and on cutting edge, with broad pink-white saddle between nostrils; iris dark brown; narrow eye-ring black; legs and feet pale grey, with dark centres to scales giving dark bands on front of tarsus (Marchant & Higgins 1993).

MEASUREMENTS Wing of 61 males 90-104 (95.5) and 43 females 87-97.5 (92.7) (Java and Sumatra), of 9 males 93-102 (98.2) and 10 females 92-102 (97.7) (Sulawesi), of 5 males 90-102 (93.7) and 10 females 89-98 (91.5) (Moluccas), of 9 males 88-101 (94, SD 4.4) and 11 females 88-99 (93.4, SD 2.5) (Philippines), of 9 males 88-101 (94, SD 2.3) and 11 females 88-99 (93.4, SD 2.5) (Australia and New Guinea); tail of 61 males 41-52 (46.6) and 43 females 39-54 (45.2) (Java and Sumatra), of 9 males 41-48 (44.9) and 10 females 39-47 (43.8) (Sulawesi), of 5 males 43-54 (47.25) and 10 females 39.5-44 (41.75) (Moluccas), of 6 males 39-48 (42.7, SD 3.3) and 7 females 35-48 (42.2, SD 4.1) (Philippines), of 5 males 40-51 (45.6, SD 4.5) and 10 females 40-54 (46.5, SD 2.5) (Australia and New Guinea); exposed culmen of 61 males 18-22.5 (20.8) and 43 females 17-22 (20) (Java and Sumatra), of 9 males 20-24 (22.3) and of 10 females 19-23 (21.5) (Sulawesi), of 5 males 20-21.5 (21) and 10 females 17-21.5 (18.75) (Moluccas); culmen to base of 6 males 25-28 (26.3, SD 1.2) and 7 females 23-25.5 (24.4, SD 0.8) (Philippines), of 3 males, 24, 24.5, 25, and 4 females 22.5-24 (23.5, SD 0.7) (Australia and New Guinea; in BMNH); tarsus of 61 males 32-39 (35.4) and 43 females 30-37 (33.7) (Java and Sumatra), of 9 males 36-38 (37.1) and 10 females 33-39 (36.2) (Sulawesi), of 5 males 32-35.5 (34.3) and 10 females 28-34.5 (32.5) (Moluccas), of 6 males 36.5-41 (39.25, SD 1.7) and 7 females 35-40 (37.1, SD 1.9) (Philippines), of 5 males 34-36 (34.6, SD 0.9) and 11 females 30.5-33.5 (31.7, SD 1.1) (Australia and New Guinea). Weight of 11 unsexed adults (Australia) 43-62.5 (52.1, SD 6.8), of 1 female from ship at sea 32, of 1 adult from Booby I 40 (Marchant & Higgins 1993); of 2 males (New Guinea) 60 (Ripley 1964).

GEOGRAPHICAL VARIATION As in many rails which are widely distributed on islands, geographical variation is not well defined. Until recently several races were commonly recognised, notably *brevipes*, *ocularis* (including *collingwoodi*), *micronesiae*, *leucophrys* (including *minimus*), *meeki*, *moluccanus* and *tannensis*, but differences between them are slight, not well defined, in most cases subject to overlap and in some affected by wear, and the advisability of recognising subspecies is questionable (e.g. Mees 1982, White & Bruce 1986, Marchant & Higgins 1993). Characters formerly used to separate subspecies include: amount of grey on top of head; colour of feather fringes, and extent of dark streaking, on upperparts; overall colour of upperparts and underparts; length of tarsus; and proportions of bill (shorter and heavier in the extinct putative race *brevipes*) – see e.g. Hartert (1924b), Mayr (1949), Ripley (1977).

MOULT Primary moult is simultaneous and is recorded from the Philippines (Luzon) in Oct. In Australian specimens, wear suggests that moult often occurs in the wet season: birds with very worn wings were collected in Nov-Jan, and with fresh or slightly worn wings in Jan-Feb; birds with worn wings in May and fresh wings in Jul suggest that moult timing may vary; body moult is recorded in Jan (Marchant & Higgins 1993). Postjuvenile moult probably begins shortly after fledging, and is partial, beginning on the upperparts, flanks and breast and ending on hindcrown and nape (Marchant & Higgins 1993).

DISTRIBUTION AND STATUS C and S Thailand; peninsular Malaysia (including Singapore and Penang); Cambodia; Borneo, Sumatra, Java (including Kangean Is) and Bali; Philippines (Basilan, Batan, Bohol, Bongao, Calayan, Catanduanes, Cebu, Dinagat, Guimaras, Leyte, Luzon, Marinduque, Mindanao, Mindoro, Negros, Palawan, Panay, Sanga Sanga, Siargao, Sibutu, Sibuyan, Simunul, Siquijor, Tawitawi, Ticao); Sulawesi including Talaud Is, Tanahjampea and Sula Is (Taliabu and

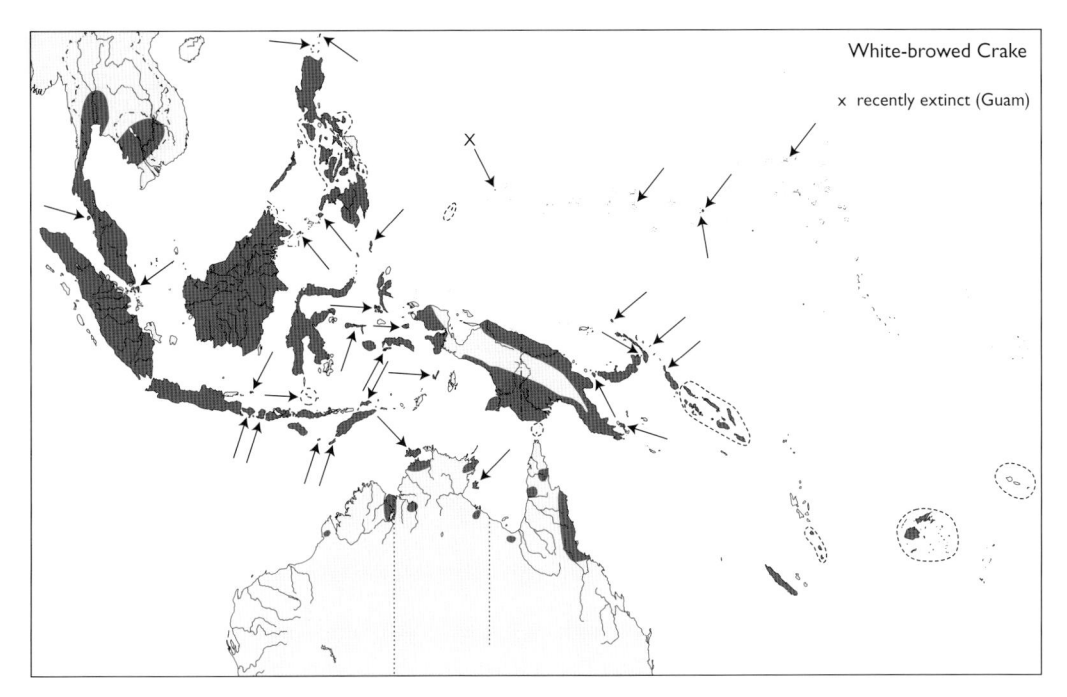

Mangola); Moluccas, from Ambon, Bacan, Buru, Halmahera, Kai Is (Kai Kecil) and Seram; Lesser Sundas (Flores, Lombok, Roti, Sawu, Sumba, Sumbawa, Timor, Wetar); New Guinea (virtually throughout in lowlands, locally at higher altitudes); Misool I; N Australia, from NE Western Australia (W to Kimberleys) to Queensland SE to Townsville; Bismarck Archipelago (Umboi, New Britain, New Ireland, Duke of York, New Hanover, Emira, Mussau, and Lihir Is); D'Entrecasteaux Archipelago (Fergusson); Bougainville and Buka; Solomon Is; Mariana Is (Guam, where recently extinct); Palau Archipelago (Koror, Babelthuap, Peleliu); Caroline Is (Yap, Ulithi[?], Truk, Pohnpei); Marshall Is (Bikini); New Caledonia; Vanuatu (Erromango, Tanna, Gaua); Fiji Is (Viti Levu, Vanua Levu, Taveuni, Kandavu, Gau, Ovalau); Samoa Is (Savai'i, Upolu); formerly also Iwo (= Volcano) Islands (Iwo-jima, Minami-iwo-jima).

Not globally threatened. It is less shy and elusive than most rails and was formerly regarded as locally fairly common to fairly abundant throughout most of its range, with the exception of Micronesia (Marianas, Palau, Caroline and Marshall Is), New Caledonia and Lifu, Vanuatu, Fiji and Samoa, where it was uncommon to rare (Ripley 1977, Pratt et al. 1980; Reed 1980). The putative form *brevipes*, confined to the Volcano Is (S of Japan), is generally regarded as having been extinct since 1911 (Greenway 1967), although it was said to have been observed in 1924/25 (Momiyama 1930); extinction was probably caused by introduced rats and cats (Momiyama 1930, Day 1989). The species apparently disappeared from Guam in the 1970s concurrent with the draining and development of many of the island's freshwater wetland habitats (Jenkins 1983). It is currently widespread and locally common in the Greater Sundas, Philippines, Sulawesi and Flores, and local and uncommon in Vanuatu (Dickinson et al. 1991, Bregulla 1992, MacKinnon & Phillipps 1993, Coates & Bishop 1997). Its status is uncertain in Australia: it is local and apparently not generally common, but it will occupy artificial wetlands, and areas grazed by livestock (see Habitat section) (Marchant & Higgins 1993). It is uncommon in Thailand (Lekagul & Round 1991), and locally common in peninsular Malaysia and Singapore (C. R. Robson *in litt.*) and in Cambodia (Mundkur et al. 1995). Its current status elsewhere is unclear but has possibly remained largely unchanged. It is accidental to the Marshall Is (Bikini) whence there are no recent records (Amerson 1969).

MOVEMENTS Imperfectly known. It is considered resident over much of its range but is probably partially migratory in some regions. It is regarded as a wet-season migrant in N Australia: although recorded throughout the year, more records occur in the summer months and birds are apparently absent from some areas in the winter; a regular migration along the wetlands of Cape York has been suggested, and it occurs on islands in the Torres Strait, where 112 birds struck Booby Island lighthouse from Dec to Jun in 1975-76 (Draffan et al. 1983, Stokes 1983, Marchant & Higgins 1993). On Viti Levu, Fiji Is, the remains of 2 birds were found at a Peregrine *Falco peregrinus* feeding post in Dec, the crakes possibly having been taken over the sea which would indicate local migration in that season (F. Clunie in Ripley 1977). It is a vagrant to Bikini, Marshall Is.

HABITAT Well-vegetated coastal and terrestrial wetlands, both freshwater and saline, especially those with abundant floating vegetation (e.g. waterlilies); the habitat may be permanent, seasonal or ephemeral. It inhabits swamps, marshes, wet grasslands, creeks, rivers, pools, inundations, lakes, dams, sewage ponds, and rank vegetation bordering streams. Occupied vegetation includes sedges (e.g. *Eleocharis*), reeds and rushes, *Sesbania* and mangroves. It uses flooded areas, and has been observed among flood debris (Marchant & Higgins 1993). It also occurs in grasslands and agricultural areas, including grazed land, rice fields, taro *Colocasia* patches, overgrown ditches, and thickets; it is recorded from mature or degraded forest, rank secondary vegetation away from water, and palm groves; it is found in areas grazed by livestock and buffalo *Bubalus bubalis*, and it has bred near a dam used for recreation (Marchant & Higgins 1993). It may occur in habitat favoured by snipe *Gallinago* (Glenister 1951). The extinct Volcano Is population occupied grass, thickets, dense bush and forest, as well as brackish swamps and marshes (Brazil 1991). Breeding habitat includes vegetated watercourses and swamps, agricultural areas and sewage farms (Marchant & Higgins 1993). It is normally recorded from lowlands; in New Guinea it occurs up to 1,350m, and was once found at 1,850m at which altitude it may have been a straggler (Diamond 1972b).

FOOD AND FEEDING Earthworms, leeches, gastropods, slugs, adult and larval insects (including Coleoptera), water spiders, frog spawn, and small fish; also seeds and leaves of aquatic plants. It is described as an omnivorous feeder, taking mainly vegetation and insects (Khobkhet 1984). It forages at mud patches and along the margins of watercourses, both in and out of cover; it frequently forages on floating vegetation, dashing around in a stop-start manner; it also catches flying insects. It takes food items from the water surface while walking in flooded reeds and often while swimming, and it also floats quietly with neck extended, picking up insects with short, sudden thrusts of the bill. Occasionally it runs up marsh vegetation or along branches. One bird emerged from adjacent dense cover for periods of 30-60 s to forage at an adjacent 3-m wide ditch, walking on floating waterlily leaves; it sometimes vigorously trampled a leaf for 1-2s, causing the leaf to submerge, when the bird pecked intensively at prey items which appeared in the swirling water above the leaf (Möller 1992).

HABITS Birds are active throughout the day, particularly so in the early morning and evening, and during overcast weather, when they are often fairly conspicuous as they forage in the open. They are often vocal and, although they are not as shy as most other crakes, they are more often heard than seen. They are wary, and run to cover, flicking the tail, at the slightest hint of danger; they will fly rather than swim; they often run up branches and roots, watch, and call (North 1913); they perch on the broad leaves of taro (Pratt *et al.* 1987). One pair, when approached, submerged in water up to their necks and then dived and swam or crawled along the vegetation under water (Boekel 1980). They walk and run, jacana-like, on lily-pads and other floating aquatic vegetation, sometimes fluttering agilely when the vegetation cannot support their weight. They swim well and frequently forage while swimming. They have been observed roosting during the day on platform-like *Sesbania* vegetation with Baillon's Crakes (94) and Buff-banded Rails (46), venturing out to feed at dusk and in the early morning (Mason & Wolfe 1975).

SOCIAL ORGANISATION Monogamous; may remain paired outside breeding season, and are possibly permanently territorial. On Bougainville, 8 pairs were counted in c. 1km of lagoon edge (Givens 1948).

SOCIAL AND SEXUAL BEHAVIOUR Copulation has been observed once; 1 bird chased another for 30-60s, then both faced each other, bill to bill, in a semi-upright position (both birds were very noisy throughout), 1 bird then mounted the other and made movements suggesting copulation for c. 5s, followed by 10-20s of cowering, and what appeared to be mutual pecking or biting (T. Aumann in Marchant & Higgins 1993). Both parents will give a broken-wing distraction display at the nest, and another distraction display in which they splash water with the wings (Marchant & Higgins 1993).

BREEDING AND SURVIVAL Season Malay Peninsula, Sep, downy young Feb and Aug; Borneo, Apr-Jun; Philippines, Jul-Aug, and downy young Oct; New Guinea, Dec-Mar, May, chicks Jun and Oct-Nov; Solomon Is, Jan, Jun-Jul, Sep; Vanuatu, mostly Nov-Feb, possibly Sep-Mar; Australia, Jan-May, also small young Sep-Nov, N Queensland; Volcano Is, May 1904. **Nest** A saucer-shaped platform of rushes, coarse grass or herbage, lined grass and other fine material; built on, or up to 1m above, ground, on trampled blades of tussock, or in grassy marsh vegetation 13-45cm above water which may be 50cm to 1.2m deep; once in fork of mangrove tree; growing vegetation often woven into a thin canopy. External diameter c. 15cm, internal diameter c. 7cm, depth c. 3.5cm; built up to 1m above ground; runway of reed stems often leads from ground to nest rim. **Eggs** 3-7 (usually 4); oval, rounded oval or elliptical; smooth and glossy; greyish-white, olive-buff, or brownish-cream to creamy buff, thickly marked with reddish-brown or chestnut and dull purplish-brown spots and blotches, particularly around larger end; size of 22 eggs (Australia) 26.9-31 x 21.1-22.9 (28.7 [SD 1.1] x 21.9 [SD 0.5]); of 20 eggs (Malaysia, Sumatra, Java, Borneo, Lesser Sundas) 28-30.1 x 21-23.7 (29.4 x 22.5), calculated weight 8.1; of 11 eggs (Philippines, Sulawesi, Mariana Is, Caroline Is, Palau Is) 29.5-32.5 x 22.2-23.1 (30.1 x 22.5), calculated weight 8.3 (Schönwetter 1961-62, Marchant & Higgins 1993). Eggs laid at daily intervals. **Incubation** Period c. 18 days, by both sexes; starts when clutch complete. **Chicks** Fed and cared for by both parents for c. 4 weeks. Postjuvenile moult probably begins shortly after fledging. In Australia has 1, sometimes 2, broods per year; possibly able to breed on assumption of first basic plumage, i.e. after postjuvenile moult, at less than 1 year old (Marchant & Higgins 1993).

AENIGMATOLIMNAS

One species, the Striped Crake, endemic to Africa. It is very close to *Porzana*, and is often merged with that genus, but it has a deeper bill with a very broad, almost vertical, nasal bar and a smaller bony nostril (Olson 1973b). It is one of the very few rails which is markedly sexually dimorphic in plumage and is also unusual in having (at least in captivity) a sequentially polyandrous mating system. It frequents seasonally inundated, grass-dominated habitats which are often ephemeral, and it is migratory over most of its range. It is not globally threatened.

109 STRIPED CRAKE
Aenigmatolimnas marginalis Plate 27

Porzana marginalis Hartlaub, 1857, Gabon.

Often retained in *Porzana* on basis of plumage characters, but differs in skull structure and in having longer legs and toes; further skeletal comparison is required. Monotypic.

Synonyms: *Ortygometra/Crex/Porzana/Limnobaenus marginalis*; *Crex/Limnobaenus suahelensis*.

IDENTIFICATION Length 18-21cm; wingspan 36-39cm. A medium-small, short-tailed, often rather slender-looking crake, with short, heavy bill and very long toes. Mantle to tail, and upperwings, dark brown, prominently streaked white except from rump to tail; lower flanks to undertail-coverts cinnamon. Male has forehead to hindneck dark brown; sides of head dusky cinnamon, and underparts to breast pale cinnamon; flanks olive-brown, streaked white. Female has forehead to upper mantle dark grey; sides of head to breast and flanks grey with whitish streaks and faint bars; both sexes have white chin and throat. Narrow white leading edge to wing. Iris golden-brown to dark brown; eye-ring pale green or yellowish, orange in non-breeding birds; bill apple-green to grey-green, with greyer tip and darker culmen; legs and feet jade-green to olive-green. Immature resembles adult, with less prominent upperpart streaking and dull bare parts; male paler and more rufous on head, neck, breast and flanks; female has only vague pale markings on neck and breast, and a brownish tinge to grey of head, neck and sides of breast. Juvenile plumage duller and often almost plain; male darker brown, female duller grey; bare parts dull. Inhabits seasonally inundated grasslands, pans and marsh edges.
Similar species Large area of reddish-cinnamon from lower flanks to undertail-coverts, usually well visible in flight, serves to separate this species from all sympatric rails, including all *Porzana* species; legs and feet also project more in flight. Also easily distinguished from Corncrake (69) and African Crake (68) by smaller size and white stripes on upperparts. In some circumstances, especially in brief view, confusion possible with sympatric *Porzana* crakes. Spotted Crake (96) somewhat larger and bulkier, with slightly longer wings, deeper wingbeats, and thinner bill which is yellow with orange-red spot at base; has buff undertail-coverts, predominantly spotted plumage, barred flanks, and broad, prominent white leading edge of wing. Baillon's Crake (94) is smaller and more dumpy, with shorter, more rounded wings, richer brown upperparts with spots and shorter streaks, darker slate-blue face and anterior underparts (not in juvenile), prominently barred flanks and undertail-coverts. Little Crake (93) is also smaller, but long and slender, with proportionally long wings; upperparts paler and with very limited white streaking; rear flanks to undertail-coverts barred; and red base to green bill (adults).

VOICE The advertising call of the female is a series of rapidly repeated ticking notes, lasting up to 1 min (Urban *et al.* 1986). The answering call, presumably of the male, is a short series of rapidly repeated, high-pitched, grating *graa* notes (W. R. Tarboton *in litt.*). This grating call appears to be the same as that uttered by both members of captive pairs during the mating and egg-laying period; any sudden noise, such as the calls of Helmeted Guineafowl *Numida meleagris*, started a pair calling together (Wintle & Taylor 1993). Advertising calls are given early and late in the day, especially in overcast weather, and also during the night; captive females called throughout the day and night, most frequently in the late evening, calling for several minutes and repeating the call c. 30 min later; thunder and lightning often stimulated the birds to call (Wintle & Taylor 1993, P. B. Taylor unpubl.). Members of a family party frequently uttered *chup* and *yup* contact calls (R. K. Brooke in Pitman 1965). Birds also have grunts and growls of alarm, which are also given when chicks are being protected, and may grunt and squeal when handled (Aspinwall 1978, Urban *et al.* 1986). When attacking a human intruder, breeding males made loud hissing noises (Wintle & Taylor 1993).

DESCRIPTION
Note that some early publications contain errors arising from incorrectly specified sexual and age-related plumage differences (Wintle & Taylor 1993).
Adult male Forehead to hindneck dark brown, often washed cinnamon, shading to dusky cinnamon on sides of head, including supercilium and lores, darkest on ear-coverts and more tawny-olive on sides of neck; malar region, chin and throat white; foreneck and breast pale cinnamon, breast sometimes with pale streaking (whitish feather edges). Mantle, back, upper rump, scapulars, tertials and upperwing-coverts dark rich brown, somewhat paler on upperwing-coverts; feathers edged white and broadly tipped tawny-olive; mantle sometimes washed cinnamon and white feather edges strongly cinnamon-tinged. White edges produce strongly striped effect. Lower rump and uppertail-coverts blackish-brown, feathers edged cinnamon to pinkish-cinnamon; rectrices dark brown, edged pale pinkish-cinnamon. Greater coverts, primaries and secondaries dark olive-brown, narrowly tipped pinkish-cinnamon except on coverts; outer primary and outermost alula feather have outer web white, others narrowly edged buffy on outer webs; marginal coverts white. Tertials as upperwing-coverts but centres darker, like mantle and back; axillaries whitish to very pale olive-brown; underwing-coverts pale olive-brown, lessers and medians often broadly tipped white to almost entirely white. Flanks olive-brown, feathers edged white and often tipped and/or washed pale cinnamon-buff; belly whitish, often tinged pale cinnamon; lower flanks, vent and undertail-coverts cinnamon; thighs pale olive-brown, sometimes washed pale cinnamon. Iris golden-brown to darkish brown; eye-ring pale green or yellow, orange in non-breeding birds, more prominent when breeding; bill apple-green to grey-green, probably brighter when breeding, with greyer tip and often with darker culmen; legs and feet jade-green to olive-green, brighter when breeding.
Adult female Upperparts similar to male but dark brown often less rich, more olive-tinged, and feather tips often colder in tone than in male; forehead to hindneck and upper mantle dark grey (may sometimes be washed dark brown from forehead to hindneck), shading to pale grey on sides of head, malar region and sides of neck; chin and throat white to very pale grey; sides of neck, and breast and flanks, grey with variably occurring whitish scalloping (pale feather tips) and with white streaks formed by white feather edges; streaks are especially prominent on sides of breast and flanks, but tend to be less obvious, and shorter, on centre breast, where grey palest. Flanks usually

lack cinnamon-buff wash of male. Eye-ring often pale green when breeding.
Immature Resembles adult, with less prominent upperpart streaking and dull bare parts; male paler and more rufous on head, neck, breast and flanks; female has only vague pale markings on neck and breast, and has a brownish tinge to grey of head, neck and sides of breast. Iris brown; bill greenish; legs and feet blue-grey to olive-green.
Juvenile Duller than immature, male being darker brown, female duller grey; less patterned than immature, often looking unstreaked on upperparts; 1 specimen has belly and undertail-coverts off-white. Iris dull brown; bill dull greenish (horn to yellowish also recorded); legs and feet blue-grey to green-grey.
Downy young Black downy chick has long dark legs; bill cream or pinky-white, dark band on culmen from base to mid-region and running down bill to lower mandible at border of cere-like sheath which extends from nostrils to base of bill; bill becomes predominantly dull greenish after 10 days, with small whitish saddle near base of upper mandible. Small white egg-tooth and wing-claw present; egg-tooth lost after 2-3 days.

MEASUREMENTS Wing of 9 males 103-109 (105.8, SD 2.15), of 11 females 103-113 (106.7, SD 3.2); tail of 9 males 38-49 (43.95, SD 3.3), of 11 females 41-52 (46.1, SD 4.0); culmen to base of 6 males 20-21 (20.5, SD 0.3), of 8 females 20-22 (20.9, SD 0.65); tarsus of 9 males 32-37 (35.3, SD 1.6), of 11 females 32-38 (34.6, SD 2.0). Weight of 2 females 41.5, 61.

GEOGRAPHICAL VARIATION None.

MOULT Most information is from captive birds (Wintle & Taylor 1993). In captivity remex moult is simultaneous and birds are flightless for c. 3 weeks. However an adult female caught in Zambia (Jan), possibly with chicks, had the outer 6 primaries old and the inners new (Aspinwall 1978). Observations indicate that 2 complete moults per year occur in captive adults, 1 after the last brood is fully grown and 1 before or during the breeding season; females normally moult in Mar-Apr and Sep, males in Apr and Dec; the males's Dec moult often occurred between successive nesting attempts but on at least 3 occasions males with half-grown remiges mated with females and nested. Postjuvenile moult begins at 13-15 weeks and is complete at 21 weeks.

DISTRIBUTION AND STATUS The distribution of the Striped Crake is imperfectly known. It is recorded patchily in West Africa, from N Ivory Coast (Korhogo), Ghana (Cape Coast E to Keta), Togo, Nigeria, W and C Cameroon, Gabon and coastal Congo (Bas-Kouilou). It is known more continuously from E Zaïre, Rwanda, Uganda (Rwenzori NP and Kampala), W and S Kenya, Tanzania, Malawi (Zomba, Mangoche, Nkhata Bay, Nkhotakota, Kasungu, Lilongwe, Blantyre), Zambia, Zimbabwe, N Botswana and extreme NE Namibia. It probably occurs throughout S Mozambique in suitable habitat (Clancey 1996) and in extreme SE Angola (as mapped by Urban *et al.* 1986). It is also reported from scattered localities in South Africa, and it may breed in NE South Africa in years of high rainfall (Taylor 1996b, 1997c). Not globally threatened. It is a highly secretive and poorly known species, undoubtedly overlooked but apparently generally uncommon throughout its range, which may be more extensive and continuous than is known. There are only 1 certain and 2 possible records from Ivory Coast (Thiollay 1985). In Ghana it occurs on the coast and there is 1 probable sighting from the N (Vea Dam) in Dec; it is possibly overlooked, at least in the S (Grimes 1987). It is uncommon in Nigeria (Elgood *et al.* 1994) and there are only a few records from Cameroon (Louette 1981); it is locally common in NE Gabon in some years (Brosset & Erard 1986). It is rarely recorded in East Africa (Britton 1980). In Zimbabwe it is of irregular occurrence but may attain a relatively high local density (Irwin 1981). In N Namibia it bred at Odonga in the 19th century and has recently been recorded very occasionally, including at Makuri Pan, Bushmanland, in Jan (Andersson & Gurney 1872, Brown *et al.* 1987, Taylor *et al.* 1994, Taylor 1997c). In Botswana it is rarely recorded: Clancey (1964) states that it was collected in the N but provides no details; 2 atlas records from the Okavango (Apr 1988 and May 1989) are rejected by Penry (1994) but the May record is accepted, and further Okavango records added, in the Southern African Atlas (Taylor 1997c).

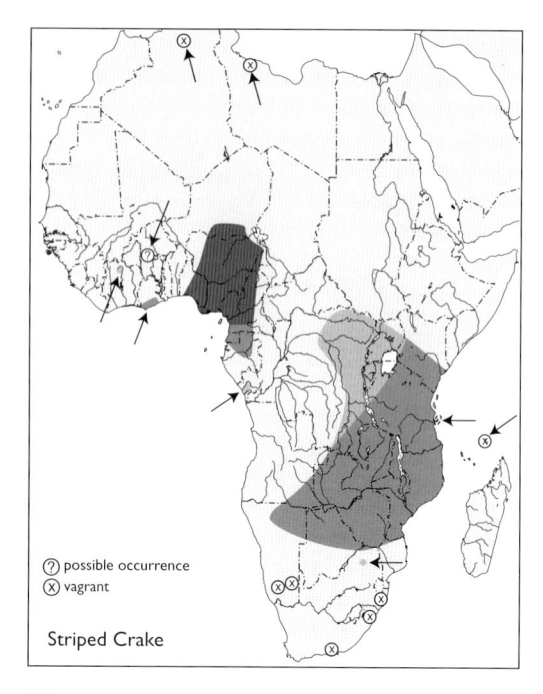

Much of this bird's breeding habitat is ephemeral, so its occurrences are often irregular and its periods of residence short (Taylor 1996b). Like the Streaky-breasted Flufftail (5) it may sometimes be locally common in large areas of breeding habitat in seasons of good rainfall but, like that species, its numbers must have been adversely affected by habitat loss resulting from overgrazing and disturbance, and the damming, draining and cultivation of seasonal and ephemeral wetlands. In South Africa, where it is rarely recorded, the maximum number known from any site is 5 on the Nyl R floodplain, where it occurred in 1988 and 1996 (Taylor *et al.* 1994); however, potentially suitable habitat could hold up to 270 birds, especially on the N KwaZulu-Natal coastal plain (where it has not yet been recorded) (Taylor 1997a). In view of the threats facing its breeding habitats, the possible presence of even a small breeding population in South Africa could be of some significance and should be investigated (Taylor 1997a). It appears to be particularly susceptible to nest predation following human disturbance (Hopkinson & Masterson 1975), while it probably tolerates only very limited grazing pressure (e.g. Mallalieu 1995). Predators include African Marsh Owl *Asio capensis* and the bullfrog *Pyxicephalus adspersus* (Urban *et al.* 1986).

MOVEMENTS Imperfectly known; the seasonality of occurrence shown on the distribution map is conjectural. In West Africa it is recorded from Ivory coast in Mar-Jul (Thiollay 1985) and it is known from Ghana only in the rains (Jun) but its status is uncertain (Grimes 1987). It is apparently resident in Nigeria but breeds in the N only in the rains and latitudinal movements probably occur, night migrants being recorded in early Dec, when 1 was taken at a light (Elgood *et al.* 1973, 1994); in the semi-arid zone of the extreme NE, at Maiduguri, a female caught alive on 30 Jun was near to breeding (Bates 1927). In N Gabon it occurs irregularly in Nov-Mar and night migrants fly into buildings (Brosset & Erard 1986). Its status elsewhere in West Africa is uncertain. In East Africa it is recorded from Kenya in May-Sep and Nov, and night migrants (in breeding condition) have hit lighted windows in Nairobi in May while coastal movements were observed in May-Jul at Mombasa (Urban *et al.* 1986, Lewis & Pomeroy 1989, Taylor *et al.* 1994). A bird on Zanzibar in Nov 1991 (1993, *Scopus* 15 3: 160) was also presumably a migrant. From Tanzania southwards it is recorded in breeding areas from Dec to Apr; much breeding habitat is ephemeral so its occurrences are irregular and its periods of residence may be short; odd birds may remain the during dry season if suitable habitat persists (Urban *et al.* 1986, Taylor *et al.* 1994, Taylor 1997c). In Zimbabwe it only appears in any numbers if there have been good rains by the end of Dec, staying if these continue into Jan-Feb and moving elsewhere if conditions become unsuitable (Irwin 1981). In Zambia and Malawi it occurs from mid-Dec to Mar, with a Sep record for Zambia and unsubstantiated claims for Apr and Jul in Malawi (Benson *et al.* 1971, Benson & Benson 1977); in Zimbabwe it occurs from Dec to Apr, with 1 May specimen (Benson 1964b, Irwin 1981, Taylor 1997c). There is a recent unsubstantiated May record from N Botswana (Maun) (Taylor 1997c). Calling birds have been recorded on the Nyl R floodplain, South Africa, in Jan-Feb (Taylor 1997c). Night migrants are recorded at Mufindi, S Tanzania, May and Dec (1986, *Scopus* 8 5: 107), in N Zambia (Mbala), Mar (Benson 1956), and in S Namibia (Kalahari Gemsbok NP), May (Taylor *et al.* 1994). A Sep 1958 record from S Namibia, and 7 scattered records from South Africa (Mar-Jun), are attributed to vagrancy, possible coastal and inland migration, and birds displaced from normal wintering areas by drought conditions; however, Jan-Feb 1988 records of calling birds from the Nyl floodplain, former Transvaal, suggest that the breeding range may extend to South Africa in years of good rainfall (Taylor *et al.* 1994, Taylor 1997c). In southern Africa it is assumed to be a wet-season breeding visitor which retreats towards equatorial regions after breeding. Vagrants are recorded from North Africa in Algeria (Biskra, Jan) and Libya (Wadi Turghat, Feb), during the northern tropics dry season; Aldabra in Dec 1904 (Benson 1964b, Cramp & Simmons 1980). As breeding habitats dry out it is also reported to retreat to more permanently wet areas locally, such as streams and marshes (Urban *et al.* 1986). In captive birds (Wintle & Taylor 1993), at the end of each breeding season (usually in May) juveniles repeatedly attempted to fly out of aviaries, always at night.

HABITAT Striped Crakes avoid dense, tall vegetation in permanent marshes, preferring areas with shallow pools, muddy patches and fine grasses, and much of their preferred habitat is ephemeral (Taylor *et al.* 1994). They breed in seasonally inundated grasslands which dry out and are often burned during the dry season; typical breeding habitats include temporary pans, short-grassed dambos, river floodplains, old ricefields and the edges of marshes, ponds and ditches (Urban *et al.* 1986, Taylor *et al.* 1994, Taylor 1997a). In Zimbabwe this species typically occurs in shallowly flooded grassland 50-100cm tall with small grassy or open pools, occupied sites being on relatively high ground above the level of the floods which follow heavy storms (Hopkinson & Masterson 1975). It nests in fine annual grasses such as *Eragrostis* (possibly *curvula*), *Setaria* (possibly *anceps*), *Sporobolus pyramidalis*, *Leersia hexandra* and *Aristida* (possibly *junciformis*), and sometimes in coarser grass such as *Paspalum urvallii* (Hopkinson & Masterson 1975). Similar habitat is occupied in Zambia, where the birds occur in dense tussocky or continuous grass cover which is 0.3-1.2m tall, interspersed with shallow pools and muddy patches, sometimes has scattered *Acacia* bushes, and may be flooded to a depth of 20-30cm (Aspinwall 1978, Taylor 1987). In Malawi the species has occurred in tussocky *Setaria* and *Eragrostis* with sedges *Cyperus* sp. and forbs, on ground flooded to a depth of 10cm and with pools up to 30cm deep (Mallalieu 1995). In Ghana it is recorded from marshy savanna containing scattered thickets to which birds retreated when flushed (Cramp & Simmons 1980), and in coastal Congo it was found in seasonal sedge and *Jardinea* grass swamp with water 20-30cm deep (Dowsett-Lemaire *et al.* 1993). It may also occur on the banks of shallow drainage ditches through rank grassland near marshes, and on flooded pans where the dominant vegetation is waterlilies, sedges and *Polygonum* (Cramp & Simmons 1980, P. B. Taylor unpubl.). It often favours muddy ground for foraging, even venturing into areas with relatively short, sparse cover where it will forage close to Great Snipe *Gallinago media* (P. B. Taylor unpubl.). The habitat of migrating and non-breeding birds is less well known; descriptions include recently inundated grassland and pool edges (Mombasa, Kenya); fine reeds near a stream (Namibia); short sedge and grass clumps fringing shallow water (Luangwa V, Zambia); rank grass and sedges adjacent to *Typha* in the centre of a dambo (Malawi); temporarily flooded, *Leersia*-dominated grassland with *Typha* and sedge patches adjacent to a river (Durban, South Africa); and puddles on a dirt road through moist grassland (Kruger NP, South Africa) (Urban *et al.* 1986, Taylor *et al.* 1994, Mallalieu 1995, P. Funston in Taylor 1997d). Migrants are also recorded from airfields in Gabon (Brosset 1968). It occurs from sea level to at least 1,900m.

FOOD AND FEEDING Food comprises earthworms, small snails, spiders, beetles, grasshoppers, flies, moths, insect larvae, small fish, and frog tadpoles. Captive birds did not eat seeds of millet or sorghum (Wintle & Taylor 1993). It forages on grass and mud, at pool edges and in shallow water; it walks deliberately, searching the ground and short vegetation; it probes into mud, immerses the head in water and chases flying insects; it also forages on floating vegetation such as waterlily pads. It feeds very actively from late afternoon to dusk. Chicks are fed on insects and other small invertebrates.

HABITS Although this species may call at night in the breeding season it is apparently mainly diurnal. It is more secretive than other sympatric crakes, is seldom seen in the open and is usually very difficult to observe on the ground; even captive-bred birds remain secretive. It walks with deliberate strides, and when not wary it stands and walks in a fairly upright position; at other times it walks with a crouched stance, flicking the tail frequently. It is usually hard to flush (but not particularly difficult with a trained dog), and prefers to run or slink away when disturbed, assuming a very elongated shape with head lowered and stretched forward, and either moving rapidly off or creeping through the vegetation; it is more difficult to flush a second time. It sometimes 'freezes' when pursued, and may be caught by hand (Hopkinson & Masterson 1975, Aspinwall 1978); birds pursued by a dog sometimes submerge in shallow muddy water and may also be caught (P. B. Taylor unpubl.). Escape flights are usually short and, on landing, the birds will often stretch up the head to watch the observer before crouching and then running or flying off (P. B. Taylor unpubl.). Birds may bob the tail vigorously while feeding, even when confident. Captive birds swam freely, with head up and tail slightly raised, jerking the head back and forth like a moorhen, climbed to the top of vegetation clumps to sunbathe in the early morning, and roosted under clumps of grass at night (Hopkinson & Masterson 1975, Wintle & Taylor 1993). Males and females build roosting platforms in grass clumps (Wintle & Taylor 1993).

SOCIAL ORGANISATION All information is from Wintle & Taylor (1993) unless otherwise specified. Captive birds showed sequential polyandry whenever sufficient numbers of males were present, the female mating with 2 or more males in succession. Polyandry is possibly also prevalent in the wild, where 6 of 9 records of incubating birds refer to males and 2 to probable males (see Table 3 in Wintle & Taylor 1993); occasional monogamy is also possible. In captivity, birds were territorial only during the breeding season; in an aviary of 125m^2 with 2 males and 2 females, one half was occupied by each male to the exclusion of the other, 1 female established a territory covering the whole aviary and monopolised both males, and the other female did not breed. In an aviary of 30 x 8m with 3 males and 2 females, 2 males set up breeding territories, one at each end of the enclosure and both females established territories, again 1 at each end. In Zimbabwe, 7 nests were found in an area of c. 24ha, the closest nests being only 45m apart (Hopkinson & Masterson 1975). No evidence was obtained in captivity for juveniles helping to rear subsequent broods, but young 26 days old fed 8-day-old flufftail chicks which approached them for food.

SOCIAL AND SEXUAL BEHAVIOUR All observations are from Wintle & Taylor (1993) unless otherwise specified. Captive birds coexisted amicably throughout the year with other small rails such as Red-chested (3) and Streaky-breasted Flufftails, but were harassed by larger species such as Black Crake (90) and African Rail (62). In defence of young chicks, captive males would attack a human intruder, running with open wings, splashing the water and hissing loudly.

The female always initiated breeding activity, uttering the advertising call to attract the male, and ceasing to call when paired; a subordinate, non-territorial female usually did not call and, if she did, she was chased by the territorial female. During the mating and egg-laying period, each pair was constantly together; paired birds allopreened, and also walked around with an upright stance, placing 1 foot very slowly in front of the other; the female solicited food from the male by crouching slightly, with head down and neck outstretched. The female always initiated copulation, crouching to encourage the male to mount; copulation was brief and occurred either on land or in shallow water; after dismounting, the male shook himself before moving away.

The pair bond broke down as soon as incubation started, when the female sought another territorial male, calling again to attract a mate; having laid for this male, she would call for a third mate. Males did not tolerate the close proximity of a female late in incubation and at hatching time; however a female would be accepted into a family group when the young were large; females and large young allopreened.

BREEDING AND SURVIVAL Season Few records; breeds at beginning of, or during, rains: S Ghana, oviduct eggs Jun; N Nigeria, Aug, breeding condition Jun; Cameroon, breeding condition Jun; Gabon, breeding condition Nov; Kenya, Jun, breeding condition May, nest-building Nov; C Tanzania, breeding condition Jan; Malawi, Jan-Feb; Zambia, Jan-Feb, breeding condition Dec; Zimbabwe, Dec-Mar, captive birds Sep-Jan, occasionally to Feb/Mar (season prolonged, and initiated by first rain in Sep); N Namibia Feb-Mar. **Nest** A strongly built shallow bowl or platform of grass (*Setaria, Eragrostis, Sporobolus, Leersia, Aristida, Paspalum*), sometimes rushes or sedges, c. 8cm in diameter; typically well concealed in a tuft of grass or sometimes sedge, vegetation sometimes green but often old and dry; usually 10-25cm above water 10-30cm deep, but occasionally floating or above damp ground, rarely on ground; surrounding vegetation pulled down to form roof over nest; 1 nest was a shallow depression in ground among grasses, lined with dry grasses (Serle 1939), 1 was under broad leaves of *Crinum macowanii* lily (Hopkinson & Masterson 1975). May nest in very short new dambo grass, 1 nest being merely a platform of trodden grass (Aspinwall 1978). In captivity, nest always built by male; male often adds material during incubation. Sometimes builds dummy nests which are abandoned before completion. **Eggs** Usually 4-5 (mean 4.35, n = 17 clutches), in captivity 2-6; laid at 1-2 day intervals; oval; smooth and somewhat glossy; often described as very handsome; pink, yellow, buff or cream, with variable overlay of red-brown spots and blotches (occasionally some lilac or purple-chocolate blotches), usually coalescing to form broad band around larger end; some also with grey or violet under-markings; size of eggs 28.7-33 x 19.7-23 (29.6 x 21.4, n = 31), calculated weight 7.3. **Incubation** In captivity by male only, period 17-18 days; in wild, apparently usually by male only (Wintle & Taylor 1993); hatching synchronous. **Chicks** Semiprecocial; parental care in captivity exclusively by male; chicks fed in nest for first 4-5 days; then leave nest and begin to pick up some food for themselves; male holds food in bill and chicks peck upwards to seize food items; chicks can feed themselves at 14 days but still approach male for food. In captivity feathers begin to grow at 8 days, when chicks swim well; body well feathered at 28 days, when may be sexed on plumage colour; young fully grown and able to fly at 6-7 weeks. Age of first breeding 1 year. Clutch losses from predation often high; in 4 of 6 nests, Zimbabwe, eggs broken, possibly by rat (*Otomys* or *Pelomys*) or mongoose and probably as result of prior human disturbance (Hopkinson & Masterson 1975, Aspinwall 1978); clutch losses from rising water levels may also be high (e.g. Aspinwall 1978). In captivity nest predators included mice and egg-eating snake *Dasypeltis scabra* (Wintle & Taylor 1993). In captivity normally raises 1-2 (1-3) broods per season; up to 4 nesting attempts per male after failures (female will lay again quickly if clutch lost before late stage in incubation, when she is not tolerated by incubating male and may be paired with another male); female can lay more than 6 clutches for 2 males per season (Wintle & Taylor 1993). A pair of siblings had a low fertility rate (16 eggs hatched from 30 fully incubated; 11 of the other 14 were infertile) and chicks were weak, 5 of the 16 not surviving; a wild-caught female paired with the same male laid 18 eggs, 17 of which hatched, and all chicks were reared successfully (Wintle & Taylor 1993). **Survival** Greatest age reached by captive-bred bird 8 years 113 days (Wintle & Taylor 1993).

Cyanolimnas

One species; a relict, almost flightless, medium-sized rail with olive-brown upperparts, slate-grey underparts, faint barring on the belly, and white undertail-coverts; its brightly coloured bill has a narrow frontal plate. It is restricted to the Zapata Swamp in WC Cuba, and it is considered Critically Endangered because of its restricted distribution and because it has been recorded very infrequently in recent years.

110 ZAPATA RAIL
Cyanolimnas cerverai   Plate 33

Cyanolimnas cerverai Barbour & J. L. Peters, 1927, Santo Tomás, Península de Zapata, Cuba.

Monotypic.

IDENTIFICATION Length 29cm. Apparently almost flightless or capable only of weak flight; wings very short and rounded; tail short and decomposed; legs shortish and relatively stout. A medium-sized, dark rail without streaks or spots, faintly barred on lower belly and with prominent white undertail-coverts. Upperparts, including upperwings, olive-brown, browner from back to uppertail-coverts; tail brown-black. Forehead, forecrown, sides of head, and underparts from neck to upper belly, slate-grey, with whitish supraloral streak and white chin and upper throat; flanks greyish-brown, narrowly barred white; lower belly, vent and thighs dusky slate with very narrow whitish bars; short undertail-coverts buffy (usually not visible), longer ones white. Iris red; bill yellowish-green towards tip, greenish in centre and red at base, with narrow, pointed frontal plate; legs and feet red. Sexes alike. Young bird slightly duller than adult. Inhabits dense, tangled, bushy swamp.

Similar species Easily separable on plumage from sympatric rallids but resembles the smaller Colombian Crake (111) and Paint-billed Crake (112) of C and S America; former differs in having unbarred pale cinnamon flanks and undertail-coverts; latter has much heavier barring from flanks to undertail-coverts. Considered intermediate in plumage characters between Colombian Crake and the larger Plumbeous Rail (115) of South America, which has unbarred dusky brown to blackish rear flanks and undertail-coverts, and a much longer, slightly decurved, greenish-yellow bill with blue and red at base.

VOICE Alarm call a loud *kwowk*, much like that of the Limpkin *Aramus guarauna* (Ripley 1977). Hardy *et al.* (1996) give a very rapid, low-pitched, throbbing, drumming call, apparently made by more than 1 bird at the same time.

DESCRIPTION

Adult Forehead and forecrown slate-grey, shading into olive-brown on crown; nape to hindneck, and rest of upperparts, including scapulars and upperwing-coverts, olive-brown, becoming browner from back to uppertail-coverts; rectrices fuscous-black. Remiges described as fuscous-black, but in all illustrations at least tertials are olive-brown like coverts; axillaries and underwing-coverts dull fuscous-black, narrowly tipped white. Narrow supraloral streak whitish; sides of head slate-grey, darker on lores; chin and upper throat white; flanks olive-sepia, feathers narrowly tipped white; rest of underparts, to upper belly, slate-grey; lower belly, vent and thighs dusky slate, faintly tinged fuscous and with very narrow white feather tips; short undertail-coverts buffy, longer ones white. Plumage rather loose, denser below than above. Iris red; bill yellowish-green towards tip, greenish in centre and red at base, which is slightly swollen; base of culmen extends into narrow, pointed frontal plate; legs and feet red.

Immature 'Juvenal female' like adult but colours slightly duller (Ridgway & Friedmann 1941).
Juvenile See immature.
Downy young Down brownish-black.

MEASUREMENTS Wing of male 109.5, of 3 females 96, 99, 103; tail of male 43, 46, of 3 females 36.5, 40, 43; exposed culmen of male 47.5, of 4 females 36-45 (39.9, SD 4.4); tarsus of male 46, 49, of 4 females 39-45 (42.1, SD 2.8) (Ridgway & Friedmann 1941, Hellmayr & Conover 1942, Ripley 1977).

GEOGRAPHICAL VARIATION None.

MOULT No information available.

DISTRIBUTION AND STATUS Zapata Swamp, WC Cuba. This species is CRITICALLY ENDANGERED. It is known from only two sites, c. 65km apart, in the 4,500km² Zapata Swamp. Four were collected near Santo Tomás in 1927, and the species was easily found in 1931 (Bond 1971). Subsequently there were no records, despite occasional searches, until the 1970s, when the voice was recorded and birds were found at a second locality, Laguna del Tesoro (Clements 1979, Garrido 1985, Collar *et al.* 1992). Despite its likely occurrence in other areas of the swamp, the very few records in recent decades suggest that its population is very small; although regarded as rare, it is said to be heard regularly in the Santo Tomás area (Sulley & Sulley 1992). Extensive grass-cutting formerly occurred for roof thatch, but the most serious threats appear to be

Zapata Rail

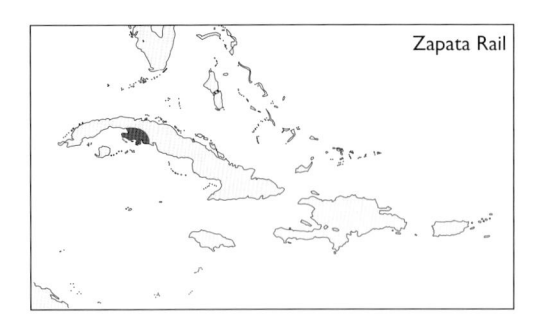

dry-season burning of habitat, which is potentially devastating for the species and is still occurring, and also the existence of introduced predators, namely a mongoose *Herpestes* sp and rats *Rattus* spp (Collar *et al.* 1992). Although some relatively small areas have been drained, this should cause no harm to the bird's habitat and the swamp has so far escaped serious drainage (King 1981, Collar *et al.* 1992). The bird has been afforded protection with an area of 10,000ha in the Corral de Santo Tomás Faunal Refuge, while the Laguna del Tesoro is within a Nature Tourism Area; the benefits of this protection are unknown (Collar *et al.* 1992). A survey of the bird's distribution is urgently needed, and dry-season burning of its habitats must be investigated and controlled. The bird's former distribution was wider, fossil bones attributable to this species having been found in cave deposits in Pinar del Río and on the Isle of Pines (Olson 1974b, Garrido 1985). This rail shares its restricted habitat with 2 other threatened endemic birds, the Zapata Wren *Ferminia cerverai* and the Zapata Sparrow *Torreornis i. inexpectata* (Collar *et al.* 1992).

MOVEMENTS None.

HABITAT Dense, tangled, bush-covered swamp with low trees, where 'arraigán' *Myrica cerifera* brush and sawgrass *Cladium jamaicense* are common, near higher ground (King 1981, Regalado Ruíz 1981).

FOOD AND FEEDING No information available.

HABITS Bond (1971) states that it is not wary and is easy to find. When disturbed, it usually runs rapidly for a short distance before stopping motionless, with tail raised and the white undertail-coverts very conspicuous; it may crouch at the approach of observers (Clements 1979), and it is adept at moving through tussocks of sawgrass without being seen (Garrido 1985).

SOCIAL ORGANISATION No information available.

SOCIAL AND SEXUAL BEHAVIOUR Not described.

BREEDING AND SURVIVAL Season Sept; 2 males with "somewhat enlarged testes", Jan (Bond 1973). **Nest and eggs** One nest found, situated c. 60cm above water level in a hummock of sawgrass; contained 3 eggs (Bond 1984).

NEOCREX

Two medium-small crakes from Central and South America, largely allopatric but occurring together at least in Panama. They both have dull brown upperparts, slate-grey underparts, red legs, and a greenish-yellow bill with a red base, but the Colombian Crake has unbarred pale cinnamon posterior underparts and unbarred white underwing-coverts, while the Paint-billed Crake is barred in these regions, and also has a relatively wider nostril. In general appearance they resemble the larger Zapata Rail of Cuba. The poorly known Colombian Crake is apparently rare and is considered Near-threatened, but the Paint-billed Crake is more widespread, has recently colonised the Galapagos Islands, and may be expanding its range in Central America.

111 COLOMBIAN CRAKE
Neocrex colombianus   Plate 33

Neocrex colombianus Bangs, 1898, Palomino, Santa Marta, Colombia.

Sometimes placed in *Porzana*. Forms superspecies with *N. erythrops*, and sometimes considered conspecific, but differs in plumage and bill structure. Two subspecies recognised.

Synonyms: *Neocrex columbianus/uniformis*; *Porzana columbiana*.

IDENTIFICATION Length 18-21cm. A medium-small, plain grey and olive-brown rail, with unbarred underparts. Top of head dark slate (nominate) to blackish grey (race *ripleyi*); upperparts and upperwings olive-brown (darker in *ripleyi*); underwing-coverts white. Chin and throat white; sides of head, sides of neck, and breast to belly pale slate-grey (darker in *ripleyi*), washed pale olive-brown on sides of neck and sides of breast; flanks, and lower belly to undertail-coverts, pale cinnamon. Iris red; bill greenish to yellowish, with black tip and reddish base; legs and feet red, orange or reddish-brown. Sexes alike. Immature undescribed; one juvenile was similar to adult but bill dull pink. Inhabits marshes, swamps, wet savanna and pasture, and overgrown forest edges; sometimes on dry ground.
Similar species The generally allopatric Paint-billed Crake (112) differs in having forehead grey, crown olive-brown like rest of upperparts, lower flanks to undertail-coverts barred blackish and white, underwing-coverts barred dusky brown and white, and bill more slender but with wider nostril, and greener with bright red base. Larger Zapata Rail (110) of Cuba similar in both plumage and bare part colours, but has dark slate flanks with faint white bars, and white undertail-coverts.

VOICE Not recorded.

DESCRIPTION *N. c. colombianus*
Adult Forehead to occiput dark slate; upperparts, including upperwing-coverts, secondaries and tertials, uniform olive-brown or slightly richer brown (almost cinnamon-brown); rectrices darker basally. Primaries

buffy-brown, narrowly edged paler; axillaries buffy-brown; underwing-coverts white, unbarred but with vague streaks or spots of olive-brown on some feathers; underside of remiges pale grey-brown. Sides of head pale slate-grey; chin and throat white; sides of neck, and breast to belly, pale slate-grey, washed pale olive-brown on sides of neck and sides of breast; flanks, and lower belly to undertail-coverts, uniform dull pale cinnamon; centre of belly white. Iris red; bill greenish, greenish-yellow or yellow, with black tip and orange to reddish base; legs and feet red, orange or reddish-brown. According to Blake (1977) bare part colours are: iris yellowish-brown; bill yellow basally, lower mandible greenish; legs and feet flesh; colours of iris, legs and feet may indicate sexual or age difference. Fjeldså & Krabbe (1990) show darkish iris and state that bill does not always have red base.

Immature Not described; see note on adult bare part colours.

Juvenile One (race uncertain) from E Panama had similar plumage to adult; primaries brownish (not fully grown); bill dull pink; legs and feet red (Behrstock 1983).

Downy young Not described.

MEASUREMENTS Wing of 6 males 100-106 (102.8), of 6 females 98-104 (101.5); tail of 1 male 30, of 1 female 29; exposed culmen of 5 males 19-21 (20), of 6 females 19-20 (19.2), culmen to base of 1 male 22.5; tarsus of 6 males 29-31 (29.9), of 6 females 26-30 (27.5).

GEOGRAPHICAL VARIATION Races differ in overall colour.

N. c. ripleyi Wetmore, 1967 – C and E Panama (W Colón, and Tocumen marsh in E Panamá Province) and adjacent NW Colombia (Acandi in NW Chocó). Without a specimen it is not certain that E Panama birds are referable to this race rather than to nominate, the latter possibility being suggested by photographs (Behrstock 1983, Ridgely & Gwynne 1989). Markedly darker overall than nominate: forehead to occiput blackish-grey; upperparts darker brown; remiges olive-grey; underparts darker slate-grey; undertail-coverts pale cinnamon-buff with white bases. Male (holotype): wing 98; tail 29.5; culmen from base 21.5; tarsus 31.5.

N. c. colombianus Bangs, 1898 – N and W Colombia, from Santa Marta Mts, E Guajira (Serranía de Macuira), the W Andes (above Calle in Valle), SW Cauca (Guapí), and W Nariño (Barbacoas); and NW Ecuador S to R Chimbo. See Description.

MOULT Primary moult is sequential and is recorded from Ecuador, Apr and Aug.

DISTRIBUTION AND STATUS C and E Panama, N and W Colombia, and NW Ecuador. Very little information is available on this crake's status and distribution, and it is currently regarded as NEAR-THREATENED. The nominate race is regarded as rare throughout its range, with scattered records in N and W Colombia and S on both slopes of the W Andes to W Ecuador, where there are very few records, the last in 1991 (Hilty & Brown 1986, Ridgely & Greenfield in prep.). The race *ripleyi* is rare in C Panama, where it was apparently known from only one specimen (the holotype of the race), collected in 1965 at Achiote Road on the NW boundary of the Canal Zone in W Colón, before several were seen in Feb 1982 at the Tocumen marsh in E Panama, where they have also been seen subsequently (Behrstock 1983, Ridgely & Gwynne 1989). It is not certain whether the species is resident at Tocumen marsh, and whether it has recently appeared in the area or has been overlooked (Ridgely & Gwynne 1989). In NW Colombia it is known from one specimen, collected at Acandi, Gulf of Urabá, near the Panama border (Hilty & Brown 1986). It possibly occurs in E Costa Rica, where a sighting near Hitoy Cerere in Mar 1985 was claimed to be either this species or Paint-billed Crake (Stiles & Skutch 1989). An investigation of its current status and distribution, and a study of its natural history, are urgently required.

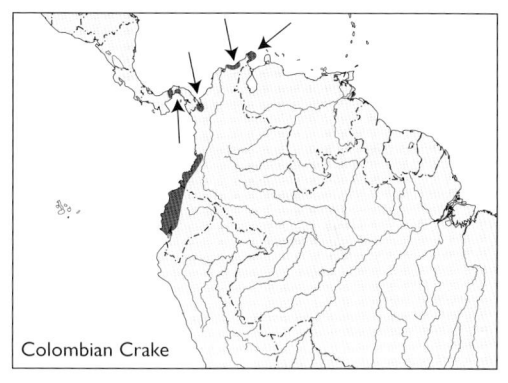

Colombian Crake

MOVEMENTS Regarded as sedentary. Three birds, all near breeding condition, which flew into lighted windows in Colombian W Andes, Dec and Jan, were believed to be resident (Miller 1963).

HABITAT Freshwater marshes (including small marshy areas), swamps, wet savannas, wet grass, pastures, and overgrown forest edges. At Tocumen marsh, E Panama, an area of vast rice fields interspersed with savanna-like habitat, it has been recorded in the dry season alongside other rails (including the generally allopatric Paint-billed Crake) at shallow drying rice field ditches and pools which were virtually the only remaining wet areas and at which prey animals were concentrated by falling water levels (Behrstock 1983, Ridgely & Gwynne 1989). It is not restricted to areas with water, and may be more of a grassland crake (Hilty & Brown 1986). It occurs up to 2,100m.

FOOD AND FEEDING There is no information on food, which is presumably similar to that of the Paint-billed Crake. It pecks at objects in mud (Behrstock 1983).

HABITS Birds at Tocumen marsh were invariably seen in the early morning as they skulked along the edges of channels and pools with some standing water; they rarely stayed in the open for very long, and they usually ran rapidly along grassy water margins or across open muddy areas, making only the briefest stops (Behrstock 1983, Ridgely & Gwynne 1989).

SOCIAL ORGANISATION No information available.

SOCIAL AND SEXUAL BEHAVIOUR No information available.

BREEDING AND SURVIVAL Season Colombia, male in breeding condition Jan, 3 birds near breeding condition Dec-Jan. E Panama, an apparently non-flying juvenile recorded 28 Feb 1982 (Behrstock 1983).

112 PAINT-BILLED CRAKE
Neocrex erythrops　　　　　　Plate 33

Porzana erythrops P. L. Sclater, 1867, Lima, Peru.

Sometimes retained in *Porzana*. Forms superspecies with *N. colombianus*, and sometimes considered conspecific, but differs in plumage and in bill structure. Two subspecies recognised.

Synonyms: *Porzana schomburgki*; *Aramides erythrops*.

IDENTIFICATION Length 18-20cm. A medium-small, predominantly olive-brown and slaty-grey crake: crown, nape and entire upperparts olive-brown (darker in race *olivascens*); tail brown; underwing-coverts barred dusky brown and white; forehead, sides of head, and underparts to upper flanks and upper belly, slaty-grey (darker in *olivascens*); lores and feathers below eye dusky; faint supraloral streak isabelline; chin and throat white (greyish in *olivascens*); lower belly dusky brown with narrow white bars; lower flanks to undertail-coverts blackish, barred white. Iris red; bill pale green, with bright red base and black tip; legs and feet red, orange or red-brown. Sexes alike. One immature had duller bill, grey iris, and pale brown legs and feet. Juvenile has pale grey, faintly barred belly, and darker bill with no red at base. Inhabits reedbeds, grassy marshes, wet to dry pastures, and thickets and damp woodland.

Similar species Very similar to generally allopatric Colombian Crake (111), which differs in having entire top of head dark grey, flanks and undertail-coverts unbarred pale cinnamon, underwing-coverts plain white, and bill greenish to yellowish with orange to reddish base and narrower nostril. Larger Zapata Rail (110) of Cuba similar in plumage but has narrow white barring on grey-brown flanks and dark slate belly and vent, and white undertail-coverts. Sympatric Grey-breasted Crake (32) smaller, with chestnut from hindcrown to upper mantle, variable amount of narrow white barring on upperwing-coverts, dusky bill with pale green at base and on lower mandible, and pale brown to yellowish legs and feet.

VOICE The song is a long, gradually accelerating and descending series of up to 36 staccato notes followed by 3-4 short churring notes which fall in pitch, the last being a 3-s flat trill (Fjeldså & Krabbe 1990). The call is also described as loud, frog-like, guttural, buzzy, single notes rendered *qurrrk* and *auuk*, sometimes given in a series (Hilty & Brown 1986, Hardy *et al.* 1996); when feeding it is said to call *tchur-ur-ur-ee* (Harris 1974). The alarm is a sharp *twack*.

DESCRIPTION *N. e. erythrops*
Adult Forehead and sides of crown clear slaty-grey to pale bluish-slate; centre of crown, nape and hindneck olive-brown, often washed darkish slate; rest of upperparts, including upperwing-coverts, secondaries and tertials, olive-brown, slightly darker and browner from lower back to uppertail-coverts; tail brown. Alula, primary coverts and primaries more ashy-brown; marginal coverts dusky brown, spotted white; axillaries and underwing-coverts barred dusky brown and white. One bird (BMNH) has narrow white bar, bordered blackish-brown, or small white spots, towards tip of some median upperwing-coverts. Lores and feathers below eye dusky slate; faint supraloral streak pale olive-brown to isabelline; chin and throat white; sides of head, sides of neck, foreneck, breast, upper flanks and upper belly clear slate-grey to pale bluish-slate, often washed olive-brown on sides of neck and sides of breast. Lower belly and thighs dusky brown with narrow white bars; lower flanks to undertail-coverts blackish, barred white, white bars broader on undertail-coverts. Iris red; bill olive-green to pale green, with bright red base and black tip; legs and feet coral-red. Sexes alike.
Immature Not properly described but one (*olivascens*) had grey and brown bill, grey iris, and pale brown legs and feet (Meyer de Schauensee 1962).
Juvenile Belly pale grey, faintly barred; bill darker and greener, with no red at base.
Downy young Not described.

MEASUREMENTS Wing of 7 males 96-111 (102.7), of 7 females 102-112 (105.5); tail of 1 female 35, of 3 unsexed 32, 35, 36; exposed culmen of 7 males 19-23 (21.1), of 7 females 17-21 (19.6), culmen to base of 3 unsexed 22, 22.5, 23; tarsus of 7 males 28-30 (28.7), of 7 females 27-30 (28.5). Weight of 2 males 51, 67, of 1 female 61.

GEOGRAPHICAL VARIATION Races separated on depth of overall colouring, and whiteness of throat.

N. e. olivascens Chubb, 1917 – E Colombia from Boyacá (Lagos de Fúenque and Tota) S to Cundinamarca (Sabana de Bogotá) and also W Meta (Villavicencio) and Vaupés (Mitú); Venezuela (Mérida, Portuguesa and Carabobo E to Miranda and Monagas, and in E Amazonas and Bolívar); Surinam and the Guianas; Brazil in N and SW Pará, E Amazonas, Paraíba, Pernambuco, Alagoas, Bahia, E Minas Gerais (Belo Horizonte), Espírito Santo, São Paulo, SW and NE Mato Grosso, and also historically in Rondônia (Forrester 1993); N, W and E Bolivia in La Paz and Santa Cruz (but see *N. e. erythrops*) and recently recorded Pando, Beni and Chuquisaca (Armonía 1995, 1997; race not given); N Argentina (Salta, Jujuy, Tucumán and Chaco) and Paraguay; also NC Costa Rica at Río Frío (Sarapiquí lowlands) and recently on the Osa Peninsula (see Distribution and Status), and W Panama (Caribbean slope in Bocas del Toro, and Tocumen marsh), where status unclear. Recently recorded from SW Ecuador (Manabí, Guayas and S Loja), the 1 specimen (Manabí) apparently being referable to this race (Ridgely & Greenfield in prep.). Upperparts darker brown than in nominate; underparts darker, slatier; throat greyer; legs and feet orange, salmon, reddish-brown or red. Wing of 8 males 99-105 (101.8), of 9 females 96-106 (100.6); tail of 6 unsexed 30-34 (32, SD 1.4); exposed culmen of 7 males 19-23 (20.8), of 9 females 19-22 (20.1), culmen to base of 5 unsexed 19.5-22 (21.9, SD 2.0); tarsus of 8 males 24-30 (27.1), of 9 females 25-29 (27.4). Weight of 1 female 43, of 3 unsexed 55-70 (62.1); of 1 unsexed probable migrant, Costa Rica, 83 (D. Westcott *in litt.*).

N. e. erythrops (P. L. Sclater, 1867) – N coastal Peru (Lambayeque S to Lima), and Galapagos Is (Santa Cruz and Santa María). One bird from NW Bolivia (La Paz) may belong here (Remsen & Traylor 1983), and racial affinities of birds from Beni, Pando and Chuquisaca (Armonía 1995, 1997) are not clear. See Description.

MOULT Remex moult is presumably sequential, as in Colombian Crake.

DISTRIBUTION AND STATUS The known distribution is patchy: Costa Rica, Panama, E Colombia, SW Ecuador, Venezuela, Surinam, the Guianas, coastal Peru, Brazil (patchily), N and E Bolivia, Paraguay, N Argentina, and the Galapagos Is. Although this species is not regarded as globally threatened, the difficulty involved in finding and observing it make its status in some areas difficult to assess; for example, in Central America it could be accidental, a migrant or a local resident, perhaps overlooked, and its range could be expanding (Behrstock 1983, Ridgely & Gwynne 1989). The race *olivascens* has been regarded as locally common in W Panama (Behrstock 1983, Ridgely & Gwynne 1989), Colombia (Olivares 1974), Venezuela (Friedmann & Smith 1950) and Argentina (Tucumán) (Wetmore 1926); however, it is currently regarded as uncommon in Argentina (Canevari *et al.* 1991), there is only 1 record from Chaco, N Argentina (Contreras *et al.* 1990), and it is rare in Paraguay (Hayes *et al.* 1994). It is probably less rare in French Guiana than the 1 existing record (Aug 1981) suggests (Tostain *et al.* 1992). There is only 1 verifiable record from Guyana, in Aug 1977 (Osborne & Beissinger 1979), it is rather rare in Surinam (Haverschmidt & Mees 1995), and it is recorded from scattered localities in Amazonian, C, E and S Brazil. In SW Ecuador it is apparently rare and local but may be overlooked; it has been recorded from Bahía de Caráquez, Chone, Macará and the Manglares-Churute Ecological Reserve (Ridgely & Greenfield in prep.). The only records (presumably *olivascens*) from Costa Rica are from Río Frío in Aug 1987, and Corcovado NP on the Osa Peninsula (M. Swartz *in litt.*: see Movements); it may also occur in E Costa Rica, where a sighting near Hitoy Cerere in Mar 1985 was claimed to be either this species or Colombian Crake (Stiles & Skutch 1989). The nominate race is possibly rare in Peru but is abundant on Santa Cruz, Galapagos. Its establishment on the Galapagos is apparently recent: it was first recorded in 1953, but was possibly overlooked for some time before that, and could occur on San Cristóbal and Isabela (Harris 1973). Its current status and distribution need investigation, as do its movements and breeding.

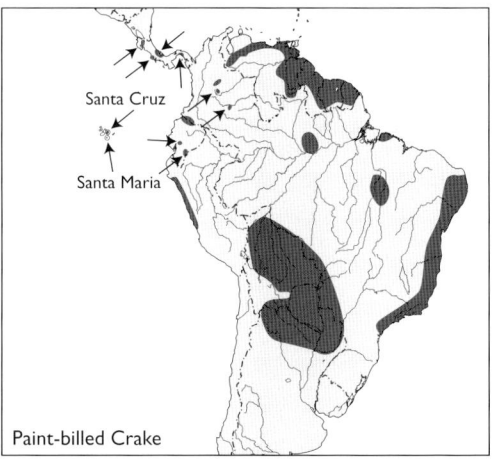

Paint-billed Crake

MOVEMENTS There is no evidence for regular movements, but many records outside the normal range are indicative of either migration or vagrant habits. Two *olivascens* were collected W Panama (Changuinola, Bocas del Toro) in Nov 1981 but the species was not noted in Apr or Jul and is believed to be of seasonal occurrence (Ripley & Beehler 1985). The bird from the Osa Peninsula, Costa Rica (date unknown), flew into a bedroom at Sirena Station, Corcovado NP, during a night in early 1990 after a period of stormy weather (M. Swartz *in litt.*, D. Westcott *in litt.*). Several records at c. 2,600m on the savannas of Bogotá and Ubaté, Colombia, are mostly in Mar-Apr and including one from the streets of Bogota; an emaciated male was also found on 10 Oct 1979 at 3,375m in a suburban garden at La Paz, Bolivia (Remsen & Traylor 1983). In N Venezuela it has been attracted to lighted windows in May-Jun at Rancho Grande (Henri Pittier National Park), and it was reported by hunters to be of seasonal occurrence at Caicara, appearing in numbers during Aug (Friedmann & Smith 1950). In Brazil 1 was found in museum gardens at Belo Horizonte, Minas Gerais (Sick 1993), and Forrester (1993) regards it as a migrant in SW Mato Grosso, accidental in E Amazonas (Manaus) but resident in coastal N Pará (Belém) and Alagoas. In Paraguay it is regarded as both resident and migrant, being less abundant or absent during the austral winter; extreme dates for the Chaco and N Orient are 24 Aug and 12 Apr (Hayes *et al.* 1994, Hayes 1995). Single dead birds have been found on Tower I, Galapagos, in Dec 1972 (Harris 1973), and on Peninsula Valdés, SE Argentina (Canevari *et al.* 1991). It is accidental in the USA: a nominate bird from Texas in Feb 1972 (Arnold 1978), and a probable *olivascens* from Virginia in Dec 1978 (Blem 1980); there is no evidence that these were captive birds. It is said also to have occurred in Nov (Stiles & Skutch 1989, no details available).

HABITAT Reedbeds, lagoons, coastal marshes (Peru), grassy marshes, rank grass, wet to dry pastures, corn and rice fields, gardens, drainage ditches, humid woodlands and savanna, and overgrown bushy areas. In South America it is described as possibly preferring dense swamp vegetation and damp secondary woodland in flooded savanna (Fjeldså & Krabbe 1990), and as being mainly a crake of grass and thickets (Hilty & Brown 1986); in Surinam it occurs in thickly overgrown shrubby, dry places, similar habitat to that occupied by the Russet-crowned Crake (26) (Haverschmidt & Mees 1995), while in Brazil it occurs at forest edges (Sick 1993). It is known to frequent puddles on dirt roads (Friedmann & Smith 1950) and the muddy margins of ponds and ditches (Ridgely & Greenfield in prep.), and in Panama it has been recorded alongside the Colombian Crake at shallow drying rice field ditches and pools, and alongside the White-throated Crake (31) in tall grass and ditches surrounding rice fields (Behrstock 1983). It occurs from tropical to temperate zones, normally in lowland areas, but presumed migrants or vagrants are recorded up to 3,375m.

FOOD AND FEEDING Very little information is available. It is known to eat Diplopoda, insects (including Coleoptera) and seeds (including grass seeds). It takes invertebrates from soil and leaf-litter; at dusk and dawn it forages at puddles, sometimes along dirt roads, and in open patches adjacent to dense vegetation.

HABITS This crake is possibly mainly crepuscular, being difficult to observe during the day but readily seen in the open at the edge of dense cover at dusk and dawn (Behrstock 1983). It is very elusive and furtive, and is reluctant to fly unless disturbed, when it often flies for a short distance with weak, rapid wingbeats and dangling legs, to land in the first available cover; this willingness to fly contrasts with the behaviour of *Laterallus* species such as the Black Rail (33) and the Galapagos Rail (34), which run when disturbed. It apparently roosts above the ground (Sick 1993).

SOCIAL ORGANISATION Apparently monogamous: birds are frequently seen in pairs (e.g. Behrstock 1983).

SOCIAL AND SEXUAL BEHAVIOUR No information available.

BREEDING AND SURVIVAL Season Venezuela, male in breeding condition Jul; Surinam, male in breeding condition Jun; Guyana, female with enlarged ovary Aug; Galapagos, Nov-Feb. **Nest** A bowl of green grass, placed on or near ground in grassy vegetation; recorded sites include "agricultural fields" and, in Venezuela, cornfields (Friedmann & Smith 1955). **Eggs** 3-7; creamy, spotted and blotched reddish-brown and grey, markings concentrated at larger end. Size of eggs: nominate (Argentina), 28.9-30.4 x 21.1-23.2 (29.7 x 23.2, n = 4), calculated weight 8.0; (3 from Galapagos) 32.6 x 23.7, 32.5 x 23.8, 33.9 x 24 (Franklin *et al.* 1979); *olivascens*, 28.7-31.2 x 22.2-23.5 (30.3 x 23, n = 5), calculated weight 8.7. **Incubation** At least 23-25 days.

PARDIRALLUS

Three distinctive, medium to medium-large rails of Central and South America, with long, greenish-yellow bills and reddish legs. Their superficially similar appearance to *Rallus* rails has often led to their being mistakenly placed in that genus, but skeletally they are quite distinct (Olson 1973b). Two species, the Blackish and Plumbeous Rails, have predominantly olive-brown upperparts and dark grey underparts, with ashy-brown to black from lower belly to undertail-coverts. The Spotted Rail differs from all other rails, including its congeners, in being entirely densely and boldly spotted, streaked and barred with white, but it does have a dark morph juvenile plumage which looks very similar to the adults of its congeners. Its variegated plumage is evidently a recently evolved condition derived from a plain plumage, and the separation of the other two species into a different genus (*Ortygonax*) is not justified (Olson 1973b). All three species inhabit marshes. Both the Plumbeous and Spotted Rails are thought to be migratory in parts of their ranges, while the latter may also show widespread dispersal or vagrancy. No species is globally threatened, but all must have been adversely affected by habitat loss and the Spotted and Blackish Rails are generally assumed to be local and uncommon. However all species are unobtrusive and often overlooked, and the Spotted Rail may even be expanding its range in Mexico.

113 SPOTTED RAIL
Pardirallus maculatus Plate 34

Rallus maculatus Boddaert, 1783, Cayenne, French Guiana. Sometimes retained in *Rallus*. Two subspecies recognised.

Synonyms: *Rallus/Pardirallus/Limnopardalus variegatus*; *Aramides/Aramus/Limnopardalus maculatus*; *Porzana variegata*; *Rallus nivosus*.

IDENTIFICATION Length 25-32cm. A medium-sized, long-billed rail with strikingly variegated plumage. Head and neck blackish-brown, heavily streaked white except on crown and nape; upperparts blackish with broad brown feather fringes and white streaks (white spots in race *insolitus*), becoming spots from lower back to upperail-coverts; tail has no white markings; upperwing-coverts browner and with fewer white markings (virtually absent from scapulars and tertials in *insolitus*); Breast striped black and white; flanks and belly barred black and white; undertail-coverts white. Iris red; bill yellowish-olive, with orange-red spot at base; legs and feet pale orange-red or pink-red. Sexes alike; female probably slightly smaller. Immature probably similar to adult. Juvenile has buff-tipped undertail-coverts (grey-tipped in *insolitus*), brownish iris, and duller bill and legs; plumage variable, in 3 colour morphs: dark morph with almost plain, dark brown upperparts and sooty, dark-tipped ventral feathers with no white bars; pale morph with throat and breast pale greyish-brown, and breast weakly barred white; barred morph with throat grey, spotted white, and breast and belly sharply barred white. Inhabits marshes, swamps, wet grassland and irrigated fields.

Similar species Adult easily distinguished from all other rallids by combination of strikingly variegated plumage and long greenish-yellow bill with red spot at base. Dark morph juvenile has almost plain plumage and is similar to Blackish Rail (114) and Plumbeous Rail (115), both of which, however, have paler upperparts and bright red legs and feet; Plumbeous Rail is larger, with chin and throat slate-grey like rest of anterior underparts, and with contrastingly grey-brown lower belly to undertail-coverts; Blackish Rail has black uppertail-coverts, tail, and lower belly to undertail-coverts.

VOICE A loud, repeated, rasping, groaning screech, usually preceded by a grunt or pop: *g'reech*, *kr'krreih*, *pum-kreep* or simply *grrr* or *kehrr*; probably a territorial or aggressive call. The species also has an accelerating series of deep, gruff, pumping notes *wuh-wuh*, like a distant motor starting up and not unlike muffled calls of the American Bittern *Botaurus lentiginosus*, and a sharp, repeated *gek* when disturbed. Birds sometimes call at night. Another type of call, a 4-note whistle of 1 long, high-pitched note and 3 rapid, short, lower notes, is also recorded (Birkenholz & Jenni 1964). There is also mention of a call "much like that of a King Rail" (56) apparently made by this species (e.g. Dickerman & Parkes 1968) and presumably referring to the sharp *gek* described above. Birds also make *tick* and *chick* calls, and adults with young give high-pitched rising *weeep-weep* notes (Hardy *et al.* 1996).

DESCRIPTION *P. m. maculatus*
Adult Forehead to nape blackish-brown, feathers of forehead and forecrown with very fine, short, terminal white shaft-streaks; hindneck and mantle dark blackish-brown to black, each feather with two elongated white spots, one at edge of each web, giving streaked effect; rest of upperparts, including scapulars, dark blackish-brown to black, feathers broadly tipped brown and edged white, forming white streak along each side of blackish central area; white streaks become fewer and shorter, degenerating to spots, on lower back and rump, which are very broadly fringed brown; uppertail-coverts and rectrices blackish-brown, fringed dull brown, coverts with few to many white spots on edges. Lesser and median upperwing-coverts patterned like scapulars and mantle but with much more extensive brown fringes, this brown becoming even more extensive from medians to greaters, which are quite uniform brown with a few broken white spots at edges, spots often bordered blackish-brown. Alula, primary coverts, primaries and secondaries dark brown to blackish-brown; tertials blackish-brown, broadly fringed brown and with variable white streaks, like back. Axillaries and underwing-coverts blackish-brown, spotted and barred white; underside of remiges dusky-brown. Lores dusky blackish-brown; sides of head dark blackish-brown, thickly marked with tiny white spots; ear-coverts sometimes browner and relatively unpatterned; chin and upper throat white, sparingly streaked blackish-brown; lower throat, foreneck, sides of neck, and breast, dark blackish-brown to black, with broad, longitudinal white spots on feather margins giving almost striped effect; flanks and belly barred blackish and white (black bars usually broader); thighs white externally, blackish behind; centre of belly to undertail-coverts white, some coverts occasionally tipped dusky. Iris bright to dark red; bill yellowish-olive, or olive with apple-green base; orange to orange-red spot at base of lower mandible and cutting edge of base of upper mandible; legs and feet pale orange-red to pinkish-red or pink.
Immature Not properly described; probably similar to adult.
Juvenile Undertail-coverts white to grey, tipped buff. Has brownish iris, and duller bill and legs. Plumage variable; 3 colour morphs described, with intermediates also recorded (Dickerman & Haverschmidt 1971). **Dark morph** has almost plain, dark brown upperparts, and sooty, dark-tipped ventral feathers with no white bars (Plate 34). **Pale morph** has throat and breast pale greyish-brown, and breast weakly barred white. **Barred morph** has throat grey, spotted white, and breast and belly dull sooty-brown, sharply barred white (Plate 34). In barred morphs, lower belly and flanks more strongly barred than breast and upper belly.
Downy young Downy chick said to be black, but 1 specimen in BMNH has down of upperparts blackish (darkest from forehead to hindneck) and rich dark brownish-black elsewhere; bill black with paler tip; legs and feet dark, possibly brownish-black.
MEASUREMENTS Wing of 14 males 121-134 (126.4), of 9 females 116-124 (119.4); tail of 10 males 46-54 (49.6, SD 2.5), of 8 females 36.5-51 (45, SD 6.0); culmen to base of 8 males 47.5-50 (49.2, SD 1.0), of 9 females 42.5-48.5 (45.3, SD 1.8); tarsus of 10 males 37-42.5 (39.3, SD 1.5), of 3 females 36, 40, 42. Weight of 7 males 148-198 (169), of 9 females 130-190 (150) (Watson 1962, Belton 1984, Haverschmidt & Mees 1995); of 16 unsexed (both races) 140-198 (171, SD 18.2) (Dunning 1993).

GEOGRAPHICAL VARIATION Possible race *inoptatus* of W Cuba included in nominate: see Watson (1962). Races separated on pattern of upperparts, and on colour of juvenile undertail-coverts.
P. m. insolitus (Bangs & Peck, 1908) – Mexico (Nayarit, Michoacán, Guerrero, Puebla, Veracruz, Quintana Roo, Oaxaca, Chiapas and Yucatán), Belize (Ycacos Lagoon), El Salvador and Costa Rica (Guanacaste and Cartago). Upperparts darker and richer brown than in nominate, and spotted rather than streaked white; inner secondaries, secondary coverts and scapulars with white spotting reduced or absent. Like nominate, juvenile has three colour morphs, but has undertail-coverts tipped sooty-grey, not buff. Measurements of 2 unsexed: wing (chord) 121, (max) 135; tail 42, 44; exposed culmen 48, culmen to base 51.5; tarsus 38, 40. Weight of 4 males 187-219 (206.25, SD 13.6), of 1 female 161 (Dickerman & Warner 1961; Dickerman 1971).
P. m. maculatus (Boddaert, 1783) – Cuba (probably including I of Pines), Hispaniola (Dominican Republic), Jamaica, Trinidad and Tobago, C and E Colombia (middle and upper Cauca V in Valle and Cauca, Magdalena V, "Bogotá" and Meta), W and N Venezuela (Mérida, Portuguesa, Carabobo and Aragua), Surinam, French Guiana, E Brazil (S from Amapá, Pará and Ceará), coastal W Ecuador (W Esmeraldas, SE Guayas and El Oro), E Bolivia (Santa Cruz and Tarija), NW and SE Peru (Piura, Lambayeque, La Libertad and Madre de Dios), Paraguay, Uruguay (Montevideo, Canelones, Maldonado, Treinta y Tres, Cerro Largo, San José, Colonia, Rocha and Lavalleja) and N Argentina (S to Tucumán, Córdoba and Buenos Aires). See Description.
Also occurs in Panama (Canal area, Tocumen marsh in E Panamá Province, and San Blas), subspecies unknown but possibly nominate.

MOULT Remiges and rectrices simultaneous. The following information is from Cuban birds (Watson 1962). Complete (prebasic) moult begins on the head (nape to back, and throat to flanks); the remiges, alula, wing-coverts and rectrices are then shed simultaneously; the crown then moults; the wings and tail grow quickly and the last tract to be moulted is the uppertail-coverts. Complete moult probably occurs Aug-Dec and has been recorded in the latter part of the breeding season, for example a laying female was in the process of replacing the alula, all the remiges, the greater coverts and the rectrices. Probable partial pre-alternate moult is recorded in Feb and May. Body moult is recorded in Mexico, Aug (Dickerman 1971).

DISTRIBUTION AND STATUS Mexico and C America; Greater Antilles; Trinidad and Tobago; Colombia E to Guianas; W Ecuador, E Bolivia and SE Peru, and from E Brazil S to Paraguay, Uruguay and N Argentina. Not globally threatened. It is locally distributed, and was formerly often regarded as uncommon to rare, although locally common in the Zapata Swamp (Cuba) and in Costa Rica, Colombia and Surinam (Orians & Paulson 1969, Ripley 1977, Bond 1979, Clements 1979, Hilty & Brown 1986). Its current status is unclear in many areas, but it is locally frequent to uncommon in Mexico, where it is more widespread than was formerly known and may be increasing – or previously overlooked (Howell & Webb 1995); it is local in French Guiana (Tostain et al 1992); there are few records from Panama, where it was only confirmed as a breeding resident in 1978 (Wetmore 1965, Ridgely & Gwynne 1989); it is rather common in Surinam (Haverschmidt & Mees 1995), scarce in Brazil (e.g. Forrester 1993, Mauricio & Dias 1996), rare and local (but probably much overlooked) in Ecuador (Ridgely & Greenfield in prep.), rare in parts of Argentina (Contreras et al. 1990, Narosky & Di Giácomo 1993), scarce in Uruguay (Arballo & Cravino in press), and a rare breeder throughout Paraguay (Hayes et al. 1994). In Bolivia it was first recorded from Tarija Dpt in May 1997 (Armonía 1997). Its total population is possibly <10,000 birds (Rose & Scott 1994). It is undoubtedly overlooked, especially when breeding, and is probably more widespread within its range than existing records suggest, although it must have suffered from habitat loss. It was thought to have been extirpated from Jamaica, but there is at least one fairly recent sight record, from the Black R marshes, in 1979 (Bond 1980). Predation by the introduced mongoose *Herpestes auropunctatus* was thought to have drastically reduced the population in drained agricultural lands of Cuba by the 1960s, while the disappearance of the bird from Jamaica was also attributed to mongoose predation (Watson 1962).

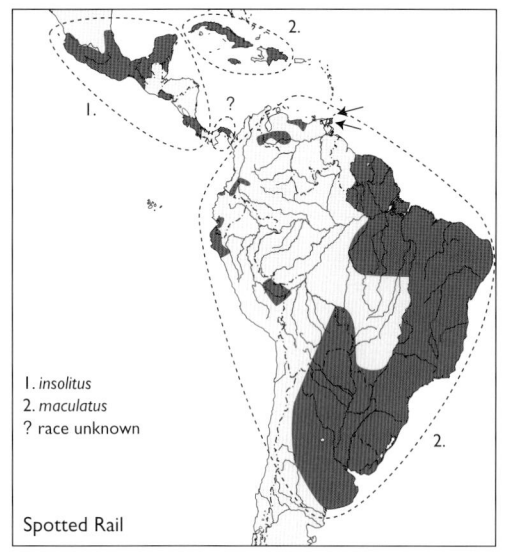

Spotted Rail
1. *insolitus*
2. *maculatus*
? race unknown

MOVEMENTS None definitely recorded, but it is regarded as a northern austral migrant in Paraguay (not shown as such on the distribution map), extreme dates for 27 specimens and 3 sight records being 30 Aug and 13 Apr (Hayes et al. 1994). It moves locally in response to adverse conditions such as drought in Surinam (Haverschmidt 1968) and possibly in response to changing water levels in Costa Rica (Stiles & Skutch 1989). In Surinam, 1 entered a house at night at Paramaribo, far from suitable habitat, in Jan (Haverschmidt 1968). It may also show widespread dispersal or vagrancy: an immature in mangroves in Panama, Oct, was almost certainly dispersing (Ridgely & Gwynne 1989), and it is accidental in North America, with records from Pennsylvania and Texas, both possibly man-assisted, and the Juan Fernandez Islands 650km from the Chilean coast (AOU 1983, Remsen & Parker 1990). One was found 500km off the coast of Espírito Santo, Brazil, in Sep (Sick 1993).

HABITAT Marshes, including those dominated by *Typha* or *Polygonum* and those with tangled secondary growth; also wooded swamps, rice paddies and other irrigated fields, wet, overgrown grassy ditches, and wet grasslands and savanna. One site in Costa Rica had *Panicum purpurescens* 45cm to 1m tall on a wet, muddy substrate (Birkenholz & Jenni 1964); in Cuba it occurs in *Cladium jamaicense*, rush and *Myrica* bush habitat (Watson 1962); in Mexico it is recorded from a large area of *Carex* taller than the surrounding grassy meadow (Dickerman & Warner 1961). In Rio Grande do Sul, Brazil, it occurred and bred (see Breeding) in marshes with *Scirpus giganteus* and *S. californicus* (Mauricio & Dias 1996). It is also recorded from tall grass on an abandoned airfield. It requires dense cover of emergent plants, grass or tangled second growth. One in mangroves in Panama was probably dispersing (Ridgely & Gwynne 1989). It occurs in lowlands, but up to 2,000m in Colombia.

FOOD AND FEEDING Earthworms; gastropods (*Ampullaria*); adult and larval insects, including Coleoptera, Hemiptera (Naucoridae), Diptera (Tabanidae) and Odonata (Aeshnidae); also other invertebrates; small fish (Cichlidae); also pondweed *Potamogeton epihydrus* and the seeds of aquatic plants. It may feed in fairly open situations at any time of day, especially in the early morning and late evening and on overcast days. Birds forage at the water's edge or while wading, and probe in mud for worms. An individual was observed capturing *Pomacea* snails which it carried to a site with a small flat rock (below a bridge); each snail was held in the bill and repeatedly smacked on the rock before being consumed (Arballo & Cravino in press). One, possibly 2, birds foraged over an area of 100 x 50m after heavy rain (Birkenholz & Jenni 1964). A captive bird ate chopped meat.

HABITS Generally retiring and difficult to see, being difficult to flush and usually remaining within dense cover except when feeding; it may fly strongly when flushed. However in Mexico it may often appear unconcerned about human presence and will feed in fairly open situations at any time of the day (Howell & Webb 1995). In Costa Rica, birds flushed more often when vegetation was short (15-45cm tall), and also ran and called freely when disturbed, but they were impossible to flush, did not run, and rarely called, when the vegetation had grown to a height of 75-100cm (Birkenholz & Jenni 1964). Haverschmidt (1968) states that during extremely dry months in Surinam it occurs in numbers around the few remaining waterholes; he also notes that it resembles the Water Rail (61) in its habits and its way of running through the rushes and along the water's edge. A captive bird frequently climbed in vegetation, using its wings for balance, and remained hidden during most of the day perched high in a shrub, but sometimes flew down, chiefly in the morning and late afternoon, to feed and bathe in a shallow pan (Friedmann 1949). It may be active and vocal at night (Stiles & Skutch 1989).

SOCIAL ORGANISATION Monogamous; apparently territorial, at least during breeding season, when 8-9 birds occurred in a 21-ha marsh (Birkenholz & Jenni 1964).

SOCIAL AND SEXUAL BEHAVIOUR The only information available is that a bird was seen fighting with a

Plumbeous Rail in a marsh in Rio Grande do Sul (Mauricio & Dias 1996).

BREEDING AND SURVIVAL Season Cuba, late summer to early autumn (at least to Sep); Hispaniola, breeding condition Jun; Trinidad, Jun-Aug; Costa Rica, breeding condition Jul; Panama, adult with small chick mid-Jan; Colombia, 2 young Dec; Venezuela, 2 young birds 15 Dec; SE Brazil (Rio Grande do Sul), adult with 1 half-grown young, 16 Nov 1992. **Nest** Cupped platform or bowl of grass or dead rushes, built low down in wet grass or other marsh vegetation, often just above shallow water (recorded depth 10-45cm); diameter 15cm. **Eggs** 2-7; oval-pyriform; creamy-white, pale buff or brownish-buff, marked and spotted brown, cinnamon and lavender at larger end. Size of 25 eggs (nominate) 35-42 x 25.5-31 (37.9 x 28.2); calculated weight 16.7. No other information available.

114 BLACKISH RAIL
Pardirallus nigricans Plate 34

Rallus nigricans Vieillot, 1819, Paraguay.

Sometimes retained in *Rallus*, or referred to genus *Ortygonax*; formerly sometimes regarded as conspecific with *P. sanguinolentus*. Two subspecies recognised.

Synonyms: *Limnopardalus/Aramides/Aramus/Ortygonax nigricans*; *Rallus/Aramides immaculatus/caesius*; *Gallinula caesia*; *Rallus bicolor*.

IDENTIFICATION Length 27-32cm. Medium-sized, plain, dark rail; upperparts and upperwings darkish olive-brown; uppertail-coverts and tail black; chin and throat white to grey (white more extensive in race *caucae*); forehead, forecrown, sides of head and neck, breast, flanks and upper belly dark bluish-slate (paler, more blue-grey, in *caucae*); lower belly, vent and undertail-coverts black (vent paler in *caucae*); bill long and slightly decurved, green with yellowish base; eye, legs and feet red. Sexes alike. Immature browner than adult, especially on posterior upperparts; entire underparts smoke-grey. Juvenile like immature but has more brown-tinged underparts; iris and bill black; legs and feet brown. Inhabits marshes, swamps and wet grassy areas.

Similar species Closely resembles partially sympatric Plumbeous Rail (115), which differs in being larger and having slate-grey chin and throat, broad dusky centres to feathers of scapulars, tertials, lower back and rump in all but the 2 southernmost races (but note that Blackish Rail often has dark centres to tertials), olive-brown uppertail-coverts and tail, grey-brown lower belly, vent and thighs, ashy-brown undertail-coverts, and more decurved, greenish-yellow bill with red and blue at base. Upperparts of Blackish Rail tend to appear more olivaceous in tone (but 2 races of Plumbeous Rail are similarly olivaceous). Bill colour and voice are the most useful field characters. May look like a small Slaty-breasted Wood-rail (78), which sometimes occurs alongside it in Brazilian marshes, but this wood-rail has upperparts tinged rufous, especially from nape to mantle and sides of upper breast, russet primaries, and paler overall cast.

VOICE Descriptions indicate that the warning call is a high whistle *tirit, kirk, PEEoo* (Sick 1993), or a very fast, metallic *tii'd'dit* and a complaining *keeeeaaa*, the latter call being similar to that of the Roadside Hawk *Buteo magnirostris*; the song (presumed to be given by the male) is described as a very sharp, loud, penetrating, repeated *whuueeee*, or *wheeee wheeee wheeee chee chee chee chee che che chechch* (Belton 1984), which may be accompanied by a low, almost inaudible *bubububu* given presumably by the female (Parker *et al.* 1991); the song is also described as beginning with an ascending series of pig-like grunts, with bellows in the background, and ending with a clear, descending tremolo (Sick 1993). Birds out of cover give a rapid, repeated *chchchee*, and the species has a sharp double squeak of anxiety (Belton 1984). The *keeeeaaa* call is also given by nesting adults in distraction and threat displays (Naranjo 1991). This species often calls in the late afternoon.

DESCRIPTION *P. n. nigricans*
Adult Forehead and forecrown dull, dark grey; hindcrown to rump, and upperwings, including scapulars and tertials, uniform darkish olive-brown; tertials often with blackish-brown centres; uppertail-coverts and tail black; primary coverts, alula, primaries and secondaries blackish-brown, primary coverts with more olive-brown outer webs; secondaries sometimes narrowly edged olive-brown; axillaries and underwing-coverts blackish-brown. Chin and throat variably white to slate-grey; sides of head, sides of neck, foreneck, breast, flanks and upper belly dark bluish-slate; lower belly, vent and undertail-coverts black. Iris brilliant carmine; bill bright green, yellowish or greenish-yellow at base and greyer at tip; legs and feet bright coral-red to orange-red. Sexes alike.
Immature Similar to adult but browner, especially on posterior upperparts. Crown and nape blackish-olive; rest of upperparts brownish-olive, becoming warmer and more sepia posteriorly; lower throat pale brownish-olive; chin, upper throat and underparts smoke-grey.
Juvenile One bird, not fully grown (BMNH), similar to immature but with underparts slightly more brown-tinged; iris black; bill blackish; legs and feet brown.
Downy young Black; 1 chick had bill and legs coral-red (Naranjo 1991). Possible chicks of this species, with black bare parts and with greyish-white filaments scattered among black down (Belton 1984) are now assigned to Plumbeous Rail (Naranjo 1991, Belton 1994).

MEASUREMENTS Wing of 9 males 136-142 (139.6), of 10 females 128-139 (132.5); tail of 5 males 54-58 (56, SD 1.6), of 8 females 49-66 (60.9); culmen to base of 5 males 50-58 (54.4, SD 3.1), of 2 females 49.5, 53.5; tarsus of 9 males 43-49 (46.8), of 10 females 40-47 (42.8). Weight of 1 male 217 (Belton 1984).

GEOGRAPHICAL VARIATION Races separated on size and on shade of underparts. Supposed race *humilis* included in nominate (Ripley 1977).
P. n. caucae (Conover, 1949) – SW Colombia (Upper Cauca V from vicinity of Medellín S to head of valley at El Tambo, and in W Nariño). Supposedly larger than nominate (but see measurements); has paler, more bluish-grey underparts; vent less black, more fuscous; throat patch usually whiter and more extensive. Bill greenish-yellow, tip darker; legs and feet bright brick-red. Wing of 10 males 121-136 (130.1), of 8 females 117-131 (127); exposed culmen of 8 males 48-57 (52.5), of 8 females 46-52 (49.5), culmen to base of 2 males 57, 58; tarsus of 10 males 40-50 (46.1), of 8 females 43-47 (45).

P. n. nigricans (Vieillot, 1819) – E Ecuador (W Sucumbios S to Zamora-Chinchipe), C and E Peru S to Madre de Dios, NW Bolivia (Pando, La Paz), W, SC and E Brazil (Pernambuco and Goiás S to Rio Grande do Sul), E Paraguay (E of Paraguay R) and extreme NE Argentina (Misiones). See Description.

MOULT Moult of primaries, at least, is simultaneous.

DISTRIBUTION AND STATUS From SW Colombia to Ecuador, Peru, Bolivia, Brazil, Paraguay and NE Argentina. Not globally threatened. Past and present status are difficult to assess, as the species is usually secretive and difficult to observe, but both races are often regarded as uncommon and patchily distributed. Its total population is estimated at 10,000-100,000 birds (Rose & Scott 1994). Although the race *caucae* is regarded as uncommon in Colombia (Hilty & Brown 1986), it is locally common in the N Cauca Valley (Naranjo 1991). The nominate race has been recorded as rare in E Peru (Parker *et al.* 1982); it is known only from 2 sites in Bolivia, Pando, and Alto Madidi, La Paz (Parker & Remsen 1987, Parker *et al.* 1991); and it is rare in Paraguay, having been recorded only E of the R Paraguay, (Hayes 1995, Lowen *et al.* in press). However it is uncommon to locally fairly common in Ecuador (Ridgely & Greenfield in prep.); in SE Brazil it is locally fairly common in Rio Grande do Sul and Santa Catarina (Belton 1984, do Rosário 1996) and it is common in suitable habitat throughout São Paulo (E. O. Willis *in litt.*); in Brazil it is regarded by Sick (1993) as one of the most abundant rails in the areas which it inhabits.

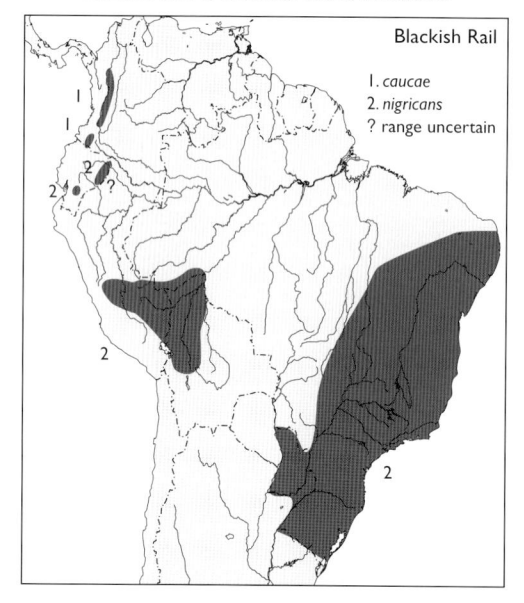

Blackish Rail
1. *caucae*
2. *nigricans*
? range uncertain

MOVEMENTS None recorded, but there is a stray record from Junín, NC Peru, at 4,080m (Fjeldså & Krabbe 1990), and on rainy nights in Brazil birds overfly cities such as Rio de Janeiro, where their voices can be heard (Sick 1993).

HABITAT Marshes of almost any type (Sick 1993), vegetation-choked waterways, flooded rice fields, tall damp grass, and swampy, lightly wooded areas. In the Santos-Cubatão region of São Paulo State, SE Brazil, this species occurs, with the Grey-necked Wood-rail (75) and the Rufous-sided Crake (28), along rivers wherever there are swamps rich in *Typha*, sedges etc; it is largely absent from mangroves proper but is found in transitional areas which have some freshwater input and are dominated by tangles of *Hibiscus tiliaceus*, and in surrounding grassy wetlands (F. Olmos *in litt.*). In Bolivia this species inhabited a marshy area of tall grass *Paspalum* in severalcm of water (Parker *et al.* 1991). It is predominantly a lowland species, but it occurs from 800 to 2,200m in the Andes, exceptionally 4080m (see Movements). In Brazil it occurs alongside the Plumbeous Rail at both lowland and upland marshes from S Minas Gerais to São Paulo and Rio de Janeiro (F. Olmos *in litt.*, J. F. Pacheco *in litt.*) and the two species are widely sympatric in Rio Grande do Sul (W. Belton *in litt.*). In Paraguay, however, A. Madroño (*in litt.*) has not found them together. For a comparison of the two species' habitat preferences, see Plumbeous Rail, Habitat section.

FOOD AND FEEDING Almost no information is available on food; it apparently eats insects and other invertebrates, and small vertebrates, and it is recorded taking small watersnakes such as *Helicops* (Canevari *et al.* 1991, Sick 1993). It occasionally forages in the open on muddy shores with other waterbirds or in small clear areas in marsh vegetation, and it searches for prey along flooded shorelines. It is recorded venturing 5m from a marsh into a restaurant garden for food scraps (Belton 1984).

HABITS This species is furtive and not easily seen unless flushed, when it flies only a short distance. Occasionally it walks in the open on muddy shores with other waterbirds or near marshy vegetation, but it seldom ventures far from marshy habitat and when alarmed it runs to cover in dense vegetation. In Rio Grande do Sul, birds permanently occupied a small marsh only 100m from a house (Belton 1984), and the species frequently occurs close to houses (Sick 1993).

SOCIAL ORGANISATION Birds permanently occupied a marsh of only 0.25ha in Rio Grande do Sul (Belton 1984). In Bolivia, at least 1 pair occupied a 1ha marsh (Parker *et al.* 1991).

SOCIAL AND SEXUAL BEHAVIOUR A nesting adult ran in tight circles around a human intruder, uttering the *keeeeaaa* call c. every 5s; when the chicks had hatched an adult also threatened the intruder with the wings partly open and the bill gaping, before jumping on the observer's back for a moment (Naranjo 1991).

BREEDING AND SURVIVAL Season Colombia, May-Jun, male 'with eggs' Sep (BMNH); Rio Grande do Sul, Brazil, male with somewhat enlarged testes Jul. **Nest** Of aquatic grasses, placed on or near the ground in tall damp grass. One nest was an elaborate shallow cup of wild rice stems woven in concentric rings and anchored to the base of several *Typha* leaves; external diameter 23cm; depth 2.5cm (Naranjo 1991). Possible nest of this species from Rio Grande do Sul, Brazil (Belton 1984), is now assigned to Plumbeous Rail (Naranjo 1991, Belton 1994). In captivity, 1 pair built nest of ivy leaves and iris leaves (Pryor 1969). **Eggs** 3 (2 in captivity); ellipsoid; pale buff, creamy-white or stone-coloured, with sparse brown or chestnut spots sometimes concentrated at larger end. Size of 40 eggs 37.7-44.6 x 28.7-33 (40.3 x 30.7); 1 egg larger, measuring 50 x 31.5 (Naranjo 1991); calculated weight 20.7. **Incubation** At least 23 days (Naranjo 1991); in captivity 18-21 days, apparently by female (Pryor 1969). **Chicks** Young leave nest soon after hatching; fed and cared for by both parents.

115 PLUMBEOUS RAIL
Pardirallus sanguinolentus   Plate 34

Rallus sanguinolentus Swainson, 1838, Brazil and Chile.

Sometimes retained in *Rallus*, or referred to genus *Ortygonax*; formerly sometimes regarded as conspecific with *P. nigricans*, or listed as *P. rytirhynchos*. Six subspecies recognised.

Synonyms: *Pardirallus/Rallus/Limnopardalus/Aramides/Ortygonax rytirhynchu(o)s*; *Rallus setosus/caesius/bicolor/ricordi/antarcticus*; *Rallina caesia*; *Aramides/ Rallus zelebori*; *Limnopardalus vigilantis*; *Ortygonax sanguinolentus*.

IDENTIFICATION Length 28-38cm. A medium-large, stocky, plain, dark rail with long, slightly decurved bill. Head, sides of neck, foreneck and entire underparts to flanks and upper belly slate-grey (darker in race *zelebori*), darker on head and washed brownish on flanks; upperparts, upperwings and tail olive-brown (slightly richer brown in *tschudii*, *zelebori* and *luridus*), with darker mottling on scapulars, tertials, lower back and rump (and sometimes on mantle) except in southernmost races *luridus* and *landbecki*; lower belly, vent and thighs ashy-grey; undertail-coverts ashy-brown. Iris, legs and feet red; bill deep green with bright blue base to upper mandible and bright red base to lower mandible. Sexes alike. Immature duskier brown than adult; throat and (sometimes) sides of head whitish; underparts ashy-grey, tinged brown; bill greenish, sometimes with darker base; iris brown; legs red-brown. Juvenile similar on upperparts; underparts brown, with buffy-white throat and belly; bill, legs and feet black. Inhabits marshes, waterside thickets and wet grassy areas.
Similar species Very similar to partially sympatric Blackish Rail (114), which differs in being smaller and having whitish chin and throat, paler and more olivaceous upperparts with no dark feather centres, dark inner remiges (sometimes with narrow olive-brown edges), black vent and tail, and straighter, greenish bill with yellowish base. Bill colour is most useful field character (but note that, in race *luridus*, red spot at base is indistinct or absent); voice also distinctive.

VOICE The male's song is a series of high, penetrating, rolling squeals *rruet'e*, *pu-rueet*, *huyr* or *tsewWIT* etc., which is often accompanied by simultaneous low, deep, hollow-sounding *hoo* notes given by the female as a duet (Fjeldså & Krabbe 1990, Ridgely & Greenfield in prep.). The species sings both day and night, especially during twilight, one song often giving rise to a chorus. The calls are repeated *giyp* or *wit* notes (Fjeldså & Krabbe 1990). When disturbed, the birds are said to repeat an almost painfully sharp note (Hudson 1920), but Sick (1993) describes the warning note as loud and coarse, similar to that of the Helmeted Guineafowl *Numida meleagris*.

DESCRIPTION *P. s. simonsi*
Adult Forehead and crown dark slate, feathers of forehead with blackish shafts; nape, hindneck, and entire upperparts and upperwings, olive-brown, sometimes rather darker and richer brown from lower back to uppertail-coverts, and with rather indistinct blackish-brown centres and bases to scapulars and feathers of lower back and rump giving mottled effect; some birds may also have obscure darker centres to mantle feathers; rectrices blackish-brown, fringed olive-brown; alula, primary coverts, primaries and secondaries dark brown to blackish-brown, narrowly edged paler on outer webs of primaries and secondaries; tertials more olive-brown, with blackish-brown centres, like lower back and rump; underside of remiges, axillaries and underwing-coverts dusky brown, coverts variably and narrowly tipped whitish. Lores dusky; sides of head dark slate, like crown; chin, throat, sides of neck, foreneck, and underparts to flanks and upper belly, slate-grey, flanks washed dusky or brownish; lower belly, vent and thighs ashy-grey; undertail-coverts blackish, tipped ashy-brown, appearing ashy-brown. Iris bright red; bill deep green, with bright blue base to upper mandible and bright scarlet base to lower mandible; legs and feet bright to dark red. Both sexes have equally bright bare part colours, which apparently are similar in both spring and autumn (Wetmore 1926).
Immature Duskier brown than adult, with slightly darker tail; chin (possibly also cheeks) and throat whitish; underparts ashy-grey, overshaded with brown and darkest on flanks; undertail-coverts isabelline-brown with black feather centres. See *P. s. sanguinolentus*.
Juvenile Similar to immature on upperparts; underparts all brown, paler than upperparts, with buffy-white throat and belly; bill, legs and feet black (also see Sick 1993).
Downy young Down blackish-brown, darker on upperparts and wings; bill black; legs and feet pale brown; small white egg-tooth present. Chicks of this species, with black bare parts and with greyish-white filaments scattered among black down, were previously assigned to Blackish Rail (Belton 1984, Naranjo 1991, Belton 1994).

MEASUREMENTS Wing of 16 males 129-142 (135), of 8 females 121-132 (128.1); tail of 7 males 52-68 (60.7, SD 4.9), of 1 female 54; culmen to base of 5 males 49-58 (53, SD 3.0), of 1 female 50, exposed culmen of 7 females 42-50 (44.8); tarsus of 16 males 41-51 (45.8), of 8 females 40-43 (41.6). Weight of 4 males 130-185 (148.7, SD 25.9), of 1 female 125 (Marín 1996).

GEOGRAPHICAL VARIATION Moderate variation exists, races being separated on size, and also on colour and pattern of upperparts, and shade of underparts. Race *luridus* includes proposed form *vigilantis* (Ripley 1977).
P. s. simonsi Chubb, 1917 – Extreme S Ecuador (Zamora Chinchipe), arid Pacific slope of Peru, from Lambayeque S to N Chile in Tarapacá and possibly Antofagasta (Johnson 1965, Blake 1977). See Description.
P. s. tschudii Chubb, 1919 – Temperate zone of Peru S from upper Marañón V through Junín to C and SE Bolivia in La Paz, Oruro and Cochabamba; this race also in extreme S Bolivia (Tarija), and in higher-altitude regions of NW Argentina (Catamarca to Jujuy) (Fjeldså & Krabbe 1990). Larger than *simonsi*; like *luridus* and *simonsi* has relatively pale upperparts, more reddish-brown than in *simonsi*. Wing of 11 males 136-153 (142.6), of 9 females 130-140 (133.6); culmen to base of 4 males 55-59 (56.75, SD 1.7), of 2 females 48, 52, exposed culmen of 7 females 44-50 (47); tarsus of 11 male 42-51 (46.5), of 9 females 39-45 (41.5).
P. s. zelebori (Pelzeln, 1865) – SE Brazil in Rio de Janeiro, "and probably also in the neighbouring provinces" (Ripley 1977): see Distribution and Status. The smallest race; very similar to nominate in having heavily black-centred scapulars, tertials, and feathers of lower back and rump; upperparts darker, more rufescent, and underparts from chin to upper belly

deeper slate; bill more slender, and paler yellowish-green. Measurements of 2 adults: wing 110, 114; tarsus 55; culmen 41, 45 (Hellmayr & Conover 1942).

P. s. sanguinolentus (Swainson, 1838) – Bolivia in Santa Cruz, and also in Tarija (Blake 1977) which is unlikely, as this appears to be within the range of *tschudii* (Fjeldså & Krabbe 1990); Paraguay (mostly E of R Paraguay); extreme SE Brazil in S Minas Gerais (J. F. Pacheco *in litt.*), São Paulo and Rio Grande do Sul (Blake 1977, Fjeldså & Krabbe 1990), and possibly also in the Paraíba do Sul V, S Rio de Janeiro (J. F. Pacheco *in litt.*); Uruguay; also in "N Argentina" SW to Mendoza and S to Río Negro (Contreras 1980, Nores & Yzurieta 1980) – this presumably refers to lower-altitude areas such as Formosa, Chaco, E Salta etc, not higher-altitude areas from Catamarca to Jujuy, where *tschudii* occurs (J. Mazar Barnett *in litt.*). Birds from Santa Catarina are possibly of this race. Larger than *zelebori*, smaller than *luridus* and *landbecki*; scapulars, tertials, and feathers of lower back and rump with dark centres and bases. Iris orange-red to carmine; bill pale green, darkening to grey or black at tip; base of upper mandible blue; base of lower mandible and cutting edge of base of upper mandible, red (Belton 1984). Immature less olivaceous on upperparts and much paler below, being almost white on throat and greyish to brownish-white on breast and belly; iris brown; bill greenish-grey with black base; tarsus reddish-grey to reddish-brown (Belton 1984). Wing of 16 males 127-144 (132.6), of 11 females 118-129 (124.8); tail of 10 males 55-72, of 5 females 54-63; culmen to base of 6 males 54-60 (56.7, SD 2.3), of 2 females 52, 53.5, exposed culmen of 9 females 45-53 (47.7); tarsus of 16 males 42-49 (46.1), of 11 females 40-46 (41.2). Weight of 3 males 170, 208, 213 (Belton 1984); of 9 adults 190-318 (232, SD 37.5) (Marín 1996).

P. s. landbecki (Hellmayr, 1932) – C Chile from Atacama S to Aisén, and SW Argentina (W Chubut and W Santa Cruz). Larger than all races except *luridus*, like which it has no dark mottling on upperparts; like *simonsi*, upperparts paler and more olive-brown than other races; red spot at base of bill small, sometimes indistinct. Wing of 11 males 144-155 (150), of 9 females 134-146 (139.3); tail of 4 males 63-79 (70.8, SD 6.7), of 4 females 66-71 (68.8, SD 2.2); culmen to base of 2 males 63, 65, of 2 females 56, 58, exposed culmen of 9 males 59-68 (62.5), of 7 females 51-60 (54); tarsus of 11 males 48-55 (52), of 9 females 41-51 (46.3). Weight of 3 males 230, 230, 318, of 6 females 190-255 (218.3, SD 23.3) (Marín 1996).

P. s. luridus (Peale, 1848) – S Chile (Blake 1977) and S Argentina to Tierra del Fuego (including Isla Grande); one doubtful record from Falkland Is. Markedly larger than other races, with heavy bill, feet and toes. Like *landbecki*, upperparts have no dark mottling, but colour more rufous-brown, and feathers may show faint buff tips; red spot at base of bill vestigial or absent. Iris golden-brown in 1 female; bill brown, sides grass-green; legs and feet pinkish-brown (Ripley 1977). Wing of 5 males 149-160 (152.4, SD 4.7), of 2 females 148, 156; tail of 5 males 67-81 (74, SD 5.8), of 1 female 78; culmen to base of 5 males 62-72 (67.2, SD 3.8), of 1 female 66; tarsus of 5 males 50-56 (52.3, SD 2.3), of 2 females 47, 49.5. Weight of 1 female 233 (Ripley 1977); of 1 male 355 and 1 female 255 (Marín 1966), these being weights given for *Rallus antarcticus* but referable to present species (M. Marín *in litt.*).

MOULT No information available.

DISTRIBUTION AND STATUS S Ecuador, Peru, Chile, Bolivia, SE Brazil, Paraguay, Uruguay and Argentina. Not globally threatened. The nominate race, *simonsi* and *landbecki* were formerly regarded as locally common (e.g. Olrog 1959, Belton 1984, Parker *et al.* 1982); the former status of others is not clear. Its current status is difficult to assess in many areas, but *tschudii* is abundant on the N Altiplano around Junín, Peru; *sanguinolentus* is rare in Paraguay but common to locally abundant in Buenos Aires Province, Argentina, and abundant in Uruguay; *luridus* is common on Isla Grande, Tierra del Fuego (Klimaitis & Moschione 1987, Fjeldså & Krabbe 1990, Narosky & Di Giácomo 1993, Hayes 1995, Arballo & Cravino in press). The race *simonsi* was recently discovered in extreme S Ecuador, at Vilcabamba (Zamora Chinchipe), where it is apparently fairly common in the quite limited amount of suitable habitat (Rasmussen *et al.* 1996), and it is also recorded from Cariamanga, S Loja (Ridgely & Greenfield in prep.).

The race *zelebori* appears to have a restricted range, occurring in the lowlands of Rio de Janeiro from the Lagoa Feia complex (c. 22°S) S to the Rio de Janeiro City area including Sepitiba and Piratininga (c. 23°S); beyond these limits suitable habitat is restricted because the mountains and tablelands reach the sea (J. F. Pacheco *in litt.*). The species (race unspecified) is locally frequent in E Santa Catarina, Brazil (do Rosário 1996).

Its total population is estimated at >1,000,000 birds (Rose & Scott 1994), and it is probably more widespread than is currently known.

Plumbeous Rail

MOVEMENTS It is thought to be migratory over part of its range but the extent of movements is unclear and is not shown on the distribution map. Birds of inland marshes on the SW pampas are thought to move N in winter, while those of Atlantic seaboard marshes, where shelter and temperature remain suitable in winter, are sedentary (Hudson 1920). Around L Junín, it was common in May but absent in Oct (Harris 1981). At marshes at Cape San Antonio, Buenos Aires, Argentina, it was seen in large numbers in Sep but fewer were present in Apr (Weller 1967). In Isla Grande it is a summer breeding visitor, occurring from Nov to Feb (Humphrey *et al.* 1970). Near Tunuyan, Mendoza, W Argentina, Wetmore (1926) found it very common in late Mar, when it was evidently on migration from colder regions in Patagonia. Marshes and *ciénagas* were filled with the birds, while others were encountered in heavy growths of weeds, at the borders of hemp fields and along irrigation ditches; they flushed easily from very heavy cover and the flight was relatively swift and high.

HABITAT Reedmarsh, sometimes of a small area, and especially around muddy creeks and ponds with much floating vegetation (e.g. *Ranunculus* and *Hydrocotyle*), and at the borders of lakes, rivers and streams. It also occurs in grass-grown marshes, swamps, and locally in waterside thickets (sometimes dense, tangled and flooded to <1 m), irrigated areas such as alfalfa crops, ditches through rushy pasture in the cultivated parts of semiarid valleys, and oases in arid regions. In S Ecuador it survives close to human habitation, being recorded on a main street near a riverbed and moist sugarcane fields, and in wet pasture near a small *Scirpus* marsh (Rasmussen *et al.* 1996). It may be found wherever there is water and sufficient vegetation cover, and it often occurs where the water is almost 1m deep (Wetmore 1926). In Rio Grande do Sul, Brazil, it was found, with Speckled Rail (15), in marshy habitat characterised by a mixture of sedges (notably *Scirpus giganteus* and *Cladium jamaicensis*) and many grasses, with a few small forbs, on a muddy substrate with little water (Mauricio & Dias 1996).

In Rio de Janeiro state, the race *zelebori* inhabits mainly open marshes of the littoral, where it often occurs alongside Blackish Rail (J. F. Pacheco *in litt.*), while at higher altitudes in São Paulo and Minas Gerais the nominate race occurs alongside Blackish Rail (F. Olmos *in litt.*, J. F. Pacheco *in litt.*). Both species are also extensively sympatric in Rio Grande do Sul, usually occupying relatively treeless marshes (W. Belton *in litt.*). However in Paraguay, A. Madroño (*in litt.*) has never found the two species side by side, although they may both occur in the same area (e.g. both occur at San Rafael NP, Itapua); Blackish Rail occurs in the Reserva Natural del Bosque Mbaracayu (Canindeyu), where Plumbeous Rail is apparently absent. In N Argentina, J Mazar Barnett (*in litt.*) has never found both species together, although they occupy similar habitats - the edges of reed and *Typha* marshes, and on open swamps with dense floating vegetation, entering taller grass areas and woodland edges. Some indication of differing habitat preferences is given by E. O. Willis (*in litt.*) for São Paulo, where the Plumbeous Rail occupies the S marshes and rice fields of the Paraíba R, not the small wood-edge marshes in which the Blackish Rail is common throughout the state, although Blackish Rails do occur in the rice fields, at least at the edges. In Paraguay, A Madrono (*in litt.*) suggests that Blackish Rails may occur more in habitat surrounded by forest, while Plumbeous Rails prefer larger wetlands with less forest. This topic would repay further study.

The Plumbeous Rail occurs in lowlands, but also locally up to 2,500m, and patchily up to c. 4,000m (e.g. Armonía 1995). Birds on migration may occur in unusual habitats (see Movements).

FOOD AND FEEDING Freshwater molluscs, small crabs, larvae, worms and insects are recorded; also Coleoptera larvae (M. Marín *in litt.*), small watersnakes such as *Helicops* (Sick 1993), and seeds (Canevari *et al.* 1991). Although mainly crepuscular, it also forages during the day and at night, when it leaves cover and wanders out singly or in small groups into cultivated fields in search of grubs, worms and insects (Johnson 1965). Migrating birds ate hemp seeds (Wetmore 1926). It has been seen to wash crabs in a stream to remove mud before eating them (Canevari *et al.* 1991).

HABITS This rail is mainly crepuscular, although it can also be seen foraging during the day; apparently it also forages during the night (Johnson 1965). It is often shy and retiring, but is also said to occur fairly often out of cover compared with other rails, and sometimes to be curious (Fjeldså & Krabbe 1990). It sometimes swims, and is normally reluctant to fly, preferring to escape by running. However, it can fly strongly if necessary; migrating birds flushed easily from dense cover and flew rather swiftly, at times rising 3-4m in the air (Wetmore 1926). When approaching an observer to investigate taped playback, birds will fly across open areas more than 3-4m wide (Belton 1984). While foraging, the birds walk with the tail raised and flicked at intervals, and they move deliberately; however, they traverse runs in the grass more rapidly, pausing at openings to peer about. It is unusual for them to venture more than 2m from cover, and they climb about like gallinules in dense, heavy tangles of shrubs up to 2m above the water (Wetmore 1926). When fleeing from danger, birds hold the tail vertical.

SOCIAL ORGANISATION Occurs singly, in pairs and family groups.

SOCIAL AND SEXUAL BEHAVIOUR No information available.

BREEDING AND SURVIVAL Season Peru, Oct (and chicks Jan); SE Brazil, breeding condition Nov; Uruguay, Sep-Feb; Argentina, Oct-Dec; Chile, Oct-Jan, including Tierra del Fuego, Nov. **Nest** Poorly constructed, a platform 15-20cm in diameter, of dry grasses; built on ground among bushes, reeds and long grass bordering water; well concealed. **Eggs** 4-6; broad oval, with little gloss; pinkish-cream to beige, with unevenly distributed rufous, reddish-brown and pale purple spots and small blotches (sometimes concentrated at larger end); 1 clutch was dull white, unspotted (Johnson 1965). Size of eggs: *sanguinolentus* 36-44 x 28-32 (40.9 x 30, n = 10), calculated weight 20.7; *landbecki* 40.2-46.4 x 30-33 (43.5 x 31.8, n = 58), calculated weight 24.1; *luridus* (3 eggs) 44.4 x 32.2, 41.4 x 32.3, 42.5 x 32.9 (Humphrey *et al.* 1970). Note that Oates (1901) describes 8 eggs of Austral Rail (60) from C Chile (Berkeley James Collection) which are too large for that species, measuring 41.2-44.5 x 30.5-31.75, and have been re-identified as Plumbeous Rail (race *landbecki*), while 4 eggs from Chile assigned by Oates to Plumbeous Rail have been re-identified as Austral Rail (M. Walters *in litt.*).

Nest of this species from Rio Grande do Sul, Brazil, on 22 Jan, previously assigned to Blackish Rail, was cup-shaped and placed on low stump among heavy second-growth shoots in swampy scrub and *Eryngium*; contained 2 chicks and 2 eggs; chicks had left nest the next day (Belton 1984, Naranjo 1991, Belton 1994).

EULABEORNIS

This genus contains only one species, a very large rail with a restricted distribution in N Australia and a race on the Aru islands. It has a longish, robust bill and a well-developed tail, and appears to be the geographical counterpart of the allopatric single species in the genera *Megacrex* and *Habroptila* (Olson 1973b). It inhabits mangroves and bears a striking superficial resemblance in plumage to the *Aramides* rails of Central and South America, but it lacks the barred underwings and the slender tarsi of that genus. It is not globally threatened, and its habitat in Australia is not under threat at present.

116 CHESTNUT RAIL
Eulabeornis castaneoventris   Plate 35

Eulabeornis castaneoventris Gould, 1844, North coast of Australia (= Flinders River, Gulf of Carpentaria).

Genus sometimes enlarged to include *Aramides*, *Amaurolimnas* and *Gymnocrex*. Two subspecies recognised.

Synonyms: *Rallina/Eulabeornis castaneoventris*.

Alternative names: Chestnut-bellied/Chestnut-breasted Rail.

IDENTIFICATION Length of male 52cm, of female 44-52cm. Largest rallid in its habitat; a large (female) to very large (male), distinctive thickset species with longish tail and long, heavy bill. Head to hindneck grey; chin white; underparts pinkish-brown varying tinged chestnut. Iris red; bill yellowish-green with whitish tip; legs and feet pale yellow or green. Polymorphic, varying mainly in colour of upperparts: dark chestnut-brown in chestnut morph (Northern Territory), olive in olive morph (Western Australia), olive-brown in olive-brown morph (Northern Territory and Queensland). Sexes alike but female smaller. Immature, and juveniles older than 4-7 months, apparently indistinguishable from adult; younger juveniles separable from adult only on colours of bare parts: iris brownish-orange; bill olive-yellow with pink-brown tip; legs and feet pale yellow, tinged pink. Race *sharpei*, of Aru Is (Irian Jaya), is like chestnut morph; bill deeper and heavier, green with yellow tip and red around nostrils. Inhabits dense mangroves.
Similar species Easily separated from all sympatric rails on combination of large size, chestnut, olive and grey plumage and mangrove habitat. Smaller Bare-eyed Rail (82) has head to mantle vinous-chestnut, upperparts olive-brown, hindpart of body black, breast to belly grey, and legs and feet red; also has pink bare skin on face.

VOICE Information is taken from Johnstone (1990) and Marchant & Higgins (1993) unless otherwise specified. The territorial call is a harsh screech, preceded by a grunt and usually repeated c. 12 times; the sequence is also described as beginning with a deep drumming followed by loud pig-like squealing, or as barking, trumpeting or donkey-like notes (Ripley & Beehler 1985, Beehler *et al.* 1986, Johnstone 1990); it is often followed by a few grunts. It is normally uttered by a single bird, but is said always to sound as if it is made by 2 birds either in unison or antiphonally; it lasts for 10-15s, and it is heard throughout the year at all times of day (but mostly during the morning); it may be answered by neighbouring birds. The contact call is a loud, repeated *chuck* or *check*, with occasional grunts. The grunt is like the drumming of an Emu *Dromaius novaehollandiae*, and is given quietly at intervals when the birds are disturbed.

DESCRIPTION *E. c. castaneoventris*
Adult chestnut morph (Northern Territory) Forehead to nape, and sides of head, grey; hindneck grey, grading to chestnut-brown towards mantle; rest of upperparts, from mantle to tail, and most upperwing-coverts, dark chestnut-brown, with faint olive gloss sometimes visible. Primaries, primary coverts and alula dark brown, narrowly fringed brownish-olive; median and greater secondary coverts and outer lesser secondary coverts with chestnut spot at tip; axillaries and underwing-coverts pale chestnut-brown, more rufous on axillaries and lessers; underside of remiges dark brown. Chin white, grading to pink-brown, tinged chestnut, on throat; foreneck more chestnut; breast and flanks pink-brown, varyingly tinged chestnut (darker when worn); belly pink-brown; undertail-coverts chestnut-brown. Iris bright red to red; bill pale green or yellowish-green, with whitish to greyish tip; legs and feet yellow or yellowish-green. **Olive morph** (Western Australia) has upperparts and upperwing-coverts uniform greyish-olive to olive (darker, less greyish, when worn), variably grading to grey on hindneck; rectrices olive-brown with broad olive fringes; remiges olive-brown, fringed olive on upper surface. **Olive-brown morph** (Northern Territory and Queensland) has upperparts, upperwing-coverts and tail olive-brown, feathers browner in centre and more olive at fringes.

Female smaller, but sexes otherwise alike, *contra* statement by Mathews & Iredale (1921) that female has upper hindneck ashy-grey like head, rather than olive: this character variable, and applies to both sexes (Plate 35).
Immature Iris said to be yellow, slightly marked with brown (Mathews & Iredale 1921). According to Franklin & Barnes (in press), captive young indistinguishable from adults, in colours of plumage and bare parts, at 121-220 days of age.
Juvenile Details from Franklin & Barnes (in press), of a bird 59 days old, from Northern Territory. Crown to nape slaty-grey, grading to brown on mantle; back and rump brown, richly tinged hazel; tail chestnut-brown; primaries, secondaries, inner primary upperwing-coverts and greater coverts chocolate-brown; outer primary upperwing-coverts and alula chestnut-brown; lesser and median upperwing-coverts brown, strongly tinged hazel; underwing-coverts pink-chestnut; underside of primaries and secondaries chocolate-brown. Sides of head slaty-grey; throat pale pink-chestnut; rest of underparts, including undertail-coverts, uniformly rich pink-chestnut. This plumage resembles olive-brown morph, differing only in rich hazel tinge to back and pale pink-chestnut of throat; there was no indication that these characters were confined to juveniles. Iris pale brownish-orange; bill olive-yellow for basal two-thirds, pale pink-brown at tip, entire bill tinged white; legs and feet pale yellow tinged pink, with indistinct darker markings, especially on toes; back of lower tibia and of tibio-tarsal joint brown.

Downy young Dark sooty-brown; bill blackish, with pale tip (egg-tooth?). (Marchant & Higgins 1993). At 10 days, down dark brown; face, crown, nape and throat brown, pale bases conveying overall pale brown appearance, especially evident around lores; iris dark brown, but smoky at 2-8 days of age; bill black; white egg-tooth lost at c. 10 days; legs and feet dark brown (Franklin & Barnes in press).

MEASUREMENTS Wing of 7 males 219-244 (224.3, SD 9.4), of 8 females 207-224 (214.6, SD 6.0); tail of 6 males 117-144 (124.7, SD 11.3), of 8 females 118-129 (124.3, SD 3.3); culmen to base of 5 males 59-66 (63.6, SD 2.7), of 7 females 56-65 (60.4); exposed culmen of 4 males 57.2-59.7 (58.2, SD 0.9), of 6 females 50.7-59.6 (54.7, SD 2.6); tarsus of 4 males 64.5-75 (71.7, SD 4.1), of 8 females 61.5-74 (69.4, SD 4.0). Weight of 5 males 626-910 (745.8, SD 103.1), of 7 females 550-710 (628) (Johnstone 1990).

GEOGRAPHICAL VARIATION Slight, involving only the size and colour of the bill (see Johnstone 1990). In Australia, chestnut morph is only known from Northern Territory, while olive morph is known from Western Australia and olive-brown morph from Northern Territory and Queensland (see Marchant & Higgins 1993).

E. c. castaneoventris Gould, 1844 – N coasts of Australia from N Western Australia (W to Kimberleys) through Northern Territory to SE Gulf of Carpentaria, and historically to NW Queensland (W shore of Cape York Peninsula). See Description.

E. c. sharpei Rothschild, 1906 – Aru Islands, including outlying islands of Karang and Enu (Diamond & Bishop 1994). Like chestnut morph of nominate; head and nape grey; mantle to tail, and upperwings, chestnut; bill deep and heavy, green (sometimes blue at base), with red around nostrils and yellow tip. Wing of 3 males 217, 220, 226, of 3 females 212, 217, 221; tail of 2 males 124, 125, of 3 females 118, 121, 137; culmen to base of 2 males 61, 62, of 3 females 54, 60, 64; tarsus of 3 males 69, 73, 75, of 3 females 67, 69, 73 (Rand & Gilliard 1967, BMNH).

MOULT Moult of primaries and secondaries is simultaneous. In Australia, a complete moult (assumed postbreeding) occurs in Feb-Mar (Johnstone 1990). Body moult is recorded in Feb, May, Aug, Oct and Dec (Marchant & Higgins 1993). In Northern Territory it has recently been recorded from 33 of 62 islands in the Wessel and English Company groups, mainly from the larger islands but also on smaller islands down to 1.3ha in area, and often not in association with mangroves (Woinarski *et al.* 1998). It is apparently fairly common in the mangrove communities of Darwin Harbour, persisting even within a few hundred m of suburban developments (D. Franklin *in litt.*). In May, postjuvenile moult of chest, back and all tail-coverts was evident in a captive bird aged 163 days (hatched Dec); another captive bird, hatched Oct, was moulting numerous body feathers, and also remiges and rectrices, in Apr, when c. 180 days old; these observations suggest that postjuvenile moult occurs at about the end of the tropical wet season and that early-hatched young may undergo a complete moult while late-hatched young have only a partial moult (Franklin & Barnes in press).

DISTRIBUTION AND STATUS N Australia, from W Kimberley Division to SW Gulf of Carpentaria (formerly also Cape York Peninsula); also Aru Is. Not globally threatened. The nominate race is patchily recorded, with several apparent gaps in its distribution which may be due to incomplete observer coverage. Its status is uncertain because it is very shy and difficult to see in dense habitat (but see Habits); it is more often located by its characteristic loud, raucous call. It was formerly regarded as moderately common to common in large areas of good habitat (e.g. Storr 1973, 1980), and its status has probably not changed in recent years (e.g. Johnstone 1990, Marchant & Higgins 1993). In Western Australia the density at one site was c. 8 birds/km, and at another 10 birds were counted along 3km of a tidal creek (Johnstone 1990). There are no known threats to the species or its habitat, but the lack of recent records from the S coast of the Gulf of Carpentaria (Blakers *et al.* 1984) is puzzling and warrants more detailed investigation (Garnett 1992). It may become accustomed to disturbance from power boats (Blakers *et al.* 1984). The race *sharpei* has a very restricted distribution and its status is unclear.

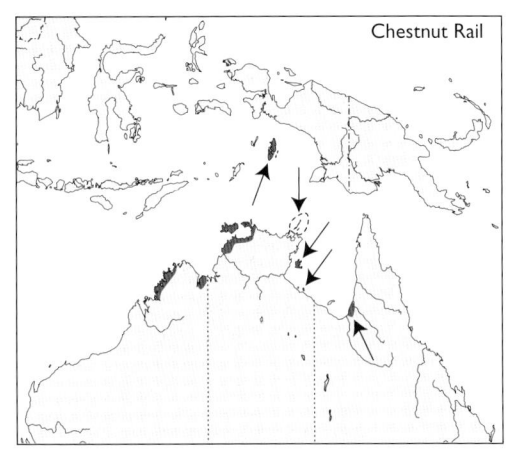

MOVEMENTS It is probably sedentary; no long-distance movements are recorded. When feeding, birds move to higher ground as the tide rises and back to creeks as the water recedes.

HABITAT Tropical coasts and estuaries, favouring large blocks of mangroves and seaward zones of dense mangroves (especially areas dominated by *Camptostemon schultzii*, *Sonneratia alba*, mixed *Rhizophora stylosa/Bruguiera* forest and *Aegiceras corniculatus*), tidal creeks and channels with shelving mudbanks, and tidal flats, especially seaward mudflats (Marchant & Higgins 1993). It prefers areas of tall trees with a dense canopy (Serventy 1985). It may prefer broad bands of dense mangrove forest, especially areas along seaward margins or those dissected by small channels; low-lying areas may be less suitable because of possible tidal damage to nests, which may preferentially be sited on higher ground close to low-lying moist foraging areas (Barnes & Franklin in press). It is uncommon along creeks with only a narrow fringe of mangroves, and it occurs occasionally in adjacent grassy flats, reedy swamps and open woodlands, sometimes because its normal mangrove habitat is flooded (Marchant & Higgins 1993). However, on islands in the Wessel and English Company groups, Northern Territory, Woinarski *et al.* (1998) recorded the nominate race not only from mangroves but also very extensively in zones of unvegetated dissected sandstone platforms and boulders between strand vegetation and the low tide mark. The vegetation adjacent to this intertidal rocky habitat varied from tussock grassland to eucalypt woodland and heath, and in very many cases the birds were seen in such rocky areas on islands with no mangroves or far from extensive mangroves. They were also recorded occasionally in dense thickets of *Pemphis acidula* which grows as a narrow strip at the upper tidal zone of many islands, while 2 birds were found roosting in an extensive *Eucalyptus tetrodonta* forest c. 400m inland of mangroves, and 1 was found at night in a coastal monsoon vine thicket c. 200m inland from the high tide mark.

FOOD AND FEEDING Information is from Johnstone (1990) and Marchant & Higgins (1993) unless otherwise stated. It takes mainly crabs (which may form up to 90% of its diet) and other crustaceans, especially fiddler crabs *Uca* spp. but also marsh crabs *Sesarma* spp. and ghost crabs *Ocypode* spp. It also takes small molluscs, insects (including beetles) and centipedes. Birds forage in soft mud or shallow water within the mangrove zone, along tidal channels and on tidal flats as the tide falls, and along the edges of saltwater at low tide. They move slowly, often flicking the tail and sometimes running a few short steps to pursue prey. Most food is taken from the ground or the water; birds often probe crab burrows, often with the bill and forehead immersed in mud or water up to the eyes; they also glean and peck at the bases of trees, take molluscs from the prop-roots of *Rhizophora*, and probe (e.g. into crab burrows). Small crabs may be tapped against mangrove roots before being eaten. Foraging takes place both day and night.

Evidence has recently been obtained for a very different foraging method, used by birds inhabiting rocky intertidal habitat on the Wessel and English Company Is (Woinarski *et al.* 1998). Here, the birds were said by local people to use isolated flat-topped stones and small rocks (diameter 5-25cm) as anvils on which to break shells, most (but not all) of which probably contained hermit crabs. These shells were 1-3.5cm in diameter and included both land and marine gastropods of the genera *Monodonta*, *Lunella*, *Turbo*, *Nerita*, *Thais* and *Xanthomelonspheroidea*; 50% of c. 200 shells examined were of *Nerita chamaeleon*.

HABITS This species is active both diurnally and nocturnally, its activity periods being influenced by tidal cycles. It is normally extremely shy, alert and secretive, but when foraging on mudflats in the open it may be less wary, sometimes almost confiding. At Darwin Harbour it is fairly readily observed from a boat at low tide, when it feeds on exposed mudflats (D. Franklin *in litt.*). It has a strutting walk, with the tail either depressed or carried high and flicked. It seldom flies (the flight is apparently weak), and when alarmed it runs at great speed through the mangroves for 20-30m before feeding again; it climbs well to its nest in mangrove trees. One bird bathed in salt water, dipping its head under the water and then raising itself up to let the water run over its back (J. L. McKean in Ripley & Beehler 1985). It has been recorded roosting (presumably on the ground) at night in an extensive *Eucalyptus tetrodonta* forest (Woinarski *et al.* 1998).

SOCIAL ORGANISATION Monogamous; territorial when breeding. Normally seen singly, in pairs or in family groups. The breeding territory size is estimated at c. 10ha, and 1 pair or group occupied 4-5ha (Marchant & Higgins 1993). In an 18ha mangrove forest along 2 rivers at Darwin Harbour, Barnes & Franklin (in press) found 5 active nests, 6 recent but inactive nests and 6 nests in various stages of degeneration, giving a minimum density of 1 pair per 0.4km of river bank or 0.33 birds/ha (1 pair/6ha).

SOCIAL AND SEXUAL BEHAVIOUR Pairs are said to feed close together (Johnstone 1990). Two fighting birds faced each other and began grunting, leaping up and striking each other with their feet; the combat lasted for c. 4 min, when 1 bird retreated (Johnstone 1990). An incubating bird left a nest when an observer approached to within c. 2 m; when an intruder neared a nest an adult approached and gave a distraction display, spreading its wings, stretching out its neck and opening its bill before running up the root of a mangrove to within 60cm of the observer and then running away (Marchant & Higgins 1993).

BREEDING AND SURVIVAL Breeds Sep-Feb. **Nest** A large, loose platform of dead sticks, grass, leaves, bark and seaweed; unlined or lined with sticks, leaves or grass; 14 nests composed predominantly or entirely of mangrove sticks, diameter 2-20mm (mostly 5-10mm) and length 15-40cm; sticks shortest at base and cup, sticks in body of nest often forked and interlocked (Barnes & Franklin in press); nest rather like flat-topped pyramid, built 0.6-3m above ground in mangrove tree (leaning tree, upturned roots, depression in trunk, horizontal branches, or butt of fallen tree); birds often choose slanting tree for ease of access; height of 14 nests 1.2-2.2m (median 1.85m) (Barnes & Franklin in press); platform 35-50cm across, 17-50cm deep; cavity diameter 13-25cm, depth 1-6cm. Some nests have gangways of sticks for access; 2 nests 1.5m up between vertical branches of mangrove tree had gangways running up over prop-roots, from ground to nest. Nest may be refurbished and used for several years. Foliage cover above nests 30-100% (Barnes & Franklin in press). **Eggs** 4-5 (1 clutch of 2 recorded); almost elliptical, being lengthened oval, swollen about upper quarter and somewhat pointed at both ends; coarse and variably glossy to matt; pinkish-white or cream, spotted and blotched chestnut, brown, tan or dull purplish-brown, with small underlying blotches of lavender. Size of eggs: *castaneoventris* 46-55.8 x 35-36.9 (52.7 x 36.0, n = 22), calculated weight c. 40; *sharpei* 46-51.4 x 34-35.6 (49.2 x 34.8, n = 7), calculated weight 32.8. Eggs laid at daily intervals. **Incubation** Not recorded. **Chicks** Precocial; leave nest soon after hatching. The following information is from Franklin & Barnes (in press). Chicks were self-feeding when caught at 2 days old; in captivity they preened, ran fast and chased insects at 12 days; they climbed and jumped efficiently at 17 days. Pin feathers began to appear on ear-coverts and wings at 24 days; young fully feathered at 46 days except on wing, where some down remaining and some feathers not fully emerged; all down lost by 59 days, when birds fully feathered except for growing primaries. At c. 230 days, 1 bird first gave adult-type calls. **Survival** No information available.

HABROPTILA

One species, endemic to Halmahera in the Moluccas. A medium-large, robust, flightless rail with dark slate and dark brown plumage, and bright red bare parts including the frontal shield. Olson (1973b) regarded it as structurally very similar to *Megacrex inepta*, the New Guinea Flightless Rail, which he included in *Habroptila*, and probably also close to *Eulabeornis castaneoventris*, the Chestnut Rail of Australia and the Aru Is. However Mees (1982) justifies the separation of *Habroptila* and *Megacrex* on the basis of differences in the structure of the bill and frontal shield and because of the feathered tibiotarsus of *Habroptila*. The Invisible Rail inhabits dense swampy thickets and marsh edges, and is very poorly known. It is considered Vulnerable, being both scarce and threatened by habitat destruction.

117 INVISIBLE RAIL
Habroptila wallacii Plate 35

Habroptila wallacii G. R. Gray, 1860, Halmahera.

Sometimes placed in *Rallus*, for no good reason, but probably derived from *Amaurornis* (Olson 1973b); may be close to *Megacrex*, which is its ecological counterpart in New Guinea. Monotypic.

Synonyms: *Habroptila/Rallina wallacei*; *Rallus wallacii*.

Alternative names: Wallace's/Halmahera/Drummer Rail

IDENTIFICATION Length 35-40cm. Medium-large, robust, flightless rail of striking appearance. Head, neck, mantle, upper back and upper breast dark slate-grey; lower back to rump dark rich brown; uppertail-coverts and tail blackish; upperwings dark brown to dark rich brown. Lower breast to vent dark slate-grey, dark brown feather centres sometimes giving vaguely barred appearance; undertail-coverts dark brown to blackish. Bare parts, including eye-ring and frontal shield, bright red. Sexes alike. Immature and juvenile undescribed. Inhabits dense swampy thickets, particularly sago swamp, and marsh edges.
Similar species Easily distinguished from all sympatric rails on size, plumage and bare part colours. Confusion may have arisen with Rufous-tailed Bush-hen (87) as a result of misidentification of calls (see Distribution and Status); the latter species is quite different in appearance, being smaller, with (in nominate race from Moluccas) olive-brown upperparts, slaty-grey underparts, brown flanks, vinous-isabelline lower belly, and sandy-buff undertail-coverts; it also has a short green bill with no frontal shield, and yellow or olive legs and feet. Possibility of confusion with Purple Swamphen (120), recently recorded from Halmahera, also exists (F. Lambert *in litt.*); swamphen larger, with short, massive red bill and red frontal shield, purple underparts, and white undertail-coverts, inhabits marshy wetlands, occasionally forest edges. Local people, who know Invisible Rail by its drumming call, say that it appears black on the body, with blue on the shoulder (M. Poulsen *in litt.*), a feature which is not visible in skins.

VOICE P. Jepson (*in litt.* to BirdLife) notes that local people describe the drumming call as a *wak wak wak*, which is given at the same time as a *tuk tuk tuk* noise made with the wings; birds call in the early morning or late evening, and calling apparently peaks during the rainy season in May, when the sago is flooded. De Haan (1950) heard most calling in the evening, during and just after twilight. Drumming with a machete on the bases of a sago leaf may elicit an answering drumming which is attributed by local people to this rail (M. Poulsen *in litt.*). Other described calls, including a loud scream (Ripley 1977) and a soft, hen-like clucking (de Haan 1950), require confirmation because they resemble calls of the Rufous-tailed Bush-hen (e.g. Coates & Bishop 1997).

DESCRIPTION
Adult Head, neck, mantle, upper back and upper breast dark slaty, darkest on head and neck; lower back to tail dark rich brown, washed olive-brown on scapulars, upperwing-coverts, back and rump, becoming more blackish-brown on uppertail-coverts and rectrices. Upperwing-coverts tend to have slaty wash on outer edges; greater coverts darker, richer brown with dusky centres; primary coverts, primaries and secondaries very dark brown; tertials slightly paler and brighter; axillaries and underwing-coverts dark slaty with brown feather centres. Lower breast to vent dark slaty, feathers with dark brown feather centres sometimes giving vaguely barred appearance; thighs and undertail-coverts very dark brown to blackish. Iris red or red-orange; eye-ring, bill, frontal shield, legs and feet bright red. Small wing-claw present. Tibiotarsus feathered almost to base (Mees 1982). Sexes alike, but P. Jepson (*in litt.* to BirdLife) reports that bill and legs are red in male, orange in female.
Immature Undescribed.
Juvenile Undescribed.
Downy young Said to be striped, but confirmation necessary.

MEASUREMENTS (2 males, 1 female, 2 unsexed) Wing of male 173, 182, of female 185, of unsexed 179, 185; tail of male 67, 71, of female 62, of unsexed 55, 65; culmen to base of male 70, 76, of female 84, of unsexed 72, 73; tarsus of male 82, 83, of female 99, of unsexed 79, 81.

GEOGRAPHICAL VARIATION None.

MOULT No information available.

DISTRIBUTION AND STATUS Confined to Halmahera, N Moluccas. It is very rarely recorded and is regarded as VULNERABLE. A report (de Haan 1950) that it was locally common in grassland is believed to have resulted from confusion with the common Rufous-tailed Bush-hen (Collar *et al.* 1994). The only recent records are of specimens collected in the early 1980s and early 1990s (Collar *et al.* 1994), possible sound records from sago swamp in the Ake Tajawe proposed protected area in Jul 1994 and from the Kao and Lalobata areas in Dec 1994 (Anon 1995f) and a sighting in a sago swamp in 1995 (Anon 1995e). However, it was not found in the Kao sago swamp, N Halmahera, in April 1996 (Anon 1997). Recent information from P. M. Taylor suggests that the species may not be as rare as the few existing records indicate (N. J. Collar *in litt.*). This species is threatened by habitat destruction: sago palm swamps have been extensively

Invisible Rail

including swamp forest and particularly areas of heavy sago palm *Metroxylon sagu* swamp, and areas with a mixture of pandan and sago at mangrove/sago transition zones; also the edges of marshes, preferring peninsulas of land jutting out into marshy expanses (Ripley & Beehler 1985, P. Jepson *in litt.* to BirdLife). A report that it occurs in alang-alang *Imperata cylindrica* grass (de Haan 1950) is thought to be incorrect (Collar *et al.* 1994) – see Distribution and Status. Its reportedly more widespread occurrence in forest, secondary growth and at forest edge (Coates & Bishop 1997) appears to be unsubstantiated.

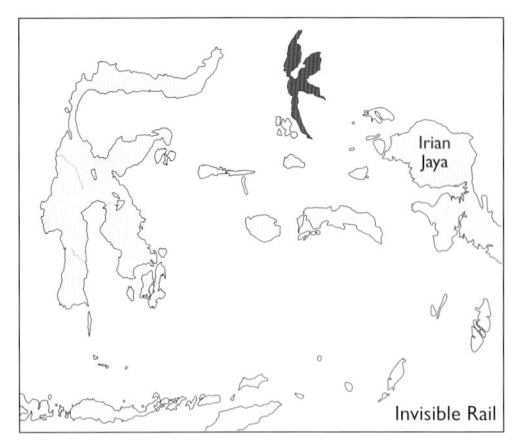

destroyed on Halmahera, where more and more areas are being given over to rice cultivation (Collar *et al.* 1994, Anon 1997). Local people regard its flesh as a delicacy, and snare it for food and catch it with dogs (de Haan 1950, N. J. Collar *in litt.*). It is also likely to be vulnerable to other introduced predators.

MOVEMENTS P. Jepson (*in litt.* to BirdLife) notes that local people seem to think that the birds leave Ako Jilolo Sago forest in the dry season.

HABITAT Dense, impenetrable, swampy thickets,

FOOD AND FEEDING Its diet and feeding habits are virtually unknown; mainly young plant shoots and beetles are recorded. Birds also feed at the open trunks of cut sago trees, either taking decaying sago or searching for other food items in it (M. Poulsen *in litt.*, P. Jepson *in litt.* to BirdLife).

HABITS Although normally keeping well concealed, these birds are reportedly not shy (de Haan 1950), but they are inconspicuous and very difficult to observe.

SOCIAL ORGANISATION Unknown.

SOCIAL AND SEXUAL BEHAVIOUR Unknown.

BREEDING AND SURVIVAL The only information is that an adult was reported with 4-5 striped chicks (Ripley & Beehler 1985); if correct, this would be an atypical downy plumage for the family. Calling activity peaks in May, during the rains (see Voice).

MEGACREX

One species, endemic to lowland areas of New Guinea; a medium-large flightless rail with a heavy greenish-yellow bill, a blackish frontal shield, very large heavy legs and a very reduced tail. It is predominantly brownish in colour, with white underparts. On the basis of structural similarities to the Invisible Rail of Halmahera, Olson (1973b) places it in the same genus, *Habroptila*; however Mees (1982) justifies its retention in *Megacrex* on the basis of differences in the structure of the bill and frontal shield and because its lower tibiotarsus is bare, whereas that of *Habroptila* is feathered. Olson (1973b) considers that it may be derived from *Amaurornis* stock; however, recent investigations using mtDNA sequence data (Trewick 1997) indicate that it falls within the *Gallirallus/Rallus* group. It inhabits dense cover in mangroves, wet forest and thickets, is very poorly known and is considered Data Deficient, but is apparently not under any immediate threat.

118 NEW GUINEA FLIGHTLESS RAIL
Megacrex inepta   Plate 35

Megacrex inepta D'Albertis & Salvadori, 1879, Fly River, New Guinea.

Sometimes placed in *Habroptila*, or in *Amaurornis*. Two subspecies recognised.

Synonym: *Rallus/Amaurornis ineptus*; *Habroptila inepta*.

Alternative names: Papuan Flightless/Grey-faced Rail.

IDENTIFICATION Length 35-38cm. A medium-large, powerful, flightless rail, with long heavy legs, robust bill with frontal shield, and very short, decomposed tail; eye large; wings constantly flicked upwards over back while walking. Crown to hindneck dull reddish-brown; mantle, back and scapulars greyish-olive; rest of upperparts, and upperwings, brownish-olive to sepia, darkest on uppertail-coverts, tail and remiges. Forehead and sides of head grey (ear-coverts often browner); lores dusky; chin to centre of breast, and belly, white, often tinged rufous on breast and belly; sides of neck, sides of breast (sometimes entire breast) and flanks brownish or vinaceous-brown in nominate race, paler and buffier in *pallida*; sides of belly to undertail-coverts brown, tinged rufous. Iris reddish; bill apple-green (may appear yellow), frontal shield blackish; legs and feet dark brown to greyish, with orange blotches on tibia. Bill may appear almost luminous in poor light (I. Burrows *in litt.*). Sexes alike; female apparently slightly smaller. Immature less brightly coloured on sides of breast. Juvenile has brownish-black hairy plumes on crown, sides of head and throat; greyish-olive of upperparts replaced by brown. Inhabits lowland mangroves, wet thickets and forest, and riparian bamboo.

Similar species Easily distinguished from sympatric rallids on structure, size, plumage and behaviour.

VOICE A harsh but shrill call *aaah-aaah*, reminiscent of the squeal of a baby pig (Ripley 1964); also a short complaining whistle (Rand & Gilliard 1967). Adults foraging with almost fully grown immatures gave quiet deep-toned notes, almost braying in quality and tone, and sometimes quite drawn-out, while one immature, fed by an adult, kept up a high-pitched cheeping like a baby chicken (Gregory 1996).

DESCRIPTION *M. i. inepta*
Adult Forehead grey; crown to hindneck dusky, dull reddish-brown; mantle, back and scapulars dull greyish-olive, tinged umber on upper back, and feathers variably edged grey to buff; rump to tail, upperwing-coverts and remiges brownish-olive to sepia, darkest on uppertail-coverts, tail and remiges. Lores dusky; sides of head grey but ear-coverts often washed pale vinous; chin, throat to centre of breast, and belly, white, often slightly tinged rufous on breast and belly; sides of neck, sides of breast (sometimes entire breast) and upper flanks tawny-olive to dull ochraceous-tawny, flanks often washed olive; lower belly and thighs dull greyish-cinnamon; lower flanks, sides of belly to undertail-coverts bright brown, tending to chestnut. Eye large; iris red to reddish-brown; bill apple-green (may appear yellow in field), upper mandible sometimes dark except at tip; frontal shield blackish to blackish-olive; legs and feet dark brown, sometimes grey-tinged, with orange blotches on tibia. No bare orbital ring; frontal shield may be soft and swollen in life (Mees 1982) but does not appear so in photographs of captive bird. Sexes alike; female apparently slightly smaller.
Immature Almost fully grown birds have plumage much as adult but less rusty wash on sides of breast; bill slightly shorter than in adult, and appears yellow with both mandibles black on central part (Gregory 1996).
Juvenile Has brownish-black hairy plumes on crown and in lines on greyish sides of head and on white throat; mantle, back and scapulars browner than in adult. Iris darker and duller; lacks orange markings on tibia.
Downy young Down black on head, and brownish-black on body, darker on upperparts than on underparts.

MEASUREMENTS (2 males, 3 females) Wing of male 184, 187, of female 180, 187, 200; tail of male 32, 37, of 2 females 27, 36; culmen (including shield) of male 71, 77, of female 71, 72.5, 80; tarsus of male 87, 92, of female 88, 94, 103. Unsexed: wing 175-190; tail 40; bill with shield 71-77; tarsus 90-97 (Ripley 1977, sample sizes and means not given). Weight c. 1,200 (Mees 1982).

GEOGRAPHICAL VARIATION Races separated on colour of neck and underparts.
M. i. pallida Rand, 1938 – NC New Guinea, from Taritatu R (Idenburg), Humboldt Bay and Sepik R. Compared with nominate, sides of neck and sides of breast paler, buffier, not tawny-olive or ochraceous-tawny; flanks much paler, feathers tipped and washed olive or pale brown; rump to tail, and remiges, slightly paler brown. Unsexed: wing 174-185; tail 31; bill with shield 70-79; tarsus 86-94 (Ripley 1977, sample sizes and means not given).
M. i. inepta D'Albertis & Salvadori, 1879 – SC New Guinea, from Setekwa R, Noord R (Lorentz), Digul R and Fly R; recently also reported from Kikori, Gulf Province (Gregory 1996). See Description.

MOULT Remex moult is sequential, and is recorded in Mar and Nov.

DISTRIBUTION AND STATUS Confined to the lowlands of NC and SC New Guinea. This rail is not globally threatened, but is regarded as DATA DEFICIENT. It was formerly regarded as locally common (Ripley 1977). It is very rarely seen, but is regarded by local people as not uncommon near Kiunga on the upper Fly R (Collar *et al.* 1994, Gregory 1996), and common in the mangrove swamps of the Kikori R delta (I. Burrows *in litt.*). Its range is probably more extensive than is known. Apart from habitat destruction by logging (e.g. Gregory 1996), it faces no obvious threats except perhaps from feral pigs, and from occasional hunting by local people, especially at sites where sago is being prepared; birds are killed by arrows and will remain even when one has been slain as long as no noise is made by the hunter (Collar *et al.* 1994, Gregory 1996). It can repel predators such as dogs by kicking, and by stabbing with its bill (see Habits). Initial populations were probably small because of the species's specialised habitat requirements (Diamond 1975a).

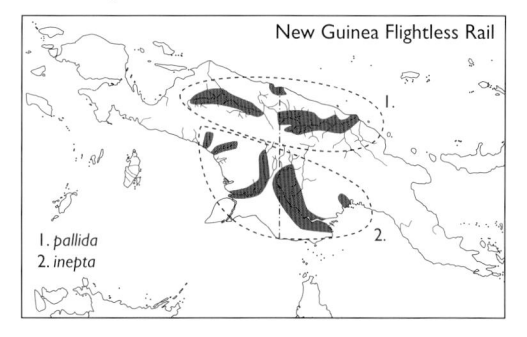

New Guinea Flightless Rail
1. *pallida*
2. *inepta*

MOVEMENTS None recorded.

HABITAT Mangrove forests, wet thickets, swamp forest and riverine bamboo thickets in lowlands. It is often associated with swamps of the sago palm *Metroxylon sagu*. In the dry season it may frequent watery ditches (Rand & Gilliard 1967).

FOOD AND FEEDING The stomach contents of 1 female were "dry pulp in which remains of beetles and vegetable fibre (sago?) could be recognised" (Mees 1982). It is attracted to places where sago is being prepared, and feeds in sago slurry; an adult gave a "white grub" to a young bird (Gregory 1996). One bird was attracted by a rotting and very smelly sago palm trunk which had been split open; it walked along the trunk and pulled off small pieces of the trunk, apparently in search of buried prey although nothing was seen to be eaten (I. Burrows *in litt.*).

HABITS This large rail is wary and runs strongly and swiftly to escape danger; while moving around it holds the wings fully open and raised above the back (Fig. 11), flicking them at each step taken; when foraging it walks with the head lowered (Rand 1942, Gregory 1996). Even though flightless, it is capable of defending itself effectively from predators such as dogs, kicking and stabbing powerfully enough to frighten off attackers and to allow it to escape into the surrounding thickets or up into trees, where it moves from branch to branch like a cat (Ripley 1964). At night it roosts on tree branches, usually > 60cm above the ground (Ripley 1964).

Figure 11: New Guinea Flightless Rail walking with raised wings. [After photograph by P. Gregory].

SOCIAL ORGANISATION Undescribed, but two adults were seen accompanying immatures (Gregory 1996), suggesting monogamy.

SOCIAL AND SEXUAL BEHAVIOUR Undescribed.

BREEDING AND SURVIVAL Undescribed; 3 downy young (nominate race) obtained at L Daviumbu, near Fly R, in Sep, during latter part of dry season (Rand 1942a); downy young, May 1929, Marienberg, Sepik R (museum specimen, BirdLife files); 3 small young with adults on 10 Jun 1995, and on 2 Jul 1995 2 young, possibly of same brood, almost fully grown (Gregory 1996, P. Gregory *in litt.*). Female *pallida* in breeding condition, Mar (Rand 1942b). Female nominate, May, had oocytes 2-4mm in diameter (Mees 1982), but this probably not indicative of breeding (Murton & Westwood 1977), especially as bird is in primary moult. Almost fully grown young still fed by adults (Gregory 1996).

GALLICREX

One species, the Watercock, a large and very distinctive rail of southern and south-eastern Asia which is unusual among the Rallidae in showing marked sexual dimorphism in both size and plumage. The male is larger and in breeding plumage predominantly black, with a prominent red 'horn' on the frontal shield; in non-breeding plumage it resembles the female and immature, which are buffy-brown with dark streaks, spots and bars. Skeletally it appears to bridge the differences between *Amaurornis* and the gallinules (Olson 1973b). It is a bird of marshes, flooded cultivation and the overgrown edges of open waters, and is migratory or dispersive over part of its range. When breeding it is noisy, pugnacious, and well known to local people. It is not globally threatened and is still regarded as common over much of its range, although possibly in decline.

119 WATERCOCK
Gallicrex cinerea Plate 38

Fulica cinerea Gmelin, 1789, China.

Monotypic.

Synonyms: *Gallinula cristata/plumbea/lugubris/cinerea/gularis/naevia/porphyrioides*; *Gallicrex cristata/cinereus*; *Rallus rufescens*; *Crex lugubris*; *Hypnodes cristatus/cinereus*.

Alternative name: Kora.

IDENTIFICATION Length of male 41-43cm, of female 31-36cm; wingspan 68-86cm. Medium-large (female) to large (male), somewhat moorhen-like rail with slim, long-necked appearance and rather long legs and toes. Male in breeding plumage distinctive: blackish, with grey to buff or ochraceous fringes of upperpart feathers giving scaled appearance; white leading edge to wing; flanks barred greyish; undertail-coverts buff with narrow black bars; bill yellow, fairly stout; iris, base of upper mandible, frontal shield, posterior-pointing horn, legs and feet bright red. Female markedly smaller; dark brown cap on crown extends as line down nape and hindneck; upperparts darkish brown with buff to ochraceous scalloping; white leading edge to wing; sides of head and neck buffy-brown; vague broad pale supercilium; chin and throat whitish; underparts pale buffy-brown with fine wavy dark barring; iris yellow; bill and small shield yellowish; legs and feet greenish-brown. Non-breeding male as female; shield sometimes absent in non-breeding female. Immature and juvenile like adult female but more buffy or tawny overall; juvenile less barred below, except on flanks and sides of belly, where bars broader, and has dark brown iris, blackish-brown frontal shield and culmen, and pale ochre bill. Inhabits reedy or grassy marshes, flooded cultivation and the overgrown edges of open waters.

Similar species Breeding male unmistakable, with combination of large size, blackish plumage, buffy undertail-coverts, red legs and feet, and mostly yellow bill with bright red base, frontal shield and horn. In non-breeding male, adult female, immature and juvenile, combination of large size, dark brown upperparts with broad buffy-brown scalloping, and pale buffy-brown underparts with fine, wavy dark barring, unlike any other sympatric rail.

VOICE A noisy bird in the breeding season, with a call of 3 components, produced fairly continuously and rhythmically with a short silence between each series: 10-12 *kok* notes with the head raised, then 10-12 deeper, more rapid, hollow, metallic, booming *utumb* notes with the head lowered, and finally 5-6 *kluck* notes with the head raised. During this performance the neck is swollen and the neck feathers raised. Calling may continue for 30 min or more, and birds call in the morning and evening, at night and also sometimes throughout the day (Neelakantan 1991b, Roberts 1991). Roberts (1991) describes how the calling bird's whole body distends with the call. It is generally silent in the non-breeding season. A captive juvenile gave a series of loud, harsh, nasal *krey* calls when handled (Neelakantan 1991b).

DESCRIPTION

Adult male, breeding Head and neck black to greyish-black; upperparts, including scapulars, tertials and most upperwing-coverts black, greyish-black or very dark blackish-brown, feathers broadly fringed grey to ashy-grey, and warm buff to ochraceous-buff or even ochraceous-tawny, giving markedly scalloped effect to upperparts; relative extent of different colours of fringes very variable, but many birds have predominantly grey fringes on mantle, back and lesser upperwing-coverts, and predominantly buff to tawny fringes from back to uppertail-coverts (very few have grey in this region) and on scapulars and other upperwing-coverts; tertials in all birds examined are broadly fringed buff to ochraceous-tawny; rectrices blackish-brown, edged brownish to greyish. Marginal coverts white, forming moderately broad line along leading edge of wing. Alula, primary coverts, primaries and secondaries blackish-brown, slightly greyish along outer edges; outer web of outermost primary, and of outermost alula feather, largely or entirely white, and shafts also white; axillaries and underwing-coverts dark brown to blackish-brown, underwing-coverts barred and fringed white. Underparts slaty-black, with fine pale grey barring (feather tips) from foreneck to belly and thighs which is lost with wear, and broader, more permanent barring on flanks; lower belly often more extensively whitish; vent and undertail-coverts predominantly buff, with narrow broken black barring. Iris bright red to reddish-brown; frontal shield, base of upper mandible, and spot on each side of base of lower mandible bright red, paling to yellow on rest of bill and duskier at tip; bright red fleshy 'horn' continues backward from frontal shield, projecting above crown; legs and feet dull to bright red.

Adult female, breeding Considerably smaller than male. Crown, and centre of nape and hindneck, dark brown to blackish-brown, giving diffuse dark cap extending as dark line down centre of nape and hindneck. Rest of upperparts, and upperwings, dark brown to blackish-brown, sometimes with slight ashy shade, feather-shafts blackish; feathers broadly fringed buff to ochraceous-buff or ochraceous-tawny, giving scalloped effect, but lower back to uppertail-coverts with narrower fringes; rectrices dark brown, edged paler brown. Upperwing-coverts slightly paler than rest of upperparts, darkish brown, sometimes tinged ashy-grey; greaters browner and with more distinct buff to ochraceous fringes; marginal coverts white, forming narrow line along leading edge of wing; alula ashy-brown, outer feathers edge white. Primaries and secondaries dark brown, ashier along outer edges; outer web of outermost primary white; tertials dark brown, fringed buff to ochraceous-buff or ochraceous-tawny, and with dusky freckling on inner webs; axillaries, underwing-coverts and underside of remiges ashy-brown, tipped whitish to buff. Sides of head and neck dusky ochraceous-buff or darker, mottled whitish in many birds, grading to whitish on chin and throat; diffuse dark brown patch from base of bill to below eye, becoming paler on ear-coverts; lores very pale buffy; vague, broad, pale buffy-brown supercilium extends above eye over darker ear-coverts. Underparts pale buffy-brown, darker on sides of breast and whitish in centre of belly; underparts (including thighs) with fine, wavy, dark bars which sometimes extend to lower foreneck and sides of neck; lower flanks rather more uniform ashy-brown; undertail-coverts pale buff to tawny, with narrow dusky-brown barring. Iris yellow or yellow-brown; bill, and small triangular frontal shield, yellowish; legs and feet dull greenish-brown.

Adult, non-breeding Both sexes like adult female breeding, including bare parts. Non-breeding female has very small frontal shield, sometimes absent.

Immature Like adult female, but may be more buffy or tawny overall than adult.

Juvenile Like adult female but more tawny overall, especially on sides of head, neck and underparts; scapulars and all upperwing-coverts dark brown with broad fulvous fringes; remiges dull blackish-brown; marginal coverts and outer web of outermost primary white. Underparts less distinctly barred than in adult, having narrow, indistinct bars of dusky brown, but bars on flanks, sides of belly, and thighs broader and darker; centre of belly very pale buff, unbarred. Iris dark brown; frontal shield, culmen and bill-tip dull blackish-brown; rest of bill pale ochre, becoming yellowish-pink; legs and feet as female. See Neelakantan (1991b).

Downy young Down of upperparts blackish, with bottle-green sheen; down of underparts dark brown. Bill blackish, with pale horn (in skins) spot at top rear corner of nostril and proximal third of both mandibles; legs and feet in skins appear darker than those of fully grown birds. Small wing-claw present.

MEASUREMENTS (20 males, 20 females) Wing of male 175-224 (207.2, SD 12.1), of female 163-192 (174.8, SD 7.2); tail of male 66-83 (75.5, SD 4.7), of female 53-70 (62.8, SD 4.6); culmen (including shield) of male 41-65 (53.1, SD 6.1), of female 32-43 (38.8, SD 2.5); tarsus of male 64-78 (71.2, SD 3.8), of female 53-67 (60.8, SD 3.3). Weight of 7 males (Philippines) 476-650 (546.1), of male 300-650, of 3 females (Philippines) 298, 336, 434, of female 200-434 (Rand & Rabor 1960, Ripley 1977, Marchant & Higgins 1993).

GEOGRAPHICAL VARIATION None.

MOULT Primary moult is possibly simultaneous; heavy postbreeding body moult recorded Nov (Qingtao, Shandong), when all remiges worn (Stresemann & Stresemann 1966).

DISTRIBUTION AND STATUS Pakistan (Sind, Punjab), Nepal, Bangladesh, India S of Himalayas, Sri Lanka, Maldive Is, Burma, Thailand, peninsular Malaysia, Andaman and Nicobar Is, Laos, Cambodia, Vietnam (Tonkin, Annam, Cochinchina), C and E China (Liaoning and Hebei S to Guangdong, W to SW Shaanxi, SE Sichuan, Guizhou and Yunnan), S Ussuriland, Hainan, Taiwan, Korea, S Ryukyu Is (Okinawa, Ishigaki-jima, Kohama-jima, Iriomote-jima), and the Philippines (Bantayan, Basilan, Batan, Bohol, Carabao, Catanduanes, Cebu, Dinagat, Jolo, Luzon, Marinduque, Masbate, Mindanao, Mindoro, Negros, Palawan, Panay, Polillo, Samar, Siargao, Sibuyan, Tablas and Ticao). It winters S to the Greater and W Lesser Sundas: Borneo, Sumatra (throughout the mainland from NE Aceh to Lampung, and also Riau Archipelago and Belitung), Java, Sumbawa and Flores, and Sulawesi (including Tana Keke I) and Banggai Is (Peleng), and also rarely in Japan and the Izu and Iwo Is (Hokkaido, Honshu, Shikoku, Kyushu, Hegura-jima, Hachijo-jima, Kita-iwo-jima). Note that Perennou *et al.* (1994) also map out-of-range occurrences in NW Pakistan and in W Xinjiang, China, with no supporting documentation. It is not globally threatened. It was formerly regarded as widespread and common over much of its range (e.g.

Ripley 1977, Perennou *et al.* 1994), although in Malaysia it became rare in localities where people killed it for food (Glenister 1951), but is now regarded as being in decline in E and SE Asia and of unknown status in S Asia (Perennou *et al.* 1994). It is a breeding visitor to Pakistan, largely confined to lower Sind but also occurring in extreme NE Punjab; it was formerly rare but has become common in lower Sind due to the great increase in rice cultivation (Roberts 1991). In Nepal it is a scarce monsoon visitor but is possibly under-recorded (Inskipp & Inskipp 1991). It is resident in India (Ali & Ripley 1980) and Bangladesh (Harvey 1990), and also in Sri Lanka where it is now described as rare (Kotagama & Fernando 1994). It is common in Thailand (Lekagul & Round 1991), and fairly common in C and E China (Cheng 1987). It is a very uncommon breeder in S Ussuriland, the Khanka lowlands and the S Kuril Is (Knystautas 1993; see details in Potapov & Flint 1987). It is regarded as a common winter visitor to Sumatra, perhaps breeding only exceptionally (van Marle & Voous 1988), but uncommon in the rest of the Greater Sundas; recorded breeding in Sabah in Jul (S. Harrap verbally). There are very few records from Wallacea, where it is assumed to be a winter visitor (White & Bruce 1986, Coates & Bishop 1997); it is a fairly common resident in the Philippines (Dickinson *et al.* 1991); it is locally numerous on Flores (Coates & Bishop 1997). It is an uncommon resident in the Yaeyama Islands (S Ryukyu Is), and a rare but annual visitor to the main islands of Japan, mostly in autumn and winter (Brazil 1991). It is apparently hunted extensively in parts of its range.

MOVEMENTS Over much of its range in continental E Asia the extent of its occurrence only in the breeding season is unclear, and the division between resident and summer breeding areas on the distribution map is largely conjectural. In Pakistan, birds arrive from mid-May and become more widespread from early July as habitat increases; after breeding they move SE to India, and are gone by late Sep (Roberts 1991). In Nepal, birds are recorded only from Jun-Sep (Inskipp & Inskipp 1991). In India and Pakistan it is resident in well-watered areas but disperses widely during the monsoon with the creation of marshy conditions in otherwise dry low-lying areas (Ali & Ripley 1980). Most birds from N China and Korea apparently migrate or disperse S for the winter (e.g. Cheng

1987), but the extent of such movements is unclear, and evidence is somewhat conflicting, with the species also reported to be resident even in Liaoning, NE China (Meyer de Schauensee 1984); migrants are recorded in Hebei (at Beidaihe) in May (Williams 1986). It is absent from Hong Kong in winter, and apparently also from Korea, where it is recorded only from Apr-Nov (Austin 1948). In Ussuriland it arrives from late May to early Jun and leaves by the middle of Oct; there are isolated records from the Kamchatka Peninsula, Schumschi, Jankitsch, Iturup and Askold Is, and the Bay of Ternay (Flint et al. 1984, Potapov & Flint 1987). In Burma there are few winter records (Smythies 1986); in Thailand it is a summer breeding visitor to the N and is resident in the S (Lekagul & Round 1991); in peninsular Malaysia it is a non-breeding visitor S of Kedah, with records of night-flying migrants, birds at lighthouses, and others presumed to be on passage, in Oct-Jan, Mar and May-Jun (Medway & Wells 1976). It is regarded as largely a visitor to the S part of its range, with winter occurrences from Borneo, Java, Sumatra, Flores and Sulawesi; however it is also recorded from Sumatra in May-July (van Marle & Voous 1988), and it has occurred in Flores in May (Paynter 1963). In Wallacea it is recorded from Dec-May (Coates & Bishop 1997). Migrants are present in Japan mainly from Sep to mid-Jan (most Oct-Nov) with occasional spring records (Brazil 1991). It is a vagrant to Christmas Island (Indian Ocean) where there are several records in Dec-Jan which are consistent with S seasonal movements (Marchant & Higgins 1993), and also possibly once to the Palau Is, W Micronesia (Pratt et al. 1987).

HABITAT Reedy or grassy swamps, marshes, flooded pasture, rice fields, irrigated sugarcane, and emergent and fringing vegetation of channels, rivers, ditches, ponds and lakes; it sometimes inhabits brackish swamps. In Vietnam it occurs on small, isolated, vegetation-fringed ponds in scrub (Delacour & Jabouille 1931). It is also known to breed in dry areas (Phillips 1978) and it occurs rarely in mangroves on migration (C. R. Robson in litt.). In India it may occupy breeding areas before the rice paddy has grown tall enough to hide the males (Neelakantan 1991b), and in Pakistan it occurs at the reedy margins of open waters and seepage zones before moving into paddyfields and low-lying flooded areas in the rice-growing season (Roberts 1991). It is a lowland species, occurring up to 1,230m on migration (Medway & Wells 1976).

FOOD AND FEEDING It feeds largely on the seeds and shoots of 'green crops' and wild and cultivated rice, and water weeds, but it also takes worms, small molluscs (*Planaxia, Neretina*), Crustacea, aquatic insects and their larvae, grasshoppers, and tadpoles. One fed on grass seeds by running its bill along the seed-head with a quick nibbling motion, taking the whole seed head at once (Smythies 1981). Neelakantan (1991b) kept a juvenile in captivity for over 3 weeks and found that it preferred small freshwater fish; if offered fish more than 50mm long it would eat the front half fully and leave the rest. It also readily consumed small aquatic snails, and would eat paddy and fresh cucumber seeds if fish was not available; however the seed-heads and tender shoots of wild grasses were ignored.

HABITS This species is largely crepuscular and is normally skulking, but is occasionally more conspicuous, especially when it emerges from dense waterside cover to feed at dawn and dusk and in overcast or rainy weather. However in India it is reported to arrive in rice paddies when the vegetation is so short that the males are often easily visible until the vegetation grows tall enough to hide them (Neelakantan 1991b). It walks with a slow, high-stepping gait, with frequent jerking of the tail, never moving far from cover; when alarmed it stretches up its neck and jerks the tail rapidly before running to cover. The flight normally appears weak, with rapid wingbeats and dangling legs, but long flights are stronger and more rapid; it is often reluctant to fly but will do so when pressed, and 1 bird perched in a bush when flushed (Glenister 1951). It swims well and will cross open water, riding quite high and looking rather duck-like (Deignan 1945). A captive juvenile often rested during the day, standing on one leg or squatting with the legs folded under its body; when approached it would raise its head and neck, lower the head almost to the ground and then raise it again, repeating this action several times before attempting to escape (Neelakantan 1991b).

SOCIAL ORGANISATION Probably monogamous, and territorial when breeding.

SOCIAL AND SEXUAL BEHAVIOUR Baker (1929) noted territorial disputes in this species. The males are very pugnacious during the breeding season, indulging in furious battles with rivals, jumping and clawing at each other trying to get a hold on the opponent's neck in the bill and to hold him down; however, little damage is usually suffered apart from scratches and the loss of neck feathers (Ripley 1977). In Bangladesh it was highly prized as a fighting bird (Ripley 1977). For further information, see Potapov & Flint (1987).

BREEDING AND SURVIVAL Season India and Pakistan, mostly Jun-Sep (monsoon months); Sri Lanka, May, also Jan-Feb and possibly Jul-Aug; Bangladesh, May-Jul; Maldives, Jun-Jul; Korea, Jul; Ussuriland, Jul; Malay Peninsula, May; Ryukyu Is, Aug; Philippines, May-Jul and Sep; Sumatra, Dec? (female and eggs collected, eggs cannot be traced). **Nest** Large, concave or deep cup-shaped, untidy pad of sedges, rushes, rice blades or grass, built low down in reeds or rice over water, or on clump of coarse grass; stems and seedheads turned down to form platform on which more material is added; sometimes domed over to form bower; occasionally builds on heap of vegetable rubbish. Nest normally in flooded areas, water depth of 70-95cm recorded (Roberts 1991), but also known to breed in dry areas. **Eggs** 3-10 (usually 3-6); rather long oval; glossy; whitish to buff, yellowish-stone or deep brick-pink, with longitudinal blotches or spots of bright reddish-brown and underlying pale purple, often more densely marked at larger end; size 39-46.6 x 28.1-33.1 (42.3 x 31.1, n = 120), calculated weight 22.5. **Incubation** Period 24-25 days, by female; hatching synchronous. Fighting birds are raised from eggs collected in the wild and incubated by being placed in half-shell of coconut and kept tied against belly of person for c. 24 days (Ripley 1977). **Chicks** Male recorded accompanying 2 almost fledged juveniles with growing remiges and no rectrices (Neelakantan 1991b). Chicks seen with adults in Jul, Sabah (S. Harrap verbally). Normally rears 2 broods (Ripley 1977).

PORPHYRIO

A genus of worldwide distribution, comprising six species of medium to very large rails, including one recently extinct. Commonly termed purple gallinules, four of the five extant species have predominantly purple and blue plumage, usually with contrastingly coloured upperparts and upperwings of green, black or brown; the undertail-coverts are white. The smallest species, the Azure Gallinule of South America, is markedly different in appearance, its plumage looking pale and washed-out and resembling that of the juveniles of its congeners. In general these gallinules have brightly coloured bare parts, including the bill and frontal shield; the legs and toes are long in the four species which inhabit palustrine wetlands and are adept at walking and climbing in marsh plants and on floating vegetation. Although sometimes split into three genera, these gallinules are obviously monophyletic: the smaller species, sometimes placed in *Porphyrula*, differ very insignificantly from *Porphyrio* but share some specialized characters with it, while the flightless Takahe of New Zealand, formerly placed in *Notornis*, is but a very recent derivative of *Porphyrio* (Olson 1973b, Trewick 1997). The widespread Purple Swamphen occurs in southern Europe and Asia, and in Africa, Australasia and the W Pacific islands; it shows great regional variation in plumage and recent phylogenetic studies (Trewick 1997) confirm that at least three races could be considered as separate species. The Takahe is the largest living rail and is endangered, with only about 200 birds remaining in the mountains of SW North Island and on offshore islands to which they have been introduced; unlike the other extant species it inhabits grassland, scrub and forest. The extinct flightless White Gallinule was endemic to Lord Howe Island and was similar in appearance to the Purple Swamphen but was completely or largely white, with red or yellow legs; it occupied wooded areas and was very quickly exterminated by the first visitors to the island. *Porphyrio* species are unique among Rallidae in using their feet to grasp and manipulate food items.

120 PURPLE SWAMPHEN
Porphyrio porphyrio   Plate 36

Fulica porphyrio Linnaeus, 1758, Asia, America (= lands bordering the western Mediterranean Sea).

Taxonomy complex and inadequately studied; races *madagascariensis*, *pulverulentus* and *poliocephalus* (incorporating all remaining races except nominate) have at times been considered separate species. Recent phylogenetic studies using mtDNA sequence data (Trewick 1997) suggest that at least *madagascariensis*, *pulverulentus* and *melanotus* could be redefined as species; Sangster (1998) also suggested, without providing any new evidence, that *poliocephalus* and *indicus* should also be raised to specific status. Numerous other races described; most forms listed as separate species until c. 1920s. Thirteen subspecies recognised.

Synonyms: *Fulica caerulea/porphyrio*; *Gallinula madagascariensis/poliocephala/porphyrio*; *Porphyrio caeruleus/hyacinthus/veterum/caesius/smarag(don)notus/aegyptiacus/poliocephalus/madagascariensis/pulverulentus/viridis/coelestis/indicus/smaragdinus/melanopterus/mertoni/bellus/chathamensis/melanotus/stanleyi/vitiensis/ellioti/neobritannicus/calvus/aneiteumensis/bemmeleni/edwardsi/samoensis/pelewensis*.

Alternative names: Purple Gallinule; Pukeko.

IDENTIFICATION Length 38-50cm; wingspan 90-100cm (nominate), 70-86cm (*melanotus*, *bellus*). Large to very large, ponderous rail, with massive triangular red bill, red shield, red to pink-red legs with long, slender toes, and red iris. Plumage very variable; most races predominantly deep blue, purple or violet on head and body, with contrastingly blackish or greenish back and upperwing-coverts; some are cerulean-blue from sides of head to foreneck and upper breast; all have prominent pure white undertail-coverts. Sexes similar; female smaller, with smaller frontal shield. Races separated on plumage and size, with 6 subspecies groups: (1) nominate *porphyrio* of Europe and N Africa around W Mediterranean has dark violet-blue upperparts and upperwings; (2) *madagascariensis* of Africa and Madagascar has lower mantle, back, scapulars and tertials bronze-green to blue-green; (3) *poliocephalus* group (3 races, Caspian Sea through Indian subregion to N Thailand, Nicobars and N Sumatra) has variably cerulean-blue throat, upper breast, sides of head, scapulars and upperwing-coverts, dark blue back, and silver-grey tinge to head feathers, with size declining south-eastwards, *caspius* largest, *seistanicus* smaller and *poliocephalus* smallest; (4) *indicus* group (2 races, of SE Asia and Greater Sundas) often has large shield with lateral ridges, upperparts and upperwing-coverts black with green tinge (less green in *indicus*), and throat and breast turquoise-green to cerulean-blue, with side of head blackish in *indicus* and cerulean-blue in *viridis*; (5) *pulverulentus* of Philippines has olive-chestnut mantle, scapulars and tertials, and whole body strongly tinged ashy-grey; (6) *melanotus* group (5 races, Australia and Oceania, E from Palau Is and Moluccas) can have relatively small shield, and has short toes, black to brown upperparts, and variably cobalt to violet throat and breast: *melanotus* has purple throat and breast, *bellus* is cerulean-blue in these regions, *samoensis* similar to *bellus* but smaller and often has greenish-brown tinge to back, *melanopterus* very variable and differs from *melanotus* and *bellus* in having blue lesser upperwing-coverts and being smaller, *pelewensis* similar to *melanopterus* but has greener gloss on upperparts and more purple tinge to lesser upperwing-coverts and breast. Immature similar to adult but may be duller; some juvenile body feathers often retained. Juvenile duller than adult; face, foreneck and breast washed grey and throat almost white (nominate and *madagascariensis*); wings as adult; bare parts duller than adult. Inhabits densely vegetated marshes and the fringes of open water; also rough, damp, more open grassy areas (e.g. in New Zealand).

Similar species Combination of large size – larger than all congeners except flightless Takahe (122) – purple and blue plumage, white undertail-coverts, massive red bill, red shield and red legs distinguish it from all sympatric rails. Takahe much larger, with much shorter legs and larger bill, and also differs from sympatric race (*melanotus*) of Purple Swamphen in having olive-green upperparts and upperwings. In Africa, vagrant American Purple Gallinule (124) considerably smaller than Purple Swamphen (race *madagascariensis*), with red and yellow bill, blue-white shield and yellow legs; Allen's Gallinule (123) much smaller and more elegant, with blue or green frontal shield and no pale blue on face or throat. Juvenile also readily separable from congeners on size, structure (e.g. massive bill and long legs), plumage characters, and colour of bare parts (red, dusky red or red-brown). On Halmahera (N Moluccas), possibility of confusion with Invisible Rail (117) also exists (F. R. Lambert *in litt.*); rail smaller, dark slate-grey with dark rich brown upperwings, lower back and rump, long red bill and smaller frontal shield; inhabits dense swampy thickets and marsh edges; call a loud drumming.

VOICE A very vocal species, with a rich and variable repertoire; the function of many calls is not known. Most information is from Cramp & Simmons (1980), Urban *et al.* (1986) and Marchant & Higgins (1993). The male has low, sonorous calls, sometimes ending with hoarse trumpet notes; the female has shriller, softer calls (Cramp & Simmons 1980: but see below). The song, possibly territorial in function, is a long (8-15s), powerful series of plaintive nasal rattles, in crescendo, rendered *quinquin-krkrrquinquinquinkrrkrr...* etc. Territorial calls include crowing notes, often preceded by squawks; in *melanotus* the female's calls are harsher and more guttural than the male's (Clapperton 1983); birds also give harsh, grating calls, e.g. *kree* or *kree-ik*, in aggressive displays. Contact and rallying calls are *cuk*, *chuck* and *n'yip* notes, and a hiccuping call comprising a harsh note, several staccato notes and a *yip*; there are also nasal rallying calls, and a deep, hollow call may also be a contact call. Other calls include an explosive trumpeting alarm call *gooweh*; repeated deep, grunting *unk* notes; short *ank* or *aak* notes; a reedy groan or wail; a dry, screechy, heron-like *wraah*; a low hoot given every few s; and a precopulatory humming note. Interacting birds give a short *tok*, a high, liquid *wik* and a low, short, reedy cackle, while a disyllabic *n'yick* is given by submissive birds. A call apparently given in a song-flight is a loud shrieking series of cackling notes, while the normal flight call is a metallic *krr* (Hollom *et al.* 1988); the flight call of the race *melanotus* is a short, repeated squawk (Marchant & Higgins 1993). The displaying male utters a quiet, guttural cluck continuously, and at the climax a very loud, sonorous chuckle or cackle. This species also has various other grunts, whistles, and bell-like notes. Calling occurs at night in chorus, as well as during the day.

The male calls chicks with a subdued mewing sound; the normal call of the young is a sparrow-like chirp, and the distress call an incessant, rather loud *peep*; an adult leading a juvenile to cover when disturbed gave a quiet *hon*, and the juvenile answered with low grunts. Juveniles also give cat-like calls and whistles.

DESCRIPTION *P. p. porphyrio*
Adult Crown to hindneck, feathers at base of bill, sides of neck, entire upperparts and upperwing-coverts dark violet-blue or hyacinth-blue, tinged plumbeous in some birds, dullest on crown and nape and purest violet on marginal coverts; rectrices black with dull violet-blue outer webs; flight feathers black, with greyish-violet outer webs; axillaries and underwing-coverts plumbeous-black, tinged violet-blue on all except on greaters and lower primary coverts. Sides of head, chin, throat, foreneck and upper breast cerulean-blue or cobalt, sometimes violet-blue; rest of underparts dark plumbeous-grey to plumbeous-black, tinged violet-blue most noticeably on breast and flanks; undertail-coverts white. Iris carmine, orange-red, or red-brown; bill and frontal shield scarlet or dark blood-red, cutting edges and tip (and sometimes either around nostrils or in line from nostril to bill tip) slightly paler and more orange; legs and feet fleshy-red, or dull coral-red to pink-red, slightly duskier at joints. Sexes alike, female slightly smaller.
Immature (first basic) Like adult but head and upperparts slightly duller violet-blue; centre of breast, belly, vent and undertail-coverts mixed dark grey and white. Bare parts like adult or slightly duller.
Juvenile Duller and paler than adult. Crown to upper mantle, and sides of head, dark plumbeous-grey, slightly tinged violet-blue; lower mantle, back and upperwing-coverts (including marginals) dull violet-blue; rump and uppertail-coverts dull grey-brown; rectrices black, tinged violet-blue towards tips and on edges of outer webs. Flight feathers blackish, outer webs dull violet-blue; axillaries and underwing-coverts apparently grey, tinged plumbeous-violet. Chin whitish; throat, foreneck, sides of neck, breast, flanks and rest of underparts grey, tinged plumbeous-violet, and feathers washed whitish on lower neck, breast to vent and thighs; undertail-coverts mixed grey and white. At c. 3 months, bare parts duller than adult: iris red-brown; bill and shield bright red but duskier at sides of bill; legs and feet dull brick-red or ochre-red.
Downy young Deep velvety black with off-white filoplumes on sides of head, throat, neck, mantle and wings; skin of hindcrown, nape and distal wing-pads red, of eyelids purple; iris dark slate; bill white, faintly tinged blue, some with purple-red at nostril and base; legs and feet rosy-flesh. At 1 month, down dull black, mixed off-white on underparts; iris olive-brown; bill black; frontal shield small, red, separated from bill by dense down; legs and feet dull grey-brown.

MEASUREMENTS Wing of 13 males 250-275 (265, SD 7.4), of 12 females 245-264 (259, SD 6.0); tail of 12 males 93-107 (100, SD 4.7), of 12 females 87-103 (92.9, SD 4.5); bill (tip to feathers at side behind nostril) of 13 males 43-47 (44.6, SD 1.5), of 12 females 38-44 (40.9, SD 1.7); tarsus of 13 males 98-107 (102, SD 2.6), of 13 females 90-100 (94.4, SD 2.7) (Cramp & Simmons 1980); bill (including shield) of 6 males 70.5-75 (72.9), of 5 females 66-69 (67.6) (Glutz von Blotzheim *et al.* 1973). Weight of 37 males 720-1,000 (869, SD 71.3), of 35 females 520-870 (724, SD 78.7). Hatching weight 30.

GEOGRAPHICAL VARIATION Great regional and individual variation exist (e.g. Mayr 1949); races differ in the colour of the head, foreneck and upperparts, and in size. The race *melanopterus* includes *chathamensis* (Marchant & Higgins 1993). Note that White & Bruce (1986) consider *melanopterus* synonymous with *samoensis*, and discuss the status of 3 other putative races, *mertoni*, *steini* and *palliatus*.

P. p. porphyrio (Linnaeus, 1758) – S and E Spain (breeds Marismas of Guadalquivir R, also Huelva, Cádiz, S Córdoba, Jaén, Castrejón reservoir), and introduced to Cataluña, Valencia and Mallorca; S

Portugal (breeds Algarve); S France (Perpignan); Sardinia; N Africa in Morocco (breeds lower R Loukos and mouth of R Moulouya, occurs Casablanca), Algeria (breeds L Tonga, L Boughzolu, Oran and Algiers) and Tunisia (breeds L Keliba and L Ichkeul). See Description.

P. p. madagascariensis (Latham, 1801) – Egypt (Nile Delta, Nile V to Aswan, Suez Canal area, L Qarun, and Wadi el Rayan); in West Africa in SW Mauritania, Senegal, Gambia, Sierra Leone, Liberia, SE Mali (Niger Delta), Burkina Faso, S Niger, Ivory Coast, S Ghana, S Togo, extreme N and S Nigeria, W Chad (L Fittri) and S Cameroon; more continuously in E, C and S Africa, in EC Sudan (Kosti), Ethiopia, Central African Republic, highlands of E and S Zaïre (L Albert S to Shaba), Rwanda, Burundi, Kenya, Uganda, Tanzania, Zambia, SW Angola (Moçâmedes, Luanda, and Huila N to Capelongo), Malawi, Mozambique, Zimbabwe, Namibia, Botswana, South Africa (excluding the arid W and SW), Swaziland and Lesotho; also Madagascar. Its former presence on Mauritius was possibly the result of an introduction from Madagascar before 1812 (Lever 1987). Hindcrown to hindneck, sides of neck and upper mantle, dull purple; lower mantle, back, scapulars and tertials bronze-green (E to S African birds), blue-green (birds from Egypt and West Africa) or intermediate (Madagascar and some from S Africa); bronze colour often most prominent on feather centres, especially of tertials and largest scapulars, which are bronze in centre and broadly edged green; rump to tail slightly darker, more dark olive, with less iridescence; rectrices often duller and browner. Most upperwing-coverts dark purple-blue; lessers and marginals may be paler and brighter blue; inner greater coverts (and, variably, some inner medians) green like scapulars and edges of tertials; demarcation between green and purple-blue on coverts often not precise, with some feathers mixed green and purple; primaries and secondaries black, with dark purple-blue outer webs; axillaries and underwing-coverts black, some coverts washed purple-blue. At least some birds have variable narrow cerulean-blue border between green and purple areas of mantle and upperwings, including along edge of scapulars and along outer webs of outermost green upperwing-coverts and tertials. Sides of head, chin, throat, foreneck and upper breast cerulean-blue; lower breast, flanks, thighs and sides of belly dull purple; centre of belly and vent variably duller, blacker. Southern African birds larger than East African; others intermediate.

As in nominate, immatures often have whitish mottling or barring on feathers of thighs and belly. Juvenile has crown to upper mantle dark grey, tinged purple; sides of head grey, washed pale blue; chin and throat white; underparts dull violet-grey, palest on belly. Juveniles from southern Africa are more brownish on head and body: crown to upper mantle pale olive-brown; rest of upperparts, and scapulars, pale olive-brown washed yellowish-green; tail olive-brown; sides of head, and chin, grey-brown with pale blue wash; throat whitish; rest of underparts pale buffy-brown, palest from foreneck to centre of belly (see Plate 36). Bill and shield dark grey; bill becomes dull red before adult-type plumage attained. Wing of 15 males 229-268 (251), of 14 females 226-258 (243); tail of 4 males 85-88 (86.5), of 5 females 81-94 (87.2); bill (method of measurement unknown) of 15 males 38-43 (40.5), of 14 females 37-41 (38.5); tarsus of 15 males 81-99 (92.1), of 14 females 84-91 (86.8) (Urban *et al.* 1986). Weight of 11 males 528-687 (636), of 8 females 480-737 (556) (Urban *et al.* 1986).

P. p. caspius Hartert, 1917 – Caspian Sea, NW Iran, Turkey and Syria; Azerbaijan (?). Like *seistanicus*, but larger. Wing of 9 males 272-295 (285, SD 6.7), of 10 females 262-275 (267, SD 3.8); bill (tip to feathers at side behind nostril) of 9 males 43-47 (44.9, SD 1.3), of 10 females 39-42 (40.8, SD 1.5); tarsus of 9 males 99-111 (105, SD 3.8), of 10 females 91-104 (97.6, SD 4.0) (Cramp & Simmons 1980).

P. p. seistanicus Zarudny & Härms, 1911 – Azerbaijan (Patrikeev 1995) (referable to *caspius*?), Iraq and SW Iran (Khuzistan), coast of Talych E through coastal areas of Caspian to lower Atrek R on Turkmenia border, E Iran and adjacent SW Afghanistan (Sistan), and Pakistan (Baluchistan in Quetta and Kalat Ds and in valleys of Mashkai and Hingol Rs: Vaurie 1965). Partly migratory: reported from Mangyshlak Peninsula, and Ashkhabad and Kabul regions. Lores and crown pale dingy grey-brown; hindcrown, nape, hindneck and upperparts deep purple-blue, tinged green; upperwings largely greenish-blue, more cerulean-blue on lesser coverts and scapulars, and sometimes more greenish on secondaries; tertials olive-brown, washed and edged greenish; rectrices blackish, edged blue. Axillaries and underwing-coverts dark greenish-blue, greaters and lower primary coverts blackish. Sides of head hoary grey tinged cerulean-blue; chin to foreneck, and upper breast, cerulean-blue, sometimes hoary grey tinged cerulean-blue on chin to foreneck; lower breast, flanks and belly purple-blue; thighs dull greenish-blue to purple-blue; vent blackish-brown. Iris deep blood-red, browner in female; bill and shield brownish blood-red, sometimes duller on anterior half of bill; legs and feet pale dingy red to dull red, dusky at joints. Immature like adult but duller, with lower breast and belly ashy-grey. Smaller than *caspius*, larger than *poliocephalus*. Wing of 9 males 262-276 (269, SD 5.0), of 7 females 248-262 (255, SD 4.6); bill (tip to feathers at side behind nostril) of 11 males 39-44 (42.1, SD 1.5), of 9 females 35-40 (37.7, SD 1.7); tarsus of 12 males 92-105 (98.5, SD 4.5), of 9 females 87-98 (90.6, SD 3.8) (Cramp & Simmons 1980). Bill (including shield) of 3 adults 63, 67, 72 (Ripley 1977).

P. p. poliocephalus (Latham, 1801) – Pakistan (E Baluchistan, Sind and Punjab); Nepal; India (from Kashmir and plains at foot of Himalayas) and Sri Lanka through Bangladesh, Andaman and Nicobar Is and Burma (throughout) to SC China (SW Yunnan and possibly W Yunnan: Cheng 1987) and N Thailand. A very similar form also occurs in N Sumatra ('*bemmeleni*'; see *indicus*). Like *seistanicus* but smaller. Female as large as male (Ali & Ripley 1980). Bill of chick blackish with green tip (Ripley 1977). Wing of male 225-271, of female 242-254; bill (including shield) of male 41-49; tarsus of male 82-90, of female 88-98 (Ripley 1977); tail of adults 82-108 (Ali & Ripley 1980). Weight of 6 adults 510-785 (662.5).

P. p. viridis Begbie, 1834 – S Burma, S Thailand and Peninsular Malaysia through Indochina; possibly to S China, where recorded accidentally in Guangxi

Zhuang Autonomous Region, probably an escaped cagebird (Cheng 1987). However, Meyer de Schauensee (1984) and Sibley & Monroe (1990) give its distribution in S China as Guizhou, Guangxi, Guangdong, S Fukien and Hainan; other recent authorities do not mention this distribution in S China, although Ripley (1977) maps a similar range and mentions that the species is recorded from S Fukien, Canton and Amoy (Swinhoe 1868, La Touche 1931-34). Caldwell & Caldwell (1931) record it from Fukien and Guangdong but say that neither they nor their collectors have ever seen it, nor does any museum (presumably in China) have a specimen (M. Walters *in litt.*). This distribution therefore requires confirmation. Possibly migrates to Sumatra (Serdang; van Marle & Voous 1988). Intermediate in plumage between *poliocephalus/seistanicus* and *indicus*, resembling the former in colour of head and neck, and latter in other plumage. Crown, lores and upper sides of head hoary grey; nape to upper mantle, and sides of neck, purple-blue; rest of upperparts blackish with green tinge; upperwing-coverts green. Lower sides of head cerulean-blue; chin grey; throat to foreneck, and upper breast, cerulean-blue, possibly sometimes tinged turquoise-green. Shield relatively large, with pronounced lateral ridges. Bare part colours as in *seistanicus/poliocephalus*. Wing of 1 male 243, of 1 female 244; tail of 1 male 87, of 1 female 86; bill (method of measurement not given) of 1 male 60, of 1 female 65; tarsus of 1 male 84, of 1 female 83 (Ripley 1977).

P. p. indicus Horsfield, 1821 – Sumatra (including Nias and Belitung), Java (including Kangean Is), Bali, S Borneo and Sulawesi. Note that White & Bruce (1986) place birds from N and SE Sulawesi in *samoensis*, while van Marle & Voous (1988) recognise the form *bemmeleni* from N Sumatra, this being like *poliocephalus* but smaller (see also Sangster 1998). Like *viridis* (but some specimens from Malay Peninsula indistinguishable from *viridis*: Ripley 1977), but crown and sides of head almost black; nape, hindneck and sides of neck deep purplish-blue, tinged blackish on hindneck; upperparts dusky blackish, less green-tinged than in *viridis* and glossed purplish-blue on mantle and less markedly on lower back and rump; scapulars and wings olive-greenish; lesser upperwing-coverts cobalt-blue, medians and greaters olive-green with bluish bases; alula, primary coverts and flight feathers dusky blackish, with dull cobalt-blue outer webs; tertials dusky olive-green; tail blackish, washed olive. Underwing-coverts greenish-blue, but lower greaters and primary coverts black (Sharpe 1894); axillaries black, edged greenish-blue. Chin dusky, spotted greenish-blue; throat to breast turquoise-green to cerulean-blue; rest of underparts rich purplish-blue; belly and vent blackish; thighs and sides of lower flanks greenish-blue. Birds from Sulawesi tend to have even less green on back. Primaries bluer, less purple, than in *melanopterus* and *samoensis*. Bare part colours as in *seistanicus/poliocephalus*. Wing of male 229-250, of female 218-235; tail of male 86; bill (including shield) of male 62; tarsus of male 79-93, of female 75-83 (Ripley 1977).

P. p. pulverulentus Temminck, 1826 – Philippines (Basilan, Bohol, Luzon, Mindanao, Mindoro, Panay),

Talaud Is (Karakelong: straggler?). A very distinctive race. Head, neck and upper mantle bluish-grey (sometimes with pale violet tinge); mantle to uppertail-coverts, and scapulars and tertials, olive-chestnut to dull orange-rufous, washed blue on edges of mantle feathers. Upperwings pale greyish-blue (bluer, less violet-tinged, than neck and underparts, especially on lesser coverts); some coverts tinged pale greenish; some medians variably olive-chestnut like tertials; flight feathers brown to blackish-brown, outer webs greyish-blue to turquoise towards bases and brownish towards tips; tail dark olive-brown. Underwing-coverts bluish-grey, greaters and some primary coverts dusky blackish. Entire underparts bluish-grey, duskier on belly and thighs. Legs and feet pale to dull red. Unsexed birds: wing 236-245; tail 88-95; bill (including shield) 67-74; tarsus 83-93 (Ripley 1977).

P. p. pelewensis Hartlaub & Finsch, 1872 – Palau Is. Similar to *melanopterus*, but upperparts more glossed green; patch on breast, and lesser upperwing-coverts, less blue, contrasting less with purple belly. Wing of 1 male 227, of 2 females 212, 227; tail of 1 male 81, of 2 females 77, 86; bill (method of measurement not given) of 1 male 62, of 2 females 57, 64; tarsus of 1 male 77, of 2 females 75, 77 (Ripley 1977).

P. p. melanopterus Bonaparte, 1856 – Moluccas (Buru, Halmahera, Seram); Lesser Sundas (Flores, Roti, Sumba, Sumbawa, Tanimbar, Timor, Wetar); Kai Is (Kai Besar) and W Papuan Is to N New Guinea (Vogelkop and N Irian Jaya) E to some islands in Milne Bay P (Goodenough, Misima, Tagula, Trobriands, Woodlarks). Note that White & Bruce (1986) place birds from Moluccas and Lesser Sundas in *samoensis*; see that work for further discussion of subspecies limits. Apparently very variable. Differs from *melanotus* in having lesser upperwing-coverts blue, and patch on breast blue (colour sometimes extending to foreneck, or even throat and malar region as in *bellus*), both areas contrasting with purple of belly and sides of breast; some birds lack blue patch on breast. Upperparts brownish-black to almost black; sides of neck brownish-black, like hindneck; thighs blackish. Wing of male 223-240, of female 221-248; tail of male 89-98, of female 80; bill (method of measurement not given) of male 63-72, of female 60-69 (Ripley 1977).

P. p. bellus Gould, 1841 – SW Australia (Moora SE to Mt Le Grand). Differs from *melanotus* in having upperparts paler, more black-brown; fringes of upper-part feathers (especially scapulars and upperwing-coverts) varyingly tinged yellowish-olive; broader purplish-blue outer edges to primaries and greater primary coverts; alula feathers more extensively blue; upper row of lesser upperwing-coverts, marginal coverts, malar region, chin, throat, foreneck and upper breast cobalt-blue; also has shorter frontal shield with squarer base. Colour of legs and feet variable: often like *melanotus*, but may also be greenish or hazel-brown. Juvenile has breast and flanks light blue, with broader pale feather tips than in *melanotus*. Measurements of 5 males, 3 females: wing of male 268-290 (281, SD 8.9), of female 266, 270, 278; tail of male 101-120 (108.8, SD 7.5), of female 105, 110, 115; bill (including shield) of male 62.73 (68.4, SD 4.0), of female 60, 63, 65.5; tarsus of male 91-95 (93.1, SD 1.6), of female 79.5, 92.5, 99. Weight of 1 male 828, of 1 female 629.

461

P. p. melanotus Temminck, 1820 – S and WC New Guinea, E to Port Moresby region; N & E Australia: Queensland (but sporadic in Cape York Peninsula and in W); throughout Victoria, New South Wales and Tasmania; South Australia (mainly S of 34°S and E of 138°E); Western Australia (Pilbara, and Kimberley Division); Northern Territory (mostly Top End); also New Zealand (including many offshore islands), Kermadec Is (Raoul), Chatham and Pitt Is, Lord Howe I and Norfolk I; migrates to S New Guinea. Head blackish; neck and upper mantle purplish-blue, becoming darker with wear; rest of upperparts, most of upperwings, and tail, blackish; marginal coverts and upper rows of lesser coverts purplish-blue; remiges black, becoming browner with wear; alula, primary coverts and P6-P9 have purplish-blue outer edges. Underside of remiges, and underwing-coverts, grey-black; all except greaters with purplish-blue fringes. Breast and flanks purplish-blue; belly and thighs blackish. Bill and shield red; legs and feet pink to pink-red, with dark grey joints. Immature separable only by scattered remnant juvenile feathers on underparts, mantle, rump and tail; not retained in some, when iris colour (brown, olive, or red with inner ring of olive) may be sole distinguishing character. Juvenile has head, neck and mantle brownish-black; lower hindneck to mantle tinged bluish; rest of upperparts blackish-brown; wings as adult; tail blackish. Breast and flanks dark blue, tinged greyish and mottled white, when worn becoming dark grey-brown with slight blue wash; belly white, mottled grey, becoming greyer with wear. Iris dark brown; bill grey-black; shield grey-black, becoming light brown at rear; legs and feet light pinkish-brown, joints grey. Chick has brown tinge to down of underparts; frontal shield mauve to violet, becoming light pink. Wing of 16 males 251-294 (275.1, SD 7.9), of 23 females 254-285 (268.1, SD 5.6); tail of 12 males 95-121 (106.7, SD 4.5), of 14 females 87-113 (100.6, SD 5.6); bill (including shield) of 16 males 56-77 (69.2, SD 4.8), of 23 females 56-72.5 (65.3, SD 4.2); tarsus of 12 males 87.5-102.5 (97.0, SD 3.8), of 15 females 81.5-101 (93.3, SD 4.6). Weight of 148 males 785-1,310 (1,091.2, SD 94.9), of 97 females 679-1,252 (885.4, SD 98.5).

P. p. samoensis Peale, 1848 – Admiralty Is (Manus, Lou, Rambutyo); Bismarck Archipelago (New Britain, New Hanover, Tabar, Umboi, Watom); Solomon Is (Bougainville, Buka); Santa Cruz Is (Ndendi); Fiji (mongoose-free islands and Rotuma); Samoa (all major islands); Tonga; Niue; Vanuatu (probably throughout – see Bregulla 1992); and New Caledonia and Loyalty Is. See also *P. p. indicus* and *P. p. melanopterus*. Extremely variable (see Mayr 1949 for detailed descriptions). Similar to *melanopterus*, differing principally in having brighter blue to greenish-blue breast patch and more richly coloured upperparts (black, tinged greenish or brownish or washed blue, or bronze-brown to bright rich brown). Iris red to red-brown; bill red; shield scarlet; legs and feet dull pink, paler, more greyish-rose in female. Wing of male 213-267 (222.1), of female 206-234 (213.8); tail of male 72-85, of female 68-85; bill (method of measurement not given) of male 63-75 (68.8), of female 54-72 (61.5); tarsus of male 72-90, of female 67-72 (Ripley 1977). Weight of 2 males 840, of 3 females 690, 690, 820 (New Caledonia: Dunning 1993).

MOULT In nominate (Cramp & Simmons 1980), postbreeding moult is complete and flight feathers are moulted simultaneously; the rectrices are usually moulted almost simultaneously, with, or very soon after, the remiges. The sides of the head and the underparts are moulted first, followed by the wings, then by the rectrices, and lastly by the mantle to the rump; moult is recorded mainly in Jun-Aug; flight feathers are sometimes bitten off just before the moult. Prebreeding moult may involve the renewal of plumage on the head, neck, or front part of the body in spring. Postjuvenile moult is partial, beginning with the head and upperparts, then the breast, flanks and underparts; it is usually completed by 4 months of age; some juvenile feathers are occasionally retained on the central belly, vent, rump, tail-coverts and tail (wings are not moulted); some birds are in completely adult-type plumage by midwinter, others only in spring.

In *madagascariensis* from the former Transvaal, South Africa, remex moult was observed in Oct-Dec and body moult in Sep-Dec and in Apr-May, 1 female with remiges and rectrices two-thirds grown had a developed egg (Fagan *et al.* 1976). Remex moult of SE Australian *melanotus* (Marchant & Higgins 1993) may occur in late summer or autumn but its relationship to breeding is uncertain; covert moult is most often recorded Sep-Jan, and adult body moult is recorded in all months, beginning with the head, then the breast, belly and nape; the mantle and rump moult in association, as do the flanks, tail-coverts and tail (tail moult is usually centrifugal). The extent of a pre-alternate (prebreeding) moult is uncertain. Postjuvenile moult is partial, is apparently similar to that in nominate, and may be complete at 5 months of age.

DISTRIBUTION AND STATUS SW and S Europe; Africa; Madagascar; Persian Gulf and Caspian Sea E across Indian subcontinent, SE Asia and S China; Philippines; Greater and Lesser Sundas; Moluccas; Australia and New Zealand; and New Guinea through W Pacific islands S to New Caledonia and E to Samoa. Not globally threatened.

The nominate race has suffered a marked decrease in its already restricted range, especially in the 20th century, because of wetland degradation and drainage and also through hunting, disturbance and the effects of pesticides (Cramp & Simmons 1980, Sánchez-Lafuente *et al.* 1992, Máñez 1994). It was formerly widespread in nearly all wetlands of the Mediterranean and S Atlantic coasts of the Iberian peninsula, and it also occurred in S Greece, inland Italy, Sicily and N Portugal (Cramp & Simmons 1980, Sánchez-Lafuente *et al.* 1992). The Guadalquivir R population has probably been adversely affected by a series of droughts since 1992, when it was estimated to comprise c. 3,000 breeding pairs; it also occurs in 4 marshy areas in Huelva and at 5 groups of lagoons in Cádiz, while the upper Guadalquivir held 300-350 individuals in 1990; its recent breeding at Castrejón reservoir is a notable range extension (Hagemeijer & Blair 1997). The Portuguese population is small (6-9 breeding pairs) and unstable, while the Sardinian population may be recovering from decline and numbers 240-300 breeding pairs in coastal wetlands, especially at the Sinis peninsula and the Golfo di Oristano, and around Cagliari (Máñez 1994, Hagemeijer & Blair 1997). In France it bred for the first time, at Étang du Canet, near Perpignan, in 1996 (Aleman 1996). It is very sparsely distributed in North Africa (Urban *et al.* 1986), but in Algeria it was reported as abundant at L Tonga (Ledant *et al.* 1981) although later only 20 pairs were reported breeding there (Mayaud 1983). Its European population is estimated at 3,274-3,777 (geometric mean 3,516) pairs and its total population at 10,000-25,000 birds (Rose & Scott 1994, Hagemeijer & Blair 1997). It requires strict protection and habitat management: the enforcement of protection measures in Spain led to an immediate recovery and a reversal of range contraction (Máñez 1994, Hagemeijer & Blair 1997). Reintroduction programmes in 3 nature reserves on the Mediterranean littoral, using birds from the Guadalquivir marismas, have produced populations in Cataluña since 1990, S'Albufera, Mallorca, since 1992 and Albufera, Valencia, since 1995, with evidence of further natural spread (Hagemeijer & Blair 1997). The 19th century occurrences in Greece may be referable to this race (Vaurie 1965).

The races *caspius* and *seistanicus* were formerly local and decreasing in numbers, while *poliocephalus* was common (but rare in Kashmir) (e.g. Ripley 1977); all may be currently generally common (Perennou 1994); *poliocephalus* is locally fairly common in Nepal, where it is chiefly a winter visitor and passage migrant (Inskipp & Inskipp 1991), a scarce resident in Bangladesh (Harvey 1990), very common in Sri Lanka (Kotagama & Fernando 1994) and locally common in Myanmar (Smythies 1986); the species is common in Pakistan (Roberts 1991). In India and Pakistan it is relished as a delicacy and often suffers considerable local persecution (Ali & Ripley 1980). Numbers of *caspius* are poorly known in Russia (100-1,000 breeding pairs) and a small range expansion may have occurred in the Volga Delta and on the Terek R (Hagemeijer & Blair 1997). The Azerbaijan breeding population has extended its range this century and now numbers at least 10,000-15,000 pairs, with 8,000-12,000 birds estimated at L Aggel; the wintering population may be 15,000-20,000 birds in mild winters (Perennou 1994, Patrikeev 1995). In Turkey, *caspius* has a very small population (100-200 pairs) and is declining (Máñez 1994). Its status in Syria is uncertain: it is recorded at Bahrat Homs, but is unlikely to breed. During the first half of the 20th century, extensive drainage of wetlands in S Turkey/N Syria, including L Antioch, resulted in the almost total destruction of the breeding population in these areas (Shirihai 1996).

The race *madagascariensis* has extended its range in Egypt in recent years, colonising the entire Nile V and also L Qarun and Wadi el Rayan; this is correlated with the recent spread of reedbeds in these regions (Goodman *et al.* 1989). It is locally common to uncommon, sometimes abundant, in E, C and S Africa, but has a fragmented distribution in W Africa (Urban *et al.* 1986). In S Mauritania, counts of 250 birds at Lac de Mal, and 996 at Lac d'Aleg, were recorded in Jan 1996 (Dodman & Taylor 1996), and in Niger 592 birds were counted at Kokoro in Feb 1997 (Dodman *et al.* 1997). It is locally common in Mali and Sierra Leone, uncommon in Liberia and apparently uncommon around L Chad, but was formerly present in large numbers at L Fittri, 200km to the E (Lamarche 1980, Urban *et al.* 1986, Gatter 1988, Field 1995). In Ivory Coast it is abundant in the few remaining reedbeds (Thiollay 1985). It is very local and generally uncommon in S Ghana, and in extreme N Nigeria and it was recently recorded from the Lekki Peninsula (E of Lagos) in the S (Grimes 1987, Elgood *et al.* 1994). It has been recorded 3 times in Cameroon since 1990, at Limbé and twice near Yaoundé, where the species bred (Sala 1991, Manners *et al.* 1993). It occurs throughout Ethiopia in suitable habitat, and may be locally common, e.g. at Lakes Awasa, Ziway, Ardibu and Ellen (M. Wondafrash pers. comm.); the only record for Sudan is of c. 25 birds at Kosti in Mar 1983, where it may not be resident (Lambert 1987). It is common in Rwanda and Burundi, and common in Uganda around L Victoria, and L Kyoga and nearby lakes, but rare elsewhere; in Kenya it is local and uncommon, occurring in the W and C highlands, extending N to L Turkana and S to L Jipe, and (rarely) on the coast at Mombasa, and it has decreased greatly in recent years due to habitat loss and the introduction of the Coypu *Myocasta coypu* to many wetlands; it is widespread in Tanzania but apparently uncommon in the N (Britton 1980, Urban *et al.* 1986, Zimmerman *et al.*1996). It is relatively common in SW Angola, locally common in Malawi and Mozambique, and not very common in Zimbabwe (Benson & Benson 1977, Irwin 1981, Pinto 1983, Clancey 1996). Southern Africa Atlas data (Taylor 1997c) show that it occurs in N and E Botswana, at scattered localities throughout much of Namibia, and widely in South Africa; it is locally common in these countries but is largely absent from the arid Karoo and Kalahari biomes (for abundance, see also Koen 1988, Hockey *et al.* 1989, Hines 1993, Maclean 1993). Its South African population is estimated at c. 50,000-60,000 birds, more than half of these (33,850) in KwaZulu-Natal, while the Nyl R floodplain may hold 5,000 breeding pairs in years of high rainfall (Tarboton *et al.* 1987, Taylor 1997a). In Madagascar it is locally rather abundant, except on the high plateau (absent) and in the S (rare), but is heavily hunted and has consequently virtually disappeared from some areas (Salvan 1972b, Langrand 1990).

The race *viridis* is apparently common (e.g. Lekagul & Round 1991). The race *indicus* was formerly locally common to very common (Ripley 1977, van Marle & Voous 1988); its present status is probably common, and 3,970 birds were estimated at Tempe L, Sulawesi (MacKinnon & Phillipps 1993, Perennou 1994). It is apparently a local resident in Kalimantan (Borneo) and it has wandered to Brunei where a single bird was seen in Apr 1991 (C. Mann verbally). Up to 5 birds have been seen at Kota Belud, Sabah on several occasions in Jul/Aug 1987-1997 where there may be a small resident population (N. Redman *in litt.*, S. Harrap verbally). The species (races *indicus* and *melanotus*) is probably very local in Wallacea, unless it has been overlooked in many places (White & Bruce 1986), but may be locally common (Coates & Bishop 1997); *pulverulentus* is uncommon and local in the Philippines (Dickinson *et al.* 1991).

The race *melanotus* is widespread and locally common in Australia and New Zealand, and in both countries its range has expanded; in Australia the construction of artificial lakes has allowed colonisation of parts of the interior, while in New Zealand, which it probably colonised c. 300 years ago (Millener 1981), the conversion of forest to open pasture has greatly increased the amount of suitable habitat used by the species (Craig & Jamieson 1990); it has recently extended into W Tasmania; it is uncommon on Norfolk I, has recently become established on Lord Howe I, and has been recorded regularly from the Kermadec Is only since 1954 (Marchant & Higgins 1993). Its total population in New Zealand is estimated at 600,000 birds (Marchant & Higgins 1993). It is sometimes considered a pest because it may damage fruit, vegetables, grain crops and pastures, and it sometimes raids chicken houses for eggs; on Norfolk I and in New Zealand it may be shot in season (Marchant & Higgins 1993), and the annual bag is estimated at 50,000-80,000 (Robertson

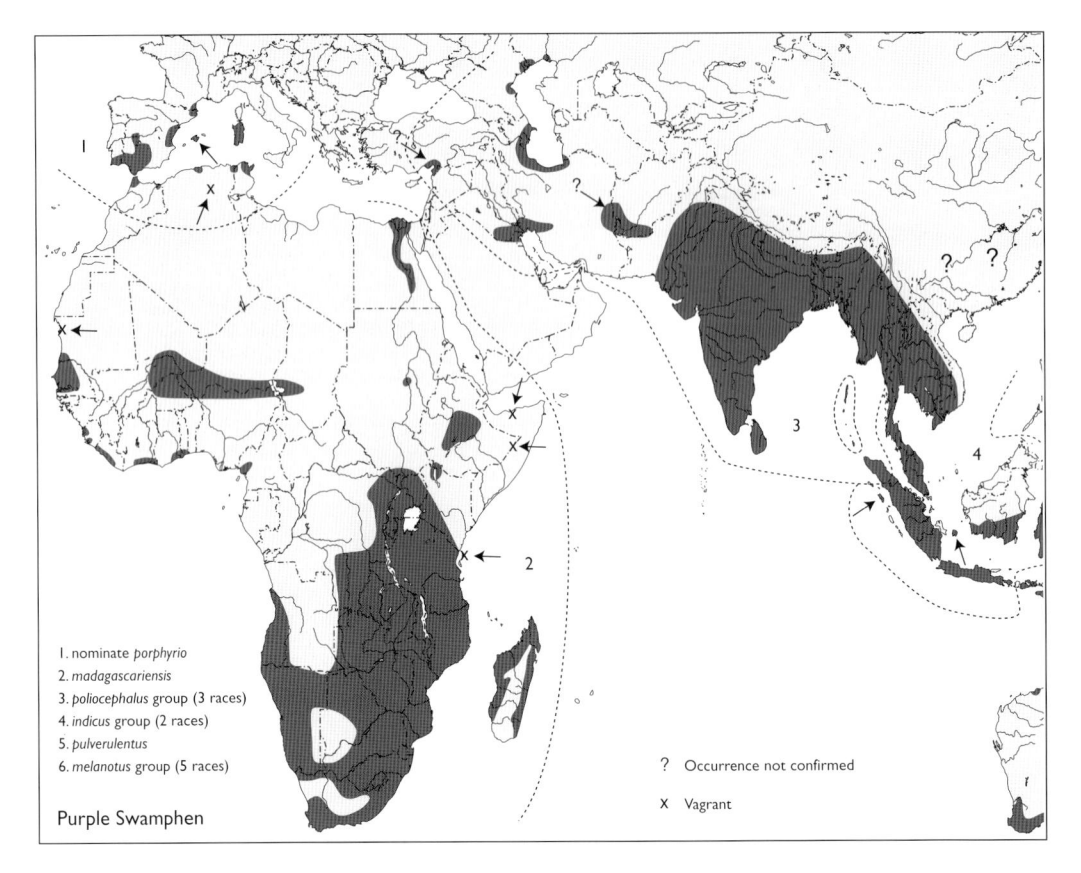

1. nominate *porphyrio*
2. *madagascariensis*
3. *poliocephalus* group (3 races)
4. *indicus* group (2 races)
5. *pulverulentus*
6. *melanotus* group (5 races)

? Occurrence not confirmed
X Vagrant

Purple Swamphen

1985). The current status of the race *bellus* is not given but is presumably similar to that of *melanotus*.

The race *melanopterus* has only recently been recorded from Halmahera, N Moluccas (Anon. 1997). From New Guinea E to the Solomon Is the species (races *melanopterus*, *melanotus*, *samoensis*) is widely distributed but local; in the highlands of New Guinea it is probably very scarce and local, but it is not uncommon along the Tebi R and is locally abundant in the middle Sepik and Port Moresby areas but is surprisingly absent from much of the Trans-Fly (Coates 1985). The race *pelewensis* is apparently quite rare, with a small population, and enjoys no legal protection; it is recommended that it be classified as Endangered (Pratt *et al.* 1980, Engbring & Pratt 1985). On Niue I, *samoensis* was formerly locally common and was considered a pest because it consumed planted yams (Kinsky & Yaldwin 1981); elsewhere in the tropical Pacific it is uncommon to rare on most islands (Pratt *et al.* 1987) but is common on most islands of Vanuatu (Bregulla 1992) and some of Tonga (Rinke 1986, 1987).

This species is vulnerable to wetland destruction, degradation and modification, and to frequent disturbance; its numbers have been reduced locally throughout its range by the removal of marginal wetland vegetation and by drainage (e.g. Marchant & Higgins 1993); domestic stock also adversely affect its habitats by grazing and trampling. Some artificially created habitat has been provided by forest clearance, swamp drainage and the creation of impoundments; on Norfolk and Lord Howe Is it has become established only since vegetation clearance has produced suitable habitats (Marchant & Higgins 1993). It responds well to protection, will occupy artificially maintained wetlands, and will tolerate a limited amount of human disturbance (Taylor 1997a).

Predators include Tawny and Imperial Eagles *Aquila rapax* and *A. heliaca*, Pallas's Fish Eagle *Haliaeetus leucoryphus* (India and Pakistan), Swamp Harrier *Circus approximans* (Australia), and Red Foxes *Vulpes vulpes* and Golden Jackals *Canis aureus* (Azerbaijan) (Ripley 1977, Roberts 1991, Patrikeev 1995).

MOVEMENTS No regular long-distance migrations are recorded, but local seasonal movements, in response to seasonally changing habitat conditions, principally the drying out of marshes, are reported from many regions. When the Guadalquivir marismas dry out each summer the birds move locally to the nearest wet habitat, and local movements in this region are believed to be made largely on foot and at night (Cramp & Simmons 1980). The Russian population moves S in winter to escape the ice, boosting numbers in Azerbaijan (Hagemeijer & Blair 1997); most birds on the Volga Delta move S in winter), and numbers increase on the S Caspian in winter (Cramp & Simmons 1980). In Azerbaijan, spring movement occurs in Mar-Apr and autumn movement in Aug-Nov (Patrikeev 1995). In India and Pakistan *seistanicus* is partly migratory, while *poliocephalus* is possibly a local migrant, and birds are especially prone to move during and after the monsoon when seasonal rains provide new feeding areas and shelter (Ali & Ripley 1980, Roberts 1991).

In N Senegal it is scarce until Jan but is then recorded in flocks of up to 550 at Grand Lac (Rodwell *et al.* 1996),

Purple Swamphen

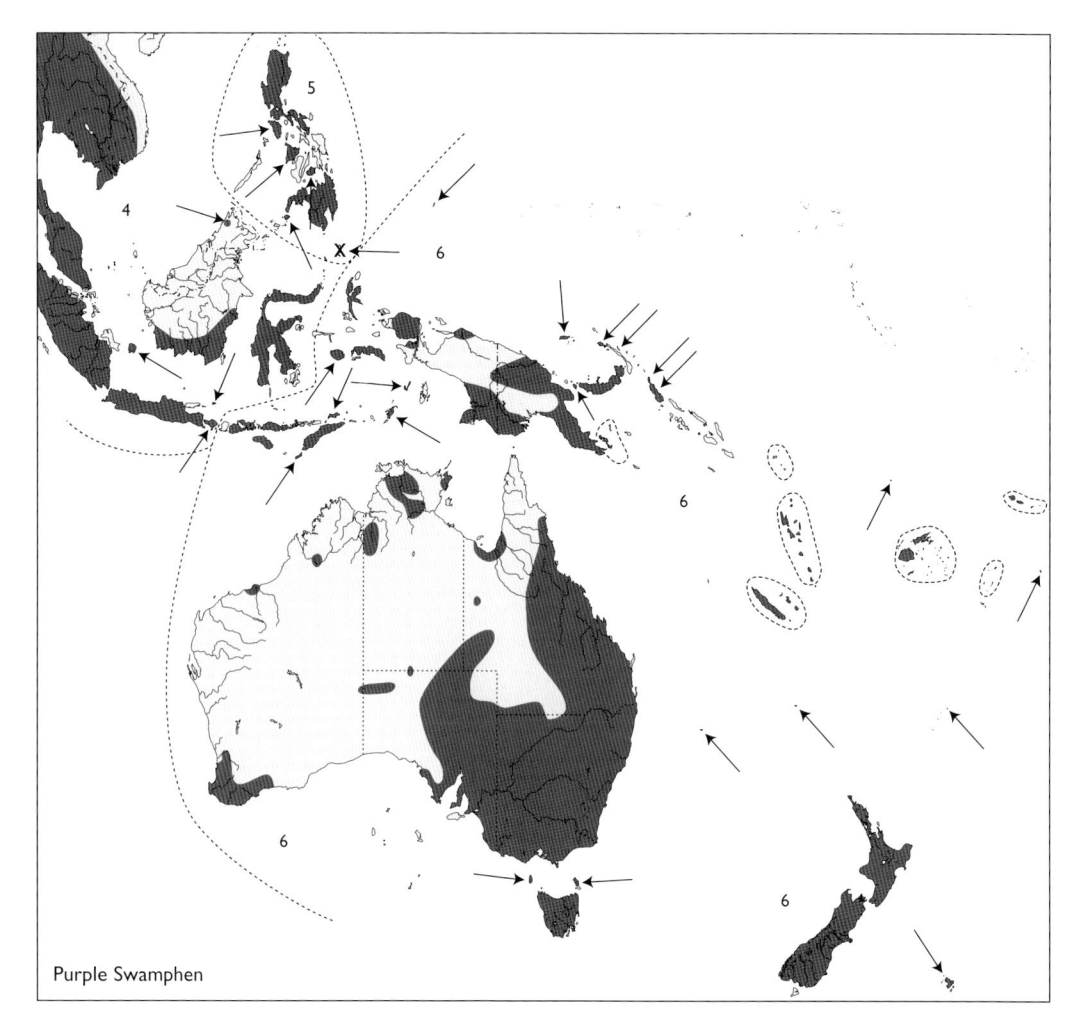

while in Gambia it is only a rare dry-season non-breeding visitor, in Dec-Sep (Gore 1990). Records from the Banc d'Arguin, N Mauritania, in Apr-May suggest occasional N movements from sub-Saharan Africa (Meininger *et al.* 1990). In East Africa it disperses to temporary ponds in seasons of unusually heavy rains (Brown & Britton 1980), while some local movements are also apparent in C Kenya in normal seasons (P. B. Taylor unpubl.). At the Nyl R floodplain, South Africa, it occurs only during the rains, is an abundant breeder in good rainfall years and is virtually absent in seasons of low rainfall (Tarboton *et al.* 1987, Taylor 1997a).

It is possibly prone to irruptive movements in some parts of Australia and New Zealand, but fluctuations in numbers may be partly associated with modifications to habitat, and there is no evidence of large-scale seasonal movement, although in some areas numbers fluctuate seasonally (Marchant & Higgins 1993). It may migrate regularly across the Torres Strait, mainly in the wet season, and birds ringed in Queensland have been recovered in Papua New Guinea, where large fluctuations in numbers on ricefields have been attributed to migration (Mees 1982, Marchant & Higgins 1993). In New Zealand it may cross the main mountain ranges, as birds have been found alive and dead on glaciers and snowfields, but such individuals may merely have been blown off course (Ripley 1977, Tunnicliffe 1985). Most ringing recoveries suggest that birds are sedentary, and most reported movements are <50km, but some birds move at least 240km and 1 from Townsville, Queensland, was recovered, 3 years later, 1,600km away in New Guinea (Marchant & Higgins 1993).

Some wandering is reported, which in Europe is confused by escapes, but it is a genuine vagrant (or very rare visitor) to France, Germany, the former Czechoslovakia, Austria, Hungary, the former Yugoslavia, Lebanon, Israel, Jordan, Kuwait, Arabia and Cyprus (Cramp & Simmons 1980, Shirihai 1996); Lewington *et al.* (1991) also record it from Belgium, the Netherlands, Norway, Poland, Switzerland, Greece and Italy. In Africa 1 was found on the beach at Berbera, NW Somalia (Archer & Godman 1937), and it has wandered to Pemba I (Jul 1942: Pakenham 1979) and to scattered desert localities, e.g. Touggourt oasis (Algeria) (Urban *et al.* 1986). The origin and racial identity of a bird from Cumbria, NW England, in Oct 1997 (Palmer 1998) are currently being evaluated. In Syria it is recorded at Bahrat Homs, where it is unlikely to breed, but it may breed in the Euphrates V (Taylor 1996b). In Israel, the race *madagascariensis* is an

465

occasional or very rare visitor in some years, being recorded throughout the year but most in Jun-Nov and especially Jul-Sep; before the 1950s, *caspius* was a scarce to uncommon winter visitor to N Israel, but since the extensive drainage of wetlands in S Turkey/N Syria (see Distribution and Status) it has not been recorded in Israel (Shirihai 1996). It is an occasional visitor to Campbell I, S of New Zealand (Marchant & Higgins 1993).

HABITAT Fresh or brackish, sheltered open waters, still or slow-flowing, densely fringed or overgrown by *Phragmites*, *Typha*, *Carex*, *Cyperus* (including *C. papyrus*), *Scirpus*, *Eleocharis* etc, and especially those which also have floating vegetation such as waterlilies. It sometimes occurs at saline, eutrophic or turbid wetlands (Marchant & Higgins 1993). Inhabited wetlands usually have fairly tall, dense vegetation and are often extensive, but in Africa birds are also found on small waters, sometimes <2ha in extent (e.g. Taylor 1985b, 1997d). In Sierra Leone it occurs in thick riverine *Vossia* grass (Field 1995). The habitat is often permanent, and in Europe it apparently prefers wetlands with stable water levels (Hagemeijer & Blair 1997) while in Australia and New Zealand it sometimes occurs at seasonal or ephemeral wetlands (Marchant & Higgins 1993). In Africa it also occupies, and breeds in, seasonal and temporary wetlands, and in South Africa non-breeding birds also occur at pans and other irregularly flooded sites in dense emergent vegetation such as *Eleocharis* which is not tall enough to conceal the birds when they stand upright (Taylor 1997d, P. B. Taylor unpubl.). It inhabits ponds, lakes, dam margins, marshes, swamps, rivers, floodplains, artesian and seismic bores, sewage farms, ricefields, and islands on the R Nile (Egypt); in Australia and New Zealand it often occurs at town lakes. It extends into open habitats adjacent to wetlands, such as grassland, agricultural land, parks, gardens, sports fields, golf courses, roadside verges, hedgerows, forest margins and even chicken runs, and occasionally occurs some distance from water (Marchant & Higgins 1993); it has been recorded from ricefields and canefields, and it may forage in burnt grass areas (e.g. Storr 1973, MacKinnon & Phillipps 1993). In New Zealand environmental conditions, especially the presence of water and green grass, are positively correlated with reproductive success, while cover is also important for the survival of eggs and chicks (Craig & Jamieson 1990). On Manus, Admiralty Is, it has been seen climbing about in the canopy of low trees in rainforest (E. Lindgren in Coates 1985). This species often occurs in lowland areas but ranges from sea level up to c. 2,350m (New Zealand) and 2,500m (Africa); once at 3,000m (Mau Narok, Kenya: Urban *et al.* 1986).

FOOD AND FEEDING Most information is from Cramp & Simmons (1980) and Marchant & Higgins (1993). This species is omnivorous, but primarily vegetarian, taking shoots, leaves, roots, stems, flowers and seeds of aquatic and semi-aquatic plants (including Lemnaceae, Juncaceae, Cyperaceae, Hydrocharitaceae, Potamogetonaceae, Zanichelliaceae and Poaceae). Principal food items include *Typha* (especially leaf bases and pith); *Scirpus* and *Bolboschoenus* (pith from stems); *Eleocharis* (stem bases); grasses (*Poa*, *Glyceria*, *Anthoxanthum*); young rice plants; seeds of grasses, rice, sedges, and *Rumex*, *Ranunculus*, *Polygonum*, *Sparganium* etc; and vegetative parts (especially tubers) of waterlilies; it also grazes clover *Trifolium*, takes fern *Salvinia natans* fronds (New Zealand), and eats bananas, tapioca roots and yam *Dioscorea* bulbs (Niue I: Kinsky & Yaldwin 1981). Animal foods normally form only a small proportion of the diet, and include onchyphorans, earthworms and leeches, molluscs (*Theba*, *Bythynia*, *Planorbis*, *Valvata*, *Acroloxus*), crustaceans (Isopoda, Amphipoda and small crabs), insects and their larvae (including beetles *Dytiscus*, *Donacia*, *Carabus*, grasshoppers Acrididae/Tettigoniidae (including *Deinacrida rugosa* on Mana I, New Zealand), Hemiptera, Diptera and Lepidoptera), spiders, fish and their eggs, frogs and their eggs, lizards, water snakes *Natrix maura*, birds, their eggs and nestlings, and small rodents. In Spain, 141 stomachs contained only 9.3% animal material (fish 2.1%, insects 7.8%); 98% of stomachs contained vegetative parts of *Typha*, and 80% seeds of *Sparganium*. It has been observed climbing a tree in a heronry to eat the eggs of Cattle Egret *Bubulcus ibis* and Yellow-billed Egret *Egretta intermedia* (Urban *et al.* 1986). It takes carrion occasionally, recorded items including dead fish, crabs and the carcass of a Red-knobbed Coot (135) (Madden & Schmitt 1976, Cramp & Simmons 1980). It feeds in cover, at the edge of cover on muddy, sandy or hard shoreline (including on sludge at sewage ponds in New Guinea), in shallow water and on floating vegetation, walking deliberately while foraging. It swims relatively little when feeding, but will dabble near shallow margins like a Mallard *Anas platyrhynchos*; it takes insects from the water surface. It climbs freely to strip flowers and seeds from reeds etc, and to take bird eggs and nestlings. It uses the bill to cut, dig up or pull out surface or submerged plants and their roots, tubers and rhizomes, to move stones and gravel, to turn over matted vegetation, and to tear and dismember food items, and the foot to grasp and manipulate food. It pulls down tall stems of reeds and other plants with the bill and cuts off sections to eat. When pulling up well-rooted plants with the bill it may splay the legs for support. Older stems of *Typha*, etc, are stripped of their outer covering with the bill and only the pith is eaten, but younger stems and leaves are eaten whole. It is an adaptable species and will forage in pasture, scrub and mown fields (race *melanotus*); it grazes clover and grass, and it also feeds in ricefields. Food may be washed before being eaten. It builds platforms for feeding. Feeding takes place mainly in the early morning and late evening, sometimes also at night.

Adults pull up *Typha* plants with the bill and feed chicks <1 week old on pieces of the lower, soft part of the stems (P. B. Taylor unpubl.); chicks also fed on tender plant shoots (Urban *et al.* 1986). Young up to 1.5 months old eat many invertebrates (68-70% of all food value), including Hemiptera, Odonata and Acrididae (Patrikeev 1995).

HABITS It is mainly diurnal and crepuscular, and is normally quite shy but is bolder when most active early in the morning and late in the evening; it also sometimes emerges from cover at night (Cramp & Simmons 1980). It may be quite shy in remote wetlands and is very secretive when harassed, but when accustomed to human proximity without interference (e.g. in urban habitats) it becomes tame; it is apparently shy and cryptic during the flightless wing moult (Marchant & Higgins 1993). It normally moves with a slow, rather dignified gait, but has a lumbering, ungainly run, often accompanied by wing-flapping. It flicks the tail frequently, usually when uneasy and in response to the approach of a predator, but also as a signal for the young to follow. Tail flicking in this species is directed towards potential predators rather than towards conspecifics, and is an alertness signal with a pursuit deterrent function rather than a warning to conspecific individuals (Woodland *et al.* 1980, Alvarez 1993), although it also occurs in response to inter- and intra-specific aggression (Marchant & Higgins 1993). When disturbed in the open it runs to cover, often paddling over the water surface and flapping the wings; it runs fast through tangled vegetation. It also flies fairly readily; short flights are heavy and unwieldy, with dangling legs, but in longer flights the legs are extended; take-off is laboured, and is preceded by a run. It normally swims infrequently, though buoyantly and strongly, with a flat-backed profile rising to a high stern, neck upstretched and head jerked back and forth (like a large moorhen); in Madagascar it frequently swims to cross open water (Langrand 1990). It may dive to escape danger, progressing under water partly or fully submerged (Marchant & Higgins 1993). It builds platforms for feeding and roosting, and may roost in trees (Marchant & Higgins 1993).

Birds bathe while standing in shallow water, ducking the head and flapping the wings, then leave the water to preen and oil the plumage; oil is applied to the body plumage with the bill, but the wingtips are rubbed over the preen gland, apparently to oil these inaccessible parts of the body; preening is often followed by sunbathing, the bird standing with partly spread wings extended sideways and/or pointed downwards; birds occasionally sunbathe by sitting on the ground with the wings closed (Holyoak 1970). In the early morning and the evening it will often climb to the top of tall plants to sunbathe, and it also sunbathes on platforms. Allopreening at the nest is often concerned with plumage care (Urban *et al.* 1986). During light rain, 2 captive birds in Vanuatu frequently pranced about their aviary, walking with quick steps with wings half open and alternately leaping high in the air; this activity was reminiscent of the dancing display of cranes and was continued until the plumage was dampened by the rain (Bregulla 1992).

SOCIAL ORGANISATION Information is mainly from the summary in Marchant & Higgins (1993) studies in New Zealand by J. L. Craig and I. G. Jamieson (see References). This species is mainly territorial, although birds also occur in non-territorial flocks; in New Zealand many territories are defended all year but some break down to a variable extent after breeding; good-quality habitat appears to be at a premium and habitat saturation means that continual defence of territories is often necessary (Craig & Jamieson 1990). Its social structure and mating system are complex. Monogamy apparently prevails in W Palearctic races and in *madagascariensis*, but elsewhere at least the races *poliocephalus* and *melanotus* often live in communal groups, *poliocephalus* being known to do so in captivity and usually only as pairs with young of the previous year which act as helpers (Harrison 1970, Holyoak 1970). In *melanotus*, stable communally breeding groups usually contain 2-7 breeding males, 1-2 (sometimes 3) breeding females, and up to 7 non-breeding helpers which are offspring from previous matings; mate-sharing occurs, all breeding adults court and copulate with each other (homosexuality is frequent and incestuous matings are common) and, if 2-3 females breed, all lay in a communal nest; all group members defend the territory and care for the young. Adult nonbreeders (usually young of the previous year) may also form part of a group, and all types of group (including monogamous pairs) may include young hatched earlier in the same season. Unstable breeding groups form in suboptimal territories, are promiscuous, usually unsuccessful, and characterised by much aggression and many males. The mating system adopted, and the size of the breeding group, depend on the size of the territory, the nature of the habitat, the stability of the group and the size of surrounding groups. Pairs are common in pasture and communal groups in swamps; monogamy occurs only where the defended boundary is short and neighbours are paired (pairs are not able to defend territories against groups): it only occurs in the most aggressive birds, and 10-22% of populations breed as pairs. Simultaneous polyandry has been recorded once (Wettin 1984).

In Spain, typical breeding densities are 1.5 nests/ha in a 4ha *Scirpus maritimus* marsh and 3.3 nests/ha in an adjacent 4ha marsh with a transition to shrubby glasswort *Arthrocnemum macrostachyum* (Hagemeijer & Blair 1997). In Azerbaijan, a breeding density of 10-15 pairs/km of lake shoreline is recorded (Patrikeev 1995). In Nigeria, 11 nests with eggs were found in 3.2ha (Serle 1939). In New Zealand, territories are 0.7-3.0ha in area.

SOCIAL AND SEXUAL BEHAVIOUR In Africa and Madagascar it is normally seen in pairs or in groups of 12 or more birds (Urban *et al.* 1986, Langrand 1990), but no proper information is available on the size of non-territorial flocks. In India it may occur in concentrations of 50+, and birds fly in large numbers to ricefields in the evenings (Ali & Ripley 1980). In New Zealand, non-territorial flocks of up to 300 birds form in summer and autumn, remain together throughout the non-breeding season and then disperse before breeding, to leave some small flocks of nonbreeders; flocks include a high proportion of young birds and males, and apparent family groups may also be absorbed into the flock (Marchant & Higgins 1993). Hierarchies occur in all social units and territorial groups (Marchant & Higgins 1993), males being dominant over females, and adults dominant over yearlings and juveniles; within each sex- and age-class, higher status is achieved with increasing weight and with increasing size of the frontal shield. Aggressive encounters between feeding birds are often observed.

This species is aggressively territorial and has many boundary squabbles, but in restricted habitats pairs sometimes associate in loose colonies (Cramp & Simmons 1980). In the W Palearctic, the territory is defended mainly by the male, but also by the female and the helpers (Cramp & Simmons 1980), and Serle (1939) noted no antagonism by breeding birds towards Allen's Gallinule, Common Moorhen (129), Lesser Moorhen (131) or African Jacana *Actophilornis africanus*.

Most information on behaviour is from observations of the race *melanotus* (Marchant & Higgins 1993) and of *poliocephalus* in captivity (Holyoak 1970). The territorial threat posture involves an upright stance (Fig. 12A), with head and bill pointing downward and neck held vertically or forward (and often extended), tail flicks to display the white undertail-coverts, and harsh, grating calls. If the opponent turns away in an appeasement gesture it may be allopreened instead, and sexual behaviour may follow (Holyoak 1970). In aggressive encounters a horizontal posture is commonly used (Fig. 12B); in a more intense version the head is lowered and the body tilted forward (Fig. 12C); if the opponent fails to retreat, the aggressor runs at it and pecks it hard on the head and neck. Fighting

A. 'Aggressive' upright.
B. Horizontal forward.
C. Depressed forward.
D. Head-bow.
E. Full bow.
F. Body bow.
G. Erect hunch.
H. Crouch.
I. Pre-copulatory hunch.
J. Pre-copulatory stance.
K. Copulation.
L. Dismount after copulation.

Figure 12: Purple Swamphen displays. [After Craig 1977 and Marchant & Higgins 1993].

may follow if the intruder does not retreat, the birds pecking each other, grappling with the feet and ripping with the claws, sometimes in mid-air; up to 4 birds may be involved in a fight; opponents in the aggressive upright display occasionally spar slowly with 1 foot. A vanquished bird flees with the head up and the tail horizontal, pursued by the victor. Aggressive bowing displays (Fig. 12D, E, F) are also recorded (see Marchant & Higgins 1993 for details), characterised by the bill pointing vertically down or resting on the ground; in the less intense forms the bird usually faces or turns away from the opponent. Hunches (Fig. 12G), with depressed head, bill and tail, and hunched neck, can precede either threat or escape displays. Submissive birds crouch in a horizontal posture with sleeked plumage (Fig. 12H); facing the head away from the opponent (thus removing the bill from view) in any display posture almost always precedes turning away in appeasement, which signifies an increased tendency to escape. During or after aggressive behaviour, displacement activities such as violent pecking at food objects, and bill-wiping, may occur.

In defence of a mate or of young against conspecifics or other species, a bird adopts a gaping-threat posture: with sleeked plumage, it holds the neck stretched horizontal and gapes at the adversary, often chasing it as well; in response to disturbance near the nest or the chicks, it makes a wing-arch or swanning display, fluffing out the plumage and raising the partly spread wings over the rump (Holyoak 1970). To distract a predator, an adult will clap its wings above its back and lead the predator away, appearing to be losing its balance but not feigning any specific injury (Marchant & Higgins 1993).

In captive *poliocephalus*, courtship behaviour usually starts with allopreening, which the female invites by turning her head away and fluffing her neck feathers; this ritualised activity serves to reduce aggression (Holyoak 1970); allopreening is also common in *melanotus*. Courtship feeding may also occur; the male approaches the female with water weeds in his bill, bows to her while calling quietly, then raises and flaps his wings, stands up straight, continues bowing, and brings the wings forwards, quivers them and calls loudly (Cramp & Simmons 1980). In *melanotus*, two birds face each other with bowed heads, and 1 bird (usually the male) passes food (usually small pieces of *Lemna*) to the other.

Allopreening may lead to a marking-time display (Holyoak 1970), in which the birds walk on the spot without closing the toes, and this may lead to copulation; the female solicits copulation by adopting an exaggerated bowing posture (arch-bow); the male mounts and copulates for 2-3s, flapping his wings to aid balance. Other birds present may allopreen the female before and during copulation. After copulation the male walks away and both birds preen themselves. In *melanotus*, courtship behaviour begins with a courtship chase, when the female is pursued by up to 6 males, holding her neck forward, with neck feathers expanded, and repeatedly dipping her head. The male approaches the female in an upright posture with the bill depressed, neck extended and wings and tail down, making humming calls, and follows the female; this posture is usually followed by a more horizontal one, with head and neck retracted. The female adopts a pre-copulatory hunched position (Fig 12I), an upright stance with neck retracted and bill pointing down; then she extends her neck downwards and partly opens her wings; as the male mounts, the female's body becomes horizontal; the male then treads the female and copulates (Fig. 12J, K), waving his wings and wagging his tail from side to side. He then dismounts over the female's head, with wings and tail raised (Fig. 12L).

BREEDING AND SURVIVAL Season Spain, Jan-Mar; Mediterranean, mainly Mar-Jun; North Africa, Mar-May; Mauritania, Nov (see Bengtsson 1997); Senegal and Gambia, Nov-Feb, Senegal, chicks Jan-Apr; Ghana, Jul-Aug; Nigeria, Jan-Feb, Jul-Oct; Cameroon, Jan; Zaïre, 2 seasons, Mar-Apr and Aug-Oct (Verheyen 1953); Ethiopia, Mar, Oct-Dec; Uganda, Feb-Apr, Aug-Oct; Kenya and NE Tanzania, May-Jul, Oct-Nov; rest of Tanzania, Apr; Malawi, Feb-Mar, Jun-Aug, Dec; Zambia, Dec-May, Jul; Zimbabwe, Jan-Oct; Botswana, Jan-Apr, Jul-Oct; South Africa, Jul-Mar (Transvaal), Oct-Feb, Aug-Apr (KwaZulu-Natal), Aug-Feb (SW Cape); in sub-Saharan Africa breeds during or late in rainy season, possibly two breeding seasons in areas with bimodal rainfall regime; Madagascar, Jan, females with enlarged ovaries Mar; Azerbaijan, Apr-Jun; India and Pakistan, mainly Jun-Sep (SW Monsoon), also Nov-Feb, Apr, chicks (Pakistan) May-Oct; Sri Lanka, Jan-May, occasionally Jul-Aug; Bangladesh, Apr-Jul; Myanmar, Aug-Sep; Peninsular Malaysia, Mar-May; Halmahera, chick Apr; Java, Feb-Jul, Nov; Borneo, Apr; Lesser Sundas (Timor), Mar; New Guinea, May, young Jun, Oct; Australia, all months, main breeding correlated with peak rainfall + 2-3 months, temperature and increase in photoperiod (Halse & Jaensch 1989); New Zealand, Aug-Feb (mainly Sept-Dec); Solomon Is (Bougainville), Sep; Vanuatu, mainly Aug-Jan; Niue I, probably May-Sep. **Nest** Nest in shallow water (depth 30-120cm), occasionally up to 180m from water; concealed in thick emergent vegetation; large, substantial structure of interwoven dead stems and leaves of water plants (reeds, rushes, coarse grass etc); sometimes lined with papyrus heads, grass blades or ferns; with shallow cup; some nests with multiple bowls (New Zealand); nest usually built on platform of beaten-down vegetation; normally just above water (occasionally up to 2m above water) but sometimes floating; may be built up from water surface; surrounding stems often pulled over to form canopy; 1-2 access ramps provided. Outside diameter of nest c. 30cm, depth 10-20cm; cavity diameter 19-30cm, depth 3-10cm; material often added by sitting bird during incubation. Both sexes build, sometimes with assistance from helpers; male usually brings material and female builds; helpers bring material and sometimes help build. Trial nests and brood nests also constructed (New Zealand). **Eggs** 2-7; clutch size 3-5 (W Palearctic, *caspius*), 3-7 (*poliocephalus*), 2-6 (*madagascariensis*); in *melanotus*, 2-6 (4.2, n = 51, Australia), clutches from 1-2 females (New Zealand) 4-13 (7.4, SD 2.4, n = 37). Eggs laid at daily intervals; in communal breeders all breeding females in group lay eggs in same nest, sometimes in double bowl, laying highly synchronised. Lays up to 3 replacement clutches after losses. Maximum of 56 days between laying of 1st and 2nd clutches, Australia (Brown & Brown 1977). Eggs long to broad oval, sometimes elliptical; smooth or slightly rough; somewhat glossy; buff to creamy-white, yellow-stone, pale green, or pink to pale red-brown or pale brown, variably spotted and blotched red, maroon, chocolate-brown, purple, pale violet or grey; markings sometimes larger and more numerous at blunt end. Size of eggs: nominate 49-59.6 x 33-39.5 (54.6 x 37.3, n = 80), calculated weight 41.5; *madagascariensis* 49-59.9 x 34.6-40 (54.2 x 37.3, n = 102), weight 40.5-42; *seistanicus* 48.2-57 x

33.7-38.6 (52.2 x 36.2, n = 13), calculated weight 37.6; *poliocephalus* 45-54.6 x 32-37.8 (49.2 x 35.5, n = 166), calculated weight 33.8; *indicus* 45.5-50 x 31.4-35.4 (47.6 x 33, n = 9), calculated weight 31; *melanopterus* 46.4-50.5 x 33.6-37.3 (48.2 x 34.6, n = 13), calculated weight c. 31.7; *bellus/melanotus* 46-58.4 x 33.5-41.9 (50.8 [SD 2.9] x 36.6 [SD 2.0]), calculated weight c. 37.5; *samoensis* 45.3-55 x 31.5-36.2 (50.4 x 34.9, n = 36), calculated weight c. 33.5.

Incubation Begins when clutch complete, or (New Zealand) in mid-clutch. Period 23-27 days, by both sexes (female takes larger share) and helpers; in New Zealand, male incubates at night; hatching synchronous (Africa) or asynchronous (New Zealand). **Chicks** Precocial or semi-precocial, and nidifugous; can leave nest soon after hatching but often remain, or are brooded, on nest for first few days, when do not receive much food; fed and cared for by parents and helpers. Chicks beg by crouching, with head and neck extended, rapidly twirling outstretched wings, sometimes moving head, and calling. Young can walk and swim soon after hatching, and can submerge to escape danger, keeping only culmen exposed; begin feeding themselves at 2-14 days but fed for 25-40 days (at least 2 months in New Zealand); attempt to hold food in feet from 2-3 weeks, only achieving success when at least half grown; fledged at 60 days or more; independent at 6-8 weeks. Breeding birds will hatch Takahe eggs and foster the chicks (Bunin & Jamieson 1996a). In *melanotus*, breeding success highest in mono-gamous birds, and is also correlated with environmental conditions (see Habitat). In New Zealand, for all groups and pairs, of 276 eggs laid, 206 (75%) hatched, 59 (21%) young (= 2.1/territory and 0.5/bird) survived to 4 months; survival higher in wet summers than in dry, and early clutches more successful than later ones; in Australia, 42 (61%) of 69 eggs hatched; of 91 eggs lost before hatching, 38% predated, 22% deserted, 16% ejected, last eggs deserted when chicks led from nest (Marchant & Higgins 1993). Nest predators include rats, crows and harriers; chicks taken by Black Kite *Milvus migrans* and harriers (Ripley 1977, Urban *et al.* 1986, Marchant & Higgins 1993, Patrikeev 1995). Age of first breeding 1-2 years; however, within breeding groups in New Zealand, birds 3 years old or less do not usually breed. May raise two broods per season in some regions.

121 WHITE GALLINULE
Porphyrio albus   Plate 37

Fulica alba White, 1790, Lord Howe Island.

Sometimes considered conspecific with *P. porphyrio* (e.g. Greenway 1967), but it is distinct by virtue of some skeletal characters, the soft rectrices, and the relative length of the secondaries and wing-coverts in relation to the primaries; in these respects it appears to be intermediate between *P. porphyrio* and *P. mantelli* (Ripley 1977, Fuller 1987). Monotypic.

Synonyms: *Gallinula/Notornis alba*; *Porphyrio porphyrio/stanleyi/melanotus* var. *alba*.

Alternative name: White/Lord Howe Swamphen.

IDENTIFICATION EXTINCT. Length 36cm (Wagstaffe 1978). Flightless. Some birds were all-white, others blue and white and some all blue; birds with blue may have been hybrids with Purple Swamphen (120) (Hindwood 1940, Hutton 1991); bill and shield red; legs and feet red or yellow. Some accounts describe males as having plumage tinged azure-blue or as having some blue feathers in wings. Young birds said to be bluish-grey. Inhabited woodland in lowlands.

Similar species Very similar in overall appearance to Purple Swamphen, but distinguished by its largely or completely white plumage. Liverpool Museum specimen differs from Purple Swamphen in having shorter and more robust legs and toes, a less massive bill, and softer remiges and rectrices (M. Largen *in litt.*). However, species also said to be considerably larger than Purple Swamphen, with very stout bill (Fuller 1987). New Zealand populations (*melanotus*) of Purple Swamphen are prone to albinism, many having pale brown, yellowish-white or white feathers mixed with normal plumage, while some individuals are buff-coloured, and others have white wings or may be completely white (Ripley 1977, Marchant & Higgins 1993); partly white birds also occur on certain Pacific islands (Mayr 1941). See Description.

VOICE Undescribed.

DESCRIPTION
Adult Liverpool specimen almost completely white, with a few scattered feathers of scapulars and mid-back regions sooty-brown at base and sooty-blue distally; central rectrices sooty-brown with faint bluish suffusion; legs and bill painted red, but no indication of original colour exists (Wagstaffe 1978, M. Largen *in litt.*, contra Ripley 1977). Vienna specimen white, with red bill and legs (Ripley 1977); both specimens have soft rectrices. Contemporary illustrations and descriptions (see Hindwood 1940) indicate that plumage could be all white, white with blue patches (e.g. on back, neck, breast and tail), white with blue feathers in the wings, and all blue or "normal coloured" as Mayr (1941) describes them; 1 illustration showed a bird with most of wings dark and face blackish. Birds with blue feathers in wings, or with plumage tinged azure-blue, were sometimes described as males. Because typical Purple Swamphens (race *melanotus*) are known to occur on Lord Howe I, it is not certain whether some descriptions might apply to this species, or to hybrids between the two species, but adult White Gallinules are thought to have been white, albinism having been dominant (Hindwood 1940, Fuller 1987). Some illustrations and descriptions refer to legs and feet as being red, but Hindwood (1940) also gives a description of yellow legs and feet, while the illustration in Salvin (1873) also shows yellow legs and feet.

Immature and juvenile Young birds said to progress from an all-black plumage to a blush-grey plumage and thence to white (Hindwood 1940).

Downy young Black; no other information recorded.

MEASUREMENTS Liverpool specimen: wing 228; tail c. 88; culmen (from base of shield) 63 ; tarsus 77 (Wagstaffe 1978). Vienna specimen: wing 228; tail 75; culmen (including shield) 79; tarsus 92 (Fuller 1987).

GEOGRAPHICAL VARIATION None.

MOULT No information available.

DISTRIBUTION AND STATUS Lord Howe Island, 480km E of the Australian mainland. Although there are claims that it occurred on Norfolk I, and on Ball's Pyramid (off

Lord Howe I), it never did so (Hindwood 1940, 1965, Garnett 1992). EXTINCT. It is known only from 2 specimens (Vienna and Liverpool Museums), for neither of which is the full provenance known (Fuller 1987), and from contemporary illustrations, some written accounts and some subfossil bones (Hindwood 1940, Hutton 1991). This gallinule was not uncommon on its discovery in 1788, but it was killed indiscriminately by the first visitors to the island, and it became extinct very quickly, either by 1844 (10 years after the first settlement in 1834), or even before settlement, almost certainly through being hunted for food by whalers and sailors who landed for supplies, and not by the destruction of its habitat (Hindwood 1940, Marchant & Higgins 1993). Predators such as rats and cats did not appear until later (Hindwood 1932, 1940).

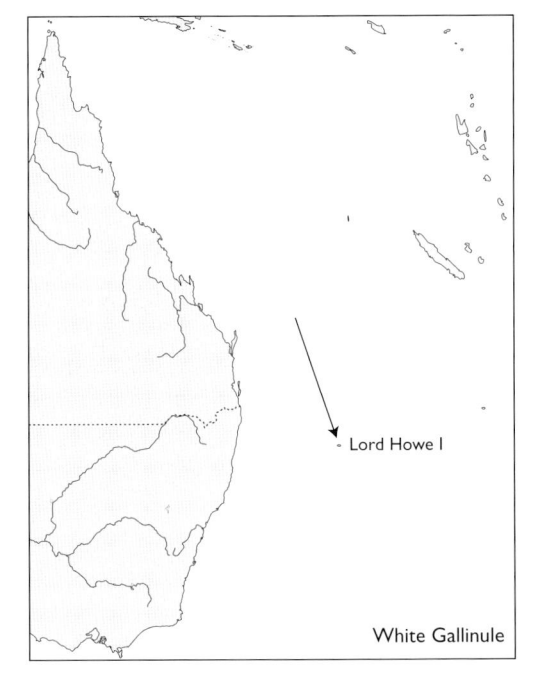

MOVEMENTS Presumably none.

HABITAT It occupied wooded lowland areas (Hindwood 1932, 1965).

FOOD AND FEEDING One visitor to the island thought that the birds were carnivorous, and said that they "hold their food between the thumb or hind claw and the bottom of the foot and lift it to the mouth without stooping so much as a parrot" (Hindwood 1940).

HABITS These birds were exceptionally tame and inquisitive, and were easily knocked down with sticks by the men who visited the island; some attempted to run away but were chased and killed (Hindwood 1940).

SOCIAL ORGANISATION Unknown.

SOCIAL AND SEXUAL BEHAVIOUR Nothing recorded.

BREEDING Not described.

122 TAKAHE
Porphyrio mantelli   Plate 37

Notornis Mantelli Owen, 1848, Waingongoro, North Island, New Zealand. Occasionally still placed in genus *Notornis*. Nominate race of North Island, New Zealand, recently extinct; may merit specific status. One extant subspecies recognised.

Synonyms: *Notornis Hochstetteri/parkeri*.

Alternative name: Notornis.

IDENTIFICATION Length 63cm; stands c. 50cm tall. Largest living rail; thickset, flightless species with reduced wings, short tail, massive, deep, laterally compressed bill and powerful, rather short, legs and feet; plumage loose and has silken sheen. Head and neck iridescent dark blue-purple; mantle paler; rest of upperparts, and most inner wing-coverts, olive-green with varying blue iridescence; flight feathers and rest of upperwing-coverts dark blue-purple. Underparts iridescent dark blue-purple; undertail-coverts white. Iris brown to reddish-brown; frontal shield and base of bill red, rest of bill more pinkish-red; legs and feet pinkish to pink-red. Sexes alike; female slightly smaller. Immature like adult, but slightly duller on back and mantle; iris attains adult colour by 5-6 months; from 6-8 months bill and shield pinkish; legs and feet dull red or pale brownish-orange, becoming like adult by 10 months. Juvenile predominantly brownish-grey on head and neck; face, chin and throat mottled white; upperparts olive-green with no iridescence; wings like adult; breast and upper flanks dull purplish-blue; rear flanks to vent cream-buff; undertail-coverts white; iris dark brown to grey; bill and shield almost black, fading to light brown or pinkish; legs and feet horn to dark purplish-brown. Second downy plumage is grey to grey-brown; Plate 37 shows young bird developing juvenile plumage but still with down on feather-tips. Inhabits alpine tussock grassland, also scrub and beech forest in winter; on islands to which introduced, occurs in pastures.

Similar species Unmistakable: much larger and more robust than Purple Swamphen (120), sympatric race of which (*melanotus*) has black upperparts and upperwings.

VOICE Information is from Marchant & Higgins (1993). The contact call is a single rising squawk *klowp*, which may be repeated in antiphonal duet following disturbance by an intruder or separation of the pair. The alarm call is a low, resonant *boomp* or percussive *oomp*, repeated slowly; a quieter version is used between the pair when together and undisturbed; when used with a superimposed soft rhythmical *kau* it serves as a recall note for chicks. Birds also give various clucking calls when feeding undisturbed, and a loud screech or hiss when suddenly alarmed, threatened, chased or caught. The contact call is easily confused with that of the Weka (39), but is generally deeper and more resonant; that of the Weka is disyllabic and more flute-like. The alarm note is also similar to that of the Weka, which is more staccato, continuous and frequent. Takahe often answer Wekas, and will respond aggressively to their calls; they also often respond to calls of Kiwis *Apteryx* at night. Chicks make cheeping calls when young; at 6 weeks they give a slow *wee-a* or a continuous repeated *weedle-weedle-weedle*, a continuous hoarse whistle may be given at 3-4 months, and captured young make a repeated, screaming distress call *chi-ching*. The adult contact call develops at c. 4.5 months and the alarm call at c. 7 months.

DESCRIPTION *P.m. hochstetteri*
Adult Head and neck dark blue-purple, tinged blue when fresh and dull black when worn; mantle glossy blue-purple; back to uppertail-coverts olive-green, back and scapulars with varying blue iridescence caused by light blue-green feather tips, feathers also having indigo bar at base of olive-green area; tail olive-green, darker and bluer than uppertail-coverts. All upperparts feathers have concealed dark grey or grey bases. Median, lesser and marginal upperwing-coverts dark blue-purple, some with pale blue iridescence; primaries, primary coverts, alula, outer secondaries and outer greater secondary coverts dark blue-purple on tips, outer webs and fringes of inner webs, elsewhere dark brown-grey; inner secondaries and their coverts similar but with olive-green or pale blue-green iridescence. Axillaries olive-green, with blue-purple iridescence; underwing-coverts dark grey with narrow blue-purple tips; underside of remiges dark grey. Underparts dark blue-purple, paler on sides of upper breast; dark grey feather bases give variable dark grey tinge; undertail-coverts white. Iris brown to reddish-brown; frontal shield and basal fifth of bill red, rest of bill pinkish-red to reddish with varying horn-yellow tinge; legs and feet pinkish to pink-red. Sharp and quite robust carpal spur, c. 10mm long, present.
Immature Like adult, but slightly duller on back and mantle. Iris attains adult colour by 5-6 months; from 6-8 months bill becomes pinkish with bluish cast, shield pinkish; legs and feet dull red or pale brownish-orange, adult colour attained by 10 months.
Juvenile From 8 weeks, head and neck predominantly brownish-grey, face, chin and throat mottled white; upperparts olive-green, with no iridescence, attaining brownish tinge with wear; breast and upper flanks dull purplish-blue, with narrow, sparse pale brown feather-tips; rear flanks, thighs and vent cream, tinged greyish; undertail-coverts white (appear from 14 weeks); tail olive-green; wing feathers grow last (fully grown at 16-20 weeks) and resemble those of adult. Iris dark brown to grey; bill and developing frontal shield almost black, fading to light brown or pinkish from 3-4 months; legs and feet horn to dark purplish-brown. Bird in Plate 37 based on illustration (Plate 47) in Marchant & Higgins (1993); shows bird developing juvenile plumage, but with down still adhering to feather-tips, giving overall grey-brown colour without proper 'woolly' appearance (D. I. Rogers *in litt.*).
Downy young Chick has black fur-like down; iris black-brown, becoming grey; bill white with black base, cutting edges and developing frontal shield, becoming black with white tip on culmen; legs and feet pale pink, becoming purple-brown; down fades to black-brown (pale brown to whitish from face to breast) and is replaced by grey to grey-brown second down at 4-5 weeks. Egg-tooth remains for c. 4 weeks; wing-claw quite conspicuous.

MEASUREMENTS Wing of 7 males 245-265 (255.1, SD 7.5), of 6 females 226-246 (232, SD 7.2); tail of 6 males 110-126 (120, SD 5.5), of 5 females 110-122 (115.6, SD 6.2); culmen to base of shield of 40 males 83.5-95 (88.2, SD 2.0), of 39 females 77-90 (83, SD 2.7); tarsus of 27 males 92.5-101.5 (97.2, SD 2.7), of 26 females 84-96 (91.1, SD 3.3). Weight: birds from suboptimal habitat in Takahe V, Fiordland, male 2,250-3,250 (2,673, SD 257.9, n = 13), female 1,850-2,600 (2,268; SD 193.8, n = 14); birds from Eyles-Wisley and Miller Peak, male mean 2936 (SD 241.6, n = 18), female mean 2,555 (SD 204.4, n = 19); hand-reared, captive and island birds, male 2,150-4,150 (3,100, n = 18), female 1,780-3,500 (2,650, n = 18) (Marchant & Higgins 1993). Hatching weight: mean 61 (SD 4.4, n = 10).

GEOGRAPHICAL VARIATION The North I and South I forms have been generally regarded as subspecifically distinct, but recent osteometric analysis indicates that the North I form *mantelli* is distinct enough from *hochstetteri* to be considered a separate species independently derived from a volant *Porphyrio* ancestor, *hochstetteri* having arisen from an earlier colonisation event (Trewick 1996b, 1997).
   P.m. mantelli Owen, 1848 – North I, New Zealand. EXTINCT. Known only from subfossils; formerly widespread in North I. Larger than *hochstetteri* in skeletal measurements: tarsometatarsus of nominate 113-129, of *hochstetteri* 90-108; femur of nominate 96-122, of *hochstetteri* 103-111; tibiotarsus of nominate 162-200, of *hochstetteri* 145-165; distance between temporal fossae at narrowest point in nominate 13.5-16, in *hochstetteri* 14.5-23.5 (Williams 1960, Greenway 1967).
   P.m. hochstetteri (A. B. Meyer, 1883) – SW South I, New Zealand. Formerly widespread in South I; now restricted to Fiordland, mainly in Murchison Mts (Middle Fiord of L Te Anau), possibly N to Stuart Mts and S to Kepler Mts, and between George and Caswell Sounds in the W. Also introduced to Tiritiri Matangi, Kapiti, Mana and Maud Is. See Description.

MOULT Information is from Marchant & Higgins (1993). Postbreeding moult is complete, and moult of primaries and secondaries is simultaneous. Moult begins about mid-Jan and continues through Mar; there is some overlap of moult and rearing of chicks. It apparently bites off old primaries before the moult, and preens out body feathers. Postjuvenile moult is partial, involving the replacement of all head, neck and body feathers, and occurs at c. 4 months. Reports of recently fledged juveniles moulting only mantle and back feathers into first-winter (first basic) plumage (Williams 1960) may be incorrect and are not confirmed by specimen and photographic evidence.

DISTRIBUTION AND STATUS Most information is from Marchant & Higgins (1993) and Clout & Craig (1994). Confined to Fiordland, SW South I, New Zealand. ENDANGERED. It was formerly widespread on both North and South Is, but has declined in the recent past, the widely accepted explanation for this decline being reduction in its grassland habitat by the spread of forest in the post-glacial Pleistocene-Holocene, and subsequent hunting of the vulnerable remaining population by Polynesian colonists who arrived c. 1000 years ago (e.g. Williams 1960, Mills *et al.* 1984). However it has also been suggested that the Takahe was not a specialist alpine grassland species and was not adversely affected by habitat loss during climatic change, but that hunting was the major cause of its decline (Beauchamp & Worthy 1988). The Takahe was probably once widespread in both forest and grassland down to sea level, and its modern distribution is probably in suboptimal habitat because of low hunting pressure there. In the 19th century the race *hochstetteri* possibly occupied c. 4000km of Fiordland, SW South I, extending to sea level in some areas, but it was thought to be extinct by the 1930s. It was rediscovered W of L Te Anau in the Murchison Mts in 1948, when its population was estimated at c. 260 pairs and its range 750-800km² (Reid 1971). Its range then contracted, mainly in the W, and by 1974 only 200 pairs occupied c. 650km² of Fiordland NP. Between 1981 and 1994 the Murchison Mts population fluctuated between 95 and 150 birds (mean c. 120), the lowest numbers being recorded in 1992 and 1994 after particularly cold winters in 1991 and 1992; Takahe survivorship in the wild, particularly of young birds, appears to have been adversely affected during these very cold winters (Maxwell & Jamieson 1997). The population may be quite stable but below the carrying capacity of the area (I. G. Jamieson *in litt.*). It was introduced to 4 mainly predator-free nearshore islands from 1984 to 1992, and has bred successfully on all 4; in 1995 its island population was 52 birds (Bunin & Jamieson 1996a).

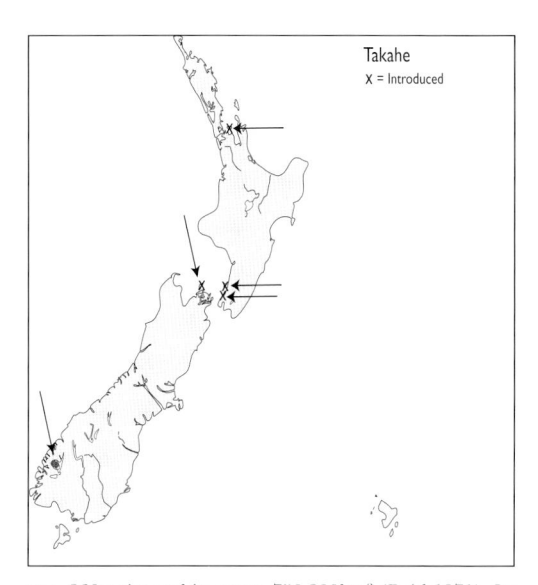

There are thought to be 2 main causes of the more recent decline since the Takahe's rediscovery. First, severe competition for food from introduced Red Deer *Cervus elaphus*, which preferentially take the same grass species and which reduce food availability so that Takahe can no longer survive in the same areas, and which have also modified habitat by overgrazing, thus eliminating the most nutritious plants and preventing some grasses from seeding; in contrast, the feeding technique of Takahe appears to have little detrimental effect on the tussocks (Mills *et al.* 1989). However Bunin & Jamieson (1995) point out that, despite intensive deer culling and the fertilization of tussock lands, which have allowed the vegetation to recover, the Takahe population has fallen to its lowest since population estimates were first recorded. They stress that a second factor was also adversely affecting the population, i.e. predation by introduced stoats *Mustela erminea*, which take eggs and all ages of birds, and they suggest that Takahe lack appropriate behavioural responses to cope with mammalian predators such as mustelids which have been introduced relatively recently (see also Bunin & Jamieson 1996b). Stoats are especially plentiful following periodic mouse irruptions linked to seeding of beech *Nothofagus* trees, and the Takahe decline in the 1960s and 1970s is thought to be at least partly due to high stoat numbers. However, stoat numbers are relatively low in the alpine zone of Fiordland compared to lowland forested areas, which might account for why Takahe have persisted in Fiordland for the last hundred years (I. Jamieson *in litt.*).

Initial attempts at captive breeding, begun in the 1950s, met with little success: fertile eggs were not produced until 1972, and few chicks survived to independence. However, since the establishment of Burwood Bush Takahe Rearing Unit in Te Anau in the 1980s methods have improved, with 90% of viable eggs hatching and 74% of chicks surviving to maturity (n = 117 viable eggs, 1982-91). Captive rearing significantly improves (up to 60%) the survival of young up to 6 months of age (Eason 1992). A recent study (Maxwell & Jamieson 1997) indicates that in Fiordland the survival of captive-reared Takahe was at least as great as wild-reared birds over a 5-year period, and showed that more captive-reared females than males formed pairs after release, this suggesting that there may be a shortage of females in the wild population (or that the males were unsuccessful because of low social status or young age).

Brood manipulation was introduced in the 1980s to ensure that all nesting pairs had at least 1 viable egg: birds with 2 viable eggs are relieved of 1, which is given to birds with infertile eggs, while non-viable eggs are removed from clutches with 1 viable and 1 non-viable egg, so that adults do not waste effort on incubating a non-viable egg after the 1 chick hatches; refined techniques now ensure that chick production is maximized each year. A rearing programme, involving artificial incubation, feeding using puppets and recorded sounds to ensure appropriate imprinting, and subsequent rearing of juveniles in groups with experienced adult birds, has enabled yearlings to be released both in Fiordland and on islands. A study to assess the feasibility of cross-fostering Takahe eggs to Purple Swamphen nests on Mana I, to increase the numbers of juveniles produced by each pair of Takahe, achieved only limited success: 8 of 12 (67%) cross-fostered eggs hatched and 2/8 (25%) of the juveniles survived to 1 year of age; for Takahe-reared eggs from the same clutches, 5 of 12 (42%) eggs hatched and 2 of 5 (40%) of young fledged successfully. Takahe eggs had low hatching success whether reared by Swamphens or Takahe. However, Swamphens exhibit greater anti-predator behaviour than Takahe and therefore a possible advantage of cross-fostering is that Takahe chicks would be influenced by the foster parents to exhibit higher levels of alertness and predator avoidance than parent-reared birds, which could be significant if Takahe are released at mainland sites where terrestrial mammalian predators occur (Bunin 1995, Bunin & Jamieson 1996a, b).

Hatching success and chick production are lower on islands than in Fiordland (see Breeding and Survival), but yearling and adult survival are much higher, and most island pairs are able to produce replacement clutches every season. In island birds, third clutches were much more successful (6 juveniles from 13 clutches) than first clutches (4 juveniles from 43 clutches), and it is thought that island productivity will improve over time as the number of inter-island transfers decreases and the proportion of breeding birds raised in the island environment increases (Bunin *et al.* 1997). The causes of low reproductive success are currently unknown but, despite the low hatching success, island populations continue to grow and are expected to reach their estimated carrying capacity before 2005 (Ryan & Jamieson 1998).

Population trends are still uncertain, but recovery plans aim to establish a self-sustaining population of over

500 birds in Fiordland, to boost island populations, to introduce birds to other island or mainland sites, and to promote public awareness. The current major causes of mortality in the Fiordland population are bad winters and predation.

The nominate race is known only from subfossils on North I, where its remains are widely distributed (e.g. Bunin *et al.* 1997).

MOVEMENTS Information is from Marchant & Higgins (1993), concerning Fiordland birds. Sedentary and flightless. It holds grassland territories until snow prevents feeding, when it descends into forest or scrub-grassland, some wandering 5-10km (possibly up to 30km) from their territories. Birds move within the territory, occupying low- and mid-altitude zones (1,000-1,100m) in Oct-Dec when breeding, and ascending in mid-Dec to higher-altitude zones (up to 1,500m) where preferred foods grow. Long movements occur, usually in winter and early spring when food is scarce. Temporary immigration of adults was recorded in 2 areas in 1972-73, and birds are known to move across and between valleys. They may move further as chicks become mobile, and have been known to move 400-800m within a few days of hatching. The maximum recorded movement from a natal or breeding territory is 21km, after the loss of a male. Young may move greater distances than adults, in search of a territory or a mate, and movements of up to 3.2km are recorded in birds up to 3 years old.

HABITAT Most information is from Marchant & Higgins (1993). Alpine tussock grassland; also subalpine scrub and beech forest in winter, when snow covers the grassland. Before the population declined it may also have inhabited coastal sand ridges and open shrubland. Its alpine grassland habitat is dominated by snow tussock grass *Chionochloa* spp c. 1m high, with sedges (*Carex* and *Schoenus*), short grass (*Festuca* and *Poa*), herbs (*Celmisia* and *Aciphylla*) and shrubs (*Hebe buxifolia, Coprosma, Olearia moschata* and *Dracophyllum uniflorum*) (Williams 1960, Marchant & Higgins 1993). Some occupied areas are dominated by *Chionochloa rubra*, and above the tree-line *C. rigida amara, C. pallens* and *C. crassiuscula* predominate. Forest habitat is dominated by beech *Nothofagus solandi* and *N. menziesii*, with an understorey of *Coprosma* shrubs, ferns (*Hypolepis* and *Blechnum*) and hookgrass *Uncinia*. In Fiordland NP it occurs in an area of heavy snows and very high rainfall (2,500-4,800mm per annum), commonly above the tree-line at 1,050-1,520m. It frequents mountain lakes, rivers, streams and bogs, where vegetation includes sedges and rushes (e.g. *Juncus gregiflorus*); it is often by fast-flowing streams; some occupied areas are prone to flooding. Territories contains tussock, and usually forest and bog, and may have above-average soil fertility; water is important for drinking, bathing and wetting food. Where introduced on islands it occurs in pastures of exotic grasses such as *Bromus willldenowii, Holcus lanatus* and *Dactylus glomerata*, and clovers *Trifolium*. Kean (1956) found a correlation between territory quality and breeding success.

FOOD AND FEEDING Most information is from Marchant & Higgins (1993). It eats predominantly leaf bases of *Chionochloa* tussocks and other alpine grass species. Leaf bases of Cyperaceae, grass seeds and fern *Hypolepis* rhizomes are also taken seasonally, mainly in winter, the latter in beech forest where they constitute 60-80% of the winter diet. It rarely takes invertebrates and small lizards. Plants eaten include: Poaceae (*Chionochloa rigida amara, C. pallens, C. crassiuscula, C. teretifolia, C. rubra*, and seeds of *Poa colensoi, P. novaezelandiae, Rhytidosperma setifolia, Festuca matthewsi* and *Anthoxanthum odoratum*); Apiaceae (*Aciphylla takahea* vegetative parts, flowers and seeds); and Asteraceae (*Celmisia petrei*) (Williams 1960, Marchant & Higgins 1993). In winter it also eats *Hypolepis millefolium* rhizomes, Juncaceae (*Juncus gregiflorus* bases), Cyperaceae (*Uncinia affinis, U. calvata, Schoenus pauciflorus* and *Carex coriacea* leaf bases and rhizomes), Poaceae (*Chionochloa conspicua* leaf bases and seeds). On islands it eats leaf blades, leaf bases and seeds of many introduced grasses, including *Dactylis glomerata, Bromus catharticus (= wildenowii/ unioloides), Phleum pratense, Poa pratensis, Holcus lanatus* and *Agrostis capillaris*, and also leaves and bases of clovers (especially *Trifolium rubra*), all parts of Chickweed *Stellaria media*, and occasionally dead sticks, grass and flax stalks. It detaches tussock tillers with the bill, by grasping them in the foot and nipping them off, or by pulling them off with the bill (an average force of up to 15.5kg is needed to detach a *Chionochloa* tiller); it then picks up the material with the bill and transfers it to the foot, holding it parrot-fashion while stripping off dead material before eating the chosen portion (Williams 1960). It strips seeds from seed-heads by running the partly open bill along the stalk, obtaining seed-heads beyond reach by first biting off the stem and holding it in the foot; it digs up fern rhizomes with the bill. It feeds selectively, taking the bases of plants richest in nitrogen, phosphorus, calcium, sodium, potassium and soluble sugars. It is thought to be essentially an extractor of plant juices, as food passes quickly through the system with little fibrous material digested and plant tissue structurally unaltered; however, it has recently been suggested (Suttie & Fennessy 1992) that its relatively long pyloric caeca may enable it to digest more fibre than was previously thought. Its energy requirement is c. 2.0-2.2 kcal/g/day (Reid 1974b). On islands its food is very similar to that of the Purple Swamphen (e.g. Bunin 1995, Trewick 1996a). Chicks are fed on plant material (e.g. *Chionochloa* bases), while invertebrates such as worms (Annelida), Arachnida, Diptera (e.g. *Calliphora quadrimaculata*), Lepidoptera (especially moths *Crambus* sp) and Odonata form a large part of their diet for the first 4-6 weeks; they become predominantly vegetarian by 6-8 weeks (Williams 1960).

HABITS Most information is from Marchant & Higgins (1993). Birds are active throughout the day, in winter to 22:00h, but there is also some activity at night. Normally they are difficult to observe, being wary and inhabiting thick vegetation in isolated areas. The normal gait is a slow, deliberate walk, and the tail is flicked continuously when the bird is nervous. When alarmed, birds break into a surprisingly fast run with the head and neck lowered and the wings used to gain impetus, and scatter in different directions, quickly disappearing into cover. They rarely venture into water, but they do wade, and they can swim and dive; 1 was seen to dive into a stream and hold onto underwater vegetation with its feet (Williams 1960). They carefully watch harriers and other flying birds, and slink into cover if these potential predators approach too close. They are said to be secretive and quiet during the postbreeding moult, and they leave large numbers of feathers under the shelter of rocks, tussocks or shrubs, or by tarns, where they rest for long periods preening out old feathers. They roost at night under the shelter of tussocks, rocks and scrub; some sites are used regularly but some only once (wherever the birds happen to be at nightfall); pairs roost standing next to each other. They sunbathe with the wings spread, and bathe in tarns. Allopreening occurs regularly between group members of all ages; the face and chin are preened, and the bird being preened stretches and twists its head upside down.

SOCIAL ORGANISATION Most information is from Marchant & Higgins (1993). Monogamous pairs are most common but a few trios and one foursome have been observed (I. Jamieson *in litt.*). Aggressive mate competition and pair switching are not uncommon on islands, with two cases resulting in lethal injuries to the losing bird (I. Jamieson *in litt.*) In Fiordland the pair-bond is permanent, at least for 12 years and probably for life; birds are territorial within loose colonies and the territory is maintained during and after breeding (Oct to May); thereafter pairs remain in semi-communal and overlapping home ranges; c. 82% of adults appear paired, and c. 70% of pairs breed, but these figures vary between years. A pair holds the same territory each year, and nonbreeding pairs also hold territories in the breeding season. It breeds in family groups; juveniles are sometimes ejected but first-year birds may assist with incubation and care of the young; multiple male helpers are reported in captivity. Many yearlings are forced out of the natal territory after the spring; occasionally a juvenile, expelled in spring, remains at the edge of the territory and is accepted back later in the season.

The territory size has been estimated as 2-60ha in Fiordland; on islands it averages much smaller (Ryan & Jamieson 1998); the actual area is not fixed, the boundary expands and contracts, and the centre may change when chicks become mobile and as young grow. Interactions between neighbours affect the size of territories and of neutral areas between them. When birds meet in such a neutral area, status is at least partly determined by the distance of each from its territory centre and the stage of the breeding cycle of each bird. By Mar, after breeding, territories enlarge, overlap and merge into semi-communal home ranges, and it is possible that, when juveniles have become independent, territories are no longer defended. The home range size may thus be larger in winter, even >200ha.

SOCIAL AND SEXUAL BEHAVIOUR Most information is from Williams (1960) and Marchant & Higgins (1993). On Mana I, where Takahe are introduced and Purple Swamphens are very recent, and very successful, colonists, there appears to be little interspecific aggression between these highly territorial species, despite overlap in habitat use; Takahe dominated in most of the infrequent aggressive encounters observed, and 1 yearling male Takahe was even adopted into a Purple Swamphen family group (Bunin 1995). Observations indicate that the Takahe's behavioural repertoire is similar enough to that of its smaller congener to elicit appropriate responses, and that Takahe behaviour is flexible and can be influenced by the behaviour of associates (Bunin 1995). There is little intraspecific conflict in the non-breeding season, but territorial disputes and fighting occur as breeding approaches. A resident rushes at an intruder with the wings held high and arched, the neck feathers ruffled and the upperwing-coverts prominent; this display is also used against avian predators such as Wekas and New Zealand Falcons *Falco novaeseelandiae*. Intruders are usually chased away, but fighting may occur, when antagonists strike with the feet, bill and neck. Birds fight intruders only of the same sex, and fighting can result in wounds and loss of feathers.

An incubating bird will cover itself with surrounding vegetation, and often responds to a close predator by tucking its head out of sight; if threatened at the nest, some birds will hiss, scream and bite rather than leave the eggs.

The male displays to the female by raising his wings close to his body, holding his neck erect, and lifting and fanning his tail to display the white undertail-coverts. He will also stand erect, hunch his neck, lower his head and fan out his wings so that the primaries trail on the ground; this posture shows the bright feathers of the upperparts. Paired birds will face each other for several seconds in a crouched stance, with bills almost touching and necks upstretched, and 1 will then crouch, droop its wings, fluff out its flank feathers and move round its partner with the white undertail-coverts prominent; similar displays, with mutual nibbling, are probably a precursor to copulation. Before copulating, birds will often approach each other with arched wings, then quickly straighten up with necks outstretched and bills touching. When attempting to mount the female, the male places a foot on the female's back, and fans and depresses his tail; the female crouches but remains standing. Overall, copulation rates appear to be low (Ryan 1997). Invertebrates are used for courtship feeding.

BREEDING AND SURVIVAL Season Mainly early Oct to late Dec; eggs occasionally to mid-Feb and small young late Mar (probably from re-nesting after failure); breeds at least 1 month earlier and 1 month later on islands (I. Jamieson *in litt.*). In Fiordland, flowering of snow tussock grass influences breeding, more pairs nesting in good flowering years, when breeding can be up to 1 month earlier than in other years (Clout & Craig 1994). **Nest** Built on well-drained ground under or between *Chionochloa rubra* and *C. rigida amara* tussocks or shrubs such as *Dracophyllum*; typically has 2 entrances connecting with runways; has latrine c. 2m away. Nest a deep bowl of fine grass and tussock leaves, placed in saucer-like scrape in ground; height 7.5-15cm, diameter 31-38cm, thickness in centre 7.5cm; in captivity both sexes build but female mainly constructs bowl. Builds new nest for re-nesting. May build trial nests; brood nests built 2-5 days after young hatch. **Eggs** Usually 2; 1-3 (1.7, SD 0.5, n = 48 clutches); rounded elliptical; pale buff, irregularly and quite sparingly blotched mauve and brown; size 68.3-78.9 x 46.1-51.0 (73.5 x 48.7, n = 87); weight 81.3-109.9 (96.5, n = 87). Laying interval between 1st and 2nd egg 48h, between 2nd and 3rd (in captivity) up to 72h. **Incubation** Begins after laying of 1st egg; duration 29-31 days, by both parents, female usually in morning and early afternoon and male for rest of day (possibly also at night); juveniles may make small contribution; see Ryan (1997). Hatching asynchronous, interval c. 48h; parents occasionally desert remaining egg when first hatches, but more often remain at nest for 3-5 days with first chick. **Chicks** Precocial and nidifugous; active soon after hatching; usually leave nest soon after hatching but may remain for up to 5 days. Young fed and brooded by both parents; brooded at night and during bad weather; dependent on adults for food for c. 4 months; young beg for food, parent passes food item to chick in bill, and chick reaches up to take it; yearlings in family group often play major role in feeding chicks. Chicks hide by burrowing into thick grass and remaining silent. Young develop adult calls at c. 4.5 months; usually remain with parents for winter and disperse in following spring, some remaining for up to 2 years. In captivity, hand-reared chicks are fostered by other adults, and second-year birds will foster groups (6-10) of first-year birds. Develops greyish-brown second down at 4-5 weeks; body feathers develop at 5-8 weeks, tail at 14-18 weeks, wings at 9-20 weeks. Egg fertility 70-80% (Bunin & Jamieson 1996a); mainland hatching success from 46 nests 67-76%; 17 (46%) of 37 pairs that hatched chicks had a chick still alive at 3-4 months; from 36 pairs, 15 chicks (0.42 chicks/pair) left nest; average productivity per breeding pair 1.78 eggs laid, 0.97 chicks hatched, 0.88 chicks survived to c. 6 weeks (Marchant & Higgins 1993). Reproductive success on islands lower than in Fiordland, figures for islands (1991-95) and Fiordland (1989-94) being (means/pair/year): eggs on islands 3.5 (SD 1.5, n = 43), in Fiordland 2.0 (SD 0.5, n = 122); chicks/egg on islands 0.30 (SD 0.31, n = 43), in Fiordland 0.60 (SD 0.41, n = 110); juveniles/egg on islands 0.18 (SD 0.19, n = 43), in Fiordland 0.37 (SD 0.39, n = 130); juveniles/year on islands 0.65 (SD 0.72, n = 43), in Fiordland 0.85 (SD 0.59, n = 171); a temporary supplementary feeding programme had no effect on island breeding success (Bunin et al. 1997). Survival to 1 year 27-71%, dependent on weather and locality; usually only 1 young raised to independence (Marchant & Higgins 1993); yearling survival on islands much higher, apparently 89% (Clout & Craig 1994). Eggs and chicks taken by Wekas and stoats. Age of first breeding 2 years, but also known to breed in first year (Williams 1960). If clutch lost will lay replacement, usually after 2-4 weeks; second brood recorded only once. Re-nesting after failure more likely in good years of snow tussock flowering, when breeding starts earlier; however, most island pairs can produce replacement clutches every season (Clout & Craig 1994). **Survival** Adult survival 73-97% per annum; on islands 76-100% (Eason & Rasch 1993, Bunin et al. 1997).

123 ALLEN'S GALLINULE
Porphyrio alleni   Plate 38

Porphyrio Alleni Thompson, 1842, Idda, Niger River.

Sometimes placed in *Gallinula* or *Porphyrula*. Forms superspecies with *P. martinica*. Monotypic.

Synonyms: *Gallinula/Porphyrula/Porphyriola/Hydrornia/Caesarornis alleni*; *Porphyrio/Porphyrula/Porphyriola chloronotus*; *Gallinula mutabilis/porphyrio*; *Porphyrio madagascariensis/variegatus*; *Hydrornia porphyrio*.

Alternative name: Lesser Gallinule.

IDENTIFICATION Length 22-26cm; wingspan 48-52cm. A medium-sized, rather slim, graceful and delicate gallinule, with bright, markedly iridescent plumage. Head black or blue-black; neck bright purplish-blue; mantle, back, scapulars, inner upperwing-coverts and tertials olive-green (darker green when worn); outer coverts, and outer webs of flight feathers, blue-green; marginal and primary coverts bright blue. Rump to tail greenish black; underparts violet-blue; lower belly blackish; undertail-coverts black and white, giving inverted heart-shaped white patch under tail. Sexes alike in plumage. Early in breeding season shield of male turquoise-blue and of female apple-green; later, both adults have blue shield which becomes dark grey or dark blue after breeding; iris coral-red, of non-breeding birds brown; legs and feet red, duller in non-breeding birds. In flight often looks almost black; very long, red legs and feet, and red bill, easily visible. Immature very like adult, with slightly duller bare parts. Juvenile has dark brown crown, hindneck, upperparts, upperwing-coverts and tertials, upperparts and coverts feathers broadly fringed pale brown giving markedly scalloped effect (fringes more olive-green on coverts); flight feathers and primary coverts as adult, but duller; chin to foreneck white to buffy; sides of head, and underparts, cinnamon-buff; central belly whiter; undertail-coverts rich buff. Iris grey-brown to olive-brown; bill ochre-brown, becoming red-brown; shield olive-brown, becoming greyer; legs and feet pale brown to red-brown. Inhabits marshes, rice fields, inundated grasslands and floodplains, and rank vegetation by open water; prefers sites with floating vegetation and often uses seasonal habitats.

Similar species Superficially resembles Purple Swamphen (120) but much smaller, with less robust bill, blue or green frontal shield and no pale blue on face or throat. Similar to American Purple Gallinule (125) but latter noticeably larger, with yellow legs and feet, yellow-tipped red bill and slightly paler frontal shield; also has purple-blue head; paler blue neck, breast and flanks; upperwing-coverts often largely blue-green, especially when worn; lacks contrastingly very dark rump to tail of Allen's; undertail-coverts wholly white. Juvenile American Purple Gallinule lacks pale feather edges on upperparts, is less white on underparts, and has pure white undertail-coverts. For further details, see American Purple Gallinule species account. Lacks white line along flanks of Common (129) and Lesser (131) Moorhens but has similar undertail-covert pattern; for further distinctions, see Lesser Moorhen species account. Normal calls are predominantly sharp *kek*, *kleek* or quacking notes, while those of Lesser Moorhen are normally more subdued clucking or *bup* notes (although there is also a sharp alarm call), a useful distinction when both species occur together and are invisible in dense emergent vegetation.

VOICE Most information is from Taylor (1985b) and Urban et al. (1986). This species gives a variety of harsh, often nasal, calls, including a dry *keck* and a drawn-out *kerk* (often repeated); also sharp, repeated *klip*, *kleep*, *kik* and *kerrr* notes given by adults probably in an alarm or aggressive context and including a series of sharp notes ending in a churring *kik-kik-kik-kik-kik-kyer-kyer-kyer-kiurr-kiurr-kurr-kurr*; and a harsh duck-like quacking. A very loud, raucous series of hooting notes has also been heard (P. B. Taylor unpubl.). The normal contact note, given by both adults and juveniles, is a subdued *kup*. A high-pitched *kli* note, repeated 6-8 times/s, is given in flight (apparently an alarm call), and there is also a sharp *click* of alarm; Serle (1939) also records a series of harsh *chuck* notes, increasing in volume and rate, rising in pitch, and often lasting for several s, which is given in alarm. A querulous anxiety call is given if an intruder approaches a nest when the eggs are hatching (Serle 1939). A half-grown chick gave a quiet *tack*, probably a contact note.

DESCRIPTION

Adult Head and nape black, sometimes tinged purple; hindneck and sides of neck bright purplish-blue to blue-violet; mantle, back, scapulars, inner upperwing-coverts and tertials olive-green with distinct rufous tinge when fresh, becoming dark bottle-green when worn; rump to tail dull black, washed green. Outer greater and median upperwing-coverts, and outer webs of flight feathers, blue-green; inner webs of flight feathers dull black; marginal and primary coverts bright blue. Axillaries and underwing-coverts dark grey to dull black, coverts adjacent to marginals tinged blue. Chin, throat and underparts violet-blue or lilac-purple; when fresh, feathers often narrowly fringed whitish; lower belly, vent, thighs and central undertail-coverts dark slate to blackish; outer undertail-coverts white, giving inverted heart-shaped white patch under tail. Sexes alike in plumage. Iris coral-red to red (sometimes yellow?) when breeding, brown, red-brown or yellow-brown in non-breeding birds; bill dark red, slightly duller in non-breeding season; early in breeding season shield of male bright turquoise-blue and of female apple-green; after chicks hatch, both adults have blue shield which becomes dark blue to dark grey in non-breeding season; legs and feet red, duller in non-breeding birds.

Immature (first basic) Like adult and not always separable, but may be duller in plumage; some have more extensive black bases to feathers of upperparts, narrow buff fringes to longer scapulars, tertials or uppertail-coverts, and grey or buff wash to tips of underpart feathers. Bare parts like non-breeding adult, possibly slightly duller. Occasionally retains some off-white or buff juvenile feathers on sides of head and breast or on centre of belly and vent. Like adult, grey shield becomes blue from edges inwards; many attain blue shield before postjuvenile moult to adult-type body plumage complete.

Juvenile Crown to hindneck dark brown to sepia; rest of upperparts, tail, upperwing-coverts and tertials dark brown to sepia, feathers broadly fringed ochre-brown to olive-brown on most of upperparts and olive-green with buff tips on upperwing-coverts; outer upperwing-coverts tinged blue basally. Flight feathers and primary coverts as adult, but blue-green slightly more olive and inner webs browner; underwing-coverts and axillaries tipped white or ochre. Sides of head and sides of neck cinnamon-buff; chin to foreneck white to pale buff; breast, flanks and sides of belly cinnamon-buff; central belly and vent white to pale buff; undertail-coverts rich buff. Iris grey-brown or olive-brown to orange or red-brown; bill ochre-brown, becoming red-brown from base; shield olive-brown, becoming grey-brown to grey; legs and feet pale brown to red-brown.

Downy young Chick has black down, browner on underparts, with silvery tips around face and chin. Half-grown chick had dark brown iris, grey bill and shield, and pinkish legs (Taylor 1985b).

MEASUREMENTS Wing of 10 males 148-162 (156, SD 3.7), of 13 females 141-164 (152, SD 6.1); tail of 11 males 60-68 (65.2, SD 2.8), of 13 females 61-73 (66.0, SD 3.6); exposed culmen of 11 males 23-25 (24.4, SD 0.7), of 13 females 22-25 (23.3, SD 0.9); tarsus of 11 males 64-72 (67.8, SD 2.8), of 13 females 61-72 (66.6, SD 3.6) (Cramp & Simmons 1980); culmen (including shield) 35-45 (Maclean 1993). Weight of 6 males 132-172 (154.2, SD 17.5), of 3 females 112, 117, 145.

GEOGRAPHICAL VARIATION None.

MOULT Postbreeding moult is complete, with simultaneous moult of the remiges (preceded by the rectrices), and mainly occurs shortly after breeding. It is recorded in Zaïre, Apr-Aug, in the dry season and the first half of the rainy season (Cramp & Simmons 1980), and in Cameroon in Dec (Urban *et al.* 1986). Moult sequence is as in the American Purple Gallinule, as also apparently is the sequence of postjuvenile moult (Cramp & Simmons 1980). Postjuvenile moult evident on upperparts before underparts; in South Africa, most if not all birds attain adult shield colour (blue) before body plumage fully moulted (P. B. Taylor unpubl.). In Kenya, postjuvenile moult of body plumage is apparently completed in c. 2 months (Taylor 1985b).

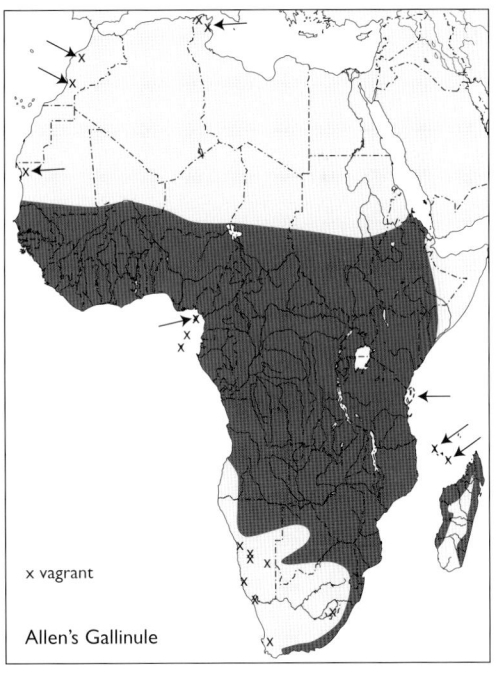

x vagrant

Allen's Gallinule

DISTRIBUTION AND STATUS Senegal and Gambia E to Ethiopia and Somalia and S to South Africa, excluding the arid SW African region; also Madagascar, the Comoro Is, and (probably introduced) Mauritius. Not globally threatened. The erratic nature of its occurrence in very seasonal habitats makes its populations difficult to assess, but it may be locally very numerous in suitable breeding habitat during seasons of good rainfall. In Senegal it is recorded from the extreme N, including the Senegal R delta, and N of Dakar, in Jan, Mar and May-Jun; in Gambia it is recorded in Jan, Feb and Apr-Oct and is apparently rare but possibly overlooked (Jensen & Kirkeby 1980, Gore 1990, Rodwell *et al.* 1996). In Sierra Leone it is locally common but is absent from the extreme N (Field 1995); it is uncommon in Liberia (Gatter 1988), not uncommon in Ghana and Nigeria (Grimes 1987, Elgood *et al.* 1994), locally common in Sierra Leone (Thiollay 1985), and rare in Benin (Claffey 1995). There are no records from Mauritania, but Urban *et al.* (1986) assume that it must occur in the extreme S. It is resident but rarely seen in S Mali (Lamarche 1980), resident in Burkina Faso (Thonnerieux *et al.* 1989), and recorded from SW Niger in Jan and Apr-Jul and at Saga in Nov (Giraudoux *et al.*

1988; Anon 1995a). In Guinea it is apparently recorded only from Macenta Prefecture in Jun (Halleux 1994), and its status in the Central African Republic is unclear (Carroll 1988, Germain & Cornet 1994). It is regular but not abundant in NE Gabon (Brosset & Erard 1986), it is recorded from Congo only at Pointe-Noire (Jan) and Ngabé (Apr) (Dowsett & Dowsett-Lemaire 1989). It is recorded as a migrant in Ivory Coast, Togo and Chad (Dowsett & Forbes-Watson 1993).

In Zaïre it is local (Chapin 1939, Lippens & Wille 1976), and in Angola it is recorded from Cabinda, Luanda, Huila and Cunene provinces but may be more widespread (Traylor 1963, Pinto 1983). In Somalia it is recorded only in the S, where it has occurred on 3 occasions in Aug and Dec (Ash & Miskell 1983). Urban & Brown (1971) regarded it as an uncommon to rare resident in Ethiopia, and in Sudan it is seasonally common to very common but local (Nikolaus 1987, 1989), while it is listed only as a vagrant to Eritrea and as resident in Rwanda and Burundi (Dowsett & Forbes-Watson 1993). It is local and usually uncommon in East Africa, being recorded from Uganda E and N of Mengo and Bunyoro, in the Rift V, on the Kenya coast, and in E and SW Tanzania with a few records from the interior (Short et al. 1990); it is common on Zanzibar and Pemba (Pakenham 1979).

It is widespread in Malawi and Zambia (Benson & Benson 1977, Benson et al. 1971). It is widespread and locally plentiful in Zimbabwe, especially at pans on the Sabi R (Irwin 1981), and in S Mozambique it is probably widespread during the rains, though largely overlooked (Clancey 1996). It occurs occasionally in N Namibia (e.g. Brooke 1967, Snow 1978, Taylor 1997c), but is a locally common breeder in the NE, on temporary pans in Bushmanland and Kavango (Hines 1993), and is a frequently observed resident at rivers in the E Caprivi (Koen 1988). It is said to be sparse and uncommon in Botswana (Penry 1994). In South Africa it occurs regularly only as far S as the Nyl R floodplain (where it is abundant in seasons of good rainfall) and the N KwaZulu-Natal coast (where it may occur in good numbers although it has not yet been proved to do so), and there are 3 old records from Lesotho (Bonde 1993, Taylor 1997a, c). In Madagascar it is uncommon, but is widespread except on the high plateau and in the S (Langrand 1990).

In favourable conditions it may occur at high densities and in very large numbers. The maximum density of a non-breeding population on a small dam at Mombasa, Kenya, was 25-40 birds/ha, and up to 35 birds roosted and sheltered in 0.55 ha of *Typha* beds (a density of 64 birds/ha) (Taylor 1985b). At least 5,000 pairs were thought to be present on the Nyl floodplain, former N Transvaal, in early 1996, during an exceptionally wet season, and this wetland is the most important known breeding site in southern Africa; there is also potential habitat for over 7,000 birds in N KwaZulu-Natal (Taylor 1997a). The destruction and modification of wetlands throughout the bird's range, and especially the loss of suitable seasonally flooded habitats, must have affected its numbers adversely, but it may have become more widespread in Zimbabwe (Irwin 1991, Taylor 1997a).

MOVEMENTS Its movements are complex and poorly known, and are not shown on the distribution map. Some birds are resident throughout the year in permanent wetlands (e.g. the Okavango Delta), but with the onset of the rains in the N tropics most migrate N (Nigeria, Cameroon and Chad), while most in the S tropics move S to breed in the rains (Urban et al. 1986, Taylor 1997c). In Ivory Coast it breeds in the N, and increases in the dry season on coastal marshes (Thiollay 1985); in Ghana it is resident in coastal areas but elsewhere is probably a migrant, occurring in the N only from Jan to Aug (Grimes 1987). In Dec birds disappear from L Chad (Elgood et al. 1973), and migrants arrive in Cameroon, where the species is absent in Aug-Nov (Serle 1954). Return movements in West Africa begin in Apr-May (Cramp & Simmons 1980). At least local movements are recorded from Sierra Leone in response to seasonal habitat changes (Field 1995). In Sudan it may be only a rainy season visitor and it is not clear whether most birds leave the S during the dry season or move to permanent swamps along the large rivers (Nikolaus 1987, 1989). It is mainly resident in Zaïre but 3 in Kivu (Jun) were possibly migrating; 1 was perched in a treetop far from suitable habitat (Urban et al. 1986). It is resident in some parts of E Africa, but seasonal occurrences are also reported. For example, at Mombasa, Kenya, numbers of non-breeding birds were much higher in Sep-Mar, with a small influx also in Jun; Sep arrivals were at a time of reduced habitat at the site, and involved birds which had just bred (including pairs with 90% grown juveniles), and departures in Mar were at the start of the long rains when habitat was increasing but was not suitable for breeding (Taylor 1985b; see this paper for a discussion of factors possibly influencing movements). Also in Kenya, numbers arrive to breed at L Baringo in May-Jun and leave in Aug-Oct, and large influxes were also noted in Jul-Aug 1977 (Britton 1980, Lewis & Pomeroy 1989); it occurs and breeds in SW Tanzania during Dec-Jan floods (Britton 1980); it also wanders to desert areas, e.g. to L Turkana, N Kenya, in Jun (Urban et al. 1986). It may have some movements in Malawi (Benson & Benson 1977); some are resident in Zambia and Zimbabwe but many appear in Nov-Dec, breed during the rains in Dec-Apr and depart in Mar-May (Benson et al. 1971, Taylor 1979, Irwin 1981). It is mainly a rains-breeding visitor to N and E Botswana from Oct to Apr, with a few remaining to Jun, and to South Africa (Sep, and Dec-May), but it is also resident on permanent waters of N Botswana and NE Namibia; there are scattered southern African occurrences during the dry season (Urban et al. 1986, Maclean 1993, Penry 1994, Taylor 1997c).

Vagrancy is widely reported, and this species is unique as an Afrotropical species which straggles to Europe (Hudson 1974). It is recorded from North Africa in Morocco (4 records), Algeria (2), Tunisia (1) and Egypt (1), from Europe in Britain (1), France (2), Spain (2), Denmark (1), Germany (1), Italy (5), Sicily and Cyprus (1), and from the Azores (4) and possibly Madeira (Heim de Balsac & Mayaud 1962, Cramp & Simmons 1980, Urban et al. 1986). Of 20 W Palearctic records, 17 are dated and all but 1 (in May) occurred in Oct to early Feb with a peak (11 records) between 1 Dec and 1 Jan; this pattern fits well with the timing of movements in West Africa, and all 4 birds found in N and C Europe occurred during or just after anticyclonic conditions with S winds from NW Africa (Cramp & Simmons 1980). It is also a vagrant to the Banc d'Arguin (Mauritania), and to the Gulf of Guinea islands of Bioko, São Tomé and Pagalu (Urban et al. 1986, Eccles 1988). In the S Atlantic it is recorded from Ascension I (May 1920 and possibly Jul 1836) and St Helena (dated record Jul 1938), 1,600km and 1,900km from the African continent (Olson 1973a, Cramp & Simmons 1980). A fairly

478

long-dead corpse found in Dec 1984 on South Georgia in the S Atlantic has been identified as this species (Prince & Croxall 1996); the locality is c. 4,800km WSW of Cape Town, where the species is also a vagrant. In the Indian Ocean it is recorded from the Comoro Is (Mayotte, Grand Comoro), and from Rodrigues I, 1,500km E of Madagascar (Benson 1960, Cramp & Simmons 1980). Stragglers are also recorded to the SW of its normal range, near Cape Town (Mar and Jun) and in coastal Namibia (Hockey et al. 1989, Maclean 1993).

HABITAT Information is from Urban et al. (1986) and Taylor (1997a, c). Freshwater marshes, reedbeds, inundated grassland and floodplains, especially those with flooded *Oryza longistaminata* and *Cyperus fastigiatus*; papyrus swamps, rice fields, and thick vegetation such as sedges, reeds and rank grass beside lakes, rivers, ponds and temporary pools. It normally prefers wetlands with *Nymphaea*, *Nymphoides*, *Ottelia* and other floating-leaved vegetation (including *Pistia stratiotes*: Taylor 1985b), and it frequently occupies and breeds in seasonal or temporary habitats. In contrast with the Lesser Moorhen it often avoids temporarily inundated grassland and grassy pans, and in the Kruger NP, South Africa, most occurrences are apparently on permanent waters in the N, while the Lesser Moorhen is more widespread and numerous in the S on temporary grass pans. It occurs up to 1,900m in E Africa.

FOOD AND FEEDING Flowers and seeds of reeds and sedges; seeds, stems and leaves of grasses and other marsh plants; unripe seedheads of waterlilies; fruits of the thorn bush *Drepanocarpus lunatus*; it also takes earthworms, molluscs, crustaceans, aquatic and terrestrial insects, spiders, fish eggs and small fish (Taylor 1985b, Urban et al. 1986). It sometimes kleptoparasitises other species, e.g. it robs Pygmy Geese *Nettapus auritus* of waterlily seedheads (Taylor 1985b). It feeds most actively in the early morning and late afternoon but also sometimes in the middle of the day and on moonlit nights. It forages while swimming, taking insects and plant material from water; also while walking on floating vegetation, and occasionally in short grass bordering water; it makes short rapid runs to catch moving prey (Taylor 1985b). It turns over floating waterlily leaves with the bill and gleans food from the exposed underside while holding down the turned leaf with its feet (e.g. Serle 1939, Fry 1966); it uses the bill to shift aside floating *Pistia* plants and to turn over dead vegetation when searching for invertebrates (Taylor 1985b). It breaks off developing waterlily seedheads with the bill and holds them down with the foot while pulling them apart with the bill; it holds *Drepanocarpus* fruit in the toes and breaks pieces off with the bill, and it also carries food in the toes to the bill (Wood 1977, Taylor 1985b, Urban et al. 1986). It climbs bushes and creepers to feed on fruits, and climbs up to 2m on reed stems, both to feed and to preen (Taylor 1985b, P. B. Taylor unpubl.). It is said to construct platforms of plaited reed stems for feeding on flowers and seeds of reeds high above the water (Mackworth-Praed & Grant 1970).

HABITS These birds are often quite shy and retiring, but they feed in the open, walking on floating vegetation up to 35m from cover on sheltered and undisturbed waters or at sites where they become accustomed to the proximity of people who do not interfere with them (e.g. Serle 1939, Taylor 1985b). They are predominantly crepuscular and diurnal, and most activity takes place during the early morning and evening; at Mombasa, Kenya, birds were active from just after dawn (06:30h) to at least 09:30h and again from 17:45h to dusk (18:30h) (Taylor 1985b); at other times the birds usually remain in dense cover such as reedbeds, papyrus or bushes, but they are sometimes active in the middle of the day and on moonlit nights (Cramp & Simmons 1980, Urban et al. 1986). No justification is given for the questionable comment in Elgood et al. (1994) that this species "may become much more nocturnal in the dry season and escape detection". Birds often jerk the tail while walking; when disturbed in the open, they raise the tail, displaying the white undertail-coverts, and then either fly to cover or lower the head and run rapidly, with long strides, into cover; when disturbed close to cover they flick the tail, adopt an upright pose with raised head, and move quietly into cover with short steps (Taylor 1985b). Short flights are laboured and clumsy, with dangling legs, but the birds can make strong, long-distance flights; they swim well and can dive, and they often climb high into grass, reeds and bushes to forage and roost (Chapin 1939, Urban et al. 1996).

SOCIAL ORGANISATION Monogamous; territorial when breeding; nests are well spaced even where birds are common (Urban et al. 1986). It also occurs in pairs in non-breeding areas, such pairs sometimes being accompanied by 1 or more juveniles which are usually tolerated until they have moulted into adult-type plumage (Taylor 1985b, P. B. Taylor unpubl.). Its breeding density in rank vegetation around lakes in the Hadejia wetlands, N Nigeria, has been estimated as 1-2 pairs/ha, and in emergent vegetation on the Nyl R floodplain, South Africa, as at least 3 pairs/ha (Taylor 1997d). In Zambia it breeds on ponds as small as 0.5ha (Taylor 1985b).

SOCIAL AND SEXUAL BEHAVIOUR Birds are often aggressive towards conspecifics when feeding, and non-breeding pairs at Mombasa actively defended their immediate foraging area, and that of any accompanying juveniles, chasing away any other Allen's Gallinules which approached closer than 3m, and they were also aggressive to Pygmy Geese, which they sometimes kleptoparasitised; however, they fed amicably alongside Purple Swamphens, Common Moorhens and African Jacanas *Actophilornis africana* (Taylor 1985b); in Sierra Leone they also consort freely with African Jacanas but are aggressive towards Lesser Moorhens (Field 1995).

When about to breed, adults chase away juveniles and immatures which may previously have been tolerated (P. B. Taylor unpubl.). Courtship and mating are undescribed. The incubating bird sits tight at the approach of an intruder and when flushed may sneak off through the vegetation but may also run away in full view across floating vegetation, or may even fly (Urban et al. 1986). When eggs are hatching, the incubating bird displays anxiety, walking nervously to and fro a fewm from the nest and uttering a querulous alarm note (Serle 1939).

BREEDING AND SURVIVAL Season Senegal, breeding condition, Aug; Sierra Leone, Sep, Nov; Nigeria, May-Oct, especially at height of rains; Ghana, Jun; Central African Republic, breeding condition May; Cameroon, Aug; Gabon, May (from young in Jun); Ethiopia, Apr, Jun, Sep-Oct; Kenya and NE Tanzania, Apr-Oct; Zanzibar & Pemba Is, May, Jul-Aug; in E Africa prefers dry months following long rains; SE Zaïre, probably begins Jan-Feb (second half of rains); Zambia, Dec-Apr; Malawi, Feb-Apr, Jun, Sept-

Dec; Zimbabwe, mainly Dec-Apr, also May, Sep; Namibia, Apr; Botswana, Nov-Dec; South Africa, Dec-Apr, peaks Jan-Feb at Nyls Vlei (former Transvaal), when habitat most flooded; Madagascar, Jan. Breeding record from Mombasa (Brown & Britton 1980; EANHS Nest record Scheme) erroneous, and 1 from Lamu (Jackson & Slater 1938) also unlikely: both based on observations of flying juveniles, which do not prove local breeding (Taylor 1985b). Newly arrived population at Nyl R floodplain, Feb, contained birds of all ages from full juvenile to breeding adult, suggesting that species may breed all year round (Taylor 1997a). **Nest** Typically in reeds, grasses or tangled vegetation at edge of water; also in open marshes and rice fields or on floating vegetation; quite flimsy; loosely constructed of reed stems and blades, dry sedges, grass and other plants, with deep or shallow cup; placed just above water; sometimes woven into surrounding vegetation, which may be bent down as foundation and pulled over to form dome. **Eggs** Eggs 3-8 (mean 4.4, n = 33 clutches); oval, occasionally oval-pyriform; smooth and slightly glossy; dirty white, pinkish-cream, pale brown or light red-brown, covered with small, sharp, distinct spots and specks of red-brown over pale purple or ashy. Size of eggs: (n = 41, Nigeria) 31.8-39.2 x 23.6-27.5 (36.2 x 26.1); (n = 14, Malawi) 35.0-39.5 x 25.5-27.0 (36.8 x 26.3); (n = 24, Zimbabwe) 34.5-39.0 x 24.7-28.0 (36.5 x 26.2); calculated weight 13.5. **Incubation** Begins with first egg; period c. 15 days; both sexes incubate. Hatching asynchronous. **Chicks** Precocial; fed and cared for by both parents. Juveniles fly well even before they are fully grown, and birds only 80-90% grown migrate with adults in Kenya (Taylor 1985b) and South Africa (P. B. Taylor unpubl.).

124 AMERICAN PURPLE GALLINULE
Porphyrio martinica   Plate 38

Fulica martinica Linnaeus, 1766, Martinique, West Indies.

Sometimes placed in *Gallinula* or *Porphyrula*. Forms a superspecies with *P. alleni*. Monotypic.

Synonyms: *Porphyriola/Crex/Hydrogallina/Gallinula/ Porphyrula martinica*; *Fulica/Gallinula/Ionornis martinicensis*; *Porphyrio americanus/cyan(e)icollis/martinicus/tavou(e)a*; *Porphyrula georgica*; *Gallinula/Fulica porphyrio*; *Ionornis martinicus/martinica*; *Fulica flavirostris*; *Parra viridis*.

Alternative name: Purple Gallinule.

IDENTIFICATION Length 27-36cm; wingspan 50-55cm. Medium-sized gallinule, about size of Common Moorhen (129) but more slender, with longer neck and legs. Moves quite gracefully, but bulging foreneck and slightly stooping posture detract from appearance of grace (Slud 1964). Adult plumage colourful and iridescent: head, neck, breast, flanks and upper belly purplish-blue; hindneck, sides of mantle and sides of breast pale blue; rest of upperparts olive-brown to olive-green, darker towards tail; upperwing-coverts olive-green and pale blue; outer webs of remiges pale blue. Lower belly, vent and thighs slate-grey to dull black; undertail-coverts white. When worn, upperparts become darker blue-green and underparts duller and slightly slaty; when fresh, narrow white feather edges visible on underparts. Iris red; bill red with broad yellow tip; frontal shield bright pale blue; legs and feet yellow. Sexes alike; female smaller. Bare parts duller in non-breeding season. Immature like adult. Juvenile has upperparts brown with bronze-green tinge on back, scapulars and upperwing coverts; outer upperwing-coverts like adult but less blue; chin and throat cream; sides of head, and neck and underparts, buff-brown; undertail-coverts white; iris brown to orange; legs and feet pale yellow or brownish; bill yellow-green with pale tip and brown-pink base, and shield brown or grey. Inhabits palustrine wetlands with lush emergent, fringing, and especially floating, vegetation.

Similar species Wholly white undertail-coverts provide good distinction from Common Moorhen and Allen's Gallinule (123), both of which have dark central wedge extending from base of coverts; white area in Allen's Gallinule has inverted heart shape. Lacks white flank line of Common Moorhen. Noticeably larger than Allen's Gallinule; similar in plumage but has purple (not black) head; paler blue neck, breast and flanks; upperwing-coverts often largely green, especially when worn (inner coverts of Allen's green, marginal and primary coverts contrastingly light blue); lacks contrastingly very dark rump to tail of Allen's; has yellow tip to bill, and brighter, paler frontal shield; legs and feet yellow (red in Allen's). However, note that legs and feet may occasionally be reddish (e.g. McLaren 1996). Juvenile Allen's Gallinule has pale edges to upperpart feathers and upperwing-coverts, giving markedly scalloped appearance; also has more extensively white underparts, and buff (not white) undertail-coverts. Also see Azure Gallinule (125).

VOICE A noisy species, with a variety of sharp, harsh, cackling and guttural notes. The commonest call is described as a harsh, shrill, rapid, laughing *hiddy-hiddy-hiddy, hit-up, hit-up, hit-up* the latter part delivered slowly (Ripley 1977). It also has a sharp, high-pitched *kyik* or *kr-lik*, sometimes with a booming undertone, and a loud *kur*, often preceded by a series of *cook* notes when it is often modified to *cu-KUR-cu*, not unlike the protest of a domestic chicken (Urban *et al.* 1986). Other typical calls include a wailing scream *whiehrrr*, likened by several authors to the call of the Limpkin *Aramus guarauna*; a rapid, clucking series of *kahw, ka* or *keh* notes, sometimes accelerating and fading, and given when disturbed (Howell & Webb 1995b); a gruff, repeated *kruk*; a cackling *kek kek kek* in flight; a low ticking; and a low, reedy buzz (Urban *et al.* 1986). Calls given by Hardy *et al.* (1996) appear to be a variant of the rapid, clucking *ka* calls, which slow down or become lower-pitched *kur* notes. Some calls may be accompanied by bill-snapping (Ripley 1977). Chicks make peeping food-soliciting calls (Krekorian 1978).

DESCRIPTION
Adult Head, nape, sides of neck, foreneck, breast, upper belly and flanks deep purplish-blue, slightly blackish adjacent to base of bill and frontal shield; when worn, underparts slightly slaty, less glossy purple-blue. Hindneck, sides of mantle and sides of breast pale blue, often washed green. Mantle to tail, scapulars, inner greater and median upperwing-coverts, and tertials, olive-brown to olive-green, sometimes more blue-green on mantle and brown-tinged on scapulars and back; purer green on upperwing-coverts, but in worn plumage often more blue-green, like edges of primaries; darker olive-green (more dark brown-tinged) towards tail. Lesser upperwing-coverts, outer median and greater coverts, and outer webs of remiges and primary coverts, pale blue, with olive-green tinge on inner

primaries (P1-P6/P7) and towards central median and greater coverts; inner webs of remiges, and axillaries and larger underwing-coverts, dark brown to dark grey-brown, slightly tinged bronze or blue especially on outer webs; smaller underwing-coverts pale blue to blue-green; larger underwing-coverts sometimes narrowly tipped white. Lower belly, vent and thighs slate-grey to dull black with slight violet tinge; undertail-coverts white. When worn, upperparts become darker blue-green, with olive tinge only on scapulars and tertials and no olive tinge on upperwing-coverts. In fresh plumage, narrow white feather edges visible on underparts; in worn plumage, underparts less glossy purple-blue, slightly slaty. Iris blood-red or carmine when breeding; at other times, red-brown, brown and hazel also recorded; bill bright carmine to dark blood-red with contrasting broad yellow tip; frontal shield bright pale blue, but bluish-white also recorded; during moult bill and shield become dark brown (Helm 1994), or shield may become pale slate-blue and bill tip yellow-green or dusky green (Cramp & Simmons 1980); legs and feet lemon-yellow or ochre-yellow, sometimes tinged green, outside breeding season becoming slightly duller. Tarsal colour variation, from green through yellow-green to yellow, may be age-related as in American Coot (138) (Crawford 1978). Sexes alike. Complete albinism is recorded (Haverschmidt & Mees 1995).

Immature Like adult; usually not separable when no worn juvenile remiges, rectrices or body feathers retained; in some, fresh violet-blue feathers of upperparts washed pale grey to buff at tips, and upperparts sometimes slightly more rufous-brown than adult. Shield gradually attains adult colours during first year of life. One bird from Nova Scotia, Feb, had reddish legs and feet; after some months in a freezer, colour became yellowish (McLaren 1996).

Juvenile Dark olive-brown from crown to hindneck; hindneck often washed olive-green; neck and sides of head pale buff-brown; chin and throat cream to whitish; upperparts dark olive-brown with bronze-green tinge on upper back, scapulars and upperwing-coverts; lower back darker olive-brown; rump to tail darkish brown, rectrices edged paler. Wings like adult but less blue and olive, more light green with blue tinge to feather bases; upperwing-coverts variably tipped pale buff; most underwing-coverts pale brown-grey, broadly fringed white, but lessers pale blue-green. Sides of head and neck, and underparts, buff-brown, ashier on breast, more olive on flanks and thighs, and palest (sometimes almost white) on rear flanks, belly and vent; undertail-coverts white. Iris brown to pale orange; bill dark yellow-green with pale tip and brown-pink base; shield smaller than in adult, dark brown tinged olive, or grey; legs and feet olive-brown to ochre-yellow, often tinged olive.

Downy young Down glossy black with silver tips on head, neck, upperwings and back; underparts more brownish-black; iris brown; bill red with narrow median black band, black tip, and white spot near tip of upper mandible (Ridgway & Friedmann 1941, Stiles & Skutch 1989); at c. 10 days, bill described as dull red from base to nostrils, followed by a 5 mm-wide black band, a narrower sub-terminal band of pale dusky blue, and a black tip (Dickey & van Rossem 1938); subterminal band described as pink by Gross & van Tyne (1929); shield flesh; legs and feet dull light brownish. White egg-tooth and small wing-claw present.

MEASUREMENTS (20 males, 20 females) Wing of male 171-195 (181.2, SD 5.8), of female 167-182 (175.1, SD 4.0); tail of male 64-78 (70, SD 4.0), of female 58-80 (68.1, SD 5.3); culmen (including shield) of male 45-52 (48.9, SD 1.9), of female 43-53 (46.5, SD 3.0); tarsus of male 57-65 (60.9, SD 2.4), of female 56-62 (59.2, SD 1.7). Weight of 10 males 203-305 (231, SD 32.4), of 10 females 142-291 (205.6, SD 43.1); of 4 emaciated vagrants, South Africa, 119-160 (141.5, SD 17.7) (Cramp & Simmons 1980, Silbernagl 1982). Hatching weight 10.1-12.2 (10.9, SD 0.8, n = 5) (Gross & van Tyne 1929, Trautman & Glines 1964).

GEOGRAPHICAL VARIATION None. Adults from El Salvador were said to differ in overall tone and depth of colour from those in Surinam (Dickey & van Rossem 1938), but there is no evidence that such variation is consistent. The 1943 specimen from South Georgia was described as a distinct species *Porphyrula georgica* by Pereyra (1944), despite the editors' conclusion (Anon. 1944) that the bird was a juvenile *P. martinica*.

MOULT Information is from Cramp & Simmons (1980) and Helm (1994). Adult postbreeding moult is complete, and flight feathers are lost simultaneously, birds remaining flightless for 3-4 weeks. Moult usually starts with a few small body feathers shortly after breeding, followed by flight feathers, then tail. Heavy moult of body and wing-coverts occurs when the remiges and rectrices are fully grown. Moult takes 6 weeks to complete, and is recorded in the USA mainly Aug-Oct, and in Surinam Jun-Oct. During moult, the bill and shield become dark brown. Adult prebreeding moult is partial, probably involving the smaller body feathers. Postjuvenile moult is apparently complete, the upperparts and sides of the body moulting first, followed by the sides of the head, the breast, flanks and upperwing-coverts, and then the underparts, underwing-coverts, remiges and rectrices. Some scattered juvenile feathers are retained for a long time. In nonmigratory populations, postjuvenile moult usually starts as soon as the young are able to fly, e.g. in Surinam, moult of body feathers is recorded Jul-Nov, and of remiges and rectrices Oct-Nov; in migratory populations body feathers are moulted in the winter quarters between late Sep and Mar.

DISTRIBUTION AND STATUS In the USA the breeding range comprises the E and SE region from SW Pennsylvania, C Ohio, S Indiana, S Illinois (formerly), extreme SE Missouri, S and E Arkansas and S and E Texas, to the Atlantic coast S from Maryland and Delaware; the highest breeding densities occur near the Gulf and lower Atlantic coasts of Florida, Georgia, Louisiana, South Carolina and Texas (Helm 1994). It ranges S through Central America from Mexico to Panama; in Mexico it occurs along the Gulf slope S from Tamaulipas, and the Pacific slope S from Nayarit, extending inland to S Guanajuato and N Michoacán, and E to Chiapas and Quintana Roo, while reports N to Sonora appear to be erroneous and in the N Yucatán Peninsula it is mapped as occurring only in the winter (Howell & Webb 1995b). In Panama it occurs in the lowlands of both slopes, ranges to the lower highlands in W Chiriquí, and also occurs on Coiba I (Ridgely & Gwynne 1989). It also occurs throughout the Greater Antilles, in the Lesser Antilles only S of Guadeloupe (including Curaçao and Aruba), and on Trinidad and Tobago. In South America it occurs in Colombia, Venezuela, the Guianas, Brazil (throughout), Ecuador, Bolivia (Cochabamba, Beni, La Paz and Santa Cruz), Peru (exact distribution uncertain), Paraguay (throughout), Uruguay (except Río Negro, Soriano, Flores and Colonia) and N Argentina (Jujuy, Tucumán, Santiago del Estero, NE Córdoba, Misiones, Santa Fe, Entre Ríos and extreme N Buenos Aires). USA populations winter S from S Texas, Louisiana and Florida, S throughout the remainder of the breeding range, and in winter the species is uncommon in the USA except in S Florida (AOU 1983, Root 1988, Helm 1994); birds from the extreme S of the range migrate N in the non-breeding season.

It is not globally threatened. In the USA it appears to be uncommon to relatively common, but no detailed information on population status is available (Helm 1994). In Mexico it is a frequent to uncommon and local resident, more widespread in winter when N migrants occur (Howell & Webb 1995b). In Puerto Rico it is now uncommon to rare over most of the coast but is regular at Cartagena Lagoon (Raffaele 1989). It is rare and local in the lowlands of Guatemala (Land 1970, deGraaf & Rappole 1995), and was formerly very local in lowland El Salvador (Dickey & van Rossem 1938) and uncommon in Honduras (Monroe 1968). It is locally common throughout Costa Rica (Stiles & Skutch 1989), and fairly common to common in Panama, where it has recently colonised Coiba I, probably because of the great expansion of rice cultivation there (Ridgely & Gwynne 1989). In Colombia it was regarded as locally common by Hilty & Brown (1986), but was rarely seen on L Tota, N Colombia, in 1982 (Varty et al. 1986); in Meta ricefields breeding birds occurred at a density of 20-27/ha (McKay 1981). It is locally common in the littoral region of French Guiana (Tostain et al. 1992), locally numerous in Surinam (Haverschmidt & Mees 1995) and Brazil (Sick 1993), and locally common in Ecuador (Ridgely et al. in press). There is only 1 record from Chile, from Tarapacá in 1943 (Johnson 1965b). In Paraguay it is an uncommon breeder W of the R Paraguay and rare to the E (Hayes et al. 1994), it is uncommon in Uruguay (Arballo & Cravino in press), and it is locally common to rare in Argentina (e.g. Contreras et al. 1990, Navas 1991, Narosky & Di Giácomo 1993). Its total population is estimated at 100,000-1,000,000 (Rose & Scott 1994). Little information exists on current populations and population trends but, as the quality and quantity of wetland habitat are keys to population stability, populations are probably decreasing throughout the bird's range as a result of freshwater wetland loss in the USA and throughout South and Central America (Helm 1994). In the USA it is a game species (31 of the contiguous 48 states selected a moorhen/gallinule hunting season in 1992) but the harvest is likely to be small because of low hunter interest, the secretive nature of the species, and the fact that the birds migrate S prior to the late autumn hunting season in many states (Helm 1994). In the USA, the conservation and management of freshwater wetlands along the lower Atlantic and Gulf Coast states are critical for the species, and relevant management practices should be developed (Helm 1994). There is also an urgent need to monitor population status, trends and harvest, and to provide baseline information on population status, annual productivity, hunting and non-hunting mortality, recruitment and survival rates, as well as on habitat use and preferences (Helm 1994).

Damage to rice by nesting birds has been a problem in Louisiana, USA, but usually causes insignificant losses (Helm 1982), but a few depredation permits, which allow the birds to be shot in ricefields, have been issued (Eddleman et al. 1988). In Surinam and some other Neotropical areas it is regarded as a pest in ricefields, where it breaks stems by sitting on them and bends plants into nests and feeding platforms, and in Surinam and Colombia birds have been destroyed by aerial spraying with the pesticide Endrin (Haverschmidt 1968, McKay 1981, Haverschmidt & Mees 1995). Some of the insects consumed by the birds in ricefields are serious pests (e.g. Noctuidae larvae) and thus the presence of the gallinule is to some extent beneficial; little is known about the effect of pesticides on gallinule population dynamics, and in view of the potential for crop and environmental contamination, studies integrating damage analysis with feeding habits are needed to assess accurately the effect of the gallinule in tropical ricefields (McKay 1981). In Brazil it is valued as a game bird, especially in the NE where it is an important supply of necessary protein to the human population and the destruction of eggs and birds is almost total; it is also accused of destroying rice when the plants sprout (Sick 1993). In Maranhão, Brazil, where many nest in Apr-Jun, the species is heavily hunted throughout its period of occurrence (Mar-Nov), adults being particularly vulnerable from Jul, when they are fat and unable to fly during postbreeding moult; a close season from Apr to Jun has been recommended (Sick 1993). Predators include alligators *Alligator mississippiensis* and snapping turtles *Chelyda serpentina* (Cramp & Simmons 1980, Hunter 1987a).

MOVEMENTS North American birds migrate S over the Gulf of Mexico in Oct-Nov, a few remaining in coastal states until Dec, where the species is generally uncommon from Dec-Mar (Helm 1994). Returning birds arrive in the breeding areas in Georgia, Texas and Louisiana in mid-April (Helm 1994). South American populations are not normally regarded as migratory, except at the S end of the range, where birds from many regions apparently move

N into the tropics for the austral winter. However in W Meta, Colombia, it is abundant from late Mar to Oct but occurs in much smaller numbers for the rest of the year and is believed to be at least locally migratory (McKay 1981; Hilty & Brown 1986), while in Maranhão, NE Brazil, it occurs in large numbers from Mar to Nov, breeding in Apr-Jun (Sick 1993). Further S, it is present in Rio Grande do Sul, SE Brazil, in May-Sep, and it disappears completely from S Brazil in the winter (Belton 1984, Sick 1993), while it is a northern austral migrant in Paraguay, extreme dates being 5 Sep and 30 May (Hayes *et al.* 1994). Despite this evidence of seasonality in the S, the species is often not considered a migrant in Argentina (e.g. Nores *et al.* 1983, Contreras *et al.* 1990, Navas 1991), and insufficient information is available to allow the extent of its seasonal occurrence in South America to be plotted on the distribution map. Night migration is recorded, and birds may come to lights on foggy nights (Belton 1984).

This gallinule is well known for numerous instances of long-distance vagrancy, both N and S of the equator; migrants meeting cyclonic storms are especially prone to being blown well beyond the normal range, while individuals often land on ships' decks and may be carried to ports outside the normal range (Ripley 1977, Remsen & Parker 1990, Sick 1993). It is thought to be able to rest on the surface of a calm sea (Sick 1993). It wanders widely but irregularly throughout much of the USA, W and N to S California (San Diego), S Nevada, Utah, Colorado, South Dakota and Minnesota, and into Canada in S Ontario, S Quebec, New Brunswick, Nova Scotia, Labrador and Newfoundland (AOU 1983). It also wanders to the Bahamas and N Lesser Antilles (N to Barbuda) (AOU 1983). Low pressure systems moving up coastal USA carry birds almost annually to Bermuda and New England, and sometimes beyond (Cramp & Simmons 1980). It is a vagrant to S Greenland (Apr), Azores (Apr and Nov), Europe (Britain, Norway, Switzerland, Malta), and to islands in the Atlantic and Pacific Oceans: Falkland Is (3 records); South Georgia (1,850km E of Tierra del Fuego, 1 in 1943 and 1 in Apr 1978), Tristan da Cunha (Mar-Aug, up to 40 birds annually), Galapagos (Santa Cruz and Hood Is, Feb, 1 on a ship between Galapagos and Ecuador, Feb), Ascension I (c. 2,500km E of South America) and St Helena (c. 6,400km from South America) (both islands May-Sep) (Olson 1972, Harris 1973, Cramp & Simmons 1980, Richardson 1984, Hockey *et al.* 1989, Remsen & Parker 1990, McLaren 1996, Prince & Croxall 1996). It has been recorded over 20 times, and may occur annually, in the SW Cape region of South Africa (6,700-8,000km from South America) in Mar-Aug (most Apr-Jul), when birds are migrating N from Argentina, Uruguay and S Brazil, whence African vagrants may originate (Siegfried & Frost 1973, Silbernagl 1982); in West Africa it is recorded off the Liberian coast in Jun (Urban *et al.* 1986, Hockey *et al.* 1989). S vagrants are nearly all immatures. All examples caught on Tristan de Cunha by Elliott (1957) were emaciated, and Elliott considered it unlikely to be capable of maintaining a viable local population, but Hockey *et al.* (1989) suggest that it may breed there occasionally. Most birds arrive in South Africa exhausted and starving, often occur in urban areas, and usually die soon after arrival, so the possibility of birds surviving long enough to establish a viable local population is remote (Silbernagl 1982, Taylor 1997c).

In Costa Rica, helpers dispersed from the natal territories when they became adult, and most disappeared from the study site to return after several months, when they usually became floaters (Hunter 1987a).

HABITAT It inhabits lush palustrine wetlands, notably grassy marshes, flooded fields and overgrown swamps (presumably not wooded), and the fringing and emergent vegetation of lagoons, pools, ponds, river mouths, and channels of slow-moving and still watercourses from rivers down to the size of roadside ditches. The following information is from Helm (1994) unless otherwise stated. Its breeding habitat is primarily freshwater wetlands, but also, in the USA, wetlands of intermediate salinity (<5ppt. salt content), and includes deepwater marshes with water 25-100cm deep, lakes, and impoundments (primarily coastal but also inland), with stable water levels and a good interspersion of dense stands of floating vegetation including *Pontederia, Alternathera, Eichhornia* and *Nuphar*, emergents such as *Zizaniopsis* and *Typha*, and submerged vegetation such as *Ceratophyllum* and *Potamogeton*. Preferred breeding habitats also include a large amount of edge created by the interspersion of robust emergents with open water areas, but it seldom uses open water free of vegetation. Closed habitat (<25% open water) was used significantly more than open habitat (50-75% open water) during the Jun nesting season, but this difference was not detected in Sep (Mulholland 1983). Breeding habitat at a Costa Rica site was characterised by floating-leaved *Nymphaea ampla*, and by emergents such as *Panicum purpurascens, Leersia hexandra, Eleocharis geniculata, Ludwigia decurrens* and *Cyperus papyrus* (Krekorian 1978, Hunter 1987a), and in Colombia some birds nested in dense growths of *Thalia geniculata* along streams and drainage canals in ricefields (McKay 1981). This gallinule is dependent on floating vegetation, and also submergents, for brood rearing; such vegetation provides food, cover and protection from aquatic predators. Ricefields are also an important nesting habitat, providing a dependable source of food and cover; after rice harvest and before migration, birds use either adjacent unharvested fields or the numerous canals associated with rice cultivation (McKay 1981). It probably has more specific habitat requirements than does the Common Moorhen. At lake edges in SE Peru it preferred taller, larger-leaved vegetation than its smaller congener the Azure Gallinule (Parker 1982). Little is known of habitat use during migration and in wintering areas S of the USA. It occurs mainly in subtropical and tropical lowlands and coastal fringes, normally ranging up to c. 1,000m, but to at least 1,500m in Costa Rica (Stiles & Skutch 1989); during movements birds frequent wetlands at higher altitudes, in the E Andes of Colombia up to at least 3,020m, and it is casual at 4,080m in Junín, Peru (Fjeldså & Krabbe 1990). Vagrants are recorded from atypical habitats such as wild yams on a small stream (St Helena: Olson 1972).

FOOD AND FEEDING Predominantly plant material, including pondweed *Potamogeton*, sedges, willows, the fruits of waterlilies, trees and other aquatic and terrestrial plants, the seeds of grasses (e.g. *Paspalum*) and sedges and of floating and submerged vegetation, flowers of *Eichhornia crassipes*, and cultivated rice grains (Helm 1994). It also eats insects, including Odonata adults and nymphs, borer moth (Noctuidae) larvae and pupae, Coleoptera, Hemiptera (including Aphididae), Hymenoptera (including stingless bees *Trigona sylvestriana* and *T. corvina*, and ants *Ponera opaciceps*), Orthoptera (grasshoppers), and fly larvae and pupae (Corydalidae, Cyclorrhapha), as well

483

as worms, molluscs, crustaceans, ticks, spiders, small frogs and fishes, and occasionally the eggs and young of herons and jacanas (Krekorian 1978, Stiles & Skutch 1989, Navas 1991, Helm 1994); it may also take carrion (McKay 1981). In a Florida study, its diet comprised 71% plant material (mostly seeds) and 29% animal matter (Mulholland & Percival 1982); proportions by volume in Colombian ricefields were very similar (68% rice grains, 5% weed seeds, 27% animal matter) (McKay 1981); however, in another Florida study plant material comprised 58% and animal matter 42% (Sprunt 1954). In a Costa Rica study, 71% of food items eaten by breeding birds were *Nymphaea* fruits (Krekorian 1978). It walks on floating vegetation to feed; it turns over lilypads with the bill to glean food from the undersides, holding the leaf down with a foot; and it swims readily. It climbs easily to feed in bushes and trees for up to 20m from the ground (Wetmore 1965). Birds visit ricefields and fields of ripe grain in the autumn, climbing the stalks to feed on seedheads (Cramp & Simmons 1980), and they perch on reed stalks, weighing them down until they break off and the seedheads can be reached and stripped (Haverschmidt & Mees 1995). To dismember frogs, two or more birds often pull at the frog's body (Krekorian 1978). It is an opportunistic species, taking advantage of locally abundant foods (Helm 1994). Birds usually remain close to cover, but feed in the open by day; when encouraged they become tame, visit gardens and will pick up food dropped by humans. Chicks eat molluscs, crayfish, spiders, and insects and insect larvae (including stingless bees *Trigona* spp.), as well as *Nymphaea* fruits and grass seeds; adults break up *Nymphaea* fruits to feed to young chicks (Krekorian 1978, Helm 1994).

HABITS This species is apparently diurnal. It is usually found close to or within cover, but is less retiring than some congeners; it feeds in the open during the day and, when unmolested, it becomes very confident and ventures onto banks and beyond them to adjacent meadows and even to gardens and lawns (Slud 1964, Urban *et al.* 1986). The presence of alligators in some favoured waters tends to inhibit incursions into exposed situations (Cramp & Simmons 1980). It jerks the tail while walking, and it climbs easily to the tops of marsh plants, where it is often seen in the evening. It often perches on bushes, low branches or even fence posts near water (Hilty & Brown 1986). When disturbed or alarmed it frequently runs rapidly to cover, often flapping its wings, but it will often fly; when flushed, it flies rather slowly but directly, with rapid wingbeats and dangling legs, and often lands on floating or emergent vegetation; on longer flights the legs are raised (Slud 1964, Hilty & Brown 1986). Running birds, when chased, lowered themselves 2-4cm under water, flattened out and remained completely submerged with eyes closed; such submerged birds may be captured by hand (McKay 1981); birds may dive and swim underwater with only the bill showing (Audubon 1840). It swims freely at times but appears on open water far less than does the Common Moorhen.

SOCIAL ORGANISATION Monogamous; territorial during the breeding season. Migrants either arrive paired on the breeding grounds or pair immediately after arrival, and territories are established prior to the nesting season and maintained throughout the brood-rearing period (Helm 1994). In Colombian ricefields, the observed inter-nest distance was usually at least 40m, but 2 nests were only 11m apart (McKay 1981). The average home range for 4 nesting birds in Louisiana was 1.03ha (Matthews 1983). In Costa Rica, non-migratory birds live in extended family groups in which immatures and juveniles help with feeding and defending chicks, and with territory defence (Krekorian 1978, Hunter 1987b). Two of 11 cooperative breeding groups contained 2 adult females and 1 male; the male mated with both females but only 1 female produced eggs (Hunter 1987a). In the breeding group, all birds over 2 months of age feed chicks, but helpers spend significantly less time feeding chicks than do breeders (Hunter 1987b). After dispersal, former helpers returned to the study site to become floaters, which formed new territories when breeding habitat at the site was increased; helping and floating are viewed as age-related responses to the ecological constraints of limited habitat availability in this nonmigratory population (Hunter 1987a).

SOCIAL AND SEXUAL BEHAVIOUR Birds are recorded mingling freely with Nothern Jacanas *Jacana spinosa* and Common Moorhens (Slud 1964); adults and juveniles defend the breeding territory against these 2 species and against conspecific individuals (Krekorian 1978). In Costa Rica, non-breeding floaters at a study site were solitary and did not flock together, but used open areas of the wetland as a communal feeding ground (Hunter 1987a). Intraspecific territorial displays include charging, chasing and bowing (Helm 1994). Noisy fights between occupants of adjacent territories sometimes occur, involving 2 or more individuals from each of the family groups; fighting birds flapped their wings and kicked each other, and were watched by other members of both families (Krekorian 1978). Pairing activities and courtship displays include billing, bowing and nibbling, swaying, and the squat arch (Meanley 1963, Helm 1982).

BREEDING AND SURVIVAL Season USA, usually May-Aug, sometimes from Apr, in some habitats, e.g. ricefields, nesting delayed until plant density adequate (Helm 1994); Puerto Rico, probably all year but peaking late Aug to Oct and Apr-May; Trinidad and Tobago, Jun-Dec; El Salvador, Jul-Aug; Costa Rica, season prolonged at sites with permanent water, wet season in Guanacaste; Panama, Mar-Nov; Colombia, May-Oct (peak in W Meta Jun-Jul); Surinam, Jan-Jun; SE Brazil (Rio Grande do Sul), young one-third grown, 30 Mar; Argentina, Dec; Uruguay, juveniles Dec, Mar; southern hemisphere populations, Nov-Dec. **Nest** Nest initiation period 72-95 days (Louisiana: Helm 1982). Nest built on floating mats of vegetation such as *Eichhornia* and *Alternanthera*, or in emergent vegetation such as *Typha*, *Thalia* and *Zizaniopsis*, occasionally on pile of drift weeds; nests in emergent vegetation normally near edge, where interface occurs with open water areas containing submergents such as *Ceratophyllum* and *Potamogeton*, or floating vegetation; also nests beside water. Nest bulky, made from readily available plant materials, e.g. leaves of *Typha*, grass or rice plants, usually obtained very close to nest site; 26 of 28 nests built from plants <1 m from nest site (Helm *et al.* 1987); nest a shallow cup usually lined with green vegetation; in *Typha*, much of structure often made of blades pulled down and intricately woven together. External diameter 16-28 cm, depth 6-20 cm; cup depth 2-9 cm; nest 0-120cm (often 20-60 cm) above water; depth of water at nest sites in rice, Colombia, 6-26 (14.7) cm, height of vegetation 55-85 (65) cm. Partial canopy of growing vegetation often made; extensive and substantial ramp up to nest usually constructed, 5-15cm wide and extending upward from water surface. Both sexes build. Trial nests and elevated roosting platforms also constructed. **Eggs** 6-10 (Florida); 3-12 (West Indies); 4-8 (Mexico); 3-7 (Trinidad, Tobago, Costa Rica); 4-5 (Panama, Surinam). Mean clutch size 4.5 (n = 120), 5.8 (n = 32) and 8.6 (n = 60) (all Louisiana); 6.5 (n = 87, Texas). Eggs blunt oval, sometimes biconical; slightly glossy; pale pink to pinkish-cream, buff or creamy white, closely or unevenly speckled with minute dots and/or spots of chocolate-brown and underlying pale purple, markings sometimes denser at larger end. Size of 120 eggs 34.6-44 x 26.2-30.7 (40.0 x 28.7); calculated weight 18 (Schönwetter 1961-62); weight of fresh eggs given erroneously as 27-28.5 (Haverschmidt & Mees 1995); weight of 7 eggs, Panama, 14.95-16.35 (15.7, SD 0.5) (Gross & van Tyne 1929), of 2 eggs (age unknown) 15.2 and 14.8 (McKay 1981). Eggs laid at daily intervals. **Incubation** 18-20 days (c. 22 recorded in Panama), by both sexes; begins before clutch complete; hatching asynchronous. **Chicks** Precocial or semi-precocial; fed in nest for 1-4 days after hatching; fed and cared for by both parents, and also by juvenile helpers (in Costa Rica); up to 3 weeks of age young return to nest for brooding; separate brood nest sometimes constructed, especially if egg nest destroyed or flooded. Chicks fed bill-to-bill; beg by stretching the neck forward, lowering or waving the head, waving the wings vigorously, and making peeping calls (Gross & van Tyne 1929, Krekorian 1978); begin to feed themselves at 7-10 days; become self-feeding at c. 21 days. Downy chicks can climb actively, using small wing-claw (Gross & van Tyne 1929); can dive and swim under water to escape danger (Olson 1972c). First feathers appear at c. 2 weeks; juvenile plumage attained, and young capable of flight, at 5-7 weeks; independent at c. 9 weeks. Parent bird observed carrying downy chick in flight, holding young by nape or back in bill (Olson 1974c). Nest success rate variable, but usually high: 91% (n = 87) and 49% (n = 87) in a 2-year Texas study (Cottam & Glazener 1959), and 85% (n = 39) and 50% (n = 28) in a 2-year Louisiana study (Helm 1982). Some nests lost to flooding (e.g. Gross & van Tyne 1929). Average brood size at fledging 1.5-3.1; average production rate (SW Louisiana) 0.6-2.8 (1.6) immatures per adult, 1979-92 (Helm 1994). Chicks in groups with helpers received more food and were accompanied for longer periods than those in groups without helpers; both factors may have contributed to increased chick survival (Hunter 1987b). North American birds sometimes renest if first attempt fails (Matthews 1983), but extent of double brood production unknown; in Costa Rica pairs produce a clutch every 2-4 months throughout year (Hunter 1987b). Predators of chicks include snapping turtles (Hunter 1987b). Age of first breeding probably 1 year; however, Dickey & van Rossem (1938) report that a female "mostly in juvenal plumage with only a few brighter-colored feathers here and there about the body" had laid a short time before being collected. **Survival** No information available on adult survival.

125 AZURE GALLINULE
Porphyrio flavirostris   Plate 37

Fulica flavirostris Gmelin, 1789, Cayenne.

Formerly listed as *P. parvus*, but this name preoccupied. Sometimes placed in *Gallinula* or *Porphyrula*. Monotypic.

Synonyms: *Porphyriola/Fulica/Gallinula/Ionornis parva*; *Glaucestes flavirostris/parvus*; *Porphyrio parvus/cyanoleucus*; *Ionornis flavirostris*; *Gallinula/Porphyrula flavirostris*.

Alternative name: Little Gallinule.

IDENTIFICATION Length 23-26cm. Smallest *Porphyrio*, medium-sized, delicately proportioned gallinule, markedly different in appearance to most congeneric adults, its plumage looking washed-out and resembling more that of juveniles. Crown to back pale brownish-olive with indistinct dusky feather centres on back; lower back to tail dark brown, darker on tail; uppertail-coverts and tail feathers pale-tipped; upperwing-coverts predominantly azure or greenish-blue, as are outer webs of primaries and secondaries. Sides of head, neck and breast pale blue-grey; chin, throat and underparts pure white. Iris reddish-brown; bill and frontal shield pale greenish-yellow; legs and feet yellow. Sexes alike. Immature browner above, with contrasting dark rump and tail; sides of head, neck and breast buffy; frontal shield and culmen green; legs and feet yellowish-orange; iris deep yellow. Juvenile like immature but has tawny to buff tips on feathers from mantle to uppertail-coverts, and on upperwing-coverts (which are brownish-olive, tinged pale greenish-blue) and remiges; has only washed-out greenish-blue on outer webs of primaries and secondaries. Inhabits marshes with dense emergent grass, marshy margins of open waters, and rice fields.

Similar species Similar to juvenile of larger American Purple Gallinule (124), alongside which it sometimes occurs, but much more slender, with longer tail; is pale blue-grey (not buff-brown) on sides of head, neck and breast; has more blue in wings, whiter underparts, and different bare part colours. Juvenile American Purple Gallinule has upperparts brown with bronze-green tinge on back, scapulars and upperwing-coverts, and turquoise wash on outer upperwing-coverts; underparts buff-brown; undertail-coverts white; iris brown to orange; bill yellow-green with pale tip and brown-pink base, and shield brown or grey; legs and feet pale yellow or brownish. Juvenile Azure Gallinule also differs in having mantle to uppertail-coverts, scapulars, upperwing-coverts and remiges tipped bright buff to tawny, giving patterned appearance.

VOICE It is usually a rather quiet species; it sometimes utters a short trill (Hilty & Brown 1986). It also gives *krrrr*, *krrra*, and shorter *kra* or *ka* calls of typical *Porphyrio* quality (see Hardy *et al.* 1996).

DESCRIPTION
Adult Crown, nape, and hindneck to back pale brownish-olive, with darker brown to sepia feather centres giving indistinct dusky pattern on back; lower back to uppertail-coverts dark brown, rump feathers often very narrowly tipped tawny, uppertail-coverts tipped whitish; rectrices blackish-brown, edged pale greyish-blue and tipped white. Scapulars dark brown, edged olive; upperwing-coverts azure or pale greenish-blue, inner webs more fuscous; tertials pale brownish-olive with dusky feather centres;

alula, primaries and secondaries dusky brown, outer edges grey-blue to turquoise-blue. Underwing-coverts and axillaries white, lessers and marginals washed very pale blue-grey. Lores, supercilium, ear-coverts, and sides of neck and breast pale blue-grey; malar region, chin, throat, foreneck, centre of breast and rest of underparts pure white. Iris reddish-brown, hazel or orange-brown; bill and frontal shield pale green to pale greenish-yellow or olive, frontal shield also described as dark olive-grey; legs and feet dull to bright yellow. Sexes alike.

Immature Browner above than adult, with contrasting dark rump and tail, buffy sides of head, neck and breast; bill as adult but frontal shield and culmen green; legs and feet yellowish-orange; iris amber-yellow.

Juvenile Blue in plumage confined to upperwings, being replaced by pale cinnamon-buff to pale tawny-olive on sides of head and breast; feathers of mantle to uppertail-coverts, including scapulars, with pale cinnamon-tawny to buff tips, especially bright on uppertail-coverts. Upperwing-coverts pale brownish-olive, tipped bright buff, and tinged pale greenish-blue on lessers and medians; alula and remiges like adult, but with only washed-out greenish-blue on outer webs, and tipped bright buff.

Downy young Undescribed.

MEASUREMENTS Wing of 18 males 128-142 (135.5), of 17 females 119-139 (132.1); tail of 5 males 67-71 (69, SD 1.9), of 6 females 65-69 (67.3, SD 1.9); culmen (including shield) of 5 males 29-34 (31.6, SD 2.4), of 6 females 30-32.5 (31, SD 1.2); tarsus of 11 males 38-42 (40.1), of 13 females 38-44 (40). Weight of 9 males 73-111 (93.3), of 8 females 79-107 (92), of 1 apparently healthy female in breeding condition 57 (Dunning 1993, Haverschmidt & Mees 1995).

GEOGRAPHICAL VARIATION None.

MOULT Nothing recorded.

DISTRIBUTION AND STATUS C and S Colombia (E of Andes from C and W Meta, W Caquetá and Amazonas, and probably N to Arauca; also once from Boyacá), Trinidad, C and S Venezuela (Delta Amacuro, Apure, Bolívar, Amazonas) to the Guianas, S to N and NW Brazil (N Roraima, Amapá, Pará, Amazonas) and NE Ecuador (Napo and Pastaza) and NE Peru (Loreto); SE Peru (Madre de Dios), N Bolivia (Beni, Santa Cruz), W and C Brazil (Goiás, W São Paulo, Mato Grosso, Mato Grosso do Sul) and Paraguay (mostly E of Paraguay R) to extreme N Argentina (Corrientes, Misiones); also E Brazil (Minas Gerais, Rio de Janeiro) and extreme SE Brazil (Rio Grande do Sul). Not globally threatened. It was formerly regarded as fairly common to abundant at least in some parts of its range, which has been conflictingly regarded as either fragmented or continuous (e.g. Blake 1977, Ripley 1977, Remsen & Parker 1990). It was apparently fairly common in Guyana but spottily distributed in Venezuela and uncommon and local in Colombia (Snyder 1966, Meyer de Schauensee & Phelps 1978, Hilty & Brown 1986), and uncommon to rare in E Peru and N Bolivia (Pearson 1975, Remsen & Parker 1990); it is probably more widespread in Amazonia than records indicate (Remsen & Parker 1990). It is currently regarded as locally common on the littoral plain in French Guiana and common in Surinam (Tostain *et al.* 1992, Haverschmidt & Mees 1995). It is probably regular in summer in the marshes of W São Paulo, Brazil, but is likely to disappear as rivers become reservoirs (Willis & Oniki 1993). It is uncommon in E Ecuador (Ridgely *et al.* in press). It is recorded from only 2 areas of Peru: in the NE it occurs on the Amazon and Napo Rs near Iquitos in Loreto, while in the SE it was fairly common at the Tambopata Reserve, near Puerto Maldonado in Madre de Dios, and possibly occurs throughout lowland E Peru, at least seasonally (Parker 1982, Remsen & Parker 1990, J. V. Remsen *in litt.*). Hayes *et al.* (1994) and Hayes (1995) give its status in Paraguay as a rare breeder, although there are many records E of the R Paraguay, but there is only 1 certain record (Jan 1937) from the Chaco (W of the R Paraguay); however, during Project Yacutinga in 1995, 5 and 3 birds were seen in the Chaco at Reserva Natural Privada La Golondrina, Dpto. Presidente Hayes, on 7 Nov (R. P. Clay *in litt.*). In E and SE Brazil, there are 2 specimens from Minas Gerais (Remsen & Parker 1990), it is recorded from Rio de Janeiro (Sick 1993), and there is an extralimital record from Rio Grande do Sul (see Movements). It is regarded as rare in N Argentina, where a published occurrence from Formosa is actually from Paraguay, and it is recorded once from Misiones (at Posadas) and 3 times from Corrientes (Navas 1991, Chebez 1994); however, R. M. Fraga (*in litt.*) has found that it occurs regularly at Estancia San Juan Poriahu, N Corrientes, up to 10-12 birds being seen daily, and considers that it probably breeds locally. There is also a considerable number of records from the Nariva Swamp area in Trinidad, all since 1978, and this may represent a recent range expansion (ffrench 1985). Seasonality of occurrence is discussed in the following section.

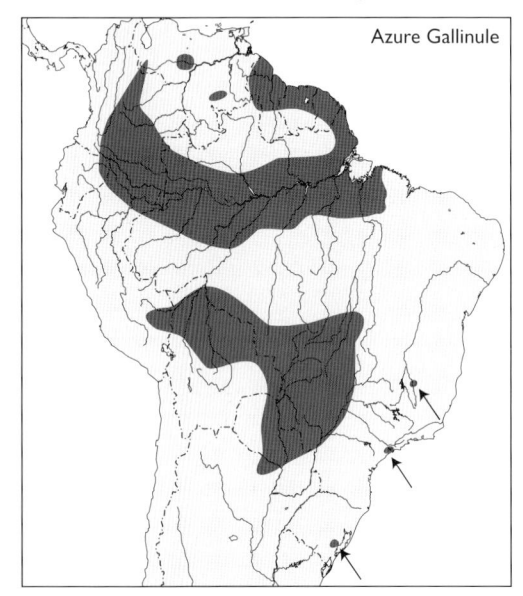

MOVEMENTS Unless otherwise stated, information is from the summary in Remsen & Parker (1990). Movements are not clearly understood, and possible seasonality of occurrence is not shown on the distribution map, although this gallinule is apparently seasonal in parts of its range. It is highly seasonal in the Guianas, where almost all records are in Apr-Aug (the wet season), and in SW Amazonia, where most records are in Oct-Jan with a few of immatures from May to Aug. It is a northern austral migrant in Paraguay, extreme dates being Oct and 28 Jan, with a sighting on the Brazilian side of the R Paraguay on

17 Aug indicating that birds may arrive as early as Aug (Hayes *et al.* 1994, Hayes 1995). It is assumed to be migratory in Amapá and Roraima, N Brazil, and Mato Grosso and Mato Grosso do Sul, SC Brazil (Forrester 1993). Closer to the equator, seasonality is not so pronounced but is still evident from specimen records: 82% of the specimens from E Amazonian Brazil are from Feb-Aug, while specimens from W Amazonia peak in Jan and observations along the upper Amazon (Iquitos and Leticia) indicated that birds are present there only in Jan-Jul when river levels are high. It is regarded as resident in NE Amazonas at Manaus (Forrester 1993).

Its occurrences may be timed to coincide with wet or high-water seasons, and it is apparently present seasonally in areas at the periphery of its range; however, at Estancia San Juan Poriahu, Corrientes, N Argentina, it occurs regularly and probably breeds, and it is of regular occurrence elsewhere in the region (R. M. Fraga *in litt.*). In SE Peru, arrival was noted at the Tambopata Reserve, Madre de Dios, in early Nov and birds were present until at least mid-Jan (Parker 1982); it is possibly absent elsewhere in Madre de Dios in Jun-Dec. It is recorded from a locality in the *llanos* of Venezuela only in Aug-Dec (Thomas 1979). It is apparently resident in Ecuador and S Colombia (Parker 1982, Ridgely & Greenfield in prep.), but in C Colombia (W Meta) it is recorded only from late Mar to mid-July, and migrants are present from Mar to May in the savanna of Bogotá (Nicéforo & Olivares 1965, Hilty & Brown 1986). After breeding in May-Jul, most of the population in N South America may disperse S c. 1,800km to the S periphery of Amazonia, returning N in Feb-Mar.

Vagrancy, and occurrence in atypical habitats, are also recorded. It has occurred at 3,000m on L Tota, Boyaca, N Colombia, in late Jul (Varty *et al.* 1986) and at Escorial in the Venezuelan Andes at 3,000m in Oct. It is recorded twice from Bogotá city at 2,600m in Dec (Nicéforo & Olivares 1965, Olivares 1969), on a rocky escarpment above forest in S Venezuela in Jan, and in Rio Grande do Sul, SE Brazil, in Nov. A specimen from New York, USA, (Dec) could be a wild bird as its occurrence falls within the period during which vagrants have been detected and during the presumed non-breeding season.

HABITAT Most information is from Remsen & Parker (1990). Freshwater marshes, especially those with fairly deep water and a thick cover of marsh grass (e.g. *Paspalum* up to 1m tall) or other emergent or floating vegetation (sedges etc.); also rice paddies, wet savannas, riverine marshes (including swampy stream and river margins) and the marshy shores of lakes (including oxbow lakes: Parker 1982, Ridgely & Greenfield in prep.), lagoons and permanent or seasonal ponds. It occurs alongside the American Purple Gallinule but prefers shorter, smaller-leaved vegetation and avoids areas with bushes and *Heliconia*; it is often flushed from emergent grass 15-30cm tall. When rivers are at or near flood stage, Azure Gallinules congregate in the *Paspalum* marshes along the edges and lower ends of river islands. Specimen records suggest that it is commoner in forested regions than in the wet savanna regions bordering Amazonia, in spite of the abundance of marsh habitats in these savannas. It normally occurs in lowlands, up to 500m, but it is recorded casually on the savanna of Bogotá in the E Andes at 2,600m (Fjeldså & Krabbe 1990) and at 3,000m in Colombia and Venezuela (Varty *et al.* 1986, Remsen & Parker 1990).

FOOD AND FEEDING Grass seeds, insects (including Hemiptera and Coleoptera) and Arachnoidea are recorded. It forages in cover and on floating vegetation, and it climbs on grass stalks to bend them to the water and eat their seeds (Willis & Oniki 1993).

HABITS It is said to be less skulking than other rails (Ripley 1977), but it apparently remains within cover much more than does American Purple Gallinule, and it seldom swims (only rarely in open water) (Hilty & Brown 1986). However, it flushes readily and flies for a short distance with dangling legs; it typically holds its wings up momentarily when alighting, and it sometimes perches in exposed places on emergent or floating vegetation (Hilty & Brown 1986, Remsen & Parker 1990).

SOCIAL ORGANISATION Monogamous; probably territorial when breeding.

SOCIAL AND SEXUAL BEHAVIOUR At a change-over during incubation, a relieving bird was seen to present a piece of straw to its mate (Haverschmidt 1968). Nothing else is recorded.

BREEDING AND SURVIVAL Season Surinam, May-Aug, half-grown young mid-Apr; Brazil, E Amazonian Brazil probably May-Jun, breeding condition Amazonas, and Pará, Mar; Ecuador, oviduct egg Jun, breeding condition Jul, Oct. Season may extend from Mar to Aug, possibly Oct, and non-breeding season may be Sep-Dec (Remsen & Parker 1990). **Nest** An open cup of dead leaves, grasses or rushes, concealed in marsh vegetation. **Eggs** 4-5; creamy with numerous small reddish-brown spots; size of 3 eggs 31.8 x 24.8, 32 x 24.7, 33.3 x 24.7 (Schönwetter 1961-62); size 30.5-34 x 23.3-25.6, weight 9.5-11.5 (Haverschmidt & Mees 1995) . **Incubation** Both sexes incubate. No other information available.

GALLINULA

This genus is of worldwide distribution and contains nine species of medium-small to very large gallinules. Most have dark, plain plumage, the exception being the Spot-flanked Gallinule of South America which is paler and brighter, with bold white flank spots; all have a brightly coloured bill and frontal shield. Several species have a prominent white line along the dark flanks and all the essentially aquatic species (plus the Tristan Moorhen) have white on the undertail-coverts. The distinctive Samoan and San Cristobal Moorhens are sometimes placed in their own genus, *Pareudiastes*, on skeletal characters, a separation which has some merit. However, it is no longer regarded as correct to place the Spot-flanked Gallinule in the monotypic genus *Porphyriops*, or the Tasmanian and Black-tailed Native Hens in the genus *Tribonyx*: the skeletons of all three species show no differences of generic significance when compared to *Gallinula* (Olson 1973b). *Gallinula* species occur in a very wide range of habitats, ranging from wetlands with open fresh water and emergent vegetation, which are occupied by the 'typical' moorhens, to the tussock grassland and bush occupied by the Tristan Moorhen and the dense primary forest habitat of the Samoan and San Cristobal Moorhens. The Common Moorhen of Eurasia, Africa and the Americas is the most widely distributed rail in the world, and its northern populations are strongly migratory. Most other species are sedentary, but the Black-tailed Native-hen of Australia is markedly dispersive and eruptive, and the Lesser Moorhen of Africa is a rains migrant over much of its range. Of the two forest-dwelling species, the flightless Samoan Moorhen is almost certainly extinct, while the flightless and critically endangered San Cristobal Moorhen is reliably known from only one specimen, collected in 1929. Although it is also flightless, the very large Tasmanian Native-hen is maintaining its numbers successfully and is sometimes regarded as an agricultural pest. The nominate race of the flightless Tristan Moorhen became extinct on Tristan da Cunha before 1900 and the island is now populated by the race *comeri*, introduced from Gough I, which is classed as Vulnerable. Two races of the Common Moorhen, *guami* from the Mariana Islands, and *sandvicensis* from the Hawaiian Islands, are endangered as a result of habitat loss.

126 SAMOAN MOORHEN
Gallinula pacifica   Plate 38

Pareudiastes pacificus Hartlaub & Finsch, 1871. Savaii, Samoa Islands.

Sometimes retained in *Pareudiastes*, with the San Cristobal Moorhen *G. silvestris*, on basis of construction of bill, skull and tarsometatarsus, and on plumage pattern (Olson 1973b, 1975b), an arrangement which may well be valid. Monotypic.

Alternative names: Samoan Wood Rail/Gallinule/Woodhen.

IDENTIFICATION Length 23-25cm. A small, almost flightless gallinule, PROBABLY EXTINCT. Head, neck and breast dark bluish-slate, blackish on face, chin and throat; upperparts very dark olive-brown, tinged greenish; lower rump to tail black; rest of underparts dark olive-greenish, but flanks dark olive-brown and undertail-coverts black. Eye-ring red; iris probably dark brown; bill possibly orange-red, shading to yellow at base of upper mandible and on shield, or red; legs and feet probably red. Eyes apparently very large. Sexes alike. Immature and juvenile undescribed. Inhabited montane forest.
Similar species Unlike its close relative the San Cristobal Moorhen (127), this species bears a strong superficial resemblance to other *Gallinula* rails. However, no congener occurs in its habitat, but confusion is possible with juvenile Purple Swamphens (120) and young domestic chickens (Pratt *et al.* 1987).

VOICE Not described.

DESCRIPTION
Adult Head and neck dark bluish-slate, but lores, supercilium, ear-coverts, chin and throat blackish, slightly tinged ashy; upperparts, including upperwing-coverts, very dark olive-brown, tinged greenish; remiges blackish, somewhat greenish on outer webs; axillaries, underwing-coverts, and underside of remiges black; lower rump, uppertail-coverts and tail black. Breast bluish-slate, like neck; rest of underparts dark slate, strongly washed olive-greenish but flanks darker olive-brown; undertail-coverts black. Eye-ring red; iris probably dark brown; bill possibly red, or orange-red, shading to yellow at base of upper mandible and on shield; legs and feet probably red. Eyes apparently very large.
Immature Unknown.
Juvenile Unknown.
Downy young Unknown.

MEASUREMENTS (2 unsexed). Wing 118, 135; tail 32, 35; culmen (with shield) 37, 37; tarsus 39, 42.

GEOGRAPHICAL VARIATION None.

MOULT Unknown.

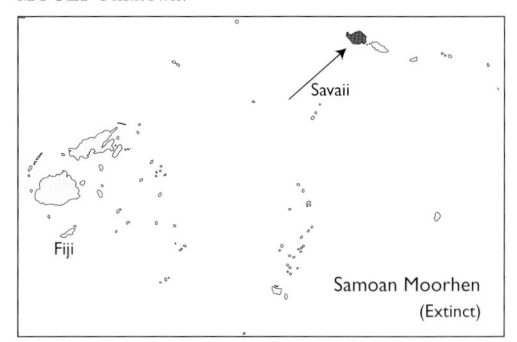

DISTRIBUTION AND STATUS PROBABLY EXTINCT. It was confined to Savaii I, but may also have occurred on Opolu I (Whitmee 1874). It was first collected in 1869 and was last recorded in 1873, when 2 specimens were probably given to the Challenger expedition but bear no dates (Greenway 1967). It was possibly extinct by 1907 (Reed 1980), and the Whitney Expedition failed to find it in 1926. While the local people esteemed it as food, its apparent demise is probably attributable to rats, cats and other introduced predators, although it had survived large populations of rats and cats for many years (Greenway 1967, Ripley 1977, Fuller 1987): see also Mees (1977). However, there were 2 possible sightings in upland forest W of Mt Elietoga in Aug 1984 (Hay 1986, Bellingham & Davies 1988) and it may yet survive (Collar *et al.* 1994).

MOVEMENTS None recorded.

HABITAT Montane forest, where it is reported to have lived in burrows in the ground (see Habits).

FOOD AND FEEDING Apparently entirely animal matter, including insects; captive birds fared poorly on a vegetable diet (Pritchard 1866, Whitmee 1874).

HABITS Its exceptionally large eyes suggest that it may have been crepuscular or even nocturnal (Fuller 1987, Day 1989). Its local name *puna'e*, apparently means "springer-up", and both Pritchard (1866) and Whitmee (1874) state that local people consistently maintained that the birds lived in burrows in the ground. Apparently the bird, when startled from its burrow, made a long spring upwards from the ground, but having very small wings could not fly (Pritchard 1866). Alternatively, it is possible that it may have only taken refuge in burrows, as did other species such as the Laysan Crake (95) and the Hawaiian Crake (101) (Greenway 1967), and the only breeding record is of a bird taken with 2 eggs from a nest on the ground (see Breeding and Survival).

SOCIAL ORGANISATION Unknown.

SOCIAL AND SEXUAL BEHAVIOUR Unknown.

BREEDING AND SURVIVAL Nest 1 nest described: situated on ground; of a few twigs and some grass; found in Oct 1873 (Knox & Walters 1994). **Eggs** 2 eggs found in this nest; 1 is in BMNH; narrow oval; creamy-white with numerous spots of reddish-brown and purplish-brown, and some pale purple underlying blotches; markings more numerous at larger end; size 41.7 x 31.8; calculated weight 25. Other local reports say that the species nested in burrows (Fuller 1987).

127 SAN CRISTOBAL MOORHEN
Gallinula silvestris Plate 40

Edithornis silvestris Mayr, 1933, San Cristobal Island.

Sometimes placed in genus *Pareudiastes*, with extinct Samoan Moorhen *G. pacifica*, on basis of construction of bill, skull and tarsometatarsus, and on plumage pattern (Olson 1973b, 1975b), an arrangement which may well be valid. Monotypic.

Synonym: *Pareudiastes silvestris*.

Alternative name: San Cristobal Gallinule.

IDENTIFICATION Length of 1 male 26.5cm. Male entirely plain: head, neck and breast dark bluish-slate, blacker on chin and face; scapulars, mantle and upperwings brown-black, tinged olive; rest of body dull brownish-black; iris chocolate; legs, feet and bill scarlet; shield dark grey-blue; bare skin on face yellow. Tail very short. Flies very little, if at all. Female, immature and juvenile unknown. Inhabits dense undergrowth of primeval mountain forest.
Similar species None within its range. Because of its distinctive appearance, it can scarcely be confused with any other rail (Olson 1975b).

VOICE Unknown.

DESCRIPTION
Adult male Head, neck and breast dark bluish-slate, almost blackish on chin and sides of head; scapulars, mantle, upperwing-coverts and secondaries brown-black, tinged olive; rest of body, including undertail-coverts, dull brownish-black; iris chocolate; legs, feet and bill bright scarlet; frontal shield dark grey-blue, posterior edge straight; eyelids, lores, and bare skin from gape under eyes to postocular region, yellow; lores and lower and posterior parts of circumocular area covered with short, scattered, brushlike feathers. Tail very short, with hairlike rectrices; secondaries soft and decomposed. Bill strong and laterally compressed; legs and feet slender and long.
Immature Unknown.
Juvenile Unknown.
Downy young Unknown.

MEASUREMENTS Type specimen (male): wing 149; tail 40; culmen (including shield) 56; tarsus 60. Weight of 1 male 450.

GEOGRAPHICAL VARIATION None.

MOULT Unknown.

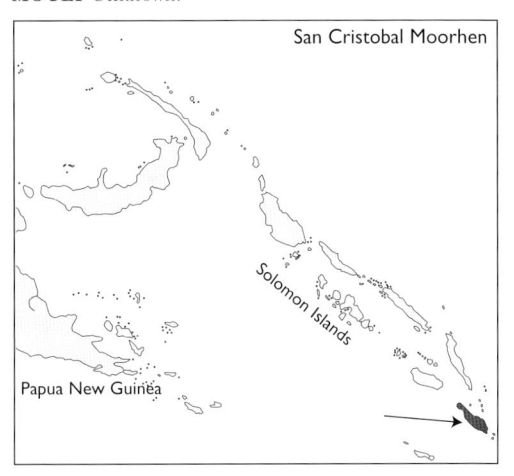

DISTRIBUTION AND STATUS San Cristobal, Solomon Islands. CRITICALLY ENDANGERED. It is known only from the holotype, collected by a local hunter for the Whitney South Sea Expedition on 4 December 1929 near the village of Hunogaraha in the central mountains of San Cristobal. Even then it was apparently rare, and was hunted by local people with dogs. A search for more specimens was unsuccessful, despite inducements offered to local people. The only subsequent observations are of 1 seen by a member of the Oxford University Expedition in 1953, when it was reported to be not uncommon in rocky valleys below Wuranakumau, on Naghasi ridge (Cain & Galbraith 1956), and a report by local people of its presence in 1974 (Collar *et al.* 1994); however, the forests where it was originally found have not been visited by ornithologists since the 1950s (Collar *et al.* 1994). It is likely to have been affected by introduced mammalian predators, such as cats, which have wiped out most native terrestrial mammals on nearby Guadalcanal (Collar *et al.* 1994). Information on the bird's status is urgently needed in view of potential threats to the island's forests (Hay 1986).

MOVEMENTS Unknown; presumably none.

HABITAT Dense undergrowth of primeval mountain forest on steep slopes; the type specimen was obtained at 580m, but mountains in the vicinity ascend to 1,200m. The area has many brooks and creeks cutting deep into the mountain slopes, but no standing water, and the whole region was covered with forest intermingled with plantations and some secondary growth. This species is also reported as occurring in rocky valleys below 425m. Other rails found in the area were the Bush-hen (86) and the Buff-banded Rail (46), both of which frequently occur far from water in the Solomons and are particularly fond of dense undergrowth in secondary vegetation, but the San Cristobal Moorhen is probably more of a true forest bird than either of these species (Mayr 1933a).

FOOD AND FEEDING Unknown. The diet of the extinct Samoan Moorhen (126), which was found in similar habitat, was apparently almost entirely animal matter.

HABITS Local people maintain that the species flies very little, if at all. It runs fast in the undergrowth and is very elusive, but it can be treed by trained dogs, when it flaps clumsily up into bushes like a domestic hen (Cain & Galbraith 1953).

SOCIAL ORGANISATION Unknown.

SOCIAL AND SEXUAL BEHAVIOUR Unknown.

BREEDING AND SURVIVAL Unknown.

128 TRISTAN MOORHEN
Gallinula nesiotis   Plate 39

Gallinula nesiotis P. L. Sclater, 1861, Tristan da Cunha.

Formerly sometimes placed in genus *Porphyriornis*, apparently solely on characters relating to loss of flight (Greenway 1973). Nominate race of Tristan da Cunha recently extinct. One extant subspecies recognised.

Synonyms: *Porphyriornis nesiotis/comeri*.

Alternative names: (Tristan) Island Cock/Hen; Gough (Island) Moorhen.

IDENTIFICATION Length 25cm. A medium-sized rail, small and thickset for a moorhen; almost flightless. Head to mantle, and entire underparts, black; rest of upperparts, including upperwings, very dark olive-brown; flight feathers darker; leading edge of wing whitish; flanks with a few white streaks; lateral undertail-coverts white. Bill and frontal shield bright red, bill tipped yellow; legs and feet heavy, red with greenish-yellow blotches, and red 'garter'. Sexes alike. Immature very dark olive-brown on head, neck and flanks; rest of upperparts paler than in adult; rest of underparts dark slate-grey; belly paler grey, with cream feather-tips; shield red; bill dull red with yellow-green tip; iris red-brown; legs and feet green. Juvenile olive-brown on head, neck, upperparts and flanks; lower sides of head, and rest of underparts, paler olive-brown, palest on belly (which also has cream feather-tips); lateral undertail-coverts brownish-white; medians very dark brown; shield and bill dark greenish, bill with whitish tip; iris grey-green; legs and feet dark olive-green. Inhabits tussock grass, bushes and fern-bush.

Similar species Superficially resembles Common Moorhen (129) but is smaller, much more strongly built, short-winged, and almost completely flightless. Also differs in having black neck and underparts (most races of Common Moorhen are grey in these regions), white on flanks confined to a few streaks, and red legs heavily blotched greenish-yellow.

VOICE The voice is metallic, and louder than that of the Common Moorhen. Information is taken from Richardson (1984) and Watkins & Furness (1986) unless otherwise stated. The common territorial or display call is a harsh, staccato, high-pitched and far-carrying *koo-ik* or *chek-a-kek*, often repeated and taken up by surrounding birds, and occasionally shortened to a monosyllabic *eek*; there is also a loud, far-carrying series of *koo-ik* notes (Richardson 1984), and a low-pitched, monotonous, whispered, repeated *ik* or *ook* given by pair members and audible only at close range. The alarm call is like the display call but more rapid and strident (Watkins & Furness 1986). The female gives a harsh repeated *tcherk aaa kak krak-krrak* near the nest (Ripley 1977). When foraging apart, pair members maintain vocal contact by giving a quiet locating call, *kek-kek*, every c. 10 min. A soft and rapid *ha-ha* call is given during change-over at the nest.

DESCRIPTION *G. n. nesiotis*
Adult Head to mantle, and entire underparts, black, sometimes faintly washed dark olive-brown. Rest of upperparts, including upperwing-coverts and tertials, very dark olive-brown. Alula, primaries and secondaries blackish-brown, outer web of alula feathers, and of outermost primary, edged white; marginal coverts mixed blackish-brown and white; underwing-coverts and axillaries very dark brown to blackish, coverts narrowly tipped whitish. Flanks with a few white streaks; lateral undertail-coverts white. Iris dark brown; bill and shield bright red, tip of bill yellow; tarsi and feet red with greenish-yellow blotches, tibiae red. Wing-claw present.
Immature Black areas of adult very dark olive-brown, becoming very dark slate-grey; rest of upperparts as adult but paler and brighter olive-brown. Remiges and rectrices as adult. Underparts dark slate-grey with pale, often creamy, feather-tips on belly, at least in fresh plumage; flanks very dark olive-brown, as upperparts. Underwings as adult. Yellow of bill tinged greenish; legs and feet green; iris reddish-brown with grey tinge.
Juvenile Entire upperparts bright olive-brown, duller on head, hindneck and mantle; remiges and rectrices very dark brown, narrowly edged pale brown; axillaries, underwing-coverts and underside of remiges blackish-brown; marginal coverts mixed dark brown and white. Underparts (and sides of head below eye) paler olive-brown, tending to buffy-brown, darker on sides of neck and breast and palest in centre of belly; flanks darker olive-brown, like upperparts; lateral undertail-coverts brownish-white; median undertail-coverts very dark brown. Iris grey-green to dark brown-grey; bill dark olive with creamy tip to dark greenish-black with black tip to upper mandible and pale horn tip to lower mandible; legs and feet dark green to dark olive-green.
Downy young Downy chick black, with fringe of long silky hairs on throat; legs and feet black. Two conflicting descriptions of bill colour exist, given without comment by Ripley (1977): bill crimson at base, blue-grey towards tip, with white spot on top of upper mandible (Wilson & Swales 1958; from captive *comeri* and presumably accurate); and bill horny-yellow, upper mandible with black tip and median black band, lower mandible with black tip and basal half of gonys (Watson 1975; source unknown).

MEASUREMENTS 2 unsexed: wing 136, 142; tail 55, 63; culmen (including shield) 39, 42; tarsus 49, 51. Weight: unsexed mean 400 (Ripley 1977).

GEOGRAPHICAL VARIATION Races differ only in skeletal measurements, nominate being slightly smaller and lighter (Beintema 1972).
G. n. nesiotis Sclater, 1861 – Tristan da Cunha. EXTINCT. See Description.
G. n. comeri (Allen, 1892) – Gough Island, S Atlantic. Recently introduced to Tristan da Cunha. Slightly larger than nominate. Measurements of 7 males, 10 females: wing of male 141-152 (148.7, SD 3.8), of female 134-148 (141.6, SD 4.1); tail of male 57-63 (60.4, SD 2.0), of female 48-62 (55.5, SD 3.7); culmen (including shield) of male 44-47 (45.7, SD 1.0), of

female 39-43 (41.6, SD 1.3); tarsus of male 46-52 (48.7, SD 2.0), of female 44-50 (465.4, SD 1.7). Weight of 3 adults 505, 505, 530 (Watkins & Furness 1986).

MOULT Moult is "in progress" in Feb (Watson 1975). Immature plumage is retained into the second year (Richardson 1984).

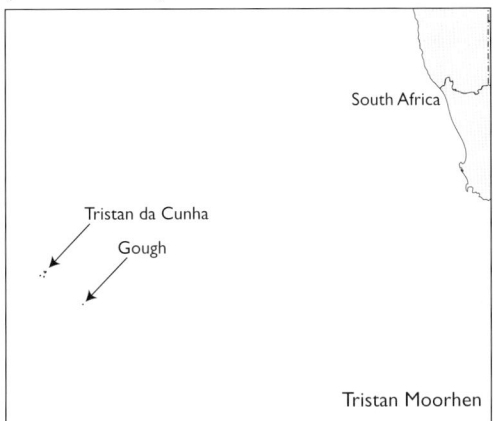

DISTRIBUTION AND STATUS Tristan da Cunha and Gough Is, S Atlantic. The nominate race of Tristan da Cunha was abundant up to 1852 but rare by 1873, when it was not found by the Challenger Expedition, and was apparently extinct by the end of the 19th century (Beintema 1972). Its extinction has been attributed to rats and feral pigs, but feral cats, hunting by islanders with dogs (it was considered excellent eating), and habitat destruction by fire, probably all contributed to its demise (Beintema 1972, Collar & Stuart 1985). The race *comeri* of Gough I is VULNERABLE. Its numbers are difficult to assess because of the bird's secretive nature, but it is common on Gough I, where the total population is probably 2,000-3,000 pairs (Collar & Stuart 1985, Collar *et al.* 1994), although Rose & Scott (1994) give a population estimate of 6,000-9,000 birds. It has no avian competitors. In May 1956, 7 were released on Tristan da Cunha and the population in 1984 was c. 250 pairs, and increasing (Collar & Stuart 1985). Both populations appear secure, and the habitats are unlikely to change significantly, but there is a permanent risk of mammalian predators becoming established on Gough I (Collar & Stuart 1985); however, the chances of this happening have been minimised (Cooper & Ryan 1994). The re-establishment on Tristan da Cunha was successful despite the presence of a large population of black rats *Rattus rattus*, and this suggests that the cause of the bird's previous extinction on Tristan was most likely to have been hunting (Collar & Stuart 1985). Gough I is a wildlife reserve, and measures have been taken to eradicate cats from Tristan (Collar & Stuart 1985). Captive stock exists in 3 European zoos (Collar *et al.* 1992), and the species breeds successfully in captivity, but hybridisation with Common Moorhen occurs and must be avoided (Wace & Holdgate 1976).

MOVEMENTS None. One captured in Cape Town in 1893 was certainly an escape (Hockey *et al.* 1989).

HABITAT The race *comeri* occurs throughout Gough I in very dense vegetation of tussock grass (*Spartina arundinacea* and *Poa flabellata*) and bushes in the shrub and tree-fern *Blechnum* zones, up to 450m where the open mountain begins; it is especially common along the shore, in boggy areas, and in valley bottoms covered with bushes and dense undergrowth including grass and tree-fern undergrowth along streams (Collar & Stuart 1985, Watkins & Furness 1986). On Tristan da Cunha, where it is introduced, it inhabits rugged, luxuriant, inaccessible *Blechnum* bush, with dense *Phylica arborea* trees, at 300-900m (Collar & Stuart 1985). Its absence from open areas has been considered a defence against predation by the Brown Skua *Catharacta antarctica* (Collar & Stuart 1985).

FOOD AND FEEDING Gough I birds appear to feed as much on vegetable matter, seeds and carrion as on invertebrates. They generally remain in cover but will feed in the open if undisturbed. They eat grass, and take grassheads with a scythe-like motion of bill. They scavenge from carcasses of Soft-plumaged Petrels *Pterodroma mollis* and Broad-billed Prions *Pachyptila vittata* which have been partly eaten by Brown Skuas; they also feed on garbage, such as fish-paste bait and fruit, at the meteorological station (Clancey 1981). They enter petrel burrows, apparently in search of food, and have been seen foraging in abandoned albatross nests, presumably for invertebrates; they also scavenge around active albatross nests. Foraging takes place within the territory, and pair members may forage together or independently (Watkins & Furness 1986).

HABITS These birds are very shy and generally stay hidden in vegetation, being more often heard than seen. At the slightest noise or disturbance they dart into cover, showing the white undertail-coverts; however, they will appear in the open when there is no danger. They are reported to take refuge in holes (Greenway 1967), but this may refer to their foraging habits (see Food & Feeding); they apparently construct tunnels in dense vegetation as hiding places (Watkins & Furness 1986). Although almost completely flightless, they are able to climb well (e.g. in *Phylica* trees) and partially fly over obstacles (Ripley 1977, Collar & Stuart 1985), and the wings are used for balance when the birds run swiftly. The white undertail-coverts are displayed when the birds are alarmed. In captivity they are "active, aggressive and extremely intelligent birds, finding their way in and out of pens with ease" (Ripley 1977). Birds bathe in fresh water but avoid salt water; they bathe with a dipping, flexing motion of the tarsus and slight opening of the wings to flush water over the back (Watson 1975, Ripley 1977).

SOCIAL ORGANISATION Monogamous and territorial; the pair bond appears to be permanent. In a study area of 7 ha, Watkins & Furness (1986) found 16 territories, which suggests a nesting density of 230 pairs/km^2; the distance between nearest neighbours or territorial foci was 20-60m (mostly 30-40m), and territories were c. 0.5ha in area. Young of the first brood help to feed those of the second brood (Watkins & Furness 1986).

SOCIAL AND SEXUAL BEHAVIOUR In captivity males or females often fight, mostly "mock fights", uttering shrill *kek kek* calls at the same time (Ripley 1977), and the species is antagonistic towards Common Moorhens in captivity (Eber 1961). Both sexes defend the territory against conspecific individuals. Copulation in captive birds is described by Wilson & Swales (1958). The male courtship-feeds the female with insect larvae, both birds make low clucking calls and the female walks slowly away with the male walking alongside her, their bodies touching. The female exposes her white undertail-coverts by tilting her tail towards the male, then bows away from the male, placing the tip of her bill on the ground, when the male mounts with outspread wings. After copulation, the male flutters off over the female's head. For further details see Eber (1961).

BREEDING AND SURVIVAL Season On Gough lays Sep-Mar, with peak activity Oct-Dec, also seen copulating in Jun and Aug; on Tristan probably breeds Dec-Mar. **Nest** Circular and cup-shaped; well concealed in dense cover; made of dry tussock leaves, or sticks; bowl diameter 14-20cm; rim 10-15cm above ground (in captivity up to 75cm above ground); built in shallow scrape in ground, or in grass *Poa flabellata* tussock with access tunnel 5-100cm long from edge of tussock to nest. Both sexes build. **Eggs** 2-6 in captivity, 2-5 in wild; narrow oval; almost without gloss; very pale buff to pinkish-cream, marked with spots and blotches of reddish-brown, sometimes concentrated at larger end, and with some evenly distributed pale purple underlying blotches; size of 1 egg (nominate) 49.5 x 33, calculated weight 28.5; of 12 eggs (*comeri*) 49-54.9 x 34.1-37.2 (51.3 [SD 1.5] x 35.4 [SD 0.9]), weight (most eggs well incubated) 22.5-37 (31.1, SD 4.4, n = 11). **Incubation** 21 days (in captivity), by both sexes. Wild birds incubated for periods of 53-200+ min; incubating bird fed by its mate every c. 10 min (Watkins & Furness 1986). **Chicks** Young fed by both parents; seen to be given fresh flesh from bird carcasses (Clancey 1981). 2 broods per season recorded, with c. 14 weeks between clutches; young of first brood help to feed second brood. Egg predation by skuas is recorded. **Survival** The longevity record in captivity is 13 years and a few months; at Amsterdam Zoo between 1964 and 1983 at least 54 young were produced, 17 surviving more than a year (Collar & Stuart 1985).

129 COMMON MOORHEN
Gallinula chloropus   Plate 39

Fulica Chloropus Linnaeus, 1758, Europe; restricted to type locality, England.

Sometimes considered conspecific with *G. tenebrosa*, or to form superspecies, but the two are sympatric in Wallacea (see *G. tenebrosa* species account). Many races described. Twelve subspecies recognised.

Synonyms: *Gallinula parvifrons/pyrrhorrhoa/sandvicensis/galatea/garmani/orientalis/fusca/fistulans/minor/sandwichensis/burnsei*; *Crex galatea/chloropus*; *Stagnicola meridionalis/brachyptera/chloropus/parvifrons/minor*, *Fulica fusca*; *Rallus chloropus*.

Alternative names: Common/Florida Gallinule; Waterhen.

IDENTIFICATION Length 30-38cm; wingspan 50-55cm. A medium-large gallinule, plumage appearing essentially black at any distance, with prominent yellow-tipped red bill and red frontal shield, prominent white line along top of flanks, and white lateral undertail-coverts (buff in Madagascar race *pyrrhorrhoa*) which are rendered obvious by constant flicking of tail. In worn plumage most or all of flank line may disappear, and flank line could also be lost temporarily in moulting birds. Head, neck and upper mantle grey-black; rest of upperparts, and upperwings, dark olive-brown in nominate race, brighter in races from N and C America and Antilles (*cachinnans, cerceris, barbadensis*), or mantle and upperwing-coverts darker and greyer, lacking brown (*meridionalis* of Africa and E races *orientalis, guami* and *sandvicensis*), or entire upperparts dark grey (South American races *pauxilla, garmani* and *galatea*); underparts dark slate; centre of lower belly and vent variably marked with white. Iris red; legs and feet bright yellow-green to yellow, upper half of tibia orange. Sexes similar but female smaller. Old World birds have elliptical shield, widest in middle and with rounded top, New World birds mostly have truncated shield, top almost square and widest near top. Immature similar to juvenile, looking more like adult when plumage worn; gradually attains adult bare part colours. Juvenile has crown, hindneck and upperparts dark brown to grey (greyer in races where adults have greyer upperparts), duller than in adult; underparts pale grey-brown to brown, with white chin and throat and whiter belly; flank line white or buff-white, less obvious than in adult; wings and undertail-coverts like adult; iris brownish, bill and shield greenish-brown, olive-grey or red-brown; legs and feet olive-grey. Birds in postjuvenile moult may lose most or all of flank line. Inhabits freshwater wetlands with mixture of open water and fringing/emergent vegetation.

Similar species For differences from adult and juvenile Dusky Moorhen (130) and Lesser Moorhen (131), see relevant species accounts. Tristan Moorhen (128) superficially similar but smaller, much more strongly built, short-winged, and almost completely flightless. Also differs in having black neck and underparts, white on flanks confined to a few streaks, and red legs heavily blotched greenish-yellow. All ages separable from all sympatric coots by presence of white flank line; also differs from Common (136) and Red-knobbed (135) Coots in having white lateral undertail-coverts; for general differences from coots in structure, appearance and behaviour, see Common Coot.

VOICE Information is from Cramp & Simmons (1980). Common Moorhens have a wide variety of clucking and chattering calls, often harsh and metallic, common to both sexes and sometimes repeated for long periods in American races (but not in Old World races). The most familiar call is the advertising call (crowing), a loud, explosive, single *krrrruk*; it also has similar *kurr-ik* and *kark* calls. The advertising call is often given from dense cover, and calling occurs most frequently during the breeding season, and at sunrise and sunset; in spring, birds may call throughout the night. During hostile encounters this call is often associated with distress calls *keh-keh*. An annoyance call is a loud, harsh, repeated *kik*. It has a soft, musical, repeated *kook* call (the murmur-call) used in various displays of paired birds, when nest-building, and when brooding or calling young. Moderate alarm is indicated by a *cuk* which is lower and louder than the murmur-call and is sometimes rapidly repeated; great alarm or distress (e.g. when captured, fighting or severely frightened) elicits a very loud *keh-keh*. Mild alarm is indicated by the well-known *kittick* call. When defending eggs or young, intense alarm or distress is indicated by a repeated *chuck* call (male) or a weaker *kuck* or *kick* (female), which may be given more loudly (an explosive *kuk* or *berk*); the explosive call softens to a faint *took* as danger recedes. The flight call is described as a soft *tit-a-tit* or *kek-kek-kek*. The contact/food call of the chicks is a shrill *chew* or *keep*; in fright this changes to a *kee-ip* or a harsh, tittering *kik-ik-ik*; a squealing *pitcheese* is given in distress.

DESCRIPTION *G. c. chloropus*
Adult Head, neck and upper mantle dark slate-grey, almost black on head and tinged olive-brown on nape and upper mantle; chin and throat duller. Lower mantle to uppertail-coverts, scapulars, upperwing-coverts and tertials dark olive-brown, tinged dark rufous when fresh and olive-green when worn; tail black, faintly tinged olive-brown and glossed green. Outer upperwing-coverts black, tinged olive at tips and glossed green; flight feathers, primary coverts and alula dull black, tinged olive-brown on outer webs when fresh; narrow white or pale buff line along bend of wing and outer edge of outer primary and outer alula feather (sometimes absent on primary); axillaries dark olive-grey, sometimes tipped or marked white; underwing-coverts dark grey, smaller feathers tinged olive and narrowly edged white. Underparts dark slate, tinged olive-brown on lower flanks and sides of rump; upper flanks with line of white formed by white streaks along upper webs of feathers; central lower belly and vent slate-grey, variably tipped white, white tips may disappear in worn plumage (male). Longer and outer undertail-coverts white, variably pale buff towards base; central shorter coverts black. When fresh, feathers of chin and underparts sometimes narrowly tipped white; when worn, scapulars tipped olive-grey, head and upper mantle slate-black with no olive-brown tinge (see Plate 39). Sexes similar, but female smaller; has feathers of central lower belly and vent more broadly tipped white, usually appearing white or mottled grey and white; when fresh, white tips to lower breast and belly feathers average broader, sometimes not disappearing with wear. Iris crimson; bill and frontal shield bright red, tip of upper mandible and distal half of lower mandible bright yellow; legs bright yellow-green, front of tarsus, upper surface of toes, and lower tibia yellow, upper tibia orange-red, joints and back of tarsus olive-grey; legs and feet sometimes all yellow except for red upper tibia and olive-grey tinge on joints. Outside breeding season iris red-brown; bill and shield dark brown-red and olive-green (late summer to early winter), shield wrinkled and reduced in size; legs duller yellow and olive-green (Aug-Feb). Toes somewhat emarginated.
First breeding Very similar to adult (sometimes inseparable), because of new feathers and because of abrasion of white and brown feather edges of underparts. Crown to upper mantle, lores and sides of head tinged olive-brown, less deep slate than in adult; underparts usually less uniformly slate, often washed brown on breast, with some white mottling on chin or white fringes to some underpart feathers; usually more white on underwing-coverts; white streaks on flanks often shorter and narrower. Bare parts as adult.
Immature Upperparts, wings, tail and tail-coverts as juvenile; underparts dark grey, variably mottled and washed brown and white, sometimes with hardly any brown or white and, when worn, very similar to adult. Chin and upper throat white, mottled grey; sides of head, neck and breast slate-grey, feathers tipped brown to buff; breast to vent slate-grey, feathers broadly tipped white and variably tinged brownish; central belly and vent mainly white; flanks dark olive-brown, feathers tipped buff; white streaks on upper flanks narrower and more irregular than in adult. Iris red-brown; bill and shield red-brown, tip of bill yellow (some birds may show adult bill colour while still in immature plumage); legs and feet assume adult colours by spring.
Juvenile Crown, hindneck, upperparts, tertials and upperwing-coverts dark rufous-brown (less olive than in adult), slightly more olive on hindneck; tail and all tail-coverts like adult; wing like adult but smaller underwing-coverts more irregularly and extensively marked white; white on wing bend irregularly extending to marginal underwing-coverts. Chin and throat off-white; sides of head and neck, and most of breast and flanks, pale grey-brown, feathers often buff at tips; some long flank feathers with indistinct buff to whitish marks or streaks, giving irregular, variably prominent, buff to whitish flank line; centre of lower breast and belly, and vent, off-white or pale buff. Whole underparts appear uniform pale brown with chin and throat white, surrounded by pale buff, and belly and vent pale buff to whitish. With wear, upperparts may become duller olive-green, and underparts mottled pale brown and off-white. Iris olive-grey, becoming grey-brown before shading to hazel and then red-brown; bill dark olive-grey, tip yellow-green; base and shield becoming red-brown and tip yellow; legs and feet dark olive-grey, paler and yellower on front of tarsus and tibia, upper tibia becoming yellow (orange in some males).
Downy young Down black, sometimes with slight green gloss on upperparts; underparts more slate-grey; some white bristles on throat and sides of neck; bare skin of crown rose-red and blue; skin of throat yellow and of wing pink; iris grey-brown, soon becoming dull olive-grey; bill and shield blood-red with orange to yellow distal part and perhaps an indication of dark transverse lines; legs and feet blackish-brown to black, becoming dull olive-grey. Well-developed wing-claw present.

MEASUREMENTS Wing of 13 males 178-194 (185, SD 3.6), of 13 females 169-184 (176, SD 4.0); tail of 13 males 68-79 (74.7, SD 3.7), of 11 females 68-80 (74.1, SD 3.7); bill (including shield) of male 37-44, of female 36-40 (Feb/Mar-Jul/Aug), of male 35-39, of female 31-36 (Jul/Aug-Feb); tarsus of 60 males 49-55 (51.3, SD 1.9), of 78 females 44-51 (47.6, SD 1.6) (Cramp & Simmons 1980). Weight of 74 males 249-493 (339), of 80 females 192-343 (271). Hatching weight 13-17 (Fjeldså 1977) (but see *meridionalis*).

GEOGRAPHICAL VARIATION Racial variation is mostly clinal, mainly involving size, the colour of the upperparts and upperwing coverts, and the size and shape of the frontal shield. Old World birds have an elliptical shield, widest in the middle and with a rounded top, New World birds mostly have a truncated shield, widest near the top which is almost square. The possible races *correiana* (Azores) and *indica* (Arabia E to Japan) are included in the nominate, and *brachyptera* in *meridionalis* (Cramp & Simmons 1980, Urban *et al.* 1986). The possible race *hypomelaena*, here included in *garmani*, may be a *galeata/garmani* intergrade (Fjeldså & Krabbe 1990). See White & Bruce (1986) for a discussion of *orientalis*.

G. c. chloropus (Linnaeus, 1758) – Europe N to Fenno-Scandia (62°-64°N), E across Russia from Leningrad via Cherepovets Region, S Kirov and S Gorkiy to 56°N in Tatar Republic, and in Bashkiria around Ufa region, extending (probably) through Omsk, Kainsk and Tomsk, and to SW Russian Altai Mts. Ranges S to Mediterranean and its larger islands, including Cyprus; breeds throughout Europe except Faeroes, Iceland, Svalbard and Madeira (Hagemeijer & Blair 1997); Azores (Terceira and São Miguel), Canary Is (formerly bred) and Cape Verde Is (Boa Vista and São Tiago); North Africa in Morocco (S to edge of desert), N Algeria, Tunisia, Libya (Fezzan, possibly also Brak and coastal Tripoli), N Chad (Tibesti) and Egypt (Nile Delta, Nile Valley S to Aswan, also Suez Canal, Wadi Natrun and Faiyum); Israel; Arabia (locally); Iraq and Iranian region E and N through Russian Turkestan, Transcaspia and Kazakhstan to SW Altai (see above); also E to W Chinese Turkestan (Xinjiang Uygur region, from Kashi in W, N to Yining and E to Ruoqiang). In Far East, in S Sakhalin, S Ussuriland, Manchuria and Korea; Japan (Hokkaido, Honshu, Shikoku, Kyushu, and throughout Nansei Shoto); S through China (Heilongjiang, Hebei, Shanxi, Henan and Shaanxi, S to Chang Jiang R and through S China W to Sichuan, Yunnan and S Xizang) to Taiwan, Hainan and Indochina, Thailand and the C Malay Peninsula, and W via Myanmar, Bangladesh and Nepal to India, Sri Lanka, Pakistan and Afghanistan. Possibly also breeds from N Mongolia to SW Transbaikalia (Vaurie 1965). N populations winter S to the Mediterranean region (and occasionally Madeira, Canaries and Cape Verde Is); also North Africa in N Morocco, N and S Algeria, Tunisia, Libya (coast, and inland at Hon), Egypt (Nile Delta and Valley S to Aswan, and Kharga oasis), Israel and Arabia; sub-Saharan Africa in Senegal, Gambia, Guinea (Koundara), C Mali (Niger basin), S and N Niger, extreme N Nigeria (Kazaure), C Chad (L Chad to Abéché) and Sudan S to 12°N. Also winters in S Asian range, but wintering areas poorly known and distribution map somewhat conjectural. In SE and E China, populations N of the Chang Jiang R are present only in summer (Cheng 1987). Possibly also winters in N Borneo (Smythies 1968, MacKinnon & Phillipps 1993; but see Marin & Sheldon 1987 and Distribution and Status section); it has also occurred twice in the Philippines, on Mindoro and Luzon (Dickinson *et al.* 1991). There is a clinal decrease in size from NW to SE, birds from Britain and NW Europe being largest and those from India, China and Malaysia smallest. Birds from Britain and from India (putative *indica*) overlap only slightly in wing length (Cramp & Simmons 1980). See Description.

G. c. meridionalis (C. L. Brehm, 1831) – Sub-Saharan Africa: N Senegal (Djoudj NP), Gambia (this race?), C Mali (Niger basin), S Niger, Guinea-Bissau (Bafatá), Guinea (Gbé outfall, this race?), C and N Sierra Leone, Liberia, Ivory Coast, Burkina Faso, Ghana, Benin, Togo, Nigeria (mostly in N), Cameroon, N Central African Republic (presumably this race), Congo (Ile M'Bamou), Zaïre, Rwanda, Burundi, Ethiopian highlands and Eritrea, NW and S Somalia, Uganda, Kenya (except NE), Tanzania, Zanzibar and Pemba Is, Malawi, Zambia, Mozambique, Angola and throughout southern Africa (except some of the dry W and most of the Kalahari); also St Helena (self-introduced c. 1930); formerly bred Pagalu (Annobon) I, now possibly extinct. Smaller than nominate; mantle dark slate; upperwing-coverts slaty blue-grey without olive wash; back, rump and tertials duller and darker olive-brown than in nominate. In young, bill and shield pinkish, legs and feet grey-black (days 19-30); bill, shield and legs become dark olive-green (days 31-45) (Urban *et al.* 1986). Wing of 12 males 156-169 (163), of 5 females 145-170 (161); tail of 12 males 62-77 (70.4), of 5 females 63-73 (70.4); bill (including shield) of 27 unsexed 36-46 (40.4); tarsus of 12 males 44-51 (46.8), of 5 females 45-48 (46.4) (Urban *et al.* 1986, Maclean 1993). Weight of 143 unsexed 173-347 (247). Hatching weight c. 25 (Urban *et al.* 1986).

G. c. pyrrhorrhoa A. Newton, 1861 – Madagascar, Réunion, Mauritius and Comoros Is. Undertail-coverts buff. Lower mantle and upperwing-coverts dark olive-brown (some Comoros birds plain dark slate in these regions). Wing of 32 unsexed 157-190 (170.7) (Benson 1960).

G. c. orientalis Horsfield, 1821 – Seychelles (granitic islands), Andamans, S Malaysia, Sulawesi and Siau, Borneo, Sumatra, Bali, Java (including Kangean Is), W Lesser Sunda Is (Flores, Lombok, Sumba and Sumbawa), Philippines (Basilan, Bohol, Calayan, Catanduanes, Cebu, Guimaras, Leyte, Luzon, Mindanao, Mindoro, Negros, Palawan, Panay and Samar) and Palau Is (Angaur, formerly Peleliu). Smaller than nominate and darker overall, like *meridionalis*; shield relatively large. Wing of male 145-175, of female 140-166; tail of male 65-80; bill (including shield) of male 40-47, of female 37-45; tarsus of male 44-52, of female 42-49 (Ripley 1977, sample presumably includes some *meridionalis*). Weight of 1 male 370, of 1 female 250 (Ripley 1977).

G. c. guami Hartert, 1917 – Mariana Is (Tinian, Saipan and Guam). Similar to nominate but upperwing-coverts darker, almost olivaceous-black; back, rump and scapulars darker and less olivaceous-brown, but paler than in *orientalis*. Frontal shield rounded at top. Unsexed birds: wing 156-180; bill 44; tarsus 47-56 (Ripley 1977).

G. c. sandvicensis Streets, 1877 – Hawaiian Is (Oahu and Kauai; re-introduced to Molokai). Similar to *galeata*, but darker and blacker, less bluish-grey, with shorter wing, larger frontal shield, and more robust tarsus. Lacks white on belly and underwing. Differs from all other races in having crimson blush on front of tarsus, deepening at sides; tibia with bright crimson ring; feet green. Unsexed birds: wing 150-158; tarsus 52-56 (Ripley 1977).

G. c. cachinnans Bangs, 1915 – SE Canada (SE Ontario, SW Quebec, S New Brunswick, and Nova Scotia); E USA (Maine W to SE Minnesota and S to S and E Texas); SW USA (C to S California, Arizona, parts of Nevada, Utah, New Mexico and W Texas); Mexico (throughout); Guatemala and Belize S to W Panama (resident in W Bocas del Toro: Wetmore 1965); also Bermuda and Galápagos Is (San Cristóbal, Santa Cruz, S Isabela, Floreana, and Fernandina). Recorded from Bahamas (Cat I), race not specified (Buden 1987). Most of E population migratory, wintering along coast from North Carolina to Texas, through Mexico and S to Panama, West Indies and possibly South America. Very similar to nominate, but has different shield shape (truncated at top) and relatively long bill, tarsus and toes. In contrast to *galeata*, has middle of back and rump extensively rufescent-brown, this colour also often extending to upperwing-coverts and secondaries. Wing (chord) of 16 males 167-181 (174.5), of 10 females 152-174 (164.2); tail of 16 males 64-86 (72.2), of 10 females 62-71 (66); culmen (including shield) of adults 36-42.5 (39.9); tarsus of adults 45-51 (48.5) (Blake 1977).

G. c. cerceris Bangs, 1910 – Greater and Lesser Antilles; also Tobago. Similar to *cachinnans* but with rufescent-brown of upperparts not usually extending to upperwing-coverts; tarsus longer. Wing (chord) of 15 males 166-186 (174), of 18 females 154-183 (166.1); tail of male 62-75.5 (70.8), of female 63.5-73.5 (67.2); culmen (including shield) of 8 males 46.5-49 (48.2); tarsus of male 52-61 (58), of female 49-59 (52.9) (Blake 1977).

G. c. barbadensis Bond, 1954 – Barbados. Differs from *cerceris* in having much brighter body plumage and paler, less blackish, head and neck; no difference in size apparent (Blake 1977).

G. c. pauxilla Bangs, 1915 – C and E Panama (R Chagres and throughout Canal area); N and W Colombia (Cauca V and lower Magdalena V, and locally in E Andes); W Ecuador (W Esmeraldas to El Oro) and N Ecuador (Ibarra, and formerly Pichincha); and coastal NW Peru S at least to Mejía area of Arequipa. Slightly smaller than *cachinnans*, with upperparts greyish-brown, more a deep plumbeous with at most only slight suffusion of olive-brown or deep olive on middle of back and rump; shield widely expanded, and truncated posteriorly. Similar to *galeata* but has shorter and more slender tarsus and toes. Wing (chord) of 12 males 160-192 (169.5), of 8 females 155-175 (162.3); tail of 8 males 64-70 (68), of 2 females 65, 67; tarsus of 12 males 48-59 (53.2), of 8 females 45-56 (48.5) (Wetmore 1965, Blake 1977, Ripley 1977).

G. c. garmani Allen, 1876 – Peru (Arequipa, Cuzco, Huancavelica, and Junín to Ancash), N Chile (Tarapacá and Tocopilla), Bolivia (La Paz, Potosí and Cochabamba), and NW Argentina (Jujuy and Salta). Larger and darker than *cachinnans*, almost uniformly dark plumbeous, sometimes slightly tinged dull olivaceous on lower back and rump; head and neck slaty-black. Wing (chord) of 10 males 207-225 (219), of 8 females 195-214 (203.7); tarsus of 10 males 57-66 (60.5), of 8 females 52-59 (55.1) (Blake 1977). Bill (including shield) of unsexed 42-50 (Ripley 1977). Weight of 3 males 400, 540, 550 (Marín 1996). Birds from Vacas in Cochabamba, NC Bolivia (proposed race *hypomelaena*) are smaller (Blake 1977).

G. c. galeata (Lichtenstein, 1818) – N Venezuela (Zulia, Carabobo, Aragua, Monagas), Trinidad, Guyana, Surinam, French Guiana (Farez); C, E and S Brazil (Pará [Belém], Ceará, Goiás, Alagoas, Bahia, Minas Gerais, Espírito Santo, São Paulo, Rio de Janeiro, Mato Grosso, Mato Grosso do Sul, Paraná, Santa Catarina and Rio Grande do Sul); Paraguay; Uruguay (throughout); and N Argentina in S Jujuy, N Salta (at Orán), Formosa, Chaco, Tucumán, Catamarca, Mendoza, Córdoba, Santa Fe, Corrientes, Misiones, Entre Ríos and Buenos Aires. Also Bolivia (outside range of *garmani*, at lower altitudes), and presumably this race in lowland NE Ecuador at Limoncocha (Ridgely & Greenfield in prep.). Like *garmani* in having deep plumbeous upperparts, but smaller; middle of back and rump sometimes faintly tinged dull olive-brown or deep olive (Blake 1977). Wing (chord) of 10 males 155-193 (174.5), of 10 females 156-179 (164.9); tarsus of 10 males 42-55 (50.4), of 10 females 45-52 (47.8) (Blake 1977). Bill (including shield) of male 44-48 (46.1), of female 40-50 (44.2) (Navas 1991).

MOULT Postbreeding moult is complete and moult of flight feathers is simultaneous. In nominate (Cramp & Simmons 1980), head, neck, upper mantle, breast and flanks start moulting from late May to late Jul, then simultaneous moult and regrowth of flight feathers, tertials and all wing-coverts occurs from late Jun to late Aug (some in May, others not until mid-Sep). The remaining body feathers are moulted when the wings are fully grown, usually before late Sep but in Oct in some migrants. The tips of some flight feathers and rectrices are sometimes bitten off, or feathers even pulled out, from May onwards. Adult prebreeding moult is partial, occurring in Mar-Jun and involving variable areas of the head, neck, mantle, breast and belly; it does not occur in all birds. Postjuvenile moult is partial, involving the head and body only; it starts before the wings are fully grown, with the head, breast and flanks, and is usually finished in Sep. The first prebreeding moult is also partial and very variable: some birds moult direct from juvenile to first breeding, but usually the immature non-breeding plumage is interposed, followed by the partial moult to first breeding. It sometimes starts in Sep-Oct, but often not until Mar-Jun after the break in moult during migration and winter; it is usually limited to a variable number of scattered feathers on the head, neck, upper mantle, breast and flanks; some birds also moult other areas, but not the central belly, vent, and some wing-coverts and tertials. This moult gives rise to a first breeding ('second immature') plumage which is variably distinct from adult breeding plumage.

In *meridionalis* (Fagan *et al.* 1976, Urban *et al.* 1986), the remiges and rectrices are moulted simultaneously at the peak of breeding: flightless birds are reported from Pemba (Sep), Namibia (Jan), and in South Africa from the former Transvaal (Sep-Mar) and the W Cape (Jan-Apr). Primaries grow 5.0-5.3 mm/day and moult is completed in 20-27 days. An adult with half-grown remiges had a 2-week-old chick; a moulting (continuously breeding) female laid and incubated eggs and at the same time tended young of a previous brood (Siegfried & Frost 1975). In the Transvaal, body feathers are moulted twice a year, in Aug-May, this moult lasting over 6 weeks.

DISTRIBUTION AND STATUS British Isles through Europe, Russia and S Siberia to Ussuriland and Japan; S through Africa and Madagascar, the Middle East and Arabia, C, S and SE Asia to the Philippines; also Hawaii, N and S America and the West Indies. N populations winter S to Mediterranean, N and W Africa, Middle East, S Asia, and S USA to Central America (possibly South America). Not globally threatened; most races are at least locally common. The nominate race is generally common throughout its range, including in NW Europe, where populations fluctuate markedly due to hard winters; an overall increase has occurred in Great Britain, and in Fenno-Scandia where it first bred in Denmark, Norway, Sweden and Finland in the mid-19th century (Cramp & Simmons 1980); it remains rare in Finland (c. 150 pairs in S and SE), is more numerous in SE Norway, and is now widespread in Denmark (Hagemeijer & Blair 1997). Its distribution is limited altitudinally and latitudinally by its intolerance of freezing conditions (Hagemeijer & Blair 1997). It has a total European population of 904,000-1,161,000 (geometric mean 992,800) in 38 countries, a total Russian population of 10,000-100,000 (31,600) and a European Turkish population of 5,000-20,000 (10,000), with the highest numbers in the United Kingdom (250,000) and France (100,000-300,000, geometric mean 175,000); c. 70% of its European population occurs in

Britain, Ireland, the Netherlands, France and Spain; Great Britain is a particularly important area for this bird (Hagemeijer & Blair 1997). Overall in Europe it is secure, with stable or fluctuating populations (1970-1990) in 26 countries, declines in 8 and increases in 5; as it is multibrooded its numbers can build up rapidly following reductions due to severe winter weather (Hagemeijer & Blair 1997). The possible race *correiana*, of Azores (Terceira, possibly Fayal and São Miguel), is limited to 20-30 breeding pairs (Hagemeijer & Blair 1997).

Its breeding population in Israel is c. 1,000-2,000 pairs, and midwinter numbers may exceed 5,000 birds (Shirihai 1996). In Arabia it is locally common and is becoming more widespread as it expands its range to occupy newly created waters (Jennings 1981, Hollom *et al.* 1988). In N Africa it is widespread, but it is rather scarce in Tunisia; in Egypt it is common in the Nile Delta (several thousand breeding pairs), and is abundant in winter, with 20,500-25,000 being shot annually on L Manzla; it is uncommon on passage in Mauritania (mainly coastal), common in winter in Senegal, common in Mali (where it occurs alongside the race *meridionalis*, with 2,500 at L Fati in Jan), uncommon in Gambia, and rare in N and C Sudan (Cramp & Simmons 1980, Lamarche 1980, Gee 1984, Urban *et al.* 1986, Nikolaus 1987, Goodman *et al.* 1989, Gore 1990).

In Azerbaijan it is a common breeding, wintering and migrant bird, but total numbers have not been estimated (Patrikeev 1995). It is abundant in Kashmir, locally common in Nepal, locally abundant in Pakistan, local in Bangladesh and common in Sri Lanka, while in the Malay Peninsula it was formerly locally distributed (Medway & Wells 1976, Ali & Ripley 1980, Harvey 1990, Inskipp & Inskipp 1991, Roberts 1991, Kotagama & Fernando 1994). In Myanmar it is widespread but local, and breeds in C and S regions (Smythies 1986). In Thailand it is a locally common resident (Lekagul & Round 1991). It is common in Vietnam (Wildash 1968; Stusák & Vo Quy 1986) but is apparently only a straggler to Korea (Austin 1948). In Japan it is fairly common throughout the Nansei Shoto, Kyushu, Shikoku and S Honshu, and a summer visitor to N Honshu (common) and Hokkaido (uncommon) (Brazil 1991). Data from the Asian Waterfowl Census (Perennou *et al.* 1994) suggest that the nominate race (including putative *indica*) is widespread and generally common in SW, S, E and SE Asia; population trends are unclear but S Asian populations may be stable.

The race *meridionalis* is patchily distributed and uncommon to locally abundant in sub-Saharan Africa. It is recorded as locally common to abundant in Mali and S and E Zaïre (E highland lakes), locally common in East Africa (mainly E Uganda, W and C Kenya and N Tanzania) and coastal Angola (but scarce in the interior), frequent in the Ethiopian highlands, not uncommon in Nigeria, widespread in Ivory Coast, uncommon in Sierra Leone, Liberia, Ghana and Zambia, scarce and scattered in NW and S Somalia, and rare in Benin (Lamarche 1980, Ash & Miskell 1983, Pinto 1983, Thiollay 1985, Urban *et al.* 1986, Gatter 1988, Lewis & Pomeroy 1989, Elgood *et al.* 1994, Claffey 1995, Field 1995). It is supposedly rare in Cameroon (Louette 1980) but is not uncommon at Yaoundé (Fotso 1990). It is recorded as common in Bamingui-Bangoran NP, N Central African Republic, in Feb-Mar, race not given (Green 1983), and as frequent at Bafatá, C Guinea Bissau, in Mar (Rodwell 1996). Its status in Guinea is unclear: a record from the Gbé outfall, SE Guinea, in Apr (Walsh 1987) may be this race, but Morel & Morel (1988) give a record of 3 nominate birds in Jan at Koundara. In N Senegal it has bred at Djoudj NP (Rodwell *et al.* 1996). It was formerly locally abundant in Rwanda (Chapin 1939). In southern Africa it is widespread and locally common in Zimbabwe, S Mozambique, N Botswana, Namibia and South Africa; 6,000+ breeding pairs may occur on the Nyl R floodplain in years of good rainfall (Urban *et al.* 1986, Tarboton *et al.* 1987, Clancey 1996, Taylor 1997c). It is increasing locally as a result of artificial dams and ponds, e.g. in Malawi (Benson & Benson 1977). It is apparently self-introduced to St Helena, where it is well established as a breeding bird (Olson 1973a). The race *pyrrhorrhoa* is abundant throughout Madagascar (Langrand 1990).

The race *orientalis* is common in the Philippines, on Flores and elsewhere in Wallacea, and in Sumatra (where it is still sold for meat in street markets in Medan), Sabah, Java and Bali, but rare and endangered in Palau (Pratt *et al.* 1980, White & Bruce 1986, Marín & Sheldon 1987, van Marle & Voous 1988, Dickinson *et al.* 1991, MacKinnon & Phillipps 1993, Coates & Bishop 1997). In Borneo there is an old breeding record from Bangkau L in the S, and the species probably also breeds in the Kelabit uplands (Smythies 1981). Immature specimens from Sarawak and Satang I in Nov may be migrants from mainland Asia (Smythies 1968, MacKinnon & Phillipps 1993), but the species was common in Sabah, N Borneo, in 1983 and a nest was found (Marín & Sheldon 1987).

In the USA (Greij 1994) *cachinnans* is widespread and locally rare to abundant; in the E it is most abundant in coastal areas from S Texas to North Carolina, and from Maryland to Maine, and common elsewhere except in centrally located E states, where it is generally rare to uncommon; in the W, it is locally common in parts of Arizona, rare to locally common in New Mexico, rare in Nevada and Utah, and rare to uncommon in Texas. In North American ricefields birds do not significantly damage the crop but rice harvesting is harmful to nests and young broods, especially in early maturing strains of rice (Greij 1994). The estimated USA gallinule harvest – most data refer to Common Moorhen rather than to American Purple Gallinule (122) averages 44,500 per year (1977-1992), 77.5% of these birds being taken in Louisiana. Habitat loss and degradation significantly affect this species, and management priorities include identifying key habitat types for breeding and wintering birds, and implementing conservation procedures for large and small sites; it is also necessary to obtain better data on moorhen population sizes and trends, and more accurate harvest estimates, and to investigate the origin and distribution of harvested birds; a reduction in bag limit should be imposed if some populations are found to be under intense pressure (Greij 1994). This race is common and widespread in Mexico (Howell & Webb 1995b), uncommon and local in Costa Rica (Stiles & Skutch 1989) and was formerly common in Honduras (Monroe 1968); in Guatemala it was formerly uncommon and local in winter, mainly on the Pacific slope (Land 1970).

The race *sandvicensis* of Hawaii was formerly present on all 5 major islands but is now restricted to a few hundred birds on Oahu and Kauai (Byrd & Zeillemaker 1981) and is ENDANGERED; in 1983 6 were released on Molokai (where the last sighting had been in 1973) but the status of these birds is not known and the lack of good habitat on Molokai precludes the development of a significant

Common Moorhen

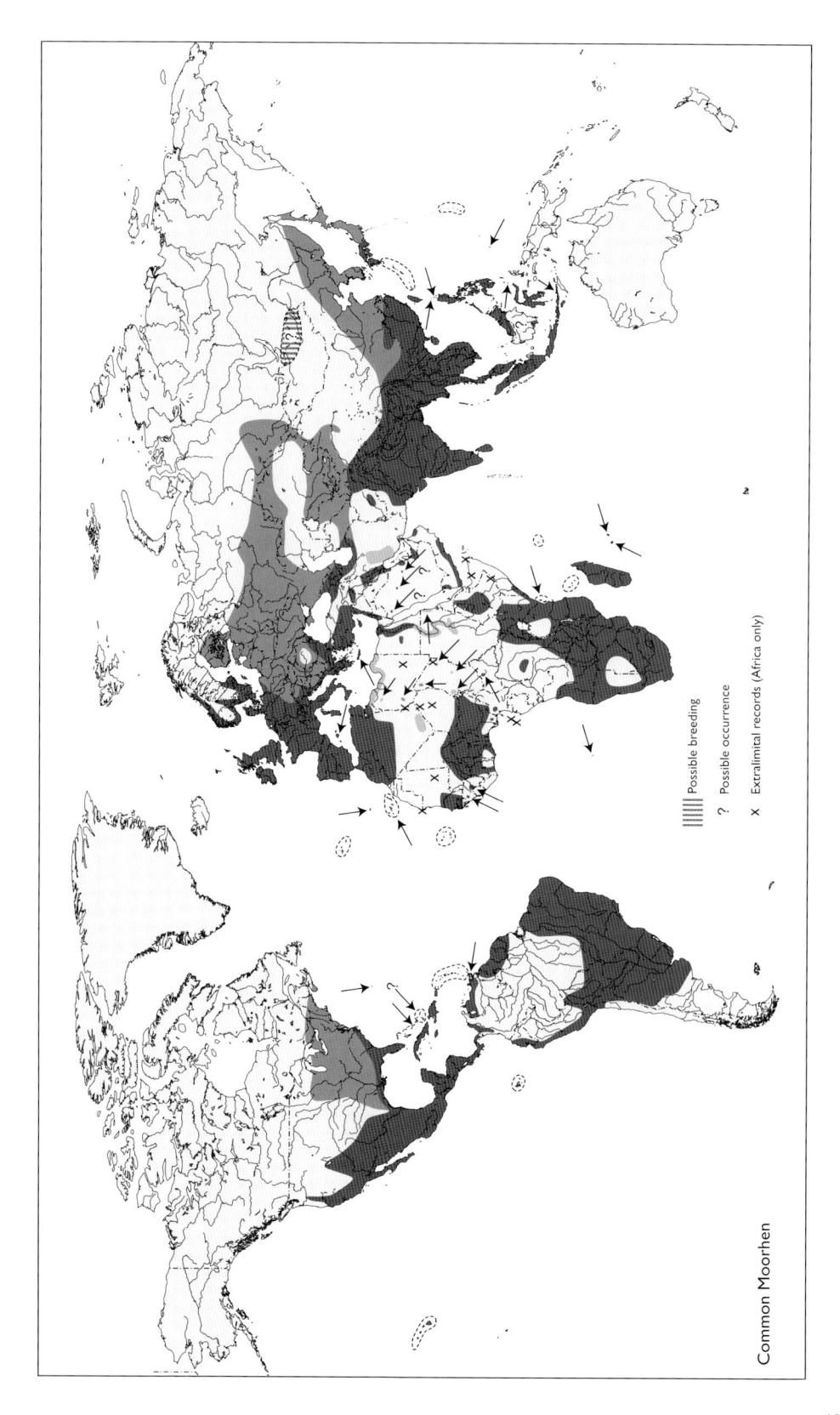

population (Greij 1994). The Mariana Is race *guami* originally occurred on Tinian, Saipan, Guam, Pagan and Rota but is now confined to Tinian (20-125 birds), Saipan (60-120) and Guam (100-200) in greatly reduced wetland habitats; it is also ENDANGERED (Anon. 1992b, Greij 1994). Its decline, which has occurred since 1945, has been caused principally by habitat loss and degradation, and its remaining habitats are still subjected to these threats, as well as potential threats from introduced predators and possibly from poaching; recovery plans include the protection and management of 240ha of freshwater wetland habitat on Guam, 120 ha on Saipan and 30ha on Tinian (this includes improving secondary habitats and developing additional habitats), and the attainment of bird densities of at least 2.5/ha in these wetlands for 5 consecutive years (Stinson *et al.* 1991, Anon. 1992b). Encouragingly, a land developer recently agreed to build small wetlands on a tourist development property in SC Guam, and a 600-m^2 wetland, built in Jan 1992, was quickly occupied by 2 moorhens which produced chicks in Aug of the same year (Ritter & Sweet 1993).

The race *cerceris* is common in Puerto Rico, but less so in the Virgin Is (Raffaele 1989); its current status elsewhere is unclear. The race *pauxilla* is locally common in Panama, where it is probably increasing in the Canal area (Ridgely & Gwynne 1989), local in Colombia but fairly common in the Cauca V (Hilty & Brown 1986), and locally common in W Ecuador (Ridgely *et al.* in press). The race *garmani* was formerly abundant at high-altitude lakes in the Peruvian Andes at Junín (Harris 1981) and was recently discovered at Chacance, Tocopilla province, Chile (Howell & Webb 1995a). The race *galeata* is rather rare and local in Surinam, and rare in French Guiana where it occurs on the coast (Tostain *et al.* 1992, Haverschmidt & Mees 1995); it was formerly local in Venezuela (Meyer de Schauensee & Phelps 1978). It is common in the Alto Chaco of W Paraguay, but uncommon elsewhere in the country (Hayes 1995), and in Brazil it is generally abundant (but scarce in the extreme N) in Rio Grande do Sul (Belton 1984), and common throughout Santa Catarina (Rosário 1996). It is abundant throughout Uruguay (Arballo & Cravino in press), and is locally common to abundant in Argentina (e.g. Contreras *et al.* 1990, Navas 1991). In South America it is very numerous in certain wetlands (Fjeldså & Krabbe 1990).

This species readily exploits newly created habitats and is tenacious of occupied areas, not being displaced easily by changes or human disturbance; it adapts well to situations at urban sites, where total acceptance of human presence is necessary in order to occupy suitable habitat of tree-lined ornamental waters and lawns (Cramp & Simmons 1980). In South Africa it has benefited from the construction of artificial waterbodies, especially in KwaZulu-Natal and the SW Cape (Taylor 1997c).

Predators of adults are not specifically recorded, but presumably include most of those which take the eggs and young (see Breeding and Survival). Bad weather may cause significant mortality: in Azerbaijan, L Aggel lost 95% of its moorhen population after 40 days of chill and snowstorms during the severe winter of 1963/64 (Patrikeev 1995).

MOVEMENTS It is normally a resident species, but in the N parts of its range it is forced to migrate because of its vulnerability to freezing. In the W Palearctic (Cramp & Simmons 1980) it is resident or dispersive in the S and the extreme W and is partially migratory to migratory elsewhere, the extent of its movements increasing further to the N and E. It is almost entirely a summer visitor to Finland and the former USSR, while relatively few remain in N Scandinavia, N Germany and Poland in the winter. N European birds winter S to Iberia, Italy, the Balkans and N Africa, with most ringing recoveries from France and Spain; 1 bird ringed in E Germany in Sep moved 1,520km to Spain in 5 days. Autumn movements are mainly SW from N and W Europe and S to SE from C Europe, the main passage being in Sep-Dec, after the flightless moult; there is no evidence for a separate moult migration. Juveniles disperse from Jul and movements of up to 570km are recorded in that month. Swedish and Danish birds winter mainly in W maritime countries such as Ireland, Britain, and NW Germany to Spain, while birds from Bavaria move to SE France, Italy, the former Yugoslavia, and Greece. Winter flocks break up in Feb, and return passage occurs in Mar-May. British and French populations are resident but both countries receive winter migrants from NW Europe.

In Israel, autumn movements occur from mid-Jul to end-Nov (mostly mid-Oct to mid-Nov) and return movements from end-Feb to end-Jun (mainly mid-Mar to end-Apr) with wintering birds present from Oct-Nov to Feb-Mar (Shirihai 1996). It is a passage and winter visitor to Oman in Jul-Apr, most birds occurring in autumn (Gallagher & Woodcock 1980). In Egypt, passage and wintering birds occur from mid-Sep to mid-Apr (Goodman *et al.* 1989). Migrant nominate birds arrive in North Africa in Sep-Dec, many overwinter but some continue down the Mauritanian coast (Port Etienne and Banc d'Arguin, Oct-Nov, Feb and Apr) or down the Nile to Sudan, others cross the Sahara on a broad front (migration is recorded at oases in Algeria, Libya, Niger and Chad); return passage through North Africa occurs in Mar-May (Gee 1984, Urban *et al.* 1986). Claimed nominate migrants are recorded from the N Aïr NP, C Niger, in Aug-Oct (Newby *et al.* 1987) but the period appears to be rather early; other presumed migrants occur in the N of Niger in Oct-Apr (Giraudoux *et al.* 1986). Palearctic birds winter in Subsaharan Africa in Senegal (Feb-Mar, many being immatures), Gambia (Nov-Apr) (Gore 1990), Mali (arrivals Nov-Dec), N Nigeria, C Chad (L Chad basin, Nov-Mar) and N Sudan (Oct-Mar) (Lamarche 1980, Urban *et al.* 1986). Danish-ringed birds have been recovered in Algeria (Feb) and Morocco (Nov), and a Dutch bird in SW Morocco (Jan); regular spring and autumn Mediterranean crossings are indicated by the bird's fairly common occurrence during those periods on Malta and Cyprus (Cramp & Simmons 1980, Urban *et al.* 1986). Some dispersal of locally breeding birds occurs in winter in at least Tunisia and Algeria (Urban *et al.* 1986).

The winter range of emigrant Russian birds, and the extent of dispersal by Near East birds, are unknown, and E birds are relatively secretive and thus more easily overlooked than are those in the W Palearctic; however, numbers apparently increase in winter in wetlands of the Near East and N Middle East; breeding birds apparently leave the C plateau of Turkey in winter (Cramp & Simmons 1980). Passage is noted in Azerbaijan in Mar-Apr and Oct-Nov (Patrikeev 1995). Numbers in India and Pakistan are greatly augmented by winter visitors, and passage is noted in Oct and Mar-May (Ali & Ripley 1980), while migrants occur in Nepal in Oct-May and passage is also recorded in Sep (Inskipp & Inskipp 1991). Migrants occur in Japan (Hokkaido and Honshu) from Mar-Oct (Brazil 1991), and birds from Japan are known to reach the Philippines (Dickinson *et al.* 1991).

The African race *meridionalis* is mainly sedentary, with at least local movements in response to changing conditions, e.g in Sierra Leone and East Africa (Urban *et al.* 1986, Field 1995). At a site near Mombasa, Kenya, numbers of non-breeding birds were high in Oct-Jan, with a increase in Feb-Mar followed by the departure of most birds in Apr before the breeding period at the site (Taylor 1985b). In Zimbabwe some may disperse to the major river systems and the SE lowveld in the rains, and it breeds at temporary wetlands e.g. on the Nyl R floodplain, former Transvaal, South Africa (6,000+ pairs recorded) and in NE Namibia (Tarboton *et al.* 1987, Taylor 1997c).

In North America (Greij 1994) migrants arrive in the N parts of the US breeding grounds (Wisconsin to Pennsylvania) in Apr-May and in Michigan until early Jun; late arrivals are thought to be young birds. Departures are recorded in Sep-Nov (e.g. Chapman 1966). Some seasonal altitudinal movements occur in W populations (Arizona and New Mexico), with birds wintering below 1,000m, and also some migration, the extent of which is not known (they are reported to winter in Baja California and W Mexico: Ripley 1977). Migrants occur in Mexico from Oct to Mar (Howell & Webb 1995b), possibly in Costa Rica (Stiles & Skutch 1989), and in Colombia from Oct-Apr (Hilty & Brown 1986). In South America a presumed winter influx from the S was noted in Rio Grande do Sul, Brazil, where many birds were present in Jul; 1 ringed bird was recovered 100km N in Santa Catarina (Belton 1984).

The Common Moorhen's success in colonising a high proportion of suitable, often scattered, habitats, indicates effective prospecting on the wing, largely at night and in lower airspace (Cramp & Simmons 1980). The nominate race is accidental to Iceland and Spitsbergen; *meridionalis* is a vagrant to NE Chad (Cramp & Simmons 1980, Urban *et al.* 1986) and, although it is regarded as a vagrant to São Tomé, 7 birds were seen there at 5 localities in early Apr 1987 (Eccles 1988); *cachinnans* is accidental to Greenland and the Commander Is (AOU 1983). This species also reaches Tristan da Cunha, aided by W winds from South America (Sick 1993), and a juvenile has been captured on Ascension I, where information from islanders suggests that this species, as well as Allen's Gallinule (123) and American Purple Gallinule, may be of fairly regular occurrence (Olson 1973a).

HABITAT Most information is from Cramp & Simmons (1980) and Greij (1994). It exploits a wide range of natural and manmade eutrophic freshwater wetlands with fringing (often emergent) vegetation, occurring on both still and moving water; the species of plants available may not be as important as having a robust growth of emergents. It is tolerant of wide range of climatic conditions but is vulnerable to freezing; however, in Pakistan it has been observed at -12°C in mid-Feb (Roberts 1991). It occurs on rivers (normally slow-flowing), oxbow lakes, streams, canals, ditches, lakes, dams, pools, ponds, disused gravel pits, ricefields, swamps and marshes, and at the seepage zones of irrigation barrages (Roberts 1991); it also occurs at sewage ponds and at seasonally flooded sites such as floodplains (e.g. Taylor 1997d). It normally avoids oligotrophic or saline situations, but it inhabits brackish waters on the Namibian coast (Cramp & Simmons 1980), in Saudi Arabia (Jennings 1981) and in the Galapagos (Harris 1973), and it occurs in mangroves on Puerto Rico (Raffaele 1989). Larger, more open lakes, even those with suitable marginal cover, tend to be relatively less frequented than small to medium-sized waters (Ripley 1977). It requires ready access to some open fresh water with adequate plant cover, but it readily occupies sites with only small areas of open water, such as ponds and pools only a few metres across, and narrow waters such as canals and ditches. In the USA, preferred breeding habitat is characterised by dense stands of robust and persistent emergents with openings giving an approximately equal interspersion of cover and open water; lowest breeding densities occur in wetlands with nonpersistent emergents, or with mixed scrub/shrub and persistent emergents, and seasonally flooded or other sites with a high proportion of open water. Plant species with which it is associated in the USA include *Typha*, *Phragmites*, *Spartina*, *Zizaniopsis*, *Panicum*, *Paspalum*, *Sagittaria*, *Sparganium*, *Pontederia*, *Alternanthera*, *Hydrocotyl*, *Brasenia* and *Nuphar*; in S and C Africa it is associated with *Typha*, *Phragmites*, *Polygonum*, *Scirpus*, *Schoenoplectus*, *Eleocharis*, *Bolboschoenus*, the larger *Cyperus* species such as *C. dives* and *C. fastigiatus*, the alien *Sagittaria*, and robust emergent grasses such as *Oryza*, *Panicum* and *Echinochloa* (P. B. Taylor unpubl). In India it also occurs in lotus *Nelumbo* (Ali & Ripley 1980). It makes good use of waterside trees and bushes (e.g. *Tamarix*) for resting, roosting and sometimes nesting. It prefers waters sheltered by woodland, bushes or tall emergent plants, and avoids very open sites, especially those exposed to wind and wave action.

It occurs alongside the Common Coot at wetland margins but is less prone to venture far out into open water. However outside the breeding season it may congregate, sometimes in large numbers, on sheltered lakes and ponds (Ripley 1977). When foraging it freely ranges into adjacent drier areas such as meadows, agricultural land or grassland. On migration and in winter it also occurs in damp fields away from water (Shirihai 1996); in Azerbaijan migrants occur along the seashore, and at channels and small ponds, and wintering birds can occur in flooded blackberry *Rubus* thickets (Patrikeev 1995). It normally occurs most commonly in lowlands, but reaches 1,700m in Switzerland, 2,300m in Madagascar, 2,400m in Kashmir, 3,000m in Kenya, 4,200m in Argentina, Bolivia and N Chile, and 4,575m in Nepal (on passage).

FOOD AND FEEDING Most information is from Cramp & Simmons (1980) and Greij (1994), with some data on animal food items from Beltzer *et al.* (1991). Omnivorous; the proportions of plant and animal foods in the diet vary. Plant foods include filamentous algae; moss; vegetative parts of *Lemna*, *Wolffia*, *Potamogeton*, *Juncus*, *Phragmites*, *Bolboschoenus*, *Veronica*, *Hydrilla*, *Ceratophyllum* and grasses; seeds of *Typha*, *Sparganium*, *Potamogeton*, *Carex*, *Scirpus*, *Bolboschoenus*, *Rumex*, *Polygonum*, *Nymphaea*, *Nuphar*, *Ranunculus*, *Ulmus*, *Zizaniopsis* and various cereals; flowers of *Eichhornia*; berries of *Taxus*, *Rubus*, *Sorbus*, *Rosa*, *Crataegus*, *Rhamnus*, *Hedera*, *Sambucus* and *Hippophae*; and various orchard fruits (plums, apples, pears). Animal foods include earthworms; molluscs (*Limnaea*, *Planorbis*, *Helix*, *Arion*, *Agriolimnax*, *Hygromia*, *Planorbella*, *Asolene*, *Marisa*); crustaceans; adult and larval insects, especially Odonata, Ephemeroptera, Hemiptera (including Belistostomatidae, Nepidae and Gerridae), Trichoptera, Orthoptera, Coleoptera (including Curculionidae, Hydrophilidae and Dytiscidae), Lepidoptera and Diptera (including Chironomidae); spiders and harvestmen; small fish;

tadpoles; and occasionally eggs of birds. It also takes carrion (birds and fish), rubbish (vegetable scraps), duck food and fish food. In Britain, 10 stomachs contained 75% plant food and 25% animal food by volume; stomachs of 8 winter birds feeding in grassland contained 38% grain, 25% seeds, 24% annelids and 11% molluscs, and of 11 away from the grassland 32% grain, 30% seeds, 13% annelids and 22% molluscs. In the USA, plant/animal percentages were 80/20-93/7 (Florida: Greij 1994) and 75/25 (Ripley 1977); in Puerto Rico 97/3 (Wetmore 1916). Two studies in Argentina (Beltzer *et al.* 1991, Lajmanovich & Beltzer 1993) showed that seeds constituted the great majority of food items in stomachs throughout the year, percentages of food items by season being as follows: spring, seeds 69.5% and 88%, molluscs 21% and 7.5%, insects 9.5% and 4.5%; summer, seeds 89%, molluscs 10%, insects 1%; autumn, seeds 98%, insects 2%; winter, seeds 95% and 89%, molluscs 1% and 8.5%, insects 4% and 2.5%; the species was found to be an opportunistic feeder.

It feeds while swimming and while walking on floating vegetation, either in cover or in the open; it also feeds on land, grazing and gleaning over open grass or on damp earth in arable farmland, usually near cover. In water it obtains food by dipping the head, surface sifting, upending and rarely by diving. It gleans insects, seeds and fruits from the ground and from plants; it often clambers over leaves and stems, and it climbs and perches well. It may be opportunistic, feeding on the most abundant food types (e.g. Mulholland 1983). It kleptoparasitises other birds, such as Great Crested Grebe *Podiceps cristatus*. In Britain 6 nestling stomachs contained 61% plant material, 23% annelids and 16% insects.

HABITS Most information is from Ripley (1977) and Cramp & Simmons (1980). Although normally diurnal, going to roost at dusk, it is sometimes active on moonlit nights. It has a rather high-stepping walk, with the head raised and the tail horizontal or slightly depressed. The tail is cocked and flicked frequently when the bird is nervous or excited, usually being jerked up strongly and lowered more slowly. Tail flicking in this species, as in the Purple Swamphen (120), is directed towards potential predators rather than towards conspecifics, and is an alertness signal with a pursuit deterrent function rather than a warning to conspecific individuals (Alvarez 1993). It climbs freely in reeds, swamp vegetation and woody plants, walking and balancing adeptly on thin branches and twigs; it perches freely and often roosts in bushes and low trees. It swims well and buoyantly, with the head raised and moved backwards and forwards, and the tail often flicked, and it will dive to evade pursuit. Under water it uses the legs for propulsion, with assistance from the wings, and it can remain underwater by holding onto vegetation with the bill (Bent 1926). When danger threatens birds may remain submerged, usually below banks or among plants, with only the bill visible, and may then slowly expose the head or the back while waiting for the danger to pass. When disturbed on land it runs for cover with lowered head and neck, but it is normally far less secretive than are crakes and rails and it swims around freely in open water, although usually within easy reach of cover. On urban waters it becomes indifferent to human presence, but it can be extremely secretive if harassed or persecuted. It normally takes off with an effort, and short flights appear laboured, with the legs dangling before being held straight out to project beyond the tail. It often flies for only a short distance before dropping into cover or onto water, but is capable of strong and sustained flight and at night may make flights, to some height, which are not associated with migration (Ripley 1977).

During heavy rain an incubating bird was seen to cover itself repeatedly with a sheet of polythene, in the manner of a cape, and to remove it when the rain stopped; this appeared to be customary behaviour for the bird in such circumstances (Hawkins 1970). Birds roost in bushes and trees, on the ground in reedbeds, or (when breeding) on old display platform or nest; winter flocks may roost in meadows if undisturbed.

SOCIAL ORGANISATION Most information is from Cramp & Simmons (1980) and recent studies by S. McRae (McRae 1994, 1995, 1996a, b, 1997, McRae & Burke 1996). Monogamous and territorial, although young in summer, and young and adults in winter, may form feeding groups of up to c. 30 birds, especially in hard weather. The territory may be permanent, and the pair bond may persist only during the breeding season but is sometimes maintained for several years. Pair formation in nominate and *cachinnans* normally takes place after winter flocks break up or after arrival on the breeding grounds (Cramp & Simmons 1980, Greij 1994). Polyandrous trios of 1 female and 2 males also occur at a low frequency, while cooperative nesting of 2 or more females (always mother and daughter) mated to 1 male is also recorded, in both nominate and *meridionalis*, and intraspecific brood parasitism regularly occurs (Gibbons 1986). In a recent study by S. McRae, most cooperative breeding groups consisted of a core pair and their previous offspring; in many cases a daughter laid in a communal nest with her mother and, if there was only one male in the group, he gained full paternity of his daughter's offspring, which had a low survival rate because of inbreeding (McRae 1996a). Furthermore, brood parasitism regularly occurred: females laid eggs in neighbours' nests, often in addition to laying their own clutches. Parasitic eggs laid before the host had started to lay were destroyed, those laid on day 1 or 2 of the host's laying period sometimes resulted in desertion, and those laid after the host's 4th egg were accepted, the costs of desertion being too high at this stage (McRae 1995). It was considered that (a) cooperative breeding might be the only breeding option for first-year females that hatch late and overwinter on their natal territories, and (b) that brood parasitism was a highly opportunistic reproductive strategy used by females who could produce more eggs than they could rear, or by those whose nests had been destroyed, or by "professional parasites" who achieved little reproductive success (McRae 1994, 1996a). The rate of brood parasitism almost doubled in the third year of the study, when rodent nest predators were not controlled and the nest predation rate increased from 10-13% to 65%; these results, with other studies, indicate that high rates of brood parasitism can occur out of constraint under unfavourable conditions (low nest success rates) and as a bonus under very favourable conditions (high nest availability, from studies of other species) (McRae 1997).

Nests may be as little as 8m apart, and the territory as little as 122m^2 in area. In artificial habitats up to 5 territories/ha are recorded but more normal densities across farmland in Britain average c. 0.03 territories/ha (Gibbons 1993). In the winter, birds defend only a core area of the former breeding territory, often feeding beyond the boundary. On streams through agricultural grassland in England breeding territories comprised 80-220m of waterway and the immediate area of the banks; in winter, c. 40 (33-47) m of waterway were defended, plus adjacent land up to 1m from the banks (Wood 1974). In the USA (Greij 1994) nesting densities vary from 0.2-4.6 pairs/ha, with 10 nests/ha also recorded; home ranges of nesting birds were estimated at 0.21-3.2ha (1.22, n = 12) and of non-nesting birds at 5.2-6.01ha (5.61, n = 2), while juveniles used 0.61-17.75ha (6.76, n = 6). Petrie (1984) found that the weight of male moorhens was highly correlated with territory size, suggesting that heavier males have a greater resource-holding potential, i.e. are better able to defend a territory; this is supported by the observation that heavier moorhens were more likely to win contests in winter flocks.

SOCIAL AND SEXUAL BEHAVIOUR Most information is from Cramp & Simmons (1980). In flocks a social hierarchy may be seen, males being dominant to females and adults dominant to immatures; for further information on feeding hierarchies, see Drost (1968). The charging attack (see below) is used by flocking birds as a spacing mechanism and in the establishment of the social hierarchy, and higher-ranking birds have larger shields than others. Displays are very similar to those of the Common and American (138) Coots. The low posture, with head and neck depressed and showing the red shield against the black head feathers, and swanning (half opening the wings), are important in aggression. In courtship, lowering the head to make the shield inconspicuous (bowing) occurs. Tail-raising is accompanied by fanning of the white undertail-coverts; when alarmed, the head and neck are stretched vertically.

An intruder is usually expelled from the territory by the charging attack, the defender swimming or running towards the intruder with head and neck low, and with shield-showing. This may develop into a splattering attack, with the wings flapping; the trespasser flees in the splattering retreat, a similar posture but with head and neck raised. The charging attack is also used against other species, e.g. Redheads *Aythya americana* and garter snakes *Thamnophis* sp. (Fredrickson 1971). At the territorial boundary neighbours may perform the hunched display (see Common and American Coots) with mutual retreat. Fighting is common, but usually occurs only during the establishment or expansion of a territory; birds usually fight only others of the same sex, but a pair may combine to attack a single intruder. Contestants face each other in an upright posture (upright challenge display), with neck raised, wings closed and tail level with the body; if neither gives way, breast to breast combat ensues, each bird striking violently with the feet, clawing at the other's breast and grappling, often while sitting back on the tail with the wings open; on land, the wings are often flapped, raising the combatants up to 1 m above the ground. On water, each appears to be attempting to drag the other below the surface. Fighting is often followed by the hunched display with mutual retreat.

Parental aggression to offspring is recorded: adults "tousle" chicks by grabbing the chick's head or neck in the bill and shaking the chick from side to side; larger chicks, which can monopolise feedings by reaching parents ahead of smaller siblings, are preferentially tousled, after which small chicks are fed more often (Leonard *et al*. 1988). Unlike the Common Coot, which apparently tousles chicks to discourage them from accompanying the aggressive parent (Horsfall 1984), the Common Moorhen does not use the behaviour to maintain brood division; its effect is to reduce demands by chicks for feeding and it appears to reduce sibling competition and to encourage chick independence (Leonard *et al*. 1988).

In response to the intrusion of a species too large to attack (e.g. man), a pair with a nest or young often flee and hide in vegetation, or adopt a defensive swanning display (see American Coot), which is like the hunched display but with the tail not raised and the undertail-coverts not fanned; the birds also sometimes perform the water-churning display, splashing water up by slapping the feet against the surface (see American Coot); both displays are frequently accompanied by the distress call (see Voice), usually with continuous tail-flicking. Terrestrial foot-slapping is also reported as a nest defence by *meridionalis* and *cachinnans*, on a branch or a brood-nest, producing a sound audible over 10m.

This species is unusual in that females compete for males rather than vice versa. In a study of nominate birds, Petrie (1983) found that the heaviest females win most agonistic encounters and select small males with large fat reserves; females paired to fat males initiate more clutches in a season because fat males can incubate for longer than can thinner males.

During pair formation the male may swim towards the female, dipping his bill rhythmically into the water, then swimming alongside her and uttering the murmur call (see Voice); the female may face the male and bill-dip with him. There is also a mock-nesting ceremony (see Glutz von Blotzheim *et al*. 1973). The bowing and nibbling ceremony (see Common and American Coots) is performed by paired birds, sometimes on special display platforms. Throughout the year the pair greet each other after separation with a meeting and passing ceremony, approaching slowly then hurrying past each other with the head lowered and the tail raised or lowered before relaxing to feed; the undertail-coverts may be spread very briefly at the end of the ceremony. In courtship chasing, which in spring sometimes occurs after the meeting and passing ceremony, the male pursues the female closely for up to 1 min, usually on land, each running with the head stretched forward and the body horizontal. Either sex may initiate the chase, and either or both may spread the tail; the male may peck the female. When the female stops running, she may initiate bowing and nibbling, or may solicit copulation by performing the standing arch display with the head bent down and the bill pointing towards the toes, and then crouching in the squat arch (see American Coot), when the male mounts with flapping wings and copulates. After mating, both birds may perform a post-copulatory bowing ceremony in which the female turns her head to the side of her body; the male may adopt an arch bow position, which sometimes leads to reverse mounting. Soliciting and mating never occur on water. The male is reported to feed the female on the nest.

BREEDING AND SURVIVAL Season Europe, Mar-Aug; Morocco, Apr-Jun, Aug; Algeria and Libya, May-Jun; Tunisia, May; Egypt, May; Israel, Mar-Oct; Mali, Nov-Jan; Sierra Leone, Jan (see Field 1995); Ghana, Jun, but all year in S except at peak of wet season; Nigeria, Jul-Aug; Uganda and W Kenya, Feb-Mar, May-Jul, Nov; rest of

Kenya, and NE Tanzania, all months except Feb; rest of Tanzania, Apr-Aug, Oct; Pemba, Jun-Aug, Dec; Malawi, Mar, May-Sep; Zambia, Feb-Mar, Jul, Nov-Dec; Zimbabwe, all months (54% Jan-Apr, 32% Jun-Aug, n = 155); southern Africa, all months, peaking Jan-Apr and Jun-Aug in Zimbabwe, in autumn in Namibia, during rains (Dec-Mar) in former Transvaal and in early summer (Sep-Jan) in SW Cape, breeding probably opportunistic in response to varying wetland conditions; Madagascar, Oct-Mar; Mauritius, Sep-Oct, second brood Jan-Feb; Russia, Apr-Jul; Azerbaijan, May; Pakistan, May-Jul; Kashmir, May-Aug (chiefly Jun-Jul); Indian Peninsula, SW monsoon (Jul-Sep); Sri Lanka, Mar-Aug; Bangladesh, Apr, Jun-Sep; Malay Peninsula, Jan, Apr, Jul, Nov; Japan, Apr-Aug; Philippines, Jul-Sep; Sulawesi, chick Mar; S Borneo, Apr; Sabah, Jun; Sumatra, Dec-Feb; Java, Jan-Jul, Nov; Flores, May; Guam, Mar, Jun, Jul, Dec; Hawaii, all year, possibly peaking Mar-Aug; N America, Apr to early Jul (season 5-6 weeks longer in S than in N), also downy chicks Florida, Dec and Feb; Panama, mostly Dec-Jan; Trinidad, Jul-Dec; Peruvian Andes, Jul-Sep and in rainy season; S Brazil (Rio Grande do Sul), downy young mid-Jan; Uruguay, Oct-Dec, chicks mid-Jan; Argentina, Dec. In Ohio, USA, peak nest initiation occurred when height of *Typha* was 45-100cm and growth rate was greatest (Brackney 1979); in Louisiana, birds began using ricefields c. 7 weeks after planting, when rice was 80-90cm tall (Helm 1982). In tropical regions, breeding may occur at any time of year. **Nest** Saucer-shaped or more substantial, of dead or living twigs, reeds, rushes and sedges, with shallow to deep cup, lined with grasses and finer material; early nests may be entirely of dead material; usually uses closest available material; nest on, or up to 1m above, water; often built in emergent vegetation (especially in *Typha*) or on solid platform of branches or matted vegetation in water; also in grass tussocks; sometimes floating (e.g. on *Eichhornia, Alternanthera, Hydrocotyl* or *Brasenia*), less often in ground vegetation or low bush on bank, usually within a few m of water, and occasionally in bushes and trees up to 8 m above ground. Floating nests often have entrance ramp. External diameter 24-30cm, depth 10-20cm; cup diameter 12-17 cm, depth 3-7cm. Both sexes build, helped by other group members; male usually brings material while female builds (Krauth 1972, Urban *et al.* 1986); material often added during incubation; builds brood nests and display platforms. Also uses old nests of Rook *Corvus frugilegus*, Jay *Garrulus glandarius*, Magpie *Pica pica*, Woodpigeon *Columba palumbus* and other birds. **Eggs** Clutch size 2-17, mostly 5-9 (6.6, n = 2,278 clutches, nominate, Britain), 3-9 (5.7, n = 134 clutches, *meridionalis*), 7-9 (*cachinnans*), 3-7 (*galeata*, Trinidad and Brazil); clutches larger than 13-14 probably from 2 females. Clutch size (nominate) increases to peak in late Apr, then declines; younger birds lay later and smaller clutches than older birds (Cramp & Simmons 1980). Up to 4 replacement clutches laid after egg loss. Eggs short oval; smooth and glossy; whitish-grey to buff, pink-buff, pale brown or greenish, with spots and blotches of red-brown, purple, slate or black over lead-blue, grey or pale purple; markings often concentrated at larger end. Size of eggs: nominate (Europe to W Asia) 37-51 x 27-34.2 (43 x 30.6, n = 200), weight 22-28.5 (24.9, n = 88); nominate (India to Japan) 38-44 x 28-32 (40.6 x 29.6, n = 200), weight of 160, Russia, 15.9-25.5 (20.4); *orientalis* 38.6-48.3 x 27.7-29.3 (42.3, n = 26), calculated weight c. 21; *meridionalis* 38.7-46.4 x 28.1-32.6 (42.4 x 29.9, n = 60), calculated weight 20.8; *cachinnans* 40-49.5 x 28-33 (44.2 x 31.2, n = 117), calculated weight 23.6; *galeata* 41.8-50.1 x 29.6-35 (46 x 32.4, n = 25), calculated weight 25.5; *garmani* 43-53 x 32-38 (48.9 x 33.9, n = 10), calculated weight 30.4. Eggs laid at daily intervals. **Incubation** Period 17-22 days, by both sexes, sometimes helped by other group members; may start with laying of first or later eggs, or not until clutch complete (normally in first clutch); in *meridionalis*, female incubates in day, male incubates for 72% of 24-h period, including during night (Siegfried & Frost 1975); hatching usually synchronous for first clutches and asynchronous for replacements and later clutches, but asynchronous for all clutches in USA. **Chicks** Precocial and nidifugous; remain in nest for 1-2 days (1-4 in *cachinnans*); swim well by 3rd day; can dive, and swim up to 3m under water, at 8 days; fed and cared for by both parents, by immatures of previous broods and by other adults in group; brood division may occur (Wood 1974, *contra* Leonard *et al.* 1988); first-hatched chicks may be led away by 1 adult while mate continues to incubate. Carrying newly hatched young in bill from nest 2m above water reported (Cramp & Simmons 1980). Chicks beg by sitting upright, waving wings and swaying head; by 2 weeks, chicks beg while squatting with neck low, rear end elevated and head raised; begging chicks persistently utter high-pitched food call (see Voice) and snatch food from adult's bill. Adults may be aggressive to chicks (see Social and Sexual Behaviour), thus apparently reducing sibling competition and encouraging chick independence (Leonard *et al.* 1988). Chicks use spur on alula to climb and grasp vegetation and to enter brood nests; brooded frequently until c. 14 days old; totally dependent on adults for 7-10 days; become self-feeding at 21-25 days, but fed for up to 45 days. Body feathers and tail growing, days 19-30, and wings growing, days 31-45 (*meridionalis*); fledge at 42-50 days, rarely 70 days (nominate and *cachinnans*), fly at 60-65 days (*meridionalis*); independent (leave territory) at 52-99 days (mean 72 days).

Intraspecific parasitism (dumping) occurs (see Social Organisation); parasitism by Black-headed Duck *Heteronetta atricapilla* occasional (Arballo & Cravino in press) and by Ruddy Duck *Oxyura jamaicensis* recorded once (Frederickson 1971); in Uruguay, 1 nest with 6 eggs contained 1 egg of Red-gartered Coot (140) (Arballo & Cravino in press). Of 1,766 nests, Britain, 1,154 (65.3%) hatched 1 or more chicks (but unknown number of clutches lost before hatching not included in total); recorded losses due to predation (8%), flooding (13.7%), human predation (23.5%); of 36 1st clutches, 18 (50%) hatched; of 267 eggs laid in 53 nests, 47 (17.8%) hatched in 11 (20.1%) nests and 45 of these young were reared to 70 days (Cramp & Simmons 1980); nest success increased during season as vegetation developed (Huxley & Wood 1976); another study gave hatching success 53.4% (n = 470) (Hornbuckle 1981). In *cachinnans* (Greij 1974), nest success 53-75% (64.1, n = 968), hatching success 51-82% (80.2, n = 2,685); survivorship difficult to ascertain but mortality in first 10 days must exceed 40%; mean brood size at 1-10 days 4.7, at 3.5-5.5 weeks 2.6; in Hawaii, mean brood size at fledging 2.3 (n = 55). Egg infertility usually low (Greij 1994) but in England Hornbuckle (1981) noted infertility rate of 14.4%, attributed to shortage of males. Predators of nests and young include turtles (Urban *et al.* 1986), pythons *Python sebae* (Giraudoux *et al.* 1986), the large frogs *Rana catesbeina* (Viernes 1995) and *R. ridibunda* (Patrikeev 1995), largemouth bass *Micropterus salmonides* (takes chicks), water moccasins *Ancistrodon piscivorus*, raccoons *Procyon lotor*, grey foxes *Urocyon cinereoargenteus*, alligators *Alligator mississippiensis*, Boat-tailed Grackles *Quiscalus major* and probably rats *Oryzomys palustris*, while cattle trampling also causes nest losses (Greij 1994); other nest predators recorded are Rooks *Corvus frugilegus* and brown rats *Rattus norvegicus* (McRae 1997).

Age of first breeding 1 year. May rear up to 3 (occasionally 4) broods per season (e.g. in Europe: Gibbons 1989); interval between broods 20-30 days. In South Africa 2 pairs nested continuously in suitable conditions when supplementary food available, producing 40 and 37 egg sets respectively in 48 months, and raised 33 and 32 broods (means 4.2 young from 7 eggs and 3.5 young from 6.4 eggs) to full independence and flying; male's larger share in nest-building, incubation, and tending of young, gives female more time for feeding and conservation of energy, potentially contributing to her ability to produce rapidly repeated clutches (Siegfried & Frost 1975). **Survival** Of 90 ringed Germany, 62 (69%) died in 1st year and 21 (23%) in 2nd year; oldest ringed bird 11 years 3 months (Cramp & Simmons 1980). In USA, from 201 recoveries or controls of 5,470 birds ringed, 149 (74%) were obtained within 2 years of ringing, and oldest birds were recaptured 8 and 9 years after ringing (Greij 1994).

130 DUSKY MOORHEN
Gallinula tenebrosa   Plate 39

Gallinula tenebrosa Gould, 1846, South Australia.

Sometimes considered conspecific with *G. chloropus*, or to form superspecies, but the two are sympatric in Wallacea. Three subspecies recognised.

Synonyms: *Gallinula frontata/haematopus/chloropus tenebrosa*.

Alternative names: Black Moorhen; Dusky Gallinule.

IDENTIFICATION Length of New Guinea birds 25-32cm, of nominate 35-40cm; wingspan of nominate 55-65cm. A medium to medium-large moorhen, with very dark plumage. Head and neck grey-black; upperparts and upperwings dark brownish-olive, greyer on mantle; tail black. Underparts dark grey; some birds have broken or continuous white stripe along flanks; undertail-coverts black in centre and white at edges. Inconspicuous narrow white leading edge to wing in some. Iris red-brown; bill red, yellow towards tip; frontal shield orange-red to orange; front of tarsus and top of toes orange-red to orange, narrowly edged yellowish or lime-green at sides; rear of tarsus, soles and joints dark olive-grey, forming broad band at each joint; tibia red to orange-red. Sexes similar. Non-breeding adult has bill dull red (most often in males) to olive-black, shrunken shield olive-green or darker, red on tarsus and toes fading to olive-green or yellow. Immature similar to adult but duller; iris dark brown, bill, shield and tibia olive-black or olive-brown, and tarsus olive-green; becomes more like adult in first breeding season. Juvenile duller than adult, browner (grey-brown) on upperparts, paler on underparts, with feathers of centre breast speckled white or fringed pale grey-brown; bill and shrunken shield pale red, becoming dark olive-brown; iris dark brown; legs and feet dark olive-green or olive-brown. Inhabits open waters with fringing, and often floating, vegetation.

Similar species Markedly darker and more uniform in plumage than Common Moorhen (nominate race), but differs from marginally sympatric dark race *orientalis* of Common Moorhen only in absence of prominent white line along top of flanks (infrequently shows narrow or broken line), and in orange to red legs and feet with dark soles and joints. However, note that worn or moulting Common Moorhens of all ages may temporarily lack white flank line (see Common Moorhen). Juvenile apparently separable from juvenile Common on darker overall appearance, with darker, duller brown upperparts and more grey-brown underparts (not clear if these distinctions apply in comparison with juvenile Common of race *orientalis*), and by juvenile Common's obvious buff-white line along flanks. Also markedly larger (especially nominate and *frontata*) than Common Moorhen. Separated from other sympatric congeners by dark plumage, white lateral undertail-coverts, and colours of bare parts.

VOICE Information is from Marchant & Higgins (1993). The territorial advertising call is a raucous, crowing *kurk*, sometimes repeated or run together as *kurruk-uk*; it is given at all times of the day, and is taken up by birds in neighbouring territories so that responses can be heard >2km from the original call. There are also various short, sharp alarm calls, including harsh squawks and shrieks, and a series of widely spaced staccato calls made by swimming or preening birds (function unclear). A soft mewing is made by either sex before or during sexual pursuit, and a short, soft *kook* is also recorded in this context. Birds make a soft hissing call when eggs are handled. Adults give a short click when separated from chicks, and chicks give a series of descending whistles in reply. Chicks make a repeated, shrill piping when an adult approaches with food and when begging, and also when separated from adults; this call is given until young are 3 months old.

DESCRIPTION *G. t. tenebrosa*
Adult Head and neck grey-black, duller sooty-grey when worn; chin and throat may be slightly paler (whitish feather fringes) when fresh; mantle dark grey, becoming duller sooty-grey with wear; lower mantle tinged dark olive; scapulars, and back to uppertail-coverts, dark brownish-olive, more rufous-tinged when fresh; tail black. Tertials and most upperwing-coverts as scapulars; primary coverts, alula and remiges black-brown, becoming dark grey-brown when very worn; leading edge of P10 whitish but inconspicuous; P11 minute; few birds have single row of white marginal coverts along leading edge of wing, occurrence apparently correlated with occurrence of white flank stripes; axillaries and underwing-coverts dark grey-brown, coverts occasionally narrowly tipped white. Breast, flanks and belly uniform dark grey, with paler, browner feather fringes when worn; 10-15% of birds have short, narrow white streaks adjacent to shafts of c. 5 central flank feathers, forming broken (sometimes continuous) stripe, visible when flank feathers folded over wing; a few have narrow white fringes to belly feathers when fresh; vent grey, feathers variably tipped white to give scaly effect; undertail-coverts black in centre and white at edges. Iris red-brown; bill bright red to crimson, with yellow distal quarter to upper mandible and distal third to lower mandible; frontal shield slightly swollen, orange-red to orange, this colour extending broadly to base of culmen and narrowly round base of rest of bill; front of tarsus and top of toes orange-

red to orange, narrowly edged olive-yellow or lime-green along sides; rear of tarsus, soles and joints dark olive-grey, forming broad banding at joints; tibia dark red to orange-red. Sexes similar. Non-breeding plumage as breeding, but new contour feathers contrast faintly with faded, unmoulted feathers; iris duller; bill olive-black, often tinged dull red (more often in males), shrunken shield dark olive to olive-black; red on tarsus and toes fades to olive-green or yellow. Some males (probably older birds) retain breeding colours throughout year.

Immature breeding (second immature) As immature non-breeding, but juvenile wing appears more worn; head, neck and most of body feathers new, very similar to adult. Bare part colours as adult breeding.

Immature non-breeding (first immature) Similar to adult but duller. Head, neck, and varying area of body (usually mantle, breast, flanks and undertail-coverts) very similar to adult; retained juvenile feathers faded and contrastingly browner or grey-brown; new breast and belly feathers typically have light grey-brown fringes. Iris dark brown; bill, shield and tibia olive-black or dark olive-brown, tarsus olive-green.

Juvenile Duller and slightly paler than adult. Top of head dark brown; sides of head, chin, throat and neck dark grey-brown; sometimes faintly mottled white on lower sides of head (pale feather bases); upperparts dark brown, less olive- or rufous-tinged than in adult; feathers strongly prone to fade patchily to grey-brown; tail and wings like adult, but coverts and tertials browner, like upperparts; some have very narrow white fringes to greater and median coverts; some have narrow white leading edge to wing, like some adults. Underparts brownish-grey; central breast often finely speckled white when fresh, and feathers develop broad pale grey-brown fringes with wear; white flank stripes present in similar proportion of juveniles as adults; vent whiter than in adult; undertail-coverts like adult. Iris dark brown; bill and shrunken shield pale red, becoming dark olive-brown when moult almost complete; legs and feet dark olive-green or olive-brown.

Downy young Chick has thin and wispy black down on upperparts and dense and woolly black-brown down on underparts, tipped silver on throat and neck; skin of crown orange-brown or red; skin round eye bright blue; skin on leading edge of wing orange-yellow. Iris dark brown; shield and basal third of upper mandible red, tip of upper mandible yellow, lower mandible pink-orange fading to pale yellow at tip; legs and feet black. Black ring quickly develops between yellow and red on bill; in 10 days, red of shield and bill fades to pale red or pink. White egg-tooth and ivory-white wing-claw present.

MEASUREMENTS Adults and immatures combined. Wing of 19 males 197-233 (208.5, SD 7.1), of 16 females 189-213 (199.6, SD 7.0); tail of 16 males 64-84 (73.9, SD 4.2), of 16 females 63-77 (70.6, SD 4.0); culmen (including shield) of 17 males 40-52 (46.2, SD 3.6), of 16 females 39-48 (43.1, SD 3.2); tarsus of 19 males 57.5-70 (63.3, SD 3.1), of 16 females 54.5-68 (60.2, SD 2.9). Weight of adults and immatures: male 490-720 (570, SD 73.6, n = 5), female 336-684 (493, SD 121, n = 7).

GEOGRAPHICAL VARIATION Races separated on size and colour.

G. t. frontata Wallace, 1863 — SE Borneo (Bankau Lakes), Sulawesi, Sula Is (Mangole), S Moluccas (Ambon, Buru, Kelang, Seram), Lesser Sundas (Flores, Roti, Sumba, Sumbawa, Timor), and W and SE New Guinea (Vogelkop; Hall Sound and Lakoli R, E to Port Moresby region); also Trans-Fly region (Coates 1985). Sight records (J. S. Ash) from Bali require confirmation (White & Bruce 1986). Slightly smaller than nominate, darker on underside, with front of tarsus and top of toes bright red, joints olive-green (described as grey in Sulawesi birds); leg colour may not change in non-breeding season, and immature may have brownish-yellow tarsi and vermilion tibiae (Eskell & Garnett 1979). Leg and foot colour may not be reliable for differentiating this race because 2 specimens of nominate described as having "red" legs, while seasonal changes in leg colour not described for *frontata* (White 1976). However, "bright" red tarsus and toes, if occurring in nominate, would be exceptional, and further investigation needed (Marchant & Higgins 1993). Wing of 8 males 188-206, of 4 females 187-201; tail c. 70; tarsus c. 57 (Ripley 1977, no further details given).

G. t. neumanni Hartert, 1930 – N New Guinea (L Sentani and middle Sepik R, E to Ramu R). Smaller and darker than other races; upperparts blackish, with little or no olivaceous tinge. Legs of adult orange, yellow-red or vermilion-and-brown; bill of immature yellow-green, legs grass-green (Eskell & Garnett 1979). Wing of male 170-182, of female 161-174; tail of male 61, of female 61; culmen (including shield) of male 42-44, of female 40-43; tarsus of male 51, of female 53 (Ripley 1977). Weight of 7 unsexed: 290-370 (332.9, SD 35.5).

G. t. tenebrosa Gould, 1846 – Australia, mostly E of 135°E and in SW Western Australia. Widespread from Cooktown to E South Australia, mainly in E Queensland, E of W slopes in New South Wales, throughout Victoria, and in E South Australia; in Western Australia, mainly in SW from Bremer Bay N to Jurien; also Dampier area; scattered records N and NW Australia. Also N and SE Tasmania, Flinders I and King I. See Description.

MOULT Information is for Australian birds, from Marchant & Higgins (1993). Moult is apparently very similar to that of the Common Moorhen but few details are available. Postbreeding moult is complete, with simultaneous moult of the flight feathers, the primary coverts moulting with the remiges. Head and body moult begins anteriorly and is protracted, but it is not known if it is interrupted during wing moult; it is recorded Nov-Jan and in Apr. Prebreeding moult of adults is restricted to head and body feathers but varies in extent; it is recorded from Jun to Sep. Post-juvenile moult is partial, excluding the flight feathers, and varies in extent: the head, neck and undertail-coverts are usually replaced, often the mantle and upper breast, but rarely all of the underparts; it possibly occurs during the autumn of the first year. The immature prebreeding moult (first pre-alternate) is partial, like the adult prebreeding, and probably occurs at the same time; it gives rise to a very similar immature breeding plumage (second immature), after which postbreeding moult produces adult plumage.

DISTRIBUTION AND STATUS Borneo, Sulawesi, Sula Is, S Moluccas, Lesser Sundas, New Guinea, SW and E Australia, and Tasmania. Not globally threatened. Australian information is from Marchant & Higgins (1993). In Australia, the nominate race is widespread and common in the E and its inland range is expanding: a

1. *frontata*
2. *neumanni*
3. *tenebrosa*
× vagrant

Dusky Moorhen

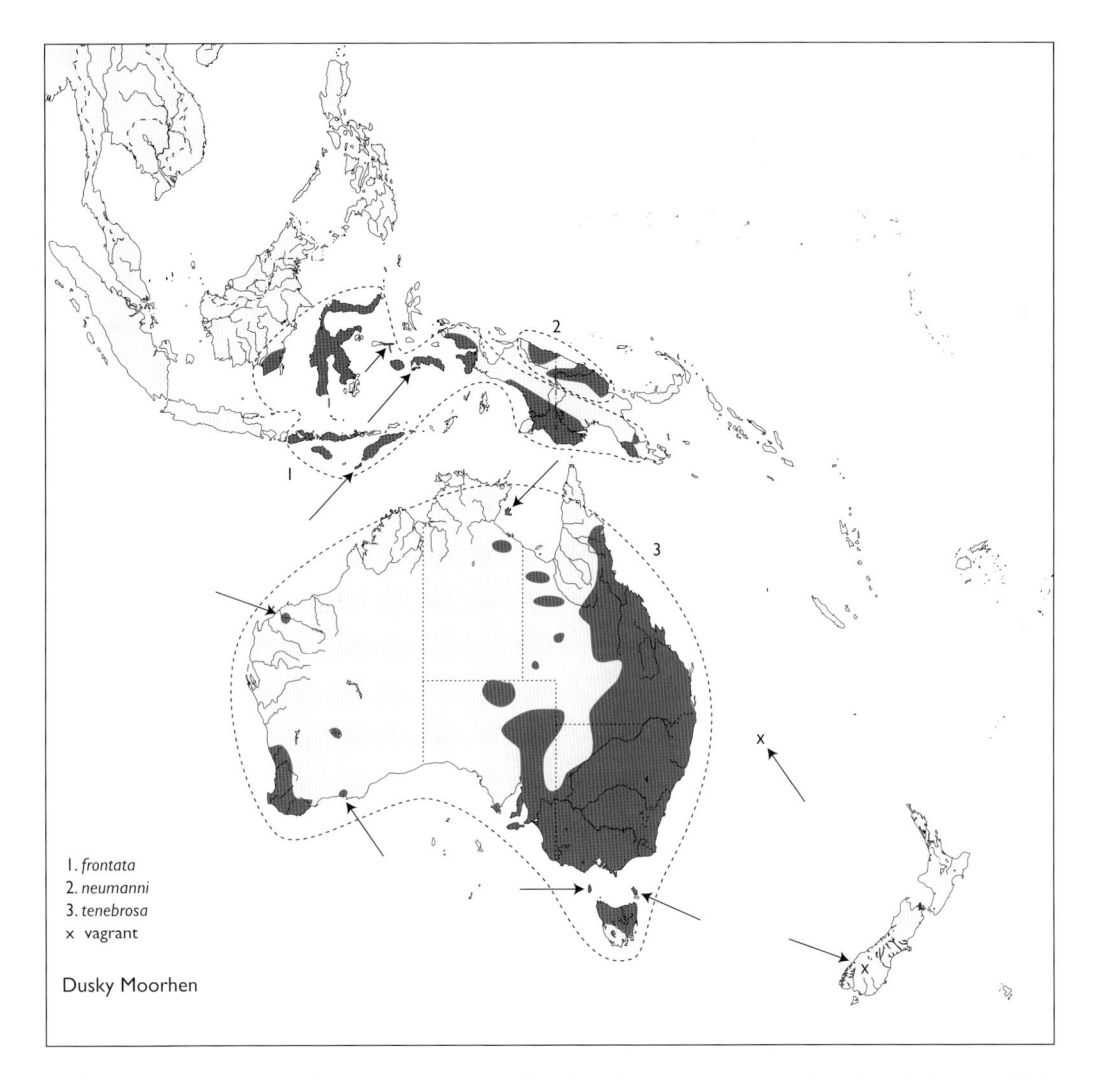

small population has recently become established in N Western Australia and the first breeding was recorded recently in some parts of W Queensland and E South Australia. It first bred on Flinders I in 1935 and King I in the 1960s, and it has colonised N Tasmania since 1976. It may be locally numerous, even on small waters, and 97 birds were recorded on a 2-ha lake, while in SW Western Australia 232 birds were counted on c. 3.3km of the Canning R (Jaensch *et al.* 1988). It is favoured by the construction of artificial wetlands, and its inland expansion has possibly been assisted by the construction of dams, but advantages are offset by habitat losses occurring through wetland drainage and modification.

The races *frontata* and *neumanni* were formerly regarded as either common over most of their ranges or widespread and locally common to scarce (Rand & Gilliard 1967, Ripley 1977, Coates 1985); their present overall status is uncertain, but *frontata* is locally moderately common to abundant, otherwise scarce to absent, in Wallacea (Coates & Bishop 1997). Dusky Moorhen is apparently sympatric with Common Moorhen in parts of Wallacea, and formerly in SE Borneo, although sympatric breeding is not recorded (Marchant & Higgins 1993). Considering the 5 islands where sympatry is recorded, on Sulawesi Common Moorhen is apparently expanding N (Watling 1983) and replacing Dusky Moorhen, while Common Moorhen was first reported from Sumba in 1949, and Dusky Moorhen has not been recorded from Borneo and Flores since the 19th century and from Sumbawa since early this century (White & Bruce 1986, MacKinnon & Phillipps 1993). However, Coates & Bishop (1997) record that on Sulawesi it coexists with Common Moorhen, at Lakes Tondano, Bone and Tolitoli (N), L Lindu, Palu V and Polewali (NC) and Tempe Ls (S), and Baltzer (1990) reports that it is far commoner than Common Moorhen at L Tempe. Baltzer's estimate of 8,900 birds at L Tempe could indicate that this site is of international significance, but the figure is derived by extrapolation from sample counts in a relatively small area and should be treated with caution (Perennou *et al.* 1994). Potential predators include domestic dogs and cats, and red foxes *Vulpes vulpes* (Garnett 1978): also see Social and Sexual Behaviour. Birds may be disturbed by dogs and duck-shooters, and are occasionally shot (Marchant & Higgins 1993).

MOVEMENTS In Australia (Marchant & Higgins 1993) it is sedentary, nomadic or dispersive, possibly partly migratory although reporting rates do not suggest regular long-distance movements. It may stay in one place for over 8 years, but apparently occurs seasonally in some areas (e.g. parts of E Australia), and also shows seasonal fluctuations in numbers, but atlas reporting rates do not suggest a large-scale seasonal pattern of movement. The intermittent occurrence of birds at isolated or temporary wetlands, the recent colonisation of Tasmania, and records of vagrancy, are evidence of some long-distance movement. It possibly moves to areas of high rainfall or surface water, to flood-waters or abundant food sources, and away from flooded areas where vegetation has been destroyed and from wetlands covered with *Eichhornia*. In SW Australia movements are possibly regular, and birds arrive in S swamps after the first rains, but they do not flock in the non-breeding season. Unlike the Black-tailed Native-hen (133), it does not undergo large-scale irruptive movements. Immatures disperse either in autumn or in spring after wintering in flocks, and may occur in habitats rarely frequented by adults. Movements are made at night, when flying birds may be heard calling (e.g. Watson 1955). In New Guinea it is regarded as locally nomadic, and a ringing recovery shows a movement of 60km NW (Coates 1985), while it has been recorded once (in Sep 1993) at L Wangbin in the Ok Tedi area, C New Guinea (Gregory 1995). It is a vagrant to New Zealand (Arrowtown, Aug-Oct 1968) and Lord Howe I (Apr 1975) (Marchant & Higgins 1993, Heather & Robertson 1997).

HABITAT It inhabits permanent or ephemeral wetlands, usually freshwater but sometimes brackish or saline: swamps, creeks, rivers, lagoons and estuaries. It also occupies artificial wetlands such as reservoirs, farm dams, and ornamental ponds and lakes in parks and gardens (Marchant & Higgins 1993). It requires open water, which usually has fringing cover such as reeds, rushes and grass, and often has floating, emergent or aquatic vegetation; however, waters choked with water hyacinth *Eichhornia* are avoided. It occasionally frequents rubbish tips and polluted water. It is seldom found far from the wetland edge, except when foraging in surrounding short vegetation. It is uncommon on saline and ephemeral waters, and occurs rarely in mangroves. It usually occurs in lowlands, but is found up to 1,000m in Sulawesi and a pair, presumed vagrants, was shot at 1,580m in New Guinea (Gyldenstolpe 1955).

FOOD AND FEEDING Most information is from Marchant & Higgins (1993). Vegetable food includes algae, and the vegetative parts, seeds and fruits of plants of the following taxa: Azollaceae, Hydrocharitaceae, Potomogetonaceae, Lemnaceae, Poaceae, Typhaceae, Polygonaceae, Portulacaceae, Solanaceae and Nymphaeaceae. Animal food includes worms, molluscs, arachnids, insects and their larvae (Odonata, Orthoptera, Hemiptera, Coleoptera, Lepidoptera and Hymenoptera), amphibians and fish. Birds also take carrion, bread, and the droppings of gulls and ducks. They forage in open water and among floating vegetation, usually within 100m of cover; also on adjacent land, often on grass and herbfields near water but rarely among tall terrestrial vegetation. They glean and peck on the ground or low vegetation, taking seeds and the tips of grass and shrubs at up to 120 pecks/min (Garnett 1978); they prefer feeding in shallow water with much vegetation; they feed swimming, gleaning from the water surface and up-ending for 2-7s with the tail and legs above water, taking food up to 30cm below the surface; they do not dive to feed. They sometimes chase insects, they rarely stretch up on the toes to take food, and they sometimes pin a food item to ground with the foot. In gardens and on golf courses they often graze on lawns near water (Serventy 1985). They drink regularly, but only from the water's edge (Garnett 1978). Chicks are fed mainly on molluscs and annelids for the first few weeks, the proportion of vegetable matter being gradually increased (Garnett 1978).

HABITS Most information is from Marchant & Higgins (1993). Diurnal. Birds are generally wary and secretive, but are easy to observe where artificially fed and at artificial sites where they are accustomed to human proximity (Serventy 1985); they are shy, retiring and elusive when moulting. They swim well and buoyantly, with the neck upstretched, the back flat but the stern higher than the front; with each swimming thrust of a foot the head is jerked and the tail often flicked. When alarmed, they raise the head, flick the tail and fan the white undertail-coverts, and may walk, run or fly to cover. A recent study (Ryan *et al.* 1996) suggests that tail-flicking represents both an interspecific signal of alertness and an intraspecific signal of social status. They may dive to escape danger and can apparently remain submerged without holding onto anything with the feet; 1 reportedly remained submerged for c. 4 min (Serventy 1985). The gait on land is an uneven, high-stepping walk, and a slightly twisting run with lowered head and often with flapping wings. It can climb well, sometimes using the wings for support. The flight is strong and direct, with rapid shallow wingbeats, extended head and neck, and legs dangling or trailing behind, but short flights appear more laboured. Birds roost at night, singly or in breeding groups, or in non-breeding flocks; roosting platforms (each holding 1 bird) of trampled vegetation are often constructed in reeds, 0.5-2m above the water and at least 0.5m apart; they also roost on branches over water, and rarely on the ground among reeds. They rest at roosting sites during the day, and also loaf on floating vegetation, banks, and emergent logs and stones; in hot weather they may perch high in trees. Thorough preening occurs up to 5 times/day in spring and summer, and bathing is frequent, particularly in hot weather: the bird dips the head under water and then raises it, lowering the body and ruffling the wing and body feathers to allow water to flow over the upperparts (Garnett 1978). Sunbathing occurs on a platform, the bird holding the wings partly open and drooped.

SOCIAL ORGANISATION Most information is from Marchant & Higgins (1993). It is territorial when breeding; outside the breeding season birds gather in loose flocks, sometimes exceeding 100 birds. This species is simultaneously promiscuous, forming breeding groups of 2-7 apparently unrelated birds; individuals sometimes switch groups between seasons. Within the group, the sex-ratio usually favours males, and all males copulate with all females. All group members defend the territory, build nests, incubate, and care for young; older siblings sometimes help to care for the young.

Territory size correlates with group size, and all-adult groups hold larger territories: in all-adult groups, total area 1,810-3,330m^2, area of reeds 375-785m^2; in groups with immatures, total area 1,818-2,450m^2, area of reeds 60-175m^2 (Garnett 1980). At Canberra, a breeding density of 53-89 birds/ha has been recorded (Garnett 1980).

A. Alert.

C. Retreat.

B. Pursuit.

D. Retreat.

E. Males chasing female during pre-copulatory display.

F. Female posture before copulation.

G. Copulation.

H. Post-copulatory pose.

I. Begging.

Figure 13: Dusky Moorhen displays. [After Garnett 1978].

SOCIAL AND SEXUAL BEHAVIOUR Most information is from Marchant & Higgins (1993). The displays are conspicuous and are identical to those of the Common Moorhen, although the social structure differs, and behaviour is similar to that of the Tasmanian Native-hen (134), although simpler. Tail-flicking represents a signal of social status (juveniles flick significantly more often than adults) as well as of alertness (Ryan *et al.* 1996). Aggression is apparent in non-breeding flocks but rarely occurs within the breeding group except when the territory is being established, when birds may give a meeting display with arched neck, bill pointed downwards and wings partly raised, slowly moving past one another before resuming normal behaviour. All group members defend the territory but only 1 member confronts an intruder, and birds generally defend only boundaries in water. When sighting an intruder, a resident raises the head and neck and remains rigid for several seconds; the neck is then extended forwards with the head just above the water, the tail is lowered and the bird swims towards the intruder, lowering the head and tail still further. The intruder retreats with raised head, flicking the tail and fanning the white undertail-coverts. The attacker may run across the water surface, flapping its wings, when the intruder also runs, usually taking flight; if the pursuer catches up, the intruder dives. Pursuit stops at the territory boundary, when the resident swims slowly away with head lowered, wings partly raised and white undertail-coverts spread ('mutual retreat'). Fighting may occur between adjacent residents at a territory boundary; birds first attack with the bill and if 1 obtains a grip it may force the other underwater, when the submerged bird retreats; otherwise both birds kick at each other's breast while flapping the wings and rising up to 0.5m above the water; the kicks eventually force the birds apart, when they mutually retreat. Agonistic behaviour is also directed at other bird species (see Marchant & Higgins 1993), and Purple Swamphens (120) elicit the full threat display, including pursuit, mutual retreat and fighting.

Adults have 2 distraction displays. The body is held low and the wings are half-opened and beaten on the water while alarm calls are given and the bird alternately approaches and retreats; if this is not successful the swimming bird lowers the head, half-raises the wings and tail, and shakes them. Breeding groups will attack some predators (cats, water rats *Hydromys leucogaster*, Brown Goshawk *Accipiter fasciatus*, Collared Sparrowhawk *A. cirrocephalus* and Australian Raven *Corvus coronoides*) using their bills, but only to defend chicks, not eggs.

A female usually initiates copulation by passing in front of a male or by calling briefly, then running or swimming away with her neck extended forward, and her undertail-coverts fanned beneath the lowered tail. 1-3 males follow, often with their bills on the female's back; after up to 3 min only 1 male remains and the chase ends on land or vegetation; the female crouches with bill pointing down and tail depressed; the male mounts, maintains balance by flapping his wings and moving his feet, copulates and then raises his tail and descends; he may briefly adopt a post-copulatory pose, moving slowly on raised toes, with lowered head, wings arched over the back, tail raised and undertail-coverts fanned. The female may copulate with another male immediately afterwards. Reverse mounting and homosexual copulations between males are rare. Allopreening occurs among all group members, and possibly also among prebreeding immatures.

BREEDING AND SURVIVAL Season Sulawesi, possibly Apr, downy young Mar; Seram, juvenile, May; New Guinea, small young May-Jun; Australia, Aug-Mar, little variation throughout range, in SW Western Australia laying correlates with peak rainfall + 3 months (Halse & Jaensch 1989). The following information is from Marchant & Higgins (1993). **Nest** Usually built up to 180cm above water (occasionally on ground) in grass tussocks, reeds (especially *Typha*, *Phragmites*), *Eragrostis australasica*, *Triglochin procera*, rushes *Juncus*, lignum *Muehlenbeckia cunninghamii* bushes, or trees; also on waterlilies floating in clear water, and in stumps or hollow logs; also recorded building inside metal drum, and in wire netting lying in water. Nest a bulky platform or shallow cup of reeds, rushes, lignum shoots, twigs, bark, leaves and waterweed; external diameter 25-30cm; bowl 15-20cm wide, 6-8cm deep. Most nests have runway leading to water; nests in reeds may have foundation of pressed-down reeds, and may have passageways leading to them. Will build nest up to 1 month before using it. In addition to egg-nest, may build 1-2 false nests, abandoned before completion; 1-2 brooding nests built after eggs hatch. All group members build; collect material close to nest, and sometimes up to 30m away. **Eggs** 5-18 (mean 7.8, n = 85 clutches); large clutches probably from 2 or more females; eggs oval; slightly lustrous; pale creamy-brown to brownish-white, irregularly spotted and blotched dull red-brown and purple-brown with few faint purple-brown and grey undermarkings, blotches often angular. Size of eggs: nominate, 45-58.4 (51.3, SD 2.3) x 32.8-37.8 (36.7, SD 1.1, n = 26); calculated weight c. 34; *neumanni*, 45.1-46.7 x 31.7-32.4 (46.1 x 32.2, n = 4), calculated weight 26. Laying irregular for first few days, then daily; in Australian study, one egg laid per day by each female in group at night or in early morning until clutch of 5-8 per female was complete, all females starting to lay within 3 days of each other (Garnett 1978). **Incubation** 19-24 days; usually begins after laying of last or penultimate egg; all members of group thought to incubate; hatching asynchronous, within 1-5 days, usually within 48h. **Chicks** Semi-precocial and nidifugous; leave nest after 3-4 days; fed intensively and brooded until c. 4 weeks old, then fed less frequently until c. 9 weeks old; begin to feed independently after 3-4 weeks. Brood sometimes divided between group members. Fed bill-to-bill; chicks beg by extending neck, waving wings and giving high piping call. When danger threatens, chicks swim to branches, climb them and hide in foliage, or lie still in mud at edge of water, or dive and swim under water to cover. Feathers appear after a few days, on head and body before wings; fully feathered on body at 1 month; fly at c. 8 weeks. Young remain in territory for 5-8 months; then independent; most birds enter breeding groups when more than 20 months old.

In Australian study, early nesting groups lost 13.4% eggs and fledged 1.97 young per group, while late nesting groups lost 38.5% eggs and fledged 0.5 young per group; Australian nest record cards show overall hatching success of 55.5%. Nests with least cover suffer greatest predation; some nests flooded by rising water; mortality of young high; eggs taken by rats *Hydromys leucogaster*; eggs and young probably taken by harriers *Circus* (for other predators, see Social and Sexual Behaviour). Will relay within 2 weeks of failure; same nest used. Normally single-brooded; may lay twice per season in Australia. One nest contained an egg of Musk Duck *Biziura lobata*; 2 nests contained eggs of Common Coot (136). **Survival** No information available.

131 LESSER MOORHEN
Gallinula angulata   Plate 39

Gallinula angulata Sundevall, 1851, Lower Caffraria (= Natal), South Africa.

Monotypic.

Synonyms: *Gallinula pumila/minor.*

IDENTIFICATION Length 22-24cm. A medium-small gallinule; male has head blackish; neck and upper mantle dark slate-grey; rest of upperparts, and upperwings, dark olive-brown; underparts dark slate-grey; line of broad white streaks along flanks; undertail-coverts black in centre, white at sides. Narrow white leading edge to entire wing. Legs and feet very variable in colour: often green or yellow-green, sometimes orange or pinkish. Female has paler, browner upperparts, light grey face with blackish only round base of bill, silvery-grey throat, and paler grey underparts, especially on belly; frontal shield smaller, duller, and orange next to feathers; female appears relatively pale greyish in flight, with darker wings. Immature paler and browner than adult on upperparts; scapulars and tertials edged pale buff; some black around base of bill; sides of head, chin and throat largely greyish; rest of underparts pale brownish-grey, washed buffy-brown on breast; white streaks on flanks less distinct; undertail-coverts as adult; bare parts usually duller than in adult; bill yellowish; shield orange to orange-red; may breed in this plumage. Juvenile has upperparts more olive-brown than in immature; sides of head, neck and breast buffy-brown, contrasting with dark top of head; chin to lower breast and belly creamy-white, shading to light grey on flanks, which lack white streaks; undertail-coverts as adult; bill brownish-yellow with dusky base to culmen; legs and feet greyish-green to dull yellow-green. Inhabits seasonal, ephemeral and permanent wetlands with abundant cover of emergent grasses, sedges etc.

Similar species Like a small version of Common Moorhen (129), but female much paler than Common, in flight appearing predominantly grey with darker wings; both sexes distinguishable by bright yellow bill (red with yellow tip in Common), scarlet frontal shield and culmen, and pointed frontal shield (rounded in Common); profile very different (see Plate 39), with stubbier, deeper bill which tapers evenly throughout its length to give almost conical shape, and culmen and shield in almost straight line (apart from slight bulge in shield when breeding) to give flat forehead (Common has bill longer and thinner, almost straight for half its length, and shield at an angle to culmen, giving steeper forehead); legs and feet may be orange or pink; lacks red 'garter' on tibia; narrower and less obvious white streaks on flanks. Common Moorhen has white markings on belly, but these difficult to see in field. Immature differs from immature Common in bare part colours, shield shape and profile; upperparts similar in colour to Common, but may be paler, and are less olive-green, more russet-tinged; scapulars and tertials pale-edged; sides of head paler, with darker area around base of bill; underparts may be paler and browner. Immature Common has sides of head, and neck and underparts, predominantly grey, browner on flanks and belly. Juvenile differs from juvenile Common in having brownish-yellow bill, dusky culmen, pointed shield, and no orange on tibia; upperparts paler, slightly warmer brown than in juvenile Common; sides of head, neck and breast buffy-brown, paler and greyer on flanks, tertials and scapulars edged buff (juvenile Common largely darkish brown to grey-brown on underparts, with buffy feather fringes, whitish chin and throat, and buff-white flank line). Although both species can occur together they are often separated by habitat, Lesser favouring temporary waters with emergent vegetation and Common permanent waters with fringing vegetation. Often occurs alongside similar-sized Allen's Gallinule (123), which differs markedly in shape and plumage and is distinguishable in flight by long, projecting red legs and feet, red bill, and overall blacker appearance. Allen's also lacks white line along leading edge of wing; prefers waters with floating vegetation.

VOICE The calls resemble those of the Common Moorhen. They include a rapid series of clucking notes, a series of subdued chuckling or pumping notes, and sharp clicking and squeaky calls. A soft *pyup*, possibly a contact note, is often given. A high-pitched rattle is also recorded (P. B. Taylor unpubl.). The alarm call near the nest is a sharp *tik* or *tek* (Urban *et al.* 1986).

DESCRIPTION
Adult male Crown, sides of head, chin and throat blackish; neck and upper mantle dark slate-grey; rest of upperparts, including scapulars, tertials and upperwing-coverts, dark olive-brown, slightly less olive on coverts, tail blackish-brown. Primary coverts, alula, primaries and secondaries dark brownish-grey; outer web of outermost primary white; alula feathers also edged white; marginal coverts mixed white and grey, giving narrow line along leading edge of wing; underwing-coverts and axillaries grey, narrowly tipped white. Entire underparts dark slate-grey; line of broad white streaks along flanks; vent black; undertail-coverts black in centre, white at sides. Iris red; bill bright yellow, culmen and pointed frontal shield red, with yellow at extreme tip of bill, rarely with narrow orange line around edge of shield. Legs and feet very variable in colour: green or yellow-green, sometimes olive-yellow, orange, flesh, pinkish-red or pinkish-brown. Compared with Common Moorhen, bill relatively short, deep and stubby, decreasing evenly in depth throughout its length to give almost conical shape, and culmen and shield in almost straight line to give flat forehead. Culmen only slightly curved over entire length; breeding birds have slight bulge in shield, extending to base of culmen.
Adult female Has paler, browner upperparts, light grey face with black only around base of bill, silvery-grey chin and throat, and paler grey underparts, especially on belly; frontal shield smaller, duller, and orange next to feathers.
Immature Crown to hindneck, upperparts and upperwing-coverts dark olive-brown, often tinged rufous but greyer on mantle; scapulars and tertials edged pale buff; rest of wing as in adult. Variable amount of black around base of bill; sides of head, chin and throat largely greyish (paler on chin and throat in female?); rest of underparts pale brownish-grey, washed buffy-brown on breast; flanks streaked whitish, less prominently than in adult; undertail-coverts as adult. Bare parts usually duller than in adult: iris brown to red-brown; bill yellowish, culmen and shield orange to orange-red; legs and feet dirty yellow-green. May breed in this plumage (Andersson & Gurney 1872, White 1945). Some immature females have yellow-horn bill and blackish shield; others have greenish-yellow bill, and shield only tinged red (White 1945). It is not clear to what extent (if any) the first breeding plumage may differ from the

509

full adult plumage, as it does in e.g. Common Moorhen and Dusky Moorhen (130).

Juvenile Crown to hindneck olive-brown to darker brown; upperparts and upperwing-coverts olive-brown, slightly greyer on mantle; scapulars and tertials edged pale buff; rest of wing as in adult. Sides of head buffy-brown, paler on lores and over eyes, sometimes contrasting sharply with dark top of head to give capped effect; sides of neck and breast buffy-brown, this colour sometimes extending more faintly across centre of breast; chin, throat, lower breast and belly creamy-white, shading to pale greyish on flanks; flanks lack white streaks; undertail-coverts as in immature. Iris brown; bill brownish-yellow, with blackish culmen or dusky base to culmen; legs and feet greyish-green to dull yellow-green.

Downy young Black; iris dark brown; bill black with white distal half, black cutting edges and pink base; frontal shield and base of culmen pale red-brown, pale purple next to forehead; legs and feet bluish-grey to dark grey; skin on crown pale red-brown.

MEASUREMENTS Wing of 9 males 125-145 (137), of 5 females 130-135 (132.2, SD 2.3); tail of 9 males 52-62 (56.3), of 3 females 55, 57, 59; culmen of 9 males 19-21.5 (20.3), of 3 females 19; culmen (with shield) of 2 males 30, 36, of 2 females 24, 25; tarsus of 9 males 35-39 (36.6), of 3 females 36, 37, 37. Weight of 4 males 145-164 (154.7, SD 8.8), of 3 females 92, 99, 137, of 2 unsexed 149, 150.

GEOGRAPHICAL VARIATION None.

MOULT Nothing is recorded. The moult pattern is presumably similar to that of the Common Moorhen.

DISTRIBUTION AND STATUS Senegal and Gambia E to Ethiopia and S to SC Namibia, N and E Botswana and E South Africa. Not globally threatened. It is widespread and locally common over much of its range but, as with Allen's Gallinule, the erratic nature of its occurrences in seasonal habitats makes its numbers difficult to assess; it is sometimes locally very numerous in suitable breeding habitat during seasons of good rainfall. It is widespread in Senegal, and there are few records from The Gambia (Jan, Feb, Apr, and Jul-Sep) but it is probably overlooked (Urban *et al.* 1986, Gore 1990). It is an uncommon resident in Sierra Leone, at least in the NW (Field 1995); in Ivory Coast it is widespread, and it outnumbers Common Moorhen in the Korhogo marshes (Thiollay 1985;, Urban *et al.* 1986); it is not uncommon in Ghana, at least in the S (Grimes 1987). It is frequent in C Benin (Claffey 1995), not uncommon in Nigeria except in the SE (Elgood *et al.* 1994), rarely seen but possibly common in C and S Mali (Lamarche 1980), recorded in Burkina Faso only at Ouagadougou (Thonnerieux *et al.* 1980), numerous and probably widespread in S Cameroon (Louette 1981), and uncommon in Chad (Urban *et al.* 1996). It is apparently resident on Gulf of Guinea islands, and is also recorded as a migrant in Liberia, Togo and the Central African Republic and resident in Niger (Gatter 1988, Dowsett & Forbes-Watson 1993). It is regular in NE Gabon and locally numerous seasonally (Brosset & Erard 1986), but has been recorded only once from Congo, in Dec (Dowsett & Dowsett-Lemaire 1989).

In Sudan it occurs locally N to Wad Medani, and also to Darfur in the rains, and it is rare in Ethiopia (Urban *et al.* 1986); there is only 1 old record from Somalia, in the S, in May (Ash & Miskell 1983). In Zaïre it is widespread but uncommon except in the E and SE (Urban *et al.* 1986).

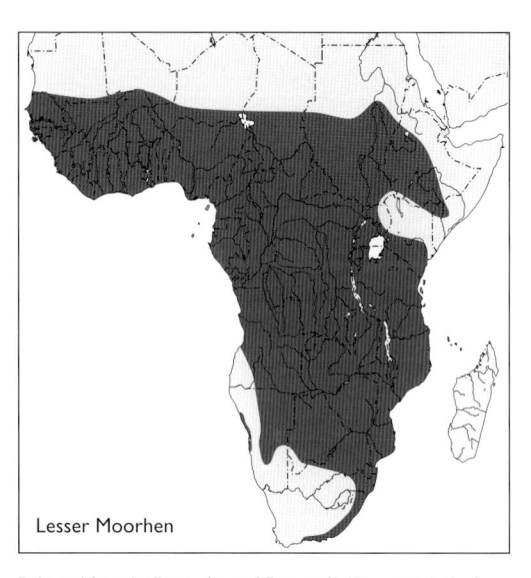

Lesser Moorhen

It is resident in Rwanda and Burundi (Dowsett & Forbes-Watson 1993); in Uganda it is resident at Awoja and Entebbe, and also occurs to the extreme N (Kidepo V NP); it ranges S from C Kenya to Tanzania, where it occurs widely except in E Tanzania S of Morogoro and Dar es Salaam (Britton 1980, Short *et al.* 1990), and it has occurred in great numbers, and nested, in the lower Tana R valley during extensive floods (Jackson & Sclater 1938). It is locally very common to abundant during the rains in Zambia, and in Zimbabwe (mostly on the C Plateau) (Benson *et al.* 1971, Irwin 1981, Urban *et al.* 1986) and a concentration of 1,000 was recorded at Lochinvar on the Kafue Flats, Zambia, in Jun (Taylor 1979). Its status is unclear in Malawi, whence there are few specimens (Benson & Benson 1977). It is generally distributed throughout most of Angola except the extreme SW (Traylor 1963, Pinto 1983), and it is probably widespread in S Mozambique during the rains (Clancey 1996). At Ondonga, N Namibia, it was formerly recorded as "literally swarming in all the vleis of this country, where it breeds most abundantly" (Andersson & Gurney 1872); it occurs most widely in the N of Namibia but also extends to SC regions (Taylor 1997c); it is a locally common breeder in the NE, on temporary pans in Bushmanland and Kavango (Hines 1993), and is a frequently observed resident at pools, swamps and lakes in the E Caprivi (Koen 1988). In Botswana it is sparse to uncommon and occurs mainly in the N and E (Penry 1994, Taylor 1997c). In South Africa it is fairly widespread in the N and E (in the former Transvaal and N KwaZulu-Natal), rarely extending to the S Cape coast as far as Swellendam and Oudtshoorn (Taylor 1997a); it is uncommon in Swaziland and has possibly occurred once in Lesotho (Bonde 1993, Parker 1994). Its numbers fluctuate widely from year to year in South Africa and it can be locally abundant, e.g. at the Nyl R floodplain, N Transvaal, 8,000-50,000 pairs have been estimated to occur in years of high rainfall (few if any occur in dry years), while the Kruger NP may hold up to 3000 birds, mostly in the S, and N KwaZulu-Natal has potential habitat for at least 7,600 birds (Tarboton *et al.* 1987, Taylor 1997a). Habitat loss must have occurred in recent years as a result of the destruction and degradation of its wetland habitats, especially the loss of suitable seasonally flooded sites, but the overall effect of this is unclear. The Nyl R floodplain is the most important known breeding site in southern Africa (Taylor 1997a).

The Lesser Moorhen's flesh is moderately palatable (Cott & Benson 1970) and in N Namibia was formerly much esteemed by local people, who made up great hunting parties to chase the birds out of the water onto dry land where they were easily secured (see Habits), while the eggs were also taken in large numbers (Andersson & Gurney 1872); it is not clear whether the species is still hunted to any significant extent in this area. No information is available on predation.

MOVEMENTS Its movements are complex, and are not shown on the distribution map. Some birds are resident throughout the year in permanently suitable habitat, but many are rains migrants. In W Africa some are resident in wet S areas but in dry N areas numbers increase during the rains, when birds breed, and decrease as seasonal habitats dry out. It occurs in Senegal and Gambia mainly in the wet season (Urban et al. 1986); it has some seasonal movements in Ivory Coast (Thiollay 1985); it occurs in Ghana mainly in Jun-Sep (Grimes 1987), and in C Benin from Sep to May (Claffey 1995), and almost all records from Nigeria are in Mar-Sep (Elgood et al. 1994). In Burkina Faso it is recorded only from Jul to Sep (Thonnerieux et al. 1980), and in Niger only in Mar, May and Jun (Giraudoux et al. 1988). It is resident in Chad except in the Sahel zone, where it is purely a rains migrant (Urban et al. 1996). Probable nocturnal migrants have been found in Cameroon, Nov, and in Zaïre (Kivu), May-Jun, while in Gabon birds have been taken around buildings between 15 and 20 Nov, and enter buildings in Nov-Dec when many arrive in the NE (Brosset & Erard 1986, Urban et al. 1986). There are some movements in Sudan (Urban et al. 1986). It is present all year near Entebbe, Uganda, and it occurs in Kenya and NE Tanzania in most months but is commonest from Apr-Jul, this possibly representing a postbreeding influx from the S (Britton 1980). However, in Kenya it has occurred in very large numbers, and has bred, on the lower Tana R in Jun, and has also nested at Nairobi in Jun (Jackson & Sclater 1938); during and after the Apr-Jul rains it occurs sporadically in temporary habitat outside its normal range, and night migrants have been attracted to lights at Ngulia in Dec-Jan (Urban et al. 1986, Lewis & Pomeroy 1989). In Zambia (Benson & Irwin 1965, Benson et al. 1971, Taylor 1979) there are records for all months except Sep, but it is largely absent from Jun to Nov, and a possible migrant was killed against a house in Apr; arrivals are noted from late Nov to Dec and departures by mid-May, and numbers build up at the end of the rains in May-Jun, especially on the Kafue Flats. It is largely a rainy season visitor in southern Africa, although a few remain in suitable habitat during the non-breeding season: 88% of specimens from SE Zaïre to South Africa were taken in Dec-Apr (38% in Jan alone) (Benson & Irwin 1965). In Zimbabwe and Botswana it breeds in semi-arid areas at pans and other temporary waters which disappear in the dry season (Urban et al. 1986); in Zimbabwe most arrive in Dec and leave in Apr, but there are records from Oct to May, and migrants pass through the Middle Zambezi and Limpopo Valleys (Irwin 1981); it occurs in Botswana from Oct-Jun, mainly in Dec-Apr (Penry 1994). In South Africa it occurs in the former Transvaal in Dec-May, and a very fat migrant in breeding condition flew into a wall in KwaZulu-Natal in Feb (Taylor 1997a, P. B. Taylor unpubl.). One was taken at night, 16km off the coast of Mozambique, near Inhambane, in Jan (Benson & Irwin 1965).

HABITAT Permanent and temporary freshwater wetlands, such as floodplains and pans with emergent grass or sedge cover and often with floating plants; marshes with rushes and open water; papyrus swamps, reedbeds, and ponds with waterlilies; also rank fringing vegetation on ponds, dams, rivers and forest streams; rice fields, flooded farmland, seasonally inundated grassland, and sewage ponds. In Namibia it bred in an ephemeral grassy marsh, dominated by *Diplachne* spp, at the edge of an unvegetated alkaline pan (Jamieson et al. in prep.). It occurs at "coastal lagoons" in Ghana (Grimes 1987), and has bred at roadside gravel pits in Zimbabwe (Hopkinson & Masterson 1984). It often occupies different habitat to that utilised by Common Moorhen, preferring temporary waters (often in savanna) with abundant cover of emergent vegetation as opposed to permanent waters with fringing vegetation, but both species sometimes occur together (Taylor 1997a). It occurs alongside Allen's Gallinule but, unlike that species, often frequents temporarily inundated grassland and grassy pans with dense emergent cover but little or no floating vegetation (Taylor 1997a), and at the Nyl R floodplain it is more widespread than Allen's Gallinule in flooded grassland, and least numerous in the gallinule's most favoured habitats (see Allen's Gallinule species account), seeming to prefer areas with relatively dense emergent grass and few or no open patches (P. B. Taylor unpubl. Breeding birds at Sokoto in N Nigeria were restricted to rice fields and to lakes and flooded farmlands with close-growing emergent grasses, while Allen's Gallinules favoured permanent lakes with a profuse growth of waterlilies (Serle 1939). Ii occurs down to sea level and, in East Africa, up to 2,000m.

FOOD AND FEEDING Molluscs, insects (especially Coleoptera), vegetable matter including seeds and flowers of reeds; in captivity also termites (Urban et al. 1986). It forages while swimming or while walking at the edge of water, on floating vegetation such as lilypads, or on open mud (Urban et al. 1986, P. B. Taylor unpubl.). During extensive floods, birds were seen swimming around to feed on small beetles and other insects which had sought refuge on the stems and leaves of flooded grass (Jackson & Sclater 1938).

HABITS It is usually much shyer than the Common Moorhen, remaining in cover even while foraging, but sometimes it tolerates the proximity of man and may even become fairly tame: it sometimes comes close to houses and footpaths at Dar es Salaam and Nairobi (Urban et al. 1986). Although it is said to swim less readily than the Common Moorhen (Urban et al. 1986), it swims frequently, and often occurs in emergent vegetation far from the shore; it is less inclined to fly than Common Moorhen but the flight action is similar. It sometimes walks on lilypads and other floating vegetation, and on mud and grass at the edge of water. Andersson & Gurney (1872) report that it is reluctant to fly when pursued, and when chased onto dry land birds try to conceal themselves in bushes and grass.

SOCIAL ORGANISATION Monogamous; territorial when breeding. Nests are usually well separated, but in Nigeria 4 pairs nested in a radius of 20m (Serle 1939) and a breeding density of 1-2 pairs/ha is recorded on lakes in the Nguru area (Elgood *et al.* 1994). On the Nyl R floodplain early in 1996, birds bred at an estimated density of 3 pairs/ha (Taylor 1997a). In a recent study in Namibia, Jamieson *et al.* (in prep.) found a high rate of conspecific brood parasitism (in 21-36% of 28 nests) in a population breeding in an ephemeral wetland where there may have been a shortage of available nesting sites and where the nesting period covered only 3 weeks. Hosts rejected 57% of parasitic eggs, 85% of which were buried in the nest lining; in the other cases the nest was deserted.

SOCIAL AND SEXUAL BEHAVIOUR It is not aggressive hen feeding, when it tolerates the proximity of its own or other species; although it is dominant over Spotted Crake (96), both species feed close together (Urban *et al.* 1986, Taylor 1987). Incubating birds usually slip off the nest before they are seen by an approaching intruder, but may occasionally sit until the observer is very close (Serle 1939). Nothing else is recorded.

BREEDING AND SURVIVAL Season Senegal and Gambia, breeding condition Aug; Sierra Leone, Nov; N Nigeria, Jul-Sep; Ghana, Jun; Gabon, Feb; Chad, Aug; Sudan (Darfur), Aug; Somalia, May; Zaïre, high plateau of SE Jan-Mar (second half of rains), at lower altitudes possibly Apr-May; Angola, Jan-Mar; Kenya and NE Tanzania, Mar, May-Jun; Tanzania, Dec-Mar; Zambia and Malawi, Jan-Mar; Zimbabwe, Dec-Apr, once May; Namibia, Feb-Mar, probably Sep; Botswana, Feb; South Africa, Dec-May, peaks Jan-Feb at Nyl R floodplain (former Transvaal), when habitat most flooded. **Nest** Smaller and more compact than that of Common Moorhen; pad of grass or sedges with shallow cup; either floating on water or placed up to 5cm above water (recorded water depth 20-100cm) in emergent grasses or sedges up to 1.5m tall, with surrounding stems often bent down over nest and bound together to form canopy; entrance ramp of grass and sedge stems sometimes built from water to lip of cup. External diameter 15-20cm; thickness 6-10cm; depth of cup 2.5-5cm; height of rim above water 10-15cm. **Eggs** 3-9 (5.0, n = 55 clutches); oval; slightly glossy; pale cream, pale buff, pale greyish or yellow-white, with few to many small spots and few large angular blotches of pale brown, red-brown or dark brown, sometimes with pale lilac undermarkings; markings often chiefly at larger end. Size of eggs: (n = 82, Nigeria) 31-37.2 x 23-27 (34.1 x 24.8); (n = 161, southern Africa) 29.9-38.5 x 22.7-27 (34.2 x 24.8); calculated weight 11.2. **Incubation** Period 19-20 days; probably by both sexes; begins before clutch complete. **Chicks** No information on parental care. Fledging 35-38 days. Young independent at 5-6 weeks (W. R. Tarboton pers. comm.). May lay twice in a season (Andersson & Gurney 1872). Conspecific brood parasitism recorded (see Social Organisation), and nest predation rates were high (58% of 33 nests lost) in an ephemeral wetland (Jamieson *et al.* in prep.).

132 SPOT-FLANKED GALLINULE
Gallinula melanops Plate 40

Rallus melanops Vieillot, 1819, Paraguay.

Sometimes placed in *Porphyriops*. Three subspecies recognised.

Synonyms: *Crex/Ortygometra/Amaurornis femoralis*; *Fulica/Gallinula/Porphyriops crassirostris*; *Hydrocicca/Ortygometra melanops*; *Porphyriops leucopterus/guttatus/melanops*.

Alternative names: Little/Spotted Waterhen.

IDENTIFICATION Length 22-30. Distinctive medium-sized gallinule, in shape resembling large *Porzana* crake more than *Gallinula*; on water appears flat-backed with scarcely raised rear end. Forepart of head, and often crown to nape, blackish; sides of head, neck, upper mantle, breast and upper belly slate-grey; rest of upperparts warm olive-brown, often washed golden-olive, more chestnut on scapulars and upperwing-coverts especially in race *bogotensis*; tail blackish. Flanks grey to pale brownish, heavily spotted white; lower belly whitish; undertail-coverts white, prominent when tail raised. Iris red; stout bill and narrow frontal shield lime-green (contrasting with blackish face); shield bluish-green when breeding; legs and feet greenish; toes lobed. Sexes alike. Immature washed chestnut on head; mantle and sides of head olive-brown; chin whitish; underparts paler than in adult, with sandy wash. Juvenile like immature, but paler and browner on sides of head; chin, throat and sometimes foreneck white (may be mottled grey); flank spotting rather faint. Inhabits waters with fringing, emergent and floating-leaved vegetation.

Similar species Although rather *Porzana*-like in shape, this species is easily distinguishable from sympatric crakes on size, plumage and green bill and frontal shield. It is much more conspicuous and easy to observe than are crakes, and is normally seen swimming. Spotted flanks, shown by all ages, easily distinguish it from gallinules. Similar-sized Azure Gallinule (125) has greenish-yellow bill and shield, but differs in plumage; has yellow legs and feet, and skulks in vegetation, rarely swimming.

VOICE Information is from Varty *et al.* (1986). Vocalisations are described as a variety of tapping, clicking and clucking calls. In intraspecific conflicts, birds give a gentle *tap-tap-tap...* growing to a loud, laughing *huh-huh-huh...* which ends abruptly. This call is probably also used to advertise occupancy of a territory, and presumably is the call referred to by Ripley (1977) as a loud, hollow cackling, resembling a sudden burst of hysterical laughter, the notes beginning loud and long and becoming brief and hurried as they die away. Birds call from within and outside the reedbeds, beginning just after dawn and continuing throughout the day until dusk. The territorial call is frequently answered immediately by a neighbour, which is then answered by the next bird, and so on along the shore. Birds also called in response to a braying donkey. When foraging, the birds normally give quieter *tap-tap-tap* or *tuh-tuh-tuh* calls and a mixture of soft clucks, these calls being inaudible beyond 15-20m. All calls are apparently given by both sexes.

DESCRIPTION *G. m. melanops*
Adult Forehead, and often also crown to nape, blackish; hindneck to upper mantle, and sometimes also crown to nape, slate-grey, variably washed dark olive-brown; rest of upperparts warm olive-brown, often washed golden-olive; scapulars tinged chestnut and washed golden-olive; upperwing-coverts strongly tinged chestnut, sometimes washed golden-olive; rectrices blackish-brown. Alula, primary coverts, primaries and secondaries dark brown, alula feathers tipped white and narrowly edged white on outer webs; outer web of outermost primary edged white; tertials dark brown, washed golden-olive; all remiges

narrowly fringed buff to tawny. Axillaries regularly barred brownish-olive and white; underwing-coverts white, spotted and barred brownish-olive. Lores, chin and forepart of face blackish; sides of head, throat, sides of neck, foreneck, breast and upper belly slate-grey, palest on upper belly; foreneck and breast sometimes washed sandy; flanks pale brownish to grey, with prominent white spots (often somewhat elongated); thighs greyish; lower belly whitish, sometimes with greyish mottling; vent whitish, sometimes with vague darker barring at sides; undertail-coverts white. Iris red; bill stout and lime-green to greenish-yellow; frontal shield pale bluish-green (when breeding) to green; shield may become duller in non-breeding-season; legs and feet olive-green to dull greenish. Unique in genus in having lobed toes. Ripley (1977) states that bill has reddish base, a feature not described elsewhere and probably an error.

Immature Similar to adult, but forehead to nape warm olive-brown, washed chestnut; hindneck often washed olive-brown; mantle dull olive-brown; sides of head dark olive-brown where adult is black and olive-brown (sometimes tinged grey) where adult is grey; chin whitish; underparts paler than in adult, strongly washed sandy and more extensively white on belly and vent.

Juvenile Similar to immature, but sides of head paler and browner; white on chin extends to throat, and sometimes foreneck, and may be mottled greyish; sides of neck sometimes mottled whitish; underparts sometimes more extensively white (variable).

Downy young Described (Fjeldså & Krabbe 1990) as blackish-brown to black, with almost bare pink and blue crown; bill banded black and white; legs and feet greyish; their illustration shows bill rather green-tinged with 2 black bands, and small white wing-claw. Navas (1991) describes bill as black, with transverse bands of greyish-cream, and tip of upper mandible white; legs and feet black.

MEASUREMENTS (7 males, 10 females) Wing of male 115-134 (123.9, SD 6.5), of female 112-123 (118.4); tail of male 46-56 (50, SD 4.3), of 2 females 49, 50; culmen (including shield) of male 25-33 (29.7, SD 3.5), of female 24-28 (26.4); tarsus of male 32-41 (35.7, SD 3.6), of females 31-36 (33.4). Weight of 1 male 195, of 1 female 225, of 1 unsexed 154 (Belton 1984, Marín 1996).

GEOGRAPHICAL VARIATION Slight; races separated on colour of axillaries, and on size.

G. m. bogotensis (Chapman, 1914) – E Andes of Colombia (Santander, Boyacá, and Bogotá savanna, Cundinamarca). Possibly R Tullumayo, Junín, Peru, but identity questionable – see comments in Blake (1977). Similar to nominate, but axillaries wholly white or dusky at base, not barred; mantle usually washed chestnut (variable), and upperwing-coverts and scapulars often more markedly washed chestnut than in other races. Wing of 6 males 117-140 (129.1), of 10 females 121-132 (125.2); tail of unsexed birds 52-57; culmen (including shield) of 6 males 21-33 (28.3), of 10 females 21-31 (27.1); tarsus of 6 males 34-44 (40), of 10 females 35-38 (36.5).

G. m. melanops (Vieillot, 1819) – Bolivia (La Paz, Santa Cruz, Tarija and Oruro) and Paraguay (W of Paraguay R), E and S Brazil (Ceará, Pernambuco, Bahia, Rio de Janeiro, São Paulo, Santa Catarina and Rio Grande do Sul) to N and C Argentina (W to Tucumán, La Rioja and Mendoza and S to Córdoba, Santa Fe and Buenos Aires) and throughout Uruguay (except Flores and Rivera). See Description.

G. m. crassirostris (J. E. Gray, 1829) – Chile (Atacama S to Aisén) and S Argentina (W Río Negro, Chubut and Santa Cruz). Larger than nominate; bill thicker. 14 males, 4 females: wing of male 126-140 (134.8, SD 3.8), of female 121-132 (126.8, SD 4.6); tail of male 52-61 (57.4, SD 2.7), of female 50-56 (53.8, SD 2.9); culmen (including shield) of male 29-35 (31.8, SD 1.8), of female 27-32 (29.75, SD 2.7); tarsus of male 36-44 (41.1, SD 2.5), of female 37-42 (38.8, SD 2.4).

MOULT Remex moult appears to be simultaneous (1 specimen, BMNH), Chile, Feb.

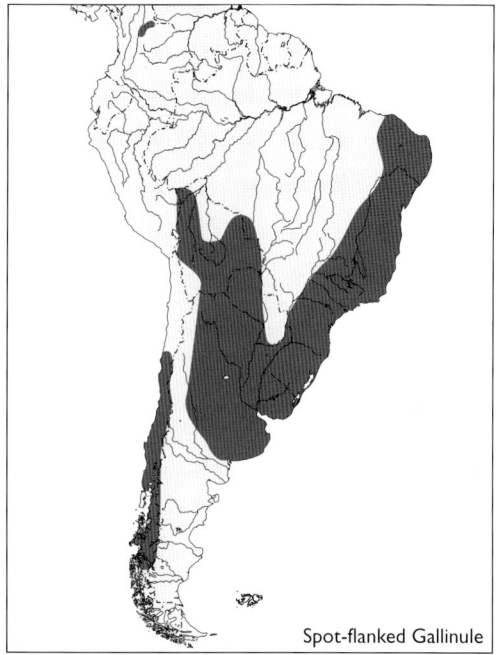

DISTRIBUTION AND STATUS NC Colombia, Bolivia, Paraguay, E and S Brazil, Uruguay, Argentina and C Chile. Not globally threatened. The nominate race was formerly regarded as fairly common, at least locally, as was *crassirostris*, and the status of *bogotensis* was unknown (e.g. Ripley 1977, Nores *et al.* 1983, Belton 1984). It is currently regarded as locally common in SE Brazil (e.g. Rosário 1996), widespread in Uruguay and common in the coastal zones (Arballo & Cravino in press), and widespread in SW Brazil, W Paraguay, E Bolivia and the lowlands of N and C Argentina (Fjeldså & Krabbe 1990, Hayes 1995), and the nominate race is thought to be spreading southwards (e.g. Fjeldså & Krabbe 1990, do Rosário 1996); however, it is relatively rare in Chaco and scarce in Buenos Aires (Contreras *et al.* 1990, Narosky & Di Giácomo 1993). The race *bogotensis* has a small range in Colombia but within this it was widespread and locally common in the 1980s, e.g. at Parque La Florida, and at L Tota where an estimated 40-50 pairs were resident in Jul-Aug 1982 (Hilty & Brown 1986, Varty 1986). In Paraguay, where the nominate race is uncommon, there are many records from W of the Paraguay R (Hayes 1995) but only 1 from E of the river, at Lagunita in the Reserva Natural del Bosque Mbaracayú (Canindeyú) in Sep 1995 (Lowen *et al.* 1997).

The total population of this species is estimated at 100,000-1,000,000 birds (Rose & Scott 1994). Its flesh is said to be bitter and unpalatable, which affords it some protection from hunters (Johnson 1965). Varty et al. (1986) mention that suitable habitat for this species could be encouraged by creating open areas amongst reeds, possibly by some form of reedbed management whereby small areas could be harvested periodically to create patches of different ages; this would diversify the habitat, lead to healthier reedbeds and provide food for a variety of reed-dwelling species.

MOVEMENTS It is said to be a summer visitor in S Argentina and S Chile (Ripley 1977) but the extent of its seasonal occurrence is not clear and is not shown on the distribution map. In Argentina it makes regional movements in response to changing water levels (Canevari et al. 1991), while at Cabo San Antonio birds were first seen in Sep and were gone by late May (Weller 1967). It is regarded as resident in SE Brazil (e.g. Belton 1984).

HABITAT Ponds, ditches, marshes, lagoons, and lake margins with emergent, fringing and (often extensive) floating-leaved vegetation; it also occurs in wet savannas, and is said to frequent dense reeds and rushes along rivers (Ripley 1977). It does not usually frequent open water areas. At L Tota, Colombia, Varty et al. (1986) found it among *Typha* and *Scirpus* beds along the shoreline, particularly in floating vegetation such as *Azolla* and *Lemna* bordering the reedbeds, and it was commonest in the wider stretches of reedbeds although it also bred along steep shorelines with only a thin band of *Scirpus*. In Argentina, it may prefer temporary waters for breeding (Canevari et al. 1991). It occurs in temperate zones; in Argentina it inhabits lowlands up to 750m but in the E Andes of Colombia it occurs at 2,000-3,100m, and in Bolivia it is recorded at 3,700m (Armonía 1995).

FOOD AND FEEDING Stomach contents are described as seeds of water plants. It is also said to take small molluscs, crustaceans, insects and other aquatic invertebrates (Navas 1991). The following information is taken from Varty et al. (1986) unless otherwise specified. It is omnivorous, taking dead fish, insects, and plants such as *Elodea* and *Lemna*; it also pecks at floating wood and reed. *Lemna* is a preferred food, and birds gorge themselves when they find a patch of it. Birds feed mostly by swimming, picking food such as insects from floating vegetation and from the water surface, and taking other floating material washed up against the reeds; they seldom forage on marsh vegetation or on land. They eat pieces of *Elodea* which float to the surface (usually breaking them up first), and they also up-end occasionally to reach submerged plants. They turn over *Azolla* mats with the bill and peck at the roots, but it is not clear whether they take insect larvae or the roots themselves. When foraging, they swim close to the shore, weaving in and out of the reeds and making frequent stops. Pair members often travel in opposite directions, crossing around the middle of the stretch being searched. It is unusual to see more than 1 bird feeding in the same area, unless chicks are present, but at 1 site up to 7 juveniles were regularly found feeding together in a large area of floating vegetation.

HABITS The habits of this species essentially resemble those of the Common Moorhen (129). It is normally seen swimming, which it does well, with its head jerking backwards and forwards in a typically moorhen-like manner. It walks with agility over floating vegetation, continually jerking its tail, and dives easily (Canevari et al. 1991); Varty et al. (1986) also describe it as quite at home on land. It is not shy and is normally conspicuous and easily seen, being tolerant of nearby human activities; it rarely takes flight, even when disturbed; it seldom ranges more than a few metres away from reedbed cover; and it is active throughout the day (Varty et al. 1986); however, Navas (1991) describes it as relatively shy. When danger threatens, it flutters across the water for a short distance; it also submerges to escape, hiding among floating or rooted vegetation with only its head above the water (Arballo & Cravino in press).

SOCIAL ORGANISATION Occurs singly, in pairs or family parties, sometimes in flocks. Six pairs observed at L Tota, Colombia, by Varty et al. (1986) held breeding territories along the lakeshore. These territories were essentially linear, as the birds foraged along the edge of the reedbeds, and ranged from c. 100-130m to c. 300-350m in length, averaging c. 160-190m. They were fairly exclusive to the occupants, although there was some overlap between neighbouring pairs.

SOCIAL AND SEXUAL BEHAVIOUR These birds are quarrelsome, often chasing each other pattering across the water (Fjeldså & Krabbe 1990). At L Tota, Colombia, Varty et al. (1986) noted that an intruder on a breeding territory would usually be met by one of the occupants, whose calls were usually sufficient to drive the intruder away. Occasionally the two birds met, when they swam side by side almost touching each other, with their bodies stiffened, throats swollen and heads bent down to the water as they called (see Voice); the intruder then retreated. This display was also seen between pair members, although the birds usually ignored each other. Territory holders showed no interspecific aggression, tolerating all other bird species in the territory.

BREEDING AND SURVIVAL Season Colombia, Feb and Jun, and downy young Jul-Aug, but also breeds in other seasons; Chile, Oct-Nov (spring); SE Brazil, season prolonged, feathered young seen with adults from mid-Oct to Mar; Uruguay, Nov, chicks mid-Dec, independent immatures end Dec (Arballo & Cravino in press); Argentina, Sep-Mar (Canevari et al. 1991). **Nest** Built among reeds or on damp ground, slightly above water or almost floating; a deep cup of dry rushes, or of strips of dead *Typha* woven among living reeds, with half-dome of same material; diameter 15-28cm, depth of cup 3-5cm. **Eggs** 4-8; blunt oval and glossy; brownish-buff, with numerous rich brown to reddish-brown spots and blotches, usually concentrated at larger end, where many eggs also have twisted and knotted lines; also a few small pale purple undermarkings. Size of eggs: *melanops* 37-43.2 x 26-31 (39 x 28.6, n = 17), calculated weight 18; *crassirostris* 37-46 x 27-31 (41.5 x 29.2, n = 65), calculated weight 19.8. **Incubation** Nothing recorded. **Chicks** Only 1 parent seen to feed chicks, but both may do so; juveniles are also fed occasionally by adults (Varty et al. 1986). Birds of prey and certain mammals take eggs and young (Johnson 1965).

133 BLACK-TAILED NATIVE-HEN
Gallinula ventralis   Plate 40

Gallinula ventralis Gould, 1837, Swan River, West Australia.

Formerly sometimes placed in genus *Tribonyx*. Although most appropriately retained in *Gallinula*, may merit separate subgenus along with *G. mortierii* on basis of morphological features. Monotypic.

Synonyms: *Tribonyx/Microtribonyx ventralis*.

Alternative names: Black-tailed Waterhen/Swamphen/ Gallinule/Native Bantam.

IDENTIFICATION Length 30-38cm; wingspan 55-66cm. Large, dark, thickset, fleet-footed rail with upright stance, long wings and vertically fanned black tail; overall appearance rather like a bantam-hen. Top of head to hindneck olive-brown; entire upperparts, and upperwings, uniform olive, with dark brown feather centres varyingly visible on scapulars and some upperwing-coverts; remiges slightly darker; tail blackish. Forehead to chin blackish, giving indistinct band round base of bill; sides of head olive-brown to dark grey; breast and flanks dark grey, browner on rear flanks and with large, long white spots on lower foreflanks; belly blacker; undertail-coverts blackish. Narrow white leading edge to outer wing. Iris orange-yellow; bill and shield light green, lower mandible orange-red towards base; legs and feet pinkish-red. Sexes similar, but female slightly smaller, and slightly duller and paler overall; dark band round base of bill reduced or absent and white spots on sides of body smaller. Immature very similar to adult, but some retain smaller juvenile white flank spots; bare part colours identical to adult. Juvenile resembles adult female but is paler on head and neck, and in worn plumage has off-white lores, face, chin and throat; has paler underparts contrasting more with blackish vent and undertail-coverts; white flank spots smaller; bill greenish-yellow with dusky tip; iris blackish; legs and feet brownish-pink. Inhabits permanent and temporary wetlands in low rainfall areas; may occur on almost any open dry area close to water and cover.

Similar species Overall appearance, size, gait and behaviour, and preference for foraging in groups in drier habitats, distinguish it from all sympatric rallids. All ages of similar-sized Dusky Moorhen (130) have conspicuous white lateral undertail-coverts, slimmer build, finer bill which is never mostly green, differently shaped tail, and shorter wing-point, which extends only a little beyond tertials in folded wing (wing-point of Black-tailed Native-hen extends well beyond tertials). Vagrant to Tasmania, where may be confused with Tasmanian Native-hen (134), which is similar in appearance but much larger, flightless and appears much shorter-winged; lacks black around base of bill; has white markings on scapulars and upperwing-coverts and large, conspicuous white patch on flanks; has red iris, olive-yellow bill and grey or grey-olive legs and feet.

VOICE Information is from Marchant & Higgins (1993). The voice is virtually unknown. The birds are usually silent, but both sexes have a sharp *kak* of alarm, a harsh cackling, and a rapid, harsh, metallic *yapyapyapyap*. They occasionally utter a peculiar sharp cry when feeding, but flocks are quiet. This species has a very small vocal repertoire compared with the Tasmanian Native-hen.

DESCRIPTION
Adult male Forehead, lores and chin black-brown, forming indistinct dark band round base of bill; crown to hindneck olive-brown to dark olive-brown; upperparts, including scapulars, tertials and all secondary coverts, uniform olive, feathers with dark brown centres which may be visible on longest scapulars and are varyingly exposed on greater and outer median coverts and less exposed on inner medians; primary coverts and alula dark brown to black-brown, outer webs more olive-brown. Primaries and secondaries as primary coverts, but most have narrow brownish-olive outer edge becoming more olive-grey on outer primaries; P10 has broad white outer edge; axillaries olive-brown, sometimes tipped white; underside of remiges, and all greater coverts and median secondary coverts, grey, coverts with broad white tips giving stripes across underwing; rest of coverts dark brown, with broad white tips and black-brown subterminal bands. Tail black to black-brown, central feathers edged dark olive basally. Sides of head grade from olive-brown of crown to black-brown or grey-black of chin and throat; breast dark grey or dark leaden-grey (when fresh, having blackish feather-tips which may make it look blackish), grading to blackish on vent and undertail-coverts; anterior flanks as breast; thighs and posterior flanks dark grey-brown; long feathers on sides of belly/lower foreflanks, which may cover hindflanks, grey-black, each with large, elongated, white subterminal spot. Iris orange-yellow, sometimes yellow; bill and shield light green, sometimes becoming duller green or pale bluish-grey at tip of lower mandible; basal half to two-thirds of lower mandible orange-red to red; legs and feet pinkish-red.
Adult female Very like male. Blackish-brown feathering round base of bill narrower, sometimes absent; crown to hindneck olive-brown, usually lighter than in male; upperparts olive to grey-olive, paler and greyer than in male; tertials and secondary coverts paler and greyer than in male; grey of throat and foreneck lighter than in male; breast paler and faintly bluer than in male, and grey colour often extends over much of belly; white flank spots usually slightly smaller than in male. Bare parts as male.
Immature Very similar to adult; some separable on narrower shape of primaries, and some also retain a few juvenile feathers on flanks (where smaller, narrower spots conspicuous), midline of underparts (feathers paler and white-tipped) and tail. Bare parts as adult.
Juvenile Head and neck similar to adult female, but paler olive-brown or olive-grey, grading to pale grey or off-white on face, chin and throat; when worn, white bases to feathers of lores and chin visible. Wings, tail, upperparts and underparts as adult female, but breast paler and greyer, this colour extending over whole belly so that blackish vent and undertail-coverts contrast sharply. Flank spots much smaller, narrower and less conspicuous, comprising white shaft-streaks with 2-3 small spots near feather-tips, distal spots bordered dark brown. With wear, breast and belly fade to grey-brown, feathers sometimes narrowly tipped white; thighs and rear flanks fade to grey-brown. Primaries narrower than in adult. Iris black-brown; bill greenish-yellow with dusky tip; legs and feet dirty pink. Adult bare part colours attained before postjuvenile moult.
Downy young Black downy chick has greenish sheen; colour fades to black-brown on body as chick grows. Iris blackish; bill black with dark pink basal third of upper mandible and narrow, white or pale grey, black-bordered

x vagrant

Black-tailed Native-hen

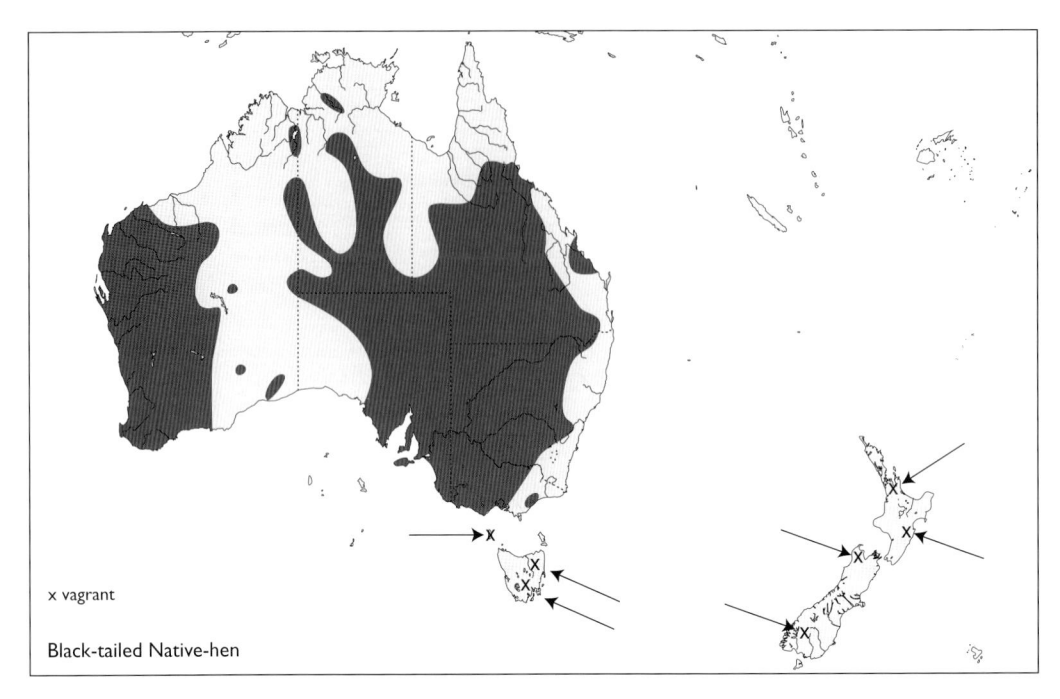

central saddle on upper mandible; legs and feet grey-black. Large white egg-tooth present.

MEASUREMENTS Wing of 18 males 205-231 (218.3, SD 6.5), of 15 females 200-223 (210.2, SD 6.8); tail of 18 males 76-88 (82.7, SD 3.5), of 15 females 72-87 (80.1, SD 4.5); bill (tip to junction of upper tomium with feathers) of 18 males 30-39 (33.0, SD 2.4), of 15 females 27.5-31.5 (29.2, SD 1.2); tarsus of 15 males 59-66 (61.7, SD 1.8), of 14 females 50.5-62.5 (56.5, SD 3.6) (Marchant & Higgins 1993); culmen (including shield) of both sexes 28-32 (Ripley 1977). Weight of 8 males 250-530 (410, SD 82.8), of 5 females 322-405 (364, SD 33.9).

GEOGRAPHICAL VARIATION None; putative races *whitei* and *territorii* not distinct (Peters 1934, Greenway 1973).

MOULT Information is from Marchant & Higgins (1993). Adult postbreeding moult is complete; there are no records of birds in remex moult and they are assumed to be cryptic during a simultaneous moult of remiges. Wear of primaries indicates that most birds moult the remiges in late summer or autumn: most Sep-Oct birds have worn wings but some have completed the moult. A few still have worn plumage in May. There are too few data to rule out the possibility of a pre-alternate (prebreeding) moult. Postjuvenile moult is partial and does not include the remiges, which are renewed during the second pre-basic (postbreeding) moult; some birds retain a few juvenile feathers in tail or on underparts. The date of postjuvenile moult presumably depends on the hatching date; it is recorded in Jan-Apr, but birds with fresh plumage and no moult have been collected even in May and Jul.

DISTRIBUTION AND STATUS Most information is from Marchant & Higgins (1993). This species is endemic to mainland Australia, where it is generally widespread S of 20°S and W of the Great Dividing Range, with only scattered records from coastal and subcoastal regions in Queensland and New South Wales. It is absent from the Cape York Peninsula, present mainly in the interior of South Australia, widespread but erratic in the SW and W of Western Australia, and widespread but localised and sporadic in Northern Territory. The Australian Atlas (Blakers *et al.* 1984) records no breeding N of 22°S, and most breeding records come from the Murray-Darling and L Eyre drainage basins. Not globally threatened. It is locally common, and wide fluctuations in numbers are recorded, large concentrations of 10,000-20,000 birds occurring during irruptions. Annual indices of relative abundance from aerial surveys (transect counts), obtained in 1983-89 and covering wetlands in c. 12% of the land area in E Australia, were between 2,222 and 25,424 (61-100% of the total numbers were counted). Irruptions were more frequent 50-100 years ago, and the lower frequency in recent years is possibly because of the reduced extent of the bird's breeding grounds following extensive drainage of swamps and control of flooding on inland rivers. During irruptions it may cause damage to crops and vegetable gardens by trampling and eating plants, and may pollute water supplies. Artificial habitats are often used, especially during irruptions. Possible predators include Grey Falcon *Falco hypoleucos* and Wedge-tailed Eagle *Aquila audax* (Hobbs 1973, Brooker *et al.* 1979).

MOVEMENTS Most information is from Marchant & Higgins (1993). It may make regular seasonal movements in some regions: it is reported as a regular visitor in the extreme SW; reporting rates N of 26° S increase in the wet summer season; and in Victoria reporting rates increase in spring and summer and decrease in winter. However, in some areas (e.g. Mannum region of South Australia, swamps in SW New South Wales, and Milloo District, Victoria) it is present at most times of the year. It is dispersive and highly irruptive, being able to exploit the erratic floods which are so characteristic of the arid interior of Australia (Serventy 1985) and since 1833

516

irruptions have occurred on average once every 2.7 years. Irruptions are apparently associated with favourable breeding conditions in the interior, allowing a buildup of numbers, followed by harsh conditions, drought and decreasing food supplies which cause birds to leave the interior and disperse often towards the coasts although possibly in all directions. Irruptions are sometimes characterised by the sudden appearance and disappearance of large numbers of birds over a period of as little as 12h, e.g. in NW New South Wales c. 10,000 birds moved into the Manara area within 12-14h of rain. Widely separated regions may experience simultaneous influxes. Movements are often linked with rain: it often appears after rain and after cyclonic disturbances, and it is also claimed to appear before rain actually falls; it possibly moves in response to cloud banks. In South Australia it arrives at wetlands partly filled by rain, and a flock arrived at a pond which had been formed by rain on the previous day. It may move into flooded areas and areas with lush green growth, and away from areas subjected to drying-out, drought or hot weather; however, it is also known to move into regions affected by droughts. It possibly moves back to its normal range after rainfall, but after irruptions in some districts (e.g. N Western Australia) it may remain during the winter. Birds may move at night, and can fly long distances. It is a vagrant to King I in the Bass Strait (3 birds, Mar-Jul 1985), Tasmania (Jun 1916, Mar-Apr 1988) and New Zealand (North I, May 1957 to mid-1958, May 1986; South I, Jun 1923, Aug 1984); these and mainland vagrant occurrences often coincide with irruptions. One was also taken at sea between Tasmania and the Auckland Is (specimen in BMNH, listed by Sharpe 1894).

HABITAT Information is from Marchant & Higgins (1993). Opportunistic, especially during influxes. It normally occurs in low rainfall areas at permanent or ephemeral terrestrial wetlands, including shallow lakes, swamps, pools, floodplains and flats of rivers and creeks, and inundated depressions; it favours fresh or brackish waters but is often on shallow and more saline ephemeral wetlands. It rarely occurs in dry *Banksia* woodland, and at mangroves and tidal pools. Its habitat is often characterised by dense clumps of lignum *Muehlenbeckia cunninghamii*, canegrass *Eragrostis australasica*, bluebush (*Chenopodium*/*Maireana*?) or saltbush (*Atriplex* sp?), and is sometimes sparsely wooded. This species also occurs at artificial wetlands and other habitats (especially during irruptions), e.g. dams, margins of reservoirs, sewage ponds, and in pasture, crops or fallow lands, and in urban areas, e.g. streets, gardens, parks, golf courses and racecourses. It also occurs in unusual situations such as arid country, sandhills, undulating hills and valleys, coastal flats, and among samphire (Chenopodiaceae; presumably *Halosarcia* spp) and coastal shrubs. It usually breeds near water, in swamps and marshes, around waterbodies, and in farmland and on plains, often in areas with receding floodwaters; at other times it may occur far from water.

FOOD AND FEEDING Information is from Marchant & Higgins (1993). It takes seeds, plant material and invertebrates (especially insects). Plant food includes seeds (e.g. Poaceae, *Polygonum*, *Triticum*, *Hordeum*); young vegetables, fodder and cereal crops (e.g. *Triticum*, *Avena*, *Hordeum*, *Zea*, *Medicago*, *Pisum*); fresh green growth of annuals; aquatic plants; and drying apricots; it also eats thistles (Asteraceae). Invertebrate food includes molluscs, and adult and larval insects: Orthoptera (Gryllacrididae, Tettigoniidae, Acrididae), Coleoptera (Carabidae, Chrysomelidae, Scarabaeidae), Lepidoptera, and Hymenoptera (Formicidae). It feeds at grassy or muddy wetland margins, on open ground near wetlands, in pastures and crops, and in scrub and sand dunes. It sometimes feeds with domestic fowl and at piggeries. It gleans from the ground, alternately running and stopping in order to disturb insects. It also feeds from the surface of water and submerges its head and shoulders.

HABITS Most information is from Marchant & Higgins (1993). Diurnal. The pose is upright and the gait a high-stepping walk, with the tail held erect and folded like that of a bantam-hen. It swims readily, with buoyant carriage and erect tail, but seldom wades. When disturbed, birds spread the tail and flick it rapidly; they run fast and nimbly, sometimes zigzagging towards cover or, if pressed, they will fly or take to water; they hide by tussocks and tall reeds, and even by railway sleepers. A flock of 4,000 escaped from a Grey Falcon *Falco hypoleucos*, by running to hide under nearby trees, each tree having >1,000 birds packed tightly beneath it (Hobbs 1973). The flight is strong, with rapid shallow wingbeats on long, narrow rounded wings, with the feet trailing; birds take off and land with a short run. They roost at night in long grass and dense vegetation, and during the day groups sleep and rest round roots of trees; they will perch in trees.

SOCIAL ORGANISATION Information is from Marchant & Higgins (1993). Unclear. On existing evidence, it appears monogamous, and it is territorial when breeding. It often nests in colonies of 5 pairs to 500+ birds, but also breeds as isolated pairs. Nests were 7-10m apart in New South Wales.

SOCIAL AND SEXUAL BEHAVIOUR Information is from Marchant & Higgins (1993). It is highly sociable; it forages gregariously or in pairs, and very large numbers gather when conditions are suitable. During irruptions it is recorded feeding in flocks of up to 20,000 birds, and it was formerly reported gathering in hundreds waiting for domestic fowl to be fed. No intraspecific aggression is reported among captive birds, but it is recorded injuring a domestic chicken. In large feeding flocks, a bird was occasionally seen to run in circles, flapping its wings, and then continue to run with outstretched wings and with the neck stretched forward; it stopped abruptly near another bird, which usually continued the performance (Christian 1909); the function of this display is not clear. A breeding colony may also include other bird species, Black-winged Stilt *Himantopus himantopus*, Red-kneed Dotterel *Erythrogonys cinctus*, Australian Crake (97) and White-fronted Chat *Ephthianura albifrons* being recorded.

Its aggressive behaviour is not well developed, possibly because it is nomadic, and it is thus not typical of moorhens and gallinules. Birds with chicks chase away non-breeding birds which enter the territory. An incubating bird behaves cryptically on the approach of a predator, turning its head away, burying the bill in the feathers and hiding the eyes; it will fly from a terrestrial predator, or creep from the nest; a brooding adult and chicks will leave the nest together.

Nothing is known about the bird's sexual behaviour.

BREEDING AND SURVIVAL Season E and S Australia, usually Aug-Dec, but timing influenced by rainfall, especially after drought, when may breed any month; in

SW Western Australia, June-Nov; in SW New South Wales breeds after large influxes, continuously through winter and summer. Breeds opportunistically, often soon after arrival following heavy rain, or as water dries up after flooding; laying time correlated with peak rainfall + 2-3 months (Halse & Jaensch 1989). **Nest** Breeds in swamps of redgum, canegrass, lignum or occasionally *Polygonum*, in grass beside water, and in inundated grass. Nest usually well concealed low in dense vegetation in shallow water or near water, e.g. in clumps of grass, reeds, bushes and herbage; also recorded nesting on tree-stump, in tree-fork or branches, on debris on log in water, in shallow depression in ground, in wooden box in tree, and under wire fence enclosure. Nest cup-shaped, of any available vegetation, including lignum stalks and leaves, reeds, twigs, leaves, grass and bark; lined grass, leaves, waterweed, feathers; often with partial roof of woven stems; sometimes with approach ramp of broken reeds; external diameter 26cm, depth 10cm; height above substrate 0-80cm (0.3, SD 0.2, n = 39); nests recorded over water 20-70cm deep (North 1913, Jaensch 1994). **Eggs** Usually 5-7, range 14-12 (5.9, n = 13 clutches); larger clutches possibly from 2 females; oval, somewhat compressed towards smaller end; lustrous; dull light green to pale sage-green, pale green-blue or dark green, with rounded blotches and minute spots of chestnut and purplish-brown, or sparsely and evenly spotted fine purplish-brown (sometimes also with a few hair-lines) with faint undermarkings of same colour or dull violet-grey. Size of eggs 43-49.5 x 29-34.7 (45.0 x 31.4, n = 60), calculated weight 24. Laying probably synchronous in large colonies where arrival virtually simultaneous. **Incubation** Probably begins before clutch complete; period 19-20 days; hatching asynchronous but all eggs probably hatch within 24-48h. **Chicks** Said to be precocial and nidifugous; remain in nest for unknown time after hatching. Both adults appear to accompany and feed young; timing of development not recorded. When in open, threatened chicks crouch in depressions (e.g. cattle hoofprints), and run or swim to escape pursuit; at the nest they run or climb to cover. May have 1-3 broods per year; may not breed at all in severe drought years (Serventy 1985). **Survival** No information.

134 TASMANIAN NATIVE-HEN
Gallinula mortierii   Plate 40

Tribonyx Mortierii Du Bus de Gisignes, 1840, Tasmania.

Although most appropriately retained in *Gallinula*, may merit separate subgenus along with *G. ventralis* on basis of morphological features. Monotypic, but distinct form from mainland Australia recently extinct.

Synonyms: *Brachyptrallus ralloides*; *Tribonyx gouldi*.

Alternative names: Narkie; Waterhen.

IDENTIFICATION Length 42-51cm. Very large, thickset, flightless rail, rather bantam-like, with long, narrow tail often held erect, and stout bill and legs. Conspicuous, active, noisy, demonstrative and aggressive; walks with short wings partly drooped and tail constantly jerked up and down or carried erect; runs very swiftly with tail erect, opening wings to balance. Crown to hindneck, and upperparts, brown to dark olive-brown, more olive-tinged on mantle and greyish-olive on outer scapulars and upperwing-coverts; remiges black-brown, olive on outer webs; tail black-brown. Outer scapulars, and most upperwing-coverts, have rather small white markings. Sides of head olive-brown; sides of neck greyish-olive; chin to foreneck grey to leaden-grey; breast bluish-grey, darkening to black-brown on belly and undertail-coverts; thighs and flanks pale brownish-grey with narrow white feather tips; large white blaze on foreflanks. Iris red; bill olive-yellow; legs and feet dark greyish. Sexes alike; immature like adult. Juvenile duller and paler overall; as plumage becomes worn, white patch develops on lores and buff feather tips show from back to uppertail-coverts; feathers of breast and belly have whitish tips; bare parts like adult but bill initially dusky. Inhabits grassland and paddocks near wetlands with abundant cover.

Similar species Only likely to be confused with Black-tailed Native-hen (133), which is vagrant to Tasmania and much smaller and longer-winged, with primaries extending well beyond tertials in closed wing, and has dark feathering round base of bill, elongated white spots on foreflanks, no pale tips to upperwing-coverts, orange-yellow iris, pale green bill with orange-red base to lower mandible, and pinkish-red legs and feet.

VOICE Information is from Marchant & Higgins (1993). The characteristic aggressive call of pairs (sometimes trios) is a loud, rasping *see-saw*, given in antiphonal duet by male and female, and at times by 3 birds. Birds also give various grunts, which include contact notes between adults, calls to chicks, calls during aggressive behaviour, and alarm grunts which are soft, low-pitched and repetitive, becoming shorter, louder and higher-pitched in extreme alarm. There is also a low, rattling click call, the notes often in pairs, in response to the distant approach of an intruder; sharp and harsh alarm calls; and screaming calls, including cackles, given when a bird is chased or attacked. The parental alarm call is quite loud and long, resembling a creaking door-hinge. In the male, aggressive and alarm calls may be higher-pitched than in the female. This species is noisy early in the morning and in the evening, and also calls at night. Immatures give a high-pitched cat-like whining *miaow*. Chicks cheep for food.

DESCRIPTION
Adult Crown, nape and hindneck brown to dark olive-brown; mantle, upper back and inner scapulars olive-brown to olive-grey; lower back to uppertail-coverts olive-brown to dark brown; outer scapulars greyish-olive with narrow white tips; tail black-brown. Median upperwing-coverts, longest lessers and greaters greyish-olive, with white spots and streaks towards tips; marginals and shortest lessers olive; primary coverts, alula, primaries and secondaries blackish-brown with outer half of outer webs dark olive-brown, and paler outer edges; alula and P10 (sometimes also P9) have small white tips; tertials have blackish-brown inner webs, olive-brown outer webs and olive fringes. Underwing-coverts dark brown, broadly tipped white; underside of remiges blackish-brown. Sides of head paler olive-brown than crown; sides of neck greyish-olive; chin, throat and foreneck grey to leaden-grey, sometimes tinged olive-grey; breast bluish-grey with varying grey-olive feather tips giving faint olive tinge towards foreneck; lower breast darker. Belly to undertail-coverts black-brown, belly feathers narrowly tipped white when fresh; thighs and flanks pale brownish-grey to pale grey, with narrow white tips which can be lost with wear;

broad white tips to feathers of foreflanks form large white blaze. When very worn, breast may be almost as pale as in juvenile. Iris red; bill olive-yellow, olive tinge strongest on frontal shield; legs and feet dark grey to dark grey-olive, some having dirty pink tinge on inside of tarsus. Sexes alike.
Immature Very similar to adult, but some retain some juvenile feathers on belly; narrow juvenile remiges retained. Bare parts as adult.
Juvenile Similar to adult but top of head to hindneck browner and paler; mantle and scapulars olive; back to uppertail-coverts like adult but olive-brown when worn, with buff feather tips; whitish patch on lores when worn; sides of neck olive-brown; some have narrow white line down middle of chin and upper throat; throat grey (lacks bluish tinge of adult); breast grey, grading to dark grey on belly, feathers with whitish tips; belly becomes dark brown when worn; undertail-coverts black-brown; flanks and thighs more brownish-grey than in adult. Bill initially dusky; adult bare part colours attained before postjuvenile moult.
Downy young Downy chick black with greenish or silvery sheen, colour fading to dark brown. Iris black; bill black, with large white egg-tooth, pink basal half to upper mandible, pink mandibular rami, and narrow lavender saddle half-way across centre of upper mandible; legs and feet grey-black. In older chicks, pink areas on bill fade to white and decrease in size.

MEASUREMENTS Wing of 19 males 194-213 (202.4, SD 5.8), of 17 females 188-211 (198.1, SD 6.1); tail of 18 males 88-102 (95.9, SD 3.1), of 15 females 83-100 (93.7, SD 4.8); bill (tip to junction of upper tomium with feathers) of 19 males 30-35.5 (32.0, SD 1.3), of 14 females 30-33 (31.2, SD 0.9); tarsus of 20 males 76.5-89 (83.9, SD 2.8), of 17 females 74-85.5 (81.1, SD 2.9). Mean weight of male 1,334 (SD 109.3, n = 152), of female 1,251 (SD 103.4, n = 120) (Ridpath 1972a).

GEOGRAPHICAL VARIATION The recently extinct mainland Australian form (see Distribution and Status) was slightly smaller and merits separation as a distinct subspecies *T.m. reperta*, for which the name *Porphyrio reperta* De Vis has priority; the species is believed to have evolved on the mainland and spread to Tasmania when the two land masses were connected (Olson 1975a). The Tasmanian form therefore should perhaps be referred to as *T.m. mortierii* (Olson 1975a).

MOULT Information is from Marchant & Higgins (1993). Postbreeding moult is complete, and primary moult is usually simultaneous; however, 1 adult with very worn remiges had 1 almost fully grown primary in each wing. Remiges are sometimes bitten off just before the moult. The pattern of tail moult is irregular. Moult occurs mainly in late summer and autumn, primary moult being recorded in Jan, Mar and Apr. Post-juvenile moult is usually partial, not involving the remiges, but it is not known if some birds also moult these; it occurs before the first winter and, in at least some birds, the juvenile remiges are retained until the second complete (pre-basic) moult when birds are c. 1 year old.

DISTRIBUTION AND STATUS Tasmania; also introduced to Maria I, off E Tasmania. Not globally threatened. Most information is from Marchant & Higgins (1993). It is endemic to the Tasmanian mainland (but generally absent from the W and SW) and nearby islands, where its distribution has not changed substantially in the recent

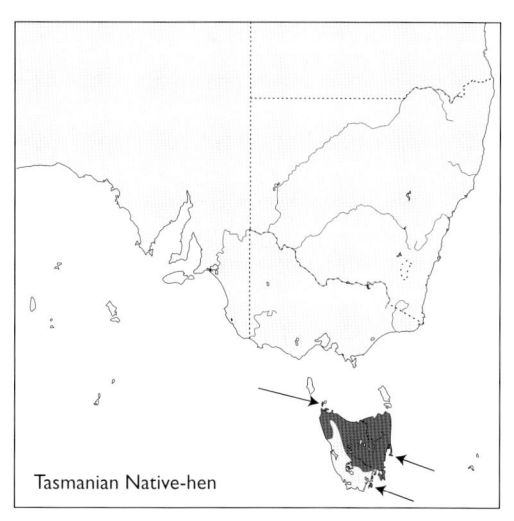

Tasmanian Native-hen

past, although it has expanded into newly opened land. Road construction may provide corridors for range expansion. It was introduced to Maria I in 1969 and has bred there since 1973; it is absent from King and Flinders Is. It has benefited from the introduction of rabbits as well as from the clearance of wooded land for grazing and crops, and the construction of artificial wetlands, but such benefits are offset by wetland drainage. This species is very numerous, despite some local declines. A drastic decline in the Geeveston area, S Tasmania, in 1989, was thought to be the result of disease (Goldizen *et al.* 1993). A decline was reported in the Midlands after the introduction of myxomatosis and the consequent reduction of rabbits (Green 1965); the rabbit calcivirus (RCD), which has spread on the Australian mainland, was introduced to Tasmania in late 1997 (H. Phillipps *in litt.*) and could have a similar effect. The Tasmanian Native-hen is not protected, and was traditionally regarded as an agricultural pest, being reputed to trample, graze and foul pasture and crops; most claims are unsubstantiated, but it may reduce sprouting oats by up to 8% by grazing activity. It was declared vermin in 1950, and was deliberately shot, poisoned and trapped; thousands were killed in 1955-58. Apparently it is still controlled by shooting and round-ups (Serventy 1985). Birds are attacked by feral cats and dogs, and are also killed by road traffic: they appear particularly susceptible on elevated sections of roads (Taylor & Mooney 1991).

This species formerly occurred on mainland Australia (SE Queensland to Victoria and SE South Australia), and is thought to have become extinct there 12,000-20,000 years ago during a period of severe aridity; however, the recent finding of remains only 4,700 years old suggests that its final disappearance coincided roughly with the arrival of the Dingo *Canis familiaris dingo* (Baird 1984, 1986, 1991). In Tasmania it survives predation by feral cats as well as by native species such as the Tasmanian Devil *Sarcophilus harrisi* (Blakers *et al.* 1984). Other possible predators include the Tiger Quoll *Dasyurus maculatus* (Goldizen *et al.* 1993) and the Wedge-tailed Eagle *Aquila audax*; see the Social and Sexual Behaviour section for details of potential predators attacked by Native-hens.

MOVEMENTS Information is from Marchant & Higgins (1993). Normally sedentary, but some groups are recorded as disappearing in summer. Many young disperse from the natal area at the end of their first year (or when up to 17 months old), and trespass into neighbours' territories just before the start of breeding. However, on Maria I 20-30% of young were still in the natal territories at the end of the 2nd year. Movements of up to 15km are recorded for young birds, once 40km, but 72 (75%) of 96 ringing recoveries were obtained <4km from the ringing site.

HABITAT Open pasture, grassland, crops (newly sown cereals or legumes) and other cleared land round permanent or seasonal freshwater wetlands which include marshes, lakes, farm dams, creeks, rivers and streams; rarely round saline wetlands. Grassland may be lightly timbered or infested with alien weeds. Birds usually frequent wetland areas with abundant cover of rushes, reeds, sedges, tussocks, bracken or willows for shelter and nesting. This species occasionally enters adjacent woodland, or penetrates forest along tracks and in clearings. It occurs throughout agricultural areas, especially where farming is less intense; it is absent from heath dominated by button-grass in the south-west. It requires short-grazed pasture all year for foraging, and is currently dependent on swards maintained by introduced species (rabbits, sheep and cattle); such swards were formerly maintained by grazing marsupials such as *Macropus rufogriseus*, *Thylogale billardierii* and *Vombatus ursinus*, and by fire. It always breeds near water. It is most abundant below 700m, but is recorded rarely up to 1,700m.

FOOD AND FEEDING Information is from Marchant & Higgins (1993). Mostly seeds and leaves of plants of: Poaceae, Cyperaceae, Restionaceae, Juncaceae, Polygonaceae, Portulaceae, Caryophyllaceae, Brassicaceae, Rosaceae, Fabaceae, Geraniaceae, Thymelaceae, Apiaceae, Epacridaceae, Primulaceae, Convolvulaceae, Boraginaceae, Scrophulariaceae, Plantaginaceae, Rubiaceae, Asteraceae and Hydrocharitaceae; also cereal crops; the diet varies with the availability of plant species. It also takes a few insects, including Orthoptera, Coleoptera (Scarabeidae, Curculionidae), Diptera, and Lepidopteran adults and larvae; in captivity it eats meat and fish. In 447 gizzards, herbage constituted 93.8% by volume, seeds 4.6%, and animal matter 1.6%. Herbage is eaten most in winter and spring, seeds and insects in summer. Cellulose is not digested well, and starch from young plants is broken down in 2 well-developed intestinal caecae. This species forages on the ground in pastures and paddocks, and also on exposed mud. It gleans and pecks at the ground and seedheads; it grasps herbage and pulls seeds off with the tip of the bill; in orchards it takes fruit from the ground or low branches, sometimes hopping onto branches; it chases flying moths. A secondary grazer, it requires primary grazers to maintain suitable habitat (see Habitat section). Young chicks are given equal quantities of plant material and invertebrate food, the latter including worms, insects (Coleoptera, Lepidoptera) and tadpoles; the proportion of plant food is increased, and by 14 days of age the diet is like that of the adult.

HABITS Most information is from Marchant & Higgins (1993). Diurnal; it is generally in the open during the day, unless alarmed. It walks with the short wings partly drooped and the slightly rounded tail constantly jerked up and down or carried erect; it runs very swiftly with the tail erect, opening the wings to balance; a running speed of 48km/h has been recorded (Serventy 1985). It can also swim and dive. When alarmed, birds stand still with the head slightly raised, and may flick the tail. In response to danger birds also flick the wings outwards and upwards, and they give a harsh alarm call when a predator attacks; the distress scream (see Voice), given when birds are attacked or handled, may attract other Native-hens which sometimes help to repel the attacker; they may threaten and fight potential predators, using intraspecific aggressive behaviour (see Social and Sexual Behaviour). Birds walk or run briskly to cover with the tail cocked and the head raised, watching the predator. To escape danger they may submerge in water, with only the bill showing, and may run under water along the bottom of a stream. Birds roost at night, as a group, on open ground near water; in cold winter weather they may roost among vegetation such as *Juncus australis*. They sleep standing on both legs with the head tucked over the back, and often huddle together. They stand in the shade beneath vegetation on hot days, especially from c. 11:00-16:00. Birds bathe vigorously in creeks and temporary puddles, and preen regularly.

SOCIAL ORGANISATION Most information is from Marchant & Higgins (1993). Monogamous or polygamous, usually polyandrous; rarely polygynous. It is permanently territorial in groups of 2-17 birds consisting of 2-5 (usually 2-3) breeding adults, plus young usually up to 2 years old. The breeding unit is usually an adult pair, or a trio of 2 males (usually brothers) and 1 female; it rarely includes 2 females (always sisters). Adults normally associate for life. All males perform copulations, and all adults share all duties approximately equally, including territorial defence, nest-building, incubation and care of the young. The sex-ratio is biased towards males: at 4-6 months 50:18; in adults 111:74 (Ridpath 1972b); on Maria I it is only slightly biased towards males. Maynard Smith & Ridpath (1972) analysed the data from Ridpath's original study (Ridpath 1972a, b, c) and showed that the selective forces responsible for the tolerance of one male by the other male in the trio depend on the fact that the 2 males are brothers, i.e. that this is an example of kin selection.

When 1-year-old birds leave natal territories and form new groups just before the breeding season, birds often enter other territories and court other 1-year-old birds; they are usually driven out by the parents. When a 1-year-old group survives the first 2-3 days of territorial establishment, its members almost always stay together, and usually breed in the same year.

At Geeveston, S Tasmania, where the population appeared to be in precipitous decline in 1989, the group composition was much less stable than normal: adult male/female relations were unstable, females having 3-6 mates (singly or in pairs) in 1 season and males having an average of 2.3 mates; and juveniles changed groups unusually early and often (Goldizen *et al.* 1993).

Territorial boundaries are usually well-defined; territory sizes of 0.4-1.5ha are recorded and can vary, e.g. when the population density increases, territory size decreases (especially in first-year birds). Larger and older groups hold the largest territories, first-year groups the smallest, usually only half or less the size of adult group territories.

SOCIAL AND SEXUAL BEHAVIOUR Most information is from Marchant & Higgins (1993). Group members usually move around together and feed within 10m of each other, and conflict between members of a breeding group is very rare; there is little evidence of a dominance hierarchy within groups under natural conditions.

Aggression within a group usually ends with 1 bird assuming a submissive crouch, with sleeked body feathers, lowered tail and open bill, generally giving the immature miaow call or the screaming cackle (see Voice). Territorial birds will attack other bird species feeding in the territory, and birds will also sometimes attack potential predators or intruders such as Swamp Harriers *Circus approximans*, goshawks *Accipiter* spp., Brown Falcons *Falco berigora*, Forest Ravens *Corvus tasmanicus*, domestic cats, Red-necked Wallaby *Macropus rufgriseus*, Brush-tailed Possum *Trichosurus vulpecula*, Echidna *Tachyglossus aculeatus*, sheep, rabbits and people. A captive pair defending a land territory excluded wild Common Moorhens (129) sharing the same pond area, and the moorhens defended only water territories (Holyoak & Sager 1970).

Adults share all territorial and breeding duties; young of the previous year sometimes help with breeding duties, and first-brood young sometimes feed and brood second-brood chicks. Territorial behaviour is intensified during the breeding season. Detection of the approach of an intruder is signalled by the click-call (see Voice); the territorial bird then stands still, with body stretched forwards, gives aggressive grunts, and then, usually with companions, lowers the head, rouses the body plumage and runs towards the intruder. When near the intruder, group members strut or trot closer in the "aggressive parade", with arched wings and ruffled neck and body feathers (the white flank patch being prominent), and make the see-saw duet (see Voice). All then abruptly stop calling and adopt a pre-fighting, half-crouched pose, in which they may stand still or stalk around or towards the intruder; conflict usually ends here but, if the intruder does not retreat, fighting occurs. Fights are most common between groups at territorial boundaries, or between groups and first-year birds attempting to enter the group or form a new group; opposing birds jump at each other, grasping and jabbing with the bill and flapping the wings, progressing to a series of vertical leaps 75-150cm high, where they kick and rip with the feet; opponents may push each other apart or may lock together and roll on the ground; most fights last 1-2 min (up to 15 min recorded) and injuries occur rarely, but even death is possible. After disengaging, a chase may ensue, the pursued bird giving the screaming-cackle call; chases usually end at the territory boundary.

Parents will distract predators by making a crouched run on a zigzagged course, with legs bent and head, body and tail flattened. Holyoak & Sager (1970) describe how a captive pair of Native-hens defending a land territory from Common Moorhens, which occupied only water territories, fed 3 very young moorhen chicks on land and usually drove the chicks' parents into the water when the moorhens attempted to feed their young on land. However, on the occasions when the moorhens managed to reach the chicks the Native-hens maintained a passive indifference a short way away, and this failure to attack an 'intruder' close to the chicks was thought to have occurred because of the risk of injury to the adults in a situation where there was little chance of the chicks being saved from a 'predator' which had already reached them.

In the breeding season, birds approaching each other give a greeting display, prancing on the toes with head down and body feathers fluffed, then giving a half-bow with partly arched wings for c. 30 s, and sometimes calling. The female normally initiates sexual display by trotting, sometimes on her toes, near the male, with neck feathers expanded, wing partly spread, tail arched and head raised and bobbed; the male follows with a similar high-stepping gait, neck sleek and extended; if he catches the female he often pecks her nape. The trot develops into a rapid prancing run; if the female holds her head high, the display ceases; otherwise the female slows down and bows, pointing her head down, then stops and crouches with tail depressed and fanned; the male mounts, slipping his legs on either side of the female, lays his head along her back and pecks her neck, then raises his head and breast and copulates, maintaining balance by clasping with his legs and flapping his wings; he finally raises his wings above his back for 1-2s and then dismounts. Both sexes then ruffle the feathers and preen, and the male adopts a postcopulatory pose in which he points his head down, fluffs his feathers, partly raises his wings, half crouches and depresses his tail.

Group members aged 2 weeks and over often allopreen on the head and neck. Behaviour designated as play is recorded: young often chase each other, starting and stopping unpredictably and indulging in harmless fights; at c. 5 months, birds run faster and more vigorously, and also leap, this behaviour often resulting in chasing and harmless fighting which spreads throughout the group; at this time (early in the breeding season), the parents are often involved in territorial fighting.

BREEDING AND SURVIVAL Season Jul-Jan (usually Aug-Nov); determined by rainfall, as dependent on fresh young plant growth. **Nest** A bulky woven cup of grass, reeds or herbage, lined with soft dry grass; well concealed in tall thick vegetation such as rushes, reeds, sedges, tussocks, ferns, prickly bushes (e.g. blackberry *Rubus* or gorse), occasionally thistles. Built 10-120cm above ground or over water, almost always on edge of stream, dam, lagoon or pond; sometimes under overhang or in hole in creek bank; c. 20% of nests concealed overhead with woven stems; entrance leads towards water. Nest may be used for second clutch. Also constructs up to 7 brood nests after eggs hatch (sometimes before); usually in exposed location, e.g. short grass, isolated tussocks, flat rocks, exposed edges of river bank, bare soil, on log, or <150m from cover in open pasture; often untidier and bulkier than egg-nests. **Eggs** Clutch 3-9 (usually 5-8), mean 6.4 (SD 1.8, n = 90), clutch size smaller in drier years, larger in adults than in first-year birds and in trios than in pairs; larger clutches (12-16) probably from 2 females. Eggs oval or elongate oval; smooth and slightly glossy; dull yellowish-stone to light buffy-brown, sparsely but uniformly spotted, blotched and streaked chestnut-brown with purplish-brown under-markings; on Maria I, 2+ family groups laid pure white eggs. Size of eggs 53-61 x 36.3-40.4 (56.0 x 38.7, n = 26), calculated weight 45. Eggs laid at daily intervals. **Incubation** All adult group members incubate, occasionally assisted by first-year birds; incubation stints 5-180 min (usually 20-28); period 19-25 days (usually 22); eggs normally hatch within 48 hours. **Chicks** Precocial and nidifugous; leave nest 1-2 days after hatching; fed and cared for by all adult members of group, occasionally by younger birds. Brood sometimes divided between 2-3 adults for brooding; adult about to brood grunts softly, crouches slightly, ruffles breast and flank feathers, bows head and waves it from side to side; young run under adult, which settles over them; chicks brooded during day for first 1-2 weeks, then at night or in cold weather until 4-5 weeks old. Chicks become self-feeding after 1-2 weeks but fed for 8 weeks;

adult calls chick with grunts, lowers and stretches head forwards, and feeds chick from tip of bill with aid of saliva (not by regurgitation). Young tended closely for 3-4 weeks and then less frequently until 6 weeks old; remain with parents until 9-15 months old. Chicks <6-10 weeks old which stray into neighbouring territory are fostered by occupying group; older young usually attacked. When threatened, chicks hide head downwards in tussocks round nest, and run to squat against irregularities in ground; after 8 weeks, climb tussocks, bushes and rocks up to 3m above ground; can run as fast as adults after 6 weeks. Hatching success 89% (n = 88), not affected by age of female or size of group; survival of young to 4 months 46% in adult pairs and 61% in adult trios; survival higher in adult groups (6.5/trio, 5/5/pair) than first-year groups (3.1/trio, 1.1/pair). Eggs and chicks taken by Forest Ravens, chicks by Brown Falcons and Swamp Harriers. Able to breed in first year. Often rears 2 broods per season, sometimes 3; older groups more often multiple-brooded; more groups lay second clutches in wetter years. Second clutch laid when first brood 1-15 weeks old, possibly sooner. **Survival** Birds aged 7-19 months suffer higher mortality if they leave breeding area; birds older than 19 months have mean annual mortality of 6.8%.

FULICA

Coots are large to very large rails of open water, with entirely black or dark grey-black plumage although most have some white on the undertail-coverts (Fig. 14). The genus contains 11 species and is of worldwide distribution in tropical and temperate regions, but the centre of species abundance and diversity is in South America, where eight species occur, and it seems likely that the genus originated there and spread to the Old World (Olson 1973b). Coots were formerly accorded their own subfamily, but they are very similar to *Gallinula*, both skeletally and in their adult and juvenile plumages, and are derived from *Gallinula*-like stock which has become adapted for diving, e.g. Coots have a relatively long, narrow pelvis. The genus is well defined and has diverged relatively little from its ancestral stock. The Hawaiian Coot has only recently been treated as a species distinct from the American Coot, while the Caribbean Coot is sometimes regarded as a morph of the American Coot.

Coots are predominantly vegetarian and have stout bills with shearing edges. There is normally a prominent frontal shield, but the Horned Coot of South America has an extensible and erectile black proboscis, which normally rests on the ridge of the bill. Bill, shield and leg colours vary markedly between species, and the Slate-coloured Coot has two morphs which differ in the colours of the bare parts. The diet of most species is predominantly submerged aquatic plants, which the birds frequently dive to obtain; the toes of all species are lobed, but the lobes are least developed in the Red-fronted Coot which is primarily a surface feeder. The wings are rather short and rounded, and adult Giant Coots are normally too heavy to fly although the immatures are smaller and fly readily. The two largest species, the Giant and Horned Coots, which inhabit barren, high-altitude lakes in the Andes, build enormous nests, that of the Horned Coot sometimes resting on a conical mound of up to 1.5 tonnes of stones which are collected by the birds. Some species are widespread and numerous but two are classed as Vulnerable: the Hawaiian Coot has declined drastically because of habitat loss and introduced predators, so that only a small population survives, and the Horned Coot is probably in decline and some of the lakes at which it occurs suffer from water removal and habitat degradation, while the species is also hunted.

135 RED-KNOBBED COOT
Fulica cristata   Plate 41

Fulica cristata Gmelin, 1789, Madagascar.

Forms superspecies with *F. atra*, *F. alai*, *F. americana*, *F. caribaea*, *F. leucoptera* and *F. ardesiaca*. Monotypic.

Synonyms: *Fulica mitrata*; *Lupha/Lophopalaris cristata*.

Alternative name: Crested Coot.

IDENTIFICATION Length 35-42cm; wingspan 75-85cm. Dark slate-grey; head and neck black; underparts dark grey; lower flanks to undertail-coverts and tail black. Has very small amount of white at bend of wing, rarely has white on outermost primary; usually has no white on tips of secondaries; undertail-coverts rarely with some white tips. Iris red; bill and shield white, sometimes tinged blue; 2 large red knobs on top of shield; legs and feet greenish. Sexes alike, but female smaller. Non-breeding adults have small knobs (often very hard to see), red-brown iris, duskier tinge at sides of bill, and predominantly dull slate legs and feet. Immature like adult but feathers of chin, throat, breast and underparts fringed white; upperparts often have dusky olive tinge. Juvenile has crown and upperparts dark brown; sides of head, neck and flanks dark olive-brown mottled off-white; lores, chin and throat white; underparts pale ash-grey; iris grey-brown to dark brown; bill dull grey; legs and feet dark grey. Inhabits lakes, lagoons, ponds, dams, floodplains and swamps, sometimes large rivers; breeds at waters with fringing or emergent vegetation.

Similar species Common Coot (136) very similar in appearance; confusion possible. Red-knobbed overall slightly darker in appearance with plumbeous-grey rather than slate-blue tinge; only slightly bulkier but wings c. 10% longer. Characters most useful in separating the 2 species are: Red-knobbed usually has no white tips to secondaries and hardly any white at wing-bend; has rounded projection of loral feathering between bill and shield (Common has

F. cristata

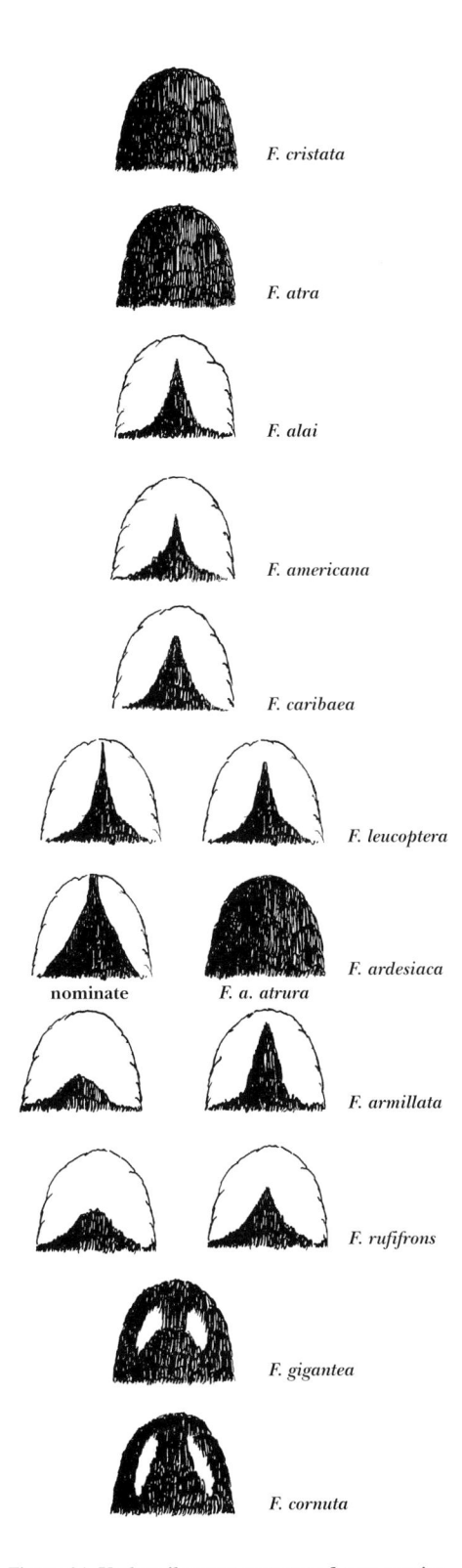

F. atra

F. alai

F. americana

F. caribaea

F. leucoptera

F. ardesiaca
nominate   *F. a. atrura*

F. armillata

F. rufifrons

F. gigantea

F. cornuta

Figure 14: Undertail-covert patterns of coot species.

more extensive and pointed projection); often has bluish tinge to bill (Common has pinkish tinge); has red knobs at top of frontal shield. Referring to additional field characters for separating these species proposed by Forsman (1991), Keijl *et al.* (1993) point out the following: in the field, greater body size of Red-knobbed not obvious; both species when swimming may look either flat-backed or round-backed; slimmer neck of Red-knobbed usually not noticeable; more triangular head shape of Red-knobbed Coot due to presence of knobs and size of shield – in birds with no knobs and small shield, head shape very like Common; differences in shield shape (see Description) hard to see in frontal view. Juvenile much darker than juvenile Common but may be hard to separate from subadult Common, which may be very similar in colour. Larger and bulkier than Common Moorhen (129), lacking white on flanks and undertail-coverts and with different bare parts colours; may appear round-backed, and swims with head closer to body and nodded much less; in flight looks much bulkier, more duck-like, than Common Moorhen; wingbeats seem fast for actual progress made; when not breeding, often occurs far from cover and forms large flocks.

VOICE Information is from Urban *et al.* (1986) and Cramp & Simmons (1980). This species has a wide vocabulary and the function of many calls is uncertain. The commonest calls include a sharp, shrill *kik* or *krik*; a low, reedy *kek*; a shrill, trilled *krrt*; a double clucking note *clukuk* or *k-kek*; and a deep *koop* or *kup*. Other calls include: a hollow *hoo*; a loud *kwon* or *cronk*; a deep, nasal, frog-like croak; a drawn-out *ker*; a *coo-dooc* or *coo* followed (possibly from another bird) by a deep *poof* (often heard, Morocco, early Mar); a metallic ringing *croo-oo-k* or snorting *tcholf* of alarm; and a high-pitched, nasal, repeated *hue* group alarm/distress call. When approaching the nest, birds give a low, repeated *kiow* which is rather different to the fuller *cauu* call given when feeding. A strange, breathy, humming *vvvv* is also recorded. Chicks make peeping noises (van Heerden 1972).

DESCRIPTION
Adult Head and neck black, in some lights with slight green gloss; upperparts and upperwing-coverts dark slate-grey, browner on outer wing and sometimes slightly tinged olive on back and rump; remiges dark grey-brown, outer webs tinged slate, primaries black towards tips; a few white-tipped feathers among marginal coverts at bend of wing (D. Allan *in litt.*); rarely has white on outer edge of outer alula or outermost primary; usually has no white on tips of secondaries. Axillaries and underwing-coverts dark grey. Underparts dark grey, tinged slate on sides of breast and flanks, in fresh plumage with faint and narrow white feather tips, duller and brown-tinged when worn; lower flanks to undertail-coverts and tail black; undertail-coverts rarely with some white tips. Iris pale crimson to bright red; bill and shield white, sometimes tinged blue; 2 large knobs on top of shield, rose-red to deep red; legs and feet olive-green to dark green. Shield has broader base, more parallel sides, and relatively greater length, than in Common Coot. Sexes alike: birds cannot be sexed, even in the hand (Dean & Skead 1978). In non-breeding birds, iris brown-red; sides of bill duskier slate and knobs much smaller; rear of tarsus, and joints and toes, dull slate.
Immature Like adult, but upperparts often slightly more extensively tinged dusky olive, appearing less uniform grey than adult; feathers of chin, throat, breast and central

underparts narrowly fringed white; sometimes retains a few off-white juvenile feathers on centre of breast and belly. Bare parts as adult. In worn plumage, olive tinge on upperparts and white feather edges on underparts abrade, birds then becoming indistinguishable from adults.

Juvenile Crown and upperparts of body dark brown, tinged olive; lores, chin and throat white to grey-buff; sides of head, neck, and flanks dark olive-brown, mottled off-white; rest of underparts pale ash-grey, feathers of belly and vent with ill-defined white tips giving whitish centre of belly. Iris pale grey-brown to dark brown; bill dull grey; legs and feet dark grey, slightly tinged olive-green on front and sides of tarsus.

Downy young Downy chick ashy or grey-black, paler grey below, tinged silvery or blue-grey; golden-yellow hair-like down on neck and collar; longer filaments on mantle and back white at tips; bare skin of crown pink and blue; iris brown; bill red, terminal part orange-pink, with narrow white subterminal band and black extreme tip; legs and feet pale grey-green with pink tinge; small egg-tooth present.

MEASUREMENTS Wing of 15 males 219-239 (227, SD 5.0), of 13 females 208-224 (217, SD 5.2); tail of 15 males 56-66 (59.4, SD 3.5), of 17 females 54-65 (58.7, SD 2.7); exposed culmen of 16 males 30-36 (32.9, SD 1.8), of 20 females 29-33 (31, SD 1.0); tarsus of 14 males 68-75 (71.8, SD 2.3), of 19 females 61-70 (66, SD 2.9) (Cramp & Simmons 1980). Weight of 10 males 770-910, of 10 females 455-790 (Ripley 1977; means not given), of 4,016 unsexed (Barberspan, South Africa) 363-1236 (737) with annual peak in Mar, just before first peak in flightless moult, followed by gradual decline to Aug and then gradual increase (Dean & Skead 1979). Mean hatching weight 24.0 (SD 1.66) (Fairall 1981).

GEOGRAPHICAL VARIATION None.

MOULT Information is from Dean & Skead (1979) and Urban *et al.* (1986). Remex moult is simultaneous and (presumably) occurs once per year; it is recorded on the Kafue Flats (Zambia) in Jun-Jul and at Barberspan (South Africa) throughout the year with peaks in Apr and Oct-Nov. The remiges are shed over 4-6 days, growth takes 40-48 days and remex hardening takes c. 5 days; the flightless period is lengthy, c. 54 days (49-59 days), which may be adaptive and energetically economical, resulting from the absence of pressure to moult quickly; flightless birds occur singly or in small groups among full-winged birds well out on open water (see Habitat). At Barberspan the Apr peak in flightless moult falls soon after the peak in mean adult mass and the growth peak in the bird's main food source (*Potamogeton pectinatus*), which both occur in Mar; moult of the tail and body plumage is not seasonal, and there is continual replacement of feathers in these tracts.

DISTRIBUTION AND STATUS S Spain and N Morocco; Ethiopia, Eritrea, Kenya, Uganda, Rwanda, Burundi, E Zaïre, Tanzania, Zambia, Malawi, S Angola, S Mozambique, Zimbabwe, Botswana, Namibia and South Africa (including Lesotho and Swaziland); also Madagascar. Not globally threatened. Its range has decreased in both Europe and N Africa, and it is now close to extinction in Europe; reasons for these declines are not clear (Cramp & Simmons 1980), although factors such as loss and degradation of habitat are probably involved (Fernández-Palacios 1994). Around 1900 most of the abundant Iberian population was concentrated in the Guadalquivir marshes and surrounding lagoons, and the La Janda wetland system, both in Andalucía; La Janda was destroyed in 1960 and the species now remains only in the Guadalquivir area, where c. 50 adults survive, giving 10-20 breeding pairs in the marshes and up to 14 breeding pairs in the lagoons (Hagemeijer & Blair 1997). Between 1992 and 1994 51 captive-born birds were released in the area, and the viability of the population is probably reinforced by Moroccan recruits (Hagemeijer & Blair 1997). Breeding has recently been confirmed at Pantano del Honde in Alicante, but numbers are not known, and the species has been recorded from several other localities (see Hagemeijer & Blair 1997 for details). It also bred in Portugal early in the 20th century but is now only accidental there (Fernández-Palacios 1994). In Morocco it was formerly regarded as locally common (e.g. 2,100 on a winter count in 1972) and it still winters on Middle Atlas lakes; in Algeria it bred in the 19th century at L Halloula, and it may have nested at L Fetzara; there have been no records in Tunisia for many years but it was not rare in the N in the 19th century (Cramp & Simmons 1980). Its total population in Iberia and N Africa is estimated at <10,000 (Rose & Scott 1994). The main factor which limits its distribution in Europe is the lack of suitable habitat, and the main cause of mortality is accidental hunting because of confusion with the Common Coot (Hagemeijer & Blair 1997).

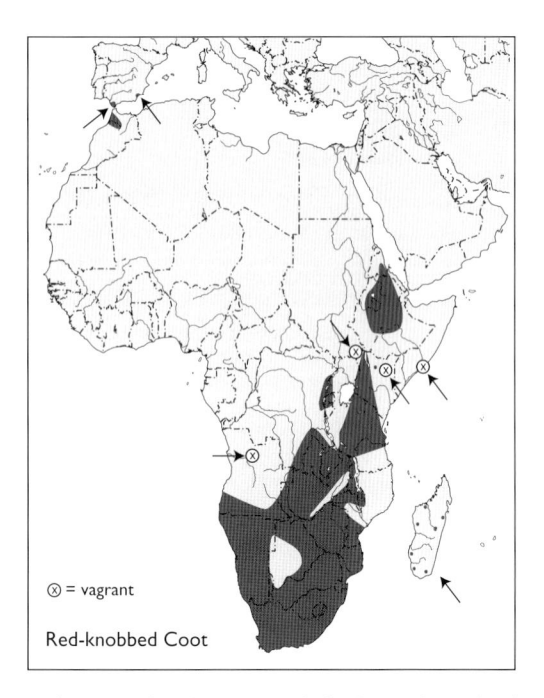

It is locally common to abundant in sub-Saharan Africa, including Ethiopia (mainly in the highlands and Rift Valley), Uganda (N to Kidepo V NP), Kenya (S from L Turkana and Moyale, especially common on L Naivasha and many smaller highland dams and lakes, generally absent from the arid N and the E, and on the coast only at the Tana R), and Tanzania (S to Rukwa, Njombe and Dar es Salaam); in Burundi it is known from the Ruzizi R delta in very small numbers (Britton 1980, Urban *et al.* 1986, Lewis & Pomeroy 1989, Short *et al.* 1990). Although reportedly common at L Turkana, where flocks of 500 have been reported at Ferguson's Gulf in Oct-Dec (Lewis & Pomeroy 1989), a Feb 1992 aerial count gave an estimate of only 1,600 birds for the entire lake (Bennun & Fasola 1996). In Zaïre it is mainly confined to the lakes of the E highlands (Kivu) and L Albert but it was formerly numerous near Lusinga Station, Upemba NP, and presumably occurs in extreme SE Zaïre between Upemba and the Zambian border (Lippens & Wille 1976, Urban et al. 1986). It is irregular in small numbers in Zambia, but common on the Kafue Flats (maximum count 1,200), and in Malawi it is widespread up to 2,500m but not on L Malawi (Benson et al. 1971, Benson & Benson 1977). It is uncommon in Angola, where it occurs mainly on the coastal plain of Mossâmedes and Benguela and N to Pungo Andongo, and it is very local and uncommon in S Mozambique, where it may be only a non-breeding visitor from the W (Urban et al. 1986, Clancey 1996). It is the most numerous and frequently reported rallid in southern Africa, distributed throughout the region except in low-rainfall areas lacking open waters, but it is relatively scarce on the KwaZulu-Natal coastal plain and in Swaziland, and remarkably absent from most of the Okavango (Taylor 1997c). In Zimbabwe it was originally found only on a few semi-permanent pans on the C plateau but is now common and widespread on artificial waterbodies (Irwin 1981). It is locally common in N, E and SE Botswana, common in lowland areas of Lesotho, and only a vagrant in Swaziland outside the highveld and middleveld areas (Bonde 1993, Parker 1994, Penry 1994). It is locally common to rare in Madagascar, where it has a fragmented distribution on suitable waterbodies (Langrand 1990). It occurs virtually throughout South Africa (except in the arid extreme N of the Northern Cape region), where there are sometimes up to 15,500 at the Bot River and Kleinmond Lagoon, 27,000 at Barberspan and over 30,000 at de Hoop Vlei (Urban et al. 1986, Tarboton et al. 1987, Hockey et al. 1989, Taylor 1997c); numbers at individual wetlands vary enormously, e.g. minimum and maximum counts of 1,800 and 18,700 were made at the Wilderness Lakes system over 4 years (Boshoff et al. 1991). The largest 1995-96 counts in Africa include 12,100 at L Tlawi (Tanzania) in Jan 1996, and 2,100 (Botswana), 38,200 (South Africa) and 4,300 (Zimbabwe) in Jul 1995 (Dodman & Taylor 1996). Its total population in sub-Saharan Africa is estimated at 100,000 - 1,000,000 (Rose & Scott 1994). In some areas, especially in South Africa and Zimbabwe, it has benefited from the proliferation of farm dams and other artificial waterbodies, especially as it can breed on very small waters, even on those <1ha in extent (Irwin 1981, Hockey et al. 1989, Taylor 1997c). It appears to be less tolerant of human disturbance than is the Common Coot, at least locally (Cramp & Simmons 1980), and in KwaZulu-Natal it is sometimes considered a nuisance where it overgrazes green crops close to dams (Urban et al. 1986).

The African Fish Eagle *Haliaetus vocifer* is an important predator of adults and young, while other predators are Tawny Eagle *Aquila rapax*, Marsh Harrier *Circus aeruginosus* and Grey Heron *Ardea cinerea* (Wood 1975, Urban et al. 1986). In Kenya, a Serval *Felis serval* pounced from cover into a group of coot in shallow water and, although the birds dived immediately, the cat dipped its head under water and caught one of them (Jackson & Sclater 1938).

MOVEMENTS Mainly sedentary; also nomadic and opportunistic, with many indications of wandering noted throughout its range. Local movements are evident from the degree of non-breeding-season flocking, e.g. in Morocco, and from fluctuations in numbers on permanent waters. Erratic winter wandering beyond the normal range was reported in the past from Morocco and Spain; in the 19th century, when the bird's range extended to Algeria and Tunisia, it wandered in winter to Portugal, S France, Sardinia, Sicily, Italy, and Malta (Cramp & Simmons 1980). Flocks in Africa are often sedentary on suitable waters but birds may move considerable distances outside the breeding season. At Barberspan, South Africa, peak numbers are recorded in Apr and Oct/Nov, and factors influencing fluctuations include rainfall, water level and the availability of the birds' favourite food (Dean & Skead 1979). Ringing recoveries (recovery rate for 16,500 birds ringed 1955-78 was c. 1%) give a mean distance travelled from Barberspan of 270km, 70% of recoveries being within 300km; the longest distances travelled are in Jan-Apr, when the population drops during the rainy season, and include birds recovered from S Zambia, Namibia, Botswana, S Mozambique and (1,072km) at Rondevlei near Cape Town (Skead 1981, Urban et al. 1986). Birds ringed at Rondevlei disperse up to 400km (Winterbottom 1966). Birds ringed at Ngorongoro Crater, Tanzania, have been recovered at Eldoret and Naivasha, Kenya (Urban et al. 1986). In Kenya, non-breeding visitors occur on dams near Nairobi in Sep-Mar, and numbers at L Baringo increase in Nov-May (Lewis & Pomeroy 1989). At Lochinvar NP on the Kafue Flats, Zambia, this species was present in Feb-Oct, with peak abundance in Mar-June and a maximum of 900 birds in May (Douthwaite 1978). In South Africa, concentrations occur on large permanent waters in the dry months and birds disperse to smaller permanent and temporary waterbodies during the rains (Urban et al. 1986, Earlé & Grobler 1987, Tarboton et al. 1987); during the summer in the SW Cape large flocks congregate at estuaries and coastal lagoons (Hockey et al. 1989). It wanders to S Somalia (2 recent records, from Dannow, Mar and Libsoma, Dec), N Angola, extreme N Uganda and NE Kenya (Wajir), and is accidental in France, Italy, Portugal and Malta (Cramp & Simmons 1980, Ash 1983, Urban et al. 1986, Lewis & Pomeroy 1989).

HABITAT Information is from Cramp & Simmons (1980) and Urban et al. (1986) unless otherwise stated. It chiefly frequents open fresh water of lakes, lagoons, ponds, dams, permanent and temporary dams and vleis, and floodplains. It also occurs in swamps with reeds and papyrus, and at sewage ponds; in S Spain it occurs in natural marshes and marshes transformed into rice-fields, at lagunas and among dunes and reedy channels of marismas. It is sometimes found on rivers and tidal lagoons but generally prefers still water (Taylor 1997c). When breeding, it frequents waters with fringing or emergent vegetation, but at other times it may occur on completely open waters; it requires submerged aquatic vegetation for food, and the quality and quantity of submerged macrophytes are probably key requirements (Hagemeijer & Blair 1997). In Namibia it bred in an ephemeral grassy marsh, dominated by *Diplachne* spp., at the edge of an unvegetated alkaline pan (Jamieson et al. in prep.). Its habitat is very similar to that of the Common Coot, which it replaces over most of its range, but it may rely more on cover and may show itself less openly, both far out on the water and on exposed land; it appears to be less hardy, requiring warmer, more sheltered habitat. It occurs up to 1,800m in Madagascar and 3,000m in E Africa. During the flightless moult it prefers permanent waters rich in submerged aquatic vegetation, and it tends to stay out on the open water among full-winged birds rather than close to the shore where food is more accessible and vegetation cover is available, possibly to avoid predation (Dean & Skead 1979).

FOOD AND FEEDING Information is from Urban *et al.* (1986) and Cramp & Simmons (1980) unless otherwise stated. Omnivorous; it takes mainly aquatic vegetation, especially non-rooting submerged or floating plants (Watson *et al.* 1970) and including filamentous algae (Chlorophyceae), water plants such as *Marsilia, Aeschynomene fluitans* (stems, flowers, fruits and aerial roots), *Polygonum limbatum* (stems, leaves and fruits), *Najas pectinata* (stems and leaves), *Potamogeton pectinatus* (the principal food at Barberspan and Swartvlei, South Africa: Dean & Skead 1979, Fairall 1980), *Ruppia maritima, Chara globuaris, Eichhornia crassipes*, possibly *Zostera* (Bot River/Kleinmond Lagoon, South Africa: Hockey *et al.* 1989), grass (e.g. leaves of *Panicum repens*) and seeds. It also eats molluscs (Gastropoda), crustaceans and arthropods (including insects), and occasionally carrion (e.g. ducks washed up on the shore), and it will come ashore readily to pick up almost any type of food, including campers' scraps. At L Naivasha, Kenya, it fed mainly on unrooted submerged or floating plants; its diet, which showed very little overlap with that of ducks at the lake, comprised seeds and fruit (7.3%), other plant material (90.1%), Arthropoda (2.5%), birds (0.4%); all 23 stomachs contained the waterweed *Najas pectinata* (mean proportion 81.6%) (Watson *et al.* 1970). Feeding techniques are similar to those of the Common Coot, with the emphasis on aquatic modes such as diving down and pulling up underwater vegetation; its diving ability is better developed than in the Common Coot, and S of the Sahara it feeds mostly in water. It also feeds from the surface, either while swimming or when standing in shallow water. It feeds much less often on land than does the Common Coot, but it grazes short grass near water, especially when food is scarce (the bill has a shearing edge adapted to grazing). On the Kafue Flats, Zambia, it grazes where the vegetation has been trampled by Kafue Lechwe *Kobus lechwe*; in KwaZulu-Natal it overgrazes green crops planted near dams. It also feeds on floating algae while standing or swimming, by repeated opening and closing of the vertical bill partly immersed in water (P. B. Taylor in Urban *et al.* 1996). Stewart & Bally (1985) found that the daily metabolised energy of captive birds of mean mass 720g was 516kJ per bird, probably less than that of wild birds, for which 691kJ was estimated; the daily macrophyte consumption of birds at the Bot R estuary was estimated at 82.4g dry weight per bird (of *Potamogeton* and *Ruppia*), and the annual consumption of the estuary coot population at 759.5 tonnes dry weight; this equates to between 9.9 and 12.5% of the estuary's annual macrophyte production.

HABITS Not described in detail, but similar to those of the Common Coot. In the W Palearctic it is much shyer than the Common Coot (Cramp & Simmons 1980). When wounded or suddenly surprised at close quarters it will dive, and if amongst submerged or floating vegetation will remain stationary with its bill above the water (Jackson & Sclater 1938).

SOCIAL ORGANISATION This species is gregarious outside the breeding season, and nonbreeders also flock during the breeding season. It is monogamous and territorial when breeding; apparently some birds are also paired in winter flocks (e.g. Morocco, Dec) (Cramp & Simmons 1980). In a recent study in Namibia, Jamieson *et al.* (in prep.) found a very high rate of conspecific brood parasitism (in 43-75% of 42 nests) in a population breeding in an ephemeral wetland where there may have been a shortage of available nesting sites. Parasites laid up to 6 eggs, mostly during the hosts' laying period, and 6 of 15 parasitic females subsequently laid their own clutches. Hosts rejected 69% of parasitic eggs, 87% of which were buried in the nest lining.

SOCIAL AND SEXUAL BEHAVIOUR Very little information is available. At Lagune de Mehdia, Morocco, in Nov-Dec, Red-knobbed and Common Coots occurred together and on the few occasions when they intermingled no interspecific intolerance was seen; the Red-knobbed Coots similarly were not aggressive to Common Moorhens (129) or Marbled Teal *Marmaronetta angustirostris* (Wood 1975). The group defence of the flock in response to an aerial predator is similar to that of the Common Coot; in defence against a Marsh Harrier, grazing birds fled into the water, formed a tight pack, and stretched their necks forwards and upwards, giving the group alarm call (Wood 1975); a separated bird escaped by repeated diving and rejoining the flock.

This species, like other coots, is a pugnacious bird, at least in the breeding season, and it has typical charging-attack, splattering-attack and water churning displays, which are performed in similar circumstances to the Common Coot (see that species account) (Cramp & Simmons 1980); conspecific individuals and also the smaller duck species are driven away from the nesting and feeding area (Vincent 1945). However, during the winter it is not as aggressive as the Common Coot (Wood 1975). The water-churning display is also given in response to the presence of man in the non-breeding season (Wood 1975). Paired birds allopreen (Urban *et al.* 1980). Incubating birds may sit very tight (Urban *et al.* 1986) but often leave the nest quickly when an intruder approaches (van Heerden 1972, P. B. Taylor unpubl.).

BREEDING AND SURVIVAL Season Spain and N Africa, May; Morocco, Feb-Sep; Ethiopia, Apr-Jul, Sep-Dec; Kenya and NE Tanzania, all months, peaks May-Jul and Sep-Oct; W Kenya and Uganda, all months except Feb, peak Apr; Zambia, Apr-Jul; Malawi, Jun-Jul; Zimbabwe, Jan-Sep; Namibia, Feb-Mar; South Africa, all months, peaks Barberspan (former Transvaal) Feb/Mar and Jul; Madagascar, Dec-May. **Nest** A bulky platform of leaves and stems of water plants (rushes, grass etc); cup lined with finer material or papyrus inflorescences; usually has ramp at one side; in shallow water (usually >1m deep; wading is necessary to reach it), or floating and anchored to vegetation; built either in open water or within emergent vegetation (reeds, rushes, sedges, coarse grass etc), sometimes on waterlilies and often on raft of fresh green reeds; if within vegetation, nest often built on foundation of bent and trampled stems and always sited to give quick and easy access to open water; no attempt made to conceal nest; nest diameter c. 30cm. Both sexes build, occasionally with help of immatures; building renewed if platform settles or water rises; material also added during incubation, off-duty bird bringing material and sitting bird incorporating it. Builds many 'false nests' and rafts, used as resting platforms. **Eggs** 3-11 (normally 5-7); mean clutch sizes, South Africa, 6.0 (SW Cape, n = 90 clutches), 4.6 (E Cape, n = 11), Zimbabwe 5.0 (n = 4); oval to long oval; smooth and slightly glossy; cream, pale to rich buff or pale grey-stone, evenly spotted and speckled dark brown and sometimes underlying pale purple. Size of eggs 50-60 x 35.4-40 (54.2 x 37.3, n = 70), calculated weight 41. Eggs laid at daily intervals; 2 females known to have laid in 1 nest. **Incubation** Period 18-25 days, by both sexes, with frequent nest-reliefs; relieving bird turns eggs with bill; hatching asynchronous; chicks hatch at 1-day intervals. **Chicks** Precocial; leave nest after 1 day; able to dive soon after hatching, suggesting capability for independent feeding at early age. Fed and cared for by both parents, sometimes with help of immatures or juveniles from earlier brood, particularly when food and nesting habitat abnormally abundant; broods often divided unequally between parents, particularly when young small and when juvenile helpers exist, and remain separate at least until fledged; little intra-family hostility in broods (Dean 1980). Fledging probably 55-60 days; breast feathers appear first, wing feathers last. Conspecific brood parasitism recorded (see Social Organisation) in an ephemeral wetland where nest predation thought to be high (Jamieson et al. in prep.). In East Africa, death rate among chicks very high (Urban et al. 1986); in South Africa, 175 (34%) of 506 eggs produced chicks; 53 (42.1%) of 126 nests produced fledged young; causes of loss of 56 nests were: predators 20, wave action 18, rise in water level 18 (Winterbottom 1966). Often only 1 brood, unless conditions very favourable; possibly double-brooded in Morocco.

136 COMMON COOT
Fulica atra   Plate 41

Fulica atra Linnaeus, 1758, Europe (= Sweden).

Forms superspecies with *F. cristata, F. alai, F. americana, F. caribaea, F. leucoptera* and *F. ardesiaca*. Four subspecies recognised.

Synonyms: *Fulica aterrima/lugubris/australis/fuliginosa/ albiventris/aethiops/atrata/pullata/tasmanica*. Race *novaeguineae* incorrectly spelt *novaeguinea* (Ripley 1977).

Alternative names: Australian/Eurasian/Black Coot.

IDENTIFICATION Length 36-45 cm; wingspan 68-80cm in nominate, 56-64cm in race *australis*. Head and neck black, slightly glossy; rest of plumage slate-grey, somewhat paler on flanks and underbody (underparts relatively pale in *australis*), and variably washed olive-brown on upperparts; narrow white line along leading edge of wing and on outer edge of alula and outermost primary (sometimes faint); tips of secondaries white (white much reduced or absent in *novaeguineae* and *australis*); undertail-coverts black. Iris red; bill and frontal shield white, bill sometimes tinged pink or yellow (or blue-grey in *australis*); tarsus yellow-green to yellow or orange-red at sides, white at front, olive-grey at back; tibia orange-yellow to red; toes bluegrey (legs and feet all grey in *australis*). Non-breeding birds have iris red-brown, sides of bill dusky, legs duller and paler. Sexes alike; female averages smaller. Immature like adult but upperparts more washed olive-brown; throat, cheeks and underparts variably mottled whitish; iris brown to red-brown; adult colours attained on other bare parts between autumn and spring. Juvenile has crown and hindneck black to dark brown; rest of head and neck mixed dark grey and white (throat whiter); upperparts dark olivebrown; tail and wings like adult but upperwing-coverts more extensively washed olive-brown; breast mainly white, with some pale slate-grey; central underparts white with grey suffusion; flanks dark olive-brown. Iris dark grey to pale brown; bill grey, tinged yellow or pink; small shield grey; tarsus and toes grey; tibia yellow to orange-yellow (grey in *australis*). Inhabits mainly large open waters, with submerged vegetation and (in breeding season) fringing and emergent vegetation.

Similar species Distinguished from Red-knobbed Coot (135) by greater contrast between black head and neck and paler body, white tips to secondaries, lack of red knobs at top of frontal shield, and pointed projection of loral feathering between bill and shield. Differs from American Coot (138) in colours of bill, shield, legs and feet, absence of white on undertail-coverts, and less contrast between dark head and paler body. Larger and bulkier than Common Moorhen (129), lacking white in plumage (Common Moorhen has white flank-line and lateral undertail-coverts) and with different bare part colours; has rounded back, and swims with head closer to body and nodding much less; in flight looks much bulkier, more duck-like, than Common Moorhen; wingbeats seem fast for actual progress made; prefers larger areas of open water (especially in non-breeding season) than does Common Moorhen; often occurs far from cover and forms large flocks. Race *australis* similar in size to sympatric Dusky Moorhen (130), which differs in bare part colours and general appearance and behaviour (as for Common Moorhen), and has white lateral undertail-coverts but usually no white flank-line.

VOICE Most information is from Cramp & Simmons (1980). Often a noisy species. As in the American Coot, there are clear sexual differences in the calls. The combat call of the male is a sharp, explosive *pssi*, and of the female a short, croaking *ai* which becomes *u* in less tense situations. The male's alarm call is a sharp variant of the combat call, while the female's is a rapid sequence of *aioeu* sounds. The typical contact call is a single, short *kow, kowk, kut,* or sharper *kick*; sometimes 2 notes are combined, e.g. *kick-kowp*, the male also has mechanical *p* or *ta* calls, and a cork-popping sound, and the female a high, short, falsetto *oeu*. In the race *australis*, the typical contact call is described as a loud, harsh *crark* (Marchant & Higgins 1993). The male's courtship call is a rapid series of low, hissing *phsi* calls, that of the female a soft, resonant *oeu*; in New Zealand, 1 bird gave a soft *pitt-pitt-pitt* before allopreening or as a greeting (Marchant & Higgins 1993). Calls are often metallic, resonant, querulous or explosive and are particularly sharp and high when birds are agitated. Adults call young with soft *kt* notes. The young make a rasping *creer* call, and later a less hoarse *quee-ip*, which become shriller in alarm; in *australis* this call is described as a plaintive *weeee*, becoming a disyllabic *whee-eep* when older (Marchant & Higgins 1993).

DESCRIPTION *F. a. atra*
Adult Head and neck black, with slight green gloss; upperparts, upperwing-coverts, tertials and tail dark slategrey; tips of longer tertials and of rectrices black, glossed green; tips of other upperparts feathers olive-brown to brown when fresh, colour retained on lower inner scapulars, inner tertials, and lower mantle to uppertail-

coverts. Narrow line along leading edge of wing white; also white on outer edge of outer alula feather, and on outer edge of outermost primary (sometimes faint); alula, primary coverts and flight feathers dark brown-grey, paler on inner webs of flight feathers; tips of secondaries white; axillaries and underwing-coverts grey, some small coverts narrowly fringed white (variable). Sides of breast and flanks darkish slate-grey (duller and paler than upperparts); rest of underparts dark grey; undertail-coverts black. When fresh, feathers of underparts faintly fringed white. Iris bright red; bill and frontal shield milk-white, bill sometimes tinged pink or yellow; sides of tarsus yellow-green to yellow or (older birds?) orange-red, front of tarsus white, tinged pale blue or yellow; back of tarsus and ankle joint dark olive-grey; tibia orange-yellow to orange-red, sometimes dark red; toes and lobes pale blue-grey. Bare parts duller in non-breeding season: iris red-brown; sides of bill dusky; legs duller and often paler. Sexes alike; male larger. Pure and partial albinos are recorded in India.

Immature Like adult but upperparts more extensively washed olive-brown; feathers of sides of head, throat and underparts more broadly and irregularly tipped whitish, central underparts appearing dark grey with white mottling; some off-white juvenile feathers of central belly and vent retained; wing as juvenile. As olive-brown on upperparts and white on underparts abrades, some become inseparable from adult but upperparts often still duller and browner, and underparts often still tinged white or brown. In first breeding season, many indistinguishable from adult but some show mixture of worn and fresh feathers, and wings old. Iris becomes dark brown to red-brown (first autumn and winter); bill becomes white from Aug-Dec; tarsus becomes green at sides from autumn, then yellow-green to yellow by spring; front of tarsus pale blue-grey.

Juvenile Crown and hindneck black, feathers sometimes narrowly tipped white; rest of head and neck mixed dark grey and white (throat whiter); upperparts dark olive-brown; tail, tail-coverts and wings like adult but upperwing-coverts more extensively washed olive-brown and underwing-coverts more extensively fringed white. Breast, belly and vent pale slate-grey, feathers broadly tipped white, breast mainly white and central underparts white with grey suffusion; flanks dark olive-brown. Iris dark grey to pale brown; bill dusky grey, tinged olive or pink; tarsus and toes dark grey, lobes paler, all tinged pink when young; tibia yellow, often more orange at top.

Downy young Black downy chick has orange-red to yellow-buff tips to down of neck, sides of head, chin, throat, wings and mantle, and red tips on lores and around shield; iris hazel to grey-brown; bare skin of crown red, but blue above eyes and on nape; bill and shield red, bill white distally and black at tip; legs and feet slate-grey.

MEASUREMENTS Wing of 21 males 211-229 (219, SD 4.5), of 23 females 197-213 (205, SD 4.1); tail of 18 males 50-61 (55.4, SD 3.1), of 22 females 49-60 (53.4, SD 2.8); culmen (without shield) of 143 males 28-36 (31.1), of 151 females 21-31 (29.1); shield of 146 males 19-37 (26.4), of 152 females 16-33 (23.0); bill (including shield) of male 45-58, of female 40-60; tarsus of 37 males 59-65 (61.7, SD 1.7), of 38 females 54-60 (56.8, SD 1.6) (Glutz von Blotzheim *et al.* 1973, Ripley 1977, Cramp & Simmons 1980). Weight of 215 males 610-1,200 (902), of 214 females 610-1,150 (770). In Azerbaijan, wintering birds in Big Bay, Kizil Agach Reserve, have mean weight of 680 after arrival but increase in weight to average 970 (up to 1,450) by Nov; by Feb they are so fat that 70-80% of the population become flightless, but when N migration begins, mean weight is only 750 (Patrikeev 1995). Hatching weight 21-26.

GEOGRAPHICAL VARIATION Races separated on size, in which there is considerable variation, and on slight variations in colour, namely amount of white on tips of secondaries and colour of underparts. Nominate race includes *turkestanica*, and *australis* includes *ingrami* (Peters 1934). See White & Bruce (1986) for a discussion of the races *lugubris*, *australis* and *novaeguineae*, and the putative race *anggiensis*.

F. a. atra Linnaeus, 1758 – Azores through Europe S of 66°N, E across C Russia to 61°N in Leningrad area, 59°N at Kama R, 57°N in Urals and 59°N E of Urals; via Angara R to upper Lena R and Aldan R mouth (64°N); SE to S coast of Sea of Okhotsk; on Amurland, Manchuria, Ussuriland, Sakhalin and Japan (Hokkaido to N and C Honshu). Occurs S to Mediterranean and its larger islands (including Cyprus); North Africa, in N and C Morocco (including Middle Atlas), N Algeria (also C Algeria at El Goléa), and Tunisia S to L Affial (near Feriana), Sidi Mansour, Gabès and Bou Grara; occasionally breeds on temporary waters further S; also Egypt (formerly Nile Delta, possibly now Nile Valley); Arabia S to Oman, and Yemen; Iraq and E to Caucasus, Transcaucasia and Azerbaijan; and Iranian region to Pakistan, India (from Kashmir and Ladakh S to Sri Lanka), Bangladesh, N China (S to Heilongjiang, NE and C Inner Mongolia, Jilin, Liaoning, Hebei, Henan, Hupeh, Ningxia, Gansu, and N, W and SW Xinjiang) and Korea. In the W Palearctic, N populations winter from the North Sea, Baltic, EC Europe, the Black Sea and Caspian Sea, Iraq and Arabia, S to the Canaries and Madeira; in Africa from Morocco to Tunisia, coastal Libya, S into desert oases of Morocco and Algeria, along Nile in Egypt and Sudan S at least to Khartoum, on lakes in Egyptian W desert, and in C and S Sudan; also Senegal, Mali (C Niger delta to Gao, and also E of Timbuktu), C Burkina Faso (Ouagadougou and Koubri), S and NC Niger, Nigeria (extreme N, also once at Ibadan), Chad (Mao and Ouinanga Kebir). There are unconfirmed sight records from the Ethiopian highlands and from Ngorongoro, N Tanzania (Moreau 1972, Urban *et al.* 1986). Further E it winters in Afghanistan, Russian Turkestan and S Japan, Nepal, India, Burma, Thailand, Indochina (Cambodia, Vietnam, Laos), S China (C and SW Sichuan, middle and lower Chang Jiang R, and all provinces to S, W to S Xizang and S through Yunnan, Guangxi, Guangdong, Fukien, Chekiang, and S Anhui and Kiangsu), Hainan, Taiwan, S Japan (S from S Honshu) and the Nansei-shoto (Okinawa to Yaeyama Is), Izu-shoto, W Java, Philippines (Luzon, Mindoro, Negros, Palawan and Sanga Sanga), and Mariana Is. Old records from N Sulawesi are unconfirmed (White & Bruce 1986). Vaurie (1965) includes the Malay Peninsula and Sumatra in its winter range, but this is not confirmed by later works. See Description.

F. a. lugubris (Müller, 1847) – NW New Guinea (Anggi Lakes, Arfak Mts); it also formerly bred in E Java (Hoogerwerf 1949). Birds recorded from Flores in recent years (Coates & Bishop 1997) possibly this race.

Smallest race; shield relatively larger than in *australis* and nominate; plumage like nominate, underparts mouse-grey, paler than in *novaeguineae*. Wing of 3 males 181-190 (186), of 9 females 170-185 (178); tail of unsexed 52; culmen (including shield) of unsexed 42-49; tarsus of unsexed 56-60 (Ripley 1977).

F. a. novaeguineae Rand, 1940 – C New Guinea (Mt Wilhelmina region, Snow Mts); birds from Carstenz Massif probably also this race (Schodde *et al.* 1974); racial affinity of birds from L Iviva (Enga P) and L Wangbin (Western P), Papua New Guinea, not clear. Larger than *australis*; underparts darker slate-grey; shield relatively larger than in *australis* and nominate, reaching crown behind rear of eyes. White on tips of secondaries narrow or absent. Wing of 4 males 200-212 (207.3, SD 5.5), of 6 females 190-202 (194.7, SD 5.1); bill (including shield) of male 56-62, of female 46-58.

F. a. australis Gould, 1845 – Australia, Tasmania and New Zealand. Birds from Port Moresby area, New Guinea, possibly of this race (Coates 1985). White on tips of secondaries narrow or absent. Iris red, but orange-red and maroon also recorded; bill white, tinged blue-grey; legs and feet grey to blackish-grey, usually darker at joints. Juvenile has top of head dark brown; upperparts dark brown, tinged greyish; bill of young birds has blackish saddle near tip, dark grey base and pinkish-white distal third; shield and basal third of bill mostly pink; colours become yellower with age, and saddle fades to grey; legs and feet grey. Chick has bristles round face, and down of back of neck and wings, tipped yellow; bill and shield red, bill becoming white towards tip and black at tip; legs and feet dark olive-grey. Wing of 6 males 173-194 (185.7, SD 7.0), of 9 females 169-181 (176.5, SD 4.0); tail of 6 males 48-51 (49.8, SD 1.1), of 10 females 44-53 (47.4, SD 2.6); bill (including shield) of male 42-55, of female 40-50; tarsus of 6 males 53.5-58.5 (56.2, SD 1.7), of 10 females 50.5-58 (54.2, SD 2.8). Weight of 6 males 481-660 (568, SD 74.6), of 5 females 476-609 (552, SD 51.8), of unsexed 305-725 (511) (Ripley 1977, Marchant & Higgins 1993).

MOULT Postbreeding moult is complete; flight feathers and all wing-coverts are moulted simultaneously. In nominate birds (Cramp & Simmons 1980), moult starts in late May with the head, neck and anterior body, but is slow when birds are nesting and with young; wing feathers are shed in non-breeding birds from mid-Jun; later in breeders (but highly variable), late Jun to early Sep, in males slightly earlier than females. Rectrices are lost soon after the flight feathers; birds are in heavy body moult in Jul, and are mostly finished by late Aug; the head, back, vent and tail-coverts are moulted by mid-Sep; some flight feathers are partly bitten off before the wing-moult. In Baluchistan (Pakistan) birds are flightless at the end of Jun, and in Rajasthan (NW India) in early to mid-Oct (Ali & Ripley 1980). Adult prebreeding moult is partial, involving the head and neck, and sometimes also a few body feathers, and occurs Dec-May. Postjuvenile moult is partial, involves the head and body, starts before the wing is fully grown, and is usually finished before Oct (sometimes only by Dec); the head and upperparts are moulted first, then the flanks, breast and most underparts, finally the centre of the breast and belly. The first prebreeding moult is partial, occurring Dec-May; it is like that of the adult but often involves more feathers.

One *australis* was moulting primaries but only a few coverts (lessers) in Feb; body moult begins on head and upperparts well before wing moult is complete; the complete moult probably occurs in late summer or autumn but its relationship with breeding is not known (Marchant & Higgins 1993). Prebreeding moult involves the head and body, and is recorded in Jul-Aug; postjuvenile moult, also of the head and body, starts with the head and upperparts, followed by the breast, flanks and the rest of the upperparts, and is recorded in Jan-May; this moult, especially of the underparts, may be interrupted (Marchant & Higgins 1993).

DISTRIBUTION AND STATUS Widespread across Eurasia, from the Azores and Canaries, through Europe S and E to Sakhalin and N Japan, S to N Africa, Turkey, the Middle East, Iraq, Iranian region, India and Sri Lanka, N China, and SE Asia to the Sundas, New Guinea, Australia and New Zealand. N populations winter S to WC Africa, SE Asia, Indonesia, Philippines, and Mariana Islands (stragglers). Not globally threatened. The nominate race suffered a considerable decrease in numbers in many European countries during the 19th century, and a similar decline has taken place in the former USSR since the mid-1950s; such declines are attributed to severe winters and to the interacting effects of varying winter mortality, and delayed density-dependence relative to territorial behaviour (Cavé & Vissier 1985, Hagemeijer & Blair 1997). It expanded its range in Europe from the late 19th century to the 1930s, and during 1970-1990 at least 15 European countries recorded some increase in numbers and range, while 6 recorded declining populations; there are marked population fluctuations in many areas due to hard winters but it has probably increased generally, aided by eutrophication, new manmade habitats, and adaptation to urban environments, and populations are quite stable (Cramp & Simmons 1980, Hagemeijer & Blair 1997). Recent estimates (Hagemeijer & Blair 1997) give a total European population of 1,116,000-1,289,500 (geometric mean 1,188,700) in 37 countries, a total Russian population of 130,000-230,000 (172,900) and a European Turkish population of 5,000-50,000 (15,800), with the highest densities in Poland, Hungary and the Netherlands, and the largest population (300,000-400,000) in Poland; densities rapidly decline with increasing latitude and altitude, presumably because of a lack of suitable vegetation and the preponderance of oligotrophic and deep waters with unsuitable margins (e.g. Cramp & Simmons 1980). In the 1970s, the W Palearctic area of the former USSR held over 580,000 pairs in summer and 804,000 birds in winter; the Black Sea and Mediterranean area held 1,035,000-1,296,000 birds in the 1969-1971 winters (Cramp & Simmons 1980).

It is a locally common breeder in North Africa, and in winter it formerly occurred in large numbers, especially in Morocco (100,000) and Tunisia (311,000 in 1973) (Urban *et al.* 1986). In the Middle East and Arabia its range is expanding as a result of newly created waterbodies, and it is both a resident and a passage/wintering bird in the area (e.g. Hollom *et al.* 1988, Shirihai 1996). In Israel it breeds rarely but is an abundant passage migrant and winter visitor, daily counts of migrants being up to 6,700 per locality and the national midwinter total in the 1980s being 12,000-54,000 (mean 30,000) (Shirihai 1996). In Egypt it probably no longer breeds in the Nile Delta

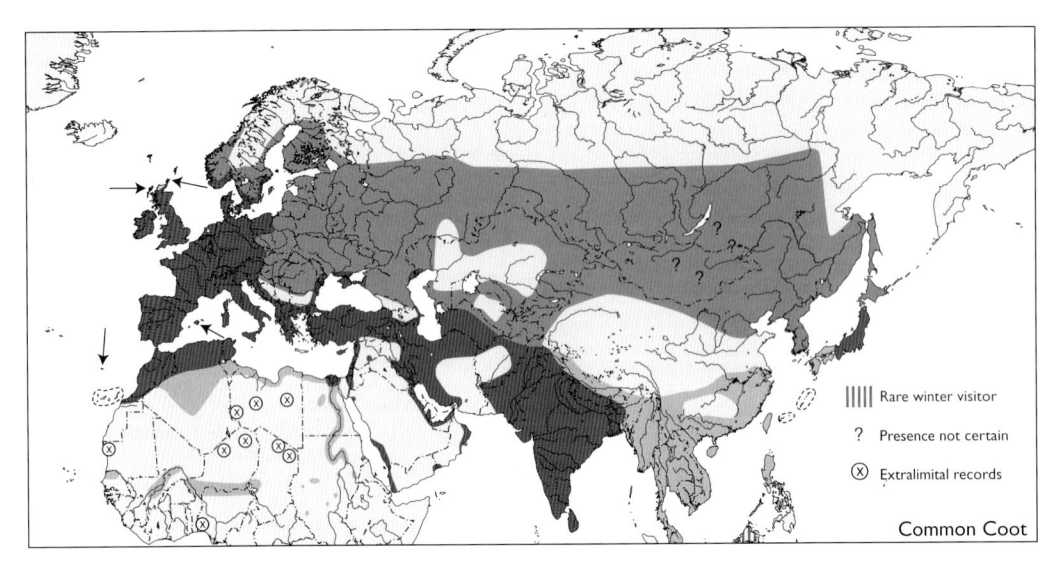

|||| Rare winter visitor
? Presence not certain
⊗ Extralimital records

Common Coot

although a few may now breed in the Nile Valley, but it is abundant in winter along the Nile S to Aswan (in flocks of up to 1,000) and at lakes in the Delta and the W desert, e.g. 154,000 at L Burullus, 51,300 at L Manzla and 18,000 at L Qarun (Goodman *et al.* 1989). In Oman it is an irregular passage migrant and winter visitor (Gallagher & Woodcock 1980) and in Saudi Arabia it is common in the winter (Jennings 1981). Relatively very few birds move as far S as the sub-Saharan region, where it is a scarce to very locally common winter visitor, several hundreds having been recorded from Senegal, small numbers in Mali (but several thousand E of Timbuktu in Jan 1980), occasional small numbers in Niger, 500+ at 1 locality (Kazaure) in Nigeria, and small numbers in C and S Sudan, while there are 2 recorded occurrences in Burkina Faso (Lamarche 1980, Newby *et al.* 1987, Nikolaus 1987, Giraudoux *et al.* 1988, Thonnerieux *et al.* 1989, Elgood *et al.* 1994, Rodwell *et al.* 1996).

Much hunted in the past, it is still shot in Mediterranean countries today, for both sport and food (Cramp & Simmons 1980). In Egypt it is the commonest bird for sale in the Port Said market, with up to 11,500 birds in 18 days in the 1982-83 winter (Goodman *et al.* 1989).

It is a numerous breeding species in Azerbaijan, where the total breeding population is probably more than 15,000-20,000 pairs, but it is severely hunted during the winter and is declining due to hunting and oil pollution (Patrikeev 1995); see below for winter census counts. In C Siberia its populations have suffered a great deal and urgently require protection (Rogacheva 1992). It is common to very abundant in India and Pakistan during the winter; in Pakistan it is only an occasional breeder (Sind and Baluchistan), the entire population normally departing in spring to breeding grounds in the former USSR (Ali & Ripley 1980, Roberts 1991). In Nepal it is mainly an uncommon winter visitor and passage migrant (Inskipp & Inskipp 1991), and in Bangladesh it is a scarce breeding resident (Harvey 1990). It first appeared in Sri Lanka in 1924 (Henry 1978) and is currently regarded as very rare (Kotagama & Fernando 1994). In India, Pakistan, Afghanistan and neighbouring countries these coots are highly esteemed as food by lake dwellers, and are killed in vast numbers, including when flightless during the postbreeding moult (Ali & Ripley 1980). Although it is still an abundant visitor to Pakistan, loss of habitat to agricultural drainage schemes have greatly reduced its population (Roberts 1991). In C Asia its range has expanded considerably due to the construction of many artificial reservoirs, which it colonises once vegetation appears (Potapov *et al.* 1987). In the N Gobi it formerly bred in very large numbers at L Orok Nor (Vaurie 1965). It is fairly common in China throughout the year but is probably rare in the lower Changjiang R and on Hainan (Cheng 1987). Its status in Myanmar is unclear: it is common on most suitable jheels in the cold weather, is thought to breed, and is reportedly present all year (Smythies 1986). It is uncommon in Thailand (Lekagul & Round 1991). In Japan it is an uncommon and localised summer visitor in Hokkaido, mainly to the SE and SW, and a localised and uncommon resident in N and C Honshu, but it is spreading, and has bred in W Honshu and N Kyushu; further S it is a winter visitor in small numbers (Brazil 1991). It was formerly not uncommon in Korea (Austin 1948), where it presumably breeds, and rare in S Vietnam (Wildash 1968). It occurs only rarely as a winter visitor in Borneo and Java (MacKinnon & Phillipps 1993), and in Bali (3 specimens, Jan 1938: White & Bruce 1986).

The Asian Waterfowl Census (Perennou *et al.* 1994) gives population estimates for the 3 main wintering groups in Asia: SW Asia (Iran and central Asian republics) 2,000,000 birds (sum of 5-year means for all sites 877,000, but 1,453,000 with 1970s data); S Asia (to Bangladesh) 1,500,000 (sum of 5-year means 734,000); E/SE Asia (S to Thailand) sum of 5-year means 269,000. Populations may be declining in SW Asia and may be stable in S Asia; sites of international importance in the region (with 5-year means) include Kirov Bay and Aggel L, Azerbaijan (243,250 and 50,000); Yancheng shore, Jiangsu, China (89,400); Chilka L, Orissa, India (26,700); Gomishan Marsh and Torkman Sahra, Mazandaran, Iran (47,600 and 45,400); in Pakistan, Chashma barrage reservoir, Punjab (93,515), Haleji L, Sind (61,994), Hub Dam, Sind (34,870) and Keenjhar L, Sind (94,331); Krasnovodsk and N Cheleken bays, and L Sarakamysh, Turkmenistan (166,500 and 87,651).

1. *atra*
2. *lugubris*
3. *novaeguineae*
4. *australis*

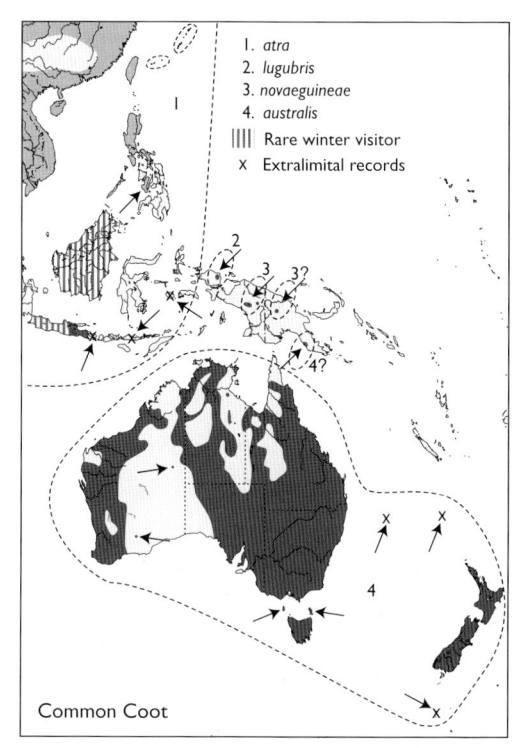

|||| Rare winter visitor
× Extralimital records

Common Coot

The race *lugubris* was formerly regarded as not uncommon to common, and the race *novaeguineae* as locally common (Rand & Gilliard 1967, Ripley 1977, Coates 1985, Beehler *et al.* 1986); their current status is not clear. Elsewhere in New Guinea this species was not known from lowland areas before 1978, after which a few birds were present in the Port Moresby area until 1981, while at L Wangbin the species was apparently exterminated when shotguns were introduced into the area (Coates 1985); however it was rediscovered at L Wangbin in 1992 (Gregory 1995). In E Java, *lugubris* formerly bred at mountain lakes of the Yang Plateau, but there are no recent records (MacKinnon & Phillipps 1993); up to 3 birds regularly recorded at Tiwu Bowu, Flores, in Apr-Nov in recent years (Coates & Bishop 1997) could be this race. The race *australis* (all details from Marchant & Higgins 1993) is widespread in Australia, mainly E of a line from Streaky Bay to Argyle and W of a line from Esperance to Port Headland; it is most widespread in E Queensland, throughout New South Wales and Victoria, in E and S South Australia, and SW Western Australia; it occurs throughout Tasmania. It is locally common and populations are stable but are characterised by large changes in local abundance; annual indices of abundance from aerial surveys in c. 12% of E Australia, 1983-89, ranged from 2,800 to 263,300, and counts in Victoria in 1991-92 totalled 131,000 and 110,000 birds. In New Zealand it was rare before 1957, when an invasion occurred, and it is now widespread but scattered, with an estimated total population of 1,655-1,955 birds, and it is increasing in range and abundance (for further details see Marchant & Higgins 1993); it is displaced by wetland drainage but rapidly colonises suitable artificial habitats such as dams, while in arid areas artificial wetlands act as refuges.

Predators include the Black-necked Stork *Ephippiorhynchus asiaticus*, which hunts down diving coots in shallow water and swallows them whole, and can even catch them in flight (Ripley 1977), Pallas's Fish Eagle *Haliaeetus leucoryphus*, White-tailed Eagle *H. albicilla* and Spotted Eagle *Aquila clanga* (Ali & Ripley 1980, Roberts 1991), and also (in Australia) raptors and water-rats *Hydromys chrysogaster* (Marchant & Higgins 1993).

MOVEMENTS This species is present all year in warm and temperate regions, although birds are not necessarily resident, but it is mainly migratory in N Eurasia under the influence of a continental climate. It is also partly nomadic, breeding opportunistically on temporary waters (Urban *et al.* 1986). Birds normally migrate at night (e.g. Patrikeev 1995). In Europe, populations from Fennoscandia and E of the Czech Republic are mostly migratory (Hagemeijer & Blair 1997); autumn movement occurs through continental Europe on a broad front, W to S, and coastal movement via the Baltic brings birds from as far E as Moscow into the North Sea area and even to Iberia or NW Africa; wintering concentrations on the Black Sea and Caspian Sea, and in Turkey and Iraq, are probably from the former USSR (Cramp & Simmons 1980). Movements occur from mid-Aug to Nov, but some birds remain in C Europe until Jan; return passage begins in late Feb, birds appear in C Europe from early Mar, and migration continues until May (Cramp & Simmons 1980, Hagemeijer & Blair 1997). In Egypt, migrants arrive from mid-Sep, numbers increase in Nov-Dec, and most birds have left by mid-Mar, some flocks remaining until early Apr (Goodman *et al.* 1989). Migrants arrive in Israel from end-Aug to end-Dec, mainly from mid-Oct to mid-Dec, and depart in Feb-May, mainly end-Feb to end-Mar (Shirihai 1996); in Oman it is recorded from Oct to Apr/May (Gallagher & Woodcock 1980). It reaches Morocco and Libya in Sep, and passes through SW Libya in Oct-Nov (Urban *et al.* 1986). Birds cross the Sahara on a broad front, and are recorded in Senegal and Mali from Nov-Apr, Niger and Nigeria Nov-Mar, Burkina Faso Jan and Apr, and Chad Sep-Apr; it is present in N Africa until Apr and in Libya until early May (exceptionally mid-Jun) (Lamarche 1980, Urban *et al.* 1986, Giradoux *et al.* 1988, Thonnerieux *et al.* 1989, Elgood *et al.* 1994, Rodwell *et al.* 1996). In Mauritania it occurs on passage, and dead birds have been found at Banc d'Arguin in Nov (Gee 1984). Birds wintering from the Black and Caspian Seas S to Iraq are presumably from the former USSR; movements occur in Azerbaijan in Sep-Dec and Feb-Apr. Birds ringed in England, Belgium and Spain have been recovered in Morocco; birds from France, Switzerland, Poland and former Yugoslavia in Algeria; and birds from Germany, Hungary and Poland in Tunisia (Urban *et al.* 1986). Birds ringed in Israel have been recovered in the former USSR (41-49°N and 32-49°E), and vice versa (Shirihai 1996), and a bird ringed near Odessa (former USSR) was recovered in Egypt (Goodman *et al.* 1989).

Moult migrations occur; these are little studied but at least some involve non-breeding adults, and moulting concentrations occur in Jun-Sep in Denmark, Bavaria, the Bodensee, the Ukrainian coast of the Black Sea, and probably in England.

Further E in Asia, N passage is observed in Afghanistan and Pakistan, Feb-May, and passage is recorded in Nepal in Nov, Mar, May and Jun (Ali & Ripley 1980, Inskipp & Inskipp 1991, Roberts 1991). In China this species is classed as a migrant in Shandong, S Shaanxi, Qinghai, and part of Xizang (Cheng 1987). In the Philippines it is recorded in Nov-Mar. A bird ringed in Indore, N India, was recovered in Uzbekistan, and 1 from Kazakhstan in Kashmir (Ali & Ripley 1980). The race *australis* (details from Marchant & Higgins 1993) is dispersive in Australia, where large changes in abundance occur, possibly in response to weather conditions and changes in water levels, as birds move to flooded areas, breed and then depart. Such patterns may not be seasonal, but seasonal changes do occur in some areas, e.g. in SW New South Wales large flocks occur in late summer and winter; in Rockingham D, Queensland, flocks occur in the summer; in SW Australia, numbers peak in summer-autumn at 2 sites and in the dry season at another; at 2 sites in Victoria, peaks occur in winter. In New Zealand numbers may fluctuate seasonally, being highest in some areas in Mar-Aug and lowest in Dec-Mar. It probably often crosses the Bass Strait and Tasman Sea, and it is recorded occasionally from Norfolk I (2 records), Lord Howe I (2 records) and Macquarie I (2 prolonged occurrences). The average distance travelled by 65 birds ringed in Victoria was 295±230km, only 4 moving >500km. Birds seen irregularly in the New Guinea lowlands may be vagrants from Australia.

It is accidental in Alaska (St Paul, Pribilof Is), Labrador, Newfoundland, Greenland, Iceland, Spitsbergen and the Faeroes. It is recorded as a rare visitor in the Mariana Is (Pratt *et al.* 1987), and *australis* was once collected from Buru (Moluccas) in May, presumably a vagrant from Australia (Hartert 1924. White & Bruce 1986). It is a vagrant to Singapore (C. R. Robson *in litt.*).

HABITAT Most information is from Cramp & Simmons (1980) and Marchant & Higgins (1993). It inhabits mainly large, still or slow-moving waters, occurring on eutrophic and mesotrophic lakes, pools, ponds, dams, reservoirs, barrages, gravel pits, canals, drainage channels, dykes, rivers and river deltas, creeks, oxbows, open marshes, freshwater meadows, floodlands and lagoons; also lakes and pools in towns, sewage ponds, and saltpans and claypans. It exploits temporary pools and seasonal marshes for breeding, and also in the winter quarters. It prefers fairly shallow waters with open and deeper water (>2m) for diving and with a muddy bottom well furnished with marginal, emergent, floating or submerged vegetation. It always requires some open water, and tolerates some exposure and wave action, but when breeding it does not normally occur far from banks or from emergent or floating vegetation, cover availability being a limiting factor during this period. In Europe it prefers mosaic biotopes where emergent vegetation is interspersed with wet grassy mounds, banks and islets with bushes or tree clumps (Hagemeijer & Blair 1997). In winter it will resort to quiet estuarine or inshore seawaters, and often occurs on the open water of lakes and reservoirs. It usually avoids closely overgrown, narrowly confined or very shallow waters, and those overshadowed by trees, cliffs etc, but it sometimes occurs on fast rivers where suitable vegetation flourishes. In contrast, the Common Moorhen prefers smaller waterbodies with more cover. In the winter it keeps to deeper areas of more open water than does the Red-knobbed Coot (Wood 1975). It normally occurs in lowlands, but up to 1,000m in Europe (locally to 1,800m in the Swiss Alps), 2,000m in Tajikistan and Iran, and 2,500m in Kashmir; it occurs up to 3,700m in New Guinea, where it is resident at montane lakes.

FOOD AND FEEDING Omnivorous, but primarily vegetarian. Most information is from Cramp & Simmons (1980). It eats mainly the vegetative parts and seeds of aquatic and sometimes terrestrial plants: algae (*Chara, Vaucheria, Cladophora, Spirogyra, Ectocarpus* and *Nostoc*); the vegetative parts of *Potamogeton, Ruppia, Zannichella, Elodea, Zostera, Vallisnaria, Lemna, Ceratophyllum, Ranunculus, Polygonum, Myriophyllum, Najas, Carex, Cyperus, Scirpus, Typha, Phragmites, Phalaris, Sparganium* and grasses (e.g. *Poa, Pennisetum* and *Paspalum*); and seeds of *Ruppia, Sparganium, Potamogeton, Carex, Scirpus, Cladium, Nymphaea, Rumex, Ceratophyllum,* grasses and cereals. It also eats clubmoss *Selaginella* and aquatic fungus *Leptomitus*. In Myanmar it eats "young crops" (Smythies 1986). Animal food is chiefly molluscs and insects, and includes: worms and leeches; molluscs (*Limnaea, Planorbis, Sphaerium, Cardium, Unio, Dreissena, Arion, Limax, Agriolimax, Hydrobia, Patella, Nerita, Littorina, Rissoa, Purpura, Tellina, Donax, Cardium, Mya, Mytilus* etc); shrimps; adult and larval insects such as Diptera, Trichoptera, Odonata, Lepidoptera, Coleoptera and Hemiptera (including *Melolontha, Notonecta* and *Corixa*); spiders; small fish (e.g. *Gasterosteus, Rutilus, Gobio, Phoxinus*) and fish eggs; frogs; birds and their eggs; and small mammals. It also takes fish food and duck food, bread, and potatoes, and it is known to eat goose and other waterbird droppings (Kear *et al.* 1980, Phillips 1991). In 157 stomachs from Britain, 84.1% by volume was plant material and 15.9% animal (Witherby *et al.* 1941); plants are most important in Dec-Feb and least in May; similar proportions are recorded from other regions. In W Bengal, 36 stomachs contained 63.5% vegetable matter, 13% freshwater gastropods, 6.2% annelid worms, 7% aquatic insects and 6.7% small fish (date of sample not given) (Roberts 1991). At L Sempach, Switzerland, which has few underwater plants, the main summer food is *Phragmites* (leaves, shoots, pith and roots), while grasses and cereal seeds predominate in winter.

It feeds both in water and on land, employing a variety of foraging techniques. It scrapes algae off stems, stones and tree stumps under water (Marchant & Higgins 1993), takes food (including plant debris) from the water surface while swimming, breaks off young emergent shoots, feeds among vegetation stirred up by other waterfowl, leaps upwards to bring down leaves, seed-heads and insects, up-ends in water up to 40cm deep, occasionally immersing the whole body, and dives to bring up plant material (occasionally also invertebrates and fish), which it eats on the surface. Dives are usually short, no more than 20s (up to 27s recorded), and normally to only 1-2m, but a depth of 6.5m is recorded. It also fishes for shrimps in flocks of up to 100, sometimes with ducks such as Hardheads *Aythya australis* (Marchant & Higgins 1993). It grazes on land in flocks, or sometimes solitarily, particularly from late autumn to spring when winds cause high waves on the water. It kleptoparasitises conspecifics as well as swans and both surface-feeding and diving ducks. Pacific Black Ducks *Anas superciliosa* steal food from Coots (Marchant & Higgins 1993). Large food items are shaken to break them up, or may be chopped up by the bill on the ground.

There are conflicting accounts of the proportion of invertebrates fed to young, but at least some are fed chiefly on invertebrates for the first 10 days, major food items including emerging aquatic insects (Horsfall 1984c).

HABITS It is diurnal, but is often active at night on moonlit nights or on floodlit waters. It swims with the neck looking fairly short and thick, and nods the head less than does the Common Moorhen. When taking off it rises with difficulty and makes a long run along the surface before becoming airborne. It flies with the neck extended and the large feet stretched out behind; the wingbeats seem rapid for the rate of progress made. When landing, the feet touch the water first and the bird then strikes the water with its breast. It dives well (see Food and Feeding) for short periods, pressing the air out of the feathers, making a short upward jump, tilting forwards and diving almost vertically with the neck extended; it uses the feet for swimming under water. On land it has a twisted walk and an awkward, waddling run assisted by the flapping wings. It is not shy and will allow very close approach when accustomed to human proximity, even feeding from the hand in town parks; it often walks or rests on land and grazes on lawns etc, within easy reach of the water to which it retreats if disturbed.

Birds normally roost at sunset; roosting and loafing sites are variable: small islets, mudbanks, sandbanks, floating mats of vegetation, rocks in water, floating logs, branches of trees over water etc; the race *australis* also roosts in disused nests of other birds, and has even been seen removing eggs from Hoary-headed Grebe *Poliocephalus poliocephalus* from a nest to do so (Marchant & Higgins 1993). Flocks roost on the open water, in shore vegetation or in adjacent meadows.

SOCIAL ORGANISATION Most information is from Cramp & Simmons (1980). It is gregarious, forming large flocks outside the breeding season, often in association with ducks, but is monogamous, territorial and pugnacious when breeding; males are also reported to exhibit successive polygyny, mating with two or more females during a breeding season (Horsfall 1984a). Given favourable conditions some birds are territorial throughout the year; pair bonds sometimes may be retained in flocking and migratory populations. In the nominate race pair formation usually takes place in the winter and spring, before the acquisition of a territory (which can take place as early as Dec-Jan when climate and food availability permit); unpaired birds may not be able to hold a breeding territory when neighbouring pairs are present.

Nests are well separated within fixed, stringently defended, all-purpose breeding territories. The typical territory size is 0.1-0.5ha, each territory having 40-50m of shoreline, and the mean distance between nests is 45-265 m (minimum 8-22m, but sometimes only 0.75-2.5m); marginal territories are often <0.05-0.1ha, with less defended shoreline. In a Danish marsh, pairs which hatched chicks early in the season withdrew into neighbouring reed swamp with their broods, and their former territories were immediately occupied by new pairs from previously non-breeding members of the flock (Fjeldså 1973). Maximum breeding densities in Europe are 13 pairs/ha (Volga Delta) and 6-7 pairs/ha (Moldova and Latvia) (Hagemeijer & Blair 1997). In Australia, 6-10 pairs/ha are recorded (Marchant & Higgins 1993).

SOCIAL AND SEXUAL BEHAVIOUR Most information is taken from Cramp & Simmons (1980). Displays are very similar to those of the American Coot (see Fig. 15). In aggression a low posture, with head and neck depressed, is common; showing the white shield against the dark head feathers, erecting the neck feathers in a ruff, and arching the wings are all important. In courtship a bowing posture, with lowered head and bill pointing towards the substrate (or even pointing backwards), serves to make the shield inconspicuous. Tail-raising, and expanding the undertail-coverts, also occur but are not as conspicuous as in the American Coot or the Common Moorhen, which have white undertail-coverts.

Intraspecific aggression within flocks is probably important as a spacing mechanism, and becomes more prominent when flocks are breaking up to breed. Shield-showing is much used in this context, while charging attacks are probably used to establish social rank-order and to steal food from subordinates; an intense form of shield-showing involves stretching the neck forward, and the bird may also raise itself on the water momentarily by treading movements with the feet. Tail-raising, practised by the American Coot, is not seen in the Common Coot, probably being replaced by the warning call, at which the flock moves to cover. On the approach of a gull or a bird of prey, a flock will form a tight pack and splash up water. When approached by a Marsh Harrier *Circus aeruginosus*, a flock of coots will "start screaming, gather in a thick mass, turn on their backs and use their legs for defence" (Patrikeev 1995). Similar kicking behaviour is recorded in Australia (Weston 1993).

Antagonistic behaviour (for details see American Coot species account) includes patrolling (with conspicuous shield-showing), and the charging attack, splattering attack and splattering retreat. At territory boundaries, adversaries typically perform a hunched display, as in the Common Moorhen, with a low posture, high wing-arch, tail raising and tail fanning; displaying birds usually turn away from each other and move off in opposite directions. Fighting generally follows charging attacks when neither bird gives way, and is as described for the American Coot. Birds do not normally fight individuals of the opposite sex, but both members of a pair often fight an opposing pair; juveniles also assist occasionally. Interspecific aggression is greatest towards Common Moorhens.

Pairs are also aggressive towards chicks of neighbouring pairs, plunging them under the water and shaking them violently by the head; attacked chicks usually feign death and thus inhibit the attacker. Parents will also attack, and sometimes kill, their own chicks by shaking them violently by the head; this behaviour, termed "tousling", apparently regulates which chicks are most often closest to the parent and are thus most often fed by that parent (chicks which are tousled frequently by a parent are seldom fed by that parent), and serves to maintain brood division (Horsfall 1984a).

In defence of nests and young the swanning display is used, often followed by the water-churning display (see American Coot). Some incubating birds may peck intruders, or may be noisy, while others may slip away from the nest quietly (Marchant & Higgins 1993). In a distraction display the incubating bird crouched at the side of the nest with wings spread and neck outstretched while its partner jumped up and down on the surface of the water (Jackson & Lyall 1964). An incubating bird will fly up to attack a Swamp Harrier *Circus approximans*, afterwards dropping down to cover the eggs (Brown & Brown 1980), and will pull vegetation over itself; adults and young, if disturbed, may move quietly to another brood-nest (Marchant & Higgins 1993).

During pair formation the female, while still within the flock, calls continually and often carries plant material in her bill; the male follows her almost ceaselessly, intermittently puffing up his feathers (and thereby provoking threats and attack from other flock members). After repeated courtship the female approaches the male and adopts the bowing posture, when the male allopreens ('nibbles') her neck feathers (see American Coot). After the territory is selected, the bowing and nibbling ceremony is still performed regularly, probably to strengthen the pair bond, until incubation starts, and continues less frequently thereafter. Pairs usually copulate 3-5 times daily but the male attempts copulation more frequently, at intervals of 1-2h. A courtship chase may precede copulation, both birds holding their necks stretched forward; the chase usually ends on land, when the female eventually stops and solicits copulation by performing the arch-bow display ('standing arch') – see American Coot. The female then crouches in a squat bow, the male steps onto her back from behind, often pattering his feet to stimulate the female to erect her tail vertically. Copulation then takes place, with the mail vibrating his tail; he then flaps his wings and, with neck stretched forward, rises steeply before falling off the female to the front. Self-preening often occurs after copulation.

BREEDING AND SURVIVAL Season Europe, Feb-Sep (most Mar-Jul); Azerbaijan, Apr-Jun; N Africa, Mar-Jun; Israel Apr-May; Kashmir and N India, May-Sep; S India, Nov-Dec; Japan, Mar-Aug; Java, Mar and May-Jun (formerly); New Guinea, Mar, breeding condition Jul-Aug; Australia, mainly Aug-Feb, also all other months, main breeding correlated with rainfall +2 months (Halse & Jaensch 1989); New Zealand, Sep-Feb, young in early Jun. **Nest** Almost always in shallow water, normally concealed in emergent vegetation but sometimes in open; often resting on bottom or trampled foundation of vegetation, occasionally floating; also builds in fork of tree or in bushes up to 3m above water (Australia); artificial platforms, rafts, tree stumps or islands sometimes used. Where no cover exists, may build among rocks and boulders, or on boulders emerging from water. Receding water level can leave nest many m from water. Nest bulky, of dead and live stems and leaves (*Typha, Scirpus, Bolboschoenus* etc.), sometimes also twigs of bushes (tamarisk *Tamarix*, lignum *Muehlenbeckia* etc), bark and roots; lined with finer material. Typical measurements: external diameter 25-55cm, height 8-35cm above water; cup diameter 16-30cm, depth 4-13cm; nest may be built up to height of 45cm if water level rises; ramp often constructed. Both sexes build, but male brings most material while female incorporates it; 1-5 brood platforms also built by male. **Eggs** Eggs 1-14 (usually 6-10); mean (1,131 clutches, nominate) 7.5, (151 clutches, *australis*) 5.8; larger clutches may be laid by 2 females. Clutch size declines during season due to replacement layings and because older birds lay before younger and lay larger clutches. Eggs oval, sometimes elliptical or biconical; smooth and slightly glossy; pale buff to stone or white-brown, evenly spotted black, dark brown or purplish-brown, sometimes with a few violet-grey undermarkings. Size of eggs: nominate 44-61 x 33-40 (53 x 36, n = 485), mean weight 38 (n = 283), weight 79 eggs from Azerbaijan 26-43.1 (34.9); *australis* 48.3-52.6 x 31.2-35.1 (50.0 [SD 1.4] x 33.9 [SD 0.9], n = 13). Eggs laid daily, sometimes at intervals of 1-2 days. Up to 3 replacement clutches laid after egg loss. **Incubation** Period 21-26 days; by both parents; begins with 1st-4th egg; hatching asynchronous, over several days. **Chicks** Precocial and nidifugous; brooded on nest by female for 3-4 days while male brings food; later fed and cared for by both parents, which may divide brood temporarily or permanently; larger chicks fed by females, which feed young at higher rate than males (Horsfall 1984a). Chicks beg either by drawing head back onto body, or by stretching head up and forward, with wings held out from body, and pecking at bill of parent; older young beg by stretching head forward and often twisting neck round. Cases of male killing chicks reported, and later-hatched chicks may starve within 4-5 days of hatching; early-hatched chicks gain feeding advantage over those which hatch later, while parents regulate which chicks accompany them on foraging trips and thus maintain feeding differences within brood (Horsfall 1984a, 1984b; see Social and Sexual Behaviour). Older young help feed chicks of later broods. Parents with young <2 weeks old may tolerate or adopt strange young of similar age; thereafter drive strange chicks off (Alley & Boyd 1950); young follow and beg from any adult, even if attacked, for first 8-11 days; thereafter avoid adults in attitude of attack; recognise parents at 3 weeks. Young begin to dive at 3-5 weeks; become self-feeding at 30 days; fledged at 55-60 days; fed by parents for up to 2 months; independent at 6-8 weeks; fly at 8-11 weeks; display territoriality at 9 weeks; remain in parental territory for up to 14 weeks, possibly helping in territory defence; also said to remain with parents until next spring (Jackson & Lyall 1964).

In Australia, 2 instances of egg-dumping recorded (Marchant & Higgins 1993). In England, of 343 eggs laid by 70 pairs, and of a further 121 laid, 116 (34%) and 42 (35%) respectively hatched, and 71 (21%) and 28 (23%) chicks fledged; main losses from flooding (33% of first sample) and predation (58% of second sample) (Cramp & Simmons 1980). Comparative figures for Sweden are 115 laid, 56 (49%) hatched, 26 (23%) chicks lived to 8 weeks; of 268 clutches, Czechoslovakia, 31 (11.5%) destroyed, mainly by flooding or wave action (Cramp & Simmons 1980). Hatching success greater with first clutches (37.5%) than with replacements (8-28%) (Lelek 1958). In Australia, of 472 eggs laid, 182 (39%) young fledged; young hatched from 51 (53%) of 96 clutches laid; from 66 eggs in successful nests, 55 (83%) hatched; from 96 nests, 12 (12.5%) lost through flooding and 33 (34.5%) from predation (Marchant & Higgins 1993). Predators of nests and young include Hooded Crow *Corvus corone* and diced snake *Natrix tesselata* (Patrikeev 1995), Australian Raven *Corvus coronoides*, Swamp Harrier *Circus approximans* and other raptors, and black rats *Rattus rattus*, while Purple Swamphens (120) usurp nests for brood nests (Marchant & Higgins 1993).

Age of first breeding 1-2 years; may not first breed for several years. May have 2 broods per season. **Survival** Mortality in NW Europe in first year of life 76-87%, in second year 48-72%; lower limits probably more accurate (Glutz von Blotzheim *et al.* 1973). Oldest ringed bird 18 years 3 months (Cramp & Simmons 1980).

137 HAWAIIAN COOT
Fulica alai
Plate 41

Fulica alai Peale, 1848, Hawaiian Islands.

Sometimes regarded as a race of *F. americana*. Forms superspecies with *F. cristata, F. atra, F. americana, F. caribaea, F. leucoptera* and *F. ardesiaca*. Monotypic.

Synonyms: *Fulica atra/americana alai*.

IDENTIFICATION Length 38-39cm; wingspan 65cm. Dark slate-grey, paler on underparts; head and neck blackish. White undertail-coverts form extensive patch (inverted heart-shape) but are often not visible in swimming birds; leading edge of wing white; secondaries broadly tipped white. Frontal shield markedly swollen; extends to crown and is often high enough to be visible from rear; bill and shield commonly white, but colour varies from bluish-white to cream, yellow or dark blood-red; in some birds shield red, and bill has broken, reddish-brown to black subterminal ring. Iris red; legs and feet pale grey or bluish, sometimes yellowish or greenish. Sexes alike, but female smaller, with narrower frontal shield. Immature has white throat and breast, dark bill and small frontal shield. Juvenile undescribed. Inhabits flooded wetlands, both fresh and saline.

Similar species American Coot (138), which occasionally occurs in Hawaiian Is (e.g. Pratt 1987), very similar but has longer wing, white bill with subterminal broken chestnut or maroon ring (but some Hawaiian Coots also show this feature), rather small white frontal shield with dark chestnut or red-brown horny callus at top (in breeding birds sometimes covering entire shield), yellow to orange legs and feet, red tibiae and pale blue-grey toe lobes.

VOICE Chicken-like *keck-keck* calls and other clucks and creaks.

DESCRIPTION
Adult Head and neck blackish; rest of plumage slate-grey. Upperparts, including tertials and tail, dark slate-grey, sometimes tinged brownish; alula, primaries and secondaries browner; marginal coverts white; outer edge of outermost alula feather, and of outermost primary, white; secondaries with relatively broad white tips; axillaries and underwing-coverts slate-grey; underside of remiges glossy grey. Underparts dark ash-grey, paler on centre of lower belly and vent; lateral and terminal undertail-coverts white. Swollen frontal shield extends back to crown; bill and shield white in most birds, but colour varies from bluish-white to cream, yellow or dark blood-red; in c. 15% of birds shield is red and bill has broken, reddish-brown to black subterminal ring. Iris red; legs and feet pale grey or bluish, sometimes yellowish or greenish (Munro 1944). Sexes alike, but female smaller, with narrower frontal shield.

Immature Not properly described; presumably similar to American Coot.

Juvenile Chin, throat and breast white; bill dark, frontal shield small.

Downy young Black downy chick has reddish-orange bristles on head, neck and throat; down short or absent on forehead and crown; bill crimson to orange-red, with black tip.

MEASUREMENTS Wing of 2 males 175, 186, of 6 females 170-178; tail of 2 males 35, 46, of 6 females 45-49; culmen (with shield) of 2 males 45, 53, of 6 females 47-53; tarsus of 1 male 55, of 6 females 50-53 (Schwartz & Schwartz 1952, Ripley 1977).

GEOGRAPHICAL VARIATION None.

MOULT Details are presumably as for the American Coot.

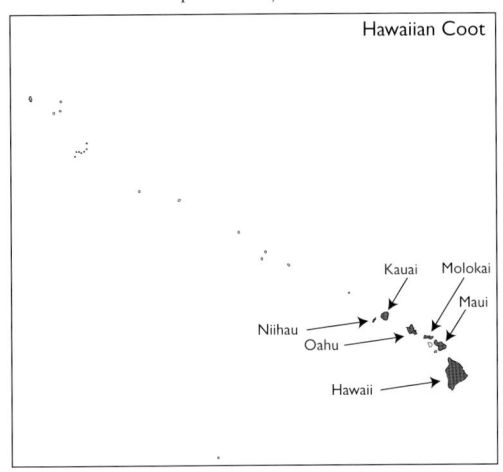

DISTRIBUTION AND STATUS Hawaiian Is. VULNERABLE. It is resident on all the main Hawaiian islands from Niihau eastward, except Lanai, with stragglers reaching as far W as Kure Atoll in the NW Hawaiian Is. It can be expected to occur on virtually any body of water but these are now scattered and very limited in area. This once plentiful species has declined drastically as a result of wetland destruction by drainage for cultivation and other development this century, as well as having suffered from the depredations of introduced predators (Berger 1981, King 1981, Collar *et al.* 1994). Insecticides, and also herbicides used to clear canals and drainage ditches serving sugarcane plantations, may have poisoned the birds directly or indirectly (Berger 1981). When nesting, this species is also vulnerable to introduced predators such as dogs, cats and mongooses (Berger 1981). Only c. 15 regular nesting locations were known in the early 1980s, on Oahu, Maui, Hawaii, Niihau and Molokai (Byrd *et al.* 1985).

The total population probably fluctuates between 2,000 and 4,000 birds (also see Movements section), 80% of which probably occur on Kauai, Oahu and Maui; between 1977 and 1986, winter counts were 422-2,823 (mean 1,447) and summer counts 915-4,466 (mean 1872), and populations were related inversely to rainfall (Pratt 1987, Engilis & Pratt 1993, Collar *et al.* 1994). Wide annual fluctuations could result from irruptive movements in the resident population, from variations in breeding success or from errors in censusing methods (Pratt 1987). Hunting was prohibited in 1939 and the species is fully protected, although Berger (1981) recorded that some illegal killing was occurring. Key wetland areas have been, or are being, acquired as refuges or sanctuaries, and other areas are protected by cooperative agreements (King 1981). Management recommendations (Byrd *et al.* 1985) include: habitat management to promote optimum conditions, i.e. a 50:50 ratio of open water and sparse emergent cover, and a water depth of c. 30cm; a better study of the seasonality of breeding so that any periods of reduced activity may be used for marsh maintenance and improvement;

more detailed studies of food requirements and breeding success; improved control of mortality factors (especially chick predation); and a study of local movements.

MOVEMENTS To some extent nomadic and irruptive (Pratt 1987), wandering between islands in response to the availability of waterbodies (e.g. Munro 1960). Engilis & Pratt (1993) found that all islands show similar population fluctuations, "indicating no annual, corresponding shift of numbers between islands that might indicate regular population movements". One bird found dead on Tern I in 1965 and several in NW Hawaiian Is in the summer of 1983 were apparently wanderers from the main islands (Pratt 1987).

HABITAT All types of wetland, including fresh and saltwater ponds, estuaries and marshes, irrigation ditches and flooded agricultural lands. Breeding sites are characterised by robust emergent plants interspersed with open, fresh water which is usually <1m deep, and interconnecting waterways make smaller open water areas more useful – at 1 pond, nesting declined drastically in a *Batis maritima* area after interconnecting waterways became overgrown (Byrd *et al.* 1985). Coots prefer fresh water for nesting.

FOOD AND FEEDING Mainly vegetarian; its food is apparently similar to that of the American Coot. Stems, leaves and seeds of water plants are recorded; also lagoon molluscs (Munro 1960), guava *Psidium guajava* seeds, fibrous plant stems, and seeds taken from the ripening heads of grass growing near open water (Schwartz & Schwartz 1952). Birds feed from the water surface and also by diving.

HABITS No detailed information available; habits are presumably very similar to those of the American Coot. When unmolested the birds become very tame.

SOCIAL ORGANISATION Monogamous; territorial when breeding. The following information is from Byrd *et al.* (1985). Nesting density varies widely between sites, recorded averages being 1.6, 2.5 and 12.5 nests/ha. Average distances between active nests at 1 site were 27 and 25.5m, and nests were found as close together as 7 and 13.5m.

SOCIAL AND SEXUAL BEHAVIOUR Very similar to that of the American Coot: for details, see Shallenberger (1978).

BREEDING AND SURVIVAL Most information is from Byrd *et al.* (1985). **Season** Breeds throughout year, mostly Apr-Aug (but peaks may only indicate peaks of observation). **Nest** Sometimes floating, usually built among robust emergent aquatic vegetation within 1m of open water and anchored to surrounding vegetation; sometimes in open water, anchored to dense algal mats. Semi-floating nests are most frequently anchored beside or within clumps of *Scirpus*, *Typha*, *Pluchia indica* or *Batis maritima*, usually adjacent to open water. Also nests on mats of *Bacopa monnieri* and *Paspalum*; twice recorded nesting on islands near water. Avoids very dense clumps of robust emergents, preferring sites with relatively low stem density; height of cover may vary considerably. Nest platform substantial, made from buoyant stems of nearby emergent plants, particularly *Scirpus*, but also *Batis*, *Brachiaria* and *Pluchia*; outside diameter up to 31cm, inside diameter 17-18cm, depth 5-6cm; broader base of nests built in open water probably serves to improve stability and reduce effects of wave action. Sometimes builds resting platforms close to breeding nest. Mean water depth at nest sites 33.5 and 60.8cm. **Eggs** 1-10 (mean 4.9, SE 0.3, n = 62 clutches); ovoid; greyish to pale brown or buffy-tan, thickly covered with small spots of brown or black, and sometimes also of purple; mean size of 146 eggs 48.3 (SE 0.2) x 33.7 (SE 0.1); size of 15 eggs 44.2-49.4 x 31.5-36 (46.9 x 33.8), calculated weight 28; eggs laid at daily intervals. **Incubation** Probably 23-27 days, by both parents; may begin after first egg laid, or may begin later; hatching asynchronous. **Chicks** Precocial; can swim shortly after hatching; details of care and development presumably similar to those for American Coot. Nesting success high (92% recorded); hatching success averages 80%, most egg losses apparently due to infertility rather than predation; chick mortality high (average 76% over first 14 days), largely due to predation by introduced mammals such as dogs, cats and mongooses; Night Herons *Nycticorax nycticorax*, fish (bass) and bullfrogs also take young chicks (Berger 1981). Two broods per year occasional.

138 AMERICAN COOT
Fulica americana   Plate 42

Fulica americana Gmelin, 1789, North America.

Sometimes considered to include *F. alai*, *F. caribaea* and *F. ardesiaca* as races; forms superspecies with these, and with *F. cristata*, *F. atra* and *F. leucoptera*. Two subspecies recognized.

Synonyms: *Fulica wilsoni/leucopyga*; *Colymbus parvus* (fossil material).

IDENTIFICATION Length 34-43cm; wingspan 60-70cm. Head and neck black; rest of plumage slate-grey; some white on leading edge of outer wing and on wing-bend; secondaries tipped white; undertail-coverts white, with black wedge at base giving rather heart-shaped white patch. Iris red; bill white with subterminal broken chestnut or maroon ring; frontal shield rather small, white, with red-brown horny callus at top (sometimes covering entire shield in breeding birds); legs and feet yellow to orange-red, tibia red. Colour and shape of shield variable: in a few birds red-brown callus reduced or lacking; shield may be washed yellow, may have reddish markings, and may be bulbous. Sexes alike, but female smaller. Immature like adult; legs and feet become yellow-green, and tibia red, by spring; tarsal colour becomes orange-red after 4 years. Juvenile olive-brown on crown and upperparts, pale ash-grey on sides of head and body, flanks and underparts; may appear mottled on throat and forehead; bill and shield ivory-grey with small, pale red callus, most birds soon attaining adult colours; legs and feet blue-grey or grey-green. Inhabits reed-fringed open waters, open marshes and sluggish rivers; also estuaries and bays in winter.

Similar species Differs from American congeners in colour of bill, shield, legs and feet, and in having red-brown horny callus above shield. Very similar to Caribbean Coot (139), which differs in lacking callus and in having swollen shield which extends well onto crown and is sometimes tinged ivory or yellow; however, a few American Coots have callus much reduced or lacking, and may have shield almost entirely white or washed yellow, sometimes with variable and irregular reddish markings; shield of such atypical birds may be bulbous (Roberson & Baptista 1988). Extent of white on outer web of outermost primary and tips of secondaries very variable in both species and apparently not a distinguishing character (Roberson & Baptista 1988, contra Blake 1977). Hawaiian Coot (137) very similar but has shorter wing; secondaries broadly tipped white; frontal shield markedly swollen, extending to crown and often visible from rear; bill and shield commonly white (some may have broken chestnut or maroon ring on bill), but may be bluish-white to cream, yellow or dark blood-red; legs and feet pale grey or bluish, sometimes yellowish or greenish. Common Coot (136) lacks white undertail-coverts; shows less contrast between blackish head and neck and paler areas of body; white tips to secondaries may be rather less prominent.

VOICE A loquacious species, with a variety of cackling and clucking notes which are often repeated; sexual differences in the calls of adults make it relatively easy to sex birds in the field (Gullion 1950). Information is from Gullion (1952). The alarm call of the male is rendered *puhlk* and that of the female *poonk*, both notes being given with a vigorous forward thrust of the head; under stress the male also gives a plaintive, crowing *puhk-cowah* or *pow-ur* and the female a simpler *cooah*. Warning notes to conspecifics are highly variable but in the male are commonly a quick *puhk-ut* or similar call, and in the female a nasal *punk-unk* or variations. The male's intimidation call is a three-part, crowing *puhk-kuh-kuh* or *cook-uk-ook*; this often leads to aggressive behaviour between neighbours. The female equivalent is a hollow, crowing *kaw-pow* or *kra-kow*. The aggressive male gives an explosive *hic*, and the courting male a cough which may become a repeated, sharp *perk* or *kerk*. When chased during courtship, an unreceptive female will face the male and give a cackling, repeated *tack-tack* call, while the female calls the male to a display platform with a repeated call which varies from a low nasal *punt* or *put* to a sharp, clear *tuk*. Recognition notes between adults are a high clear *puhk* from the male and a low, nasal *punk* from the female. To young, the male gives a clear *puht* and the female a nasal *punt*; the young are sent to shelter by an explosive *chuck* or *chook* from the male or a quick, nasal *punt-unt* from the female. Hardy *et al.* (1996) also give low-pitched gargling or rattling calls, and rapid, varied group interaction calls including *kow* and *krrr* notes.

DESCRIPTION *F. a. americana*
Adult Head and neck dark slate-black to black, with slight bluish or greenish gloss; upperparts, tail, upperwing-coverts and tertials dark slate-grey, often tinged olive from lower mantle and tertials to uppertail-coverts; leading edge of wing at wing-bend white; outer web of outermost primary (P10) white for variable distance from base (white sometimes absent) and not reaching tip; outer web of outermost alula feather also white. Primaries and secondaries dark sepia-grey, slightly paler on inner webs; outer webs tinged ash-grey when fresh; tips of secondaries broadly white; axillaries and underwing-coverts slate-grey. Underparts dark ash-grey, tinged slate on breast and flanks, paler on centre of lower belly and vent; undertail-coverts white, shorter central feathers black. When fresh, some birds have narrow white tips to feathers of underparts and underwings; in late spring and summer, tips of feathers on body and upperwings appear pale grey to off-white through wear. Iris deep red to red-brown; bill white, sometimes with pink or yellow-green tinge at border with shield; narrow band round bill half-way between nostril and tip maroon or dark red-brown, bordered in front by darker line; shield white, tinged pink or yellow-green, with red-brown or chestnut horny callus at top; callus may cover entire shield in territorial adults but, except for old birds, often shrinks during autumn and winter. Some birds may have callus much reduced or lacking, and may have shield almost entirely white or washed yellow, sometimes with variable and irregular reddish spots, streaks or splotches; shield of such atypical birds may be bulbous (Roberson & Baptista 1988). Most or all white-shielded birds may be male; shield size and degree of swelling increase in both sexes when pair-bonded and territorial, and when nesting; some old birds may have swollen shield throughout year (Gullion 1951, Roberson & Baptista 1988). Legs and feet yellow-green to bright yellow, tibia red above and behind, toe lobes pale blue-grey. Tarsal colour progresses through yellow-green and yellow to become orange-red after 4 years (Crawford 1978).
Immature After postjuvenile moult, usually inseparable from adult on plumage, but worn juvenile remiges retained. When fresh, white edges to feathers of underparts broader and less distinctly defined; upperparts slightly duller and more olive, less slate. Legs and feet become yellow-green, and tibia red, by spring.
Juvenile Crown and upperparts pale to dark olive-brown; rectrices slate-grey, washed brownish; sides of head and body, flanks, and underparts pale ash-grey, feather tips tinged white to pale olive-buff on underparts and olive-brown on flanks and vent; may appear mottled on throat and foreneck; undertail-coverts as adult. Wing similar to adult. Iris red-brown to deep red, with grey outer ring in some; at 2.5 months bill and shield ivory-grey and small callus pale red; adult colours attained at c. 4-5 months; legs and feet blue-grey or grey-green.
Downy young Downy chick black, greyer on underside; orange to red bare skin on top of head, with 2 median lines of black down; dense yellow to red bristles on forehead, lores and chin; bill red, shading to pink distally and with black tip; legs and feet greenish. See race *columbiana*.

MEASUREMENTS Wing (chord) of 10 males 186-212 (199.5, SD 9.0), of 10 females 185-209 (194.4, SD 9.8); tail of 31 males 44.5-61 (51.4), of 26 females 41-55 (49.0); culmen (including shield) of 10 males 40.5-52.5 (45.3, SD 3.8), of 10 females 36.5-54.5 (43.6, SD 4.9); tarsus of 10 males 55.5-63.5 (59.1, SD 2.3), of 10 females 54.5-65 (57.7, SD 4.0) (Ridgway & Friedmann 1941, Roberson & Baptista 1988). Cramp & Simmons (1980) give wing (max) of 15 males 192-210 (199, SD 5.9), of 8 females 177-200 (188, SD 7.8); tail of 12 males 50-54 (52.4, SD 1.4), of 7 females 48-56 (51.4, SD 2.5). Weight of 27 males 576-848 (742), of 20 females 427-628 (560) (Dunning 1993); mean spring mass of 98 males 670 (SE 8.0), of 99 females 530 (SE 5.7) (Arnold 1990).

GEOGRAPHICAL VARIATION Races separated on bare part measurements and plumage. Proposed race *grenadensis* is included in nominate (Ripley 1977).

F. a. americana Gmelin, 1789 – N and C America, from SE Alaska, S Yukon, S Mackenzie, NW and C Saskatchewan, C Manitoba, W and S Ontario, SW Quebec, S New Brunswick, Prince Edward I and Nova Scotia, locally through USA and S to N West Indies in Cuba (including I of Pines), Jamaica, Hispaniola,

Virgin Is (St John), Grand Cayman and Bahamas, and locally through Mexico (to S Baja California, Nayarit, Jalisco, Guadalajara and Michoacán) to Nicaragua and NW Costa Rica (Guanacaste); breeds locally Guatemala (L Atitlán; Dueñas). N populations winter from SE Alaska and British Columbia S through Pacific states of USA and from N Arizona, N New Mexico, C Texas, lower Mississippi and Ohio Vs, and Maryland (casually N of Canadian border E of Rocky Mts), W to Hawaii and S to SE USA, West Indies (E to Puerto Rico and St Croix, S to Grenada), and through Central America to C Panama (Canal Zone) and (possibly) N Colombia (Cauca V). See Description.

F. a. columbiana Chapman, 1914 – Colombia, in Cauca V (resident breeders?), in E Andes from C Boyacá (L Tota) S to Cundinamarca (Sabana de Bogotá), and in C Andes at Parque Nacional Puracé; also N Ecuador (Ibarra) but now extirpated (Ridgely & Greenfield in prep., Ridgely *et al.* in press). Its recorded occurrence in Nariño (Hilty & Brown 1986) may be the result of confusion with red-fronted Slate-coloured Coots (141) (Fjeldså & Krabbe 1990). Tarsus and toes longer; bill heavier and longer, with yellow base when breeding, and larger, higher, more rounded frontal shield; plumage darker, slatier, especially on underparts; underwing-coverts darker with little or no white edges; less white at tips of secondaries. Immature has distal part of bill pinkish or smoke-grey; chick has red skin on crown, blue skin above eyes, blood-red shield, bill scarlet basally, becoming orange-pink and then black at tip, and legs and feet black (Fjeldså 1983c). Wing of 7 males 186-210 (196.2), of 5 females 182-194 (187.5); tail of male 50; culmen (including shield) of unsexed birds 45-57.5; tarsus of 7 males 53-59 (55.6), of 5 females 50-55 (52.4) (Blake 1977, Ripley 1977, Fjeldså 1983c).

MOULT The adult postbreeding moult is complete, occurring from late Jul to early Sep (nominate); moult of flight feathers and all wing-coverts is simultaneous (Alisauskas & Arnold 1994), and birds are apparently flightless for c. 4 weeks; the rectrices are dropped and replaced 1 or 2 at a time throughout the period of wing moult (Gullion 1953a). Moult details are apparently as for the Common Coot (Cramp & Simmons 1980), in which moult starts with the head, neck and the front of the body. The prebreeding moult is partial, involving head and neck, sometimes also some body feathers. The postjuvenile moult is partial, involving the head and body. The first prebreeding (pre-alternate) moult is partial, and is variable in extent.

DISTRIBUTION AND STATUS SE Alaska E to Nova Scotia and S to West Indies, Nicaragua and Costa Rica, wintering W to Hawaii and S to Panama; it also occurs in Colombia but has disappeared from Ecuador. Not globally threatened. Information for North America is from Alisauskas & Arnold (1994) unless otherwise stated. The main breeding areas of the nominate race are in the pothole country of the US Great Plains states (especially North and South Dakota and Oregon) and the prairie provinces (particularly Saskatchewan) of Canada; lower densities occur in the W USA, and few now occur in the E USA where they were abundant before extensive wetland drainage occurred (Cramp & Simmons 1980, Lang 1991, Alisauskas & Arnold 1994); detailed distribution patterns are given by Alisauskas & Arnold (1994). It winters in the

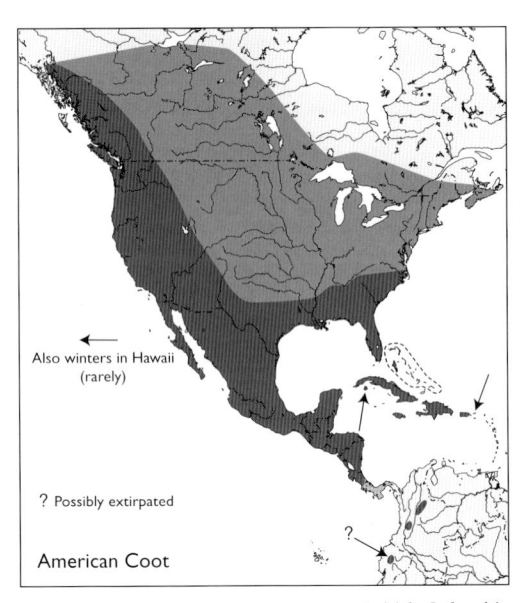

Also winters in Hawaii (rarely)

? Possibly extirpated

American Coot

Pacific and Atlantic coastal states N to British Columbia and Maryland, not normally occurring in the hinterland above the frost-line; almost all birds E of 100°W occur in regions with a mean minimum Jan temperature $>-7°C$ (Root 1988). Most birds winter in the S USA (California, Mississippi, Florida, Louisiana and Texas) and in Mexico, but it is occasional in winter N to Colorado, Nebraska, Michigan, New York, Connecticut, Rhode I and Massachusetts (Ripley 1977, Alisauskas & Arnold 1994). In North America it is considered abundant, and estimated breeding-season numbers in the prairie-parkland region have increased in the last 3 decades, sometimes reaching 1.5-2 million birds in the Canadian prairies and >1 million in the E Dakotas, but this apparent increase may be partly explained by a tendency for surveyors to count coots more accurately in recent years. USA midwinter indices over the last 40 years range from 1-3 million birds; numbers increased after 1955 and decreased after the mid-1970s, and these trends mostly reflect the patterns of abundance in the Mississippi flyway. Winter distribution has shifted during the period, with more birds wintering in the Mississippi flyway and a substantial decline in the Pacific flyway states; this decline may be caused by wetland destruction or drought in California.

It is a game species; the annual harvest over the last 3-4 decades has averaged c. 8,000 birds in Canada, and 880,000 in the USA, where it is hunted in 48 states (Eddleman *et al.* 1988); the harvest has declined in recent years in both countries. Local populations of coots are extremely vulnerable to heavy shooting. Major population management requirements include the evaluation of breeding-ground and wintering-ground surveys; the establishment of surveys in Mexico; updated estimates of survival and recovery rates; and an assessment of possible bias in harvest surveys. Traditionally, wetland habitats have not been managed for coots but the birds appear to benefit from many waterfowl management activities; however, wetland loss has reduced the potential breeding population in regions such as Iowa and Minnesota. The species's ability to pioneer new habitats should enable it to take advantage of wetland restoration efforts. It is sometimes regarded as an urban and agricultural pest in winter, e.g. on golf courses and in ricefields.

It is fairly common in Puerto Rico and uncommon in the Virgin Is in most months except during the summer (Raffaele 1989). In the winter it is locally common in Panama (but it may have declined recently in the Canal area) and Guatemala (on volcanic lakes), and widespread in Costa Rica, and was formerly common in Honduras (where it was also a locally uncommon resident) and El Salvador, and occurred in small numbers in Belize (Dickey & van Rossem 1938, Russell 1964, Monroe 1968, Ridgely et al. 1989, Stiles & Skutch 1989, deGraaf & Rapploe 1995). It has been recorded occasionally in Haiti, in Nov-Jun (Wetmore & Swales 1931). It is a rare and irregular visitor to the Hawaiian Is in winter (Pratt 1987, Pratt et al. 1987).

The race *columbiana* has been regarded as common to locally common in Colombia (e.g. Ripley 1977, Hilty & Brown 1986). However, in recent years its numbers have declined, decreases in Cundinamarca and Boyacá being attributed to intense shooting and the deterioration of wetland habitats: for example, the apparently suitable 4,500ha L Fuquene held only 1 coot in 1981, being so turbid due to sediment from soil erosion that submerged vegetation was almost absent, while the Colombian hunting ban was ignored at the lake (Fjeldså 1983c). The most recent estimate of its total numbers, by Fjeldså (1983c), is c. 2,000 adults, with the main population (800 birds) at L Tota; in 1982 Varty et al. (1986) estimated 500-600 coots at L Tota.

In North America predators of adults and juveniles include alligator *Alligator mississippiensis*, mink *Mustela vison*, Great Horned Owl *Bubo virginianus*, Hen Harrier *Circus cyaneus*, Bald Eagle *Haliaeetus leucocephalus* and Great Black-backed Gull *Larus marinus*; extensive mortality sometimes occurs during severe spring weather (Alisauskas & Arnold 1994).

MOVEMENTS Most information is from Cramp & Simmons (1980) and Alisauskas & Arnold (1994). This species is mainly migratory, especially in N America E of the Rocky Mountains, but many birds in the S USA, Mexico and Central America are probably sedentary (Gullion 1954). Birds will moult on breeding wetlands if water levels remain high, but many birds apparently migrate to large N wetlands for the postbreeding moult (Alisauskas & Arnold 1994). Birds ringed in Alberta winter mainly in California and W Mexico, with a few recorded to the SE on passage (Great Lakes) and in winter (Florida); those from Saskatchewan and Manitoba winter mainly from E Mexico to Louisiana, some reaching the Atlantic coast states and the Caribbean islands. All populations mix in the Gulf states in winter, and wintering birds ringed in Louisiana occur in summer over a vast area from Yukon to Ontario. Birds nesting more to the S move furthest S in winter, e.g. 32% of recoveries of those ringed in Iowa have been found in Cuba, compared with only 4% of recoveries from the prairie provinces. It winters throughout C America to Panama and N Colombia; small numbers reach the larger Hawaiian islands in winter (Pratt 1987). It occurs in Panama from Oct to late Apr (Ridgely et al. 1989) and in Costa Rica from Oct-Apr (Stiles & Skutch 1989).

It migrates at night, singly or in loose flocks, probably on a broad front; birds often hit powerlines (Cramp & Simmons 1980, Alisauskas & Arnold 1994). Males and nonbreeders congregate in late summer and move S ahead of females and juveniles (Ryder 1963). Autumn passage occurs from late Aug to Dec, and peaks in the Mississippi flyway in late Oct; wintering birds are present in Louisiana and Mexico Oct-Mar, Costa Rica Oct-Apr, and Panama Oct-late Apr (Cramp & Simmons 1980, Alisauskas & Arnold 1994). Spring migration occurs from late Feb through mid-May; males and older birds migrate first; birds arrive on N breeding grounds from late Mar through Apr (Cramp & Simmons 1980, Alisauskas & Arnold 1994). In British Columbia, S movements occur in Aug-Nov (peaking mid-Sep to mid-Oct), and the main N movement occurs on the S coast in Apr-May while in the interior it starts in Mar and peaks in Apr (Campbell et al. 1990). According to Bent (1926), birds move N as fast as the advancing spring melts the ice on the waterbodies, often arriving while some ice remains.

There are many cases of birds wandering N of the breeding range in summer and autumn, as far as Newfoundland, Labrador, Franklin and S and W Greenland; it is also casual W to the E Aleutians and N to W Alaska (Seward Peninsula). It is a vagrant to Iceland (Cramp & Simmons 1980). Numbers increase in Colombia Oct-Apr, but the origins of the birds involved are not clear (Hilty & Brown 1986).

HABITAT Most information is from Alisauskas & Arnold (1994). This species inhabits reed-fringed lakes and ponds, open marshes and sluggish rivers; also estuaries and bays in winter. The nominate race has maximum nesting densities on well-flooded, semi-permanent and persistent wetlands with a good interspersion of emergent vegetation and open water; nest densities are usually highest on semi-permanent wetlands, and nests are usually found where foundations (e.g. mats of algae) are available. It uses seasonal wetlands when water levels are high. See the Breeding section for details of nesting habitat. The race *columbianus* (details from Fjeldså 1983c) occurs mainly in eutrophic marshes and bays of lakes, where open water is fringed by tall *Scirpus californicus*, *Juncus bogotensis* and *Typha latifolia*; the highest concentrations occur in areas with mosaics of floating low *Bidens laevis* and *Limnobium stoloniferum* outside the main reed border. Oligotrophic coasts with tall *Cortaderia bifida* vegetation are avoided. At 3,250m on L Chingaza, the birds nested in *Carex* beds outside a delta with *Chusquea* bogs. Colombian breeding localities have dense submerged vegetation such as *Nitella*, *Elodea*, *Potamogeton* or *Myriophyllum elatinoides*, or at least areas of floating *Azolla*, *Limnobium* or *Myriophyllum brasiliense*.

This species prefers fresh water but in winter it is sometimes forced to shift temporarily to salt water and salt marshes on sheltered coasts or estuaries; in winter it also uses manmade wetlands including brackish impoundments, crayfish ponds, catfish ponds and coalmine sediment ponds. In coastal Texas, birds fed on average in areas with water 90cm deep, 20% cover of emergent vegetation and 40% cover of floating plus submerged vegetation (White & James 1978). *Hydrilla* cover may be an important component of wintering habitat in the S USA. It occupies a similar niche in the Nearctic to that of the Common Coot in the W Palearctic. A wide variety of habitats is occupied during migration; in Oklahoma, birds staying >6 days used feeding habitats differing from random sites in having greater vegetative cover, shallower water, lower pH, higher alkalinity and higher conductivity (Eddleman 1983). In N Colombia this species occurs up to 1,000m and in the Andes at 2,100-3,400m (Hilty & Brown 1986).

FOOD AND FEEDING Most information is from Alisauskas & Arnold (1994). It eats principally aquatic vegetation, mostly submerged plants, especially *Chara* and other algae, and *Potamogeton*, *Lemna*, *Elodea* and *Myriophyllum*, and also *Scirpus*; it takes seeds of *Potamogeton* and *Sparganium* (Bent 1926). It feeds on land, taking grass, sprouting grain and waste grain. It also takes aquatic insects and molluscs, more frequently during the breeding season (Jones 1940). Its primary foraging methods are diving in shallow water and dabbling at or near the water surface; it will immerse the head and neck and will sometimes upend; it also grazes on land or on floating matted vegetation, and will fly-catch. It will also eat the eggs of other birds, including Franklin's Gull *Larus pipixcan*, Pied-billed Grebe *Podylimbus podiceps* and Red-winged Blackbird *Agelaius phoeniceus* (Burger 1973). It sometimes scavenges, and cannibalism is also recorded (Paullin 1987). Ripley (1977) records that in winter birds swim along in close-packed large flocks ('rafts'), frequently submerging to take invertebrate prey which are driven forwards and upwards by the moving flock. Aquatic invertebrates may constitute a large proportion (45-85%) of the diet of young chicks, important taxa including Trichoptera and Diptera larvae, Coleoptera and Gastropoda (Driver 1988).

Sometimes this species feeds in association with Canvasbacks *Aythya valisinaria*, taking rootstocks of *Potamogeton pectinatus* disturbed by the diving ducks (Anderson 1974). Associations with other waterfowl species are also recorded. Kleptoparasitism is recorded: it robs Canvasbacks, Redheads *A. americana* and Ring-necked Ducks *A. collaris*, and it is kleptoparasitised by American Wigeon *Anas americana* and Gadwall *A. strepera* and relinquishes food passively to them, apparently to gain time to dive a second time and eat food unmolested (Bent 1926, Ryan 1981, Bergan & Smith 1986). Leshack & Hepp (1995) suggest that kleptoparasitism is an alternative feeding method used by subordinate Gadwalls that do not have access to good feeding areas.

HABITS Most information is from Bent (1926). It is apparently diurnal. On land it walks actively, often with a hunch-backed posture. It is a strong swimmer, floating high in the water with a level back; when swimming it jerks the head back and forth like a moorhen. To take off from water, birds run along the surface, beating the water with their wings and feet and making spray fly until they become airborne. The flight is strong and direct, with neck extended and the conspicuous bill pointed slightly downwards and the feet stretched out behind with the toes pointing slightly upwards; most flights are made low over the water, up to a height of 3-5m. It is more likely to escape danger by swimming, diving or scurrying across the surface than by flying. When danger threatens, the tail is erected to show the white undertail-coverts, which are not spread laterally (Fig. 15F).

SOCIAL ORGANISATION Gregarious; but monogamous, territorial and pugnacious when breeding. In resident birds with suitable territory the pair bond probably lasts for life, and territorial behaviour may occur throughout the year but is at its lowest ebb during the winter, when only a core area of the territory is defended (Oct-Feb); winter migrants also defend territories (Gullion 1953b, 1954). In the nominate race, pair formation in migrant birds probably occurs at terminal staging areas and/or after arrival at the breeding grounds (Alisauskas & Arnold 1994).

In a 4-year study of a population of American Coots in British Columbia, Canada, Lyon (1993a, b) found that conspecific brood parasitism occurred commonly, with over 40% of nests parasitised and 13% of all eggs laid parasitically. Parasitism occurred in several ecological contexts, each involving different constraints and trade-offs. A quarter of the parasitic eggs were attributed to floater females which had no nest or territory; parasitism for these birds was a low-paying alternative to not breeding (the annual reproductive success of territorial, nesting females was 16 times greater than that of floaters). Nest loss during laying accounted for a few cases of parasitism, but most instances involved nesting females which usually laid parasitically in neighbours' nests prior to laying their own full-sized clutches. It was suggested that laying surplus eggs parasitically allowed females to bypass the constraints of parental care and increase their total production of offspring: brood reduction through starvation was prevalent and indicated that parental care normally limits the number of offspring that pairs can raise. Only 8% of parasitic eggs (n = 571 eggs) produced independent offspring, compared to 35% of non-parasitic eggs (n = 2183 eggs from 237 pairs); most mortality was due to egg rejection by hosts, while hatching success and post-hatching survival decreased with the stage of the host's nesting cycle; no egg laid >6 days after clutch completion ever hatched. Hosts are capable of egg discrimination, and 80% of rejected eggs were buried in the nest material while 20% were ejected from the nest.

Territory sizes in California (from Gullion 1953b) were (in ha): winter, 0.035-0.22 (0.1, SD 0.075, n = 5); spring, 0.16-0.51 (0.29, SD 0.14, n = 7); summer, 0.22-0.56 (0.43, SD 0.15, n = 5); 4 spring and all 5 summer territories contained emergent vegetation (0.012-0.26ha), comprising 4-30% (mean 13%) of the territory area in spring and 6-46% (mean 21.5%) in summer; the mean area of vegetation regularly used in summer territories was 0.065ha. Other authors give figures of 1 nest/0.28ha in a 91-ha slough or 1 nest/0.12ha for the 44.1ha of emergent vegetation, and 1 nest/ha in *Typha* and *Carex* (Gullion 1953b). The defended area does not include adjacent dry land (Gullion 1953b). Nudds (1981) gives a breeding density of 13.6, 17.8 and 32.6 nests/km^2 in Saskatchewan pothole country, and Gorenzel et al. (1982) recorded breeding densities of 10.2-33.1 successful nests/ha of *Typha* and *Scirpus* in Colorado.

SOCIAL AND SEXUAL BEHAVIOUR Both sexes assist in territory defence, and territories are vigorously defended against conspecifics, and sometimes against other species of wetland birds, such as ducks (e.g. Ruddy Duck *Oxyura jamaicensis* and Mallard *Anas platyrhynchos*) and grebes; all species of small birds, and most species of small vertebrates (including garter snakes *Thamnophis elegans* and mud turtles *Clemmys marmorata*) are attacked when coots have young chicks (Bent 1926, Gullion 1953b, Alisauskas & Arnold 1994). Large waterbirds such as swans and Muscovy Duck *Cairina moschata* are avoided (Gullion 1953b). Coots compete intensively with Franklin's Gulls for nest sites; in most cases the coots succeed in displacing the gulls, and may nest within a gull colony area (Burger 1973).

Gullion (1952) describes displays in detail (see Fig. 15). Holding the head and neck depressed signifies aggression, while a bowed head is the basic form of courtship and mating displays; wing-arching is important in most displays; and the neck feathers are erected to form

A. Normal posture.

G. Bowing and nibbling.

B. Patrolling.

H. Bracing.

C. Charging (note ruff is erected).

I. Arching.

D. Paired display (note neck ruff).

J. Splattering.

E. Swanning (note neck ruff).

K. Fighting.

F. Warning.

Figure 15: American Coot displays. [After Gullion 1952].

a ruff in aggressive displays; the position of the tail, and the consequent prominence of the white undertail-coverts, are also important. Patrolling, with head lowered, ruff erected and tail slightly depressed, occurs when a conspecific approaches the territory. If an intruder enters the territory, the occupant charges with its neck extended forwards on the water, the ruff erected and the wings and tail in a normal position. A more intense charge is termed the splattering attack, when the bird runs over the water with flapping wings, the neck extended horizontally and the ruff erected; the attacked bird often flees in a like manner (splattering retreat) but with the head and neck raised, and it may also dive to escape pursuit. After charging and/or splattering at conspecifics, the territory occupants may indulge in a paired display (the hunched display) with their heads held fairly low, the wings arched high over the back (often with tips crossing), the tail held vertically and the white undertail-coverts expanded prominently. This display is often interspersed with fighting, usually follows a fight, and is often followed by displacement feeding and preening.

Fighting is described by Bent (1926) and Gullion (1952). Antagonists swim rapidly towards each other with heads extended and wingtips elevated, and when they meet they strike viciously at each other with their bills before lying back and striking with the feet (the claws are long and sharp); the birds' feet often become interlocked, when the fight is continued with savage thrusts of the bills. Birds may grasp the breast of the adversary with the claws of one foot, leaving the other foot free for striking; frequent quick jabs with the bill are made to knock the opposing bird off balance. The weaker bird may be forced onto its back and then held under water, while the victor leisuredly plucks out feathers; vanquished birds often escape by swimming for long distances under water. Females frequently participate in fights, and several birds may be involved. Fights are often vicious and may result in injury or even death. Defenders will hit other coots directly from flight, and underwater fighting certainly occurs but its nature is unknown (Gullion 1952).

Parental aggression towards chicks is common (Lyon 1993a) and involves "tousling" behaviour, chicks being shaken and pecked by their parents until they feign death. It may result in chicks ceasing to beg and thus dying of starvation. Furthermore, late-hatching chicks were often found dead in the nest 2-3 days after hatching, and many had peck marks and bruises on the top of the head, indicating that they had been killed by the parents.

Swanning (Fig. 15E) is used in defence of nests and young: the wings are arched over the back and also spread laterally, so that the primaries touch the water; the tail is not lifted, but the neck is extended, the head held low and the ruff displayed: the bird appears about twice its normal size (Gullion 1952). If swanning fails, e.g. in defence of the nest against a human intruder, the male will *churn*, foot-slapping against the surface to create a turmoil; by so doing, the bird's body is raised backwards out of the water (Bent 1926). This display is also used with other aggressive displays, when it appears to be a displacement activity in situations where rapid success has been achieved in aggressive action (Gullion 1952).

During pair formation birds touch bills when meeting; bowing (Fig. 15G) follows billing, one bird bowing its head to present its head and neck for allopreening; the other bird then nibbles through the feathers, often burying its entire bill among the breast and back feathers (Gullion 1952). Bracing (Fig. 15H) occurs when paired birds meet after a period of separation and its function is not understood: the swimming bird raises the forepart of its body high in the water with the ruff erected and the head stiffly erect. In the courtship chase (Bent 1926) the male frequently rushes after the female, both birds paddling over the water surface with flapping wings; the female may dive if too closely pressed. The male displays by swimming towards the female, with his head and neck resting on the water, wingtips raised high above the tail, and the tail elevated and spread to display the white undertail-coverts laterally; the female usually assumes the same attitude as the male approaches; when <1m away, the male turns away from the female, swims off and then repeats the performance (Bent 1926). Paired birds often swim towards each other with heads extended on the water and making *kuk* calls; as they meet they assume a more erect attitude, brush against each other and turn about, sometimes dabbling in the water with the open bill; the female often allopreens the male's head and then lowers her head for him to reciprocate (Bent 1926). The female displays to the male by swimming with her tail raised and her undertail-coverts spread; she often leads the male to a display platform, on which she performs a standing arch display, with her head lowered, her tail elevated and her undertail-coverts expanded; she often slaps the platform with one foot, and gives low *tuk* or *punt* calls (Gullion 1952). When copulation is imminent, the female squats with her tail erect and her head lowered in the squat arch (Fig. 15I) (Gullion 1952). The male mounts, the female raises her head, and the male rears back as copulation takes place; after copulation, both birds may fluff out their feathers (Gullion 1954).

BREEDING AND SURVIVAL Season N America, Apr-Jul (peak late Apr-May); winter breeding reported in Florida, eggs Jan (Woolfenden 1979); Colombia, laying peaks Jun-Jul but also lays Sep, building reported Mar, breeds nonseasonally at L Tota (Varty *et al.* 1986). Usually begins nesting within 10-14 days of arrival on breeding grounds (Alisauskas & Arnold 1994). **Nest** A large, bulky floating pile of dead or rotted aquatic vegetation (e.g. *Typha*, *Scirpus*, *Sparganium*, *Potamogeton*, reeds or grasses), preferably dry; structures built early and late in season composed almost entirely of dead material, but growing material often used at other times; cup lined with dry, smooth material; may use most readily available materials for nest; access ramp often constructed. Nest usually in, and anchored to, residual emergent vegetation, preferably *Typha* or *Scirpus* although *Carex*, *Scholochloa* or *Salix* are sometimes used; late-season nests sometimes in new-growth emergents; as nest is floating, material must be added constantly, to prevent structure settling below surface. Nest usually built within 1-5m of open water, sometimes well concealed in dense reedbed but sometimes in open and resting on mass of old matted reeds etc; typically floating in water 0.3-1.2m deep; average external diameter 35-46cm, internal diameter c. 18cm. Both sexes build; female may do most building, male bringing material to her. May build up to 9 platform structures associated with display, nesting and brooding during a season, most such building by male. **Eggs** Eggs 6-12 (nominate), 3-5 (*columbiana*); average clutch size declines seasonally, and with altitude, in most populations (Arnold 1990); clutches in excess of 12 (up to 22 recorded) probably laid by 2 or more females. Eggs oval, sometimes quite pointed, to subelliptical; smooth and slightly glossy; pinkish-buff, pale buffish-stone or tan, thickly and finely spotted blackish-brown. Size of 145 eggs (nominate) 41.5-53.2 x 30.0-36.0 (48.5 x 33.0), weight 26.1-32.8 (28.5, n = 12); 3 eggs of *columbiana* measured 56.3 x 35.4, 56.4 x 36.4 and 57.3 x 36.1 (Fjeldså 1983c). Eggs laid at daily intervals. Some females are intraspecific brood parasites, laying eggs in other nests before producing their own clutches; this coot is capable of both conspecific and interspecific egg discrimination (Arnold 1987, Lyon 1991; see Social Organisation). **Incubation** Both sexes incubate; incubation starts with 3rd-6th egg; period 21-27 days; hatching asynchronous (Alisauskas & Arnold 1994). **Chicks** Precocial, leave nest soon after hatching, when they can swim and dive well; chicks 1 day old remained under water for almost 3 min (Bent 1926); return to nest or brood platforms for frequent brooding; cared for by both parents; usually brooded by female and fed by male for first 3-4 days; later, brood may be divided between parents; female plays greater part in feeding young than does male (Ryan & Dinsmore 1979). Chicks beg by raising hindquarters, depressing neck, turning head up, and spreading and waving wings. Aggression by parents towards young is common (Lyon 1993a) and may indirectly result in death by starvation; late-hatched chicks are often found dead in nests (see Social and Sexual Behaviour). Fed almost entirely by parents for first 2 weeks; gradually become independent over next 3-10 weeks. Juvenile plumage acquired from c. 3 weeks of age; fledged at 60-70 days. Nest parasitism by Ruddy Ducks and Redheads occurs rarely (Arnold 1987). Age of first breeding 1 year, but many yearlings are nonbreeders (Alisauskas 1987); clutch size increases with age; breeding performance increases with age, especially between 1 and 2 years of age; older coots begin nesting earlier in the season than younger ones (Crawford 1980, Alisauskas & Arnold 1994). Nest success usually over 80% (68.6-84.9% in Colorado), may be lower in drought years; hatching success 80-100% (85.2-92.4% in Colorado); fledging success averages just over 50%; renesting may occur up to 4 times after clutch or brood loss; predators caused most nest (64.3%) and egg (45.9%) loss in Colorado (Gorenzel et al. 1982, Alisauskas & Arnold 1994); in a study in British Columbia, 70% of chicks which died before independence died by the age of 10 days, mortality usually being due to starvation (Lyon 1993a). Two sequential broods occasionally raised in S parts of breeding range; elsewhere, 2nd clutch may be laid while 1st clutch still being incubated (Hill 1986, Alisauskas & Arnold 1994). **Survival** Average survival of adults 49% per year, 44% for juveniles and 45% overall (Ryder 1963); adult survival also calculated at 43% (Burton 1959).

139 CARIBBEAN COOT
Fulica caribaea   Plate 42

Fulica caribaea Ridgway, 1884, St. John, Virgin Islands.

Sometimes regarded as conspecific with *F. americana*, but West Indian populations normally mate assortatively without evidence of crossing (Sibley & Monroe 1990). Forms superspecies with *F. cristata, F. atra, F. alai, F. americana, F. leucoptera* and *F. ardesiaca*. Monotypic.

Synonyms: *Fulica mexicana/atra/americana caribaea*.

IDENTIFICATION Length 33-38cm. Predominantly slate-grey, underparts slightly paler; head and neck blackish; leading edge of wing white; secondaries variably and narrowly tipped whitish; lateral undertail-coverts white, forming 2 broad bands which meet under tail tip. Frontal shield white, sometimes tinged ivory or yellow; broad, bulbous, and oval or elliptical, extending well up onto crown; bill white, sometimes with reddish-brown spot or band near tip. Iris red; legs and feet dull olive to yellowish. Sexes alike, but female averages smaller. Immature grey; it and juvenile presumably similar to those of American Coot. Inhabits freshwater lakes, ponds and marshes; sometimes brackish lagoons.
Similar species American Coot is very similar in size and overall appearance, but normally has reddish-brown callus at top of white shield, and different profile (shield not swollen). However, a small percentage of American Coots have reddish callus much reduced or lacking, and may have shield almost entirely white or washed yellow, sometimes with variable and irregular reddish patches; shield of such atypical birds may be bulbous (Roberson & Baptista 1988). Extent of white on outer web of outermost primary and tips of secondaries very variable in both species and apparently not a distinguishing character (Roberson & Baptista 1988, *contra* Blake 1977).

VOICE Similar to that of the American Coot: a variety of cackling, croaking and clucking notes (Raffaele 1989). Hardy *et al.* (1996) give grating, clucking calls, and a low-pitched disturbance call *krr-rrrh* which rises in pitch at the end.

DESCRIPTION
Adult Head and neck blackish-slate; remainder of plumage slate-grey, darker on tertials and tail and slightly paler on underparts; remiges browner; variably extensive narrow white edge to outer web of outermost primary and outermost alula feather; variable narrow whitish tips to some or all secondaries; marginal coverts white; lateral undertail-coverts white, forming two broad bands which meet under tip of tail. Frontal shield entirely white (sometimes tinged ivory or yellow), broad, bulbous and oval or elliptical; bill entirely white or with reddish-brown spot near tip of both mandibles, sometimes forming narrow band. Iris red; legs and feet dull olive to yellowish. Sexes alike, but female averages smaller.
Immature Described only as 'grey'; presumably similar to that of American Coot.
Juvenile Not described; presumably similar to that of American Coot.
Downy young Not described.

MEASUREMENTS Wing (chord) of 10 males 171-202 (188.7, SD 9.1), of 9 females 164-197 (176.9, SD 10.1); tail of 11 males 42-56 (49.1), of 8 females 42.5-51.5 (47.2); culmen (including shield) of 11 males 46.5-58 (52.4, SD 3.4), of 9 females 43-51.5 (47.2, SD 3.0); tarsus of 11 males 56-65.5 (62.5, SD 4.3), of 9 females 54-65.5 (59.4, SD 4.0) (Ridgway & Friedmann 1941, Roberson & Baptista 1988).

GEOGRAPHICAL VARIATION None. Putative race *major* from Puerto Rico is not recognised (Ripley 1977).

MOULT No information available.

DISTRIBUTION AND STATUS S Bahamas (N Caicos), Greater Antilles (except I of Pines but including Puerto Rico and Virgin Is), Lesser Antilles (S to Grenada and Barbados), Trinidad, Curaçao, and NW Venezuela (Zulia and Aragua). Records of birds from North American localities all refer to American Coot (Roberson & Baptista 1988). It is not globally threatened, but it was formerly regarded as locally uncommon to rare over much of its range, but apparently locally common on Hispaniola (Wetmore & Swales 1931). In Puerto Rico it was formerly abundant but has diminished greatly as a result of overhunting and habitat destruction, so that it is now decidedly uncommon (Biaggi 1983, Raffaele 1989). In other areas its current status is unclear, and research is required.

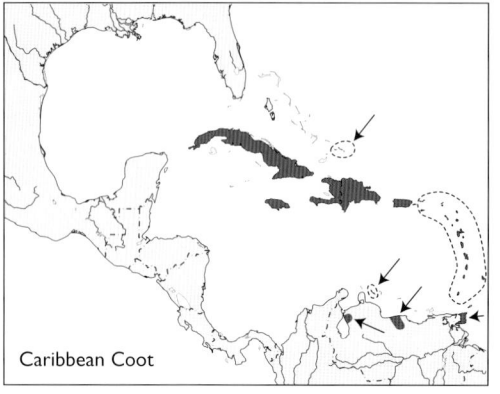

Caribbean Coot

MOVEMENTS None recorded.

HABITAT Freshwater lakes, ponds and marshes; rivers, streams or swampy channels with emergent vegetation; swamps with floating vegetation; also less frequently in coastal brackish lagoons. It inhabits lowlands, occurring up to 500m in Venezuela.

FOOD AND FEEDING Aquatic plants and small invertebrates; foraging methods are presumably similar to those of the American Coot.

HABITS Little is recorded. Birds are not shy, and are usually seen in groups, sometimes walking on mud in swampy channels, but when hunted they take shelter in emergent vegetation.

SOCIAL ORGANISATION Presumably as for American Coot.

SOCIAL AND SEXUAL BEHAVIOUR Presumably similar to that of American Coot.

BREEDING AND SURVIVAL Poorly known; details presumably similar to American Coot. **Season** Puerto Rico, throughout year, with peaks in spring and autumn (Raffaele 1989); Virgin Is, May-Jun and Sep-Oct. **Nest** Three nests were among rushes growing in a few inches of water; well-cupped, with detached canopy; built up to c. 30cm above water; other nests described as floating. **Eggs** 4-8; ovate; smooth with slight gloss; white, stone-coloured or pale olive-buff, with small spots and speckles of brownish-black to blackish-slate and a few underlying ones of pale lilac-grey, markings fairly evenly distributed; size 47-51.4 x 33.9-35 (49.8 x 34.6, n = 12, 2 clutches). Although West Indian populations normally mate assortatively without evidence of crossing (Sibley & Monroe 1990), cross-breeding with American Coot has been observed on St John, Virgin Is (Raffaele 1989) and at St Maarten, Netherlands Antilles (Roberson & Baptista 1988).

140 WHITE-WINGED COOT
Fulica leucoptera   Plate 42

Fulica leucoptera Vieillot, 1817, Paraguay.

Forms a superspecies with *F. cristata, F. atra, F. alai, F. americana, F. caribaea* and *F. ardesiaca*. Monotypic.

Synonyms: *Fulica gallinuloides/leucopyga/stricklandi/ chloropoides*.

IDENTIFICATION Length 35-43cm. Medium-large coot; dark slate-grey, paler on underparts and with black head and neck; lateral undertail-coverts white, very conspicuous as bird swims away with raised tail; white line formed by broad white tips to secondaries conspicuous in flight; leading edge of wing white. Bill yellow to greenish-yellow; frontal shield small, pale yellow to orange-yellow, rounded at top giving round-headed appearance. Iris red; legs and feet olive-green to yellow-green, with grey to blackish lobes and joints. Some birds have predominantly bluish-pink or pinkish-grey bill, others have pink to pale reddish shield; colour variations in Brazilian birds could be age-related (see Description for other colour variations). Sexes alike. Immature has grey-brown upperparts and paler underparts; chin to breast may be whitish; iris brown, bill olivaceous; shield pinkish; legs and feet greenish-grey. Juvenile dark greyish, paler on underparts; head and neck white, mottled dusky from forehead to hindneck; shield pinkish. Inhabits lagoons, river backwaters, ponds and marshes.
Similar species Differs from congeners in having broad white tips to secondaries, yellow to greenish-yellow bill and small yellow frontal shield (but see Description for colour variations), and round-headed appearance. Andean Coot (141) averages larger, with white only at sides of undertail-coverts (White-winged has lateral white patches meeting under tip of tail). Red-gartered Coot (142) is larger, with pointed frontal shield giving angular profile; has no white on secondaries, and yellow legs and feet with red 'garter'. Red-fronted Coot (143) has almost straight-line profile from bill tip to knob-like top of elongated frontal shield, more white on undertail-coverts (inverted heart-shape), and often no white on outermost primary. Darker than American Coot (138) on body.

VOICE Vocal, with a variety of loud, hollow cackling calls, clucking and grating notes, including typical coot calls such as *kow, kowk, kut, kak, oeu* etc; some calls resemble peals of laughter (see Hardy *et al.* 1996). The voice is similar to that of the Red-gartered Coot but higher-pitched and more tremulous (Canevari *et al.* 1991). When concealed in emergent vegetation, and also sometimes when in the open, White-winged Coots call frequently to each other. Breeding birds give *wurt* or *wyt* calls and aggressive males give an explosive *huc* (Fjeldså & Krabbe 1990).

DESCRIPTION
Adult Head and neck black; rest of plumage dark slate-grey, slightly blacker on tertials and tail, and paler grey on underparts; secondaries broadly tipped white; outer web of outermost alula feather and outermost primary edged white; marginal coverts white; axillaries and underwing-coverts slate-grey, narrowly tipped whitish; underside of remiges glossy grey; lateral undertail-coverts white (sometimes with dark subterminal spots), forming

two broad bands under tail; median undertail-coverts black to dark neutral-grey. Iris red, orange or magenta; bill chrome-yellow to greenish-yellow, occasionally with dark spot near tip, or tip greenish; frontal shield small, pale yellow to orange-yellow, rounded at top giving round-headed appearance; legs and feet olive-green, or pale sea-green to yellow-green, with grey to blackish lobes and joints. Some birds have predominantly bluish-pink or pinkish grey bill (sometimes with orange-yellow spots or yellowish-green subterminal band), others have bill dull green or green-yellow shading to pale vinaceous-fawn on proximal half, and sometimes to whitish at extreme base; some have pink to light reddish shield; colour variations of shield in Brazilian birds not sex-related but could be age-related (Belton 1984). One bird with all-white bill and shield recorded (Weller 1967). Sexes alike.

Immature Upperparts grey-brown; underparts paler; chin to breast whitish to grey; iris brown; bill dusky olivaceous, becoming yellower; shield small, rather pinkish; legs and feet greenish-grey.

Juvenile Dark drab grey, paler on underparts, with white head and neck mottled dusky from forecrown to hindneck; bill may be blackish; small shield rather pinkish.

Downy young Black downy chick has partly naked pink and blue crown, orange tips to down of head and dense yellow bristles around neck, blood-red bill with black tip, and greenish-brown legs and feet.

MEASUREMENTS Wing of 14 males 178-227 (191.9), of 11 females 173-191 (186.5, SD 4.9); tail of 5 males 49-57 (53.7, SD 3.9), of 9 females 46-56 (50.6, SD 3.5); culmen (including shield) of 4 males 42-52 (47.5, SD 4.2), of 9 females 38-48 (43.8, SD 3.7) ; tarsus of 14 males 50-67 (56.4), of 9 females 48-60 (53.6, SD 3.6). Weight of 6 females 400-500 and 607 (Ripley 1977, Belton 1984, Marín 1996); of 5 unsexed 469-616 (554) (Navas 1991).

GEOGRAPHICAL VARIATION None.

MOULT No information available.

DISTRIBUTION AND STATUS Chile (S from Arica), E and SE Bolivia (Santa Cruz, Chuqisaca, Tarija, also occasional on L Alalay, Cochabamba (Fjeldså & Krabbe 1990, Armonía 1997), Uruguay (throughout, except in Río Negro, Paysandú and Salto), Paraguay (mostly W of R Paraguay) and extreme SE Brazil (Rio Grande do Sul), and Argentina S to Tierra del Fuego. Not globally threatened. It was formerly regarded as widespread, and common to locally abundant (e.g. Humphrey et al. 1970); it has been recorded in flocks of thousands on large marshy lagoons of the Argentine pampas (Ripley 1977). At present it is regarded as generally common to abundant, with a total population in excess of 1,000,000 birds (Rose & Scott 1994), but it is relatively uncommon to rare in Paraguay (Hayes 1995).

MOVEMENTS Although the species is present throughout the year in Rio Grande do Sul it makes considerable local movements, most or all birds being absent from some breeding areas in autumn or winter (Belton 1984). It is regarded as a migrant in N Argentina (Nores et al. 1983) and it migrates to E Bolivia (Fjeldså & Krabbe 1990, Armonía 1995). At Cabo San Antonio (Buenos Aires), where it apparently occurred all year, it was present in small numbers in Sep and increased in Oct (Weller 1967). On Isla Grande (Tierra del Fuego) it is apparently a summer breeding visitor, some individuals remaining throughout the year (Humphrey et al. 1970).

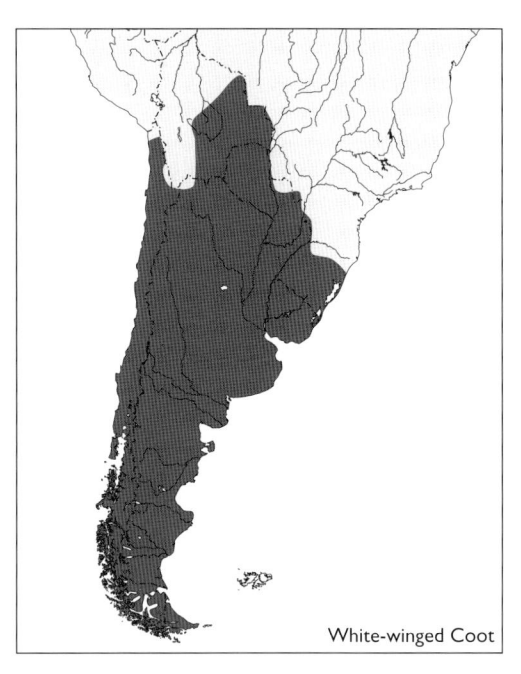

White-winged Coot

HABITAT Freshwater lagoons, river backwaters, ponds and marshes; sometimes saline waters. In some regions it prefers grassy or vegetation-free shores, waters with many submerged plants, or *Lemna* duckweed-covered waters (Weller 1967, Fjeldså & Krabbe 1990), but it often occurs on waters with much fringing and/or emergent vegetation. Upland breeding sites are usually barren, with no fringing reeds but with much floating *Myriophyllum*; it may occupy temporary waters when nesting (Fjeldså & Krabbe 1990). It occasionally occurs on the sea, close to the shore. It occurs over a wide altitudinal range (up to 4,900m), and is apparently more of a generalist than its sympatric congeners, being remarkably tolerant of widely differing environmental conditions; in Chile it is equally at home in the sheltered subtropical valleys of Arica, the bleak and windswept country in the extreme S, the saline marshes and lagoons bordering the Atlantic and Pacific coasts, and lakes at the foot of the Andean peaks where the temperature range between 05:00h and 11:00h may be 35°C (Johnson 1965). It sometimes occurs alongside the larger Red-gartered and Red-fronted Coots; in a Chilean study, Cody (1970) found that all species eat the same water plants, but Red-fronted is commonest at pool edges and in weed-beds, and Red-gartered in open water, while White-winged utilises all zones. In a study of White-winged and Red-gartered Coots at an artificial pond in Buenos Aires, Heimsath et al. (1993) found that both species occurred and foraged more in open water than in emergent vegetation or along muddy, vegetated shorelines, which they used more for resting. However, White-winged Coot also made some use of *Typha* reedbeds when foraging and resting, while Red-gartered never occurred in this vegetation type. The authors note that White-winged can make more use of the vegetated zones than can Red-gartered and conclude that it is effectively forced to do so by its larger, more specialised congener's aggressive behaviour and superior numbers – a conclusion which may be debatable on the existing evidence, especially as Cody (1970) states that these coots normally interact without regard to species identity (see Social and Sexual Behaviour); however, see Red-gartered Coot (Social and Sexual Behaviour section) for other examples of interspecific aggression.

FOOD AND FEEDING Mostly aquatic weeds. A study of birds from the Paraná R, Santa Fe, Argentina (Mosso & Beltzer 1993) showed that the grass *Paspalum repens* formed the basic diet, with *Polygonum acuminatum* and a small amount of *Myriophyllum* sp. also being eaten, while insects formed a very small proportion of the diet; the species also takes molluscs and various seeds. It is chiefly a surface feeder, but it dives occasionally; it pecks at the surface, and upends frequently to drag up submerged waterweeds with the bill. Birds also graze on land, sometimes over 100m from marsh edges, and sometimes pull seeds from bent-over grass heads (Weller 1967). In Chile, Cody (1970) found that surface feeding, up-ending and diving were equally used foraging methods, and that dives averaged 1.5s (see Red-gartered and Red-fronted Coot species accounts for comparisons).

HABITS When chased, these coots rise noisily, splashing the water surface with their feet for 20-30m before becoming airborne (Ripley 1977). They are more inclined to taxi along the water surface and to fly short distances than are sympatric congeners. When alarmed, they swim into emergent cover and call frequently (Wetmore 1926).

SOCIAL ORGANISATION Monogamous; strongly territorial when breeding. It is gregarious outside the breeding season, normally occurring in large flocks, often with other coot species.

SOCIAL AND SEXUAL BEHAVIOUR In the non-breeding season this species occurs in mixed flocks with Red-gartered and Red-fronted Coots. In such flocks all species will tolerate the presence of other feeding individuals down to c. 1m; closer approach results in a brief but spectacular scuffle and a tolerable spacing is regained; because such interactions occur within and between species in every combination, the coots behave effectively as a single species (Cody 1970).

The White-winged Coot is very aggressive when breeding, and it excludes conspecific individuals and other bird species from the territory (Canevari *et al.* 1991). Birds threaten with the head low over the water, and often with expanded body plumage, the tail cocked and spread, and the wings partly tilted rear-edge up (but not spread); the birds repeatedly face each other and turn away, displaying the undertail pattern (Fjeldså 1983c). Territorial disputes are common, usually consisting of noisy chases which often end in fights, when the combatants rise up in the water, or float on their backs with spread wings, and strike with the feet (Canevari *et al.* 1991). Fighting males also grasp one another by the feet and strike savage blows with their bills (Wetmore 1926).

BREEDING AND SURVIVAL Season Lowlands of C Argentina, mainly Apr-Nov, also Dec; uplands, Nov-Jan; season very prolonged in Rio Grande do Sul, Brazil, where the species nests throughout spring and summer; Chile, Oct-Jan; Uruguay, Oct, nest-building starts mid-Sep. **Nest** A floating platform of rushes, green grass or other aquatic plants, often among emergent vegetation (water depths of 35-50cm recorded) but (especially at high altitudes) also floating offshore among patches of aquatic plants to which it is anchored below the surface; also in *Solanum malacoxylon* bushes; may have access ramp; slight canopy sometimes built over nest. External diameter c. 40cm, internal diameter c. 17cm, height 10-24cm. May nest alongside Red-gartered Coot. **Eggs** 4-12 (3 also recorded); oval, fairly smooth and with almost no gloss; cream, creamy-buff or olive-cream, finely spotted blackish or chocolate-brown, and purplish, over whole surface; size 43-54 x 29-34 (48.8 x 33.2, n = 28); calculated weight 29.3; Belton (1984) found eggs weighing 33. **Chicks** Young fed by both parents. Occasional parasitism by Black-headed Duck *Heteronetta atricapilla* recorded (Arballo & Cravino in press).

141 ANDEAN COOT
Fulica ardesiaca   Plate 42

Fulica ardesiaca Tschudi, 1843, Lake Junín, Peru.

Chestnut- and white-fronted birds formerly treated as separate species, *F. americana peruviana* and *F. ardesiaca*, and later lumped as *F. americana ardesiaca*. Forms superspecies with *F. cristata*, *F. atra*, *F. alai*, *F. americana*, *F. caribaea* and *F. leucoptera*. Two subspecies recognised.

Synonyms: *Fulica chilensis*; *Lysca ardesiac(e)a*; *Fulica americana peruviana/ardesiaca*.

Alternative name: Slate-coloured/Peruvian Coot.

IDENTIFICATION Length 40-44cm. Medium-large, stocky, slate-grey coot, with black head and neck; marginal coverts and leading edge of outermost primary white; secondaries usually have small white tips; lateral undertail-coverts white (nominate race), giving 2 white lines under tail, or almost black (race *atrura*); large, rounded frontal shield gives round head shape. Iris red to reddish-yellow. Two colour morphs: **red-fronted**, with more solid, deep chestnut frontal shield, chrome-yellow bill becoming pale yellow to green (sometimes white) near tip, and green legs and feet; and **white-fronted**, with white bill, white to orange-yellow frontal shield, and slaty legs and feet; for other variants, see Description. Sexes alike. Immature not described. Juvenile dark drab grey, washed dull brown, with paler underparts and mainly white face and chin to foreneck; iris dark brown; bill dark greyish-horn with paler tip; legs and feet greyish-horn with darker joints. Inhabits lakes, ponds, rivers and marshes, with or without emergent vegetation and often at high altitudes.
Similar species Stocky, but smaller and much less heavy-bodied than Giant Coot (144), with rounder head and different bare part colours. Smaller than Red-gartered Coot (142), which has yellow frontal shield, yellow legs and feet with red 'garter', and no white on secondaries. Similar in size to Red-fronted Coot (143), with (in red-fronted morph) similarly deep chestnut frontal shield, yellow bill and olive to greenish legs and feet. However, Red-fronted has chestnut extending to ridge of bill, and different profile, with almost straight line from bill tip to knob-like top of elongated frontal shield; also has more extensive white on undertail-coverts, and no white on secondaries. White-winged Coot (140) averages slightly smaller, with entirely yellow bill and shield (but colour variable: see White-winged species account), and more extensive white on secondaries and undertail-coverts. Larger than American Coot (138) with sometimes paler

slate-coloured body plumage contrasting more with blackish head and neck, and different bare part colours (race *columbiana* 'Colombian Coot' of American Coot has smaller, chestnut frontal shield with pointed posterior outline, chestnut-red subterminal spots on both mandibles, and yellow only at base of upper mandible; legs and feet olive-green, with yellow and orange areas).

VOICE Details are from Fjeldså (1982). The usual call is a low *churrrr* or (in males?) a harder *chrrp*, *hrr* or *hp* often repeated (see Hardy *et al.* 1996); such calls are given in agitated, frightened and aggressive situations, including fights and paired displays. The male also uses a short *hrp* at meeting, or while chasing a bracing female (see Social and Sexual Behaviour) when the female gives a low chittering *phyji phyji ...* call. A male once gave an explosive *phx* call.

DESCRIPTION *F. a. ardesiaca*
Adult Head and neck velvety black; rest of plumage slate-grey, with slight olive wash on upperparts; tertials and rectrices slightly darker; marginal coverts white; primaries and secondaries blackish-brown; outer web of outermost primary prominently edged white; secondaries usually with small white tips; underwing-coverts variably edged white (white sometimes absent); lateral undertail-coverts white. Large, rounded frontal shield giving round head shape. Iris red to reddish-yellow. Two colour morphs: red-fronted, with larger, rounded, more solid, deep chestnut frontal shield, chrome-yellow bill becoming pale yellow to green (sometimes white) near tip, and green legs and feet; and white-fronted, with white bill, white to orange-yellow frontal shield, and slaty legs and feet. Other variants also exist: bill may be flesh coloured, or (rarely) may have chestnut subterminal spot on upper mandible; shield variably pale to primrose-yellow; legs and feet sometimes grey or lavender. Sexes alike.
Immature May be slightly paler and greyer overall than adult.
Juvenile Dark grey, washed dull brown; underparts paler, especially on breast; sides of head below eye, and chin, throat, sides of neck and foreneck, mainly white; iris dark brown, tinged reddish; bill dark greyish-horn with paler tip and distal cutting edge; legs and feet greyish-horn, more greenish on tibiae, and with darker joints.
Downy young Black downy chick has red and blue skin on crown, and short orange bristles on throat; bill red at base, lemon-yellow distally, colours separated by double black lines of variable development.

MEASUREMENTS (9 males, 7 females) Wing of male 223-241 (233.3, SD 6.6), of female 206-237 (220.4, SD 9.6); tail of male 52-61 (56.6, SD 2.8), of female 47-58 (53.15, SD 4.1); culmen (including shield) of male 54-68 (60.8, SD 4.3), of female 55-64 (59.2, SD 3.3); tarsus of male 66-76 (70.7, SD 3.3), of female 63-71 (67.4, SD 2.8). Weight of 5 adults 900-1,100 (988, SD 77.9) (Marín 1996).

GEOGRAPHICAL VARIATION Races separated on amount of white in undertail-coverts. For a discussion of the taxonomy of this species and American Coot, see Fjeldså (1982, 1983a).

F. a. atrura Fjeldså, 1983 – S Colombia (Nariño, W Putumayo), N and C Ecuador (Carchi S to Chimborazo and Azuay, also Santa Elena Peninsula in W Guayas), and coastal Peru S to Lima. Undertail-coverts have black inner webs and extensive black streaking and freckling on outer webs. Measurements as for nominate. Weight of 1 female 670 (FMNH; D. E. Willard *in litt.*).

F. a. ardesiaca Tschudi, 1843 – Peru (S from Ancash, at Mollendo in Arequipa, Pisco (once) in Ica, occasionally to sea level), to C and W Bolivia (La Paz, Cochabamba, Oruro, Potosí and Tarija), NW Chile (Tarapacá, Antofagasta) and NW Argentina (Jujuy, Salta and N Catamarca). See Description.

MOULT No information available.

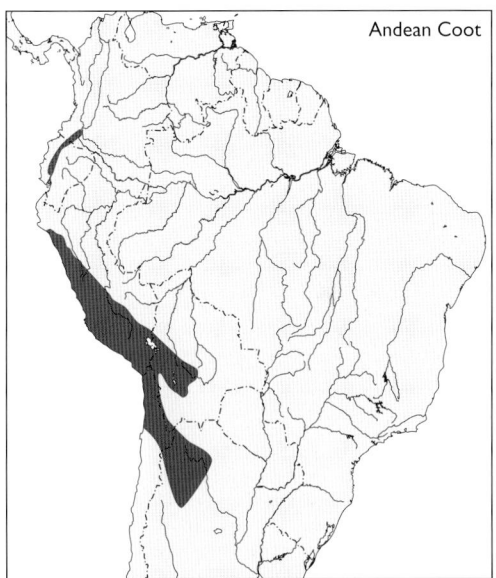

Andean Coot

DISTRIBUTION AND STATUS S Colombia, Ecuador, Peru, Bolivia, N Chile and NW Argentina. Not globally threatened. Both colour morphs occur from S Colombia to S Peru, and white-fronted dominates at L Junín, but only red-fronted morph occurs further S, from Cuzco and L Titicaca in Peru and to Argentina and Chile (Gill 1964, Fjeldså & Krabbe 1990). Both types are known to nest together only on Yaguarcocha and Calta (= Cotta) in Ecuador, and at Laguna de Paca and L Junín in Peru; in most of the overlap zone, one or other type may occur alone locally (Fjeldså 1982). See Fjeldså (1982) for further details of the distribution of the 2 forms.

This coot was formerly regarded as locally frequent to common (e.g. Parker *et al.* 1982, Hilty & Brown 1986). Few data are available on its current status but it is locally common to abundant, sometimes assembling in thousands (Fjeldså & Krabbe 1990), while Ridgely & Greenfield (in prep.) regard it as uncommon to locally fairly common in Ecuador, although a record of a flock from the Guayaquil area (Fjeldså & Krabbe 1990) was an isolated occurrence.

MOVEMENTS Some seasonal population movements occur, but these are not properly described.

HABITAT Ponds, lakes, rivers and marshes. Information is taken from Fjeldså (1982) and Fjeldså & Krabbe (1990). It occurs mostly on fairly large, reed-fringed lakes with extensive shallows having dense submerged vegetation (*Myriophyllum*, *Potamogeton*, *Chara* and *Elodea*), but it also breeds on barren lakes lacking reeds. Except furthest S, there is a tendency for red-fronted birds to dominate at well vegetated lakes and white-fronted birds to dominate at barren high-altitude lakes where *Chara* is the principal submerged plant. At L Junín, both morphs were abundant in mosaics of rushes and shallow open areas with marl bottoms and varied submerged growth, but the best habitat for red-fronted was in *Scirpus* zones with mixed floating vegetation (see Food & Feeding), while white-fronted dominated in shallows and offshore areas where *Chara* covered the lake bed. In the Titicaca area, birds nested throughout the enormous *Scirpus* zones of the main lake, and on smaller lakes with *Scirpus* beds, except where growth was very scantly due to intense human harvesting. Non-breeders assembled far offshore, especially where dense weeds reached the surface. The nominate race occurs from 1,120 to 4,700m; *atrura* occurs in coastal Peru, through Ecuador and in paramos and some lowland swamps in S Colombia, being found up to 3,600m in Colombia. See Food & Feeding for further habitat details.

FOOD AND FEEDING Most information is taken from Fjeldså (1982). Eats mainly aquatic vegetation; in white-fronted birds especially *Chara*, and in red-fronted birds *Myriophyllum* and *Elodea*. Birds feed in shallow water areas with dense submerged or floating vegetation, generally gregariously and sometimes in mixed species flocks. They dive for food at depths of 2-5m, at L Titicaca mostly at 4-5m in places with scattered *Myriophyllum elatinoides* and *Potamogeton strictus* on the surface and *Chara* at the bottom, and in an area with dense *Elodea potamogeton* at 2-4m depth. On L Lagunillas, which has no fringing reedbeds, some birds grazed along the shoreline in *Stylites andecolus*, *Elodea*, *Zanichellia palustris*, *Lilaeopsis* spp, *Hypsella reniformis* etc), but most dived where uniform carpets of *Chara* appeared, outside the zone of floating *Myriophyllum*, *Potamogeton* and *Zanichellia* which was the main habitat of Giant Coots (from which species they were thus ecologically isolated, as was also seen at L Shegue). They also feed by walking on floating vegetation and on beaches. On L Junín birds fed when walking on dense, floating tracts of *Hydrocotyle bonariensis*, *Ranunculus limoselloides* and *Myriophyllum* near the shore, within dense tall zones of *Scirpus*, while at other lakes they fed by walking on floating *Chara* and on clay beaches with *Chara*.

HABITS These birds are usually seen swimming in the open water of reed-fringed waterbodies but they are often quite shy, pattering into cover if approached in a boat. They loaf and swim in deep water in monospecific flocks (Fjeldså & Krabbe 1990). See Food & Feeding.

SOCIAL ORGANISATION Generally gregarious, sometimes assembling in thousands when feeding; monogamous, territorial and pugnacious when breeding. Mixed pairs of red-fronted and white-fronted birds often occur but there may be a tendency to assortative mating (Fjeldså 1982). Juveniles have been seen feeding downy chicks.

SOCIAL AND SEXUAL BEHAVIOUR Information is taken from Fjeldså (1982). As in other coots, lifting the tail but not spreading it seems to be a sign of alarm. Birds threaten with the head and neck held low, the back rounded and the tail horizontal; in intense threat, which is often the precursor of an attack, the bird swims forwards strongly and is flat-backed with the tail lowered, the neck extended forwards and the crown feathers sleeked so that the shield is prominent. On land, the bird runs with extended neck bent low. The species has a typical splattering charge, running across the water with flapping wings and extended low neck and lowered tail. A similar posture is used in retreat, but the tail is raised and the neck looks thin. When close to the opponent the attacking bird raises its body and makes a long kick with one foot, simultaneously throwing its body around to the opposite side; it often turns away while in the air, and some fights look like a series of such jumps with a kick and a somersault. In sustained fights the birds face each other upright in the water, flap the wings and kick and splash water for several seconds. An attacking bird may brake some way from its opponent by 'churning': paddling both feet rapidly and backing water with a boiling sound, the body raised at an angle, the neck bent down and the wings closed. In defensive situations, pairs threaten with the swollen neck proudly raised and arched, and the wings often raised over the back and tilted rear-edge up.

During pair formation males often circle females in defensive threat postures, which easily provoke attacks from other males. At some point the female bows her neck or turns her head away, which stimulates nibbling of her nape feathers by the male. Paired birds both nibble each other's neck, this action sometimes lasting for several minutes and being preceded by meeting with bill-touching. When nest-building, one bird may ascend the platform and stand rigid, with lowered head, slapping one foot on the substrate several times. If the mate approaches, the displaying bird may adopt the 'standing arch' position, with body raised and head bent low, before walking off in this posture and then swimming away with erect, swollen neck, raised nape-ruff, body sloping backwards and undertail-coverts hidden (the 'bracing' posture). A bracing posture is also adopted by the female prior to copulation, when the male follows her closely towards the nest, sometimes even charging, giving aggressive calls, and pecking at her neck. The female ascends the nest, adopts a 'squat-arch' invitation posture, and copulation follows.

BREEDING AND SURVIVAL Season Colombia, Feb; N Chile, Nov-Jan; Peru, Jan, Apr and Jun-Nov; Bolivia, young Jun, adults feeding small chicks Aug; laying peaks in dry season (Jul-Aug) but may nest in any season. **Nest** Builds large platform nest, placed among reeds or open to view in floating waterweeds. **Eggs** 4-7; ovate; cream to pale buffy-brown, spotted reddish-brown and blackish or dark brown; size of 10 eggs 55.3-64.2 x 39.3-41.6 (59.2 x 40.1), calculated weight 51.9. Breeding behaviour poorly known; apparently close to that of American Coot. Second clutches likely to occur, as juveniles seen feeding downy chicks.

142 RED-GARTERED COOT
Fulica armillata   Plate 43

Fulica armillata Vieillot, 1817, Paraguay.

Monotypic.

Synonyms: *Fulica galeata/chilensis/frontata/chloropoides/ leucopygia*.

IDENTIFICATION Length 43-51cm. Large, dark slate-grey coot, shading to black on head and neck; leading edge of wing white; no white on secondaries; white lateral undertail-coverts form 2 broad white stripes under tail (sometimes coverts may appear entirely white). Has pointed frontal shield, giving somewhat angular profile, and distinct red mark along base of culmen, making yellow bill and shield appear unconnected; yellow may look almost white from a distance. Bill and shield colours variable: shield may be paler yellow than bill; bill sometimes reddish. Iris reddish, sometimes yellow; legs and feet orange-yellow to yellow, with red 'garter' above ankle joint; when swimming away, heel and occasionally red garter are visible alternately on each leg. Sexes alike. Immature plumage not described; has horn-coloured bill but soon acquires yellow shield; legs and feet olive to greenish-yellow. Juvenile dull drab grey-brown, paler on underparts (whitest on centre of breast and belly); head and neck white, heavily mottled grey-brown from crown to hindneck and to below eye, lightly mottled on malar region and sides of neck; bare parts olivaceous (iris may be yellower). Inhabits lakes, large ponds, rivers and marshes, mainly in lowland areas.

Similar species Smaller White-winged Coot (140) has white tips to secondaries, yellow bill with no red markings, small yellow frontal shield which is rounded at top giving round-headed appearance, and pale greenish legs and feet with dark toes and joints; some birds have predominantly bluish-pink or pinkish-grey bill, or pink to pale reddish shield (see White-winged species account). Smaller Red-fronted Coot (143) often has no white on outermost primary, has prominent white undertail-coverts of inverted heart-shape (but present species sometimes shows similar pattern), and characteristic profile, with almost straight line from bill tip to knob-like top of elongated frontal shield; iris reddish-brown; bill yellow, ridge and frontal shield dark chestnut-red, base reddish; legs and feet olive. Smaller Andean Coot (141) has rounded head shape, with chestnut frontal shield, yellow bill and green legs and feet (red-fronted morph) or white to yellow frontal shield, white bill and slaty legs and feet (white-fronted morph) as well as other variants, and usually has some white on secondaries.

VOICE Information is from Ripley (1977) and Fjeldså & Krabbe (1990). The male's alarm call is a short, whistled *huit* or *juit*; he also has an explosive *puit* or *pit* and a repeated *wuw* of aggression, and a repeated soft *cuit* repeated 6-7 times when near the female. She has a *yec* of alarm (equivalent to the male's *juit*) and a loud *terr* which is repeated 7-9 times. See also Hardy *et al.* (1996) for *kow* notes and rapid *krr-krr-krrik* calls. The voice is similar to that of the White-winged Coot but lower-pitched and less tremulous (Canevari *et al.* 1991).

DESCRIPTION
Adult Predominantly dark slate-grey with black head and neck; remiges slightly browner; outer web of outermost primary edged white; marginal coverts white; greater underwing-coverts, and underside of remiges, more silvery-grey; lateral, and usually terminal, undertail-coverts white, sometimes almost all undertail-coverts white. Bill and frontal shield yellow, with distinct red mark along base of culmen below shield. Frontal shield pointed, giving somewhat angular profile. Colours of bill and shield may vary, possibly with age and/or sex: shield may be paler yellow than bill; bill sometimes greenish-yellow or reddish; in Brazil, some birds may have red spot at base of lower mandible, and migrants usually have only a scarlet line separating bill and shield (Sick 1993). Iris usually reddish but can be yellow; legs and feet greenish orange-yellow to yellow, with pale to deep red 'garter' above ankle joint. Feet very large, much larger than those of congeners with which this species normally associates, and with proportionately broader lobes (Wetmore 1926). Sexes alike.

Immature Bill horn-coloured, but soon acquires yellow shield; 1 bird had brownish-black bill with yellow mottling; legs and feet olive; in second or third year legs and feet duller, more olive-green or olivaceous, than in adult. Presumed young bird in captivity had iris crimson with narrow dark brown inner ring; bill and shield whitish-horn, with yellow-orange patches towards base, and red-brown towards tip of upper mandible; legs and feet dull yellow-green (Ripley & Beehler 1985).
Juvenile Dull, drab grey-brown with white head and neck, heavily mottled grey-brown from crown to hindneck and on sides of head to below eye, and lightly mottled on malar region and sides of neck; has distinct whitish demarcation towards frontal shield (Fjeldså & Krabbe 1990); upperparts slightly cinnamon-tinged in 1 specimen, most markedly from back to rump; underparts paler and whiter than upperparts, whitest on centre of breast and belly. Older birds become progressively greyer on head, neck and underparts. Iris dull olivaceous to dull yellowish; bill and shield dark olivaceous grey-brown; legs and feet dark olive-greyish. For further information, see Navas (1970).
Downy young Black downy chick has partly naked pink and blue crown, small orange bristles on chin, throat, foreneck and sides of neck; iris grey-brown; bill black with orange band and small red shield; legs and feet dark grey-brown.

MEASUREMENTS (8 males, 9 females) Wing of male 189-218 (207.5, SD 9.5), of female 185-218 (202.5, SD 9.7); tail of male 44-53 (49.3, SD 3.2), of female 42-53 (47.1, SD 3.1); culmen (including shield) of male 47-53 (50.2, SD 2.1), of female 43-54 (48.8, SD 3.7); tarsus of male 58-67 (63.8, SD 2.8), of female 58-63 (59.8, SD 2.0). Weight of 7 unsexed 744-1,079 (918) (Navas 1991).

GEOGRAPHICAL VARIATION None.

MOULT Postjuvenile moult is complete at c. 5 months (Navas 1970) – see this publication for further information.

DISTRIBUTION AND STATUS C and S Chile (N to Coquimbo), and SE Brazil (Rio de Janeiro, São Paulo and Santa Catarina to Rio Grande do Sul) and Uruguay (except Artigas, Flores, Durazno, Rivera and Tacuarembó) S through Argentina (except altiplano) from Jujuy to Tierra del Fuego; possibly also Paraguay and Falkland Is (1 fully-grown female from Falkland Is, May 1923, in BMNH). Not globally threatened. It was formerly regarded as generally abundant (Ripley 1977), although in Rio Grande do Sul, Brazil, Belton (1984) found it uncommon along most of the littoral but commoner S of 32°S It is still regarded as common to locally abundant (e.g. Nores *et al.* 1983, Fjeldså & Krabbe 1990, Narosky & Di Giácomo 1993, do Rosário 1996, Arballo & Cravino in press), but few data are available for most areas; however, its total population is estimated to be in excess of 1,000,000 birds (Rose & Scott 1994). A concentration of c. 10,400 birds was counted in a coastal inlet at Puerto Natales, S Chile, in Aug 1992, in company with small numbers of White-winged Coots (Taylor 1996b). There are no recent records from Paraguay. In Uruguay, birds occurring on coastal waters suffer mortality from spills of fuel and other toxic substances, and some also die in storms of wind and heavy seas which occur on the Rio de la Plata coast (Arballo & Cravino in press).

MOVEMENTS Apparently sedentary when conditions permit. Weller (1967) notes that in Cabo San Antonio,

? Possible occurrence

Red-gartered Coot

vegetated shorelines, which they used for resting, the Red-gartered using shorelines for resting much more frequently than emergent vegetation. Red-gartered never occurred in *Typha* reedbeds, which the White-winged Coot made some use of when foraging and resting. It occurs mainly in lowland areas, but ascends to 1,200m on the barren plateaux of inland Patagonia and up to 1,000m on lakes of the S Andes, and it breeds at 2,100m in L Volcán, Jujuy, NW Argentina. In the winter it can form large agglomerations on sheltered marine bays.

FOOD AND FEEDING This species eats primarily aquatic plants, especially *Myriophyllum elatinoides* and also algae (*Oedogonium*, *Spirogyra*, *Zygnema* and Characeae), with some seeds (of reeds and other plants) (Navas 1991). A skilled diver, it often obtains most of its food in open, fairly deep water (Ripley 1977) by upending and diving: in Chile, Cody (1970) found that up-ending was the most frequently used foraging method (used by 50% of observed individuals) while diving was used by 37% and surface feeding by only 13% (see White-winged and Red-fronted Coot species accounts for comparisons). It occasionally raids waterweeds pulled about by Rosy-billed Pochards *Netta peposaca* and Black-headed Ducks *Heteronetta atricapilla* (Ripley 1977).

HABITS Gregarious, sometimes swimming or loafing in large rafts; they feed in scattered company but when alarmed gather in a close flock (Wetmore 1926). When duck species with which they are associated take flight at the approach of danger, the coots splatter over the water with flapping wings to shelter in the emergent marginal vegetation, from which their characteristic calls soon come (Johnson 1965). The species may sometimes be found walking along the shores of waterways (Arballo & Cravino in press).

SOCIAL ORGANISATION Normally gregarious, assembling in very large flocks when not breeding, but monogamous, strongly territorial and pugnacious when breeding.

SOCIAL AND SEXUAL BEHAVIOUR In the non-breeding season this species occurs in mixed flocks with the smaller White-winged and Red-fronted Coots. In such flocks all species will tolerate the presence of other feeding individuals down to c. 1m; closer approach results in a brief but spectacular scuffle and a tolerable spacing is regained; because such interactions occur within and between species in every combination, the coots behave effectively as a single species (Cody 1970). It is described as aggressive in all seasons, chasing all other waterbirds except the Coscoroba Swan *Coscoroba coscoroba* and the Black-necked Swan *Cygnus melanocoryphus* (Weller 1967), and in a study of White-winged and Red-gartered Coots at an artificial pond in Buenos Aires, Heimsath *et al.* (1993) proposed that the larger Red-gartered, as a result of its aggressive behaviour and superior numbers, forced the smaller species to forage more in vegetated parts of the pond (see Habitat section, White-winged Coot). Weller (1967) suggested that the aggressiveness of Red-gartered Coots towards Red-fronted Coots in one area may have been a factor in the delayed nesting of the latter species compared to Red-fronted Coots in other areas of the marsh.

The threat and attack behaviour resembles that of other coots. There is a typical threat posture, with the head and neck held low, the back rounded and the tail horizontal, an intense threat in which the bird swims forwards

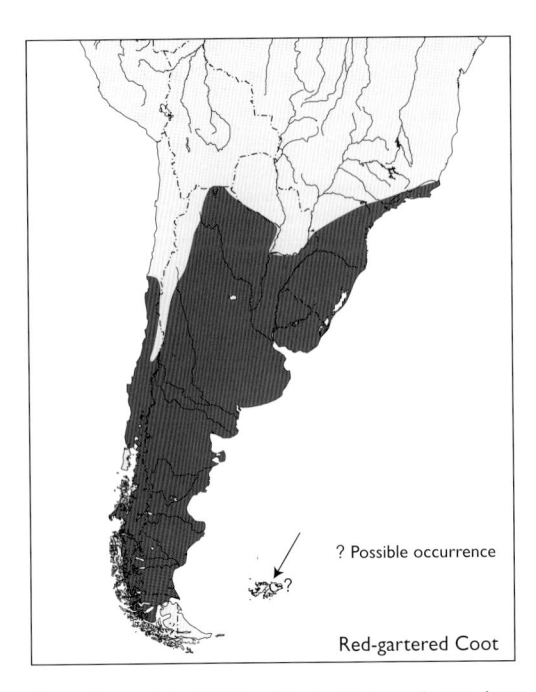

Argentina, relatively little of its habitat was lost to late summer drought so that birds remained well into the autumn and reduced populations remained all year; in some parts of Uruguay small numbers may be seen throughout the year (Arballo & Cravino in press). In the Rio Grande do Sul it is present all year but is given to considerable movement and local concentration (Belton 1984): near Santa Isabel it was common in Nov 1972 and May 1972 but only 2 were seen in Feb 1974, while at a reservoir 50km away several hundred were present in Jan 1975 – see also White-winged Coot. It is a migrant in NE Argentina (where it winters at least in Chaco) and in Paraguay and S Brazil (Nores *et al.* 1983, Contreras *et al.* 1990, Fjeldså & Krabbe 1990, do Rosário 1996), and in Brazil it is known from São Paulo in Jun and Rio de Janeiro in Feb-Aug (Sick 1993); it may be only a breeding visitor to Tierra del Fuego (Ripley 1977). The extent of these movements is not clear and seasonal occurrences are not shown on the distribution map. A "quite young" coot from Tristan da Cunha before 1909 is probably this species (Watson 1975).

HABITAT Lakes, large ponds, rivers, marshes and deep, clear roadside ditches; it is rarely seen on small pools but occasionally feeds in protected waters associated with lakes. It usually occurs on open waters with abundant water-weeds, and often with fringing emergent vegetation; on the Patagonian uplands it breeds mainly on exposed shallow lakes with extensive carpets of floating *Myriophyllum* (Fjeldså & Krabbe 1990). It sometimes occurs alongside the smaller White-winged and Red-fronted Coots; in a Chilean study, Cody (1970) found that all species eat the same water plants, but Red-fronted is commonest at pool edges and in weed-beds, and Red-gartered in open water, while White-winged utilises all zones. In a study of Red-gartered and White-winged Coots at an artificial pond in Buenos Aires, Heimsath *et al.* (1993) found that both species occurred and foraged more in open water than in emergent vegetation or along muddy, in the same posture, and a splattering charge (Navas 1960). Birds fight by confronting the opponent, rising up in the water, flapping the wings and striking with the bill and feet, sometimes grappling with the feet while pecking (Navas 1960). Birds apparently also fight using the kick-and-somersault method, as described for the Andean Coot (141) (Fjeldså 1982, 1983c). The paired display given by territorial birds in defensive situations closely resembles that of the Andean Coot: pairs threaten with the swollen neck proudly raised and arched, and the wings raised over the back and tilted rear-edge up, the wings sometimes being even more apart and the bill rising (Fjeldså 1982). For further information, see Navas (1960).

BREEDING AND SURVIVAL Season Chile, Oct-Nov, with replacement clutches until Jan when eggs harvested by local people; Brazil, Nov and three-quarter-grown young, Feb; Uruguay, nest-building from mid-Jun, dependent immatures mid-Jan (Arballo & Cravino in press); Cabo San Antonio, Argentina, Sep-Dec, 2 peaks, mid-Sep and mid-Oct, latter possibly due to renesting. **Nest** A loosely built platform of dried rushes (e.g. *Scirpus californicus*) with built-up rim forming damp, cup-shaped deposit; usually in tules of moderate density near open water; sometimes floating (water depth of 40cm recorded); often with access ramp; nest area usually free of floating vegetation. External diameter 30-42cm, height 11-19cm, depth of cup 5-10cm; ramp 140cm long recorded. May nest alongside White-winged Coot. **Eggs** 2-8 (mean 5.3, n = 22 clutches); ovate; pale brown, sepia or reddish-brown, freely spotted darker brown, sepia and reddish-brown; size 54-65 x 36-46 (58 x 39.4, n = 45), calculated weight 49.6; Belton (1984) found eggs of length 67 and mass 48. After chicks hatch, nest built up to serve as brood site, or brood nests or platforms often constructed. **Incubation** Period 24-25 days, by both sexes. **Chicks** Young fed by both parents; able to dive quite early in life. Lays replacement clutches after egg loss. In Cabo San Antonio, Argentina, nest success low because of flooding losses, and only 1 nest destroyed by an aerial predator (Weller 1967). Nest parasitism by Black-headed Duck *Heteronetta atricapilla* frequent (Johnson 1965), but coots usually do not raise ducklings (Weller 1967); in Uruguay, nest of Common Moorhen (129) with 6 eggs contained 1 egg of Red-gartered Coot (Arballo & Cravino in press).

143 RED-FRONTED COOT
Fulica rufifrons   Plate 43

Tulica [sic] *rufifrons* Philippi & Landbeck, 1861, Chile.

Monotypic.

Synonyms: *Fulica leucopygia/leucopyga/chloropoides*.

IDENTIFICATION Length 36-43cm. Dark slate-grey, head and neck blacker; prominent white undertail-coverts of inverted heart shape; often swims with tail raised; back very flat. Outer web of outermost primary often lacks white, but sometimes has prominent white edge; marginal coverts white, forming narrow white leading edge to inner wing. Characteristic profile, with almost straight line from bill tip to knob-like top of elongated frontal shield. Bill yellow, with reddish base and green tip; ridge and frontal shield dark chestnut-red. Iris reddish-brown; legs and feet olive.

Sexes alike. Immature browner than adult, with duller bill; some white mottling on malar region, chin and throat; more white on belly than adult; becomes very similar to adult during first winter. Juvenile grey-brown above, darker from forehead to hindneck; sides of head, and chin and throat, mottled grey-brown and white; underparts pale grey, washed brown on breast and flanks; white undertail-coverts mottled grey-brown; bill olivaceous or blackish; has diagnostic pointed shield. Inhabits marshes and reedy lakes, lagoons and ponds, normally at low altitudes.
Similar species Similar in size to White-winged Coot (140) but lacks white on secondaries and has more extensive white patch on undertail-coverts (White-winged has white more confined to lateral coverts). White-winged has yellowish bill, small yellow to orange-yellow shield which is rounded at top giving round-headed appearance, and pale greenish legs and feet with dark toes and joints; some birds have predominantly bluish-pink or pinkish-grey bill, or pink to light reddish shield (see White-winged species account for variations in bill colour). Slightly larger Red-gartered Coot (142) has white edge to outermost primary, white lateral undertail-coverts forming 2 broad white stripes under tail, pointed frontal shield, giving somewhat angular profile, distinct red mark along base of culmen, between yellow of bill and shield, and yellow to olive legs and feet, with red 'garter' above ankle joint. See also Andean Coot (141). Juvenile rather uniform grey-brown as in somewhat smaller juvenile of Common Moorhen (129) but has diagnostic pointed shield and large white area on undertail-coverts, and lacks white line along flanks.

VOICE A long, chattering series of calls, described as *togo togo togo...*, *cu cu cu...*, *puHUH puHUH puHUH...* etc; the alarm note of the male is *tuc* or *toc*, of the female *toec* (Ripley 1977, Fjeldså & Krabbe 1990); on Hardy *et al.* (1996) the calls are very rapid, incorporate a variety of notes, including *ker ker*, *co co*, *wakawaka* etc, are given in chorus and have a laughing and sometimes a gobbling quality. The *toc* call is given when birds are patrolling; the warning call of the male is a *go...go...goo*, and that of the female a repeated *taeg*; and pair members call in an "intimidating duet" (Escalante 1991). It is generally less vocal than other coots (Arballo & Cravino in press), and is silent during the non-breeding season and during nest-building and incubation (Escalante 1991). Also see Navas (1991).

DESCRIPTION
Adult Dark slate-grey, blacker on head and neck; often some white on centre of belly (usually more than in White-winged and Red-gartered Coots), and sometimes over whole of belly to centre of lower breast; prominent white undertail-coverts of inverted heart shape. Primaries fuscous, outer web of outermost primary often with no white or very faint white border, but sometimes with prominent white edge; secondaries uniform fuscous; marginal coverts white. Iris reddish-brown to brown; bill yellow, tip green, sides at base reddish, basal half of culmen, and narrow frontal shield, dark chestnut-red, rest of culmen orange; legs and feet olive to greenish-olive (sometimes yellowish, or with yellow spot at garter), with greyer joints and lobes. Characteristic profile, with almost straight line from bill tip to knob-like top of frontal shield; beyond this, shield often continues as thin projection to centre of crown. May have less enlarged lobes on toes than sympatric congeners. Sexes alike. In non-breeding season, red at base of bill reduced or absent; shield dull red and slightly wrinkled.

551

Immature One fully grown bird intermediate between juvenile and adult: head, neck and upperparts browner than in adult; some white mottling on malar region, chin and throat; underparts more brown-tinged than in adult, and centre of lower breast to centre of belly more extensively whitish. Frontal shield small; bill colours apparently like adult but duller. At 120 days of age ('first-winter plumage'), similar to adult in size and plumage but shield only half adult size and vinous-red (Escalante 1991).

Juvenile Rather uniform grey-brown on upperparts, darker from forehead to hindneck; sides of head dark grey-brown, mottled white; chin to foreneck white, mottled dark grey-brown; breast pale grey, upper region washed darker grey-brown; flanks pale grey, often tinged brown; rest of underparts pale greyish, mottled white (most white on belly and vent); undertail-coverts as adult but white feathers tipped dark grey-brown. Diagnostic pointed shield; bill said to be blackish, but 1 had bill olivaceous with paler tip and lower mandible; legs and feet pale olivaceous. At 80 days, bill yellow-orange, washed dark; no red at base; shield has trace of dull red at base (Escalante 1991).

Downy young Down of upperparts black, with bottle-green sheen; down of underparts blackish-brown; crown partly naked, with pink and blue skin; chin and throat with some broad, flat, orange bristles; down of foreneck (1 specimen) with white tips; iris dark brown-grey; bill red with 2-3 black bands; shield red; legs and feet dark greyish; wing has 2 red stripes until c. 35 days of age (Escalante 1991).

MEASUREMENTS Wing of 5 males 171-184 (179.8, SD 5.2), of 10 females 162-180 (169.4, SD 6.3); tail of 3 males 55, 56, 66, of 10 females 50-68 (56.3, SD 5.5); culmen (including shield) of 3 males 48, 49, 52, of 10 females 40-57 (49.9, SD 5.0); tarsus of 5 males 50-60 (57, SD 4.2), of 10 females 52-57 (54.6, SD 1.8). Weight of 1 male 685, of 1 female 550; of 20 unsexed 533-872 (644) (Navas 1991).

GEOGRAPHICAL VARIATION None.

MOULT No information available.

DISTRIBUTION AND STATUS Coastal S Peru (Mejía in Arequipa); extreme SE Brazil (São Paulo, Santa Catarina and Rio Grande do Sul); coastal C Chile (Atacama S to Osorno) E to W Paraguay, Uruguay (Montevideo, Río Negro, Colonia, Canelones, Maldonado, Rocha, San José, Florida, Soriano, Treinta y Tres, Cerro Largo and Artigas), and NE Argentina (Chaco, Córdoba, Santa Fe and Entre Ríos) W to foot of Andes in San Juan and Mendoza, and S to Chubut, apparently more casual to Tierra del Fuego (1 specimen, Dec) (Humphrey et al. 1970, Fjeldså & Krabbe 1990); also Falkland Is (formerly?). Erroneous records from N Argentina (Tucumán and Jujuy) are based on confusion with Red-gartered Coot (Blake 1977), but the species is still recorded as occurring in these provinces (Nores et al. 1983, Fjeldså & Krabbe 1990). Not globally threatened. The species was formerly regarded as not uncommon over most of its range, and abundant in pampas marshes (e.g. Wetmore 1926, Weller 1967, Ripley 1977, Nores & Yzurieta 1980), but scarce in Rio Grande do Sul (Belton 1984). In Peru, its population was thought to be increasing in the 1970s (Tallman et al. 1978). It is still regarded as common to abundant in some areas (e.g. Narosky & Di Giácomo 1993, Arballo & Cravino in press) but is rare in Paraguay, where it breeds sporadically in the chaco (Short 1975, Hayes 1995). Its population is apparently stable, and is estimated to be in excess of 1,000,000 birds (Rose & Scott 1994). It has not been recorded from the Falkland Is since 1924, where perhaps it may never have been more than a vagrant.

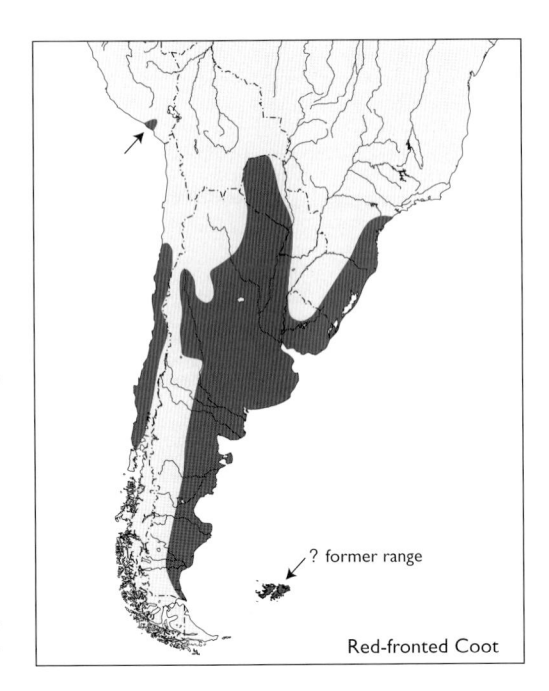

? former range

Red-fronted Coot

MOVEMENTS Some regular seasonal variations in numbers are recorded, and records of casual birds and occasional influxes also indicate some wandering or dispersal. In Brazil it is apparently a migrant (or vagrant?) in São Paulo (Blake 1977), and in Santa Catarina it has movements in winter (do Rosário 1996); at Cabo San Antonio, Argentina, most birds left during the winter when marshes dried up, the remaining birds being mostly immatures (Weller 1967). In Chaco, Argentina, there are very few records in the hot season and migrants arrive in the winter (Contreras et al. 1990)

HABITAT Densely vegetated to semi-open marshes and reedy lakes, lagoons and ponds, especially with much floating duckweed and water-ferns and often with very shallow water; it prefers places where emergent reed cover is relatively dense. It sometimes occurs alongside the White-winged Coot and the larger Red-fronted Coot; in a Chilean study, Cody (1970) found that all species eat the same water plants, but Red-fronted is commonest at pool edges and in weed-beds, hardly ever occurring away from edge or weed-bed habitat, and Red-gartered is commonest in open water, while White-winged utilises all zones. In coastal Peru it inhabits freshwater coastal marshes with *Scirpus* sedges (Tallman et al. 1978). In Uruguay it occurs at waters with *Scirpus californicus*, *Typha* and *Potamogeton pectinatus* (Arballo & Cravino in press). It inhabits lowlands up to 800m, but is casual up to 2,100m in NW Argentina.

FOOD AND FEEDING Primarily a surface feeder, pecking at *Azolla* and duckweed mats, but it also dives skilfully and grazes in upland areas adjacent to water when marshes dry up seasonally (Weller 1967). In Chile, Cody (1970) found that 78% of individuals fed from the water surface, 11% by up-ending and 11% by diving (see White-winged

552

and Red-fronted Coot species accounts for comparisons). An important food in Uruguay is *Potamogeton pectinatus* (Arballo & Criado in press).

HABITS Birds are usually found among vegetation, rarely venturing far from cover, and are regarded as being shy and less visible than other coots (Wetmore 1926, Friedmann 1927, Arballo & Cravino in press). This species generally occurs in loose flocks, sometimes with Red-gartered Coots. It swims with somewhat less nodding than the Common Moorhen, and with the tail usually raised. It flies less readily than other coot species, swimming into cover when danger threatens.

SOCIAL ORGANISATION Gregarious when not breeding, assembling in flocks of various sizes. Monogamous, territorial and aggressive when breeding. In 1 pair, territory size in *Scirpus* at start of breeding c. 30m², increasing to c. 100m²; male later moved family into *Typha*, where young fed in an area of c. 800m² (Escalante 1991).

SOCIAL AND SEXUAL BEHAVIOUR Aggressive when breeding, both to conspecific individuals and to birds of other species, but apparently less aggressive than the larger Red-gartered Coot. Weller (1967) suggested that the aggressiveness of Red-gartered Coots in one area may have been a factor in the delayed nesting of Red-fronted Coots in that area compared to elsewhere in the marsh. This species has agonistic behaviour similar to that of congeners such as the American Coot (141); for further information, see Navas (1956) and Escalante (1991).

BREEDING AND SURVIVAL Season In Chile lays Sep-Oct; Argentina, May-Nov, peaking Oct-Nov in Chaco and Sep-Oct at Cabo San Antonio; Uruguay, breeds from late Sep, chicks Dec, juvenile Apr; Peru, Sep-Jan. **Nest** A floating platform of reeds and aquatic vegetation; generally well hidden; anchored in emergent vegetation (e.g. *Scirpus californicus*) or built on floating vegetation; usually low in water; depth of water may be <50cm; often bends or weaves reeds over nest (Arballo & Cravino in press); may have access ramp; nest rather small for genus, external diameter 25cm, height 25cm (15cm above water), depth of cup 6cm. Species very adaptable in nest-site selection, and will use dense vegetation adjacent to small pools or sparse vegetation which has few openings (Weller 1967). **Eggs** 2-9 (mean 5.6, n = 67 clutches); late clutches smaller than early ones; oval; smooth and slightly glossy; pale creamy-buff, tinged green, spotted and blotched reddish to chocolate-brown, with underlying markings few, small, and pale purple; markings tend to be concentrated at larger end; size 51-60 x 34.5-38.7 (55.5 x 37.2, n = 40); calculated weight 42. **Incubation** Period calculated as c. 25 days (Escalante 1991). **Chicks** Duration of parental care c. 150 days; at 60 days almost fully fledged but coverts growing; reaches c. 80% adult size at 80 days (Escalante 1991) After chicks hatch, nest often relined with more delicate material and serves as brood platform for at least 4 days. Nest success high (83%) at Cabo San Antonio, E Argentina, and second broods common, but deserts readily, mostly as a result of human intrusion, even when clutch well incubated (Weller 1967). This species apparently is main host of parasitic Black-headed Duck *Heteronetta atricapilla*; few desertions associated with parasitism (Weller 1967).

144 GIANT COOT
Fulica gigantea   Plate 43

Fulcia [sic] *gigantea* Eydoux & Souleyet, 1841, Peru.

Monotypic.

Synonyms: *Phalaria gigas*; *Fulica maxima*.

IDENTIFICATION Length 48-59cm; young birds smaller. Adult very large and heavy-bodied, with relatively small head, concave forehead and high knobs formed by enlarged orbital rims. Adult deep slate-grey, black on head, neck and undertail-coverts (but some birds completely black); variable white streaking on lateral undertail-coverts, not usually very prominent. Bill dark red with white at tip and from posterior ridge to centre of shield; sides of upper mandible, and sides of shield, yellow; bill colour variable (see Description); crown has pair of feathered knobs at top of shield above eyes. Iris red-brown; legs and feet dark red. Sexes alike. Adult normally too heavy to fly but smaller immature can fly readily. Immature paler than adult; head and neck darkest, and underparts paler grey than upperparts; bill duller; legs and feet pale greyish-red. Juvenile dark dull grey, paler on underparts and blacker from top of head to below eyes; foreneck and sides of head below eyes whitish, becoming greyer with age; bill black; legs and feet dark grey. Inhabits highland ponds and lakes in Andean *puna* zone.

Similar species Adults distinguishable by very heavy-bodied and small-headed appearance, combined with head shape (concave forehead and high knobs). Also distinguished from all other coots by dark red legs and feet; shows less white on undertail-coverts than sympatric congeners. Young birds not much larger than Andean Coot (141) but very heavy bodied, with relatively small head which has characteristic profile as in adult.

VOICE Information is from Fjeldså (1981) unless otherwise stated. The male gives gobbling or laughing *houehouhouhouhoou* calls alternating with low, growling *hrr* or *horr* sounds (see Hardy *et al.* 1996); it sometimes utters only growls, mainly from the nest. The female has a low, crackling *chu-jrrrh*, low squeaking or cracking sounds sometimes becoming a soft chittering, and a soft *hi-hirr hirrr hirrr...* given before copulation and during courtship feeding. A contact or welcoming call is a rasping *grrrrrr* (Hardy *et al.* 1996). Begging chicks call *chic* repeatedly, and give soft, repeated *phi* at other times.

DESCRIPTION
Adult Plumage dense, thick and rather fur-like. Head and neck velvety black; general colour of upperparts dark slate-grey; remiges and rectrices blackish-brown to black; outermost primary with white edge (often absent); axillaries and underwing-coverts dark slate-grey; underside of remiges dark grey. Underparts slaty-grey; undertail-coverts black, lateral coverts with very variable amount of white on outer webs. Some individuals entirely black, with body only slightly paler than head. Iris reddish-brown; bill dark red with white tip; upper mandible with white on ridge from nostril back to shield, and laterally yellow towards base (above dark red cutting edge); shield white in centre, yellow at sides; legs and feet dark red. Bill colour variable, possibly with age or season; some birds have reddish spot on culmen at base of frontal shield (Ripley 1977) or reddish area bordering whitish centre of shield

553

(Plate 43); some have upper mandible black distally and white basally (McFarlane 1975). Crown has pair of feathered knobs at apex of frontal shield above eyes, formed by enlarged orbital rims. Sexes alike.

Immature Head and neck dark blackish-grey; body dark grey, darkest on upperparts and wings, paler on underparts; bill colours duller than in adult; legs and feet pale greyish-red.

Juvenile Forehead, crown and ocular region more blackish-grey, becoming medium grey on nape and hindneck; foreneck, and sides of head below eyes, initially white or very pale grey, becoming darker grey possibly through feather wear; rest of body dark grey, somewhat lighter on underparts; bill black, legs and feet dark grey.

Downy young Sooty-black with steel-like lustre; duller and browner on underparts with few orange-yellow bristles (and some white bristles) on throat; some white sheaths on tips of down on crown and nape; small, orange-scarlet club-like sheaths around base of bill; pink and blue skin of crown hardly visible through dense down; iris brown; frontal shield pink, mauve at sides; bill with magenta base, yellow tip and pinkish-buff centre separated by black bands; egg-tooth white; legs and feet greenish-brown.

MEASUREMENTS Wing of 10 males 241-287 (271.6), of 11 females 252-300 (274.6); tail of 4 males 60-71 (66, SD 4.5), of 6 females 61-66 (63.7, SD 2.0); culmen (including shield) of 4 males 62-70 (65.8, SD 3.9), of 6 females 60-69 (64.7, SD 3.4); tarsus of 10 males 82-103 (93.6), of 11 females 82-103 (92.5). Weight of 1 male 2,700, of 2 females 2,020, 2,440 (Navas 1991, Marín 1996). Weight of chick c. 40 at hatching, 100 at 17 days (Fjeldså 1981).

GEOGRAPHICAL VARIATION None.

MOULT Wing moult has been recorded in Feb (Fjeldså 1981).

DISTRIBUTION AND STATUS C Peru (N to L Junín, where occasional, and L Paron in Ancash) through W and SW Bolivia (La Paz, Cochabamba, Tarija, Oruro, Potosí) to N Chile (Tarapacá); it also extends to extreme NW Argentina (altiplano of Jujuy, and Anconquiza Mts,

Tucumán) (Chebez 1994). This species is not globally threatened. It was formerly regarded as scarce to rare and local through most of its range (e.g. Ripley 1977), but large populations exist on some lakes, such as Yaurihuiri lakes in Ayacucho, Conococha in Ancash, and Lauca National Park, Arica, N Chile, where c. 12,000 birds occur (Fjeldså & Krabbe 1990). Populations seem to have exploded in Chile and Peru, thanks to control on the use of firearms (Fjeldså & Krabbe 1990). However, it is still considered Vulnerable in Chile and it is scarce and localised in extreme N Argentina, where it is less numerous than the Horned Coot (145) (Chebez 1994). Its total population is estimated at 10,000-100,000 birds (Rose & Scott 1994).

MOVEMENTS Immatures probably disperse widely by flight at night but adults are normally permanently attached to the territory; when small breeding ponds freeze, adult birds may walk to larger ice-free lakes. In Peru, it is an occasional visitor to L Junín and L Titicaca and a few birds occur on waters near breeding sites, most such birds probably being immature (Fjeldså 1981). Although normally confined to high altitudes, it straggles to lower elevations, occasionally to the Pacific coast, where 1 was found at sea level on Mejía Lagoons, Peru, from Dec 1979 to Jan 1980 (Hughes 1980; Fjeldså & Krabbe 1990).

HABITAT Ponds and lakes in the barren highlands of the Andean *puna* zone, where it is most numerous on lakes with extensive weedy shallows and nests in metre-deep water with dense weeds such as *Myriophyllum*, *Potamogeton* and *Ruppia* growing to the surface (Fjeldså & Krabbe 1990). Immatures visit lower lakes with emergent vegetation but always stay outside cover (Fjeldså & Krabbe 1990). Its requirement of wide shallows with abundant weeds for nesting restricts the number of good breeding sites (Fjeldså 1981). It occurs chiefly at 3,600-5,000m, also up to 6,540m.

FOOD AND FEEDING A vegetarian species: it takes mostly aquatic vegetation, especially *Myriophyllum*, *Potamogeton*, *Zanichellia* and *Ruppia*; it also eats some filamentous algae, and grazes grass on the shore. Unlike the sympatric Andean Coot it does not eat *Chara*, and on L Lagunillas, Peru, it was ecologically isolated from its congener, which grazed along the shoreline in *Stylites*, *Elodea*, *Zanichellia*, *Lilaeopsis*, *Hypsella* etc, and dived where uniform carpets of *Chara* appeared, outside the zone of floating *Myriophyllum*, *Potamogeton* and *Zanichellia* which was the main habitat of the Giant Coots (Fjeldså 1982). Birds feed from the surface of the water with a sideways throwing movement; they occasionally up-end like a dabbling duck, and will also dive. Young are given weed fragments and invertebrates such as amphipods *Hyalella*; captive young ate plants (*Elodea*, *Oxalis*), small fish *Orestias*, tadpoles of *Telmatobius marmoratus* and a small frog *Pleurodema marmorata*, and also pieces of meat, spaghetti and breadcrumbs.

HABITS This species is normally quite confident, but becomes shy when persecuted. Alarmed birds float low in the water, with flattened backs and tail raised but not spread (Fjeldså 1981). Although adults are normally too heavy for flight, a captive pair in France managed to become airborne in a high wind and disappeared over the enclosure fence (Ripley 1977), while Taczanowski (1886) claimed that the species could fly. Birds preen while standing on the shore or on the nest platform.

SOCIAL ORGANISATION Monogamous and permanently territorial, probably being attached to the territory and nest for life. The defended territory extends from the nest to the adjacent shoreline and includes a small area of open water towards the centre of the lake (McFarlane 1975). In Peru, at L Shegue, nests were situated 40-80m apart along suitable shorelines; in wide weed zones at L Lagunillas territories were semicircular units, platforms being 40-80m apart (Fjeldså 1981). Immatures from previous broods often feed young.

SOCIAL AND SEXUAL BEHAVIOUR Information is from McFarlane (1975) and Fjeldså (1981). These coots usually occur in pairs or family groups, often with 2 generations of young together. They are gregarious when together in the centre of a lake, but are very aggressive at the nest, defending the territory vigorously. They attack any duck, gull, cormorant or coot which approaches the nest too closely, and Andean Coots are attacked particularly violently. Birds are often seen making splattering attacks, running across the water with heavy wingbeats, neck extended forward, head held low and tail down; both sexes will attack intruders. Few attacks end in fighting: the splattering bird usually closes its wings, adopts a hunch-backed posture with head down and neck sloping forward, and brakes and treads water vigorously; it may then swim with upright thin neck and the threatened bird moves away with thin, erect neck and raised tail.

In the threat posture the bird protrudes the breast, highly arches the neck, keeps the tail horizontal, does not tilt the wings, and swims towards the intruder. This display can lead to fighting, after antagonists have circled one another and have made short sorties with beating wings. Fights can be quite vicious, antagonists facing one another, kicking with the feet, sometimes with an aerial leap, whirling around and landing partly facing away from the opponent; at times they fight with toes interlocked and bodies so upright that the birds topple over. Fleeing birds make a splattering retreat, with thin neck and slightly raised head and tail; the pursuer may remain low in the water with outstretched head and flapping wings. Nesting birds may make splattering attacks on a human intruder. On the nest, either sex may stand and call with inflated, extended and raised neck, presumably denoting territorial possession.

The greeting display at the nest involves two birds arching their swollen neck towards each other, making slight undulating or side to side movements and giving low chittering calls; when meeting, the birds may bill briefly, and may indulge in allopreening of the neck; the male may courtship feed by offering weed to the female. A gentle courtship chase may occur, the male holding his neck high and curved, and sometimes pecking the female; the female making herself look small, with tail down. In full display the female chitters and adopts a 'squat-arch' position with bent legs and bill pointing towards her feet, which is followed by copulation, the male flapping his wings to maintain balance. After copulation the female may perform foot and wing stretching, and the male may bathe and preen.

BREEDING AND SURVIVAL Season Breeds at any season but laying peaks in local winter, Jun-Jul, often with second clutch Nov-Feb. **Nest** Nests in water c. 1m deep with dense waterweeds growing to surface, sometimes on miry clay; small new nests usually float but large old ones rest on bottom; usually built farther offshore than aquatic plant beds which supply construction material, probably as protection from terrestrial predators (McFarlane 1975). Nest enormous: of aquatic vegetation, forming platform; large platforms become elongated, parallel to shore, rather than circular, being c. 1m wide and up to 3m long at waterline (and much larger at base) and projecting 20-50cm above water. Nest has high rim, and cup c. 75cm in diameter and up to 25cm deep; structure maintained and enlarged over several years, original one having compacted into peat in centre; pairs hold same platform for several years, probably often for life; new cup built for each clutch. Both adults build; to obtain nest material, bird grasps growing weeds in bill and swims backwards, churning water with considerable force, to break stems or pull up plants from substrate (McFarlane 1975). Fresh weed constantly added to rim while young on raft; high rim hides young, provides shelter in strong winds, and acts as food source (Fjeldså 1981). Heat produced by decomposition of plant material in nest raft should be considerable and may be important in reproduction (McFarlane 1975). Territory may also contain 1-4 smaller low platforms for loafing (Fjeldså 1981). **Eggs** 3-7 (mean 4.4, n = 10 clutches); blunt oval or elliptical; shell coarse with no gloss; pale cream, pale mouse-grey or pale olive-grey, with few to many fine spots and large blotches of dark vinaceous-brown or reddish-brown and a few underlying areas of pale purple; may have overlying areas of pale brown, probably nest stain; size 63-71 x 43.7-46.5 (67 x 45.1, n = 6) (Schönwetter 1961-62), Johnson (1965b) gives size as 62.4-72.5 x 43.7-46.1 (66.4 ± 0.82 x 44.6 ± 0.19); calculated weight 74.3. **Incubation** Both parents incubate; second clutch may be incubated by 1 adult, the other remaining with young of previous brood. **Chicks** Most information from H Macedo in Fjeldså (1981). Newly hatched, still partly poikilo-thermic, captive chicks hardly ate. Young fed until 2 months old; in captive juvenile, plumage complete after 67 days, fully fledged at 4 months, postjuvenile moult occurred some time later. In fine weather young a few days old are fed in water but because of frequent strong, cold winds, chicks spend much time in nest, feeding on fresh nest material (and possibly also invertebrates) from high rim. Chicks beg by raising bill with jerks of neck and head waggles; larger young extend neck forward, keep head low but raise and waggle bill. Adults often feed chicks in nest; juveniles beg by walking with neck bent low and bill forward. Immatures from previous broods often feed young. Many eggs taken by people at some sites, but survival of young quite high: pre-fledging mortality estimated at 36% (Fjeldså 1981). Broods of 8-9 young may be indicative of 2 female parents, or may be creches.

145 HORNED COOT
Fulica cornuta   Plate 43

Fulica cornuta Bonaparte, 1853, Potosí, Bolivia.

Monotypic.

Synonym: *Lycornis cornuta*.

IDENTIFICATION Length 46-53cm. Adult very large and heavy-bodied, with small head; plumage slate-grey, blacker on head and neck; undertail-coverts black with lateral white stripes. Bill greenish-yellow with dull orange base, black ridge, and long, extensible and erectile black proboscis which normally rests on ridge and has 2 black tufts at base. Iris orange-brown; legs and feet olive, joints grey. Sexes alike. Immature like juvenile; iris brown. Juvenile greyer, with extensively white chin and throat; bill black, tinged greenish; proboscis reduced or absent. Inhabits high-altitude Andean lakes with dense submerged aquatic plants.

Similar species Shape and size like Giant Coot (144), but differs strikingly in colours of bare parts and in structure of forecrown.

VOICE Apparently similar to that of the Giant Coot, although only a low grunting series of 3-5 syllables, and loud sharp grunts, are described (Fjeldså & Krabbe 1990).

DESCRIPTION
Adult Slate-grey to ashy-black, darkest and dullest on head and neck, palest on underparts; undertail-coverts mixed black and white, forming 2 white stripes. Iris orange-brown; bill greenish-yellow with dull orange base, black ridge, and long (30-42mm), extensible and erectile black proboscis, which normally rests on ridge of bill and has 2 black wattles at its base; all 3 protuberances terminate in tuft of thick papillae; also small, fleshy white patch below proboscis, at base of upper mandible. Legs and feet olive to greenish-brown, with dark grey joints. Sexes alike; feet of female may be considerably smaller.
Immature Apparently similar to juvenile; iris brown, gradually attaining adult colour; legs and feet like adult.
Juvenile Paler and greyer than adult, lacking blackish head; chin and throat extensively white; iris brown; bill greenish-black to black, tinged brownish; proboscis reduced or absent; legs and feet olive-green.
Downy young Black, with down also on crown; small, bare whitish area on chin; bill light grey to pink, with partly blackish upper mandible, and yellow at base and tip; egg-tooth yellow or white; legs and feet black.

MEASUREMENTS Wing of 9 males 285-309 (295.6), of 6 females 280-300 (929.1); tail of 2 males 78, 78, of 2 females 67, 75; bill of 4 males 44, 49, 50, 53, of 2 females 44, 50; tarsus of 9 males 77-86 (82.1), of 6 females 69-103 (79.8). Weight of 1 male 2,100, of 3 females 1,660, 1,900, 2,010, of 2 juveniles 1,360, 1,500 (Ripley 1977, Marín 1996); can reach 2,290 (Chebez 1994).

GEOGRAPHICAL VARIATION None.

MOULT No information available.

DISTRIBUTION AND STATUS N Chile (Tarapacá, Antofagasta and Atacama), SW Bolivia (Oruro in 1903, and Potosí) and NW Argentina (Jujuy, Salta, N Catamarca in 1918, Tucumán and extreme N San Juan). VULNERABLE, and probably in decline. It normally occurs at low densities: 1-10 nesting pairs at some sites with up to 70-80 at a few (for full details, see Collar et al. 1992). However, concentrations of over 100 birds are reported occasionally, with 550 and 782 in the Eduardo Avaroa National Faunal Reserve, Potosí, Bolivia, in Oct-Nov 1989, while 2,800 birds were reported on Laguna Pelada in that reserve in Nov 1989 (Collar et al. 1992). The Chilean population is estimated at 620 birds (Collar et al. 1992). Total numbers are not accurately known but are estimated at 5,000 (Collar et al. 1994). Since these estimates were made, in Chile 180-200 birds and 70-80 nests were counted at Laguna Menique in Nov 1993, with 30-40 birds at the adjacent Laguna Miscanti (Howell & Webb 1995a). Very significantly, 8,988 birds and 180 active nests were counted in Oct 1995 at a little known lagoon complex in the Vilama and Pululos areas around Sierra Aconquija, Tucumán, Argentina (Caziani & Derlindati 1996); "large numbers" were also recorded at Pululos before 1941 (Collar et al. 1992). Little is known about population trends, but local populations fluctuate greatly between periods of drought and inundation (Fjeldså & Krabbe 1990), and the species was badly affected by a drought in the Vilama/Pululos areas in 1995 (Caziani & Derlindati 1996).

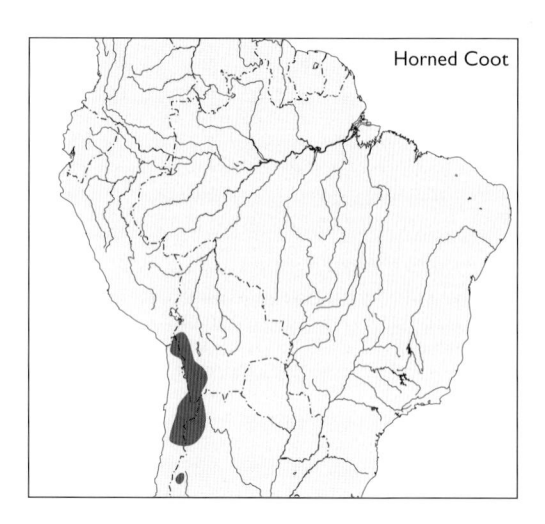

Lakes at which the species occurs, although remotely situated, remain vulnerable to contamination and trampling by cattle, while from some, water is pumped to coastal cities and towns and to mines; this coot also suffers from hunting and egg-harvesting (Collar et al. 1992). A recent decline in Tucumán is attributed to unpredictable changes such as droughts and floods (Vides-Almonacid 1988) but, as the species has evolved in such extreme ecosystems, these changes should not be a threat: however, in association with threats such as hunting, egg-harvesting and water removal (by pipeline to coastal cities and towns or mining centres), such changes may prove fatal (Collar et al. 1992). Predation by the Andean Gull *Larus serranus* is sometimes reported as a threat (e.g. Chebez 1994). Some lakes at which this coot occurs have been protected (see Collar et al. 1992), but only 6 of its known sites of occurrence are classed as key areas for threatened birds and 4 of these enjoy some form of protection, including the important Laguna Pelada, and the very important Vilama/Pululos lagoon complex which lies within the Altoandina de la Chinchilla Provincial Reserve and for which better protection has been proposed; more areas need to be protected, and existing protection laws need to be enforced (Collar et al. 1992, Wege & Long 1995, Caziani & Derlindati 1996). This species is officially considered 'vulnerable' in Chile, where censuses, surveillance and other studies are being carried out. Studies are needed to obtain overall breeding population estimates and to clarify habitat requirements, feeding and ecology, and seasonal movements, in order to prepare a global conservation strategy for the species (Collar et al. 1992).

MOVEMENTS Birds may fly from one feeding ground to another, and flocks have been seen arriving after dusk at a lake, staying throughout the next day, and flying out during the next night (Johnson 1965b). Local populations fluctuate greatly between periods of drought and inundation, while altitudinal movements or displacements, some over long distances, may occur in harsh weather, especially in the winter when food becomes locally unavailable as some lakes freeze (e.g. Collar *et al.* 1992); at such times birds are recorded as low as 2,000m. Movements take place at night.

HABITAT Barren Andean high-altitude altiplano lakes, both freshwater and brackish, with dense submerged aquatic plants, especially *Myriophyllum*, *Potamogeton* and *Ruppia*, and also aquatic grasses. These plants are apparently absent from saltmarshes, which may explain the bird's avoidance of the highly saline environments commonly found in the puna. This coot occurs chiefly at 3,000-5,200m in desert puna, but may wander as low as 2,000m in winter or during harsh weather (Collar *et al.* 1992).

FOOD AND FEEDING Poorly known; it eats mainly aquatic plants, especially *Myriophyllum*, *Potamogeton* and *Ruppia*, and also aquatic grasses and some seeds. *Myriophyllum* is apparently preferred to *Ruppia* when both plants are available (Behn & Millie 1959).

HABITS Hardly anything is recorded apart from the behaviour associated with nesting. These birds are reluctant to fly during the day, preferring to skim along the surface treading the water, but are apparently strong fliers when necessary, e.g. when moving between feeding grounds at night. The function of the bird's proboscis not clear, but it is sometimes used to assist in transportation of weed (Johnson 1965b).

SOCIAL ORGANISATION Monogamous; may breed in colonies of up to 80 pairs.

SOCIAL AND SEXUAL BEHAVIOUR Nothing is recorded except that the proboscis can be raised during display or in moments of excitement.

BREEDING AND SURVIVAL Season Breeds mainly Oct-Feb, although nest-building recorded as early as Sep. **Nest** Information taken from Ripley (1957a) and Johnson (1965b). Nest enormous; resembles that of Giant Coot but usually built on conical mound of up to 1.5t of stones, up to 4m in basal diameter and built up from lake bed to a height of c. 60cm, ending just below water surface in a platform c. 1m on which birds build nest of waterweed in shape of truncated cone. Nest sometimes built on large stone just above water level; in lakes with very dense vegetation, may build floating platform of waterweeds; in very shallow water nest may be built on top of natural hillocks on muddy bottom. Both adults build; stones (weight up to c. 450g) picked up one by one from lake shore or from bottom in shallow water and carried in bill; fresh plant material (usually *Ruppia* or *Myriophyllum*) constantly added to rim while nest in use. Nest height 30-60cm; diameter c. 2m at base and 35-60cm across cavity at top. Wall of cup continually renovated by addition of fresh weeds, pulled up from below water surface and carried to nest in bill, weed partly lying on bird's back and partly floating in water. Nest often used over many years, and also serves as nest-site for other waterbirds after coots have bred. **Eggs** 3-5 (Chebez [1994] gives 1-7); stone-grey to buff, speckled or blotched dark grey-brown; size 61-78 x 43.9-58 (66.2 x 47.1, n = 18); calculated weight c. 80. **Incubation** Period not recorded. **Chicks** Young fed by both parents; when adults bring fresh material to nest, seed pods are torn off and given to young. Chicks roost in nest, leaving it each morning when temperature rises and water surface thaws (Canevari *et al* 1991). Chick mortality estimated at c. 25% (Collar *et al.* 1992).

BIBLIOGRAPHY

ABBOTT, C. C. 1940. Notes from the Salton Sea, California. *Condor* 42: 264-265.
ABDULALI, H. 1970. A catalogue of the birds in the collection of the Bombay Natural History Society, 5. *J. Bombay Nat. Hist. Soc.* 66: 544-599.
ABDULALI, H. 1978. The birds of Great and Car Nicobars with some notes on wildlife conservation in the islands. *J. Bombay Nat. Hist. Soc.* 75: 744-772.
ABREUS, E. & CRIADO, J. 1994. *Distribution and conservation approach to Zapata Rail* Cyanolimnas cerverai *and Zapata Wren* Ferminia cerverai, *Zapata Swamp, Cuba.* BP Conservation Expedition Award 1994. BirdLife International & the Flora and Fauna Preservation Society.
ADAMS, D. A. & QUAY, T. L. 1958. Ecology of the Clapper Rail in southeastern North Carolina. *J. Wildl. Manage.* 22: 149-156.
AGUILAR, H. A. & KOWALINSKI, E. A. 1996. Nidificación del burrito común (*Laterallus melanophaius*) en Hudson, Buenos Aires. *Nuestras Aves* 35: 33.
AGUON, C. F. 1983. Survey and inventory of native landbirds on Guam. In *Research Project Report, Guam Aquatic and Wildlife Resources, FY 1983 Annual Report.* Guam: Dept. of Agriculture.
AIGNER, P. A., TECKLIN, J. & KOEHLER, C. E. 1995. Probable breeding population of Black Rail in Yuba County, California. *Western Birds* 26: 157-160.
ALBIGNAC, R. 1970. Mammifères et oiseaux du Massif du Tsaratanana (Madagascar Nord). *Mém. ORSTOM* 37: 223-229.
ALEMAN, Y. 1996. [Purple Gallinule, a new breeding bird for France.] *Ornithos* 3: 176-177.
ALEXANDER, W. B. 1923. A week on the Upper Barcoo, central Queensland. *Emu* 23: 82-95.
ALI, S. 1977. *Field guide to the birds of the eastern Himalayas.* Delhi: Oxford University Press.
ALI, S. & RIPLEY, S. D. 1980. *Handbook of the birds of India and Pakistan.* Vol. 2. Delhi: Oxford University Press.
ALISAUSKAS, R. T. 1987. Morphological correlates of age and breeding status in American coots. *Auk* 104: 640-646.
ALISAUSKAS, R. T. & ARNOLD, T. W. 1994. American Coot. In TACHA, T. C. & BRAUN, C. E. (eds.). *Management of migratory shore and upland game birds in North America.* Washington D.C.: International Assoc. Fish & Wildlife Agencies.
ALLEY, R. & BOYD, H. 1950. Parent-young recognition in the Coot *Fulica atra. Ibis* 92: 46-51.
ALLOUCHE, L. 1988. Strategies d'hivernage comparées du Canard Chipeau et de la Foulque Macroule pour un partage spatio-temporal des milieux humides de Carmargue. Montpellier: Université des Sciences et Techniques du Languedoc (PhD thesis).
ALSTRÖM, P. & OLSSON, U. 1982. Bill colour of the Crested Coot. *British Birds* 75: 287.
ALVAREZ, F. 1993. Alertness signalling in two rail species. *Animal Behav.* 46: 1229-1231.
AMADON, D. & DUPONT, J. E. 1970. Notes on Philippine birds. *Nemouria* 1: 1-14.
AMERSON, A. B. 1969. Ornithology of the Marshall and Gilbert Islands. *Atoll Res. Bull.* 127: 205-206, 229-230.
AMERSON, A. B., WHISTLER, W. A. & SCHWANER, T. D. 1995. Breeding pattern in the Banded Rail (*Gallirallus philippensis*) in Western Samoa. *Notornis* 42: 46-48.
AMOS, E. J. R. 1991. *A Gguide to the birds of Bermuda.* Warwick, Bermuda: E. J. R. Amos.
ANDERSON, A. 1975. A method of sexing moorhens. *Waterfowl* 26: 77-82.

ANDERSON, B. W. & OHMART, R. D. 1985. Habitat use by Clapper Rails in the lower Colorado River valley. *Condor* 87: 116-126.
ANDERSON, J. M. 1977. Yellow Rail (*Coturnicops noveboracensis*). In SANDERSON, G. C. (ed.). *Management of migratory shore and upland game birds in North America.* Washington D.C.: International Assoc. Fish & Wildlife Agencies.
ANDERSON, M. G. 1974. American Coots feeding in association with Canvasbacks. *Wilson Bull.* 86: 462-463.
ANDERSSON, C. J. & GURNEY, J. H. 1872. *Notes on the birds of Damara Land and the adjacent countries of South-West Africa.* London: John van Voorst.
ANDREAS, U. 1996. Brutverhalten der Wasserralle *Rallus aquaticus*: Ergebnisse von Volierenbeobachtungen. *J. Orn.* 137: 77-90.
ANDREW, P. 1992. *The birds of Indonesia. A checklist (Peters' sequence).* Kukila Checklist 1. Jakarta: Indonesian Orn. Soc.
ANDREW, P. & HOLMES, D. A. 1990. Sulawesi Bird Report. *Kukila* 5: 4-26.
ANDREWS, D. A. 1973. Habitat utilization by Sora, Virginia Rails and King Rails near southwestern Lake Erie. Columbus: Ohio State University (MS thesis).
ANDRIAMAMPIANINA, J. 1981. Les réserves naturelles et la protection de la nature à Madagascar. Pp. 105-111 in OBERLÉ, P. (ed.). *Madagascar, un sanctuaire de la nature.* Paris: Lechevalier.
ANON. 1888. Note on *Zapornia spilonota. Ibis* 5(6): 285.
ANON. 1944. Editorial comment on *Porphyrula georgica. Hornero* 8: 489-490.
ANON. 1983. *Yuma Clapper Rail Recovery Plan.* Albuquerque, New Mexico: US Fish & Wildlife Service.
ANON. 1984. *Salt Marsh Harvest Mouse and California Clapper Rail Recovery Plan.* Region One, Portland, Oregon: US Fish & Wildlife Service.
ANON. 1985. *Light-footed Clapper Rail Recovery Plan (Revised).* Region One, Portland, Oregon: US Fish & Wildlife Service.
ANON. 1992a. East African Bird Report 1990. *Scopus* 14: 129-158.
ANON. 1992b. *Recovery Plan: Mariana Common Moorhen* Gallinula chloropus guami. Region One, Portland, Oregon: US Fish & Wildlife Service.
ANON. 1993. *Estudio geográfico integral de la Ciénaga de Zapata.* Havana: Academia de Ciencias de Cuba, ICGC.
ANON. 1995a. Recent Reports. Niger. *Bull. African Bird Club* 2: 63.
ANON. 1995b. Recent Reports. South Africa. *Bull. African Bird Club* 2: 64.
ANON. 1995c. Corncrakes: better news from Scottish islands. *Birdwatch* 41: 7.
ANON. 1995d. New records for the Gambia; and Inaccessible Island. *Bull. African Bird Club* 2: 10-11.
ANON. 1995e. The Invisible Rail. *World Birdwatch* 17(3): 3.
ANON. 1995f. Invisible Rails heard on Halmahera? *Bull. Oriental Bird Club* 21: 17.
ANON. 1996. Birds get worst of two worlds. *BBC Wildlife* 14(1): 60.
ANON. 1997a. Progress report on the BOU-sponsored Maluku Programme, Indonesia. *Ibis* 139: 437-439.
ANON. 1997b. Threatened crake found in Bolivia. *World Birdwatch* 19(3): 4.
AOU 1983. *Check-list of North American birds.* 6th. edn. Lawrence, Kansas: American Ornithologists' Union.
ARBALLO, E. 1990. Nuevos registros para avifauna Uruguaya. *Hornero* 13: 179-187.
ARBALLO, E. & CRAVINO, J. 1987. Nido de Burrito de Patas Rojas, en Tacuarembó (Uruguay). *Nuestras Aves* 5(14): 21.

ARBALLO, E & CRAVINO, J. (in press). *Aves del Uruguay, manual ornitológico.* Vol. 1. Montevideo: Editorial Hemisfero Sur.

ARCHER, G. & GODMAN, E. M. 1937. *The birds of British Somaliland and the Gulf of Aden.* Vol. 2. London: Gurney & Jackson.

ARMONÍA. 1995. *Lista de las aves de Bolivia.* Santa Cruz, Bolivia: Associación Armonía.

ARMONÍA. 1997. Database information on Bolivian Rallidae. Santa Cruz, Bolivia: Associación Armonía (unpublished).

ARMSTRONG, R. H. 1983. *A new, expanded guide to the birds of Alaska.* Anchorage, Alaska: Alaska Northwest Publishing Company.

ARNOLD, K. A. 1978. First United States record of Paint-billed Crake (*Neocrex erythrops*). *Auk* 95: 745-746.

ARNOLD, T. W. 1987. Conspecific egg discrimination in American Coots. *Condor* 89: 675-676.

ARNOLD, T. W. 1990. Food limitation and the adaptive significance of clutch size in American coots (*Fulica americana*). London, Ontario: University of Western Ontario (PhD thesis).

ARNOLD, T. W. 1991. Intraclutch variation in egg size of American Coots. *Condor* 93: 19-27.

ARNOLD, T. W. 1992a. Continuous laying by American coots in response to partial clutch removal and total clutch loss. *Auk* 109: 407- 421.

ARNOLD, T. W. 1992b. The adaptive significance of eggshell removal by nesting birds: testing the egg-capping hypothesis. *Condor* 94: 547-548.

ARNOLD, T. W. 1993. Factors affecting renesting in American Coots. *Condor* 95: 273-281.

ARTMANN, J. W. & MARTIN, E. M. 1975. Incidence of ingested shot in Sora Rails. *J. Wildl. Manage.* 39: 514-519.

ASH, J. S. 1978. *Sarothrura* crakes in Ethiopia. *Bull. Brit. Orn. Club* 98: 26-29.

ASH, J. S. 1983. Over fifty additions of birds to the Somalia list including two hybrids, together with notes from Ethiopia and Kenya. *Scopus* 7: 54-79.

ASH, J. S. 1990. Additions to the avifauna of Nigeria, with notes on distributional changes and breeding. *Malimbus* 11: 104-116.

ASH, J. S. & GULLICK, T. M. 1989. The present situation regarding the endemic breeding birds of Ethiopia. *Scopus* 13: 90-96.

ASH, J. S. & MISKELL, J. E. 1983. Birds of Somalia, their habitat, status and distribution. *Scopus Special Suppl.* 1: 1-97.

ASKANER, T. 1959. Några iakttagelser över häckningsbeteende och häckningsresultat hos sothönan (*Fulica atra*). *Vår Fågelvärld* 18: 285-310.

ASPINWALL, D. R. 1978. Striped Crakes and some other dambo birds near Lusaka. *Bull. Zamb. Orn. Soc.* 10: 52-56.

ASTLEY MABERLY, C. T. 1935a. Notes on *Sarothrura elegans. Ostrich* 6: 39-42.

ASTLEY MABERLY, C. T. 1935b. Further notes upon *Sarothrura elegans. Ostrich* 6: 101-104.

ATKINSON, I. 1989. Introduced animals and extinctions. In WESTERN, D. & PEARL, M. C. (eds.) *Conservation for the Twenty-first Century.* New York: Oxford University Press.

ATKINSON, P., ROBERTSON, P., DELLELEGN, Y., WONDAFRASH, M. & ATKINS, J. 1996. The recent rediscovery of White-winged Flufftails in Ethiopia. *Bull. African Bird Club* 3: 34-36.

AUDUBON, J. J. 1842. *The birds of America.* Vol. 5. New York: J. J. Audubon.

AUMANN, T. 1991. Notes on the birds of the upper and middle reaches of Kimberley rivers during the dry season 1989. *Aust. Bird Watcher* 14: 51-67.

AUSTIN, O. L. 1948. The birds of Korea. *Bull. Mus. Comp. Zool.* 101.

AUSTIN, O. L. & KURODA, N. 1953. Birds of Japan, their status and distribution. *Bull. Mus. Comp. Zool.* 109: 279-637.

AVERY, G., BROOKE, R. K. & KOMEN, J. 1988. Records of the African Crake *Crex egregia* in western southern Africa. *Ostrich* 59: 25-29.

AVISE, J. C. & ZINK, R. M. 1988. Molecular genetic divergence between avian sibling species: King and Clapper Rails. *Auk* 105: 516-528.

BADMAN, F. J. 1979. Birds of southern and western Lake Eyre drainage. *S. Aust. Orn.* 28: 29-54.

BAHA EL DIN, S. 1993. *The catching of Corncrakes* Crex crex *and other birds in northern Egypt.* Cambridge, UK: BirdLife International (Study Report 55).

BAHA EL DIN, S. 1996. Trapping and shooting of Corncrakes *Crex crex* on the Mediterranean coast of Egypt. *Bird. Conserv. Internatn.* 6: 213-227.

BAILEY, A. M. 1956. *Birds of Midway and Laysan Islands.* Mus. Pictorial no. 12, Denver Mus. Nat. Hist.

BAIRD, K. E. 1974. A field study of the King, Sora and Virginia Rails at Cheyenne Bottoms in west-central Kansas. Fort Hayes: Kansas State College (MS thesis).

BAIRD, R. F. 1984. The Pleistocene distribution of the Tasmanian Native Hen *Gallinula mortierii mortierii. Emu* 84: 119-123.

BAIRD, R. F. 1986. Tasmanian Native-hen *Gallinula mortierii*: the first late Pleistocene record from Queensland. *Emu* 86: 121-122.

BAIRD, R. F. 1991. The dingo as a possible factor in the disappearance of *Gallinula mortierii* from the Australian mainland. *Emu* 91: 121-122.

BAKER, E. C. S. 1929. *The Fauna of British India.* Vol. 6. London: Taylor & Francis.

BAKER, E. C. S. 1935. *The Nidification of birds of the Indian Empire.* Vol. 4. London: Taylor & Francis.

BAKER, N.E., BEAKBANE, A.J. & BOSWELL, E.M. 1984. Streaky-breasted Pygmy Crake *Sarothrura boehmi*: first documented records for Tanzania. *Scopus* 8: 64-66.

BAKER, R. H. 1951. The avifauna of Micronesia, its origin, evolution, and distribution. *Univ. Kansas Pubs. Mus. Nat. Hist.* 3: 1-359.

BAKKER, B. J. & FORDHAM, R. A. 1993. Diving behaviour of the Australian Coot in a New Zealand lake. *Notornis* 40: 131-136.

BALDWIN, M. 1975. Birds of Inverell District, New South Wales. *Emu* 75: 113-120.

BALDWIN, P. H. 1945. The fate of the Laysan Rail. *Audubon Mag.* 47: 343-348.

BALDWIN, P. H. 1947. The life history of the Laysan Rail. *Condor* 49: 14-21.

BALTZER, M. C. 1990. A report on the wetland avifauna of South Sulawesi. *Kukila* 5: 27-55.

BANG, B. G. 1968. Olfaction in Rallidae (Gruiformes), a morphological study of thirteen species. *J. Zool. Soc. Lond.* 156: 97-107.

BANGS, O. 1907. On the wood rails, genus *Aramides,* occurring north of Panama. *Amer. Nat.* 41: 177-187.

BANKS, R. C. 1984. Bird specimens from American Samoa. *Pacific Science* 38: 150-169.

BANKS, R. C. & TOMLINSON, R. E. 1974. Taxonomic studies of certain Clapper Rails of southwestern United States and northwestern Mexico. *Wilson Bull.* 86: 325-335.

BANNERMAN, D. A. 1911. On three new species of birds from south-western Abyssinia (*Anomalospiza macmillani, Eremomela elegans abyssinicus* and *Ortygops macmillani*). *Bull. Brit. Orn. Club* 29: 37-39.

BANNERMAN, D. A. 1935. Notes on the birds of Nigeria. *Bull. Brit. Orn. Club* 55: 170-172.

BANNERMAN, D. A. & BANNERMAN, W. M. 1965. *The birds of the Atlantic islands.* Vol. 2. Edinburgh: Oliver & Boyd.

BARBOUR, T. 1923. The birds of Cuba. *Mem. Nuttall Orn. Club* 6: 1-141.
BARBOUR, T. & PETERS, J. L. 1927. Two more remarkable new birds from Cuba. *Proc. New England Zool. Club* 9: 95-97.
BARLOW, M. & SUTTON, R. R. 1975. Nest of a Marsh Crake. *Notornis* 22: 178-180.
BARNARD, H. G. 1911. Field notes from Cape York. *Emu* 11: 17-32.
BARNARD, H. G. 1914. Northern Territory birds. *Emu* 14: 39-57.
BARNES, K. N. (Ed.) (in prep.) *Important Bird Areas in Southern Africa - a first inventory*. Johannesburg: BirdLife South Africa.
BARNES, T. A. & FRANKLIN, D. C. (in press). Notes on nests of the Chestnut Rail *Eulabeornis castaneoventris*. *Northern Territory Naturalist*.
BARROW, K. M. 1910. *Three years in Tristan de Cunha*. London: Skeffington & Son.
BART, J. R., STEHN, A., HERRICK, J. A., HEASLIP, N. A., BOOKHOUT, T. A. & STENZEL, J. R. 1984. Survey methods for breeding Yellow Rails. *J. Wildl. Manage*. 48: 1382-1386.
BASILIO, A. 1963. *Aves de la Isla de Fernando Poo*. Madrid: Coculsa.
BATES, G. L. 1927. Notes on some birds of Cameroon and the Lake Chad region; their status and breeding-times. *Ibis* (12)3: 1-64.
BATES, J. M., PARKER, T. A., CAPPARELLA, A. P. & DAVIS, T. J. 1992. Observations on the campo, cerrado and forest avifaunas of eastern Dpto. Santa Cruz, Bolivia, including 21 species new to the country. *Bull. Brit. Orn. Club* 112: 86-98.
BATES, R. S. P. & LOWTHER, E. H. N. 1952. *Breeding birds of Kashmir*. Bombay: Oxford University Press.
BAUER, K. 1960a. Variabilität und Rassengliederung des Haselhuhnes (*Tetrastes bonasia*) in Mitteleuropa. *Bonn zoöl. Beitr*. 11:1-18.
BAUER, K. 1960b. Studies of less familiar birds 108. Little Crake. *British Birds* 53: 518-524.
BEAUCHAMP, A. J. 1986. A case of co-operative rearing in Wekas. *Notornis* 33: 51-52.
BEAUCHAMP, A. J. 1987a. A population study of the Weka *Gallirallus australis* on Kapiti Island. Victoria University (PhD thesis).
BEAUCHAMP, A. J. 1987b. The social structure of the Weka (*Gallirallus australis*) at Double Cove, Marlborough Sound. *Notornis* 34: 317-325.
BEAUCHAMP, A. J. 1988. Status of the Weka *Gallirallus australis* on Cape Brett, Bay of Islands. *Notornis* 35: 282-284.
BEAUCHAMP, A. J. 1996. Weka (*Gallirallus australis*) and *Leiopelma* frogs – a risk assessment. *Notornis* 43: 59-65.
BEAUCHAMP, A. J., CHAMBERS, R. & KENDRICK, J. L. 1993. North Island Weka on Rakitu Island. *Notornis* 40: 309-312.
BEAUCHAMP, A. J. & WORTHY, T. H. 1988. Decline in the distribution of the Takahe *Porphyrio* (=*Notornis*) *mantelli*. A re-examination. *J. Roy. Soc. N. Z*. 18: 103-112.
BECK, R. E. 1988. Survey and inventory of the native landbirds on Guam. In *Research Project Report, Guam Aquatic and Wildlife Resources, FY 1988 Annual Report*. Guam: Dept. of Agriculture
BECK, R. E. & SAVIDGE, J. A. 1990. *Recovery Plan for the native forest birds of Guam and Rota of the Northern Mariana Islands*. Honolulu, Hawaii: U. S. Fish & Wildlife Service.
BECKER, P. 1983. Zum Brutvorkommen des Zwergsumpfhuhns (*Porzana pusilla*) in Niedersachsens. *Beitr. Naturkd. Niedersachsens* 36: 193-203.
BECKER, P. 1990. Kennzeichen und kleider der europäischen kleinen Rallen und Sumpfhühner *Rallus* und *Porzana*. *Limicola* 4: 93-144.

BECKER, P. 1995. Identification of Water Rail and *Porzana* crakes in Europe. *Dutch Birding* 17: 181-211.
BEEBE, W., HARTLEY, G. I. & HOWES, P. G. 1917. *Tropical wildlife of British Guiana*. New York: New York Zoological Society.
BEEHLER, B. M. 1978. *Upland birds of northeastern New Guinea*. Wau Ecology Institute Handbook 4. Wau, Papua New Guinea.
BEEHLER, B. M., PRATT, T. K. & ZIMMERMAN, D. A. 1986. *Birds of New Guinea*. Princeton, New Jersey: Princeton University Press.
BEHN, F. & MILLIE, G. 1959. Beitrag zur Kenntnis des Rüsselblässhuhns (*Fulica cornuta* Bonaparte). *J. Orn*. 100: 119-131.
BEHRSTOCK, R. A. 1983. Colombian Crake (*Neocrex colombianus*) and Paint-billed Crake (*N. erythrops*): first breeding records for Central America. *Amer. Birds* 37: 956-957.
BEHRSTOCK, R. A. 1996. Voices of Stripe-backed Bittern *Ixobrychus involucris*, Least Bittern *I. exilis*, and Zigzag Heron *Zebrilus undulatus*, with notes on distribution. *Cotinga* 5: 55-61.
BEINTEMA, A. J. 1972. The history of the Island Hen (*Gallinula nesiotis*), the extinct flightless gallinule of Tristan da Cunha. *Bull. Brit. Orn. Club* 92: 106-113.
BELCHER, C. & SMOOKER, G. D. 1935. Birds of the colony of Trinidad and Tobago. Part II. *Ibis* (13)5: 279-297.
BELL, H. L. 1970. Field notes on the birds of Amazon Bay, Papua. *Emu* 70: 23-26.
BELL, H. L. 1982. A bird community of lowland rainforest in New Guinea. 1. Composition and density of the avifauna. *Emu* 82: 24-41.
BELLINGHAM, M. & DAVIES, A. 1988. Forest bird communities in Western Samoa. *Notornis* 35: 117-128.
BELTON, W. 1984. Birds of Rio Grande do Sul, Brazil. Part 1. Rheidae through Furnariidae. *Bull. Amer. Mus. Nat. Hist*. 178: 371-631.
BELTON, W. 1994. *Aves do Rio Grande do Sul: distribuição e biologia*. São Leopoldo, RS: Editora Unisinos.
BELTZER, A. H., SABATTINI, R. A., & MARTA, M. C. 1991. Ecología alimentaria de la Polla de Agua Negra *Gallinula chloropus galeata* (Aves: Rallidae) en un ambiente lenítico del río Paranámedio, Argentina. *Orn. Neotropical* 2: 29-36.
VAN BEMMEL, A. C. V. & VOOUS, K. H. 1951. On the birds of the islands of Muna and Buton, S.E. Celebes. *Treubia* 21: 27-104.
BENGTSON, S.-A. 1962. Småfläckige sumphönans (*Porzana porzana*) förekomst och häckningsbiologi i nordöstra Skåne. *Vår Fågelvärld* 21: 253-266.
BENGTSON, S.-A. 1967. Revirförhällenden hos vattenrall (*Rallus aquaticus*) tidigt på våren. *Vår Fågelvärld* 26: 6-18.
BENGTSSON, K. 1997. Some interesting observations in Mauritania and Senegal. *Malimbus* 19: 96-97.
BENITO-ESPINAL, E. & HAUTCASTEL, P. 1988. Les oiseaux menacés de Guadeloupe et de Martinique. Pp 37-60 in THIBAULT, J.-C. & GUYOT, I. (eds.). *Livre rouge des oiseaux menacés des régions françaises d'Outre-mer*. Saint-Cloud: Conseil International pour la Protection des Oiseaux (Monogr. 5).
BENNETT, W. W. & OHMART, R. D. 1978. Habitat requirements and population characteristics of the Clapper Rail (*Rallus longirostris yumanensis*) in the Imperial Valley of California. Unpubl. report. Univ. of California, Lawrence Livermore Laboratory, Livermore, CA. Unpublished.
BENNUN, L. & FASOLA, M. (eds.). 1996. Resident and migrant waterbirds at Lake Turkana, February 1992. *Quad. Civ. Staz. Idrobiol*. 21: 7-62.
BENSON, C. W. 1956. New or unusual birds from Northern Rhodesia. *Ibis* 98: 595-605.
BENSON, C. W. 1960. The birds of the Comoro Islands. Results

of the British Ornithologists' Union Centenary Expedition 1958. *Ibis* 103B.

BENSON, C. W. 1964a. The European and African races of Baillon's Crake, *Porzana pusilla*. *Bull. Brit. Orn. Club* 84: 2-5.

BENSON, C. W. 1964b. Some intra-African migratory birds. *Puku* 2: 53-56.

BENSON, C. W. 1967. The birds of Aldabra and their status. *Atoll Res. Bull.* 118: 68-111.

BENSON, C. W. & BENSON, F. M. 1977. *The birds of Malawi*. Limbe, Malawi: Montfort Press.

BENSON, C. W., BROOKE, R. K., DOWSETT, R. J. & IRWIN, M. P. S. 1971. *The birds of Zambia*. London: Collins.

BENSON, C. W., COLEBROOK-ROBJENT, J. F. R. & WILLIAMS, A. 1976. Contribution à l'ornithologie de Madagascar. *Oiseau et R.F.O.* 2: 1-96.

BENSON, C. W. & HOLLIDAY, C. S. 1964. *Sarothrura affinis* and some other species on the Nyika Plateau. *Bull. Brit. Orn. Club* 84: 131-132.

BENSON, C. W. & IRWIN, M. P. S. 1965. Some intra-African migratory birds II. *Puku* 3: 45-55.

BENSON, C. W. & IRWIN, M. P. S. 1971. A South African male of *Sarothrura ayresi* and other specimens of the genus in the Leiden Museum. *Ostrich* 42: 227-228.

BENSON, C. W. & IRWIN, M. P. S. 1974. On a specimen of *Sarothrura ayresi* from the Transvaal in the Leiden Museum. *Ostrich* 42: 227-228.

BENSON, C. W. & PITMAN, C. R. S. 1964. Further breeding records from Northern Rhodesia. *Bull. Brit. Orn. Club* 84: 54-60.

BENSON, C. W. & PITMAN, C. R. S. 1966. On the breeding of Baillon's Crake *Porzana pusilla* (Pallas) in Africa and Madagascar. *Bull. Brit. Orn. Club* 86: 141-143.

BENSON, C. W. & WAGSTAFFE, R. 1972. *Porzana olivieri* and *Limnocorax flavirostris*; a likely affinity. *Bull. Brit. Orn. Club* 92: 160-164.

BENSON, C. W. & WINTERBOTTOM, J. M. 1968. The relationship of the Striped Crake *Crecopsis egregia* Peters and the White-throated Crake *Porzana albicollis* (Vieillot). *Ostrich* 39: 177-179.

BENT, A. C. 1926. Life histories of North American marsh birds. *Bull. U.S. Natn. Mus.* 135: 1-490.

BERGAN, J. F. & SMITH, L. M. 1986. Food robbery of wintering Ring-necked Ducks by American Coots. *Wilson Bull.* 98: 306-308.

BERGER, A. J. 1951. Nesting density of Virginia and Sora Rails in Michigan. *Condor* 53: 203.

BERGER, A. J. 1981. *Hawaiian birdlife*. Honolulu: University of Hawaii Press.

BERGH, L. van den. 1991. Status, distribution and research on Corncrakes in the Netherlands. *Vogelwelt* 112: 78-83.

BERGMAN, R. D. 1973. Use of southern boreal lakes by postbreeding canvasbacks and redheads. *J. Wildl. Manage.* 37: 160-170.

BERLIOZ, J. & PFEFFER, P. 1966. Étude d'une collection oiseaux d'Amboine (Iles Moluques). *Bull. Mus. Natn. d'Hist. Nat.* (2)37: 907-915.

BERNEY, F. L. 1907. Field notes on birds of the Richmond District, North Queensland. IV. *Emu* 6: 106-115.

BERULDSEN, G. 1975. The bush-hen in south-eastern Queensland. *Aust. Bird Watcher* 6: 75-76.

BERULDSEN, G. R. 1976. Further notes on the Bush-hen in south-eastern Queensland. *Sunbird* 7: 53-58.

BEST, B. J., CLARKE, C. T., CHECKER, M., BROOM, A. L., THEWLIS, R. M., DUCKWORTH, W. & MCNAB, A. 1993. Distributional records, natural history notes, and conservation of some poorly known birds from southwestern Ecuador and northwestern Peru. *Bull. Brit. Orn. Club* 113: 108-119.

BEZZEL, E. & SCHÖPF, H. 1991. Der Wachtelkönig im Murnauer Moos: Artenschutzerfolg durch Ausweisung eines Naturschutzgebietes? *Vogelwelt* 112: 83-90.

BIAGGI, V. 1983. *Las aves de Puerto Rico*. Rio Piedras, Puerto Rico: Editorial de la Universidad de Puerto Rico.

BILLARD, R. S. 1948. An ecological study of the Virginia Rail (*Rallus limicola*) and the Sora (*Porzana carolina*) in some Connecticut swamps. Ames, Iowa: Iowa State College (MS thesis).

BINFORD, L. C. 1973. Virginia Rail and Cape May Warbler in Chiapas, Mexico. *Condor* 75: 350-351.

BIRKENHOLZ, D. E. & JENNI, D. A. 1964. Observations on the Spotted Rail and Pinnated Bittern in Costa Rica. *Auk* 81: 558-559.

BISHOP, K. D. 1983. Some notes on non-passerine birds of West New Britain. *Emu* 83: 235-241.

BLABER, S. J. M. 1990. A checklist and notes on the current status of birds of New Georgia, Western Province, Solomon Islands. *Emu* 90: 205-214.

BLACK, J. 1919. Pajingo notes. *Emu* 18: 206-207.

BLACKBURN, A. 1968. The birdlife of Codfish Island. *Notornis* 15: 51-65.

BLACKBURN, A. 1971. Some notes on Fijian birds. *Notornis* 18: 147-174.

BLAKE, E. R. 1959. New and rare Colombian birds. *Lozania* 11: 1-10.

BLAKE, E. R. 1977. *Manual of Neotropical Bbirds*. Vol. 1. Chicago: Chicago University Press.

BLAKERS, M., DAVIES, S. J. J. F. & REILLY, P. N. 1984. *The atlas of Australian birds*. Victoria: Royal Australian Ornithologists' Union.

BLANCO, J. C. & GONZALES, J. L. (eds.) 1992. *Libro rojo de los vertebrados de España*. Madrid: Colección Técnica, ICONA.

BLASIUS, W. 1897. Neuer Beitrag zur Kenntniss der Vogelfauna von Celebes. Fests. Herzogl. Technischen Hochschule Carol. Wilhelm. Gelegenheit 69 Vers. D. ntf. Aerzte in Braunschweig 1897: 275-395.

BLEDSOE, A. H. 1988. Status and hybridization of Clapper and King Rails in Connecticut. *Connecticut Warbler* 8: 61-65.

BLEM, C. R. 1980. A Paint-billed Crake in Virginia. *Wilson Bull.* 92: 393-394.

BOCAGE, J. V. B. 1877. *Ornithologie d'Angola*. Lisbon: Imprimerie Nationale.

BOEKEL, C. 1980. Birds of Victoria River Downs Station and of Varradin, Northern Territory. Part 1. *Aust. Bird Watcher* 8: 171-193.

BOESMAN, P. 1997. Recent observations of the Rusty-flanked Crake *Laterallus levraudi*. *Cotinga* 7: 39-42.

BOND, J. 1971. *Sixteenth supplement to the Check-list of birds of the West Indies (1956)*. Philadelphia: Academy of Natural Sciences.

BOND, J. 1973. *Eighteenth supplement to the Check-list of birds of the West Indies (1956)*. Philadelphia: Academy of Natural Sciences.

BOND, J. 1979. *Birds of the West Indies*. 4th. edn. London: Collins.

BOND, J. 1980. *Twenty-third supplement to the Check-list of birds of the West Indies (1956)*. Philadelphia: Academy of Natural Sciences.

BOND, J. 1984. *Twenty-fifth supplement to the Check-list of birds of the West Indies (1956)*. Philadelphia: Academy of Natural Sciences.

BOND, J. 1985. *Birds of the West Indies*. 5th. edn. London: Collins.

BONDE, K. 1993. *Birds of Lesotho*. Pietermaritzburg, South Africa: University of Natal Press.

BOOKHOUT, T. A. 1995. Yellow Rail. In POOLE, A. & GILL, F. (eds.). *The birds of North America*, No. 139. Washington, D.C.: American Ornithologists' Union.

BOOKHOUT, T. A. & STENZEL, J. R. 1987. Habitat and movements of breeding Yellow Rails. *Wilson Bull.* 99: 441-447.

BOOTH, D. F. 1983. Classified summarised notes, 1 July 1981 to 30 June 1982. *Notornis* 30: 34-38.

BOPP, R. 1959. Das Blesshuhn (*Fulica atra*). Wittenberg-Lutherstadt: Neue Brehm-Bücherei.

BORNSCHEIN, M. R. & REINERT, B. L. 1996. On the diagnosis of *Aramides cajanea aviceniae* Stotz, 1992. *Bull. Brit. Orn. Club* 116: 272.

BORNSCHEIN, M. R., REINERT, B. L. & PICHORIM, M. 1997. Notas sobre algumas aves novas ou pouco conhecidas no sul do Brasil. *Ararajuba* 5: 53-59.

BOSCHERT, M. 1995. [High population density of the Moorhen (*Gallinula chloropus*).] *J. Orn. Jahrb. Bad.-Württ.* 11: 159-165.

BOSHOFF, A., PALMER, N. G. & PIPER, S. E. 1991. Spatial and temporal abundance patterns of waterbirds in the southern Cape Province. Part 1: Diving and surface predators. *Ostrich* 62: 156-177.

BOURNE, W. R. P. & DAVID, A. C. F. 1983. Henderson Island, central south Pacific, and its birds. *Notornis* 30: 233-243.

BOWMAN, R. I. 1960. *Report on a biological reconnaissance of the Galápagos Islands during 1957.* Paris: UNESCO.

BOWLER, J. & TAYLOR, J. 1989. An annotated checklist of the birds of Manusela National Park, Seram. Birds recorded on the Operation Raleigh Expedition. *Kukila* 4: 3-29.

BOYD, H. J. & ALLEY, R. 1948. The function of the head-coloration of the nestling Coot and other nestling Rallidae. *Ibis* 90: 582-593.

BOYLE, W. J., PAXTON, R. O. & CUTLER, D. A. 1987. The winter season, Hudson-Delaware Region. *Amer. Birds* 41: 260-263.

BRAAKSMA, S. 1962. Voorkomen en levensgewoonten van de Kwartelkoning (*Crex crex* L.). *Limosa* 35: 230-259.

BRACE, R. C., HORNBUCKLE, J. & PEARCE-HIGGINS, J. W. 1997. The avifauna of the Beni Biological Station, Bolivia. *Bird Conserv. Internatn.* 7: 117-159.

BRACE, R., HORNBUCKLE, J. & ST PIERRE, P. (1998). Rufous-faced Crake *Laterallus xenopterus*: a new species to Bolivia. *Cotinga*. 9: 76-80.

BRACKENBURY, J. 1984. Physiological responses of birds to flight and running. *Biol. Rev. Cambr. Philos. Soc.* 59: 559-575.

BRACKNEY, A. W. 1979. Population ecology of common gallinules in southwestern Lake Erie marshes. Columbus: Ohio State University (PhD thesis).

BRACKNEY, A. W. & BOOKHOUT, T. A. 1982. Population ecology of common gallinules in southwestern Lake Erie marshes. *Ohio J. Sci.* 82: 229-237.

BRADSHAW, C. 1993. Separating juvenile Little and Baillon's Crakes in the field. *British Birds* 86: 303-311.

BRAINE, S. 1988. Vagrants and range extensions found in and adjacent to the Skeleton Coast Park. *Lanioturdus* 24: 4-15.

BRANFIELD, A. 1989. New bird records for the East Caprivi. *Lanioturdus* 25: 4-21.

BRAVERY, J. A. 1970. The birds of Atherton Shire, Queensland. *Emu* 70: 49-63.

BRAZIL, M. A. 1984. Observations on the behaviour and vocalizations of the Okinawa Rail *Rallus okinawae*. *J. Coll. Dairying (Ebetsu)* 10: 437-449.

BRAZIL, M. A. 1985a. Notes on the Okinawa Rail *Rallus okinawae*: observations at night and at dawn. *Tori* 33: 125-127.

BRAZIL, M. A. 1985b. The endemic birds of the Nansei Shoto. Pp 11-35 in ANON. 1985. *Conservation of the Nansei Shoto.* Part II. WWF Japan Scientific Committee.

BRAZIL, M. A. 1991. *The birds of Japan.* Washington D.C.: Smithsonian Institution Press.

BREGULLA, H. L. 1992. *Birds of Vanuatu.* Oswestry, UK: Anthony Nelson.

BRIGGS, S. V. 1989. Food addition, clutch size, and the timing of laying in American Coots. *Condor* 91: 493-494.

BRIGHT, J. & TAYSOM, A. R. 1932. Birds of Lake Cooper, Victoria, and surroundings. *Emu* 32: 42-48.

BRINKHOF, M. W. G. (in press). Seasonal decline in body size of coot chicks. *J. Avian Biol.* 27.

BRINKHOF, M. W. G., CAVÉ, A. J., HAGE, F. J. & VERHULST, S. 1993. Timing of reproduction and fledging success in the coot *Fulica atra*: evidence for a causal relationship. *J. Animal Ecology* 62: 577-587.

BRINKHOF, M. W. G., CAVÉ, A. J. & PERDECK, A. C. 1997. The seasonal decline in the first-year survival of juvenile coots: an experimental approach. *J. Animal Ecology* 66: 73-82.

BRITTON, P. L. (ed.). 1980. *Birds of East Africa.* Nairobi: East African Natural History Society.

BRODKORB, P. 1943. Birds from the Gulf Lowlands of southern Mexico. *Misc. Publ. Mus. Zool. Univ. Michigan* 55: 1-88.

BRODKORB, P. 1967. Catalogue of fossil birds: Part 3 (Ralliformes, Ichthyornithiformes, Charadriiformes). *Bull. Florida State Mus.* 11: 99-220.

BROEKHUYSEN, G. J., LESTRANGE, G. K. & MYBURGH, N. 1964. The nest of the Red-chested Flufftail (*Sarothrura rufa* Vieillot). *Ostrich* 35: 117-120.

BROEKHUYSEN, G. J. & MACNAE, W. 1949. Observations on the birds of Tristan da Cunha Islands and Gough Island in February and early March, 1948. *Ardea* 37: 97-113.

BROOK, B. W., LIM, L., HARDEN, R. & FRANKHAM, R. 1997. How secure is the Lord Howe Island Woodhen? A population viability analysis using VORTEX. *Pacific Conserv. Biol* 3: 125-133.

BROOKE, R. K. 1964. Avian observations on a journey across Central Africa and additional information on some of the species seen. *Ostrich* 35: 277-292.

BROOKE, R. K. 1968. On the distribution, movements and breeding of the Lesser Reedhen *Porphyrio alleni* in southern Africa. *Ostrich* 39: 259-262.

BROOKE, R. K. 1974. The Spotted Crake *Porzana porzana* (Aves: Rallidae) in south-central and southern Africa. *Durban Mus. Novit.* 10: 43-52.

BROOKE, R. K. 1975. Cooperative breeding, duetting, allopreening and swimming in the Black Crake. *Ostrich* 46: 190-191.

BROOKE, R. K. 1984. *South African Red Data Book – birds.* Pretoria: Council for Scientific and Industrial Research.

BROOKE, R. K. 1992. Apparent reverse mounting in the African Rail *Rallus caerulescens*. *Ostrich* 63: 185.

BROOKE, R. K. & PARKER, S. A. 1984. The African bird collections of S. A. White of South Australia. *Scopus* 8: 33-36.

BROOKER, M. G., RIDPATH, M. G., ESTBERGS, A. J., BYWATER, J., HART, D. S. & JONES, M. S. 1979. Bird observations on the north-western Nullarbor Plain and neighbouring regions, 1967-1978. *Emu* 79: 176-190.

BROOKS, T. & DUTSON, G. 1997. Twenty-nine new island records of birds from the Philippines. *Bull. Brit. Orn. Club* 117: 32-37.

BROSSET, A. 1963. La reproduction des oiseaux de mer des Iles Galápagos en 1962. *Alauda* 31: 81-109.

BROSSET, A. 1968. Localisation écologique des oiseaux migrateurs dans la forêt équatoriale du Gabon. *Biol. Gabon* 4: 211-226.

BROSSET, A. & ERARD, C. 1986. *Les oiseaux des régions forestières du nord-est du Gabon.* Vol. 1. Paris: Société Nationale de Protection de la Nature.

BROTHERS, N. P. 1979. Further notes on the birds of Maatsuyker Island, Tasmania. *Emu* 79: 89-91.

BROTHERS, N. P. & SKIRA, I. J. 1984. Wekas on Macquarie Island. *Notornis* 31: 145-154.

BROUWER, J. & GARNETT, S. T. (eds.). 1990. *Threatened birds of Australia. An annotated list.* RAOU Report 68. Melbourne: Royal Australasian Ornithologists' Union.

BROWN, C. J. & S.W.A. RARE BIRD COMMITTEE. 1985. South West Africa/Namibia rare bird report for 1984/1985. *Lanioturdus* 21(4).

BROWN, C. J. & S.W.A. RARE BIRD COMMITTEE. 1987. South West Africa/Namibia rare bird report for 1985/1986. *Lanioturdus* 23: 66-71.

BROWN, L. H. & BRITTON, P. L. 1980. *The breeding seasons of East African birds.* Nairobi: East Africa Natural History Society.

BROWN, P. B. 1974. Breeding the Ypecaha Wood Rail at Harewood Bird Garden. *Avicult. Mag.* 80: 11-13.

BROWN, R. H. 1938. Notes on the Land-Rail. *British Birds* 32: 13-16.

BROWN, R. J. & BROWN, M. N. 1977. Observations on swamphens breeding near Manjimup, Western Australia. *Corella* 1: 82-83.

BROWN, R. J. & BROWN, M. N. 1980. Eurasian Coots breeding on irrigation dams near Manjimup, Western Australia. *Corella* 4: 33-36.

BROWN, M. & DINSMORE, J. D. 1986. Implications of marsh size and isolation for marsh bird management. *J. Wildl. Manage.* 50: 392-397.

BROWNE, M. M. & POST, W. 1972. Black rails hit television tower at Raleigh, North Carolina. *Wilson Bull.* 84: 491-492.

BROYER, J. 1987. L'habitat du Râle de Genêts *Crex crex* en France. *Alauda* 3: 161-186.

BROYER, J. 1991. Situation des Wachtelkönigs in Frankreich. *Vogelwelt* 122: 71-77.

BROYER, J. 1994. La régression du Râle de genêts *Crex crex* en France et la gestation des milieux prairiaux. *Alauda* 62: 1-7.

BROYER, J. & ROCAMORA, G. 1994. Enquête nationale Râle de genêts 1991-92. Principaux résultats. *Ornithos* 1: 55-56.

BRUCE, M. D. 1981. Final supplement to "A field list of the birds of the Philippines". In BRUCE, M. D. (ed.) *The Palawan Expedition, Stage II.* Sydney: Assoc. Res. Expl. Aid.

BRUNER, P. L. 1972. *The birds of French Polynesia.* Honolulu: Pacific Scientific Information Center, Bernice P. Bishop Museum.

BRYANT, C. E. 1940. Photography in the swamps: the Eastern Swamphen. *Emu* 39: 236-239.

BRYANT, C. E. 1942. Photography in the swamps: the Marsh Crake. *Emu* 42: 31-35.

BRYANT, C. E. & AMOS, B. 1949. Notes on the crakes of the genus *Porzana* around Melbourne, Victoria. *Emu* 48: 249-275.

BUDDLE, G. A. 1941a. The birds of the Poor Knights. *Emu* 41: 56-68.

BUDDLE, G. A. 1941b. Photographing the Spotless Crake. *Emu* 41: 130-134.

BUDEN, D. W. 1987. The birds of Cat Island, Bahamas. *Wilson Bull.* 99: 579-600.

BULL, P. C., GAZE, P. B. & ROBERTSON, C. J. R. 1985. *The atlas of bird Ddistribution in New Zealand.* Wellington: Ornithological Society of New Zealand.

BULLER, W. L. 1873. *A history of the birds of New Zealand.* London: van Voorst.

BULLER, W. L. 1905. *Supplement to the birds of New Zealand.* London: author.

BUNIN, J. S. 1995. Preliminary observations of behavioural interactions between Takahe (*Porphyrio mantelli*) and Pukeko (*P. porphyrio*) on Mana Island. *Notornis* 42: 140-143.

BUNIN, J. S. & JAMIESON, I. G. 1995. New approaches towards a better understanding of the decline of the takahe (*Porphyrio mantelli*) in New Zealand. *Conserv. Biol.* 9: 100-106.

BUNIN, J. S. & JAMIESON, I. G. 1996a. A cross-fostering experiment between the endangered Takahe (*Porphyrio mantelli*) and its closest relative, the Pukeko (*P. porphyrio*). *N.Z. J. Ecol.* 20: 207-213.

BUNIN, J. S. & JAMIESON, I. G. 1996b. Responses to a model predator of New Zealand's endangered Takahe and its closest relative, the Pukeko. *Conserv. Biol.* 10: 1463-1466.

BUNIN, J. S., JAMIESON, I. G. & EASON, D. 1997. Low reproductive success of the endangered Takahe *Porphyrio mantelli* on offshore island refuges in New Zealand. *Ibis* 139: 144-151.

BURGER, J. 1973. Competition for nest sites between Franklin's Gull and the American Coot. *Wilson Bull.* 85: 449-451.

BURTON, J. H. 1959. Some population mechanics of the American coot. *J. Wildl. Manage.* 23: 203-210.

BUTLER, A. L. 1899. The birds of the Andaman and Nicobar Islands. *J. Bombay Nat. Hist. Soc.* 12: 684-696.

BUTLER, T. Y. 1979. *The birds of Ecuador and the Galapagos Archipelago.* Portsmouth, N.H., USA: The Ramphastos Agency.

BÜTTIKOFER, J. 1893. Description of a new genus of crakes. *Notes Leyden Mus.* 15(39): 274-275.

BUXTON, A. 1948. *Travelling naturalist.* London: Collins.

BYRD, G. V., COLEMAN, R. A., SHALLENBERGER, R. J. & ARUME, C. S. 1985. Notes on the breeding biology of the Hawaiian race of the American Coot. *Elepaio* 45: 57-63.

BYRD, G. V. & ZEILLEMAKER, G. F. 1981. Ecology of nesting Hawaiian Common Gallinules at Hanali, Hawaii. *Western Birds* 12: 105-116.

CAIN, A. J. & GALBRAITH, I. C. J. 1956. Field notes on birds of the eastern Solomon Islands. *Ibis* 98: 100-134.

CAIRNS, J. 1953. Studies of Malayan birds. *Malayan Nature J.* 7: 173-179.

CALDWELL, H. R. & CALDWELL, J. C. 1931. *South China birds.* Shanghai: Vanderburgh.

CAMPBELL, A. J. 1906. Oological notes and further description of a new fruit-pigeon. *Emu* 5: 195-199.

CAMPBELL, A. J. & BARNARD, H. G. 1917. Birds of the Rockingham Bay District, north Queensland. *Emu* 17: 2-38.

CAMPBELL, E. G. & WOLF, G. A. 1977. Great Egret depredation on a Virginia Rail. *Western Birds* 8: 64.

CAMPBELL, R. W., DAWE, N. K., MCTAGGART-COWAN, I., COOPER, J. M., KAISER, G. W. & MCNALL, M. C. E. 1990. *The birds of British Columbia.* Vol. 2. Victoria: Royal British Columbia Museum.

CANEVARI, M., CANEVARI, P., CARRIZO, B. R., HARRIS, G., MATA, J. R. & STRANECK, R. J. 1991. *Nueva guia de las aves Argentinas.* Buenos Aires: Fundación Acindar.

CARDIFF, S. W. & SMALLEY, G. B. 1989. Birds in the rice country of southwest Louisiana. *Birding* 21: 232-240.

CARPENTER, R. E. & STAFFORD, M. A. 1970. The secretory rates and the chemical stimulus for secretion of the nasal salt glands in the Rallidae. *Condor* 72: 316-324.

CARROLL, A. L. K. 1963a. Food habits of the North Island Weka. *Notornis* 10: 289-300.

CARROLL, A. L. K. 1963b. Breeding cycle of the North Island Weka. *Notornis* 10: 300-302.

CARROLL, A. L. K. 1963c. Sexing of Wekas. *Notornis* 10: 302-303.

CARROLL, A. L. K. 1966. Food habits of the Pukeko (*Porphyrio melanotus*, Temminck). *Notornis* 13: 133-144.

CARROLL, A. L. K. 1969. The Pukeko (*Porphyrio melanotus*) in New Zealand. *Notornis* 16: 101-120.

CARROLL, R. W. 1988. Birds of the Central African Republic. *Malimbus* 10: 177-200.

CARTER, T. 1904. Birds occurring in the region of the north-west Cape. *Emu* 3: 171-177.

CATESBY, M. 1731. *The natural history of Carolina, Florida and the Bahama Islands.* Vol. 1. London: privately published.

CAVÉ, A. J. & VISSER, J. 1985. Winter severity and breeding numbers in a Coot population. *Ardea* 73: 129-138.

CAVE, F. O. & MACDONALD, J. D. 1955. *Birds of the Sudan.* Edinburgh: Oliver & Boyd.

CAZIANI, S. M. & DERLINDATI, E. 1996. *Fulica cornuta* en la laguna de Pululos y otros cercanas, puna arida del noroeste de Argentina. *TWSG News* 9: 34-39.

CEMPULIK, P. 1991. Bestandsentwicklung, Schutzstatus und aktuelle Untersuchungen am Wachtelkönig in Polen. *Vogelwelt* 112: 40-45.

CEMPULIK, P. 1992. [Wintering of the Moorhen (*Gallinula chloropus*) in Upper Silesia.] *Notatki Ornitol.* 33: 275-283.

CHACÓN, G. 1993. El Guión de Codornices en la peninsula Ibérica y Baleares. *Quercus* 83: 26-29.

CHAFFER, N. 1940. Photographing the Avocet – and other birds – at Lake Midgeon. *Emu* 40: 126-128.

CHAMBERS, S. 1989. *Birds of New Zealand. Locality guide.* Hamilton, N.Z.: Arun Books.

CHAPIN, J. P. 1939. The birds of the Belgian Congo. Part II. *Bull. Amer. Mus. Nat. Hist.* 75: 1-632.

CHAPIN, J. P. 1948. The mystery of the Mabira Banshee. *Audubon Mag.* 50: 341-349.

CHAPMAN, F. M. 1966. *Handbook of birds of Eastern North America.* New York: Dover.

CHAPPUIS, C. 1975. Illustration sonore de problèmes bioacoustiques posés par les oiseaux de la zone éthiopienne. *Alauda*, Suppl. Sonore, 43: 427-474.

CHASEN, F. N. & HOOGERWERF, A. 1941. The birds of the Netherlands Indian Mt Leuser Expedition 1937 to North Sumatra. *Treubia* 18, Suppl.: 1-125.

CHEBEZ, J. C. 1994. *Los que se van. Especies argentinas en peligro.* Albatros.

CHEESMAN, R. E. & SCLATER, W. L. 1935. On a collection of birds from north-western Abyssinia – Part II. *Ibis* 13(5): 297-329.

CHEKE, R. A. & WALSH, J. F. 1980. Bird records from the Republic of Togo. *Malimbus* 2: 112-120

CHENG, TSO-HIN 1987. *A synopsis of the avifauna of China.* Beijing: Science Press & Hamburg: Paul Parey.

CHILD, H. 1972. A survey of mixed heronries in the Okavango delta, Botswana. *Ostrich* 43: 60-61.

CHRISTIAN, E. J. 1909. Notes on the Black-tailed Native Hen (*Microtribonyx ventralis*). *Emu* 9: 95-97.

CHUBB, C. 1916. *The birds of British Guiana.* Vol. 1. London: Bernard Quaritch.

CLAFFEY, P. M. 1995. Notes on the avifauna of the Bétérou area, Borgou Province, Republic of Benin. *Malimbus* 17: 63-84.

CLANCEY, P. A. 1964. *The birds of Natal and Zululand.* Edinburgh: Oliver & Boyd.

CLANCEY, P. A. 1971. *A handlist of the birds of southern Moçambique.* Lourenço Marques: Instituto de Investigação Científica de Moçambique.

CLANCEY, P. A. 1980. *S.A.O.S. checklist of southern African birds.* Johannesburg: Southern African Ornithological Society.

CLANCEY, P. A. 1981. On birds from Gough Island, central south Atlantic. *Durban Mus. Novit.* 12: 187-200.

CLANCEY, P. A. 1996. *The birds of southern Mozambique.* KwaZulu-Natal, South Africa: African Bird Book Publishing.

CLAPPERTON, K. 1983. Sexual differences in Pukeko calls. *Notornis* 30: 69-70.

CLAPPERTON, K. & JENKINS, P. F. 1984. Vocal repertoire of the Pukeko (Aves: Rallidae). *N. Z. J. Zool.* 11: 71-84.

CLAPPERTON, K. & JENKINS, P. F. 1987. Individuality in contact calls of the Pukeko (Aves: Rallidae). *N. Z. J. Zool.* 14: 19-28.

CLARK, C. T. 1985. Caribbean Coot? *Birding* 17: 84-86.

CLARK, J. H. 1975. Observations on the Bush-hen at Camp Mountain, south-east Queensland. *Sunbird* 6: 15-21.

CLARK, R. 1986. *Aves de Tierra del Fuego y Cabo de Hornos, guía de campo.* Buenos Aires: Literature of Latin America.

CLARKE, G. 1985. Bird observations from northwest Somalia. *Scopus* 9: 24-42.

CLAY, R. P., JACK, S. R. & VINCENT, J. P. 1994. A survey of the birds and large mammals of the proposed Jatun Sacha Bilsa Biological Reserve, north-western Ecuador. Project Esmeraldas '94 preliminary report.

CLEMENTS, J. F. 1979. Viva Zapata! *Birding* 11: 2-6.

CLOUT, M. N. & CRAIG, J. L. 1995. The conservation of critically endangered flightless birds in New Zealand. *Ibis* 137 (Suppl. 1): S181-S190.

COATES, B. J. 1977. *Birds in Papua New Guinea.* Port Moresby: Robert Brown & Associates.

COATES, B. J. 1985. *The birds of Papua New Guinea.* Vol. 1. Alderley, Queensland: Dove Publications.

COATES, B. J. & BISHOP, K. D. 1997. *A guide to the birds of Wallacea.* Alderley, Queensland: Dove Publications.

CODY, M. L. 1970. Chilean bird distribution. *Ecology* 51: 455-464.

COLEMAN, J. D., WARBURTON, B. & GREEN, W. Q. 1983. Some population statistics and movements of the Western Weka. *Notornis* 30: 93-107.

COLEMAN, R. A. 1978. Coots prosper at Kakahaia refuge. *Elepaio* 38: 130.

COLLAR, N. J. 1987. Rising sun, falling trees. *World Birdwatch* 9(1): 6-7.

COLLAR, N. J. 1993. The conservation status in 1982 of the Aldabra White-throated Rail *Dryolimnas cuvieri aldabranus*. *Bird Conserv. Internatn.* 3: 299-305.

COLLAR, N. J. & ANDREW, P. 1988. *Birds to watch: the ICBP world checklist of threatened birds.* ICBP Tech. Pub. 8. Cambridge, UK: International Council for Bird Preservation.

COLLAR, N. J., CROSBY, M. J. & STATTERSFIELD, A. J. 1994. *Birds to watch 2: the world list of threatened birds.* Cambridge, UK: BirdLife International.

COLLAR, N. J., GONZAGA, L. P., KRABBE, N., MADROÑO NIETO, A., NARANJO, L. G., PARKER, T. A. & WEGE, D. C. 1992. *Threatened birds of the Americas.* Cambridge, UK: International Council for Bird Preservation.

COLLAR, N. J. & STUART, S. N. 1985. *Threatened birds of Africa and related islands.* Cambridge, UK: International Council for Bird Preservation.

COLLINGE, W. E. 1936. The food and feeding habits of the Coot (*Fulica atra* Linn.). *Ibis* (13)6: 35-39.

COLSTON, P. R. & CURRY-LINDAHL, K. 1986. *The birds of Mount Nimba, Liberia.* London: British Museum (Natural History).

CONOVER, H. B. 1934. A new species of rail from Paraguay. *Auk* 51: 365-366.

CONTRERAS, A. 1992. Winter status of the Sora in the Pacific northwest. *Western Birds* 23: 137-142.

CONTRERAS, J. R. 1980. Nota acerca del limite Rionegrino de la distribución de las subespecies de *Rallus sanguinolentus* en la Argentina (Aves, Rallidae). *Historia Natural* 1: 149-152.

CONTRERAS, J. R. 1988. Acerca de la biología reproductiva del burrito silbón, *Laterallus melanophaius* (Aves, Rallidae). *Nót. Faun.* 9: 1-2.

CONTRERAS, J. R., BERRY, L. M., CONTRERAS, A. O., BERTONATTI, C. C. & UTGES, E. E. 1990. *Atlas ornitogeográfico de la provincia del Chaco – Républica Argentina.* Vol. 1. No passeriformes. Cuadernos Técnicos "Felix de Azara".

CONTRERAS, J. R. & CONTRERAS, A. O. 1994. Acerca de *Laterallus exilis* (Temminck, 1831) y de *Calidris bairdii* (Coues, 1861) en la República del Paraguay (Aves: Rallidae, Scolopacidae). *Nót. Faun.* 51: 1-4.

CONWAY, C. J. 1990. Seasonal changes in movements and habitat use by three sympatric species of rails. Laramie: University of Wyoming (MS thesis).

CONWAY, C. J. 1995. Virginia Rail. In POOLE, A. & GILL, F. (eds.). *The birds of North America,* No. 173. Washington, D.C.: American Ornithologists' Union.

CONWAY, C. J. & EDDLEMAN, W. R. 1994. Virginia Rail. In TACHA, T. C. & BRAUN, C. E. (eds.). *Management of migratory shore and upland game birds in North America.* Washington D.C. International Assoc. Fish & Wildlife Agencies.

CONWAY, C. J., EDDLEMAN, W. R. & ANDERSON, S. H. 1994. Nesting success and survival of Virginia Rails and Soras. *Wilson Bull.* 106: 466-473.

CONWAY, C. J., EDDLEMAN, W. R., ANDERSON, S. H. & HANEBURY, L. R. 1993. Seasonal changes in Yuma Clapper Rail vocalization rate and habitat use. *J. Wildl. Manage.* 57: 282-290.

COOMANS DE RUITER, L. 1947a. Over de wederontdekking van *Aramidopsis plateni* in de Minahasa en het voorkomen van *Gymnocrex rosenbergii* aldaar. *Limosa* 19: 65-75.

COOMANS DE RUITER, L. 1947b. Roofvogel waarnemingen in Zuid-Celebes. *Limosa* 20: 213-229.

COOMANS DE RUITER, L. 1951. Vogels van het dal van de Bodjo-rivier (Zuid-Celebes). *Ardea* 39: 261-318.

COOPER, J. 1970. Nest of the African Crake. *Honeyguide* 62: 34.

COOPER, J. & RYAN, P. G. 1994. *Management plan for the Gough Island Wildlife Reserve.* Edinburgh, Tristan da Cunha: Government of Tristan da Cunha.

COOPER, W. J., MISKELLY, C. M., MORRISON, K. & PEACOCK, R. J. 1986. Birds of the Solander Islands. *Notornis* 33: 77-89.

COOPER, W. & MORRISON, K. 1984. Solander Island birds. *Notornis* 31: 182-183.

CORDONNIER, P. 1983. Hand-rearing and growth of the Purple Gallinule *Porphyrio p. porphyrio* at the Park of Villares-les-Dombes, France. *Avicult. Mag.* 89: 205-209.

COSENS, S. E. 1981. Development of vocalizations in the American Coot *Fulica americana. Can. J. Zool.* 59: 1921-1928.

COTT, H. B. & BENSON, C. W. 1970. The palatability of birds, mainly based upon observations of a tasting panel in Zambia. *Ostrich* Suppl. 8: 357-384.

COTTAM, C. & GLAZENER, W. C. 1959. Late nesting of water birds in south Texas. *Trans. N. Amer. Wildl. Conf.* 24: 383-395.

COTTRELL, C. B. 1949. Notes on some nesting habits of the Buff-spotted Pigmy Rail. *Ostrich* 20: 168-170.

COX, J. B. & PEDLER, L. P. 1977. Birds recorded during three visits to the far north-east of South Australia. *S. Aust. Orn.* 27: 231-250.

CRACRAFT, J. 1973. Systematics and evolution of the Gruiformes (Class Aves). 3. Phylogeny of the Suborder Grues. *Bull. Amer. Mus. Nat. Hist.* 151: 1-127.

CRAIG, J. L. 1976. An interterritorial hierarchy: an advantage for a subordinate in a communal territory. *Z. Tierpsychol.* 42: 200-205.

CRAIG, J. L. 1977. The behaviour of the Pukeko, *Porphyrio porphyrio melanotus. N. Z. J. Zool.* 4: 413-433.

CRAIG, J. L. 1979. Habitat variation in the social organization of a communal gallinule, the Pukeko *Porphyrio porphyrio melanotus. Behav. Ecol. Sociobiol.* 5: 331-358.

CRAIG, J. L. 1980a. Pair and group breeding behavior of a communal gallinule, the Pukeko, *Porphyrio porphyrio melanotus. Animal Behav.* 28: 593-603.

CRAIG, J. L. 1980b. Breeding success of a communal gallinule. *Behav. Ecol. Sociobiol.* 6: 289-295.

CRAIG, J. L. 1982. On the evidence for a "pursuit deterrent" function of alarm signals of swamphens. *Amer. Nat.* 119: 753-755.

CRAIG, J. L. 1984. Are communal Pukeko caught in the prisoner's dilemma? *Behav. Ecol. Sociobiol.* 14: 147-150.

CRAIG, J. L. & JAMIESON, I. G. 1985. The relationship between presumed gamete contribution and parental investment in a communal breeder. *Behav. Ecol. Sociobiol.* 17: 207-211.

CRAIG, J. L. & JAMIESON, I. G. 1988. Incestuous mating in a communal bird: a family affair. *Amer. Nat.* 131: 58-70.

CRAIG, J. L. & JAMIESON, I. G. 1990. Pukeko: different approaches and some different answers. In STACEY, P. B. & KOENIG, W. D. (eds.). *Co-operative breeding in birds: long-term studies of ecology and behaviour.* Cambridge, UK: Cambridge University Press.

CRAIG, J. L., MCARDLE, B. H. & WETTIN, P. D. 1980. Sex determination of the Pukeko or Purple Swamphen. *Notornis* 27: 287-291.

CRAMP, S. & SIMMONS, K. E. L. (eds.). 1980. *The birds of the Western Palearctic.* Vol. 2. Oxford: Oxford University Press.

CRAWFORD, D. N. 1972. Birds of Darwin area, with some records from other parts of Northern Territory. *Emu* 72: 131-148.

CRAWFORD, R. D. 1978. Tarsal colour of American Coots in relation to age. *Wilson Bull.* 90: 536-543.

CRAWFORD, R. D. 1980. Effects of age on reproduction in American Coots. *J. Wildl. Manage.* 44: 183-189.

CRAWFORD, R. L., OLSON, S. L. & TAYLOR, W. K. 1983. Winter distribution of subspecies of Clapper Rails (*Rallus longirostris*) in Florida with evidence for long-distance and overland movements. *Auk* 100: 198-200.

CRIADO, J., ABREUS, E., GÓMEZ, T., ARDÁ, M., FIAIO, J. & COTAYO, L. 1995. La Ciénaga de Zapata, maravilla natural cubana. *Garcilla* 93: 32-37.

CRICK, H. Q. P. & MARSHALL, P. J. 1981. The birds of Yankari Game Reserve, Nigeria: their abundance and seasonal occurrence. *Malimbus* 3: 103-114.

CROUCHLEY, D. 1994. *Takahe Recovery Plan.* Wellington: Dept. of Conservation.

CROUS, R. & TEBELE, P. 1995. A Buffspotted Flufftail *Sarothrura elegans* in the central Kalahari. *Babbler* 29: 38-40.

CROUTHER, M. M. 1994. Spotless Crake grounded at Eungella, North Queensland. *Sunbird* 24: 28-29.

CROWLEY, S. K. 1994. Habitat use and population monitoring of secretive waterbirds in Massachusetts. Amherst: University of Massachusetts (MS thesis).

CURRY-LINDAHL. K. 1981. *Bird migration in Africa.* Vols. 1 & 2. London: Academic Press.

CYRUS, D. & ROBSON, N. 1980. *Bird atlas of Natal.* Pietermaritzburg, South Africa: University of Natal Press.

CZECHURA, G. V. 1983. The rails of the Blackall-Conondale Range region with additional comments on Latham's Snipe *Gallinago hardwickii. Sunbird* 13: 31-35.

DAHM, A. G. 1969. A Corn-crake, *Crex crex* L., trapped in Kumasi, Ghana. *Bull. Brit. Orn. Club* 89: 76-78.

DANIELSEN, F., BALETE, D. S., CHRISTENSEN, T. D., HEEGAARD, M., JAKOBSEN, O. F., JENSEN, A., LUND, T. & POULSEN, M. K. 1994. *Conservation of biological diversity in the Sierra Madre mountains of Isabela and southern Cagayan province, the Philippines.* Manila & Copenhagen: Department of Environment & Natural Resources, BirdLife International & Danish Ornithological Society.

DARLINGTON, P. J. 1931. Notes on the birds of the Río Frío (near Santa Marta), Magdalena, Colombia. *Bull. Mus. Comp. Zool.* 71: 349-421.

DAVID-BEAULIEU, A. 1939. Liste complémentaire des oiseaux du Tranninh. *Oiseau et R.F.O.* 9: 183-186.

DAVID-BEAULIEU, A. 1941. Deuxième liste complémentaire des oiseaux du Tranninh. *Oiseau et R.F.O.* 10: 78-97.

DAVIS, L. I. 1972. *A field guide to the birds of Mexico and Central America.* Austin & London: University of Texas Press.

DAY, D. 1989. *Vanished species.* 2nd edn. New York: Gallery Books.

DE HAAN, G. A. L. 1950. Notes on the Invisible Flightless Rail of Halmahera (*Habroptila wallacii* Gray). *Amsterdam Nat.* 1: 57-60.

DEAN, W. R. J. 1976. Breeding records of *Crex egregia, Myrmecocichla nigra* and *Cichladusa ruficauda* from Angola. *Bull. Brit. Orn. Club* 96: 48-49.

DEAN, W. R. J. 1980. Brood division by Redknobbed Coot. *Ostrich* 51: 125-127.

DEAN, W. R. J. & MACDONALD, I. A. W. 1981. A review of African birds feeding in association with mammals. *Ostrich* 52: 135-155.

DEAN, W. R. J. & SKEAD, D. M. 1978. Problems in sexing Red-knobbed Coots. *Safring News* 7: 9-11.

DEAN, W. R. J. & SKEAD, D. M. 1979. Moult and mass of the Redknobbed Coot. *Ostrich* 50: 199-202.

DECEUNINCK, B. 1995. Programme LIFE–Râle des genêts. Bilan des opérations d'études et de conservation menées en France. *Ornithos* 2: 188-189.

DEE, T. J. 1986. *The endemic birds of Madagascar.* Cambridge: International Council for Bird Preservation.

DEGRAAF, R. & RAPPOLE, J. H. 1995. *Neotropical migratory birds.* Ithaca: Cornell University Press.

DEIGNAN, H. G. 1945. *The birds of northern Thailand.* Washington, D.C.: Smithsonian Institution.

DEIGNAN, H. G. 1963. Checklist of the birds of Thailand. *Bull. U.S. Natn. Mus.* 226.

DEL HOYO, J., ELLIOTT, A. & SARGATAL, J. (eds.). 1996. *Handbook of birds of the world.* Vol. 3: Hoatzin to Auks. Barcelona: Lynx Edicions.

DELACOUR, J. 1932. Les oiseaux de la Mission Franco-Anglo-Américaine à Madagascar. *Oiseau et R.F.O.* 2: 1-96.

DELACOUR, J. 1947. *Birds of Malaysia.* New York: Macmillan.

DELACOUR, J. 1966. *Guide des oiseaux de la Nouvelle-Calédonie.* Neuchâtel: Delachaux et Niestlé.

DELACOUR, J. & JABOUILLE, P. 1931. *Les oiseaux de l'Indochine française.* Vol. 1. Paris: Exposition Coloniale Internationale.

DELACOUR, J. & MAYR, E. 1946. *Birds of the Philippines.* New York: Macmillan.

DELAP, E. 1979. Breeding of King Rail in Washington County, Oklahoma. *Bull. Oklahoma Orn. Soc.* 12: 14.

DEMENT'EV, G. P., GLADKOV, N. A. & SPANGENBERG, E. P. 1969. *Birds of the Soviet Union.* (English translation.) Vol. 3. Jerusalem: Israel Program for Scientific Translations.

DEMEY, R. & FISHPOOL, L. D. C. 1991. Additions and annotations to the avifauna of Côte d'Ivoire. *Malimbus* 12: 61-86.

DEMEY, R. & FISHPOOL, L. D. C. 1994. The birds of Yapo Forest, Ivory Coast. *Malimbus* 16: 100-122.

DEMPSEY, J. 1991. A King Rail observation. *Loon* 63: 73-74.

DERRICKSON, S. 1996. Reintroduced Guam Rail breeds on Rota Island. *Endangered Species Update* 13: 15.

DESROSCHERS, B. A. & ANKNEY, C. D. 1986. Effect of brood size and age on the feeding behaviour of adult and juvenile American coots (*Fulica americana*). *Can. J. Zool.* 64: 1400-1406.

DEVITT, O. E. 1939. The Yellow Rail breeding in Ontario. *Auk* 56: 238-243.

DIAMOND, J. M. 1969. Preliminary results of an ornithological exploration of the North Coastal Range, New Guinea. *Amer. Mus. Novit.* 2362.

DIAMOND, J. M. 1972a. Further examples of dual singing by southwest Pacific birds. *Auk* 89: 180-183.

DIAMOND, J. M. 1972b. Avifauna of the eastern highlands of New Guinea. *Publ. Nuttall Orn. Club* 12: 1-438.

DIAMOND, J. M. 1975a. The island dilemma: lessons of modern biogeographical studies for the design of natural reserves. *Biol. Conserv.* 7: 129-146.

DIAMOND, J. M. 1975b. Distributional ecology and habits of some Bougainville birds (Solomon Islands). *Condor* 77: 14-23.

DIAMOND, J. M. 1981. Flightlessness and fear of flying in island species. *Nature* 293: 507-508.

DIAMOND, J. M. 1985. New distributional records and taxa from the outlying mountain ranges of New Guinea. *Emu* 85: 65-91.

DIAMOND, J. M. 1987. Extant unless proven extinct? Or, extinct unless proven extant? *Conserv. Biol.* 1: 77-79.

DIAMOND, J. M. 1991. A new species of rail from the Solomon Islands and convergent evolution of insular flightlessness. *Auk* 108: 461-470.

DIAMOND, J. M. & BISHOP, K. D. 1994. New records and observations from the Aru Islands, New Guinea region. *Emu* 94: 41-45.

DIAMOND, J. M. & TERBORGH, J. W. 1968. Dual singing by New Guinea birds. *Auk* 85: 62-82.

DICKERMAN, R. W. 1966. A new subspecies of the Virginia Rail from Mexico. *Condor* 68: 215-216.

DICKERMAN, R. W. 1968a. Notes on the Red Rail (*Laterallus ruber*). *Wilson Bull.* 80: 94-99.

DICKERMAN, R. W. 1968b. Notes on the Ocellated Rail (*Micropygia schomburgkii*) with first record from Central America. *Bull. Brit. Orn. Club* 88: 25-30.

DICKERMAN, R. W. 1971. Notes on various rails in Mexico. *Wilson Bull.* 83: 49-56.

DICKERMAN, R. W. 1975. Nine new specimen records from Guatemala. *Wilson Bull.* 87: 412-413.

DICKERMAN, R. W. & HAVERSCHMIDT, F. 1971. Further notes on the juvenal plumage of the Spotted Rail (*Rallus maculatus*). *Wilson Bull.* 83: 444-446.

DICKERMAN, R. W. & PARKES, K. C. 1969. Juvenal plumage of the Spotted Rail. *Wilson Bull.* 81: 207-209.

DICKERMAN, R. W. & WARNER, C. W. 1961. Distribution records from Tecolutla, Veracruz, with the first record of *Porzana flaviventer* for Mexico. *Wilson Bull.* 73: 336-340.

DICKEY, D. R. & VAN ROSSEM, A. J. 1938. *The birds of El Salvador.* Chicago: Field Museum of Natural History Press.

DICKINSON, E. C. 1984. Notes on Philippine birds, 1. The status of *Porzana paykullii* in the Philippines. *Bull. Brit. Orn. Club* 104: 71-72.

DICKINSON, E. C., KENNEDY, R. S. & PARKES, K. C. 1991. *The birds of the Philippines.* BOU Checklist No. 12. London: British Ornithologists' Union.

DICKINSON, E. C., KENNEDY, R. S., READ, D. K. & ROZENDAAL, F. G. 1989. Notes on the birds collected in the Philippines during the Steere expedition of 1887-1888. *Nemouria* 32: 1-19.

DISNEY, H. J. de S. 1974a. Woodhen. *Aust. Nat. Hist.* 18: 70-73.

DISNEY, H. J. de S. 1974b. Survey of the Woodhen. Appendix G, pp. 73-76. RECHER, H. F. & CLARK, S. S. (eds.). *Environmental survey of Lord Howe Island.* Sydney: Lord Howe Island Board.

DISNEY, H. J. de S. 1976. Report on the Woodhen of Lord Howe Island. Canberra: Australian National Parks and Wildlife Service, unpublished ms.

DISNEY, H. J. de S. & FULLAGAR, P. J. 1984. Saved from extinction: Lord Howe Island Woodhen. *Aust. Nat. Hist.* 21: 259.

DISNEY, H. J. de S. & SMITHERS, C. N. 1972. The distribution of terrestrial and freshwater birds on Lord Howe Island, in comparison with Norfolk Island. *Aust. Zool.* 17: 1-11.

DOD, A. S. 1978. *Aves de la República Dominicana.* Santo Domingo: Museo Nacional de Historia Natural.

DOD, A. S. 1980. First records of Spotted Rail *Pardirallus maculatus* on the island of Hispaniola. *Auk* 97: 407.

DOD, A. S. 1992. *Endangered and endemic birds of the Dominican Republic.* Fort Bragg, California: Cypress House.

DODMAN, T., DE VAAN, C., HUBERT, E. & NIVET, C. 1997. *African Waterfowl Census, 1997.* Amsterdam: Wetlands International.

DODMAN, T. & TAYLOR, V. 1996. *African waterfowl census 1996.* Wageningen, Netherlands: Wetlands International.

DONAHUE, P. 1994. *Birds of Tambopata. A checklist.* London: Tambopata Reserve Society.

DORST, J. & ROUX, F. 1973. L'avifaune des fôrets de *Podocarpus* de la province de l'Arussie, Ethiopie. *Oiseau et R.F.O.* 43: 269-304.

DOTT, H. E. M. 1984. Range extensions, one new record, and notes on winter breeding birds in Bolivia. *Bull. Brit. Orn. Club* 104: 104-109.

DOUTHWAITE, R. J. 1978. Geese and Red-knobbed Coot on the Kafue Flats in Zambia, 1970-1974. *E. Afr. Wildl. J.* 16: 29-47.

DOWSETT, R. J. & DOWSETT-LEMAIRE, F. 1980. The systematic status of some Zambian birds. *Gerfaut* 70: 151-200.

DOWSETT, R. J. & DOWSETT-LEMAIRE, F. 1989. Liste preliminaire des oiseaux du Congo. *Tauraco Res. Rep.* 2: 29-51.

DOWSETT, R. J. & DOWSETT-LEMAIRE, F. (eds.). 1991. Flore et faune du bassin du Kouilou (Congo) et leur exploitation. *Tauraco Res. Rep.* 4.

DOWSETT, R. J. & DOWSETT-LEMAIRE, F. (eds.). 1993. A contribution to the distribution and taxonomy of Afrotropical and Malagasy birds. *Tauraco Res. Rep.* 5.

DOWSETT, R. J. & FORBES-WATSON, A. 1993. *Checklist of the birds of the Afrotropical and Malagasy regions, Vol. 1. Species limits and distribution.* Liège: Tauraco Press.

DOWSETT-LEMAIRE, F. 1990. Eco-ethology, distribution and status of Nyungwe Forest birds. Pp. 31-85 in DOWSETT, R. J. (ed.) *Enquête faunistique et floristique dans la Forêt de Nyungwe, Rwanda.* Ely, UK: Tauraco Press (*Tauraco Res. Rep.* 3).

DOWSETT-LEMAIRE, F., DOWSETT, R. J. & BULENS, P. 1993. Additions and corrections to the avifauna of Congo. *Malimbus* 15: 68-80.

DRAFFAN, R. D. W., GARNETT, S. T. & MALONE, G. J. 1983. Birds of the Torres Strait: an annotated list and biogeographical analysis. *Emu* 83: 207-234.

DRIVER, E. A. 1988. Diet and behaviour of young American coots. *Wildfowl* 39: 34-42.

DROST, R. 1968. Auf dem Lebenslauf eines Teichhuhns (*Gallinula chloropus*). *Bonn. zool. Beitr.* 19: 346-349.

DUNLOP, R. R. 1970. Behaviour of the Banded Rail, *Rallus philippensis*. *Sunbird* 1: 3-15.

DUNNING, J. B. (ed.) 1993. *CRC handbook of avian body masses.* Boca Raton, Florida: CRC Press.

DUPONT, J. E. 1971. *Philippine birds.* Greenville, Delaware: Delaware Mus. Nat. Hist. Monogr. No. 2.

DURNFORD, H. 1877. Notes on the birds of the province of Buenos Aires. *Ibis* (4)1: 166-203.

DVORAK, M., RANNER, A. & BERG, H.-M. (eds.). 1993. *Atlas der Brutvögel Österreichs.* Wien: Umweltbundesamt.

DYER, P. K. 1992. Other occupants of Wedge-tailed Shearwater burrows. *Sunbird* 22: 38-40.

DWIGHT, J. 1900. The sequence of plumages and moults of the passerine birds of New York. *Annals New York Acad. Sci.* 13: 73-360.

EAMES, J. C. & ERICSON, P. G. P. 1996. The Björkegren expeditions to French Indochina: a collection of birds from Vietnam and Cambodia. *Nat. Hist. Bull. Siam Soc.* 44: 75-111.

EARLÉ, R. A. & GROBLER, N. 1987. *First atlas of bird distribution in the Orange Free State.* Bloemfontein, South Africa: National Museum.

EASON, D. 1992. Takahe, *Notornis mantelli*, artificial incubation of eggs and methods to determine sex. University of Otago, Dunedin (Diploma of Wildlife Management thesis).

EASON, D. & RASCH, G. 1993. Takahe management 1992/93. Wellington: Department of Conservation (internal report).

EASTERLA, D. A. 1962. Some foods of the Yellow Rail in Missouri. *Wilson Bull.* 74: 94-95.

EBER, G. 1961. Vergleichende Untersuchungen am flugfähigen Teichhuhn *Gallinula chl. chloropus* und an der flugunfähigen Inselralle *Gallinula nesiotis*. *Bonn. zool. Beitr.* 12: 247-315.

ECCLES, S. D. 1988. The birds of São Tomé – record of a visit, April 1987, with notes on the rediscovery of Bocage's Longbill. *Malimbus* 10: 207-217.

ECK, S. 1976. Die Vögel der Banggai-Inseln, insbesondere Pelengs. *Zool. Abh. Staatl. Mus. Tierk. Dresden* 34: 53-100.

EDEN, S. F. 1987. When do helpers help? Food availability and helping in the Moorhen, *Gallinula chloropus*. *Behav. Ecol. Sociobiol.* 21: 191-195.

EDDLEMAN, W. R. 1983. A study of migratory American coots, *Fulica americana*, in Oklahoma. Stillwater, Oklahoma: Oklahoma State University (PhD thesis).

EDDLEMAN, W. R. & CONWAY, C. J. 1994. Clapper Rail. In TACHA, T. C. & BRAUN, C. E. (eds.). *Migratory shore and upland game bird management in North America.* Washington, D.C.: International Assoc. Fish & Wildlife Agencies.

EDDLEMAN, W. R. & CONWAY, C. J. 1998. Clapper Rail. In POOLE, A. & GILL, F. (eds.). *The birds of North America*, No. 340. Washington, D.C.: American Ornithologists' Union.

EDDLEMAN, W. R., FLORES, R. E. & LEGARE, M. L. 1994. Black Rail. In POOLE, A. & GILL, F. (eds.). *The birds of North America*, No. 123. Washington, D.C.: American Ornithologists' Union.

EDDLEMAN, W. R. & KNOPF, F. L. 1985. Determining age and sex of American coots. *J. Field Orn.* 56: 41-55.

EDDLEMAN, W. R., KNOPF, F. L. & PATTERSON, C. T. 1985. Chronology of migration by American Coots in Oklahoma. *J. Wildl. Manage.* 49: 241-246.

EDDLEMAN, W. R., KNOPF, F. L., MEANLEY, B., REID, F. A. & ZEMBAL, R. 1988. Conservation of North American rallids. *Wilson Bull.* 100: 458-475.

EDWARDS, E. P. 1989. *A field guide to the birds of Mexico.* Sweet Briar, Virginia: E. P. Edwards.

EHRLICH, P. R., DOBKIN, D. S. & WHEYE, D. 1992. *Birds in jeopardy – the imperiled and extinct birds of the United States and Canada, including Hawaii and Puerto Rico.* Stanford: Stanford University Press.

ELGOOD, J. H., FRY, C. H. & DOWSETT, R. J. 1973. African migrants in Nigeria. *Ibis* 115: 1-45.

ELGOOD, J. H., HEIGHAM, J. B., MOORE, A. M., NASON, A. M., SHARLAND, R. E. & SKINNER, N. J. 1994. *The birds of Nigeria.* BOU Checklist No. 4 (Second Edition). London: British Ornithologists' Union.

ELLIOTT, G. P. 1983. The distribution and habitat requirements of the Banded Rail (*Rallus philippensis*) in Nelson and Marlborough. Wellington: Victoria University (MSc thesis).

ELLIOTT, G. P. 1987. Habitat use by the Banded Rail. *N. Z. J. Ecol.* 10: 109-115.

ELLIOTT, G. P. 1989. The distribution of Banded Rails and Marsh Crakes in coastal Nelson and Marlborough Sounds. *Notornis* 36: 117-123.

ELLIOTT, G., WALKER, K. & BUCKINGHAM, R. 1991. The Auckland Island Rail. *Notornis* 38: 199-209.

ELLIOTT, H. F. I. 1953. The fauna of Tristan da Cunha. *Oryx* 2: 41-53.

ELLIOTT, H. F. I. 1957. A contribution to the ornithology of the Tristan da Cunha group. *Ibis* 99: 545-586.

ELLIOT, R. D. & MORRISON, R. I. G. 1979. The incubation period of the Yellow Rail. *Auk* 96: 422-423.

EMANUEL, V. L. 1980. First documented Panama record of Spotted Rail (*Pardirallus maculatus*). *Amer. Birds* 34: 214-215.

EMISON, W. B., BEARDSELL, C. M., NORMAN, F. I., LOYN, R. H. & BENNETT, S. C. 1987. *Atlas of Victorian birds.* Melbourne: Dept. Forests and Lands and RAOU.

ENGBRING, J. & ENGILIS, A. 1988. Rediscovery of the Sooty Rail (*Porzana tabuensis*) in American Samoa. *Auk* 105: 391.

ENGBRING, J. & PRATT, H. D. 1985. Endangered birds in Micronesia; their history, status and future prospects. *Bird Conserv.* 2: 71-105.

ENGELBACH, P. 1948. Liste complémentaire aux oiseaux du Cambodge (1948). *Oiseau et R.F.O.* 18: 5-26.

ENGILIS, A. & PRATT, T. K. 1993. Status and population trends of Hawaii's native waterbirds, 1977-1987. *Wilson Bull.* 105: 142-158.

ERARD, C. & VIELLIARD, J. 1977. *Sarothrura rufa* au Togo. *Oiseau et R.F.O.* 47: 309-310.

ERLANGER VON, C. 1905. Beiträge zur Vogelfauna Nordostafrikas. *J. Orn.* 53: 42-158.

ESCALANTE, R. 1983. *Catálogo de las aves uruguayas. 3a parte, Galliformes y Gruiformes.* Intendencia Municipal de Montevideo: Museo Dámaso A. Larrañaga.

ESCALANTE, R. 1991. Notas sobre la biología de la reproducción de la Gallareta de Escudete Rojo (*Fulica rufifrons*, Rallidae). *Commun. Zool. Mus. Hist. Nat. Montevideo* 12 (177): 1-15.

ESCOTT, C. J. & HOLMES, D. A. 1980. The avifauna of Sulawesi, Indonesia: faunistic notes and additions. *Bull. Brit. Orn. Club* 100: 189-194.

ESKELL, R. & GARNETT, S. 1979. Notes on the colours of the legs, wings and flanks of the Dusky Moorhen *Gallinula tenebrosa*. *Emu* 79: 143-146.

ESSENBERG, G. 1984. De Braziliaanse dwergral. *Onze Vogels* 45: 8-9.

ETCHÉCOPAR, R. D. & HÜE, F. 1964. *Les oiseaux du Nord de l'Afrique de la Mer Rouge aux Canaries.* Paris: Éditions N. Boubée.

ETCHÉCOPAR, R. D. & HÜE, F. 1978. *Les oiseaux de Chine, de Mongolie et de la Corée. Non Passereaux.* Papeete, Tahiti: les Éditions du Pacifique.

EVANS, M. I. (ed.). 1994. *Important bird areas in the Middle East.* BirdLife Conservation Series 2. Cambridge: BirdLife International.

EVENS, J. & PAGE, G. W. 1986. Predation on Black Rails during high tides in salt marshes. *Condor* 88: 107-109.

EVENS, J. G., PAGE, G. W., LAYMON, S. A. & STALLCUP, R. W. 1991. Distribution, relative abundance and status of the California Black Rail in western North America. *Condor* 93: 952-966.

EVENS, J. G., PAGE, G. W., STENZEL, L. & WARNOCK, N. 1986. Distribution, abundance and habitat of California Black Rails in tidal marshes of Marin and Sonoma counties, California. Point Reyes Bird Observatory, contribution no. 336 unpublished report.

EVERITT, C. 1962. Breeding the Red-legged Water-Rail. *Avicult. Mag.* 68: 179-181.

EWNHS 1996. *Important bird areas of Ethiopia: a first inventory.* Addis Ababa: Ethiopian Wildlife and Natural History Society.

FAGAN, M. J., SCHMITT, M. B. & WHITEHOUSE, P. J. 1976. Moult of Purple Gallinule and Moorhen in the southern Transvaal. *Ostrich* 47: 226-227.

FAIRALL, N. 1981. A study of the bioenergetics of the Red-knobbed Coot *Fulica cristata* on a South African estuarine lake. *S. Afr. J. Wildl. Res.* 11: 1-4.

FAIRON, J. 1971. Exploration ornithologique au Kaouar (hiver 1970). *Gerfaut* 61: 146-161.

FALLA, R. A. 1949. *Notornis* rediscovered. *Emu* 48: 316-322.

FALLA, R. A. 1951. The nesting season of *Notornis*. *Notornis* 4: 97-100.

FALLA, R. A. 1967. An Auckland Island rail. *Notornis* 14: 107-113.

FALLA, R. A., SIBSON, R. B. & TURBOTT, E. G. 1978. *The new guide to the birds of New Zealand and outlying islands.* London: Collins.

FANNUCCHI, W. A., FANNUCCHI, G. T. & NAUMAN, L. E. 1986. Effects of harvesting wild rice, *Zizania aquatica*, on Soras, *Porzana carolina*. *Can. Field Nat.* 100: 533-536.

FARMER, R. 1979. Checklist of birds of the Ile-Ife area, Nigeria. *Malimbus* 1: 56-64.

FAYAD, V. C. & FAYAD, C. C. 1980. *An ecological survey of the Nguruman Forest, Kenya.* Nairobi: privately published.

FEDUCCIA, J. A. 1968. Pliocene rails of North America. *Auk* 85: 441-453.

FEDUCCIA, J. A. 1996. *The origin and evolution of birds.* New Haven: Yale University Press.

FEINDT, P. 1968. *Vier europaïsche Rallenarten.* (Record). Hildesheim.

FERNÁNDEZ-PALACIOS, J. 1994. Crested Coot *Fulica cristata*. Pp. 232-233 in TUCKER, G. M. & HEATH, M. F. 1994. *Birds in Europe: their conservation status.* BirdLife Conservation Series No. 3. Cambridge, UK: BirdLife International.

FERNÁNDEZ-PALACIOS, J. M. & RAYA, C. 1991. Biología de la focha cornuda (*Fulica cristata*) en Cádiz y otros humedales del Bajo Guadalquivir. Pp. 97-117 in FERNÁNDEZ-PALACIOS, J. M. & MARTOS, M. J. (eds.). *Plan rector de uso y gestión de las Reservas Naturales de las Lagunas de Cádiz.* Sevilla: Junta de Andalucía. Agencia de Medio Ambiente.

FFRENCH, R. P. (1973) *A guide to the birds of Trinidad and Tobago.* Wynnewood, Pa.: Livingston.

FFRENCH, R. 1985. Changes in the avifauna of Trinidad. In BUCKLEY, P. A., FOSTER, M. S., MORTON, E. S., RIDGLEY, R. S. & BUCKLEY, F. G. (eds.) *Neotropical ornithology.* Ornithol. Monogr. No. 36. Washington, D. C.: American Ornithologists' Union.

FIELD, G. 1995. Unpublished notes on birds of Sierra Leone. Ms.

FINCH, B. W. 1985. Noteworthy observations in Papua New Guinea and the Solomons. *Papua New Guinea Bird Soc. Newsletter.* 215: 6-10.

FISHER, W. K. 1906. Birds of Laysan and the Leeward Islands. *Bull. U.S. Fish Comm.* 23 (for 1903), pt. 3: 767-807.

FLEGG, J. & MADGE, S. 1995. *Photographic field guide, birds of Australia.* London: New Holland.

FJELDSÅ, J. 1973. Territorial regulation of the progress of breeding in a population of Coots *Fulica atra*. *Dansk. orn. Foren. Tidsskr.* 67: 115-127.

FJELDSÅ, J. 1977. *Guide to the young of European precocial birds.* Tisvildeleje: Skarv.

FJELDSÅ, J. 1981. Biological notes on the Giant Coot *Fulica gigantea*. *Ibis* 123: 423-437.

FJELDSÅ, J. 1982. Biology and systematic relations of the Andean Coot *Fulica americana ardesiaca* (Aves, Rallidae). *Steenstrupia* 8: 1-21.

FJELDSÅ, J. 1983a. Geographic variation in the Andean Coot *Fulica ardesiaca*. *Bull. Brit. Orn. Club* 103: 18-22.

FJELDSÅ, J. 1983b. A Black Rail from Junín, Central Peru: *Laterallus jamaicensis tuerosi* ssp. n. (Aves, Rallidae). *Steenstrupia* 8: 277-282.

FJELDSÅ, J. 1983c. Systematic and biological notes on the Colombian Coot *Fulica americana columbiana* (Aves: Rallidae). *Steenstrupia* 9: 209-215.

FJELDSÅ, J. 1990. Systematic relations of an assembly of allopatric rails from western South America (Aves: Rallidae) *Steenstrupia* 16: 109-116.

FJELDSÅ, J. & KRABBE, N. 1990. *Birds of the high Andes.* [Copenhagen and] Svendborg: Zoological Museum, University of Copenhagen and Apollo Books.

FLADE, M. 1991. Die Habitate des Wachtelkönig *Crex crex*: eine kurze biologische Charakterisierung. *Vogelwelt* 112: 16-40.

FLEISCHER, R. C., FULLER, G. & LEDIG, D. B. 1995. Genetic structure of endangered Clapper Rail (*Rallus longirostris*) pop-ulations in southern California. *Conserv. Biol.* 9: 1234-1243.

FLEMING, A. 1976. Multiple feeding by Dusky Moorhens. *Aust. Bird Watcher* 6: 325-326.

FLEMING, C. A. 1951. *Notornis* in February 1950. Some general reflections on *Notornis*. *Notornis* 4: 101-106.

FLETCHER, J. A. 1909. Bird notes from Cleveland, Tasmania. *Emu* 8: 210-214.
FLETCHER, J. A. 1912. Notes on the Native Hen (*Tribonyx mortierii*). *Emu* 11: 250-252.
FLETCHER, J. A. 1913. Field notes on some Rallinae. *Emu* 13: 45-47.
FLETCHER, J. A. 1914. Field notes on the Spotless Crake (*Porzana immaculata*). *Emu* 13: 197-202.
FLETCHER, J. A. 1916a. Spotless Crake. *Emu* 15: 188.
FLETCHER, J. A. 1916b. Further notes on the Spotless Crake (*Porzana immaculata*). *Emu* 16: 46-48.
FLINT, J. H. 1967. Conservation problems on Tristan da Cunha. *Oryx* 9: 28-32.
FLINT, V. E., BOEHME, R. L., KOSTIN, Y. V. & KUZNETSOV, A. A. 1984. *A field guide to the birds of the USSR.* Princeton, New Jersey: Princeton University Press.
FLORES, R. E. 1991. Ecology of the California Black Rail in southwestern Arizona. Kingston, Rhode Island: University of Rhode Island (MSc thesis).
FLORES, R. E. & EDDLEMAN, W. R. 1991. Ecology of the California Black Rail in southwestern Arizona. Final Rept., U.S. Bur. Reclam., Yuma Proj. Off. & Arizona Dept. Game & Fish. Yuma, Arizona.
FLORES, R. E. & EDDLEMAN, W. R. 1993. Nesting biology of the California Black Rail in southwestern Arizona. *Western Birds* 24: 81-88.
FLORES, R. E. & EDDLEMAN, W. R. 1995. California Black Rail use of habitat in southwestern Arizona. *J. Wildl. Manage.* 59: 357-363.
FORBES, H. O. 1893. A list of the birds inhabiting the Chatham Islands. *Ibis* (6)5: 521-546.
FORBUSH, E. H. 1925. *Birds of Massachusetts and other New England states.* Part 1. Boston: Commonwealth of Massachusetts, Massachusetts Department of Agriculture.
FORD, F. B. C. 1906. Longbreach (Q.) notes. *Emu* 5: 158.
FORD, J. R. 1962. Northern extension for the ranges of the Spotless and Spotted Crakes. *Emu* 62: 61-62.
FORDHAM, R. A. 1983. Seasonal dispersion and activity of the Pukeko *Porphyrio p. melanotus* (Rallidae) in swamp and pasture. *N. Z. J. Ecol.* 6: 133-142.
FORDHAM, R. A. 1985. The mineral content of the faeces of Pukeko *Porphyrio p. melanotus*. *Notornis* 32: 74-79.
FORRESTER, B. C. 1993. *Birding Brazil. A check-list and site guide.* Rankinston, UK: B. C. Forrester.
FORSMAN, D. 1991. Aspects of identification of Crested Coot. *Dutch Birding* 13: 121-125.
FOSBERG, F. R., SACHET, M. H. & STODDART, D. R. 1983. Henderson Island (southeastern Polynesia): summary of current knowledge. *Atoll Res. Bull.* 272.
FOTSO, R. C. 1990. Notes sur les oiseaux d'eau de la région de Yaoundé. *Malimbus* 12: 25-30.
FRANKLIN, A. B., CLARK, D. A. & CLARK, D. B. 1979. Ecology and behaviour of the Galapagos Rail. *Wilson Bull.* 91: 202-221.
FRANKLIN, D. C. & BARNES, T. A. (in press). The downy young and juvenile of the Chestnut Rail, with notes on development. *Corella.*
FRASER, E. 1972. Some notes on the Spotless Crake. *Notornis* 19: 87-88.
FRASER, G. C. & MENDEL, G. J. 1976. The Bush-Hen in New South Wales. *Aust. Birds* 11: 25-27.
FRASER, M. 1989. The Inaccessible Island Rail: smallest flightless bird in the world. *African Wildlife* 43(1): 14-19.
FRASER, M. W., DEAN, W. R. J. & BEST, I. C. 1992. Observations on the Inaccessible Island Rail *Atlantisia rogersi*: the world's smallest flightless bird. *Bull. Brit. Orn. Club* 112: 12-22.
FRASER, M. W. & FRASER, J. G. 1984. Allopreening by Corncrakes. *British Birds.* 77: 567.
FREDRICKSON, L. H. 1970. Breeding biology of American Coots in Iowa. *Wilson Bull.* 82: 445-457.

FREDRICKSON, L. H. 1971. Common Gallinule breeding biology and development. *Auk* 88: 914-919.
FREDRICKSON, L. H. 1977. American coot (*Fulica americana*). In SANDERSON, G. C. (ed.). *Management of migratory shore and upland game birds in North America.* Washington, D.C.: International Assoc. Fish and Wildl. Agencies.
FREETHY, R. 1980. Moorhens' rapid construction of brood nest. *British Birds* 73: 35.
FRIEDMANN, H. 1927. Notes on some Argentina birds. *Bull. Mus. Comp. Zool.* 68(4): 141-236.
FRIEDMANN, H. 1949. The status of the Spotted Rail, *Pardirallus maculatus*, in Chiapas. *Auk* 66: 86-87.
FRIEDMANN, H. & SMITH, F. D. 1950. A contribution to the ornithology of northeastern Venezuela. *Proc. U.S. Natn. Mus.* 100: 411-538.
FRIEDMANN, H. & SMITH, F. D. 1955. A further contribution to the ornithology of northeastern Venezuela. *Proc. U.S. Natn. Mus.* 104: 463-524.
FRIEDMANN, H. & WILLIAMS, J. G. 1969. The birds of the Sango Bay forests, Buddu County, Masaka District, Uganda. *Los Angeles County Mus. Contrib. Sci.* 162: 1-48.
FRITH, C. & FRITH, D. 1988. The Chestnut Forest-Rail, *Rallina rubra* (Rallidae), at Tari Gap, Southern Highlands Province, Papua New Guinea and its vocalizations. *Muruk* 3: 48-50.
FRITH, C. B. 1977. Life history notes on some Aldabra land birds. *Atoll Res. Bull.* 201: 3-5.
FRITH, C. B. & FRITH, D. W. 1990. Nidification of the Chestnut Forest-Rail *Rallina rubra* (Rallidae) in Papua New Guinea and a review of *Rallina* nesting biology. *Emu* 90: 254-259.
FROHAWK, F. W. 1892. Description of a new species of rail from Laysan Island. *Ann. Mag. Nat. Hist.* 9: 247-249.
FRY, C. H. 1966. On the feeding of Allen's Gallinule. *Bull. Nig. Orn. Soc.* 3: 97.
FRYER, J. C. F. 1911. The structure and formation of Aldabra and neighbouring islands – with notes on their flora and fauna. *Trans. Linn. Soc. Lond.*, Ser. 2, Zool. 14 (Percy Sladen Expedition Reports, 3): 397-422.
FULLAGAR, P. J. 1985. The Woodhens of Lord Howe Island. *Avicult. Mag.* 91: 15-30.
FULLAGAR, P. J. & DISNEY, H. J. de S. 1975. The birds of Lord Howe Island: a report on the rare and endangered species. *ICBP Bull.* XII: 187-202.
FULLAGAR, P. J. & DISNEY, H. J. de S. 1981. Discriminant functions for sexing woodhens. *Corella* 5: 106-108.
FULLAGAR, P. J., DISNEY, H. J. de S. & DE NAUROIS, R. 1982. Additional specimens of two rare rails and comments on the genus *Tricholimnas* of New Caledonia and Lord Howe Island. *Emu* 82: 131-136.
FULLER, E. 1987. *Extinct birds.* London: Viking/Rainbird.
GAINES, D. 1988. *Birds of Yosemite and the East Slope.* Lee Vining, Ca: Artemisia Press.
GAJDACS, M. & KEVE, A. 1968. Beiträge zur Vogelfauna des mittleren Äthiopien. *Stuttgarter Beitr. Naturkd.* 182: 1-13.
GALLAGHER, M. D. & ROGERS, T. D. 1980. On some birds of Dhofar and other parts of Oman. *J. Oman Studies, Special Report* 22: 347-385.
GALLAGHER, M. D. & WOODCOCK, M. W. 1980. *The birds of Oman.* London: Quartet.
GARNETT, S. (ed.). 1993. *Threatened and extinct birds of Australia.* 2nd corrected edition. RAOU Report 82. Victoria: Royal Australasian Ornithologists' Union.
GARNETT, S. T. 1978. The behaviour patterns of the Dusky Moorhen *Gallinula tenebrosa* Gould (Aves: Rallidae). *Aust. Wildl. Res.* 5: 363-384.
GARNETT, S. T. 1980. The social organization of the Dusky Moorhen, *Gallinula tenebrosa* Gould (Aves: Rallidae). *Aust. Wildl. Res.* 7: 103-112.

GARRIDO, O. H. 1985. Cuban endangered birds. In BUCKLEY, P. A., FOSTER, M. S., MORTON, E. S., RIDGELY, R. S. & BUCKLEY, F. G. (eds.) *Neotropical ornithology*. Ornithol. Monog. 36. Washington, D.C.: American Ornithologists' Union.

GARRIDO, O. & KIRKCONNELL, A. 1993. *Checklist of Cuban birds*. Published by the authors.

GATTER, W. 1988. The birds of Liberia: a preliminary list with status and open questions. *Verh. orn. Ges. Bayern* 24: 689-723.

GEARY, N. 1922. Spotted Crake in a grass crop. *Emu* 21: 312.

GEE, J. P. 1984. The birds of Mauritania. *Malimbus* 6: 31-66.

GERMAIN, M. & CORNET, J.-P. 1994. Oiseaux nouveaux pour la République Centrafricaine ou dont les notifications de ce pays sont peu nombreuses. *Malimbus* 16: 30-51.

GIBBON, G. 1989. Long-tailed (*sic*) Flufftail at Dzalanyama: first record for Malawi. *Nyala* 14(1): 46-47.

GIBBON, G. 1991. *Southern African bird sounds*. Tape recording (6 cassettes). Durban, South Africa: Southern African Birding.

GIBBONS, D. W. 1986. Brood parasitism and cooperative nesting in the moorhen, *Gallinula chloropus*. *Behav. Ecol. Sociobiol.* 19: 221-232.

GIBBONS, D. W. 1987. Juvenile helping in the Moorhen, *Gallinula chloropus*. *Animal Behav.* 35: 170-181.

GIBBONS, D. W. 1989. Seasonal reproductive success of the Moorhen *Gallinula chloropus*: the importance of male weight. *Ibis* 131: 57-68.

GIBBONS, D. W. 1993. Moorhen *Gallinula chloropus*. In GIBBONS, D. W., REID, J. B. & CHAPMAN, R. A. (eds.). *The new atlas of breeding birds in Britain and Ireland: 1988-1991*. London: Poyser.

GIBBS, D. 1996. Notes on Solomon Island birds. *Bull. Brit. Orn. Club* 116: 18-25.

GIBBS, J. P. & MELVIN, S. M. 1993. Call-response surveys for monitoring breeding waterbirds. *J. Wildl. Manage.* 57: 27-34.

GIBBS, J. P., SHRIVER, W. G. & MELVIN, S. M. 1991. Spring and summer records of the Yellow Rail in Maine. *J. Field Orn.* 62: 509-516.

GIBSON, E. 1920. Further ornithological notes from the neighbourhood of Cape San Antonio, province of Buenos Aires. *Ibis* (11)2: 1-97.

GIBSON, L. 1979. Breeding *Laterallus leucopyrrhus*. *Avicult. Mag.* 85: 63-67.

GIBSON-HILL, C. A. 1949. The birds of the Cocos-Keeling Islands. *Ibis* 91: 221-243.

GIFFORD, E. W. 1913. Expedition of the California Academy of Sciences to the Galápagos Islands, 1905-1906. VIII: The birds of the Galápagos Islands, with observations on the birds of Cocos and Clipperton Islands (Columbiformes to Pelecaniformes). *Proc. Calif. Acad. Sci.* (4) 2: 1-132.

GILBERT, P. A. 1936. Field notes on *Rallus pectoralis*. *Emu* 36: 72-73.

GILL, F. B. 1964. The shield color and relationships of certain Andean coots. *Condor* 66: 209-211.

GILL, H. B. 1970. Birds of Innisfail and hinterland. *Emu* 70: 105-116.

GILL, R. G. 1965. Some observations on the Red-necked Rail. *Emu* 64: 321-322.

GILLIARD, E. T. & LECROY, M. 1961. Birds of the Victor Emanuel and Hindenburg Mountains, New Guinea. Results of the American Museum of Natural History expedition to New Guinea in 1954. *Bull. Amer. Mus. Nat. Hist.* 125: 1-86.

GILLIARD, E. T. & LECROY, M. 1966. Birds of the Middle Sepik Region, New Guinea. *Bull. Amer. Mus. Nat. Hist.* 132: 247-275.

GILLIARD, E. T. & LECROY, M. 1967. Results of the 1958-1959 Gilliard New Britain expedition. *Bull. Amer. Mus. Nat. Hist.* 135: 175-216.

GILMOUR, J. G. 1972. Corncrakes breeding in Stirlingshire. *Scottish Birds* 7: 52.

GIRAUDOUX, P., DEGAUQUIER, R., JONES, P. J., WEIGEL, J. & ISENMANN, P. 1988. Avifaune du Niger: état des connaissances en 1986. *Malimbus* 10: 1-140.

GIVENS, T. V. 1948. Notes on the White-browed Crake. *Emu* 48: 141-147.

GLAHN, J. F. 1974. Study of breeding rails with recorded calls in north-central Colorado. *Wilson Bull.* 86: 206-214.

GLAYRE, D. & MAGNENAT, D. 1977. Nidifications de la Marouette de Baillon et de la Marouette poussin à Chavornay. *Nos Oiseaux* 34: 3-22.

GLENISTER, A. G. 1951. *The birds of the Malay Peninsula, Singapore and Penang*. London: Oxford University Press.

GLUTZ VON BLOTZHEIM, U. N., BAUER, K. M. & BEZZEL, E. 1973. *Handbuch der Vögel Mitteleuropas*. Vol. 5. Frankfurt am Main: Akad. Verlag.

GOCHFELD, M. 1972. Observations on the status, ecology and behaviour of Soras wintering in Trinidad, West Indies. *Wilson Bull.* 84: 200-201.

GOCHFELD, M. 1973. Observations on new or unusual birds from Trinidad, West Indies, and comments on the genus *Plegadis* in Venezuela. *Condor* 75: 474-478.

GODFREY, W. E. 1966. *The birds of Canada*. Ottawa: National Museum of Canada.

GODFREY, W. E. 1986. *The birds of Canada*. Revised Edition. Ottawa: National Museum of Canada.

GOLDIZEN, A. W., GOLDIZEN, A. R. & DEVLIN, T. 1993. Unstable social structure associated with a population crash in the Tasmanian Native Hen *Tribonyx mortierii*. *Animal Behav.* 46:1013-1016.

GONZALES, P. C. & KENNEDY, R. S. 1989. Notes on Philippine birds, 14. Additional records for the island of Palawan. *Bull. Brit. Orn. Club* 109: 126-130.

GOODMAN, S. M. & GONZALES, P. C. 1989. Notes on Philippine birds, 12. Seven species new to Catanduanes Island. *Bull. Brit. Orn. Club* 109: 48-50.

GOODMAN, S. M., MEININGER, P. L., BAHA EL DIN, S. M., HOBBS, J. J. & MULLIÉ, W. C. (eds.) 1989. *The birds of Egypt*. Oxford: Oxford University Press.

GOODMAN, S. M., MEININGER, P. L. & MULLIÉ, W. C. 1986. The birds of the Egyptian Western Desert. *Misc. Pub. Mus. Zool. Univ. Michigan* no. 172.

GORE, M. E. J. 1990. *The birds of the Gambia*. BOU Checklist No. 3. London: British Ornithologists' Union.

GORE, M. E. J. & GEPP, A. R. M. 1978. *Las aves del Uruguay*. Montevideo: Mosca Hermanos.

GORENZEL, W. P., RYDER, R. A. & BRAUN, C. E. 1981a. American coot distribution and migration in Colorado. *Wilson Bull.* 93: 115-118.

GORENZEL, W. P., RYDER, R. A. & BRAUN, C. E. 1981b. American Coot response to habitat changes in a Colorado marsh. *Southwestern Naturalist* 26: 59-65.

GORENZEL, W. P., RYDER, R. A. & BRAUN, C. E. 1982. Reproduction and nest site characteristics of American Coots at different altitudes in Colorado. *Condor* 84: 59-65.

GOSSE, P. H. 1847. *The birds of Jamaica*. London: John van Voorst.

GRAHAM, G. L., GRAVES, G. R., SCHULENBERG, T. S. & O'NEILL, J. P. 1980. Seventeen bird species new to Peru from the Pampas de Heath. *Auk* 97: 366-370.

GRANDIDIER, G. & BERLIOZ, J. 1929. Description d'une espèce nouvelle de Madagascar de la famille des Rallidés. *Bull. Acad. Malagache* n. s. 10: 83-85.

GRAVES, G. R. 1982. First record of Brown Wood Rail (*Aramides wolfi*) for Peru. *Gerfaut* 72: 237-238

GRAVES, G. R. 1992. The endemic land birds of Henderson Island, southeastern Polynesia; notes on natural history and conservation. *Wilson Bull.* 104: 32-43.

GREEN, A. A. 1983. The birds of Bamingui-Bangoran National Park, Central African Republic. *Malimbus* 5: 17-30.

GREEN, A. A. 1984. Additional bird records from Bamingui-Bangoran National Park, Central African Republic. *Malimbus* 6: 70-72.

GREEN, A. A. 1990. The avifauna of the southern sector of the Gashaka-Gumti Game Reserve, Nigeria. *Malimbus* 12: 31-51.

GREEN, A. A. & CARROLL, R. W. 1991. The avifauna of Dzanga-Ndoki National Park and Dzanga-Sangha Rainforest Reserve, Central African Republic. *Malimbus* 13: 49-66

GREEN, A. A. & RODEWALD, P. G. 1996. New bird records from Korup National Park and environs, Cameroon. *Malimbus* 18: 122-133.

GREEN, R. E. & RAYMENT, M. D. 1996. Geographical variation in the abundance of the Corncrake *Crex crex* in Europe in relation to the intensity of agriculture. *Bird Conserv. Internatn.* 6: 201-211.

GREEN, R. E. & STOWE, T. J. 1993. The decline of the Corncrake *Crex crex* in Britain and Ireland in relation to habitat change. *J. Appl. Ecol.* 30: 689-695.

GREEN, R. E. & WILLIAMS, G. 1994. The ecology of the Corncrake *Crex crex* and action for its conservation in Britain and Ireland. Pp. 69-74 in BIGNAL, E. M., MCCRACKEN, D. I. & CURTIS, D. J. (eds.). *Nature conservation and pastoralism in Europe*. Peterborough, UK: JNCC.

GREEN, R. H. 1963. A forgotten record. *Emu* 62: 240.

GREEN, R. H. 1965. Impact of rabbits on predators and prey. *Tasmanian Naturalist* 3: 1-2.

GREEN, R. H. 1989. *The birds of Tasmania*. Launceston: Potaroo Press.

GREENLAW, J. S. & MILLER, R. F. 1982. Breeding Soras on a Long Island salt marsh. *Kingbird* 32: 78-84.

GREENLAW, J. S. & MILLER, R. F. 1983. Calculating incubation periods of species that sometimes neglect their last eggs: the case of the Sora. *Wilson Bull.* 95: 459-461.

GREENWAY, J. C. 1952. *Tricholimnas conditicius* is probably a synonym of *Tricholimnas sylvestris*. *Breviora* 5: 1-4.

GREENWAY, J. C. 1967. *Extinct and vanishing birds of the world*. New York: Dover Publications.

GREENWAY, J. C. 1973. Type specimens of birds in the American Museum of Natural History. Part 1. Tinamidae – Rallidae. *Bull. Amer. Mus. Nat. Hist.* 150: 207-346.

GREGORY, P. 1995. *The birds of the Ok Tedi area*. Tabubil, Papua New Guinea: National Library of Papua New Guinea.

GREGORY, P. 1996. The New Guinea Flightless Rail (*Megacrex inepta*) in Gulf Province. *Muruk* 8: 38-39.

GREGORY, P., HALSE, S. A., JAENSCH, R. P., KAY, W. R., KULMOI, P., PEARSON, G. B. & STOREY, A. W. 1996. The middle Fly waterbird survey 1994-95. *Muruk* 8: 1-7.

GREIJ, E. D. 1994. Common Moorhen. In TACHA, T. C. & BRAUN, C. E. (eds.). *Management of migratory shore and upland game birds in North America*. Washington D.C. International Assoc. Fish & Wildlife Agencies.

GRIESE, H. J., RYDER, R. A. & BRAUN, C. E. 1980. Spatial and temporal distribution of rails in Colorado. *Wilson Bull.* 92: 96-102

GRIMES, L. G. 1987. *The birds of Ghana*. BOU Checklist No. 9. London: British Ornithologists' Union.

GRIMMETT, R. F. A. & JONES, T. A. 1989. *Important bird areas in Europe*. Cambridge, UK: International Council for Bird Preservation (Tech. Publ. 9).

GRONOW, R. W. 1969. Notes on breeding the Rufous-sided Crake (*Laterallus melanophaius*). *Avicult. Mag.* 75:100.

GROSS, A. O & VAN TYNE, J. 1929. The Purple Gallinule (*Ionornis martinicus*) of Barro Colorado Island, Canal Zone. *Auk* 46: 431-446.

GUICHARD, K. M. 1948. Notes on *Sarothrura ayresi* and three birds new to Abyssinia. *Bull. Brit. Orn. Club* 68: 102-104.

GUICHARD, K. M. 1950. A summary of the birds of the Addis Ababa region, Ethiopia. *J. E. Afr. Nat. Hist. Soc.* 19(5): 154-178.

GULLION, G. W. 1950. Voice differences between sexes in the American coot. *Condor* 52: 272-273.

GULLION, G. W. 1951. The frontal shield of the American Coot. *Wilson Bull.* 63: 157-166.

GULLION, G. W. 1952. The displays and calls of the American coot. *Wilson Bull.* 64: 83-97.

GULLION, G. W. 1953a. Observation on molting of the American coot. *Condor* 55: 102-103.

GULLION, G. W. 1953b. Territorial behaviour in the American coot. *Condor* 55: 169-186.

GULLION, G. W. 1954. The reproductive cycle of American coots in California. *Auk* 71: 366-412.

GUTHRIE-SMITH, H. 1910. *Birds of wood, water and waste*. Wellington: Whitcombe & Tombs.

GUTHRIE-SMITH, H. 1914. *Mutton birds and other birds*. Christchurch: Whitcombe & Tombs.

GUTHRIE-SMITH, H. 1925. *Bird life on island and shore*. Edinburgh: Blackwood.

GWIADZA, R. 1992. [A new record of the Baillon's Crake (*Porzana pusilla*) in Poland.] *Notatki Ornitol.* 33: 164-165.

GYLDENSTOLPE, N. 1945. The fauna of Rio Jurua in western Brazil. *Kung. Svenska Vet. Handl.* 22: 1-338.

GYLDENSTOLPE, N. 1955. Notes on a collection of birds made in the Western Highlands, central New Guinea, 1951. *Ark. Zool.*, (2)8: 1-181.

HAAGNER, G. V. & REYNOLDS, D. S. 1988. Notes on the nesting of the African Crake at Manyeleti Game Reserve, eastern Transvaal. *Ostrich* 59: 45.

HADDEN, F. C. 1941. Midway Islands. *Hawaiian Planters' Record* 45: 179-221.

HADDEN, D. 1970. Notes on the Spotless Crake in the Waingaro District. *Notornis* 17: 200-213.

HADDEN, D, 1972. Further notes on the Spotless Crake. *Notornis* 19: 323-329.

HADDEN, D. 1981. *Birds of the North Solomons*. Wau, Papua New Guinea: Wau Ecological Institute (Handbook 8).

HADDEN, D. 1993. First Spotless Crake's nest for the South Island. *Notornis* 40: 231-232.

HAGEMEIJER, E. J. M. & BLAIR, M. J. 1997. *The EBC atlas of European breeding birds: their distribution and abundance*. London: T & A.D. Poyser.

HAGEN, Y. 1952. *Birds of Tristan da Cunha*. Oslo: Det Norske Videnskaps-Akademi.

HAIG, S. M., BALLOU, J. D. & DERRICKSON, S. R. 1990. Management options for preserving genetic diversity: reintroduction of Guam Rails to the wild. *Conserv. Biol.* 4: 290-300.

HAIG, S. M., BALLOU, J. D. & DERRICKSON, S. R. 1993. Genetic considerations for the Guam Rail. *Re-introduction News* 7: 11-12.

HALLEUX, D. 1994. Annotated bird list of Macenta Prefecture, Guinea. *Malimbus* 16: 10-29.

HALSE, S. A. & JAENSCH, R. P. 1989. Breeding seasons of waterbirds in south-western Australia – the importance of rainfall. *Emu* 89: 232-249.

HAMBLER, C., NEWING, J. & HAMBLER, K. 1993. Population monitoring for the flightless rail *Dryolimnas cuvieri aldabranus*. *Bird Conserv. Internatn.* 3: 307-318.

HAMLING, H. H. 1949. King Reed-Hen or Purple Gallinule. *Ostrich* 20: 91-94.

HANAWA, S. & MORISHITA, E. 1986. The Okinawa Rail's distribution and estimated number of individuals. *Survey on special birds for protection*. Environment Agency. (In Japanese).

HANCOCK, J. 1985. Moorhens eating apples. *British Birds* 78: 453.

HANNECART, F. 1988. Les oiseaux menacés de la Nouvelle Calédonie et des îles proches. Pp. in THIBAULT, J.-C. & GUYOT, I. (eds.). *Livre rouge des oiseaux menacés des régions françaises d'Outre-mer*. Saint-Cloud: Conseil International pour la Protection des Oiseaux. Monogr. 5.

HANNECART, F. & LETOCART, Y. 1980-1983. *Oiseaux de Nouvelle Calédonie et des Loyautés*. Vols I & II. Nouméa, Nouvelle Calédonie: Editions Cardinalis.

HARATO, T. & OZAKI, K. 1993. Roosting behaviour of the Okinawa Rail. *J. Yamashina Inst. Orn.* 25: 40-53.

HARDEN, R. H. & ROBERTSHAW, J. D. 1987. Lord Howe Island Woodhen Census 1986. Sydney: New South Wales National Parks and Wildlife Service. Unpubl. Report.

HARDEN, R. H. & ROBERTSHAW, J. D. 1988. Lord Howe Island Woodhen Census 1987. Sydney: New South Wales National Parks and Wildlife Service. Unpubl. Report.

HARDY, J. W. & DICKERMAN, R. W. 1965. Relationships between two forms of the Red-winged Blackbird in Mexico. *Living Bird* 4: 107-129.

HARDY, J. W., PARKER, T. A., REYNARD, G. B. & TAYLOR, T. 1996. *Voices of the New World rails*. Gainesville, Florida: ARA Records.

HARRIS, M. P. 1973. The Galápagos avifauna. *Condor* 75: 265-278.

HARRIS, M. P. 1974. *A field guide to the birds of Galápagos*. New York: Taplinger.

HARRIS, M. P. 1981. The waterbirds of Lake Junin, central Peru. *Wildfowl* 32: 137-145.

HARRIS, M. P. 1982. *A field guide to the birds of Galapagos*. London: Collins.

HARRISON, C. 1978. *A field guide to the nests, eggs and nestlings of North American birds*. New York: Collins.

HARRISON, C. J. O. 1970. Helpers at the nest in the Purple Gallinule (*Porphyrio porphyrio*). *Avicult. Mag.* 76: 2-4.

HARRISON, C. J. O. 1975. The Australian subspecies of Lewin's Rail. *Emu* 75: 39-40.

HARRISON, C. J. O. & PARKER, S. A. 1967. The eggs of Woodford's Rail, Rouget's Rail and the Malayan Banded Crake. *Bull. Brit. Orn. Club* 87: 14-16.

HARRISON, H. H. 1975. *A field guide to birds' nests*. Boston: Houghton Mifflin.

HARTERT, E. 1924a. Notes on some birds from Buru. *Novit. Zool.* 31: 104-111.

HARTERT, E. 1924b. Birds of St. Matthias Island. *Novit. Zool.* 31: 261-275.

HARTERT, E. 1927. Types of birds in the Tring Museum. B. Types in the general collection. VIII Columbae-Struthionidae. *Novit. Zool.* 34: 1-38.

HARTERT, E. & VENTURI, S. 1909. Notes sur les oiseaux de la République Argentine. *Novit. Zool.* 16: 159-267.

HARTLAUB, G. 1893. Vier seltene Rallen. *Abh. naturw. Ver. Bremen* 12: 389-402.

HARTY, S. T. 1964. The discovery of the Black Rail (*Laterallus jamaicensis*) in Panama and the first breeding record. *Cassinia* 48: 19-20.

HARVEY, T. E. 1988. Breeding biology of the California Clapper Rail in south San Francisco Bay. *Trans. West. Sect. Wildl. Soc.* 24: 98-104.

HARVEY, W. G. 1990. *Birds in Bangladesh*. Dhaka: University Press Ltd.

HASHMI, D. 1991. Bestand und Verbreitung des Wachtelkönigs in der Bundesrepublik Deutschland vor 1990. *Vogelwelt* 112: 66-71.

HAVERSCHMIDT, F. 1968. *Birds of Surinam*. London: Oliver & Boyd.

HAVERSCHMIDT, F. 1974. Notes on the Grey-breasted Crake *Laterallus exilis*. *Bull. Brit. Orn. Club* 94: 2-3.

HAVERSCHMIDT, F. & MEES, G. F. 1995. *Birds of Suriname*. Paramaribo: Vaco.

HAWKINS, A. F. 1970. Incubating Moorhen repeatedly pulling cover over itself in rain. *British Birds* 63: 33-34.

HAY, R. 1986. *Bird conservation in the Pacific islands*. ICBP Study Report 7. Cambridge: International Council for Bird Preservation.

HAYES, F. E. 1995. *Status, distribution and biogeography of the birds of Paraguay*. Monographs in Field Ornithology 1. American Birding Association.

HAYES, F. E., SCHARF, P. A. & RIDGELY, R. S. 1994. Austral bird migrants in Paraguay. *Condor* 96: 83-97.

HEATHER, B. D. & ROBERTSON, H. A. 1997. *The field guide to the birds of New Zealand*. Oxford: Oxford University Press.

VAN HEERDEN, J. 1972. Observations on a coot. *Bokmakierie* 24: 88-89.

HEIM DE BALSAC, H. & MAYAUD, N. 1962. *Les oiseaux du nord-est de l'Afrique*. Paris: Editions Paul Lechevalier.

HEIMSATH, S. F., LÓPEZ, J., CUETO, V. R. & CITTADINO, E. A. 1993. Uso de habitat en *Fulica armillata*, *F. leucoptera* y *Gallinula chloropus* durante la primavera. *Hornero* 13: 286-289.

HEINRICH, G. 1932. *Der Vögel Schnarch: zwei Jahre Rallenfang und Urwaldforschung in Celebes*. Berlin: D. Reimer.

HEINRICH, G. 1956. Biologische Aufzeichnungen über Vögel von Halmahera und Batjan. *J. Orn.* 97: 31-40.

HEINROTH, O. & HEINROTH, M. 1928. *Die Vögel Mitteleuropas*. Vol. 3. Berlin: Hugo Bermuhler.

HEITMEYER, M. E. 1980. Characteristics of wetland habitats and waterfowl populations in Oklahoma. Stillwater: Oklahoma State University (MS thesis).

HELLEBREKERS, W. P. J. & HOOGERWERF, A. 1967. A further contribution to our oological knowledge of the island of Java (Indonesia). *Zool. Verhand.* 88: 1-164.

HELLMAYR, C. E. 1929. A contribution to the ornithology of north-eastern Brazil. *Chicago Field Mus. Nat. Hist. Ser.* 12: 235-501.

HELLMAYR, C. E. & CONOVER, B. 1942. Catalog of birds of the Americas and adjacent islands. Part 1, No. 1. *Chicago Field Mus. Nat. Hist., Zool. Ser.* 13: 1-636.

HELM, R. N. 1982. Chronological nesting study of common and purple gallinules in the marshlands and rice fields of southwest Louisiana. Baton Rouge: Louisiana State University (MS thesis).

HELM, R. N. 1994. Purple Gallinule. In TACHA, T. C. & BRAUN, C. E. (eds.). *Management of migratory shore and upland game birds in North America*. Washington D.C.: International Assoc. Fish & Wildlife Agencies.

HELM, R. N., PASHLEY, D. N. & ZWANK, P. J. 1987. Notes on the nesting of the Common Moorhen and Purple Gallinule in southwestern Louisiana. *J. Field Orn.* 58: 55-61.

HENRY, G. M. 1978. *A guide to the birds of Ceylon*. Kandy: De Silva.

HERKLOTS, G. A. C. 1961. *The birds of Trinidad and Tobago*. London: Collins.

HILL, W. L. 1986. Clutch overlap in American Coots. *Condor* 88: 96-97.

HILL, W. L. 1988. The effect of food abundance on the reproductive patterns of coots. *Condor* 90: 324-331.

HILTY, S. L. 1985. Distributional changes in the Colombian avifauna: a preliminary blue list. In BUCKLEY, P. A., FOSTER, M. S., MORTON, E. S., RIDGELY, R. S. & BUCKLEY, F. G. (eds.) *Neotropical ornithology*. Ornithol. Monogr. No. 36. Washington, D. C.: American Ornithologists' Union.

HILTY, S. L. & BROWN, W. L. 1986. *A guide to the birds of Colombia*. Princeton, New Jersey: Princeton University Press.

HINES, C. J. H. 1993. Temporary wetlands of Bushmanland and Kavango, northeast Namibia. *Madoqua* 18: 57-69.

HINDWOOD, K. A. 1932. An historic diary. *Emu* 32: 17-29.

HINDWOOD, K. A. 1940. Birds of Lord Howe Island. *Emu* 40: 1-86.

HINDWOOD, K. A. 1953. River pollution and birds. *Emu* 53: 90-91.

HINDWOOD, K. A. 1965. John Hunter, a naturalist and artist of the first fleet. *Emu* 65: 83-96.

HOBBS, J. N. 1961. The birds of south-west New South Wales. *Emu* 61: 21-55.

HOBBS, J. N. 1967. Distraction display by two species of crakes. *Emu* 66: 299-300.

HOBBS, J. N. 1973. Reactions to predators by Black-tailed Native Hen. *Aust. Bird Watcher* 5: 29-30.

HOCKEY, P. A. R., UNDERHILL, L. G. & NEATHERWAY, M. 1989. *Atlas of the birds of the southwestern Cape.* Cape Town: Cape Bird Club.

HOESCH, W. & NIETHAMMER, G. 1940. *Die Vogelwelt Deutsch-Südwestafrikas.* Deutschen Ornithologischen Gesellschaft, Sonderheft, Berlin.

HOFF, J. G. 1975. Clapper Rail feeding on water snake. *Wilson Bull.* 87: 112.

HOGG, P., DARE, P. J. & RINTOUL, J. V. 1984. Palaearctic migrants in the central Sudan. *Ibis* 126: 307-331.

HOLDGATE, M. W. 1957. Gough Island – a possible sanctuary. *Oryx* 4: 168-175.

HOLDGATE, M. W. & WACE, N. M. 1961. The influence of man on the floras of southern islands. *Polar Record* 10: 475-493.

HOLLIS, E. & BEDDING, J. 1994. Can we stop the wetlands from drying up? *New Scientist* 143 (1932): 30-35.

HOLLOM, P. A. D., PORTER, R. F., CHRISTENSEN, S. & WILLIS, I. 1988. *Birds of the Middle East and North Africa.* Calton, UK: Poyser.

HOLMES, D. A. & VAN BALEN, S. 1996. The birds of Tinjil and Deli Islands, West Java. *Kukila* 8: 117-126

HOLMES, D. A. & NASH, S. 1991. *The birds of Java and Bali.* Singapore: Oxford University Press.

HOLYOAK, D. T. 1970. The behaviour of captive Purple Gallinules. *Avicult. Mag.* 76: 98-109.

HOLYOAK, D. T. 1979. Notes on the birds of Viti Levu and Taveuni, Fiji. *Emu* 79: 7-18.

HOLYOAK, D. T. & SAGER, D. 1970. Observations on captive Tasmanian Native Hens and their interactions with wild Moorhens. *Avicult. Mag.* 76: 56-57.

HOLYOAK, D. T. & THIBAULT, J. C. 1984. *Contribution à l'étude des oiseaux de Polynésie Orientale.* Memoirs du Muséum National D'Histoire Naturelle Série A. Zoologie 127. Paris: Éditions du Muséum d' Histoire Naturelle.

HON, T., ODUM, R. R. & BELCHER, D. P. 1977. Results of Georgia's Clapper Rail banding program. *Proc. Ann. Conf. S. E. Assoc. Fish Wildl. Agencies* 31: 72-76.

HONEGGER, R. E. 1966. Ornithologische Beobachtungen von das Seychellen. *Natur und Museum* 96: 481-490.

HOOGERWERF, A. 1949. Bijdrage tot de oölogie van Java. *Limosa* 22: 1-279.

HOOGERWERF, A. 1964. On birds new for New Guinea or with a larger range than previously known. *Bull. Brit. Orn. Club* 84: 70-77.

HOPKINSON, G. & MASTERSON, A. N. B. 1975. Notes on the Striped Crake. *Honeyguide* 84: 12-21.

HOPKINSON, G. & MASTERSON, A. N. B. 1977. On the occurrence near Salisbury of the White-winged Flufftail. *Honeyguide* 91: 25-28.

HOPKINSON, G. & MASTERSON, A. N. B. 1984. The occurrence and ecological preferences of certain Rallidae near Salisbury, Zimbabwe. *Proc. V Pan-Afr. Orn. Congr.* Johannesburg: Southern African Ornithological Society.

HORAK, G. J. 1970. A comparative study of the foods of the Sora and Virginia Rail. *Wilson Bull.* 82: 206-213.

HORNBUCKLE, J. 1981. Some aspects of the breeding biology of the Moorhen. *Magpie* 2: 45-56.

HORNBUCKLE, J. 1994. *Birdwatching in the Philippines.* Privately published.

HORSFALL, J. A. 1984a. Brood reduction and brood division in Coots. *Animal Behav.* 32: 216-225.

HORSFALL, J. A. 1984b. The "dawn chorus" and incubation in the coot (*Fulica atra* L.). *Behav. Ecol. Sociobiol.* 15: 69-71.

HORSFALL, J. A. 1984c. Food supply and egg mass variation in the European Coot. *Ecology* 65: 89-95.

HORSFALL, J. A. 1985. Cranes and rails. Pp. 142-151. in PERRINS, C. M. & MIDDLETON, A. L. A. (eds.). *The encyclopaedia of birds,* London: Allen & Unwin.

HORTON, W. 1975. The birds of Mt. Isa. *Sunbird* 6: 49-69.

HOVEL, H. 1987. *Check-list of the birds of Israel.* Tel Aviv: Society for the Protection of Nature in Israel.

HOWARD, H. E. 1940. *A waterhen's world.* Cambridge: Cambridge University Press.

HOWARD, R. & MOORE, A. 1991. *A complete checklist of the birds of the world.* 2nd. edn. London: Academic Press.

HOWELL, L. 1986. Classified summarised notes, 1 July 1984 to 30 June 1985, North Island. *Notornis* 33: 95-119.

HOWELL, L. 1987. Classified summarised notes, 1 July 1985 to 30 June 1986, North Island. *Notornis* 34: 117-147.

HOWELL, S. N. G., DOWELL, B. A., JAMES, D. A., BEHRSTOCK, R. A. & ROBBINS, C. S. 1992. New and noteworthy bird records from Belize. *Bull. Brit. Orn. Club* 112: 235-244.

HOWELL, S. N. G. & WEBB, S. 1995a. Noteworthy bird observations from Chile. *Bull. Brit. Orn. Club* 115: 57-66.

HOWELL, S. N. G. & WEBB, S. 1995b. *A guide to the birds of Mexico and northern Central America.* Oxford: Oxford University Press.

HUDSON, A. V., STOWE, T. J. & ASPINALL, S. J. 1990. The status and distribution of Corncrakes in Britain, 1988. *British Birds* 83: 173-186.

HUDSON, R. 1974. Allen's Gallinule in Britain and the Palaearctic. *British Birds* 67: 405-413.

HUDSON, W. H. 1920. *Birds of La Plata.* Vol. 2. New York: Dutton.

HÜE, F. & ETCHÉCOPAR, R. D. 1970. *Les oiseaux du Proche et du Moyen Orient de la Méditerranée aux contreforts de l'Himalaya.* Paris: Éditions N. Boubée.

HUGHES, R. A. 1980. Additional puna zone bird species on the coast of Peru. *Condor* 82: 475.

HUMPHREY, P. S., BRIDGE, D., REYNOLDS, P. D. & PETERSON, R. T. 1970. *Birds of Isla Grande, Tierra del Fuego.* Washington, D.C.: Smithsonian Institution.

HUMPHREY, P. S. & PARKES, K. C. 1959. An approach to the study of molts and plumages. *Auk* 76: 1-31.

HUNTER, L. A. 1985. The effects of helpers in cooperatively breeding purple gallinules. *Behav. Ecol. Sociobiol.* 18: 147-153.

HUNTER, L. A. 1987a. Acquisition of territories by floaters in cooperatively breeding purple gallinules. *Animal Behav.* 35: 402-410.

HUNTER, L. A. 1987b. Cooperative breeding in purple gallinules: the role of helpers in feeding chicks. *Behav. Ecol. Sociobiol.* 20: 171-177.

HURTER, H.-U. 1972. Nahrung und Ernährungsweise des Blässhuhns *Fulica atra* am Sempachersee. *Orn. Beob.* 69: 125-149.

HUTTON, F. W. 1874. On a new genus of Rallidae. *Trans. and Proc. New Zealand Inst.* 6: 108-110.

HUTTON, I. 1991. *Birds of Lord Howe Island.* Coffs Harbour, Australia: privately published.

HUXLEY, C. R. 1976. Gonad weight and food supply in captive Moorhens *Gallinula chloropus. Ibis* 118: 411-413.

HUXLEY, C. R. 1979. The tortoise and the rail. *Phil. Trans. R. Soc. Lond B* 286: 225-238.

HUXLEY, C. R. & WILKINSON, R. 1977. Vocalizations of the Aldabra White-throated Rail *Dryolimnas cuvieri aldabranus. Proc. R. Soc. Lond. B.* 197: 315-331.

HUXLEY, C. R. & WILKINSON, R. 1979. Duetting and vocal recognition by Aldabra White-throated Rails *Dryolimnas cuvieri aldabranus. Ibis* 121: 265-273.

HUXLEY, C. R. & WOOD, N. A. 1976. Aspects of the breeding of the moorhen in Britain. *Bird Study* 23: 1-10.

IKENAGA, H. 1983. Appearance time and behaviour of the Okinawa Rail *Rallus okinawae* at a water site in the late afternoon. *Strix* 2:11.

IKENAGA, H. & GIMA, T. 1993. Vocal repertoire and duetting in the Okinawa Rail *Rallus okinawae*. *J. Yamashina Inst. Orn.* 25: 28-39.

VAN IMPE, J. & LIECKENS, H. 1993. [Aspects of the breeding biology of the Coot *Fulica atra* in an uncommon habitat.] *Oriolus* 59: 3-13.

INGOLD, N. 1930. Weiters vom Wachtelkönig. *Orn. Beob.* 28: 23-24.

INSKIPP, C. & INSKIPP, T. 1991. *A guide to the birds of Nepal.* London: Croom Helm.

INSKIPP, T., LINDSEY, N. & DUCKWORTH, W. 1996. *An annotated checklist of the birds of the Oriental Region.* Sandy, UK: Oriental Bird Club.

INSKIPP. T. P. & ROUND, P. D. 1989. A review of the Black-tailed Crake *Porzana bicolor*. *Forktail* 5: 3-15.

IRISH, J. 1974. Postbreeding territorial behaviour of Soras and Virginia Rails in several Michigan marshes. *Jack-Pine Warbler* 52: 115-124.

IRWIN, M. P. S. 1981. *The birds of Zimbabwe.* Harare, Zimbabwe: Quest Publishing.

JACKSON, F. J. & SCLATER, W. L. 1938. *The birds of Kenya Colony and the Uganda Protectorate.* Vol. 1. London: Gurney & Jackson.

JACKSON, H. D. 1989. Weights of birds collected in the Mutare Municipal area, Zimbabwe. *Bull. Brit. Orn. Club* 109: 100-106.

JACKSON, R. & LYALL, H. 1964. An account of the establishment of the Australian Coot in the Rotorua District with some notes on its nesting habits. *Notornis* 11: 82-86.

JAENSCH, R. P. 1989. *Birds of wetlands and grasslands in the Kimberley Division, Western Australia: some records of interest, 1981-88.* RAOU Report 61. Victoria: Royal Australasian Ornithologists' Union.

JAENSCH, R. P. 1994. *An inventory of wetlands of the sub-humid tropics of the Northern Territory.* Palmerston, NT: Conservation Commission of the Northern Territory.

JAENSCH, R. P. & BELLCHAMBERS, K. 1997. *Waterbird conservation values of ephemeral wetlands of the Barkly Tablelands, Northern Territory.* Australian Heritage Commission and Parks & Wildlife Commission of the Northern Territory.

JAENSCH, R. P., VERVEST, R. M. & HEWISH, M. J. 1988. Waterbirds in nature reserves of south-western Australia 1981-1985: Reserve Accounts. RAOU Report 30. Victoria: Royal Australasian Ornithologists' Union.

JAFFE, M. 1997. *And no birds sing.* New York: Barricade Books.

JAMES, D. A. 1987. A Yellow Rail (*Coturnicops noveboracensis*) with dark plumage from Arkansas. *Proc. Arkansas Acad. Sci.* 41: 107-108.

JAMES, H. W. 1970. *Catalogue of the birds' eggs in the collection of the National Museums of Rhodesia.* Salisbury: National Museums of Rhodesia.

JAMIESON, I. G. & CRAIG, J. L. 1987a. Dominance and mating in a communal polygynandrous bird: cooperation or indifference towards mating competitors? *Ethology* 75: 317-327.

JAMIESON, I. G. & CRAIG, J. L. 1987b. Male-male and female-female courtship and copulation behaviour in a communally breeding bird. *Animal Behav.* 35: 1251-1252.

JAMIESON, I. G., MCRAE, S. B., SIMMONS, R. B. & TREWBY, M. (in prep.). High rates of conspecific brood parasitism and egg rejection in a species of coot and moorhen breeding in an ephemeral wetland in Namibia.

JAMIESON, I. G. & QUINN, J. S. 1997. Problems with removal experiments designed to test the relationship between paternity and parental effort in a socially polyandrous bird. *Auk* 114: 291-295.

JENKINS, A. R. & MEIKLEJOHN, M. F. M. 1960. Song of the Water Rail. *Scottish Birds* 1: 187-188.

JENKINS, J. M. 1979. Natural history of the Guam Rail. *Condor* 81: 404-408.

JENKINS, J. M. 1983. The native forest birds of Guam. *Orn. Monogr.* 31: 1-61.

JENNI, D. A. 1996. Family Jacanidae (Jacanas). Pp. 276-291 in: DEL HOYO, J., ELLIOTT, A., & SARGATAL, J. (eds). *Handbook of birds of the world.* Vol. 3: Hoatzin to Auks. Barcelona: Lynx Edicions.

JENNINGS, M. C. 1981. *The birds of Saudi Arabia: a check-list.* Cambridge: Privately published.

JENSEN, J. V. & KIRKEBY, J. 1980. *The birds of the Gambia.* Århus: Aros Nature Guides.

JOHNSON, A. W. 1964. The Giant Coot *Fulica gigantea* Eydoux and Souleyet. *Bull. Brit. Orn. Club* 84: 170-172.

JOHNSON, A. W. 1965a. The Horned Coot, *Fulica cornuta* Bonaparte. *Bull. Brit. Orn. Club* 85: 84-88.

JOHNSON, A. W. 1965b. *The birds of Chile and adjacent regions of Argentina, Bolivia and Peru,* vol. 1. Buenos Aires: Platt Establecimientos Gráficos.

JOHNSON, A. W. 1972. *Supplement to the birds of Chile and adjacent regions of Argentina, Bolivia and Peru.* Buenos Aires: Platt Establecimientos Gráficos.

JOHNSON, R. R. 1984. Breeding habitat use and postbreeding movements by Soras and Virginia Rails. Ames: Iowa State University (MS thesis).

JOHNSON, R. R. & DINSMORE, J. J. 1985. Brood-rearing and postbreeding habitat use by Virginia Rails and Soras. *Wilson Bull.* 97: 551-554.

JOHNSON, R. R. & DINSMORE, J. J. 1986a. Habitat use by breeding Virginia Rails and Soras. *J. Wildl. Manage.* 50: 387-392.

JOHNSON, R. R. & DINSMORE, J. J. 1986b. The use of tape-recorded calls to count Virginia Rails and Soras. *Wilson Bull.* 98: 303-306.

JOHNSON, R. W. 1973. Observations on the ecology and management of the Northern Clapper Rail, *Rallus longirostris crepitans* Gmelin, in Nassau County, N. Y. Ithaca, New York: Cornell University (PhD thesis).

JOHNSON, T. H. & STATTERSFIELD, A. J. 1990. A global review of island endemic birds. *Ibis* 132: 167-180.

JOHNSTONE, G. W. 1985. Threats to birds on Subantarctic Islands. Pp. 101-121 in MOORS, P. J. 1985. *Conservation of island birds.* ICBP Tech. Pub. 3. Cambridge: International Council for Bird Preservation.

JOHNSTONE, R. E. 1990. Mangroves and mangrove birds of Western Australia. *Rec. West Aust. Mus.* Suppl. 32.

JOHNSTONE, R. E., VAN BALEN, S. & DEKKER, R. W. R. J. 1993. New bird records for the island of Lombok. *Kukila* 6: 124-127.

JONES, J. C. 1940. Food habits of the American coot with notes on distribution. *U.S. Dep. Inter. Wildl. Res. Bull.* 2: 1-52.

JONES, P., SCHUBEL, S., JOLLY, J., BROOKE, M. de L. & VICKERY, J. 1995. Behaviour, natural history and annual cycle of the Henderson Island Rail *Porzana atra* (Aves: Rallidae). *Biol. J. Linn. Soc.* 56: 167-183.

JONSSON, L. 1993. *Birds of Europe with North Africa and the Middle East.* Princeton, N.J.: Princeton University Press.

JORGENSEN, P. D. 1975. Habitat preference of the Light-footed Clapper Rail in Tijuana Marsh, California. San Diego: San Diego State University (MSc thesis).

JORGENSEN, P. D. & FERGUSON, H. L. 1982. Clapper rail preys on Savannah Sparrow. *Wilson Bull.* 94: 215.

JOSEPH, L. & DRUMMOND, R. 1982. Food item of the Black Butcherbird. *Sunbird* 12: 49-50.

JUNGE, G. C. A. 1953. Zoological results of the Dutch New Guinea Expedition 1939. (5). The birds. *Zool. Verhand.* 20: 1-77.

KAESTNER, P. 1987. Some observations from lowland swamp forest in south Bougainville. *Muruk* 2: 34-38.

KAKEBEEKE, B. 1993. Striped Flufftail found breeding in Somerset West. *Birding in Southern Africa* 45: 9-11.

KARHU, S. 1973. On the development stages of chicks and adult moorhens *Gallinula chloropus* at the end of a breeding season. *Ornis Fenn.* 50: 1-17.

KAUFMANN, G. W. 1971. Behaviour and ecology of the Sora (*Porzana carolina*) and Virginia Rail (*Rallus limicola*). University of Minnesota (PhD dissertation).

KAUFMANN, G. W. 1977. Breeding requirements of the Virginia Rail and the Sora in captivity. *Avicult. Mag.* 83: 135-141.

KAUFMANN, G. W. 1983. Displays and vocalizations of the Sora and the Virginia Rail. *Wilson Bull.* 95: 42-59.

KAUFMANN, G. W. 1987a. Swamp habitat use by Spotless Crakes and Marsh Crakes at Pukepuke Lagoon. *Notornis* 34: 207-216.

KAUFMANN, G. W. 1987b. Growth and development of Sora and Virginia Rail chicks. *Wilson Bull.* 99: 432-440.

KAUFMANN, G. W. 1988a. The usefulness of taped Spotless Crake calls as a census technique. *Wilson Bull.* 100: 682-686.

KAUFMANN, G. W. 1988b. Development of Spotless Crake chicks. *Notornis* 35: 324-327.

KAUFMANN, G. W. 1988c. Social preening in Soras and in Virginia Rails. *Loon* 60: 59-63.

KAUFMANN, G. W. 1989. Breeding ecology of the Sora *Porzana carolina* and the Virginia Rail *Rallus limicola*. *Can. Field Nat.* 103: 270-282.

KAUFMANN, G. & LAVERS, R. 1987. Observations of breeding behaviour of Spotless Crake (*Porzana tabuensis*) and Marsh Crake (*P. pusilla*) at Pukepuke lagoon. *Notornis* 34: 193-205.

KAUTESK, B. M. 1985. "Caribbean Coots" in British Columbia? *Discovery* 14(2): 49-51.

KEAN, R. I. 1956. *Notornis* faeces in evidence on foods as a factor in chick rearing success. *Notornis* 6: 229-231.

KEAR, J., HILGARTH, N. & ANDERS, S. 1980. Coots eating goose droppings. *British Birds* 73: 410.

KEELEY, B. R. 1988. Classified summarised notes, 1 July 1986 to 30 June 1987, North Island. *Notornis* 35: 285-310.

KEELEY, B. R. 1989. Classified summarised notes, 1 July 1987 to 30 June 1988, North Island. *Notornis* 36: 197-222.

KEIJL, G. O., EGGENHUIZEN, A. H. V. & RUITERS, P. S. 1993. Identification of Red-knobbed Coot. *Dutch Birding* 15: 22-25.

KEITH, G. S. 1973. The voice of *Sarothrura insularis*, with further notes on members of the genus. *Bull. Brit. Orn. Club* 93: 130-136.

KEITH, S. 1978. Review: *Rails of the world*. *Wilson Bull.* 90: 322-325.

KEITH, S., BENSON, C. W. & IRWIN, M. P. S. 1970. The genus *Sarothrura* (Aves, Rallidae). *Bull. Amer. Mus. Nat. Hist.* 143: 1-84.

KENNEDY, R. S., GLASS, P. O., GLASS, E. J., GONZALES, P. C. & DICKINSON, E. C. 1986. Notes on Philippine birds, II. New or important records for the island of Palawan. *Bull. Brit. Orn. Club* 106: 173-179.

KENNEDY, R. S. & ROSS, C. A. 1987. A new subspecies of *Rallina eurizonoides* (Aves: Rallidae) from the Batan Islands, Philippines. *Proc. Biol. Soc. Washington* 100: 459-461.

KERLINGER, P. & WIEDNER, D. S. 1990. Vocal behaviour and habitat use of Black Rails in south Jersey. *Records of New Jersey Birds* 16: 58-62.

KHAN, M. A. R. 1982. *Wildlife of Bangladesh, a checklist*. Dhaka: University of Dhaka.

KHOBKHET, O. 1984. Ecology of jacanas and some rallids in Thailand. Pp. 113-120 in COTO, Z. & SUMARDJA, E. A. (eds.) *Wildlife Ecology in Southeast Asia*. Biotrop Special Publication No. 21: Southeast Asian Regional Center for Tropical Biology.

KIEL, W. H. 1955. Nesting studies of the coot in southwestern Manitoba. *J. Wildl. Manage.* 19: 189-198.

KIFF, L. F. 1975. Notes on southwestern Costa Rican birds. *Condor* 77: 101-103.

KIFF, L. F. & HOUGH, D. J. 1985. *Inventory of bird egg collections of North America*. Norman, Oklahoma: American Ornithologists' Union & Oklahoma Biological Survey.

KIKKAWA, J. 1976. The birds of Cape York Peninsula. *Sunbird* 7:81-105.

KILHAM, L. A. 1979. Snake and pond snails as food of Grey-necked Wood-Rails. *Condor* 81: 100-101.

KING, B. 1980. Individual recognition and winter behaviour of Water Rails. *British Birds* 73: 33-35.

KING, B. F., DICKINSON, E. C. & WOODCOCK, M. W. 1975. *A field guide to the birds of South-East Asia*. London: Collins.

KING, W. B. 1981. *Endangered birds of the world: the ICBP bird Red Data Book*. Washington, D.C.: Smithsonian Institution Press.

KINGSFORD, R. 1991. *Australian waterbirds*. Sydney: Kangaroo Press.

KINSKY, F. C. & YALDWYN, J. C. 1981. The bird fauna of Niue Island, southwest Pacific, with special notes on the White-tailed Tropicbird and Golden Plover. *Nat. Hist. Mus. New Zealand Misc. Series* 2: 1-49.

KIRWAN, G. 1994. First record of White-breasted Waterhen *Amaurornis phoenicurus* in Yemen. *Sandgrouse* 16: 55.

KLAAS, E. E., OHLENDORF, H. M. & CROMARTIE, E. 1980. Organochlorine residues and shell thicknesses in eggs of the clapper rail, common gallinule, purple gallinule and limpkin (Class Aves), eastern and southern United States, 1972-74. *Pestic. Monitor. J.* 14: 90-94.

KLEEFISCH, T. 1984. Die Weiß-brust alle (*Laterallus leucopyrrhus*). *Gefiederte Welt* 108: 100-103.

KLIMAITIS, J. F. & MOSCHIONE, F. N. 1987. *Aves de la Reserva Integral de Selva Marginal de Punta Lara y sus alrededores*. Dirección de Servicios Generales del Ministerio, Argentina.

KNOX, A. & WALTERS, M. 1994. *Extinct and endangered birds in the collections of the Natural History Museum*. British Ornithologists' Club Occasional Publications No. 1.

KNYSTAUTAS, A. 1993. *Birds of Russia*. London: Harper Collins.

KOCHAN, Z. 1992. Wintering of the Coot (*Fulica atra*) on the Gulf of Gdansk during the seasons of 1984/1985-1986/1987. *Notatki Ornitol.* 34: 125-130.

KOEN, J. H. 1988. Birds of the Eastern Caprivi. *Southern Birds* 15: 1-73.

KOENIG, O. 1943. *Rallen und Bartmeisen*. Wien: Niederdonau Nat. Kult., H. 25.

KOEPCKE, M. 1964. *Las aves del Departmento de Lima*. Lima: privately published.

KORNOWSKI, G. 1957. Beiträge zur Ethologie des Bläßhuhns (*Fulica atra* L.). *J. Orn.* 98: 318-355.

KOSHELEV, A. I. 1994. Baillon's Crake *Porzana pusilla*. Pp. 226-227 in TUCKER, G. M. & HEATH, M. F. 1994. *Birds in Europe: their conservation status*. BirdLife Conservation Series No. 3. Cambridge, UK: BirdLife International.

KOSKIMIES, P. 1989. *Distribution and numbers of Finnish breeding birds*. Appendix to Suomen Lintuatlas. Helsinki: Lintutieto Oy.

KOSKIMIES, P. 1993. Population sizes and recent trends of breeding and wintering birds in Finland. *Linnut* 28(2): 6-15.

KOSTER, S. H. & GRETTENBERGER, J. F. 1983. A preliminary survey of birds in Park W, Niger. *Malimbus* 5: 62-72.

KOTAGAMA, S. & FERNANDO, P. 1994. *A field guide to the birds of Sri Lanka*. Colombo: Wildlife Heritage Trust of Sri Lanka.

KOUDIJS, N. 1991. Broedresultaat met Braziliaanse dwergral. *Onze Vogels* 51: 508.

KOZICKY, E. L. & SCHMIDT, F. V. 1949. Nesting habits of the Clapper Rail in New Jersey. *Auk* 66: 355-364.

KRAMER, P. & BLACK, J. 1970. *Scientific and conservation report No. 21*. Galápagos, Ecuador: Charles Darwin Research Station.

KRAUS, C., BERNATH, O., ZBINDEN, K. & PILLERI, G. 1975. Zum Verhalten und Vokalisation von *Tribonyx mortierii* du Bois, 1840 (Aves, Rallidae). *Rev. Suisse Zool.* 82: 6-13.

KRAUS, M. & LISCHKA, W. 1956. Zum Vorkommen der *Porzana*-Arten im Fränkischen Weihergebiet. *J. Orn.* 97: 190-201.

KRAUTH, S. 1972. The breeding biology of the Common Gallinule. University of Wisconsin, Oshkosh (MS thesis).

KREKORIAN, C. O. 1978. Alloparental care in the Purple Gallinule. *Condor* 80: 382-390.

KURODA, N. 1993. Morpho-anatomy of the Okinawa Rail. *J. Yamashina Inst. Orn.* 25: 12-27.

KURODA, N., MANO, T. & OZAKI, K. 1984. The Rallidae, their insular distribution and conservation, with special note around discovery of *Rallus okinawae*. *50 years' Retrospect of Yamashina Inst. Orn.* 14: 36-57.

LAJMANOVICH, R. C. & BELTZER, A. H. 1993. Aporte al conocimiento de la biología alimentaria de la pollona negra *Gallinula chloropus* en el parana medio, Argentina. *Hornero* 13: 289-291.

LAMARCHE, B. 1980. Liste commentée des oiseaux du Mali. 1ère partie: non-passereaux. *Malimbus* 2: 121-158.

LAMARCHE, B. 1987. *Liste commentée des oiseaux de Mauretanie*. Nouakchott: Etudes Sahariennes et Ouest-Africaines 1,4 et spécial.

LAMARCHE, B. 1988. *Liste commentée des oiseaux de Mauretanie. Supplément no. 1*. Nouakchott: privately published.

LAMBERT, F. R. 1987. New and unusual bird records from the Sudan. *Bull. Brit. Orn. Club* 107: 17-19.

LAMBERT, F. R. 1989. Some field observations of the endemic Sulawesi rails. *Kukila* 4: 34-36.

LAMBERT, F. R. 1998a. A new species of *Gymnocrex* from the Talaud Islands, Indonesia. *Forktail* 13: 1-6.

LAMBERT. F. R. 1998b. A new species of *Amaurornis* from the Talaud Islands, Indonesia, and a review of taxonomy of bush hens occurring from the Philippines to Australasia. *Bull. Brit. Orn. Club* 118: 67-82.

LAND, H. C. 1970. *Birds of Guatemala*. Wynnewood, Pa.: Livingston.

LANG, A. L. 1991. Status of the American coot in Canada. *Cana. Field Nat.* 105: 530-541.

LANGRAND, O. 1990. *Guide to the birds of Madagascar*. New Haven: Yale University Press.

LA TOUCHE, J. D. D. 1931-34. *A handbook of the birds of eastern China*. Vol. 2. London: Taylor & Francis.

LAVERS, R. & MILLS, J. 1984. *Takahe*. Dunedin: John McIndoe and New Zealand Wildlife Service.

LAYARD, E. L. 1875. Notes on Fijian birds. *Proc. Zool. Soc. Lond.* 1875: 423-442.

LAYARD, E. L. 1876. Notes on some little known birds of the Fiji Islands. *Ibis* (3)6: 137-156.

LAYARD, E. L. & LAYARD, E. C. L. 1882. Notes on the avifauna of New Caledonia. *Ibis* (4)6: 493-550.

LECK, C. F. 1979. Avian extinction in an isolated tropical wet forest preserve, Ecuador. *Auk* 96: 343-352.

LEDANT, J. P., JACOB, J. P., JACOBS, P., MAHLER, F., OCHANDO, B. & ROCHE, J. 1981. Mise à jour de l'avifaune Algérienne. *Gerfaut* 71: 295-398.

LEES, A. 1991. *A protected forests system for the Solomon Islands*. Nelson, New Zealand: Maruia Society.

LEIBAK, E., LILLELEHT, V. & VEROMANN, H. 1994. *Birds in Estonia: status, distribution and numbers*. Tallinn: Estonian Academy Publishers.

LEICESTER, M. 1960. Some notes on the Lewin Rail. *Emu* 60: 20-24.

LEKAGUL, B. & ROUND, P. D. 1991. *A guide to the birds of Thailand*. Bangkok: Saha Karn Bahet.

LELEK, A. 1958. Contribution to the bionomy of the Coot (*Fulica atra* L.). *Folia Zoologica* 7: 143-168.

LEONARD, M. L., HORN, A. G. & EDEN, S. F. 1988. Parent-offspring aggression in moorhens. *Behav. Ecol. Sociobiol.* 23: 265-270.

LEONARD, M. L., HORN, A. G. & EDEN, S. F. 1989. Does juvenile helping enhance breeder reproductive success? A removal experiment on moorhens. *Behav. Ecol. Sociobiol.* 25: 357-362.

LESHACK, C. R. & HEPP, G. R. 1995. Kleptoparasitism of American Coots by Gadwalls and its relationship to social dominance and food abundance. *Auk* 112: 429-435.

LE SOUËF, D. 1903. Descriptions of birds' eggs from the Port Darwin district, northern Australia. *Emu* 2: 139-159.

LEVER, C. 1987. *Naturalized birds of the world*. Harlow, UK: Longman Scientific & Technical.

LEVI, P. J. 1966. Observations on the Southern White-breasted Crake in captivity. *Avicult. Mag.* 72: 24-26.

LEWINGTON, I. ALSTRÖM, P. & COLSTON, P. 1991. *A field guide to the rare birds of Britain and Europe*. London: HarperCollins.

LEWIS, A. D. & POMEROY, D. 1989. *A bird atlas of Kenya*. Rotterdam: Balkema.

LEWIS, J. C. & GARRISON, R. L. 1983. Habitat suitability index models: Clapper Rail. *U.S. Fish & Wildl. Serv.* FWS/OBS-82/10.51.

LINDSAY, C. J., PHILLIPPS, W. J. & WATTERS, W. A. 1959. Birds of Chatham Island and Pitt Island. *Notornis* 8: 99-106.

LINDSEY, T. R. *Encyclopedia of Australian animals*. Birds. Sydney: Angus & Robertson.

LINT, K. C. 1968. A rail of Guam. *Zoo Nooz* 41(5): 16-17.

LIPPENS, L. & WILLE, H. 1976. *Les oiseaux du Zaïre*. Lannoo: Tielt.

LITTLER, F. M. 1910. *A handbook of the birds of Tasmania and its dependencies*. Launceston, Tasmania: privately published.

LIVERSIDGE, R. 1968. The first plumage of *Sarothrura rufa*. *Ostrich* 39: 200.

LIVEZEY, B. C. (in press). A phylogenetic analysis of the Gruiformes (Aves) based on morphological characters, with an emphasis on the rails (Rallidae). *Phil. Trans. R. Soc. Lond. B*.

LLANDRES, C. & URDIALES, C. 1991. *Las aves de Doñana*. Barcelona: Lynx Ediciones.

LOERY, G. 1993. The ups and downs of a Virginia Rail population. *Connecticut Warbler* 13: 54-56.

LORD, E. A. R. 1936. Notes on the Dusky Moorhen. *Emu* 36: 128-129.

LORD, E. A. R. 1956. The birds of Murphy's Creek District, southern Queensland. *Emu* 56: 100-128.

LOUETTE, M. 1981. *The birds of Cameroon: an annotated checklist*. Brussels: Paleis der Academiën.

LOURIE-FRASER, G. 1982. Captive breeding of the Lord Howe Island Woodhen, an endangered rail. *AFA Watchbird* 10(2): 30-44.

LOWE, P. R. 1928. A description of *Atlantisia rogersi*, the diminutive and flightless rail of Inaccessible Island (Southern Atlantic), with some notes on flightless rails. *Ibis* (12)4: 99-131.

LOWEN, J. C., CLAY, R. P., BROOKS, T. M., ESQUIVEL, E. Z., BARTRINA, L., BARNES, R., BUTCHART, S. H. M. & ETCHEVERRY, N. I. 1995. Bird conservation in the Paraguayan atlantic forest. *Cotinga* 4: 58-64.

LOWEN, J., BARTRINA, L., BROOKS, T., CLAY, R. P. & TOBIAS, J. 1996a. Project YACUTINGA '95: bird surveys and conservation priorities in eastern Paraguay. *Cotinga* 5: 14-19.

LOWEN, J., BARTRINA, L., CLAY, R. P. & TOBIAS, J. 1996b. *Biological surveys and conservation priorities in eastern Paraguay*. Cambridge, UK: CSB Conservation Publications.

LOWEN, J. C., CLAY, R. P., MAZAR BARNETT, J., MADROÑO N., A., PEARMAN, M., LÓPEZ LANÚS, B., TOBIAS, J. A., LILEY, D. C., BROOKS, T. M., ESQUIVEL, E. Z. & REID, J. M. 1997. New and noteworthy observations on the Paraguayan avifauna. *Bull. Brit. Orn. Club* 117: 275-293.

LOWEN, J., MAZAR BARNETT, J., PEARMAN, M., CLAY, R. P. & LÓPEZ LANÚS, B. (in press). New geographical information for 25 species in eastern Paraguay. *Ararajuba*.

LOWERY, G. H. & DALQUEST, W. W. 1951. *Birds from the state of Veracruz, Mexico*. Lawrence, Kansas: Museum of Natural History, University of Kansas Publications 3.

LOWTHER, J. K. 1977. Nesting biology of the Sora at Vermilion, Alberta. *Can. Field Nat.* 91: 63-67.

LOZANO, I. E. 1993. Observaciones sobre la ecología y el comportamiento de *Rallus semiplumbeus* en el Humedal de la Florida, Sabana de Bogota. Unpublished.

LUDLOW, F. & KINNEAR, N. 1937. The birds of Bhutan and adjacent territories of Sikkim and Tibet. Part 3. *Ibis* (14)1: 467-504.

LUDLOW, F. & KINNEAR, N. 1944. The birds of south-eastern Tibet (3). *Ibis* 86: 348-389.

LYON, B. E. 1991. Brood parasitism in American coots: avoiding the constraints of parental care. *Proc. Internatn. Orn. Congr.* 20: 1023-1030.

LYON, B. E. 1993a. Conspecific brood parasitism as a flexible reproductive tactic in American Coots. *Animal Behav.* 46: 911-928.

LYON, B. E. 1993b. Tactics of parasitic American Coots: host choice and the pattern of egg dispersion among host nests. *Behav. Ecol. Sociobiol.* 33: 87-100.

MACDONALD, J. D. 1988. *Birds of Australia*. French's Forest, Australia: Reed.

MACDONALD, R. 1966. Australian Coots on Virginia Lake, Wanganui. *Notornis* 13: 165.

MACDONALD, R. 1968. The Australian Coot established on Virginia Lake, Wanganui. *Notornis* 15: 234-237.

MACGILLIVRAY, C. W. 1914. Notes on some North Queensland birds. *Emu* 13: 132-186.

MACGILLIVRAY, C. W. 1917. Ornithologists in North Queensland. *Emu* 17: 63-87.

MACGILLIVRAY, C. W. 1928. Bird life of the Bunker and Capricorn Islands. *Emu* 27: 230-239.

MACGILLIVRAY, W. D. K. 1914. Notes on some North Queensland birds. *Emu* 13: 132-186.

MACKAY, R. D. 1970. *The birds of Port Moresby and district*. Melbourne: Nelson.

MACKINNON, J. & PHILLIPPS, K. 1993. *A field guide to the birds of Borneo, Java, Sumatra and Bali*. Oxford: Oxford University Press.

MACKWORTH-PRAED, C. W. & GRANT, C. H. B. 1937. A survey of the *Sarothrura* crakes. *Ibis* (14)1: 626-634.

MACKWORTH-PRAED, C. W. & GRANT, C. H. B. 1957. *Birds of eastern and north eastern Africa*. Vol. 1. London: Longmans.

MACKWORTH-PRAED, C. W. & GRANT, C. H. B. 1962. *Birds of the southern third of Africa*. Vol. 1. London: Longmans.

MACKWORTH-PRAED, C. W. & GRANT, C. H. B. 1970. *Birds of west central and western Africa*. Vol. 1. London: Longmans.

MACLEAN, G. L. 1993. *Roberts' birds of Ssouthern Africa*. 6th. edn. Cape Town: John Voelcker Bird Book Fund.

MACMILLAN, B. W. H. 1990. Attempts to re-establish Wekas, Brown Kiwis and Red-crowned Parakeets in the Waitakere Ranges. *Notornis* 37: 45-51.

MADDEN, S. T. & SCHMITT, M. B. 1976. Scavenging by some Rallidae in the Transvaal. *Ostrich* 47: 68.

MADOC, G. C. 1976. *An introduction to Malayan birds*. Kuala Lumpur: Malayan Nature Society.

MAGARRY, D. 1991. Red-necked Crakes in Cairns. *Queensland Orn. Soc. Newsl.* 22(6): 5.

MAJNEP, I. S. & BULMER, R. 1977. *Birds of My Kalam country*. Auckland: Auckland University Press.

MALLALIEU, M. 1995. Crakes in the Lilongwe area: results of fieldwork in 1991-1993. *Nyala* 18: 1-10.

MÁÑEZ, M. 1994. Purple Gallinule *Porphyrio porphyrio*. Pp. 230-231 in TUCKER, G. M. & HEATH, M. F. 1994. *Birds in Europe: their conservation status*. BirdLife Conservation Series No. 3. Cambridge, UK: BirdLife International.

MANNERS, G. R., BURTCH, P., BOWDEN, C. G. R., BOWDEN, E. M. & WILLIAMS, E. 1993. Purple Gallinule *Porphyrio porphyrio*, further sightings in Cameroon. *Malimbus* 14: 59.

MANOLIS, T. 1978. Status of the Black Rail in central California. *Western Birds* 9: 151-157.

MANSON, A. J. 1986. Notes on the breeding of the Buff-spotted Flufftail at Seldomseen, Vumba. *Honeyguide* 32: 137-142.

MARCHANT, S. & HIGGINS, P. J. (eds.). 1993. *Handbook of Australian, New Zealand and Antarctic birds*. Vol. 2. Melbourne: Oxford University Press.

MARÍN, M. 1996. Pesos corporales de aves chilenas: Tinamiformes a Charadriiformes. *Noticiario Mensual* No. 326, July 1996. Santiago, Chile: Museo Nacional de Historia Natural.

MARÍN, M. A. & SHELDON, F. H. 1987. Some new nesting records of padi-dwelling birds in Sabah, East Malaysia (North Borneo). *Bull. Brit. Orn. Club* 107: 23-25.

VAN MARLE, J. G. 1940. Aanteekeningen omtrent de vogels van de Minahasa (N.O.-Celebes). II. *Limosa* 13: 119-124.

VAN MARLE, J. G. & VOOUS, K. H. 1988. *The birds of Sumatra*. BOU Checklist No. 10. London: British Ornithologists' Union.

MARCHANT, J. H., HUDSON, R., CARTER, S. P. & WHITTINGTON, P. 1990. *Population trends in British breeding birds*. Tring, UK: British Trust for Ornithology.

MARTIN, A. C., ZIM, H. S. & NELSON, A. L. 1951. *American wildlife and plants*. New York: Dover.

MARTIN, E. M. & PERRY, M. C. 1981. Yellow Rail collected in Maryland. *Maryland Birdlife* 37: 15-16.

MARTIN, P. R., THOMPSON, B. G. & WITTS, S. J. 1979. Niche separation in three species of water birds. *Corella* 3: 1-5.

MARTINEZ, M. M., BO, M. S. & ISACCH, J. P. 1997. Habitat and abundance of Speckled Crake (*Coturnicops notata*) and Dot-winged Crake (*Porzana spiloptera*) in Mar Chiquita, Buenos Aires province, Argentina. *Hornero* 14: 274-277.

MASON, A. G. 1940. On some experiments with Corncrakes. *Irish Nat. J.* 7: 226-237.

MASON, A. G. 1941. Further experiments with Corncrakes. *Irish Nat. J.* 7: 321-332.

MASON, A. G. 1944. Combat display of Corncrake. *Irish Nat. J.* 8: 200-202.

MASON, A. G. 1945. The display of the Corn-Crake. *British Birds* 38: 351-352.

MASON, A. G. 1947. Aggressive behaviour of Corn-Crake and Ringed Plover. *British Birds* 40: 191-192.

MASON, A. G. 1950. The behaviour of Corn-Crakes. *British Birds* 43: 70-78.

MASON, A.G. 1951. Aggressive display of the Corncrake. *British Birds* 44: 162-166.

MASON, C. G. & MAXWELL-LEFROY, H. 1912. The food of birds in India. *Mem. Agr. Dept. India, Entomol. Ser.* 3.

MASON, I. J., GILL, H. B. & YOUNG, J. H. 1981. Observations on the Red-necked Crake *Rallina tricolor*. *Aust. Bird Watcher* 9: 69-77.

MASON, I. J. & WOLFE, T. O. 1975. First record of Marsh Crake for the Northern Territory. *Emu* 75: 235.

MASON, P. F. 1940. A brief faunal survey of north-western Benin. *Nigerian Field* 9: 68-80.

MASON, V. 1996. Baillon's Crake, a new species for Bali. *Kukila* 8: 157-158.

MASSEY, B. W. & ZEMBAL, R. 1987. Vocalizations of the Light-footed Clapper Rail. *J. Field Orn.* 58: 32-40.

MASSEY, B. W., ZEMBAL, R. & JORGENSEN, P. D. 1984. Nesting habitat of the Light-footed Clapper Rail in southern California. *J. Field Orn.* 55: 67-80.

MASSOLI-NOVELLI, R. 1988. Segnalazione di Schiribilla Alibianche, *Sarothrura ayresi*, in Etiopia. *Riv. Ital. Orn.* 58: 40-42.

MASTERS, J. R. & MILHINCH, A. L. 1974. Birds of the shire of Northam, about 100 km east of Perth, W. A. *Emu* 74: 228-244.

MASTERSON, A. N. B. 1991. Notes on the African Crake *Crecopsis egregia*. Unpublished.

MATHESON, W. E. 1974. The irruption of Native Hens in South Australia in 1972-73. *S. Aust. Orn.* 26: 151-156.

MATHESON, W. E. 1978. A further irruption of Native Hens in 1975. *S. Aust. Orn.* 27: 270-273.

MATHEWS, G. M. 1911. *The birds of Australia*. Vol. 1, pt. 4. London: Witherby.

MATHEWS, G. M. 1915. Diggles' Ornithology of Australia, and other works. *Austral Avian Record* 2: 137-153.

MATHEWS, G. M. 1928. *The birds of Norfolk and Lord Howe islands*. London: Witherby.

MATHEWS, G. M. & IREDALE, T. 1921. *A manual of the birds of Australia*. London: Witherby.

MATTHEWS, W. C. 1983. Home range, movements and habitat selection of nesting gallinules in a Louisiana freshwater marsh. Baton Rouge: Louisiana State University (MS thesis).

MAURICIO, G. N. & DIAS, R. A. 1996. Novos registros e extensos de distribuicão de aves palustres e costeiras no litoral sul do Rio Grande do Sul. *Ararajuba* 4: 47-51.

MAURO, I. 1994. Enkele bemerkingen bij nieuwe gegevens over het Porseleinhoen *Porzana porzana* te Molsbroek-Lokeren (Oost-Vlaanderen) tijdens het broedseizoen 1992. *Oriolus* 60: 30-33.

MAXWELL, J. M. & JAMIESON, I. G. 1997. Survival and recruitment of captive-reared and wild-reared Takahe in Fiordland, New Zealand. *Conserv. Biol.* 11: 683-691.

MAY, L. 1994. Individually distinctive corncrake *Crex crex* calls: a pilot study. *Bioacoustics* 6: 25-32.

MAYAUD, M. 1983. Les oiseaux du nord-ouest de l'Afrique. Notes complémentaires. *Alauda* 51: 271-301.

MAYES, E. 1993. The decline of the Corncrake *Crex crex* in Britain and Ireland in relation to habitat. *J. Appl. Ecol.* 30: 53-62.

MAYES, E. & STOWE, T. J. 1989. The status and distribution of the Corncrake in Ireland in 1988. *Irish Birds* 4: 1-12.

MAYNARD SMITH, J. & RIDPATH, M. G. 1972. Wife sharing in the Tasmanian Native Hen, *Tribonyx mortierii*: a case of kin selection. *Amer. Nat.* 106: 447-452.

MAYAUD, N. 1982. Les oiseaux du nord-ouest de l'Afrique. Notes complémentaires. *Alauda* 50: 45-67.

MAYR, E. 1933a. Birds collected during the Whitney South Sea expedition. XXII. *Amer. Mus. Novit.* 590.

MAYR, E. 1933b. On a collection of birds, supposedly from the Solomon Islands. *Ibis* (13) 3: 549-552.

MAYR, E. 1933b. Birds collected during the Whitney South Sea expedition. XXIII. *Amer. Mus. Novit.* 609.

MAYR, E. 1938. Birds collected during the Whitney South Sea Expedition. XL. *Amer. Mus. Novit.* 1007.

MAYR, E. 1941. Taxonomic notes on the birds of Lord Howe Island. *Emu* 40: 321-322.

MAYR, E, 1944. The birds of Timor and Sumba. *Bull. Amer. Mus. Nat. Hist.* 83: 127-194.

MAYR, E. 1945. *Birds of the Southwest Pacific*. New York: Macmillan.

MAYR, E. 1949. Notes on the birds of northern Melanesia, 2. *Amer. Mus. Novit.* 1417.

MAYR, E. & COTTRELL, G. W. (eds.). 1979. *Check-list of birds of the world*. Cambridge, Mass.: Museum of Comparative Zoology.

MAYR, E. & GILLIARD, E. T. 1951. New species and subspecies of birds from the highlands of New Guinea. *Amer. Mus. Novit.* 1524: 1-15.

MAYR, E. & GILLIARD, E. T. 1954. Birds of central New Guinea. Results of the American Museum of Natural History expedition to New Guinea in 1950 and 1952. *Bull. Amer. Mus. Nat. Hist.* 103.

MAYR, E. & RAND, A. L. 1937. Results of the Archbold Expeditions. 14, The birds of the 1933-1934 Papuan Expeditions. *Bull. Amer. Mus. Nat. Hist.* 73: 1-248.

MCALLAN, I. A. W. & BRUCE, M. D. 1988. *The birds of New South Wales. A working list*. Turramurra, Australia: Biocon Research Group.

MCCLURE, H. E. 1974. *Migration and survival of the birds of Asia*. Bangkok, Thailand: U.S. Army Medical Component, SEATO Medical Project.

MCCLURE, H. E. & LEELAVIT, L. P. 1972. *Birds banded in Asia during the MAPS Program, by locality, 1963-1971*. U.S. Army Research & Development Group, Far East, Report No. FE-315-7.

MCFARLANE, R. W. 1975. Notes on the Giant Coot (*Fulica gigantea*). *Condor* 77: 324-327.

MCGILP, J. N. 1923. Birds of Lake Frome district, South Australia. *Emu* 22: 237-243.

MCKAY, W. D. 1981. Notes on Purple Gallinules in Colombian rice-fields. *Wilson Bull.* 93: 267-271.

MCKEAN, J. L., EVANS, O. & LEWIS, J. H. 1976. Notes on the birds of Norfolk Island. *Notornis* 23: 299-301.

MCKEAN, J. L. & HINDWOOD, K. A. 1965. Additional notes on the birds of Lord Howe Island. *Emu* 64: 79-97.

MCKEAN, J. L. & READ, M. 1979. Further sightings of the Bush-Hen *Gallinula olivacea* from sub-coastal Northern Territory. *Sunbird* 10: 73-74.

MCLAREN, I. 1996. A reddish-legged Purple Gallinule. *Birders J.* 5: 22-24.

MCNAB, B. K. 1994. Energy conservation and the evolution of flightlessness in birds. *Amer. Nat.* 144: 628-642.

MCRAE, S. B. 1994. An ecological and genetic analysis of breeding strategies in the moorhen, *Gallinula chloropus*. University of Cambridge (PhD thesis).

MCRAE, S. B. 1995. Temporal variation in responses to intraspecific brood parasitism in the moorhen. *Animal Behav.* 49: 1073-1088.

MCRAE, S. B. 1996a. Family values: costs and benefits of communal nesting in the Moorhen. *Animal Behav.* 52: 225-245.

MCRAE, S. B. 1996b. Brood parasitism in the moorhen: brief encounters between parasites and hosts and the significance of an evening laying hour. *J. Avian Biol.* 27: 311-320.

MCRAE, S. B. 1997. A rise in nest predation enhances the frequency of intraspecific brood parasitism in a moorhen population. *J. Animal Ecol.* 66: 143-153.

MCRAE, S. B. & BURKE, T. 1996. Intraspecific brood parasitism in the moorhen: parentage and parasite-host relationships determined by DNA fingerprinting. *Behav. Ecol. Sociobiol.* 38: 115-129.

MEANLEY, B. 1953. Nesting of the King Rail in the Arkansas ricefields. *Auk* 70: 262-269.

MEANLEY, B. 1956. Food habits of the King Rail in the Arkansas ricefields. *Auk* 73: 252-258.

MEANLEY, B. 1957. Notes on the courtship behavior of the King Rail. *Auk* 74: 433-440.

MEANLEY, B. 1960. Fall food of the Sora Rail in the Arkansas ricefields. *J. Wildl. Manage.* 24: 339.

MEANLEY, B. 1963. Pre-nesting activity of the purple gallinule near Savannah, Georgia.

MEANLEY, B. 1965a. Early fall food and habitat of the Sora in the Patuxent River Marsh, Maryland. *Chesapeake Science* 6: 226-237.

MEANLEY, B. 1965b. King and Clapper Rails of Broadway Meadows. *Delaware Conservationist*, winter issue: 3-7.

MEANLEY, B. 1969. Natural history of the King Rail. *Bureau of Sports, Fisheries and Wildlife: North American Fauna*, 67: 1-108.

MEANLEY, B. 1985. *The marsh hen: a natural history of the Clapper Rail of the Atlantic coast salt marsh*. Centreville, Md.: Tidewater Publications.
MEANLEY, B. 1992. King Rail. In POOLE, A., STETTENHAM, P. & GILL, F. (eds.). *The birds of North America*, No. 3. Washington, D.C.: American Ornithologists' Union.
MEANLEY, B. & MEANLEY, A. G. 1958. Growth and development of the King Rail. *Auk* 75: 381-386.
MEANLEY, B. & WETHERBEE, D. K. 1962. Ecological notes on mixed populations of King Rails and Clapper Rails in Delaware Bay marshes. *Auk* 79: 453-457.
MEDWAY, LORD & WELLS, D. R. 1976. *The birds of the Malay Peninsula*. Vol. V. Conclusions and Survey of Every Species. London: Witherby, & Kuala Lumpur: Penerbit Universiti Malaya.
MEES, G. F. 1965. The avifauna of Misool. *Nova Guinea, Zoology* 31: 139-203.
MEES, G. F. 1970. Birds of the Inyanga National Park, Rhodesia. *Zool. Verhand.* No. 109.
MEES, G. F. 1977. Enige gegevens over de uitgestorven ral *Pareudiastes pacificus* Hartlaub & Finsch. *Zool. Meded.* 50: 231-242.
MEES, G. F. 1982. Birds from the lowlands of Southern New Guinea (Merauke and Koembe). *Zool. Verhand.* 191.
MEES, G. F. 1986. A list of the birds recorded from Bangka Island, Indonesia. *Zool. Verhand.* 232.
MEINERTZHAGEN, R. 1930. *Nicoll's birds of Egypt*. London: H. Rees.
MEININGER, P. L., DUIVEN, P., MARTEIJN, E. C. L. & VAN SPANJE, T. M. 1990. Notable bird observations from Mauritania. *Malimbus* 12: 19-24.
MEISE, W. 1934. Zur Brutbiologie der Ralle *Laterallus leucopyrrhus* (Vieill.). *J. Orn.* 82: 257-268.
MELVIN, S. M. & GIBBS, J. P. 1994. Sora. In TACHA, T. C. & BRAUN, C. E. (eds.). *Management of migratory shore and upland game birds in North America*. Washington, D.C.: International Assoc. Fish & Wildlife Agencies.
MELVIN, S. M. & GIBBS, J. P. 1996. Sora. In POOLE, A., & GILL, F. (eds.). *The birds of North America*, No. 250. Washington, D.C.: American Ornithologists' Union.
MENDELSOHN, J. M., SINCLAIR, J. C. & TARBOTON, W. R. 1983. Flushing flufftails out of vleis. *Bokmakierie* 35: 9-11.
MERTON, D. V. 1970. Kermadec Islands Expedition reports: a general account of birdlife. *Notornis* 17: 147-199.
MEYER, P. O. 1936. *Die Vögel des Bismarckarchipel*. Kokopo, New Britain: Catholic Mission.
MEYER DE SCHAUENSEE, R. 1962. Notes on Venezuelan birds, with a history of the rail *Coturnicops notata*. *Notulae Naturae* 357.
MEYER DE SCHAUENSEE, R. 1966. *The species of birds of South America and their distribution*. Narbeth, Pa.: Livingston.
MEYER DE SCHAUENSEE, R. 1970. *A guide to the birds of South America*. Wynnewood, Pa.: Livingston.
MEYER DE SCHAUENSEE, R. 1982. *A guide to the birds of South America*. (Reprint with additions and addenda) Narberth, Pa: Livingston.
MEYER DE SCHAUENSEE, R. 1984. *The birds of China*. Washington, D.C.: Smithsonian Institution Press.
MEYER DE SCHAUENSEE, R. & PHELPS, W. H. Jr. 1978. *A guide to the birds of Venezuela*. Princeton, New Jersey: Princeton University Press.
MILLEDGE, D. 1972. The birds of Maatsuyker Island, Tasmania. *Emu* 72: 167-170.
MILLENER, P. R. 1981. The quaternary avifauna of New Zealand. University of Auckland, New Zealand (PhD thesis).
MILLER, A. H. 1960. Additional data on the distribution of some Colombian birds. *Novedas Colombianas* 1: 235-237.
MILLER, A. H. 1963. Seasonal activity and ecology of the avifauna of an American equatorial cloud forest. *Univ. Calif. Publ. Zool.* 66: 1-74.
MILLER, B. & KINGSTON, T. 1980. Lord Howe Island Woodhen. *Parks & Wildlife*. Aug. 1980: 17-26.
MILLER, B. & MULLETTE, K. J. 1985. Rehabilitation of an endangered Australian bird: the Lord Howe Island Woodhen *Tricholimnas sylvestris* (Sclater). *Biol. Conserv.* 34: 55-95.
MILLER, R. F. 1928. Virginia Rail lays in Sora's nest. *Oologist* 45: 132.
MILLER, R. F. 1946. The Florida gallinule: breeding birds of the Philadelphia region. *Cassinia* 36: 1-15.
MILLS, J. A. 1973. Takahe research. *Wildlife - A Review*. New Zealand Wildlife Service, 4: 24-27.
MILLS, J. A. 1975. Population studies on Takahe, *Notornis mantelli* in Fiordland, New Zealand. *ICBP Bull.* 12: 140-146.
MILLS, J. A. 1976. Takahe - red deer. *Wildlife - A Review*. New Zealand Wildlife Service, 7: 24-30.
MILLS, J. A. & LAVERS, R. B. 1974. Preliminary results of research into the present status of Takahe (*Notornis mantelli*) in the Murchison Mountains. *Notornis* 21: 312-317.
MILLS, J. A., LAVERS, R. B. & LEE, W. G. 1984. The Takahe – a relict of the Pleistocene grassland avifauna of New Zealand. *N. Z. J. Ecol.* 7: 57-70.
MILLS, J. A., LEE, W. G. & LAVERS, R. B. 1989. Experimental investigations of the effect of Takahe and deer grazing on *Chinochloa pallens* grasslands, Fiordland, New Zealand. *J. Appl. Ecol.* 26: 397-417.
MILLS, J. A., LEE, W. G., MARK, A. F. & LAVERS, R. B.. 1980. Winter use by Takahe (*Notornis mantelli*) of the Summergreen Fern *Hypolepis milleifolium*) in relation to its annual cycle of carbohydrates and minerals. *N. Z. J. Ecol.* 3: 131-137.
MILLS, J. A. & MARK, A. F. 1977. Food preferences of Takahe in Fiordland National Park, New Zealand, and the effect of competition from introduced Red Deer. *J. Anim. Ecol.* 46: 939-958.
MILON, P., PETTER, J.-J. & RANDRIANASOLO, G. 1973. *Faune de Madagascar, 35, Oiseaux*. Tananarive & Paris: ORSTOM & CNRS.
MISKELLY, C. M. 1981. Leg colour and dominance in Buff Wekas. *Notornis* 28: 47-48.
MISKELLY, C. M. 1987. The identity of the Hakawai. *Notornis* 34: 95-116.
MITCHELL, M. H. 1957. *Observations on birds of southeastern Brazil*. Toronto: University of Toronto Press.
MIYAGI, K. 1989. Conservation of Okinawa Rail and Okinawa Woodpecker. *Proc. Joint Meeting of International Council for Bird Preservation Asian Section and East Asia Bird Protection Conference, Bangkok, Thailand*. (Unpublished).
MÖLLER, E. 1992. Feeding technique of a White-browed Crake *Porzana cinerea*. *Forktail* 7: 154-155.
MOLTONI, E. & RUSCONE, G. G. 1944. *Gli uccelli dell'Africa Orientale Italiana, 3*. Milan: Museo Civico di Storia Naturale.
MOMIYAMA, T. T. 1930. On the birds of Bonin and the Iwo-Islands. *Bull. Bio-Geogr. Soc. Japan* 1: 89-186.
MONCRIEFF, P. 1928. Bird migration in New Zealand. *Emu* 28: 138-149.
MONROE, B. L. 1968. *A distributional survey of the birds of Honduras*. Orn. Monogr. 7. American Ornithologists' Union.
MOODY, A. F. 1932. *Water-fowl and game-birds in captivity*. London: Witherby.
MOON, G. J. H. 1960. *Focus on New Zealand birds*. Wellington.
MOORE, P. J. 1983. Marsh Crake at Lake Wairarapa. *Notornis* 30: 246-249.
MOREAU, R. E. 1972. *The Palaearctic-African bird migration systems*. London & New York: Academic Press.
MOREL, G. J. 1972. *Liste commentée des oiseaux du Sénégal*. Dakar: ORSTOM.
MOREL, G. J. & MOREL, M.-Y. 1988. Les oiseaux de Guinée. *Malimbus* 10: 143-176.

MOREL, G. J. & MOREL, M.-Y. 1990. *Les oiseaux de Sénégambie.* Paris: ORSTOM.

MOREL, G. J. & ROUX, F. 1966. Les migrateurs paléarctiques au Sénégal. I. Non Passereaux. *Terre et Vie* 20: 19-72.

MOREL, G. J. & ROUX, F. 1973. Les migrateurs paléarctiques au Sénégal: notes complémentaires. *Terre et Vie* 27: 523-550.

MORENO, A. 1953. Consideration about the systematic value of *Laterallus jamaicensis jamaicensis* (Gmelin) and *Laterallus jamaicensis pygmaeus* (Blackwell). *Torreia* 20.

MORGAN, B. & MORGAN, J. 1968. The Bush-hen in southeastern Queensland. *Emu* 68: 150.

MORIARTY, T. K. 1972. Birds of Wanjarri, Western Australia. *Emu* 72: 1-7.

MORRIS, R. B. 1977. Observations of Takahe nesting behaviour at Mt Bruce. *Notornis* 24: 54-58.

MORRISON, A. 1939a. Notes on birds of Lake Junin, central Peru. *Ibis* (14)3: 643-654.

MORRISON, A. 1939b. A new coot from Peru. *Bull. Brit. Orn. Club* 59: 56-57.

MORTON, E. S. 1979. Status of endemic birds of the Zapata Swamp, Cuba. Memorandum to S. D. Ripley, 6 November (unpublished).

MORTON, H. J. 1953. Notes on Spotless Crake (*Porzana plumbea*). *S. Aust. Orn.* 21: 5.

MOSSO, E. D. & BELTZER, A. H. 1993. Nota sobre la dieta de *Fulica leucoptera* en el valle aluvial del Río Paraná medio, Argentina. *Orn. Neotropical* 4: 91-93.

MOUNTFORT, G. 1988. *Rare birds of the world.* London: Collins.

MOUSLEY, H. 1937. A study of Virginia and Sora Rails at their nests. *Wilson Bull.* 49: 80-84.

MOUSLEY, H. 1940. Further notes on the nesting habits of the Virginia Rail. *Wilson Bull.* 52: 87-90.

MULHOLLAND, R. 1983. Feeding ecology of the common moorhen and purple gallinule on Orange Lake, Florida. Gainsville, Fla: University of Florida (MS thesis).

MULHOLLAND, R. & PERCIVAL, H. F. 1982. Food habits of the common moorhen and purple gallinule in north-central Florida. *Proc. Southeast Assoc. Fish and Wildl. Agencies* 36: 527-536.

MUNDKUR, T., CARR, P., SUN HEAN & CHHIM SOMEAN. 1995. *Survey for large waterbirds in Cambodia, March-April 1994.* Gland, Switzerland: IUCN.

MUNRO, G. C. 1944. *Birds of Hawaii.* Honolulu: Tongg.

MUNRO, G. C. 1947. Notes on the Laysan rail. *Elepaio* 8: 24-25.

MUNRO, G. C. 1960. *Birds of Hawaii.* Rutland, Vermont: Charles E. Tuttle Co.

MURANYI, M. & BAVERSTOCK, P. R. 1996. The distribution and habitat preferences of the Bush-hen *Amaurornis olivacea* in north-eastern New South Wales. *Emu* 96: 285-287.

MURPHY, R. C. 1924. Birds collected during the Whitney South Sea Expedition. II. *Amer. Mus. Novit.* 124.

MURTON, R. K. & WESTWOOD, N. J. 1977. *Avian breeding cycles.* Oxford: Clarendon Press.

MYERS, J. G. 1923. The present position of the endemic birds of New Zealand. *N. Z. J. Sci. & Tech.* 6: 65-99.

MYERS, P. & HANSEN, R. L. 1980. Rediscovery of the Rufous-faced Crake (*Laterallus xenopterus*). *Auk* 97: 901-902.

NAPIER, J. R. 1969. Birds of the Break O'Day Valley, Tasmania. *Aust. Bird Watcher* 3: 179-192.

NARANJO, L. G. 1991. Nest, eggs and young of the Blackish Rail. *Orn. Neotropical* 2: 47-48.

NAROSKY, T. 1985. *Aves Argentinas: guía para el reconocimiento de la avifauna bonaerense.* Buenos Aires: Editorial Albatros.

NAROSKY, T. & DI GIÁCOMO, A. G. 1993. *Las aves de la provincia de Buenos Aires: distribución y estatus.* Asoc. Ornit. del Plata. Buenos Aires.

NAROSKY, T. & YZURIETA, D. 1989. *Birds of Argentina and Uruguay: a field guide.* Buenos Aires: Vázquez Mazzini.

NAUMBURG, E. M. B. 1930. The birds of Matto Grosso, Brazil. *Bull. Amer. Mus. Nat. Hist.* 60: 1-432.

DE NAUROIS, R. 1994. *Les oiseaux des îles du Golfe de Guinée (São Tomé, Príncipe et Annobon).* Lisbon: Instituto de Investigação Científica Tropical.

NAVAS, J. R. 1956. Manifestaciones vocales de las gallaretas. *Hornero* 10: 119-135.

NAVAS, J. R. 1960. Comportamiento aggressivo de *Fulica armillata* Vieillot. *Revta. Mus. Argent. Cienc. Nat. "Bernadino Ricadavio" Inst. Nac. Invest. Cienc. Nat.* 6: 103-129.

NAVAS, J. R. 1962. Reciente hallazgo de *Rallus limicola antarcticus* King (Aves, Rallidae). *Neotropica* 8: 73-76.

NAVAS, J. R. 1970. Desarrollo y secuencias de plumajes en *Fulica armillata* Vieillot (Aves, Rallidae). *Revta. Mus. Argent. Cienc. Nat. "Bernadino Ricadavio" Inst. Nac. Invest. Cienc. Nat.* 16: 65-85.

NAVAS, J. R. 1991. *Fauna de agua dulce de la Republica Argentina.* Vol. XLIII. CONICET.

NEELAKANTAN, K. K. 1958. The voice of the Kora *Gallicrex cinerea* (Gmelin). *J. Bombay Nat. Hist. Soc.* 55: 560-561.

NEELAKANTAN, K. K. 1991a. Bluebreasted Banded Rail *Rallus striatus* Linn. nesting in Kerala. *J. Bombay Nat. Hist. Soc.* 88: 448-450.

NEELAKANTAN, K. K. 1991b. Breeding of the Kora or Watercock *Gallicrex cinerea* in Kerala. *J. Bombay Nat. Hist. Soc.* 88: 450-451.

NEGERE, E. 1980. The effects of religious belief on conservation of birds in Ethiopia. *Proc. IV Pan-Afr. Orn. Congr.*: 361-365.

NEGRET, A. & TEIXEIRA, D. M. 1984a. Notas sobre duas espécies de aves raras: *Micropygia schomburgkii* e *Laterallus xenopterus* (Rallidae) na região de Brasilia-DF. P. 337 in *Resumos, XI Congresso Brasiliero de Zoologia.* Imprensa Universitária, Belo Horizonte.

NEGRET, A. & TEIXEIRA, D. M. 1984b. The Ocellated Crake (*Micropygia schomburgkii*) of central Brazil. *Condor* 86: 220.

NEUBY VARTY, B. V. 1953. Reichenow Striped Flufftail. *Bokmakierie* 5: 52.

NEUFELDT, I. A. 1978. [Extinct birds in the collection of the Zoological Institute of the Academy of sciences of the USSR]. Pp. 101-110 in SKARLATO, O. A. (ed.) 1978. [*Systematics and life history of rare and little known birds*]. Proceedings of the Zoological Institute 76. (In Russian).

NEWBY, J., GRETTENBERGER, J. & WATKINS, J. 1987. The birds of the northern Aïr, Niger. *Malimbus* 9: 4-16

NICE, M. M. 1962. Development of behavior in precocial birds. *Trans. Linn. Soc. N.Y.* 8: 1-211.

NICÉFORO, M. H. & OLIVARES, A. 1964. Adiciónes a la avifauna Colombiana, I (Tinamidae-Falconidae). *Bol. Inst. La Salle* 204: 5-27.

NICÉFORO, M. H. & OLIVARES, A. 1965. Adiciónes a la avifauna Colombiana, II (Cracidae-Rynchopidae). *Bol. Soc. Venezolana Cienc. Nat.* 110: 370-393.

NICHOLLS, E. B. 1905. A trip to the west. *Emu* 5: 78-82.

NICOLL, M. J. 1906. On the birds collected and observed during the voyage of the 'Valhalla', R.Y.S., from November 1905 to May 1906. *Ibis* (8)6: 666-712.

NICOLL, M. J. 1908. *Three voyages of a naturalist, being an account of many little-known islands in three oceans visited by the 'Valhalla' R.Y.S.* London.

NIEMANN, S. 1995. *Habitat management for Corncrakes.* Sandy, UK: Royal Society for the Protection of Birds.

NIKOLAUS, G. 1981. Palaearctic migrants new to the North Sudan. *Scopus* 5: 121-124.

NIKOLAUS, G. 1987. *Distribution atlas of Sudan's birds with notes on habitat and status.* Bonn: Zoologischesforschungs Institut und Museum Alexander Koenig (Bonner Zoologische Monographien 25).

NIKOLAUS, G. 1989. Birds of South Sudan. *Scopus Special Supplement* No. 3.

NOLL, H. 1924. *Deutscher Verlag für Jugend und Volk.* Vienna.

NORES, M. & YZURIETA, D. 1980. *Aves de ambientes acuáticos de Córdoba y centro de Argentina*. Córdoba, Argentina: Sec. Estado Agric. y Granaderia.

NORES, M., YZURIETA, D. & MIATELLO, R. 1983. Lista y distribución de las aves de Córdoba, Argentina. *Bol. Acad. Nac. Cienc.* 56 (1-2): 1-114.

NORMAN, F. I. & MUMFORD, L. 1985. Studies on the Purple Swamphen, *Porphyrio porphyrio*, in Victoria. *Aust. Wildl. Res.* 12: 263-278.

NORRIS, C. A. 1947. Report on the distribution and status of the Corncrake. *British Birds* 40: 226-244.

NORTH, A. J. 1913. *Nests and eggs of birds found breeding in Australia and Tasmania*. Sydney: F. W. White.

NORTON, D. W. 1965. Notes on some non-passerine birds from eastern Ecuador. *Breviora* 230: 1-11.

NOSKE, R. A. 1996. Abundance, zonation and foraging ecology of birds in mangroves of Darwin Harbour, Northern Territory. *Aust. Wildl. Res.* 23: 443-474.

NUDDS, T. D. 1981. The effects of coots on duck densities in Saskatchewan parkland. *Wildfowl* 32: 19-22.

NUSSER, J. A., GOTO, R. M., LEDIG, D. B., FLEISCHER, R. C. & MILLER, M .M. 1996. RAPD analysis reveals low genetic variability in the endangered light-footed clapper rail. *Molecular Ecology* 5: 463-472.

OAKE, K. & HERREMANS, M. 1992. The Buffspotted Flufftail *Sarothrura elegans* new to Botswana. *Babbler* 24: 18-19.

OATES, E. W. 1901. *Catalogue of the collection of birds' eggs in the British Museum*. Vol. 1. London: Taylor & Francis.

OBERHOLSER, H. C. 1937. A revision of the Clapper Rails (*Rallus longirostris* Boddaert). *Proc. U.S. Natn. Mus.* 84: 313-354.

OBERHOLSER, H. C. 1974. *The bird life of Texas*. Austin: University of Texas Press.

O'DONNELL, C. F. J. 1994. Distribution and habitats of Spotless Crakes in Canterbury. *Notornis* 41: 211-213.

OGILVIE-GRANT, W. R. 1913. [Description of syntypes of *Rallicula klossi*, sp. n.]. *Bull. Brit. Orn. Club* 31: 104.

OHMART, R. D. & TOMLINSON, R. E. 1977. Foods of western Clapper Rails. *Wilson Bull.* 89: 332-336.

OLIVARES, A. 1957. Aves de la costa del Pacifico, Municipio de Guápi, Cauca, Colombia, II. *Caldasia* 8: 217-251.

OLIVARES, A. 1959. *Cinco aves que aparentemente no habian sido registradas en Colombia*. Lozania (Acta Zool. Colombiana) 12: 51-56.

OLIVARES, A. 1969. *Aves de Cundinamarca*. Bogotá: Universidad Nacional de Colombia.

OLIVARES, A. 1974. Aves de la Orinoquia Colombiana. *Inst. Cienc. Nat. Ornithologia Univ. Nac. Colombia*.

OLIVER, W. R. B. 1974. *New Zealand birds*. Wellington: Reed.

OLROG, C. C. 1959. *Las aves Argentinas: una guia de campo*. Tucumán: Universidad Nacional de Tucumán.

OLROG, C. C. 1984. *Las aves Argentinas*. Buenos Aires: Administración de Parques Nacionales.

OLSON, S. L. 1970. The relationships of *Porzana flaviventer*. *Auk* 87: 805-808.

OLSON, S. L. 1972. The American Purple Gallinule, *Porphyrula martinica*, on Ascension and St. Helena Islands. *Bull. Brit. Orn. Club* 92: 92-93.

OLSON, S. L. 1973a. Evolution of the rails of the South Atlantic Islands (Aves: Rallidae). *Smithsonian Contrib. Zool.* 152: 1-53.

OLSON, S. L. 1973b. A classification of the Rallidae. *Wilson Bull.* 85: 381-446.

OLSON, S. L. 1974a. The pleistocene rails of North America. *Condor* 76: 169-175.

OLSON, S. L. 1974b. A new species of *Nesotrochis* from Hispaniola, with notes on other fossil rails from the West Indies (Aves: Rallidae). *Proc. Biol. Soc. Washington* 87: 439-450.

OLSON, S. L. 1974c. Purple Gallinule carrying young. *Florida Field Nat.* 2: 15-16.

OLSON, S. L. 1975a. The fossil rails of C. W. de Vis, being mainly an extinct form of *Tribonyx mortierii* from Queensland. *Emu* 75: 49-54.

OLSON, S. L. 1975b. The south Pacific gallinules of the genus *Pareudiastes*. *Wilson Bull.* 87: 1-5.

OLSON, S. L. 1975c. A review of the extinct rails of the New Zealand region. *Natn. Mus. New Zealand Rec.* 1(3): 63-79.

OLSON, S. L. 1977. A synopsis of the fossil Rallidae. Pp. 339-373 in RIPLEY, S. D. *Rails of the world: a monograph of the family Rallidae*. Boston: Godine.

OLSON, S. L. 1986. *Gallirallus sharpei* (Büttikofer) nov. comb. A valid species of rail (Rallidae) of unknown origin. *Gerfaut* 76: 263-269.

OLSON, S. L. 1989. Extinction on islands: man as a catastrophe. In WESTERN, D. & PEARL, M.C. (eds.) *Conservation for the Twenty-first Century*. New York: Oxford University Press.

OLSON, S. L. 1991. Requiescat for *Tricholimnas conditicius*, a rail that never was. *Bull. Brit. Orn. Club* 112: 174-179.

OLSON, S. L. 1997. Towards a less imperfect understanding of the systematics and biogeography of the Clapper and King Rail complex (*Rallus longirostris* and *R. elegans*). Pp. 93-111 in DICKERMAN, R. W. (compiler). *The era of Allan R. Phillips: a festschrift*. Albuquerque, New Mexico: Horizon Communications.

OLSON, S. L. & JAMES, H. F. 1982. Fossil birds from the Hawaiian Islands: evidence for wholesale extinction by man before Western contact. *Science* 21: 633-635.

OLSON, S. L. & JAMES, H. F. 1991. Descriptions of thirty-two new species of birds from the Hawaiian Islands. Part I, non-Passeriformes. *Orn. Monogr.* 45: 1-88.

OLSON, S. L. & STEADMAN, D. W. 1987. Comments on the proposed suppression of *Rallus nigra* Miller 1784, Z. N. (S.) 2276, and *Columba r. forsteri* Wagler, 1829, Z. N. (S.) 2277. *Bull. Zool. Nomenclature* 44: 126-127.

ONLEY, D. 1982a. The Spotless Crake (*Porzana tabuensis*) on Aorangi, Poor Knights Island. *Notornis* 29: 9-21.

ONLEY, D. 1982b. The nomenclature of the Spotless Crake (*Porzana tabuensis*). *Notornis* 29: 75-79.

ORIANS, G. H. & PAULSON, D. R. 1969. Notes on Costa Rican birds. *Condor* 71: 426-431.

ORTIZ, F. & CARRIÓN, J. M. 1991. *Introducción a las aves del Ecuador*. Quito: Fecodes.

O.S.N.Z. 1990. *Annotated checklist of New Zealand birds*. Wellington: Reed.

OSBORNE, D. R. & BEISSINGER, S. R. 1979. The Paint-billed Crake in Guyana. *Auk* 96: 425.

OWEN, K. L. & SELL, M. G. 1985. The birds of Waimea Inlet. *Notornis* 32: 271-309.

PACHECO, J. F. & WHITNEY, B. M. 1995. Range extensions for some birds in northeastern Brazil. *Bull. Brit. Orn. Club* 115: 157-163.

PAKENHAM, R. H. W. 1943. Field notes on the birds of Zanzibar and Pemba. *Ibis* 85: 165-189.

PAKENHAM, R. H. W. 1945. Field notes on the birds of Zanzibar and Pemba Islands. *Ibis* 87: 216-223.

PAKENHAM, R. H. W. 1979. *The birds of Zanzibar and Pemba*. BOU Checklist No. 2. London: British Ornithologists' Union.

PALLISER, T. 1992. An unknown rail sighted in west New Britain. *Muruk* 5: 62-63.

PALMER, P. 1998. The Purple Gallinule in Cumbria – a new British bird? *Birding World* 10: 463-466.

PANDE, P., KOTHARI, A. & SINGH, S. 1991. *Andaman and Nicobar Islands*. New Delhi: Indian Institute of Public Administration.

PARKER, T. A. 1982. Observations of some unusual rainforest and marsh birds in southeastern Peru. *Wilson Bull.* 94: 477-493.

PARKER, T. A., CASTILLO U., A., GELL-MANN, M. & ROCHA O., O. 1991. Records of new and unusual birds from northern Bolivia. *Bull. Brit. Orn. Club* 111: 120-138.

PARKER, T. A., PARKER, S. A. & PLENGE, M. A. 1982. *An annotated checklist of Peruvian birds.* Vermilion, S Dakota: Buteo Books.

PARKER, T. A. & REMSEN, J. V. 1987. Fifty-two Amazonian bird species new to Bolivia. *Bull. Brit. Orn. Club* 107: 94-107.

PARKER, T. A., SCHULENBERG, T. S., KESSLER, M. & WUST, W. H. 1995. Natural history and conservation of the endemic avifauna in north-west Peru. *Bird Conserv. Internatn.* 5: 201-231.

PARKES, K. C. 1949. *Rallus philippensis* on Mindoro, Philippine Islands. *Auk* 66: 200-201.

PARKES, K. C. 1971. Taxonomic and distributional notes on Philippine birds. *Nemouria* 4: 1-67.

PARKES, K. C. & AMADON, D. 1959. A new species of rail from the Philippine Islands. *Wilson Bull.* 71: 303-306.

PARKES, K. C., KIBBE, D. P. & ROTH, E. L. 1978. First records of the Spotted Rail (*Pardirallus maculatus*) for the United States, Chile, Bolivia and western Mexico. *Amer. Birds* 32: 295-299.

PARNELL, G. W. 1967. Spotted Crake at Marandellas. *Honeyguide* 50: 10.

PARROTT, J. 1979. Kaffir Rail *Rallus caerulescens* in West Africa. *Malimbus* 1: 145-146.

PATON, P. W. C., SCOTT, J. M. & BURR, T. A. 1985. American Coot and Black-necked Stilt on the island of Hawaii. *Western Birds* 16: 175-181.

PATRIKEEV, M. 1995. Birds of Azerbaijan. Draft ms.

PATTERSON, R. M. 1989. A Black-tailed Native Hen near Hobart. *Tasmanian Bird Rep.* 18: 29-30.

PAULER, K. 1968. Eine Brut des Tüpfelsumpfhuhns (*Porzana porzana*) in Gefangenschaft. *Egretta* 11: 16-19.

PAULLIN, D. C. 1987. Cannibalism in American coots induced by severe spring weather and avian cholera. *Condor* 89: 442-443.

PAYNE, R. B. & MASTER, L. L. 1983. Breeding of a mixed pair of white-shielded and red-shielded American Coots in Michigan. *Wilson Bull.* 95: 467-469.

PAYNTER, R. A. 1950. A new Clapper Rail from the territory of Quintana Roo, Mexico. *Condor* 52: 139-140.

PAYNTER, R. A. 1955. The ornithogeography of the Yucatán Peninsula. *Peabody Mus. Nat. Hist. Bull.* 9: 1-347.

PAYNTER, R. A. 1963. Birds from Flores, Lesser Sunda Islands. *Breviora* 182: 1-5.

PAZ, U. 1987. *The birds of Israel.* Bromley, Kent, UK: Christopher Helm.

PEABODY, P. B. 1922. Haunts and breeding habits of the Yellow Rail *Coturnicops noveboracensis. J. Mus. Comp. Oology* 2: 33-44.

PEAKE, T. M., MCGREGOR, P. K., SMITH, K. W., TYLER, G., GILBERT, G. & GREEN, R. E. 1998. Individuality in Corncrake *Crex crex* vocalizations. *Ibis* 140: 120-127.

PEARCE-HIGGINS, J., THOMPSON, A., HALL, S., KEALEY, I. & KEMP, P. 1995. Nottingham University Bolivia project 1995: a survey of mammals and birds in the Parque Nacional Noel Kempff Mercado. Nottingham, UK: Unpublished.

PEARSON, D. L. 1975. Range extensions and new records for bird species in Ecuador, Peru, and Bolivia. *Condor* 77: 96-99.

PECK, G. K. & JAMES, R. D. 1983. *Breeding birds of Ontario – nidology and distribution.* Vol. 1: Nonpasserines. Toronto: Royal Ontario Museum.

DE LA PEÑA, M. R. 1992. *Guia des aves Argentinas.* 2nd. edn. Vol 2. Falconiformes – Charadriiformes. Buenos Aires: Literature of Latin America.

PENNY, M. J. 1974. *The birds of Seychelles and the outlying islands.* London: Collins.

PENNY, M. J. & DIAMOND, A. W. 1971. The White-throated Rail *Dryolimnas cuvieri* on Aldabra. *Phil. Trans. Roy. Soc. Lond.* B 260: 529-548.

PENRY, E. H. 1994. *A bird atlas of Botswana.* Pietermaritzburg: University of Natal Press.

PERCY, W. 1951. *Three studies in bird character: bitterns, herons and water rails.* London: Country Life.

PERENNOU, C., MUNDKUR, T., SCOTT, D. A., FOLLESTAD, A. & KVENILD, L. 1994. *The Asian Waterfowl Census 1987-91: distribution and status of Asian waterfowl.* AWB Publication 86. IWRB Publication 24. Kuala Lumpur & Slimbridge: AWB and IWRB.

PERES, C. A. & WHITTAKER, A. 1991. Annotated checklist of the bird species of the upper Rio Urucu, Amazonas, Brazil. *Bull. Brit. Orn. Club* 111: 156-171.

PEREYRA, J. A. 1931. Los crescicus (gallinetas enanas). *Hornero* 4: 414-415.

PEREYRA, J. A. 1938. Algunos nidos poco conocidos de nuestra avifauna. *Hornero* 7: 25-30.

PEREYRA, J. A. 1944. Descripcion de un nuevo ejemplar de ralido de las Isla Georgia del Sud. *Hornero* 8: 484-489.

PEREZ, G. S. A. 1968. Notes on the breeding season of Guam rails (*Rallus owstoni*). *Micronesica* 4: 133-135.

PERKINS, R. C. L. 1903. Vertebrata. In *Fauna Hawaiiensis, or The zoology of the Sandwich (Hawaiian) Isles,* vol. 1, pt. 4. Cambridge, England.

PETERS, J. L. & GRISCOM, L. 1928. A new rail and a new dove from Micronesia. *Proc. New Eng. Zool. Club* 10: 99-106.

PETERS, J. L. 1934. *Check-list of birds of the world.* Vol. 2. Cambridge, Mass.: Harvard University Press.

PETERSON, R. T. & CHALIF, E. L. 1973. *A field guide to Mexican birds.* Boston: Houghton Mifflin.

PETRIE, M. 1983. Female Moorhens compete for small fat males. *Science* 220: 413-414.

PETRIE, M. 1984. Territory size in the Moorhen (*Gallinula chloropus*): an outcome of RHP asymmetry between neighbours. *Animal Behav.* 32: 861-870.

PETRIE, M. 1986. Reproductive strategies of male and female moorhens (*Gallinula chloropus*). Pp. 43-63 in D. I. RUBENSTEIN & R. W. WRANGHAM (eds.) *Ecological aspects of social evolution.* Princeton: Princeton University Press.

PHILLIPPS, W. J. 1959. The last (?) occurrence of *Notornis* in the North Island. *Notornis* 8: 93-94.

PHILLIPS, H. A. 1991. Eurasian Coot *Fulica atra* eating waterbird faeces. *Aust. Bird Watcher* 14: 150.

PHILLIPS, N. J. 1984. Migrant species new to the Seychelles. *Bull. Brit. Orn. Club* 104: 9-10.

PHILLIPS, W. W. A. 1978. *Annotated checklist of the birds of Ceylon.* Colombo: Wildl. & Nature Protection Soc. Sri Lanka & Ceylon Bird Club.

PIMM, S. L. 1987. The snake that ate Guam. *Trends in Ecology and Evolution* 2: 293-295.

PINTO, A. A. DA ROSA 1983. *Ornitologia de Angola.* Vol 1 (Non-passeres). Lisbon: Instituto de Investigação Cientifica Tropical.

PINTO, O. M. D'OLIVIERA 1964. *Ornitologia brasiliense.* Vol. 1. Parte Introdutória e Familias Rheidae a Cuculidae. São Paulo: Secretaria da Agricultura do Estado de São Paulo.

PITMAN, C. R. S. 1929. Notes on the breeding habits and nesting of *Limnocorax flavirostra* – the Black Rail. *Ool. Rec.* 9: 37-41.

PITMAN, C. R. S. 1965. The nest and eggs of the Striped Crake, *Porzana marginalis* Hartlaub. *Bull. Brit. Orn. Club* 85: 32-40.

PIZZEY, G. & DOYLE, R. 1980. *A field guide to the birds of Australia.* Sydney: Collins.

PORTEOUS, B. & ACEVEDO, C. 1996. Potentially important populations of Chocó Tinamou *Crypturellus kerriae* and Brown Wood-rail *Aramides wolfi* in Colombia. *Cotinga* 6: 31-32.

POSPICHAL, L. B. & MARSHALL, W. H. 1954. A field study of Sora Rail and Virginia Rail in central Minnesota. *Flicker* 26: 2-32.

POTAPOV, R. L. & FLINT, V. E. (eds.). 1987. Galliformes, Gruiformes. Vol. 4. In ILLICEV, V. D. & FLINT, V. E. 1985-1989. *Handbuch der Vögel der Sowjetunion*. Wittenberg Lutherstadt: A. Ziemsen Verlag.

POULSEN, M. K. 1995. The threatened and near-threatened birds of Luzon, Philippines, and the role of the Sierra Madre mountains in their conservation. *Bird Conserv. Internatn.* 5: 79-115.

PRACY, L. T. 1969. Weka liberations in the Palliser Bay region. *Notornis* 26: 212-213.

VAN PRAET, L. 1980. Breeding *Laterallus leucopyrrhus*. *Avicult. Mag.* 86: 60.

PRATT, H. D. 1978. Do mainland coots occur in Hawaii? *Elepaio* 38: 73.

PRATT, H. D. 1987. Occurrence of the North American Coot (*Fulica americana americana*) in the Hawaiian Islands, with comments on the taxonomy of the Hawaiian Coot. *Elepaio* 47: 25-28.

PRATT, H. D. 1993. *Enjoying birds in Hawaii: a birdfinding guide to the Fiftieth State*. Honolulu: Mutual Publishing.

PRATT, H. D., BRUNER, P. L. & BERRETT, D. G. 1979. America's unknown avifauna: the birds of the Marianas Islands. *Amer. Birds* 33: 227-235.

PRATT, H. D., BRUNER, P. L. & BERRETT, D. G. 1987. *A field guide to the birds of Hawaii and the tropical Pacific*. Princeton, New Jersey: Princeton University Press.

PRATT, H. D., ENGBRING, J., BRUNER, P. L. & BERRETT, D. G. 1980. Notes on the taxonomy, natural history and status of the resident birds of Palau. *Condor* 82: 117-131.

PRATT, P. D. 1992. Black Rail: new to Ontario and Canada. *Ontario Birds* 10: 90-92.

PRATT, T. K. 1982. Additions to the avifauna of the Adelbert Range, Papua New Guinea. *Emu* 82: 117-125.

PRESTWICK, A. A. 1964. News and views. *Avicult. Mag.* 70: 150.

PRILL, A. G. 1931. A land migration of coots. *Wilson Bull.* 43: 148-149.

PRINCE, P. A. & CROXALL, J. P. 1996. The birds of South Georgia. *Bull. Brit. Orn. Club* 116: 81-104.

PRITCHARD, W. T. 1866. *Polynesian reminiscences; or life in the South Pacific islands*. London: Chapman & Hall.

PRYOR, G. R. 1969. Breeding the Blackish Rail (*Rallus nigricans*). *Avicult. Mag.* 75: 164-165.

PULICH, W. M. 1961. A record of the Yellow Rail from Dallas County, Texas. *Auk* 78: 639-640.

PULIDO, V. 1991. *El libro rojo de la fauna silvestre del Perú*. Lima: INIAA, WWF & US Fish & Wildlife Service.

PYE-SMITH, G. 1950. The nest and eggs of the Uganda White-spotted Crake (*Sarothrura pulchra centralis*). *Ool. Rec.* 24: 48-49.

RABOR, D. S. 1977. *Philippine birds and mammals*. Quezon City: University of Philippines Press.

RAFFAELE, H. A. 1989. *A guide to the birds of Puerto Rico and the Virgin Islands*. Princeton, New Jersey: Princeton University Press.

RAGLESS, G. B. 1977. The Chestnut Rail at Darwin. *S. Aust. Orn.* 27: 254-255.

RAMANAMPAMONJY, J. R. 1995. Rencontre inattendue avec le Râle d'Olivier (*Amaurornis olivieri*) au Lac Bemamba. *Newlsetter, Working Group on Birds in the Madagascar Region* 5(2): 5-7.

RAND, A. L. 1936. The distribution and habits of Madagascar birds. *Bull. Amer. Mus. Nat. Hist.* 72: 143-499.

RAND, A. L. 1938. Results of the Archbold Expeditions. No. 19. *Amer. Mus. Novit.* 990.

RAND, A. L. 1942a. Results of the Archbold Expeditions. No. 42, Birds of the 1936-37 New Guinea Expedition. *Bull. Amer. Mus. Nat. Hist.* 79: 289-366.

RAND, A. L. 1942b. Results of the Archbold Expeditions. No. 43, Birds of the 1938-39 expedition. *Bull. Amer. Mus. Nat. Hist.* 79: 425-515.

RAND, A. L. 1951. Birds from Liberia, with a discussion of barriers between Upper and Lower Guinea sub-species. *Fieldiana Zool.* 32: 561-653.

RAND, A. L. & GILLIARD, E. T. 1967. *Handbook of New Guinea birds*. London: Weidenfeld & Nicholson.

RAND, A. L. & RABOR, D. S. 1960. Birds of the Philippine Islands: Siquijor, Mount Malindang, Bohol, and Samar. *Fieldiana Zool.* 35: 225-441.

RASMUSSEN, J. F., RAHBEK, C., POULSEN, B. O., POULSEN, M. K. & BLOCH, H. 1996. Distributional records and natural history notes on threatened and little known birds of southern Ecuador. *Bull. Brit. Orn. Club* 116: 26-46.

REAGAN, W. W. 1977. Resource partitioning in the North American gallinules in southern Texas. Logan, Utah: Utah State University (MS thesis).

RECHER, H. F. & CLARK, S. S. 1974. A biological survey of Lord Howe Island, with recommendations for the conservation of the island's wildlife. *Biol. Conserv.* 6: 263-273.

REED, C. 1965. *North American birds' eggs*. New York: Dover Publications.

REED, C. S. 1941. Notas referentes a *Laterallus jamaicensis salinasi*, Phil. *Publicación Oficial No. 14 del Jardín Zoológico Nacional de Chile* 2: 7-21.

REED, S. 1980. The birds of Savai'i, Western Samoa. *Notornis* 27: 151-159.

REGALADO RUÍZ, P. 1981. El género *Torreornis* (Aves: Fringillidae), descripción de una nueva subespecie en Cayo Coco, Cuba. *Centro Agrícola* 2: 87-112.

REICHHOLF, J. H. 1991. Crex crex: eine kurze biologische Charakteristierung. *Vogelwelt* 112: 6-9.

REID, B. E. 1967. Some features of recent research on the Takahe (*Notornis mantelli*). *Proc. N. Z. Ecol. Soc.* 14: 79-87.

REID, B. E. 1969. Survival status of the Takahe, *Notornis mantelli*, of New Zealand. *Biol. Conserv.* 1: 237-240.

REID, B. E. 1971. Takahe – a vanishing species. *Wildlife – A Review*. Wellington: Dept. of Internal Affairs 3: 39-42.

REID, B. E. 1974a. Sightings and records of the Takahe (*Notornis mantelli*) prior to its "official rediscovery" by Dr. G. B. Orbell in 1948. *Notornis* 21: 277-295.

REID, B. E. 1974b. Faeces of Takahe (*Notornis mantelli*): a general discussion relating the quantity of faeces to the type of food and to the estimated energy requirements of the bird. *Notornis* 21: 306-311.

REID, B. E. 1977. Takahe at Mt. Bruce. *Wildlife – A Review*. Wellington: Dept. of Internal Affairs 8: 56-73.

REID, F. A. 1989. Differential habitat use by waterbirds in a managed wetland complex. Columbia: University of Missouri (PhD dissertation).

REID, F. A., MEANLEY, B. & FREDRICKSON, L. H. 1994. King Rail. In TACHA, T. C. & BRAUN, C. E. (eds.). *Management of migratory shore and upland game birds in North America*. Washington D.C.: International Assoc. Fish & Wildlife Agencies.

REITHMÜLLER, E. 1931. Nesting notes from Willis Island. *Emu* 31: 142-246.

REMSEN, J. V. & PARKER, T. A. 1990. Seasonal distribution of the Azure Gallinule (*Porphyrula flavirostris*), with comments on vagrancy in rails and gallinules. *Wilson Bull.* 102: 380-399.

REMSEN, J. V. & TRAYLOR, M. A. 1983. Additions to the avifauna of Bolivia, part 2. *Condor* 85: 95-98.

REMSEN, J. V. & TRAYLOR, M. A. 1989. *An annotated list of the birds of Bolivia*. Vermilion, S Dakota: Buteo Books.
REMSEN, J. V., TRAYLOR, M. A. & PARKES, K. C. 1985. Range extensions for some Bolivian birds, 1 (Tinamiformes to Charadriiformes). *Bull. Brit. Orn. Club* 105: 124-130.
RENSCH, B. 1931. Die Vogelwelt von Lombok, Sumbawa und Flores. *Mitt. Zool. Mus. Berlin* 17: 451-637.
REPKING, C. F. 1975. Distribution and habitat requirements of the Black Rail (*Laterallus jamaicensis*) along the lower Colorado River. Tempe, Arizona: Arizona State University (MSc thesis).
REPKING, C. F. & OHMART, R. D. 1977. Distribution and density of Black Rail populations along the lower Colorado River. *Condor* 79: 486-489.
REYNARD, G. B. 1974. Some vocalizations of the Black, Yellow and Virginia Rails. *Auk* 91: 747-756.
RICH, E. 1973. Black-tailed Native Hen in desert country. *Aust. Bird Watcher* 5: 28-29.
RICHARDSON, M. E. 1984. Aspects of the ornithology of the Tristan da Cunha group and Gough Island, 1972-1974. *Cormorant* 12: 123-201.
RIDGELY, R. S. 1981. *A guide to the birds of Panama*. Princeton, New Jersey: Princeton University Press.
RIDGELY, R. S. & GAULIN, S. J. C. 1980. The birds of Finca Merenberg, Huila Department, Colombia. *Condor* 82: 379-391.
RIDGELY, R. S. & GWYNNE, J. A. 1989. *A guide to the birds of Panama with Costa Rica, Nicaragua and Honduras*. Princeton, New Jersey: Princeton Univ. Press.
RIDGELY, R. S., GREENFIELD, P. J. & GUERRERO, M. (in press). *Una lista Anotada de las aves de Ecuador continental*. Quito: CECIA.
RIDGELY, R. S., GREENFIELD, P. J. (in prep.) *The birds of Ecuador*. 2 Vols. Cornell University Press.
RIDGWAY, R. 1895. On birds collected by Dr W. L. Abbott in the Seychelles, Amirantes, Gloriosa, Assumption, Aldabra and adjacent islands, with notes on habits etc. by the collector. *Proc. U.S. Natn. Mus.* 18: 509-539.
RIDGWAY, R. 1912. *Color standards and color nomenclature*. Washington, DC: Privately published.
RIDGWAY, R. & FRIEDMANN, H. 1941. The birds of North and Middle America. *Bull. U.S. Natn. Mus.* 50: 1-254.
RIDPATH, M. G. 1964. The Tasmanian Native Hen. *Aust. Nat. Hist.* 14: 346-350.
RIDPATH, M. G. 1972a. The Tasmanian Native Hen, *Tribonyx mortierii*. I Patterns of behaviour. *CSIRO Wildl. Res.* 17: 1-51.
RIDPATH, M. G. 1972b. The Tasmanian Native Hen, *Tribonyx mortierii*. II The individual, the group and the population. *CSIRO Wildl. Res.* 17: 53-90.
RIDPATH, M. G. 1972c. The Tasmanian Native Hen, *Tribonyx mortierii*. III Ecology. *CSIRO Wildl. Res.* 17: 91-118.
RIDPATH, M. G. & MELDRUM, G. K. 1968a. Damage to pastures by the Tasmanian Native Hen *Tribonyx mortierii*. *CSIRO Wildl. Res.* 13: 11-24.
RIDPATH, M. G. & MELDRUM, G. K. 1968b. Damage to oat crops by the Tasmanian Native Hen *Tribonyx mortierii*. *CSIRO Wildl. Res.* 13: 25-54.
RILEY, J. H. 1924. A collection of birds from north and north-central Celebes. *Proc. U.S. Natn. Mus.* 64(16): 1-118.
RILEY, J. H. 1938. Three birds from Banka and Borneo. *Proc. Biol. Soc. Washington* 51: 95-96.
RINEY, T. A. & MIERS, K. H. 1956. Initial banding of *Notornis mantelli*. *Notornis* 6: 181-184.
RINKE, D. 1986. Notes on the avifauna of Niuafo'ou Island, Kingdom of Tonga. *Emu* 86: 82-86.
RINKE, D. 1987. The avifauna of 'Eua and its off-shore islet Kalau, Kingdom of Tonga. *Emu* 87: 26-34.
RIPLEY, S. D. 1957a. Notes on the Horned Coot, *Fulica cornuta* Bonaparte. *Postilla* 30: 1-8.

RIPLEY, S. D. 1957b. Additional notes on the Horned Coot, *Fulica cornuta* Bonaparte. *Postilla* 32: 1-2.
RIPLEY, S. D. 1961. *A synopsis of the birds of India and Pakistan*. Bombay: Bombay Natural History Society.
RIPLEY, S. D. 1964. A systematic and ecological study of birds of New Guinea. *Peabody Mus. of Nat. Hist. Yale Univ. Bull.* 19: 1-87.
RIPLEY, S. D. 1977. *Rails of the world*. Boston: Godine.
RIPLEY, S. D. 1982. *A synopsis of the birds of India and Pakistan*. Bombay: Bombay Natural History Society.
RIPLEY, S. D. & BEEHLER, B. M. 1985. Rails of the world, a compilation of new information, 1975-1983 (Aves: Rallidae). *Smithsonian Contrib. Zool.* 417: 1-28.
RIPLEY, S. D. & HEINRICH, G. H. 1960. Additions to the avifauna of northern Angola. II. *Postilla* No. 95: 1-29.
RIPLEY, S. D. & OLSON, S. L. 1973. Re-identification of *Rallus pectoralis deignani*. *Bull. Brit. Orn. Club* 93: 115.
RISEN, K. W. 1992. Unusual feeding behaviour observed for King Rail. *Loon* 64: 162-163.
RITTER, M. W. 1994. Notes on nesting and growth of Mariana Common Moorhens on Guam. *Micronesica* 27: 127-132.
RITTER, M. W. & SWEET, T. M. 1993. Rapid colonization of a human-made wetland by Mariana Common Moorhen on Guam. *Wilson Bull.* 105: 685-687.
ROBBINS, M. B. & EASTERLA, A. 1992. *Birds of Missouri, their distribution and abundance*. Columbia: University of Missouri Press.
ROBERSON, D. & BAPTISTA, L. F. 1988. White-shielded coots in North America: a critical evaluation. *Amer. Birds* 42: 1241-1246.
ROBERT, M. 1996. Yellow Rail (*Coturnicops noveboracensis*). In GAUTHIER, J. & AUBRY, Y. (eds.) *Atlas of the breeding birds of southern Quebec*. Montreal: Canadian Wildlife Service.
ROBERT, M. 1997. A closer look: Yellow Rail. *Birding* 29: 283-290.
ROBERT, M., CLOUTIER, L. & LAPORTE, P. 1997. Summer diet of the Yellow Rail in southern Quebec. *Wilson Bull.* 109: 702-710.
ROBERT, M. & LAPORTE, P. 1993. Le râle jaune (*Coturnicops noveboracensis*) à l'Ile aux Grues et à Cacouna. Rapport d'activités estivales – 1993. Sainte-Foy, Quebec: Service Canadien de la Faune.
ROBERT, M. & LAPORTE, P. 1996. *Le Râle Jaune dans le sud du Québec: inventaires, habitats et nidification*. Technical Report Series No. 247. Sainte-Foy, Quebec: Canadian Wildlife Service.
ROBERT, M. & LAPORTE, P. 1997. Field techniques for studying breeding Yellow Rails. *J. Field Orn.* 68: 56-63.
ROBERT, M., LAPORTE, P. & SHAFFER, F. 1995. Plan d'action pour le rétablissement du Râle Jaune (*Coturnicops noveboracensis*) au Québec. Sainte-Foy, Quebec: Canadian Wildlife Service.
ROBERTS, G. J. 1975. Observations of water birds in south west Queensland. *Sunbird* 6: 69-75.
ROBERTS, G. J. 1980. Records of interest from the Alice Springs region. *S. Aust. Orn.* 28: 99-102.
ROBERTS, G. J. 1981. Observations of water-birds at the Alice Springs sewage ponds. *S. Aust. Orn.* 28: 175-179.
ROBERTS, T. J. 1991. *The birds of Pakistan*. Vol. 1. Non-passeriformes. Karachi: Oxford University Press.
ROBERTS, T. S. 1932. *The birds of Minnesota*. Minneapolis: University of Minnesota Press.
ROBERTSON, C. J. R. 1985. *Reader's Digest complete book of New Zealand birds*. Sydney: Reader's Digest.
ROBERTSON, D. B. 1976. Weka liberation in Northland. *Notornis* 23: 213-219.
ROBINSON, A. C. 1995. Breeding pattern in the Banded Rail (*Gallirallus philippensis*) in Western Samoa. *Notornis* 42: 46-48.

ROBSON, C. 1989. Recent reports. *Bull. Oriental Bird Club* 10: 44.
ROBSON, C. 1996. From the field. *Bull. Oriental Bird Club* 23: 53.
ROBSON, N. F. & HORNER, R. 1996. The birds of the Ozabeni Section of the Greater St Lucia Wetland Park. Pietermaritzburg, South Africa: Unpublished report.
ROCHA, O. & PEÑARANDA, E. 1995. Avifauna de Huaraco: una localidad de la puna semiárida del altiplano central, Departamento de La Paz, Bolivia. *Cotinga* 3: 17-25.
RODEWALD, P. G., DEJAIFVE, P.-A. & GREEN, A. A. 1994. The birds of Korup National Park and Korup Project Area, Southwest Province, Cameroon. *Bird Conserv. Internatn.* 4: 1-68.
RODRIGUEZ, R. & HIRALDO, F. 1975. [Dietary requirements of *Porphyrio porphyrio* in the Guadalquivir marshes.] *Doñana Acta Vertebrata* 2: 201-213.
RODWELL, S. P. 1996. Notes on the distribution and abundance of birds observed in Guinea-Bissau. *Malimbus* 18: 25-43.
RODWELL, S. P., SAUVAGE, A., RUMSEY, S. J. R. & BRÄUNLICH, A. 1996. An annotated check-list of birds occurring at the Parc National des Oiseaux du Djoudj in Senegal, 1984-1994. *Malimbus* 18: 74-111.
ROE, N. A. & REES, W. E. 1979. Notes on the Puna avifauna of Azángaro Province, Department of Puno, southern Peru. *Auk* 96: 475-482.
ROGACHEVA, H. 1992. *The birds of central Siberia*. Husum, Germany: Husum-Druck und Verlagsgesellschaft.
ROOT, T. 1988. *Atlas of wintering North American birds*. Chicago: University of Chicago Press.
DO ROSÁRIO, L. A. 1996. *As aves em Santa Catarina: distribuição geográfica e meio ambiente*. Florianópolis, Santa Catarina: Fundação do Meio Ambiente.
ROSE, P. M. & SCOTT, D. A. 1994. *Waterfowl population estimates*. Slimbridge, UK: International Waterfowl and Wetlands Research Bureau (IWRB Special Publication 29).
ROSENBERG, D. K. 1990. The impact of introduced herbivores on the Galapagos Rail (*Laterallus spilonotus*). *Monogr. Syst. Bot. Missouri Bot. Gard.* 32: 169-178.
ROSS, C. A. 1988. Weights of some New Caledonian birds. *Bull. Brit. Orn. Club* 108: 91-93.
ROST, V. F. 1995. Der Brutbestand von Blaßhuhn (*Fulica atra*) und Teichhuhn (*Gallinula chloropus*) in Thüringen 1994. *Anz. Ver. Thüring. Orn.* 2: 145-157.
ROTH, R. R., NEWSOM, J. D., JOANEN, T. & MCNEASE, L. L. 1972. The daily and seasonal behaviour patterns of the clapper rail (*Rallus longirostris*) in the Louisiana coastal marshes. *Proc. Southeast Assoc. Game & Fish Comm.* 26: 136-159.
ROTHSCHILD, LORD. 1928. [The hitherto unknown egg of the Flightless Rail (*Atlantisia rogersi* Lowe) of Inaccessible Island, Tristan d'Acunha group.] *Bull. Brit. Orn. Club* 48: 121-122.
ROTHSCHILD, W. 1893. *The avifauna of Laysan and the neighbouring islands*. Part 1. London: Porter.
ROTHSCHILD, W. 1903. [Original description of *Rallus wakensis*.] *Bull. Brit. Orn. Club* 13: 78.
ROTHSCHILD, W. 1907. *Extinct birds*. London: Hutchinson & Co.
ROTHSCHILD, W. & HARTERT, E. 1915. The birds of Dampier Island. *Novit. Zool.* 22: 26-37.
ROUX, F. & BENSON, C. W. 1969. A note on *Sarothrura lugens*. *Bull. Brit. Orn. Club* 89: 67-68.
ROUX, F. & MOREL, G. 1964. Le Sénégal, région privilégiée pour les migrateurs paléarctiques. *Ostrich Suppl.* 6: 249-254.
RSPB. 1995. *Species Action Plan 0421 Corncrake Crex crex a Red Data Bird*. Sandy, UK: Royal Society for the Protection of Birds.

ROZENDAAL, F. G. & DEKKER, R. W. R. J. 1989. Annotated checklist of the birds of the Dumoga-Bone National Park, North Sulawesi. *Kukila* 4: 85-109.
RUNDLE, W. D. & FREDERICKSON, L. H. 1981. Managing seasonally flooded impoundments for migrant rails and shorebirds. *Wildl. Soc. Bull.* 9: 80-87.
RUNDLE, W. D. & SAYRE, M. W. 1983. Feeding ecology of migrant Soras in southeastern Missouri. *J. Wildl. Manage.* 47: 1153-1159.
RUSCHI, A. 1979. *Aves do Brasil*. São Paulo: Editora Rios.
RUSSELL, S. M. 1964. *A distributional survey of the birds of British Honduras*. Orn. Monogr. 1. American Ornithologists' Union.
RUSSELL, S. M. 1966. Status of the Black Rail and Gray-breasted Crake in British Honduras. *Condor* 68: 105-107.
RUTGERS, A. & NORRIS, K. A. 1970. *Encyclopaedia of Aviculture*. London: Blandford Press.
RYAN, C. 1997. Observations on the breeding behaviour of the Takahe (*Porphyrio mantelli*) on Mana Island. *Notornis* 44: 233-240.
RYAN, C. & JAMIESON, I. G. (in press). Estimating the home range and carrying capacity for Takahe on predator-free offshore islands: implications for future management. *N. Z. J. Ecol.* 22.
RYAN, D. A., BAWDEN, K. M., BERMINGHAM, K. T. & ELGAR, M. A. 1996. Scanning and tail-flicking in the Australian Dusky Moorhen (*Gallinula tenebrosa*). *Auk* 113: 499-501.
RYAN, M. R. 1981. Evasive behaviour of American Coots to kleptoparasitism by waterfowl. *Wilson Bull.* 93: 274-275
RYAN, M. R. & DINSMORE, J. J. 1979. A quantitative study of the behaviour of breeding American Coots. *Auk* 96: 704-713.
RYAN, M. R. & DINSMORE, J. J. 1980. The behavioral ecology of breeding American Coots in relation to age. *Condor* 82: 320-327.
RYAN, P. G., WATKINS, B. P. & SIEGFRIED, W. R. 1989. Morphometrics, metabolic rate and body temperature of the smallest flightless bird: the Inaccessible Island Rail. *Condor* 91: 465-467.
RYDER, R. A. 1963. Migration and population dynamics of American coots in western North America. *Proc. Internatn. Orn. Congr.* 13: 441-453.
SAAB, V. A. & PETIT, D. R. 1992. Impact of pasture development on winter bird communities in Belize, Central America. *Condor* 94: 66-71.
SALA, A. 1991. La Talève poule-sultane *Porphyrio porphyrio madagascariensis* à Yaoundé, Cameroun. *Malimbus* 13: 78.
SALAMOLARD, M. 1995. Râle des genêts. Au secours du "roi des cailles". *Oiseau et R.F.O.* 39: 34-37.
SALATHÉ, T. 1991. Möglichkeiten und Probleme eines internationalen Schutzprogramms für den Wachtelkönig. *Vogelwelt* 112: 108-116.
SALVAN, J. 1968. Contribution à l'étude des oiseaux du Tchad. *Oiseau et R.F.O.* 38: 53-85.
SALVAN, J. 1970. Remarques sur l'évolution de l'avifauna malgache depuis 1945. *Alauda* 38: 191-203.
SALVAN, J. 1972a. Essai d'évaluation des densités d'oiseaux dans quelques biotopes malgaches. *Alauda* 40: 163-170.
SALVAN, J. 1972b. Statut, recensement, reproduction des oiseaux dulçaquicoles aux environs de Tan arive. *Oiseau et R.F.O.* 42: 35-51.
SALVIN, O. 1873. Note on the *Fulica alba* of White. *Ibis* (3)3: 295 & Plate X.
SALVIN, O. 1876. On the avifauna of the Galápagos Archipelago. *Trans. Zool. Soc. Lond.* 9: 447-510.
SALVIN, O. & GODMAN, F. D. 1903. *Biologia Centrali-Americana*. Aves: Vol. 3. London: Taylor & Francis.
SÁNCHEZ-LAFUENTE, A. M., REY, P., VALERA, F. & MUÑOZ-COBO, J. 1992. Past and current distribution of the purple swamphen *Porphyrio porphyrio* L. in the Iberian Peninsula. *Biol. Conserv.* 61: 23-30.

SANGSTER, G. 1998. Purple Swamp-hen is a complex of species. *Dutch Birding* 20: 13-22.

SANTORO, V., GANDARA, A. & GANDARA, E. 1988. Gallareta *Fulica cornuta* en la región de Antofagasta. *Medio Ambiente* 9: 88-93.

SATO, M., OGI, H., TANAKA, M. & SUGIYAMA, A. 1997. Band-bellied Crake (*Rallina paykullii*) take at Oshima Island, Hokkaido, Japan. *J. Yamashina Inst. Orn.* 29: 102-107.

SAUVAGE, A. 1993. Notes complémentaires sur l'avifaune du Niger. *Malimbus* 14: 44-47.

SAVIDGE, J. A. 1987. Extinction of an island forest avifauna by an introduced snake. *Ecology* 68: 660-668.

SAYRE, M. W. & RUNDLE, W. D. 1984. Comparison of habitat use by migrant Soras and Virginia Rails. *J. Wildl. Manage.* 48: 599-605.

SCARLETT, R. J. 1979. Letter dated 6 October 1978. *Notornis* 26: 99.

SCARLETT, R. J. 1972. Bones for the New Zealand archaeologist. *Canterbury Mus. Bull.* 4: 1-69.

SCHÄFFER, N. 1994. Methoden zum Nachweis von Bruten des Wachtelkönigs *Crex crex*. *Vogelwelt* 115: 69-73.

SCHÄFFER, N. 1995. Rufverhalten und Funktion des Rufens beim Wachtelkönig *Crex crex*. *Vogelwelt* 116: 141-151.

SCHÄFFER, N. & WEISSER, W. 1996. Modell für den Schutz des Wachtelkönigs *Crex crex*. *J. Orn.* 137: 53-90.

SCHALDACH, W. J. 1963. The avifauna of Colima and adjacent Jalisco, Mexico. *Proc. Western Fnd. Vert. Zool.* 1(1).

SCHMID, C. K. 1993. Birds of Nokopo. *Muruk* 6(2): 1-62.

SCHMITT, C. G. & SCHMITT, D. C. 1987. Extensions of range of some Bolivian birds. *Bull. Brit. Orn. Club* 107: 129-134.

SCHMITT, M. B. 1975. Observations on the Black Crake in the southern Transvaal. *Ostrich* 46: 129-138.

SCHMITT, M. B. 1976. Observations on the Cape Rail in the southern Transvaal. *Ostrich* 47: 16-26.

SCHMUTZ, E. 1977. Die Vögel der Manggarai (Flores). Unpublished.

SCHMUTZ, E. 1978. Addenda und corrigenda. Unpublished.

SCHNEIDER-JACOBY, M. 1991. Verbreitung und Bestand des Wachtelkönigs in Jugoslawien. *Vogelwelt* 112: 48-57.

SCHODDE, R., VAN TETS, G. F., CHAMPION, C. R. & HOPE, G. S. 1975. Observations on birds at glacial altitudes on the Carstensz Massif, western New Guinea. *Emu* 75: 65-72.

SCHODDE, R. & DE NAUROIS, R. 1982. Patterns of variation and dispersal in the Buff-banded Rail (*Gallirallus philippensis*) in the south-west Pacific, with description of a new subspecies. *Notornis* 29: 131-142.

SCHODDE, R. & TIDEMANN, S. C. (eds.) 1990. *Readers Digest complete book of Australian birds*. 2nd edn. Surrey Hills, NSW: Readers Digest Services Pty.

SCHÖNWETTER, M. 1935. Vogeleier aus Neubrittanien. *Beitr. Fortpfl. Biol. Vögel.* 11: 129-136.

SCHÖNWETTER, M. 1961-62. *Handbuch der Oologie*. Lieferung 5-6. Berlin: Akademie-Verlag.

SCHRADER, N. W. 1974. Invasion of Black-tailed Native-hens. *Aust. Bird Watcher* 5: 234.

SCHULENBERG, T. S. & REMSEN, J. V. 1982. Eleven bird species new to Bolivia. *Bull. Brit. Orn. Club* 102: 52-57.

SCHUZ, E. 1959. *Die Vogelwelt des südkaspischen Tieflandes*. Stuttgart: E. Scheizer-bart'sche Verlagsbuchhandlung.

SCHWARTZ, C. W. & SCHWARTZ, E. R. 1952. The Hawaiian Coot. *Auk* 69: 446-449.

SCLATER, P. L. & SALVIN, O. 1879. On the birds collected by T. K. Salmon in the state of Antioquia, United States of Colombia. *Proc. Zool. Soc. Lond.* 1879: 486-550.

SCLATER, W. L. 1906. *The birds of South Africa*. Vol. 4. London: R. H. Porter.

SCLATER, W. L. & MOREAU, R. E. 1932. Taxonomic and field notes on some birds of northeastern Tanganyika Territory. Part IV. *Ibis* (13)2: 487-522.

SCLATER, W. L. & MOREAU, R. E. 1933. Taxonomic and field notes on some birds of northeastern Tanganyika Territory. Part V. *Ibis* (13)3: 399-440.

SCOTT, D. A. 1989. *A directory of Asian wetlands*. Gland, Switzerland and Cambridge, UK: International Union for Conservation of Nature and Natural Resources.

SCOTT, D. A. & CARBONELL, M. 1986. *A directory of Neotropical Wetlands*. Cambridge, UK: International Union for Conservation of Nature and Natural Resources.

SCOTT, J. H. 1882. Macquarie Island. *Trans. Proc. N.Z. Inst.* 15: 484-493.

SEGRE, A., HAILMAN, J. P. & BEER, C. G. 1968. Complex interactions between Clapper Rails and Laughing Gulls. *Wilson Bull.* 80: 213-219.

SÉRIOT, J., PINEAU, O., DE SCHATZEN, R. & DUBOIS, P. J. 1986. Black-tailed Crake *Porzana bicolor*: a new species for Thailand. *Forktail* 2: 101-103.

SERLE, W. 1939. Observations on the breeding habits of some Nigerian Rallidae. *Ool. Rec.* 19: 61-70.

SERLE, W. 1954. A second contribution to the ornithology of the British Cameroons. *Ibis* 96: 47-80.

SERLE, W. 1957. A contribution to the ornithology of the eastern region of Nigeria. *Ibis* 99: 371-418.

SERLE, W. 1959. Some breeding records of birds at Ndian, British Southern Cameroons. *Nigerian Field* 24: 76-77.

SERPELL, J., COLLAR, N., DAVIS, S. & WELLS, S. 1983. Submission to the Foreign and Commonwealth Office on the future conservation of Henderson Island in the Pitcairn group. Unpublished.

SERVENTY, D. L. 1953. The southern invasion of northern birds during 1952. *W. Aust. Nat.* 3: 177-196.

SERVENTY, D. L. 1958. A new bird for the Australian list: the Malay Banded Crake. *Emu* 58: 415-418.

SERVENTY, D. L. & WHITTELL, H. M. 1976. *Birds of Western Australia*. 5th edn. Perth: Univ. of Western Australia Press.

SERVENTY, V. 1959. The birds of Willis Island. *Emu* 59: 167-176.

SERVENTY, V. N. (ed.) 1985. *The waterbirds of Australia*. North Ryde, New South Wales: Angus & Robertson.

SHALLENBERGER, R. J. 1977. *An ornithological survey of Hawaiian wetlands*. Honolulu: Ahuimanu Productions.

SHALLENBERGER, R. J. 1978. *Hawaii's birds*. Honolulu: Hawaiian Audubon Society.

SHARLAND, M. S. R. 1925. Tasmania's endemic birds. *Emu* 25: 94-103.

SHARLAND, M. S. R. 1981. *A guide to birds of Tasmania*. Hobart: Oldham, Beddome & Meredith.

SHARPE, R. B. 1894. Catalogue of the Fulicariae and Alectorides. Vol. 23 in *Catalogue of the birds in the British Museum*. London: British Museum (Natural History).

SHARPE, R. B. 1907. On further collections of birds from the Efulen District of Cameroon, West Africa. *Ibis* (9)1: 416-464.

SHARROCK, J. T. R. 1980. Rare breeding birds in the United Kingdom in 1978. *British Birds* 73: 5-26.

SHEPPARD, R. & GREEN, R. E. 1994. Status of the Corncrake in Ireland in 1993. *Irish Birds* 5: 125-138.

SHERRINGTON, P. 1994. Yellow Rail in Yoho National Park. *Brit. Columbia Birds* 4: 15-16.

SHIRIHAI, H. 1996. *The birds of Israel*. London: Academic Press.

SHORE-BAILEY, W. 1926. Brazilian rails nesting (*Aramides ypecaha* and *Aramides cayennensis*). *Avicult. Mag.* (4)4: 305-310.

SHORE-BAILEY, W. 1929. Breeding the South African Water Rail. *Avicult. Mag.* (4)7: 286-288.

SHORT, L. L. 1975. A zoogeographic analysis of the South American chaco avifauna. *Bull. Amer. Mus. Nat. Hist.* 154: 167-352.

SHORT, L. L., HORNE, J. F. M. & MURINGO-GICHUKI, C. 1990. Annotated check-list of the birds of East Africa. *Proc. Western Fnd. Vert. Zool.* 4(3).

SHUEL, R. 1938. Notes on the breeding habits of birds near Zaria, Nigeria, with descriptions of their nests and eggs. *Ibis* (14)2: 230-244.

SIBLEY, C. G. 1951. Notes on the birds of New Georgia, central Solomon Islands. *Condor* 53: 81-92.

SIBLEY, C. G. & AHLQUIST, J. E. 1972. A comparative study of the egg-white proteins of non-passerine birds. *Bull. Peabody Mus. Nat. Hist.* 39: 1-276.

SIBLEY, C. G. & AHLQUIST, J. E. 1985. The relationships of some groups of African birds, based on comparisons of the genetic material, DNA. Pp. 115-162 in *Proc. Internatn. Symp. African Vertebrates*. Bonn: Zoologisches Forschungsinstitut und Museum Alexander Koenig.

SIBLEY, C. G. & AHLQUIST, J. E. 1990. *Phylogeny and classification of birds: a study in molecular evolution*. New Haven: Yale University Press.

SIBLEY, C. G., AHLQUIST, J. E. & DE BENEDICTIS, P. 1993. The phylogenetic relationships of the rails, based on DNA comparisons. *J. Yamashina Inst. Orn.* 25: 1-11.

SIBLEY, C. G. & MONROE, B. L. Jr. 1990. *Distribution and taxonomy of birds of the world*. New Haven: Yale University Press.

SIBLEY, C. G. & MONROE, B. L. Jr. 1993. *A supplement to distribution and taxonomy of birds of the world*. New Haven: Yale University Press.

SIBSON, R. B. 1982. *Birds at risk. rare or endangered species of New Zealand*. Wellington: Reed.

SICK, H. 1930. The nesting of the Black Rail. *Avicult. Mag.* 8: 270-273.

SICK, H. 1979. Notes on some Brazilian birds. *Bull. Brit. Orn. Club* 99: 115-120.

SICK, H. 1985. *Ornitologia Brasileira, uma introdução*. Brasília: Editora Universidade de Brasília.

SICK, H. 1993. *Birds in Brazil. A Natural History*. Princeton, New Jersey: Princeton University Press.

SICK, H. 1997 *Ornitologia Brasileira*. 2nd edn. Rio de Janeiro: Editora Nova Fronteira.

SIEBERS, H. C. 1930. Fauna Buruana. Aves. *Treubia* 7, Suppl.: 165-303.

SIEGFRIED, W. R. & FROST, P. G. H. 1973. Regular occurrence of *Porphyrula martinica* in South Africa. *Bull. Brit. Orn. Club* 93: 36-38.

SIEGFRIED, W. R. & FROST, P. G. H. 1975. Continuous breeding and associated behaviour in the Moorhen *Gallinula chloropus*. *Ibis* 117: 102-109.

SIGMUND, L. 1958. Die postembryonale Entwicklung der Wasserralle (*Rallus aquaticus*). *Sylvia* 15: 85-118.

SIGMUND, L. 1959. Mechanik und anatomische Grundlagen der Fortbewegung bei Wasser-ralle (*Rallus aquaticus* L.), Teichhuhn (*Gallinula chloropus* L.) und Bläßhuhn (*Fulica atra* L.). *J. Orn.* 100: 3-24.

SILBERNAGL, H. P. 1982. Seasonal and spatial distribution of the American Purple Gallinule in South Africa. *Ostrich* 53: 236-240.

SIMEONOV, S. D., MICHEV, T. & NANKINOV, L. 1990. [*The fauna of Bulgaria*]. Vol. 20. Aves, part 1. Sofia: Bulgarian Academy of Sciences.

SIMMONS, R. 1995. Namibia's second Buffspotted Flufftail. *Birding in Southern Africa* 47: 26-27.

SIMPSON, K. & DAY, N. 1994. *Field guide to the birds of Australia*. London: Christopher Helm.

SKEAD, D. M. 1967. Ecology of birds in the eastern Cape Province. *Ostrich* Suppl. 7: 1-104.

SKEAD, D. M. 1981. Recovery distribution of Redknobbed Coots ringed at Barberspan. *Ostrich* 52: 126-128.

SKEAD, D. M. & DEAN, W. R. J. 1977. Status of the Barberspan avifauna, 1971-1975. *Ostrich* Suppl. 12: 3-42.

SKEMP, J. R. 1955. Notes on the Lewin's Rail. *Emu* 55: 169-172.

SKERRETT, A. 1996. The first report of the Seychelles Bird Records Committee. *Bull. African Bird Club* 3: 45-50.

SKINNER, J. F. 1979. Fernbird duetting with Spotless Crake. *Notornis* 26: 22.

SKUTCH, A. F. 1976. *Parent birds and their young*. Austin & London: University of Texas Press.

SKUTCH, A. F. 1994. The Gray-necked Wood-Rail: habits, food, nesting and voice. *Auk* 111: 200-204.

SLUD, P. 1964. The birds of Costa Rica. *Bull. Amer. Mus. Nat. Hist.* 128: 1-430.

SMALL, A. 1994. *California birds: their status and distribution*. Vista, California: Ibis Publ. Co.

SMITH, P. M. 1975. Habitat requirements and observations on the Clapper Rail (*Rallus longirostris yumanensis*). Tempe, Arizona: Arizona State University (MSc thesis).

SMITHE, F. B. 1975. *Naturalists' color guide*. In 3 parts. New York: American Museum of Natural History.

SMITHERS, R. H. N. 1964. *A checklist of the birds of the Bechuanaland Protectorate and the Caprivi Strip*. Salisbury: Trustees of the National Museums of Southern Rhodesia.

SMYTHIES, B. E. 1953. *The birds of Burma*. 2nd. edn. Edinburgh & London: Oliver & Boyd.

SMYTHIES, B. E. 1968. *The birds of Borneo*. 2nd. edn. Edinburgh: Oliver & Boyd.

SMYTHIES, B. E. 1981. *The birds of Borneo*. 3rd. edn. Kota Kinabalu and Kuala Lumpur: Sabah Society and Malayan Nature Society.

SMYTHIES, B. E. 1986. *The birds of Burma*. 3rd. edn. Liss, UK: Nimrod Press.

SNOW, D. W. (ed.). 1978. *An atlas of speciation in African non-passerine birds*. London: British Museum (Natural History).

SNOW, D. W. 1979. Atlas of speciation in African non-passerine birds – addenda and corrigenda. *Bull. Brit. Orn. Club* 99: 66-68.

SNYDER, D. E. 1966. *The birds of Guyana*. Salem, Mass.: Peabody Museum.

SOBKOWIAK, S. & TITMAN, R. D. 1989. Bald Eagles killing American Coots and stealing coot carcasses from Greater Black-backed Gulls. *Wilson Bull.* 101: 494-496.

VAN SOMEREN, V. D. 1947. Field notes on some Madagascar birds. *Ibis* 89: 235-267.

SONOBE, K. 1982. *A field guide to the birds of Japan*. Tokyo: Wild Bird Society of Japan.

SOPER, M. F. 1969. Kermadec Island Expedition reports: the Spotless Crake (*Porzana tabuensis plumbea*). *Notornis* 16: 219-220.

SOPER, M. F. 1976. *New Zealand birds*. Christchurch: Whitcoulls Publishers.

SORENSEN, M. D. 1995. Evidence of conspecific nest parasitism and egg discrimination in the Sora. *Condor* 97: 819-821.

SORENSEN, U. G. 1995. [Truede og sjaeldne danske ynglefugle 1976-1991. Status i relation til den generelle landskabsudvikling. *Dansk Orn. Foren. Tidsskr.* 89: 1-48.

SPEIGHT, J. A. 1981. *The status of King Rails and Clapper Rails along the Mississippi Gulf coast*. Mississippi State University (MS thesis).

SPENDELOW, J. A. & SPENDELOW, H. R. 1980. Clapper Rail kills birds in a net. *J. Field Orn.* 51: 175-176.

SPRUNT, A. 1954. *Florida bird life*. New York: Coward McKann.

SSC. 1994. *IUCN Red List categories*. Gland, Switzerland: IUCN.

ST. CLAIR, C. C. & ST. CLAIR, R. C. 1992. Weka predation on eggs and chicks of Fiordland Crested Penguins. *Notornis* 39: 60-63.

STALHEIM, P. S. 1974. Behaviour and ecology of the Yellow Rail (*Coturnicops noveboracensis*). Minneapolis: University of Minnesota (MSc thesis).

STALHEIM, P. S. 1975. Breeding and behaviour of captive Yellow Rails. *Avicult. Mag.* 81: 133-141.

STEADMAN, D. W. 1985. Fossil birds from Mangaia, southern Cook Islands. *Bull. Brit. Orn. Club* 105: 58-66.

STEADMAN, D. W. 1986. Two new species of rails (Aves: Rallidae) from Mangaia, southern Cook Islands. *Pacific Sci.* 40: 27-43.

STEADMAN, D. W. 1988. A new species of *Porphyrio* (Aves: Rallidae) from archaeological sites in the Marquesas Islands. *Proc. Biol. Soc. Washington* 101: 162-170.

STEADMAN, D. W. 1989. Extinction of birds in eastern Polynesia: a review of the record, and comparisons with other Pacific island groups. *J. Archaeol. Sci.* 16: 177-205.

STEADMAN, D. W. 1995. Prehistoric extinctions of Pacific island birds: biodiversity meets zooarchaeology. *Science* 267: 1123-1131.

STENDELL, R. C., ARTMANN, J. W. & MARTIN, E. 1980. Lead residues in Sora Rails from Maryland. *J. Wildl. Manage.* 44: 525-527.

STENZEL, J. R. 1982. Ecology of breeding yellow rails at Seney National Wildlife Refuge. Columbus, Ohio: Ohio State University (MSc thesis).

STEPHENS, J. 1974. Kanaha Pond: vital habitat for Hawaii's en-dangered coot and stilt. *Natn. Parks Conserv. Mag.* 48: 10-13.

STERN, M. A., MORAWSKI, J. F. & ROSENBERG, G. A. 1993. Rediscovery and status of a disjunct population of breeding Yellow Rails in southern Oregon. *Condor* 95: 1024-1027.

STEVENS, G. W. & TEDDER, J. L. O. 1973. *A Honiara bird guide.* Honiara: British Solomon Islands Scout Association.

STEVENS, H. 1925. Notes on the birds of the Sikkim Himalayas. Part 7. *J. Bombay Nat. Hist. Soc.* 30: 872-893.

STEWART, B. A. & BALLY, R. 1985. The ecological role of the Red-knobbed Coot *Fulica cristata* Gmelin at the Bot River Estuary, South Africa: a preliminary investigation. *Trans. R. Soc. S. Afr.* 45: 419-426.

STEWART, R. E. 1954. Migratory movements of the Northern Clapper Rail. *Bird-Banding* 25: 1-5.

STEWART, R. E. & ROBBINS, C. S. 1958. Birds of Maryland and the District of Columbia. *North American Fauna* 62. Washington.

STEYN, P. & MYBURGH, N. 1986. A tale of two flufftails. *Afr. Wildlife* 40: 22-27.

STIEFEL, A. 1991. Situation des Wachtelkönigs in Ostdeutschland (vormalige DDR). *Vogelwelt* 112: 57-66.

STIEFEL, A. & BERG, W. 1975. Geschlechtsunterschiede in einigen Rufen der Wasserralle (*Rallus aquaticus*). *Beitr. Vogelkd.* 21: 330-339.

STILES, F. G. 1981. Notes on the Uniform Crake in Costa Rica. *Wilson Bull.* 93: 107-108.

STILES, F. G. 1988. Notes on the distribution and status of certain birds in Costa Rica. *Condor* 90: 931-933.

STILES, F. G. & LEVEY, D. J. 1988. The Gray-breasted Crake (*Laterallus exilis*) in Costa Rica: vocalizations, distribution and interactions with White-throated Crakes (*L. albigularis*). *Condor* 90: 607-612.

STILES, F. G. & SKUTCH, A. F. 1989. *A guide to the birds of Costa Rica.* Ithaca, New York: Cornell University Press.

STINSON, D. W., RITTER, M. W. & REICHEL, J. D. 1991. The Mariana Common Moorhen: decline of an island endemic. *Condor* 93: 38-43.

STODDARD, H. L. & NORRIS, R. A. 1967. Bird casualties at a Leon County, Florida TV tower: an eleven-year study. *Bull. Tall Timbers Res. Stn.* no. 8.

STODDART, D. R. 1971. White-throated Rail *Dryolimnas cuvieri* on Astove Atoll. *Bull. Brit. Orn. Club* 91: 145-147.

STOKES, T. 1979. On the possible existence of the New Caledonian Wood Rail *Tricholimnas lafresnayanus*. *Bull. Brit. Orn. Club* 99: 47-54.

STOKES, T. 1983. Bird casualties in 1975-76 at the Booby Island lightstation, Torres Strait. *Sunbird* 13: 53-58.

STOKES, T., SHEILS, W. & DUNN, K. 1984. Birds of the Cocos (Keeling) Islands, Indian Ocean. *Emu* 84: 23-28.

STONE, A. C. 1912. Birds of Lake Boga, Victoria. *Emu* 12: 112-122.

STONE, W. *Bird studies at old Cape May.* Philadelphia: Academy of Natural Sciences.

STORER, R. W. 1981. The Rufous-faced Crake *Laterallus xenopterus* and its Paraguayan congeners. *Wilson Bull.* 93: 137-144.

STORER, R. W. 1989. Notes on Paraguayan birds. *Occas. Pap. Mus. Zool. Univ. Michigan* 719: 1-21.

STORR, G. M. 1973. List of Queensland birds. *Spec. Publs. West. Aust. Mus.* no. 5.

STORR, G. M. 1980. Birds of the Kimberley Division, Western Australia. *Spec. Publs. West. Aust. Mus.* no. 11.

STOTZ, D. F. 1992. A new subspecies of *Aramides cajanea* from Brazil. *Bull. Brit. Orn. Club* 112: 231-234.

STOWE, T. J. & BECKER, D. 1992. Status and conservation of Corncrakes *Crex crex* outside the breeding grounds. *Tauraco* 2: 1-23.

STOWE, T. J. & HUDSON, A. V. 1991a. Radio telemetry studies of Corncrake in Great Britain. *Vogelwelt* 112: 10-16.

STOWE, T. J. & HUDSON, A. V. 1991b. Corncrakes outside the breeding grounds, and ideas for a conservation strategy. *Vogelwelt* 112: 103-107.

STOWE, T. J., NEWTON, A. V., GREEN, R. E. & MAYES, E. 1993. The decline of the Corncrake *Crex crex* in Britain and Ireland in relation to habitat. *J. Appl. Ecol.* 30: 53-62.

STRESEMANN, E. 1931. Vorläufiges über die ornithologischen Ergebnisse der Expedition Heinrich 1930-1931. *Orn. Monatsb.* 39: 167-171.

STRESEMANN, E. 1936. A nominal list of the birds of Celebes. *Ibis* (13)6: 356-369.

STRESEMANN, E. 1941. Die Vögel von Celebes. Teil III. Systematik und Biologie. *J. Orn.* 89: 1-102.

STRESEMANN, E. 1950. Birds collected during Capt. James Cook's last expedition. *Auk* 67: 66-88.

STRESEMANN, E. & STRESEMANN, V. 1966. Die Mäuser der Vögel. *J. Orn.* 107 (Sonderheft): 1-439.

STROPHLET, J. J. 1946. Birds of Guam. *Auk* 65: 534-540.

STUSÁK, J. M. & VO QUY. 1986. *The birds of the Hanoi area.* Prague: University of Agriculture.

SUDDABY, D. & HARVEY, P. V. 1992. Baillon's Crake, Fair Isle, Shetland. *Scottish Birds* 16: 211-214.

SUGDEN, L. G. 1979. Habitat use by nesting American Coots in Saskatchewan parklands. *Wilson Bull.* 91: 599-607.

SULLEY, S. C. & SULLEY, M. E. 1992. *Birding in Cuba.* Derbyshire, UK.: Worldwide Publications.

SUTHERLAND, J. M. 1991. Effects of drought on American coot, *Fulica americana*, reproduction in Saskatchewan parklands. *Can. Field Nat.* 105: 267-273.

SUTHERLAND, J. M. & MAHER, W. J. 1987. Nest-site selection of the American Coot in the aspen parklands of Saskatchewan. *Condor* 89: 804-810.

SUTHERLAND, W. J. 1994. How to help the Corncrake. *Nature* 372: 223.

SUTTIE, J. M. & FENNESSY, P. F. 1992. Organ weight and weight relationships in Takahe and Pukeko. *Notornis* 39: 47-53.

SWINHOE, R. 1868. Ornithological notes from Amoy. *Ibis* (2)4: 52-65.

SYKES, P. W. 1975. Caribbean Coot collected in southern Florida. *Florida Field Nat.* 3: 25-27.

SZABO, L. V. 1970. Vergleichende Untersuchungen der Brutverhältnisse der drei *Porzana*-Arten in Ungarn. *Aquila* 66-67: 73-113.

SZABO, L. V. 1976. Das Nisten des Zwergsumpfhuhnes (*Porzana pusilla*) in der Puszta von Hortobagy. *Aquila* 82: 165-175.

SZEP, T. 1991. The present and historical situation of the Corncrake in Hungary. *Vogelwelt* 112: 45-48.

TACHA, T. C. & BRAUN, C. E. (eds.). *Management of migratory shore and upland game birds in North America*. Washington D.C.: International Assoc. Fish & Wildlife Agencies.

TACZANOWSKI, L. 1886. *Ornithologie du Pérou*. Vol. 3. Rennes: Oberthuz.

TALLMAN, D. A., PARKER, T. A. & LEITER, G. D. 1978. Notes on two species of birds previously unreported from Peru. *Wilson Bull.* 90: 445-446.

TALLMAN, D., TALLMAN, E. & PEARSON, D. L. 1977. *The birds of Limonocha, Napo province, Ecuador*. Quito: Instituto Linguistico de Verano.

TANNER, W. D. & HENDRICKSON, G. O. 1954. Ecology of the Virginia Rail in Clay County, Iowa. *Iowa Bird Life* 24: 65-70.

TANNER, W. D. & HENDRICKSON, G. O. 1956a. Ecology of the King Rail in Clay County, Iowa. *Iowa Bird Life* 26: 54-56.

TANNER, W. D. & HENDRICKSON, G. O. 1956b. Ecology of the Sora in Clay County, Iowa. *Iowa Bird Life* 26: 78-81.

TANO, J. 1996. More new records from Ambua, and a further sightings (sic) of a mystery bird. *Muruk* 8: 37.

TARBOTON, W. R., KEMP, M. I. & KEMP, A. C. 1987. *Birds of the Transvaal*. Pretoria: Transvaal Museum.

TAYLOR, G. A. 1990. Classified summarised notes, 1 July 1988 to 30 June 1989, North Island. *Notornis* 37: 183-235.

TAYLOR, G. A. & PARRISH, G. R. 1991. Classified summarised notes, 1 July 1989 to 30 June 1990, North Island. *Notornis* 38: 267-314.

TAYLOR, G. A. & PARRISH, G. R. 1994. Classified summarised notes, 1 July 1992 to 30 June 1993, North Island. *Notornis* 41: 235-274.

TAYLOR, P. B. 1979. Palaearctic and intra-African migratory birds in Zambia: a report for the period May 1971 to December 1976. *Zambian Orn. Soc. Occas. Pap.* 1.

TAYLOR, P. B. 1980. Little Crake *Porzana parva* at Ndola, Zambia. *Scopus* 4: 93-95.

TAYLOR, P. B. 1984. A field study of the Corncrake *Crex crex* at Ndola, Zambia. *Scopus* 8: 53-59.

TAYLOR, P. B. 1985a. Field studies of the African Crake *Crex egregia* in Zambia and Kenya. *Ostrich* 56: 170-185.

TAYLOR, P. B. 1985b. Observations of Allen's Gallinule at Mombasa, Kenya. *Malimbus* 7: 141-150.

TAYLOR, P. B. 1987. A field study of the Spotted Crake *Porzana porzana* at Ndola, Zambia. *Ostrich* 58: 107-117.

TAYLOR, P. B. 1994. The biology, ecology and conservation of four flufftail species, *Sarothrura* (Aves: Rallidae). Pietermaritzburg: University of Natal (PhD thesis).

TAYLOR, P. B. 1996a. The Whitewinged Flufftail and its wetland habitats: report on a study in Ethiopia, 11-27 August 1996. Unpublished report to the Ethiopian Wildlife and Natural History Society.

TAYLOR, P. B. 1996b. Family and species accounts (Rallidae). Pp. 108-209 in DEL HOYO, J., ELLIOTT, A. & SARGATAL, J. (eds.). *Handbook of the birds of the world*. Vol. 3: Hotazin to Auks. Barcelona: Lynx Edicions.

TAYLOR, P. B. 1997a. *The status and conservation of rallids in South Africa: results of a wetland survey in 1995/96*. ADU Research Report No. 23. Avian Demography Unit, Department of Statistical Sciences, University of Cape Town.

TAYLOR, P. B. 1997b. Whitewinged Flufftails and wetlands: results of fieldwork in Ethiopia from 29 July to 8 August 1997. Unpublished report to the Ethiopian Wildlife and Natural History Society.

TAYLOR, P. B. 1997c. Species accounts (Rallidae). Pp. in HARRISON, J. A., ALLAN, D. G., UNDERHILL, L. G., BROWN, C. J., TREE, A. J., PARKER, V. & HERREMANS, M. (eds). *The atlas of southern African birds*. Johannesburg: BirdLife South Africa.

TAYLOR, P. B. 1997d. *South African palustrine wetlands: the results of a survey in summer 1995/96*. ADU Research Report No. 24. Avian Demography Unit, Department of Statistical Sciences, University of Cape Town.

TAYLOR, P. B. 1997e. Hope for the Whitewinged Flufftail. *Africa Birds and Birding*. 2(5): 14-15.

TAYLOR, P. B. & HUSTLER, K. 1993. The southern African specimen of the Longtoed Flufftail *Sarothrura lugens*: a case of misidentification. *Ostrich* 64: 38-40.

TAYLOR, P. B., SMITH, E. P. & HERHOLDT, J. J. 1994. The Striped Crake in southern Africa. *Birding in Southern Africa* 46: 18-21.

TAYLOR, P. B. & TAYLOR, C. A. 1986. Field studies of forest flufftails (*Sarothrura* species) in Kenya. Unpublished.

TAYLOR, R. J. & MOONEY, N. J. 1991. Increased mortality of birds on an elevated section of highway in northern Tasmania. *Emu* 91: 186-188.

TEIXEIRA, D. M. 1981. Notas sobre a Saracura Três-potes, *Aramides cajanea* (Müller, 1776): a ocorrência do Ninho-Criadeira. *Bol. Mus. Paraense Emilio Goeldi, Zool.* 100.

TEIXEIRA, D. M., NACINOVIC, J. & TAVARES, M. S. 1986. Notes on some birds of northeastern Brazil. *Bull. Brit. Orn. Club* 106: 70-74.

TEIXEIRA, D. M., NACINOVIC, J. B. & LUIGI, G. 1989. Notes on some birds of northeastern Brazil (4). *Bull. Brit. Orn. Club* 109: 152-157.

TEIXEIRA, D. M. & PUGA, M. E. M. 1984. Notes on the Speckled Crake (*Coturnicops notata*) in Brazil. *Condor* 86: 342-343.

TEMME, M. 1974. New records of Philippine birds on the island of Mindoro. *Bonn. zool. Beitr.* 25: 292-296.

TENNYSON, A. J. D. & MILLENER, P. R. 1994. Bird extinctions and fossil bones from Mangere Island, Chatham Islands. *Notornis* (Suppl.) 41: 165-178.

TERRILL, L. M. 1943. Nesting habits of the yellow rail in Gaspé County, Quebec. *Auk* 60: 171-180.

THAYER, J. E. & BANGS, O. 1912. Some Chinese vertebrates: Aves. *Mem. Mus. Comp. Zool.* 40: 137-200.

THIEDE, U. 1982. "Yambaru Kuina" (*Rallus okinawae*), eine neu entdeckte Rallenart in Japan. *Vogelwelt* 103: 143-150.

THIOLLAY, J.-M. 1985. The birds of Ivory Coast: status and distribution. *Malimbus* 7: 1-59.

THOMAS, B. T. 1979. The birds of a ranch in the Venezuelan llanos. Pp. in EISENBERG, J. F. (ed.). *Vertebrate ecology in the northern Neotropics*. Washington, D.C.: Smithsonian Institution Press.

THOMAS, B. T. 1993. Birds of a northern Venezuelan secondary-scrub habitat. *Bull. Brit. Orn. Club* 113: 9-17.

THOMAS, D. G. 1979. *Tasmanian bird atlas*. Fauna of Tasmania, Handbook no. 2. Hobart: University of Tasmania.

THONNERIEUX, Y., WALSH, J. F. & BORTOLI, L. 1989. L'avifaune de la ville de Ouagadougou et ses environs (Burkina Faso). *Malimbus* 11: 7-40.

TIMMIS, W. H. 1972. Breeding and behaviour of the North Island Weka Rail at Chester Zoo. *Avicult. Mag.* 78: 53-64.

TIMMIS, W. H. 1974. The breeding and behaviour of the Blue-breasted Banded Rail. *Avicult. Mag.* 80: 125-131.

TODD, R. L. 1977. Black Rail, Little Black Rail, Black Crake, Farallon Rail (*Laterallus jamaicensis*). Pp. in SANDERSON, G. C. (ed.) *Management of migratory shore and upland game birds in North America*. Washington, D.C.: International Assoc. Fish & Wildl. Agencies.

TODD, R. L. 1986. *A saltwater marsh hen in Arizona: a history of the Yuma Clapper Rail (Rallus longirostris yumanensis)*. Completion Rept. Fed. Aid Proj. W-95-R. Phoenix, Arizona: Arizona Game & Fish Dept.

TOLHURST, L. 1992. Bush-hens *Amaurornis olivaceus* breeding in the dry season. *Muruk* 5: 64-65.

TOLHURST, L. 1996. Two new species for Misima Island, Milne Bay Province. *Muruk* 8: 34-35.

TOMIALOJC, L. 1994. Corncrake *Crex crex*. Pp. 228-229 in TUCKER, G. M. & HEATH, M. F. 1994. *Birds in Europe: their conservation status*. BirdLife Conservation Series No. 3. Cambridge, UK: BirdLife International.

TOMLINSON, R. E. & TODD, R. L. 1973. Distribution of two western Clapper Rail races as determined by responses to taped calls. *Condor* 75: 177-183.

TOSTAIN, O., DUJARDIN, J.-L., ERARD, C. & THIOLLAY, J.-M. 1992. *Oiseaux de Guyane*. Brunoy, France: Société d'Études Ornithologiques.

TRAPP, J. L., ROBUS, M., TANS, G. J. & TANS, M. M. 1981. First breeding record of the Sora and American Coot in Alaska – with comments on drought displacement. *Amer. Birds* 35: 901-902.

TRAUTMAN, B. M. & GLINES, S. J. 1964. A nesting of the Purple Gallinule (*Porphyrula martinica*) in Ohio. *Auk* 81: 224-226.

TRAYLOR, M. A. 1958. Birds of northeastern Peru. Fieldiana: Zoology 35(5). Chicago Nat. Hist. Museum.

TRAYLOR, M. A. 1963. *Check-list of Angolan birds*. Lisbon: Companhia de Diamantes de Angola, Museu do Dundo.

TREWICK, S. A. 1996a. The diet of Kakapo (*Strigops habroptilus*), Takahe (*Porphyrio mantelli*) and Pukeko (*P. porphyrio melanotus*) studied by faecal analysis. *Notornis* 43: 79-84.

TREWICK, S. A. 1996b. Morphology and evolution of two takahe: flightless rails of New Zealand. *J. Zool. Lond.* 238: 221-237.

TREWICK, S. A. 1997. Flightlessness and phylogeny amongst endemic rails (Aves: Rallidae) of the New Zealand region. *Phil. Trans. R. Soc. Lond.* B 352: 429-446.

TUCKER, G. M. & EVANS, M. I. 1997. *Habitats for birds in Europe: a conservation strategy for the wider environment*. Cambridge: BirdLife International.

TUCKER, G. M. & HEATH, M. F. 1994. *Birds in Europe: their conservation status*. BirdLife Conservation Series No. 3. Cambridge, UK: BirdLife International.

TUNNICLIFFE, G. A. 1985. High altitude records of Pukeko in the Southern Alps. *Notornis* 32: 81-82.

TURBOTT, E.G. 1951. Winter observations on *Notornis*. *Notornis* 4: 107-113.

TURBOTT, E. G. 1967. *Buller's birds of New Zealand*. Auckland: Whitcombe & Tombs.

TURNER, E. L. 1924. *Broadland birds*. London: Country Life.

UDVARDY, M. D. F. 1960. Movements and concentrations of the Hawaiian Coot on the island of Oahu. *Elepaio* 21: 20-22.

URBAN, E. K. 1980. *Ethiopia's endemic Bbirds*. Addis Ababa: Ethiopian Tourist Commission.

URBAN, E. K. & BROWN, L. H. 1971. *A checklist of the birds of Ethiopia*. Addis Ababa: Haile Sellassie 1 University.

URBAN, E. K., FRY, C. H. & KEITH, S. (eds.). 1986. *The birds of Africa*. Vol. 2: 84-131. London: Academic Press.

VANNINI, J. P. 1994. Nearctic avian migrants in coffee plantations and forest fragments of southwest Guatemala. *Bird Conserv. Internatn.* 4: 209-232.

VARTY, N., ADAMS, J., ESPIN, P. & HAMBLER, C. J. 1986. An ornithological survey of Lake Tota, Colombia, 1982. *ICBP Study Report*, 12. Cambridge: International Council for Bird Preservation.

VAURIE, C. 1965. *The birds of the Palaearctic fauna: non-passeriformes*. London: Witherby.

VAURIE, C. 1972. *Tibet and its birds*. London: Witherby.

VAN VELZEN, A. & KREITZER, J. F. 1975. The toxicity of p.p'-DDT to the Clapper Rail. *J. Wildl. Manage.* 39: 305-309.

VEIT, R. & PETERSEN, W. 1993. *Birds of Massachusetts*. Lincoln, Mass: Massachusetts Audubon Society.

VERHEYEN, R. 1953. *Exploration du Parc National de l'Upemba. Oiseaux*. Bruxelles: Institut des Parcs Nationaux du Congo Belge.

VERHEYEN, R. 1957. Contribution au demembrement de l'ordo artificiel des Gruiformes (Peters 1934) 1. -Les Ralliformes. *Bull. Inst. Roy. Sci. Nat. Belgique.* 33(21): 1-44.

VESEY-FITZGERALD, L. D. E. F. 1940. The birds of the Seychelles. I. The endemic birds (Appendix, Birds of the Aldabra Group). *Ibis* 14(4): 480-489.

VESTJENS, W. J. M. 1963. Remains of the extinct Banded Rail at Macquarie Island. *Emu* 62: 249-250.

VESTJENS, W. J. M. 1977. Status, habitats and food of vertebrates at Lake Cowal. *Tech. Mem. Div. Wildl. Res. CSIRO, Aust.* 12: 1-87.

VIDES-ALMONACID, R. 1988. Notas sobre ee estado de las poblaciones de la gallareta cornuda (*Fulica cornuta*) en la provincia de Tucumán, Argentina. *Hornero* 13: 34-38.

VIELLIARD, J. 1974. The Purple Gallinule in the marismas of the Guadalquivir. *British Birds* 67: 230-236.

VIERNES, K. J. F. 1995. Bullfrog predation on an endangered Common Moorhen chick at Hanalei National Wildlife Refuge, Kaua'i. *Elepaio* 55(6): 37.

VILLA, J. & PONCE, A. 1984. Islands for people and evolution: the Galapagos. Pp. in MCNEELY, J. A. & MILLER, K. R. (eds.). *National parks, conservation, and development*. World Congress on National Parks (1982: Bali, Indonesia). Washington D.C.: Smithsonian Institution.

VINCENT, A. W. 1945. On the breeding habits of some African birds. *Ibis* 87: 345-365.

VISSER, J. 1974. The post-embryonic development of the Coot *Fulica atra*. *Ardea* 62: 172-189.

VOISIN, J.-F. 1979. Observations ornithologiques aux îles Tristan da Cunha et Gough. *Alauda* 47: 73-82.

VOOUS, K. H. 1948. Notes on a collection of Javanese birds. *Limosa* 21: 85-100.

VOOUS, K. H. 1960. *Atlas of European birds*. London: Nelson.

VOOUS, K. H. 1961. The generic distinction of the Gough Island flightless gallinule. *Bijd. Dierkunde* 31: 75-79.

VOOUS, K. H. 1962. Notes on a collection of birds from Tristan da Cunha and Gough Island. *Beaufortia* 9: 105-114.

VUILLEUMIER, F. & GOCHFELD, M. 1976. Notes sur l'avifaune de Nouvelle-Calédonie. *Alauda* 44: 237-273.

VUILLEUMIER, F., LECROY, M. & MAYR, E. 1992. New species of birds described from 1981 to 1990. *Bull. Brit. Orn. Club* 112A: 267-309.

WACE, N. M. & HOLDGATE, M. W. 1976. *Man and nature in Tristan da Cunha*. Morges, Switzerland: IUCN (Monograph no. 6).

WACHER, T. 1993. Some new observations of forest birds in The Gambia. *Malimbus* 15: 24-37.

WAGSTAFFE, R. 1978. *Type specimens of birds in the Merseyside County Museums*. Liverpool: Merseyside County Council.

WALKINSHAW, L. H. 1937. The Virginia Rail in Michigan. *Auk* 54: 464-475.

WALKINSHAW, L. H. 1939. The Yellow Rail in Michigan. *Auk* 56 227-237.

WALKINSHAW, L. H. 1940. Summer life of the Sora Rail. *Auk* 57: 153-168.

WALKINSHAW, L. H. 1991. Yellow Rail. Pp. in BREWER, R., MCPEEK, G. A. & ADAMS, R. J. (eds.). *The atlas of breeding birds of Michigan*. East Lansing, Michigan: Michigan State University Press.

WALLACE, D. I. M. 1976. Sora Rail in Scilly and the identification of immature small crakes. *British Birds* 69: 443-447.

WALSH, J. F. 1987. Records of birds seen in north-eastern Guinea in 1984-1985. *Malimbus* 9: 105-122.

WALTERS, M. 1987. The provenance of the Gilbert Rail *Tricholimnas conditicius* (Peters & Griscom). *Bull. Brit. Orn. Club* 107: 181-184.

WALTERS, M. 1988. Probable validity of *Rallus nigra* Miller, an extinct species from Tahiti. *Notornis* 35: 265-269.

WARNER, D. W. 1947. The ornithology of New Caledonia and

the Loyalty Islands. Cornell University (PhD thesis).
WARNER, D. W. & DICKERMAN, R. W. 1959. The status of *Rallus elegans tenuirostris* in Mexico. *Condor* 61: 49-51.
WATKINS, B. P. & FURNESS, R. W. 1986. Population status, breeding and conservation of the Gough Moorhen. *Ostrich* 57: 32-36.
WATLING, D. 1981. A flying flightless rail and the Unexpected Barn Owl in Lore Lindu. *Conserv. Indon. Newsl. WWF* 5: 6.
WATLING, D. 1982a. Notes on the birds of Makogai Island, Fiji Islands. *Bull. Brit. Orn. Club* 102: 123-127.
WATLING, D. 1982b. *Birds of Fiji, Tonga and Samoa*. Wellington, New Zealand: Millwood Press.
WATLING, D. 1983. Ornithological notes from Sulawesi. *Emu* 83: 247-261.
WATSON, G. E. 1962. Notes on the Spotted Rail in Cuba. *Wilson Bull.* 74: 349-356.
WATSON, G. E. 1975. *Birds of the Antarctic and sub-Antarctic*. Washington, D.C.: American Geophysical Union.
WATSON, I. M. 1955. Some species seen at the Laverton Saltworks, Victoria, 1950-1953, with notes on seasonal changes. *Emu* 55: 224-248.
WATSON, J. L. 1969. Aviary breeding of the Black Crake. *Honeyguide* 59: 11-12.
WATSON, R. M., SINGH, T. & PARKER, I. S. C. 1970. The diet of duck and coot on Lake Naivasha. *E. Afr. Wildl. J.* 8: 131-144.
WEBB, H. P. 1992. Field observations of the birds of Santa Isabel, Solomon Islands. *Emu* 92: 52-57.
WEBER, M. J., VOHS, P. A. & FLAKE, L. D. 1982. Use of prairie wetlands by selected bird species in South Dakota. *Wilson Bull.* 94: 550-554.
WEBER, P., MUNTEANU, D. & PAPADOPOL, A. (eds.). 1994. *Atlasul provisoriu al pasarilor clocitoare din Romania*. Medias: Societatea Ornitologica.
WEBSTER, C. G. 1964. Fall foods of Soras from two habitats in Connecticut. *J. Wildl. Manage.* 28: 163-165.
WEID, R. 1991. Verhalten und Habitatansprüche des Wachtelkönigs im intensiv genutzen Grünland in Franken. *Vogelwelt* 112: 90-96.
WEGE, D. C. & LONG, A. J. 1995. *Key areas for threatened birds in the Neotropics*. BirdLife Conservation Series No. 5 Cambridge, UK: BirdLife International.
WELLER, M. W. 1967. Notes on some marsh birds of Cape San Antonio, Argentina. *Ibis* 109: 391-411.
WELLER, M. W. & FREDRICKSON, L. H. 1974. Avian ecology of a managed glacial marsh. *Living Bird* 12: 269-291.
WELLS, D. R. & MEDWAY, LORD. 1976. Taxonomic and faunistic notes on the birds of the Malay Peninsula. *Bull. Brit. Orn. Club* 96: 20-34.
WESKE, J. S. 1969. An ecological study of the Black Rail in Dorchester County, Maryland. Ithaca, New York: Cornell University (MSc thesis).
WESTERSKOV, K. E. 1970. Leg and foot colour of the Marsh Crake *Porzana pusilla*. *Notornis* 17: 324-330.
WESTON, M. 1993. Anti-predatory behaviour of the Coot. *Aust. Bird Watcher* 15: 192
WETHERBEE, D. K. & MEANLEY, B. 1965. Natal plumage characters in rails. *Auk* 82: 500-501.
WETMORE, A. 1916. Birds of Puerto Rico. *US Dept. Agric. Bull.* 326.
WETMORE, A. 1926. Observations on the birds of Argentina, Paraguay, Uruguay, and Chile. *Bull. U.S. Natn. Mus.* 133: 1-448.
WETMORE, A. 1957. A fossil rail from the Pliocene of Arizona. *Condor* 59: 267-268.
WETMORE, A. 1963. An extinct rail from the island of St Helena. *Ibis* 103b: 379-381.
WETMORE, A. 1965. The birds of the Republic of Panama. Part 1. *Smithsonian Misc. Coll.* 150: 1-483.

WETMORE, A. & SWALES, B. H. 1931. The birds of Haiti and the Dominican Republic. *Bull. U.S. Natn. Mus.* 155: 1-483.
WETTIN, P. 1984. Simultaneous polyandry in the Purple Swamphen. *Emu* 84: 111-112.
WHEELER, R. 1948. More notes from "The Bend". *Emu* 48: 1-7.
WHEELER, R. 1973. The Black-tailed Native Hen recent invasion. *Geelong Nat.* 10: 45-50.
WHITMEE, S. J. 1874. Letter to P. Sclater on birds of Samoa. *Proc. Zool. Soc. Lond.* 1874: 183-186.
WHITE, C. M. N. 1945. The ornithology of the Kaonde-Lunda Province, Northern Rhodesia. Part III, Systematic list (contd.). *Ibis* 87: 309-345.
WHITE, C. M. N. 1976. Comments on the Dusky Moorhen *Gallinula tenebrosa*. *Bull. Brit. Orn. Club* 96: 125-128.
WHITE, C. M. N. & BRUCE, M. D. 1986. *The birds of Wallacea (Sulawesi, The Moluccas & Lesser Sunda Islands, Indonesia)*. BOU Checklist No. 7. London: British Ornithologists' Union.
WHITE, D. H. & JAMES, D. 1978. Differential use of fresh water environments by wintering waterfowl of coastal Texas. *Wilson Bull.* 90: 99-111.
WHITE, H. L. 1922. A collecting trip to Cape York Peninsula. *Emu* 22: 99-116.
WHITE, S. A. 1918. Birds of Lake Victoria and the Murray River for 100 miles down stream. *Emu* 18: 8-25.
WHITE, S. R. 1946. Notes on the bird life of Australia's heaviest rainfall region. *Emu* 46: 81-122.
WHITLOCK, F. L. 1914. Notes on the Spotless Crake and Western Ground-Parrot. *Emu* 13: 202-205.
WHITTELL, H. M. 1933. The birds of Bridgetown district, south-west Australia. *Emu* 32: 182-189.
WIENS, J. A. 1966. Notes on the distraction display of the Virginia Rail. *Wilson Bull.* 78: 229-231.
WILBUR, S. R. 1974. The status of the Light-footed Clapper Rail. *Amer. Birds* 28: 868-870.
WILBUR, S. R., JORGENSEN, P. D., MASSEY, B. W. & BASHAM, V. A. 1979. The Light-footed Clapper Rail: an update. *Amer. Birds* 33: 251.
WILBUR, S. R. & TOMLINSON, R. E. 1976. The literature of the western Clapper Rails. *U.S. Fish & Wildl. Serv. Spec. Sci. Rept.* – Wildl. No. 194.
WILDASH, P. 1968. *Birds of south Vietnam*. Rutland, Vermont: Tuttle.
WILEY, R. H. & RICHARDS, D. G. 1982. Adaptations for acoustic communication in birds: sound transmission and signal detection. Pp. 131-181 in D. E. KROODSMA & E. H. MILLER (eds.) *Acoustic communication in birds*, Vol. 1. New York: Academic Press.
WILKINSON, A. S. 1927. Birds of Kapiti Island. *Emu* 26: 237-258.
WILKINSON, R., BEECROFT, R. & AIDLEY, D. J. 1982. Nigeria: a new wintering area for the Little Crake *Porzana parva*. *Bull. Brit. Orn. Club* 102: 139-140.
WILKINSON, R. & HUXLEY, C. R. 1978. Vocalizations of chicks and juveniles and the development of adult calls in the Aldabra White-throated Rail *Dryolimnas cuvieri aldabranus* (Aves: Rallidae). *J. Zool. Lond.* 186: 487-505.
WILLIAMS, G., HOLMES, J. & KIRBY, J. 1995. Action plans for United Kingdom and European rare, threatened and internationally important birds. *Ibis* 137 (Suppl. 1): S201-S213.
WILLIAMS, G. R. 1952. Notornis in March, 1951. A report of the sixth expedition. *Notornis* 4: 202-208.
WILLIAMS, G. R. 1957. Some preliminary data on the population dynamics of the Takahe (*Notornis mantelli* Owen, 1848). *Notornis* 7: 165-171.
WILLIAMS, G. R. 1960. The Takahe (*Notornis mantelli* Owen, 1848). A general survey. *Trans. Roy. Soc. N. Z.* 88: 235-258.
WILLIAMS, G. R. 1979. Review: Rails of the world. *Notornis* 26: 102-103.

WILLIAMS, G. R. & GIVEN, D. R. 1981. *The Red Data Book of New Zealand.* Wellington: Nature Conservation Council.

WILLIAMS, G. R. & MEIRS, K. H. 1958. A five year banding study of the Takahe (*Notornis mantelli* Owen). *Notornis* 8: 1-12.

WILLIAMS, M. D. (ed.). 1986. *Report on the Cambridge Ornithological Expedition to China 1985.* Scarborough, UK: Cambridge Ornithological Expedition to China 1985.

WILLIAMS, M. D., CAREY, G. J., DUFF, D. G. & XU WEISHU. 1992. Autumn bird migration at Beidaihe, China, 1986-1990. *Forktail* 7: 3-55.

WILLIAMS, R. 1995. Neotropical notebook. Chile. *Cotinga* 4: 67.

WILLIAMS, S. O. 1989. Notes on the rail *Rallus longirostris tenuirostris* in the highlands of central Mexico. *Wilson Bull.* 101: 117-120.

WILLIS, E. O. & ONIKI, Y. 1985. Bird specimens new for the state of São Paulo, Brazil. *Revta. Bras. Biol.* 45: 105-108.

WILLIS, E. O. & ONIKI, Y. 1993. New and reconfirmed birds from the state of São Paulo, Brazil, with notes on disappearing species. *Bull. Brit. Orn. Club* 113: 23-34.

WILMÉ, L. & LANGRAND, O. 1990. Rediscovery of Slender-billed Flufftail *Sarothrura watersi* (Bartlett, 1879), and notes on the genus *Sarothrura* in Madagascar. *Biol. Conserv.* 51: 211-223.

WILSON, A. E. & SWALES, M. K. 1958. Flightless Moorhens *Porphyrio c. comeri* from Gough Island breed in captivity. *Avicult. Mag.* 64: 43-45.

WINTERBOTTOM, J. M. 1966. Some notes on the Red-knobbed Coot *Fulica cristata* in South Africa. *Ostrich* 37: 92-94.

WINTERBOTTOM, J. M. 1971. *Priest's eggs of southern African birds.* Johannesburg: Winchester Press.

WINTERBOTTOM, J. M. 1976. New distributional data: Buff-spotted Flufftail *Sarothrura elegans. Ostrich* 47: 217.

WINTLE, C. C. 1988. Crakes and flufftails in captivity. Unpublished.

WINTLE, C. C. & TAYLOR, P. B. 1993. Sequential polyandry, behaviour and moult in captive Striped Crakes *Aenigmatolimnas marginalis. Ostrich* 64: 115-122.

WITHERBY, H. F., JOURDAIN, F. C. R., TICEHURST, N. F. & TUCKER, B. W. 1941. *The handbook of British birds.* Vol. 5. London: Witherby.

WITTEMAN, G. J. & BECK, R. E. 1991. Decline and conservation of the Guam Rail. In MARUYAMA, N. *et al.* (eds.). Wildlife conservation: present trends and perspectives for the 21st century. *Proc. Internatn. Symp. on Wildlife Conservation in Tsukuba and Yokohama, Japan, August 21-25 1990.* Tokyo: Japan Wildlife Research Centre.

WITTEMAN, G. J., BECK, R. E., PIMM, S. L. & DERRICKSON, S. R. 1990. The decline and restoration of the Guam Rail *Rallus owstoni. Endangered Species Update.* 8: 36-39.

WODZICKI, K. 1971. The birds of Niue Island, South Pacific: an annotated checklist. *Notornis* 18: 291-304.

WOINARSKI, J. C. Z., FISHER, A., BRENNAN, K., MORRIS, I., WILLAN, R. C. & CHATTO, R. (in press). The Chestnut Rail *Eulabeornis castaneoventris* on the Wessel and English Company Islands: notes on unusual habitats and use of anvils. *Emu* 98

WOLFF, S. W. & MILSTEIN, P. le S. 1976. Rediscovery of the Whitewinged Flufftail in South Africa. *Bokmakierie* 28: 33-36.

WOLTERS, H. E. 1982. *Die Vogelarten der Erde* (1975-1982). Hamburg & Berlin: Paul Parey.

WOOD, N. 1977. A feeding technique of Allen's Gallinule *Porphyrio alleni. Ostrich* 48: 120-121.

WOOD, N. A. 1974. The breeding biology and behaviour of the Moorhen. *British Birds* 67: 104-115, 135-158.

WOOD, N. A. 1975. Habitat preference and behaviour of Crested Coot in winter. *British Birds* 68: 116-118.

WOODALL, P. F. 1993. The distribution and abundance of the Australian Crake (*Porzana fluminea*) and Baillon's Crake (*Porzana pusilla*) and their association with lignum (*Muehlenbeckia cunningham*) at Lake Bindegolly, south-west Queensland. *Queensland Nat.* 31: 107-113.

WOODLAND, D. J., JAFFAR, Z. & KNIGHT, M. 1980. The "pursuit deterrent" function of alarm signals in the Swamphen (*Porphyrio porphyrio*). *Amer. Nat.* 115: 748-753.

WOODS, R. W. 1988. *Guide to birds of the Falkland Islands.* Oswestry, UK: Nelson.

WOOLFENDEN, G. E. 1979. Winter breeding by the American Coot at Tampa, Florida. *Florida Field Nat.* 7: 26.

VAN'T WOUDT, B. D. & DOBBS, R. A. 1988. Wetland wildlife tracks. *Notornis* 35: 27-34.

WRIGHT, A. 1981. Wekas swimming. *Notornis* 28: 28.

YAMASHINA, Y. & MANO, T. 1981. A new species of rail from Okinawa Island. *J. Yamashina Inst. Orn.* 13: 1-6.

YANAGISAWA, N., FUJIMAKI, Y. & HIGUCHI, H. 1993. Japanese data on waterbird population sizes, summarised for IWRB. In ROSE, P. M. & SCOTT, D. A. 1994. *Waterfowl population estimates.* Slimbridge, UK: International Waterfowl and Wetlands Research Bureau (Spec. Pub. 29).

YAN ANHOU & PANG BINGZHANG. 1986. [Ecology of the White-breasted Waterhen *Amaurornis phoenicurus*.] *Chinese Wildlife* 6: 31-33. (In Chinese).

YEALLAND, J. 1952. Notes on some birds of the British Cameroons forest. *Avicult. Mag.* 58: 48-59.

YEPEZ, T. G. 1964. Ornitologia de las islas Margarita, Coche y Cubagua (Venezuela). Part 2. *Mem. Soc. Cienc. Nat. La Salle* 24.

ZEMBAL, R. & FANCHER, J. M. 1988. Foraging behaviour and foods of the Light-footed Clapper Rail. *Condor* 90: 959-962.

ZEMBAL, R. & MASSEY, B. W. 1985. Function of a rail "mystery" call. *Auk* 102: 179-180.

ZEMBAL, R. & MASSEY, B. W. 1987. Seasonality of vocalizations by Light-footed Clapper Rails. *J. Field Orn.* 58: 41-48.

ZEMBAL, R., MASSEY, B. W. & FANCHER, J. M. 1989. Movements and activity patterns of the Light-footed Clapper Rail. *J. Wildl. Manage.* 53: 39-42.

ZIMMER, J. T. 1930. Birds of the Marshall Field Peruvian Expedition. *Field Mus. Nat. Hist., Zool. Ser.* 17: 1-480.

ZIMMER, J. T. & PHELPS, W. H. 1944. New species and subspecies of birds from Venezuela. *Amer. Mus. Novit.* 1270.

ZIMMERMAN, D. A., TURNER, D. & PEARSON, D. J. 1996. *Birds of Kenya and northern Tanzania.* London: Christopher Helm.

ZIMMERMAN, J. L. 1984. Distribution, habitat and status of the Sora and Virginia Rail in interior Kansas. *J. Field Orn.* 55: 38-47.

ZIPP, W. C. H. 1939. The African Moorhen. *Ostrich* 10: 54-56.

INDEX

abbotti, Dryolimnas cuvieri 313
Abyssinian Rail. *See* Rouget's Rail
admiralitatis, Gallirallus philippensis 252
Aenigmatolimnas 430
aequatorialis, Rallus limicola 286
affinis, Porzana pusilla 381
affinis, Sarothrura 64, *168*
affinis, Sarothrura affinis 168
African Crake **27** 116, *317*, 321, 431
African Rail **19** 100, 294, *299*, 317, 327, 379
akool, Amaurornis 112, *353*
akool, Amaurornis akool 354
alai, Fulica 144, *535*
alberti, Lewinia pectoralis 307
albicollis, Porzana 120, *408*
albicollis, Porzana albicollis 408
albigularis, Laterallus 80, *217*
albigularis, Laterallus albigularis 216
albiventer, Gallirallus striatus 265
albiventris, Aramides cajanea 106, *338*
albus, Porphyrio 136, *470*
Aldabra (White-throated) Rail. *See* White-throated Rail
aldabranus, Dryolimnas cuvieri 104, *313*
alleni, Porphyrio 138, *476*
Allen's Gallinule **38** 138, 459, *476*, 480, 509
alvarezi, Rallina eurizonoides 204
Amaurolimnas 346
amauroptera, Rallina eurizonoides 76, *203*
Amaurornis 353
American Coot **42** 146, 527, 535, *536*, 543, 544, 546
American Purple Gallinule **38** 138, 459, 476, *480*, 485
American Yellow Rail. *See* Yellow Rail
americana, Fulica 146, *536*
americana, Fulica americana 537
anachoretae, Gallirallus philippensis 252
Andaman Banded Crake. *See* Andaman Crake
Andaman Crake **7** 76, *200*, 265
Andean Coot **42** 146, 544, *546*, 549, 553
andrewsi, Gallirallus philippensis 250
angulata, Gallinula 140, *509*
antarcticus, Rallus 98, *293*
antonii, Sarothrura affinis 64, *169*
Anurolimnas 206
Apaiang Rail. *See* Gilbert Rail
Aphanapteryx 59
aquaticus, Rallus 100, *293*
aquaticus, Rallus aquaticus 294
Aramides 334
Aramidopsis 329
ardesiaca, Fulica 146, *546*
ardesiaca, Fulica ardesiaca 547
armillata, Fulica 148, *548*
Ascension Rail 61

Ash-throated Crake **29** 120, 218, *407*
Ashy Crake. *See* White-browed Crake
Asian Yellow Rail. *See* Swinhoe's Rail
assimilis, Gallirallus philippensis 90, *252*
Atlantisia 331
atra, Fulica 144, *527*
atra, Fulica atra 527
atra, Porzana 124, *422*
atrura, Fulica ardesiaca 547
Auckland Island Rail. *See* Auckland Rail
Auckland Rail **20** 102, 306, *310*
Austral Rail **18** 98, 289, *291*
Australian Coot. *See* Common Coot
Australian Crake **29** 120, 379, *400*, 427. *See also* Australian Crake
australis, Fulica atra 529
australis, Gallirallus 86, *233*
australis, Gallirallus australis 233
avicenniae, Aramides cajanea 339
axillaris, Aramides 106, *335*
ayresi, Sarothrura 68, *172*
Azure Gallinule **37** 136, 480, *485*, 512

Baillon's Crake **28** 118, 294, 299, 374, *379*, 396, 431
Bald-faced Rail **24** 110, 330, *349*
Band-bellied Crake **30** 122, 201, 203, 354, 411, *414*
Banded Land Rail. *See* Buff-banded Rail
Banded Rail. *See* Buff-banded Rail; Dieffenbach's Rail
bangsi, Porzana flaviventer 426
Bar-winged Rail **11** 84, 230, *231*
barbadensis, Gallinula chloropus 495
Bare-eyed Rail **24** 110, *350*, 352, 448
Bare-faced Rail. *See* Bald-faced Rail
Barred Rail **27** 88, 243, *245*, 247
batesi, Sarothrura pulchra 152
beldingi, Rallus longirostris 271
belizensis, Rallus longirostris 273
bellus, Porphyrio porphyrio 461
berliozi, Canirallus kioloides 182
bicolor, Amaurornis 114, *372*
Bismarck Rail. *See* New Britain Rail
Black Coot. *See* Common Coot
Black Crake **26** 114, *366*, 371. *See also* Black Rail
Black Moorhen. *See* Dusky Moorhen
Black North Island Woodhen. *See* Weka
Black Rail **10** 82, 184, 188, *220*, 224, 406. *See also* Black Crake; Spotless Crake
Black South Island Woodhen. *See* Weka
Black-banded Crake **8** 78, 206, 208, *209*, 211, 213
Black-tailed Crake **26** 114, 353, *372*, 411
Black-tailed Forest-rail. *See* Mayr's Forest-rail
Black-tailed Native-hen **40** 142, *515*, 518
Black-tailed Waterhen/Swamphen/Gallinule. *See* Black-tailed Native-hen

593

Blackish Rail **34** 130, 344, 440, *443*, 445
Blue-breasted (Banded) Rail. *See* Slaty-breasted Rail
Blue-faced Rail. *See* Bald-faced Rail
boehmi, Sarothrura 66, *165*
Boehm's Flufftail. *See* Streaky-breasted Flufftail
Bogotá Rail **18** 98, *289*, 293
bogotensis, Gallinula melanops 513
bonapartii, Sarothrura rufa 66, *160*
bonasia, Aphanapteryx 59
brachipus, Lewinia pectoralis 102, *308*
Brazilian Crake. *See* Rufous-sided Crake
Brown Crake **25** 112, *353*, 372
Brown Moorhen. *See* Rufous-tailed Bush-hen
Brown North Island Woodhen. *See* Weka
Brown Rail. *See* Rufous-tailed Bush-hen
Brown South Island Woodhen. *See* Weka
Brown Wood-rail **23** 108, 334, 337, *341*
Brown-backed Wood-rail. *See* Brown Wood-rail
Brown-banded Rail **20** 102, 265, *304*, 306, 310
brunnescens, Anurolimnas viridis 208
Buff-banded Rail **14** 90, 233, 239, 245, 247, *248*, 256, 258, 260, 261, 262, 263, 264, 330
Buff-spotted Crake. *See* Buff-spotted Flufftail
Buff-spotted Flufftail **1** 64, 151, *154*
Bush-hen. *See* Plain Bush-hen

cachinnans, Gallinula chloropus 494
caerulescens, Rallus 100, *299*
cajanea, Aramides 106, *337*
cajanea, Aramides cajanea 338
calopterus, Aramides 108, *346*
Canirallus 179
canningi, Rallina 76, *200*
captus, Lewinia pectoralis 102, *307*
caribaea, Fulica 146, *547*
caribaeus, Rallus longirostris 273
Caribbean Coot **42** 146, 536, *543*
carmichaeli, Rallina mayri 74, *197*
carolina, Porzana 120, *399*
Carolina Rail/Crake. *See* Sora
caspius, Porphyrio porphyrio 460
castaneiceps, Anurolimnas 78, *206*
castaneiceps, Anurolimnas castaneiceps 206
castaneoventris, Eulabeornis 132, *448*
castaneoventris, Eulabeornis castaneoventris 448
castaneus, Amaurolimnas concolor 110, *347*
caucae, Pardirallus nigricans 443
Cayenne Crake. *See* Russet-crowned Crake
Cayenne Wood-rail. *See* Grey-necked Wood-rail
celebensis, Gallirallus torquatus 88, *246*
Celebes Bush-hen. *See* Isabelline Bush-hen
Celebes Rail. *See* Snoring Rail
centralis, Sarothrura pulchra 64, *151*
cerceris, Gallinula chloropus 494
cerdaleus, Laterallus albigularis 217
cerverai, Cyanolimnas 128, *435*
chapmani, Micropygia schomburgkii 190
Chatham Island Banded Rail. *See* Dieffenbach's Rail
Chatham Islands Rail. *See* Chatham Rail

Chatham Rail **16** 94, *263*
Chestnut Forest-rail **6** 74, *192*, 194, 195, 197
Chestnut Rail **35** 132, 350, *448*
Chestnut-bellied Rail. *See* Chestnut Rail
Chestnut-breasted Crake. *See* Band-bellied Crake
Chestnut-breasted Rail. *See* Chestnut Rail
Chestnut-headed Crake **8** 78, *206*, 208, 209, 213, 347. *See also* Chestnut-headed Flufftail
Chestnut-headed Flufftail **2** 66, 159, *163*, 165
Chestnut-headed Rail. *See* Chestnut-headed Crake
Chestnut-tailed Crake. *See* Striped Flufftail
Chestnut-tailed Flufftail. *See* Striped Flufftail
Chinese Banded Crake. *See* Band-bellied Crake
chloropus, Gallinula 140, *492*
chloropus, Gallinula chloropus 493
christophori, Gallirallus philippensis 251
cinerea, Gallicrex 138, *454*
cinerea, Porzana 124, *427*
cinereiceps, Laterallus albigularis 217
Clapper Rail **17** 96, *269*, 279, 283, 284
clelandi, Lewinia pectoralis 307
coccineipes, Amaurornis akool 354
coccineipes, Anurolimnas castaneiceps 78, *207*
Colombian Crake **33** 128, 435, *437*, 438
colombianus, Neocrex 128, *437*
colombianus, Neocrex colombianus 437
columbiana, Fulica americana 538
comeri, Gallinula nesiotis 490
Common Bush-hen. *See* Rufous-tailed Bush-hen
Common Coot **41** 144, 492, *527*, 537
Common Moorhen **39** 140, 321, 476, 480, 490, 492, 504, 509, 551
Common/Florida Gallinule. *See* Common Moorhen
concolor, Amaurolimnas 110, *346*
concolor, Amaurolimnas concolor 347
conditicius, Gallirallus 86, *242*
Corncrake **27** 116, 294, 317, *320*, 431
cornuta, Fulica 148, *555*
coryi, Rallus longirostris 272
Coturnicops 183
coturniculus, Laterallus jamaicensis 82, *221*
crassirostris, Gallinula melanops 513
crassirostris, Rallus longirostris 274
crepitans, Rallus longirostris 96, *271*
Crested Coot. *See* Red-knobbed Coot
Crex 315
crex, Crex 116, *321*
Crimson-legged Crake. *See* Brown Crake
cristata, Fulica 144, *522*
cuvieri, Dryolimnas 104, *312*
cuvieri, Dryolimnas cuvieri 312
Cuvier's Rail. *See* White-throated Rail
Cyanolimnas 435
cypereti, Rallus longirostris 273

dieffenbachii, Gallirallus 94, *262*
Dieffenbach's Rail **16** 94, *262*
Dot-winged Crake **29** 120, 188, 220, *406*
Dot-winged Rail. *See* Dot-winged Crake

Dotted Crake. *See* Ocellated Crake
Drummer Rail. *See* Invisible Rail
dryas, Rallina forbesi 195
Dryolimnas 311
Durnford's Crake. *See* Dot-winged Crake
Dusky Gallinule. *See* Dusky Moorhen
Dusky Moorhen **39** 140, *503,* 515
Dusky Rail. *See* Hawaiian Crake

ecaudatus, Gallirallus philippensis 252
Ecuadorian Wood-rail. *See* Red-winged Wood-rail
edwardi, Porzana tabuensis 122, *417*
egregia, Crex 116, *317*
elegans, Rallus 96, *278*
elegans, Rallus elegans 279
elegans, Sarothrura 64, *154*
elegans, Sarothrura elegans 155
elizabethae, Sarothrura rufa 160
elpenor, Atlantisia 61
Elwes's Crake. *See* Black-tailed Crake
erythrops, Neocrex 128, *438*
erythrops, Neocrex erythrops 438
erythrothorax, Porzana fusca 122, *412*
Eulabeornis 448
Eurasian Coot. *See* Common Coot
Eurasian Spotted Crake. *See* Spotted Crake
eurizonoides, Rallina 76, *203*
eurizonoides, Rallina eurizonoides 204
European Water Rail. *See* Water Rail
exilis, Laterallus 82, *217*
exquisitus, Coturnicops 72, *183*
exsul, Lewinia pectoralis 307

fasciata, Rallina 76, *201*
fasciatus, Anurolimnas 78, *209*
Fiji Rail. *See* Bar-winged Rail
flavirostris, Amaurornis 114, *366*
flavirostris, Porphyrio 136, *485*
flaviventer, Porzana 124, *424*
flaviventer, Porzana flaviventer 426
fluminea, Porzana 120, *405*
forbesi, Rallina 74, *195*
forbesi, Rallina forbesi 196
Forbes's Chestnut Rail. *See* Forbes's Forest-rail
Forbes's Forest-rail **6** 74, 194, *195,* 197
formosana, Rallina eurizonoides 76, *204*
friedmanni, Rallus limicola 285
frontata, Gallinula tenebrosa 504
Fulica 522
fusca, Porzana 122, *411*
fusca, Porzana fusca 411

Galapagos Crake. *See* Galapagos Rail
Galapagos Rail **10** 82, 220, *224*
galeata, Gallinula chloropus 495
Gallicrex 454
Gallinula 487
Gallirallus 232
garmani, Gallinula chloropus 495

Giant Coot **43** 148, 546, *553,* 556
Giant Wood-rail **23** 108, 334, 341, *342*
gigantea, Fulica 148, *553*
Gilbert Rail **12** 86, *242*
goldmani, Coturnicops noveboracensis 185
goodsoni, Gallirallus philippensis 252
gossii, Porzana flaviventer 425
Gough (Island) Moorhen. *See* Tristan Moorhen
Great Rail. *See* Giant Wood-rail
Grey-bellied Crake. *See* White-browed Crake
Grey-breasted Crake **10** 82, 213, *217,* 227, 438
Grey-faced Rail. *See* New Guinea Flightless Rail
Grey-necked Wood-rail **22** 106, 334, 335, *337,* 341, 342
Grey-throated Rail **4** 70, *179,* 181
Grey-throated Wood-rail. *See* Madagascar Wood-rail
greyi, Gallirallus australis 86, *234*
grossi, Rallus longirostris 273
Guam Rail **13** 88, 256, *258*
guami, Gallinula chloropus 494
guatemalensis, Amaurolimnas concolor 347
gularis, Gallirallus striatus 92, *266*
Gymnocrex 348

Habroptila 451
haematopus, Himantornis 70, *177*
Halmahera Rail. *See* Invisible Rail
Hauxwell's Crake. *See* Black-banded Crake
Hawaiian Coot **41** 144, *535,* 537
Hawaiian Crake **32** 126, *409*
hectori, Gallirallus australis 86, *234*
Henderson Crake **31** 124, 422
hendersoni, Porzana flaviventer 425
hibernans, Rallus aquaticus 295
Highland Rail. *See* King Rail
Himantornis 177
hoeveni, Gymnocrex plumbeiventris 351
Horned Coot **43** 148, *555*
Horqueta Crake. *See* Rufous-faced Crake
Hutton's Rail. *See* Chatham Islands Rail

immaculatus, Nesoclopeus woodfordi 230
Inaccessible Island Rail. *See* Inaccessible Rail
Inaccessible Rail **21** 104, *331*
Indian Water Rail. *See* Water Rail
indicus, Porphyrio porphyrio 461
indicus, Rallus aquaticus 100, *295*
inepta, Megacrex 132, *452*
inepta, Megacrex inepta 453
insignis, Gallirallus 88, *247*
insolitus, Pardirallus maculatus 441
insularis, Amaurornis phoenicurus 364
insularis, Sarothrura 68, *171*
insularum, Rallus longirostris 272
insulsus, Lewinia pectoralis 307
intermedia, Porzana pusilla 380
Invisible Rail **35** 132, *451,* 459
Isabelline Bush-hen **25** 112, *355*
Isabelline Waterhen. *See* Isabelline Bush-hen

isabellinus, Amaurornis 112, *355*
Island Hen/Cock. *See* Inaccessible Rail

jamaicensis, Laterallus 82, *220*
jamaicensis, Laterallus jamaicensis 220
jouyi, Gallirallus striatus 266

Kelp Hen. *See* Weka
King Clapper Rail. *See* King Rail
King Rail **17** 96, 270, *278*, 284
kioloides, Canirallus 70, *181*
kioloides, Canirallus kioloides 181
Kioloides Rail. *See* Madagascar Wood-rail
klossi, Rallina rubra 74, *193*
korejewi, Rallus aquaticus 295
Kosrae Crake **32** 126, 422, 423
kuehni, Gallirallus torquatus 246

lacustris, Gallirallus philippensis 251
lafresnayanus, Gallirallus 86, *237*
Land Rail. *See* Buff-banded Rail
landbecki, Pardirallus sanguinolentus 446
Landrail. *See* Corncrake
Large Crake. *See* Band-bellied Crake
latens, Aramides cajanea 338
Laterallus 210
Laysan Crake **32** 126, *387*
leguati, Aphanapteryx 60
Leguat's Rail 60
lesouefi, Gallirallus philippensis 251
Lesser Gallinule. *See* Allen's Gallinule
Lesser Moorhen **39** 140, 476, *509*
Lesser Rail. *See* Virginia Rail
leucomelanus, Amaurornis phoenicurus 114, *364*
leucophaeus, Rallus longirostris 273
leucoptera, Fulica 146, *544*
leucopyrrhus, Laterallus 84, *227*
leucospila, Rallina 74, *194*
levipes, Rallus longirostris 271
levraudi, Laterallus 80, *213*
Levraud's Crake. *See* Rusty-flanked Crake
Lewinia 304
Lewin's Rail **20** 102, 304, *306*, 310, *396*
limarius, Gallirallus torquatus 246
limicola, Rallus 98, *284*
limicola, Rallus limicola 285
Little Crake **28** 118, 294, *374*, 379, *390*, 395
Little Gallinule. *See* Azure Gallinule
Little Wood-rail **22** 106, *334*, 335, 341, 346
Little (Spotted) Waterhen. *See* Spot-flanked Gallinule
Long-billed Clapper Rail. *See* Clapper Rail
Long-toed Flufftail. *See* Chestnut-headed Flufftail
longirostris, Rallus 96, *269*
longirostris, Rallus longirostris 273
Lord Howe Island Rail. *See* Lord Howe Rail
Lord Howe Island Woodhen. *See* Lord Howe Rail
Lord Howe Rail **12** 86, 237, *238*, 242
lugens, Sarothrura 66, *166*

lugens, Sarothrura lugens 167
lugubris, Fulica atra 528
luridus, Pardirallus sanguinolentus 446
Luzon Rail. *See* Brown-banded Rail
lynesi, Sarothrura lugens 164

macquariensis, Gallirallus philippensis 252
maculatus, Pardirallus 130, *440*
maculatus, Pardirallus maculatus 441
Madagascar Crake. *See* Madagascar Flufftail
Madagascar Flufftail **3** 68, 168, *171*, 176
Madagascar Rail **19** 100, *302*
Madagascar Wood-rail **4** 70, 179, *181*, 302, 312
madagascariensis, Porphyrio porphyrio 134, *460*
madagascariensis, Rallus 100, *302*
magnirostris, Amaurornis 112, *361*
Malay/Malaysian Banded Crake. *See* Red-legged Crake
Mangere Rail. *See* Chatham Islands Rail
mangle, Aramides 106, *334*
mantelli, Porphyrio 136, *471*
Maori Hen. *See* Weka
margaritae, Rallus longirostris 273
marginalis, Aenigmatolimnas 116, *431*
Marsh Crake. *See* Baillon's Crake
Marsh Hen. *See* Clapper Rail; King Rail
martinica, Porphyrio 138, *480*
Mascarene Coot 61
Mauritian Red Rail 59
mayri, Lewinia pectoralis 307
mayri, Porzana pusilla 381
mayri, Rallina 74, *196*
mayri, Rallina mayri 197
Mayr's Chestnut Rail. *See* Mayr's Forest-rail
Mayr's Forest-rail **6** 74, 192, 194, 195, *196*
Meadow Chicken. *See* Sora
Megacrex 452
melanophaius, Laterallus 80, *210*
melanophaius, Laterallus melanophaius 211
melanops, Gallinula 142, *512*
melanops, Gallinula melanops 512
melanopterus, Porphyrio porphyrio 461
melanotus, Porphyrio porphyrio 134, *462*
mellori, Gallirallus philippensis 90, *249*, 252
meridionalis, Gallinula chloropus 494
Mexican Clapper Rail. *See* King Rail
mexicanus, Aramides cajanea 337
meyerdeschauenseei, Rallus limicola 286
meyeri, Gallirallus philippensis 251
Micropygia 189
midnicobaricus, Amaurornis phoenicurus 364
Mills's Crake. *See* Hawaiian Crake
minahasa, Rallina eurizonoides 204
mira, Porzana pusilla 118, *381*
mirificus, Lewinia 102, *304*
Modest Rail. *See* Chatham Islands Rail
modestus, Gallirallus 94, *263*
Moho. *See* Hawaiian Crake
moluccanus, Amaurornis 112, *357*

moluccanus, Amaurornis moluccanus 359
monasa, Porzana 126, *422*
morrisoni, Aramides cajanea 338
mortierii, Gallinula 142, *518*
Mud Hen. *See* Clapper Rail; King Rail
muelleri, Lewinia 102, *310*
murivagans, Laterallus jamaicensis 221

Narkie. *See* Tasmanian Native-hen
Neocrex 436
nesiotis, Gallinula 140, *490*
nesiotis, Gallinula nesiotis 490
Nesoclopeus 230
neumanni, Gallinula tenebrosa 504
New Britain Rail **13** 88, 243, 245, *247*
New Caledonian Rail **12** 86, *237*, 239
New Caledonian Wood-rail. *See* New Caledonian Rail
New Guinea Chestnut Rail. *See* Chestnut Forest-rail
New Guinea Flightless Rail **34** 132, *452*
newtonii, Fulica 61
niger, Nesophylax 60
nigra, Rallus 60
nigricans, Pardirallus 130, *443*
nigricans, Pardirallus nigricans 443
nigrifrons, Amaurornis moluccanus 359
Nkulenga Rail. *See* Nkulengu Rail
Nkulengu Rail **4** 70, *177*
notatus, Coturnicops 72, *187*
novaeguineae, Fulica atra 529
noveboracensis, Coturnicops 72, *184*
noveboracensis, Coturnicops noveboracensis 185

obscurior, Gallirallus striatus 266
obsoletus, Rallus longirostris 96, *270*
Ocellated Crake **5** 72, *190*, 425
Ocellated Rail. *See* Ocellated Crake
oculeus, Canirallus 70, *179*
oenops, Laterallus melanophaius 211
Okinawa Rail **15** 92, *243*, 245, *247*
okinawae, Gallirallus 92, *243*
olivacea, Porzana albicollis 409
olivaceus, Amaurornis 112, *356*
olivascens, Neocrex erythrops 128, *438*
olivieri, Amaurornis 114, *371*
Olivier's Crake/Rail. *See* Sakalava Rail
orientalis, Gallinula chloropus 494
Ortolan. *See* Sora
owstoni, Gallirallus 88, *258*

Pacific Rail. *See* Tahiti Rail
pacifica, Gallinula 138, *488*
pacificus, Aramides cajanea 338
pacificus, Gallirallus 94, *261*
Paint-billed Crake **33** 128, 218, 224, 409, 435, *437*, *438*
Painted Rail. *See* Buff-banded Rail
pallida, Megacrex inepta 453
pallidus, Rallus longirostris 273
palmeri, Porzana 126, *387*

palustris, Porzana pusilla 118, *381*
Papuan Flightless Rail. *See* New Guinea Flightless Rail
paratermus, Gallirallus striatus 266
Pardirallus 440
parva, Porzana 118, *374*
parva, Rallina forbesi 195
pauxilla, Gallinula chloropus 495
paykullii, Porzana 122, *414*
Pectoral. *See* Buff-banded Rail
Pectoral Rail. *See* Lewin's Rail
pectoralis, Lewinia 102, *306*
pectoralis, Lewinia pectoralis 306
pelewensis, Gallirallus philippensis 251
pelewensis, Porphyrio porphyrio 461
pelodramus, Rallus longirostris 273
Peruvian Coot. *See* Andean Coot
Peruvian Rail. *See* Bogota Rail
peruvianus, Rallus semiplumbeus 290
phaeopyga, Porzana fusca 412
phelpsi, Rallus longirostris 273
philippensis, Gallirallus 90, *248*
philippensis, Gallirallus philippensis 251
Philippine Rail. *See* Barred Rail
phoenicurus, Amaurornis 114, *363*
phoenicurus, Amaurornis phoenicurus 363
Pink-legged Rail. *See* New Britain Rail
Plain Bush-hen **25** 112, *356*, 358
Plain-flanked Rail **18** 98, 270, *283*, 294
plateni, Aramidopsis 104, *329*
Platen's Celebes Rail. *See* Snoring Rail
Platen's Rail. *See* Snoring Rail
plumbeicollis, Aramides cajanea 106, *338*
plumbeiventris, Gymnocrex 110, *350*
plumbeiventris, Gymnocrex plumbeiventris 350
Plumbeous Rail **34** 130, 435, 440, 443, *445*
Plumbeous-breasted Rail. *See* Slaty-breasted Rail
poecilopterus, Nesoclopeus 84, *231*
poliocephalus, Porphyrio porphyrio 460
Porphyrio 458
porphyrio, Porphyrio 134, *458*
porphyrio, Porphyrio porphyrio 459
Porzana 373
porzana, Porzana 118, *389*
praedo, Gallirallus philippensis 252
Pukeko. *See* Purple Swamphen
pulchra, Sarothrura 64, *151*
pulchra, Sarothrura pulchra 152
pulverulentus, Porphyrio porphyrio 134, *461*
Purple Gallinule. *See* American Purple Gallinule; Purple Swamphen
Purple Swamphen **36** 134, 451, *458*, 488
pusilla, Porzana 118, *379*
pusilla, Porzana pusilla 381
pyrrhorrhoa, Gallinula chloropus 494

Rallina 192
Rallus 269
ramsdeni, Rallus elegans 280

Red Crake. *See* Ruddy Crake
Red Rail. *See* Ruddy Crake; Uniform Crake
Red-and-white Crake **11** 84, *226*, 228
Red-backed Forest-rail. *See* Forbes's Forest-rail
Red-billed Rail. *See* Tahiti Rail
Red-chested Crake. *See* Red-chested Flufftail
Red-chested Flufftail **2** 66, *159*, 163, 165
Red-fronted Coot **43** 148, 544, 546, 549, *551*
Red-gartered Coot **43** 148, 544, 546, *548*, 551
Red-knobbed Coot **41** 144, 492, *522*, 527
Red-legged Banded Crake. *See* Red-legged Crake
Red-legged Crake **7** 76, 198, *201*, 203, 354, 411, 414
Red-necked Crake **7** 76, 195, *197*, 201, 203, 350
Red-necked Rail. *See* Red-necked Crake
Red-tailed Flufftail. *See* Striped Flufftail
Red-winged Wood-rail **23** 108, 334, 341, 342, 344, *345*, 350
reductus, Gallirallus philippensis 251
reichenovi, Sarothrura elegans 155
Rice Chicken. *See* King Rail
richardsoni, Porzana tabuensis 417
ripleyi, Neocrex colombianus 128, *437*
rogersi, Atlantisia 104, *331*
rosenbergii, Gymnocrex 110, *349*
Rosenberg's (Bare-eyed) Rail. *See* Bald-faced Rail
rougetii, Rougetius 104, *327*
Rougetius 327
Rouget's Rail **21** 104, 299, *327*
Roviana Rail **14** 90, *256*, 258
rovianae, Gallirallus 90, *256*
ruber, Laterallus 80, *214*
rubra, Rallina 74, *192*
rubra, Rallina rubra 192
Ruddy Crake **9** 80, 206, *214–238*, 217. *See also* Ruddy-breasted Crake
Ruddy-breasted Crake **30** 122, 201, 203, 354, *411*, 414
rufa, Sarothrura 66, *159*
rufa, Sarothrura rufa 159
ruficrissus, Amaurornis moluccanus 112, *358*
rufifrons, Fulica 148, *553*
Rufous-backed Crake. *See* Black-tailed Crake
Rufous-crowned Wood-rail. *See* Rufous-necked Wood-rail
Rufous-faced Crake **11** 84, 211, 227, *228*
Rufous-necked Wood-rail **22** 106, 334, *335*, 341, 346
Rufous-sided Crake **9** 80, 208, 209, *210*, 213, 217, 218, 227, 228
Rufous-sided Rail. *See* Rufous-sided Crake
Rufous-tailed Bush-hen **25** 112, 198, 350, 356, *357*
Rufous-tailed Crake. *See* Rufous-tailed Bush-hen
Rufous-tailed Moorhen. *See* Rufous-tailed Bush-hen
Rufous-tailed Rail. *See* Rufous-tailed Bush-hen
Rufous-tailed Waterhen. *See* Rufous-tailed Bush-hen
Rufous-vented Crake. *See* Rufous-sided Crake
Rufous-vented Moorhen. *See* Rufous-tailed Bush-hen
Rufous-vented Rail. *See* Rufous-tailed Bush-hen
Russet-crowned Crake **8** 78, 206, *207*, 209, 213, 347
Rusty-flanked Crake **9** 80, 206, *213*, 214

Sakalava Rail **26** 114, 367, *371*, 372
salinasi, Laterallus jamaicensis 221
Salt Marsh Clapper Rail. *See* Clapper Rail
Salt Marsh-hen. *See* Clapper Rail
Samoan Moorhen **37** 138, *488*
Samoan Wood Rail/Gallinule/Woodhen. *See* Samoan Moorhen
San Cristobal Gallinule. *See* San Cristobal Moorhen
San Cristobal Moorhen **40** 142, 488, *489*
sandvicensis, Gallinula chloropus 494
Sandwich Crake/Rail. *See* Hawaiian Crake
sandwichensis, Porzana 126, *409*
sanguinolentus, Pardirallus 130, *445*
sanguinolentus, Pardirallus sanguinolentus 130, *446*
saracura, Aramides 108, *344*
Saracura Wood-rail. *See* Slaty-breasted Wood-rail
Sarothrura 151
saturatus, Rallus longirostris 272
schomburgkii, Micropygia 72, *190*
schomburgkii, Micropygia schomburgkii 190
Schomburgk's Crake. *See* Ocellated Crake
Sclater's Rail. *See* New Britain Rail
scotti, Gallirallus australis 86, *234*
scotti, Rallus longirostris 96, *272*
seistanicus, Porphyrio porphyrio 460
semiplumbeus, Rallus 98, *289*
semiplumbeus, Rallus semiplumbeus 289
sepiaria, Rallina eurizonoides 204
sethsmithi, Gallirallus philippensis 252
sharpei, Eulabeornis castaneoventris 132, *449*
sharpei, Gallirallus 94, *264*
Sharpe's Rail **16** 94, *264*
Short-toed Rail. *See* Lewin's Rail
Siberian Rail. *See* Swinhoe's Rail
Siberian Ruddy Crake. *See* Band-bellied Crake
silvestris, Gallinula 142, *489*
simonsi, Pardirallus sanguinolentus 445
Slate-breasted Rail. *See* Lewin's Rail
Slate-coloured Coot. *See* Andean Coot
Slaty-breasted Rail **15** 92, 264, *265*, 294, 305, 306, 330
Slaty-breasted Wood-rail **23** 108, 334, 341, 342, *344*, 346, 443
Slaty-legged Banded Crake. *See* Slaty-legged Crake
Slaty-legged Crake **7** 76, 200, 201, *203*, 354, 411, 414
Slaty-legged Philippine Crake. *See* Slaty-legged Crake
Slaty-legged Ryukyu Crake. *See* Slaty-legged Crake
Slender-billed Flufftail **3** 68, 171, *175*
Snoring Rail **21** 104, *329*, 349
Solomons Rail. *See* Woodford's Rail
Sooty Crake/Rail. *See* Spotless Crake
Sora **29** 120, 218, 390, *399*, 408, 425
Speckled Rail **5** 72, *187*, 220, 406

spilonotus, Laterallus 82, *224*
spiloptera, Porzana 120, *406*
Spot-flanked Gallinule **40** 142, *512*
Spotless Crake **30** 122, 396, *416*, 422, 423
Spotted Crake **28** 118, 294, 317, 374, *389*, 399, 431. *See also* Australian Crake
Spotted Rail **34** 130, 264, *440*
steini, Rallina forbesi 74, *195*
Streaky-breasted Crake. *See* Streaky-breasted Flufftail
Streaky-breasted Flufftail **2** 66, 159, 163, *165*
striatus, Gallirallus 92, 265
striatus, Gallirallus striatus 266
Striped Crake **27** 116, 317, 374, 390, *431*
Striped Flufftail **1** 64, *168*, 171, 172
sulcirostris, Gallirallus torquatus 88, *246*
swindellsi, Gallirallus philippensis 90, *252*
Swinhoe's Rail **5** 72, *183*, 184
Swinhoe's Yellow Rail. *See* Swinhoe's Rail
sylvestris, Gallirallus 86, *238*

tabuensis, Porzana 122, *416*
tabuensis, Porzana tabuensis 416
Tahiti Crake 60
Tahiti Rail **16** 94, *261*
Tahitian Rail. *See* Tahiti Rail
taiwanus, Gallirallus striatus 266
Takahe **37** 136, 459, *471*
Talaud Bush-hen **25** 112, *361*
Talaud Rail **24** 110, *352*
talaudensis, Gymnocrex 110, *352*
Tasmanian Native-hen **40** 142, 515, *518*
telefolminensis, Rallina rubra 193
telmatophila, Rallina eurizonoides 204
Temminck's Crake. *See* Grey-breasted Crake
tenebrosa, Gallinula 140, *504*
tenebrosa, Gallinula tenebrosa 503
tenuirostris, Rallus elegans 280
tertius, Nesoclopeus woodfordi 230
Tiny Crake. *See* Baillon's Crake
torquatus, Gallirallus 88, *245*
torquatus, Gallirallus torquatus 245
tounelieri, Gallirallus philippensis 252
tricolor, Rallina 76, *197*
Tristan Island Cock/Hen. *See* Tristan Moorhen
Tristan Moorhen **39** 140, *490*, 492
Tristan Rail. *See* Inaccessible Rail
Troglodyte Rail. *See* Weka
tschudii, Pardirallus sanguinolentus 445
tuerosi, Laterallus jamaicensis 82, *221*
Twopenny Chick. *See* Yellow-breasted Crake

ultimus, Amaurornis moluccanus 359
Uniform Crake **24** 110, 206, 208, 213, 335, *346*

vanrossemi, Aramides cajanea 338
ventralis, Gallinula 142, *515*
Virginia Rail **18** 98, 279, 283, *284*, 289, 293
viridis, Anurolimnas 78, *207*
viridis, Anurolimnas viridis 208

viridis, Porphyrio porphyrio 460

Wake Island Rail. *See* Wake Rail
Wake Rail **15** 92, *260*
wakensis, Gallirallus 92, 260
Wallace's Rail. *See* Invisible Rail
wallacii, Habroptila 132, *451*
Water Crake. *See* Australian Crake.
Water Rail **19** 100, 289, *293*, 299, 375, 386. *See also* Lewin's Rail
Watercock **38** 138, *454*
Waterhen. *See* Common Moorhen; Tasmanian Native-hen
Waters's Crake/Flufftail. *See* Slender-billed Flufftail
watersi, Sarothrura 68, *175*
waynei, Rallus longirostris 272
Weka **12** 86, *233*, 237, 239
wetmorei, Rallus 98, *286*
Wetmore's Rail. *See* Plain-flanked Rail
White Gallinule **37** 136, *470*
White-breasted Crake. *See* Red-and-white Crake
White-breasted Swamphen. *See* White-breasted Waterhen
White-breasted Waterhen **26** 114, *363*
White-browed Water Crake/Rail. *See* White-browed Crake
White-browed Crake **31** 124, *427*
White-necked Crake. *See* Ash-throated Crake
White-spotted Crake. *See* White-spotted Flufftail
White-spotted Flufftail **1** 64, *151*, 154
White-striped Chestnut Rail. *See* White-striped Forest-rail
White-striped Forest-rail **6** 74, 192, *194*, 195, 197
White-throated Crake **9** 80, 208, 209, 210, 211, 214, *216*, 218, 227. *See also* Ash-throated Crake
White-throated Rail **21** 104, 181, *312*. *See also* White-throated Crake
White-winged Coot **42** 146, *544*, 546, 549, 551
White-winged Crake. *See* White-winged Flufftail
White-winged Flufftail **3** 68, *172*
wilkinsoni, Gallirallus philippensis 251
wolfi, Aramides 108, *341*
Wolf's Wood-rail. *See* Brown Wood-rail
Wood Rail. *See* Madagascar Wood-rail; Uniform Crake
woodfordi, Nesoclopeus 84, *230*
woodfordi, Nesoclopeus woodfordi 230
Woodford's Rail **11** 84, *230*, 232
Woodhen. *See* Lord Howe Rail
woodi, Porzana flaviventer 426

xenopterus, Laterallus 84, *228*
xerophilus, Gallirallus philippensis 251

Yellow Rail **5** 72, 183, *184*, 425
Yellow-bellied Rail. *See* Yellow-breasted Crake
Yellow-breasted Crake **31** 124, 184, 190, *424*
Yellow-breasted Rail. *See* Yellow-breasted Crake

yorki, Gallirallus philippensis 251
Ypecaha (Wood) Rail. *See* Giant Wood-rail
ypecaha, Aramides 108, *342*
yumanensis, Rallus longirostris 271

Zapata Rail **33** 128, *435*, *437*, *438*
zelebori, Pardirallus sanguinolentus 445
zenkeri, Sarothrura pulchra 64, *152*
zeylonica, Porzana fusca 412